ENCYCLOPEDIA OF ENVIRONMENTAL SCIENCE AND ENGINEERING

Pfafflin/Ziegler

ENCYCLOPEDIA OF ENVIRONMENTAL SCIENCE AND ENGINEERING

VOLUME ONE

A–M

GORDON AND BREACH SCIENCE PUBLISHERS
New York London Paris

Copyright © 1976 by

Gordon and Breach Science Publishers Inc.
One Park Avenue
New York, N.Y. 10016

Editorial office for the United Kingdom

Gordon and Breach Science Publishers Ltd.
42 William IV Street
London W.C.2.

Editorial office for France

Gordon & Breach
7–9 rue Emile Dubois
Paris 75014

Library of Congress catalog card number 72–89286. ISBN 0 677 14650 7. Printed in east Germany.

ACKNOWLEDGMENT

THE EDITORS wish to thank their colleagues who put their knowledge on to paper and the families who endured long hours of detached work by those involved in this endeavor. One of the editors would like to acknowledge a particular debt of gratitude to the late Professor William K. Roots.

EDITORS' PREFACE

THE MAJOR purpose of the *Encyclopedia of Environmental Science and Engineering* is to bring the contributions of key environmental areas into a single readily available work. The *Encyclopedia* covers such broad subjects as Air, Water and Soil Pollution, Public Health and Ecology. Air Pollution, taken as a typical example, is then given extensive treatment by division into subjects such as Meteorology, Sources, Automotive Pollution, Effects of Atmospheric Pollutants, Energy Sources, Control of Gaseous Pollutants, Noise, Particulates, Remote Detection, Sulfur Removal, Stack Sampling, Urban Pollution Modelling and Environmental Law.

The *Encyclopedia* thus makes the information of areas related to one's own specialization readily available to the environmentalist and also provides him with a more general knowledge of his field.

The length of articles has been chosen to establish effectively the relationships among various areas within a discipline, rather than risk the segregation of concepts into overspecialized and less readable portions. In each area covered, the most significant facts and data have been sought and useful references have been provided for those wishing to pursue the field in greater depth.

Man's increased awareness of his relation to other organisms in the scheme of life has led him to examine many disciplines. Application of these disciplines to solutions of everyday problems of his surroundings has given rise to the new professions of environmental scientist and environmental engineer. It is possible that continued survival of civilization, as we know it, may depend on the energy and wisdom of the environmental worker. It is to be hoped, however, that in addition to the professional, the *Encyclopedia* will also be useful to anyone sufficiently interested in, and concerned about, the environment to seek information in the Public Library.

J. R. PFAFFLIN
Windsor, Ontario

E. N. ZIEGLER
New York, N.Y.

CONTRIBUTORS TO VOLUME 1

NEAL E. ARMSTRONG—Department of Civil Engineering, University of Texas. *Estuarine Water Quality*
BRUCE A. BELL—Evirotech Corp. *Groundwater*
J. K. BEWTRA—Department of Civil Engineering, University of Windsor. *Biological Treatment of Wastewater*
RAUL R. CARDENAS, JR.—Department of Civil Engineering, Polytechnic Institute of New York. *Composting*
P. J. DITCHETT—School of Pharmacy, University of Wales. *Disinfection and Chemosterilization*
ROBERT DRESNACK—Department of Civil Engineering, Newark College of Engineering. *Eutrophication*
FRANCIS C. EVANS—Department of Zoology, University of Michigan. *Ecology of Primary Terrestrial Consumers*
W. L. FAITH—Consulting Chemical Engineer, San Marino, California. *Effects of Atmospheric Pollutants*
ERIC R. GIDLUND—Department of Civil Engineering, Polytechnic Institute of New York. *Groundwater*
CHARLES R. GOLDMAN—Division of Environmental Studies, University of California (Davis). *Aquatic Primary Production*
HELENE N. GUTTMAN—Department of Biological Sciences, University of Illinois at Chicago Circle. *Microbiology*
WALTER J. HICKEL—Anchorage, Alaska. *Conservation Planning*
KENNETH C. HOFFMAN—Brookhaven National Laboratory. *Energy Sources—Alternatives*
JOSEPH R. JONES—Windsor Metro Health Center, Windsor, Ontario. *Effects of Chemicals on Human Ecology*
PHILIP H. JONES—Institute of Environmental Science and Engineering, University of Toronto. *Environmental Education*
PETER B. LEDERMAN—Environmental Protection Agency. *Management of Solid Waste*
CAMERON MACINNIS—Department of Civil Engineering, University of Windsor. *Municipal Wastewater*
COLIN A. MAWSON—Atomic Energy of Canada Limited. *Management of Radioactive Waste*
NELSON L. NEMEROW—Department of Civil Engineering, Syracuse University. *Liquid Wastes of Industry*
EUGENE P. ODUM—Institute of Ecology, University of Georgia. *Ecosystem Theory*
DONALD F. OTHMER—Department of Chemical Engineering, Polytechnic Institute of New York. *Desalination*
HANS A. PANOFSKY—Department of Meteorology, Pennsylvania State University. *Air Pollution Meteorology*
JAMES R. PFAFFLIN—Department of Electrical Engineering, University of Windsor. *Effects of Chemicals on Human Ecology. Municipal Wastewater*
MICHAEL C. QUICK—Department of Civil Engineering, University of British Columbia. *Hydrology*
ELROY L. RICE—Department of Botany, University of Oklahoma. *Ecology of Plants*
DONALD P. ROSEMAN—Hydronautics, Inc. *Marine Spillage—Sources and Hazards*
A. D. RUSSELL—School of Pharmacy, University of Wales. *Disinfection and Chemosterilization*
JOSEPH A. SALVATO, JR.—New York State Department of Health. *Environmental Health*
CONRAD SIMON—Environmental Protection Agency. *Air Pollution Sources*
S. SOURIRAJAN—National Research Council of Canada. *Applications of Reverse Osmosis*
MEREDITH H. THOMPSON—New York State Department of Health. *Environmental Health Administration*

STEPHEN VARRO, JR.—Ecology, Inc. *Composting*
ISRAEL WILENITZ—Quirk, Lawler and Matusky. *Modeling of River and Estuary Water Quality*
VICTOR J. YANNACONE, JR.—Yannacone and Yannacone. *Environmental Law*
EDWARD N. ZIEGLER—Department of Chemical Engineering, Polytechnic Institute of New York. *Automotive Pollution. Effects of Chemicals on Human Ecology. Gaseous Pollutant Removal*
ABEL WOLMAN—Johns Hopkins University. *Introduction*

CONTRIBUTORS TO VOLUME 2

H. I. ALI—Department of Civil Engineering, University of Windsor. *Wastewater: Physical and Chemical Treatment*

STANLEY I. AUERBACH—Oak Ridge National Laboratory. *Radiation Ecology*

JOHN E. BARDACH—Department of Zoology, University of Michigan. *Pollution Effects on Fish*

DOMINICK P. BARTENOPE—Charles Pfizer Co. *Sulfur Removal*

J. K. BEWTRA—Department of Civil Engineering, University of Windsor. *Wastewater: Physical and Chemical Treatment*

GERARD P. CANEVARI—Esso Research and Engineering. *Spillage into Water—Treatment*

RAUL R. CARDENAS, JR.—Department of Civil Engineering, Polytechnic Institute of New York. *Pollution Instrumentation—Wastewater*

LAWRENCE K. CECIL—Consulting Chemical Engineer, Champaign, Illinois. *Water Reuse*

S. P. CHEE—Department of Civil Engineering, University of Windsor. *Water Flow*

CHARLES C. COUTANT—Oak Ridge National Laboratory. *Thermal Effects on Fish Ecology*

DAVID W. ECKHOFF—Department of Civil Engineering, University of Utah. *Urban Runoff*

MORTON FELS—Battelle Memorial Institute. *Particulate Removal*

MARTIN FORSBERG—Harvard University. *Water Chemistry. Water: Properties, Structure and Occurrence in Nature*

R. BUCKMINSTER FULLER—Southern Illinois University. *The Terrestrial System*

SUZETTE GAUVAIN—London School of Hygiene and Tropical Medicine. *Occupational Health*

STEVEN GHERINI—Harvard University. *Water Chemistry. Water: Properties, Structure and Occurrence in Nature*

ERNST P. HALL—Environmental Protection Agency. *Pollution from Mine Drainage*

J. M. HARRINGTON—London School of Hygiene and Tropical Medicine. *Occupational Health*

JAMES P. HARTT—Department of Civil Engineering, University of Windsor. *Urban Planning*

KENNETH L. JOHNSON—Environmental Protection Agency. *Stack Sampling*

PHILIP H. JONES—Institute of Environmental Science and Engineering, University of Toronto. *Water Treatment*

ASHFAQ A. KHAN—Department of Chemical Engineering, Polytechnic Institute of New York. *Recycling of Waste Materials*

JEFFREY S. LIFE—Toxicology Center, University of Iowa. *Pollution Effects on Fish*

DAVID S. K. LIU—RAND Corp. *Urban Runoff*

GEORGE C. MALING, JR.—I. B. M. Corp. *Noise*

CONRAD J. MASON—Department of Meteorology and Oceanography, University of Michigan. *Urban Air Pollution Modeling*

JOHN M. MATSEN—Esso Research and Engineering. *Particulate Emissions. Particulate Removal*

J. A. MCCORQUODALE—Department of Civil Engineering, University of Windsor. *Sediment Transport and Erosion*

ROBERT L. METCALF—Department of Zoology, University of Illinois. *Pesticides*

GEORGE MICHANOWSKY—Amazonia Foundation, Inc. *Remote Detection of Air Pollutants*

HARRY MOSES—Argonne National Laboratory. *Urban Air Pollution Modeling*

JAMES R. PFAFFLIN—Department of Electrical Engineering, University of Windsor. *Recycling of Waste Appendix: Materials. Appendix: Selected Properties.*

SHEILA M. PFAFFLIN—Bell Telephone Laboratories. *Psychological Aspects of Man's Environment. Statistical Methods for Environmental Science*

SYED R. QASIM—Department of Civil Engineering, University of Texas at Arlington. *Wastewater: Treatment for Special Applications*

S. SOURIRAJAN—National Research Council of Canada. *Reverse Osmosis — Process Fundamentals*

WERNER STUMM—Swiss Federal Republic Institute of Technology. *Water Chemistry. Water: Properties, Structure and Occurrence in Nature*

ISRAEL WILENITZ—Quirk, Lawler and Matusky. *Systems Modeling for Water Pollution Control*

G. M. WILKENING—Bell Telephone Laboratories. *Non-Ionizing Radiations*

EDWARD N. ZIEGLER—Department of Chemical Engineering, Polytechnic Institute of New York. *Sulfur Removal. Appendix: Selected Properties. Table of Environmental Speeds*

INTRODUCTION

Only the historian will be able to diagnose the causes of the great upsurge of interest in environmental problems on the part of the man in the street. Until recently, this universal excitement was restricted primarily to the "developed" worlds of the East and West. Aroused interest is now penetrating into the less developed regions, inhabited by billions of people. The leaders of these people have observed somewhat cynically that they are concerned more about starvation than about automobile emissions. They hasten to add, however, that, if they can, they would like to avoid the errors which industrialization, increased urbanization and excessive population growth have engendered in the more favored countries.

It was inevitable that, as public interest was aroused, a search for scapegoats for creating actual or assumed environmental ills would be pursued. This readily took the form of a retreat to a non-existent past, to a demand for reversing economic progress, to a general anti-intellectualism and, more specifically, to an attack on the "obvious" vices of science and technology. If it were not for the anti-human machinations of the scientist and the engineer, ecologic disturbance would not have occurred, nature would not have been despoiled and Utopia would have prevailed. The polemicist's argument, of course, is not new. It has been the hallmark of the ages, when each new advance was fought on the fields of religion, law, and authority. Fortunately, science and technology have much to their credit in alleviating the ills of man, in prolonging life, in providing leisure and comfort and in protecting the amenities. On the debit side, wisdom, logic and ecologic equilibrium have on occasion been distinguished by their absence. It is undoubtedly true, however, that the intellectual forces which created a better life for man, and likewise violated some of the environment, are the same forces to which one must look for correcting past and avoiding future deteriorations.

Essentially, the scientist-technologist is now challenged to recapture a role, old in practice, but apparently lost in the present drama of revolutionizing the world. Perhaps the major stress of today is to see to it that human knowledge provides a tool for facilitating human progress, as much as it is for purely scientific or technologic advancement. To the student of progress this is merely re-stating the axiom that "engineering was an instrumentality of civilization, and not a mere physical rearrangement of physical things."*

The volumes produced here are designed to shed authoritative light upon the ingredients of a wholesome environment and how they may be assessed and developed for the benefit of the biosphere—encompassing man, plant and animal life in their multi-variate relationships. The effort to provide a source-book in so vast a field as the environment is a heroic one. Any encyclopedia poses a major challenge of coverage, of priorities and of selection, since the cumulative knowledge in the field here covered is vast, but still not great enough. Many of the contributions admit the lacunae of ignorance, as well as the depths of understanding.

These two volumes have assembled accurate information by competent and respected authorities, to which the student, the practitioner and even the interested layman might have recourse. The "television instant expert" has been avoided, not for moral or ethical reasons, but to assure the reader that the contributions are soundly based on scientific verity and not on specious propaganda.

The environment is great and coverage is likely to be deficient. The simple listing of the subjects in the table of contents serves the useful purpose of disclosing to many the tremendous spectrum of subject matter we so lightly package in the term *environment*. The assembly of topics alone should intrigue the young student threading lightly through his "earth day" exercises. In spite of the breadth and depth of the subject matter, and because of the expanding nature of the field, it may be suggested that supplements might well be on the way.

---

* Nicholas Murray Butler, Introduction, *Engineers and Engineering in the Renaissance*, William Barclay Parsons. The Williams and Wilkins Company, Baltimore, Maryland, 1939.

It may be emphasized, in conclusion, that the present assembly is not intended to provide a long historical record of the science and art of the control of the environment. That record is part of a "very ancient process, rooted in many civilizations" and one of great accomplishment.* The present and the future are the setting for these deliberations. It is certainly, to be hoped that, with appetites whetted, the reader will turn to excellent long time rehearsals of how we came to now!

Abel Wolman

The Johns Hopkins University

---

* *A History of Technology*. Edited by Charles Singer, E. J. Holmyard and A. R. Hall. Five volumes. Oxford University Press, New York and London, 1954.

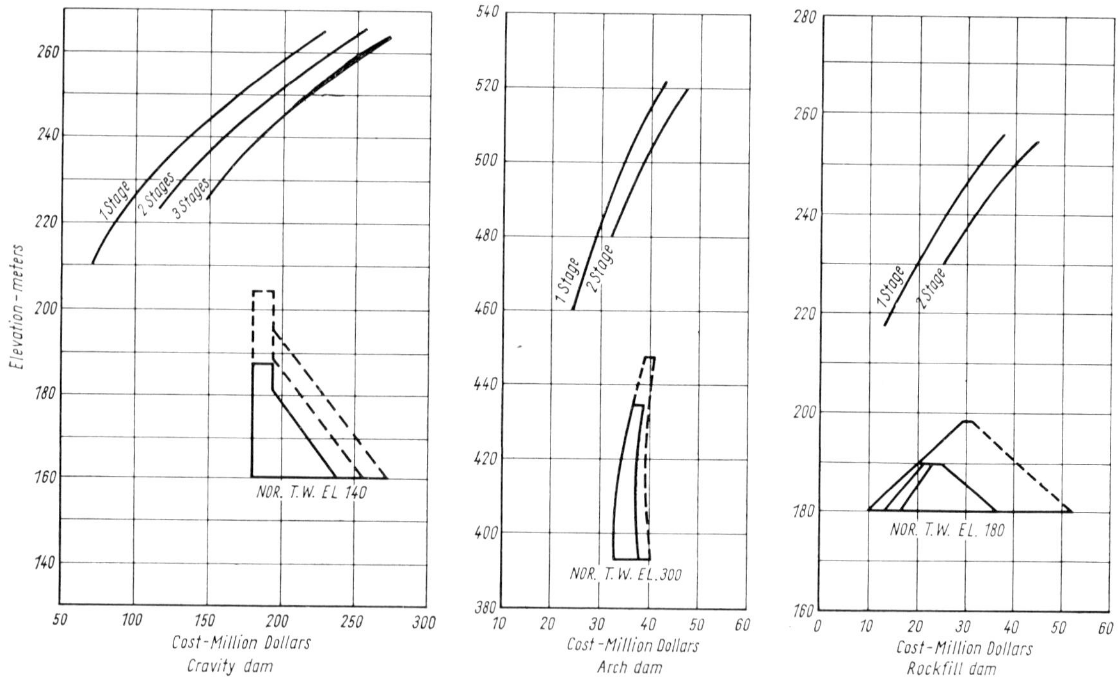

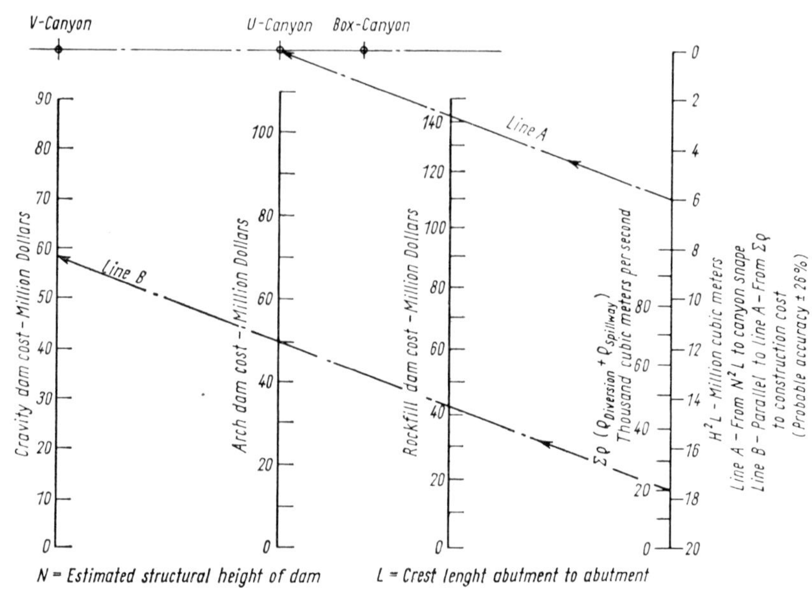

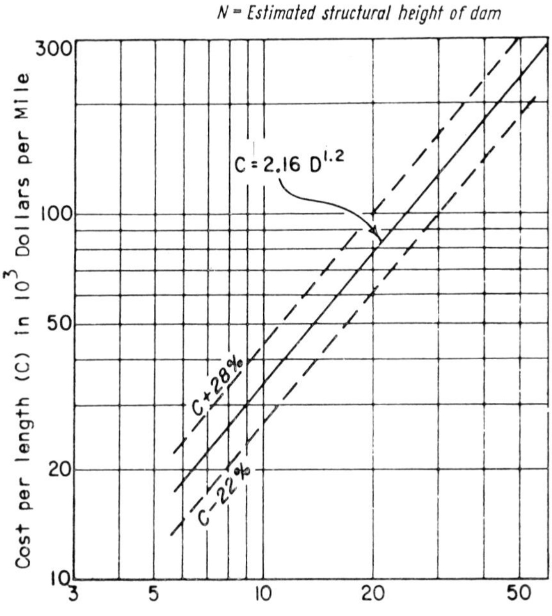

A

ACID MINE DRAINAGE: see POLLUTION FROM MINE DRAINAGE

ACOUSTICS OF THE ENVIRONMENT: see NOISE

AEROSOLS: see PARTICULATE EMISSIONS; PARTICULATE REMOVAL

AGRICULTURAL CHEMICALS: see PESTICIDES

AIR POLLUTANT REMOVAL: see GASEOUS POLLUTANT CONTROL; PARTICULATE REMOVAL; AUTOMOTIVE POLLUTION

AIR POLLUTION INSTRUMENTATION: see STACK SAMPLING; REMOTE DETECTION OF AIR POLLUTANTS

AIR POLLUTION LAW: see ENVIRONMENTAL LAW

AIR POLLUTION METEOROLOGY

I. Effects of Weather on Air Pollution

a) Introduction

As the world's population and industrialization grow, air pollution (Fig. 1) becomes a progressively more serious problem. The control of air pollution requires the involvement of scientists from many disciplines: physics, chemistry and mechanical engineering, meteorology, economics, and politics. The amount of control necessary depends on the results of medical and biological studies.

The state of the atmosphere affects, first, many types of sources of pollution. For example, on a cold day, more fuel is used for space heating. Also, solar radiation, which is affected by cloudiness, has an influence on smog production. Second, atmospheric conditions determine the behavior of pollutants after they leave the source or sources until they reach receptors, such as people, animals, or plants. The question to be answered is: given the meteorological conditions, and the characteristics of the source or sources, what will be the concentration of the pollutants at any distance from the sources? The inverse question also is important for some applications: given a region of polluted air, where does the pollution originate?

Finally, the effect of the pollution on the receptor may depend on atmospheric conditions. For example, on a humid day, sulfur dioxide is more corrosive than on a dry day.

Meteorological information is needed in three general areas of air pollution control:

1. In planning control measures, wind climatology is required. Pollution usually must be reduced to a

point where the air quality is substantially better than the existing quality. In order to assure improved quality, certain standards are set which prescribe maximum concentrations of certain pollutants.

In order to reach such standards, the points of origin of the pollution must first be located; traditionally, everybody blames everybody else for the unsatisfactory air quality. Given possible pollution sources, tracing of air trajectories coupled with estimates of atmospheric dispersion will give the required answers. Once the relative importance of different pollution sources is known, strategies have to be developed to determine the degree to which each source must reduce its effluent.

The most economical way to cut concentration of some pollutant may not be to cut the effluent of each emitter by the same amount. In order to find the best strategy, city models must be constructed, separately for each pollutant and for different meteorological conditions, which show how the air pollution climate of an urban region is affected by the existing distribution of sources, and what change would be produced when certain sources are controlled. The construction of such models will be discussed later, and requires a fairly sophisticated handling of meteorological data. The same models then also help in planning future growth of housing and industry.

Of course, not all problems of air pollution meteorology are as complex as those involving urban areas. The planning of individual plants, for example, must be based in part on the air pollution to be expected from the plant under various atmospheric conditions; meteorological calculations may show whether expensive techniques for cleaning the effluent before leaving the stack may be required.

2. Meteorological forecasts can be used to vary the effluent from day to day, or even within a 24-hour periods. This is because at different times the atmosphere is able to disperse contaminants much better than at other times; purer fuels must be used, and operation of certain industries must be stopped completely in certain areas when the mixing ability of the atmosphere is particularly bad.

Figure 1 Air pollution in New York City.

3. Meteorological factors have to be taken into account when evaluating air pollution control measures. For example, the air quality in a region may improve over a number of years—not as a result of abatement measures, but because of gradual changes in the weather characteristics. If the effects of the meteorological changes are not evaluated, efforts at abatement will be relaxed, with the result of unsupportable conditions when the weather patterns change again.

b) Effects between source and receptor

The way in which the atmospheric characteristics affect the concentration of air pollutants after they leave the source can be divided conveniently into three parts:

1. The effect on the "effective" emission height.
2. The effect on transport of the pollutants.
3. The effect on the dispersion of the pollutants.

c) Rise of effluent

To begin with the problem of effluent rise, inversion layers limit the height and cause the effluent to spread out horizontally; in unstable air, the effluent theoretically keeps on rising indefinitely—in practice, until a stable layer is reached. Also, wind reduces smoke rise.

There exist at least 20 formulae which relate the rise to the meteorological and nonmeteorological variables. Most are determined by fitting equations to smoke rise measurements. Because many such formulae are based only on limited ranges of the variables, they are not generally valid. Also, most of the formulae contain dimensional constants suggesting that not all relevant variables have been included properly.

For a concise summary of the most commonly used equations, the reader is referred to a paper by Briggs (1969). In this summary, Briggs also describes a series of smoke rise formulae based on dimensional analysis. These have the advantage of a more physical foundation than the purely empirical formulae, and appear to fit a wide range of observed smoke plumes. For example, in neutrally stable air, the theory predicts that the rise should be proportional to horizontal distance to the 2/3 power which is in good agreement with observations. Although dimensionally correct formulae are not yet in general use, they appear to be the most promising in the long run.

Given the height of effluent rise above a stack, an "effective" source is assumed for calculation of transport and dispersion. This effective source is taken to be slightly upwind of a point straight above the stack, by an amount of the excess rise calculated. If the efflux velocity is small, the excess rise may actually be negative at certain wind velocities (downwash).

d) Transport of pollutants

Pollutants travel with the wind. Hourly wind observations at the ground are available at many places, particularly airports. Unfortunately, such weather stations are normally several hundred kilometers apart, and good wind data are lacking in between. Further, wind information above 10 meters height is even less plentiful, and pollutants travel with winds at higher levels.

Because only the large-scale features of the wind patterns are known, air pollution meteorologists have spent considerable effort in studying the wind patterns between weather stations. The branch of meteorology dealing with this scale—the scale of several km to 100 km—is known as mesometeorology. The wind patterns on this scale can be quite complex, and are strongly influenced by surface characteristics. Thus, for instance, hills, mountains, lakes, large rivers, and cities cause characteristic wind patterns, both in the vertical and horizontal. Many vary in time, for example, from day to night. One of the important problems for the air pollution meteorologist is to infer the local wind pattern on the mesoscale from ordinary airport observations. Such influences are aided by theories of sea breezes, mountain-valley flow, etc.

In many areas, local wind studies have been made. A particularly useful tool is the tetroon, a tetrahedral balloon which drifts horizontally and is followed by radar. In some important cities such as New York and Chicago, the local wind features are well-known. In general, however, the wind patterns on the mesoscale are understood qualitatively, but not completely quantitatively. Much needs to be done in this area.

e) Atmospheric dispersion

Dispersion of a contaminant in the atmosphere essentially depends on two factors: on the mean wind speed, and on the characteristics of atmospheric "turbulence". To see the effect of wind speed, consider a stack which emits one puff per second. If the wind speed is 10 m/sec, the puffs will be 10 m apart; if it is 5 m/sec, the distance is 5 m. Hence, the greater the wind speed, the smaller the concentration.

Atmospheric "turbulence" consists of horizontal and vertical eddies which are able to mix the contaminated air with clean air surrounding it; hence, turbulence decreases the concentration of contaminants in the plume, and increases the concentration outside. The stronger the turbulence, the more the pollutants are dispersed.

There are two mechanisms by which "eddies" are formed in the atmosphere: heating from below and wind shear. Heating produces convection. Convection occurs whenever the temperature decreases rapidly with height—that is, whenever the lapse rate exceeds 1°C/100 m. It often penetrates into regions where the lapse rate is less. In general, convection occurs from the ground up to several hundred meters elevation on clear days and in cumulus-type clouds.

The other type of turbulence, mechanical turbulence, occurs when the wind changes with height. Because there is no wind at ground level, and there usually is some wind above the ground, mechanical turbulence just above the ground is common. This type of turbulence increases with increasing wind speed (at a given height) and is greater over rough terrain than over smooth terrain. The terrain roughness is usually characterized by a "roughness length" z_0 which varies from about 0.1 cm over smooth sand to a few meters over cities. This quantity does not measure the actual height of the roughness elements; rather it is proportional to the size of the eddies that can exist among the roughness elements. Thus, if the roughness elements are close together, z_0 is relatively small.

TABLE 1

Turbulence Characteristics with Various Richardson Numbers

$0.25 < Ri$	No vertical mixing
$0 < Ri < 0.25$	Mechanical turbulence, weakened by stratification
$Ri = 0$	Mechanical turbulence only
$-0.03 \leqq Ri < 0$	Mechanical turbulence and convection, but mixing mostly due to the former
$Ri < -0.04$	Convective mixing dominates mechanical mixing

The relative importance of heat convection and mechanical turbulence is often characterized by the Richardson number, Ri, Actually, $-Ri$ is a measure of the relative rate of production of convective and mechanical energy. For example, negative Richardson numbers of large magnitude indicate that convection predominates; in this situation, the wind are weak, and there is strong vertical motion. Smoke leaving a source spreads rapidly, both vertically and laterally (Fig. 2). As the mechanical turbulence increases, the Richardson number approaches zero, and the angular dispersion decreases. Finally, as the Richardson number becomes positive, the stratification becomes stable and damps the mechanical turbulence. For Richardson numbers above 0.25 (strong inversions, weak winds), vertical mixing effectively disappears, and only weak horizontal eddies remain.

Because the Richardson number plays such an important role in the theory of atmospheric turbulence and dispersion, Table 1 gives a qualitative summary of the implication of Richardson numbers of various magnitudes.

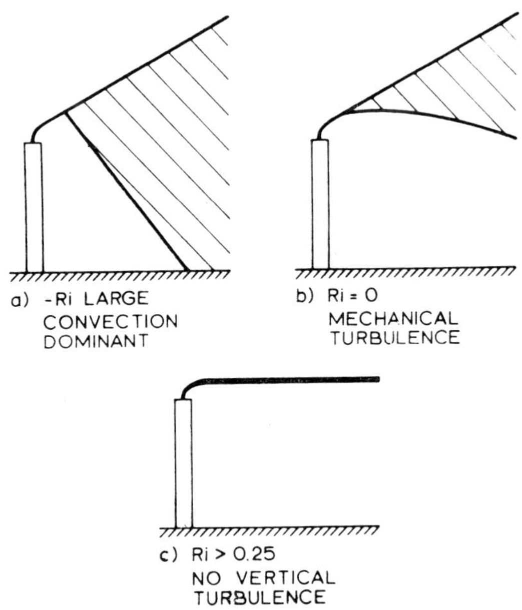

Figure 2 Average vertical spread of effluent from an elevated source under different meteorological conditions (schematic).

It has been possible to describe the effect of roughness length, wind speed, and Richardson numbers on many of the statistical characteristics of the eddies near the ground quantitatively. In particular, the standard deviation of the vertical wind direction is given by an equation of the form:

$$\sigma_\theta = \frac{f(Ri)}{ln\, z/z_0 - \psi(Ri)} \tag{1}$$

Here z is height and $f(Ri)$ and $\psi(Ri)$ are known functions of the Richardson number which increase as the Richardson number decreases. The The standard deviation of vertical wind direction plays an important role in air pollution, because it determines the initial angular apread of a plume in the vertical. If it is large, the pollution spreads rapidly in the vertical. It turns out that under such conditions, the contaminant also spreads rapidly sideways, so that the central concentration decreases rapidly downstream. If σ_θ is small, there is negligible vertical spreading.

Equation (1) states that the standard deviation of vertical wind direction does not explicitly depend on the wind speed, but at a given height, depends only on terrain roughness and Richardson number. Over rough terrain, vertical spreading is faster than over smooth terrain. The variation with Richardson number given in equation (1) gives the variation of spreading with the type of turbulence as indicated in Table 1: greatest vertical spreading with negative *Ri* with large numerical values, less spreading in mechanical turbulence ($Ri = 0$), and negligible spreading in stable temperature stratification with little wind change in the vertical.

An equation similar to equation (1) governs the standard deviation of horizontal wind direction. Generally, this is somewhat larger than σ_θ.

In summary, then, dispersion of a plume from a continuous elevated source in all directions increases with increasing roughness, and with increasing convection relative to mechanical turbulence. It would then be particularly strong on a clear day, with a large lapse rate and a weak wind, particularly weak in an inversion, and intermediate in mechanical turbulence (strong wind).

f) Estimating concentration of contaminants

Given a source of a contaminant and meteorological conditions, what is the concentration some distance away? Originally, this problem was attacked generally by attempting to solve the diffusion equation:

$$\frac{d\chi}{dt} = \frac{\partial}{\partial x} K_x \frac{\partial \chi}{\partial x} + \frac{\partial}{\partial y} K_y \frac{\partial \chi}{\partial y} + \frac{\partial}{\partial z} K_z \frac{\partial \chi}{\partial z} \quad (2)$$

Here, χ is the concentration per unit volume; x, y and z are Cartesian coordinates, and the K's are diffusion coefficients, not necessarily equal to each other.

If molecular motions produced the dispersion, the K's would be essentially constant. In the atmosphere, where the mixing is produced by eddies (molecular mixing is small enough to be neglected), the K's vary in many ways. The diffusion coefficients essentially measure the product of eddy size and eddy velocity. Eddy size increases with height; so does K. Eddy velocity varies with lapse rate, roughness length, and wind speed; so does K. Worst of all, the eddies relevant to dispersion probably vary with plume width, and depth, and therefore with distance from the source. Due to these complications, solutions of equations (2) have not been very successful with atmospheric problems except in some special cases such as continuous line sources at the ground at right angles to the wind.

The more successful methods have been largely empirical: one assumes that the character of the geometrical distribution of the effluent is known, and postulates that effluent is conserved during the diffusion process (this can be modified if there is decay or fall-out).

The usual assumption is that the distribution of effluent from a continuous source has a normal (Gaussian) distribution relative to the center line both in the vertical direction, z (measured from the ground) and the direction perpendicular to the wind, y. The rationalization for this assumption is that the distributions of observed contaminants are also nearly normal. Subject to the condition of continuity, the concentration is given by (including reflection at the ground)

$$\chi = \frac{Q}{2\pi V \sigma_y \sigma_z}\left(\exp - \frac{y^2}{2\sigma y^2}\right) \times \left(\exp - \frac{(z-H)^2}{2\sigma_z^2} + \exp \frac{-(z+H)^2}{2\sigma_z^2}\right) \quad (3)$$

Here, H is the "effective" height of the source, given by stack height plus additional rise, σ is the standard deviation of the distribution of concentration in the y and z-direction, respectively, and V is the wind speed, assumed constant. Q is the amount of contaminant emitted per unit time.

The various techniques currently in use differ in the way σ_y and σ_z are determined. Clearly, these quantities change with downwind distance x (Fig. 3) as well as with roughness and Richardson number.

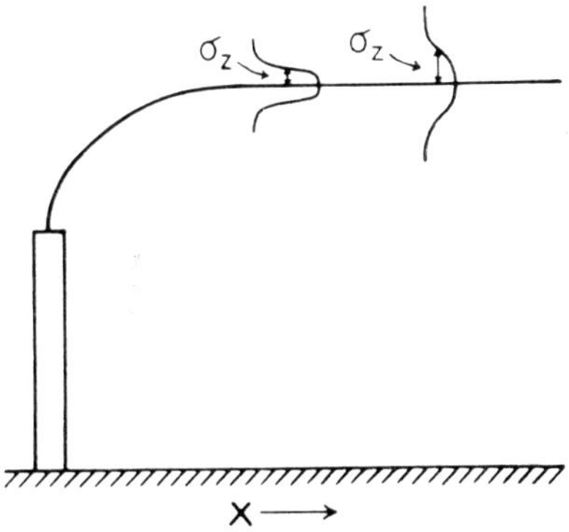

Figure 3 Change of vertical effluent distribution downstream.

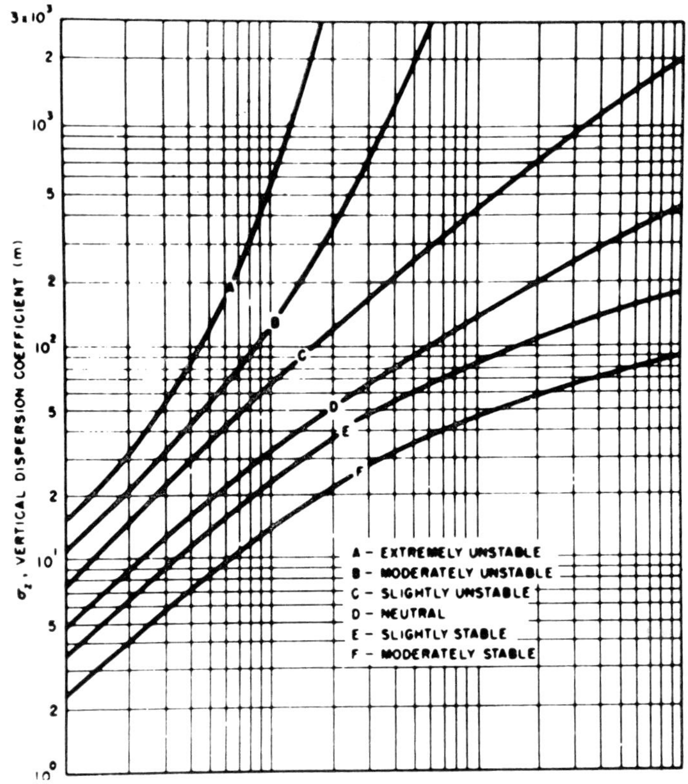

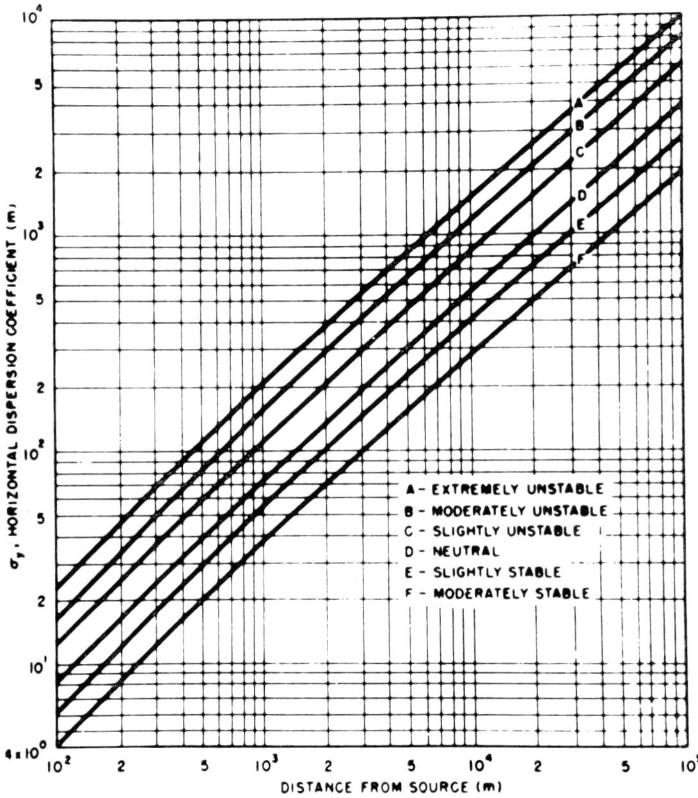

Figure 4 Pasquill-Gifford scheme for estimating vertical and lateral plume width as function of downwind distance and meteorological conditions.

Quantitative estimation of the Richardson number requires quite sophisticated instrumentation; approximately, the Richardson number can be estimated by the wind speed, the time of the day and year, and the cloudiness. Thus, for example, on a clear night with little wind, the Richardson number would be large and positive, and σ's in equation (3) are small; on the other hand, with strong winds, the Richardson numbers are near zero, and the dispersion rate as indicated by the σ would be intermediate.

For many years, standard deviations were obtained by Sutton's technique, which is based on a very arbitrary selection for the mathematical form of Lagrangian correlation functions. More popular at present is the Pasquill-Gifford method in which σ_y and σ_z as function of x are determined by empirical graphs (Fig. 4). Note that the dependence of the standard deviations on x varies with the "stability category" (from A to F). These categories are essentially Richardson number categories, judged more or less subjectively. Thus, A (large dispersion) means little wind and strong convection; D is used in strong winds, hence strong mechanical turbulence and less dispersion; F applies at night in weak winds.

The main drawback of the Pasquill-Gifford method is that it does not allow for the effect of terrain roughness; the empirical curves were actually based on experiments over smooth terrain, and therefore underestimate the dispersion over cities and other rough regions. Some users of the method suggest allowing for this by using a different system of categories over rough terrain than originally recommended.

This difficulty can be avoided if measurements of horizontal and vertical wind direction variations are available; in general, the standard deviations of wind direction are proportional to the angular dispersion of contaminants, as already mentioned, no matter what the meteorological and surface conditions. For example, Smith (1968) suggests for elevated sources equations of the forms in stable air:

$$\sigma_y = 0.15\sigma_\alpha x^{0.71}$$

$$\sigma_z = 0.15\sigma_\theta x^{0.71}$$

and in unstable air: (4)

$$\sigma_y = 0.045\sigma_\alpha x^{0.86}$$

$$\sigma_z = 0.045\sigma_\theta x^{0.86}$$

Here, σ_α and σ_θ are the standard deviations of vertical and horizontal wind directions in degrees, and lengths are in meters.

Although these equations are not in general use, this approach seems quite promising except for some controversies about the exact numbers to be used in equations (4).

As the plume expands vertically, the vertical distribution cannot remain normal indefinitely. At the bottom, the plume is limited by the ground. At the top, the plume will be limited by an elevated inversion layer. Eventually, the vertical distribution becomes uniform. In that case, the concentration is given by the equation:

$$\chi = \frac{Q}{\sqrt{2\pi}\, VD\sigma_y} \exp - \frac{y^2}{2\sigma_y^2} \tag{5}$$

where D is the height of the inversion layer, which is also the thickness of the "mixed layer." Note that the concentration is inversely proportional to VD, the "ventilation factor," which is the product of D, and V, the average wind in the mixed layer.

The lateral spread is often limited by topography. In a valley of width W, the factor $(\exp - y^2/2\sigma_y^2)/(\sqrt{2\pi}\,\sigma_y)$ in equations (3) and (5) is replaced by $1/W$, after the contaminant concentration fills the valley uniformly in the y-direction (the direction perpendicular to the wind, and also usually perpendicular to the valley). The effect of this change is that relatively large concentrations are maintained at large distances from the sources.

g) City models

The equations above give the pollutant concentrations downwind from a single source. In order to obtain the total picture of air pollution from a city, the concentrations resulting from all sources must be added together, separately for all different wind directions, different meteorological conditions, and for each contaminant. Such a procedure is expensive, even if carried out with an electronic computer, and even if, as is usually done, all small sources in square-mile areas are combined. Therefore, complete city models of air pollutant concentrations have only been constructed for very few locations. It is necessary, however, to have city models in order to understand the distribution of contaminants; only then is it possible to determine the most economical strategy to reduce the pollution, and to evaluate the effects of expansion of housing and industry.

Because the construction of a complete city model is so expensive, city models are often simplified. For example, if the city is represented by a series of parallel line sources, the computations are greatly reduced. Many other simplifications have been introduced; for a summary of many city models now in existence, see Stern (1968)

h) Diurnal variation of air pollution

Equation (5), which shows that concentrations at considerable distances from individual sources are inversely proportional to the ventilation factor (*VD*), can be used to explain some of the variations in air pollution caused by meteorological factors. First, we shall consider the diurnal variation of air pollution. Of course, the actual variation of pollution may be different if the source strength varies systematically with time of day. The diurnal variation is different in cities and in the country. Consider typical vertical temperature distributions as seen in Figure 5. During the day, both over cities and country, the ground temperature is high, giving a deep mixed layer. After sunset, the air temperature near the surface in the country falls, producing an inversion reaching down to the ground. After air moves from the country out over the relatively warmer and rougher city, a thin mixed layer is formed near the ground. The thickness of this mixed layer varies with the size of the city, and depends on how long the air has moved over the city. In New York, for example, the mixed layer is typically 300 m thick; in Johnstown, Pa., an industrial valley city with just under 100,000 population, it is only a little over 100 m.

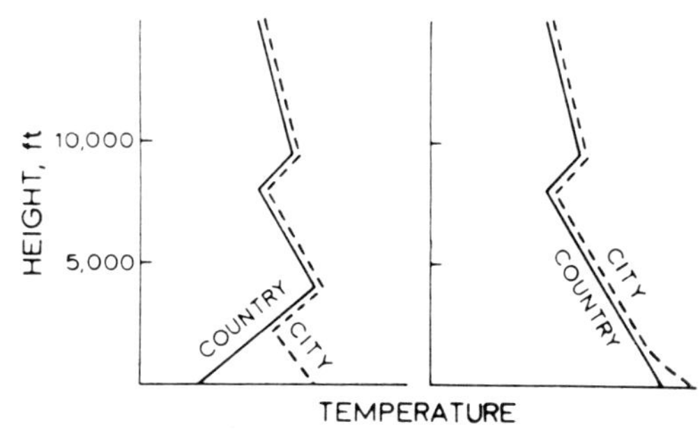

Figure 5 Vertical temperature distribution (schematic) over city and country, day and night.

Figure 6 indicates how the temperature changes shown in Figure 5 influence the diurnal variation of pollution due to an elevated source in the country; at night, vertical mixing is negligible and the air near the ground is clean. Some time shortly after sunrise, the mixed layer extends to just above the source, and the elevated polluted layer is mixed with the ground air, leading to strong pollution (also

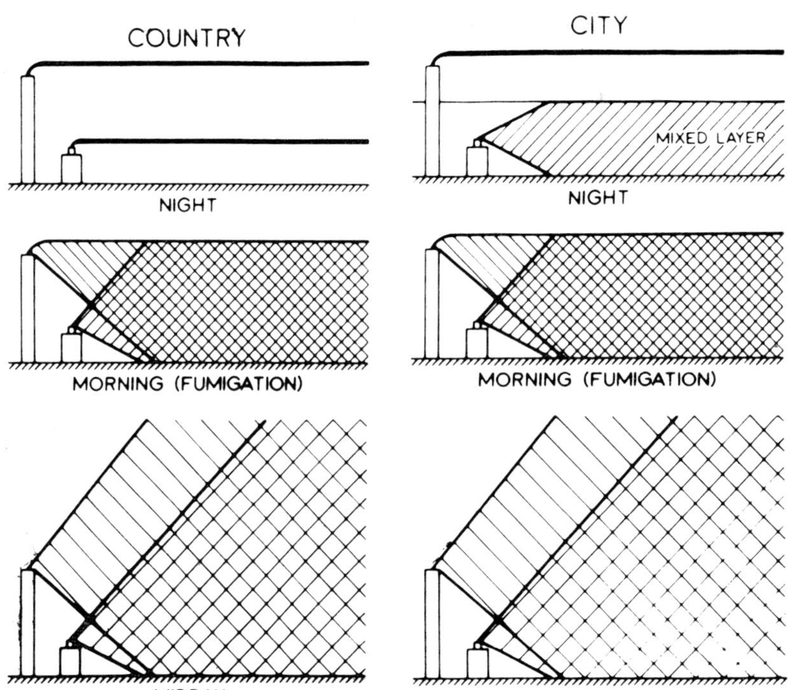

Figure 6 Concentrations of effluent (schematic) as function of time of day, over city and country.

referred to as "fumigation"), which may extend many kilometers away from the source. Later in the morning and early afternoon, the heating continues and thickens the mixed layer. Also, the wind speed typically increases, and the pollution decreases.

In the city, many sources usually exist in the thin night-time mixed layer. Since this layer is so thin, and the wind usually weak, large pollution occurs at night. Right after sunrise, the pollution at first increases somewhat, as the effluent from large, elevated sources is brought to the ground. As the mixed layer grows, the concentrations diminish, and, in the early afternoon, they are often less than the night-time concentrations (see Fig. 7).

Thus, the main difference between air pollution climates in the city and country is that country air near industrial sources is usually clean at night, whereas the city air is dirtier at night than in the middle of the day. These differences are most pronounced during clear nights and days, and can be obliterated by diurnal variations of source strengths. Figure 7 shows the characteristic behavior only because the sources of pollution at Johnstown, Pa., are fairly constant throughout.

i) Day-to-day variations in air pollution

Equation (5) shows that, other things being equal, the concentration of contaminants is inversely proportional to the wind speed. Figure 8 shows this effect on 24-hr. total particulate concentration at Johnstown, for cases where the source strengths were roughly the same, during the fall of 1964.

Conditions of particularly bad air pollution over wide areas and for extended periods are accompanied not only by light winds and calms, but also by unusually small mixing depths (D) so that the ventilation factor is usually small. Such conditions occur within large high-pressure areas (anticyclones). In such areas, air is sinking. Sinking air is warmed by compression. Thus, in an anticyclone

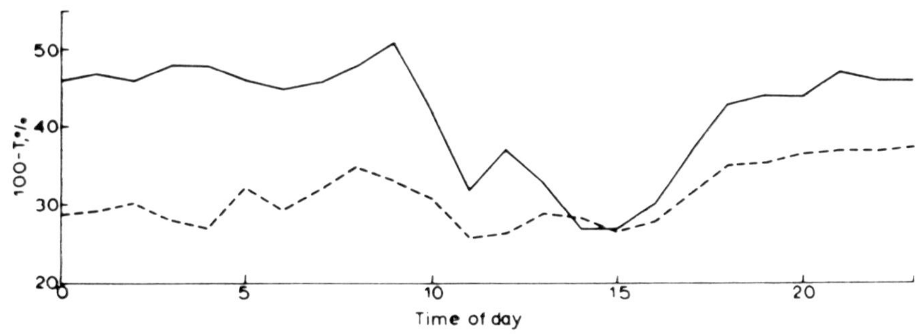

Figure 7 Concentrations of air pollution (100-T%), as function of time of day, on clear day (solid line) and cloudy day (dashed line), at Johnstown, Pa.

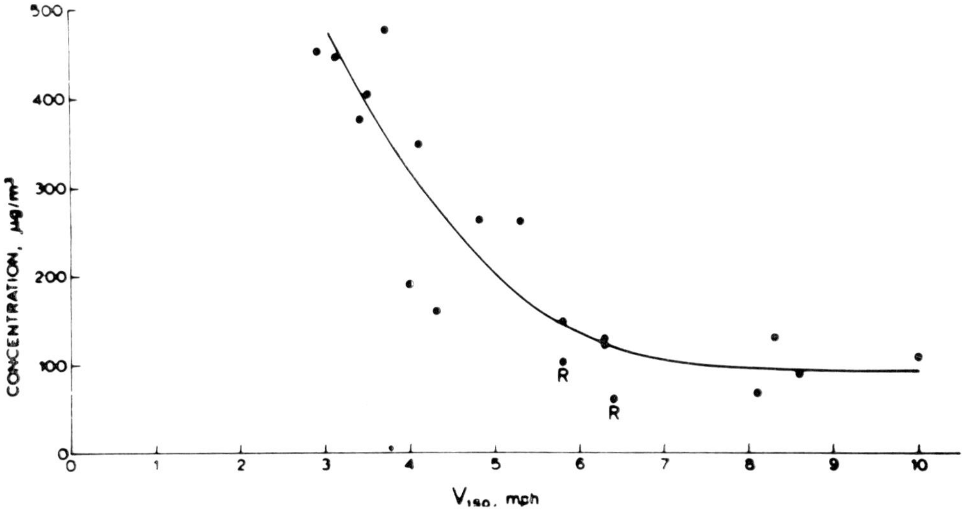

Figure 8 Dependence of 24-hour average particle concentrations at Johnstown on wind speed at 150 ft R denotes rain.

(high-pressure area), an elevated warm layer forms, below which there is room only for a relatively thin mixed layer (Fig. 9). The inversion on top of the mixed layer prevents upward spreading of the pollution, and when mountains or hills prevent sideways spreading the worst possible conditions prevail. A particularly bad situation arose in the industrial valley town of Donora, Pa., in which many people were killed by air pollution in 1948.

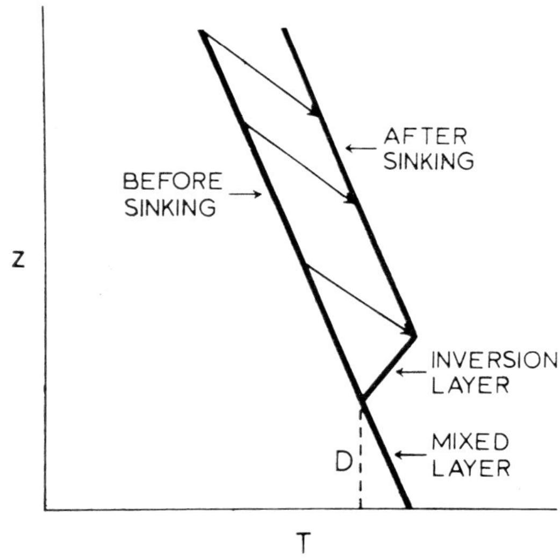

Figure 9 Effect of sinking on vertical temperature distribution (schematic).

Cities in California, like Los Angeles, are under the influence of a largescale anticyclone throughout the summer, and an elevated inversion at a few hundred meters height occurs there every day; that is why Los Angeles had air pollution problems as soon as pollutants were put into the atmosphere to any large extent. In the United States outside the West Coast, stagnant anticyclones occur only a few times per year, usually in the fall.

In order to warn of generally bad air pollution conditions, the Weather Service computes daily mixing depth D and V, the average wind speed within it, at all stations, where upper-air observations are made. These quantities are transmitted along with the usual weather data over the weather teletype network. If the ventilation factor VD remains low for over 24 hours (and when some other conditions are satisfied), an air pollution alert is declared, at which time the air pollution potential is expected to be high. On the basis of ventilation factors and air pollution alerts, industries are asked to cut their sources of contaminants.

So far, relatively little use has been made in the USA of forecast changes in air pollution potential from day to day. As air pollution problems become more severe, more use will be made of such forecasts. Already, this type of information has proved itself in air pollution management in some European countries.

Not much has been said about the influence of wind direction on air pollution. When pollution is mainly due to many, relatively small sources, as it is in New York, the pollution is surprisingly insensitive to changes in wind direction. Even in Johnstown, Pa., wind direction is unimportant except for the case of easterly winds, when a single, huge steel plant adds significantly to the contaminant concentration.

In contrast, wind direction plays a major role when most of the pollution in a given area is due to a single or a few major plants or if an industrial city is nearby. Also, there are special situations, in which wind direction is particularly important; for example, in Chicago, which has no pollution sources east of the city, east winds bring clean air.

The main difference between the effects of lapse rate, mixing depth, and wind speed on the one hand, and wind direction on the other, is that the wind direction has different effects at various sites, depending on the location of the sources; the other factors have similar effects generally.

II. Effect of Air Pollution on Local and Regional Weather

a) Visibility

The most obvious effect of air pollution is to reduce visibility. This effect has been studied frequently by comparing visibility in different parts of a city, or the visibility in a city with visivility in the country. For a summary of many such investigations, see Peterson, 1969.

To give some examples: Around London and Manchester, afternoon visibility less than $6\frac{1}{4}$ miles occur on more than 200 days; in Cornwall in SW England, the number is less than 100. In central London, there are 940 hours a year with visibilities less than $\frac{5}{8}$ mile; in SE England, only 494.

In many cities, visibilities have recently improved probably due to control of particle emissions; however, as mentioned before, some of this change may be due to changes in large-scale weather patterns.

Although decreased visibility is usually associated with industrial or automobile pollution, considerable attention has been paid recently to decreased visibilities due to the "contamination" of the atmosphere by water droplets by industry. This problem arises because many processes generate excess heat; if this is added to streams and lakes, undesirable effects ensue; hence, progressively more and more heat is used to evaporate water which is then emitted into the atmosphere, and later condenses to form water plumes.

There are many unpublished studies estimating the effect of cooling towers on visibility. This varies greatly with meteorological conditions, but is particularly serious in winter, when the air is nearly saturated and little additional vapor is required to produce liquid drops. Under those conditions, water plumes from industries produce clouds and fog which may reach over a hundred miles from the sources. Automobile accidents have been blamed on such fogs, particularly when the particles freeze and make roads slippery, adding to the visibility hazard.

b) Sunshine intensity

"Turbidity" is an indicator of the reduction of light due haze, smoke and other particles. Turbidity is now being monitored at many places in the world. It is quite clear that it is larger over cities than over the country; it has been suggested that the average decrease of sunshine over cities is 15 to 20% due pollution. The effect is even larger if only ultraviolet light is considered.

Control of smoke emission in cities such as London has caused a very noticeable increase of sunshine intensity: for example the hours of "Bright sunshine" increased by 50% after control measures had become effective. Again, for a summary of some of these studies, the reader is referred to Peterson, 1969.

c) Precipitation amount

There have now been several studies suggesting precipitation is increased downstream of industrial centers. The investigations are of a statistical nature, and it is not known whether the effects are due to increased convection (increased heat), increased condensation nuclei of increased water, vapor. Further, the reliability of the statistics has been questioned.

For example, Changnon (1969) found a large precipitation anomaly at La Porte (Indiana) just downwind of large industrial complexes of Northwestern Indiana. But change in observational techniques of rainfall and other uncertainties have thrown doubt on the results. Hobbs *et al.* (1970) have compared rainfall distribution in Western Washington before and after the construction of industries and found an increase by 30% or so; but some of this increase may have been due to "normal" climatic change. For a summary of these and other studies see Robinson (1970). It becomes quite clear from this summary that more, careful investigations of this type are needed before effects of air pollution on precipitation patterns can be definitely proven.

d) Acid rain

Most fuels contain sulphur, and therefore sulphur dioxide is a common pollutant. When rain falls through sulphur dioxide, acid can be produced which may be harmful to vegetation. Thus, recently, rain in Southern Sweden has been found to be quite acidic; but, in this case, the origin of the acid is not known. Since the prevailing wind is from the SW, the original hypothesis was that the acid originated from sulphur dioxide emitted by industrial complexes in Western Germany, but this question has by no means been settled.

The situation is more clear in the case of mine-mouth power plants in northwestern Pennsylvania, where sulphur-rich coal is burned and sulphur dioxide is emitted from 800′–1,000′ high stacks. Here, rain samples collected a few miles downwind from the stacks frequently have pH values below 4. Thus it is clear that acid rain can certainly be a local problem; but it is uncertain whether the problem may be regional as well.

III. Global Effects of Air Pollution

a) Natural climatic changes

The climate of the earth has been changing, over centuries, over thousands and millions of years. The reasons for these changes are unknown. Many theories have been constructed to explain these changes: variations in the radiation from the sun; changes in the earth's orbit; mountain building; volcanic dust; changes in the composition of the atmosphere, and many more. Each theory has its proponents, but none is generally accepted.

In this century, the temperature first rose, at an average rate of about 1°C per century. Since 1945–1950, the temperature has decreased, at a slightly slower rate. These rates of change are by no means unusual. Therefore, although industry and air pollution have increased at a growing rate in this century, there is no necessity to explain the change of climate since 1900 by air pollution. In fact, there is no way to determine whether any of the changes in this century were significantly influenced by air pollution. But, as the pollution increases, a time may come when climatic changes produced by it may become large enough to exceed "natural" changes, and produce cataclysmic events, such as freezing of the oceans, or, conversely, melting of all the ice with accompanying widespread flooding.

We will now assess the effect of some atmospheric pollutants as to their ability to change the earth's climate. In doing so, we are hampered by the fact that the present climate is produced by a multitude of interacting factors; if one factor is changed, others will too, and a complex chain reaction will ensue. These reactions can be studied by complex mathematical models of the atmosphere, which so far have been quite successful in describing the existing climate. But, as yet these models contain assumptions which make it impossible at this time to assess accurate the effects of changes in some of the factors affecting climate. Until such models are improved, then, we cannot really estimate quantitatively climatic changes produced by pollutants.

b) Carbon dioxide

The concentration of CO_2 is about 330 parts per million (ppm). According to observations at Mauna Loa in Hawaii, over the last ten years or so, it has increased at the rate of 0.7% per year. This is less than half the amount put into the atmosphere by industry. The other half goes into the ocean or into vegetation; but it is not known how much goes into each. Further, we do not know whether the same fraction can disappear out of the atmosphere in the future—e.g., the amount going into the ocean is sensitive to temperature, and the amount going into vegetation may be limited by other factors. However, a reasonable guess is that the fraction of CO_2 in the atmosphere in the year 2,000 will be about 380 ppm.

The basic effect of CO_2 on climate is due to the fact that it transmits short-wave radiation from the sun, but stops a part of the infrared radiation emitted by the earth. Hence, the more CO_2, the greater the surface temperature. This is known as the greenhouse effect. Also, since CO_2 increases the radiation into space, the high atmosphere is cooled by increasing CO_2.

The heating rate at the ground expected with a doubling of CO_2 has been calculated by many radiation specialists. The answers differ, depending on how many other variables (such as water vapor concentration) are allowed to change as the CO_2 changes. The best current estimates are that doubling CO_2 would increase the surface temperature about 2°C, and decrease the temperature aloft a little more. But these estimates omit the effect of most atmospheric motions and other factors such as changes of ice cover, which affects the fraction of solar radiation retained by the earth—atmosphere system. Still, if we expect only a 20% change in CO_2 by the end of the century, the climatic change due to this factor should be small. However, a serious problem could arise in the next century, particularly because it is difficult to see how a trend in CO_2 concentration can be reversed. It is therefore of great importance to continue monitoring CO_2 concentration accurately.

c) Particles

Particles come from many sources: deserts, agriculture, ocean spray, volcanoes, industry, and from chemical reactions between gases. It is estimated that at present 1 in five atmospheric particles is manmade; but that by the year 2,000, 50% will be manmade. Certainly, a gradual increase of particles in the atmosphere is well documented by a variety of measurements.

The interaction between particles and radiation is complex and differs between particles. Sunlight is reflected, and a portion is absorbed; and infrared radiation is absorbed from the ground and atmosphere below, and emitted again. These processes not only depend on the nature of the particle, but also on the radiative properties of the air and ground; hence, particles can produce either warming or cooling. There is some evidence that particles emitted from volcanoes have led to some surface cooling; and there is good evidence that the explosion of the volcano Agung in Bali in 1963 caused a 5°C heating in the equatorial stratosphere.

In view of the uncertainties of the radiative effects of particles, highest priorities are now being assigned to studies of such properties for different types of particles. In addition, particles can affect production of clouds and precipitation. But, so far, few quantitative statements can be made about the direction or magnitudes of climatic effects to atmosphere contamination by particles.

Of special interest is the possible effect of depositing particles in the stratosphere (above about 10 km height). Particles placed in the stratosphere are likely to stay there for a year or more, because vertical mixing is weak there. Further, the density in the stratosphere is lower than further down. Now, if large numbers of supersonic transports (SST's) are built, various types of particles will enter the stratosphere: such as sulphates produced from the sulphur dioxide in the exhaust, and "smog" formed by the interaction of hydrocarbons with the ozone in the stratosphere. Some calculations have suggested that 500 SST's, flying 7 hours a day, will produce about as many particles in the stratosphere as Agung did. Since these particles will enter the stratosphere in middle and high latitudes, there is no exact analogy to the Agung event; still considerable stratospheric changes are possible with unknown consequences at lower levels.

d) Water vapor and clouds

Water vapor in the atmosphere can also produce a greenhouse effect, in fact more so than CO_2. But water vapor is already much more plentiful than CO_2, so that increased amounts as a result of human activity are not likely to be important. Even the addition of H_2O vapor into the normally very dry stratosphere is likely to produce very minor temperature changes directly.

However, two other effects of H_2O must be considered: it might change the ozone concentration, and it can form clouds. As to the first, recent computations have shown that ozone change likely produced by H_2O insertion is likely to be much smaller than normal day-to-day changes. But the importance of "artificial" clouds is more problematical.

There is fairly good evidence that jet aircraft have increased the cloudiness at levels around 10–11 km where they normally fly; further, the increase of these clouds seems to be proportional to the increase in jet flights. But it is not at all clear if or how the clouds can change the climate. For thin clouds do reflect a little sunlight; but they also trap infrared radiation; thus the net effect on the ground temperature may be small. But this matter has not yet had as much attention as it deserves.

Also, the SST's will fly through very cold regios of the stratosphere where even small amounts of water vapor can lead to clouds. In such regions, we already sometimes see clouds, called mother-of-pearl clouds. The frequency of such clouds will almost certainly be increased by SST's but the climatic importance of such increases is not known.

Another point about clouds is that it is known that they may become dirty due to air pollution. This will change the reflecting power of the clouds, and hence the temperature distribution in the atmosphere. Again, the magnitude of this effect is unknown.

e) Summary

In summary, increasing air pollution can modify the climate in many ways. There is no evidence that any significant change has occurred so far; but eventually, large effects are likely. But we have not yet the means to estimate the magnitude—and often not even the sign of such effects. For a more complete summary, see Robinson, 1970.

The discussion of the effects of atmospheric variables on air pollution meteorology was based largely on an earlier paper by the author on Air Pollution Meteorology (1969), brought up to date and modified primarily as a result of discussions with R. A. McCormick, director of the Meteorology Division of the National Air Pollution Control Administration.

The discussion of the effects of air pollution on weather and climate was strongly infuenced by the fact that the author was permitted to attend the MIT study on Critical Global Pollution Problems, held at Williamstown, Mass., in July, 1970. A non-technical discussion of the results of this study is published in paper-back form by the MIT press in Oct., 1970; and the technical papers leading to these results published as Clep papers in 1971. See also "Study of Man's Impact on Climate". (MIT Press, 1971).

REFERENCES

Briggs, G. A. (1969), Plume Rise. USAEC critical review series, TID-25075, Clearinghouse for federal scientific and technical information.

Changnon, S. H. (1968), The LaPorte Weather Anomaly, fact or fiction, *Bull. Amer. Met. Soc.* **49**, pp. 4–11.

Hobbs, P. V., L. F. Radke, and Shumway S. E. (1970), Cloud condensation nuclei from industrial sources and their appearent influence on precipitation in Washington State. *Jour. Atmos. Sci.* **27**, pp. 81–89.

Peterson, James T. (1969), The Climate of Cities: A Survey of Recent Literature, U.S. Dept. HEW, NAPCA, Pub. No. AP-59.

Robinson, G. D. (1970), Long-Term Effects of Air Pollution, Center for Environment and Man, Hartford, CEM 4029-400.

Smith, Maynard (1968), Predication of the Dispersion of Airborne Effluents. Amer. Soc. Mech. Engr.

Stern, A. C. (1968), *Air Pollution*, Vol. 1, Academic Press, New York.

HANS A. PANOFSKY
Pennsylvania State University

AIR POLLUTION MODELING—URBAN: see URBAN AIR POLLUTION MODELING

AIR POLLUTION SOURCES

Air Pollution may be defined as the presence in the gas, aerosol or particulate atmosphere of one or more contaminants or combinations thereof in such quantities and of such duration as may be, or may tend to be injurious to human, plant, or animal life, or property, or which unreasonably interfere with the comfortable enjoyment of life, or property, and the natural environment.*

Air pollutants may be classified according to their origin, state of matter or chemical composition.

A. Classification according to origin

Air pollutants are classified according to origin as: (1) primary and (2) secondary.

(1) Primary Pollutants—primary pollutants are those emitted to the atmosphere as a result of some process. The remain for a long time in the atmosphere in the chemical form in which they are emitted.

(2) Secondary Pollutants—Secondary pollutants are those which are formed in the atmosphere as products of some reaction. This reaction may occur between two pollutants or between a single pollutant and a natural component of the atmosphere. The reaction may be photochemical or non-photochemical, e.g. hydrolysis, catalytic oxidation.

B. Classification according to the state of matter

According to state of matter, pollutants may be classified as: (1) gaseous and (2) particulate.

(1) Gaseous Pollutants—Gases which are present in the atmosphere as contaminants behave much as the air itself. Once diffused, they do not tend to settle out.

(2) Particulate Pollutants—Particulate pollutants are any finely divided solids or liquids. The largest particulates settle out quickly, and effect quality air near the source.
Moderate size particles travel further, and eventually settle out some distance from the source. The smallest particulates behave almost like a gas; they remain suspended and are readily transported by wind currents.

C. Classification according to chemical composition

According to chemical composition pollutants may be classified as: (1) organic and (2) inorganic.

(1) Organic Pollutants—Organic compounds may be defined as those which contain carbon and hydrogen, and may contain other elements. By this definition we exclude the very simple carbon monoxide and carbon dioxide. These contain carbon, but no hydrogen.

(2) Inorganic Pollutants—Inorganic materials found in contaminated atmospheres include the very simple carbon compounds (CO and CO_2), and non-carbon-containing compounds. Many of the most commonly encountered pollutants are inorganic.

Classification of pollutants according to their physical and chemical characteristics identifies the major classes and subclasses of the gases and particulates. The more important the pollutant, the more specifically it is classified. An example of such a classification is shown in table 1.

A considerable quantity of air pollution occurs naturally as a consequence of such process as soil erosion, volcanic eruptions. However, those pollutants which pose a treat to human health and cause extensive damage to property are primarily derived from activities associated with the development of community living, as well as the growth of affluence and living standards in industrial societies. These activities have included the burning of fuel for heat and power the processing of materials for food and goods, and the disposal of wastes.

TABLE 1

Classification of Pollutants

Major classes	Sub-classes	Typical members of sub-classes
Organic Gases	Hydrocarbons	Hexane, benzene, ethylene, methane, butane, butadiene
	Aldehydes and Ketones	Formaldehyde, acetone
	Other organics	Chlorinated hydrocarbons, alcohols
Inorganic Gases	Oxides of nitrogen	Nitrogen dioxide, nitric oxide
	Oxides of sulfur	Sulfur dioxide, sulfur trioxide
	Oxides of carbon	Carbon monoxide, carbon dioxide
	Other inorganics	Hydrogen sulfide, hydrogen fluoride, ammonia, chloride
Particulate	Solid particulates	Dust, smoke, fume
	Liquid particulates	Mist, spray

Much of the materials which pollute our atmosphere represent valuable resources which are being wasted by a non-conservationist industrial system. We have available today the technological means of controlling most sources of air pollution. The cost of control however has been estimated on to order of 10 to 20% of the world's gross annual product. (a) Moreover, implementation of effective control measures would require a significant change in life style in many of our large centers of population. In the face of these costs government and industry has shown great reluctance to undertake the degree of control required to restore the quality of the air we breathe.

II Classification of Sources of Air Pollutants

Sources of air pollutants may be classified according to the number, size or density, the nature of the pollutant producing activity or on the basis of the pollutant or pollutant types or source type characteristic of the specific source.

A. Classification by size and density

For convenience of analysis an pollution sources and divided into two classes (1) specific sources and (2) multiple sources.

1) Specific Sources are those associated with large plants and whose emissions quantities are large enough to cause them to have a dominant role in the pollution potential of an area. These sources are fixed (stationary) and commonly occupy a limited area relative to a community. They include:
   (a) Major industrial and manufacturing plants
   (b) Steam—Electric power plants
   (c) Industrial and Municipal Incinerators

Industrial plants constitute a highly varied and complex chemical system, each industry presenting a unique an pollution problem. The characteristics of the emission produced are directly related to the peculiarities of the operation in question i.e., on the raw materials, the fuels, the process method, the efficiency of the closenprocess method and the type of our pollution control measures applied.

2) *Multiple sources* are those which cannot be assayed practicably on a source by source basis. They many be stationary or mobile and are commonly spread throughout the community. These sources are associated with:
   (a) Combustion of fuels in residences and commercial buildings and institutions for personal comfort and convenience.
   (b) Service industries such as laundries, dry-cleaning plants repair servoces, etc.
   (c) Residential and commercial incineration (on-site).
   (d) Evaporation of petroleum products from solvent usage (surface coating, degreasing, dry–cleaning, plastics manufacture, rubber manufacture), storage and marketing (tank farms, service stations).
   (e) Combustion of fuels for transportation in motor vehicles, aircraft, ships and railroads.
   (f) Odors Producing Activities—Odors are derived from

(1) Industrial processes such as refineries, chemical plants
(2) Animal processing
(3) Manufacturing and handling of paints, lacquers and solvents
(4) Food processing in restaurants, grills, coffee roasting, etc.
(5) Combustion of wastes in dumps, incinerators, etc.
(6) Decomposition of organic waste
(7) Sewage treatment plants or leaking conducts

B. Classification by source type

Five major types of sources have been generally identified on the basis of the pollutant—producing activity.

These are: (1) Fuel combustion for transportation (Mobile sources); (2) Fuel combustion in stationary sources; (3) Industrial Processes; (4) Solid Waste Disposal; (5) Miscellaneous sources.

1) *Fuel combustion in mobile sources occurs* in motor vehicles, aircraft, railroads and ships. Fossel fuels—oil coal and natural gas-rovide the primary source of energy. Distillate or synthetic products are used in systems requiring cleaner fuels.

Emissions are in the form of combustion gases and unburned fuels in gaseous form. Contamination of the atmosphere also occurs through losses form the storage systems by escape of gases, evaporation of liquid fuels and entrainment of fine solid particle of coal.

More than 90% of mobile source pollution is derived from motor vehicles. Some 99% of the motor vehicles in the United States are powered by the gasoline engine. These engines discharge pollutants from four points: Gasoline vapors escape from the gas tank and carburetor. Large amounts of unburned or partly burned fuel are discharged from the tailpipe. Small amounts of unburnt fuel blow by the pistons to the crankcase and are vented to the atmosphere.

2) *Fuel combustion in stationary sources* occurs in steam electric powerplants and residential commercial structures. The primary functions is to produce electricity or heat for industrial or residential purposes. Fuel oil has recently been replacing coal as the primary fuel used in these systems. Combustion products of major importance in air pollution are carbon occurs particulates or soot, flyash and oxidised forms of impurities such as sulfur which form irritant acid gases.

3) *Industrial processes* (See specific sources). These sources account for less than 15% of the estimated annual emissions of major pollutants in the United States.

4) *Solid waste disposal* relates to the combustion of residential, commercial and industrial refuse in municipal, industrial and on-site residential incinerators. In the United States more than 150 million tons of wastes including paper, grass, garbage, ashes, metal and glass must be disposed of annually. Sanitary land fill operations and open dumps are used to dispose a majority of this refuse.

5) *Miscellaneous*—This category includes forest fires, structural fires,the burning of agricultural wastes and evaporative losses from storage and marketing facilities. These sources account for more than 15% of the amount of major pollutants emitted in to the atmosphere over the United States annually.

III Classification of Air Pollution Problems

The major categories used to classify air pollution problem are:

A. Community Air Pollutants
B. Industrial Air Pollutant
C. Miscellaneous Other

Community air pollutants

The Community Air Pollutants are those derived from the everyday and commercial activities of community. These pollutants are primarily derived from fuel combustion in stationary as well as mobile sources, incineration of wastes, the use of solvents in dry cleaning and architectural coatings, construction and demolition activities.

The six major community air pollutants are carbon monoxide, oxides of sulfur, particulates, hydrocarbons, oxides of nitrogen and photochemical oxidants. Other important community air pollutants include aldehydes, ammonia and various organic carcinogens.

Industrial air pollutants

Industrial Air Pollutants relate to those emissions uniquely associated with the manufacturing or processing of materials and goods. Each industry is characterized by its particular raw materials, end products, and by products. The use of caustic materials as well as such acid gases as chlorine is common in non metal industrial processes. The processing of metals and metal products has been found to be a major source of particulate and acid gas emissions.

Miscellaneous other

This group is used to include such various air environmental problems as aeroallegens, biological aerosols, odorous compounds carbon dioxide, waste heat radioactive emissions and pesticide. In many cases they are not normally characterized as air pollutants.

The tables which follow are divided into two parts.

In part I, the pollutants are listed by type and associated problem. The characteristics of each pollutant type are described, its abundance and method of abatement.

In part II, standard categories of industries are tested, the nature of the activity described and types of air pollutant problems associated with each identified.

United States Pollutant Emissions in 10^6 tons/year (1968)

Sources	Carbon monoxide (CO)	Particulates	Sulfur oxides (SO_x)	Hydro carbons (HC)	Nitrogen oxides (NO_x)
Fuel-Combustion-Transportation	63.8	1.2	0.8	16.6	8.1
Motor vehicles	59.2	0.8	0.3	15.6	7.2
Gasoline	59.0	0.5	0.2	15.2	6.6
Diesel	0.2	0.3	0.1	0.4	0.6
Aircraft	2.4	Na	Na	0.3	Na
Railroads	0.1	0.2	0.1	0.3	0.4
Ships	0.3	0.1	0.3	0.1	0.2
Non-highway use of motor fuels	1.8	0.1	0.1	0.3	0.3
Fuel combustion stationary sources	1.9	8.9	24.4	0.7	10.0
Coal	0.8	8.2	20.1	0.2	4.0
Fuel oil	0.1	0.3	4.3	0.1	1.0
Natural gas	Na	0.2	N	Na	4.8
Wood	1.0	0.2	N	0.4	0.2
Industrial processes	9.7	7.5	7.3	4.6	0.2
Solid waste disposal	7.8	1.1	0.1	1.6	0.6
Miscellaneous	16.9	9.6	0.6	8.5	1.7
Forest fires	7.2	6.7	N	2.2	1.2
Structural fires	0.2	0.1	0.6	0.1	N
Coal refuse burning	1.2	0.4	—	0.2	0.2
Agricultural burning	8.3	2.4	—	1.7	0.3
Organic sdv. evaproation	—	—	—	3.1	—
Evaporation gasoline marketing	—	—	—	1.2	—
Total	100.1	28.3	33.2	32. 0	20.6

a = Slightly less than 10% of the SO_x emissions are derived from the ore of distillate oil.
N = Negligible in terms of the units being used.
Na = Not available.

Pollutant types	Sources and abundance	Abatement and control methods
CARBON MONOXIDE A tasteless, colorless, odorless gas, slightly lighter than air. When the gas enters the bloodstream it interferes with the ability of the blood to transport oxygen which at concentrations of 100 ppm causes most people to experiance dizziness, headache, lassitude and other symptoms of poison. The mean residence time of atmospheric CO has been estimated to be between 1 month and 5 years. The average background concentration has been estimated at 0.10 ppm in the northern hemisphere.	Most atmospheric CO is produced by the incomplete combustion of carbonaceous materials used as fuels for vehicles, space heating and industrial processing or burned as refuse. Transportation activities represent the largest source category. Other major man-made sources of CO emissions are stationary heat-generating facilities, industrial processes and solid waste combustion. It also occurs in high concentrations in cigarette smoke. Natural sources of CO include volcanoes, forest fires, lighting and the photochemical degredation of various reactive organic compounds. Some CO is formed in the upper atmosphere (above 70 km) photodissociation of CO_2. Biologically CO is formed by certain by brown algae, nature leaves of green plants various microorganism and various ocean dewelling biological specimen.	Afterburners, catalytic oxydation

Estimates of Carbon Monoxide Emissions

CO Emissions in the United States in 1968

Sources	10^6 tons/year	per cent
1. Mobile sources	59.2	58
2. Miscellaneous combustion sources (forest fires)	17.3	17
3. Industrial processes	11.2	11
4. Disposal of solid waste	8.2	8
5. Stationary fuel combustion	2.0	2
6. Miscellaneous other	4.1	4
	102.0	100

Robinson and Robbins (1968) have estimated that CO from the use of Gasoline in automobiles is responsible for 193 million of the total 232 million tons of CO representing 84% of the total CO emissions.

Pollutant types	Sources and abundance	Abatement and control methods
OXIDES OF SULFUR: Chemical compounds of sulfur and oxygen. a. *Sulfur dioxide*—SO_2 a colorless gas, which has a pungent and irritating odor at concentrations above 3 ppm and which can be tated at ambient concentrations as low as 0.3 ppm. Sulfur dioxide is a gas under ambient atmospheric conditions and can act as a reducing agent or an oxidizing agent of considerable importance to the problem of air pollution is the ability of the gas to react photochemically or catalytically with materials in the atmo-	*Major sources:* On an annual basis some 220 tons of sulfur are discharged into the atmosphere on a world wide basis. Of this amount about one third comes from air pollution sources mostly in the form of SO_2. This SO_2 is derived from sulfur bearing fuels (coal, oil) in order to produce electric power and a space heating industrial processes (primarily smelting and petroleum refining). Thus sulfur diocide pollution is characteristically found in large cities in mid latitude regions. Of the total U.S. emission of 33.2 tons of sulfur oxides in 1967 approximately	1. Substitution of low sulfur coal and oil (sulfur free fuels) for high sulfur to low sulfur fuels. 2. Removal of sulfur from fuel befor use. For coal, cleaning techniques (crushing and flatation); for fuel oil, catalytic treatment with hydrogen or blending with low-sulfur distillate oils. 3. *Removal of sulfur oxides from flue gases by:* Reaction with calcined limestone, then removal by fly ash control devices or wet scrubbers; Reaction with alkalized alumina,

Pollutant types	Sources and abundance	Abatement and control methods
sphere to form sulfur trioxide, sulfuric acid, and salts of sulfuric acid. The only natural source significance for SO_2 is volcanic activity. Sulfuric acid, H_2SO_4 rather than sulfur trioxide, is thus the compound normally found in the atmosphere. Sulfur trioxide is a strong acid and will combine with other materials (basic oxides) to form sulfates. In non-industrial areas sulfates are formed from the oxidation of hydrogen sulfide H_2S. A considerable quantity of sulfates are produced by sea spray and as an aerosol can make significant contribution to ambient particulate concentrations. b. *Sulfur trioxide*—SO_3 is a gas derived from combustion sources directly or from the oxidation of atmospheric sulfur dioxide. Sulfur trioxide may exist in the air as a vapor if the water vapor concentration in the air is low enough; but if sufficient water vapor is present, the sulfur trioxide combines immediately with water to form sulfuric acid. c. *Sulfates* (SO_4)—Most of the sulfur compounds originating from natural sources are in the form of Sulfate aerosols produced from sea spray. These sulfates relate to ions which are generally contained in sulfate compounds including ammonium sulfate and sulfuric acid.	75% resulted from the combustion of coal, oil, gasoline and other petroleum products containing sulfur. Most of the remaining emissions resulted from industrial processed particularly petroleum refining and smelting. The background atmospheric concentrations of SO_2 is estimated as 0.2 ppb. (See table: Worldwide Emissions of SO_2, 1965, below.) The annual world wide production o sulfate from sea spray is estimated as sulfates are also formed from the oxidation of SO_2 and H_2S most sulfates found in industrial air or in precipitation and derived from the combustion of SO_2. The background atmospheric concentrations are estimated at 2 $\mu g/m^3$.	followed by recovery of sulfur (commercial value); Catalytic oxidation to sulfuric acid (commercial valve). 4. Dispersion by use of tall stacks subject to the influence of local meteorological and topographic conditions. Natural removal occurs through precipitation and absorption by surfaces and vegetation. Conventional particulate control devices are used to control sulfates.
PARTICULATES: Are defined as any dispered matter solid or liquid in which the industrial aggregates are larger than single small molecules (about 0.0002 μ in diameter) but smaller than 500 μ. Are composed of the incomplete combustion, metals, large ions or salts, mists, fumes, fugitive, dusts and various other liquid or solid particles. Particles usually encountered in the atmosphere range from about 0.1 μ to 10 μ. In general, the smaller ranges of particles are quite-abundant while the larger particles exist in the atmosphere in very low concentrations. Particulates can remain airborne from a few seconds to several months. The category includes some compounds which are not solid or liquid while contained which condense when they enter into the atmosphere. Included are: aerosols (solids and liquids of microscopic size which are dissolved in gas, forming smoke, fog or mist),	(1) The major sources of particulates are combustion: a) Condensables are particles with change from the vapor phase when emitted into the atmosphere and range in size from 0.1 to 1.0 μ. b) Ash and unberned fuel particles is sizes greather than 1 μ. c) Soot from partial combustion of fuel. (2) Mechanical processer such as wind erosion, grinding, spraying, demolition, industrial activity, salt. Most of these particulate are in the 1–10 μ range and generally predominate very near the source. Electric generation, forest products industry, agriculture and related operations the crushed stone industry, the cement industry, the iron and steel industry.	(1) *From stationary sources (industry power plants, fires):* (a) Use of air cleaning techniques and devices fy industry and power plants to remove particulates: — Inertial separations or gravitational setting chambers — Cyclones — Bag houses and fabric filters — Electrostatic precipitators — Scrubbers and venturi scrubbers (b) Control of construction and demolition — (including grading of earth, paving roads, and parking lots, sand blasting, spray-painting). Hooding and venting, to air pollution control equipment, wetting down working surfaces with water or oil. (c) *Disposal of solid waste by sanitary land fill, composting, shreding and grinding rather than* incineration. Improvement in burning characteristics of combustion units.

Worldwide Emissions of SO_2, 1965

Source	Emissions 10^6 tons/year
1. Combustion of coal	102
2. Petroleum combustion	
a. Gasoline	0.36
b. Kerosene	0.20
c. Distillate oil	2.00
d. Residual oil	20.3
3. Petroleum refining	5.7
4. Smelting	
a. Copper	12.9
b. Lead	1.5
c. Zinc	1.3
Total	146.2

Pollutant types	Sources and abundance	Abatement and control methods
large particles (such as fly ash, metallic particles and dust), soot (carbon particles impregnated with tar), oil and grease.		*From motor vehicle sources* The basic aim is to develop. Methods of obtaining complete combustion. It this is accomplished, then the particulates (like soot and smoke) would be reduced. Research on particulate emission control is being National Air Pollution Control Administration conducted by NAPCA. It has not yet been determined whether emission control systems (described under carbon monoxide) are effective in removing particulates from vehicles exhaust. The removal of lead from gasoline.

Combustion—stationary sources:

Incineration: Analysis of emissions from municipal incinerators indicated 20–30% by weight of the discharge to be condensable and approximately 5–15% of the condensate, to be sulfuric acid. The remaining 70–80% were solids containing silicon, lead, aluminum, calcium, iron and traces of other elements.

Fuel oil combustion: 70 to 80% of particulate emissions comprise of carbon (cenospheres) and fly ash. As much as 25% appears as sulfates. Silicon and various metallic oxides comprise the remainder.

Coal combustions: The major component of particulate emissions is generally comprised of silicon oxides. Emissions carbon range from zero to almost 40%. Alumina and iron oxide particles are common in fly ash.

In the combustion of fossil fuels to produce energy in power plants industrial or domestic sources the follow emissions are typical for each billion BTU of energy produced.

Natural gas	15– 20 lb
Distillate oil	60–105 lb
Residual oil	70–160 lb
Coal	1000–13,000 lb

Mobile sources: Particles contained in vehicle exhaust include lead compounds, carbon particles, motor oil, such non volatile reaction products as aldehydes and ketones. Exhuast from a crusing automobile has been shown to contain 40–50 μg/l of particulates. From 62–80 per cent of these had diameters less than 2 μ. The lead content averaged about 40%.

Major non-combustion sources include cement manufacturing, kraft pulp mills sulfuric acid manufacturing steel manufacturing in open-hearth and/or electric furnaces.

Several organic substances such as benzo-(a)-pyrene and inorganic substances such as metal dusts and asbestos are measured as particulates have been demonstrated to be occupationally associated with lung cancer.

Arithmetic Mean and Maximum Urban Particulate Concentrations in the United States, Biweekly Samplings, 1960 to 1965

Pollutant	Number of stations	Concentrations μg/m³	
		Arith. average[a]	Maximum
Suspended particulates	291	105	1254
Fractions:			
Benzene-soluble organics	218	6.8	[b]
Nitrates	96	2.6	39.7
Sulfates	96	10.6	101.2
Ammonium	56	1.3	75.5
Antimony	35	0.001	0.160

Pollutant	Number of stations	Concentrations $\mu g/m^3$	
		Arith. average[a]	Maximum
Arsenic	133	0.02	[b]
Beryllium	100	<0.0005	0.010
Bismuth	35	<0.0005	0.064
Cadmium	35	0.002	0.420
Chromium	103	0.015	0.330
Cobalt	35	<0.0005	0.060
Copper	103	0.09	10.00
Iron	104	1.58	22.00
Lead	104	0.79	8.60
Manganese	103	0.10	9.98
Molybdenum	35	0.005	0.78
Nickel	103	0.034	0.460
Tin	85	0.02	0.50
Titanium	104	0.04	1.10
Vanadium	99	0.050	2.200
Zinc	99	0.67	58.00
Gross beta radioactivity	323	(0.8 pci/m^3)	(12.4 pci/m^3)

[a] Arithmetic averages are presented to permit comparable expression of averages derived from quarterly composite samples; as such they are not directly comparable to geometric means calculated for previous years' data. The geometric mean for all urban stations during 1964–65 was 90 $\mu g/m^3$, for the nonurban stations, 28 $\mu g/m^3$.

[b] No individual sample analyses performed.

Pollutant types	Sources and abundance	Abatement and control methods
HYDROCARBONS: Compounds and carbon all compounds of carbon except the oxides of carbon, the carbides, and the carbonates, are organic compounds which we are concerned with. Polluting hydrocarbons exist in the atmosphere primarily as gases (including methane, ethane, ethylene and acetylene), others, including cancer—inducing agents such as benzo-a-pyrene, are solid particulates. Large quantities of terpene-type compounds emitted by vegetation are polymerized in atmospheric photochemical reactions to form those aerosols responsible for the blue haze found in many forests. Reactions of hydrocarbons in the urban atmosphere are important because they give rise to secondary contaminants and reaction intermediates that cause nearly all the detrimental effects of hydrocarbon air pollution i.e. (photochemical oxidants). The hydrocarbons found in the largest concentrations in ambient air are methane (CH_4), Ethane (C_2H_6), Propane (C_3H_8), acetylane (C_3H_4), ethylene (C_2H_4), butane and isopentane.	The American Chemical Society estimates that most (approximately 85%) of all hydrocarbons in the air are emitted from natural sources (forests, vegetation and the bacterial decomposition of organic matter). The major man-made sources of hydrocarbons are the processing, distribution storage, marketing and use of solvents or petroleum. The hydrocarbons in vehicle exhausts, which result from the incomplete burning of gasoline in the internal combustion engine, accounted for about one half of the man-made hydrocarbon emissions in 1968. Other important sources were the evaporation of industrial solvents and incineration of waste. The National Air Pollution Control Administration developed the estimates of emissions given *for the United States* in the table below. On a world wide basis it is estimated that gasoline evaporation may be responsible for as much as 34×10^6 tons per year, solvent use 10×10^6 tons per year and incinerators for some 25×10^6 tons per year. About one third of the total emissions are considered reactive. Over 50% of reactive emissions are due to gasoline consumption.	FROM MOTOR VEHICLE SOURCES: Same as for carbon monoxide. Emissions resulting from evaporation of gasoline from fuel tanks and carburators can be limited by storage of the vapors (within the engine itself or in a carbon canister which absorbs the fuel vapors) and then routing the vapors back to the engine, where they will be burned. FROM STATIONARY SOURCES: a. Design equipment to use or consumed completely the processed materials. b. Use materials which have a higher boiling point or are less photochemically reactive. c. Use control equipment to reduce emissions. d. Elimination of open burning of waste in sanitary land fills. e. Evaporation emissions control. f. Control by adsorption, absorption and condensation.

Pollutant types	Sources and abundance	Abatement and control methods

Approximately 90% by volume of these hydrocarbons is methane which is inert in photochemical reactions. The non urban background concentrations of methane range from 1.0 to 1.5 ppm. Hydrocarbon concentrations are generally determined by the flame ionization technique which in effect counts the number of organically combined carbons.

TABLE 2-6

Estimates of Hydrocarbon Emissions by Source Category, 1968

Source	Hydrocarbon emissions				
	10^6 tons/year		% of total emissions		per cent reactive
Transportation	16.6		51.9		
Motor vehicles		15.6[a]		48.7	44
Airport		0.3		1.0	18
Railroads		0.3		1.0	18
Vessels		0.1		0.2	18
Nonhighway use motor fuels		0.3		1.0	44
Fuel combustion	0.7		2.2		
Coal		0.2		0.7	15
Fuel oil		0.1		0.3	18
Natural gas		negligible		negligible	—
Wood		0.4		1.2	15
Industrial processes	4.6		14.4		14
Solid waste disposal	1.6		5.0		30
Miscellaneous	8.5		26.5		
Forest fires		2.2		6.9	21
Structural fires		0.1		0.2	21
Coal refuse		0.2		0.6	15
Organic solvent evaporation		3.1		9.7	15
Gasoline marketing		1.2		3.8	20
Agricultural burning		1.7		5.3	21
Total	32.0		100.0		

[a] Diesel use accounts for about 2.5% of the total motor vehicle production.

Natural Sources:

	10^8 tons/year	
Methane	3.0	Decomposition of organic matter
Terpenes and isoprene	4.4	vegetation

Pollutant types	Sources and abundance	Abatement and control methods

ETHYLENE (ethene), a colorless hydrocarbon gas of the olefin series, is one of the major growth petrochemicals of the United States and of the world. Ethylene is not toxic to humans or animals, but it is the only hydrocarbon that has adverse effects on vegetation at ambient concentrations of 1 ppm or less. It therefore represents a considerable air pollution problem, for two reasons:

1. it is a significant phytotoxicant, and (2) it contributes to photo-chemically produced air pollution. Ethylene is the most abundant (based on mole-

Major sources:

1. Ethylene may form as a by-product from incomplete combustion of hydrocarbons and other organic substances. Thus, ethylene has been found to be one of the components of automobile and diesel combustion emissions (exhust and blowby emissions), incinerator effluents, and agricultural waste combustion gases.

2. Incinerator EFFLUENTS:

—effluents of multiple—chamber incinerators usually contain ethylene in concentrations of as much as 11,500 μg/m^3 (10 ppm).

3. Burning of agricultural wastes.

Ethylene poses no peculiar control problem in these emissions and, this can be controlled by methods generally used for hydrocarbons. These methods include combustion techniques, adsorption methods, and vapor recovery systems.

Pollutant types	Sources and abundance	Abatement and control methods
volume) of the photoreactive hydrocarbons in the lower atmosphere-it undergoes photochemical reactions with nitrogen oxides and ozone.	NATURAL OCCURRENCE: Ethylene is not normally found in deposits of petroleum or natural gas. Ethylene is produced naturally from many types of plants. Productions sources: —1967 production was about 12 billion pounds. —1969–1971 production was about 14.1 and 17.3 billion lbs respectively.	
A. Nitrogen oxides of the various oxides 1. nitrogen the most important as air pollutants are nitric oxide (NO) nitrogen dioxide (NO_2) and nitrous oxide. Nitrous oxide is not generally considered a pollutant although there is evidence that it is involved in photochemical reactions. a. Nitric oxide, a colorless gas which is relatively harmless but can be converted photochemically in the atmosphere to the more dangerous nitrogen dioxide. It is produced in any combustion processes by the combustion of oxygen and nitrogen. The rate of production increases with temperatures $N_2 + O_2 \rightleftharpoons 2NO$. The equilibrium concentration of NO is negligible below 1000°F but significant above 3000°F. b. Nitrogen dioxide a redish brown with apungent odor gas formes primarily in the atmosphere by the photochemical reaction $$NO + {}^1/_2 O_2 \rightarrow NO_2$$ NO_2 is corrosive and physiologically irritating and toxic and a known corrosive agent. It reacts with water to form nitric acid. Nitrous oxide (N_2O) is a colorless and odorless gas used as an anesthetic and is familiarily known as "laughing gas".	The major component of world wide atmospheric nitrogen oxide (NO_2) is biologically produced nitric oxides. Natural sources produce about 50×10^5 tons NO 1 year man made sources emit about 1/10 of that value. In general, NO_x concentrations urban atmosphere are 10 to 100 times higher than those in non-urban atmospheres. Therefore, the problem associated with NO_x consequently reflect the importance of man-made sources over natural sources. These emissions are primarily associated with power generation (boilers) and space heating. Non-combustion sources play a relatively small role in NOx emissions. However, chemical process dealing directly with nitrogen (nitric acid plants) do contribute to the amount of NO_x in the atmosphere. Motor vehicle sources account for approximately 40% of the total NO_x in the atmosphere.	I. *Fuel combustion:* Possible and speculative abatement methods for stationary sources are: 1. Conversion to a lower NOr producing fuel. Emissions are highest with coal, intermediate with oil and lowest with natural gas on an equivalent BTU basis. 2. Combustion modification which has been successful in reducing NO_x emissions from stationary sources. These include a 30–50% reduction using the top stage combustion in oil and gas fired boiler, a 30–60% reduction from use of low excess air reduction in a switch from front wall to tangential firing of a boiler. 3. Injection of steam and wetes into the combustion chamber. 4. The use of fuel additive. II. Internal combustion engine: A. Diesel The following methods have been used with varying degrees of success to reduce NO_x emissions from diesels. 1. Use of the pre-chamber or turbulent chamber design. 2. Changing the fuel type and properties. The use of natural gas has decreases NO_x emission. 3. Modification of a fuel injection timing. 4. Varying the compression ratio. B. Gasoline engines: Similar results have been found with the following abatement methods. 1. Exhaust gas recyling. 2. Changing the air/fuel ratio. The most economical a/f ratio prodoes the maximum NO_x emissions. By making the operation leaner (i.e. less fuel to air) NO_x emissions are decreases, however, the power output is correspondingly reduced. 3. Low compression ratios in engines. 4. Water injection into cylinders. 5. Catalytic conversion of NO_x in exhaust gases.

Pollutant types	Sources and abundance	Abatement and control methods
		6. Use of lower polluting engines such as gas turbine engine.
PHOTOCHEMICAL OXIDANTS: Photochemical oxidants result from a complex series of atmospheric reactions initiated by sunlight. When reactive organic substances and nitrogen oxides accumulate in the atmosphere and are exposed to the ultraviolet component of sunlight, the formation of new compounds, including ozone and peroxyacyl nitrates, takes place. Absorption of ultraviolet light energy by nitrogen dioxide results in its dissociation into nitric oxide and an oxygen atom. These oxygen atoms for the most part react with air oxygen to form ozone and also with certain hydrocarbons to form free radical intermediates and various products. Photochemical oxidants are known as secondary pollutants because they are formed in the atmosphere as the result of primary pollutants reacting; in this case hydrocarbons and nitrogen dioxide. Therefore, the concentrations of photochemical oxidants in the atmosphere will vary proportional to the amount of hydrocarbons and NO_2 in the air and the intensity of sunlight.	Except for the formations of ozone from electric discharge from high voltage lines and in such devices as room air purifiers, oxidants are formed as secondary pollutants. Since secondary pollutants are not emitted from any man-made sources, the amount of oxidants will depend entirely in the amount of primary pollutants present. Sources are the same as those for hydrocarbons and nitrogen oxides. *Natural sources:* A layer of ozone exits in the atmosphere at an elevation of about 20 kilometers. It has been postulated that some of this ozone is periodically transported downwards to the surface. Maximum concentrations in non urban areas vary from 20 to 100 $\mu g/m^3$. Ozone may also be formed naturally by electric discharge but the quantity is not considered significant (less than 20 $\mu g/m^3$) in terms of urban concentrations. Ozone has been identified in the Los Angeles area as the primary component of photochemical oxidants. Peroxyacetyl nitrates (PAN) have been identified as the next highest component. Aldehydes are the major products in the photooxidation of hydrocarbons. Formaldehyde accounts for about 50% of the total aldehydes and acrolein about 5%.	Abatement is achieved through the abatement of hydrocarbons and oxides of nitrogen.
ALDEHYDES: are one of a group of organic compound with the general formula R–CHO which yield acids when oxidixed and alchodols when reduced. Are products of incomplete combustion of hydrocarbons and other organic materials. Some common aldehydes are: Formaldehyde, Acrolein–Acetaldehyde similar irriations, exposure to these causes irritation of the mucous membranes of the eyes, nose, and other portions of the upper respiratory tract.	*Major sources:* Aldehydes are produced by the combustion of fuels in motor vehicles, space heating, power generation and in other combustion activities (such as the ingineration of wastes). In addition, aldehydes are formed in photo-chemical reactions between nitrogen oxides and certain hydrocarbons. *Natural abundance:* Natural sources of aldehydes do not appear to be important contributors to air pollution. Some aldehydes are found in fruits and plants. Man-made: Formed by oxidation of various hydrocarbons. *Formaldehyde:* Production for 1968 was 2 million tons. *Acrolein*	Control methods include more effective combustion as may be obtained in direct flame and the use of catalytic afterburners.
AMMONIA: NH_3, a compound of nitrogen and hydrogen, which is colorless, pungent, and very suffocating. It is very soluble in water (aqueovs-ammonia). In terms of total air content of ammonia, urban-produced ammonia is of lesser importance, though it may be important from the air pollution standpoint in localized situations.	*Major sources:* The primary sources of ammonia air pollution in cities is the combustion of fuels in stationary and well as mobile sources and the incineration of wastes. Industrial sources emitting ammonia are chemical plants, coke ovens, and refineries. Other sources are stock yards and similar installations, where ammonia is formed by biological degradation.	1. Collection of the bulk of escaping coke-over gas, coal dust and tar by a *vacuum* system during coke-oven charging can reduce ammonia emissions reduce in half. 2. Wet scrubbers are used to remove fly ash and ammonia which is contained in the gas stream from incinerators. 3. Where ammonia occurs as a solid

<table>
<tr><th>Pollutant types</th><th>Sources and abundance</th><th>Abatement and control methods</th></tr>
<tr>
<td>At high concentrations ammonia can be harmful to humans, animals and plants.</td>
<td>NATURAL ABUNDANCE: Approximately 3.7×10^9 tons or 99.9 per cent of the ammonia released into the atmosphere annually are the product of natural biological processes. The main source is the decomposition of organic waste material.
MAN-MADE ABUNDANCE: Approximately 4.2×10^6tons of ammonia are emitted to the atmosphere as a result of industrial and ammunity associated activities.</td>
<td>(as ammonium sulfate in the fertilizer industry for instance) conventional methods for solids removal can be used such as bag filters, electrostatic precipitators, and wet scrubbers.</td>
</tr>
<tr>
<td>ORGANIC CARCINOGENS: Carbon compounds which cause cancer in experimental animals and therefore suspected of playing a role in causing humancancer, particularly cancer of the lung. There is some question as to the carcinogenicity of selected compounds. Polynulcear aromatic hydrocarbons (PAH) in our environment evolve from high-temperature reactions under pyrolytic conditions during incomplete combustion of organic matter-pyrolysis and pyrosynthesis and large quantites of Aza-heterocyclic compounds are found in compounds contained in some air pollution source effluents in low concin automative exhaust fumes, and in moderate concs. in the air. The major classes of organic carcinogens are as follows:
1. Polynuclear Aromatic Hydrocarbons(PAH) Benzo-(a)-pyrene(BAP)-substance found in cigarette smoke.
Benzo (e) pyrene
Benz (a) anthracene
Benz (e) acephenanthrylene
Benzo (b) fluoranthene
Chrysene
Other carcinogens are:
2. Polynuclear azo-heterocyclic compounds:
Dibenz (a, h) acridine
Dibenz (a, j) acridine
3. Polynuclear imino-heterocyclic compounds.
4. Polynuclear carbonyl compounds 7H-Benz (de) anthracen-7-one
5. Alkylating agents:
Aliphatic and alifinic epoxides
Peroxide
Bactones</td>
<td>*Major source:* The incomplete combustion of matter containing carbon. Heat generation (burning coal, oil and gas) accounts for more than 85 per cent. Sources of heat generation that were tested ranged in size from residential heaters to heavy industrial power plant boilers.
Refuse burning incineration of wastes in units ranging from small commercial incinerators to large municipal incinerators to large municipal incinerators produce about 5% of emissions.
Motor vehicle exhaust contributes about 5% of emissions.
Industrial processes also account for 5%.
Natural abundance: Organic carcinogens are primarily unwanted by-products of imperfect combustion. However, a few sources of organic carcinogens might be defined as naturally occurring. Bituminous coal are asbestos contain certain organic carcinogens (BAP).
Since the (BAP)'s are the most common carcinogens to contend with their emissions (annual) are given in tons for the following four:
<table>
<tr><th>Source</th><th>Annual emission (tons)</th></tr>
<tr><td>1. Heat generation</td><td>421.6</td></tr>
<tr><td>2. Refuse burning</td><td>20.2</td></tr>
<tr><td>3. Industries</td><td>18.8</td></tr>
<tr><td>4. Motor vehicles</td><td>20</td></tr>
<tr><td>Total</td><td>481</td></tr>
</table>
* Alkylating agents are primarily found in liquid fuels, such as gasoline and diesel fuel, which are used in the internal combustion engine.</td>
<td>FROM MOTOR VEHICLE SOURCES: Same as for carbon monoxide. Emissions resulting from evaporation of gasoline from fuel tanks and carburetors can be limited by storage of the vapors (within the engine itself or in a carbon canister which absorbs the fuel vapors back to the engine, where they will be burned. Federals standards controlling these emissions will be in effect beginning with 1971 vehicle models.
FROM STATIONARY SOURCES:
1. Design equipment to use or consume completely the processed materials.
2. Use materials which have a higher boiling point or are less photochemically reactive.
3. Use control equipment to reduce emissions.
4. Stop open burning of waste by use of multiple-chamber incinerators or disposing of waste in sanitary land fills.</td>
</tr>
<tr>
<td>(1) ARSENIC: a brittle, very poisonous chemical element, found widely distributed over the earth's crust. It is most often found with copper, lead, cobalt, nickel, iron, gold and silver.
Arsenic is commonly found as a sulfide, arsenide, arsenite, or arsenate.</td>
<td>Major sources:
a. Smelters processing copper, zinc, lead and gold—arsenic is recovered as by product.
b. Cotton ginning and the burning of cotton trash.
c. Use as a pesticide (DDT).</td>
<td>a. Use of air cleaning devices to remove particulates from smelters and cotton gins. Equipment must operate at temperatures low enough to condense arsenic fumes ($<100°$ centrigrade)
— Electrostatic precipitators</td>
</tr>
</table>

Pollutant types	Sources and abundance	Abatement and control methods
	d. Combustion of coal. e. Incineration. Possible sources are: manufacturing of glass—arsenic pentoxide, As_2O_5, arsenic trisulfide, As_2S_3 manufacturing of ceramics—arsenic trichloride, As_2Cl_3 mining—arsenic trioxide, As_2O_3, arsine, AsH_3 wood preservatives, pigment and painting—arsenic disulfide, As_2S_2, also found is oil and non ferrous alloys.	— Cooling flues — Bag houses, especially those using wet scrubbing vacuum pumps instead of fabric filters. b. No methods available to control emissions produced by burning cotton trash.

TABLE 13

Arsenic Emissions in the United States, 1968

Source	Emissions, tons
Copper Smelters	2,700
Zinc Smelters	1,390
Lead Smelters	376
Agricultural Reprocessing	196
Glass Reprocessing	638
Agricultural Uses	3,270
Coal	743
Cast Iron (Incineration)	97
Other	5
Total	9,415

Pollutant types	Sources and abundance	Abatement and control methods
ASBESTOS: General name given to a variety of fibrous minerals found in rock masses. The value of asbestos ensues from the indestructible nature of products fabricated from the various grades of mineral fibers. *The major asbestos minerals are:* (Pyroxenes) chrysolite (amphiboples——), crocidolite, amosite, and anthophyllite. Tremolite and actinolite are considerably less important. Over 90% of the asbestos is chrysolite.	*Major source are:* a. Asbestos mines and factories. b. The wearing away of brake linings, roofing, insulation and shingles. c. Fireproofing of buildings with sprayed asbestos applications. d. Road surfacing. e. Asbestos cement. Man-made sources abundance: In the manufacturing of asbestos linings 30% of asbestos loss through emissions. In asbestos cement products 15 to 20% of the cement is composed of asbestos. One lb of asbestos is lost per ton of cement. This usage accounts for over 50% of the U.S. asbestos production. In the grinding brakelinings and clutch facings 10 lbs of asbestos is lost per ton of asbestos processed.	a. IN MANUFACTURING: Ventilation through fabric sleeve filters carrying out some operations (such as spinning and weaving of asbestos fabrics) as wet processes to elimate dust. b. IN TRANSPORTATION: Use of plastic-coated bags to transport asbestos. c. IN CONSTRUCTION: Use of insulators to enclose the work area when asbestos fire-proofing is blown onto steel frames. This technique. is not completely effective. d. IN MINES AND MILLS: No information available.

Asbestos Emissions, 1968

Source category	Source group	tons/year	
Mining and other basic processing	Mining and milling		5,610
Reprocessing	Friction materials	312	
	Asbestos cement products	205	
	Textiles	18	
	Paper	15	
	Floor tile	100	
	Miscellaneous	28	
	Subtotal		678
Consumptive uses	Construction	61	
	Brake linings	190	
	Steel fireproofing	15	
	Insulating cement	25	
	Subtotal		291
Incineration or other disposal			NA
	Total		6,579

NA = Data not available

Pollutant types	Sources and abundance	Abatement and control methods
BARIUM: A silvery white malleable, active, divalent, metallic element occurring in combination chiefly as barite. Inhalation of barium compounds can cause Baritosis a non-malignant lung disease. Characterized by fibrous hardening.	*Major sources are:* 1. Industrial processes involved in mining, refining and production of barium and barium-base chemicals. 2. Use of barium compounds as a fuel additive for the reduction of black smoke emissions from diesel engines. (This is accomplished by the production in vehicle exhaust of micron-sized particles which have minimal effects on visibility.) Concentration of about 0.075 per cent *barium* by weight of additive is most effectively used. *Natural abundance:* Barium frequently appears as gangne in lead and zinc ore deposits. The two main minerals are barite (barium sulfate, $BaSO_4$) and witherite (barium carbonate, $BaCO_3$).	The conventional methods for removal of barium are the same as thosfor solids, and include bag filters, electrostatic precipitators, and wet scrubbers.
BORON: A non-metallic chemical element which occurs only in combination with other elements (as with sodium and oxygen in borax). Most important pollutants are boron dust and borane fuel. The boranes are the most highly toxic of the boron compounds, consists chiefly of pentaborane, decaborane, and diborane.	*Major sources are:* Rocket motor or jet engines which use borane, a compound of boron, for a high energy fuel; combination of petroleum fuels which contain boron as an additive; burning of coal containing boron; manufacturing processes employed to produced boron compounds which are used as wastes softness. *Natural abundance:* Boron is widely distributed in nature, but constitutes only an estimated in 0.001 per cent of the earth's crust. It is present in sea water, and is an essential constituent of a number of rock-forming silicate minerals, such as datolite and tourmaline. Boron occurs naturally only in combined forms, usually as air alkaline earth borate or as boric acid.	(1) Prevention of accidental spilling of fuels. (2) Reduction or elimination of boron additives in vehicle fuels.
CHLORINE: Chlorine is a dense, greenish-yellow gas with a distinctive, irritating odor. It is noted for its very	The major commercial source of chlorine is the electrolysis of alkali chloride solution. Electrolytic diaphragm cells account	To prevent air pollution from these sources, the emissions from these systems can be tied indirectly either to a

Pollutant types	Sources and abundance	Abatement and control methods
strong oxiding and bleaching properties. Because of these properties, chlorine can be extremely hazardousto all life forms, as well as corrosive to metals and other materials.	for over two-thirds of the chlorine production. In this process, a saturated aqueous solution of sodium chloride or potassium chloride is electrolized. Electrolytic mercury cells account for about one fourth of the total chlorine production. *Natural occurrence:* Natural occurrences of free chlorine gas are extremely rare, due to the high reactivity of chlorine with many substances. Volcanic gases contain very small amounts of chlorine gas. Low concentrations of chlorine may, however, be formed by atmospheric reactions. The major source of chlorine emission is the residual gases remaining after the liquification of the chlorine cell gas. These residual gases may contain from 20 to 50% chlorine by volume, depending on the conditions of liquification. Other sources of emission include the loading and cleaning of tank cars, barges, or cylinders; dechlorination of spent brine solutions; and power or equipment failure.	scrubber system or a high stack for dispersion. i.g.—water scrubbers, alkali scrubbers, carbon tetrachloride scrubbers.
FLUORIDES: Fluorine is the 13th element in order of abundance and exists in nature primarily as fluorspar and as fluorapatite. Fluorspar and fluorapatite contain 49% and 3–4% fluorine respectively. Fluorspar is the source of nearly all commercially, used fluorine. Fluorapatite is also known as phosphate rock and is used in the manufacture of phosphate fertizers and elemental phosphorus compounds comprising of fluorine. It may occur in extremely low concentrations in the atmosphere as solid particles (sodium and calcium fluoride) or highly irritant and toxic gases (for example, hydrofluoric acid). Processing of phosphate rock produces silicon tetrafluoride SiF_4.	*Major sources:* The use of processing fluor-spar and fluorspatite are predominant sources of fluorine air pollutants. Phosphate rock processing to produce fertilizers is responsible for an estimated 18,700 tons of fluoride emissions annually. From electrolytic reduction of aluminum oxide, whereby aluminum is formed, atmospheric emissions of fluoride amounted to 16,000 tons per year in 1968. Industrial plants manufacturing steel, glass, brick and tile, are among the major emitters. Other major sources are: Combustion of coal and the firing of some types of rockets. Approximately 302,000 tons of hydrogen-fluoride are produced annually from fluorspar for subsequent conversion to fluorine, fluorocarbons and other compounds. The emission of fluorine from these processes in negligible.	a. In manufacture of phosphate fertilizer and processing of ANIMAL AND POULTRY FEED: Use of scrubbers: Waste water accumulates in ponds, where lime treatment may prevent evaporation into the air. b. In ALUMINUM MANUFACTURE: Use of air cleaning devices (cyclones and electrostatic precipatators with scrubbers). c. IN STEEL MANUFACTURE: Partial control through use of air cleaning devices to remove particulates. d. IN BRICK AND TILE MANUFACTURE: No devices now used. Could use scrubbers. e. IN GLASS MANUFACTURE: Little or no control today. Some small furnaces use scrubbers and bag houses. f. IN COAL COMBUSTION: Little or no control at present. Could use electrostatic precipitation in combination with scrubbers to remove particulates.

Fluoride Emissions in the United States, 1968

Source	Emissions, tons/year	
Phosphate rock processing		18,700
Fertilizers		
Normal superphosphate	9,700	
Wet process phosphoric acid	3,000	
Triple superphosphate	300	
Diammonium phosphate	100	
Elemental phosphorus	5,500	
Defluorination animal feed	100	
Aluminum manufacture		16,000
Fluorspar fluxing		11,300
Open hearth steel furnaces	2,300	
Basic oxygen steel furnaces	500	
Electric steel furnaces	1,800	
Iron foundries	N	
Welding coatings	2,700	
Non-ferrous foundries	4,000	
Brick and tile products		18,500
Glass and frit		2,700
Coal combustion		16,000
	Total	83,200

Pollutant types	Sources and abundance	Abatement and control methods
HYDROGEN SULFIDE: Hydrogen sulfide (H_2S) is a colorless gas that has an obnoxious odor at low concentrations. In higher concents, the gas is toxic to humans and animals and corrosive to many metals.	Air pollution by hydrogen sulfide is not a widespread urban problem, but is generally localized in the vicinity of an emitter such as kraft paper mills, industrial waste disposal ponds, sewage plants, petroleum refineries, and coke oven plants. It is estimated that 40–60,000 tons of H_2S are emitted annually from pulp production all over the world. In emissions from other industrial sources it has been estimated that H_2S accounts for as much as 2% of the quantity attributed to SO_2. This would indicate a total worldwide industrial H_2S emission of about 3 million tons per year. *Natural abundance:* It is estimated the H_2S from natural processes acounts for almost one half of the total sulfur released to the atmosphere or 10×10^6 tons. Hydrogen sulfide is produced in nature primarily through decomposition of proteinaceous material (vegetable and animal) by bacteria. It develops principally in stagnant and insufficiently aerated water such as found in swamps and polluted water. Hydrogen sulfide also occurs naturally as a constituent of natural gas, petroleum, sulfur deposits, and numerous volcanic gases and sulfur springs. The background atmospheric concentrations are estimated at 0.2 ppb.	Many of the removal systems are based on scrubbing the gas streams with a suitable absorbant and then removing the absorbed gas from the absorbent for disposal by burning or disposal by burning or conversion to valuable by-product. Different types of contacting devices (wet scrubbers) have been used, including conventional and novel design spray towers, plate towers and venture scrubbers. Natural removal occurs through precipitation and absorption by surfaces and vegetation.

Pollutant types	Sources and abundance	Abatement and control methods
Hydrogen Chloride (HCl): is a hygroscopic, colorless gas with a strong, pungent, and irritating odor. Because of its high solubility in water, the gas fumes in moist air. An aqueous solution of hydrogen chloride is called hydrochloric acid. Hydrochloric acid can be extremely harmful to the skin or mucous membranes.	The major sources of emissions of hydrogen chloride or hydrochloric acid are from their commercial production and use, and from the burning of paper products or from the combustion of chlorine containing substances such as polyvinyl chloride. Many industries also produce hydrochloric acid as an unwanted by-product in the manufacture of chemicals. *Natural abundance:* Natural occurrances of hydrochloric acid are rare. Minute quantities of hydrochloric acid are present in nature in volcanic fumes and in some rivers. *Man-made production (HCl):* 1967—1.6 million tons.	Due to the high solubility of hydrogen chloride in water, its can be achieved very effectively and yet, inexpensive by operational costs through the use of absorption facilities. Different types of absorbers are used, but the system generally consist of a packed tower or a cooling absorption tower followed by a packed tail tower.—Water scrubbing systems can be used to reduce emissions to less than 0.1 pound of hydrogen chloride per ton of acid produced. Other methods: Rotary brush scrubbers and ejector venture scrubber.
PHOSPHORUS: a solid non metallic element existing in at least two allotropic forms, one yellow (poisonous, infammable, and luminous in the dark), the other red (less poisonous, and less infammable). Elemental phosphorus (yellow) is a protoplasmic poison some of its compounds, especially organic phosphates, can also be lethal to man and animals in the case of exposure to high air concentrations.	The compounds known to be emitted in appreciable quantities into the ambient air are phosphorus oxides, phosphoric acid, and the organophorous compounds, found mostly in agricultural chemicals. Other organic phosphorus compounds are very probably emitted into the ambient air by the chemical industry from processes in which phosphorous products are intermediate or final outputs. *Major sources:* 1. Oil-fired boilers—0.9% phosphorus in fly ash. 2. Iron and steel industry—phosphorus pentoxide accounts for an average of 0.2% of the total weight of fume from furnaces. 3. Transportation sources: Organophosphorus compounds used as fuel additives. *Natural abundance:* Natural phosphates are divided into three classes on the basis of the metal or metals to which is bound. The three major classes are aluminum (iron) phosphates, calcium-aluminum (iron) phosphates, and calcium phosphates.	*Major control methods:* Scrubbers cyclones, fiber mist eliminators, high-energy wire-mesh contactors and electrostatic precipitators are used in the control of phosphorus emissions.
SELENIUM: is a nonmetallic element chemically resembling sulfur and tellurium, occurring in several allotropic forms. The soils of Midwest are particularly high in selenium content. Extensive studies have been made of the people, animals, food products, and soils of the high seleniferous areas. Selenium has also been found to be an essential nutrient for animals and may be necessary for humans.	*Major sources:* The sources of atmospheric selenium are believed to be terrestrial, such as fuels and ores used by industry (copper refinery), or possibly the burning of trash, particularly paper. Natural abundances: In nature, selenium is widely distributed in the earth's crust at a concentration of about 0.09 ppm. Selenium can be also found in coal and igneous rocks. Approximately 0.001 $\mu g/m^3$ of selenium has been found in samples of rain, snow, and air analyzed.	No study has been made of the methods for control of selenium and its compounds. However, based on the properties and on the methods of recovery and purification of selenium wet scrubbers and high-voltage electrostatic precipitators should be effective.
BERYLLIUM: is a light-weight, grayish metal that has a high strength-	Major sources: Beryllium is commonly found as an atmospheric pollutant within	1. a. Use of conventional air cleaning devices: scrubbers, venturi scrub-

Pollutant types	Sources and abundance	Abatement and control methods
to-weight ratio, great stiffness and valuable nuclear properties. A hard metallic element which forms strong, hard alloys with several metals, including copper and nickel. Almost all the presently known beryllium compounds are acknowledged to be toxic in both the soluble and insoluble forms: — beryllium sulfate } soluble — beryllium chloride } — metallic beryllium } insoluble — beryllium oxide } In concentrated form, it is found in relatively few minerals, and there are basically compounds of beryllium oxide. The most important such minerals are as follows: Principal ore: Beryl—$3BeO \cdot Al_2O_3 \cdot 6SiO_2$ Beryllium is used in nuclear reactors, gas turbines, airplane brakes, optical devices, springs bellows, diaphragms electrical contacts especially in high voltage insulation.	the confines and in the proximity of industrial plants producing or using beryllium substances. Such plants engage in the extraction, refining, machining and alloying of the metal. b. Combustion of coals and oil containing of on the average 1.9 ppm and 0.08 ppm of beryllium respectively. c. Use of beryllium as additive in rocket fuels. d. During the 1930s, use of beryllium in production of fluorescent lamps was a major source of pollution. NATURAL ABUNDANCE: Beryllium makes up a small portion of the earth's crust (10 ppm) or 0.006%.	bers packed towers, organic wet collectors, wet cyclones. b. For dry processes; conventional bag collectors, reverse-jet bag collectors, electrostatic precipitators, cyclones, unit filters. 2. Discontinuance of the use of beryllium in fluorescent lamp tubes.

Beryllium Emissions in the United States 1968

Source		Emission tons/year
Metallurgical Processing		5.30
Reprocessing		
Beryllium Fabrication		0.01
Beryllium Alloys		0.01
Beryllium Ceramics		0.01
Consumptive Uses		
Coal		147.00
Oil		8.00
Cast Iron		4.00
	Total	164.33

Pollutant types	Sources and abundance	Abatement and control methods
CADMIUM: Is a relatively rare metal which is not found in a free natural state. It is obtained from zinc, lead, copper, and other ores that contain zinc minerals. Pollution exists as fumes, and vapors. The major use of Cadmium is for electroplating iron and steel. The most common cadmium compounds and their uses are: 1. Electroplating—cadmium cyanide, $Cd(CN_2)$, and Cadmium acetate, $Cd(CH_3COO)_2$. 2. Photography and dyeing—cadmium chloride, $CdCl_2$.	*Major sources are:* 1. MINING—Since no ore is mined solely for cadmium recovery, emissions of cadmium dust ore vapors are those that occur during mining and concentration of zinc-bearing ores. 2. Metallurgical processing most of the atmospheric emissions occurs during the roasting and sintering of zinc concentrates as impurities are removed. Cadmium is volatized and condensed to be collected as dust in baghouses or electrostatic precipitators. Lead and copper smelters also process concentrates containing cadmium.	General control procedures for the prevention of air pollution by dusts, fumes, and mists applicable to the metal refinery alloying, and machining industries are considered suitable to these processes in the cadmium industry. — Copper mining and smelting: addition of bag filters and cyclones added to increase the recovery of cadmium. — Use of flue systems to direct the flow gases to proper receptables.

Pollutant types	Sources and abundance	Abatement and control methods
3. Manufacture of phosphores, glass, in nuclear reactor controls. 4. Manufacture of electrodes for storage batteries—cadmium hydroxide $Cd(OH)_2$. 5. Cadmium iodide, CdI_2, electrodeposition of Cd, manufacturing of phosphors. 6. Cadmium oxide—CdO In phosphores, semi-conductors, manufacture of silver alloys. 7. Cadmium selenide—CdSe In photoconductors. 8. Cadmium sulfate—$CdSO_4$ In electrodeposition Cd, Cu, and N. 9. Dimethylcadmium—$Cd(CH_3)_2$ In organic synthesis	3. Reprocessing—emissions occur during electroplating, manufacturing. 4. Consumptive uses—include use of rubber tires motor oil, fungicides and fertilizers. 5. Incineration and ether disposal- gaseous emissions will occur when scrap metal is melted to make new steel. 6. Cadmium used in plastics and pigments. NATURAL OCCURRENCE: The concentration of cadmium is almost always in direct proportion to that of zinc. The cadmium to zinc ratio varies from about 0.0002 to 0.002. With respect to the cadmium concentration percentage of the earth's crust, it is roughly 0.000055. Small concentrations of cadmium have been estimated in soil and sea water. MAN-MADE OCCURRENCE: Mining—2 lbs of emissions are estimated occur per ton of cadmium mined in wind loss from tailings. Metallurgical processing. Total U.S. consumption in 1968 was 6,650 tons.	

Cadmium Emissions in the United States, 1968
(tons/year)

Source category	Source group	Emissions (tons/year)	
Mining	Mining		0.26
Metallurgical processing	Cd separation from ores		1050.00
Reprocessing	Pigments	10.50	
	Plastics	3.00	
	Alloys	2.50	
	Batteries	0.20	
	Miscellaneous	0.55	
	Sub-total		16.75
Consumptive uses	Rubber tires	5.70	
	Motor oil	0.91	
	Fungicides	0.25	
	Fertilizers	0.46	
	Sub-total		7.32
Incineration or other disposal	Plated metal	1000	
	Radiators	125	
	Other	95	
	Sub-total		1220
Total			2294

Pollutant types	Sources and abundance	Abatement and control methods
CHROMIUM: Chromium is a lustrous brittle metallic element usually occurring in compound form with	*Major Sources:* Chromium concentrations in urban air average 0.015 $\mu g/m^3$ and range as high as 0.350 $\mu g/m^3$. Although a	Chromium air pollution usually occurs as particulate emissions, which may be controlled by the usual dust-

Pollutant types	Sources and abundance	Abatement and control methods
other elements. Most of the chromium ore produced is used in the production of stainless and austenite steels. Chromium (Cr) is commonly known for its use as a decorative finish in chrome plating.	complete inventory of sources of ambient chromium has not been made some possible sources are metallurgical industry, chromate-producing industry, chrome plating, the burning of coal, and the use of chromium chemicals as fuel additives, corrosion inhibitors, pigments, tanning agents, etc. *Natural abundance:* Elemental chromium is not found in nature. The only important commercial chromium mineral is chromite ($FeOCr_2O_3$) which is also never found in the pure form. Most soils and rocks contain small amounts of chromium, usually as chromic oxide (Cr_2O_3). The continental crust average 0.037% by weight, of chromium. In addition, most animal and plant tissues contain small amounts of chromium.	handling equipment, such as bag filters, precipitators, and scrubbers. *Chrome-plating facilities:* Moisture-extractor vanes in hood-duct systems have been used to break up bubbles in the exhaust gases. *Mist emissions:* Mist emissions from a decorative-chrome plating tank with problems can be substantically eliminated by adding a suitable surface-active agent to the plating solution.
IRON: A ductile, malleable silver-white metallic element, scarcely known in a pure condition, but abundantly used in its crude or impure forms containing carbon. Although inhalation or iron oxide is believed to cause a benign pneumoconiosis, there is growing concern about its synergistic effects with sulfur dioxide and carcinogens. Iron particulates may also act to reduce visibility.	*Major sources:* Iron and steel industry, sintering plants, blast furnaces, gray iron cupolas (used to melt gray iron), fuel sources (coal and oil), and incineration. *Natural abundance:* Iron abounds in nature and is an essential element for both animals and plants. The iron content of the earth's crust has been calculated at 5.6%.	Control of emissions from the iron and steel industry is being accomplished through improvements in steel processing. Dust removal is accomplished by high-efficiency electrostatic precipitators, venturi type scrubbers, or filters.
LEAD: Lead is a haevy, soft malleable, bluish-gray metallic element. It is the sulfide ore, galena, soil, water, vegetation and animal. It is introduced into the body with the intake of water and in air. Most lead air pollution is in the form of aerosols, fume, and sprays. The largest use of lead 39% of the total in 1968 is in the construction of storage batteries. The second largest use is in the manufacture of gasoline and tetramethyl lead. Approximately 1.30 million tons of lead were consured in the United States in 1968.	*Major sources:* The *major source* of airborne lead in urban areas is the exhaust from gasoline powered vehicles. Other man-made sources are manufacturing of lead additives for gasoline, processing and manufacture of lead products, the combustion of coal.	(1) *From vehicles sources:* Reduction or elimination of lead in fuel; use of particulate traps on vehicle exhausts. (2) *From lead processing and the manufacture of lead products:* Control of operating conditions (temperature and timing): Use of conventional air cleaning techniques (bag house filters, scrubbers, electrostatic precipitators). (3) *From coal combustion:* Use of electrostatic precipitators. (4) *From manufacture of lead additives for gasoline:* Use of water scrubbers and bag house filters. (5) *From transfer and transportation of lead gasoline:* Use of vapor recovery systems; reduction or elimination of lead in gasoline. (6) *From use of pesticides:* Use of pesticides which do not contian lead; improved techniques of pesticide use. (7) *From incineration of refuse:* Use of conventional air cleaning techniques or sanitary land fills instead of incinerators.

TABLE 9

Lead Emissions in the United States (1968)

Source	Emissions tons/year
Leaded gasoline combustion	180,000
Secondary lead smelting	2,000
Gray iron foundries	1,400
Waste crankcase oil combustion	1,250
Coal combustion	830
Lead alkyl manufacture	810
Primary zinc smelting	250
Primary copper smelting	380
Brass Founding	520
Municipal incineration	320
Primary lead smelting	680
Open hearth steel production	150
Gasoline transfer	40
Fuel Oil Combustion	24
Total	188,654

Pollutant types	Sources and abundance	Abatement and control methods
MANGANESE: A hard, brittle grayish-white metallic element whose oxide (MnO_2) is a valuable oxidizing agent, used as allaying agent in steel to give it toughness. Although manganese (Mn) is one of the elements essential to the human body, a high atmospheric concentration may result in poisoning and disease of several types.	*Major sources:* Air pollution by manganese arises almost entirely from the manganese and steel industries. Fumes from welding rods and organic manganese compounds may also contribute to air pollution. The organic compounds that have been tested as additives in gasoline, fuel oil, and diesel oil for use in both internal combustion engines and turbine engines may become an increasingly important sources of pollution. *Natural occurrence:* Manganese is widely distributed in the combined slate, ranking 12th in abundance (1,000 $\mu g/m^3$) among the elements in the earth's crust. Almost all of the manganese in the atmosphere enters as manganese oxides, such as MnO, Mn_2O_3 or Mn_3O_4.	Abatement: Control of manganese from furnances is accomplished by various types of collectors, including electrostatic precipitators, high-efficiency scrubbers, and fabric filters.
MERCURY: A high density, silver-white metal, is liquid at normal ambient temperatures. Although it is contained in at least 25 minerals, the chief source is cinnabar, (H_gS).	*Major sources:* The combustion of coal was the largest source of mercury emissions in the United States in 1968. The paint industry was the source of more than 25% of lead emissions in the U.S. in 1968 marine anti-fouling paints contain mercurial compounds as a toxicant cant; latex paints use mercurial compounds as a preservative. The third largest source of mercury emissions is the combustion of wastes which accounted for almost 17% of all mercury emissions in 1968. Other sources of mercury include mining and processing of ore. In industrial applications mercury is used in rectifiers, mercury precision lighting, batteries (mercury cell and alkaline energy cell). Labor-	For applications which use mercury at normal temperatures: 1. Proper ventilation in work areas. 2. Cleaning up spilled mercury (sweeping with special vacuum cleaners or chemical treatment). 3. Use of non-porous material for floors, working surfaces and protective clothing. 4. Conventional control of pesticides. For applications which use mercury at high temperatures: Condensing mercury vapors by: Cold-water jacketed condensers, impregnated charcoal. Water scrubbers.

Pollutant types	Sources and abundance	Abatement and control methods
	atory equipment and instruments—such as barometers, thermometers, flow meters, pressure-sensing devices, switches and relays all contain mercury (spillage creates droplets which vaporizes). Electrolytic preparation of chlorine. Agricultural use of mercury compounds as pesticides (now declining).	

TABLE 12

Nationwide Emissions of Mercury

Source	Emission (tons)
Mining	2.6
Processing	66.0
Reprocessing	4.0
Consumptive Uses	
Paint	215.0
Agriculture	19.0
Pharmaceuticals	2.6
Electrolytic Chlorine	70.2
Instruments	2.6
Dental Preparations	1.2
General Laboratory Use	32.0
Other	3.0
Fuel Combustion	
Coal	255.0
Oil	37.0
Waste Disposal	142.0
Total	852.2

Pollutant types	Sources and abundance	Abatement and control methods
NICKEL: A grayish white metallic element–hard, rough partially magnetic resistant to oxidation and corrosion. Nickel forms a variety of alloys with other metals. It is very important in making steel alloys and particularly stainless steel. Major pollutants are nickel dust and vapors.	Major sources: The processing of nickel to produce various alloys is the major source of emissions: This includes: 1. PLANT producing nickel alloys (including stainless steel) contains anywhere between 3–65% nickel in the alloys. 2. Nickel plating facilities via, electroplating, electroless plating (chemical plating), electroforming (nickel cas on mold) etc. 3. Nickel is used extensively as a catalyst (for i.e. Raney Nickel) used in hydrogenation of organic compounds, dehydrogenation of organic compounds, aging of liquors, etc. 4. Aviation and automobile engines burning fuels containing nickel concentrations range from 1 to 10% nickel. 5. Burning coal and oil-nickel in ash varies from 3 to 10,000 μg/g?	A. Use of conventional air cleaning devices: 1. Bag filters 2. Precipitators 3. Scrubbers B. Dceomposition of gaseous emissions at high temperature forming nickel (which can be removed as a particulate) and carbon monoxide. C. No control methods currently available for vehicle engine exhausts.

Pollutant types	Sources and abundance	Abatement and control methods
	6. Incineration of nickel products. * Schrocder, H. A., J. J. Balaser, and I. H. Tipton, Abnormal Trace Metals in Man-Nickel, *J. Chronic Disease* **15**, 51 (1962).	

Nickel Emissions in the United States (1968)

Source category	Source group	tons/year	
Mining and metallurgical processing			248
	Mining and smelting	248	
			834
Reprocessing	Stainless steel	442	
	Alloy steel	147	
	Nickel alloys'	53	
	Cast iron	79	
	Copper base alloys	6	
	Electrical alloys	5	
	Other	102	
Consumptive uses			5095
	Oil	4970	
	Gasoline	25	
	Coal	100	
	Total		6177

Pollutant types	Sources and abundance	Abatement and control methods
VANADIUM: Is grayish, malleable, ductile element found in combined in many minerals. Vanadium is used primarily to form alloy. Vanadium is also found in coal and oil as an impurity. The consumption of vanadium in 1968 was reported as 5,495 tons. Of this total about 80% was used in making various steels. More than 65 vanadium-bearing minerals have been identified. The most important of these are: a. patronite (V_2S_5S) b. Bravoite (FeNi)(S_2) c. Sulvanite ($3CU_2S \cdot V_2S_5$) d. Davidite—titanium ore e. roscoelite ($CaO \cdot 3V_2S_5 \cdot 9H_2O$)	Major sources: Almost all emissions of vanadium in the United States are derived from the combustion of fuel oil and coal both of which contain small amounts of metal. Fuel oil is by far the largest contribution (almost 90% of total emissions). In oil, the concentrations, of vanadium pentoxide vary from 0.01% (Continental crude) to 0.06% (Venezuelan crude). The ash from combustion of residual oil varies from 0.002 to 0.3% (by weight). In coal, there is a small contribution of vanadium in the lignite deposit and the ash emitted. Vanadium percentages in ash can range anywhere between 0.001 to 0.11%. Other minor sources are the processing of steel, cast iron and non ferrous alloys. Some additional emissions result from the manufacture of glass and ceramics and the use of vanadium as a catalyst.	*Use of additives:* Use of magnesium oxide in oil-fired burners, resulting in the reduction of fine particulate and amounts of vanadium escaping to the atmosphere. Use of conventional devices to remove particulates: Use of cetufigual collectors to gather ash emissions. Use of efficient fly-ash control equipment such as cyclones, electrostatic precipitators.

Vanadium Emissions in the United States 1968

Source	Emissions (tons/year)
Mining and Processing	81
Metallurgical Processing	144
Reprocessing of steel	236
Miscellaneous	6
Consumptive Uses	
Coal	1,750
Oil	17,000
Total	19,219

Pollutant types	Sources and abundance	Abatement and control methods
ZINC: A bluish-white metallic element occurring combined as the sulfide, oxide, carbonate, silicate etc. resembling magnesium in its chemical reactions. Although zinc is an essential element of the human and animal body, zinc and its compounds have been found to be toxic under certain conditions.	*Natural occurrence:* Zinc, widely distributed in the earth's crust, occurs in small quantities in almost all igneous rocks. The primary sources of emissions of zinc compounds into the atmosphere are zinc, lead, and copper smelting operations, secondary processing to recover scrap zinc, and possibly the incineration of zinc-bearing materials. Zinc oxide fumes are the zinc compounds most commonly emitted from these sources.	*Zinc gores* can be collected by electrostatic (rod-curtain-type) precipitators and then further treated in cyclone scrubbers. Zinc particles—can be collected by use of electrostatic precipitators, a central cloth-bag collector system, or some efficient filtering device.
AEROALLERGENS: Aeroallergens (pollens) are airborne materials that elicit a hypersensitivity or allergic response in susceptible individuals. The most common aeroallergens are the pollens of wind-pollinated plants—especially ragweed pollen, which is the main cause of hay fever. In addition to the pollens, aeroallergens include molds, danders, house, cosmetics, and others. It has been estimated that between 10 and 15 million people in the United States are affected by seasonal allergic (hay-fever).	Major sources: Most aeroallergens are produced by natural causes, although some may be produced through man-made interferences. 1. Natural sources: The aeroallergens encompass a wide variety of materials, but pollens are the most important member of this group. a) *Ragweed*—has been found in all 50 States, it produces large quantities of pollen, and the grains are especially adapted for aerial dissemination by virtue of their size (20 μ) shape, and density. It has been estimated that an acre of giant ragweed may produce as much as 50 lbs of pollen during a single season. b) *Fungi*—(molds) usually habitating in soil and dust, can become a menace when airborne. Their concentration in the air is dependent upon the magnitude of the source, their death rate in the air, humidity, temperature and other factors. c) *Danders*—(small particulate organic materials), including feathers of fowl and hair of animals and house dust. 2. *Man-made sources:* a) Flour mills—grain dusts produced in flour-milling plants (have been identified as a cause of asthma). b) Castor bean dust-oil processing plants. Most sources of biological aerosals are in natural.	Abatement and control measures for aeroallergens have been directed primarily at the ragweed. Since ragweed grows quickly in areas where the soil has been disturbed, it is not controlled by pulling it up when noticed, since the soil is thus disturbed and the growth may be heavier the following year. Herbicide (plant killers)—are sometimes used, but they are not only to ragweed, but to all plants. For eradicating molds, a number of disinfectants have been utilized. Man made sources are subject to normal particulate control methods as well as good housekeeping practices in plants.

Pollutant types	Sources and abundance	Abatement and control methods
BIOLOGICAL AEROSOLS: Are suspensions of microorganisms in the air. These biological contaminants occur as solid or liquid particles in the air. The vary in size from virus units of less than 0.1 microns to fungal spores of 100 microns or larger. The micro-organism classes (microbes) usually of concern are, bacteria, fungi (yeasts and molds), and viruses.	NATURAL ABUNDANCE: Biological aerosols ubiquitons in nature. Nowever, microorganisms found in the air had their orginal habitat either soil, water, humans, animals, or plants. Microorganisms become airborne from the soil and plants by wind disturbances, and from water by wave and wind action. The life of biological aerosols, depends on various factors wich define and affect the organisms.	It is extremely difficult to control the spread of biological aerosols. In general, knowledge is incomplete concerning the various parameters of biological aerosol, production, the survival and transmission of the aerosol, the sampling procedures, and other factors. 1. *Immunization*—some diseases as influenza can be controlled by this method. 2. *Ultra violet light*—can control airborne infections indoors (measles). 3. *Pesidual disinfectants, and dust removers*—can stop the spread of airborne bateria in dust. 4. Air filtration devices and air masks can sometimes eliminate contact of airborne aerosols from reaching respiratory systems.
CARBON DIOXIDE: The most commonly emitted air contaminant is not typically considered a pollutant in air pollution regulations. It is however the only air pollutant that has been proven to be of global importance to man's environment. CO_2 is a strong absorber and radiator of infrared radiation. In the „greenhouse" effect it acts to trap long wave radiation in the lower layers of the atmosphere thereby exerting profound influence on the temperature balance. The reaction of fossil fuels (hydrocarbon-based) with oxygen in complete combustion produces heat and water.	*Major sources:* The combustion of carbon in fossil fuels (coal, oil and natural gas) in both stationary and mobile sources produces almost 90% of the man made CO_2 emissions throughout the world. Other combustion sources including refuse disposal systems, the burning of wood and forests account for most of the additional 10% of emissions. Several industrial processes including coke ovens and smelting operations emit a significant amount of CO_2 on a localized basis. *Natural sources:* CO_2 is a natural constituent of earth's atmosphere comprising about 0.030% or 2.4×10^{18} gm of the total gas volume. Since the industrial revolution the combustion of fossil fuel has lead to a significant rate of rise in the CO_2 in the atmosphere to a level of about 0.7 ppm per year. The current equilibrium concentration is about 0.032%. If present trends in the increase of fuels continues by the year 2000 the increase in atmospheric CO_2 is produced by the decay of vegetation and respiration in man and animals, a considerable quantity is absorbed by the oceans and by living plants in the formation of food through photosynthesis.	Use of alternate sources of energy including nuclear, solar or chemically derived energy stimulation of the growth of plant life to increase or maintain maximum rate of photosynthesis. The global consumption of CO_2 by land plants has been estimated at $2\% 10^{17}$ gm per year.

Carbon Dioxide and Heat Emissions in the United States, 1969

Source	CO_2 10^6 tons/year	Heat 10^{12} BTU/year	Heat 10^{15} calories/year
Fuel combustion			
Stationary			
Coal	1,420	13,416	3,381
Oil	610	7,552	1,903

cont.

Source	CO_2 10^6 tons/year	Heat 10^{12}BTu/year	Heat 10^{15} calories/year
Natural gas	1,340	22,000	5,544
Wood	130	756	191
Transportation			
Highway vehicles			
Gasoline	800	10,519	2,651
Diesel	71	914	230
Non-highway use of Motor fuel			
Gasoline	61	797	201
Diesel	71	914	230
Railroads	48	573	144
Vessels	36	456	115
Aircraft	155	2,058	519
Refuse disposal	114	1,645	415
Miscellaneous	297	4,790	1,209
Total	5,153	66,390	16,733

Pollutant types	Sources and abundance	Abatement and control methods
ODOROUS COMPOUNDS: Odor may be defined as the sensation of smell perceived as a result of alfortory stimulus. Odorous compounds may have offensive smells which provoke people into complaining about air pollution. Hydrogen sulfide in a major offender. For names of some common o-doronts —p. 171, table 6.	*Major sources:* (1) Industrial processes such as kraft paper mills. Stack gases from the recovery furnaces may contain H_2S, mercaptans, organic sulfide and organic di-sulfides. (2) Animal rendering plants. (3) Chemical plants—Nitrogen or sulfide compounds are associated with organic materials in high temperature processes. (4) Petroleum refineries—Malodorous gases are produced in petrochemical plants during cracking and desulfurization reactions. Other well known sources are: Metallurgical plants—in which H_2S is a common by-product, diesel engines, sewers and sewage treatment plants, coke ovens and coal, food processing and aircraft odors. *Natural abundance:* Odors are produced in nature primarily from the decomposition of proteinaceous material (vegetable and animal) by bacterial action. They develop principally in stagnant and insufficiently aerated water. Some of the odors are traceable to the rotting stench of dead organisms.	1. Incineration of odorous waste products at the source. 2. Absorption of the odorant, which is soluble in water, or using various devices (including spray towers, cyclone scrubbers trays), to bring vapor into contact with liquid. 3. Absorption—Activated charcoal commonly used. 4. Masking or counteracting odors—usually by use of synthetic compounds with a pleasant smell. 5. Dilution with air—Requires favorable weather conditions to minimize nuisance effects. 6. Source elimination—Use of low sulfur fuels. 7. Removal of particulates which carry odors. 8. Chemical control—Oxidation or combination may change odorous to non-odorous compounds. 9. Biological control—Use of organisms which control sewage odor by oxidizing the odorous gases. 10. Containment—Covers on fuel tanks, sewage ponds and other open storage areas.
PESTICIDES: Economic poisons used to control or destroy pests that cause economic losses or adverse human health effects. These chemicals can be grouped as insecticides, herbicides (weed and brush killers, defoliants, and desicants), fungicides, molluscacides, algaecides, rodenticides,	The primary source of pesticides in the air is from application processes; a certain amount of drift is unavoidable even under normal conditions. Pesticides can evaporate into the air from soil, water and treated surfaces. Pesticides contained in dust from the soil can enter the air and be transported for considerable distances before falling	1. Control of pesticide drift during application: Improved application equipment and methods; Improved formulas for pesticides (increased use of large oil-based droplets rather than dusts). Wider distribution and use of weather

Pollutant types	Sources and abundance	Abatement and control methods
iscaricides, nematocides, repellents, attractants, and plant growth regulators.	back to the earth. Plants manufacturing pesticides also produce pollutant emissions. *Natural abundances:* Pesticides do not occur naturally in the atmospheric environment. *Man-made abundances:* 1.049 Billion of lbs synthetic organic pesticides are produced annually.	data in area where pesticides are to be used. 2. Control and abatement during production: Enclosure of grinding, mixing, packing and filling operations; Venting of solid emissions through bag houses and cyclones; Venting of liquid emissions through liquid scrubbers.
RADIOACTIVE SUBSTANCES: Substances which give off radiant energy in the form of particles or rays as a result of the disintegration of atomic nuclei. Major pollutants are radioactive gases and particulates (dust, fumes, smokes and mists). Radiation is classified as gamma radiation, alpha radiation is comprised of electron magnetic rays. Serious health effects have been associated with indiscriminate exposure to X-Rays. These include the loss of hair, the occurrence of burns, as well as the development of chronic ulcers and cancers. If exposures are of concentrations high enough, certain adverse biological and genetic effects can also arise: Alpha radiation is comprised of positively charged particles that are identical with the nucleus of a helium atom. Beta radiation is comprised of electrons emitted from nucleus of an atom.	Atmospheric radiation arises from both natural sources (rocks, soils and cosmic rays and from artificial sources (nuclear explosions and the nuclear industry in general). Production sources: a. Nuclear fuel production—mining of crude uranium. b. Mining, milling and refining of uranium. c. Nuclear reactors—heat is produced by nuclear fission. Emissions include fission products and activation products which remain incorporated in the coolant. d. Fuel reprocessing—radioactive fuels taken from power reactors are reprocessed. Experience to date has shown that, although there is a potential for radiation release from all facets of the nuclear industry it is insignificant when compared with the natural radiation dose. However, because of projected expansion of nuclear industry, there is evidence that krypton-85, which is released from nuclear fuel reprocessing, may be a problem. The U.S. Atomic Energy Commission has established maximum permissible concentrations for radionuclides that can be released from nuclear plants.	MINING: Wet drilling in uranium mines, underground drainage, and by clearing away the ore as rapidly as possible. NUCLEAR REACTORS— closed —cycle coolant systems to reduce pollution. ATMOSPHERIC ABATEMENT: The air cleaning method utilized will depend upon the form of the radioactive material (particulate or gas), the particle size, and the chemical and physical properties of the atmosphere and its contaminants. *Particulates* 1. Filtration—roughing filter, absolute filters, bag filters, or combinations of these. — Electrostatic precipitators, wet scrubbers. *Gases* 1. *Chemisorption* and *Adsorption*—adsorbents like carbon and silica gels absorb gases from streams of flow gas.
ON HUMANS AND ANIMALS: Radiation causes leukemia and genetic effects. Since the genetic effects of various amounts of radiation cannot always be determined, many scientists accept the belief that there is no safe level of exposure, i.e., no threshold below which there is no danger of biological damage to humans.	NATURAL OCCURRENCE: The two major sources of natural radioactivity are the gases which emanate from minerals in the earth's crust and the interaction of cosmic radiation with gases in the atmosphere. Solids and rocks contain naturally radioactive minerals such as radium-226 and radium-229 in variable amounts. Radon and thoron emanate from the earth's crust and contribute greatly to atmospheric radioactivity. The probability of certain radioactive gases emanating from the earth's crust depends primarily on their half-lives. Radon, with a half-life 3.8 days has a higher probability of escaping from the earth's crust before it decays than thoron, which has a half-life of 54 seconds.	2. Absorption—is suitable for removing from the ventilation air gases that react chemically with the scrubbing liquid or are highly soluble in it—water and weak alkaline solution. 3. Delay in storage—the delay or retention of gases tank until the radioisotope have decayed enough to permit release.

Pollutant types	Sources and abundance	Abatement and control methods
	Interactions of cosmic rays with atmospheric gases produce a number of radioactive species, the most important of which are tritium, carbon-14 and beryllium. These interactions produce electrons, gamma rays, nucleons, and muons. At low radiation levels the muons account for 70% of the cosmic radiation. Certain fuels such as coal, oil, and natural gas contain radioactive materials in small quantities. 1 Ci = 1 curie = disintegrations of a radioactive nucleide numbering exactly to 3×10^{-10} (intensity of radioactivity). coal—1 MW contains 10.8 MCi of radium-235 and 17.2 μCi of radium-226.	

Chemical manufacturing industry	Nature of activity	Type of air pollution problems
ADIPIC ACID	Adipic Acid, COOH · $(CH_2)_4$ · COOH, is a dibasic acid used in the manufacture of synthetic fibers. Adipic acid is produced by the oxidation of cyclohexane by air over a catalyst and then purified by crystallization.	Emissions: The only significant emissions from the manufacture of adipic acid are nitrogen oxides. In oxidizing the cycloheranol, nitric acid is reduced to onrecoverable N_2O and potentially recoverable NO and NO_2 emitted into the atmosphere.
AMMONIA	The manufacture of ammonia (NH_3) is accomplished primarily by the catalytic reaction of hydrogen and nitrogen at high temperatures and pressures.	Emissions: Range from CO, HC, to NH_3 gases. Wet scrubbers and water can be utilized to reduce the atmospheric emissions.
CARBON BLACK	Carbon black is produced by reacting a hydrocarbon fuel such as oil and/or gas with a limited supply of air at temperatures of 2500–3000 °F. Part of the fuel is burned to CO_2, CO and water, thus generating heat for combustion of fresh feed. The unburnt carbon is collected as a black fluffy particle.	Emissions: A high percentage of the emissions are carbon monoxide and hydrocarbons. The particulate and hydrogen sulfide problem are not as prevalent but do occur at amounts warranting attention. NO_2 emissions are relatively low due to the lack of available oxygen in the reaction.
CHARCOAL	Charcoal is generally manufactured by means of pyrolysis, or destructive distillation of wood waste from members of the deciduous hardwood species. Four tons of hardwood are reguired to produce one ton of charcoal. In the pyrolysis of wood, all the gases, tars, oils, acids, and water are driven of leaving virtually pure carbon.	During pyrolysis of wood, carbon monoxide, hydrocarbons, particulate crude methanol, and acetic acid are emitted into the atmosphere. Some of these gases can be recovered by utilizing a chemical recovery plant.
CHLOR-ALKALI	Chlorine and caustic are produced concurrently by the electrolysis of brine in either the diaphragm or mercury cell.	Emissions from diaphragm and mercury cell chlorine plants include chlorine gas, carbon dioxide, carbon monoxide, and hydrogen. Other emissions include mercury vapor, chlorine, wet scrubbers (alkaline) can be utilized for emission reduction.
EXPLOSIVES	An explosive is a material which, under the influence of thermal or mechanical shock, de-	Emissions: Sulfur oxides and nitrogen oxides emissions from processes which produce

Chemical manufacturing industry	Nature of activity	Type of air pollution problems
	composes rapidly and spontaneously with the evolution of large amounts of heat and gas.	some of the raw materials for explosives production can be considerable.
HYDROCHLORIC ACID	Hydrochloric acid is manufactured by a number of different chemical processes. Approximately 80% of the hydrochloric acid is produced by the by-product hydrogen chloride process. By product hydrogen chloride process. By product hydrogen chloride is produced when chlorine is added to an organic compound such as benzene, touluene, and vinyl chloride.	The recovery of the hydrogen chloride from the chlorination of an organic compound is the major source of hydrogen chloride emissions. The exit gas from the absorption or scrubbing system is the actual source of the hydrogen chloride emitted.
HYDROFLUORIC ACID	All hydrofluoric acid in the United States is currently produced by reacting acid grade fluorspar with sulfuric acid for 30–60 minutes in externallys fired rotary kilns at a temperature of 400–500°F.	The exit gases from the final absorber contain small amounts of HF, silicon tetrafluoride (SiF_4), CO_2, and SO_4 and may be scrubbed with a caustic solution to further reduce emissions. Dust emissions may also result from raw fluorspar grinding and driying operations.
NITRIC ACID	The ammonia oxidation process (AOP) is the principal method of producing commercial nitric acid. It involves high temperature oxidation of ammonia with air over a platinum catalyst from nitric oxide. The nitric oxide air mixture is cooled, and additional air water added to produced nitric acid.	The main source of atmospheric emissions from the manufacture of nitric acid is the tail gas from the absorption tower, which contains urabsorbed nitrogen oxides. These oxides are largely in the form of nitric oxide and nitrogen dioxide.
PAINT AND VARNISH	The manufacture of paint involves the dispersion of a colored oil or pigment in a vehicle, usually an oil or resin, followed by the addition of an organic solvent for viscosity adjustment.	Particulate emissions amount to 0.5 to 1% of the pigment handled; 1 to 2% of the solvent are lost. Hydrocarbons are the pollutant of primary concern.
PHOSPHORIC ACID	Phosphoric acid is produced by two principal methods, the wet process and the thermal process. In the wet process findy-ground phosphate rock is fed into a reactor with sulfuric acid to form phosphoric acid and gypsum. In the thermal process phosphate rock, siliceous flux, and coke are vaporized and the place in contact with water to produced phosphoric acid.	Emissions from the wet process are primarily gas fluorides, consisting mostly of silicon tetrofluoride and hydrogen fluoride. The principal emissions from the thermal process acid are P_2O_2 acid and acid mist. Particulates are also emitted in fairly large quantities.
PHTHALIC ANHYDRIDE	Phthalic anhydride is produced primarily by oxidizing naphthaline vapors with excess air over a catalyst, usually V_2O_5. The phthalic anhydride is then purified by a chemical soak in sulfuric acid.	The major source of emissions is the excess air from the production system which contains some uncondensed phthalic anhydride, maleic anhydride, quinines, and other organics.
PLASTICS	The manufacture of most resins or plastics begins with the polymerization or linking of the basis compound (monomer) usually a gas or liquid, into high molecular weight non crystalline solids.	The air contamination from plastics manufacturing are the emissions of raw material or monomer, emissions of solvents or other volatile liquids during the reaction, emissions of sublimed solids such as phthalic anhydride in alkyd production, and emissions of solvents during storage and handling of thinned resins.
PRINTING INK	Printing ink is produced by adding dyes to water and then flushing it with and ink vehicle.	Particulate emissions result from the addition of pigments to the vehicle while gases like terpenses, carbon dioxide, and aldehydes are

Chemical manufacturing industry	Nature of activity	Type of air pollution problems
		emitted into the atmosphere, during the preliminary stages of ink production.
SOAP AND DETERGENTS	Soap is manufactured by the catalytic hydrolysis of various fatty acids with sodium or potassium hydroxide to form a glycerol-soap mixture. This mixture is separated by distillation, neutralized and blended to produced soap. In the manufacture of detergents, a fatty alcohol is sulfated neutralized, and then sprayed dry to obtain the product.	The main atmospheric pollution problem in the manufacture of soap is odor from the spray drying operation, storage on mixing tanks and particulate emissions from the spray drying tower.

Metals industry	Nature of activity	Type of air pollution problems
BRASS AND BRONZE INGOTS (COPPER ALLOYS)	Obsolete domestic and industrial copper-bearing scrap is the basic raw material of the brass and bronze ingot industry. The ingots are produced from a number of different furnaces through a combination of melting, smelting, refining and allowing of the process scrap materials.	The exit gas from the furnaces may contain fly, as, soot and smoke, and some zinc oxides other particulates emissions include the preparation of raw materials and the pouring of ingots.
GRAY IRON FOUNDRY	The major type of furnace used to produce gray iron costings is the cupola, which uses an extremely hot bed of cake to melt the iron.	Emissions from cupola furnaces include CO, dust and fumes, smoke and oil vapors.
SECONDARY LEAD SMELTING	Furnaces similar to the ones mentioned above are used to melt the impure leaded scraps into the desirable products (Hasd-lead, semi-soft lead, and pure lead).	The primary emissions from lead smelting are particulates, lead oxides, and carbon monoxide.
SECONDARY MAGNESIUM SMELTING	Magnesium smelting is arried out in crucible or pot type furnaces charged with magnesium, scrap, melted, and poured into perspective molds.	Emissions from magnesium smelting include particulate magnesium (MgO), oxides of nitrogen, sulfur dioxide, and chloride gases.
CHEMICAL INDUSTRY SODIUM CARBONATE	The Solvay process is used to manufacture over 80% of all soda ash. In this process, the basic raw materials of ammonia, cake, limestone (calcium carbonate) and salt (sodium chloride). Are purified inabsorbent using ammonia and CO_2, to produce sodium bicarbonate as a by product.	The major emissions from the manufacture of soda ash is ammonia. Small amounts of ammonia are emitted in the vent gases from the brine purification system. Traces of particulate emissions can result from drying rotaries dry solids handling and processing of lime.
SULFURIC ACID	The conctact process is responsible for producing 90% of all the sulfuric acid in the U.S. and uses. In this process sulfuric acid is produced from the contact of SO_2 and SO_3 with water.	The waste gas contains unreacted sulfur dioxide, unabsorbent sulfur trioxide, as well as sulfuric acid mist and spray. When the waste gas reaches the atmosphere, sulfur trioxide is converted to acid mist.

Food and agricultural industry	Nature of activity	Type of air pollution problems
	This section deals with the manufacture of food and agricultured products and the intermediate steps which present on air pollution problem.	

Food and agricultural industry	Nature of activity	Type of air pollution problems
ALFALFA DEHYDRATING	An alfalfa dehydrating plant produces an animal feed from alfalfa. The dehydrating and grinding of alfalfa constitute the bulk of the manufacturing process of alfalfa meal. It is a very dusty operation most commonly carried out in rural areas.	Sources of dust emissions are the primary cyclone, grindeirs and air-meal separator. Overall dust loss have been reported as high as 7 per cent by weight of the meal produced. The use of a bag house as a secondary collection system can greatly reduce emissions.
COFFEE ROASTING	Coffee, which is imported in the form of green beans, must be cleaned, blended, roasted and packaged before being sold.	Dust, chaff, coffeebean oils (as mists) smoke, and odors are the principal air contaminants emitted from coffee processing. The major source of particulate emissions and practically the only source of aldehydes, nitrogen oxides and organic acids is the roasting process.
COTTON GINNING	In separating the seed from the lint in raw seed cotton, a large amount of trash is left over. From one ton of cotton approximately one 500 pound bale of cotton can be made, the rest is discarded as trash.	The major sources of particulates from cotton ginning are the unloading fan, the cleaner and the stick and bur machine. When cyclone collectors are used emissions have been reported to be about 90 per cent less.
FEED AND GRAIN MILLS AND ELEVATORS	Grain elevators are primarily transfer and storage units of various sizes. At grain elevator locations the following operations might occur: recewing, transfer and storages, cleaning, drying and milling or gunding.	Almost all emissions emanating from grain milles are dust particulates (minute grain particulates). The emissions from grain elevators operations are dependent on the type of grain, the moisture content of the grain, amount of foreign material, and the loading and unloading areas.
FERMENTATION	Fermentation occurs when various organisms (as molds, yeast certain bacteria, etc.) agitate or excite substances into onother form. The fermentation industries include the food, beer, whisky, and wine categories.	Emissions from fermentation process are nearly all gases and primarily consists of carbon dioxide, hydrogen, oxigen, and water vapor, none of which presents an air pollution problems. However, particulate emissions can occur in handling of the grain used as raw material, while gaseous hydrocarbons can be emitted during the drying of spent grains.
FISH PROCESSING	The canning, dehydrating, smoking of fish, and the manufacture of fish oil are the important sigments of fish processing.	The biggest problem from fish processing is emissions of odors such as hydrogen sulfide and trime-thylamine. Some of the methods used to control odors include activated carbon adsorbers, scrubbing with some oxidizing solution and incineration.
MEAT SMOKEHOUSES	Smoking is a diffusion process in which food products are exposed to atmosphere of hardwood smoke, causing various organic compounds to be absorbed by the food.	Emissions from smokehouses are generated from the burning hardwood, and included particulates, carbon monoxide, hydrocarbons, (CH_4), aldehydes (HCH°) and organic acids (acetic).
NITRATE FERTILIZERS	Nitrate fertilizers the product of the reaction of nitric acid and ammonia to form ammonia nitrate solution or granules.	The main emissions from the manufacture of nitrate fertilizers are the ammonia and nitric oxides lost in the neutralization and drying operation.
PHOSPHATE FERTILIZERS	Nearly all phosphate fertilizer are made from naturally occuring phosphorous-containing minerals such as phosphate rock. The phosphorous content of these minerals is not in a from that is readily available to growing plants so that the minerals must be treated to	Emissions from manufacturing phosphate fertilizers include vent gases containing particulates ammonia, silicon tetrafluoride, carbon dioxide, stean and sulfur oxides. The sulfur oxides emissions arise from the reaction of phosphate rock and sulfuric acid.

Food and agricultural industry	Nature of activity	Type of air pollution problems
	convert the phosphorous to a plant-available form.	
STARCH MANUFACTURING	Starch is obtained through the separation of coarse starch in corn to a fine dry powder form ready for marketing.	The manufacture of starch from corn can result in significant dust emissions from cleaning grinding and screening operations.
SUGAR CANE PROCESSING	The processing of sugar cane starts with harvesting crops, then through a series of processes (washing, crushing, milling, diffusing) into the final sugar product.	The largest sources of emissions from sugra cane processing are the open burning in the harvesting of the crop and the burning of bagasse as fuel. Emissions include particulates, CO usually large. HC, and nitrogen oxides.

Wood processing industry	Nature of activity	Type of air pollution problems
WOOD PULPING INDUSTRY	Wood pulping involves the production of cellulose from wood by dissolving the lignin that binds the cellulose fiber together. The three major chemical processes for pulp production are the kraft or fulfate process, the sulfite process and the neutral sulfite semi chemical process. The kraft process involves cooking wood ships in sodium sulfide and sodium hydroxide to dissolve the lignin. The excess pulp and impurities are washed away and the remaining clean pulp pressed and dried into the finished product.	Particulate emissions from the kraft process occur primarily from the recovery furnace, the line kilm and smelt dissolving tank. Ths characteristic kraft mill odor is principally due to the presence of a variable mixture of hydrogen sulfide and dimethyl disulfide. Some sulfur dioxide emission results from the oxidation of the sulfur compounds CO emissions may occur from the recovery furnaces and kilms.
PULPBOARD	Pulpboard manufacturing includes the manufacture of fibrons boards from a pulp slurry. After the pulp is washed, it is entered into a board machine and subsequently, dried and ready for fabrication.	Emissions from the paper board machine consist of only water vapor. Little or no particulates are emitted from the dryers.

METALLURGICAL INDUSTRY

The metallurgical industries can be broadly divided into primary and secondary metal production operations. Primary metal industry includes the production of the metal from ore; among these industries are the non ferrous operations involved in aluminum ore reduction, copper smelters, lead smelters, zinc smelters, iron and steel mills, ferro alloys and metallurgical coke manufacture. The secondary metals industry includes the recovery of the metal from scrap and salvage, the production of alloys from ingot, secondary aluminum operations, gray iron foundries, lead smelting, magnesium smelting, steel foundaries, and zinc processing.

Metals industry	Nature of activity	Type of air pollution problems
ALUMINUM ORE REDUCTION	Bauxite, a hydrated oxide of aluminum associated with silicon, titanium, and iron, is the base ore for aluminum production. After preliminary purification using the (Boyer) process, the new oxide (Al_2O_3) is reduced in	During the reduction process, the affluent released contains fluorides particulate and gaseous hydrogen fluoride. Particulate matter such as aluminum and carbon from the anodes are also emitted.

Metals industry	Nature of activity	Type of air pollution problems
	the Hall-Heroult process and pure aluminum is produced. Four tons of bauxite are required to make 1 ton of aluminum.	
METALLURGICAL COKE MANUFACTURE	Coking is the process of heating coal in an atmosphere of low oxygen content, i.e. destructive distillation. During the process organic compounds in the coal break down to yield gases and a relatively non-volatile residue.	Visable smoke, hydrocarbons, carbon monoxide, sulphur dioxide, nitrogen oxide and ammonia originate from by product coking operations.
COPPER SMELTERS	Copper is produced primarily from low-grade sulfide ores, which are concentrated by gravity and subjected to melting and purifying procedures.	The raw waste gases from the process contain significant amounts of dust and sulfur oxides.
FERRO ALLOY PRODUCTION	Ferro alloys is the generic term for alloys consisting of iron and one or more other metals. The major method used to produce ferroalloy for steel making is the electric furnace process. In this process suitable oxides are reduced to the appropriate metals.	Most of the emissions of carbon monoxide and particulates (dust) are a direct result of the electric furnace, which uses carbon as the reducing agent.
BRASS AND BRONZE INGOTS (COPPER ALLOYS)	Obsolete domestic and industrial copper-bearing scrap is the basic raw material of the brass and bronze ingot industry. The ingots are produced from a number of different furnaces through a combination of melting, smelting, refining and alloying of the process scrap materials.	The exit gas f om the furnaces may contain fly ash, soot an d smoke and some zinc oxides. Other particulate emissions include the preparation of raw materials and the pouring of ingots.
GRAY IRON FOUNDRY	The major type of furnace used to produce gray iron castings is the cupola, which uses an extremely hot bed of coke to melt the iron.	Emissions from cupola furnaces include CO dust and fumes, smoke and all vapors.
SECONDARY LEAD SMELTING	Furnaces similar to the ones mentioned above are used to melt impure leaded scraps into desirable products (hard-lead, semi-soft lead, and pure lead).	The primary emissions from lead smelting are particulates, lead oxides, and carbon monoxides.
SECONDARY MAGNESIUM SMELTING	Magnesium smelting is carried out in crucible or pot type furnaces charged with magnesium scraps, melted and poured into perspective molds.	Emissions from magnesium smelting include particulate magnesium (MgO), oxides of nitrogen, sulfur dioxide and chloride gases.
IRON AND STEEL MILLS	To make steel, iron ore is reduced to pig iron, and some of its impurities are removed in a blast furnace. The pig iron is further purified in other processes (open hearth, Bessemer converters, basic oxygen furnaces, or electric furnaces).	Particulates and carbon monoxide are the major pollutant emission resulting from the virious furnace reactions.
LEAD SMELTERS	The ore from which primary lead is produced contains both lead and zinc. After melting the metals are concentrated.	Effluent gases from the various concentrating processes include considerable particulate matter and sulfur dioxide.
ZINC SMELTERS	Most domestic zinc comes from zinc and lead ores. Another important source of raw material for zinc metal has been zinc oxide from fuming furnaces, the roasted are electrolytically purified.	Dust, fumes, and sulfur dioxide are evolved from zinc concentrate roasting.
SECONDARY ALUMINUM OPERATIONS	Secondary aluminum operations involve making light weight metal alloys for industrial castings and ingots. Copper, magnesium, and	Emissions from secondary aluminum operations include fine particulate matter and small quantities of gaseous chlorides and fluorides.

Metals industry	Nature of activity	Type of air pollution problems
	silicon are the most common alloying constituents.	
STEEL FOUNDRIES	Steel foundries produce steel castings by melting steel metal and pouring it into molds. The basic melting process operations are furnace charging, melting, tapping the furnace into a ladle and pouring the steel into molds.	Particulate emissions from steel foundry operations include iron oxide fume, sand fires, graphite, and metal dust. Gaseous emissions from foundry operations include oxide of nitrogen, oxides of sulfur, and hydrocarbons.
SECONDARY ZINC PROCESSING	Zinc processing includes zinc reclaiming (separation of zinc from the scrap), zinc oxide manufacturing (distilling metallic zinc into dry air stream), and zinc galvanizing (flux cover over zinc).	A potential for particulate emissions, mainly zinc oxide, occurs, if the temperature of the furnaces are very high (1100°F). Small quantities of ammonia chloride, nitrogen oxides, and carbon monoxides are also emitted into the atmosphere.

Mineral products industry	Nature of activity	Type of air pollution problems
	This section involves the processing and production of various minerals. Mineral processing is characterized by particulate emissions in the form of dust. However, most of the emissions from the manufacturing proces-conventional in this section acn be reduced by conventional particulate control equipment such as cyclones, scrubbers and fabric filters.	
ASPHALT BATCH PLANTS	Hot-mix asphalt paving consists of a combination of aggregater, coarse or fine, uniformly mixed and coated with asphalt cement. The coarse aggregates are usually crushed stone, crushed slag or crushed gravel, while the fine aggregates are usually natural sand and finely crushed stones.	The largest pollutant type is dust, emanating from the rotary dryers and filtering systems, normally used in producing asphalt.
ASPHALT ROOFING	The manufacture of asphalt roofing felts and shingles involves saturating a fiber media with asphalt by means of dipping and/or spraying.	The major pollutants are particulate emissions from asphalt roofing plants during the asphalt blowing operations and the felt saturation. Common methods of control at asphalt saturation plants include complete enclosure of the spray area and saturation followed by good ventilation through one or more collection devices. Some traces of carbon monoxide and hydrocarbons are also present in the emissions from this asphalt process.
BRICKS AND RELATED CLAY PRODUCTS	The manufacture of brick and related products such as clay pipe, pottery and some types of refraction brick involves the grinding, screening, blending of the raw materials, forming drying or curing, firing and ferial cutting or shaping.	Particulate emissions similar to those obtained in clay processing are emitted from the materials handling process in refractory and brick manufacturing. Gaseous flourides and nitrogen oxides are also emitted from brick manufacturing operations.
CALCIUM CARBIDE	Calcium carbide is manufactured by heating a mixture of quick-lime (CaO) and carbon in an electric arc furnace when the lime is reduced by the coke to calcium carbide and carbon monoxide. About 1,900 pounds of lime and 1,300 pounds of coke yield 1 ton of calcium carbide.	Particulates, acetylens, sulfur compounds and some carbon monoxide arc emitted from calcium carbide plants.

Mineral products industry	Nature of activity	Type of air pollution problems
CASTABLE REFRACTORIES	Castable or fused—cast refractories are manufactured by carefully blending such components as alumina, zirconia, silica, chrome, and magnesium, melting the mixture, pouring into molds, and slowly cooling to the solid state.	Particulate emissions occur form drying, crushing and handling procedures while gaseous fluoride occur during melting operations.
PORTLAND CEMENT MANUFACTURING	Lime (calcareous), silica (siliceous), alumina (argillaceous) and iron (ferriferous) are the four major components used to manufactur cement. The various substances are crushed in exact proportions, fired in a film, and then ground in gypsum to be bagged for shipment as cement.	Particulate matter is the primary emission in the manufacture of portland cement and is emitted primarily from crushing operations and rotary kilms. Control systems usually include multicyclones, electrostatic precipitators or combinations of these types of control.
CERAMIC CLAY MANUFACTURE	The manufacture of ceramic clay involves the conditioning of the basic ores, coolinate and montmorillonite (aluminous-silicate materials), into dry clay products.	Emissions consist primarily of particulates, but some fluorides and acid gases are also emitted in the drying process.
CLAY AND FLY ASH SINTERING	Both the sintering of clay and fly ash involve the firing and burning off of residual matter to the desirable product. In fly ash, carbon is burned off while in clay, entrained volatile matter is driven off.	Dust is the major pollutant emitted from the screening and sintering process.
COAL CLEANING	Coal cleaning is the process by which undesirable materials are removed from both bituminous and authorite coal. The coal is screened, classified, washed and dried at coal preparation plants.	Particulates in the from of coal dust constitute the major air pollution problem from coal clearing plants.
CONCRETE BATCHING	Concrete batching involves the proportioning of sand, gravel, cement, and water by means of weight hoppers and conveyors into a mixing receiver.	Particulate emissions consist primarily of cement dust, but some sand and aggregate dust emissions do occur during batching operations.
FIBER GLASS MANUFACTURING	Fiber glass manufactured by melting various raw materials to form glass, drawing the molten glass into fibers, and coating the fibers with an organic material.	The major emissions from fiber glass manufacturing processes are particulates from the glass melting furnace and the product coaling line.
FRIT MANUFACTURING	Raw materials such as borax, feldspar, sodium fluoride and soda ash are melted and then quenched with water to produce shattered small glass particles-called frit. The frit particles are then grounded into fine particles used in enameling iron and steel or in glazing porcelain or pottery.	The major emissions from frit-smelting operations are dust and fumes (usually condensed metallic oxide fumes) from the molten charge. A small quantity of hydrogen fluoride also can be detected in the emissions.
GLASS MANUFACTURE	Nearly all glass produced commercially is either soda-lime, lead, fused silica, borasilicate, or 96 per cent silica. Soda lime glass, being of the largest type, is produced on a massive scale in large, direct fired, continuous melting furnaces in which the blended raw material are melted at 2700° to form glass.	Emissions from the glass melting operation consist primarily of particulate (only a few microns in diameter) and fluorides, if fluoride —containing fluxes are used in the process.
GYPSUM	Gypsum or hydrated calcium sulfate is a naturally occurring mineral which hardens when in contact with water to form a solid crystalline hydrate. Gypsum is an important building material, and if it loses its water of hydration, becomes plaster of paris.	Gypsum rock dust and partially calcined gypsum dust are emitted into the atmosphere from the grinding and mixing of the gypsum material.

Mineral products industry	Nature of activity	Type of air pollution problems
LIME MANUFACTURING	Lime (CaO) is the high temperature product of the calcination of limestone ($CaCO_3$). Lime is manufactured in vertical or rotary kilms fired by coal, oil, or natural gas.	Atmospheric emissions in the lime manufacturing industry include the particulate emissions from the mining, handling, crushing screening and calcining of the limestone and the combustion products from the kilm.
MINERAL WOOL	The product mineral wool is made by firing charge material (slag wool and rock wool) in a furnace with silica rock and coke, into long fibrons tails for a "blanket" of wool.	Gases such as sulfur oxides and fluorides are major emissions from cupolas or furnace stacks. Minor particulate emissions are found in existing fumes.
PERLITE MANUFACTURE	Perlite is a glassy, volcanic rock consisting of oxides of silican and aluminum combined as a natural glass by water of hydration. By a process called exfolication, the material is slowly heated to release water of hydration and thus expand the spherules into low-density particles used primarily as aggregate in plaster and concrete.	A fine dust is emitted from the outlet of the last product collector in a perlite expansion plant. In order to achieve complete control of these particulate emission a bag-house is needed.
PHOSPHATE ROCK PROCESSING	Phosphate rock preparation involves the benefication to remove impurities, drying to remove moisture, and grinding to improve reactivity.	Emissions in the form of fine rock dust may be expected from drying and grinding operations.
SYNTHETIC FIBERS	Synthetic fibers are classified into two major categories—semi synthetic, or "True synthetic." Semi-synthetics, such as viscose rayon and acetate fibers, result when natural polymeric materials such as cellulose are brought into a dissolved or dispersed state and then spun into find filaments. True synthetic polymers, such as nylon, orlon and dacron result from addition and polymerization reaction to form long chain molecules.	In the manufacture of viscose rayon, carbon disulfide are the major gaseous emissions. Some examples of minor pollutants emitted from the drying of the finished fiber are hydrocarbons and oil vapor (mist).
TEREPHTHALIC ACID	Terephthalic acid is an intermediate in the production of polyethylene terephthalate, which is used in polyester films and other miscellaneous products and by oxidying paraxylene by nitric acid.	The NO in the off gas from the reactor is the major air contaminant from the manufacture of terephthalic acid.
STONE QUARRING AND PROCESSING	Rock and gravel products are loosened by drilling and blasting from their deposit bedsand removed with the use of heavy equipment. Further processing includes crushing regrinding, and removal of fines.	Dust emissions occur from many operations in stone quarrying and processing.

Petroleum industry	Nature of activity	Type of air pollution problems
PETROLEUM REFINING	The operation of a petroleum refinery can be divided into four major steps: separation, conversion, treating, and blending. The crude oil is first separated into selected fractions (e.g., gasoline, kerosine, fuel oil, etc.). Some of the less valuable products such as heavy naphtha, are converted to products with a greater sale value such as gasoline. This is done by splitting, uniting, or rearranging the original molecules. The final step is the blending of the refined base stocks with each other and various additives to meet final product specifications.	The major pollutants emitted are sulfur oxides, nitrogen oxides, hydrocarbons, carbon monoxide, and malodorons materials. Other emissions of lesser importance include particulates, aldehydes, ammonia, and organic acids. Most of the above mentioned emissions come from boiling process heaters, and catalytic craking unit regerators.

CONRAD SIMON

Environmental Protection Agency

APPLICATIONS OF REVERSE OSMOSIS

(see also Part I, REVERSE OSMOSIS–PROCESS FUNDAMENTALS)

Introduction

As pointed out already in Part I, reverse osmosis is a general separation technique and its potential practical applicability in chemical process engineering in general, and in water treatment in particular, is limited only by the availability of suitable membranes and equipments for the particular purpose. From this point of view, reverse osmosis is still in its very early stages of development. However, a large number of potential industrial applications of the technique has been studied in laboratory scale and pilot plant units using Loeb-Sourirajan type porous cellulose acetate or similar membranes. The following is a summary of the significant results reported with respect to some typical water treatment and water pollution control applications.

Brackish Water Conversion

The most well established application of reverse osmosis is with respect to brackish water conversion using Loeb-Sourirajan type porous cellulose acetate membranes. The pilot plants in Coalinga, California[1–6], and Yotvatah in the Negev Desert of Israel[7] are among the most successful installations for the production of potable water from brackish water.

The Coalinga plant was built by the University of California, Los Angeles, and it has been in operation since June 4, 1965. It is the first practical installation of its kind whose operation has been large and long enough for its performance to be one of tremendous experience in reverse osmosis engineering.

Figure 1 is a simplified flow diagram of the Coalinga plant.[2] It uses tubular cellulose acetate membranes laid horizontally in tiers each containing 28 tubes. The plant operates at 600 psig, and produces nominally 5000 gal/day of potable water from 10,000 gal/day of a feed well-water containing ~2500 ppm of dissolved salts. The raw feed water, supplied to the plant at 50 psig, first flows through a Cuno filter which minimizes the entrance of particulate water into the other parts of the plant. The feed can be treated by injection of a treating-agent either upstream or downstream of the filter.

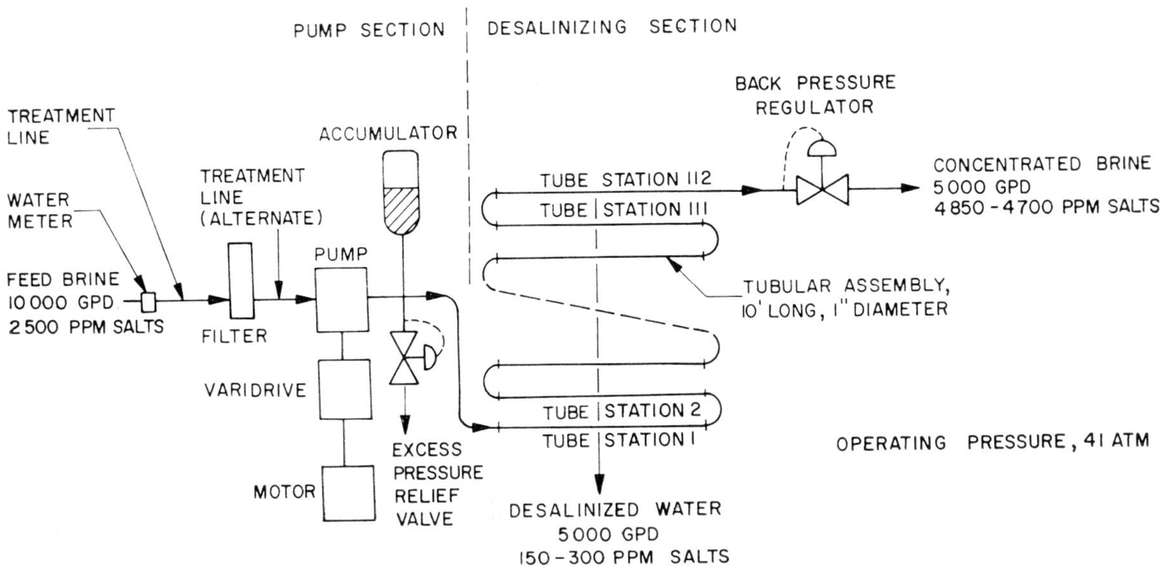

Figure 1 Flow diagram of Coalinga plant[2].

The main plant pump is a Worthington triplex reciprocating pump. The feed flow rate can be adjusted as desired during normal operation by the U.S. Motors Varidrive. A Greer accumulator of one gallon capacity is mounted on the discharge side of the pump to smooth out flow and pressure cycling due to piston reciprocation. The discharge line has also an excess pressure relief valve and a low-pressure audible alarm (not shown in Fig. 1). three layers of nylon could be included to provide more strength at the perforated areas.

The tubes are connected in series through 1-inch U-bends. The valving permits either the upper or the lower two tiers to be operated independently so that half of the plant can be kept in operation during the 10 to 15 minute period when the other half is down for a tube assembly change. There is a collection trough for each tire and each individual tube is enclosed in a plastic sleeve with all production passing through a small side outlet to a representative station in the collection trough for that tier. By this means, production from each tube can be monitored separately.

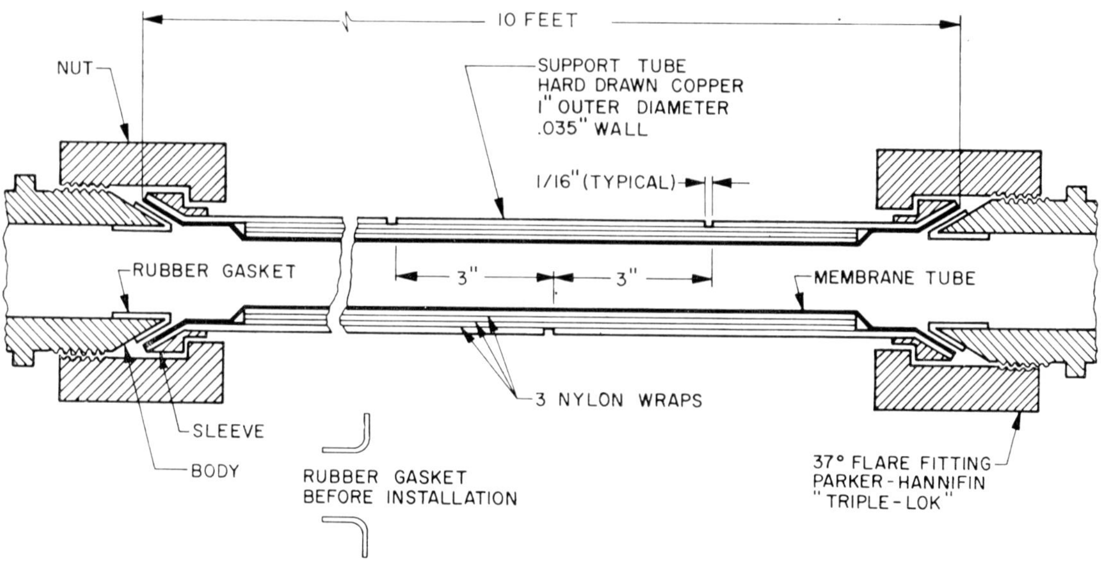

Figure 2 Tubular membrane assembly in Coalinga plant[1].

The tubular membranes were made by the gravity-drop technique[1], and assembled as shown in Figure 2. Each assembly is 10 ft long and nominally 1 inch in diameter with an effective membrane area of 2.24 ft^2. The assembly consists of a tubular membrane around which are several porous wraps to provide a low resistance path to the $\frac{1}{16}$ inch perforations in the copper support tube which assumes the expansive stresses due to internal pressure everywhere except where the tube is perforated. In this region the porous wraps support the membrane and prevent it from extruding through the perforations. It is hence important to have adequate strength in the porous wraps. Most of the assemblies used in the early test period contained membrane tubes which were wrapped with one layer of smooth filter paper around which were $1\frac{1}{2}$ layers of nylon fabric (French Fabric Co. No. 627T), the latter to provide a low resistance path and strength. However, some of the latter membrane tubes were wrapped only with nylon fabric, under which conditions

The system pressure is controlled by setting the back-pressure regulator on the concentrated brine stream. A manual bypass valve is used during start-up and shut-down to relieve pressure without changing the setting on the back pressure regulator.

Tables I to IV give some data on the performance of the Coalinga plant. Details of some problems and solutions relating to the operation and performance of the plant are given by Loeb and Selover[3]. Deposition of a ferric hydroxide slime caused membrane fouling and decline in production. Chlorination of feed minimized the decline but appeared to cause failures of the membrane and/or its nylon support wraps[2]. Production decline due to fouling could be controlled without membrane failures by feed pretreatment with catalysed sodium

TABLE I

Coalinga Pilot Plant Analyses, 17 August 1965
Data of Loeb and Johnson[2]

Test details	Feed brine	Concen-trated brine	Desali-nized water
pH	7.7	8.0	7.8
Iron, ppm	0.018	0.06	0.012
Hydroxide, ppm	nil	nil	nil
Boron, ppm	2.75	3.50	2.25
Carbonates (CO_3), ppm	nil	nil	nil
Bicarbonates (HCO_3), ppm	161.7	375.2	24.4
Chlorides (Cl), ppm	262.4	517.7	74.5
Sulfates (SO_4), ppm	1260.5	2942.9	34.2
Phosphates (PO_4), ppm	nil	nil	nil
Silica (SiO_2), ppm	49.2	92.8	8.4
Iron and Alumina (R_2O_3), ppm	1.2	2.8	0.8
Calcium (Ca), ppm	128.7	299.4	6.9
Magnesium (Mg), ppm	89.0	199.1	1.0
Sodium (Na), ppm	521.0	1168.0	66.0
Total dissolved solids, ppm	2477.6	5592.8	220.8
Total hardness (as $CaCO_3$), ppm	692.4	1577.7	25.2
Total incrustating solids, ppm	742.8	1673.3	34.4
Free carbon dioxide, ppm	5.0	6.0	1.0

TABLE II

Changes in the Number of Tubular Assemblies in Coalinga Pilot Plant
Data of McCutchan[5]

Date	Size of membrane unit	Effective membrane area, ft^2
4 June 1965	56-1″ tubes	125
11 Aug 1965	112-1″ tubes	250
2 Oct 1966	134-1″ tubes	300
12 Oct 1966	130-1″ tubes + 4-2″ tubes	309
16 Mar 1967	172-1″ tubes + 4-2″ tubes	403

sulfite to eliminate the dissolved oxygen. Slime removal can also be effected by scrubbing the membrane every 3 to 4 days by passing, under hydraulic pressure, a polyurethane ball of about 2 inch uncompressed diameter, through the tube-array. The latter technique adequately prevents production-decline and requires no continuous expenditures of chemicals. The data presented in Table IV show that the plant as a whole has maintained an outstanding performance record.

TABLE III

Antifouling Methods Used in Coalinga Plant
Data of McCutchan[5]

Method	Duration of test	Results
Chlorine (continuous dosing)	72 days	Flux decline somewhat reduced
Copper sulfate (continuous dosing)	14 days	No effect
Catalysed sodium sulfite (continuous dosing)	250 days	Flux decline noticeably lessened
Citric acid (once through flushes)	5 times ½ hr each time	No effect
Rapid brine flushes	3 times ½ hr each time	No effect
Ball flushes (once-through every 3 days)	160 days	Same as Na_2SO_3 flushing
Ball flushes (once-through every 6 days)	85 days	Poor

TABLE IV

Performance of Coalinga Pilot Plant
Data of Johnson, McCutchan, and Bennion[6]

Performance for first 3½ Years, June 4, 1965–Dec. 9, 1968	
Total production	7,605,000 gallons
Average product flux	20 gal/day · ft^2
Average TDS of feed	2450 ppm
Average TDS of product	260 ppm
Average solute separation	89.4%
Average desalination ratio	9.4%
Plant availability	98.6%

The Yotvatah plant designed by the Negev Institute for Arid Zone Research, is in operation since September 11, 1968. It was designed to produce 200 m^3/day (52840 gal/day) of potable water from brackish water containing 2400 ppm of dissolved solids (of which about half are divalent ions) at an operating pressure of 40 to 50 atm. Thus the Yotvatah plant represented a 10-fold enlargement of the Coalinga plant. Though both the plants use the same type of membrane, their design features are different in some basic respects. In the Yotvatah plant, the tubular assemblies are 20 ft long and 2.2 inches in diameter; the assemblies are in vertical array; the material of construction of the support tubes is either aluminum or glass reinforced epoxy; the entire plant stands in the open with no protective building; and, the product water collection system is protected from weather

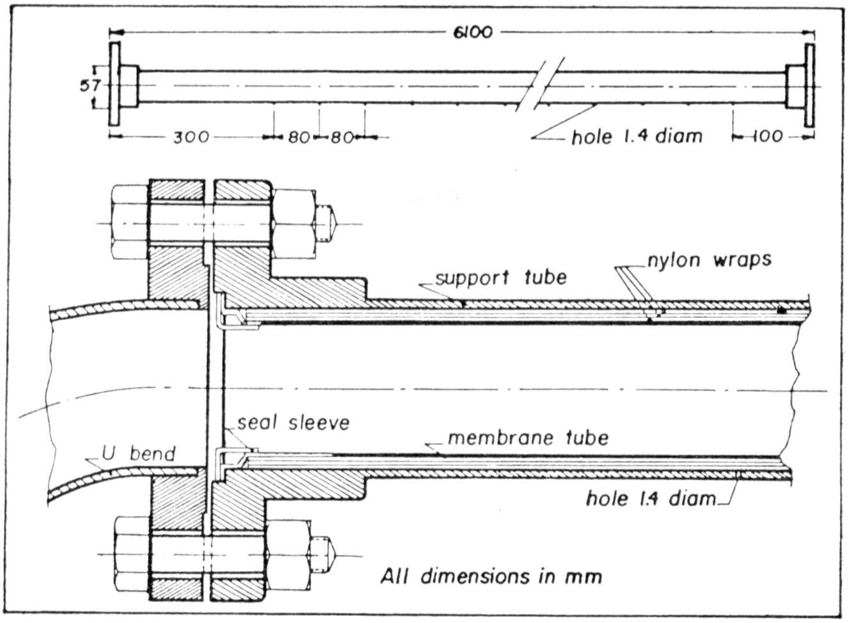

Figure 3 Tubular membrane assembly in Yotvatah plant[7].

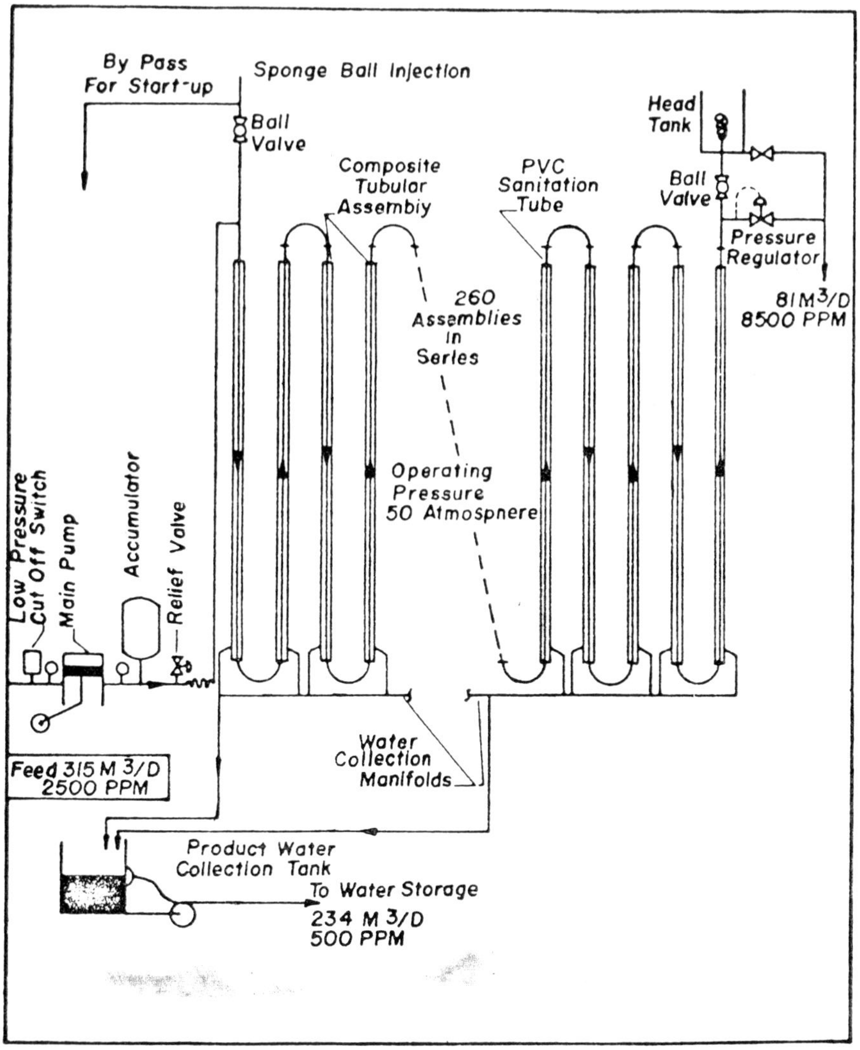

Figure 4 Flow diagram of Yotvatah plant.[7]

TABLE V

Performance of Yotvatah Plant
Data of Cohen and Loeb[7]

Sample location*	Operating hours	Film shrinkage temp., °C	Ion, ppm						TDS	Flux at 40 atm, gal/day · ft²
			Ca	Mg	Na	Cl	SO_4	HCO_3		
Feed			266	142	328	584	881	281	2482	
Product			16	5	49	89	22	24	205	15.4
Station 43	2675	88	29	11	61	97	77	43	318	14.5
Station 61	7411	84	32	14	86	159	64	43	398	14.0
Station 88	158	84	4	Trace	21	27	8	12	72	13.7
Station 115	3398	80	28	7	96	159	37	55	382	15.0
Station 138	1800	84	17	5	34	53	36	31	176	16.7
Station 182	430	84	22	7	67	115	35	37	283	20.3
Brine			431	234	515	912	1465	439	3996	

* 201 stations in operation

by PVC sanitary tubing over each of the tubular assemblies. Figures 3 and 4 give schematic diagrams of tubular assembly and the plant lay-out. A set of typical data on plant performance is given in Table V. In plant operations, it was found that cleaning of membrane surface with dilute citric acid solution at a pH of 4 was far more effective than the sponge ball cleaning method described earlier. It was also found that storage of membranes for periods up to 12 months did not affect their performance.

Several pilot plants are being field tested by the OSW (Office of Saline Water, U.S. Department of the Interior) for brackish water conversion. A typical set of their field test data[8] is given in Table VI.

In 1970, OSW initiated a program[9] on the design, construction and operational acceptance testing of a 250,000 gal/day brackish water reverse osmosis test bed plant. This program supports membrane assembly designs developed by Aerojet-General and Gulf General Atomic. The Aerojet design is based on the use of $\frac{9}{16}$ inch i.d. porous wall reverse osmosis tubes with separable membrane-lined substrate sleeves. Two different membrane assembly configurations are being considered. One is comprised of 18 tubes each having a 868-foot path length, manifolded in parallel, and containing $\frac{1}{8}$ of the total membrane area installed in the test bed. The other configuration is composed of 50 tubes each having a 868-foot path length, manifolded in parallel. The Gulf General Atomic's membrane assembly has 10 spiral-wound membrane elements contained in an 8-inch diameter, 33-foot long Schedule-40 seamless pipe conforming to ASTM material specification A 53-B. These membrane elements are similar to commercial, smaller size, spiral-wound elements described later in this article. The above OSW program is indicative of the progress of reverse osmosis technology for brackish water conversion.

TABLE VI

Reverse Osmosis Treatment of Brackish Waters
Data of Channabasappa[8]

Film type: cellulose acetate
Operating pressure: 600 psig
Product to waste ratio: 3 : 1

Test details	Solute concentration, ppm		
	Feed	Product	Waste
pH	6.7	5.3	6.8
Total dissolved solids	5170	320	20600
Total hardness as $CaCO_3$	1880	20	7350
Ca^{++}	360	7.6	1400
Mg^{++}	240	0.5	940
Na^{++}	900	110	3400
K^+	26	3.8	91
HCO_3^-	340	12	1150
SO_4^{--}	630	0	2580
Cl^-	2020	170	7850
NO_3^-	0.5	0.26	3.3
BO_3^-	0.3	0.22	0.53
SiO_2	34	76	120
Total iron	6.5	0.12	24
Total manganese	3.8	0	16
Total phosphate	10	0.18	40
Total alkalinity	280	10	940

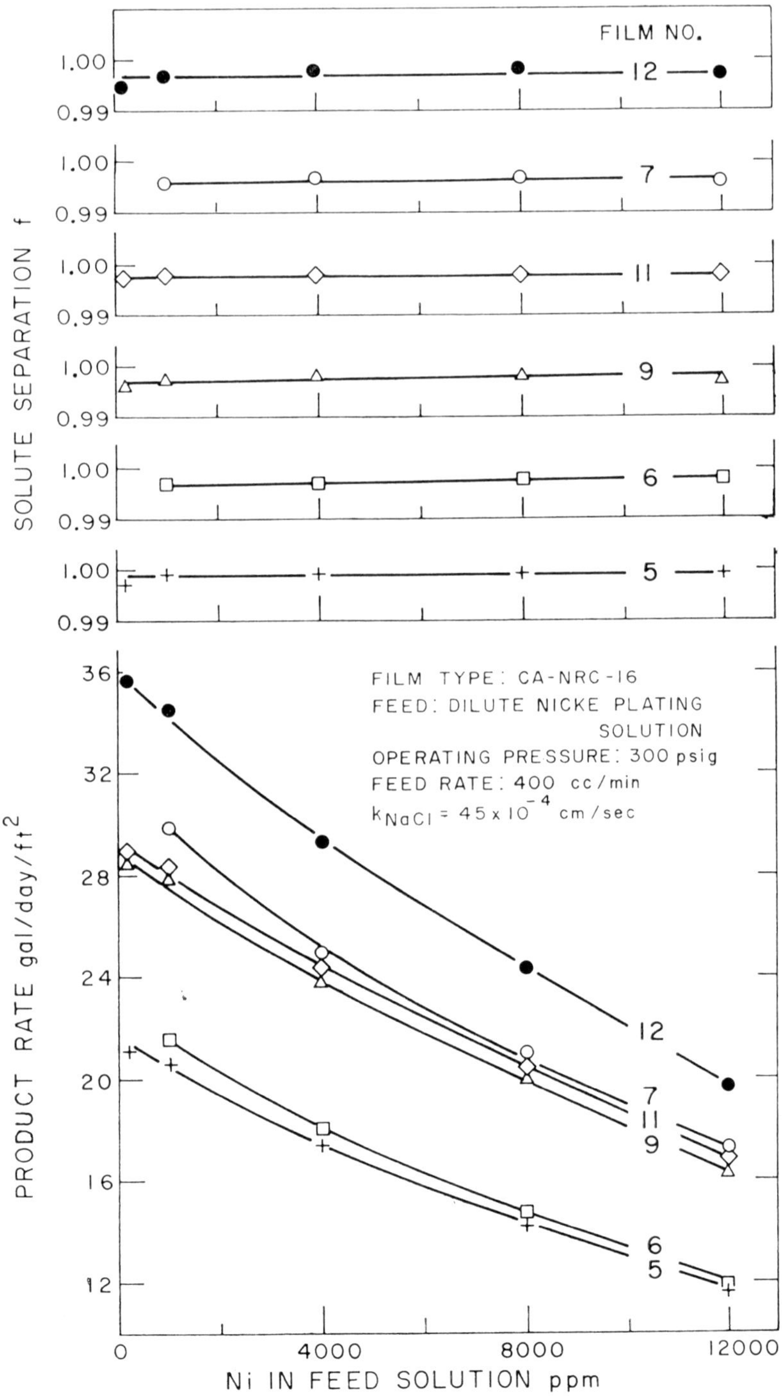

Figure 5 Effect of nickel concentration in diluted nickel plating solution on membrane performance.[66]

Sea Water Conversion

To obtain potable water from sea water in a single stage reverse osmosis operation at 1500 psig with significant product water recovery (~50%), an initial solute separation of >99% is required. Currently available cellulose acetate and other cellulose ester membranes have been shown to be capable of giving the required level of solute separation for sea water or its equivalent (3.5 wt-% $NaCl$–H_2O) feed solutions as indicated by data in Table III in Part I. The membrane materials particularly being studied by the OSW for sea water conversion[10] are given in Table VI; the use of higher acetylated (degree of substitution >2.45) cellulose acetate material for membranes is believed to contribute to higher solute separation and lower flux decline with time.

Successful single-stage sea water conversion using flat membranes made from Eastman cellulose acetate with acetyl content of 39.8%, has been demonstrated several times (see Reference 11, and also data in Part I). Using tubular membranes made from similar cellulose acetate material, single-stage sea water conversion has been reported by McCutchan and Bennion[12] (Table VII) with

TABLE VII

Some Membrane Materials for Sea Water Conversion
Data of Gillam and Podall[10]

Feed: 3.5% $NaCl-H_2O$
Operating pressure: 1500 psig

Membrane Material	Typical membrane performance: Solute sepn., %	Initial flux, gal/day · ft²	% decrease in flux in 1 year (estimated)
Commercial $CA_{2.45}$	98	10	50
$CA_{2.63}$ blend	99.5	11.5	10
$CA_{2.61}$ homopolymer	99.5	11	22
X_l–$CA_{2.45}M_{0.17}$	99.7	9	10
$CA_{2.64}$/CN–CA composite	99.5	10	22

CA = cellulose acetate; the degree of acetylation is given by the subscript
X_l = cross-linked
CAM = cellulose acetate methacrylate
CN–CA = cellulose nitrate–cellulose acetate porous support

somewhat low productivities. However, these results show that neither a higher degree of acetylation nor a change in the chemical nature of the ester group is a necessary requirement for the choice of cellulose esters as membrane materials for single-stage sea water conversion even though all compounds listed in Table VI are undoubtedly suitable membrane materials for such purpose. The suitable choice of membrane material by itself cannot ensure successful single-stage sea water conversion; the latter can be ensured only by the appropriate choice of the membrane-solution-operating system requirements as discussed in general in Part I, and by Ohya and Sourirajan[13] with particular relevance to sea water conversion.

The surface of a reverse osmosis membrane has in general a nonuniform porous structure. The

TABLE VIII

Reverse Osmosis for Sea Water Conversion
Data of McCutchan and Bennion[12]

Film type: cellulose acetate (tubular)
Feed: sea water containing 35200 ppm of total dissolved solids
Temperature of feed water: 16–25°C

Film material type[a]	Film shrinkage temp., °C	Operating pressure, psig	Solute separation, %	Membrane flux, gal/day · ft²
E-398-10	94	800	98.4	3.5
	97	800	99.2	0.97
	94	900	98.5	4.0
	97	900	99.0	1.13
	94	1000	98.6	4.6
	97	1000	99.1	1.3
	94–96	1000	99.1	2.45
	94	1100	98.5	5.1
	97	1100	99.1	1.6
	94	1200	98.7	5.1
	97	1200	99.1	1.95
E-394-45	94–96	1000	99.0	4.54
E-394-60	94–96	1000	98.8	3.51
E-400-25	94–96	800	98.9	5.4
E-398-3[b]	—	1500	99.64	2.7

[a] Eastman designation
[b] Feed: 3.5% $NaCl-H_2O$

solute separation and membrane flux are functions of number, size, and distribution of surface pores. The sizes of such pores may be broadly divided into two classes: those which are bigger than the critical pore size (which pores contribute to lower solute separations), and those which are equal to or smaller than the critical pore size (which pores already yield the maximum solute separation) for the particular reverse osmosis system under consideration. By increasing the shrinkage temperature for the membrane, the former pore sizes become smaller (which is desirable to obtain higher solute separation), and the latter ones become either still smaller or essentially closed (which is not desirable from the points of view of both solute separation and membrane flux). Consequently, beyond a certain temperature, any further increase in shrinkage temperature, while contributing necessarily to a decrease in membrane flux, affects solute separation either negligibly or negatively. From the point of view of a fuller practical utilization of the pore structure on the membrane surface and higher membrane productivity, it is hence not desirable to try to resort to higher and higher shrinkage temperatures in an attempt to reduce the size of the biggest pore to that of the critical pore size, and approach reverse osmosis operation close to 100% solute separation. With reference to sea water conversion, it is preferable (from the point of view of membrane productivity) to operate the reverse osmosis process in two stages with 90% overall solute separation in each stage. For example, using sea water feed containing 35000 ppm of total dissolved solids (TDS), the first stage may be operated at 1500 psig to yield product water containing 3500 ppm of TDS for use as feed for the second stage which can be operated at 600 psig or less, to yield the final product (potable) water containing 350 ppm of TDS. Development of such two-stage reverse osmosis operating systems for sea water conversion may be expected.

TABLE IX

Water Softening by Reverse Osmosis
Data of Hauck and Sourirajan[14]

Film type: cellulose acetate (Batch 18)
Operating pressure: 1000 psig

Film No.	System	Solute concn., ppm Feed	Solute concn., ppm Product	Product rate gal/day·ft²
		Calcd. as $CaCO_3$		
H-1	$CaCl_2$–H_2O	302	2	38.6
	$CaCl_2$–H_2O	498	2	42.3
	$CaCl_2$–H_2O	782	2	41.4
	$MgCl_2$–H_2O	315	<1	41.1
	$MgCl_2$–H_2O	516	<1	37.4
	$MgCl_2$–H_2O	800	2	36.1
	$(CaCl_2+MgCl_2)$–H_2O	300	<1	34.2
	$(CaCl_2+MgCl_2)$–H_2O	495	<1	36.7
	$(CaCl_2+MgCl_2)$–H_2O	807	<1	35.1
H-2	$CaCl_2$–H_2O	302	17	42.8
	$CaCl_2$–H_2O	504	21	42.8
	$CaCl_2$–H_2O	806	26	42.6
	$MgCl_2$–H_2O	315	24	49.2
	$MgCl_2$–H_2O	509	36	43.7
	$MgCl_2$–H_2O	829	45	41.5
	$(CaCl_2+MgCl_2)+H_2O$	304	19	43.5
	$(CaCl_2+MgCl_2)+H_2O$	504	28	43.8
	$(CaCl_2+MgCl_2)+H_2O$	805	41	42.6
	Natural hard waters			
H-1	Coalinga, Calif.	843	<1	26.5
	Webster, S.D.	610	<1	28.6
	Roswell, N.M.	641	4	29.6
	San Diego, Calif.	340	2	30.4
	Indianapolis, Ind.	247	4	30.0
H-2	Coalinga, Calif.	843	12	40.1
	Webster, S.D.	610	5	43.2
	Roswell, N.M.	641	14	42.9
	San Diego, Calif.	340	7	45.1
	Indianapolis, Ind.	247	11	45.0

Separation of Some Inorganic Solutes

The criteria for the separation of inorganic solutes in aqueous solutions using porous cellulose acetate membranes have already been discussed in Part I. Hardness is of special concern in municipal and industrial water supply. The substances responsible for hardness are primarily calcium and magnesium, and to a lesser extent, iron, manganese, strontium, and aluminum. Phosphates, nitrites, nitrates, chlorides, fluorides, sulfates, and ammonia are some of the other typical inorganic ions present as pollutants in waste waters. The data given in Tables IX to XII illustrate the possible applicability of reverse osmosis for the separation of several inorganic solutes commonly present in many waste waters.

Separation of Some Organic Solutes

The criteria for the separation of alcohols, phenols, and monocarboxylic acids in aqueous solutions using porous cellulose acetate membranes have already been discussed. Similar criteria may be expected to emerge for the separation of other classes of organic solutes. Extensive experimental data are available in the literature[17–21] for the separation of glycerol, sucrose, glucose, maltose, lactose, ethylene glycol, propylene glycol, and urea using porous cellulose acetate membranes; these data can be used for reverse osmosis process design from the points of view of waste recovery, water pollution control, or water renovation. The data of Hindin, Bennett, and Narayanan, given in Table XIII, are the most extensive available for the separation of organics with particular reference to waste water treatment. While all the above data indicate the possibility of the reverse osmosis separation of organic solutes in aqueous solutions, more detailed and fundamental studies on the subject are clearly called for in order to realize the fullest potentialities of the technique for the separation of organics present in waste waters.

TABLE X

Reverse Osmosis Separation of Some Inorganic Solutes in Aqueous Solution

Data of Matsuura and Sourirajan[15]

Film type: Cellulose acetate, Batch 316 (0/24)
Operating pressure: 250 psig
Feed rate corresponding to mass transfer coefficient, $k = 57 \times 10^{-4}$ cm/sec for 1500 ppm NaCl–H_2O feed

System	Feed concentration, ppm		Data for Film 2			Data for Film 4		
			Solute sepn., %		Product rate gal/day · ft²	Solute sepn., %		Product rate gal/day · ft²
A1–H_2O or A1–A2–H_2O	A1	A2	A1	A2		A1	A2	
NaH_2PO_4–H_2O	108	—	99.7	—	14.6	99.4	—	31.5
	422	—	99.9	—	14.4	99.5	—	30.3
	1035	—	99.4	—	14.2	98.4	—	30.2
	1540	—	99.8	—	14.0	98.5	—	30.2
NaH_2PO_4–$Na_4P_2O_7$–H_2O	51.2	53.9	98.7	99.3	14.4	99.3	99.0	31.0
	80	60	99.7	99.6	14.8	99.6	99.7	31.1
	537	536	99.9	99.3	14.2	99.5	99.2	29.3
	1086	958	99.7	99.9	14.0	99.1	99.9	29.3
NaH_2PO_4–$Na_5P_3O_{10}$–H_2O	55.6	54.9	99.3	99.3	14.7	99.3	99.3	31.2
	307	211	99.9	99.8	14.7	99.8	99.9	30.6
	505	504	99.9	99.8	14.2	99.7	99.7	29.3
	1011	1188	99.8	99.9	13.5	99.3	99.9	28.0
NaH_2PO_4–$(NaPO_3)_6$–H_2O	54.3	75.2	99.3	99.2	14.7	99.1	99.3	31.6
	288	392	99.8	99.7	14.8	99.7	99.7	30.2
	581	612	99.8	99.9	14.3	99.5	99.9	29.6
	1074	1321	99.9	99.8	14.2	99.5	99.6	29.3
$NaNO_2$–H_2O	118	—	89.7	—	15.2	83.6	—	32.1
	521	—	91.7	—	14.7	83.1	—	31.1
	1005	—	92.3	—	14.5	83.1	—	30.7
	1467	—	92.6	—	14.1	84.0	—	29.9
$NaNO_3$–H_2O	110	—	91.8	—	14.9	84.2	—	31.1
	550	—	89.8	—	14.3	80.4	—	30.4
	1065	—	91.1	—	14.3	80.5	—	30.4
	1523	—	90.9	—	13.9	81.0	—	29.6
NH_4Cl–H_2O	102	—	97.2	—	14.9	94.1	—	31.3
	501	—	95.0	—	14.3	86.5	—	30.2
	914	—	93.8	—	14.2	83.6	—	30.0
	1575	—	95.2	—	13.9	85.0	—	29.7
NaCl–H_2O	96	—	96.1	—	14.4	91.3	—	31.1
	497	—	95.8	—	14.2	88.6	—	30.7
	972	—	96.0	—	13.7	89.7	—	29.8
	1515	—	96.1	—	13.4	89.4	—	28.9
	1992	—	97.0	—	13.0	89.2	—	28.2
NaF–H_2O	102	—	94.8	—	14.9	93.7	—	31.5
	538	—	98.0	—	14.6	95.5	—	30.9
	1020	—	98.4	—	14.8	95.0	—	31.3
	1440	—	98.3	—	13.3	94.8	—	28.1
Na_2SO_4–H_2O	99	—	98.8	—	14.8	98.7	—	31.2
	508	—	99.6	—	14.8	99.4	—	31.3
	1005	—	99.8	—	14.4	99.4	—	30.4
	1488	—	99.6	—	14.2	99.2	—	29.9

Cellulose Acetate Membranes and pH of Feed Solutions

Cellulose acetate material is subject to hydrolysis when in contact with feed solutions of different pH. Hydrolysis changes the chemical nature of the membrane surface and adversely affects the performance of the membrane in reverse osmosis. Therefore with respect to the use of cellulose acetate membranes for the reverse osmosis operation, the pH of the feed solution is of critical importance. For cellulose acetate material whose acetyl content is 39.8%, hydrolysis rate is minimum at pH of ~5 which is hence the best operating pH for the reverse osmosis process using membranes

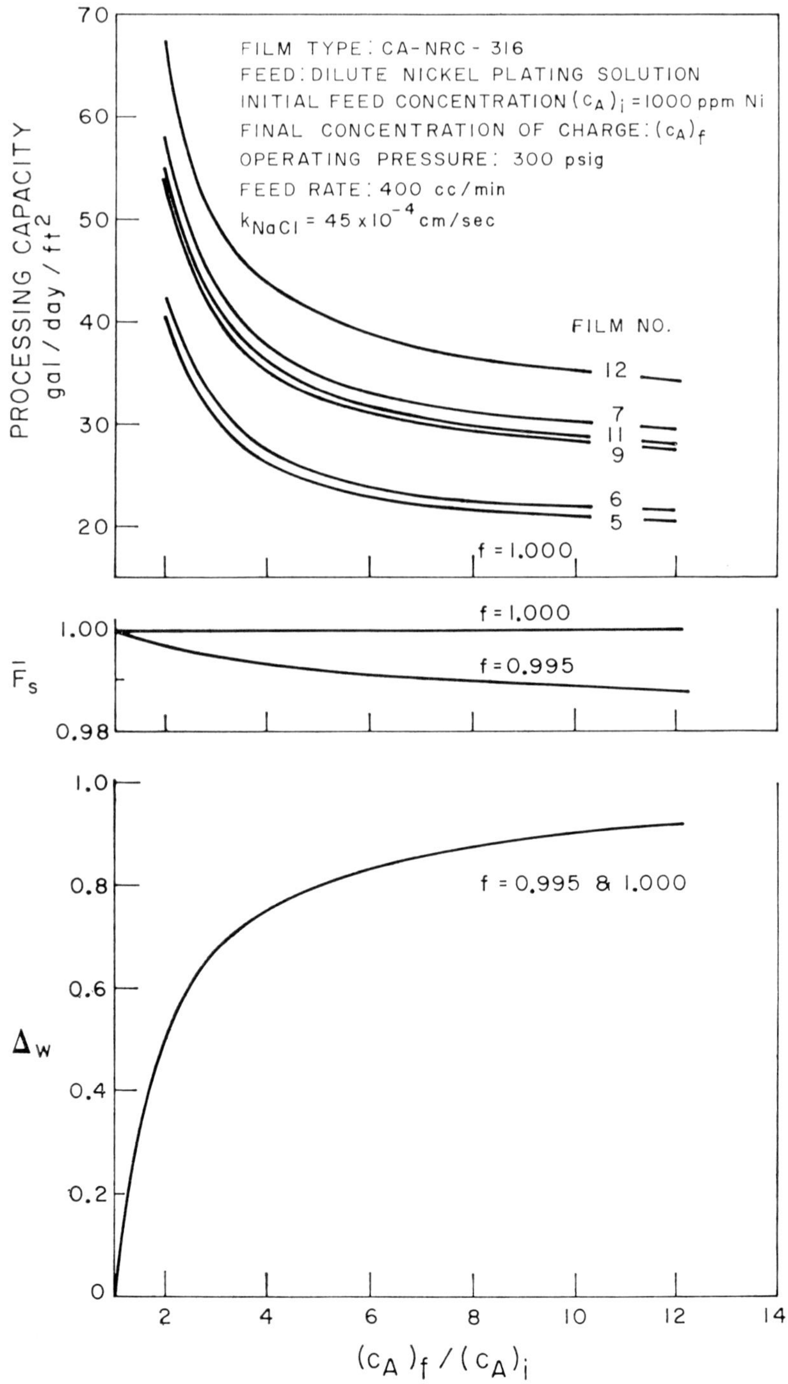

Figure 6 Water recovery, solute recovery, and processing capacities of membranes for the reverse osmosis treatment of diluted nickel plating solution.[66]

TABLE XI

Reverse Osmosis Separation of Some Inorganic Solutes in Aqueous Solutions

Data of Hindin and Bennett[16]

Film type: Cellulose acetate (Schleicher and Schuell, B-18)
Operating pressure: 100 atm

Solute	Feed solution		Solute sepn., %	Membrane flux, gal/day·ft²
	concn., mg/l	pH		
$NaCl(Na^+)$	10.7	—	86.5	14.8
	105.6	—	83.5	7.0
	1040.6	—	79.0	5.5
$KCl(K^+)$	5.5	—	90.0	22.2
	53.4	—	74.3	17.6
	521.0	—	39.4	15.5
$NH_4Cl(NH_4^+—N)$	10.8	—	97.2	6.8
	48.8	—	97.1	5.2
	250.0	—	90.4	4.5
$CaCl_2(Ca^{++})$	10.8	—	63.0	11.6
	48.8	—	59.0	11.0
	250.0	—	61.1	9.9
$MgCl_2(Mg^{++})$	11.1	—	63.9	16.4
	54.0	—	60.0	14.2
	102.2	—	61.5	12.2
	266.8	—	61.5	11.0
$CdCl_2(Cd^{++})$	9.25	—	70.0	20.8
$CuCl_2(Cu^{++})$	9.96	5.47	86.0	14.0
	100.5	5.20	82.3	13.2
$FeCl_3(Fe^{+++})$	9.55	—	97.3	12.4
	93.6	—	97.9	9.2
$AlCl_3(Al^{+++})$	10.29	4.38	95.1	13.0
	150.1	4.08	96.2	11.2
$KMnO_4(Mn^{+7})$	10.8	—	66.2	8.7
	106.8	—	55.4	7.8
$K_2CrO_4(Cr^{+6})$	5.01	—	95.0	10.5
	49.6	—	93.5	9.7
$NaCl(Cl^-)$	12.1	—	90.9	18.1
	109.8	—	83.2	15.8
	1080	—	73.9	14.4
$NaF(F^-)$	6.04	—	85.1	10.8
	58.8	—	98.3	10.5
$NaNO_2(NO_2^-—N)$	4.43	5.68	77.7	9.2
	40.1	—	72.0	8.3
$NaNO_3(NO_3^-—N)$	2.53	6.0	68.0	18.3
	25.6	6.35	68.3	18.3
	254	5.27	73.3	18.4
$NaCN(CN^-)$	7.55	9.0	85.1	17.8
	78.1	10.3	79.0	18.2
$Na_2SO_4(SO_4^{--})$	13.8	5.75	>93.0	14.9
	154	5.75	98.0	14.8
	1480	5.72	>99.9	14.8
$Na_2B_4O_7(B_4O_7^{--}—B)$	13.4	9.0	50.3	15.8
	132	9.1	51.0	15.9
$Na_2B_4O_7(SiO_2^{--})$	1.8	8.9	77.0	22.0
	17.0	10.2	91.8	23.0
	175	11.3	93.5	23.3

TABLE XI (*cont.*)

Solute	Feed solution		Solute sepn., %	Membrane flux, gal/day. ft²
	concn., mg/l	pH		
$Na_2HPO_4(PO_4^{-4})$	2.9	7.0	88.0	42.5
	29.7	7.5	88.0	40.6
	238	7.7	66.0	35.8

TABLE XII

Reverse Osmosis Separation of Ions from Sewage Effluents and Single-Solute Aqueous Feed Solutions

Data of Hindin and Bennett[16]

Solute	Calculated as	Solute separation, %			
		Trickling filter effluent feed	Activated sludge effluent feed	Single solute aqueous feed	Sepn. based on total ionic strength
Ammonia	N	96	98	97	91
Hardness	$CaCO_3$	87	94	—	—
Calcium	Ca	92	>99	61	61
Magnesium	Mg	83	90	64	62
Sodium	Na	79	91	82	80
Potassium	K	96	95	88	60
Iron	Fe	—	87	95	97
Total alkalinity	$CaCO_3$	84	85	—	—
Chloride	Cl	81	89	91	80
Fluoride	F	97	93	88	>98
Nitrate	N	53	—	68	75
Nitrite	N	—	65	83	72
Sulfate	SO_4	90	90	93	99
Phosphate	PO_4	97	93	88	66
Silica	SiO_2	—	81	92	95
Polyphosphate	PO_4	—	95	—	—

Film type: cellulose acetate (Schleicher and Schuell, B-18)
Operating pressure: 100 atm

TABLE XIII

Reverse Osmosis Separation of Some Organic Solutes in Aqueous Solutions

Data of Hindin, Bennett, and Narayanan[22]

Film type: cellulose acetate (Schleicher and Schuell, B-18)
Operating pressure: 100 atm

Solute	Feed solution		Solute sepn., %	Membrane flux, gal/day·ft²
	Concn., mg/l*	pH		
Alkyl benzene sulfonate (ABS)	1.06	—	94.0	16
	10.6	—	82.0	18
Linear alkyl benzene sulfonate (LAS)	0.935	7.2	97.2	19
	9.6	5.0	99.1	18
Sodium oleate	38.8	—	99.9	7

TABLE XIII (*cont.*)

Solute	Feed solution Concn., mg/l*	Feed solution pH	Solute sepn., %	Membrane flux, gal/day·ft²
Sodium stearate	40.0	—	99.9	10
Sodium laurate	25.0	—	>99.9	10
Motor oil, SAE-20	10	—	>99.7	19
Tannic acid	1.11	4.7	90.0	12
	11.1	4.6	92.1	11
	56.0	3.9	94.9	11
Lindane	50 ng	—	84.0	18
	500 ng	—	73.4	16
pp′-DDT	910 ng	—	>97.3	20
	7000 ng	—	99.5	17
Technical TDE (DDD)	69 μg	—	>99.9	15
	532 μg	—	>99.9	14
Benzene hexa-chloride	638 μg	—	52	14
Casein	100	6.6	>99	18
	1000	7.8	99	17
Peptone	100	4.5	>98	17
	1000	4.0	>99	17
Cellulose	8.4	—	61	24
	47.0	—	69	25
Soluble starch	1.18	—	>74	15
	12.8	—	>97	13
Technical humic acid	1000 units†	6.5	>99	11
	2000 units†	6.5	>99	12
	11000 units†	9.2	>99	9
p-Chloro nitro-benzene	95 μg	5.5	27.3	9
Methyl formate	80	5.4	14	10
Methyl acetate	93	5.5	44	9
2,4-D Isopropyl-ester	9.4 μg	—	>99.7	6
	97.2 μg	—	95.6	12
	673 μg	—	84.9	9
Acetone	90	5.3	47	9
Mixture, Phenol	10.1	7.5	17.8	10.2
+Chlorophenol	9.5		66.3	
Mixture, Phenol	101	8.7	18	11.6
+Chlorophenol	106		78.3	
Mixture, Phenol	1031	9.0	18.1	9.4
+Chlorophenol	1165		87.0	
Ester mixture,				
Methyl formate	98	4.2	66.6	
Ethyl formate	87		67.8	9.6
Methyl propionate	92		51.0	
Ethyl propionate	92		56.5	
Amino acid mixture,				
L-Glycine	10	—	50	
L-Lysine–HCl	10		>95	12.0
dl-Phenylalanine	10		>95	
Monocarboxylate mixture,				
Sodium acetate	10.4	—	84	
Sodium butyrate	9.5	—	80	19.4
Sodium valerate	9.2	—	72	
Sodium isovalerate	9.2	—	72	
Monocarboxylate mixture,				
Sodium acetate	104.2	—	77.3	
Sodium propionate	99.1	—	74.2	
Sodium butyrate	95.1	—	72.6	14.5
Sodium valerate	92.0	—	69.4	
Sodium isovalerate	93.0	—	70.6	
Monocarboxylate mixture,				
Sodium acetate	1042	—	64.9	
Sodium propionate	991		64.2	
Sodium butyrate	951		65.2	8.2
Sodium valerate	920		68.1	
Sodium iso-valerate	930		65.0	
Mixture,				
2,4-D Isopropyl ester	65.8 μg	—	54.4	
p-Chloro nitro-benzene	82.0 μg		25.6	
Phenol	10.3		13.8	8.2
Chlorophenol	9.8		36.7	
Acetone	100.7		33.3	
Ethyl acetate	100.0		60.4	
Waste water effluent feed				
Alkyl benzene sulfonate[a]	3.24	—	93.0	16
Carbohydrate[a]	13.5	7.4	56	12
Anthrone reactive[b]	4.3	7.8	93	17
Pectin[b]	10	—	>91	10
Cellubiose[b]	10	—	>95	10
Mixture[b],				
2,4-D Iso-propyl ester	87.2 μg	—	94.1	
p-Chloronitro-benzene	118.7 μg		40.4	
Phenol	9.5		21.4	
Chlorophenol	9.6		50	9
DDT	97.5 μg		>98	
Ethyl acetate	72.6 μg		66	
Acetone	95.8 μg		39.1	
Neutral lipids[b]	500 μg	—	90.4	14

* Unless otherwise specified
† Color units
ng = nanogram (10^{-6} mg)/l; μg = microgram (10^{-3} mg)/l
[a] Trickling filter effluent
[b] Activated sludge effluent

TABLE XIV

Separation of Phenols
Data of Matsuura and Sourirajan[27]

Film type: Cellulose acetate, Batch 316 (0/24)
Operating pressure: 500 psig

Expt. No.	Feed solution: Alkali used to change pH	Alkali concn., ppm	pH	Solute concn., ppm	Membrane permeated product solution: pH	Solute sepn., %	Product rate, gal/day · ft²	pH	Solute sepn., %	Product rate, gal/day · ft²
		Solute: Phenol				Data for Film 1			Data for Film 3	
7	—	0	6.3	91.5	—	−6.4	30.1	—	−6.4	62.1
12	NaOH	6.9	9.7	94.4	—	15.3	30.9	8.6	11.4	63.1
8	NaOH	23	10.6	97.4	—	60.4	30.2	8.7	54.8	61.0
10	NaOH	46	11.2	95.4	7.6	84.5	31.5	9.5	76.5	65.0
14	NaOH	69	11.5	92.3	7.5	88.0	31.5	10.1	79.9	64.6
18	NaOH	230	12.0	97.6	9.9	90.9	32.7	11.5	80.3	68.3
		Solute: Phenol				Data for Film 51			Data for Film 55	
61	NaOH	23	11.4	10.2	7.5	75.9	28.9	8.7	62.0	53.7
62	NaOH	24.6	11.4	19.5	7.5	78.7	29.3	9.0	67.6	54.1
63	NaOH	32.2	11.4	48.3	7.5	78.9	28.7	9.0	69.8	53.1
64	NaOH	38.6	11.4	64.4	7.4	83.9	28.2	9.0	76.3	51.9
65	NaOH	53.0	11.2	108.4	7.9	85.2	28.2	10.2	76.6	52.6
		Solute: p-Chlorophenol				Data for Film 21			Data for Film 23	
38	—	0	6.4	111.1	—	2.1	33.1	—	−0.3	65.7
41	NaOH	5.8	9.1	116.2	6.7	17.8	33.4	7.2	10.0	65.8
39	NaOH	12	9.6	113.0	6.9	40.1	36.3	7.5	30.5	68.2
40	NaOH	23	10.2	117.8	6.7	66.4	37.0	7.5	57.3	67.6
		Solute: p-Chlorophenol				Data for Film 61			Data for Film 65	
68	NaOH	46	11.2	128.6	6.7	93.7	27.8	9.4	81.7	62.4
69	$Ca(OH)_2$	40	10.6	128.6	5.6	98.5	28.1	8.5	83.0	63.9
70	$Ca(OH)_2$	40	10.3	128.6	5.8	88.1	29.4	8.3	77.5	66.8

Membrane performance with 1500 ppm NaCl–H_2O feed solution under the same operating conditions of pressure and feed flow rate:

Film No.	1	3	21	23	51	55	61	65
Solute sepn., %	94.9–92.5	86.1–83.8	96.1	83.8	95.9–93.7	86.1–85.6	95.7–89.9	86.8–77.9
Product rate, gal/day · ft²	29.5–32.6	60.6–63.9	38.2	69.4	27.4–27.5	51.0–51.6	26.7–28.6	59.9–66.1

made of the above material[23]. In the alkaline range, the hydrolytic degradation is particularly rapid for cellulose acetate membranes which are hence practically unsuitable for feed solutions whose pH is greater than 7. The hydrolysis rate can also be minimized by operating at lower temperatures (<25°C).

Separation of Phenols

Many industrial waste waters, such as those associated with the manufacture of plastics, dyes, disinfectants, herbicides, and petrochemicals are contaminated with phenol and substituted phenols usually in the concentration range of 10 to 100 ppm[24]. For purposes of water pollution control, it is usually required that the concentration of phenols in waste water streams be reduced to 1 ppm or less[25]. Therefore the separation of phenols from waste waters is an important industrial problem.

Using the Loeb-Sourirajan type porous cellulose acetate membranes and aqueous solutions feed containing 70 to 80 ppm of phenol, Lonsdale,

Merten and Tagami[26] reported negative separations (i.e., phenol enrichment in permeate) of 10 to 20% in the operating pressure range of 34 to 102 atm at 30°C; the negative separations increased with increase in operating pressure. The data given in Fig. 6 of Part I also show either negligible or slightly negative separations for phenol at 250 psig and 25°C with the particular films used.

On the basis of the physicochemical criteria of reverse osmosis separation discussed earlier, one may conclude that low or negative separation for phenol is due to its high hydrogen bonding ability and extremely small dissociation constant ($pK_a = \sim 10$), and its separation can be increased by ionizing it in the feed solution. The latter possibility is illustrated by the data given in Table XIV. In these experiments, cellulose acetate membranes were used in conjunction with feed solutions containing 10 to 118 ppm of phenol or p-chlorophenol. The phenols were ionized by increasing the pH of the feed solution by the addition of appropriate amounts of caustic soda or lime. Data are given for two sets of membranes to emphasise the fact that the levels of solute separation and product rate obtained do not represent limiting values. The membranes used were affected to some extent (as indicated by data on sodium chloride separations for the feed system NaCl–H_2O) by the high pH feed solutions indicating the need for more suitable membranes for the practical utilization of the process. The data presented, however, illustrate effectively the principle involved.

The results show that phenol separation increased from —6.4 to +90% for Film 1 when the pH of the feed solution was increased by the addition of NaOH, from 3.6 to 12.0. Under the conditions given for the experiments, 7, 12, 8, 10, 14, 18, 61, 62, 63, 64, and 65, the degree of dissociation of phenol in the feed solution was experimentally found to be ~0, 16.5, 65.9, 87.3, 94.9, 99.6, 87, 87.4, 90.2, 90.3, and 91.5% respectively. These data show that an increase in the degree of dissociation results in an increase in the level of solute separation. The results obtained with the other films for phenol feed solutions, and those obtained for p-chlorophenol feed solutions are similar; in the latter case, solute separations were actually increased from 2.1% to 98.5% by changing the pH of the feed solution and the nature of the alkali used for changing the pH. These results illustrate the potential applicability of reverse osmosis for the separation of phenols in waste waters.

Separation of Microorganisms

Field tests conducted with a Gulf General Atomic reverse osmosis unit using cellulose acetate membranes showed nearly complete elimination of bacteria from the Potamic River water feed[28]. A typical bacteriological test showed a plate-count reduction from the river feed of 4200 per cc to 7 per cc in the product water; the coliforms were reduced from a most probable number of 4600 per cc (positive) to less than 3 (negative) in the product.

TABLE XV

Reverse Osmosis Treatment of Some Polluted Waters
Data of Ironside and Sourirajan[29]

Film type: Cellulose acetate (Batch 18)
Operating pressure: 1000 psig

Film shrinkage temp., °C	Feed solution	Analysis of: Feed solution	Analysis of: Product water	Product water rate, gal/day · ft^2
88	Sewage water, unchlorinated	B. coli* av. 100,000/cm^3	sterile	31.1
88	Sewage water, chlorinated	B. coli* av. 400/cm^3	sterile	29.3
90	Aqueous solution of sodium dioctyl sulfo-succinate	370 ppm[a]	<0.5 ppm[a]	24.4
92	Aqueous solution of sodium alkylbenzene sulfonate	512 ppm[a]	0.6 ppm[a]	24.0
88	Brown lignin solution (acidic)	10.4%[a]	0.035%[a]	16.2

* Standard plate count
[a] Solute concentration

Ironside and Sourirajan[29] obtained sterile product water from unchlorinated and chlorinated sewage waters in reverse osmosis experiments using cellulose acetate membranes (Table XV). Essentially the same results were obtained by Hindin and Bennett[1] using Schleicher and Schuell type cellulose acetate membranes. Sinskey, Chu and Wang[30] found that the Abcor HFA-300 membranes (capable of retaining solutes of molecular weight >50,000) and the Amicon XM-100 membranes (capable of retaining solutes of molecular weight >100,000) were able to concentrate poliovirus type 1 (attenuated, Chat strain) without any virial inactivation. The permeate through the XM-100 membrane showed either no virus activity or only an insignificant level of detectable viruses. The above results indicate the possibility of obtaining essentially complete separation of viruses also using cellulose acetate membranes.

Treatment of Sewage Water

The application of reverse osmosis for the treatment of sewage water is one of the most active fields of current research and development. Reverse osmosis is being studied for the treatment of primary-treated, secondary-treated, or carbon-treated secondary sewage waters. Extensive laboratory scale and pilot plant data are already available.

Table XVI gives the results obtained with a typical cellulose acetate membrane and a number of samples of raw sewage water obtained from the Ottawa City primary sewage treatment plant[14]. Experiments were made at two operating pressures, 1000 and 500 psig with particular reference to the removal BOD, nitrate, phosphate, ABS, and total dissolved solids. The performance of the membrane (Batch 18-type) was good with respect to the removal of all the above contaminants. The recently developed Batch 316-type membranes (see Part I) may be expected to be even better.

Table XVII gives the data of Hardwick[31] obtained with a small recirculatory reverse osmosis system using cellulose acetate membranes operating at Harwell, England. The performance of the membrane was reported to be very satisfactory even after 6000 hours of continuous operation.

Table XVIII gives the data of Aerojet[32] which show that all grades of municipal waste water may be significantly upgraded by reverse osmosis treatment using cellulose acetate membranes. The data provide comparisons on the behaviour and response of the reverse osmosis process to carbon-treated secondary sewage, alum-treated secondary sewage, secondary sewage, alum-treated primary sewage, primary sewage, raw sewage, and digester supernatant. High removals of dissolved minerals, organic substances and suspended matter have been achieved in all cases. A continuous extended test of 2.5-months duration demonstrated that a constant product water flux of about 5½ gal/day.ft²

TABLE XVI

Reverse Osmosis Treatment of Primary-Treated Sewage Water
Data of Hauck and Sourirajan[14]

Film type: Cellulose acetate (Batch 18)
Product recovery: 90%

Test details	Solute concn. in feed, ppm	Operating pressure, psig	Product	
			Solute concn., ppm	Product rate, gal/day·ft²
B.O.D.	24	1000	3	29.6
	46	1000	8	35.8
	37	1000	2	38.4
	25	1000	5	31.0
	36	1000	5	28.7
	44	500	15	14.6
	46	500	8	16.2
	24	500	5	19.6
	37	500	7	18.8
	21	500	4	18.4
NO_3^-	0.24	1000	0.1	34.2
	0.50	1000	0.25	34.0
	0.07	1000	0.03	28.7
PO_4	2.5	1000	0.01	34.2
	1.8	1000	0.01	34.0
	3.5	1000	0.02	29.7
ABS	0.7	1000	0.05	29.6
	0.8	1000	0.05	35.8
	0.4	1000	0.02	38.4
	1.2	1000	0.08	31.0
	1.5	1000	0.10	28.7
	1.8	500	0.20	14.6
	0.8	500	0.08	16.2
	1.2	500	0.01	19.6
	1.1	500	0.01	18.8
	1.3	500	0.01	18.4
Total dissolved solids	284	1000	32	29.6
	454	1000	9	35.8
	76	1000	0.1	38.4
	324	1000	9	31.0
	278	1000	7	28.7
	434	500	49	14.6
	385	500	19	16.2
	350	500	30	19.6
	265	500	23	18.8
	294	500	62	18.4

TABLE XVII

Water Quality at Various Stages of Sewage Treatment
Data of Hardwick[31]

Solutes in mg/l unless otherwise specified	Settled sewage	After treatment with lime	After aeration and neutralisation	After reverse osmosis*	Stevenage tap water
Gross properties					
Total solids	657	772	570	63	390
Color (Hazen units)	—	36	120	0	0
Turbidity (ATU)	—	18	16	0	0
Conductivity (micromhos/cm)	1380	1560	946	144	620
Inorganic constituents					
Ammonia (as N)	51.4	49.6	14.0	2.0	0
Nitrate (as N)	3	0.3	1.8	1.0	5
Nitrite (as N)	0.07	0.05	0.05	0.03	0
Borate (as B)	2.79	2.34	1.95	1.90	0.20
Chloride	84	82	85	16	17
Sulfate	69	69	65	2.1	14.8
Soluble phosphate (as P)	10.08	0.19	0.31	0.04	0.03
Bicarbonate	690	610	—	71	—
Total hardness	306	392	193	11	280
Sodium	81	83	86	13.8	—
Potassium	23	23	22	5.2	—
Calcium	60	80	50	<1.0	80
pH value	7.4	11.7	6.6	7.5	—
Organic constituents					
Permanganate value	53	25	24	1.9	<0.3
BOD	—	77	51	—	—
COD	545	300	276	4.5	3
Organic carbon	186	107	84	1.6	<0.5
Anionic detergent (as Manoxol OT)	26.6	14.6	9.4	0.3	0

* 80% recovery of water, 600 psig

of film surface could be maintained with alum treated, sand-filtered primary settled sewage at product water recovery of 95%.

Aerojet also reported[33] three other classes of membranes suitable for waste water renovation; these are (i) cross linkable cellulose acetate methacrylate membrane synthesised by methacrylation of Eastman type E-360-60 cellulose acetate having a degree of substitution (d.s.) of 2.09 to 2.12 to a total d.s. of 2.34 to 2.39; (ii) cellulose acetate with a d.s. of 2.28; and (iii) cellulose diacetate-triacetate blend. These membranes were tested on secondary sewage and were found to give satisfactory separation of organic and inorganic solutes; but the product rates sharply decreased by organic fouling which however could be reduced by cleaning the membrane surface with an enzyme detergent solution.

Smith, Masse, and Miele[34] made a comparison of spiral-wound, tubular, and flat-plate membrane configurations in pilot plant studies for the treatment of secondary-treated sewage water; a summary of their results is given in Table XIX. In all cases excellent separation of contaminants of major concern was achieved. In general, the membranes used were found capable of separating 93 to 95% of total dissolved solids, 95 to 99% of phosphate, 80 to 90% of ammonia nitrogen, 60 to 70% of nitrate nitrogen, 99 to 100% of particulate matter, 90 to 95% of total organic carbon, and >90% of COD. Control of pH of feed was found necessary to avoid precipitation of calcium carbonate. Possible precipitation of a complex calcium phosphate at 80% product recovery was noted once during the investigation and was controlled by reducing the pH of the feed solution from 5.5 to 5.0. No other instance of inorganic fouling was reported. Organic fouling of the membranes was experienced in all cases and was cited as the principal cause of flux decline.

TABLE XVIII

Average Solute Separation and Product Water Quality in Reverse Osmosis Treatment of Waste Water
Data of Aerojet[32]

Membrane: Cellulose acetate (flat)
Operating pressure: 700 psig

Feed water	EC[a]	Total COD	NH_4–N	Organic N	NO_2–N	NO_3–N	Total PO_4–P	MBAS[b]
	Solute separation, %							
Pomona waste water								
Carbon-treated secondary sewage	92.0	83.6	87.3	92.0	67.4	47.0	98.3	92.5
Secondary sewage	91.7	93.2	86.7	86.8	44.7	61.3	99.1	83.0
Primary sewage	87.6	92.7	85.0	82.1	82.8	68.0	98.2	99.0
Fountain Valley Waste Water								
Alum-treated secondary sewage	85.9	78.9	79.5	70.7	89.6	23.0	93.4	90.2
Secondary sewage	90.2	91.0	86.4	87.6	57.0	68.8	98.9	92.9
Primary sewage	90.5	94.2	82.1	89.4	89.9	92.0	99.2	93.9
Raw sewage	82.8	88.2	73.6	78.3	60.1	69.5	96.6	92.5
Digester sewage	77.6	97.4	82.9	98.2	99.4	78.1	94.4	99.4
	Product water quality, solute mg/l[c]							
Pomona waste water								
Carbon-treated secondary sewage	193	9.16	1.86	0.330	0.015	2.28	0.40	0.52
Secondary sewage	246	5.15	3.13	1.01	0.024	1.68	0.18	0.28
Primary sewage	328	16.2	5.19	1.65	0.008	0.520	0.27	0.24
Fountain Valley waste water								
Alum-treated secondary sewage	483	15.4	4.47	1.46	0.001	1.27	0.072	0.12
Secondary sewage	374	11.0	5.67	1.29	0.011	0.910	0.19	0.11
Primary sewage	298	6.88	9.70	2.03	0.003	0.047	0.097	0.072
Raw sewage	668	17.0	11.4	3.55	0.011	0.280	0.36	0.11
Digester sewage	2585	151	116	10.8	0.001	2.94	9.36	0.37

[a] Electrical conductivity, μmhos/cm at 25°C
[b] Methylene blue active substances
[c] Adjusted to product quality of total output of plant operating at 80% recovery except digester sewage which is at 50% recovery.

The problem of membrane fouling has received attention from several investigators. Aerojet-studies[32] involved over 90 controlled experiments to evaluate the effects of feed quality and pretreatment on flux decline. Several classes of feed additives as chelating agents, dispersants, and flocculating agents were evaluated on a number of waste streams including primary-, secondary-, and carbon-treated secondary effluents. In general flocculating agents were found to be the most successful cleaning agents especially when followed by rapid sand filtration as feed-pretreatment for reverse osmosis. Another approach to the problem, reported by Gulf General Atomic investigators[35], is to depressurize the unit periodically and flush the membrane surface using a dilute bioenzyme (BIZ) detergent solution; this technique arrests flux decline significantly, and its success is attributed to protein hydrolysis of the slime layer deposited on the membrane surface and the cleaning action of the detergent.

Several other investigations have also been made by Gulf General Atomic[36–38] to evaluate potential reverse osmosis membranes for tertiary treatment of secondary sewage effluent. The types of membrane materials tested included polyurethane, cellulose diacetate, cellulose 2.5-acetate, polyvinylpyrrolidone-polyisocyanate interpolymers, and polyelectrolytes. The best combination of high flux and solute separation was obtained with asymmetric membranes made from cellulose 2.5-diacetate material, preshrunk in water at 55 to 70°C.

A series of ultrathin membranes with thicknesses from 250 to 5000 Å and consisting of various polysaccharide mixed esters and ethers were tested on microporous supporting films for tertiary treat-

TABLE XIX

Pilot Plant Studies for the Renovation of Municipal Waste Water by Reverse Osmosis
Data of Smith, Masse, and Miele[34]

Membrane type: Cellulose acetate

Plant type	Gulf General Atomic		Universal Water Corp.		Havens		Aerojet	
Membrane	Cellulose acetate in all cases							
Plant location			Pomona, California				Lebanon, Ohio	
Module type	Spiral-wound		Tubular		Tubular		Flat plate	
Feed water type	Carbon treated secondary effluent				Secondary effluent			
pH of feed water	5.5		5.0		5.0		5.0	
Operating pressure, psig	400		600		700		450–485	
Duration of test	3000 hours		1000 hours		325 hours		127 days	
Flux (initial-final), gal/day · ft^2	8–5		24–18		12–10		12–8	
Recovery	80%		47–58%		40–73%		~60%	
Water quality, mg/l	Feed[a]	Product	Feed	Product	Feed	Product	Feed[b]	Product
COD	11.4	0.7	9.4	0.3	45.3	1.9		
TOC							10.6	1.0
Org N							1.8	0.5
NH_3–N	17.1	2.9	11.7	0.7	11.4	1.5	9.6	1.1
NO_2–N							1.9	1.7
NO_3–N	2.1	0.8	3.6	0.7			1.5	0.5
PO_4					30.5	0.15	1.0	0.1
PO_4–P	9.7	0.07	10.1	0.12				
TDS	552	51			607	26		
Turbidity (JTU)					12.8	0.1	1.3	<0.1

[a] Analysis for period 0–1600 hours
[b] Average values for the first 72 days

TABLE XX

Performance of Some Ultrathin Membranes for Municipal Waste Water Treatment by Reverse Osmosis
Data of North Star Res. Develop. Inst.[39]

Conditions of Expt.: Pressure 600 psig
Temperature 25°C
Effluent Filtered secondary effluent
Feed pH ~7 for salt solution and 7–8 for effluent
Feed flow rate 3500 cc/min
Support film Millipore VFWP filter
Each membrane was tested initially for 20 hours in 0.1% NaCl at pH 7

Membrane material and batch number	Thickness, Å	Flux[a], gal/day · ft^2	Solute separation, %					
			NaCl[b]	TDS	NO_3	NH_3	TOC	PO_4
Cellulose methyl sulfonate O-propyl sulfonic acid (211-96A)	570	44	90	95	81	93	93	>99
Cellulose acetate O-propyl sulfonic acid (211-95A)	370	43	96	96	93	96	89	>99
Cellulose acetate phosphate (211-94A)	260	49	74	89	64	81	89	96
Cellulose acetate adduct of methylol phenol (211-83B)[c]	740	64	50	73	43	70	89	95
Cellulose acetate (Eastman) (RO-97)[c]	Asymmetric	20	96	96	74	96	94	>99

[a] Flux after ~20 hours with salt solution
[b] With salt solution
[c] Effluent feed rate = 1000 cc/min

TABLE XXI

Reverse Osmosis Treatment of Sewage Water Using Dynamically Formed Membranes
Data of Kraus[40]

Run No.	Duration hrs	Pressure psig	Velocity ft/sec	Support	Membrane	20 hours		100 hours		400 hours	
						Sepn. %	Flux gal/day·ft²	Sepn. %	Flux gal/day·ft²	Sepn. %	Flux gal/day·ft²
					Experiments with primary effluent						
59	33	1000	44	0.45 μ A	D 50	65	71	—	—	—	—
60	140	750	44	0.45 μ A	D 50	55	110	37	76	—	—
61	128	1000	44	0.025 μ M	PAA	65	11	98.5	5	—	—
62	546	1000	44	0.45 μ A	D 50	58	179	64	105	54	92
66	108	1000	44	0.45 μ A	D 50	72	110	70	93	—	—
67	394	1000	44	0.45 μ A	D 50	41	185	46	122	35	42
68	138	1000	44	0.45 μ A	PAA + $FeCl_3$	43	95	49	53	—	—
69	484	1000	18	0.45 μ A	PAA + $FeCl_3$	80	20	74	28	65	30
70	285	1000	30	0.025 μ M	D 50	60	12	60	11	—	—
71	213	1000	30.5	0.025 μ M	PAA + Zr(IV)	69	9	74	8	—	—
72	210	1000	30	0.025 μ M	Sephadex	56	13	58	12	—	—
46	113	780	17	h	PAA	43	68	—	—	—	—
					Experiments with secondary effluent						
28	50	280	14	h	CMC	18	100	—	—	—	—
30	53	250	17	h	PAA	14	26				
31	64	450	28	0.025 μ M	11	27	20	—	—	—	—
32	144	500	16.5	h	10	23	88	—	—	—	—
33	18	600	22	0.1 μ M	10	24	97	—	—	—	—
35	68	520	14	h	7	25	68	—	—	—	—
36	49	600	45	0.45 μ A	13	50	115	—	—	—	—
37	87	630	14.5	h	11	43	56	—	—	—	—
38	22	1000	12	*	15	63	11	—	—	—	—
39	105	750	14.5	h	PAA	37	140	—	—	—	—
40	110	750	14.5	h	16	30	66	—	—	—	—
41	26	760	29	0.025 μ M	16	43	10	—	—	—	—
42	72	800	14.5	h	16	32	75	—	—	—	—
43	13	750	14.5	h	17	41	77	—	—	—	—
44	445	760	14	h	17	43	72	—	—	—	—
45	15	760	17	h	7	30	108	—	—	—	—
57	18	500	44	0.45 μ A	D 50	48	294	—	—	—	—
58	54	1000	45	0.45 μ A	D 50	53	149	—	—	—	—

Support: A = Gelman Acropor AN 450; M = Millipore filter; h = Polyester or cotton firehose jacket; * Millipore VF, = 0.025 μ nominal pore size.

Membrane: D 50 = Low cross-linked Dowex-50; PAA = Poly(acrylic acid); CMC = Carboxymethyl cellulose; 7 = Starch sulfate (~1.5 sulfate/hexose); 10 = Leonardite; 11 = Pulp mill waste; 13 = Homogenized Dowex-50, cross linked with 0.25% divinyl benzene; 15 = Homogenized Dowex-50, cross linked with 1% divinyl benzene; 16 = PAA + AquaNuchar A; 17 = PAA + AquaNuchar A + Leonardite.

All separation data are with respect to chloride.

ment of municipal waste waters by the North Star Institute[39]; some of their results are given in Table XX. Membranes prepared from cellulose methyl sulfonate o-propyl sulfonic acid, and cellulose acetate o-propyl sulfonic acid looked promising. Long term (150 hours) testing of the former membrane with secondary effluent resulted in an average water flux of 34 to 36 gal/day · ft² over the last 100 hours at 600 psig; during the latter period flux decline was virtually zero, and solute separations of 96% for total dissolved solids, 94% for ammonia, and 83% for total organic carbon were obtained.

The application of reverse osmosis for the treatment of municipal sewage effluents using dynamically formed membranes has been extensively tested by Kraus[40]. The additive substances tested as membranes included negatively charged synthetic polyelectrolytes such as polystyrene sulfonate and polyacrylic acid, positively charged polyelectrolytes such as polyvinyl amine and polyvinyl pyridine, polyelectrolytes derived from natural products such as sulfonated starches and carboxymethyl cellulose, industrial wastes such as pulp mill wastes, natural products such as humic acid and Leonardite (a highly oxidized lignite), and ground or homogenized low cross-linked ion exchange resins. Comparisons were also made with membranes formed from unidentified constituents of sewage.

The porous support structures used included porous stainless steel tubes and plates, tubular ceramic filters, various grades of Millipore and Gelman (Acropor) filters, and woven fire hose jackets precoated in situ with filter aids.

Three different techniques were used for testing the membranes: (i) the additive material was added to sewage and a membrane prepared by passing the mixture under pressure over porous support structures; the run was then continued by keeping varying amounts of this additive in the sewage; (ii) the membrane was preformed by passing tap water containing the additive over the porous support until satisfactory performance was observed; sewage containing the same additive was then passed over the preformed membrane; or (iii) membranes were preformed and the sewage being processed did not contain further additives.

Since the sewage used had a very low dissolved solids level (~100 ppm) in most cases, the solutions were "spiked" to ~0.008 M NaCl to facilitate analysis. The membrane performance was tested primarily for chloride separation and the resultant flux. These data were also supplemented with analyses for organic and inorganic carbon, alkaline earth, phosphate, and turbidity measurements. Table XXI is a summary of the data of Kraus on membrane performance; all solute separations reported are with respect to chloride.

It was found that the constituents of sewage effluents themselves could form membranes which typically separated 35% of salt, and >70% of organic carbon present at fluxes between 50 and 100 gal/day · ft^2 at 750 psig at a feed velocity of 25 ft/sec. With the constituents of secondary sewage serving as membrane formers, observed salt separations were a little lower (20 to 30%).

Most of the membrane additives listed in Table XXI were tested with continuous addition of the material with or without preforming the membranes. Cation exchange membranes were almost always superior to anion exchange membranes in performance. Additive concentrations were usually 100 ppm or less; solute separation was relatively insensitive to concentrations of additive used. In general, higher separations and lower fluxes were obtained with preformed membranes. Organic separations with secondary sewage were >70% in most of the runs.

Membranes prepared from homogenized low cross-linked Dowex-50 (poly(styrene sulfonate) cross linked with divinyl benzene), or Sephadex (carboxymethyl substituted dextran), or polyacrylic acid deposited on top of a hydrous oxide such as Zr(IV) or Fe(III), seemed to be operable without continuous addition.

A long-run experiment was carried out with both primary and secondary effluent trucked in. Salt separations were 50 to 60% irrespective whether primary or secondary effluent was used. Flux declined from 200 to 150 gal/day · ft^2 during the first 50 hours the last 20 of which were with primary. At shutdown after 78 hours, flux was 92 gal/day · ft^2, chloride separation 58%, caclium and magnesium separation 76%, and reductions in organic carbon, phosphate, and turbidity were 82, 65, and 99% respectively.

Since the fouling layer responsible for flux decline was suspected to contain negatively charged polyelectrolytes from the sewage, flux rejuvenation was attempted with one-hour rinses with 0.1 M Na_2CO_3 and 0.1 M NaOH. The carbonate wash caused a large initial decrease in flux probably because of carbonate precipitation; this was followed by an increase in flux. Fourteen hours after reexposure to primary sewage, fluxes were 84, 92, and 33 gal/day · ft^2 for the test sections operating at feed velocities of 44, 31, and 18 ft/sec respectively (using Dowex-50 preformed membranes at 1000 psig), compared to values of 43, 42, and 23 gal/day · ft^2 just prior to wash. Flux decline was small during the next 50 hours of operation. Before this run was terminated after more than 500 hours of total operation (including several sodium carbonate and sodium hydroxide washes), fluxes were still 74, 84, and 37 gal/day · ft^2 at the above three feed velocities, chloride separation was between 55 and 65%, reduction in turbidity was >95%, and reductions in organic carbon, phosphate, and calcium and magnesium were 60 to 75%, 70 to 77%, and ~85% respectively.

Extended tests with membranes prepared form polyacrylic acid, both with and without added hydrous oxides in the original preparation showed similar results. At 1000 psig, a membrane formed by depositing homogenized Sephadex on 0.45 μ Acropor gave a stabilized flux of 80 gal/day · ft^2 with salt separation of ~50% for primary sewage effluent feed spiked with ~0.01 M NaCl.

The above results indicate the technical feasibility of a sewage treatment process based on dynamically formed membranes provided suitable support structures are identified and available.

The experimental data presented in Tables XV to XXI indicate that reverse osmosis is potentially capable of upgrading sewage water to a quality practically suitable for all water uses. Therefore the process is important from the points of view of both water pollution control and water renovation. The apparent current emphasis on reverse osmosis as a possible technique for the tertiary treatment of sewage water is entirely due to problems associated with membrane fouling. More detailed studies are called for in order to obtain precise information on the physicochemical nature of membrane fouling. Such studies can lead to the development of appropriate pretreatment techniques for raw sewage water prior to reverse osmosis treatment. The object of such pretreatment would be to separate all undissolved solids (including colloidal suspensions) from the raw sewage water and render the bacteria inactive in the clear solution. Such pretreated sewage water can then be directly subjected to reverse osmosis treatment (without any prior secondary or biological treatment) in a series of units operating at different pressures in the range 250 to 1500 psig to obtain high recovery (>95%) of high quality product water.

Treatment of Paper Mill Wastes

The pulp and paper industry is a large user of water. An average size pulp mill producing 150 to 500 tons of cellulose pulp for making paper and allied products may commonly use from 5 to 50 million gallons of water daily. A large part of this water can be processed by reverse osmosis and reused again. The concentrated wastes may be suitable for waste recovery or may be disposed of more economically by conventional methods of evaporation and burning. Therefore the application of reverse osmosis for the treatment of waste waters from pulp and paper mills is an area of great

TABLE XXII

Reverse Osmosis Treatment of Paper Mill Waste Waters
Data of Wiley, Ammerlaan, and Dubey[41]

Film type: Cellulose acetate
Test unit: Gulf General Atomic A, B, C Modules (size 4)
Operating pressure: 450 psig
Feed rate: 1.8 gal/min
Product recovery: 80%

Sample	Flux, gal/day · ft^2	Concentration, mg/l			pH	OD††
		Solids	C.O.D.‡	Chloride		
Process feed†		1960	945	430	2.2	8.75
Product water						
Membrane grade A*	6.1	14	100	36	2.6	0.194
Membrane grade C	7.0	72	132	251	2.2	0.261
Membrane grade B	10.9	822	516	400	2.3	3.63
Concentrated stream						
Membrane grade A*		12120	6655	3260	2.2	80.4
Membrane grade C		6940	4366	1964	1.9	44.8
Membrane grade B		5160	3405	852	2.2	31.7

* A = Dense; C = Intermediate; B = Coarse
† First stage sulfite bleach plant effluent
‡ Chemical oxygen demand
†† Optical density at 281 μm (a measure of the lignin content)

TABLE XXIII

Reverse Osmosis Treatment of Paper Mill Waste Waters
Data of Ammerlaan, Lueck, and Wiley[42]

Film type: Cellulose acetate (tubular)
Test unit: Havens Industries
Operating pressure: 600 psig

Test detail	Separation, %							
	Feed A to F		Feed G					
	Membrane type 3	Membrane type 5	Membrane type 3			Membrane type 5		
			pH			pH		
			2.7	4.8	6.1	2.7	4.8	6.1
Solids	75–98	95–99	27	57	70	45	81	95
OD (Color)	98–99.8	99–99.8	66	55	48	79	74	80
B.O.D.	80–95	90–99	26	45	58	42	62	82
C.O.D.	85–98	90–99	19	43	57	37	62	81
Inorganics	60–95	94–99	—	—	—	—	—	—
Volatile acids			7	40	65	26	70	95

Feed water A: Diluted calcium base spent sulfite liquor
B: Calcium base pulp wash water
C: Kraft first stage bleach effluent
D: Kraft second stage bleach effluent
E: Sulfite first stage bleach effluent
F: Total effluent
G: Ammonia base liquor condensate
Product rate, Membrane type 3: 2 to 12 gal/day · ft^2
Membrane type 5: 2 to 5 gal/day · ft^2
Membrane type 3: Intermediate
Membrane type 5: Very dense
B.O.D. = Biochemical oxygen demand
C.O.D. = Chemical oxygen demand

practical importance from the points of view of waste recovery, water pollution control, and water reuse. Extensive pilot plant test data on the above application are given by Wiley and coworkers[41–44] at the Pulp Manufacturers Research League and the Institute of Paper Chemistry at Appleton, Wisconsin; their data are summarized in Tables XXII to XXV.

Table XXII gives the results obtained with the processing of first stage sulfite bleach plant effluent with three Gulf General Atomic modules using cellulose acetate membranes. The solute separations obtained with the use of the "dense" type-A membrane in terms of chemical oxygen demand (94%), chlorides (92%), and excellent color removal between the process feed and product water, as well as the recovery of these components in the concentrate stream are of significant interest to pulp and paper industry.

Table XXIII gives the results of more extensive studies on the process using the Havens tubular cellulose acetate membranes. Seven different kinds of paper mill effluent feeds were tested. The flux of product water dependend on the type of feed water used, concentration level of the feed, porous structure of the membrane, and the operating pressure and temperature. The pH of the feed solution also affected the product rate in many cases; a change in the original pH of the waste water increased the concentration of insoluble material which tended to coat the membrane surface and decrease the product rate. Such membrane fouling was identified as a major problem still to be solved with respect to this application. The product rate for feed solutions free from membrane fouling problems increased by about 2.8% for every degree centigrade increase in operating temperature.

A mobile Havens reverse osmosis unit has been

in operation in Appleton, Wisconsin, since October 1968[43]. The system, wholly contained within a standard 40-ft trailer van, comprises 390 active 18-tube modules with a total membrane area of 6580 ft^2 which at a flux of 7.5 gal/day · ft^2 give a nominal capacity of about 50,000 gal/day. The modules are arranged in 5 banks. The normal operating pressure is 600 psig although the system can be operated at pressures up to 1000 psig. The unit is equipped with safety switches which automatically shut it down when pH, temperature, or pressure go wrong. There is only a minimal need for supervision approximately $\frac{1}{2}$ man-day/day. The main pump is a triplex reciprocating Manton Gaulin pump with a direct current motor and a variable speed drive. The recycle pumps are Goulds high pressure centrifugal process pumps. The pumps are equipped with Crane packing single balanced mechanical seals which work well in combination with two flush systems, one at each side of the seals.

TABLE XXIV

Reverse Osmosis Treatment of Paper Mill Waste Waters
Data of Ammerlaan and Wiley[43]

Film type: Cellulose acetate (tubular)
Test unit: 50,000–100,000 gal/day Havens pilot plant
Operating pressure: 600 psig
Feed water: Calcium base sulfite pulp wash water*

Period, Operating hours	Recovery of product water, %	% Solids in Concentrate	Temperature, °C	Membrane flux, gal/day. ft^2
0–100	90	12	23	6.9
100–200	85	9	24	5.2
200–268	90	12	26	4.8
Removal of $CaSO_4$ in Bank V				
268–400	83	6	28	6.2
400–500	80	6	27	5.7
500–587	78	5.5	28	5.7
Washup with detergent				
587–690	82	6	28	6.4

* Feed water analysis:

Solids = 8 to 13 g/l
B.O.D. = 2.8 to 4.5 g/l
Volatile acids = 1.2 to 2.0 g/l
C.O.D = 7 to 17 g/l
Calcium = 0.4 to 0.7 g/l
pH = 3.5 to 4.5

All metal pump components in contact with the process solutions are 316 stainless steel.

In the first demonstration experiment, the unit was maintained in operation 46% of the available time for 690 actual hours of running time. A total amount of 1,260,000 gal of calcium base sulfite pulp wash water containing 1% solids was concentrated and 45 tons of solids were recovered in the form of a 6 to 10% solids concentrate sent to the evaporators. The average intake was 44,000 gal/day. Operating time was lost in the start-up phase owing to seal problems in the centrifugal pumps and persistant problems with the pH controller installed for adjusting the pH of the feed solution. But the unit operated successfully during the last $2\frac{1}{2}$ weeks of continuous operation with only the scheduled weekend shutdown. The performance of the unit is illustrated by the data in Table XXIV. During the first 268 operating hours, there was a steady decline in flux. It was then discovered that bank V, which was operating at a concentration level of 9 to 12% solids, was badly plugged up with calcium sulfate scale; when the scale was removed with a solution of ethylene diamine tetra acetic acid, the flux returned to normal. After bank V was put back in service, the average flux for the system was up again. But, after awhile, the flux for the banks I, II, and III declined due to deposition of a mixture of coniferous wood pitch and silicates on the membrane surface. However, a washup with a nonionic detergent was able to remove the deposits

TABLE XXV

Reverse Osmosis Treatment of Dilute Pulping Wastes
Data of Wiley, Dubey, Holderby, and Ammerlaan[44]

Film type: Cellulose acetate (tubular), Havens pilot plant
Operating pressure: 600 psig

Pulp wash water	Solids	BOD	COD	Base	OD*
Average concentration in feed water, g/l					
Calcium sulfite	11.6	3.9	12.8	0.56 Ca	
Ammonia sulfite	15.9	3.7	24.8	0.36 NH_3	
Neutral sulfite	8.7	2.5	9.8	1.0 Na	
% Separation by reverse osmosis					
Calcium sulfite	87	74	87	96 Ca	96
Ammonia sulfite	93	70	94	90 NH_3	99
Neutral sulfite	95	88	95	88 Na	99

* Optical density at 281 mμ—a measure of lignosulfonate content.

announcement

G+B

to all purchasers of the 'Encyclopedia of Environmental Science and Engineering', edited by E. N. Ziegler and J. M. Pfafflin.

There is a possibility that Gordon and Breach will publish an annual up-date of many of the aspects covered by the Encyclopedia, in the form of an Annual Review of Environmental Science.

It will help us to gauge the demand for a publication such as this if you would complete and return the attached form. We can then keep you up-to-date with the progress of the Review, and inform you of any special pre-publication/subscription prices for those who have already bought a copy of the Encyclopedia.

Gordon and Breach Sci. Pub. Inc.
1 Park Avenue, New York
N.Y. 10016
U.S.A.

Gordon and Breach Sci. Pub. Ltd.
41/42 William IV Street
London WC2
England

- -

ANNUAL REVIEW OF ENVIRONMENTAL SCIENCES

I have purchased a copy of the Encyclopedia of Environmental Science and Engineering, ed. by E. N. Ziegler and J. M. Pfafflin, and am interested in knowing more about the proposed Annual Review of Environmental Science.

Please send further information, as it becomes available, to:

Ms/Mr/Dr/Prof ____________________ Date: __________

____

____

____

____

and restore the flux. The average feed water analysis for the experiment is given in Table XXIV; of these components, the following were recovered in the final concentrate: solids 85 to 90%, BOD 70 to 80%, volatile acids 25 to 40%, COD 85 to 90%, calcium 96 to 98%, and color 96 to 98%. In general, recovery percentages were considered good with the exception of volatile acids.

Table XXV gives further data from the above Havens unit operated with three different pulp wash waters[44] obtained respectively from the calcium-base acid sulfite process, ammonia-base acid sulfite process, and neutral sulfite semichemical pulp washing process (also called NSSC white water effluent). As much as 25% of the total dissolved solids in the neutral sulfite water might have been in the form of sodium acetate accounting for most of its BOD content. The reverse osmosis separations obtained with respect to total dissolved solids, BOD, COD, base, and color (Table XXV) are considered excellent for the pulp and paper industry from the points of view of pollution control and water reuse. Improved membrane flux, and methods of controlling membrane fouling problems are cited[44] as the main requirements to make the process economically acceptable to the paper industry.

Calcium base sulfite liquors form dynamic membranes when circulated under pressure over porous bodies, such as porous ceramic, and hence it may be possible to use such membranes for the treatment of pulp mill sulfite wastes[45]. In some pilot experiments, a membrane was formed with 1% spent liquor on Type 02 and 03 porous ceramic tubes (average pore diameters 1.4 μ and 0.90 μ respectively) in series. After 15 hours of operation, typical solute separations were obtained with permeation rates of 7.5 and 9 gal/day · ft^2 for the 02 and 03 tubes respectively at 500 psig. The loop was drained and filled with the chlorination bleach waste. At 400 psig and 30°C, the type 03 tube gave 90% solute separation based on optical density and 22% by chloride analysis; the corresponding product rate was 7 gal/day · ft^2. Solute separation for the 02 tube rose over 4.5 hours to 67% by color and 18% of chloride with a flux of 9 gal/day · ft^2. The loop was again drained, rinsed with demineralized water, and filled with alkaline extraction bleach waste. Solute separation based on optical density for the 03 tube was 90% (flux, 8 gal/day · ft^2) and for the 02 tube was 51% (flux, 14 gal/day · ft^2). Chloride separations were 32% and 16% respectively.

Dynamic membranes were also formed on 0.1 μ Millipore supports with the chlorination bleach wastes[45]. After about 5 hours of operation, solute separation based on optical density was 95 %, and chloride separation was 34%, and the product rate was 10 gal/day · ft^2 at 500 psig. In a similar test, second stage alkaline bleach waste formed a membrane separating 94% of color, and 43% of chloride at 15 gal/day · ft^2. It was estimated that a production facility operating at 60°C could attain product rates of about 30 gal/day · ft^2 and solute separations greater than 90% at pressures less than 500 psig.

Treatment of Effluents from Food Processing Plants

In practical terms of industrial process and product development, food processing is likely to be the most profitable field for reverse osmosis application both for the concentration of dilute and/or heat sensitive liquid foods such as fruit and vegetable juices, coffee, tea, milk, etc., and for the treatment of high BOD waste water plant effluents. Among the several studies on the subject which have been reported[19,20,46–64] those giving experimental data on the treatment of whey-bearing dairy wastes, and effluents from starch manufacturing plants are particularly significant from the point of view of water pollution control and waste recovery.

Whey, the fluid portion of milk obtained by coagulation of casein during manufacture of cheese, contains about half of the milk solids, most of the lactose, about $\frac{1}{5}$ of the protein, and most of the vitamins and minerals. Typical compositions (wt-%) of cottage cheese whey (pH = 4.6) and cheddar cheese whey (pH = 6.0) are respectively: total solids 6.53 and 7.31, lactose 4.39 and 5.20, protein 0.86 and 0.87, and ash 0.61 and 0.53. About 11 million tons of whey are produced in the United States every year. Whey can be condensed or dried for use in animal feeds; it can also be processed as a food ingredient for use in bread, cheese, sherbet, candy, infant formulas, etc. Fractionation of whey yields such products as lactose for pharmaceuticals and latic acid for food-grade acidulant. Fermentation of whey yields yeast products and whey proteins[49].

Thus whey could be a valuable by-product of the cheese industry; instead, for many factories, it is a troublesome waste product because of the

TABLE XXVI

Reverse Osmosis Concentration of Whey
Data of Marshall, Dunkley and Lowe[49]

Film type: Cellulose acetate (flat)
Operating pressure: 1000 to 1500 psig

Run No.	Sample No.	Time hr	Concn. in permeate					Concn. in concentrate				
			Solids %	Lactose %	N %	Ash %	K^+ mg/g	Solids %	Lactose %	N %	Ash %	K^+ mg/g
2	6	3	2.7	1.0	0.022	0.98	1.58	23.3	18.6	0.516	—	1.78
3	4	2	0.6	—	—	0.29	1.00	28.2	24.0	0.465	—	1.65
4	3	2	2.6	—	0.026	0.51	1.40	37.0	24.4	0.820	2.0	1.06
8	5	1	1.1	0.17	0.022	0.35	0.88	27.4	19.7	0.589	2.0	1.06
8	15	19	0.7	0.05	0.019	0.28	0.94	23.6	16.8	0.468	1.8	1.04
9	1	1	1.3	0.22	—	0.52	0.62	53.8	36.0	0.941	3.2	1.03
9	5	3	1.1	0.08	—	0.48	0.60	53.2	41.0	0.978	3.2	1.11
9	10	5	1.2	0.22	—	0.51	0.69	55.2	42.2	1.034	3.4	0.54

Feed for runs 2, 3, 4, and 8 was cottage cheese whey (6.5% solids, 4.4% lactose, 0.12% nitrogen, 0.7% ash, and 1.5 mg K^+/g). Product from run 8 was processed to remove part of the lactose (by crystallization and centrifugation) and used in run 9 as the feed (26% solids).

high cost of concentrating it by conventional methods. Whey has a high BOD. Every 1000 gal/day of whey discharged into a sewage treatment plant imposes a load equal to that from 1800 people; every 1000 gal discharged into a stream requires for its oxidation the dissolved oxygen in over 4.5 million gallons of unpolluted water[49].

The applicability of reverse osmosis using porous cellulose acetate membranes for the concentration of whey has been demonstrated in several studies[49,50,55,60–62]; some typical data are given in Tables XXVI to XXIX. Table XXVI shows that solids-concentration from 6.5 to 55.2%, lactose-concentration from 4.5 to 42.2%, and nitrogen-concentration from 0.12 to 1.03% have been actually obtained using membranes of different surface porosities in the operating pressure range 1000 to 1500 psig. Table XXVII gives data obtained at 650 psig in operating the process in four stages using membranes of three different surface porosities. The product water flux naturally decreased with increasing concentration of feed whey in all cases. The titratable acidity, total solids, nitrogen, and lactose contents of the membrane permeated water indicated that practically no higher molecular weight material was lost in the permeate; ionizable salts and lactic acid passed through the membranes used to different extents. The average membrane fluxes while removing 75% of water were 7.19, 5.58, and 4.90 gal/day · ft^2 respectively for the three grades of membranes used. On the basis of these data, McDonough[50] estimated that a cheese plant processing 150,000 lb of milk/day would require 2200 to 3200 ft^2 of membrane surface to treat its whey; space requirements for such a unit would be between 10 and 20 ft^3 depending on module design. Aerojet is reported to have concentrated whey to a whipping cream consistency of 56% solids operating the process at 1500 psig with a membrane flux of 11.5 gal/day · ft^2 [50]. Tables XXVIII and XXIX show that the salt constituents of whey can be fractionated, and that the concentration of whey to about double its original total solids content could result in the elimination of almost half of NaCl and smaller proportions of other salt constituents; such elimination is often desirable since the amount of whey which can be added to food products is limited by its salt content. With proper choice of membrane porosities, both demineralization and fractionation of whey are possible yielding low salt and high protein containing food products.

While the above results indicate that reverse osmosis is promising for the concentration of whey, some technical problems still remain to be solved before the process can become an economic success. These problems are essentially concerned with the drop in membrane flux (usually obtained as a function of time and concentration of feed) due to accumulation of protein, fat, and precipitated lactose on the membrane surface, and also due to generally low mass transfer coefficient obtained on

TABLE XXVII

Reverse Osmosis Treatment of Whey
Data of McDonough[50]

Film type: Cellulose acetate (tubular, Havens unit)

Membrane No.	1	2	3
Membrane-desalting ratio	2.5 : 1	10 : 1	20 : 1
Membrane area, ft^2	55	55	55
Operating pressure, psig	650	650	650
Product rate (gal/day · ft^2)			
Stage 1[a]	11.9	7.70	5.40
Stage 2	7.07	6.26	5.44
Stage 3	5.44	5.17	5.17
Stage 4	4.35	3.81	3.67
Average	7.19	5.58	4.90
Analysis of product water			
Total solids, %	0.343	0.044	0.034
Total nitrogen, %	0.0152	0.0088	0.0066
Lactose, %	Nil	Nil	Nil
Titratable acidity, %	0.255	0.135	0.095
Dissolved solids, ppm			
Stage 1	400	135	85
Stage 2	1150	220	130
Stage 3	1500	290	170
Stage 4	2000	380	215

Analysis of	Initial whey	Concentrated whey
Total solids, %	6.3	23.0
pH at 25°C	5.4	5.7
Lactose, %	4.79	18.65
Titratable acidity	1.10	3.34
Total nitrogen	0.14	0.49

[a] Flux rates determined, and product samples taken during stages of concentration referring to:

	Total solids, %
Stage 1	6.3– 9.5
Stage 2	9.5–13.2
Stage 3	13.2–15.7
Stage 4	15.7–23.0

TABLE XXVIII

Rejection Ratio for Selected Constituents During Reverse Osmosis of Cottage Cheese Whey
Data of Dunkley[55]

Constituent	Rejection ratio
Sodium	0.14
Potassium	0.10
Calcium	0.76
Phosphate	0.44
Citrate	0.85
Lactate	0.44
Lactose	0.90

Film type: Cellulose acetate
Operating pressure: 500 psig
Rejection ratio = 1 − (concn. in permeate/concn. in concentrate)

TABLE XXIX

Losses of Salt Constituents in the Permeate During Reverse Osmosis of Cottage Cheese Whey
Data from Dunkley[55]

Constituent	% of total in the permeate
Chloride (NaCl)	45.9
Calcium	16.3
Ash	20.8
Lactose	7.3

Film type: Cellulose acetate (tubular, Havens unit)
Operating pressure: 500 psig
Concentration of whey from 6.5 to 12.6% total solids
Average total solids in permeate = 0.6%

the high pressure side of the membrane, in addition to membrane-compaction while operating at high pressures; further, appropriate sanitation techniques for the control of bacterial growth in the operating system need to be developed for the success of the process.

In the manufacture of potato-starch[63] about 6% of the weight of fresh potatoes (solids content ~19%) consists of nonstarch solids which end up in waste waters. This waste contains both insoluble and soluble materials. The industry has developed satisfactory mechanical filtering procedures to remove the insoluble materials. The problem is with regard to the treatment of waste-waters containing dissolved solids which consist mostly of proteins, free amino acids, organic acids, sugars, and inorganic ions. The output of this waste for a 30 tons/day starch plant is about 432,000 gallons of waste waters per day with a dissolved solid content of 0.5 to 1.0 wt.-%. These waste waters have a COD requirement of 9000–14000 mg/l, which is equivalent to that of municipal waste waters from a city of approximately 85,000 people. Presently, the starch factory waste waters are pumped into lagoons and treated biologically. Such treatment is inefficient and inadequate, requires large land areas and contributes nothing to waste recovery.

The proteins and amino acids in the above waste waters contain adequate amounts of lysine and methionine which have considerable market value in food industry; their recovery can be economically efficient if only the solids concentration in the effluent can be increased to 2 to 4%.

TABLE XXX

Treatment of Potato-Starch Waste Water by Reverse Osmosis
Data of Porter, Siciliano, Krulick, and Heisler[63]
Film type: Cellulose acetate (tubular, Havens unit)
Operating pressure: 600 psig
Product recovery: 75%

Test details	Total solids, %	C.O.D., ppm	Av. flux, gal/day · ft²
*Feed**			
Coarse membrane	0.43	3670	
Medium membrane	0.49	4269	
Fine membrane	0.43	3838	
Concentrate			
Coarse membrane	1.60	13364	
Medium membrane	1.93	16163	
Fine membrane	1.57	13784	
Permeate			
Coarse membrane	0.035	268	11.5
Medium membrane	0.012	91	9.5
Fine membrane	0.005	40	7.8

* Feed was simulated waste effluent produced by grinding potatoes, filtering, and washing to give a concentrated product (in a manner similar to that employed in a starch factory) which was diluted to desired solids content.

The data given in Table XXX illustrate the possible applicability of reverse osmosis for the treatment of potato-starch waste waters to give a product water good enough for reuse in the plant, and a concentrate suitable for economic recovery of valuable chemicals; the COD reduction of 94 to 99% in the product water is particularly significant from the point of view of water pollution control.

Similar results have been obtained by Besik[64] with respect to the application of reverse osmosis to waste streams from starch manufacturing plants in corn processing industry. The effluents from such starch manufacturing plants have BOD 714–1575 ppm, oil and grease 39–67 ppm, phenol <1 ppm, suspended solids 420–1690 ppm, and total solids 1474–4448 ppm. A typical effluent was tested on a pilot reverse osmosis unit containing three grades of Havens tubular cellulose acetate membrane modules at 600 psig, giving rejection ratios of 25 : 1, 15 : 1 and 4 : 1 respectively for sodium chloride. The corresponding reductions obtained with the waste water feed were 96, 90, and 75% for total dissolved solids, 99.5, 98, and 97 for BOD, 99.5, 98 and 97 for COD, and 99.5% with all modules for total carbohydrates. It was also shown that the waste waters were concentrated from 2 to 100 g/l of total dissolved solids, the average value of permeate flux/pure water flux for all the membranes used was 0.85 for product water recovery of 80 to 90%. The recovery of dissolved proteins and carbohydrates from the concentrate into marketable products adds to the possible applicability of reverse osmosis for the economic treatment of wastewaters from starch manufacturing plants. The development of appropriate sanitation techniques for the control of bacterial growth in the operating system is a common problem to be solved with respect to the application of reverse osmosis in all food processing plants.

Treatment of Electroplating Wastes

In the electroplating industry, the major pollution problem arises from the toxicity of the chemicals involved and the very large volumes of water used in the rinsing operations, which also results in a significant loss of economically valuable chemicals. Therefore the treatment of rinse waters from metal finishing plants is of interest from the points of view of both water pollution control and waste recovery. The data presented in Table XXXI illustrate the possible applicability of reverse osmosis for the treatment of a wide variety of plating wastes provided operating conditions are appropriate for the practical operation of the process. Most extensive work has been reported on the treatment of diluted nickel plating solutions[65,66] since the latter

TABLE XXXI

Separation of Some Salts Present in Plating Wastes
Data of Hauck and Sourirajan[14]

Film type: Cellulose acetate (Batch 18), film H-4
Operating pressure: 1000 psig
Product recovery: 90%

System	Solute concn., ppm		Product rate, gal/day · ft²
	Feed	Product	
$ZnSO_4$–H_2O	534	48	20.7
$Pb(CH_3COO)_2$–H_2O	504	32	20.4
$CuSO_4$–H_2O	500	8	19.2
$NiCl_2$–H_2O	500	14	19.2
CrO_3–H_2O	512	22	21.5
$SnCl_2$–H_2O	500	49	20.8
$AgNO_3$–H_2O	500	135	22.6
$Fe(SO_4)_2(NH_4)_2$–H_2O	525	19	20.1
$Ni(SO_4)_2(NH_4)_2$–H_2O	515	22	20.9
$Cr(SO_4)_3$–H_2O	500	9	22.1
$HAuCl_4$–H_2O	500	109	19.1

usually have a pH of 4 to 6 and hence are particularly amenable to successful reverse osmosis operation at the ordinary room temperature using the well-developed porous cellulose acetate membranes.

Fig. 5, 6, and 7 give some of the data obtained for the reverse osmosis treatment of diluted samples of an actual nickel plating solution whose composition was as follows: $NiSO_4 \cdot 6\frac{1}{2}H_2O$ 48.2 oz/gal, $NiCl_2$ 1 $6H_2O$ 12.2 oz/gal, boric acid 6.0 oz/gal, and organic brightners 3.2% by volume. Batch 316 type cellulose acetate membranes were used at 300 psig. The presence of brightners had no effect on the performance of the membranes. Fig. 5 gives the effect of feed concentration (100 to 12,000 ppm Ni) on the performance of six different membranes preshrunk at temperatures in the range 65 to 75°C. The results showed that the level of solute separation obtained was in the range 99.5 to >99.9% with the films used, and the per cent solute separation remained essentially constant in the entire range of nickel concentration studied. Experiments with NaCl–H_2O system showed that solute transport parameters ($D_{AM}/K\delta$) for NaCl for the films used were in the range 3×10^{-5} to 11.2×10^{-5} cm/sec which indicate the membrane pore sizes appropriate for the reverse osmosis treatment of nickel plating rinse waters using cellulose acetate membranes.

For purpose of process design it is useful to determine the processing capacity of a membrane

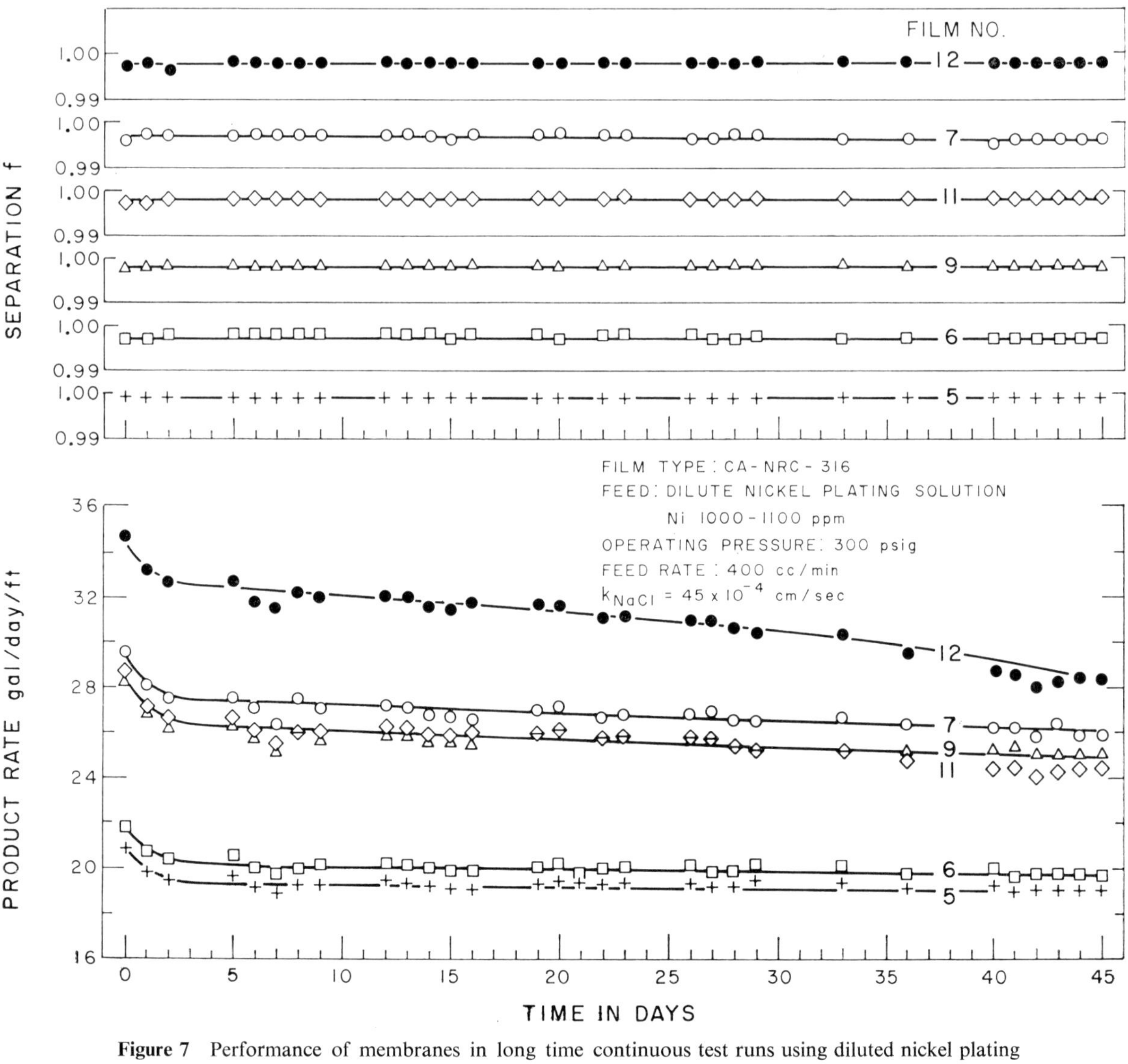

Figure 7 Performance of membranes in long time continuous test runs using diluted nickel plating solution as feed.

defined as the volume of charge (feed solution) that one ft^2 of film surface can handle per day in a batch concentration process to increase the solute concentration from $(c_A)_i$ to $(c_A)_f$ on the high pressure side of the membrane at a given operating pressure. Making use of the data given in Fig. 5, the processing capacities of the membranes used were calculated for the treatment of dilute nickel plating solutions containing 1000 ppm of Ni. The results, given in Fig. 6, show, for example, that for 90% product water recovery, the processing capacities of the films 12, 7, 11, 9, 6, and 5 are 35.0, 30.1, 28.8, 28.3, 21.7, and 20.8 gal/day · ft^2 respectively under which conditions the obtainable solute recovery is greater than 98.5%.

Figure 7 gives data on a 45-day continuous test run conducted with a diluted plating solution containing 1000 ppm Ni, to assess the performance of the membranes under such conditions. There was practically no change in solute separation during the entire test period in all cases. With respect to the product rate data for films 12, 7, 11, 9, 6, and 5, at the end of 2 days of continuous operation the rates were 94.3, 93.1, 93.1, 93.1, 93.5, and 93.1% respectively of those at the start of the run; at the end of 45 days of continuous operation, the rates were 81.7, 87.5, 84.8, 88.5, 90.0, and 91.0% respectively of those obtained at the start of the run. The effect of compaction on permeation rate is relatively more with membranes shrunk at lower temperatures. At the end of the test run it was found that there was some fine solid deposit on the membrane surfaces. This solid deposit was found to contain 56% ash whose spectrographic analysis showed the presence of Ni, Fe, Cu, Zn, Al, Si, Sn, Pb, B, Ti, Mg, and Mn; the origin of some of these metals could be the materials of construction used in the pump parts and fittings. At least a part of the drop in product rate in the long-time run could be attributed to the above solid deposit on the membrane surfaces. In any case, on the basis of the results reported in Figure 7, it may be concluded that the practically realizable processing capacities of the membranes during continuous operation for extended periods of time would be 80 to 90% of those given in Fig. 6. Even under these conditions, the performances of the membranes tested seem significant enough for consideration for their practical industrial use.

The applicability of reverse osmosis for the treatment of quite a large number of other cyanide and noncyanide electroplating waste waters awaits detailed investigations. While it is obvious that cellulose acetate membranes may not be suitable for use in conjunction with all wastewaters encountered in the electroplating industry, there can be no question about the utility of the reverse osmosis process itself under appropriate operating conditions, for the profitable treatment of a wide variety of such waste waters.

Treatment of Acid Mine Drainage Waters

Drainage from coal mines is a significant source of contamination of rivers and streams. Oxygen from the air reacts with sulfur present in coal, and sulfuric acid is formed. This acid is then dissolved by water that seeps into the mine; the acid solution also dissolves significant quantities of iron, calcium, and magnesium. After the water drains out of the mine and is further aerated, the iron is oxidized from the ferrous to the ferric state. The ferric iron is then precipitated in the stream beds. This precipitate and the dissolved solids present serious hazards to wildlife as well as to municipal and industrial water supplies. Prevention of such hazards is an important environmental problem in the solution of which reverse osmosis can be of significant service. Although the composition of the acid mine water varies from mine to mine, it is typically very acid (pH 2.5 to 3.5) and it contains about 2000 ppm of dissolved solids which are mostly sulfates of Ca, Mg and Fe which are easily separated by reverse osmosis. Table XXXII gives some typical data on the analysis of feed, product, and concentrate in a reverse osmosis operation[67]. There are at least two reports[68,69] giving details of extensive pilot plant tests with acid mine waters using Gulf General Atomic cellulose acetate membrane modules. Product water recoveries of 75 to 92% have been actually achieved in extended test runs with an average membrane flux of 10 gal/day · ft^2 at 600 psig. More than 99% separation has been obtained with respect to acid, sulfate, total hardness, Ca, Mg, and Fe. No major problems have been reported. Based on test data in a 10,000 gal/day unit, it is estimated[69] that for a 1 million gal/day (product water output plant), the cost of product water (including capital costs), based on present—day (1970) technology, would be 64.3 cents per 1000 gallons.

TABLE XXXII

Typical Chemical Analysis of Feed, Product, and Concentrate in Reverse Osmosis Test with Acid Mine Waters

Data of OSW[67]

Feed source: Coal mine, Decker No. 5, Kittanning, Pa.
Film type: Cellulose acetate

Sample	pH	Acidity (as $CaCO_3$) ppm	Total dissolved solids, ppm	Sulfates ppm	Ca (as CaO_3) ppm	Mg (as $CaCO_3$) ppm	Fe (as Fe_2O_3) ppm
Feed	2.35	695	1920	1326	389	261	279
Product	4.75	3.8	3.4	0.7	<1	<1	Trace
Concentrate	1.85	7270	19364	12650	1660	3687	2869

Treatment of Waste Streams from Petrochemical Plants

The application of reverse osmosis for the treatment of waste waters from petrochemical plants is illustrated by the studies reported by the Gulf South Research Institute on the possible recovery of glycerine from a petrochemical waste stream, The need for this study arose from a consideration of the water quality problems affecting one section of the Mississippi River which serves as a major waste water collection stream for a petrochemical complex between Baton Rouge and New Orleans, Louisiana. The river serves as the outfall collector for a refining capacity of more than 600,000 barrels of oil per day. In addition, a varied and large chemical industry based on these refined products adds to the effluent loadings. Current (1971) levels of organic extractables in the river water are between 0.5 to 1 ppm which are above the recommended maximum of 0.25 ppm. The average COD in the river is 50 ppm of which 15 to 20 ppm are attributed to petrochemical sources. With a flow of 400000 ft^3/sec, this represents 100000 tons per day of petrochemicals involving an estimated loss of $ 150,000/day in material. The recovery of useful chemicals is not only an ecological problem, but has also significant economic merit; an equally important incentive for such recovery is the development of in-plant water purification systems for recycle purposes.

A typical wastewater stream from a glycerine manufacturing plant was investigated for reverse osmosis treatment to recover glycerine. The aqueous stream contained 1.2 to 2.7% of glycerine, 14.2 to 17.3% of various polyglycerines (molecular weight >100) and 2.7 to 3.0% of mixture of inorganic (presumably sodium chloride and sodium hydro xide); the waste stream had a dark brown color, density of 1.05 g/cc, viscosity of 1.5 cp, and a pH of 8 to 9. Several commercially available cellulose acetate, cellulose acetate butyrate, and polyelectrolyte membranes were tested, and the former two types of membranes were found particularly suitable for the purpose. Some of the test data obtained are given in Tables XXXIII and XXXIV. Based on these and other data, three modes of possible reverse osmosis operations were analysed to illustrate the recoveries which could be achieved: (1) a single-stage unit capable of recovering 27% of the available glycerine, in which 92% of the inorganics and 94% of the polyglycerines available in the feed could be separated in the product; such a unit could be amortized over 1.5 years based on the current glycerine value; (2) a multistage unit to provide a product stream containing 31% of the available glycerine, 3.9% of available polyglycerines and no inorganics; the amortization period for such a design was 3.75 years; and (3) an alternative multistage unit with countercurrent feed to provide a product stream containing 31% of the available glycerine, 5.2% of the available polyglycerines and no inorganics; such a unit could be amortized over 4.5 years. While these designs could result in economic glycerine recovery and significant recycle water, acceptable methods for the treatment of final concentrates should also be developed for effective pollution control.

Many other petrochemical waste streams having potential for the application of reverse osmosis for waste recovery, water reuse, and water pollution control undoubtedly exist, and it might prove profitable to subject them to similar detailed investigations.

TABLE XXXIII

Membrane Screening for Glycerine Stream
Data of Gulf South Res. Inst.[70]

Solute concentration: 10,000 ppm

Membrane No.	NaCl		Glycerine		Polyol 400		Polyol 600	
	Flux, gal/day · ft^2	Sepn., %	Flux, gal/day · ft^2	Sepn., %	Flux, gal/day · ft^2	Sepn., %	Flux, gal/day · ft^2	Sepn., %
	CA (flat, Eastman), operating pressure, 800 psig							
RO-89	27.9	88.0	26.2	79.7	26.0	95.0	—	—
R-3	4.6	99.6	4.9	95.0	5.4	97.4	—	—
G-7	34.7	53.7	38.7	59.0	32.6	88.0	38.2	92.0
	PE Complex (flat, Amicon), operating pressure, 70 psig							
UM-05	6.3	0	6.2	11.9	5.6	88.2	6.9	91.3
	CAB (flat, UWC), operating pressure, 800 psig							
171-15	7.4	98.3	8.4	>99	8.4	>99	8.4	>99
171-25	7.9	97.2	8.0	99.0	8.7	>99	9.2	>99
IV-64-5	26.2	90.8	27.0	84.1	30.4	>99	27.0	>99
	CA (tubular, AS), operating pressure, 800 psig							
AS-1	22.3	93.5	19.9	93.0	22.3	98.4	21.7	99.1
AS-2	61.2	73.0	48.3	75.5	55.0	94.9	50.7	96.0
	CA (tubular, Eastman), operating pressure, 800 psig							
G-5	29.4	10.5	25.0	10.0	23.7	38.0	22.0	41.2
	CAB (tubular, UWC), operating pressure, 800 psig							
U-1	33.0	85.6	36.4	75.5	34.1	93.2	34.1	93.2
U-2	38.8	74.0	37.1	68.4	38.5	92.5	37.2	95.0
U-3	46.7	64.2	47.2	60.1	44.6	81.6	48.4	85.9

CA cellulose acetate
CAB cellulose acetate butyrate
PE polyelectrolyte
UWC Universal Water Corp.
AS American Standard

Treatment of Wash Waters from Photographic Processing Plants

In the photographic processing industry, the film passes through several processing solutions which are replenished continuously. Each processing solution is generally followed by a tank of wash water used for the removal of unwanted chemicals from the film. The wash water flow rate is generally 4 to 20 times that the of replenisher solution. Thus wash waters are very dilute compared to the processing solutions. If the chemicals in the wash waters could be recovered and reused, this would reduce the cost of processing chemicals needed, produce product water which might be reused for washing, and reduce the chemicals ultimately discharged into the sewer. The application of reverse osmosis for the treatment of such wash waters has been examined by Mahoney, Rowley, and West[70]; some of their test results are given in Tables XXXV to XXXVII.

Table XXXV gives the results obtained with some of the major chemicals used in the industry, tested individually for reverse osmosis separation. The concentrations used in the tests are generally those encountered in processing solutions or in wash waters. The results show that inorganic ions

TABLE XXXIV

Treatment of Waste Stream from Glycerine Manufacturing Plant

Data of Gulf South Res. Inst.[70]

Operating pressure: 800 psig

Membrane No. as in Table XXXIII	Flux, gal/day · ft²	Component	Feed Composition, wt · %	Product Composition, wt · %	Solute sepn., %
U-3	5.2	Glycerine	1.19	1.0	16.1
		Polyglycerines	14.22	3.28	76.8
		Inorganics	2.66	0.32	88.0
		Water	81.93	95.48	—
AS-2	1.43	Glycerine	1.57	0.69	56.1
		Polyglycerines	13.02	1.35	89.6
		Inorganics	2.79	0.35	87.5
		Water	82.62	97.61	—

TABLE XXXVI

Quality of Product Water from Fixing-Bath Wash Water

Data of Mahoney, Rowley, and West[71]

Film type: Cellulose acetate
Operating pressure: 400 psig

	Analyses		
	Feed pH	$Na_2S_2O_3$-$5H_2O$ mg/l	Silver mg/l
Feed	5.3	5000	
Product Fraction, % water recovered			
(1) 0–20	5.8	25	0.1
(2) 20–40	5.9	19	<0.1
(3) 40–60	6.1	23	<0.1
(4) 60–80	6.2	25	0.1
(5) 80–85	6.2	27	0.3
(6) 85–90	6.1	28	<0.1
(1)–(6) average		24	
Concentrate	5.4	33200[a]	60

[a] 99+% of that in feed

TABLE XXXV

Reverse Osmosis Treatment of Aqueous Solutions of Chemicals Used in Photographic Processing

Data of Mahoney, Rowley, and West[71]

Film type: Cellulose acetate

Solute	Film No.	Operating pressure, psig	Concentration, g/l* in feed	Concentration, g/l* in product	Solute sepn., %	Flux, gal/day · ft²
Benzyl alcohol	RO-94	600	2.5 ml/l	2.5 ml/l	0	13
Elon (with sodium sulfite)	RO-94	600	6.5	5.3	18	10
Hydroquinone (with sodium sulfite)	RO-94	600	6.0	5.5	9	11
Formalin	RO-94	600	27 ml/l	20 ml/l	25	14
Sodium sulfite	RO-94	600	7	0	99+	11
Sodium thiosulfate, hydrated	RO-89	600	5	0.02	99+	39
Sodium thiosulfate, hydrated	RO-94	1000	220	0	99+	7
Sodium ferrocyanide, dehydrated	RO-94	600	20	0	99+	20
Potassium ferricyanide	RO-94	600	50	0	99+	14
Potassium ferricyanide	RO-89	1200	200	1	99+	8.5
Sodium bromide	RO-94	600	5	0	99+	20
Potassium dichromate	RO-94	600	20	1.3	94	—
Hydroquinone monosulfonate	RO-94	600	10	1.2	88	—
Silver thiosulfate complex (as Ag)	RO-89	600	0.10	<0.001	99+	38

* g/l unless otherwise stated

are separated more completely than the organic solutes tested.

Inorganic chemicals are used almost exclusively in photographic bleaches and fixing baths; hence those baths or the washes that follow are most favorable for reverse osmosis operation. Table XXXVI gives some test results showing the quality of product water obtained by the reverse osmosis treatment of a typical fixing bath wash water. The data show that an overall solute separation of 94.5% with a 90% product water recovery was obtained in the test; it may also be noted that the concentrate contained 60 ppm of Ag whose concentration in the product water was 0.3 ppm or less.

TABLE XXXVII

Treatment of Simulated Waste Water from Photographic Processes by Reverse Osmosis
Data of Mahoney, Rowley, and West[71]

Film type: Cellulose acetate (20-module unit, 7 ft^2/module)
Operating pressure: 600 psig
Feed rate: 2680 ml/min

Process	COD (mg/l)		
	Ekta-chrome ME-4	Koda-chrome K-12	Ekta-print C
Feed	265	280	350
Product	30–45	40	65
Concentrate	1200	2485	4750
% COD removal	87	86	81
% water recovery	80	90	94

As the photographic film is processed, silver is transferred from the film to the fixing bath forming an inorganic complex ion. The recovery of silver from the fixing bath is usually accomplished by an electrolytic process which, however, is not economical for solutions containing less than 500 ppm of silver. Wash waters generally contain 10 to 100 ppm of silver. Reverse osmosis treatment of such wash waters can yield a concentrate in which silver concentration is high enough for electrolytic recovery; actual test results showed[70] that when the silver content was 30 to 40 ppm in the feed, it was reduced to not over 2 ppm in the product water and the silver in the concentrate could be increased to levels suitable for further electrolytic processing.

It is especially desirable to recover ferro- and ferricyanides from bleach washers for reuse for which concentrations of about 75 g/l are necessary. Using a tubular reverse osmosis unit, a 15 g/l solution of sodium ferrocyanide was tested. Portions were concentrated to 71 to 96 g/l at 600 and 700 psig, recovering 91% of the water. A solute separation of 95.4% was obtained at 700 psig with 89% product water recovery. At these concentrations of ferrocyanide the product water rates dropped about 10-fold from the start to the end of the tubular stack. These results indicate the feasibility of recovering bleach from wash waters, so that it can be reused.

One measure of the separation of solutes from waste waters is the COD before and after it passes through the reverse osmosis unit. Test results given

TABLE XXXVIII

Reverse Osmosis Treatment of Hospital Kitchen Waste Waters
Data of Gouveia and Hooton[72]

Film type: Cellulose acetate (Havens tubular modules)
Operating pressure: 400 to 900 psig
Operating temperature: 35°C

Test details	Solute concentration, ppm		
	Raw kitchen waste	Module concentrate[a]	Product water
Calcium (Ca)	41	984	0.1
Magnesium (Mg)	8	203	0.1
Sodium (Na)	171	7150	20
Potassium (K)	10	230	0.9
Boron (B)	—	0.2	0.1
Dissolved iron (Fe)	1.3	140	0.1
Ammonia (NH_3)	4	67	4
Silica (SiO_2)	50	60	11
Nitrate (NO_3)	—	1.0	0.1
Fluoride (F)	—	0.1	0.1
Chloride (Cl)	170	4340	28
Sulfate (SO_4)	66	10450	3
Bicarbonate (HCO_3)	181	3355	0
Carbonate (CO_3)	0	0	0
Free CO_2	—	—	193
Total phosphate (PO_4)	28	588	0.1
Orthophosphate (PO_4)	25	163	—
Oil and grease	878	15	—
Dissolved solids (calculated)	730	27567	67[b]
Dissolved solids (evaporation)	1071	37420	45[b]
Suspended solids	2498	492	—
Total solids	3552	37912	45[b]
C.O.D.	2800	10900	20
pH	6.0	4.6	3.6
Specific conductance (micromhos/cm at 25°C)	950	25500	221

[a] Mean of influent to modules and effluent from modules
[b] Excluding free CO_2

TABLE XXXIX

Removal of Tritons from Aqueous Solutions by Reverse Osmosis
Data of Sourirajan and Sirianni[73]

Film type: Cellulose acetate (Batch 18)
Operating pressure: 1500 psig
Feed concentration: 2 g Triton/100 g water

Film shrinkage temp., °C	Solute-Triton type	Solute separation, %	Product rate gal/day · ft^2
86	N-128	98.6	32.5
89	X-165	99.1	29.4
90	X-205	99.1	29.5
89	X-305	~100	26.2

in Table XXXVII show COD removals of 81 to 87% with product water recoveries of 80 to 94% were actually achieved by the reverse osmosis treatment of typical photographic processing waste waters.

Treatment of Kitchen and Laundry Waste Waters

Kitchen and laundry wastes from Williams Air Force Base, Arizona, and a synthetic shower waste were subjected to reverse osmosis treatment using Havens tubular cellulose acetate membrane modules and some of the results obtained are given in Table XXXVIII[72]. With 90% recovery, the product water obtained met U.S. Public Health Standards for potable water. Table XXXIX gives some of the results obtained on the reverse osmosis separation of several Tritons (polyoxyethylated nonionic surface active agents used as detergents) in aqueous solution[73]. These results illustrate the possible applicability of reverse osmosis for the renovation and reuse of kitchen, laundry, and similar waste waters.

Reverse Osmosis for Agriculture

The possible use of reverse osmosis treated water for agricultural purposes is an application of tremendous social significance. This application is already pioneered by the Negev Institute in Israel, who have designed and put into operation three reverse osmosis units (besides the one at Yotvatah) one of which (at Gilath) produces product water containing less than 100 ppm of chloride for citrus crops[74]. Similar and more extended applications of reverse osmosis may be expected in the future. The OSW has conducted a 1-year field test at Fresno, California, on the reverse osmosis treatment of high salinity irrigation return flows; some

TABLE XL

Reverse Osmosis Treatment of Irrigation Return Waters
Data of Channabasappa[8]

Film type: Cellulose acetate
Operating pressure: 600 psig
Product to waste ratio: 1 : 1

Test details	Solute concentration, ppm		
	Feed	Product	Waste
pH	7.9	6.0	7.4
Total dissolved solids	4890	340	7420
Total hardness	1290	15	2060
Ca^{++}	274	6	505
Mg^{++}	147	0–1	194
Na^{+}	1020	102	1620
K^{+}	4	—	5
HCO_3^-	—	—	394
SO_4^{--}	2690	43	4370
Cl^-	371	107	532
NO_3^-	43	32	54
Boron	8.5	8	11
SiO_2	44	13	62

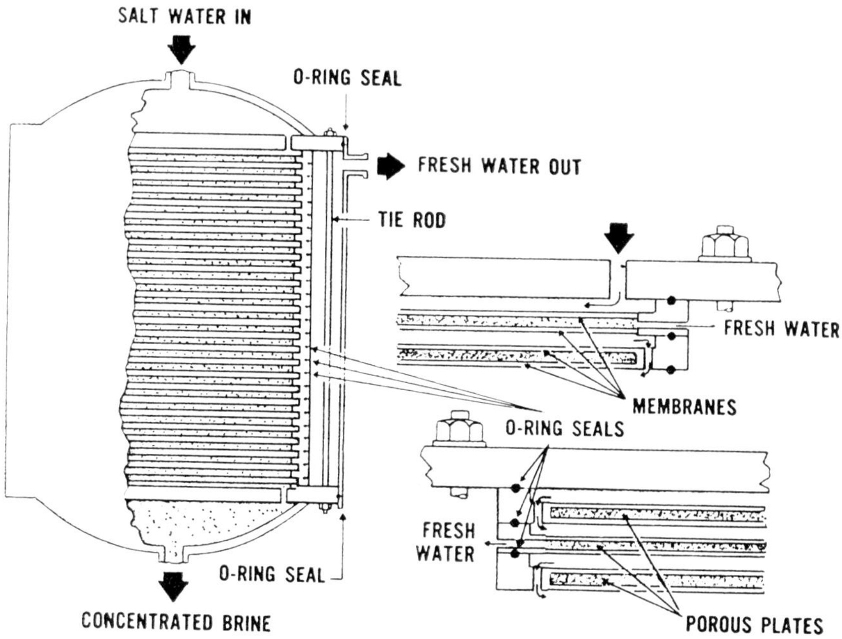

Figure 8 Schematic representation of plate and frame modular design.

of their test data are given in Table XL which shows that the quality of the product water could be suitable for agricultural reuse but for its high boron content. The boron content is required to be between 0.5 and 1 ppm for irrigation waters, which calls for further developments in reverse osmosis. The above results however point to the possible opportunities for bringing reverse osmosis into the field of agricultural practice.

Modular Designs

References have already been made to different membrane configurations and modular designs used in reverse osmosis research and development. The membranes used are in flat, tubular, or hollow-fiber forms. Flat membranes are most convenient for laboratory studies, and tubular membranes seem to be preferred type for larger scale equipments. The latter are essentially assemblies of one or more membrane modules of specific design. Four basic modular designs are currently employed, namely, plate-and-frame, spiral-wound, tubular, and shell-and-tube types[10]. The first two types use flat membranes, and the last type uses hollow fiber membranes.

The plate-and-frame design, illustrated in Fig. 8, is analogous to a filter press. It utilizes a stack of parallel plastic plates as supports for the membranes, with grooves or a porous spacer between the membrane and plate for the permeate stream to exit. Membranes are placed on each face of the plastic support, with the spacing between successive plates varying from less than 10 mils to greater than 100 mils; in the latter case, a baffle or porous spacer may be employed between the membranes on each place to enhance the feed velocity. The plate-and-frame design has typically a membrane area packing density of 150 ft²/ft³.

The spiral-wound modular design, developed by Gulf General Atomic, is illustrated in Fig. 9. It consists of a sealed envelope of two membranes

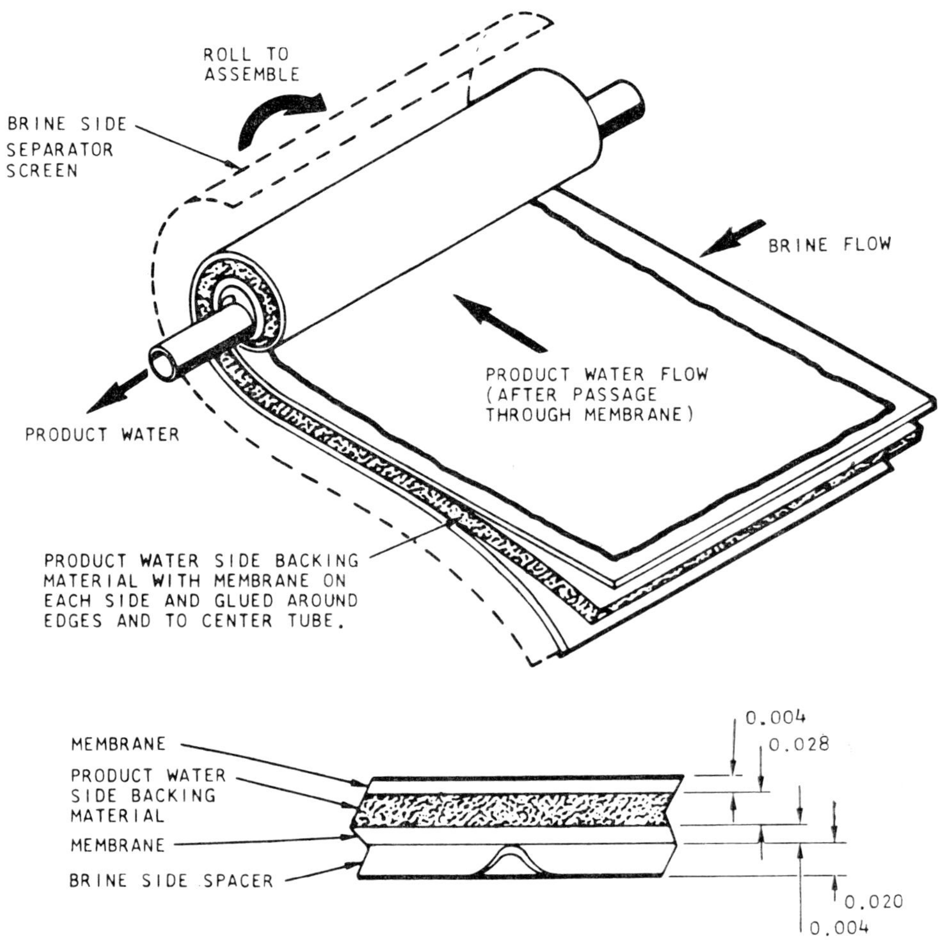

Figure 9 Schematic representation of spiral-wound modular design.

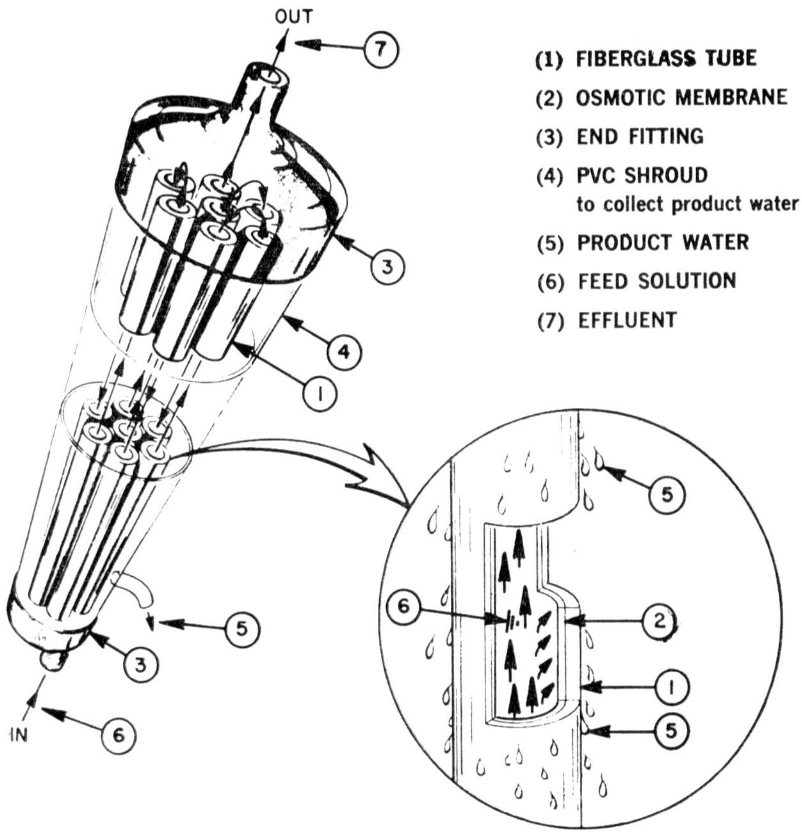

Figure 10 Schematic representation of Havens tubular module.

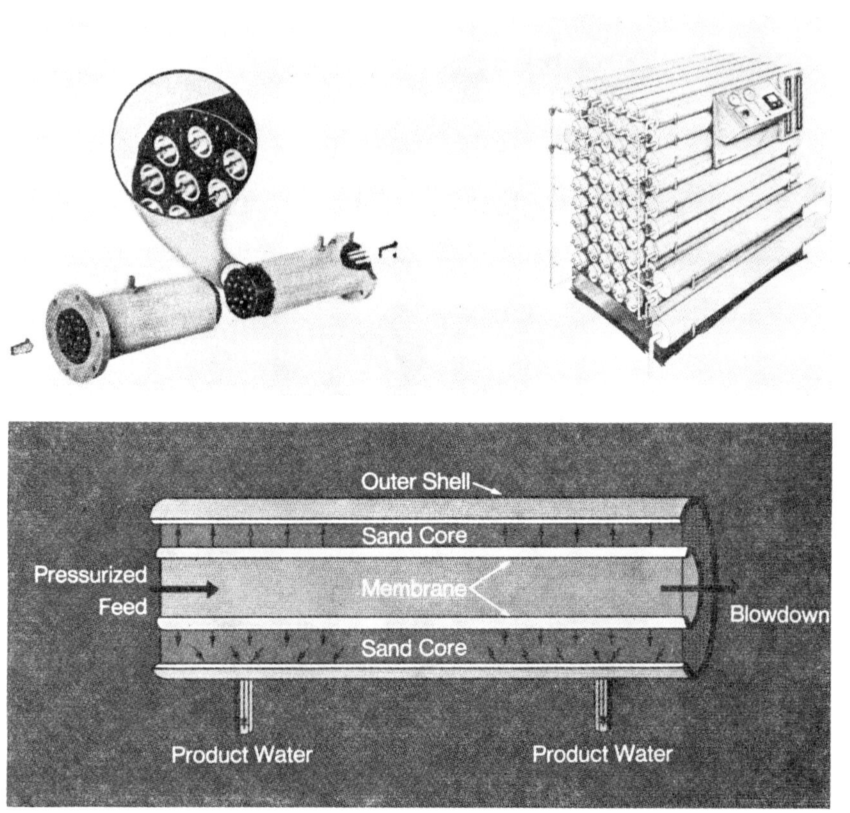

Figure 11 Schematic representation of Westinghouse tubular module.

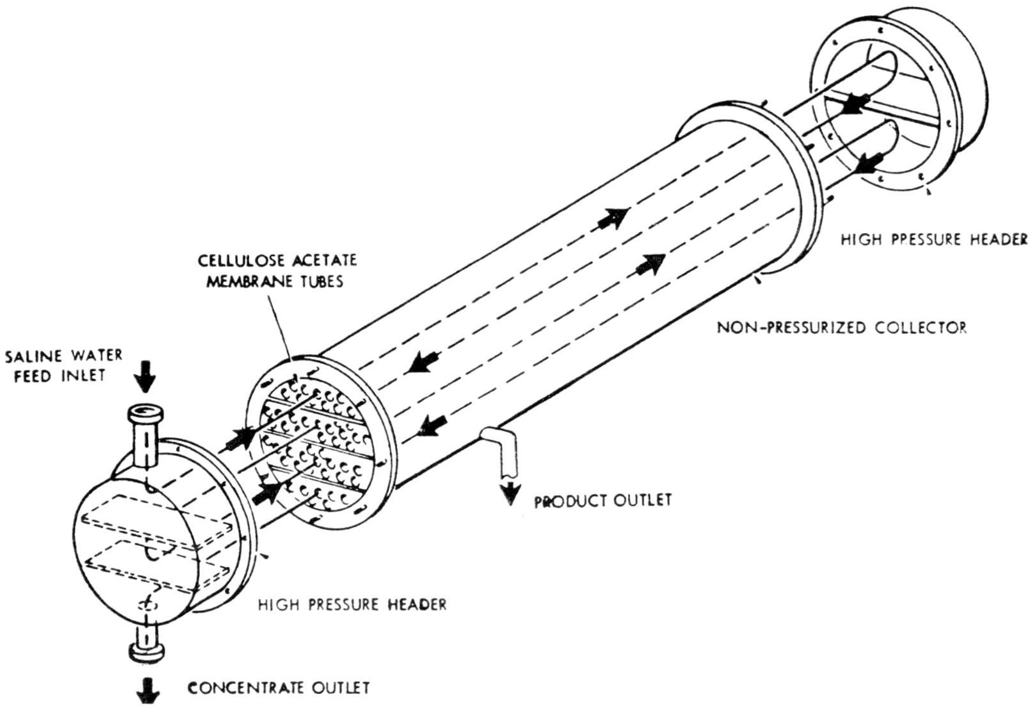

Figure 12 Schematic representation of Hydronautics tubular module.[10]

containing a product water spacer, 30–50 mil thick, separating the two membranes, and a feed-brine spacer between successive envelopes. The envelope is sealed on three edges with the fourth edge open and sealed to a central product water collection tube on which the envelope is wrapped. The spiral-wound design is compact with membrane area packing densities ranging from 200 to 300 ft^2/ft^3 of module.

Tubular membranes of diameters $\frac{1}{2}$ to 2 inches are used in tubular modules. The modular designs used in the Coalinga and Yotvatah plants (Fig. 2 and 3) have already been described. Havens Industries (presently, Calgon-Havens System) of San Diego, California, were the earliest to pioneer the commercial development of reverse osmosis units using cellulose acetate membranes integrally lined in porous fiberglass tubes. A schematic representation of Havens module is given in Fig. 10. The tubes are imbedded in an inboard fitting at each end of the module. A shroud surrounds the tubes and collects the membrane-permeated product water dripping through the porous walls of the tubes which are connected in series with U-turn connections at each end. Each end of the module has an outboard hexagonal fitting that supports the U-tube fittings, and provides a convenient method of connecting the individual modules together.

There are also several other kinds of tubular membrane modules currently available or under development. The Westinghouse tubular membranes[76] are supported on a porous resin-bonded sand matrix housed in a shell suitable for supporting the assembly and for collecting the product as shown in Fig. 11. The Hydronautics design for tubular hydrocast membranes[10,77], shown in Fig. 12, is an in-situ membrane replacement system. In this system, air is blown into the end of a tube containing the viscous cellulose acetate casting solution which is then forced out at the other end, leaving behind a fraction of the fluid coating the inner wall surface in the form of an annular layer. The annular coating is then gelled by passing a stream of cold water thereby forming a continuous tube of cellulose acetate. A subsequent hot water treatment results in an asymmetric tubular reverse osmosis membrane. Cosmodyne's reverse osmosis system utilizes a helical element consisting of a flat membrane applied to the outside of a porous tubular support; feed solution flows through an annular passage around the element, and product water is collected from within the support tube.

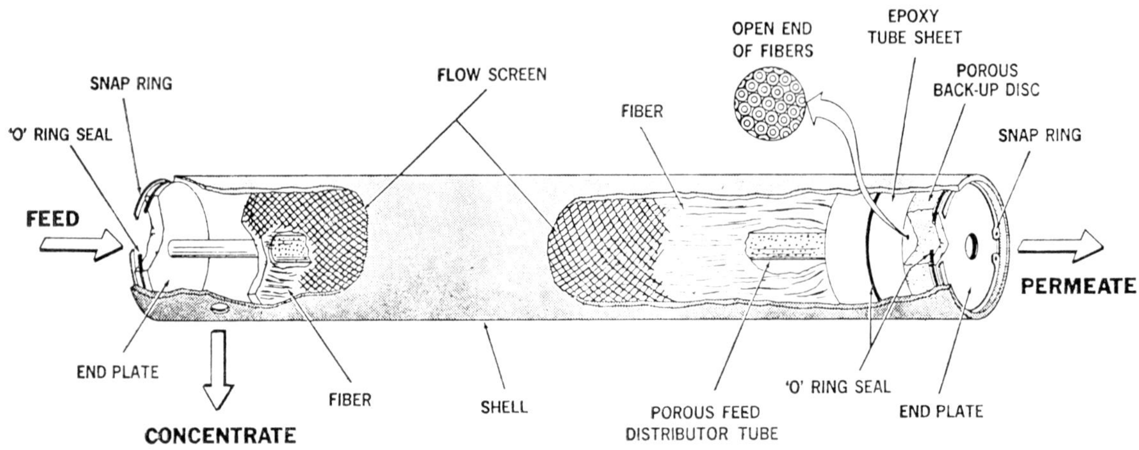

Figure 13 Schematic representation of DuPont hollow fiber module.[78]

The hollow fiber system of DuPont is illustrated in Fig. 13[78]. The system utilizes a tightly packed bundle of hollow fibers within a high pressure vessel in a shell-and-tube type configuration. The hollow fibers serve as the membrane element and as their own support against the pressure of the feed water. The open ends of each fiber loop are potted in a header where the product water from the interior of each fiber exits. An early model of the DuPont B-9 permeator has the following dimensions: shell 4.5 inches i.d., by 4 ft long, made of aluminum and pressure rated at 650 psig; active fiber length 2.5 ft; fiber size 42 microns i.d. by 84 microns o.d.; and nominal fiber surface area 1500 ft^2.

The membrane modular design is an area of activity where there is scope for infinite technological ingenuity. One may hence look forward to the continuous commercial development of reverse osmosis equipments practically suitable for different specific applications.

Conclusion

The extensive experimental data presented in the foregoing discussion on the applications of reverse osmosis must be understood and appreciated in their true context. The data represent what has been achieved, rather than what can be achieved, by reverse osmosis. Further, in most cases, the membranes and systems used are not adequately specified; consequently, the data presented represent not necessarily the best that could have been achieved with the particular membranes tested. The concepts of membrane and system specifications and the predictability of membrane and system performance discussed earlier (in Part I) are fundamental to reverse osmosis engineering. These concepts have not yet been built into the design of present day equipment manufacturers, or the parametric studies on applications using the available equipments. The physicochemical criteria for reverse osmosis separation, especially with respect to organic solutes, and the governing principles of reverse osmosis membrane science are only slowly developing. All this means that reverse osmosis is still at its very early stages of development. While what has been achieved is significant enough to illustrate the immense potential of reverse osmosis to contribute effectively to better utilization of resources and creation of healthier environments, further work in the field can be expected to contribute far more to the practical and widespread realization of the vast potentialities of this new process, and consequently contribute significantly to the economic prosperity and physical well-being of all mankind.

REFERENCES

1. Loeb, S., *Desalination* **1**, 35 (1965).
2. Loeb, S., and J. S. Johnson, *Chem. Eng. Progress* **63**, No. 1, 90 (1967).
3. Loeb, S., and E. Selover, *Desalination* **2**, 75 (1967).
4. Stevens, D., and S. Loeb, *Desalination* **2**, 56 (1967).
5. McCutchan, J. W., in Saline Water Research Progress Summary, Department of Engineering, University of California, Los Angeles, Report No. 68-1 (1968).
6. Johnson, J. S., J. W. McCutchan, and D. N. Bennion, School of Engineering and Applied Science, University of California, Los Angeles, Report No. 69-45, July 1969.
7. Cohen, H., and S. Loeb, *Proc. 3rd International Symp.*

on Fresh Water from the Sea, Dubrovnik, Vol. 2, 339 (1970).
8. Channabasappa, K. C., Chem. Eng. Progress Symp., Ser. No. 97, Vol. 65, 140 (1969).
9. Saline Water Conversion Report 1970–71, Office of Saline Water, U.S. Department of the Interior, Washington, D.C., 1971.
10. Gillam, W. S., and H. E. Podall, *Proc. 3rd International Symp. on Fresh Water from the Sea*, Dubrovnik, Vol. 2, 385 (1970).
11. Loeb, S., and S. Sourirajan, Department of Engineering, University of California, Los Angeles, Report No. 60-60 (1961).
12. McCutchan, J. W., and D. N. Bennion, in Saline Water Research Summary, School of Engineering and Applied Science, University of California, Los Angeles, Report No. 70-1, Jan. 1970.
13. Ohya, H., and S. Sourirajan, *Desalination* **6**, 153 (1969).
14. Hauck, A. R., and S. Sourirajan, *Environmental Sci. Tech.* **3**, 1269 (1969).
15. Matsuura, T., and S. Sourirajan, *Water Research* (in press).
16. Hindin, E., and P. J. Bennett, *Water and Sewage Works* **116**, 66 (1969).
17. Sourirajan, S., and S. Kimura, *Ind. Eng. Chem. Process Design Develop.* **6**, 504 (1967).
18. Ohya, H., and S. Sourirajan, *Ind. Eng. Chem. Process Design Develop.* **8**, 131 (1969).
19. Kimura, S., and S. Sourirajan, *Ind. Eng. Chem. Process Design Develop.* **7**, 548 (1968).
20. Sourirajan, S., *Ind. Eng. Chem. Process Design Develop.* **6**, 154 (1967).
21. Matsuura, T., and S. Sourirajan, *Ind. Eng. Chem. Process Design Develop.* **10**, 102 (1971) .
22. Hindin, E., P. J. Bennett, and S. S. Narayanan, *Water and Sewage Works* **116**, 466 (1969).
23. Vos, K. D., F. O. Burris, Jr., and R. L. Riley, *J. Appl. Polym. Sci.* **10**, 825 (1966).
24. Henshaw, T. B., *Chem. Eng.* **78** (12), 47 (1971).
25. Public Health Service Drinking Water Standards, U.S. Department of Health, Education, and Welfare, Public Health Service, Washington, D.C., 1962.
26. Lonsdale, H. K., U. Merten, and M. Tagami, *J. Appl. Polym. Sci.* **11**, 1807 (1967).
27. Matsuura, T., and Sourirajan, S. (unpublished).
28. OSW, Saline Water Conversion Report, U.S. Department of the Interior, Office of Saline Water, 1965.
29. Ironside, R., and S. Sourirajan, *Water Research* **1**, 179 (1967).
30. Sinskey, A. J., G. C. Y. Chu, and D. I. C. Wang, Chem. Eng. Progress Symp. Series, Vol. 67, No. 108, 75 (1971).
31. Hardwick, W. H., *Chemistry and Industry* **9**, 297 (1970).
32. Aerojet-General Corp. (Environmental Systems Div.), "Reverse Osmosis Renovation of Municipal Waste Water," Water Pollution Control Research Series ORD-17040EFQ 12/69, U. S. Department of the Interior, Federal Water Quality Administration, Washington, D.C., 1969.
33. Envirogenics Co. (Aerojet-General Corp.), "New Technology for Treatment of Waste Water by Reverse Osm osis", Water Pollution Control Research Series 17020DUDO9/70, U.S. Environmental Protection Agency, Water Quality Office, Washington, D.C., September 1970.
34. Smith, J. M., A. N. Masse, and R. P. Miele, "Renovation of Municipal Waste Water by Reverse Osmosis", Water Pollution Control Research Series 17040-05/70, U.S. Environmental Protection Agency, Water Quality Office, Washington, D.C., May 1970.
35. Nusbaum, I., J. H. Sleigh, Jr., and S. S. Kremen, "Study and Experiments in Waste Water Reclamation by Reverse Osmosis", Water Pollution Control Research Series 17040-05/70, U.S. Department of the Interior, Federal Water Quality Administration, Washington, D.C., May 1970.
36. Bray, D. T., U. Merten, and M. Augustus, "Reverse Osmosis for Water Reclamation", Gulf General Atomic, Inc., San Diego, Calif., Rept. GA-6337 (1965).
37. Sudak, R. G., and I. Nusbaum, "Pilot Plant Operation of Spiral Wound Reverse Osmosis Systems," Gulf General Atomic, Inc., San Diego, Calif., Rept. GA-8515 (1968).
38. Douglas, A. S., M. Tagami, and C. E. Milstead, "Membrane Materials for Waste Water Reclamation by Reverse Osmosis," Water Pollution Control Research Series ORD-17040EF006/70, U.S. Department of the Interior, Federal Water Quality Administration, Washington, D.C., June 1970.
39. North Star Research and Development Institute, "New and Ultrathin Membranes for Municipal Waste Water Treatment by Reverse Osmosis," Water Pollution Control Research Series 17020EFA10/70, U.S. Environmental Protection Agency, Water Quality Office, October 1970.
40. Kraus, K. A., "Application of Hyperfiltration to Treatment of Municipal Sewage Effluents", Water Pollution Control Research Series ORD-17030EOH01/70, U.S. Department of the Interior, Federal Water Quality Administration, Washington, D.C., January 1970.
41. Wiley, A. J., A. C. F. Ammerlaan, and G. A. Dubey, TAPPI **50**, 455 (1967).
42. Ammerlaan, A. C. F., B. F. Lueck, and A. J. Wiley, TAPPI **52**, 118 (1969).
43. Ammerlaan, A. C. F., and A. J. Wiley, *Chemical Eng. Prog. Symp.* Ser. No. 97, Vol. 65, 148 (1969).
44. Wiley, A. J., G. A. Dubey, J. M. Holderby, and A. C. F. Ammerlaan, J. WPCF **42** (8), R 279 (1970).
45. Perona, J. J., F. H. Butt, S. M. Fleming, S. T. Mayr, R. A. Spitz, M. K. Brown, H. D. Cochran, K. A. Kraus, and J. S. Johnson, Jr., *Environmental Sci. Tech.* **1**, 991 (1967).
46. Willits, C. O., J. C. Underwood, and U. Merten, *Food Tech.* **21**, 24 (1967).
47. Merson, R. L., and A. I. Morgan, Jr., *Food Tech.* **22**, 631 (1968).
48. Lowe, E., E. L. Durkee, and A. I. Morgan, Jr., *Food Tech.* **22**, 915 (1968).
49. Marshall, P. G., W. L. Dunkley, and E. Lowe, *Food Tech.* **22**, 969 (1968).
50. McDonough, F. E., *Food Engineering* **40**, 124 (1968).
51. Merson, R. L., L. F. Ginnette, and A. I. Morgan, Jr., *Dechema-Monographien* Band 63, 179 (1969).

52. Merson, R. L., in Albany Symp. on Reverse Osmosis in Food Processing, ARS 74-51, Agricultural Research Service, U.S. Department of Agriculture, Albany, Calif., August 1969.
53. Lowe, E., in *Albany Symp. on Reverse Osmosis in Food Processing*, ARS 74-51, Agricultural Research Service, U.S. Department of Agriculture, Albany, Calif., August 1969.
54. Underwood, J. C., in *Albany Symp. on Reverse Osmosis in Food Processing*, ARS 74-51, Agricultural Research Service, U.S. Department of Agriculture, Albany, Calif., August 1969.
55. Dunkley, W. L., in *Albany Symp. on Reverse Osmosis in Food Processing*, ARS 74-51, Agricultural Research Service, U.S. Department of Agriculture, Albany, Calif., August 1969.
56. Mercer, W. A., in *Albany Symp. on Reverse Osmosis in Food Processing*, ARS 74-51, Agricultural Research Service, U.S. Department of Agriculture, Albany, Calif., August 1969.
57. Ginnette, L. F., in *Albany Symp. on Reverse Osmosis in Food Processing*, ARS 74-51, Agricultural Research Service, U.S. Department of Agriculture, Albany, Calif., August 1969.
58. Lowe, E., E. L. Durkee, R. L. Merson, K. Ijichi, and S. L. Cimino, *Food Tech.* **23**, 753 (1969).
59. Merson, R. L., and L. F. Ginnette, in *Applied Polymer Symp.* **13**, Editor A. F. Turbak, p. 309, Interscience, New York, 1970.
60. Lim, T. H., W. L. Dunkley, and R. L. Merson, *J. Dairy Science* **54**, 306 (1971).
61. Fenton-May, R. I., and C. G. Hill, Jr., *J. Food Science* **36**, 14 (1971).
62. Peri, C., and W. L. Dunkley, *J. Food Science* **36**, 25 (1971).
63. Porter, W. L., J. Siciliano, S. Krulick, and E. G. Heisler, in *Membrane Science and Technology*, Editor J. E. Flinn, p. 220, Plenum, New York, 1970.
64. Besik, F. K., *Water and Pollution Control* **107** (10), 24 (1969); **107** (11), 47 (1969).
65. Golomb, A., *Plating* **57** (10), 1001 (1970).
66. Hauck, A. R., and S. Sourirajan, *J. WPCF* **44**, 1372 (1972).
67. OSW, Saline Water Conversion Report, p. 195, U.S. Dept. Interior, Office of Saline Water (1965).
68. Riedinger, A., and J. Schultz, "Acid Mine Water Reverse Osmosis Test at Kittanning, Pennsylvania", U.S. Dept. Interior, Office of Saline Water, Research and Development Progress Report No. 217 (1966).
69. Kremer, S. S., A. B. Riedinger, J. H. Sleigh, and R. L. Truby, "Reverse Osmosis Field Testing on Acid Mine Waters at Norton, West Virginia," U.S. Dept. Interior, Office of Saline Water, Research and Development Progress Report No. 586 (1970).
70. Gulf South Research Institute, "Polymeric Materials for Treatment and Recovery of Petrochemical Wastes," Water Pollution Control Research Series 12020 DQC 03/71, U.S. Environmental Protection Agency, Water Quality Office, Washington, D.C., March 1971.
71. Mahoney, J. G., M. E. Rowley, and L. E. West, in *Membrane Science and Technology*, Editor J. E. Flinn, p. 196, Plenum, New York, 1970.
72. Gouveia, A., and K. A. K. Hooton, *Chem. Eng. Progress Symp.* Ser. No. 90, Vol. 64, 280 (1968).
73. Sourirajan, S., and A. F. Sirianni, *Ind. Eng. Chem. Product Res. Develop.* **5**, 30 (1966).
74. Cohen, H., Report from The Negev Institute, Beersheva, Israel, Feb. 1971.
75. Roberts, R. M., and L. E. Gressingh, "Development of Economical Methods of Boron Removal from Irrigation Return Waters", U.S. Dept. Interior, Office of Saline Water, Research and Development Progress Report No. 579 (1970).
76. Coit, R. L., *Proc. 3rd International Symp. on Fresh Water from the Sea*, Dubrovnik, Vol. **2**, 351 (1970).
77. Gollan, A., and M. P. Tulin, *Proc. 3rd International Symposium on Fresh Water from the Sea*, Dubrovnik, Vol. **2**, 423 (1970).
78. Caracciolo, V. P., and W. G. Smith, "The B-9 Permeator—A Major Advance in Commercial RO," paper presented at the OSW Conference at Roswell, New Mexico, November 4–5, 1970.

S. SOURIRAJAN
National Research Council of Canada

AQUATIC PRIMARY PRODUCTION

Primary productivity in aquatic systems, like the same process in terrestrial environments, provides the base of the food web upon which all higher levels of an ecosystem depend. Biological productivity is the increase in organic material per unit of area or volume with time. This addition of organic matter is the material from which the various plant and animal communities of an ecosystem are made, and is dependent on the conversion of inorganic matter into organic matter. Conversion is accomplished by plants through the photosynthetic process. Plants are therefore considered to be the *primary producers*, and in an aquatic ecosystem these plants include algae, bacteria, and sometimes higher plants such as water grasses and water lillies. *Primary productivity*, the first level of productivity in a system, can be measured as the rate of photosynthesis, addition of biomass per unit of time (yield), or indirectly by nutrient loss or a measure of respiration of the aquatic community.

Methods of Study

Standing crop refers to the part of biological production per unit area or per unit volume that is physically present as biomass and that is not lost in respiration. Standing crop measurements over a period of time give an indirect measure of productivity in terms of yield. Plankton, microscopic floating plants and animals, can be collected in a plankton net and may be counted under a micrcosope or weighed. Aquatic biologists have used standing crop measurements to estimate productivity longer than any other method (e.g. Lohman, 1908). This method is still also used for periphyton (attached algae) or rooted plants.

Only within the past few decades have biologists progressed from merely counting numbers of organisms to calculating biomass, and more recently, to expressing biomass *yield*. Fishery biologists, like farmers, for many years have measured fish productivity in terms of tons produced per acre of water surface per year. Calculating biomass and biomass yield is an important step forward since changes in standing crop reflect the net effect of many biological and physical events and therefore are not directly proportional to productivity. For example, the standing crop of a phytoplankton community may be greatly diminished by predation and water movement, while photosynthetic rates of the survivors may remain high.

The measurement of plant pigments such as chlorophyll *a* is also a standing crop measurement that is frequently used and may now be done through remote sensing by aircraft or satellites.

Uptake of Nutrients

Another early attempt at measuring the rate of production in aquatic ecosystems was to measure the inorganic nutrients taken up in a given system and to calculate the amount of biological production required to absorb this amount. Atkins (1922, 1923) studied the decrease in carbon dioxide and phosphate in measuring production in the North Sea, and Steel (1956), also working in the North Sea, estimated the annual cycle of plant production by considering changes in the inorganic phosphate in relation to vertical mixing of the water mass. Many biologists consider phosphorus to be a difficult element to study in this respect because organisms often store it in excess of their requirements for optimum growth.

Measuring nutrient uptake is an indirect method of determining the rate of productivity in an aquatic ecosystem and is influenced by various other biological activities. Nevertheless, it has been important in the development toward more precise measurements of the dynamic aquatic ecosystem.

Measurements of Oxygen and Carbon Dioxide

The net rate at which the phytoplankton community of a given ecosystem incorporates carbon dioxide may be estimated in moderately to highly productive aquatic environments by direct measurement of the short-term fluctuations in the dissolved oxygen it produces. The calculations are based on the assumption that a mole of oxygen is released into the environment for each mole of carbon dioxide reduced in photosynthesis. This method precludes the necessity of enclosing the phytoplankton in a bottle. If measurements are made at regular hourly intervals over a 24-hour period, the average hourly decrease in oxygen during times of darkness when no photosynthesis is occurring can be determined. It is assumed that respiration removes this amount of oxygen each hour throughout the day thus giving a measure of the gross rate at which the community incorporates carbon dioxide. An analogous method exists for recording fluctuations in carbon dioxide.

The pH meter, which measures acidity, has been successfully employed to measure these carbon dioxide changes in the aquatic ecosystem since the removal of carbon dioxide from the water for photosynthesis is accompanied by a proportional rise in pH. This pH shift has been used to estimate both photosynthesis and respiration. The sea and some fresh waters are too buffered against changes in pH to make this method useful in all environments, but it has been employed with success in lakes and for continuously monitoring the growth of cultures. Carbon dioxide may also be directly measured by standard volumetric or gasometric techniques.

Although carbon dioxide and oxygen can be measured with relative precision, the overall precision of productivity measurements made by these techniques is not generally great because of uncertainties in the corrections for diffusion, water movements, or extended enclosure time. Some of the oxygen produced by higher aquatic plants may not be immediately released thus causing a lag period in the evolution of oxygen into the environment. The primary advantage this method has over the more sensitive ^{14}C method is the added benefit of an estimate of community respiration.

Some of the uncertainties of the previous method can be reduced by enclosing phytoplankton samples just long enough in glass bottles for measurable changes in the concentration of oxygen and carbon dioxide to occur, but not long enough for depletion of nutrients or the growth of bacteria on the inside bottle surface. This method is called the *light and dark bottle* method. The name is derived from the fact that identical samples are placed in a transparent "light bottle" and an opaque "dark bottle." Gross and net productivity of the plankton community from which the samples were taken can be estimated by calculating the difference in the oxygen content between the two bottles after a predetermined period of incubation and with that present initially.

Productivity determinations that are dependent on measurements of oxygen are based on some estimated photosynthetic quotient (moles O_2 liberated/moles CO_2 incorporated). For the photosynthesis of carbohydrates the ratio is unity. For the synthesis of an algal cell, however, the expected ratio is higher, and presumably varies with the physiological state of the algae and the nutrients available.

Oxygen methods in general have rather poor sensitivity and are of no use if the gross incorporation of inorganic carbon during the test period is less than about 20 mg of carbon per cubic meter. Several days may be required in many of the less productive aquatic environments for this much photosynthesis to occur and bacteria may develop on the insides of the container during this time invalidating the results.

Photosynthetic rates can be measured in light and dark bottles also by determining the amount of carbon fixed in particulate form after a short incubation. This can be done by inoculating the bottles with *radioactive carbon* ($Na_2{}^{14}CO_3$). Sensitivities with this method are much greater than the standard method and much shorter periods of incubation are possible. It is possible to obtain easily measurable amounts of ^{14}C in particulate form after only two hours by adjusting the specific activity of the inoculums. However, unlike the oxygen method, the dark bottle results do not provide an estimate of community respiration thus giving the ecologist less information with which to work.

The ^{14}C method has been widely used because it is sensitive and rapid. One outcome of its popularity is that a great deal of scrutiny has been devoted to the method itself. After 18 years of use, however, it is still not clear whether the ^{14}C is measuring gross productivity, net productivity, or something in between. The results probably most closely estimate net productivity, but it may be that this method applies only to a particular set of experimental conditions.

Already mentioned is the evidence that some of the ^{14}C that is fixed during incubation may seep out of the algal cells in the form of water-soluble organic compounds. This material is presumably utilized by bacteria rather than passed on directly to the next higher trophic level as is the remainder of the consumed primary productivity. The amount of primary production liberated extracellularly is large enough to be measured with precision and a number of workers are now routinely including quantitative studies of extracellular products of photosynthesis as part of the measurements of primary productivity.

Calibration of radioactive sources and instruments for measuring radioactivity pose a serious technical problem for the ^{14}C method. In order to calculate productivity in terms of carbon uptake it is necessary to know accurately the amount of ^{14}C added in microcuries and the number of microcuries recovered in particulate form by filtering the sample through a membrane filter.

Further it has been found that phytoplankton cells may become damaged during filtration and calculations based on these conditions will show lower productivity rates than are actually the case.

A point deserving emphasis is that those of us measuring primary productivity are still attempting to determine more precisely what is being measured, and generalizations about the transfer of energy through aquatic food-webs should be made continuously. Neither this nor any other practical technique adequately measures the change in oxidation state of the carbon that is fixed. The subsequent ecological role of newly fixed carbon is even more difficult to measure because of the various ways the photosynthate may be used.

Use of Primary Productivity Measurements in Aquatic Ecosystems

Lindeman (1942) developed a trophic-dynamic model of an aquatic ecosystem and introduced the concept of "energy flow," or the efficiency of energy transfer from one trophic level to the next, to describe its operation. A certain value derived from the measured primary productivity represented the input of energy into the next grazing level, and so forth up the food chain. It was consistent with Lindeman's purpose to express his data as energy units (calories). Subsequent workers have continued to probe the concept of energy flow. However, advances in biochemistry, physiology, and ecology require such a complex model of energy flow that it is difficult to relate it to the natural world. In an imaginary world or model of a system in which the function units are discrete trophic levels, it is not only possible but stimulating to describe the flow of energy through an ecosystem. But when the functional units of the system being investigated are conceived of as macromolecules it is difficult to translate biomass accumulation into energy units.

Besides requiring a portion of their autotrophic production for respiration, phytoplankton communities must also reserve a portion for the maintenance of community structure. In terms of information theory, energy expended for community maintenance is referred to as "*information.*" Energy information cost has never been measured directly but there is indirect evidence that it must be paid. For example, when an aquatic ecosystem is altered artificially with the aim of increasing the production of fish, zooplankton and fish may increase in greater proportion than the phytoplankton (McConnell, 1965; Goldman, 1968). Perhaps a large amount of primary production remains with the phytoplankton as information necessary for the maintenance or development of community structure. Grazers then have access only to the production in excess of this threshold level. If the magnitude of the information cost is high relative to primary production, then a small increase in the rate of growth of the primary producers will provide a relatively larger increase in the food supply of grazers and in turn the fish that consume them.

There are difficulties that must be met in the course of fitting measurements of primary productivity to the trophic-dynamic model. A highly variable yet often significant portion of primary production, as measured by ^{14}C light-and-dark bottle experiments, is not retained by the producers but instead moves into the environment in soluble form. It is difficult to measure the absolute magnitude of such excretion by a community of natural plankton because the excreta can rapidly serve as a substrate for bacterial growth and thus find its way back to particulate or inorganic form during the incubation period. Although this excrement is part of the primary productivity and eventually serves as an energy source for organisms at the higher trophic levels, the pathway along which this energy flows does not follow the usual linear sequence modeled for the transfer of energy from phytoplankton to herbivorous zooplankton. There is evidence that the amount of energy involved may sometimes be of the same order of magnitude as that recovered in particulate form in routine ^{14}C productivity studies.

The role of allochthonous material (material brought in from outside the system) in supporting the energy requirements of consumer organisms must also be considered in studies of energy flow. No natural aquatic ecosystem is entirely closed. Potential energy enters in the form of organic solutes and debris. Organic solutes undergo conversion to particulate matter through bacterial action. Sorokin (1965) in Russia found this type of production of particulate matter to be the most important in producing food for crustacean filter-feeders. Particulate and dissolved organic matter may also arise in the aquatic environment through chemosynthesis. This is a form of primary production not usually considered and therefore not usually measured. Although its magnitude may not be great in many systems, Sorokin found it to be very important in the Rybinsk reservoir and in the Black Sea.

Primary Production and Eutrophication

The process of increasing productivity of a body of water is known as eutrophication and in the idealized succession of lakes, a lake would start as oligotrophic (low productivity), becoming mesotrophic (medium productivity) eventually eutrophic (highly productive) and finally dystrophic, a bog stage in which the lake has almost been filled in by weeds and the productivity has been greatly decreased. The concept of eutrophic and oligotrophic lake types is not a new one. It was used by Naumann (1919) to indicate the difference between the more productive lakes of the cultivated lowlands and the less productive mountain lakes. The trophic state of five different aquatic environments will be discussed below.

The general progression from an oligotrophic to an eutrophic and finally to a dystrophic lake (lake succession) is as much a result of the original basin shape, climate, and such edaphic factors as soil, as it is of geologic age. It is unlikely that some shallow lakes ever passed through a stage that could be considered oligotrophic, and it is just as unlikely that the first lake to be considered here, Lake Vanda, will ever become eutrophic. It is also possible that the "progression" may be halted or reversed.

Lake Vanda, located in "dry" Wright Valley near McMurdo Sound in Antarctica, is one of the least productive lakes in the world. The lake is permanently sealed under 3 to 4 meters of very clear ice which transmits 14 to 20% of the incident radiation to the water below. This provides enough light to power the photosynthesis of a sparse phytoplankton population to a depth of 60 meters (Goldman *et al.*, 1967). Lake Vanda can be classified as ultraoligotrophic, since its mean productivity is only about 1 mg $C \cdot m^{-2} \cdot hr^{-1}$.

Lake Tahoe in the Sierra Nevada of California and Nevada is an alpine lake long esteemed for its remarkable clarity. Although it is more productive than Lake Vanda, it is still oligotrophic. The lake is characterized by a deep euphotic (lighted) zone, with photosynthesis occurring in the phytoplankton and attached plants to a depth of about 100 m. Although the production under a unit of surface area is not small, the intensity of productivity per unit of volume is extremely low. Lake Tahoe's low fertility (as inferred from its productivity per unit volume) is the result of a restricted watershed, whose granitic rocks provide a minimum of nutrient salts. This situation is rapidly being altered by human activity in the Tahoe Basin. The cultural eutrophication of the lake is accelerated by sewage disposal in the basin and by the exposure of mineral soils through road building and other construction activities. Since Lake Tahoe's water is saturated with oxygen all the way down the water column, the decomposition of dead plankton sinking slowly towards the bottom is essentially complete. This means that nutrients are returned to the system and because of a water retention time of over 600 years the increase in fertility will be cumulative.

Castle Lake, located at an elevation of 5600 feet in the Klamath Mountains of northern California, shows some of the characteristics of Lake Tahoe as well as those of more productive environments. It, therefore, is best classified as mesotrophic. Although it has a mean productivity of about 70 mg $C \cdot m^{-2} \cdot hr^{-1}$ during the growing season, it shows a depletion in oxygen in its deep water during summer stratification and also under ice cover during late winter.

Clear Lake is an extremely eutrophic shallow lake with periodic blooms of such bluegreen algae as *Aphanizomenon* and *Microcystis* and inorganic turbidity greatly reducing the transparency of the water. The photosynthetic zone is thus limited to the upper four meters with a high intensity of productivity per unit volume yielding an average of about 300 mg $C \cdot m^{-2} \cdot hr^{-1}$ during the growing season. Because Clear Lake is shallow, it does not stratify for more than a few hours at a time during the summer, and the phytoplankton which sink below the light zone are continuously returned to it by mixing.

Cedar Lake lies near Castle Lake in the Klamath Mountains. Its shallow basin is nearly filled with sediment as it nears the end of its existence as a lake. Numerous scars of similar lakes to be found in the area are prophetic of Cedar Lake's future. Terrestrial plants are already invading the lake, and higher aquatic plants reach the surface in many places. The photosynthesis beneath a unit of surface area amounts to only about 6.0 mg $C \cdot m^{-2} \cdot hr^{-1}$ during the growing season as the lake is now only about four meters in depth and may be considered a dystrophic lake. Some lakes of this type pass to a bog condition before extinction; in others, their shallow basins may go completely dry during summer and their flora and fauna become those of vernal ponds.

In examining some aspects of the productivity of these five lakes, the variation in both the intensity of photosynthesis and the depth to which it occurs is evident. The great importance of the total available light can scarcely be over emphasized. This was first made apparent to the author during studies of primary productivity and limiting factors in three oligotrophic lakes of the Alaskan Peninsula, where weather conditions imposed severe light limitations on the phytoplankton productivity. The average photosynthesis on both a cloudy and a bright day was within 10% of being proportional to the available light energy.

Nutrient limiting factors have been reviewed by Lund (1965) and examined by the author in a number of lakes. In Brooks Lake, Alaska a sequence of the most limiting factors ranged from magnesium in the spring through nitrogen in the summer to phosphorous in the fall (Goldman, 1960). In Castle Lake potassium, sulfur, and the trace element molybdenum were found to be the most limiting. In Lake Tahoe iron and nitrogen gave greatest photosynthetic response with nitrogen of particular importance. Trace elements, either singly or in combination, have been found to stimulate photosynthesis in quite a variety of lakes. In general, some component of the phytoplankton population will respond positively to almost any nutrient addition, but the community as a whole will tend to share some common deficiencies. Justus von Liebig did not intend to apply his law of the minimum as rigidly as some have interpreted it, and we can best envision nutrient limitation from the standpoint of the balance and interactions of the whole nutrient medium with the community of organisms present at any given time. Much about the nutrient requirements of phytoplankton can be gleaned from the excellent treatise of Hutchinson (1967).

It must be borne in mind that the primary productivity of a given lake may vary greatly from place to place, and measurements made at any one location may not provide a very good estimate for the lake as a whole.

Variability in productivity beneath a unit of surface area is particularly evident in Lake Tahoe, where attached algae are already becoming a nuisance in the shallow water and transparency is often markedly reduced near streams which drain disturbed watersheds. In July, 1962, the productivity of Lake Tahoe showed great increase near areas of high nutrient inflow (Goldman and Carter, 1965). This condition was even more evident in the summer of 1967 when Crystal Bay at the north end of the lake and the southern end of the lake showed different periods of high productivity. This variability in productivity may be influenced by sewage discharge and land disturbance. Were it not for the great volume of the lake (155 km^3), it would already be showing more severe signs of eutrophication.

In the foregoing paper I have attempted to sketch my impressions of aquatic primary productivity treating the subject both as a research task and as a body of information to be interpreted. I believe that biological productivity can no longer be considered a matter of simple academic interest, but of unquestioned importance for survival. The productivity and harvest of most of the world's terrestrial and aquatic environments must be increased if the world population is to have any real hope of having enough to eat. This increase is not possible unless we gain a much better understanding of both aquatic and terrestrial productivity. Only with a more sound understanding of the processes which control productivity at the level of the primary producers can we have any real hope of understanding the intricate pathways that energy moves and biomass accumulates in various links of the food chain. With this information in hand the productivity of aquatic environments can be increased or decreased for the benefit of mankind.

REFERENCES

Atkins, W. R. G. (1922), Hydrogen ion concentration of sea water in its biological relation. *J. Mar. Biol. Assoc. U.K.* **12**, 717–771.

Atkins, W. R. G. (1923), Phosphate content of waters in relationship to growth of algal plankton. *J. Mar. Biol. Assoc. U.K.* **13**, 119–150.

Goldman, C. R. (1960), Primary productivity and limiting factors in three lakes of the Alaska Peninsula. *Ecol. Monogr.* **30**, 207–230.

Goldman, C. R. (1968), Absolute activity of ^{14}C for eliminating serious errors in the measurement of primary productivity. *J. du Conseil.* **32**, 172–179.

Goldman, C. R., and R. C. Carter (1965), An investigation by rapid carbon-14 bioassay of factors affecting the cultural eutrophication of Lake Tahoe, California-Nevada. *J. Water Pollution Control Fed.* **37**, 1044–1059.

Goldman, C. R., D. T. Mason, and J. E. Hobbie (1967), Two Antarctic desert lakes. *Limnol. Oceanogr.* **12**, 295–310.

Hutchinson, G. E. (1967), *A Treatise on Limnology*. Vol. II. Introduction to lake biology and the limnoplankton. John Wiley and Sons, New York.

Lindeman, R. L. (1942), The trophic-dynamic aspect of ecology. *Ecology* **23**, 399–418.

Lohman, H. (1908), Untersuchungen zur Feststellung des vollständigen Gehaltes des Meeres an Plankton. *Wiss. Meeresunters. N.F. Abt. Kiel* **10**, 131–370.
Lund, J. W. G. (1965), The ecology of the freshwater phytoplankton. *Biol. Rev.* **40**, 231–293.
McConnell, W. J. (1965), Relationship of herbivore growth to rate of gross photosynthesis in microcosms. *Limnol. Oceanogr.* **10**, 539–543.
Naumann, E. (1919), Några synpunkter angående planktons ökologi. Med särskild hänsyn till fytoplankton. *Svensk bot. Tidskr.* **13**, 129–158.
Sorokin, Y. A. (1965), On the trophic role of chemosynthesis and bacterial biosynthesis in water bodies, p. 187–205. *In* C. R. Goldman (ed.), *Primary productivity in aquatic environments.* University of California Press, Berkeley.
Steele, J. H. (1956), Plant production on the Falden Ground. *J. Mar. Biol. Ass. U.K.* **35**, 1–33.

CHARLES R. GOLDMAN
University of California, Davis

ATMOSPHERIC: see AIR– various titles

AUDIBLE NOISE: see NOISE

AUTOMATED WATER POLLUTION INSTRUMENTATION: see POLLUTION INSTRUMENTATION—WASTEWATER

AUTOMOTIVE POLLUTION

The automobiles discovery appears to satisfactorily combine modern man's requirement for rapid transportation with his desire for independence and flexibility. However, the increasing vehicle population poses a series of threats to continued physical and psychological well-being and to the future of our environment.

As the automotive industry expands, other auxiliary industries such as petroleum production and concrete and tire manufacturing also grow, with additional potential for pollution problems. Additional roads also have to be built, with a negative impact on both ecology and landscape. A balance must be made between the right of an individual to use his own car when and where he drives and the harm brought upon society as a whole by his doing so. If it is accepted that society needs to be protected, a number of legislative and economic measures can be initiated to discourage automotive usage.

Legislation has been enacted to limit the emission of HC (hydrocarbons), NO_x (oxides of nitrogen), CO (carbon monoxide) and particulate matter including lead compounds. However, other waste materials such as tires and the automobile itself (including the repugnant abandoned cars) must be disposed of.

The broad approach to automotive pollution control is to encourage alternative means of transportation. This would include improvements in mass transit such as high-speed trains, moving sidewalks, increased links and modernization. In the city, bicycle riding and walking are low pollution, high exercise alternatives; in rural areas car pools might be formed.

Cars may be built using different energy concepts (ex. battery, turbine engine, Wankel engine) and different fuels (ex. natural gas, alcohol). Special control devices (ex. catalytic reactors, afterburners) may be placed at the exhaust of internal combustion engines to reduce emissions of various materials. The total solution to the problem will probably incorporate a few of the methods proposed above.

In the present discussion, we shall concentrate

on the major moving source of pollutants, the internal combustion engine. In the automobile, evaporative losses of pollutants occur from the fuel tank and carburetor (ca. 5%), fumes from the crankcase (ca. 20%) and the exhaust system (ca. 75%). The major offenders are unburned hydrocarbons, carbon monoxide, nitrogen oxides and HC oxidants. Positive Crankcase Ventilation (PCV) on all modern cars reduces the emissions by pulling air and fumes into the engine by maintaining a vacuum at the engine. Of course a highly efficient combustion process will eliminate the partially oxidized substances. One reason for incomplete combustion is that as the "flame front" generated from the spark moves toward the relatively cool cylinder walls, a quenching action takes place preventing further reaction, the kinetics of which are very temperature sensitive. Other factors influencing incomplete combustion are improper dilution by poor cycle timing and less than the proper excess of oxygen admitted at the carburetor. Standard engine cycles have been developed so that pollutant guidelines might be drawn. For example, a test (Table A) was set up by California for observing the maximum guidelines of 180 ppm HC and 1% CO emissions (1970) and 350 ppm, oxides of nitrogen for large engines. For smaller engines the limits are more permissive since the total volume emitted per unit time is less.

TABLE A

California Seven Mode Cycle Test

Mode	Time Reqd. sec	Weighting factor
1. Idle	20.0	0.042
2. Accel. 0–25 mph	11.5	0.244
3. Cruise 30 mph	15.0	0.118
4. Decel. 30–15 mph	11.0	0.062
5. Cruise 15 mph	15.0	0.050
6. Accel. 15–30 mph	12.5	0.455
7. Decel. 50–20 mph	25.0	0.029

The seven modes start cold and are repeated seven times on a continually warming chassis dynamometer with the weight of each time segment being determined from past observations of the seriousness of emissions. H.E.W. standards for 1975 and 1980 indicate that interest in automotive pollutant control will be forced to increase.

	CO		HC		NO_x	
	gm/mile	%	gm/mile	ppm	gm/mile	ppm
1975	11.0	0.5	0.5	40	0.9	225
1980	4.7	0.2	0.25	20	0.4	100

They used another 7 cycle (Federal Cycle) test similar to that above.[1] Companies will follow more complex Federal cycles as they come into use; these changes are made both between the cycle and the length of cycle.

The CO standard is based primarily on tests which have shown that the capacity of blood to carry oxygen is hampered by CO absorption.[2] Brain function is retarded after an exposure of 10 to 15 ppm CO for several hours. The HC and NO_x levels are mainly based on their role in photochemical smog using ambient levels. Shy *et al.*[3] found in their Chattanooga study that "NO_2 alone and exposure to suspended particulate matter alone appeared to be the most probable explanation for the observed excess in respiratory illness rates." The data are not completely convincing as pointed out by automobile industry researchers.[4] However, the U.S. government has taken a bold initiative for the welfare of the public in not waiting for all the details to be perfectly established. Certainly there are those among us (aged, with heart conditions, etc.) who will be more severely affected than the average by typical ambient pollutant concentrations and will need protection.

At least two approaches to the removal of internal combustion engine pollutants have come into wide acceptance. One is the improvement of the combustion process itself. This includes reducing manifold vacuum (and hence dilution of charge) between exhaust and intake steps; increasing the cylinder wall temperature; and designing for less maximum horsepower by minimizing the surface to volume ratio.

The other approach is controlling exhaust emissions by further reaction, either after-burning or catalytic. After-burning can be accomplished slightly downstream from the exhaust valve by additional oxygen injection converting HC and CO to CO_2 and water vapor. The afterburner has the disadvantage of not being able to remove NO_x. In fact, because of its high temperature, still more NO_x is formed. The catalytic approach is to find a catalyst or set of catalysts and temperature or set of temperatures which will completely oxidize both CO and HC, but also will reduce the NO_x. These two approaches will be discussed in detail below.

Improvement of the Combustion Process

As mentioned above the explosion of fuel accounts for chemical pollutant formation. In the Rankine cycle engines, liquids may continually be vaporized and reconditioned, and in the Sterling cycle engines a gas is repeatedly heated and cooled. Both cycles are accomplished in a sealed container and the heat used for the process comes from an open flame external to the engine. Very little pollution is generated in such a controlled rather than explosive flame.

A version of internal combustion which is being considered for pollution control is the Wankel rotary engine. It is lighter and more compact than the reciprocating engines currently in use. Although the engine may not be effective a combustor as the current type, its size permits of much additional space under the hood of the typical American auto hood in which an afterburner or reactor device may be placed.

A recent H.E.W. report[5] discusses engine modification systems, "Features shared by essentially all versions of the engine modification system include calibrated carburetors that provide (a) relatively lean air-fuel mixtures for idle and cruise operation and (b) higher engine idle speeds. Refined control of spark timing is also used, and, in some cases, retarded spark timing at idle is employed. In addition, many engines are fitted with special air cleaners and ducting designed to supply heated air at nearly constant temperature to the carburetor, to permit even leaner mixture settings. Most versions also incorporate high-temperature radiator thermostats to raise coolant temperatures and thus improve mixture distribution and promote complete combustion. In some cases, higher-capacity cooling systems are used to handle the additional cooling load at idle that results from wider throttle openings and retarded ignition timing during this operating condition. In addition, combustion chamber design attempts to avoid flame quenching zones where combustion might otherwise be incomplete, and result in high hydrocarbon emissions.

Hydrocarbon and CO emissions are reduced by adjusting the carburetor to a fuel-lean mixture during part throttle and idle operation. 'Lean surge during cruise has been largely overcome through improvement in manifolding (better mixture distribution), better carburetor fuel-metering characteristics, higher coolant temperatures, increased heating of the air-fuel mixture, and, in some cases, provision for heating the incoming air to the carburetor.

Exhaust emissions of CO and HC are praticularly difficult to control during engine idle and closed-throttle operation (deceleration). Considerable effort has gone into designing carburetor idle systems that will provide a lean air-fuel mixture and minimize emissions during these periods. To ensure that idle air-fuel mixture cannot be adjusted to be too rich (which would tend to increase CO and HC emissions appreciably), some means of limiting idle-mixture adjustment is used on most carburetors. Such devices allow idle mixture to be adjusted leaner than a predetermined value, but not richer."

The effects of charge dilution on the exhaust emission of nitric oxide (NO) from a single-cylinder engine were evaluated over a range of engine design and operating parameters.[6] Nitric oxide emission decreased as much as 70% as charge dilution fraction was increased from 0.065 to 0.164 due to increased valve overlap, external exhaust recirculation, and reduced compression ratio. No emission was strongly dependent on charge dilution fraction, but was independent of the specific method used to change charge dilution. The combined effects of increased charge dilution and 10 degree spark retard reduced NO emission 90%. However, definite limits of operation were observed on the single-cylinder engine with high charge dilution.

Emission Control Devices

a. Additional combustion

An afterburner is an additional baffled tubular reactor in which the gases are reignited and burned to completion. An air pump provides the necessary oxygen-rich mixture and the heat of reaction maintains a high temperature to speed its completion.

It is not necessary in all cases to have a separate afterburner. The aforementioned report[5] states, "Air injection systems decrease exhaust CO and HC emissions by injecting air at a controlled rate and at low pressure into each exhaust port. Here, the oxygen in the air reacts with the hot exhaust gases, resulting in further combustion of the unburned hydrocarbons and CO that would otherwise be exhausted to the atmosphere. Optimum reduction of emissions by this method depends on proper air injection rates over a wide range of engine operating conditions, carefully tailored air-fuel mixture ratios and spark advance characteristics, and in some cases the use of heated carburetor air. Some engines also provide for retarded ignition timing during closed-throttle operation.

All air injection systems use essentially the same basic air pump, a positive displacement rotary-vane type.

To guard against excessive temperature and back pressure in the exhaust system resulting from high air delivery rates at full throttle and high speeds, a pressure-relief valve is installed in the pump housing. The valve opens to bleed off some of the pump flow at a predetermined pressure setting.

Output from the air pump is directed through hoses and an air distribution manifold (or two manifolds-one for each bank on V-8 engines) to the air injection tubes located in each exhaust port. A check valve between the air distribution manifold and the air pump prevents reverse flow of hot exhaust gases in the event that pump output is interrupted.

A vacuum-controlled antibackfire valve is used to prevent flow of air to the exhaust ports during the initial stage of closed-throttle acceleration. The high vacuum that occurs during deceleration causes rapid evaporation of liquid fuel from the intake manifold walls. The resulting rich mixture creates a potentially explosive vapor in the exhaust manifold if injected air is present.

As with engine modification systems, most air injection systems also employ spark retard during idle or idle and deceleration through use of 'ported' vacuum sources or dual-diaphragm distributor-vacuum-advance mechanisms."

Besides residual gas dilution and wall quenching, engine variables have the most effect on the amount of hydrocarbons in exhaust gases. Over sixty privately owned and operated automobiles fueled with commercial leaded gasoline have been tested and seven main engine variables were found which changed the hydrocarbon concentration in the exhaust gases. These results can be briefly summarized as:

1. Air-fuel ratio: Low value for gas mixture results in higher hydrocarbon concentration in exhaust gases.
2. Ignition timing.
3. Speed: Increase in speed of engine decreases the amount of hydrocarbon.
4. Air-flow rate: The effect of air-flow rate on the total hydrocarbon concentration depends on the air-flow ratio, ignition timing combination.
5. Compression-ratio: Increasing the compression ratio, by decreasing head-to-piston distance, increases the total hydrocarbon concentration.
6. Exhaust Back Pressure: The amount of hydrocarbon in exhaust decreases with increasing exhaust pressure.
7. Coolant Temperature: Increasing the coolant temperature decreases the hydrocarbon concentration.

b. Catalytic reactions

The engine exhaust gases may be passed through a cylindrical shaped cannister packed with catalytic particles. Although this method has great potential, two problems may arise. The long term stability of the catalyst (50,000 miles) is difficult to maintain since lead and other chemicals in trace amounts poison the catalyst. The catalyst structural stability is difficult to maintain under the influence of varying gas flow rates and fairly high temperature. Also, removing three or four pollutants simultaneously can prove difficult for any single type of catalyst. The removal of NO_x, for instance, requires a reduction catalyst, whereas CO requires an oxidation catalyst. For this reason dual stage catalytic reactors been proposed. The technical problems for this method are great but the potential advantages are even greater.

(i) Nitric Oxide Removal

A review of some chemical reaction data which might be useful in automobile pollution control work have been presented by Shelef and Kummer.[7] The decomposition properties of NO are presented in Table B for various catalysts and operating conditions. The activation energies are typically less than half those of non-catalyzed reaction. The best catalysts thus far uncovered for the decomposition reaction are copper chromite,[11] copper silica[13] and cobalt oxide.[16] These unfortunately appear to have decomposition rates which at 500°C and later fall within the 10^{-9}–10^{-8} moles/m^2 (min)

TABLE B

Summary of Results on the Decomposition of Nitric Oxide on Various Catalysts[7]

Catalyst	Pressure range	NO concentration range	Temp. range °C	Activation energy kcal/mole	Order of reaction	Ref.
Quartz			<723	21.4	2	8
Ga_2O_3	atm	10%	<1000	19.2	0	9
Al_2O_3	atm	10%	<1000	31.6	0	9
CaO	atm	10%	<1000	23.0	0	9
ZrO_2	atm	10%	<1000	27.8	0	9
Cr_2O_3	atm	10%	<1000	40–60	0	9
Fe_2O_3	atm	10%	<1000	16	0	9
ZrO_2	atm	10%	<1000	31.5	0	9
Al_2O_3	atm	5–15%	700–1100	31.0	0	10
ZrO_2	atm	5–15%	700–1100	24.5	0	10
Fe Oxide*	atm	0.25%	325			11
Copper chromite*	atm	0.62%	320			11
Supported Pt*	1–15 atm	0.4%	430–540		2	12
$CuO : SiO_2$	atm	900 ppm	510	39	1	13
Al_2O_3	atm	10–50%	664–807	20.0	2	14
$CuO : Al_2O_3$ (1 : 1)	atm	720–2200 ppm	304–520	31.6	0	15
$CeO_2 : Al_2O_3$ (1 : 1)	atm	720–2200 ppm	30–520	42.2	0	15
Co_3O_4	100–200 torr	100%	250–350	29.0	1.5–1.8	16
NiO	100–300 torr	100%	450–550	20.0	2	16
CuO-I†	not given	100%	350–700	12.0	2	16
CuO-II††	not given	100%	300–450	12.0	2	16
Fe_2O_3	100–400 torr	100%	300–400	20.0	2	16
Cr_2O_3	150–270 torr	100%	350–475	20.0	2	16
ZrO_2	100–225 torr	100%	650–750	35.0	2	16

* Commercial Catalysts
† Calcined at 700° in air
†† Calcined at 450° in air

TABLE C

The Equilibrium Constants for some Reduction Reactions of Nitric Oxide[17]

Reactions	log K_p 600	700	800	900	1000 K
$2NO + 5H_2 \rightarrow 2NH_3 + 2H_2O$	49.0	48.1	32.13	26.44	21.8
$2NO + 2H_2 \rightarrow N_2 + 2H_2O$	51.7	43.4	37.1	32.2	28.3
$NO + CO \rightarrow \frac{1}{2}N_2 + CO_2$	27.3	22.6	19.7	16.46	14.29
$NO + CH_4 \rightarrow \frac{1}{2}N_2 + CO + 2H_2$	19.59	18.1	17.6	16.2	15.5
$NO + H_{at} \rightarrow HNO$	12.4	9.77	7.79	6.24	5.0
$NO + \frac{1}{2}H_2 \rightarrow HNO$	−3.7	−3.6	−3.52	−3.46	−3.42

range, far too low for practical engine exhaust use.[7]

For this reason, reactions which involve combination with rather than decomposition of NO_2 are being studied very carefully. The equilibrium constants in terms of partial pressure are given in Table C for NO combination with hydrogen, CO and methane at various exhaust temperatures. The thermodynamic conditions (large K_p values) are generally favorable for conversion (reduction). A Monel (nickel-copper alloy) catalyst has been found reasonably successful for removing NO by combining it with residual CO in the exhaust stream. The Monel dissociates the oxygen from NO and then oxides CO to the harmless dioxide.

In this system, when the engine operation is too rich (too little oxygen) nitrogen oxide reduction is found to be excellent, but carbon monoxide conversion poor. The nitrogen oxide will readily combine with Monel forming Monel oxide and nitrogen since there is little oxygen to complete with the nitrogen oxide for active sites on the Monel. However, since the CO : NO ratio is so large, there will be insufficient Monel oxide formed to give up its oxygen to the carbon monoxide.

On the other hand, if the engine operation is too lean, NO reduction will be poor and CO conversion excellent since the reaction $M + \frac{1}{2}O_2 \rightarrow MO$ becomes significant and oxygen will compete with the nitrogen oxide for sites on the Monel.

Some typical conversion data[1] at space velocities of 150,000 v/vhr indicated that conversion of NO in synthetic gas mixture (2% CO, 1000 ppm NO) was over 95% if the temperature was above 700 °C, but fell off sharply at temperature below 650 °C. Problems associated with this technique are the formation of NH_3 if any residual H_2 is present and back pressure buildups with the current catalyst structure. In addition, dusts form which are emitted as particulates. It is essential for efficient performance of this catalyst that it warm up to operating temperatures very rapidly. Lead in the fuel reduces the chemical activity and ultimately increased the rate of deterioration of the catalyst.

Another catalyst which showed promise for the same reaction but at lower temperature (200 to 350°C) has also been mentioned.[18] Despite the fact that conversions of better than 90% were reported equal amounts of CO and NO were used. Automotive exhaust have about 16 times more CO than NO. Activated carbon has been used[19] successfully with H_2 gas at 600°C. However, activated carbon lost a considerable portion of its activity after only 7 hours service.

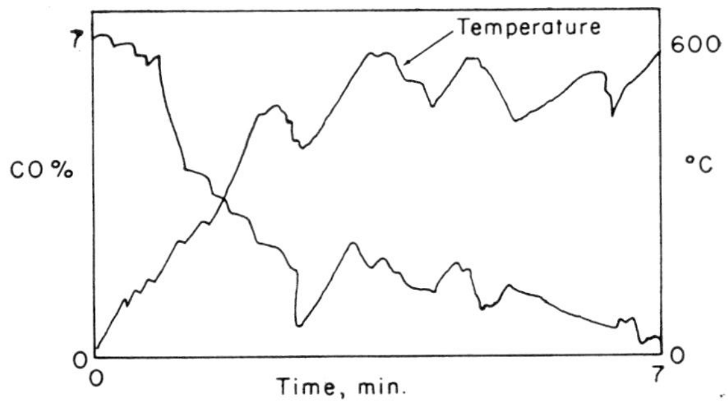

Figure 1

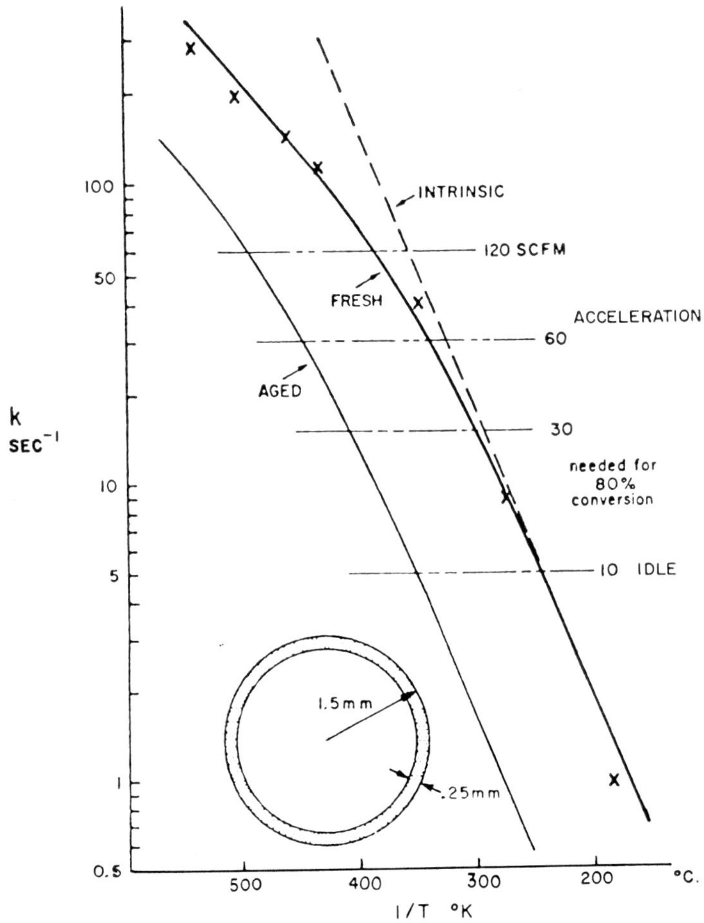

Figure 2 Carbon monoxide activity of an eggshell catalyst.[20]

(ii) CO and hydrocarbon removal

Major automotive and petroleum companies have combined efforts in the development of an inexpensive multi-thousand mile catalytic package for reducing CO and HC exit concentrations. Since the input of gasoline and hence the effluent gases are rarely at steady state, any study of a catalytic reactor must consider the dynamic situation. To give some idea of magnitude, the exhaust gas flow in standard cubic feet per minute (SCFM) is roughly twice the miles per hour equivalent of an automobile and the temperatures of the exhaust gas change from that of the ambient to well over 600°C. In Fig. 1 are typical CO and temperature levels in the exhaust stream after engine startup in a Federal cycle.

For balancing pollution problems with no catalytic or afterburner control an air-fuel ratio of about 16 is recommended. Above this level the NO_x level increases markedly and below it the amount of unburned HC and CO substantially increases. With catalytic devices this ratio may no longer be optimum, since the catalyst selectivity may be greater towards removal of one of the pollutants than any of the others.

Wei[20] has noted that the kinetics of CO oxidation over an egg-shell catalyst turn out to be first order for CO and 0.2 order for O_2 in the range of 1 to 9% oxygen. The curvature in the Arrhenius plot (Fig. 2) is believed to be caused by a pore diffusional phenomena. As the catalyst ages and activity falls the reaction rate becomes controlling and the Arrhenius plot becomes a straight line. The catalyst of the figure is best above 350 °C, but a lower operating temperature may be preferred for longer catalyst life. Wei[20] found that, "As the catalyst lost 90% of its activity, the emission rose by only 30%; but the last 10% of activity loss would result in a precipitous rise of carbon monoxide emission. A catalyst with 50% reduction in heat capacity performs much better; a reactor with 50% reduction in volume performs better when the catalyst is fresh and worse when it is aged." His philosophy is, "It is our engineering goal to produce a low-cost and convenient solution. However, any solution requires some inconvenience and cost from everyone. Quick warm-up is no problem if we are willing to sit and wait in the car for 2 min for an auxiliary heater to warm up the catalyst bed before the car moves. We have ninety million cars on the road, and a $ 100 device will cost us nine billion dollars. How much are we willing to pay for 90% cleaner air and how much more for 99% cleaner air? These decisions belong to the public, not the engineers. For the sake of everyone we hope to be able to present to the public an economical and convenient solution in the near future."

Stein *et al.*[21] have evaluated the effectiveness of possible catalysts by a microcatalytic technique based on gas chromatography. The technique which is described in detail allows a large variety of hydrocarbons and catalysts to be rapidly tested over a wide range of temperatures (100–600°C). In general oxides of cobalt, chromium, iron, manganese, and nickel are the most effective catalysts. The higher molecular wieght hydrocarbons are more easily oxidized than the lower and hydrocarbons of a given carbon number increased in reactivity according to the series: aromatic $<$ branched paraffin or alicyclic $<$ normal paraffin $<$ olefinic $<$ acetylenic. The olefinic hydrocarbons, generally considered the most undesirable, are relatively easy to remove. Other results could be summarized in the following manner:

a) With most of the catalyst tested, some cracking occurs during the oxidation of hydrocarbons. Oxidation without CH_4 formation was possible with the oxides of Co, Cr, Cu, Mn, and Ag only. Zirconium oxides are unique in that they produce only cracking products, mainly methane and smaller amounts of intermediate hydrocarbons.

b) Complete removal of all hydrocarbons was not attained with some of the catalyst, even at 600 °C. The oxidation is not a simple function of temperature.

The potential of copper oxide-alumina catalyst for air pollution control has been studied by Sourirajan and Accomazzo.[22] They stated that, "the simultaneous removal of hydrocarbons and carbon monoxide present in the auto exhaust gases has been tested making use of a six-cylinder Chevrolet engine run on leaded gasoling fuel. The hydrocarbon and carbon monoxide concentrations encountered in these studies varied in the range 170–17,000 ppm and 1–7% respectively. It was found that the minimum initial temperature of the catalyst bed required for the complete removal of both hydrocarbons and carbon monoxide, simultaneously, was 226°C under no load condition, 342°C, under an engine load of 2.5 hp, 400°C, under an engine load of 5.1 hp or higher, and

TABLE D[22]

Data on the Use of the $CuO–Al_2O_3$ (1 : 1) *Catalyst for the Simultaneous Removal of Hydrocarbons and Carbon Monoxide Present in Auto Engine Exhaust Gases*

	Engine conditions						*Reactor conditions and catalyst performance*							
Expt. No.	r.p.m.	Cylinder vacuum in. of Hg	hp	Air/fuel ratio	Hydrocarbon in exhaust ppm	Carbon monoxide in exhaust %	Hydrocarbon in entering gas ppm	Carbon monoxide in entering gas ppm	SV hr^{-1}	Stoichiometric air factor	Initial temperature of catalyst bed °C	Maximum temperature of catalyst bed °C	Extent removal of hydrocarbons %	Extent removal of carbon monoxide %
1	500	20.0	0.0	11.2	580	5.0	390	38,000	10,000	2.07	230	708	100	100
2	500	19.8	0.0	11.1	650	5.0	440	38,000	10,000	2.03	174	708	89	100
3	500	20.0	0.0	11.1	720	7.0	500	47,500	10,000	1.86	145	518	84	100
4*	500	18.5	0.0	10.7	1,300	5.0	810	34,000	10,000	2.18	204	458	94	100
5	500	19.0	0.0	10.7	1,300	5.0	730	36,000	10,000	2.08	254	770	100	100
6*	500	19.0	0.0	10.7	1,300	5.0	880	36,000	15,000	2.09	226	878	100	100
7	1,000	19.5	2.5	13.3	400	4.0	240	30,400	10,000	2.78	160	330	37	99
8	1,000	19.7	2.5	13.4	420	4.0	240	30,400	10,000	2.63	188	426	46	100
9	1,000	19.7	2.5	12.9	420	4.0	240	30,400	10,000	2.63	308	510	83	100
10*	1,000	19.8	2.5	13.3	440	3.6	260	28,800	10,000	2.27	342	536	100	100
11	1,000	18.5	2.5	13.2	620	4.0	420	30,400	10,000	2.45	254	466	69	99
12	1,000	19.8	2.5	12.6	530	4.0	330	30,400	10,000	2.53	364	622	100	100
13	1,350	18.5	5.1	14.1	240	2.0	190	16,800	10,000	3.10	360	476	79	100
14	1,350	18.7	5.1	14.1	320	2.0	210	16,800	10,000	2.92	416	494	100	100
15*	1,350	18.4	5.1	13.8	360	2.0	250	16,800	10,000	2.84	398	484	100	100
16	1,350	18.2	5.1	13.9	340	2.0	250	16,800	10,000	3.47	374	466	86	100
17	1,350	18.2	5.1	13.6	370	2.0	240	16,800	10,000	2.82	304	434	50	100
18*	1,350	18.0	5.1	13.3	350	3.0	220	24,000	15,000	2.55	400	562	100	100
19*	1,350	18.0	5.1	13.3	360	2.0	220	16,800	20,000	2.86	400	540	100	100
20	1,550	16.5	10.1	14.4	240	2.0	160	17,000	10,000	3.00	406	496	100	100
21	1,550	16.5	10.1	14.4	210	2.0	160	17,000	10,000	3.07	466	530	100	100
22	1,600	17.0	10.1	14.3	190	1.2	130	11,000	10,000	2.56	398	442	100	100
23	1,550	16.2	10.1	14.3	230	1.0	130	9,000	10,000	3.12	358	470	78	100
24	1,550	16.2	10.1	14.3	230	1.0	140	9,000	10,000	3.12	408	490	100	100
25	1,550	17.2	10.1	14.3	170	1.0	130	9,550	10,000	1.39	446	448	100	100
26	1,550	17.0	10.1	14.5	170	1.0	130	9,550	10,000	1.39	478	484	100	100
27	1,550	15.8	10.1	14.3	260	1.0	210	9,000	10,000	2.97	248	360	19	100
28*	1,550	15.5	10.1	14.5	270	1.0	160	9,500	15,000	2.71	400	480	100	100
29	1,550	15.5	10.1	14.4	270	1.0	170	9,200	20,000	2.29	404	498	82	100
30*	1,550	15.5	10.1	14.4	260	1.0	160	9,000	10,000	2.91	404	498	100	100
31	1,550	16.2	10.1	14.4	220	1.0	140	9,000	10,000	3.12	302	432	46	99
32*	1,400	24.5	DEC	11.7	16,000	4.0	4,800	12,800	10,000	2.48	252	822	99	100
33	850	24.0	DEC	11.3	11,000	5.5	2,900	17,500	10,000	3.21	258	828	97	100
34*	850	24.2	DEC	11.4	11,000	5.5	2,900	17,500	10,000	3.21	236	740	99	100

* This experiment was repeated after the 100 hour test period; the results obtained were identical with those reported herein.

236°C under deceleration conditions. The catalyst showed no deterioration in performance even after 100 hours of continuous service in conjunction with the above auto exhaust gases." In the experiments of Table D, "the concentrations of carbon monoxide in the engine exhaust gases varied in the ranges, 5–7%, 3.6–4%, 2–3%, 1–2% and 4–5.5% under the engine load conditions of 0 hp, 2.5 hp, 5.1 hp, 10.1 hp and deceleration respectively; the corresponding hydrocarbon concentrations varied in the ranges 580–1300 ppm, 400–620 ppm, 240 to 370 ppm, 170–270 ppm and 11,000–16,000 ppm under the respective engine load conditions stated; and, the initial temperature of the catalyst bed was varied from 145 to 478°C. At all conditions of engine operation tested above, it was found that carbon monoxide in the exhaust gas was completely removed by oxidation. The extent of removal of

TABLE E 1

Compound	Temperature (°C)	% of theoretical CO_2 found
	400	27
Methane	300	2
	315	12
Ethane	260	6
	315	25
Propane	260	11

TABLE E 2

Combustion of the Higher Hydrocarbons over Hopcalite

Compound	Percentage of theoretical CO_2 found		
	350°C	300°C	200°C
n-hexane	97	97	77
n-decane	96	98	89
2,2-dimethylbutane	98	84	40
Octene-1	94	89	83
Cyclohexane	91	—	—
Benzene	92	87	74
1,2,4-trimethylbenzene	100	98	55
t-butylbenzene	96	96	86

hydrocarbons from the exhaust gases was found to depend on the initial temperature of the catalyst bed and the engine load condition.

It is realized that a successful 100 hour run does not constitute a life test on the catalyst, but it does indicate the potential applicability of the catalyst in air pollution control devices. The engineering design of the suitable converter for any particular practical application of the catalyst should naturally take into account the heat liberated during oxidation. Instantaneous catalyst temperatures of the order of 900 °C have been encountered in this work with no deleterious effect on the subsequent effectiveness of the catalyst. ... the heat liberated during the reaction can be advantageously used to maintain the full effectiveness of the catalyst under all conditions of engine operation encountered in normal practice."

When contaminants are passed through a Hopcalite (unsupported coprecipitate of copper and manganese oxides) catalyst burner, the results vary from almost complete oxidation of some organics to very slight oxidation of the lower molecular weight aliphatic hydrocarbons[23] at

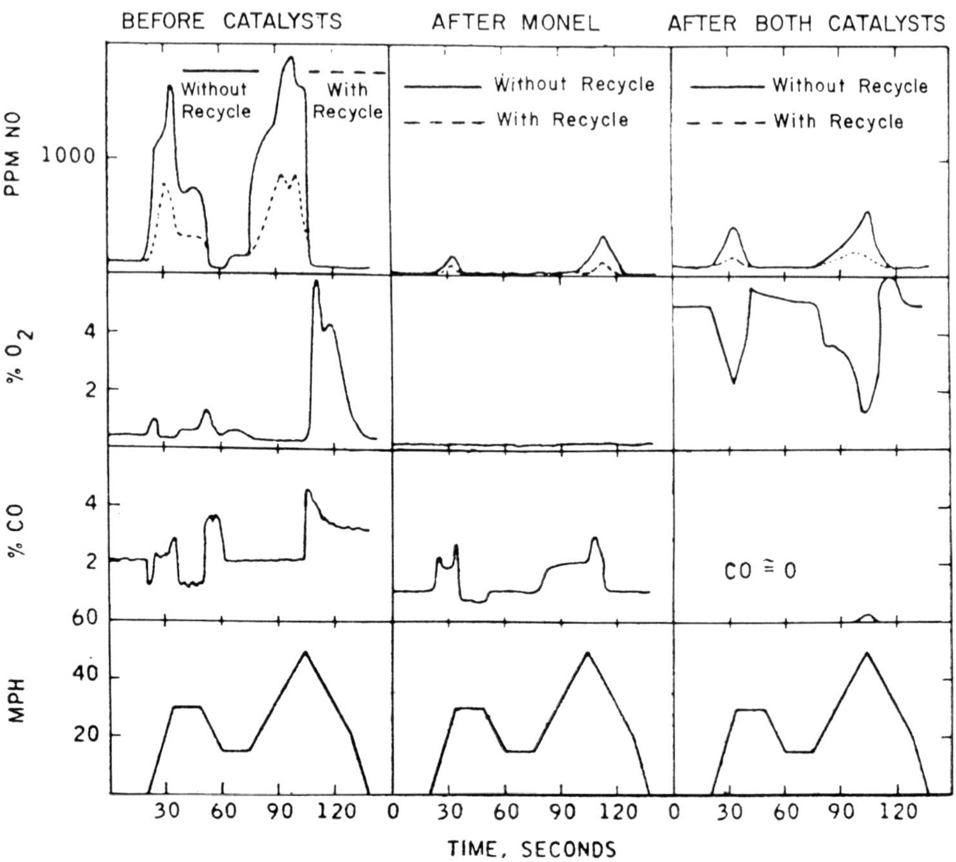

Figure 3

about 300 °C. (See Table E.) Nitrogen compounds form N_2O when oxidized and halogenated compounds indicate a strong acid reaction when the reactor effluent is tested with detector paper.

An interesting example of the use of exhaust gas recycle with catalysts has been presented by the Esso Research Group.[1] In Fig. 3, typical hot cycle traces of CO, O_2 and NO are presented for cases before the catalyst, after Monel catalyst and after a 2nd stage Platinum-alumina catalyst, for instance with and without recycle. The major beneficial effect of recycle is on the NO concentration.

Related Transportation Problems

a. Diesel exhaust odors

Diesel engines are found in buses, trucks, off-road vehicles and power applicators. Public reaction to diesel-engine exhaust odors provides the impetus for controlling effluents of that type of fuel combustion.[24] A list of oily kerosene and smokyburnst odor compound identified by A. D. Little, Inc. is included in Table F. Exhaust odor and smoke from diesel engines are more objectionable than those from spark ignition. CO emissions are generally less serious but NO_x is troublesome (4 to 10 g/mile). Improved fuel injection and afterburners are considered to be the most promising of the existing control methods. The injection differs from the internal combustion engine in that fuel does not enter the cylinder as a mixture with air, but is injected under high pressure into the chamber in exact quantities through low tolerance nozzles. For NO_x removal the basic approaches are engine refinement, fuel additives, catalytic conversion and reduction of peak combustion temperature.

TABLE F

Partial List of Odorifous Compounds

General classification	Compounds
Indans and tetralins	Methyl indan
	Tetralin
	Dimethyl indan
	Methyl tetralin
	Dimethyl tetralin
	Trimethyl tetralin
	Alkyl tetralin
Alkyl-substituted naphthalenes	Methylnaphthalenes
	Dimethylnaphthalenes
Indenes, acenaphthenes, and benzothiophenes	Alkyl-substituted indenes
	Dimethylbenzothiophenes
	Acenaphthene

General class	Carbon range
Alkenone	C_5 to C_{11}
Furan	C_6 to C_{10}
Diene	C_9 to C_{12}
Furfural	C_6 to C_7
Methoxy benzene	C_8 to C_9
Phenol	C_7 to C_{12}
Benzaldehyde	C_7 to C_{10}
Benzofuran	C_8 to C_9
Indanone	C_9 to C_{12}
Indenone	C_6 to C_{10}
Naphthol	C_{10} to C_{14}
Naphthaldehyde	C_{11}

b. Gas turbines

The exhaust of gas turbines contains perhaps less pollutants than that of any other internal combustion process. The high airport traffic density makes the problem a real one, however. Sulfur dioxide emissions are low, but the smoke and odor producing compound levels are high. For automobile use, the total mass of gas turbine exhaust is many times greater than that for the gasoline engine of equivalent power. Hydrocarbon and CO mass emissions are known to be low and difficult to reduce. The data for NO_x concentration are sparse and inconclusive, and control will be difficult. A great deal of the pollution control work underway is in the engine modification area. In aircraft the turbofan engine is largely replacing the turbojet engine. Turbofans bypass some of the air past the engine and rejoin it with the burned gases at the exhaust tailpipe.

The airplane is a much more efficient carrier (pollution wise) than the automobile, on a people x miles basis. Aircraft engine research has been concerned primarily with smoke reduction. Fuel type, fuel additives and combustion chamber designs have been the primary variables studied.

Particulate Emissions

The particulate matter emitted from automobiles has been characterized[25] as consisting of "lead salts, alkaline earth compounds, iron oxides, soot carbonaceous material, and tars. This material ranges in size from large flakes to submicron particles and varies in consistency from hard and brittle particulate to oil mists. Some of the particulate material is generated in the engine combustion chamber and nucleated and agglomerated in the vehicle exhaust system before it passes out the tail pipe. On the other hand, a large portion of the particulate material generated in the engine subsequently deposits on various surfaces of the exhaust system. At some later time, this deposited material flakes off and becomes re-entrained in the exhaust gas to be emitted from the vehicle. Obviously, opportunities exist for various types of chemical and physical processes to take place and, as a result, the overall particulate emission process for a vehicle is quite complex and difficult to define."

Much of the particulate is in the form of polynuclear aromatic (PNA) hydrocarbons and tars. As for gaseous products, standard federal cycles have been designed to test for particulate emissions.

Lead particulate matter is emitted at a rate of about 0.16 g/mile. The average particle size emitted is slightly less than a micron, but sizes are substantially distributed over the 0.10–10 micron range. In Table G, the effect of mileage on particle distribution may be observed. As the number of miles driven increases, the segment of particles greater than 9 microns also increases, probably because of deposition buildup on exhaust pipes.

Using 500 microns as a divider, about 35% of the lead burned over the lifetime of a car is in the form of fines, 40% in the form of coarse particulate.

TABLE G

Lead Particulate Emissions as a Function of Size and Mileage[25]

Average mileage	Average lead salt emissions, g/mile >9 Microns	<1.0 Microns	<0.3 Microns
5,000	0.04	0.07	0.05
16,000	0.06	0.05	0.04
21,000	0.06	0.05	0.03
28,000	0.10	0.03	0.02

Research studies[27] showed that a relationship exists between traffic volume, proximity to the highway engine acceleration *vs* constant speed, wind direction and the amount of lead in the air. Lead values can be expressed as a quadratic function of time and a linear function of traffic volume. Ter Haar[28] and coworkers found that for Federal 7-mode cycle conditions (1) cars vary widely in the amounts and composition of their particulate emissions; (2) cold cycle operation gives 2–8 times more particulate than hot engine operation; (3) lead compounds represent less than one-third of total particulates, the remainder being carbon compounds along with ammonium and nitrate ions and unknown materials; (4) carbon emission for stabilized cars using leaded gasoline varies widely but averages about 35% of the total; (5) suspended particulate emissions are nearly equal with new cars whether

TABLE H

Particulate Trapping—Federal 7-Mode Cycle[28]

Car, year, and make	Miles	Exhaust systems	Particulates, Total	g/mile Pb
1970 make D	6,000	Standard	0.169	0.046
	6,000	Anchored vortex	0.139	0.039
		% Reduction	19	16
1970 make B	6,000	Standard	0.274	0.071
	6,000	Anchored vortex	0.227	0.056
		% Reduction	17	21
1970 make A	6,000	Standard	0.471	0.114
	6,000	Anchored vortex	0.189	0.043
		% Reduction	60	66
1970 make D	0	Standard	0.116	
	0	Agglom.-inertial # 1	0.037	0.006
		% Reduction	68	
16 cars (1966–1969)	30,000+	Standard	0.339	0.060
1965 make D	24,000	Agglom.-inertial # 2	0.125	0.036
		% Reduction	63	39
1969 make D	12,000	Agglom.-inertial # 3	0.108	0.034
		% Reduction	67	43
1970 make D	12,000	Agglom.-inertial # 4	0.098	0.019
		% Reduction	71	68

or not lead is present; (6) exhausted lead varies with the condition of the exhaust system and ranges between 7 and 30% of the lead consumed by the engine; (7) fuel additivies affect the amount of emitted particulates; (8) probe sampling techniques underestimate by a large factor the amount of particulates emitted by vehicles; and (9) trapping systems offer potential for greatly reducing the emission of suspended total particulates. Non-leaded-fuel cars were found to exhaust 0.165 g of particulate per mile while while leaded-fuel cars exhausted 0.152 g/mile. The reason for this is that nonleaded fuel is more aromatic in nature and that the percentage emission for aromatic hydrocarbons is considerably greater than that for paraffinic material. An increase in the volume of aromatics

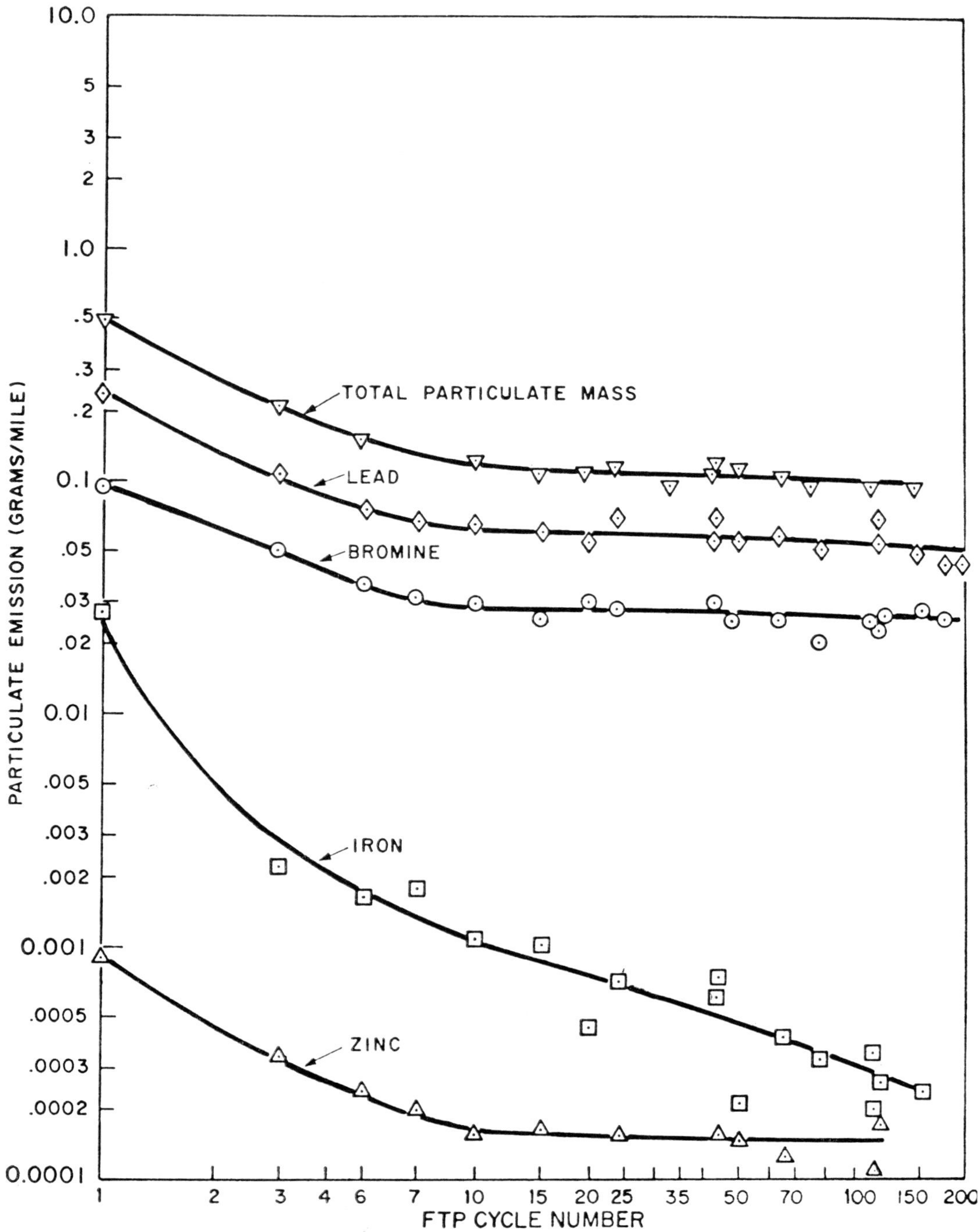

Figure 4 Particulate mass emissions from car operating on Indolene-30, as measured by the *Total Volume System*, versus FTP driving cycle number.[30]

in the fuel from 10 to 70% doubles the amount of carbon particulates emitted. It has been claimed[29] that a reduction of 80–95% of the lead emitted is possible by trapping systems.

Two simple traps have been compared for particulate removal and the results presented in Table H.

More complex traps are likely to be very successful in reducing both lead and total particulate matter.

A more detailed chemical analysis of particulates has been described under Federal emission test conditions.[30] This investigation found that a decrease in the total mass of particulates especially particles of very small size, occurred when using nonleaded fuel. The discrepancy with the aforementioned work[28] may be a result of differences in exhaust deposition for the two systems. Particles emitted in the vehicle exhaust varied in size from 0.01 to 5,000 microns, the latter for such products as rust scale.

Typical chemical compositions for the particulate exhausts are presented in Fig. 4.

The ratio of Pb to Br is relatively constant 2.1 : 1. The ratio on impactor plates was somewhat lower, indicating that the chemical composition of lead salts may be related to particle size. Considering some of the differences in the findings for different particulate matter studies it is critical that such investigations be perfected and intensified. An interesting additional discussion of the characterization of particulate lead in vehicle exhaust may be found in the works of Habibi[31] and Mueller.[32]

In addition to the fuel and exhaust line, additional sources of particulates are to be found in catalysts for gaseous pollutant removal and from the disintegration of tires. The monel catalyst, described previously[1] for NO_x and Co removal, deteriorates with time at about 0.019 g/mile. The presence of lead in the fuel substantially increases the rate of Monel deterioration.

One other source of vehicle particulate emissions which might be mentioned is that from jet aircraft. By weight, more than fifty per cent of the total particulate matter emitted from aircraft engines appears to be due to unburned and partially oxidized fuels. More detailed analysis of this constituent must await development of improved measurement techniques. About 90 per cent of the non-hydrocarbon material present is elemental carbon (i.e., smoke), while the remainder is material eroded from the engine, fuel contaminants, or dust ingested by the engine. Fuel additives containing metallic compounds are seldom used for normal aircraft operations. The emission of such metallic particles does not appear to constitute a problem.

It has been demonstrated that fuel-rich pockets in the primary combustion zone of the burner were the source of the smoke-producing carbon particles and that the smoke could be reduced by "leaning out" the fuel-rich areas. Redesign of the burner is believed to be the most effective approach to smoke reduction.

The author would like to acknowledge the assistance of George Musser and William Peterson of Esso Research and Engineering Company for their discussion on particulate matter pollution.

REFERENCES

1. Bernstein, L. S., K. K. Kearby, A. Raman, J. Vardi, and E. E. Wigg, Paper presented at S.A.E. Meeting, Detroit, Mich., Jan. 1971.
2. Beard, R. R., and G. A. Wertheim, *Amer. J. Public Health* **57**, 2012 (1967).
3. Shy, C. M., *et al.*, *J. Air Poll. Cont. Ass.* **20**, 539 (1970).
4. Heuss, J. M., G. J. Nebel, and J. M. Colucci, JAPCA **21**, 535 (1971).
5. Report AP-66, U.S. Dept. of H.E.W., March 1970.
6. Benson, J. D., and R. F. Stebar, Paper presented at S.A.E. Meeting, Detroit, Mich., Jan. 1971.
7. Shelef, M., and J. T. Kummer, *Chem. Eng. Prog. Symp. Ser.* **67**, 74 (1971).
8. Wise, H., and M. F. Frech, *J. Chem. Phys.* **20**, 22 (1952).
9. Fraser, J. P., and F. Daniels, *J. Phys. Chem.* **62**, 215 (1958).
10. Yuan, E. L., *et al.*, *J. Phys. Chem.* **63**, 952 (1959).
11. Taylor, F. R., Air Poll. Foundation Rept. No. 28 (1959).
12. Sakaida, R. R., *et al.*, *A.I.Ch.E.J.* **7**, 658 (1961).
13. Sourirajam, S., and J. L. Blumenthal, *Actes du Deuxieme Congres International de Catalyse*, Vol. 1, p. 2521 (1960).
14. Peters, M. S., AEC TID-18423 (1963).
15. Wikstrom, L. L., and K. Nobe, *J & EC, P.D. & D.* **4**, 191 (1965).
16. Yur-eva, T. M., *et al.*, *Kinet. Katal.* **6**, 1041 (1965).
17. Chkhubianishvili, M. G., and V. T. Chagunava, *Soobsheh. Akad. Nauk Gruz. S.S.R.* **41**, 591 (1966).
18. Ayen, R. J., and Y-S Ng, *A & W Poll. Int. J.* **10**, 1 (1966).
19. Bedjai, *et al.*, *I. & E.C.* **50**, 1165 (1958).
20. Wei, J., *C.E.P. Monograph Series* **65**, 14 (1969).
21. Stein *et al.*, *I. & E.C.* **52**, 671 (1960).
22. Sourirajan, S., and M. A. Accomazzo, *Can. J. of Ch. E.* 88, April 1961.
23. Christian, J. G., and J. E. Johnson, *Int. J. of A. & W. Poll.* **9**, 1 (1965).
24. Somers, J. H., and G. D. Kittredge, *J.A.P.C.A.* **21**, 764 (1971).

25. Habibi, K., *et al.*, Paper presented APCA—5th West Coast Mtg. Oct. 8–9 (1970).
26. Seizinger, D. E., and B. Dimitriades, *J.A.P.C.A.* **22**, 47 (1972).
27. Motto, H. L., *et al., Env. Sci. Tech.* **4**, 231 (1970).
28. Ter Haar, G. L., *et al., J.A.P.C.A.* **22**, 39 (1972).
29. Hirschler, D. A., and F. J. Marsee, NPRA Paper No. AM-70-5, April (1970).
30. Ninomiya, J. S., W. Berman, and B. H. Simpson, Paper EN-10G presented at 2nd Inter. Clean Air Congress, Dec. 6–11 (1970), Washington, D.C.
31. Habibi, K., *Env. Sci. and Tech.* **4**, 230 (1970).
32. Mueller, P. K., *Env. Sci. and Tech.* **4**, 248 (1970).

EDWARD N. ZIEGLER
Polytechnic Institute of New York

B

BIOLOGICAL TREATMENT OF WASTEWATER

Biological treatment is the most widely used method for removal as well as partial or complete stabilization of biologically degradable substances present in wastewaters. Most often the degradable substances are organic in nature and may be present as suspended, colloidal, or dissolved matter, the fraction of each form depending on the nature of wastewater. In operation of treatment facilities, the characteristics of wastewater are measured in terms of its Chemical Oxygen Demand, COD, Biochemical Oxygen Demand, BOD, and Volatile Suspended Solids, VSS, concepts of which have been discussed elsewhere.[1]

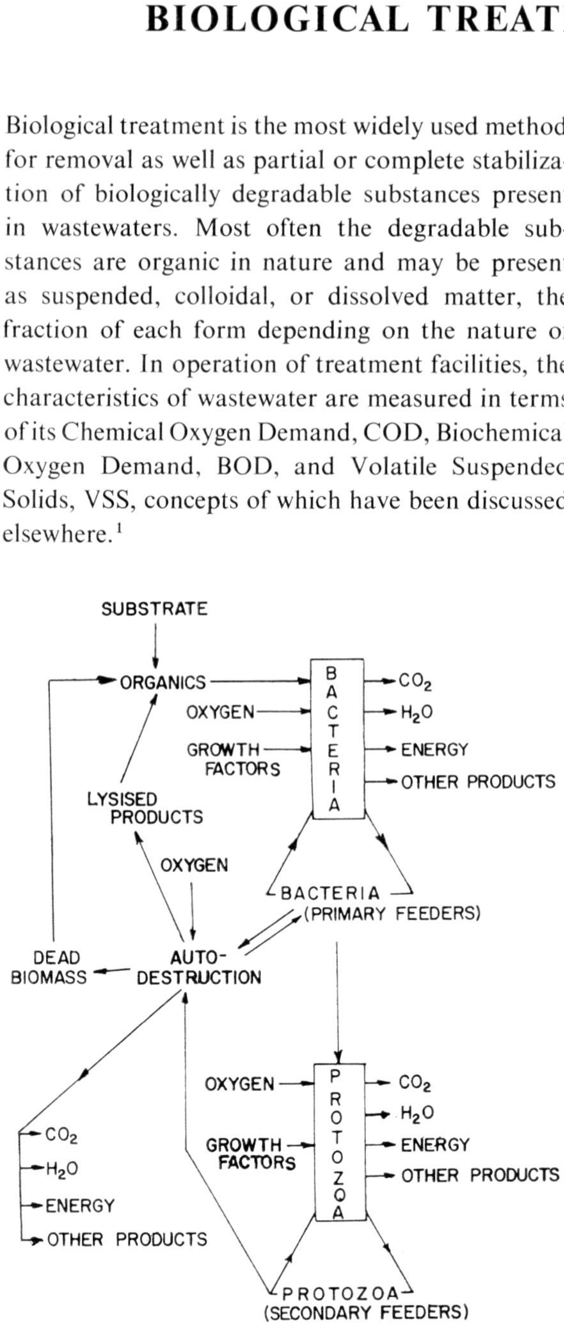

Figure 1 Biological component system existing in BOD process.[2]

Most of the biological wastewater treatmen; processes employ bacteria as the primary organismst however, certain other microorganisms may also play an important role. Gates and Ghosh[2] have presented the biological component system existing in BOD process and it is shown in Fig. 1.

The degradation of organic matter is accomplished by their use as food by bacteria or other microorganisms to produce protoplasm for new cells during growth process. When a small number of microorganisms are inoculated into a bacteriological culture medium, growth of bacteria with time follows a definite pattern as depicted in Fig. 2 by plotting viable count and mass of bacteria against time.[3] The population dynamics of bacteria in biological treatment processes depends upon various environmental factors including pH, temperature, type and concentration of substrate, hydrogen acceptor, availability and concentration of essential nutrients like nitrogen, phosphorus, sulfur etc., and essential minerals, osmotic pressure, toxicity of media or byproducts, and degree of mixing.[4]

Metabolism

The metabolic reactions occurring within a biological treatment reactor can be divided into three phases: oxidation, synthesis, and endogenous respiration. Oxidation-reduction may proceed either in the presence of free oxygen, aerobically, or in its absence, anaerobically. While the overall reactions carried out may be quite different under aerobic and anaerobic conditions, the processes of microbial growth and energy utilization are similar. The reactions in these three phases are formulated below:

Organic matter oxidation (respiration)

$$C_xH_yO_z + O_2 \rightarrow CO_2 + H_2O + \text{Energy}$$

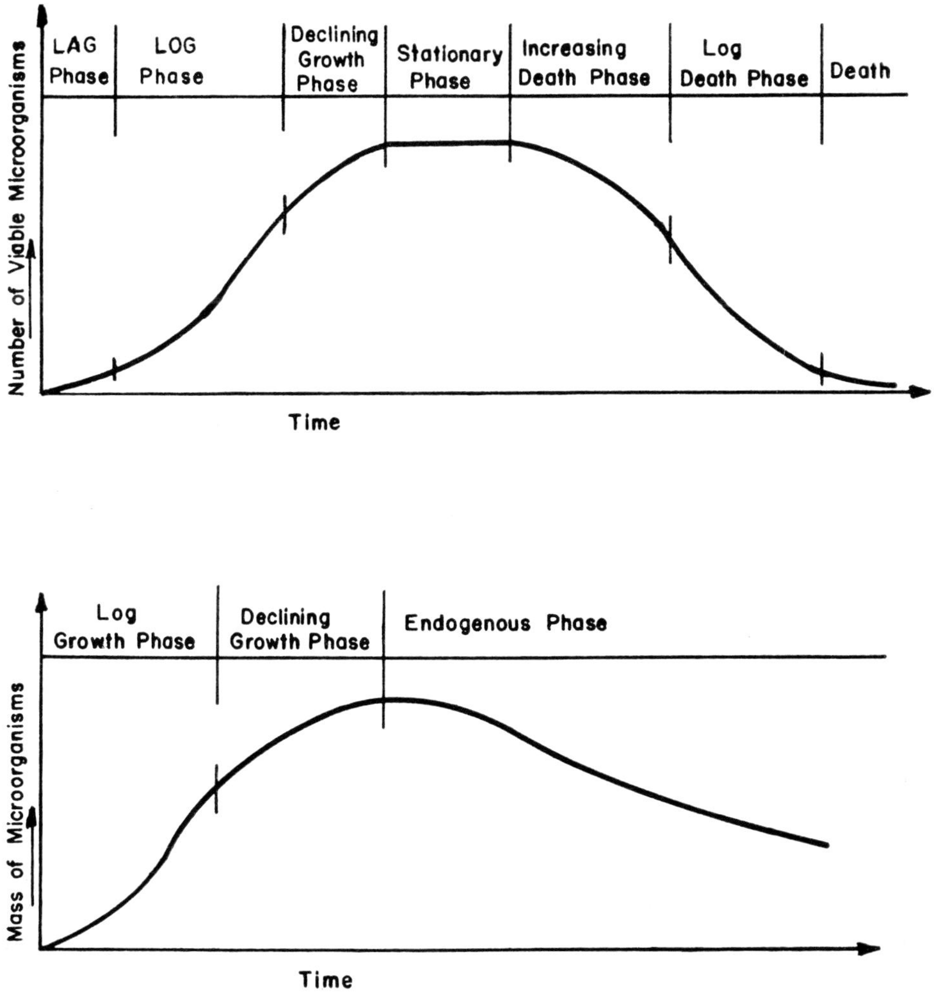

Figure 2 Growth pattern of microorganisms.[3]

Protoplasm (cell material) synthesis

$$C_xH_yO_z + NH_3 + O_2 + \text{Energy}$$
$$\rightarrow C_5H_7NO_2 + CO_2 + H_2O$$

Protoplasm (cell material) oxidation

$$C_5H_7NO_2 + 5O_2$$
$$\rightarrow 5CO_2 + 2H_2O + NH_3 + \text{Energy}$$

Bacterial respiration, is therefore, a biochemical process in living protoplasm whereby energy is made available for endothermic life processes. Being a dissimilative in nature, respiration is an important process in wastewater treatment practices. On the otherhand, endogenous respiration is the internal process in microorganisms that results in auto-digestion or self destruction of cellular material.[3] Actually, bacteria require a small amount of energy to maintain normal functions such as motion and enzyme activation and this basal-energy requirement of the bacteria has been designated as endogenous respiration. Even when nutrients are available, endogenous metabolism proceeds with the breakdown of protoplasm.[5] According to Bertalanffy's hypothesis,[6] the microbial growth is the result of competition between two opposing processes: Aufban—assimilation, and Abban—endogenous metabolism. The rate of assimilation is proportional to the mass of protoplasm in the cell and the surface area of the cell, whereas the endogenous metabolism is dependent primarily on environmental conditions.

In the presence of enzymes, produced by living microorganisms, about $\frac{1}{3}$ of the organic matter removed is oxidized to carbon-dioxide and water in order to provide energy for synthesis of the other $\frac{2}{3}$ of the organic matter removed to cell material. Metabolism and process reactions occurring in typical biological wastewater treatment processes are explained schematically by Stewart[7] as shown in Fig. 3. Thus the basic equations for biological metabolisms are:

Organic matter metabolized = Protoplasm synthesized + Energy for synthesis and Net protoplasm accumulation = Protoplasm synthesized − Endogenous respiration.

Growth Kinetics

Irvine and Schaezler[8] have developed the following expression for non-rate limited growth of microorganisms in logarithmic phase,

$$N_t = N_0 e^{k_0 t}$$

or

$$\frac{dN}{dt} = k_0 N \quad (1)$$

where

N_0 = Number of viable microorganisms per unit volume at time $t = 0$

N_t = Number of viable microorganisms per unit volume at time t

and

k_0 = Logarithmic growth rate constant, time^{-1}

In wastewater treatment practices, the growth pattern based on mass of microorganisms has received more attention than the number of viable microorganisms. If each microorganism is assumed to have an average constant weight, then N in equation (1) can be replaced with S, the mass of (biological) volatile suspended solids per unit volume to obtain

$$\frac{dS}{dt} = k_0 S \quad (2)$$

The growth of bacterial population becomes limited either due to exhaustion of available nutrients or by the accumulation of toxic substances. The growth rate of bacteria starts slowing down, and the equation (1) reduces to the form,

$$\frac{dN}{dt} = k_t N \quad (3)$$

where growth rate factor k_t varies with time and becomes a function of temperature, T, pH, substrate concentration, C_s, and concentrations of various nutrients, C_{n_1}, C_{n_2}, etc., i.e., $k_t = V_1$ (T, pH, C_s, C_{n_1}, C_{n_2}, ...). Fig. 4 shows variation in growth rate k_t with change in nutrient concentrations, assuming that T and pH are held constant and substrate concentration, C_s, is greater than the critical substrate concentration, C_s^*, above which k_t is independent of C_s. Several interesting observations are made from these curves[8]. First, the maximum value of k_t is essentially constant. Second, the shape of the curve and the limiting concentration is different for each nutrient. Third, k_t is shown to be zero when any of the nutrients is missing. Fourth, as the biological reaction proceeds all nutrients are consumed. Thus, even if all nutrients are initially in excess, the growth may eventually become limited. Finally, as the concentration drops to zero, a stationary phase is reached, i.e., dN/dt becomes zero.

In case of a substrate limited system, rate of growth is given by

$$\frac{dN}{dt} = k'(C_s)\, N \quad (4)$$

The simple relationship between growth rate, k, of microorganisms and substrate concentration, C_s,

$$k' = k_0 \left(\frac{C_s}{C_s + K_s} \right)$$

was developed by Monod[9] and has been widely accepted. K_s is a constant called half velocity coefficient and k_0 is maximum specific growth rate.

It is postulated that the same amount of substrate is incorporated in each cell formed. Therefore, the rate of increase in number of microorganisms in logarithmic growth phase, dN/dt, or dS/dt, is proportional to the rate of substrate consumption, dC_s/dt, or dL/dt (if the substrate concentration is measured in terms of BOD, L), and the following relationship can be stated:

$$\frac{dN}{dt} = \frac{dS}{dt} = -a\frac{dC_s}{dt} = -a\frac{dL}{dt}$$

or

$$\Delta S = -a\,\Delta L \quad (5)$$

where a is called the growth yield coefficient, ΔS is the cell mass synthesized per day, and ΔL is BOD removed per day. However, under conditions of rate limited growth, i.e., nutrient exhaustion or auto-oxidation, this expression becomes

$$\frac{dS}{dt} = -a\frac{dL}{dt} - bS \quad (6)$$

* It must be recognized that BOD is an operational parameter and not an actual measurement of substrate.

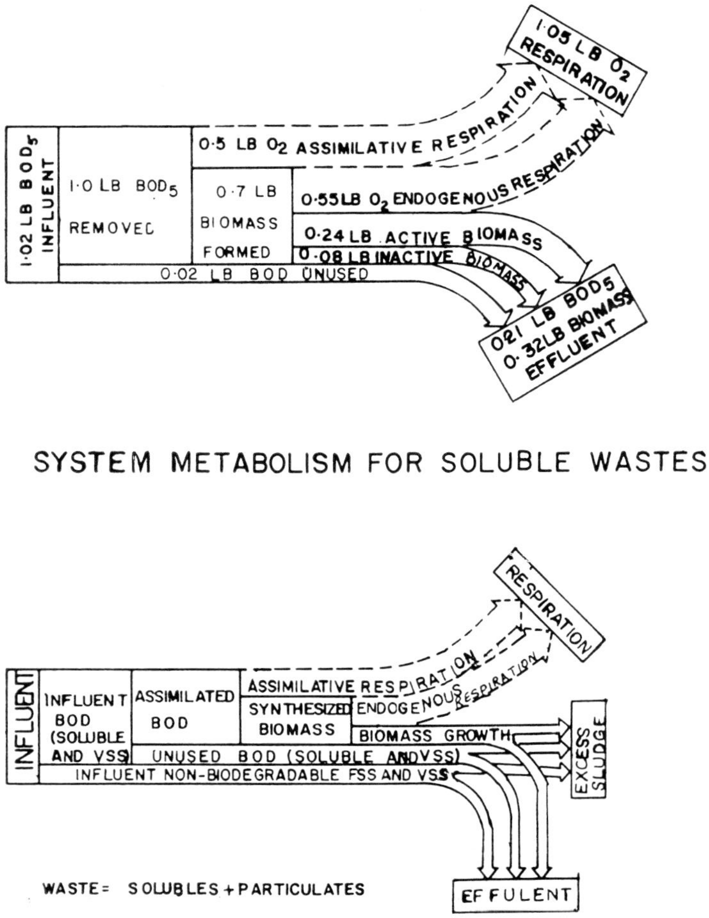

Figure 3 Metabolism and process reactions.[7]

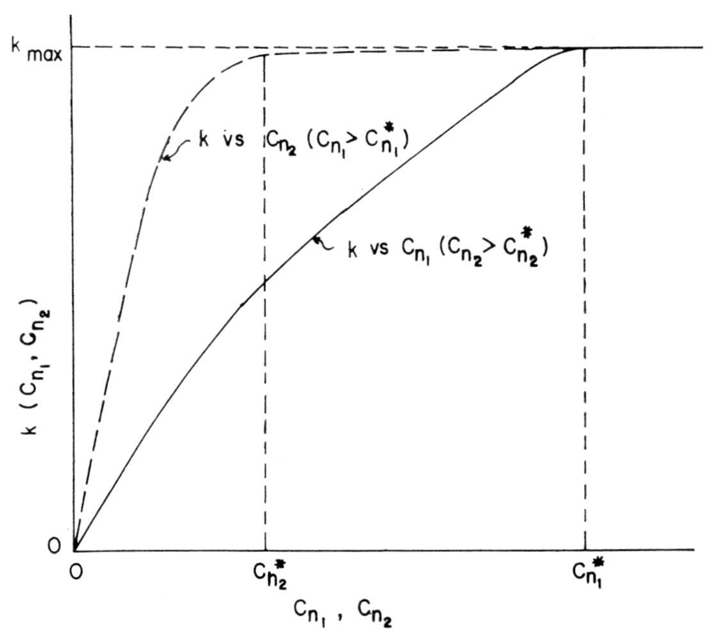

Figure 4 K vs Nutrient Concentration.[8]

where b is the auto-oxidation rate or the microbial decay coefficient. In absence of substrate, equation (6) is reduced to

$$\frac{dS}{dt} = -bS \tag{7}$$

Several kinetic equations have been suggested for analysis and design of biological wastewater treatment systems and the following have been applied frequently:[10,11,12,13]

$$\frac{dL}{dt} = -\frac{k_0 SL}{a(K_s + L)} \tag{8}$$

$$\frac{dL}{dt} = -\frac{k'SL}{a} = -KSL \tag{9}$$

$$\frac{dL}{dt} = -\frac{k'S}{a}\frac{L^2}{L_0} = -KS\frac{L^2}{L_0} \tag{10}$$

Combining equations (6) and (8) gives the specific growth rate

$$\frac{dS}{S\,dt} = \frac{k_0 L}{K_s + L} - b \tag{11}$$

where L_0 is the initial BOD.

Effect of Temperature

One of significant parameters influencing biological reaction rates is the temperature. In most of the biological treatment processes, temperature effects more than one reaction rates, and the overall influence of temperature on the process becomes important. The applicable equation for the effect of temperature on rate constants is given by

$$k_T = k_{20}\theta^{T-20} \tag{12}$$

where θ is the temperature coefficient. This equation shows that reaction rates increase with increase in temperature.

Mechanism of BOD Removal

In wastewater treatment processes, the microorganisms are not present as isolated cells, but are a collection of microorganisms such as bacteria, yeast, molds, protozoa, rotifers, worms and insect larvae in a gelatinous mass.[13] These microorganisms tend to collect in a biological floc, called biomass, which is expected to possess good settling characteristics. The biological oxidation or stabilization of organic matter by the microorganisms present in the floc is assumed to proceed in the following sequence:[13, 14]

a. An initial high rate of BOD removal from wastewater on coming in contact with active biomass by adsorption and absorption. The extent of this removal depends upon the loading rate, the type of waste, and the ecological condition of the biomass.
b. Utilization of BOD in direct proportion to biological cell growth. Substances concentrating on the surface of biomass are decomposed by the enzymes of living cells, new cells are synthesized, and end products of decomposition are washed into the water or escape to the atmosphere.
c. Oxidation of biological cell material through endogenous respiration whenever the food supply becomes limited.
d. Conversion of the biomass into settleable or otherwise removable solids.

The rates of reactions in the above mechanisms depend upon the transport rates of substrate, nutrients, and oxygen in case of aerobic treatment, first into the liquid and then into the biological cells. Any one or more of these rates of transport can become the controlling factors in obtaining the maximum efficiency for the process. However, most often the interfacial transfer or adsorption is the rate determining step.[14]

The typical progression of BOD removal with time by biomass during biological purification in a batch operation was first shown by Ruchhoft[15] as reproduced in Fig. 5. The corresponding metabolic reactions in terms of microorganisms to food ratio are shown in Fig. 6. This figure shows that the food to microorganisms ratio maintained in a biological reactor is of considerable importance in operation of the process. At a low M/F ratio, microorganisms are in the log growth phase, characterised by excess food and maximum rate of metabolism. However, under these conditions the settling characteristics of biomass is poor because of their dispersed growth; also the BOD removal efficiency is poor as the excess unused organic matter in solution passes out with the effluent. On the other hand, high M/F ratio means the operation in endogenous phase. Competition for a small amount of food available to a large mass of microorganisms results in starvation conditions within a short duration and continued aeration under these conditions results in auto-oxidation of biomass. Although the rate of metabolism is relatively low

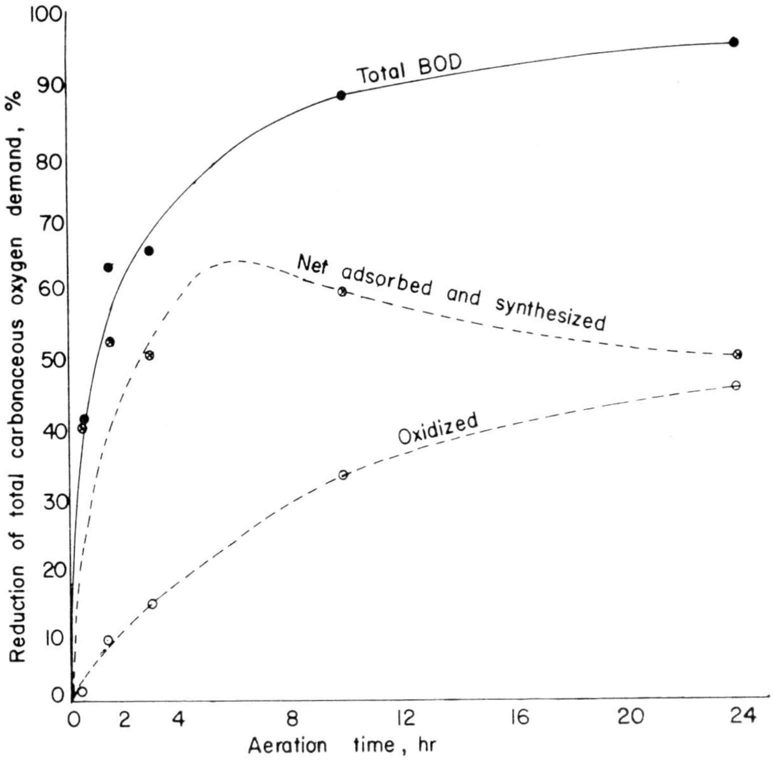

Figure 5 Removal of organic imbalance by biomasses in a batch operation.[15]

at high M/F ratio, settling characteristics of biomass are good and BOD removal efficiency is high.

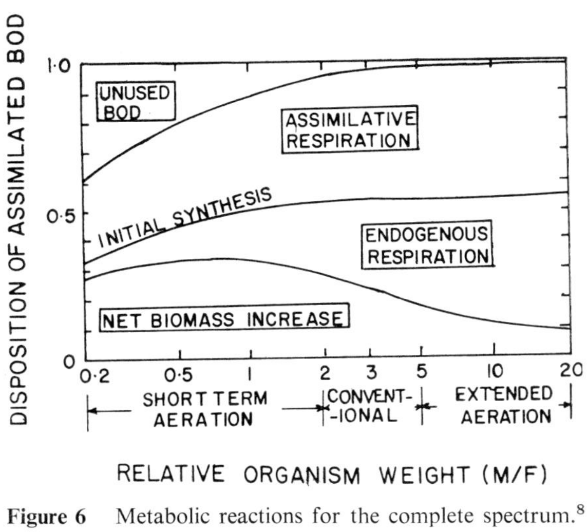

Figure 6 Metabolic reactions for the complete spectrum.[8]

Types of Reactors

Three types of reactors have been idealized for use in biological wastewater treatment processes: (a) Batch reactors in which all reactants are added at one time and composition changes with time, (b) plug flow or non-mix flow reactors in which no element of flowing fluid overtakes another element, and (c) completely mixed or back-mix reactors in which the contents are well stirred and are uniform in composition throughout. Most of the flow reactors in biological treatment are not ideal, but with negligible error, some of these can be considered ideal plug flow or back-mix flow. Others have considerable deviation due to channeling of fluid through the vessel, by the recycling of fluid through the vessel, or by the existence of stagnant regions of pockets of fluid.[16] The non-ideal flow conditions can be studied by tagging and following each and every molecule as it passes through the vessel but it is almost impossible. Instead, it is possible to measure the distribution of ages of molecules in the exit stream.

The mean retention time, t_d, for a reactor of volume V, and having volumetric feed rate of Q is given by $t_d = V/Q$. In non-ideal reactors, every molecule entering the tank has a different retention time scattered around t_d. Since all biological reactions are time dependent, knowledge on age distribution of all the molecules becomes important. The distribution of ages molecules in the exit streams of both ideal and non-ideal reactors in which a tracer is added instantaneously in the inlet

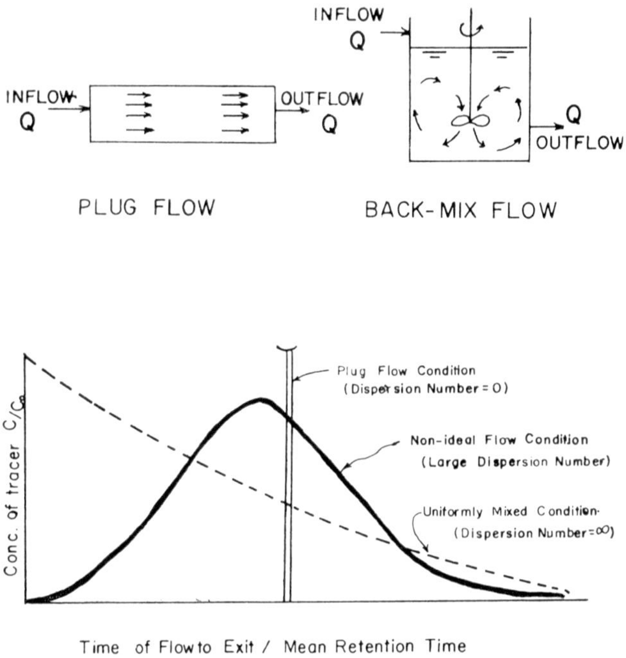

Figure 7 Hydraulic characteristics of basins.

stream is shown in Fig. 7. The spread of concentration curve around the plug flow conditions depends upon the vessel or reactor dispersion number, D/ul, where D is longitudinal or axial dispersion coefficient, u is the mean displacement velocity along the tank length and l is the length dimension.[17] In case of plug flow the dispersion number is zero whereas it becomes infinity for completely mixed tanks.

Treatment Models

Lawrence and McCarty[11] have proposed and analysed the following three models for existing continuous flow aerobic or anaerobic biological wastewater treatment configurations: (a) a completely mixed reactor without biological solids recycle, (b) a completely mixed reactor with biological solids recycle, and, (c) a plug flow reactor with biological solids recycle. These configurations are shown schematically in Fig. 8. In all these treatment models the following equations can be applied in order to evaluate kinetic constants:[18]

Solid balance equation

$$\begin{bmatrix}\Delta\text{ cells}\\ \text{reactor}\end{bmatrix} = \begin{bmatrix}\Delta\text{ cells}\\ \text{growth}\end{bmatrix} - \begin{bmatrix}\Delta\text{ cells}\\ \text{decay}\end{bmatrix} - \begin{bmatrix}\Delta\text{ cells}\\ \text{effluent loss}\end{bmatrix}$$

Substrate balance equation

$$\begin{bmatrix}\Delta\text{ substrate}\\ \text{reactor}\end{bmatrix} = \begin{bmatrix}\Delta\text{ substrate}\\ \text{influent}\end{bmatrix} - \begin{bmatrix}\Delta\text{ substrate}\\ \text{growth}\end{bmatrix} - \begin{bmatrix}\Delta\text{ substrate}\\ \text{effluent loss}\end{bmatrix}$$

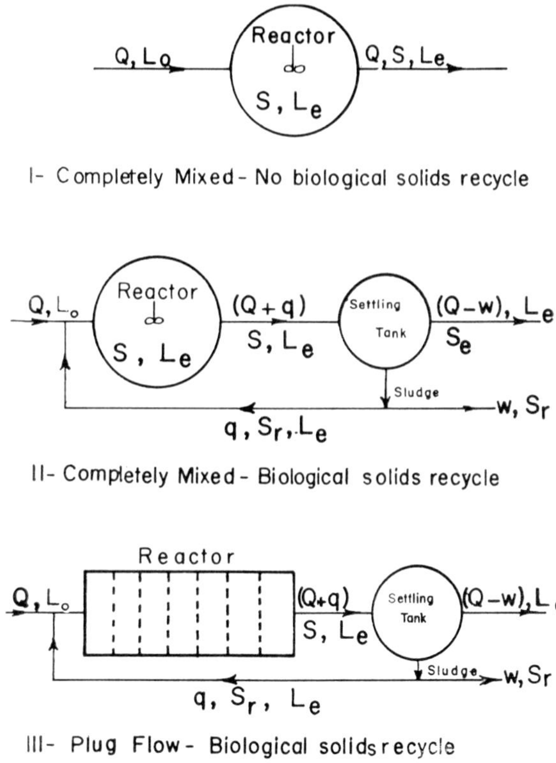

Figure 8 Treatment models.[11]

Parameters for Design and Operation

Various parameters have been developed and used in the design and operation of biological wastewater treatment processes and most significant parameters are:

t_s = Biological solids retention time, or sludge age, or mean cell retention time, is defined as the ratio between total active microbial mass in treatment system, S_T, and total quantity of active microbial mass withdrawn daily, including solids wasted purposely as well as those lost in the effluent, $\Delta X/\Delta t$.

U = Process loading factor, or substrate removal velocity, or food to microorganisms ratio, or specific utilization, is defined as the ratio between the mass of substrate utilized over a period of one day, $\Delta L/\Delta t$, and the mass of microorganisms, X_T, in the reactor.

t_d = Hydraulic retention time or detention time, or mean holding time, is defined as the ratio between the volume of reactor, V, and the volumetric feed rate, Q.

V = Volumetric loading rate or hydraulic loading rate is defined as the ratio between the mass of substrate applied over a period of one day, $L_T/\Delta t$, and the volume of the reactor, V.

E = Process treatment efficiency or process performance is defined as percentage ratio between the substrate removed, $(L_0 - L_e)$, and influent substrate concentration, L_0. A desired treatment efficiency can be obtained by control of one or more of these paramters separately or in combination.

Biological Treatment Systems

The existing biological treatment systems can be divided into the following three groups:

I. Aerobic stationary-contacts systems: Irrigation beds, irrigation sand filters, and trickling filters fall in this group. In these treatment processes the biomass remains stationary in contact with the solid supporting media like sand or rocks and the wastewater flows around it.

II. Aerobic suspended-contact systems: Activated sludge process and its various modifications and aerobic lagoons are included in this group. In these treatment processes both the biomass and the substrate are in suspension or in motion.

III. Anaerobic suspended-contact system: Anaerobic digestion of sludges and anaerobic decomposition of wastewater in anaerobic lagoons fall in this category.

A typical layout of a wastewater treatment plant incorporating biological treatment is shown in Fig. 9. Primary sedimentation separates settleable solids and the aerobic biological treatment is designed to remove the soluble BOD. The solids collected in primary sedimentation tanks and the excess sludge produced in secondary treatment are mixed together and may be digested anaerobically in digesters. Trickling filter and activated sludge processes are most common secondary treatment processes for aerobic treatment and are discussed in detail. Discussion of sludge digestion by anaerobic process has also been included.

Trickling Filters

Wastewater is applied intermittently or continuously to a fixed bed of stones or other natural or synthetic media resulting in a growth of microbial slime or biomass on the surface of this media. Wastewater is sprayed or otherwise distributed so that it trickles slowly through in contact with the air. For maximum efficiency food should be supplied continuously by recirculating, if necessary, the treated wastewater or settled sludge or both. Oxygen is provided by the dissolved oxygen in influent wastewater, recirculated water, and from the air circulating through the interstices between the media to maintain aerobic conditions.

Active microbial film, biomass, consisting primarily of bacteria, protozoa, and fungi, coats the surface of filter media. The activity in biological film is aerobic, with movement of oxygen food and end products in and out of it as shown in Fig. 10. But as the thickness of the film increases, the zone next to filter medium becomes anaerobic. Increased anaerobic activity near the surface may

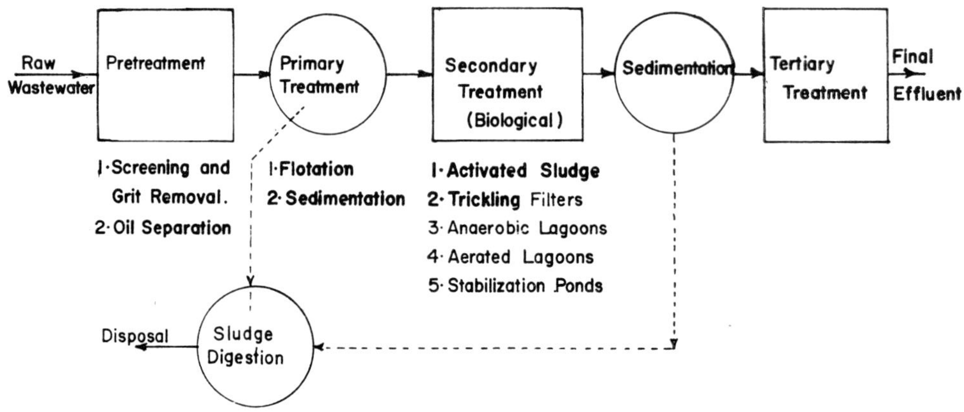

Figure 9 Typical wastewater treatment sequence.

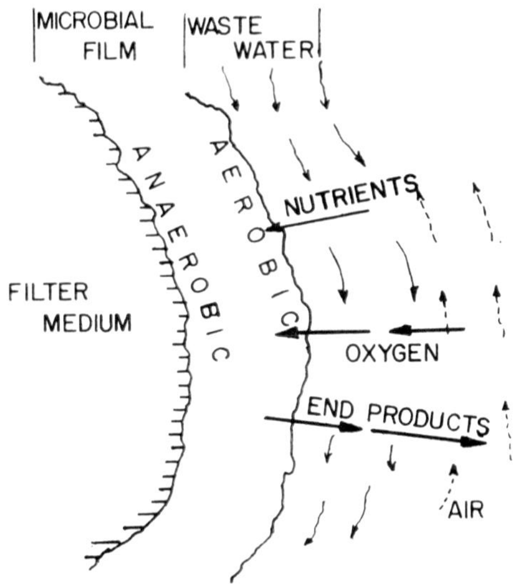

Figure 10 Process of BOD removal in trickling filters.

liquify the film in contact with the medium, resulting in sloughing or falling down of the old film and growth of a new film. The sloughed solids are separated in the secondary settling tanks and a part of these may be recirculated in the system. Two types of trickling filters are recognized, primarily on the basis of their loading and method of operation, as shown in Table I. In low rate trickling filters, the wastewater passes through only once and the effluent is then settled prior to disposal. In high-rate trickling filters wastewater applied to filters has been diluted with recirculated flow of treated effluent, settled effluent, settled sludge, or their mixture, so that it is passed through the filter more than once. Several recirculation patterns used in high rate filter systems are shown in *ASCE Manual.*[19] Sometimes two filter beds are placed in series and these are called Two-Stage Filters. The advantages and disadvantages of recirculation are listed below:

TABLE I

Comparison of Low-rate and High-rate Filters

Parameters	Low-rate filters	High-rate filters
Hydraulic loading		
US gallons per day per square foot	25 to 100	200 to 1000
Million gallons per acre per day	1.1 to 4.4	8.7 to 44
Organic loading (BOD)—V		
Pounds per 1000 cubic feet per day	5 to 25	25 to 300
Pounds per acre-foot per day	220 to 1100	1100 to 13000
Recirculation	Generally absent	Always provided $R = 0.5$ to 3
Effluent quality	Highly nitrified, lower BOD	Not fully nitrified, higher BOD

Advantages of recirculation

1. Part of organic matter in sewage feed is brought into contact with growth on filter media more than once.
2. Recirculated liquid contains active microorganisms not found in such quantity in raw sewage, thus providing seed continually. This continuous seeding with active microorganisms and enzymes stimulates hydrolysis and oxidation and increases rate of biochemical stabilization.
3. Diurnal organic load is distributed more uniformly. Thus when plant flow is low, operation is not shut off. Also stale sewage is freshened.
4. Increased flow improves uniformity of distribution, increases sloughing, and reduces clogging tendencies.
5. Higher velocities and continual scouring makes conditions less favorable for growth of filter flies.
6. Provides for more flexibility of operation.

Disadvantages

1. There is increased operating cost because of pumping. Larger settling tanks in some designs may increase capital cost.
2. Temperature is changed as a result of number of passes of liquid. In cold weather this results in decreased biochemical activity.
3. Amount of sludge solids to digesters may be increased.

The ASCE-FSIWA Manual on "*Sewage Treatment Plant Design*"[19] lists the following factors affecting the design and operation of filters:

1. Composition and characteristics of the wastewater after pretreatment,
2. Hydraulic loading applied to the filter,
3. Organic loading applied to the filter,
4. Recirculation, system, ratio and arrangement,
5. Filter beds, their volume, depth and air ventilation,
6. Size and characteristics of media,
7. Temperature of wastewater.

Assuming that the flow through the packed column could be approximated as plug flow, and if BOD removal rate occurs by first order reaction [equation (9)], then the formula to use in trickling filters will become,

$$\frac{dL}{dt} = -KLS = -k_f L$$

or

$$\frac{L}{L_0} = e^{-KSt} = e^{-k_f t} \tag{13}$$

Another equation suggested[13] for application in trickling filters is

$$\frac{L}{L_0} = \frac{1}{1 + KSt} = \frac{1}{1 + k_f t} \tag{14}$$

Where trickling filter rate coefficient, k_f, is a function of active film per unit volume and remains constant for a given specific area and uniform slime layer. Contact time, t, is related to filter depth, H, volumetric rate of flow per unit area, q, specific surface area of filter media, A_v. Sinkoff, Porges, and McDermott[20], have proposed the following relationship based on their experiments,

$$t = c_1 H \left[\frac{A_v}{q}\right]^n \tag{15}$$

c_1 is assumed to be a constant and exponent n ranges between 0.53 and 0.83 depending upon the type of filter medium and the hydraulic characteristics of the system. Substitution of this value of t in equation (13), gives

$$\frac{L}{L_0} = e^{-k_f\left[\frac{A_v}{q}\right]^n Hc_1} = e^{-k'_f H/q^n} \tag{16}$$

Eckenfelder[13] suggests that the amount of active surface film covering the filter medium decreases with depth H; therefore, combining equations (14) and (15) and substituting $c_1 \propto \frac{1}{H^m}$, gives

$$\frac{L}{L_0} = \frac{1}{1 + k_f A_v^n H^{1-m}/q^n} = \frac{1}{1 + k''_f H^{1-m}/q^n} \tag{17}$$

For treatment of domestic sewage on rock filters, Eckenfelder has obtained the values of $n = \frac{1}{2}$, $m = \frac{1}{3}$, and $k''_f = 2.5$ with H in ft and q in mgd/acre. Several emperical relationships for process efficiency in trickling filters have been proposed and successfully applied. Most significant of these are the National Research Council Formula and Rankin's Formula which have been described in detail in *ASCE Manual.*[19] Eckenfelder and O'Connort[31] have reported a value of 1.035 for overall temperature coefficient θ, in equation (12). An adjustment in process efficiency due to variation in temperature should be provided.

Activated Sludge Process

It is a biological treatment process in which biologically active mass, called activated sludge, is continuously mixed with the bio-degradable matter in an aeration basin in presence of oxygen. The combination of wastewater and activated sludge is called mixed liquor. The oxygen is supplied to the mixed liquor either by diffusing compressed air into the liquid or by mechanical aeration. The activated sludge is subsequently separated from the mixed liquor by sedimentation in a clarifier and a part of this sludge is recirculated to the aeration basin. The rest of this sludge, indicating net excess production of biological cell material, is disposed off. Activated sludge treatment plants vary in performance due to variation in unit arrangements, methods of introducing air and wastewater into the aeration basin, aeration time, concentration of active biomass, aerator volume, degree of mixing, etc. Some important types of activated sludge processes are discussed below and their operating parameters are summarized in Table II.

Kinetic rate: Depending upon the design and operating conditions, one or more of the kinetic rate equations (6), (8), (9) and (10) for BOD removal can be applied to each type of the activated sludge processes.

Oxygen requirement: Oxygen is used to provide energy for synthesis of biological cells and for endogenous respiration of the biological mass. The total oxygen requirement, ΔO_2, can be expressed with the following equation,

$$\Delta O_2 = a' \Delta L + b' S_T \tag{18}$$

where a' is the fraction of BOD removed that is oxidized for energy and b' is the oxygen used for endogenous respiration of the biological mass, per day. In conventional aeration basins, an hourly oxygen demand of 50–80 mg/l per 1000 mg/l of VSS is exerted near the beginning of the tank and is reduced to 20 mg/l per 1000 mg/l of VSS in the course of 4 to 6 hours.[14]

Excess sludge yield: By applying material balance for volatile suspended solids in activated sludge

TABLE II

Activated Sludge Process Parameters

Parameters	Conventional	Step aeration	Short term	Biosorption	Complete mixing	Extended aeration	Aerated lagoons
Organic loading lb BOD_5 per day per 1000 ft^3, V	30–40	50–150	100–400	30–70	125–180	10–20	5
Process loading factor lb BOD_5 per day per lb MLVSS, U	0.2–0.5	0.2–0.5	2–5	0.2–0.5	0.6–1.0	0.05–0.2	0.2
Sludge age, days, t_s	3–4	3–4	0.2–0.5	3–4		14–∞	3–5
Aeration time, hours, t_d	6–7.5	6–7.5	2–4	0.5–1.5 (aeration) 1.5–6 (activation)	3–5	20–30	70–120
BOD_5, removal, %, E	90–95	90–95	60–85	85–90	85–90	85–95	85–90
$\dfrac{\text{Normal return sludge}}{\text{Average design flow}} \times 100$	30 (15–75)*	50 (20–75)*	20 (10–50)*	100 (50–150)*	100 (50–150)*	100 (50–200)*	0
Primary settling required	Yes	Yes	No	Optional	Optional	No	No

* Provision in design should be made for these maximum and minimum values.

system, and using the concept shown in Fig. 3:

$$\begin{bmatrix}\text{Excess solids in activated}\\ \text{sludge system}\end{bmatrix} = \begin{bmatrix}\text{Non-biodegradable}\\ \text{suspended solids}\\ \text{in effluent}\end{bmatrix} + \begin{bmatrix}\text{Biomass synthesized}\\ \text{during BOD removal}\end{bmatrix} - \begin{bmatrix}\text{Biomass broken}\\ \text{down by endogenous}\\ \text{respiration}\end{bmatrix}$$

or

$$\Delta S = fS_0 + a\,\Delta L - bS_T \qquad (19)$$

where

ΔS = Net accumulation of volatile suspended solids, lb/day

f = Fraction of volatile suspended solids present in the influent which are non-degradable

S_0 = Influent volatile suspended solids, lb/day.

Temperature effect: According to Eckenfelder and O'Connor,[13] the value of temperature coefficient θ in equation (12) varies between 1.0 for low loading rates to 1.04 for high loading rates. Friedman and Schroeder[21] have studied in detail the

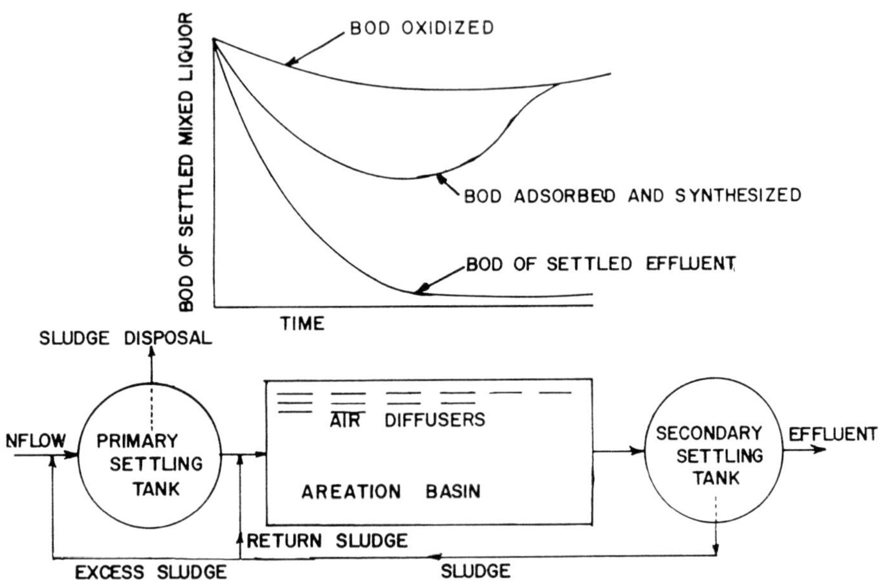

Figure 11 Conventional activated sludge.

affect of temperature on growth and cell yield for activated sludge and found that both the maximum growth and the maximum cell yield occurred at 20°C.

Elements of a conventional activated sludge system are shown in Fig. 11. In this system, the untreated wastes are mixed with the return sludge at the inlet end of the aeration tank. The microorganisms receive the full impact of any shock load and respond accordingly with sudden increase in oxygen demand during growth. By the time microorganisms leave the aeration tank, the organic matter has been stabilized and the microorganism population starts dying off. Thus, the microbial population undergoes a continual shifting and never reaches a relatively constant equilibrium.[7]

A mass of activated sludge of three to four times the weight of the daily BOD load must be kept in the system in order to consume all the new food and also acquire good settling properties. These types of plants have been used for treating domestic wastewaters of low biochemical oxygen demands. In conventional activated sludge plants that have plug flow design, high BOD in influent causes higher oxygen demand at that point in the mixed liquor and this oxygen demand diminishes as the flow passes down the aeration tank. Most of the plants designed these days are provided with tapered aeration, with highest air supply near the inlet end and lowest near the outlet end of the aeration tank.

Modifications of the Conventional Activated Sludge Process

A. Step aeration activated sludge

Step aeration process, developed by Gould[22] at New York City, offers more flexibility than the conventional activated sludge process. In this process, wastewater is introduced at four or more points along the aeration tank in order to maintain a uniformly distributed loading. In addition to evening out the oxygen demand, this also keeps sludge reaerated in presence of substrate and this remains in more active form instead of reaching endogenous phase near the end of conventional aeration tank. Step aeration system layout and fluctuations in BOD in aeration tank are shown in Fig. 12. This method has been successfully employed in the treatment of domestic wastewater and industrial wastewaters of similar nature.

B. Short term aeration or high rate or modified activated sludge

These systems have very high loading factors, both in terms of organic and volumetric loading, and a low mixed liquor volatile suspended solids, thus requiring small aeration tank capacities and reduced air requirements. Because of shorter aeration time and lower organisms weight, this process provides an intermediate degree of treatment. Organic matter is removed largely by synthesis thus exerting high rate of oxygen demand and producing a relatively large volume of sludge per

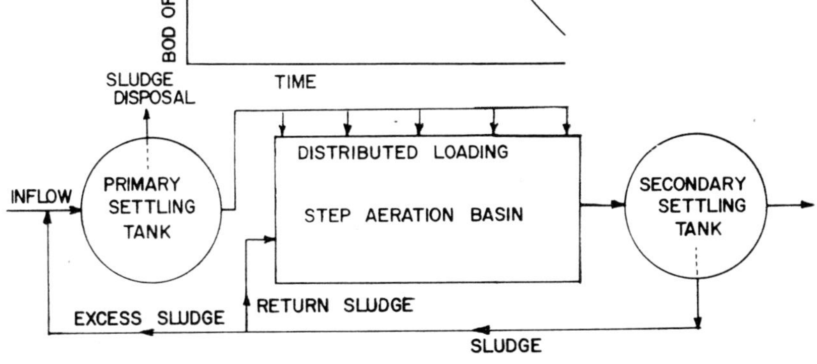

Figure 12 Step aeration activated sludge.

unit weight of BOD removed. Since the sludge still contains certain unstabilized organic matter, the settled sludge in secondary settling tanks should be removed rapidly in order to avoid its anaerobic decomposition and flotation. The flow diagram is similar to the conventional system as shown in Fig. 11.

C. Contact stabilization or biosorption

The elements of this type of plant are shown in Fig. 13. This system is ideally suited to the treatment of wastewaters in which a large portion of BOD is present in suspended or colloidal form. The suspended BOD is rapidly adsorbed in a short period, $\frac{1}{2}$ to $1\frac{1}{2}$ hours, by the well-activated organisms and a part of soluble BOD is metabolized. In the activation tank, the sludge is reaerated for bio-oxidation and stabilization of adsorbed food; and when returned to the aeration tank, it is activated for higher BOD removal as compared to conventional plant where sludge has become lean and hungry in the absence of a food supply. The additional advantage of this process is the reduced

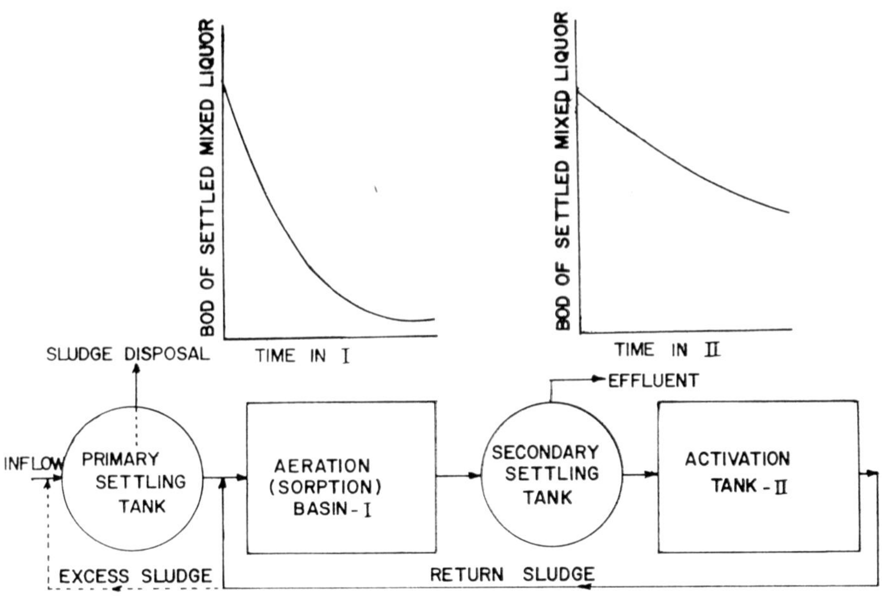

Figure 13 Biosorption (contact stabilization) activated sludge.

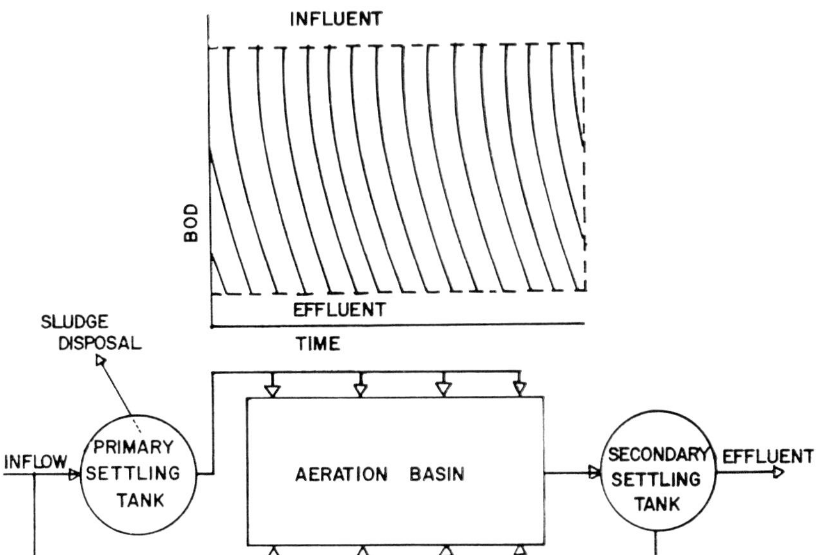

Figure 14 Complete mixing activated sludge.

overall tank volume required as compared to the conventional system.

D. Completely mixed activated sludge

"Completemix" approach is with respect to combining the return sludge and wastewater in order to maintain the entire contents of the aeration chamber in essentially a homogeneous state. Wastewater introduced into the aeration is dispersed rapidly throughout the mass and is subjected to immediate attack by fully developed organisms throughout the aeration basin. Biological stability and efficiency of the aeration basin is enhanced by this design. Layout of a completely-mixed activated sludge plant and variation in BOD are shown in Fig. 14.

In his mathematical analysis, McKinney[5] considered the complete mixing activated sludge process as the one in which the untreated wastes are instantaneously mixed throughout the aeration tank. In effect, the organic load on the aeration tank is uniform from one end to the other end and consequently a uniform oxygen demand and a uniform biological growth are produced. It is assumed to reduce the effect of variations in organic loads that produce shock loads on conventional units, retain a more biological population and hence produce a more uniform effluent, and be able to treat organic wastes of any concentration and produce an effluent of any desired concentration.[5] Using treatment model II (Fig. 7), as an example of completely mixed system Lawrence and McCarty[11] have shown analytically that although the complete-mixing will reduce the shock loads due to variations in organic loads, plug flow type conventional units (Treatment Model III) are more efficient.

Assuming that equation (9) is applicable for BOD removal rate, and since the BOD in a completely mixed aerator, L, is equal to the effluent BOD, L_e, therefore under steady state conditions,

$$\frac{dL}{dt} = -KSL_e$$

or

$$\frac{L_e}{L_i} = \frac{1}{1 + KSt} \tag{20}$$

E. Extended aeration

Extended aeration plant is the one where the net growth rate is made to approach zero, i.e., rate of growth becomes approximately equal to rate of decay. This is achieved by increasing the aeration time in order to keep the sludge in endogenous growth phase for a considerable time. In practice it is impossible to operate an extended-aeration system without sludge accumulation, because certain volatile solids, mainly polysaccharide in nature and inert to organisms in activated sludge process,

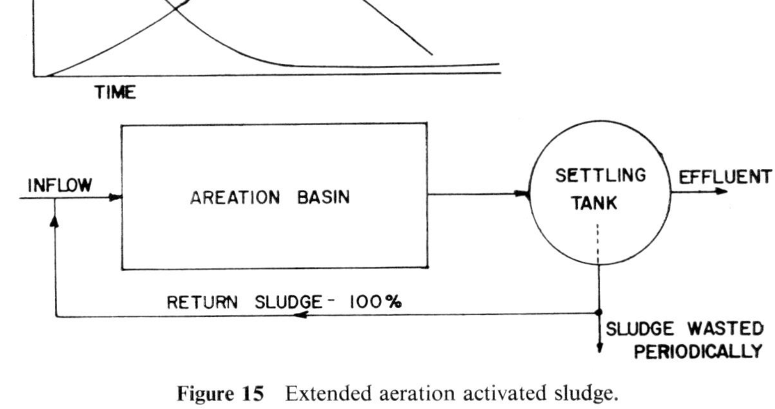

Figure 15 Extended aeration activated sludge.

accumulate in the plant. Excess sludge is not generally wasted continuously from an extended aeration, but instead, the mixed liquor is allowed to increase in suspended solids concentration and a large volume of the aeration tank contents or return sludge are periodically pumped to disposal. Oxidation ditch plants are designed and operated on this principle. Layout of a typical extended-aeration plant and variation in BOD in aeration tank are shown in Fig. 15.

F. Aerated lagoons

These are similar to activated sludge system but without recirculation of sludge. Mechanical or diffused aeration devices are used for supplying oxygen and also providing sufficient mixing. All suspended solids may or may not be kept in suspension, depending upon the degree of mixing. Deposited solids may undergo anaerobic decomposition. Mathematically, the BOD removal rate in aerated lagoons is given by equation (9) and assuming the aerated lagoon to be a completely mixed system, without recycle and maintaining sufficient turbulence, this equation becomes similar to equation (20). In practice this equation has proven to represent a generalized response function for design of most aerated lagoons.[18]

The exact solid level in an aerated lagoon can be approximated by applying a material balance around the lagoon, under equilibrium conditions:

Solids In + Net Synthesis In Basin = Solids Out

or

$$S_0 + (a\,\Delta L - bSt) = S$$

or

$$S = \frac{S_0 + a\,\Delta L}{1 + bt} \qquad (21)$$

Because of very low solid concentration, the detention time in aeration basins is very high, consequently requiring large volume of aeration basins. The temperature variation, therefore, exerts a profound effect on the rate of BOD removal. Eckenfelder and Ford[10] have given a relationship for estimating the lagoon temperature at both extreme conditions. Once this temperature is established, corrected k_T value should be obtained from equation (12), using θ equal to 1.035 and then adopted in the kinetic equation (20).

Several other modifications in activated sludge process have been discussed elsewhere;[23] but most of these modifications are similar in concepts to one of the types discussed above. For an example, in Hatfield and Kraus systems the supernatant from digestion tanks or even digested sludge are added to the reaeration tank to provide nutrients. Similarly an Activated Aeration plant is a combination of a conventional activated sludge process and the short term aeration process.

Anaerobic Treatment

In this process, anaerobic bacteria stabilize the organic matter in absence of free oxygen. Anaerobic treatment has been used widely for stabilization of sludges collected from primary and secondary settling tanks and recently is being adopted for treatment of soluble wastes in anaerobic lagoons, anaerobic filters, etc. One of the important advantages of anaerobic processes over aerobic processes is a high percentage conversion of organic matter to gases and liquid and a low percentage conversion to biological cells. McCarty[24] has mentioned that efficient anaerobic treatment of soluble wastes with BOD concentration as low as 500 mg/l is now feasible. Wastes with lower BOD can also be treated anaerobically, although the waste treatment efficiency will not be of the same magnitude expected for aerobic treatment.

Anaerobic treatment of wastewaters takes place in two stages as shown in Fig. 16. In the first stage, complex organic materials like protein, fats, carbohydrates, are converted into simple organic acids

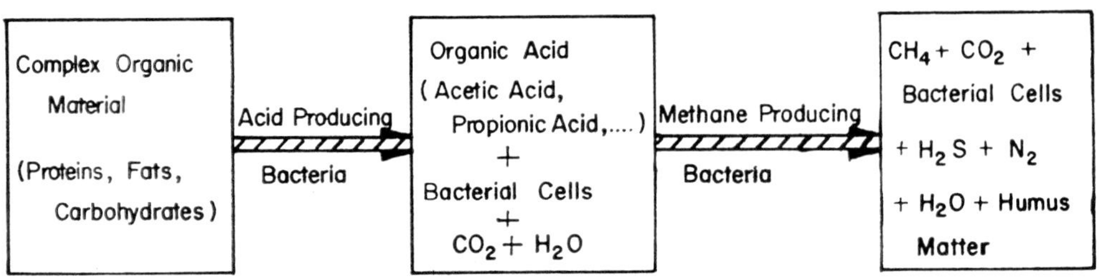

Figure 16 Sequential mechanism of anaerobic waste treatment.

by acid forming bacteria, but with little change in BOD or COD values. In the second stage, these fatty acids are converted to carbon dioxide and methane, thereby stabilizing the BOD or COD.

In a conventional anaerobic treatment process, the substrate is fed into the digester continuously or intermittently. In most of the existing digesters, the contents are mixed, mechanically or with compressed gas collected from digesters. There is no recirculation of digested sludge and system is a typical flow through system. The hydraulic detention time, t_d, in conventional process becomes equal to the solid retention time, t_c. Recently several modifications have been made in the conventional anaerobic treatment process. McCarty[24] has grouped the basic anaerobic process designs into Conventional Process, Anaerobic Activated Sludge Process, and Anaerobic Filter Process. Operating conditions of these process designs are shown in Fig. 17. It is suggested that the conventional process be used for concentrated wastes like sludges where economical treatment can be obtained by keeping hydraulic detention time, t_d, equal to the desired solid retention time, t_c. The economical treatment of diluted wastes, however, requires hydraulic detention time, t_d, much below the desired solid retention time, t_c, and thus anaerobic contact processes become more applicable.[24]

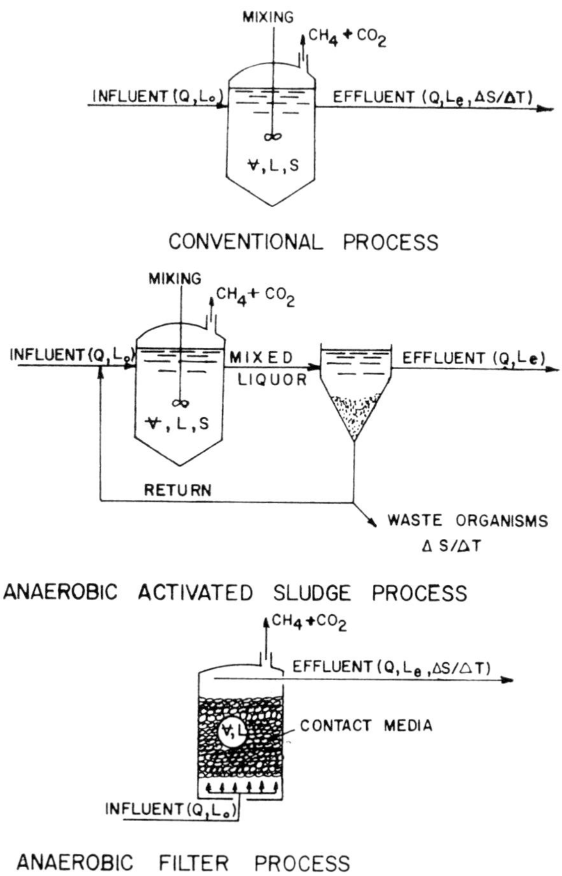

Figure 17 Basic anaerobic process designs.[24]

Anaerobic treatment processes are more sensitive to operating parameters and their environments as compared to aerobic processes. The best parameter for controlling the operation of anaerobic treatment is the biological retention time. A minimum SRT exists below which the critical methane producing bacteria are removed from the system faster than they can reproduce themselves. In practice, SRT values of two to ten times this minimum value are used. Thus the hydraulic detention time and solid retention time maintained in anaerobic treatment processes is very high and the net growth of biological solids becomes very low due to significant decay as given by equation (8).

Mixing of the digester content is becoming a common practice. The advantages of mixing are better contact between food and microorganisms, uniform temperature, reduction in scum formation, accelerated digestion, and distribution of metabolic inhibitors.

TABLE III

Operating Parameters of Conventional Anaerobic Digesters

Parameters	Unmixed	Mixed
V Loading rate, lb VSS/ft^3/day	0.02–0.05	0.1–0.2
t_d Detention time, days	30–90	10–15
E Volatile solids reduction,	50–70	50
Mixing	Absent	Present
pH	6.8–7.4	6.8–7.4
Temperature, °C	30–35	30–35

Certain cations, such as sodium, potassium, calcium, or magnesium show a toxic or inhibitory effect on anaerobic treatment when present in high concentrations. Generally cation concentrations up to 500 mg/l are considered as stimulatory to the process, whereas concentrations above 1000 mg/l start showing adverse effects. Ammonia in higher concentrations, depending upon pH, may also be detrimental.

Operating parameters of conventional anaerobic digesters are shown in Table III.

Nomenclature

A_v	Specific surface area of filter media, length^{-1}
C_s	Substrate concentration, mass per unit volume
D	Longitudinal dispersion coefficient, (length)^2 per unit time
H	Filter, depth length
K_s	Half velocity coefficient = substrate concentration at which rate of its utilization is half the maximum rate, mass per unit volume
K	k'/a = constant, (time)$^{-1}$ (mass per unit volume)$^{-1}$
L	Substrate concentration around microorganisms, measured in terms of BOD, mass per unit volume
L_e	Effluent BOD or final substrate concentration, mass per unit volume
L_0	Influent BOD or the initial substrate concentration, mass per unit volume
L_T	Total mass of substrate applied, mass
ΔL	BOD removed, mass per unit time
N_0	Number of microorganisms per unit volume at time $t = 0$
N_t	N = Number of microorganisms per unit volume at time t
ΔO_2	Amount of oxygen requirement, mass per unit time
Q	Volumetric rate of flow, volume per unit time
S	Mass of microorganisms present per unit volume
S_0	Influent volatile suspended solids, mass per unit time
S_T	Total mass of microorganisms, mass
ΔS	Cell mass synthesized, mass per unit time
T	Temperature, °C
V	Volume of the reactor, volume
a	Growth yield coefficient, dimensionless
a'	Fraction of BOD removed that is oxidized for energy
b	Microorganisms decay coefficient, time^{-1}
b'	Oxygen used for endogenous respiration of biological mass, time^{-1}
c_1	Constant
f	Fraction of volatile suspended solids present in the influent which are non-degradable
k_f, k_f', k_f''	Rate coefficient in filters, time^{-1}
k_0	Logarithmic growth rate constant, time^{-1}
k_t	Growth rate factor, time^{-1}
k'	Growth rate factor, (time)$^{-1}$ (mass per unit volume)$^{-1}$
l	Length dimension in reactor, length
m	Exponent
n	Exponent
q	Volumetric rate of flow per unit area, length per unit time
t	Contact time in filter or anyother reactor, time
t_c	Mean cell retention time, time
$t_d = \dfrac{V}{Q}$	Mean retention time, time
u	Mean displacement velocity in reactor along length, length per unit time
θ	Temperature coefficient for microbial activity
$\dfrac{dC_s}{dt}$	Rate of substrate consumption, mass per unit volume per unit time
$\dfrac{D}{ul}$	Reactor dispersion number, dimensionless
$\dfrac{dL}{dt}$	Rate of waste utilization measured in terms of BOD, mass per unit volume per unit time
$\dfrac{dN}{dt}$	Rate of growth in number of microorganisms, number per unit volume per unit time
$\dfrac{dN}{N\,dt} = \dfrac{dS}{S\,dt}$	Specific growth rate, time^{-1}
$\dfrac{dS}{dt}$	Rate of growth of mass of microorganisms, mass per unit volume per unit time

REFERENCES

1. Pfafflin, J. R., "Municipal Wastewater". *Encyclopedia of Environmental Science and Engineering*. Vol. 1. Gordon and Breach, New York, 1976.
2. Gates, W. E., and S. Ghosh, "Biokinetic Evaluation of BOD Concepts and Data," *Journal of the Sanitary Engineering Division, Proceedings of the American Society of Civil Engineers*, Vol. 97, No. SA 3, June 1971, p. 287 to 309.
3. McKinney, R. E., "*Microbiology for Sanitary Engineers*," McGraw-Hill Book Company, Inc., New York, 1962.
4. Stanier, R. Y., M. Doudoroff, and E. A. Adelberg, *The Microbial World*, Prentice-Hall, Inc., Englewood Cliffs, N.,J. 1957.
5. McKinney, R. E., "Mathematics of Complete Mixing Activated Sludge," *Journal of the Sanitary Engineering Division, Proceedings of the American Society of Civil Engineers* **88**, SA 3, May 1962, p. 87–113.
6. Tsuchiya, H. M., A. G. Fredrickson, and R. Avis, "Dynamics of Microbial Cell Populations," *Advances in Chemical Engineering*, Vol. 6, edited by T. B. Drew, J. W. Hoopes, Jr., and T. Vermeulen, Academic Press, New York, 1966.
7. Stewart, M. J., "Activated Sludge System Variations, Specific Applications, *Proceedings of the Fifteenth Ontario Industrial Waste Conference*," June 1968, p. 93 to 115.
8. Irvine, R. L., and D. J. Schaezler, "Kinetic Analysis of Data from Biological Systems," *Journal of the Sanitary Engineering Division, Proceedings of the American Society of Civil Engineers*, Vol. 97, No. SA 4, August 1971, p. 409–424.
9. Monod, J., "The Growth of Bacterial Cultures," *Annual Review of Microbiology* **3**, 371 (1949).
10. Eckenfelder, W. W., and D. L. Ford, *Water Pollution Control*, Jenkins Publishing Company, Austin, Texas, 1970.
11. Lawrence, A. W., and P. L. McCarty, "Unified Basis for Biological Treatment Design and Operation," *Journal of the Sanitary Engineering Division, Proceedings of the American Society of Civil Engineers* **96**, SA 3, June 1970, p. 757–778.
12. Pearson, E. A., "Kinetics of Biological Treatment", *Advances in Water Quality Improvement*, edited by E. F. Gloyna, and W. W. Eckenfelder, Jr., University of Texas Press, Austin, 1970.

13. Eckenfelder, W. W., and D. J. O'Connor, *Biological Waste Treatment*, Pergamon Press, New York, 1961.
14. Fair, G. M., J. C. Geyer, and D. A. Okun, *Water and Wastewater Engineering*, Vol. 2, John Wiley and Sons, Inc., 1968.
15. Ruchhoft, C. C., "Studies of Sewage Purification—IX," *Public Health Reports* **54**, 468 (1939).
16. Levenpiel, O., *Chemical Reaction Engineering*, John Wiley and Sons, Inc., New York, 1967.
17. Murphy, K. L., and B. I. Boyko, "Longitudinal Mixing in Spiral Flow Aeration Tanks," *Journal of the Sanitary Engineering Division, Proceedings of the American Society of Civil Engineers* **96**, SA2, April 1970, p. 211 to 221.
18. Parker, C. E., *Anaerobic-Aerobic Lagoon Treatment for Vegetable Tanning Wastes*, Report prepared for the Federal Water Quality Administration Environmental Protection Agency, U.S. Government Printing Office, Washington, D.C., December 1970.
19. *Sewage Treatment Plant Design*, American Society of Civil Engineers Manual of Engineering Practice No. 36, New York, 1959.
20. Sinkoff, M. D., R. Porges, and J. H. McDermott, "Mean Residence Time of a Liquid in a Trickling Filter," *Journal of the Sanitary Engineering Division*, Proceedings of the American Society of Civil Engineers **85**, SA6 (1959).
21. Friedman, A. A., and E. D. Schroeder, "Temperature Effects on Growth and Yield for Activated Sludge," Presented at 26th Purdue Industrial Waste Conference, Lafayette, Indiana, May 4–6, 1971.
22. Gould, R. H., "Tallmans Island Works Opens for World's Fair," *Municipal Sanitation*, Vol. 10, No. 4, p. 185, April 1939.
23. Skrinde, R. T., and R. F. Ward, "Activated Sludge Processes: Conventional Processes and Modifications-Applications," Presented at Short Course in Water Quality Control, Department of Civil Engineering, University of Massachusetts, Amherst, Mass., March 1970.
24. McCarty, P. L., "Anaerobic Treatment of Soluble Wastes," *Advances in Water Quality Improvement*, Edited by E. F. Gloyna, and W. W. Eckenfelder Jr., University of Texas Press, Austin, 1970.

J. K. Bewtra
University of Windsor

C

CHEMICAL TREATMENT: see WASTEWATER: PHYSICAL AND CHEMICAL TREATMENT

CITY AND REGIONAL PLANNING: see URBAN PLANNING

COMPOSTING

Composting is the biological decomposition of a semi-dry organic waste material by microorganisms. During the process certain physical-chemical changes take place which alter the character of the waste material, converting it to a humus like consistency, further biochemical decomposition ceases, rendering the material "stable" to further decomposition. The material is at this point, nuisance free, and reduced in volume by 30 to 40%.

Composting is, in fact, the way much of the vegetable material on land is recycled in nature. This process, for example, is continuously occurring wherever vegetable residues accumulate on the land surface, and the environmental conditions are appropriate, in soils, woods, under leaf piles, etc. Historically, man has utilized composting as a method of enriching his soil since the dawn of agriculture. Only during this century however, has civilized man began to understand the principles of composting. Thus only recently have we began to use composting as an engineering process.

In the United States the composting of municipal refuse to produce a soil conditioner or low grade fertilizer has not generally been successful. For example, during the period 1951–1966, six out of nine compost plants built in the U.S. were shut down mostly because of an inability to sell the compost at a realistic price.[1] A more recent report indicated that during the last seventeen years, only one of the fifteen composting plants in the U.S. operated long enough to indicate success. Of these eight shut down in failure.[2]

At present, the U.S. has a large solid waste problem. For example, almost 6 lbs. per person per day of solid waste is generated in the U.S.—more than 190 million tons per year. By 1980, some 235 million people in the U.S. will generate approximately 8 lbs. per person per day of solid waste material (340 million tons per year).[3] In light of the present environmental "crisis," the emphasis on the quality of our urban environment, and the huge quantities of solid waste material which must be disposed of, composting lends itself as a natural and obvious method of conserving, reusing, and recycling our organic containing solid waste resourees. There are, however, certain practical aspects that should be considered. Generally, however, this discussion will be limited to the composting of urban generated solid wastes.

The Process

General

Composting operations in the U.S. are carried out by two general methods—batch and continuous (or semicontinuous) operation. Either method can employ mechanization, depending on the particular circumstances and requirements—i.e., space allocation, recovery operations, output desired, etc. Two general steps are usually employed in composting which can be described as (1) refuse or sample preparation, and (2) biological stabilization or decomposition by microorganisms. In addition, further drying and product preparation is usually necessary in commercial enterprises.

Batch composting

In its easiest form, batch process composting (Fig. 1) can be achieved by periodically heaping the solid waste materials into stacks (windrows). The stacks of material may be "seeded" with microorganisms (usually sewage sludges), moistened and even have nutritional supplements added. Periodically the heaps are "turned," i.e., mixing and aeration as achieved—either manually or by mechanization. During the process, normally several months, microorganisms begin to multiply, the temperature rises well above ambient within the heap, and the material slowly undergoes a change in character. Usually the most obvious changes occur at the bottom of the heap where a dark material not unlike soil or humus is formed. Eventually the temperature will return to ambient, the volume is reduced, and the biological process can be considered to have run its course and the material "stabilized." In the batch method, few control parameters are generally necessary, however, long time periods are required for stabilization. Among the important parameters that will be discussed and which will hasten the achievement of stability are: sample preparation of the raw material (i.e., shredding, chopping or grinding), maintaining of proper moisture (45–65%), and aeration ("turning" the pile). In Europe, on a large scale, the Dutch[4,5] as well as the Germans[6]

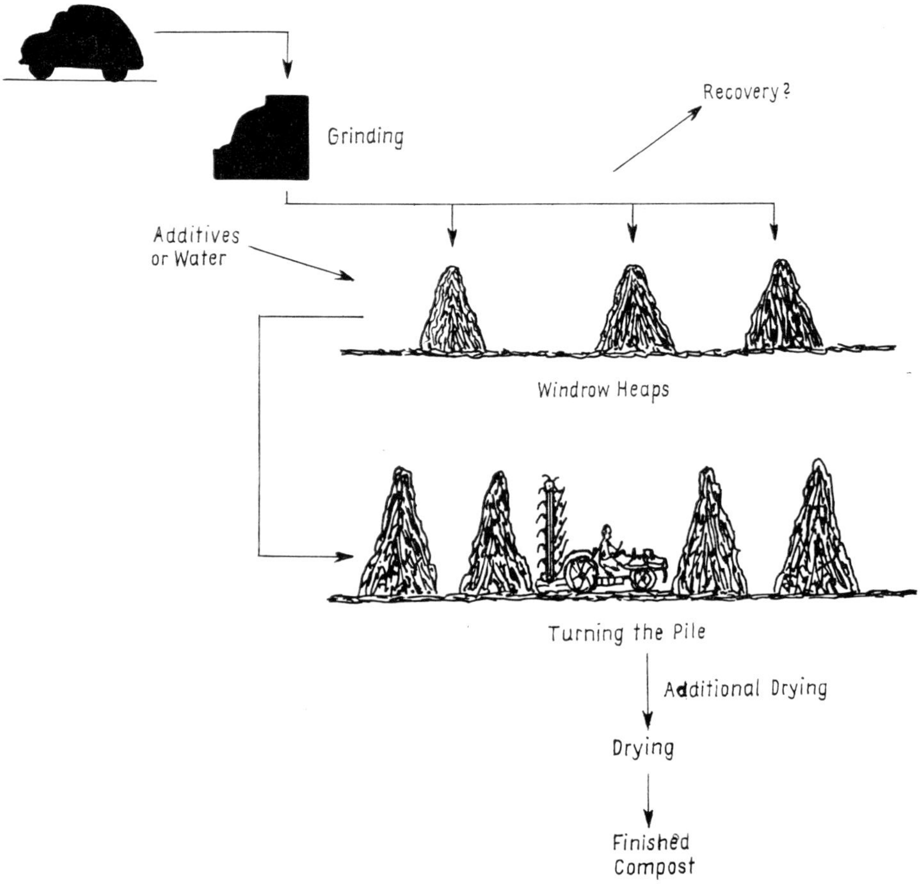

Figure 1 Batch composting using windrows.

and the British[7] have carried out such compost procedures.

Continuous-intermittent composting

Continuous flow or intermittently fed compost processes (Fig. 2) have been used as a method of processing urban wastes since the 1920's. Many of the plants now operating in Europe are of this type and were built during the 1950's. In the U.S., a number of mechanized, continuously operating procedures have also been used. As indicated only a few are still operating. Among those operating are the Altoona FAM, Inc. which has a capacity of 45 tons a day and has been operating since 1951. Perhaps the largest compost plant in the U.S. Was the plant in Houston, Texas with a capacity of 360 tons a day.[8]

Continuous flow compost operations will have a number of common unit processes, which may be broadly classified as sample preparation, storage, biochemical activation or stabilization and drying. Sample preparation generally consists of reducing the material by crushing, shredding, chopping or grinding to a small particle size of uniform consistency. A typical particle size is smaller than one inch. Segregation and recovery of ferrous metals, rags, glass or other materials may be carried out. The material is then transported to a storage tank via mechanized belt conveyors, screws or gravity fed hoppers and from there fed mechanically on a more or less continuous basis into the unit where composting is achieved, the biological reactor. Detention time in the storage tank is usually short enough to avoid nuisance conditions in the raw material but sufficient to

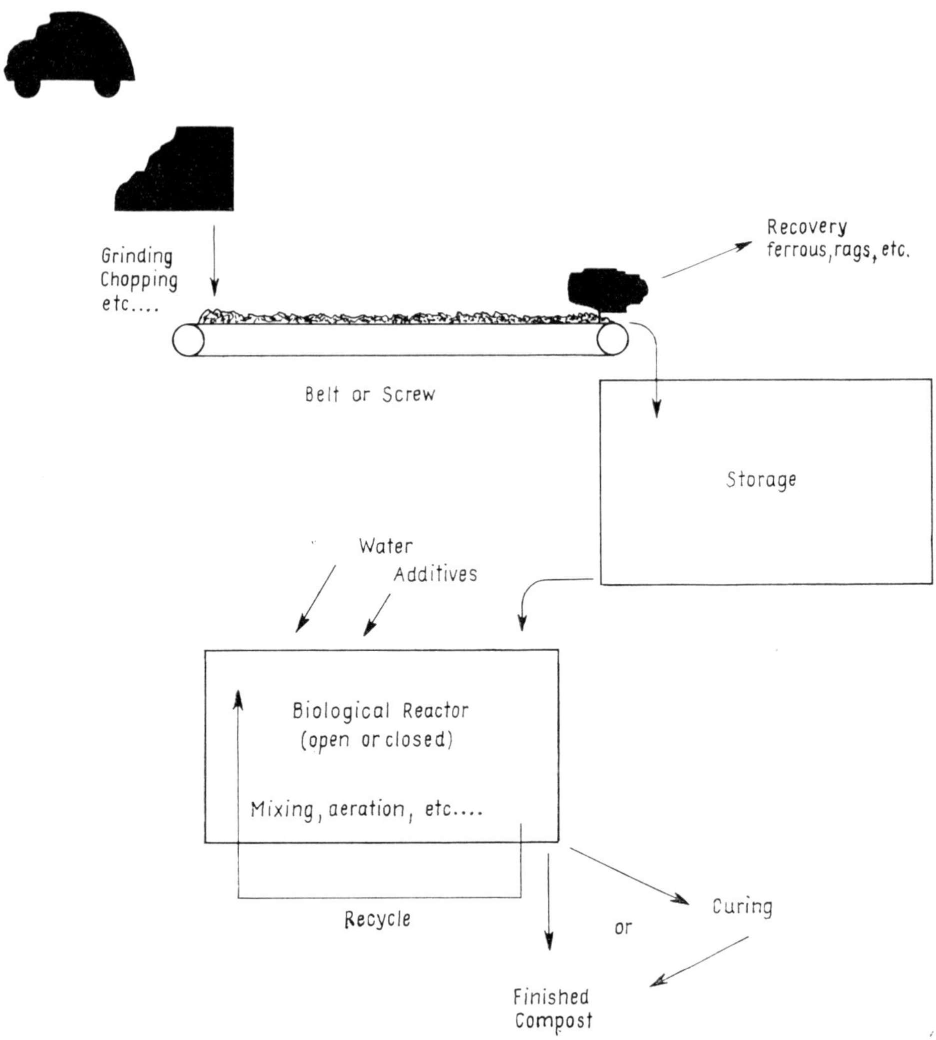

Figure 2 Continuous (or intermittently fed) composting using biological reactor.

maintain continuous feed to the reactor. At least one process (FAM) pulps the material with liquid additives prior to introduction to the biological reactor. In the biological reactor or digestor the environmental conditions are such that composting by microorganisms will take place. The reactor may be open or closed, depending on the control required. Whereas process time in windrow operations require months, a continuously operating composter can achieve stability in days. In the biological reactor the same processes occur as in the batch operation, but at a more accelerated and controlled sequence. The temperature rises, from 130 to 175°F. During this period the microorganisms multiply and the physical character of the material undergoes a change.

Among the more important control parameters which will be discussed in a later section are moisture, aeration, detention time, and seed recycle. After a period in the biological reactor the material, on drying, will become stable and further decomposition will not take place.

A number of variations in the biological reactor or digestor design can be found. Generally, the material can be continuously mixed or operated in a staged, moving sequence. There has been at least one instance in which multi stage vertical silos have been used.[9] Either open pits can be employed as in the Houston and Gainesville plants, or closed digestors can be used with varying amounts of environmental control. Mixing, agitation, or aeration is usually carried out by some mechanical means such as screws, plows, rakes, bucket lifts, etc. The Varro process for example[10] achieves both aeration and forward movement by use of plows which move along horizontal decks, vertical movement being achieved by gravity drop to the lower decks. The Houston and the Gainesville plants both used forced draft aeration. Environmental control can consist of simply the addition of water to maintain appropriate moisture levels in open pits, to the addition of nutrients such as various forms of nitrogen, phosphorus or iron, and the control of temperature in the ambient air in closed reactors. Sewage as well as sewage sludge has also been used combining nutrients and moisture. Accelerated drying may or may not be employed in the process. Some plants, for example, use a "curing" stage, where the material is heaped into windrows for an additional period of time.

Fundamentals of Composting

Raw materials

For a material to compost, organic matter must be present. Municipal refuse will have a variable composition, depending on geographical location, season, and whether the area is heavily urbanized and industrial. Table 1 will illustrate its average composition.[11,12]

TABLE 1

Municipal Refuse Composition
(as received)

Category	Weight, %	
	Oceanside, N.Y. Study*	Philadelphia, Pa. Study†
Paper	33.7–53.3	54.4
Textiles (rags)	2.4–4.0	2.6
Rubber, plastics, leather	2.5–8.8	1.7
Garbage	7.2–16.7	5.0
Wood	1.2–6.6	2.4
Glass and ceramics	7.1–11.9	9.1
Metal	7.2–10.6	8.4
Ash and dirt	—	16.4
Grass, leaves and dirt	0.3–33.3	—

* After Kaiser *et al.* 1968[11]
† After Purdom *et al.* 1966[12]

Although organic matter is present in several of the categories listed, much of the organic component is in fact cellulose—a material that does not biologically degrade at an accelerated rate. Grass leaves and related garden or lawn generated wastes, since they too are composed largely of, cellulose materials (i.e., cellulose, hemi-cellulose and lignin) will also fall into this same category. However, this material as well as some of the paper goods may also contain small amounts of available nutrients such as soluble sugars, amino acids and aliphatic acids as well as fats, oils, waxes and resins which may not be readily available. Thus garbage, as well as the small amount of readily available organic materials present in the paper and garden lawn activities, will contribute the bulk of the material which is decomposed by the microbiological action during the compost process. A better idea of the organic matter—although not necessarily that biologically available—is shown from Table 2.[11] The volatile matter (mostly paper) content is shown to be quite high. A study of the composition of garbage[13] indicates that it too is to some extent paper (about 20%). However, garbage will also contain quantities of breads, cooked table vegetables, fruits, raw vegetables, cooked meats and table residues. These materials as can be seen, are largely waste food materials, cooked and uncooked, and serve as ready and available substrate to microorganisms present in the digestor.

TABLE 2

*Municipal Refuse Composition Proximate Analysis**

Item	Weight, %
Moisture	28.0
Volatile matter	43.4
Fixed carbon	6.6
Glass, ash, metal	22.0
	100.0

* After Kaiser 1968[11]

Biochemical processes

As in aquatic systems, two general type of biochemical decomposition are carried out by microorganisms in the decomposition and degradation of organic materials—aerobic and anaerobic. No attempt will be made to discuss the details of these metabolic transformations. It is important to note, however, that there are major differences in both the efficiency and the end products of these general types of metabolism. Aerobic decomposition will generally lead to the formation of oxidized end products such as carbon dioxide, water, sulfates, nitrates, etc.—compounds considered to be stable and relatively non offensive. Anaerobic processes on the other hand will generally yield only partially oxidized compounds as well as some reduced chemicals which are, in fact, capable of continued biological decomposition, and which can result in serious nuisance conditions. The most obvious end products of anaerobic metabolism and decomposition are the formation of partially oxidized compounds such as small chain organic fatty acids, aldehydes, alcohols and other materials. Perhaps more important is the formation of the more obnoxious sulfides originating from the evolution of hydrogen sulfide gas formed from organic sulfur containing proteins or sulfates. Hydrogen sulfide is a very reactive substance and frequently combines with the partially degraded organic intermediates noted above. The resulting compounds are usually associated with the odors of decomposition. Although no biological process is odor free, aerobic processes in composting will generally have a "sweetish" odor, the odor will not be offensive.

While anaerobic composting has been used as a source of soil additives for centuries in Asia and combinations of aerobic and anaerobic composting have been used, aerobic composting is generally the method of choice in the U.S. and Europe.

Although most organic compounds can eventually be degraded biologically, it is the more readily available foodstuffs that undergo the metabolic changes described above—these materials can generally be classified as carbohydrates (or sugars and starches mostly), proteins (such as in meat scraps) and the fats. While some of the organic materials used in the paper industry are readily decomposable, cellulose is generally not degraded to any significant extent. Decomposition occurs sequentially, generally the carbohydrates disappearing first, followed by proteins and fats. Accompanying these degradation reactions, a pH change, predicable on the basis of the decomposition sequence follows—i.e., at first the pH of raw refuse will drop as carbohydrates are degraded to organic acids and carbon dioxide. As the carbohydrates are exhausted, the microorganisms begin to decompose the protein substrate and pH will rise.

Microorganisms

The microorganisms that carry out composting processes are generally those that are naturally indigeneous to the soil and refuse environment. Composting is generally carried out, however, by the bacteria, yeasts, fungi and actinomycetes. The more rapid changes are probably brought about by the bacteria since they have relatively short reproduction times, can adjust rapidly to environmental changes, and readily assimilate a wide variety or organic substrates. While protozoa, higher fungi as well as a wide variety of multi cellular forms or even virus might be recovered during various stages of the decomposition, it is doubtful that they play a significant role in degradation. During the composting process, both the kinds of microorganisms and their sequential appearance have been noted. Generally during the early phases of composting (carbohydrate decomposition) the bacteria and yeasts will be the dominant population. As the decomposition continues and the pH continues to rise, the predominant population becomes the spore forming fungi and actinomycetes although the bacteria can be seen at all stages. As with soils, the microorganisms are rarely found other than in close association with the solid materials. The solid materials are usually coated with water films or uniformly moist—an environmental condition necessary for the reproduction of microorganisms. During the active early phases of composting the total number of bacteria may reach numbers as high as 10^8–10^9 per gram of compost;[7] coliform bacteria may be approximately $10^5/g$.[14] While microorganisms can usually be conveniently classified as aerobic and anaerobic—most microorganisms active in accelerated composting processes are probably aerobic and facultatively anaerobic. Their numbers, are thus a function of the environmental conditions (macro and micro) (i.e., moisture, pH and the substrate). The majority of the microorganisms are saprophytic, capable of living on dead organic matter.

The specific microorganisms most frequently present include those common to the soil and air environments. Among the bacteria are the aerobic spore forming rods *Bacillus* spp., as well as *Achromobacter*, and numerous small gram negative rods (coliforms, etc.). Among the typical fungi are: *Aspergillus*, *Fusarium*, *Penicillium* and *Rhizopus* as well as others. Among the Actinomycetes are *Streptomyces*, *Nocardia*, *Micromonospora*, and *Streptosporangium*. Among the latter two large categories—Fungi and Actinomycetes are large numbers of spore forming microorganisms. Numerous yeasts are often seen, most often in high sugar wastes—especially during the early stages of attack.

Since composting will proceed from ambient temperatures to higher temperatures, i.e., from mesophilic to thermophilic ranges, the microorganisms are frequently designated by this terminology. Based on optimal growth at different temperature ranges, this would include those microorganisms whose optimum growth is from approximately 20°C to 45°C for mesophilic, and from 45°C to 70°C for thermophilic.

Composting operations have occasionally employed the addition of "seed microorganisms, starter" cultures, or even "additives" or enzymes, in efforts to improve or maintain the process. On the whole, the seed microorganisms employed have been commercial preparations (the Earp-Thomas "starter") or preparations grown on "special" substrates and added in the dry or liquid form to the raw compost or to the operating digester. Other efforts have included the use of sewage sludge or even sewage as a source of microbes as well as nutrient supplements. Since the microbial population that is active in the decomposition of the refuse material is one that is acclimated to the semi-dry, compost environment and which is developed sequentially with the process, it is doubtful that any of these deliberate additions of microorganisms to enhance the process is beneficial.

Seed recycle, on the other hand, since it consists of microbes grown on the same substrate under the same environmental conditions, is often employed and is beneficial. The judicious use of seed recycle will both sustain a continuous process at peak activity and serve to shorten the lag period in batch processes. The amount of recycle used can vary from as much as 20% to as little as 1%.

Nutrients

For the appropriate growth of microorganisms in composting the major nutrients, carbon and nitrogen should be present in the ratios of approximately 30 to 40/l. Although materials ranging from C/N ratios of 20 to 78/l have been composted successfully,[15,16] a recent review of this subject by Jalal[17] notes that the optimal value will vary between 26 and 31/l. The end product—stable compost or humus—may have a C/N ratio of 10/l. Implicit in this relationship is the availability of

TABLE 3

Municipal Refuse Composition Ultimate Analysis (U.S.)*

Category	Weight, %
Moisture	28.0
Carbon	25.0
Hydrogen	3.3
Oxygen	21.1
Nitrogen	0.5
Sulfur	0.1
Glass, ceramics, etc.	9.3
Metals	7.2
Ash, other inerts	5.5
	100.0

* After Kaiser 1968[11]

the organic carbon for the needs of the microorganisms, i.e., providing energy for growth and in the formation of new microbial tissue. Table 3[11] shows the ultimate analysis of a typical eastern refuse in terms of several of the principal elements. It is worth noting that the C/N ratio is about 50/1. With increased urbanization and greater cellulose usage, this ratio will probably continue to rise in the U.S. Phosphorus in the ratio of approximately $\frac{1}{6}$ to $\frac{1}{10}$ of the nitrogen should also be present. Minor nutrients or minerals should also be present in smaller amounts. This would include sulfur, iron and calcium as well as trace quantities of manganese, zinc, copper, cobalt and molybdenum. Normally, refuse will contain all of these minor mineral nutrients in appropriate quantities (Kaiser 1968).[11] Although growth factors, such as amino acids or vitamins are at times listed as nutritional requirements, composting will normally take place in the absence of these materials and such compounds are either present in the raw materials or are not required by the microorganisms (i.e., they are synthetized from the raw materials). For accelerated composting nutritional supplements such as nitrogen or phosphorus are sometimes added, either to minimize the lag period, to accelerate the process or to supply a nutrient that can be used in the end product.

Environmental conditions

Composting will not take place unless the environmental conditions are appropriate for microbial decomposition to take place. This will include the physical as well as the chemical environment.

The most important environmental parameters for the occurrence of composting are the conditions of the material—i.e., particle size, moisture aeration, temperature, and pH. Since refuse as collected is both irregular and coarse, it is not suitable for rapid microbial decomposition. Chopping, grinding or shredding of the raw materials is beneficial in a number of ways. The surface area is increase and exposed so that microbial attack can readily begin. Also the refuse is made into a uniform mass which can be managed for aerating and mixing. Proper moisture is very important in composting. All chemical reactions, of course, must occur in a water substrate. The biochemical reactions that occur in a refuse pulp, if composting is to take place, are no exception. Generally, for batch or windrow operations moisture content can be somewhat lower, but moistures below 30% are seldom compatible with rapid or satisfactory composting. It is also true that materials that are very wet—i.e., above 70% moisture, do not compost satisfactorily. At higher moistures, more rapid bacterial growth takes place and anaerobic conditions are enhanced. Generally, optimal composting in batch operations can be carried out at moisture between 40 to 65%. In continuous compost operations, moisture concentrations of 50 to 60% moisture are usually used. Optimal results are usually obtained at moistures of around 50–50%, however.

Aeration, either by forced draft (mechanically) or by "turning the pile" is usually necessary to maintain sufficient oxygen for aerobic microbiological attack to continue. Since the microorganisms actually grow in the water films or layers, the oxygen must penetrate to the site where microbial growth is occurring. Thus mechanical turning and mixing is usually employed to assure aeration. As noted, at times forced draft ventilation is also used. At least one investigator recommends 10 to 30 cubic feet per day per pound of volatile solids initial charge,[16] whereas oxygen utilization was found to be 0.0244 parts of oxygen per minute per 100 parts of dry weight.[13]

Temperature is a very important and basic parameter. Microorganisms will not grow rapidly at lower temperatures. Thus lower temperatures will result in slower biological decomposition and slower composting. Although temperatures in a batch, or continuous sequential flow (i.e., plug flow) compost process will slowly rise above ambient to as high as 175°F, unless adequate precautions are taken against heat losses, considerable heat can be dissipated. Since a slowly adapting, sequentially developed microbial population is in fact necessary for final stabilization, no efforts are usually made to force heat into the system, other than to replace heat losses. Under certain conditions artificial heating may actually be inhibitory to the process.[16] Thus the system (i.e., the digesting pulp) is permitted to establish its own temperature.

Outside of nutrient availability and temperature, pH is the most important parameter for the growth of microorganisms. Although it is commonly stated that composting will occur with the broad in pH limits of 5 to 9, it is not true that any one pH will assure the occurrence of composting. Most often no attempt is made to control the pH unless the pH remains depressed as a result of low moisture values or some other process upset. In such cases lime or caustic, carefully added, since buffers are characteristic of composting materials, is used to readjust the pH.

Control parameters

Process control in composting operations varies from the simplest methods of turning and moistening the composting pulp as used in batch (windrow) operations, to the application and use of sophisticated instrumentation such as oxygen probes or respirometers.

Among the most basic control parameters are moisture, pH and temperature. By control of the moisture and monitoring of pH and temperature, the course of composting to stabilization can be followed with relative ease. Although no attempt will be made to discuss instrumentation, it should be noted that the measurement of such parameters as pH and even moisture, since we are dealing with a non-homogeneous, semi-dry material of varying solubilities, requires special considerations.

Other more sophisticated parameters which while not always control parameters, can be used to follow or monitor the course of composting. These include the relative amounts of oxygen or carbon dioxide, oxidation-reduction potential, respirometric changes in oxygen uptake, changes in microbial populations, volatile matter, and dew points. In addition, the disappearance of various specialized substrates groups such as sugars, starches, proteins or fats will tell much of the course of the process.

Changes during composting

The changes that occur during composting of municipal refuse are summarized as Fig. 3. Refuse as received, will usually have a slightly acid or neutral pH value. As the microorganisms begin their biochemical degradation, the most available substrates are attacked first (usually the sugars or starches) and the pH value will drop, at times, to pH 5 or slightly below. The pH does not remain low for long however, and will gradually rise as the next level of substrate, the protein or nitrogen containing organic materials, is attacked by the microorganisms. This biochemical decomposition leads to a slow elevation of the pH to a value well above 7 and usually above 7.5. At times the pH can rise to 8. Usually the pH will continue to climb, but generally toward the end of the process fall to below 8.

The temperature of the pulp will typically rise rapidly from ambient to high levels—150 to 170°F—within a few hours in continuous processes—and remain at these levels, until the readily available nutrients are exhausted. The temperature will then slowly return to ambient levels. One of the tests for stability of compost, in fact, consists of removing a sample from the process and noting whether further temperature elevations will occur. If the temperature remains that of the ambient air, the material is considered to be composted and thus "stable."

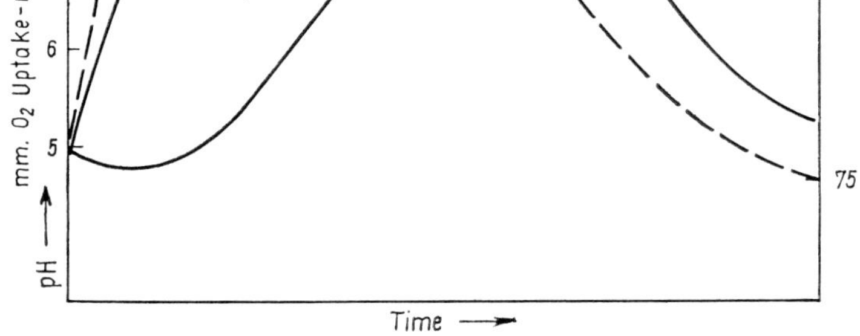

Figure 3 Changes occurring during composting. Note: mm. O_2 uptake is relative.

A parameter that has been used in research operations[18,19] as well as in pilot plant and full scale operations to optimize nutrient supplement additions for accelerated composting is the measurement of oxygen uptake by use of the Warburg respirometer modified for solid wasteapp lications.[19] A typical response is shown in Fig. 3. Typically oxygen uptake will show a more rapid rise. Thus this parameter is usually a more sensitive and rapidly fluctuating indication of the progress of the compost process than either temperature or pH. Other changes which also occur during the compost process and which have been described earlier include the sequential occurence of particular microbial groups as well as the sequential attack of particular substrates.

Public health aspects

Quite often the question arises concerning either the survival or reproduction of pathogenic microorganisms in composting operations. As a rule, pathogenic microorganisms do not generally compete well or survive in enriched environments such as soil or sewage. In addition, pathogens cannot usually withstand elevated temperatures maintained even for short periods of time. A number of literature surveys have been made on the survival of bacteria, parasitic worms, parasitic ova, cysts, and virus[20,21,22] in compost. Most authors agree that none of the pathogens noted as a class can withstand the elevated temperatures reached in a compost mass (between 140 and 160°F) for the typical retention times used in composting operations—usually two days or longer. Among the pathogenic microorganisms surveyed or studied were those typical of the most common and resistant pathogens: *Salmonella typhosa*, *Salmonella newport*, *Vibrio cholera*, *Candida albicans*, *Mycobacterium tuberculosis*, *Brucella abortus*, Endamema histolitica, Trichinella spiralis, Necator americanus, Ascaris lumbricoides, poliovirus type 1 as well as several others. While some of the pathogens—especially the parasitic cysts and eggs are especially resistant to destruction, it is apparent that properly prepared compost of municipal refuse, even when mixed with sewage sludge, does not constitute a public health hazard when the end product is used in agriculture or gardening. Proper preparation, of course, implies that the compost reaches appropriate compost temperatures and that proper mixing will subject all parts of the mass to this heat for the proper period of time.

Other public health aspects that should be noted deal with common nuisance problems associated with refuse handling operations. Flies as well as rats, which are also attracted to the refuse, however, can also serve as disease vectors. Generally, the best way to cope with these problems is to maintain compost operations at optimum conditions and practice good housekeeping.

REFERENCES

1. "Cleaning our Environment—The Chemical Basis," Amer. Chem. Soc. Washington, D. C., 1969.
2. "Solid Waste Mangement," prepared by the Ad Hoc Group, for office of Science and Technology, Exec. Office of the President, Washington, D.C., 1969.
3. Black, R., A. Muhich, A. Klee, A. Hickman, and R. Vaughn, "An Interim Report, 1968 National Survey of Community Solid Waste Practices", Presented at 1968 Annual Meeting of Institute for Solid Wastes, American Public Works Association, Miami Beach, Fla., Oct. 1968, Bureau of Solid Waste Management, U.S Dept. of Health, Education and Welfare, Washington, D. C., 1968.
4. V.A.M. Report, "The Processing of City Refuse into Compost in the Netherlands," Amsterdam, 1953.
5. Van Dommelen, Dirk, "Municipal Composting Seen Increasing in Holland," *Compost Science* **11** (5), 25.
6. Wylie, J. C., Composting. In: *Waste Treatment*, P.C.G. Isaacs Ed., Pergamon Press, London, England, 1960.
7. Gray, K. R., "Accelerated Composting," Brit. Chem. Engr. (1966).
8. Prescott, J. H., „Composting Plant Converts Refuse into Organic Soil Conditioner", *Chem. Eng.* **74** (23), 232 (1967).
9. Anon., "Compost Plant in Jersey," *Contract Rec.* **67**, 20 (1957).
10. MacDuffie, M., and S. Varro, Jr., U.S. Patent No. 3,142,557 (1964).
11. Kaiser, E. R., C. D. Zeit, and J. B. McCaffery, "Municipal Incinerator Refuse and Residue," Proceedings, 1968 National Incinerator Conference, Amer. Soc. Mech. Engrs., New York, N.Y. (1968).
12. Purdom, P. W., R. Shoenberger, A. Michaels, and A. Bergstrom, "Incinerator Residue—A Study of its Characteristics," Annual Meeting Amer. Public Works Assoc. (1966).
13. Michigan State University, "A Study of the Composting of Carbage and other Solid Organic Wastes," Interim Report No. 1, East Lansing, Mich. (1956).
14. U.S. Dept. of Health, Education and Welfare, Public Health Service, Bureau of Solid Waste Management, *Gainesville Compost Plant—An Interim Report* (1969).
15. California State Water Pollution Board, "Reclamation of Municipal Refuse by Composting," Techn. Bull. No. 9 I.E.R., Series 37, Berkeley, Univ. of California Press, 1953.

16. American Public Works Association, "Municipal Refuse Disposal," Amer. Public Works Assoc., Public Administration Service, Chicago, Illinois.
17. Jalal, K. F., "A Technological Evaluation of Composting for Community Waste Disposal in Asia," *Compost Science* **10** (1–2), 20 (1969).
18. Jeris, J., R. Regan, and R. Gasser, "Cellulose Degradation in Composting," Progress Report, USPHS Solid Waste Program, Research Grant U 100508 (Dec. 1968).
19. Cardenas, R. R., and S. Varro, Unpublished Data.
20. U.S. Dept. of Health, Education and Welfare, Public Health Service, Solid Waste/Disease Relationship—a literature survey (1967).
21. Wiley, B. B., and S. Westerberg, "Survival of Human Pathogens in Composted Sewage," *Appl. Microbiol.* **18** (6), 994–1001 (1969).
22. Wiley, J. S., "Pathogene Survival in Composting Municipal Wastes," *Jour. Water Pollution Control Federation* **34** (1), 80–89 (1962).

RAUL R. CARDENAS, JR.
Polytechnic Institute of New York
STEPHEN VARRO, JR.
President, Ecology Inc.

CONSERVATION: see ENERGY SOURCES—ALTERNATIVES

CONSERVATION PLANNING

The experience of the author in the U.S.A. is intended to serve as an introduction to the area of Conservation Planning. By 1969 pollution here was no longer a joke. This fact was made clear by the nature of the author's confirmation hearing for the post of Secretary of the Interior. The subject of environmental pollution was exasperating millions of Americans. Frustration and hostility toward government and industry were growing. The nation was desperately looking for leadership and the author decided that this leadership must be assumed.

The Department of the Interior was created in 1849 to look after the public lands, located mainly in the central and western parts of the United States. As industry began to gain importance at the turn of the century, it turned more and more of its attention to this government department. Before long, Interior was said, with some truth, to be in the pocket of big business. Public demand was for goods and services derived from the natural resources of America and the Interior Department was responsible for managing the bulk of those resources. It is not fair to label the developing Interior-industry relationship as an unholy alliance, but as the twentieth century progressed, private interests called more and more of the shots. It began to appear that those interests owned America.

The author knew differently. The problem faced as Secretary of the Interior was how to restore a just and fair balance in the management of those things owned by 200 million Americans—the public lands, the Continental Shelf, and the properties that are privately owned but that can be abused in ways that infringe upon the rights of others.

The financing of park and recreation land was perhaps the biggest piece of unfinished business found waiting upon assumption of the position of Secretary of the Interior. The American people have been terribly misled through the legislative game of authorizing a measure without appropriating the funds to carry it out. A citizen, hearing the announcement that such-and-such national park has been authorized, believe it has thereby been acquired. Often, only the concept of the park has been authorized. No money has been appropriated to pay for it.

The authorizing of parks without appropriating funds has made many land speculators rich. Once a park is authorized, the value of surrounding lands increases sharply over a period of years, reaching a point where they are ten and sometimes fifty times as expensive as when the park was initially authorized. It is the author's opinion that, when it comes to protecting the American people and protecting America's open spaces, authorization should be tied to appropriation or else long-term real estate contracts should be signed, a down payment made, and the balance paid over a period of ten or twenty years.

A perfect example of how not to help the public was the Point Reyes National Seashore in California. When it was authorized for around $ 14 million the funding was not appropriated. For this reason the federal government, over the years, ended up paying $ 38 million for the property. If appropriation had been tied to authorization, or long term real estate contracts had been used, the total payment, including interest would have been about $ 17 million, thereby saving $ 21 million. Besides, the area could have been used by the public for the past ten years.

Another scandalous example is the Cape Cod National Seashore, where land values have increased nearly a hundredfold since legislation was passed to authorize the park. In the name of conservation, the public is being raped.

The polluters of the past were providing goods and services for the public but were oblivious to the environmental consequences of their actions; one cannot blame them. Even twenty years ago the industrial and municipal effluent being discharged into public waters was, in most cases, a built-in system to break up and clear away pollution. But, today it is a different story. Many of our lakes and rivers have passed the level of tolerance and even the oceans are in danger. Today's polluters cannot avoid the issue of environmental responsibility. The additional cost to protect the environment must now and forever be included in the cost of any item.

Vast assets are being squandered. Revenue potential dwindles away. Our great inheritance, the natural beauty of our land, is fading at a frightening pace, while the government is last in the rat race of coping with one crisis after another. Time and again those things owned by the American public have been exploited due to the pressure of special interests. These interests can be anybody, large or small, an individual conglomerate or an individual rancher.

It was Abraham Lincoln's theory that government should do only those things that private enterprise cannot do or cannot do as well. This deos not mean total involvement by government in private enterprise, but it does mean guidance and regulation. The free enterprise system is the greatest incentive system ever put together on earth, but free enterprise allowed to run totally free will destroy itself.

The thrust of government must change so that actions are positive instead of producing negative reactions. An inventory must be made of all the things owned. The task of putting together such a catalogue is now possible with the technology now available, including computer systems, satellites and instant communication. Once the inventory is complete, judgments can be made about the highest and best use of all these assets. These judgments must be reached with the help of public hearings to provide a balance between the primary needs of a local area and the requirements of the nation as a whole.

Resources must be developed and used; homes and office buildings must be built. But it is inexcusable to take these actions without considering the impact they have on the heart and soul and spirit of the American people. Industry and architects have proved that projects can be accomplished with proper respect for the environment. The question is: Do we care enough about our land to see that they are done that way?

The problem of getting a clean land also involves the argument about how best to cut the timber in our forests. This is of concern to the author because much of the nation's lumber is taken from public property. The emphasis should be on "selective cutting," in which a tree is cut only after it has reached proper maturity. Selectivity does not always mean the same thing: it depends on the forest and the kinds of tree one is talking about. Certain trees, the Douglas fir for instance, are subject to "blowdown" if left unsurrounded by other trees.

This fact is used by the lumber industry to argue in favor of "clearcutting," a procedure by which, with the help of enormous machinery, a crew of men will move into a forest and level everything in a given area. But the real argument for clearcutting is that it is cheaper to cut that way and take a chance on the forest's restoration.

Undoubtedly there are situations in which the clearcutting of trees is justified, but there are many cases in which it represents an abuse of a resource. A much harder look should be taken at selective cutting, even if it does cost a bit more. Does it make any sense to cut a three year old tree, a thirty year old tree and a three hundred year old tree all at the same time just because they are standing next to each other? Is the highest and best use of a tree to make it into a telephone pole? It is interesting that clearcutting is practiced a great deal more by private firms when they use public land than when they use their own private property.

Strip mining of coal is another example of not facing the real cost of retrieving a resource—the total cost. It is possible to strip mine without desecrating the environment. It just costs a little more. What is lacking are regulations to see that such mining is done correctly. The procedure is simple: recover the top soil. When the dig is finished, the top soil is returned to its original contour. This sould be done immediately so that the machines are never more than about 1000 feet in advance of an area to be reclaimed. It is important to cover the mined quickly because when the land is open, it bleeds. Then, erosion silts the streams and acidity in the runoff, resulting from contact with sulfur, pollutes the water as well as the vegetation around the open land.

The point of returning land to its original contour is important. In a thin seam, this entire procedure costs 15 or 20 cents a ton. There is a mandate to "return it like you found it."

How will America be left? Will the heritage this generation leaves behind be an exhausted earth and a human who is degraded? Will the rugged individualism on which we have prided ourselves result in collective destruction? Not if we have a truly national approach to government.

There must be more men with the courage to accept responsibility for planning and implementing policies that will lift the nation into the next century, the century of the human. Government must be visionary in nature and be prepared to "return to us" those things that only national governments can give back, such as clear rivers, unpolluted air and a free spirit in the inner man.

At stake is the living of life and this is not a narrow or regional thing. It cannot be sustained by a government based on the principle that the blending of a conglomeration of special interests will produce a truly national policy. An individual or local government can take care of a problem in an individucal or local way, but there are very few local splashes that do not make waves across the country: local problems tend to become problems of the community called the United States. Decisions on where to put a freeway or how to dispose of municipal sewage have repercussions that are important throughout the land.

There are very few purely private decisions any more. Increasingly, every private decision related to our society must also be considered a public decision, one that cannot be undertaken without regard for its effect upon other individuals. Continued indifference to the public today can only burden more severely the public of tomorrow.

As a mode of transporting people and goods, the iron wheel and a rail still make a combination hard to beat. This was the feeling of the author while Governor of Alaska. In the first State of the State message, delivered in 1967, the Alaska House of Representatives and Senate, sitting in joint session, were asked to fund a commission to explore the future of transportation within the state. In mind, primarily, was expansion of the federally owned Alaska Railroad to points such as Prudhoe Bay, where the oil strike was beginning to come in, and to Nome on the Bering Sea.

Washington displayed it usual lack of interest in Alaska and oil companies that could have used the railroad to carry their produce in and out showed no interest at all. Had the railroad been extended, an enormous area of northern Alaska would have been opened up for its highest and best use and many of the environmental dangers mentioned in the argument about the building of an Alaska pipeline from Prudhoe Bay to Valdez would have been avoided.

Coordination of all overland transport inside "corridors" could make manageable the delivery of goods and people to their destinations. It would also provide a practical way for millions of Americans to see and enjoy the scenic grandeur of this vast area.

Although there are limitations to resources in the United States, there is still a vast potential to be explored and wisely planned for. Population is growing and urban centers especially give the impression that available space is being rapidly exhausted. In reality, there are still great regions of private and public land available both for living and for restoration of man's spirit through recreation. Anyone who crosses the country can readily see that the immediate problem is one of distribution—not only of goods but of people. The greatest lack has not been of space but rather of imagination in caring and planning for the nation's property and people. Is it right to give a cow one hundred acres of public land on which to roam, while a ghetto family is penned up in one hundred square feet?

As in other areas of the environment, the development of a comprehensive land use policy would mean jobs. And these jobs would demand the creative and imaginative talent of everyone involved, because it would be a pioneering venture. The aim would be to put a value, not measured in dollars and cents, on a mountain stream, a prairie, a forest and a wilderness. This would demand people who understand the balance between meeting the needs of the population for minerals, energy and all kinds of resources on the one hand, and for esthetics and recreation on the other.

What can be done about private decisions that are destructive? People can be alerted, educated and inspired. There is no substitute for the motivation provided by millions of people. That power, when people are informed and committed to a great cause, can do what billions of federal dollars cannot begin to do. Further, in those areas where individuals or institutions refuse to listen to the public's demand for a voice of reason, government has got to take a strong stand. For example, in dealing with the environment, a man should be free to turn out a product for sale, but his freedom ends when his factory pollutes the air, contaminates the waterways, desecrates beauty or harms those around him.

In the area of the environment, immediate survival is at stake for certain wild creatures and plant life. Man's ultimate survival is at stake as he realizes that the earth is a closed system. New resources cannot be created by man. They can only be altered. They cannot be destroyed, but they can be wasted and irretrievably lost. Environmental responsibility is a way of life, an attitude toward one's habitat, an ever-present awareness of the interrelationship and interdependence of all living things on earth.

With this sober awakening has come the joyousness of a morning. Values that were forgotten—the poetry of an untouched mountain stream, the philosophy of a lonely wilderness, the majesty of a mountain range or a great ocean's expanse—these simple satisfactions of life, which have been man's companions from the very beginning, are returning the enrich the individual and shift the world back into balance.

This awakening brings an awareness that the individual cannot live without his neighbor. Life styles of consumption and extravagance must change and the individual can no longer remain isolated. His personal decisions affect his neighbor. His neighbor's decisions affect him.

The individual, who has appeared to be becoming more and more impotent in a vast society, is now discovering that he alone is the key to change, to survival, to the improvement of life for all humanity. To accomplish a worldwide change of values, so that our resources are not squandered and our surroundings destroyed, challenges the imagination of all the people of the earth.

REFERENCE

Hickel, Walter J., *Who Owns America?* Prentice-Hall, Inc., Englewood Cliffs, New Jersey, 1971.

WALTER J. HICKEL
Anchorage, Alaska

CONTROL OF AIR QUALITY: see GASEOUS POLLUTANT REMOVAL; PARTICULATE REMOVAL

CONTROL OF WATER QUALITY: see MUNICIPAL WASTEWATER; BIOLOGICAL TREATMENT OF WASTEWATER; LIQUID WASTES OF INDUSTRY; WATER TREATMENT

CORROSION: see EFFECTS OF ATMOSPHERIC POLLUTANTS

D

DDT: see PESTICIDES

DESALINATION

Fresh water, for all of historical time, has been an uncontrollable happenstance of nature wherever and however it was found. The Bible describes "good land" as one "of brooks of water, of fountains, and depths that spring out of valleys and hills." More familiar to its writers were the opposite type of areas and Hosea complained "I did not know Thee in the wilderness, in the land of great drought."

Only fresh water allows human life; and communities developed where there was water. Values of land depend on the presence of water; some dry and worthless areas become very valuable when fresh water becomes available.

With the increase of population and the increase of living standards the demands for water increase exponentially; and wells and other sources run dry. Deepening the wells demonstrates the exhaustion of the geologic reservoirs which are being *mined* and not replenished. Increased population also increasingly pollutes streams, to reduce further available water supplies.

Containers of water have been transported, usually for longer distances than was economical for conduits to carry. Animals have carried goatskin bags, jugs, wooden barrels, or old 5 gallon oil cans. Trucks and railways have carried tanks. One very modern engineering facility not far from Baltimore has a 100,000 gallon tank filled by a truck-tank, over a 12 mile haul. Ships have carried water in ballast tanks sometimes halfway around the world; and tankers otherwise returning light from oil deliveries may make the return voyage with fresh water, which may be as precious to the oil-rich, water-poor country as the oil is to its markets. More recently, tremendous plastic and textile "sausages" have been used for both oil and water transport, towed by relatively small tugs. Towing of icebergs from polar regions has been suggested as a batch system of supply of coastal towns—and if a cheap method of undercoating the ice with an impervious plastic film could be developed, there might be some possibility for slow eating away—or drinking away—of the polar ice caps. The water given up would supply the arid lands of the earth for centuries, if transportation could be arranged. (Actually, icebergs have been towed to tropic lands for their refrigerant effect, before mechanical refrigeration was developed.)

Engineers among the ancients successfully developed dams and aqueducts to bring flowing fresh water considerable distances to growing cities, and to irrigate agricultural lands by water flowing through canals and flumes. As the skill and capabilities of engineers developed under the pressure of the increasing needs for water, conduits have become longer and longer. Usually closed pipes are now used with tunnels, inverted syphons, and pumps. New York City has studied pipelines to bring water from the Great Lakes; and Tokyo has planned many hundreds of miles of pipelines back to a supply in the mountains.

Water may be transported in large quantities for from \$0.05 to \$0.15 per 1,000 U.S. gallons per 100 miles. A system costing billions of dollars will carry very large amounts in California at about \$0.50 per 1,000 gallons for many hundreds of miles, over mountains 4,000 feet high. This may be the largest and most costly project ever to carry fresh water to the sea coast, but it adds nothing to California's supply, and steals the water from arid communities en route. Further requirements, by desalination, have been estimated in a $\frac{1}{2}$ million dollar engineering study to cost half as much, or 22 cents to 30 cents; and this notes that fresh water can often be *produced* from the sea more cheaply when all costs on the tremendous investment of impounding and transport are considered. Also desalination becomes cheaper with constantly improving technology, while that of dams and pipes can hardly be improved.

Three times as much water is used in continental U.S. per person as in 1900. In some southern cities, usage is 10 to 20 times, as much per person. The great population increase multiplies the total withdrawal. In some areas twice as much water is being drawn out of the ground as soaked into it. The water table in well populated desert areas drops almost 10 feet each year. Water is far from evenly divided, with major shortages in some very populous areas; *always* the southwest and periodically the northeast. On average, the United States will balance water supply and increasing demand in about 15 years. This is if all could be divided *equally*. Some areas will be desperately short much sooner.

Water shortages are acute in dry years, as in 1965, and in 1967 when 1,000 communities in 47 of the 48 contiguous states restricted the use of water. In Texas, with its thousands of oil wells, people lined up to buy water—at twice the price of gasoline!

Many towns normally pump treated sewage back into the ground, and wells bring the water up for reuse. In dry 1957, at least one town survived through complete re-use of its sewage. Cities near the mouth of a major river, with scores of cities, above, use water which has been through sewers upstream a dozen or more times. Two out of five cities have inadequate water supplies, even though city dwellers take much more than their share. A quarter of our population faces serious shortages.

Yet, half of our states, two-thirds of our industry, and over half of our population, border on the limitless ocean. In many places inland, there are large quantities of water too brackish to drink.

There is a tremendous requirement for chemical engineers to provide processes and equipment to separate water from one-thirtieth as much salt in sea water, or one-thousandth as much in brackish waters. Within 20 years, over one hundred billion dollars will be spent in the United States alone for plants for desalinating water, according to a recent estimate by the then head of the Office of Saline Water. This, at an average rate of \$5 billion a year will be much greater than will be spent for all equipment for the chemical industries. Many more billions will go for plants for recovering water from sewage for immediate use or discharge to streams.

The greatest, although possibly not the most dramatic, engineering achievement of this next generation will probably be the development of processes and the building of plants for producing pure *water*. Chemically, this is the most unusual compound. It is also the most important: chemically, physically, biologically, industrially, and politically. While it is the most abundant material in our world, which should indeed be called *Water* and not *Earth*, its abundant and cheap supply will be more important to the world than atomic energy. Its cheap supply is certainly needed more in the United States than a man on the moon! The late President Kennedy said that the nation developing a successful process "would get a great deal more lasting benefit than those countries that may be even first in space."

Some of the most attractive areas and beaches of the world are almost devoid of water. Not only is this living space—and space for resort hotels—lost, but, in some cases, profitable resources cannot be exploited. Thus, known minerals on Egypt's Red Sea coast cannot be mined, and fishing industries on South America's Pacific coast cannot be expanded for lack of water. These represent major losses in the world supplies of minerals and foods; there are many other possible examples.

Water usage

An average human must consume the equivalent of possibly a half gallon of water per day. However, at least several hundred gallons per day is required for producing the absolute minimal daily food ration for a vegetarian; and some thousands of gallons are required to produce one pound of protein via beef and its cereal raw materials.

Accounting for the great losses of agriculture as practiced, using much greater than minimum amounts, also the use of streams of water for waste disposal, the uses of water for industry, etc., the per capita daily usage may vary between hundreds and thousands of gallons per day per person. The higher values are those under highly developed and more complex patterns of communal life in countries with higher standards of living.

Obviously, the individual sees very little of this water, so essential to his existence; and in many places of the world which are short of water—and underdeveloped therefore—he may be rationed only a few gallons per day. In other highly industrialized, and prosperous municipalities, the withdrawal per person, including wastage, may be in the hundreds of gallons per day.

Agriculture

Over a ton of water makes the wheat grow for a loaf of bread; which contains little, if any, of this water. Vegetables require from 1,000 to 2,000 times as much water as their final water content. Irrigation loses much water:—directly, by seepage, by evaporation from open channels and the soil, and otherwise. Only a small fraction of water available is absorbed by the plants. Plastic ground covers, through which the plants protrude, minimize evaporation as well as weeds at great expense for the materials. Other and cheaper systems are necessary to deliver water into the plant system.

In several hot dry places on the worlds seacoasts many tons of vegetables are profitably grown each month by hydroponics using man made fresh water from the sea. Much water is used per unit weight of plant; but the largest losses of ordinary irrigation are saved.

Economics of desalination

Only the most specialized forms of agriculture can afford desalinated sea water under any probable costs, since irrigation water cannot now cost more than 2 to 15 cents per 1,000 gallons. Industrial water usually cannot cost more than 15 to 25 cents, but some industries may afford much higher prices.

Water to quench thirst has no upper limit of value, but generally fresh water for municipal use should not cost—excluding distribution—more than $ 0.50 to $ 1.00 per 1,000 gallons; it is less in most major cities of the world. Desalination processes will supply water possibly as low as 35 cents to 40 cents within the next five years, and present processes approach these costs now with plants of 5 to 10 million gallons per day. Thousands of pages have been written analyzing these economics under many possible conditions.

Plant costs also are going down—particularly for the larger sizes. Evaporation plants may cost $ 500 to $ 1,000 per 1,000 gallons daily capacity, Very large units, will have lower unit costs.

Materials from the Sea

The oceans are the greatest source of material available to man—about one third of a billion cubic miles. On average, there is about 35,000 parts per million of dissolved solids—more in the Red Sea, less in the Black and Baltic Seas. These dry solids represent millions of cubic miles. The four metals—sodium, magnesium, calcium and potassium—and their combining ions, chlorides, sulfates, bicarbonates, and bromides are present in amounts beyond comprehension. Sea water, as Leonardo da Vinci, the artist-engineer, said "comes from springs which leach minerals from their deposits and carry these dissolved solids to the oceans from whence they are never lifted by the clouds which produce the rivers. So the sea now would be more salty in our times than it has ever been any time previously"—and it contains a higher concentration of magnesium, copper, gold, uranium, and every other metal, than in Leonardo's day, 500 years ago. For most of these the amounts are too small to be recovered economically, as yet. As mines on land are exhausted, and the skill of the chemical engineer is sharpened in separating processes, the future may find concentrating plants for metals as by-products of fresh water production. There may be 10 tons of uranium and 70 pounds of gold for each cubic mile of sea water. These amounts of very valuable metals, multiplied by the almost infinite volume of the ocean, represent tremendous values of these materials—all safely locked away, so far, by their dilution.

Soon, world-wide about 1 billion gallons per day of fresh water from sea water will be produced—in less than a generation, this number will be 15 to 20 billion. Usually twice as much sea water is processed, so that the solids concentration of the effluent brine is doubled. Chemical recovery will be cheapened not so much by a doubled concentration as by the pumping, purification, and the presence of the sea water in a processing plant.

Materials for Desalination Plants

To produce soon this billion (10^9) gallons of fresh water at about $ 1 per 1,000 gallons, there will be spent daily about $ 1 million. These expenditures will be principally for energy, and for the capital charges, i.e., interest, depreciation, amortization, for the equipment and the materials therein. Relatively smaller items of cost will be for operating labor and for expendable materials, principally chemicals used in the processing.

The design of processes will have, as a first object, the minimization of the amounts of thermal or mechanical energy and of the amount of equipment; hence the amounts of materials used. Quite often, in such optimization studies, as energy goes down, equipment goes up, and vice versa. Thus, the most economic balance must be struck. Often this is at a point where the energy cost and the capital cost of the equipment are about equal.

Of the $ 5 billion per year to be spent for plants, materials for the equipment might be regarded as at least 50% with costs of engineering, fabrication, transportation, installation, etc. accounting for the rest. Of materials used in equipment, by far the largest amount will be for metals, particularly those metals which are least corroded by sea water. These are not the most abundant or least expensive. The specification of these metals for equipment—those which are suitable for withstanding the corrosion and other deleterious effects of this service—are of greatest importance. And their fabrication into sheets, tubes, shapes, and then finished vessels and parts and accessories will be a great test of the skill of the metallurgist, the chemical engineer, and the mechanical engineer.

Many other materials besides metals are used in equipment:—concrete, not only for foundations, but in pre-stressed shapes, also in larger and more sophisticated major vessels and shells of processing units, tanks, pipelines, etc. This will reduce the vast weight required of metals, which indeed, as a major group of man's resources, become more precious year by year, as does water itself. The next major class of materials will be plastics and high polymers in other forms:—pipes, sheets or membranes, ion exchange resins, gaskets, coatings and linings, etc.

Processing materials which are used and expendable include various liquids for solvents, heat transfer agents, refrigerants, and in other thermodynamic uses. These are all used in cycles: and in some cases, they may have only such a small loss that their original cost may almost be regarded as a fixed part of the plant itself, and their make-up as depreciation. Chemicals which are strictly expendable are those used for treating water, such as sulfuric acid and lime, algeaecides, phosphates, and other scale-prevention materials, higher alcohols to stop evaporation in storage basins, and minor amounts of many others.

These materials will go into major plants weighing many thousands of tons and processing millions of gallons of water per day, with individual units having hundreds or thousands of miles of tubes, or thousands of square feet, actually acres, of sheet plastics. However, in our modern "do-it-yourself" world, many desalination plants will come into the home, the farm, the small apartment house, or resort hotel, as major appliances, and will be priced in the range of automobiles, without their complexity, but hopefully with their dependability. As time goes on, we will have become educated in their use within the next generation or two, as we have become educated to the use and dependence upon automobiles in the last several generations.

Processes

The ingenuity and imagination of chemists and engineers have been stimulated by the problem of production of this most common and most useful and yet most unusual material,—water, by separation from one of the next most common materials—salt. The chemist realizes that if it were possible to cancel for an instant ionization effects, salt would be quite insoluble and would precipitate out of solution. And the engineer realizes that if he were able to run one of the reversible processes he likes to consider, he could separate the water for less than one-fiftieth of the thermal cost of its evaporation.

In between these idealized and quite impossible processes, hundreds of processes have been proposed based on the various properties of water and its saline solutions. Complete books, and hundreds of articles and patents each year, describe these processes on which millions of research dollars are being spent annually. The most active agency in this search and engineering development and in publications of reports of progress is the Office of Saline Water of the U.S. Department of the Interior.

The required separation may be of *water from salt*, or of *salt from water*. However, the separation of fresh water from brine is much less expensive of energy than from solid salt. Thus, sea water gives fresh water and a brine usually twice its concentration or about 7%, which is returned to the ocean. While this inexhaustible supply would allow the separation more cheaply (thermodynamically) of fresh water from a very large amount of only slightly concentrated brine, say 4%, the pumping, deaerating, and possible chemical treatment becomes correspondingly more expensive as more sea water is handled. The most economic concentration ratio is usually about 2 to 1 for sea water, about 3–6 to 1 for brackish water—and here, the disposal of the concentrated brine is often a major problem.

Evaporation

Aristotle wrote: "Salt water, when it turns into vapor, becomes sweet; and the vapor does not form salt water again when it condenses." Sailors have used simple evaporation apparatus to make drinking water for almost 400 years; and ocean going ships have traditionally used evaporators, often multiple-effect, as an accessory to steam boilers.

Solar energy for evaporation was first used on a major scale about 100 years ago in Chile, where a glass-roofed unit had over 50,000 square feet to make 6000 GPD in a mining camp. Today, many units, the largest of 93,000 square feet (24 acres) have been built by the Doctors Delyannis in Greece using black butyl-rubber sheets to make long, narrow, trough-like solar evaporating pans, which cover the former central square of a small island city. These pans are filled with sea water evaporating at a temperature much lower than the boiling point, and are covered with transparent polyvinyl fluoride sheets, through which the solar radiation passes, and on which condenses the vapors.

Daily production of each 1,000 gallons by solar evaporation may require 5,000 to 10,000 or more square feet. This requires much material—clear plastic sheets or glass for the cover-condenser, concrete or black plastic for the pans, troughs for condensate-water produced, also much piping—metal or plastic—for sea water distribution. Total costs of materials and plants are much higher than for any other type of water desalination system; and the large condensing areas are vulnerable to storms. However, energy is free—except for pumping; and operation is simple.

The usual still or single-effect evaporator uses much thermal energy; and so multiple-effects are used. Several units or effects operate at different pressures, with the vapors from one evaporation passing to condense and thus evaporate about the same amount in each successive effect. This system has been used for many years in the sugar and other process industries with from 4 to 6 effects being the usual maximum. This number has been increased to 12 to give a net evaporation—or "gain ratio"—or about 10 pounds of water per pound of boiler steam used. A great mass of metal is required, since its amount goes up almost proportionally to the gain ratio.

In the Vapor Compression Evaporator, the vapors from boiling sea water inside the tubes at, say, atmospheric pressure, are compressed to a

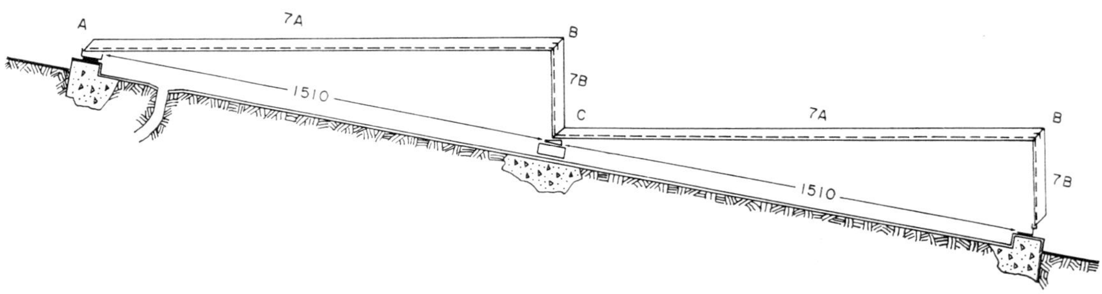

Figure 1 Solar still cross section. Delyannis type—dimensions in mm. Floor is black-rubber or other plastic to absorb energy from sun. Glass sheets as roof with 12° angle condense water on underside which drains to troughs on either end and in middle, thence to collection tanks. Courtesy Drs. A. and E. Delyannis.

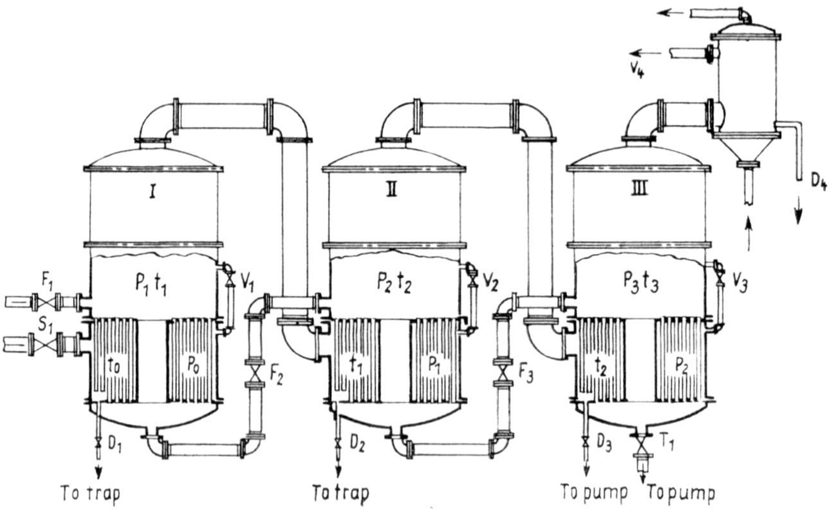

Figure 2 Mutliple effect. This simple diagram is of a vertical tube triple effect evaporator such as has been used for over a hundred years in the salt, sugar, and other process industries, and in early desalination plants. Modern units have as many as 12 effects for fresh water production, with long evaporator tubes and a total of 30 or more heat exchangers for (a) preheating feed, (b) evaporation, (c) and recovery of heat from condensate streams. Here the flows of both steam and sea water—through F_1, I, F_2, II, F_3, III and discharge—goes from left to right, from higher pressure to successively lower pressure or vacuum. Steam at pressure P_0 passes through S_1 to boil sea water flowing into I at pressure P_1. This gives vapors, passing out the top, large pipe to heating tubes of II. These vapors are condensed to boil brine coming from I through F_2 to II; and the vapors formed pass to the heating tubes of III, fed with brine through F_3 which on evaporating gives vapors passing for condensation to the tubular condenser on far right. Heat is removed here by cold sea water pumped into the bottoms of the condenser tubes and out of the tops. Steam condensing in the several bundles of the evaporator tubes is vented of noncondensible gases to the vapor space immediately above by vents V_1, V_2, and V_3. The noncondensibles are finally vented from the condenser through V_4 and removed by a vacuum pump, because of the low pressure. Boiling in the tubes is aided by the vertical movement of the liquid caused by the rising steam bubbles; and liquid from all of the tubes then recirculates down through the central well and back to the tube bottoms. Condensate from the drain line D_1 is usually returned as boiler feed, that from D_2, D_3 and D_4 is combined as the product fresh water. The gain ratio of this simple unit may be about 2.3 pounds of fresh water produced (besides that going back to the boiler) from D_1 depending on the temperature of the feed at F_1. By using 12 effects with preheaters for the feed, using steam from each effect, and heat recovery exchangers for the condensate, a gain ratio of 10 has been obtained.

higher pressure to give a sufficiently high temperature to allow them to condense on the outside of the tubes—to cause continuing boiling. Mechanical or electrical power drives the compressors. This system is used where electric power is available, or where power is generated by boiler steam which, after going through the turbine, is utilized ultimately for another evaporation process. Also, this system is useful in places where only small amounts of brackish water is available, since no condenser cooling is necessary. Newly developed additions to the process flow sheet may make possible the production of 1,000 U.S. gallons of fresh water with only 24 kWh of power. A part of this power is always used for pumping the boiling liquid through the heating tubes at a velocity which increases the rate of theat transfer.

Obviously the area of heat transfer surface, hence the amount of metal required, is reduced by increasing the coefficient of heat transfer, *U*, from the steam on one side through the surface to the boiling liquid on the other side. *U* is defined as BTU per hour per square foot per °F. temperature difference. From values of *U* below 1,000 for standard tubes, high values up to 10,000 have been reached on rotating spheroidal surfaces and 2,500 or more on special expanded surfaces.

A stream of pre-heated water may be passed through chambers at successively lower pressures in each of which it flash evaporates as it cools. The vapors from each "flash" are used to preheat the raw sea water passing through a tremendous system of tubes forming condenser—heaters in countercurrent. This is called Multi-Stage Flash

Figure 3 Vapor compression evaporator. One million gallons per day of fresh water from brackish water. Two effects, each with forced circulation of brine through vertical tubes. Large, closer black sphere is first effect vapor space, white cylinder beneath is steam heated bundle of long vertical tubes with pump below circulating brine downwardly through large pipe on left and upwardly to boil in tubes. Vapors pass from vapor space through pipe on right at angle down to heat the tube bundle of the second effect. Vapors from second effect pass down in pipe on right to vapor compressor beneath, then through dark angular pipe in center of picture to heat the tube bundle of the second effect. Vapors from second effect pass down in pipe on right to vapor compressor beneath, then through dark angular pipe in center of picture to heat the tube bundle of the first effect. Heat exchangers, and pipes for salt water in, brine out, and product water out not readily identified. Courtesy of Office of Saline Water.

or MSF Evaporation. The sea water from the high temperature condenser-heater is heated to the highest temperature by boiler steam in the "brine heater" before the first or highest pressure flashing. About 97% of present sea water desalination uses the MSF system; and plants up to 100 million gallons per day have been designed. With 40 stages, about 10 pounds of evaporation may be secured per pound of boiler steam used, a gain ratio of 10. By combining two MSF units a double effect is possible which increases this gain ratio at double the equipment cost.

By allowing the brine heater to evaporate a small amount of the feed, the vapors from which pass to another condenser—evaporator, a so-called Half Stage, the gain ratio is increased by 5% to 25%, without increasing equipment costs.

Each of the many stages of MSF should have *all* cooler brine and *all* vapor discharged in equilibrium without entrainment of brine. However,

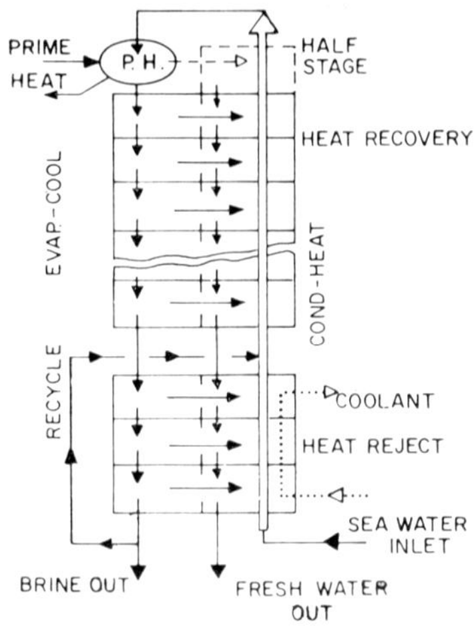

Figure 4 Conventional multistage flash evaporation—MSF evaporating and cooling of hot feed brine (vertical arrows down) on left side at successively lower pressures after heating to highest temperature in brine heater at top; vapors (horizontal arrows) from MSF, passing to preheat the sea water by condensation-heating tubes on right side; fresh water condensate passing stagewise from top to discharge at bottom; additional sea water coolant (dotted line) rejecting heat in lower stages; withdrawing of vapors from prime heater to be condensed in half-stage (dashed lines) increasing the production of fresh water.

a 2.5 million gpd fresh water plant handles a torrent of brine, mainly recycle, of 25 million gpd or 17,500 gpm, and it is impossible to obtain equilibrium in the conventional flash chamber. The driving force toward equilibrium in the few seconds in a flash chamber several feet deep may be 2 to 5°F; in lower stages a pressure drop of only 2 to 3 inches of water; and no vapor bubbles can form at any greater depth. Violent agitation does not suffice. Vapor bubbles almost explode, to carry entrained brine to the product. Demisters reduce this but give additional temperature losses up to 0.2°F per stage. Low flash temperatures per stage reduce flash violence and entrainment; but this increases number of stages to 40 or more also increases equipment costs and the inefficiencies which decrease capacity and increase heat costs. Temperature differences between brine and vapor streams leaving stages may be from $\frac{3}{4}$°F to 3°F, besides boiling point elevation.

Controlled Flash Evaporation (CFE)

More and larger MSF plants magnify inefficiencies. A surface or interfacial evaporation at

Material Balance for Fig. 5. 250°F Maximum Brine Temperature

Location	Flow, lb/hr × 10^3	Flow, gpm	Temp., °F	Dissolved solids, ppm	Specific gravity
1 seawater supply	7880	15,250	85	35,000	1.022
2 coolant seawater to evaporator	6293	12,300	85	35,000	1.022
3 seawater reject	6293	12,400	97.73	35,000	1.019
4 decarbonated make-up	2044.119	4,000	97.73	35,000	1.02
5 deaerated make-up	2044.119	4,000	97.73	35,000	1.02
6 recycle brine	6409	12,400	97.86	52,500	1.031
7 blowdown	1178.719	2,520	97.86	60,697	1.041
8 product water to storage	865.4	1,730	93.96	25 Max	1.0
9 brine heater inlet	6409	13,050	237.41	52,500	0.982
10 brine heater outlet	6409	13,000	250	52,500	0.980
11 brine heater steam	83.082		264.9		
12 brine heater condensate	83.082	177	262	2 Max	0.935
13 ejector steam	2000				1.84
14 acid, H_2SO_4	0.240				
15 antifoam	0–3 ppm				
16 caustic	0–3 ppm				
17 noncondensibles, vent to atm	0.225		210		
18 decarbonated vent CO_2 to atm	0.20		95		
19 seawater coolant to vent condenser	1100	2,150	85	35,000	1.022
20 seawater coolant from vent condenser	1100	2,160	92	35,000	1.021
21 seawater to cool steam jet air ejector (SJAE)	500	980	85	35,000	1.022
22 seawater from SJAE	500	985	95	35,000	1.02
23 seawater discharge	7880	15,300	97	35,000	1.02

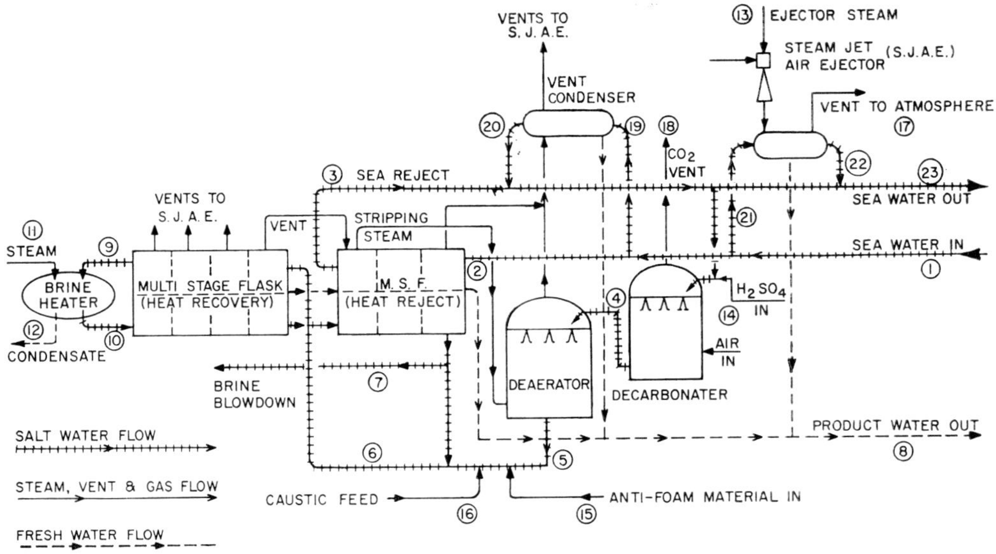

Figure 5 *Sea water in* supplies cooling to vent condensers and the heating tubes of the heat reject section of MSF evaporator; some of it, warmed now, is acid treated to eliminate CO_2 and is then deaerated. It joins the recycle brine to pass to heat recovery section of MSF. Here it is further preheated, then passed to brine heater where prime steam heats it to highest temperature in the system. From here, it flash evaporates through heat recovery and then heat reject stages; a major part is recycled; and some is blown down to waste. Product water, the condensate from the MSF (both heat recovery and heat reject) passes successively from those of high temperatures and pressures to those of lowest. In each, part evaporates, to give steam which condenses on tubes to help preheat the sea water in. The cooled condensate discharges as product water out to storage tanks, not shown. (Courtesy of Burns & Roe, Inc.)

near equilibrium is desired; and Othmer showed in 1927, its temperature drop—liquid to vapor—is at most 0.01 to 0.05 °F. Controlled Flash Evaporation (CFE) is a surface evaporation, and eliminates MSF losses due to turbulence. Many parallel chutes have long top, narrow, rectangular orifices between spacers through which heated brine from an upper stage flows downwardly as pressure decreases to go below the saturation. Cavitation quietly forms vapor practically in equilibrium, and within 0.01°F of the temperature of films of brine about 0.05 inches thick descending on either side. No bubbles form to burst and give mist as in the usual "flash." The chute cross section widens as the volume of vapors increases as they pass out laterally to the condensing tubes with practically no entrainment. Brine enters the next system of chutes.

One design uses a 24″ high chute with 3″ high expander section, and an 18″ skirt. Overall flash in one case was 14.5°F, sometimes as much as 20 °F but always much more than the conventional 3 to 5°F. The salinity of the almost pure condensate was 14.0 ppm without demisters, from a feed of sea water having 35,000 ppm solids. Higher chutes flash over 30°F per stage to make very pure water and require many fewer stages, therefore, lower capital cost.

Vapors discharge from C.F.E. stages between the skirts to condenser tubes at right angles. Parallel chutes 8 feet long are arranged in sets to discharge to an 8 foot length of condenser tubes. Six sets in line are shop fabricated with 48 foot condenser tubes adjacent both ends of the chutes.

Conventional MSF gives transport losses (non-approach to equilibrium values) from 0.05°F, for a stage at 250°F, to 2.5°F for a stage at 100°F, whereas CFE losses range from 0.1°F at 250°F to 0.5°F at 100°F—one-fifth as much. A CFE stage may have twice the flash range; thus its transport loss per °F, or overall for the entire temperature, may be only one-tenth as much.

Other CFE designs evaporate from thin films Short columns packed with shapes as in distillation columns operate with flashes up to 8°F per stage.

Plant operation

One stage with a 10°F flash range, part of a CFE plant desalination sea water without demisters, produced under test 15,000 gpd of water with only 1 to 2 ppm of salts, under the worst possible operating conditions:—pressure, below 1 psia;

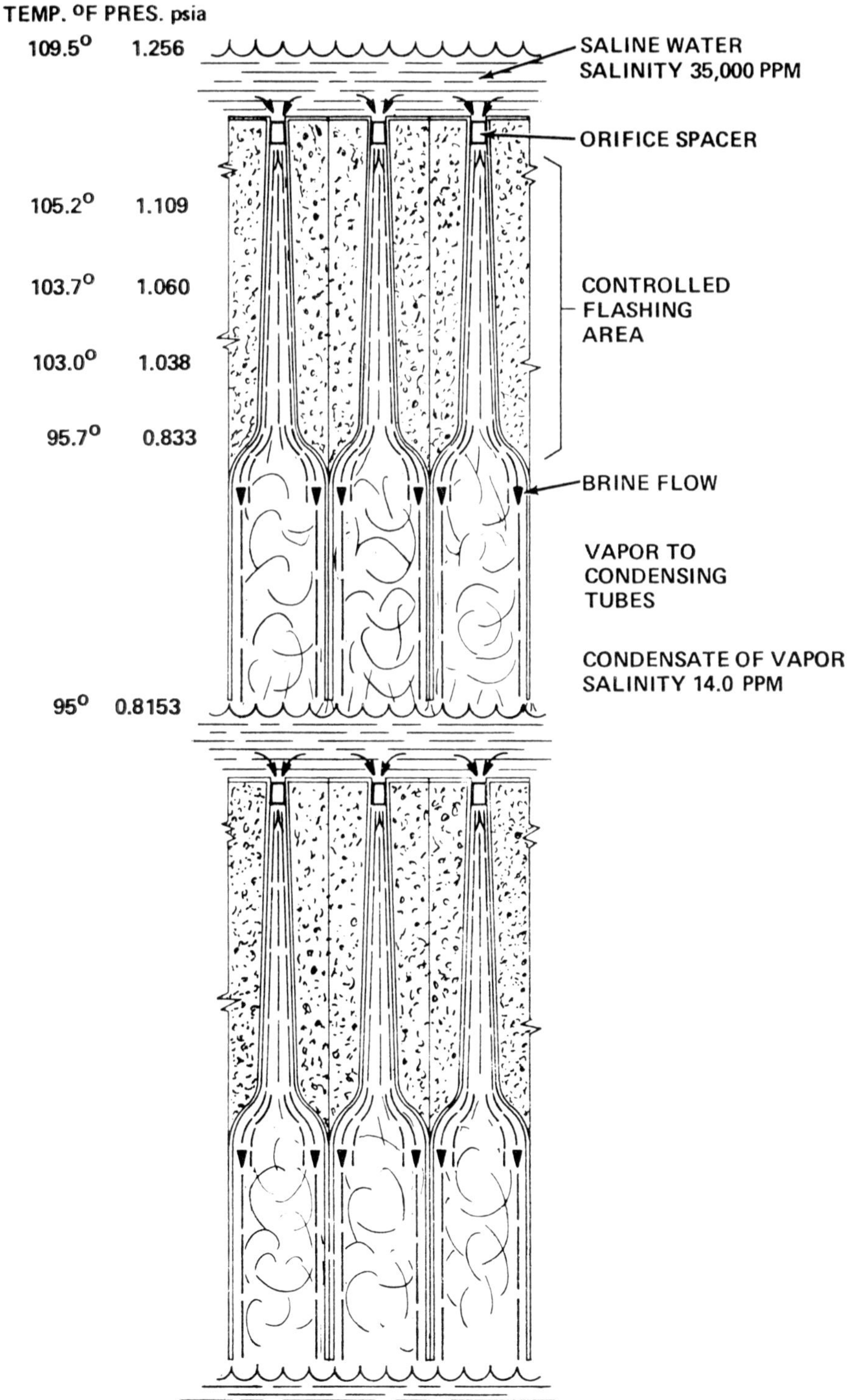

Figure 6 Controlled flash evaporation (CFE). Cross-section through flashing chutes. Sea water at top passes through orifices between orific spacer blocks. As pressure lowers, water evaporates and cavitation forms a core stream of vapor, with films of liquid descending on each side wall of chute. Vapor separates free of sea water in lower expanded section, and passes frontwards and backwards to condenser tubes. The evaporating conditions are shown at left at respective distances of passage of liquid. (Courtesy of Burns & Roe, Inc.)

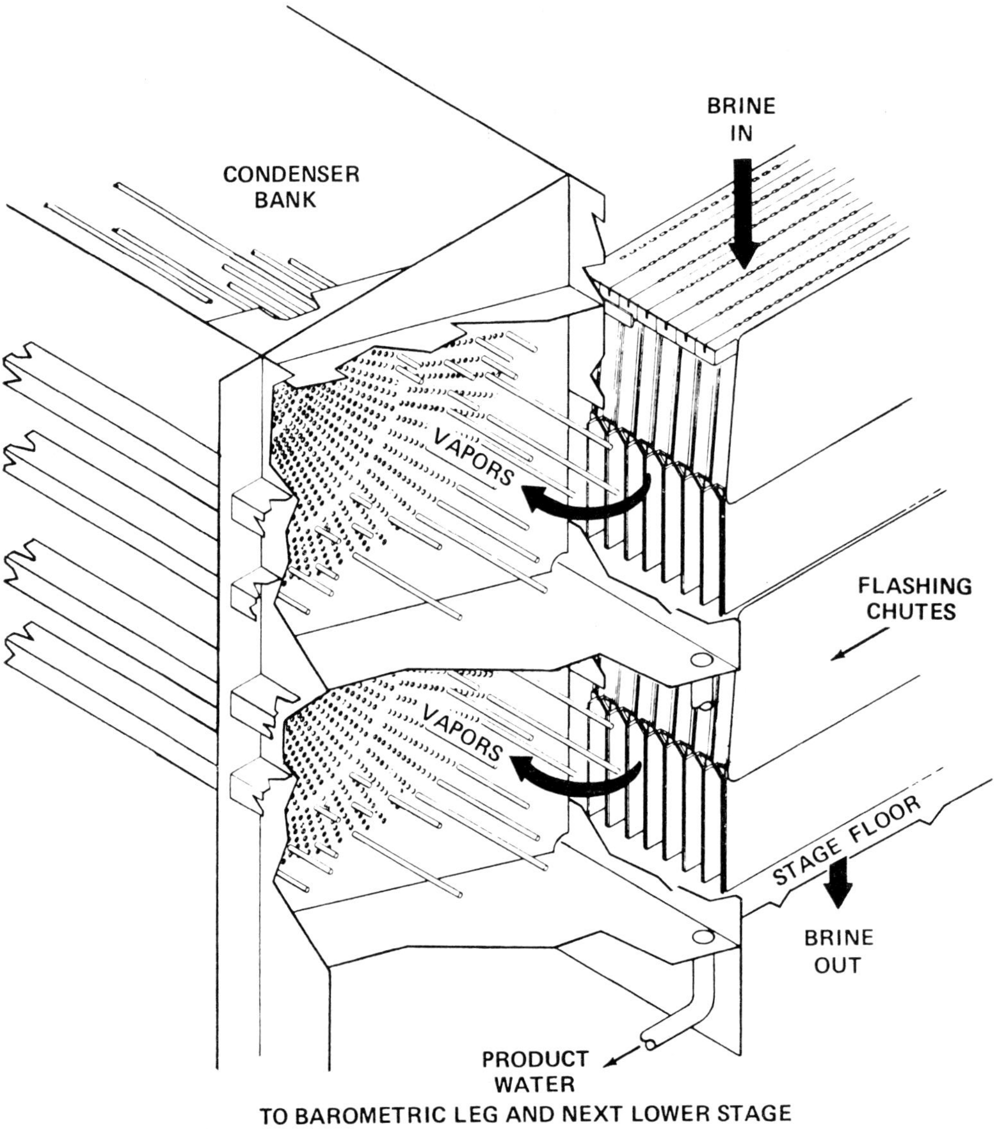

Figure 7 CFE—Flashing chutes and condensers of 2 stages. Brine from stage above descends through rectangular orifices on right formed by spacers at top of 8-foot long chutes; vapors disengage and pass, first in line with chutes, and then between the skirts to condenser tubes on left, also to right, not shown. Product water condensate goes to water side of next lower stage. Brine continues stage to stage, discharges at bottom. (Courtesy of Burns & Roe, Inc.)

temperature below 100°F; and over 350 cubic feet of vapor disengaging from the brine per pound of fresh water.

Many superimposed CFE stages have downward flow of brine evaporating from the top, hotter, to the bottom, cooler, stages. Meanwhile, brine passing through condenser tubes on each stage is heated in climbing to the top stage, where it passes to the brine heater, then back to the top stage to flash evaporate. Condensate from each stage flows to the next lower condenser through a liquid seal, to be cooled as it also flash evaporates in each successively lower pressure stage to give more vapors and more condensate.

CFE with is high thermal efficiency may give an optimum number of stages of only 13, even with high flash temperatures. Conventional MSF requires many more stages. A modular unit has 6 sets of 2 foot high flashing chutes and condenser tubes, and is about 20 feet wide, 48 feet long (tube length) and 58 feet high. Such shop-assembled components welded together in the field will produce 3 million

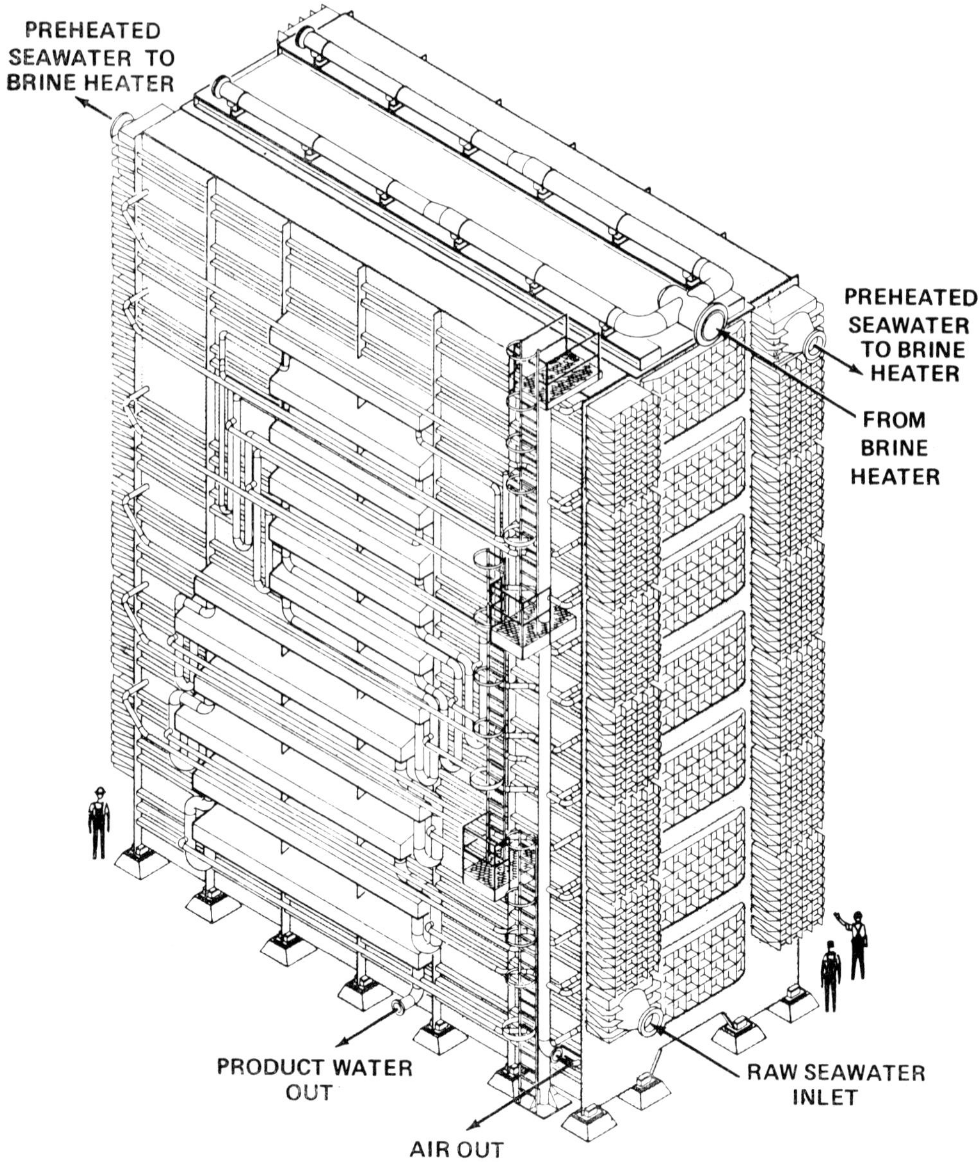

Figure 8 CFE—13-stage plant modulus, capacity 3 million GPD. Condensing-preheating tubes on each stage, about 48 feet long on either side of 6 sets of 8-foot long flashing chutes, each set about 8 feet wide; total width about 20 feet, total height about 58 feet. Cold sea water enters bottom both at nearest corner and at farthest corner, climbs back and forth through 13 superimposed tube bundles to top; the two streams join to go to the brine heater, thence back to top distribution header to inlets, one on each side of top set of flashing chutes. (Courtesy of Burns & Roe, Inc.)

gpd, water, using 80 °F sea water heated to 250°F, then cooled to 90 °F, averaging a flash of 12.4°F per stage.

High purity boiler water by waste heat from condenser water

Much thermal energy from fuel, more if nuclear, is rejected in exhaust steam to cooling water, fresh or salt, passing turbine condensers and heated 10 to 25°F. From this nuisance, a *thermal pollution*, about 40% of the weight of the boiler steam generated, is recoverable as fresh water by CFE. 100,000 gpd units are shop-assembled; larger ones are field-assembled. With no heat cost, the CFE cools the hot condenser water discharged and minimizes thermal pollution.

Deaeration of sea water before CFE

Sea water has dissolved air, also carbon dioxides if carbonates which cause scale are decomposed by

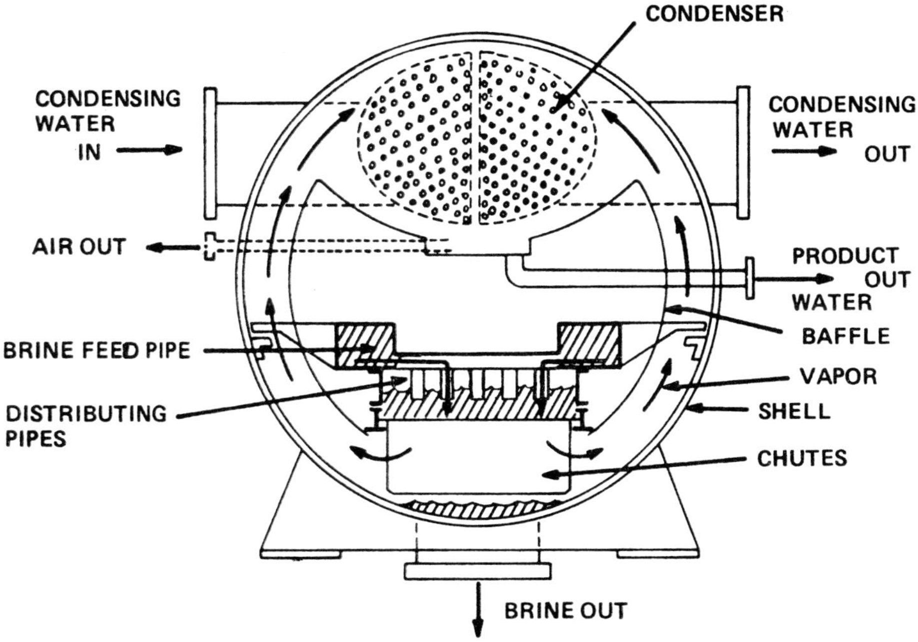

Figure 9 Cross-section of small CFE unit. Assembled in tubular steel shell to produce makeup water for boiler feed. Sea water, warmed in condenser after turbine, flows from inlet (not shown) to cross-hatched distribution system at right angles to section, and supplies the flash chutes, parallel to section, with warmed brine, which is cooled and partly vaporized, then flows out at bottom. Vapors go up to 2-pass condenser tubes cooled by cooling water-entering rear left, leaving at rear right. Condensate fresh water discharge at right center. (Courtesy of Burns & Roe, Inc.)

acids. These gases may be removed by vacuum before evaporation, at a high energy cost. The CFE deaerates with little equipment and no power cost by a patented "ingestor" using the cooler brine discharging from the chutes to entrain and dissolve these gases. They desorb when the brine discharges to the atmosphere.

Advantages of CFE compared to convential MSF

(1) Very high flashing temperature ranges without important transport losses of pressure and temperature due to turbulence.
(2) No brine passes through the stage without coming to equilibrium.
(3) Practically no entrainment of mist to add brine to product water.
(4) Fewer stages, thus lower capital cost for same heat input or gain ratio. Or, for same capital cost, the gain ratio increases greatly.
(5) Gain ratio and ease of operation are uniform over wide flow ranges.
(6) Interstage flooding cannot occur; it often may in conventional MSF.
(7) Plant site area only 25 to 50% that for conventional MSF: thus substantial reduction in real estate requirements, often in urban, high cost areas.
(8) Piping and erection costs much less; also lower operating costs.
(9) Proximity and arrangement of condensers in relation to the flash chamber of the stage reduces required condenser area.
(10) Higher stage flash temperatures gives higher temperature drops between vapors and tubes; thus higher heat transfer rates and lesser area of tube surface.
(11) No operating cost for deaeration of sea water feed.

Water costs by conventional MSF and CFE

Considerable experience in the design and costing of plants of both types shows, compared with CFE the cost for conventional MSF will be:—from 15 to 20% more for a 250,000 gpd unit, to about 8 to 10% more for a 15 million gpd unit. Similarly, water costs at the same Performance Ratio for conventional MSF plants are higher than for CFE units, varying from 15 to 20% more for 15 million gpd.

The 1969–70 Report of the Office of Saline Water states:—"The CFE cost of water is always lower than the MSF, but the percentage of difference is greater at the low temperature end, as expected. CFE water costs at 100°F are $ 0.10/1000 gallons cheaper than MSF; while at 250°F the CFE water costs are $ 0.06/1000 gallons cheaper. This comparison is based on water costs of $ 0.87 per 1000 gallons for the MSF plant."

Vapor Reheat (VR)

An evaporator has heat transfer surfaces to boil the liquid and to condense the vapors. In a vapor compression unit, and in a multiple effect, these functions are combined by condensing vapors to boil the liquid. The usual multi-flash evaporator

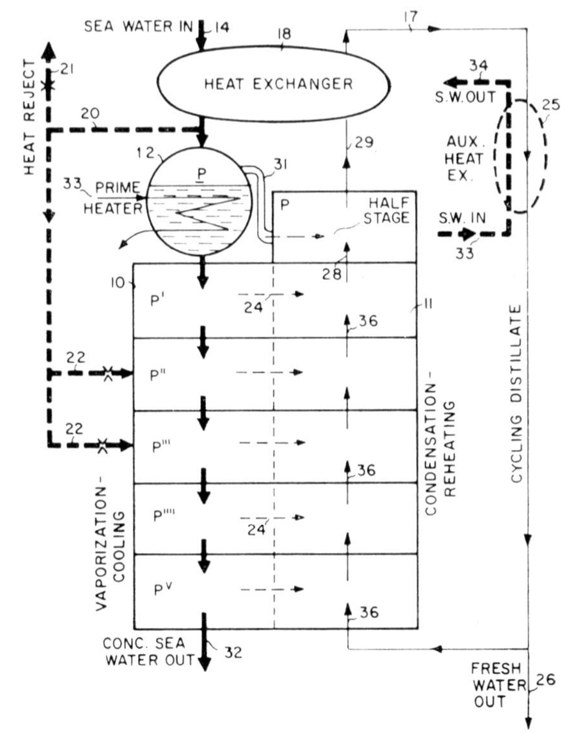

Figure 10 Vapor reheat MSF evaporation—without metallic heat transfer surface (from U.S. Patent 3,329,583). MSF evaporating and cooling of hot feed brine on left as in Fig. 4, vapors (horizontal arrows, 24) from MSF condensing by direct contact with and heating of cooler fresh water stream on right in open sprays, 36, and counter-current flow; heated fresh water being cooled in heat exchanger, 18, to preheat sea water feed and brine recycle; withdrawing fresh water product, 26, before recycle, 27, on right; increasing production of fresh water by withdrawal of vapors from prime heater in pipe, 31, to be condensed in half-stage; additional sea water coolant in auxiliary heat exchanger, 25, cooling the hot fresh water stream for rejecting heat; or by-passing some part of preheated brine around prime heater (by dotted lines, 20, 22) to upper stage of MSF as preferred method of heat reject.

eliminates the heat transfer surfaces for the evaporation, but not for the condensation. The Vapor Reheat Process is a modified multiple flash system and may use the standard or the CFE flashing system. VR eliminates the tubular heat transfer surface in both the evaporation and the condensation. These tubes may cost one half of the entire plant cost; and in the largest plants so far designed amounts to tens of thousands of miles of expensive copper-nickel tubes. VR may also use a large number of stages with heat requirements usually decreasing with increasing number.

In a conventional multi-flash unit, the liquid to be evaporated is preheated in condensing the flash vapors; in the VR evaporator, a stream of colder fresh water is directly and counter-currently contacted with these streams of vapors from the flash evaporations. The vapors condense directly on the surface of the fresh water. This heats the fresh water stream and it is cooled in preheating the cold sea water feed. Again, there is a prime heater for the sea water; and it also may evaporate a small part of the sea water to increase the gain ratio.

With a 50-stage unit, the heat requirements might be less than 40/BTU per pound, i.e., 20 pounds of evaporation per pound of steam, or twice that of a 12-effect unit, or of the most efficient multi-flash unit of the usual type yet reported.

The pre-heater for the sea water is large and requires a large area of the usual metallic heat transfer surfaces or the more recently developed polyvinyl fluoride tubes about 0.002 inches thick. The liquid-liquid-liquid type of heat exchanger (LEX) first used over 30 years ago has been suggested. The fresh water stream, after heating by the many direct condensations may be contacted directly and in counter-current, with droplets of an oil stream in a vertical vessel, the inside of which is fitted merely with spray arrangements, as in a liquid-liquid extractor. The oil so heated is contacted counter-currently with the incoming cold sea water in another identical device. The two vessels from a liquid-liquid-liquid heat exchanger, which eliminates the need for metallic heat transfer surface. About 10,000 BTU/hr/ft^3/°F can be transferred. Since there are almost no internals, the cost will be quite low compared to a usual tubular exchanger. More recently poly vinyl fluoride (du Pont's Tedlar) tubes made of 0.002 inch thick film have been used in the VR heat exchanger with good results.

Several advantages of the basic VR evaporator over conventional multi-flash unit, show low plant cost and low thermal cost:—

(a) Scaling effects may be largely disregarded. If precipitates form in the liquid-liquid-liquid exchanger or in the plastic film exchangers they are readily separable.

(b) The approach to thermal equilibrium on each side of each stage is extremely small—may be as low as 0.1°F.

(c) The low temperature difference between saturated condensing vapor and liquid, (0.1°F), as compared to that across the two liquid films and a metallic wall (3 to 10°F) means that a much larger number of stages may be used for a given overall temperature range with improved thermal economy.

(d) Equipment is much less expensive, because there are no metallic heat transfer surfaces and substantially no internals in the vessels.

(e) Lower temperatures can be utilized efficiently in the lower stages because of the low temperatures of approach, and low pressure drops.

Vapor reheat modifications

VR may be used also with some modification which decrease its energy requirements, or which makes it more adaptable to particular conditions. Thus, it may be used with the so called Half-Stage mentioned above wherein some evaporation in the brine heater is obtained without energy cost; also with a submerged combustion in the brine heater to obviate the necessity of a steam boiler.

Mechanical vapor compression may be used around the stages of flash evaporation of VR to give a low energy requirement of about 25–30 kWh per 1000 gallons as compared with about 60 in the large demonstration vapor compression of the

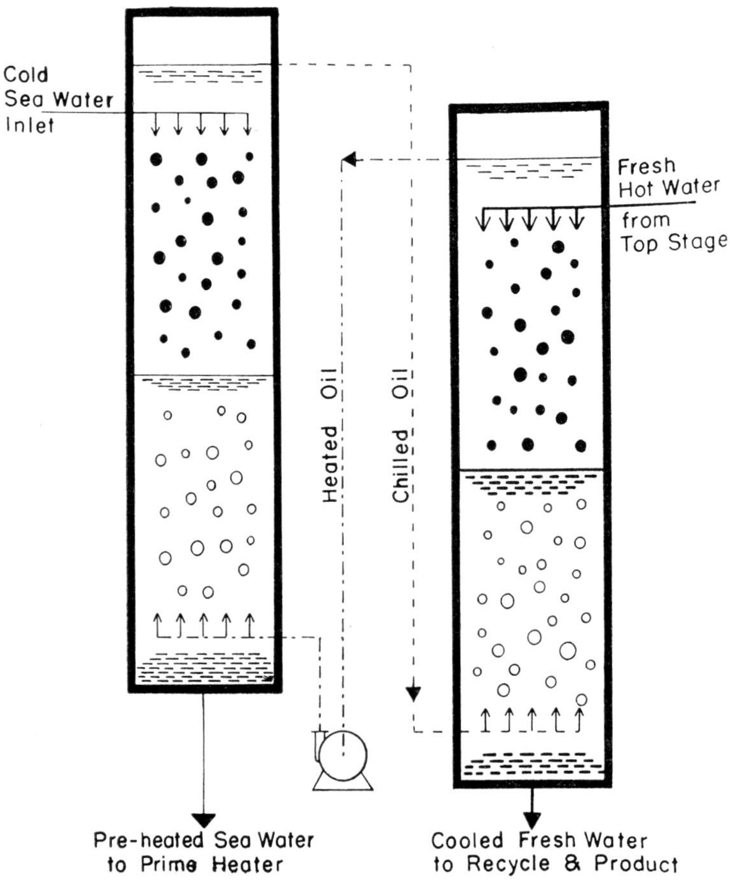

Figure 11 Liquid-liquid-liquid heat exchanger. Hot fresh water condensate from the top stage enters top of right vessel, descends as droplets against a rising stream of droplets of colder oil, which is heated thereby. The fresh water, now cooled, is removed from the bottom of the right vessel, the heated oil, as droplets, enters the bottom of the left vessel, rises against the descending stream of the droplets of cold sea water, which is preheated thereby to go to the brine heater; and the chilled oil is recycled back to the right vessel.

Office of Saline Water. Using a vapor-absorption system of vapor compression, the gain ratio of a VR system may be increased by 25–30%, while if this is combined with a multi stage absorber as a heat exchanger, the gain ratio may be 40–50% higher than for a conventional MSF system.

The VR system may be incorporated into a Hybrid Evaporator which has a multi-effect evaporator utilizing some of the heat passing from the high to the low temperature for an increase in the gain ratio—also in the cost and complications of the plant.

When combined with extraction, as will be hereinafter discussed, the gain ratio of VR again increases very considerably. Some two or three of these steps may make VR combined with Controlled Flash Evaporation the low cost evaporation system both as to equipment costs and as to energy costs.

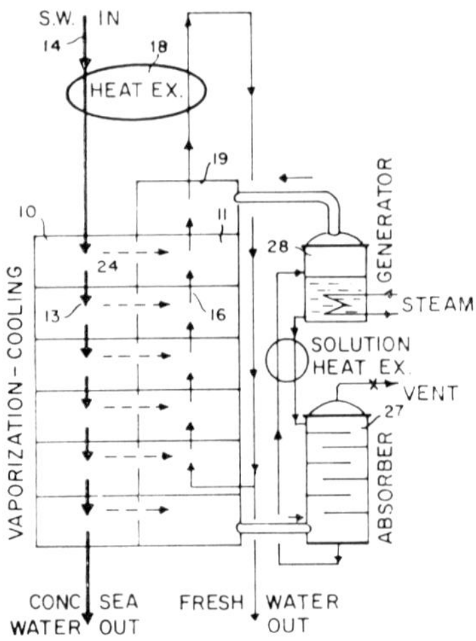

Figure 12 Vapor reheat with vapor recompression by absorption (from U.S. Patent 3,288,686). Vapor reheat MSF as in Fig. 10, vapors formed in low pressure stage pass to absorber, 27, and are absorbed in the hydrophilic liquid from solution heat exchanger, which cycles through heat exchanger o evaporator or generator, 28, for concentration at higher pressure. Vapors leave at higher pressure to half-stage, 19, to supply prime heat to fresh water stream, which in turn heats raw sea water in heat exchange, 18.

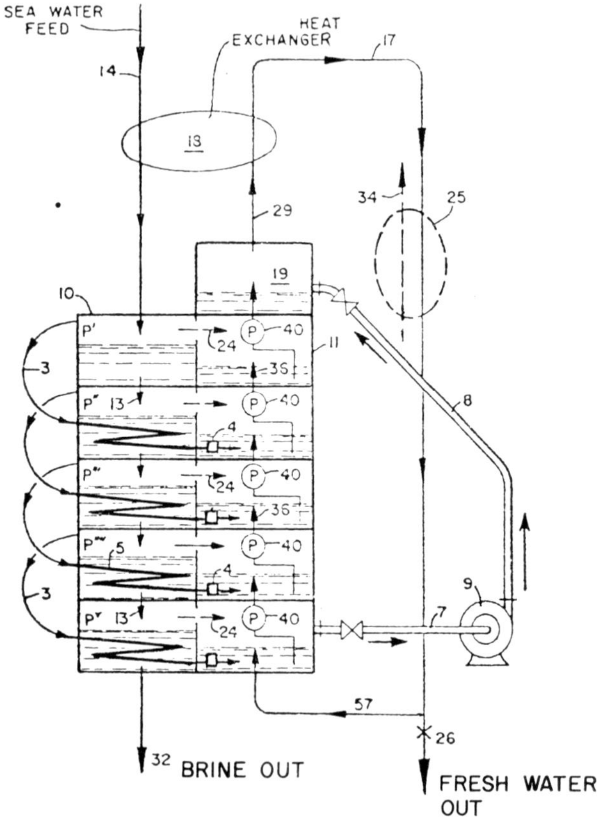

Figure 13 Hybrid evaporator—This uses a combined VR evaporator of MSF and a multieffect with evaporator surface, also vapor recompression for suppling prime energy. To the MSF and VR system, the evaporator surfaces are added as diagrammed by the tubular coils, 5, in the flashing chambers of pressure P'', P''', P'^{V} and P^{V}. By withdrawing vapors through lines 3 from a higher stage, each coil becomes one surface of a multieffect evaporator built into the VR system. The condensate passes through a steam trap, 4, and thence directly into the fresh water circulating in the VR condensate circuit by a small pump 40 lifting the condensate to each next higher stage. A brine heater of the usual type may use steam to heat the brine to the top temperature before it enters the top stage with pressure P'. As shown however, a vapor recompressor, 9, removes vapors from the bottom stage, recompresses them and passes them to the highest temperature of the half stage, 19, where they are condensed by direct contact with the preheated condensate to pass to the main heat exchanger, 18, of tubular or LEX type. The other heat exchanger, 25, is for heat reject to a separate stream of sea water, 34.

Freezing Processes

These have been given much attention, but so far, no plant data are available from commercial operations for desalination. The temperature of freezing is fixed by the concentration of the brine, while evaporation can operate over a wide temperature range. Freezing has basic advantages:—(a) much lower latent heat of fusion of water than of evaporation; (b) less heat losses (gains) because of working at temperatures closer to the ambient; (c) no scale-formation from the usual impurities; (d) less corrosion of steel at the freezing point than at the boiling point of water; and (e) cheaper materials of construction may be used.

Against these there are basic disadvantages: (a) the time required for phase transition from liquid to solid is very much greater than that required

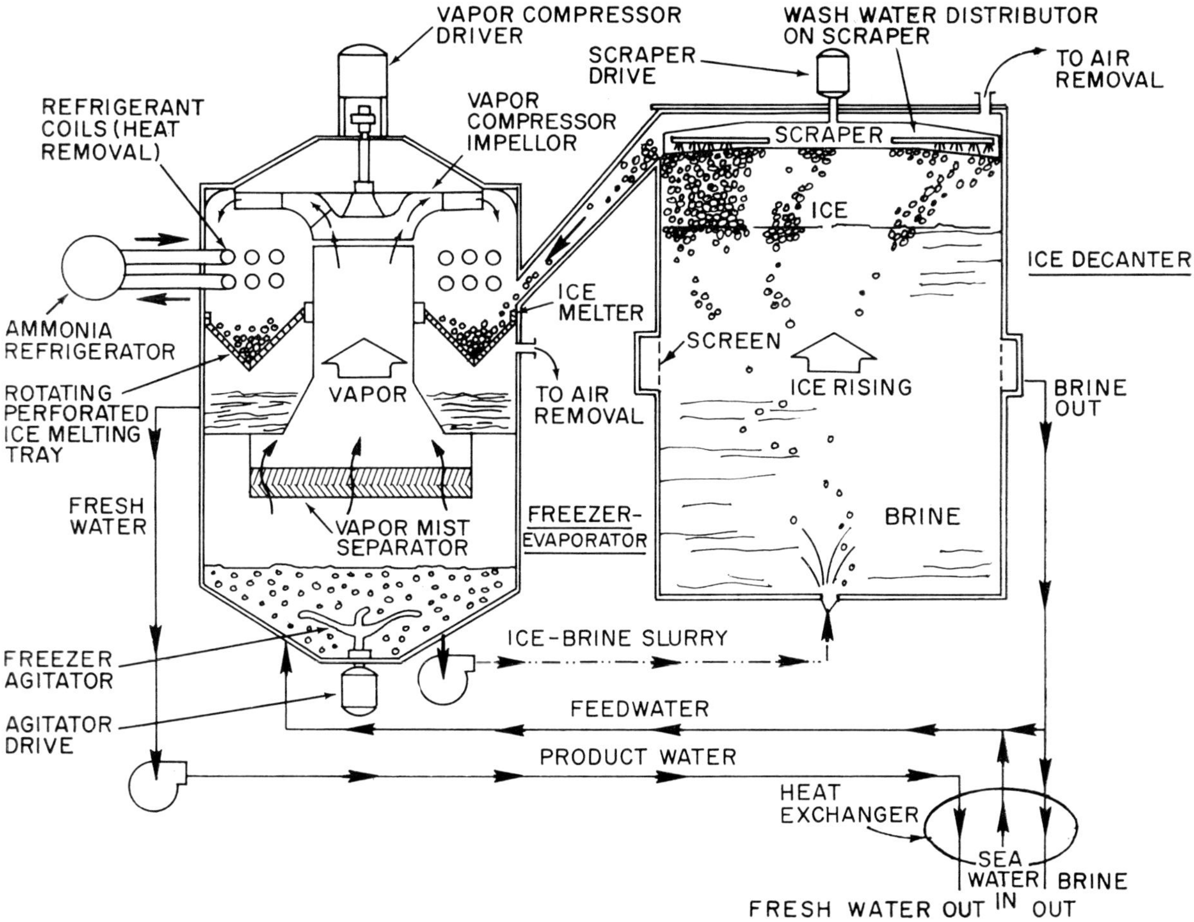

Figure 14 Zarchin-colt freezing process. Pre-chilled sea water in enters the freezer-evaporator, the left vessel—maintained at the pressure of the vapor from freezing ice. Mist is separated from the water vapors formed, which are compressed by a flexible blade compressor to above the vapor pressure of ice in equilibrium with fresh water. A brine slurry of ice crystals grown to a size approaching 1 mm passes to an ice decanter where the mass of crystals float. A small amount of fresh water at top washes off the film of brine, and a scraper discharges them into a chute aelding to a perforated melting tray in the compressed vapor space above the freezer. The condensation of water vapois and melting of ice gives pure water, augmented by condensation on an auxiliary coil supplied with external refrigeration to compensate for heat gains and inefficiencies. An additional air removal system purges the ice decanter and the freezer. A heat exchanger warms the cold product water and the brine discharged as it chills the same total amount of *Sea Water In*.

from liquid to vapor; (b) handling the crystals of ice is very much more difficult than handling the fluids in evaporation processes; (c) separation of the pure water phase (ice versus steam) is extremely difficult; (d) the cost of removing heat energy is very much more expensive than that of adding heat energy; and (e) freezing operates, practically, only at the freezing point of the final solution; and there can be no multistaging to reuse the latent heat of phase transition.

However, if energy is cheap, the freezing process may be considered. As with vapor compression evaporation, the freezing might use the power product of a dual product plant, electric power and water which could utilize the thermal energy of steam, and either the freezing process or vapor compression could use its mechanical energy.

As in all refrigeration, there are two types of processes for desalination of sea water: (a) mechanical compression of vapors, and (b) absorption of vapors by a hydrophylic liquid which then uses heat energy for desorption by evaporation. Both may use water as the refrigerating fluid; both may use other refrigerants in either direct or indirect heat transfer relation.

In the Zarchin-Colt Freezing Process, water vapor itself is the refrigerant, even though the specific volume at the freezing point is about

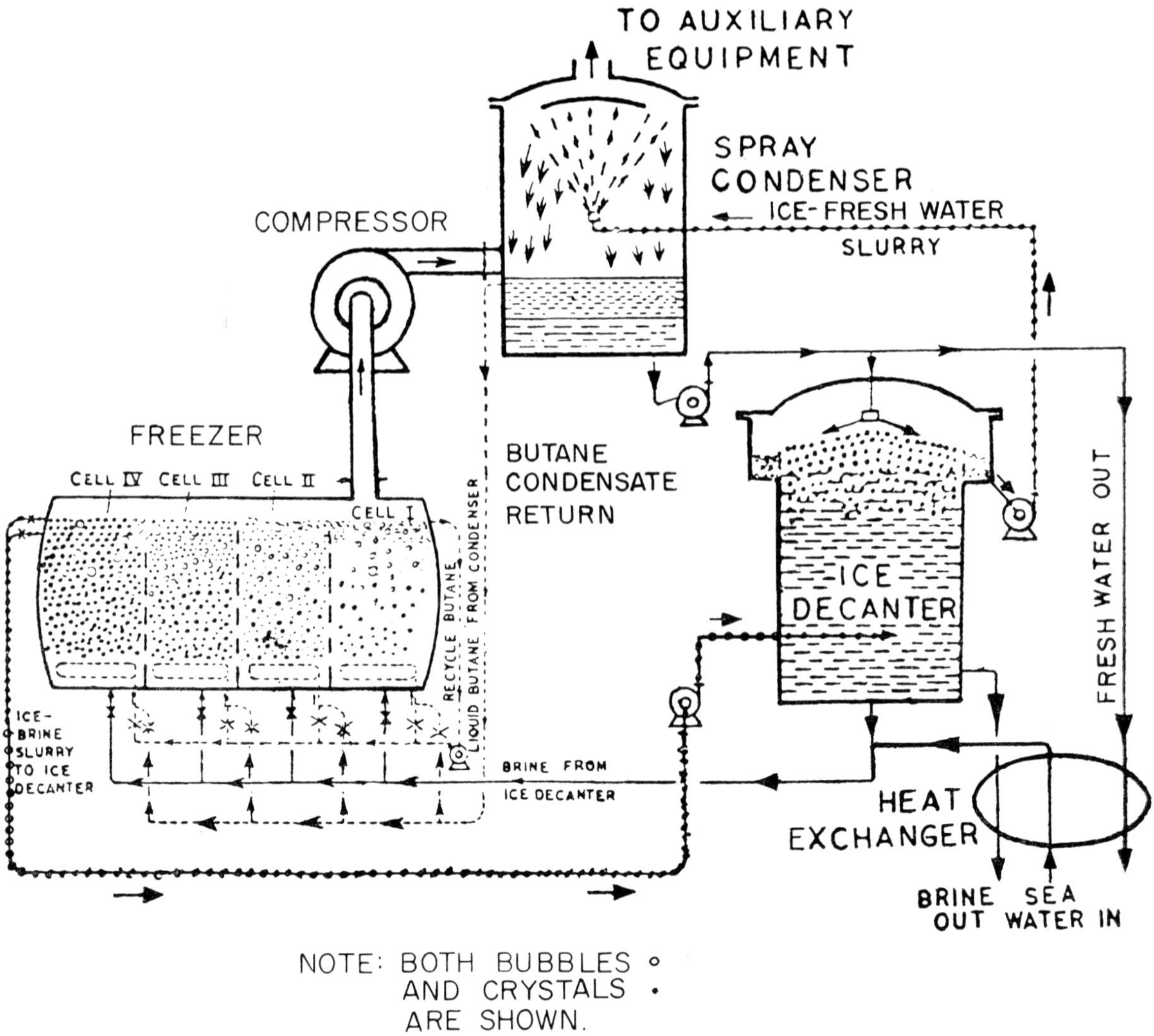

Figure 15 Freezing by direct contact with secondary refrigerant. Pre-chilled sea water is refrigerated in the freezer having four cells by evaporation of butane. Liquid butane released in controlled amounts at bottom of each cell rises and evaporates as bubbles. Brine may enter each cell, most enters cell I and passes through perforated separators to cell II, III, and IV as crystals form and increase in size. More butane is fed than is evaporated and an amount is continually withdrawn from surface, flowing from cell IV, III, II and I to the recycle pump. If a small amount of gasoline is in the butane recycle—it will not evaporate under existing conditions, flows of recycle butane and of butane from condenser may be controlled to give progressively less gasoline in I, II, and III—and none in IV—so that the effective boiling point of the butane as it becomes purer and purer from right to left is lower and lower. This control allows crystals to grow to maximum size without unwanted formation of excess seed crystals. Ice crystals in a brine slurry are removed to the ice decanter and brine is returned. Ice floats above the liquid in the decanter and is trimmed and washed off by a rotating jet spray of fresh water from which some water is intercepted as wash for the crystals. Ice-fresh-water slurry so formed passes through a spray nozzle to form a full spray in melter-condenser supplied with compressed vapors of butane which have been exhausted by the compressor from the freezer. Refrigeration—effect of cold brine and cold fresh water is recovered in pre-chilling the fact sea water, preferably by a liquid- liquid-liquid heat exchanger (LEX). Auxiliary equipment removes air and adds additional refrigeration to make up for losses.

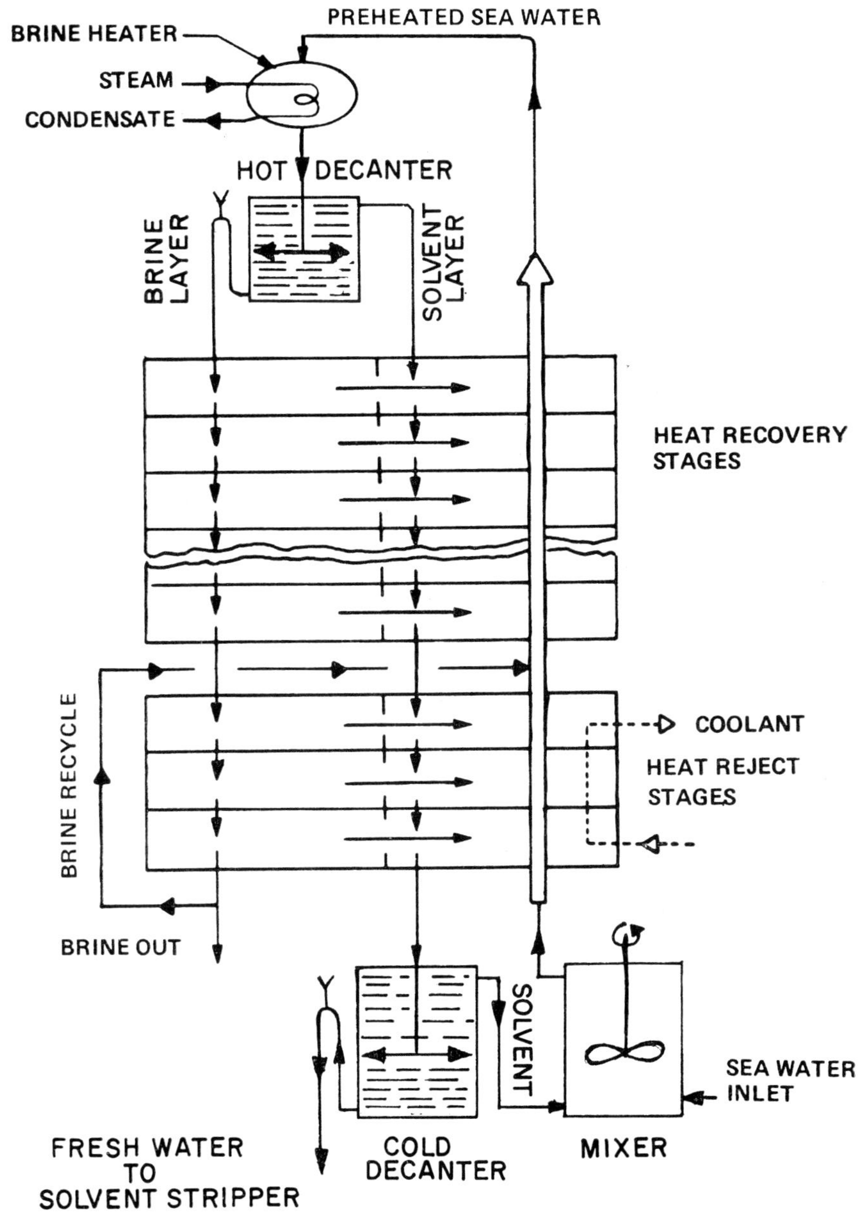

Figure 16 Extraction and MSF. Cold sea water is mixed with a solvent which has greater solubility when hot and the mixture passes through heating tubes of MSF stages, then through the brine heater. The solvent phase has extracted considerable watre at this higher temperature, and when the mixture is passed through the hot decanter, the watre saturated solvent is decanted from a brine layer. Both are passed to a sequence of different flush chambers in the same MSF stages. Condensate includes solvent; and a cold decanter separates it out and passes it back to the mixer for recycle. The water layer is saturated with solvent which is then stripped out separately from the product fresh water. Alternatively, the solvent stripping can be done in the MSF system, which then discharges fresh water directly.

150 times that at the boiling point—and a very ingenious compressor of very large capacity has been developed. With the numerous accessories, a one million gpd plant uses 27–30 kWh of power, costs about \$ 1.5 million and produces water at about \$ 0.61 per 1000 gallons with energy at 0.74 per kWh.

In some cases, an absorbent for the water vapor, rather than a compressor, is used—such as a solution of lithium chloride, an absorption refrigeration—or heat pump thus gives the fresh water by evaporating fresh water out of the solution.

In other freezing processes, liquid refrigerants

other than water are used. These materials have been, for cheapness and other advantages, hydrocarbons in the range of butane. Their make-up losses are noticeable, but do not need to add appreciably to the cost of the water.

The evaporation of a direct refrigerant such as butane in contact with water and simultaneous with the freezing of ice, with a transfer of the latent heat freezing of the ice to supply the latent heat of vaporization of the butane has long been known. One such process controls the production of seed crystals and their growth into crystals of optimum size by the presence of a controlled amount of a higher boiling hydrocarbon, e.g. a gasoline fraction, in each of several zones of the crystallizer to elevate the boiling point of the butane therein, this controls the heat removal by the temperature difference to the freezing liquid. Thus there is varying from the inlet zone to the outlet:—(a) the rate of cooling, (b) the concentration of salt and hence the freezing point of the brine, (c) the size of crystals and the mass of ice per cubic foot. Meanwhile, a slurry of crystals in brine is withdrawn from the last zone, passed to an ice decanter where crystals float, are skimmed off by a high velocity jet stream of fresh water to give a new slurry of ice in fresh water. This slurry is sprayed into an open condenser space where butane vapors are discharged by the compressor which evacuates them from the freezer. The butane condenses for recycle, while the ice evaporates to give fresh water.

Extraction

Evaporation separates fresh water from sea water as a *vapor*; freezing separates water as a *solid*; extraction may be used to separate as water a *liquid*.

No solvent has been found which will extract salt from water; but most liquids have some mutual solubility with water. Thus extraction may be used for separating fresh water from sea water, leaving a more concentrated brine behind. The difference in the solubility for water at different temperatures of the liquid may be utilized. The water is extracted by the solvent, at the temperature of maximum solubility; then the temperature is changed to that of minimum solubility to precipitate out the water. The solubility may be greater at either a high temperature or at a low temperature. Solvent recovery systems are required, and are costly to operate, in addition to the required heat exchanges. Thermodynamically, it may be shown that extraction processes are fundamentally less efficient than those using evaporation or freezing. One of several reasons for this is that the sea water always dissolves some of the solvent liquid which must then be recovered therefrom. Besides the

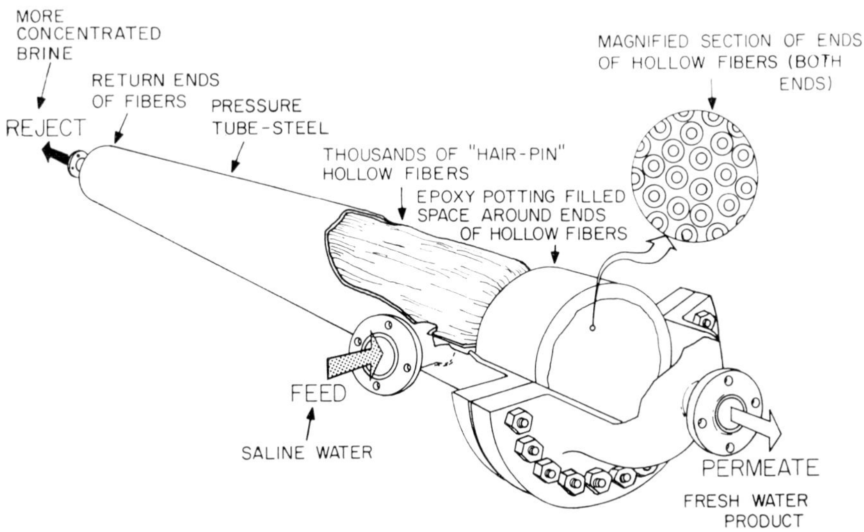

Figure 17 Reverse osmosis of brackish water using as a membrane fine nylon hollow fibers. Millions of fiber loops, like hairpins, have both ends encapsulated in an epoxy tube sheet or short cylinder enclosed in the flanged water end on right. A magnified section of the ends of these thread -fine fibers is indicated. The saline water feed passes on the shell side from the feed at the right end to reject of more concentrated brine on the left. The permeate of fresh water discharges from the right nozzle. The steel casing is designed for the pressure necessary, usually 40–50 atmospheres. (Courtesy of E.I. DuPont de Nemours.)

thermal cost, the number of processing steps in different units of equipment increases as does the plant cost.

Exotic liquids as solvents, include the substituted amines. These are probably too expensive but they have been most studied. They have a greater solubility for water at lower temperatures. Other common liquids as hydrocarbons may be used when advantage is taken of their differences of solubility for water, which for most liquids, is greater at higher temperatures.

Thus extraction processes for water depend not only on the solubility relations of the particular solvent at usual temperatures, but also on the change in these relations at different temperatures. Since two different operating temperatures are thus indicated—as also in MSF or multi-effect evaporation, and to a lesser extent in freezing processes—combined extraction and MSF processes have been studied and patented. They indicate that a very substantial increase in the gain ratio of a VR or other MSF plant is obtained when extraction with a carefully chosen liquid is combined at the high or the low temperatures, with the release of the solvent for recycle.

Various flow sheets have been developed for both conventional MSF, also VR. Because of the greater flexibility of VR, which has two cycling aqueous streams while conventional MSF has only one, there are more possible flow sheets with VR and greater opportunities for improving the gain ratio. However a relatively simple one utilizing the high and low temperatures of conventional MSF is shown in Fig. 17, with a solvent such as butanol in which the solubilities—both water in butanol and butanol in water—are greater at higher temperatures than at lower temperatures. The two liquids are completely miscible at temperatures above about 255°F, thus a suitable upper operating temperature might be taken at about 248°F where 1 pound butanol dissolves from sea water 0.90 pounds of water, and practically no salt. When this is cooled to, say, 86°F the one pound of butanol will dissolve only about 0.25 pounds of water and thus there comes out of solution about 0.65 pounds of salt-free water which is, however, saturated with butanol at this temperature. If the heats involved in heating and cooling between 248°F and 86°F can be utilized in an MSF evaporation the additional water obtained may increase the gain ratio by over 50% with a comparatively small increase in equip-

Table of Conversion Factors for Desalination

To convert	To	Multiply by
acre feet	cubic feet	43560
acre feet	cubic metres	1233.49
acre feet	gallons (U.S.)	325,851
acres	square metres	4047
BTU	kcal	0.252
BTU	kWh	2.930×10^{-4}
BTU	Watt-hours	0.2930
BTU/pound	kcal/kg	0.5556
BTU/hour/ square foot/°F	kcal/hour/ square metre/°C	4.88
centimetres	inches (U.S.)	0.3937
cubic feet	litres	28.317
cubic feet	cubic metres	0.028317
cubic feet	gallons	7.481
cubic feet water (60°F)	pounds	62.37
cubic metres	cubic feet	35.3144
cubic metres	gallons	264.173
feet	metres	0.3048
gallons (British)	litres	4.546
gallons (British)	cubic feet	0.16054
gallons (British)	gallons (U.S.)	1.20094
gallons (British)	pounds water (62°F)	10
gallons (U.S.)	cubic metres	0.0037854
gallons (U.S.)	gallons (British)	0.83268
gallons (U.S.)	litres	3.78533
1,000 gallons (U.S.)	metric tons	3.7853
gallons (U.S.)	pounds water (60°F)	8.3370
grains/gallon (British)	parts per million	14.254
grains/gallon (U.S.)	parts per million	17.118
grains	grams	0.0648
grams	grains	15.4324
grams/litre	grains/gallon (U.S.)	58.417
Horsepower hours	kWh	0.7457
kcal	BTU	3.9685
kcal	kWh	0.0011628
kcal/hour/ square metre/°C	BTU/hour/ square foot/°F	0.2049
kg	pounds	2.2046
kWh	BTU	3413
kWh	kcal	860
litres	cubic feet	0.035316
litres	gallons (British)	0.219976
litres	gallons (U.S.)	0.26418
metre	feet	3.2808
parts per million	grains/gallon (U.S.)	0.0584
parts per million	pounds/million gallons (U.S.)	8.345
pounds	kg	0.45359
pounds of water (62°F)	cubic feet	0.016033
pounds of water	gallons (U.S.)	0.1198
square feet	square metres	0.0929
square metres	square feet	10.7639
tons (metric)	pounds	2204.62
tons (metric)	tons (short)	1.10231
tons (metric)	1,000 gallons (U.S.)	0.26418

ment and processing. The flow sheet may be somewhat more complicated with VR but an even greater increase in the gain ratio may be expected.

Hyper-Filtration or Reverse Osmosis (RO)

Another system for separating *liquid* fresh water from sea water depends on what may be regarded as a size separation of water molecules from a more concentrated residual brine containing the salt molecules. This is called *hyper-filtration* or *reverse osmosis* (RO). Plastic membranes with extremely fine pores are used. Fresh water discharges through these pores in the plastic film which is supported by wire mesh or perforated plates against the 500 to 1500 pounds per square inch of the pressurized sea water; and the brine, correspondingly more concentrated, as it loses part of its water flows past the membrane and out of the high pressure side.

It is not known whether the permeation of water molecules through the membrane is an osmotic phenomenum, i.e., the molecular migration of the water molecules there—through by adsorption, diffusion, and desorption or of the large hydrated cluster around the salt ions or molecules which increases their size above that of the pores in the membrane, through which the smaller water molecules pass. Experimental and large scale operations present aspects which support theories for both mechanisms.

Cellulose acetate membranes prepared by specially developed techniques are presently favored; but membranes of other polymeric materials are also being used. Large sums of money are being spent currently to develop membranes, operable processes, and staging of processes, since it is impossible to make a salt-free water from sea water or a concentrated brine in a single hyper-filtration step. Thus several successive steps would be required to give potable water—below 500 parts per million of salts—from sea water having 35,000 parts per million of salts.

The membranes are supported on screens or perforated plates to withstand the pressure—1000 to 1500 pounds per square inch. The capital cost is high for equipment to withstand the pressure; and the maintenance cost is high because of the frequency with which the membranes must be replaced —every few months or at most every year or two. From 10 to 50 square feet of an excellent membrane may be used to produce 1,000 gallons fresh water per day from brackish water; this area increases rapidly if sediments are present to plug the pores.

The mechanical energy per 1,000 gallons required by present processes from *brackish waters* may be in the range of 25–35 kWh; and this is no lower than for the best evaporating processes or freezing processes which have been devised from *sea water*. The hope is that hyper-filtration will be reduced to some 15–20 kWh per 1,000 gallons—from brackish waters.

Ingenious systems have been developed to back up the very fragile membranes—about 0.005 inches thick—with metallic screens to accomodate the large pressure drop across them. A break-through in mechanical design has been made by some of the major manufacturers of plastic materials in forming the membrane as a capillary tube—actually finer than a hair, of about 0.002 inch outside diameter and a wall of about 0.0005 inch. The very small tube can take extremely high external pressures without collapse. Nylon is used for the tubes, as it is much more resistant than cellulose acetate to water chemicals and particularly bacteria. While basically less selective in its permeability, nylon may be made thinner in this tubular design; but between 5,000 and 10,000 square feet of fiber surface is required for each 1,000 gallons of fresh water produced per day. Millions of such tubes, going down and back like a hairpin, have both ends molded by an epoxy resin in a single tube sheet which closes one end of a heavy steel tube. Brackish water flows under high pressure the length of the shell side to discharge at the far end:—water, which permeates the walls and flows out both ends of the fine tubes through the tube sheet for collection.

Ionic Processes

Common salt, and other salts as well as acids and bases are ionized in solution into the positive (for example, sodium ions) and the negative (for example, chlorine ions). Two general methods of removing the respective ions are used, one depending on individual, separate chemical reactions with each ion, and the other depending on electrical properties. Both remove the salt, in more concentrated solutions, by separating ions from the fresh water remaining. The previously discussed processes have separated the water from the salt, i.e., in the more concentrated brine remaining. Here, however since the ionic processes remove chemically or electrically, the ions of the salt from the brackish water, these ions—or the salt represented must be discharged into another stream which may be wasted, usually a part of the original salt water. Either a chemical equivalent of the salt or usually several times this amount of chemicals must be used to remove the ions; or there must be used an electrochemical equivalent or more of the ionic charges. Thus, ionic processes may be much too expensive to use with the higher salt concentration of sea water—but may be considered for the lower concentrations—$^1/_{10}$th to $^1/_{50}$th as much—in brackish water, estuaries of rivers, or inland waters. Improvements in processing may use less expensive chemicals than are now standard; and the simplicity of the several processes is attractive. In each case, the ionic processes depend on specially developed polymeric resins in special forms.

Ion exchange

The chemical system for removal of the ions of salt, is called ion exchange and has long been used in the reverse process, in which, for example, the sodium ion of salt is used to replace calcium ions to "soften" hard water. In this case, the "bed" of ion-exchanging material consists of granules of resin which have the property of removing positive or metal ions from a circulating aqueous solution by displacing them by a chemical bond with the resin of the bed. This replaces them with ions which are loosely bound in the molecular structure of the resin. When the positive ions of the resin are completely replaced, an aqueous solution, relatively concentrated in the other positive ion, is cycled, and the process is reversed. Other beds of resin have the properties of exchanging negative ions.

Similarly, the sodium ion of a salt solution such as sea water passing through a suitable first bed may be interchanged with a hydrogen ion from the resin to leave the sodium ion in a loose combination with the resin and a resulting hydrochloric acid in the effluent solution. This solution is then passed through a second resin bed, in which a hydroxyl ion interchanges with the chlorine ion to give a resin-chloride combination and an equivalent amount of chemically formed water which is added to the aqueous stream.

When the two resin beds have interchanged all of their respective hydrogen and hydroxyl ions, their activity ceases and they must be regenerated. This is done by stopping the flow and adjusting valves to take the beds off-stream and to pass through the first bed a dilute solution of sulfuric acid and then to pass through the second bed a dilute solution of caustic soda. The hydrogen ions of the sulphuric acid displace the sodium ions on the resin of the first bed to give sodium sulphate in its effluent which is run to waste; and the hydroxyl ions of the caustic soda displace the chlorine ions on the second resin to give sodium chloride (salt) in the effluent of the second bed, which also runs to waste. Again, the ions on the beds are completely interchanged, the beds are thus regenerated; and the controlling valves are adjusted, with those allowing the sequential flow of the original sea water being opened to allow the process to be repeated.

This process thus requires both sulphuric acid and caustic soda in amounts which are chemically equivalent to the amount of salt present. Usually, at least 50 to 100 per cent excess of each is used. This will represent a very high and impossible cost for large-scale desalination of sea water. In special cases, particularly for desalinating brackish water, ion -exchange beds have been used, since only a relatively smaller amount of chemicals is required to interchange ionically with the salt in the feed solution. Thus, water containing 1,750 ppm of salt would be regarded as non-potable, but it would require only about 5 per cent of the chemicals that sea water would require. On the other hand, the energy cost for its desalination by evaporation or freezing would not be so greatly different from that required for sea water, because in these processes it is the water which is being separated.

For emergency kits, packages of ion-exchange resins have been made to have only a single use, e.g. by aviators downed at sea. Sea water is passed through these small "beds" to make a small amount of drinking water. Provision for regeneration of the resins would be complicated; and the resins are discarded when charged with sodium and chlorine, respectively.

The simplicity of ion exchange has attracted much effort to finding less expensive methods of regeneration of the ion-exchange resins. Carbon dioxide is a weak but cheap acid, as lime is a cheap alkali; and special resins and processes have been developed to use them—also to use the major differences of ion exchange equilibria at 80°C as compared to those at room temperature.

Ion exchange processes depend upon good

resins. Excellent resins are available, and they are not the limiting factor. The major cost is that of the chemicals required for the much less than stoichiometric replacement of salt by two chemicals.

Electrodialysis

The electrical mechanism of ion removal is much more complicated, and much cheaper than ion exchange since it uses electrical energy rather than chemical equivalence to replace the two ionic charges of a molecule of salt. This is called electrodialysis, since an electric current assists greatly the dialysis or movement of the ions through membranes permeable to the positive ions and to the negative ions, respectively. A membrane, which is permeable to sodium ions, forms the wall on one side of a channel of flowing saline water; and a membrane permeable to chlorine ions forms the wall on the other side. The deionized water flows between the two membranes, and the electric current may be regarded as flowing at right angles. The other aqueous streams on the other side of the respective membranes may flow out counter-currently in the other direction. Membranes of synthetic resins have been developed which are highly selective to the passage of positive ions, and others have been developed which are highly selective to the passage of negative ions. Hundreds of membranes form passages in parallel between, somewhat like a plate and frame filter press. This is called a stack.

Since alternate membranes are used, the water streams lose sodium ions through the membrane on the one side, and chloride ions through the one on the other; and the water discharges substantially less salt. The alternate channels between the ones where the water being purified passes, receive from their respective two sides the opposing ions, which unite to form salt in these streams, so that these channels discharge a more concentrated brine. Here again, it may be impossible to have a pure water separating from a concentrated brine in one step, and staging may be used. The electrical energy required is proportional to the salt concentration; and the process is economic only with brackish waters with much lower amounts of salt than sea water.

The process is governed by the distance of the flow in the compartment, the density of electric current flow perpendicular ot the channels and membranes, the permeability and other specifications of the membranes, their distance apart, and some other considerations.

Polymeric materials for electro-dialysis membranes have been developed which are more efficient materials than the earlier ones; and much effort is being given to develop even greater improvements to produce more rugged, industrially dependable materials with improved separation capabilities. At best, the frequent replacement and high cost will be an important part of the cost of electrodialysis processes.

The success in South Africa with electrodialysis plants for individual farms (1000 to 2000 gallons per day) and intermediate size units (up to 2.5 million gallons per day) when working with brackish water indicates that this type of unit may be made dependable in many circumstances. Plants in the 1 to 2.5 million gallons per day range are now regarded as feasible.

In ion-exchange and electrodialysis as well as reverse osmosis processes, sedimentary and other impurities in the water to be treated can spoil surfaces and greatly reduce capacities. Careful pretreatment of the water to remove undesirable materials usually may be necessary.

Summary

Fresh water, man's greatest necessity, becomes more and more scarce in much of the world; and this scarcity limits industry, agriculture, every aspect of our life and thus population. Numerous systems for desalination have been suggested, based on almost every possible property of water, of salt, and of saline solutions. Every one with any possibility of success is being investigated. Nothing can be overlooked, since it is firmly expected that the reduction in costs of water desalination in the next years will necessarily have to be comparable to the 90% reduction in costs made in the past fifteen years.

Present experience indicates that often approximately 50 per cent of fresh-water costs are made up of variable charges, principially energy, and 50 per cent of fixed charges, principally plant costs. The latter are reduced to some extent per unit production in larger plants.

While, theoretically, the energy costs of separating fresh water from sea water should be very small, large-scale evaporators produce a maximum of about 10 tons of water per ton of steam used; another process, vapor reheat, which has been demonstrated to be possible but has not yet been operated on a large scale, will give 25 to 30 tons of fresh water per ton of steam. If this is combined with controlled flash evaporation or, possibly, extraction the gain ratio may go to 40.

When electric power is available, vapor recompression evaporators have required 50–60 kWh per 1000 gallons. This may be cut in half by interconnecting other systems. This 25–30 kWh seems to be about the requirement per 1000 gallons by present freezing processes, although these also have possibilities of substantial improvement.

Enormous tonnages of metals are required for single plants, like tens of thousands of tons of copper-nickel tubing. Other metals and plastics are also important; but energy will be the most important cost—certainly supplies of fossil fuels will not be adequate. Indeed, fuel costs may usually be the predominating factor. Plants using nuclear fuels have not yet generated steam for power production as cheaply, when all costs are considered, as can fossil fuels. Until steam is made more cheaply by nuclear fuels, the discussion of nuclear fuels as a means of supplying the energy for water desalination is merely prophetic. When steam costs are lowered by nuclear installations, their apparent disadvantages, e.g., the large amount of materials required for shielding, and the remoteness of installations for safety, etc. may be more than balanced. However, nuclear energy or geo-thermal energy, or some other type of energy as yet untapped, must be made cheaply available if water is to continue to be plentiful, or at least sufficient for the numbers of people who are yet to be born, just in this present century.

DONALD F. OTHMER
Polytechnic Institute of New York

DESULFURIZATION: see SULFUR REMOVAL

DISINFECTION AND CHEMOSTERILIZATION

I. Introduction

Disinfection is a term which has for many years been used with different shades of meaning. It has frequently been confused with antisepsis which, although analogous to disinfection (see later), does not strictly have the same interpretation. This, particularly when considered in conjunction with other terminology, means that any article dealing with disinfection must clearly define the sense in which the term is being used. The values and uses of disinfectants were comprehensively reviewed by Finch (1958), but it is surprising that the meaning of "disinfectant" is not stated, although it is apparent from the manner in which it is used in his book that Finch considers the term to refer, in many cases (e.g. sterilization of surgical instruments, sputum, mugs, bedpans) to sporicidal substances. More recently, Davis (1968) rather vaguely defines a disinfectant as "a material having powerful germicidal activity and suitable for use as such."

Fortunately, some at least of the confusion relating to disinfection and to similar, but not identical, terms has now been resolved with the publication by the British Disinfectant Manufacturers' Association (BDMA, 1970) of a brief report which deals specifically with the terminology in sterilization and disinfection. The terms defined in this report have been classified as shown in Table 1, but only those applicable to this chapter will be defined here, and these are based on the BDMA report, together with the term "chemosterilizer" (Borick, 1968).

Sterilization is the process of destroying or removing all microbial life.

A *sterilant* (*sterilizer*) is an agent used in sterilization which destroys microbial life, including bacterial spores, and is thus distinct from a disinfectant. The term "sterilant" may itself be somewhat confusing, however, for a "chemosterilant" is sometimes used in the United States to denote a "chemical substance used to sterilize insects and render

TABLE 1

*Terminology Used in Sterilization and Disinfection**

Group		
I. Absolute Definitive Terms	II. Classified Definitive Terms	III. Terms in Common Use (but not definitive)
Sterilization	Bactericide, bacteriostat	Germicide
Sterilizer (Sterilant)	Fungicide, fungistat	Disinfectant
Sporicide	Tuberculocide	Antiseptic
	Virucide	Sanitizer
	Protozoacide	
	Amoebicide	

* Based on report by British Disinfectant Manufacturers' Association (1970)

them incapable of reproduction on mating with non-sterile partners" (Borick, 1968). Davis (1968) defines a sterilant as a disinfectant suitable for use in the food industry.

A *sporicide* is a sterilant, but is usually confined to chemical agents.

A *chemosterilizer* (Borick, 1968) refers to a chemical compound which is used to destroy all forms of microbial life, and is thus the same as a sterilant defined above. The term "chemosterilizer" will subsequently be used.

Disinfection is the destruction of micro-organisms but not usually bacterial spores. Commercially, the term applies solely to the treatment of inanimate objects, and does not necessarily imply that all micro-organisms are killed, but rather that they are reduced to a level not normally harmful to health.

Antisepsis is the destruction of micro-organisms, but not bacterial spores, on living tissues: not necessarily killing all micro-organisms, but reducing them to a level not normally harmful to health. The term is thus analogous to disinfection.

A *sanitizer* is a disinfectant with the connotation also of cleansing it is used mainly in the food and catering industries.

The suffixes "-cide" and "-stat" may be added to various words to give a precise meaning, e.g. bactericide means a substance which kills bacteria but not spores, bacteriostat (bacteristat) a substance which inhibits the growth of bacteria, thereby producing the state of bacteriostasis. Other terms which are frequently used in this context include the following: germicide, sporicide (see earlier), fungicide, fungistat, virucide, microbicide and biocide.

Not all authorities would agree with all of the definitions listed, and one term in particular which might be hotly disputed is "antisepsis." This is often used to denote a chemical agent usually applied to human skin either by destroying micro-organisms or inhibiting their growth (Olivant and Shapton, 1970).

Another term which is frequently employed is "detergent-sterilizers" or "detergent-sterilants"; these consist of two components, one of which has a cleansing action, and the other an anti-microbial activity. Unfortunately, "sterilizer" or "sterilant" has an absolute meaning (see above) and this would imply that a detergent-sterilizer (sterilant) is sporicidal as well as being lethal to other micro-organisms, whereas those compounds which comprise the "sterilizer" or "sterilant" component are usually not sproicidal. There is probably, however, need of a term which includes the word "detergent." Foster *et al.* (1953) use "sanitization" to denote the application of a bactericidal process sufficient to render dairy equipment approximately sterile, this also implying that pathogens likely to be associated with such equipment and with eating and drinking vessels will be killed. Davis (1968) has employed the name "detergent-sterilant" throughout his review, although, as pointed out earlier, his definition of a sterilant differs from that of BDMA (1970). Throughout the present chapter, "sanitizer," without quotes, will be employed rather than these other terms.

II. Kinetics

It is tempting to imagine that the application of an appropriate disinfection procedure will result in immediate elimination of all micro-organisms from the site of interest. This temptation is often fostered by various advertising interests in pursuance of their sales campaigns. However, a cursory inspection of the literature soon dispels this cosy, over-simplified view. Disinfection has been shown repeatedly to be not only a gradual or even prolonged, process, but also a complesc one.

Almost invariably, investigations into the course of disinfection processes have involved the study of purified cultures of micro-organisms (usually bacteria) under specified conditions. This has led to certain criticisms that such systems are too far removed from reality to be of practical significance. While it is true that considerable caution must be exercised in applying the results of these studies to practical situations, the experimental systems are still far from simple, and have yielded much useful information.

1. Survivor curves

The most common method of monitoring the progress of a disinfection process is by means of viable counting techniques. These suffer from certain inherent limitations. In particular, the absolute values obtained are dependent on the specific technique and the experimental conditions associated with it; and in addition, cells which have been exposed to the process, may respond quite differently from those examined prior to exposure. In order to obviate this difficulty, alternative methods of assessing "vital activity" have been suggested, usually biochemical in nature. Unfortunately, while the greater simplicity of these methods allows more precise measurements to be made, the killing of micro-organisms usually involves a whole series of complex reactions, which makes correlation of the results rather difficult. Despite their faults, viable counting methods do reflect the complexity of the killing process.

The usual scheme of events is to expose the chosen culture of micro-organisms to the disinfection process of interest, under controlled experimental conditions. Estimates of the viable population density of the system are made by performing viable counts on representative samples removed from the system. For convenience, these estimates are usually plotted graphically against time of exposure or occasionally dosage of the disinfection agent employed. While the estimated numbers of organisms may be plotted directly, they are usually converted to a proportional basis such as "surviving fraction" or "percentage survivors," since this facilitates visual comparison of the results.

The simplest graph so obtained is the arithmetic plot which invariably exhibits a curve of similar general form to Fig. 1. The main point of interest about this curve is that it indicates that the rate of disinfection varies inversely with the number of surviving organisms. This is interpreted as an indication that the individual cells of the culture exhibit differing sensitives to the process, i.e. there is a distribution of resistances. Unfortunately,

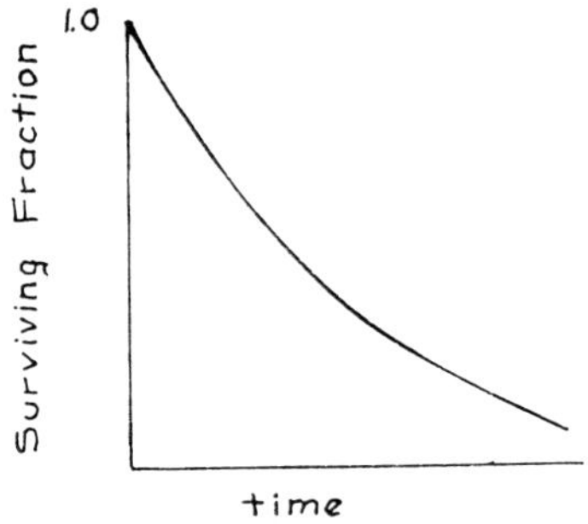

Figure 1

curves of this type are difficult to analyse or to compare visually, and so the survivors are often plotted in a logarithmic fashion. This results in a whole "family" of possible results, as shown in Fig. 2.

Figure 2a shows the simplest result, the familiar straight line which is often prized for its ease of characterization. It also possesses the sometimes dubious advantage of ease of extrapolation; a property which should be utilised only with extreme caution. This graph indicates that the rate of disinfection is inversely proportional to the logarithm of the number of surviving organisms. The similarity between this situation and the kinetics of a first-order chemical reaction has caused this type of response to be described as unimolecular or monomolecular. It is important to stress, however, that the description applies to the graphical response of the system; for it would be extremely naive to assume from this that the mode of death of the cells is attributable to a first-order chemical reaction. The straight line may be described mathematically by the equations:—

$$k = \frac{1}{t}\log\left(\frac{N_0}{N}\right)$$

where

k = rate constant or slope of line
t = time elapsed
N_0 = number of viable cells initially
N = number of viable cells at time t

or, if natural logarithms are used:—

$$\frac{N_0}{N} = \exp(kt)$$

These equations have led to the alternative and preferred descriptions of the response as being exponential or logarithmic. Since the number of

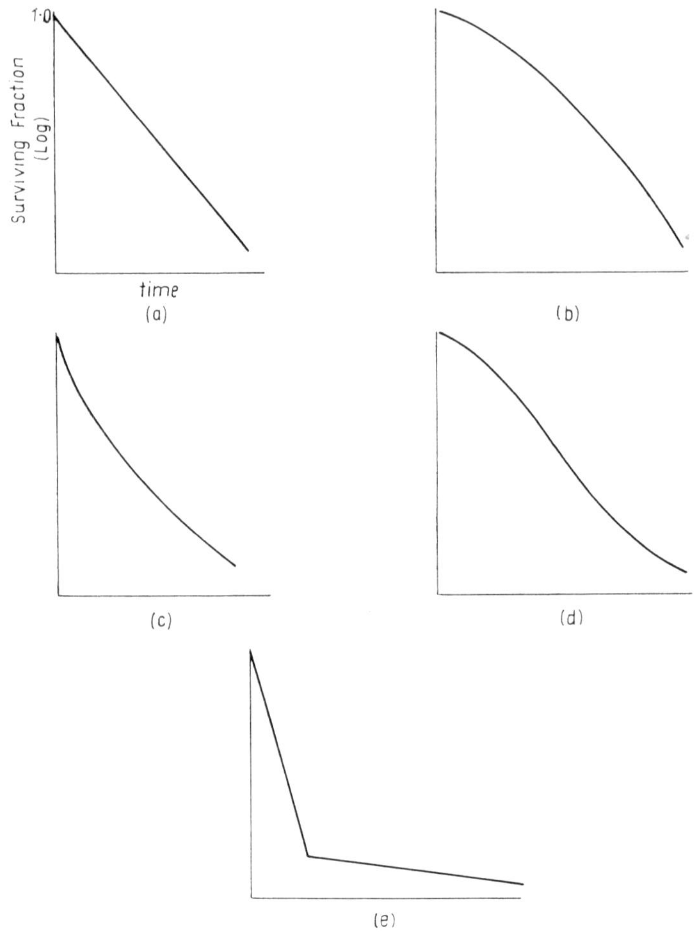

Figure 2

viable cells decreases throughout the process, then the rate constant, k, is always negative in character.

The two curves shown in Fig. 2b and 2c illustrate similar, though opposite, deviations in response from the straight line ease. These are described, according to their shape, as concave or convex, and qualified by the additional designation upward or downward, to indicate orientation. The graphs indicate that the rate of disinfection changes gradually in the early stages, but then assumes a more steady rate of change similar to that in the straight line case. In Fig. 2b, the rate of disinfection is relatively slow at first, but gradually increases to a steady, limiting value. Various interpretations have been suggested to account for this change; among them, that the distribution of resistances between the individual cells exhibits a relative deficiency of cells of low resistance; or alternatively, that the cells must pass through one or more intermediate stages before becoming sensitive to the disinfection process in question. The weakness of such interpretations is underlined by the fact that changes in the experimental conditions often result in a change in the shape of the graphical response. Figure 2c illustrates a relatively fast rate of disinfection initially, which gradually decreases to a steady, limiting value.

In Fig. 2d is illustrated the most common deviation from the straight line response, the sigmoid curve. Responses of this type are more easily demonstrated in systems employing a moderate rate of disinfection. As the rate of disinfection is increased, the limitations of viable counting techniques make it more and more difficult to monitor the progress of the process with any precision. This results in the apparent response becoming indistinguishable from the straight line case. It is sometimes suggested that the sigmoid response is the most common situation encountered, but that it is often unrecognised due to the practical difficulties experienced. The sigmoid response is usually interpreted as an indication that the distribution of resistances between the individual cells is of the log normal type. Complete agreement with this model distribution is indicated when the sigmoid curve is symmetrical.

Figure 2e illustrates a response of particular interest. The graph consists of two parts, both of which are linear but of different slope, with a fairly sharp transition between the two. This would appear to indicate a fairly rapid rate of disinfection initially, followed by a fairly sharp transition to a lower, but steady, rate. Such a sudden transition naturally engenders interest, if not suspicion. It has been suggested that this type of response indicates the presence of two distinct groups of cells, each of which exhibits its own characteristic distribution of resistances. Experiments with a mixture of two bacterial cultures of different identity, whether obtained from different species, or consisting of spores and vegetative cells of the same species, can be shown to yield this type of response. The first part of the graph corresponds to the usual response of the more sensitive component, and the second part to that of the more resistant component. However, at relatively low temperatures and humidities, exposure of nominally homogeneous cultures to ethylene oxide gas often yields this type of response. While it is sometimes suggested that this indicates the presence of two distinct groups of cells, as discussed above, it must also be considered that not only can this phenomenon be demonstrated with cultures apparently homogeneous to other sterilization methods, but also that this two-part response reverts to the exponential type on increasing the temperature or humidity of the system.

As indicated in the foregoing discussion, consideration of the shapes of survivor curves may provide useful circumstantial evidence on which to base hypotheses relating to the response of cell populations to disinfection processes. However, as also indicated in the discussion, the operative word is "circumstantial".

2. Empirical parameters

While a survivor curve illustrates the response of a cell population in terms of variation in number of survivors with time (or dose) of disinfection treatment, this is essentially a static situation since all other factors are held constant. Often, it is the variation in response of the system to changes in experimental conditions which is of particular interest. This variation in response may be monitored by constructing families of survivor curves, one curve for each level of the factor being varied. Comparison of curves within these families indicates the nature of the change in response.

The two factors most prominent in the regulation of disinfection processes are temperature and concentration of disinfectant. For both of these factors, it is found that their influence on the disinfection process varies in a regular manner over a fairly wide range of conditions. As a result, empirical parameters have been derived which enable the influence of these two factors to be characterised in a convenient form.

Temperature coefficient

In general, it is found that the activity of a disinfectant varies directly with the temperature, i.e. the higher the temperature the greater the activity. This change in activity may be quantified by expressing the disinfection rates observed at two different temperatures as a ratio. The value of this ratio is found to remain reasonably constant over a wide range of temperature. This may be expressed mathematically:—

$$\theta_{(T_2 - T_1)} = k_2/k_1$$

where k_1 and k_2 are the rate constants at temperatures T_1 °C and T_2 °C, respectively, with T_2 greater than T_1. The constant, θ, is termed the temperature coefficient, and assumes a numerical value which is characteristic of the agent and, to a certain extent, of the organism, employed. The subscript $(T_2 - T_1)$ °C is necessary to indicate the temperature difference. Since the disinfection rate is inversely related to killing or extinction time, the latter may be used instead. The expression then becomes

$$\theta_{(T_2 - T_1)} = t_1/t_2$$

where t_1 and t_2 are the extinction times at temperatures T_1 °C and T_2 °C, respectively. A high value for θ indicates that the process is relatively sensitive to temperature changes.

It is usual to find two versions of the temperature coefficient in most common use: the coefficient for a 1°C temperature change, θ, when the subscript is usually omitted; and the coefficient for a 10°C temperature change, θ_{10}, which may be sometimes expressed as Q_{10}. The popularity of the 10°C coefficient follows from its use to characterise changes in reaction rates in chemical systems. This enables interesting comparisons to be made between disinfection processes and non-living chemical reactions. Attempts have been made to deduce modes of death of the organisms by such comparisons. However, subsequent biochemical investigations have tended to disagree with these deductions.

Dilution coefficient

The activity of disinfectants under otherwise constant conditions is found to vary directly with the concentration of disinfectant employed over a considerable concentration range. As before, killing or extinction times are usually employed as a measure of disinfection rate. The effect of concentration may be expressed mathematically as:—

$$c^{\eta}t = \text{a constant}$$

or:—

$$\eta \log c + \log t = \text{a constant}$$

where

c = concentration of disinfectant
t = extinction time
η = dilution coefficient

A high value for the dilution coefficient indicates that the process is relatively sensitive to changes in concentration of disinfectant. The dilution coefficient is sometimes referred to as the concentration exponent.

Other factors influencing activity

The antimicrobial activity of several disinfectants is influenced to a considerable extent by changes in pH. For example, a rise in pH results in a decrease in the activity of phenols (Bennett, 1959), organic, acids, compounds liberating chlorine, benzoic acid and iodine, although iodine is less affected by acidity than is chlorine. An increase in pH increases the dissociation of phenols and benzoic acid (Wedderbern, 1964); chlorine in water forms HClO, which is dissociated with a rise in pH with a concomitant loss of activity. In contrast to the above, however, there is an increased microbicidal activity of the quaternary ammonium compounds (QACs) and of acridines (Foster and Russell, 1971); in the case of QACs, the effect of pH is considered by Salton (1957) to be on the cell rather than the disinfectant molecule, since the number of negatively-charged groups on the bacterial surface will be increased as the pH rises, thus influencing the number of positively charged molecules which can be attacked.

Another factor which influences the activity of certain antimicrobial agents is organic matter, e.g. the presence of blood, serum, pus, etc. In general terms, the more chemically reactive a compound, the greater the effect of organic matter on its activity. This is particularly true with the hypochlorites. Other examples are provided under individual compounds later.

3. Mathematical models

A considerable number of mathematical models have been derived at various times, in attempts to reconstruct the disinfection process. These have utilised deterministic, probabilistic, and thermodynamic approaches to the problem; and have recently been reviewed in detail by Prokop and Humphrey (1970).

In general, these models are based on attempts to reconstruct the survivor curves obtained, even though, as already discussed, these are liable to be changed by changes in experimental conditions. In addition, they usually involve preliminary assumptions concerning the nature of the disinfection process. Consequently, attempts to deduce mechanisms of killing from these models tend to be rather disappointing. They are, however, useful in terms of concise descriptions of the results of the process; and as further biochemical information on modes of killing becomes available, their usefulness and reliability should increase further.

III. Disinfectant Testing

The testing of disinfectants is a topic with a long history of controversy. Differing opinions on the merits and relevance of various methods are legion, and have been the cause of many heated exchanges between those holding them. The wide variety of methods available may most conveniently be considered under the headings of screening, standardisation, and "in-use" tests. There is, however, a degree of overlap between the three categories.

1. Screening tests

These are usually of the simplest type, since they are speculative by nature, and often involve the testing of large numbers of disinfection agents or formulations; both time and cost usually dictate this simplicity. The actual methods employed vary according to the physical characteristics of the disinfectant and the type of use envisaged. Ideally, any test should be as realistic as possible, although for screening purposes this will be subject to the requirements of simplicity discussed above.

Disinfectants which are soluble or miscible in water are often incorporated in microbiological culture media which are then inoculated with suitable micro-organisms. After incubation under appropriate conditions, the inoculated media are examined for growth of the organisms, absence of growth indicating that the inoculum has been inhibited. Where the continuous presence of the disinfectant is inappropriate, then the method is modified to include a suitable means of removing or inactivating the disinfectant after a pre-determined exposure time.

Other disinfectants may be tested in a similar manner, by arranging for inocula of suitable organisms to be subjected to standardised exposure to them. After exposure, the organisms are inoculated into suitable culture media, and incubated as before. Since the end-point in these tests is the death of all the organisms involved, this type of test is often referred to as an Extinction Test.

The phrase "suitable organisms" used above, represents one of the key factors in the testing of disinfectants. A disinfectant is expected to kill all undesirable organisms, which usually refers to organisms injurious to health (see Section I). Test organisms may be chosen, therefore, either for their resistance to disinfection or, like *Salmonella typhi*, for their medical significance. Although the designation "undesirable organisms" covers a wider range of life-forms, most disinfectant testing has employed various species of bacteria for reasons of convenience. In retrospect, this does not appear to have been a significant disadvantage, since activity against bacteria usually coincides with that against the other life-forms. However, although it is recognised that bacterial spores are much more resistant to disinfection than the vegetative forms, the latter have, with few exeptions, invariably been used in testing. This is partly due to convenience, but it should be noted that there are many disinfectants marketed which under normal conditions of use are incapable of killing bacterial spores. This has been countered, in certain quarters, by re-defining a disinfectant as a chemical agent capable of killing bacteria but not necessarily bacterial spores (Section I).

2. Standardisation tests

This is the area which has given rise to the greatest amount of contention. In most cases, this has been due to misinterpretation and misapplication of the results of testing in situations where they have little, if any, relevance. The standardisation of disinfectants usually requires the use of microbiological techniques rather than chemical assay. Even when chemical characterisation of the active agent(s) involved is possible, the activity of a disinfectant will usually be significantly influenced by factors concerned with the formulation and method of use the product. However, although the disinfectant may be standardised in terms of its intended biological effect, this is carried out under specific, controlled conditions. Depending on the similarity of their circumstances, the results from standardisation tests may, or may not, be capable of extrapolation to practical situations. This is where the controversy tends to arise.

The most popular methods of testing for standardisation purposes have been extinction tests. These are basically similar to the general method discussed in the previous section, except that the procedure and materials used are rigidly standardised in order to achieve best reproducibility of results. In addition, the most widely used tests employ the pure chemical, phenol, as a control of the resistance of the test organisms used. Since results are expressed by comparison of the activities of phenol and the disinfectant being tested, these tests are referred to as Phenol Ceofficient methods. The most popular and official versions in use are the Rideal-Walker Method (as modified by B.S.S. 541), the Chick-Martin Method (as modified by B.S.S. 808), and the Association of Official Agricultural Chemists (A.O.A.C.) Phenol Coefficient Method (1960). Details of these tests may be found in the appropriate publications.

While the use of phenol as a control on the resistance of the test organisms is extremely valuable, it has led to widespread assumptions that the ratio of activities of disinfectant and phenol indicated by the phenol coefficient will hold true in all circumstances. This, of course, is far from true. The A.O.A.C. (1960) attempted to improve this situation by introducing a Use-Dilution Method. This test is based on the hypothesis that disinfectants in use shall be at least as efficient as 5 per cent phenol, and that a dilution of twenty times the phenol coefficient should achieve this. Accordingly, this dilution is tested against standardised cultures in order to confirm this, or alternatively, in order to determine a correction factor.

Whereas the extinction methods discussed above evaluate the extinction concentration corresponding to pre-determined exposure times, Berry and Bean (1954) devised a test for evaluating extinction times for chosen disinfectant concentrations. While the test is only applicable to phenolic and other easily inactivated disinfectants, it has been claimed to be at least as reproducible as other methods in common use (Cook and Wills, 1954). The method of assessing the end-point of the reaction has been further improved by Mather (1949).

Methods other than extinction tests, for standardising disinfectants include various methods based on assessment of cessation of vital enzyme activities. The enzymes involved have generally been certain oxidases or dehydrogenases (Sykes, 1939; Knox *et al*., 1949). In a less specific manner, inhibition of respiration has been used as a method of assessment (Roberts and Rahn, 1946). The controversy surrounding these methods centres around the problem of correlating enzyme activity with viability.

Other methods which have been proposed include measurement of postincubation opacity corresponding to standard survivor levels (Needham, 1947), and also measurement of cell volume increase following postexposure incubation (Mandels and Darby, 1953).

The range of tests discussed above is by no means exhaustive, and only covers general purpose tests applicable to water-miscible disinfectants. Any number of alternative tests could be devised by appropriate selection and standardisation of the various parameters. In addition, there are numerous tests which can be, and have been, devised in order to standardise disinfectants intended for specific uses such as sporicidal or tuberculocidal duties (A.O. A.C. 1960).

Methods of standardising disinfectants other than those to which the above discussion applies are usually designed more closely around the particular use envisaged. Thus, a considerable element of actual, or simulated, in-use testing is usually involved. For convenience, they are best discussed in the following section.

3. In-use tests

The previous two sections have dealt with testing methods applied in the laboratory, which yield information primarily of value to the disinfectant manufacturer. The "consumer," however, is almost solely concerned with the performance of disinfectant materials under the conditions of use which are associated with the application envisaged. To this end, he is more interested in testing methods which resemble, as closely as possible, these practical conditions. However, the foregoing remarks should not be taken to imply that laboratory standardisation methods are completely arbitrary. In view of the greater difficulty experienced in killing organisms which are present as dried surface films, as opposed to those in fluid suspension, many standardisation tests have employed films on various surfaces as their inocula. Surfaces used have included silk thread (Koch, 1881), garnets (Kronig and Paul, 1887), glass cover-slips (Jensen and Jensen, 1933), glass cylinders (Mallmann and Hanes, 1945), glass slides (Johns, 1946), stainless steel cylinders (A.O.A.C. Use-Dilution Confirmatory Test, 1960), and glass tablet tubes (Hare, Raik and Gash, 1963). It should be noted that while these surfaces represent a step towards the practical situation, they nevertheless comprise a collection of laboratory "artifacts" when compared with real situations. A nearer approach was achieved by use of surfaces such as rubber strips (Goetchius and Botwright, 1950) and glazed, waxed, and rubber tiles (Rogers, Mather and Kaplan, 1961).

Where the physical size of articles required to be disinfected is fairly small, it is quite feasible to carry out in-use testing in the laboratory. Hence metal trays were used by Neave and Hoy (1947), 10 gallon milk churns by Hoy and Clegg (1953), and small drinking glasses by Gilcreas and O'Brien (1941). Similarly, scalpels, syringes and similar small items may conveniently be subjected to in-use testing in the laboratory. Where the physical size of the system is somewhat greater, then the laboratory must be forsaken in order to carry out on-site testing. Typical examples of such situations include hospital walls and floors, and industrial or dairy processing machinery. The outstanding value of in-use testing arises from the fact that results obtained are directly applicable to the system without the need for interpretation and extrapolation, and that practical difficulties such as short contact times or inaccessibility of certain areas can be accounted for.

The choice of test organisms usually reflects either of two main approaches. The simplest approach is to inoculate the system artificially with organisms considered to be of practical significance in the particular application. This significance may be due to the resistance to disinfection of the organism, or alternatively, to its practical effect on the system should it survive the disinfection process. In order to simulate the practical system, the organisms may be suspended in appropriate materials before inoculation. An example would be the use of milk as a suspending fluid in tests on dairy disinfection. A slight variation on this approach is to inoculate with indeterminate mixtures of likely organisms obtained from natural sources, such as low quality, raw milk. The second approach involves the use of the normal, pre-existing flora of the system which has arisen during normal use. Investigations of this flora before and after the disinfection process would provide direct evidence as to whether the process has achieved its object. One difficulty with this approach is that of providing suitable growth conditions for all of the possible species of organisms which may be present. Also, the normal flora is likely to vary with time, and so, ideally, this method is best applied in the form of a routine monitoring procedure. Both of these difficulties produce an increase in the financial cost of this type of approach, but the reliability of the results obtained is correspondingly high. The ideal method would involve a combination of these two approaches: an initial test with known resistant organisms in order to indicate the upper levels of activity which may be required; this to be followed by routine monitoring tests to guard against both unsuspected resistance amongst the normal flora, and unforeseen breakdowns in the method of application.

Any testing method which involves the use of surface films is subject to the problem of physical recovery of the organisms. In the case of small surfaces, the articles can often be placed in or on a suitable nutrient medium, and provided this allows contact between the medium and the organisms in the film, then growth takes place. Where larger or less accessible surfaces are concerned, then the simple direct method will have to be replaced by some kind of sampling technique. Sampling techniques vary widely and not all are applicable in all situations. In the case of accessible surfaces, simple methods such as wiping with sterile swabs may be used; or alternatively, flooding the surface with a sterile liquid, some or most of which is then removed by swab or pipette. An interesting alternative consisting of a sterile, agar medium, "sausage" was devised by Ten Cate (1965). The exposed transverse surface of this sausage is pressed against the surface to be sampled. After sampling a slice is cut from the sausage and incubated with the exposed side uppermost. A simple method which is used for sampling skin surfaces may also be used in other applications. This method employs adhesive cellophane tape. The adhesive surface is pressed onto the surface to be sampled, and then removed for transfer to a suitable medium. Where surfaces are inaccessible, as in pipes and processing machinery, then a rinsing technique is usually most convenient. The resulting liquid may be added directly to suitable nutrient media, or, if present in large bulk, may be filtered through membrane filters to remove the organisms. The resulting filter membrane is then transferred to a suitable medium.

One final problem which is of importance in all methods of testing which involve assessment of viability, is that of recovery and growth of the test organisms following exposure to the disinfectant. Organisms which have survived a disinfection process often show altered requirements for optimal growth. Response to physical conditions such as incubation temperature, as well as to biochemical conditions such as dependence on certain nutrients, may be completely altered from that of unexposed cells. While considerable effort has been made to derive efficient recovery methods (Flett *et al.*, 1945; Jacobs and Harris, 1954, 1960; Harris, 1963; Russell, 1964) the problem is so variable and so many combinations of factors must be considered, that it is far from being solved. There is general agreement that recovery methods should be selected which will allow maximum recovery of exposed organisms; but putting this into effect can be extremely difficult.

IV. Liquid Disinfectants and Chemosterilizers

Several chemical agents have long been employed for destroying micro-organisms, although it is frequently asserted that such substances are without effect on bacterial spores. With many agents, however, this is untrue (Sykes, 1970; Russell, 1971). The most important substances are phenols and cresols, biguanides, chlorine—releasing compounds and other halogens, aldehydes, alcohols, quaternary ammonium compounds, mercury compounds, strong acids and alkalis and hydrogen peroxide. The majority of these are considered in detail below.

1. Phenols and cresols

Although Kronig and Paul (1897) and Chick (1908) showed that phenol was active against spores, the concentrations, 5%, employed, were considerably higher than those needed to kill vegetative bacteria. More recent studies have indicated that bacterial spores are not killed even after long exposure to phenols (Sykes, 1958, 1965; Loosemore and Russell, 1963, 1964; Russell and Loosemore, 1964; Russell, 1965, 1971; Rubbo and Gardner, 1965; Briggs, 1966). Of the bacterial spores, *Bacillus stearothermophilus* is the most resistant to phenol and *B. megaterium* the most sensitive (Briggs, 1966). However, in contrast to its lack of sporicidal activity, phenol is sporostatic at low concentrations.

Several factors influence the antimicrobial activity of phenols and cresols (Bennett, 1959; Cook, 1960; Bean, 1967):

(a) Concentration. These compounds have high concentration exponents, η, which, as described above, indicates that they rapidly lose their antibacterial activity on dilution. This also means that dilution procedures can be used to prevent the carryover of inhibitory concentrations into recovery media when viable counts or sterility tests are being carried out (Russell, 1964; Russell and Gilbert, 1964). Studies from Bean's laboratory (Bean and Walters, 1955; Bean and Das, 1966; Bean, 1967) are of interest, for they show that with dilute solutions of disinfectants with high intrinsic activity, e.g. benzylchlorphenol, the high proportion taken up by the cells means that the concentration remaining is only weakly bactericidal, so that the surviving cells do not meet lethal conditions.

(b) Temperature. The bactericidal activity of the phenols and cresols increases rapidly with an increase in temperature. Examples of temperature coefficients (θ_{10}) against *E. coli* at 30–40° (Tilley, 1942) are: phenol 8.4, *o*-cresol 6.9, *p*-cresol 5.6.

Of importance, also, is the finding that these substances are sporicidal at elevated temperatures (Berry *et al.*, 1937; Russell and Loosemore, 1964). As a result of the findings of Berry *et al.* (1937), one method of sterilizing certain injections by heating them with 0.2% w/v chlorocresol is still an official method in Britain (British Pharmacopoeia, 1968).

(c) pH. The phenols are more active at an acid pH than in alkaline solution, as phenates (phenoxides) are formed at high pH. Acid pH also results in a more effective, although still slow, sporicidal action (Sykes and Hooper, 1954).

(d) Organic matter. The presence of organic matter may decrease the antimicrobial activity of these compounds; this was early recognized in the design of the Chick-Martin test for evaluating phenolic disinfectants (Chick and Martin, 1910; Garrod, 1934, 1935). The results do, however, depend on the actual phenol used and on the kind of organic matter, and the interference is less than with other disinfectants such as the quaternary ammonium compounds (Cook, 1960).

(e) Oxygen tension. Anaerobic bacteria are generally more resistant than aerobes to phenols. Moreover, facultative organisms, e.g. *E. coli*, are more resistant when grown under anaerobic conditions.

(f) Type of organism. As described above, these compounds are bactericidal and sporostatic at low concentrations, and sporostatic and not sporicidal even at high concentrations. As a group, however, they are also fungicidal to several moulds, and use is made of this in the inclusion of cresol and chlorocresol as preservatives in creams which are liable to fungal contamination (Wedderbern, 1964).

Morris and Darlow (1971) have pointed out that phenolic compounds with high R-W coefficients are effective against some viruses, but that they are generally too variable in their activity to be suitable as general virucidal agents.

(g) Chemical nature. Dihydric and trihydric phenols (Fig. 3) are generally less active than phenols, and alkylation of monohydric phenols to give cresols potentiates the antimicrobial activity. Also halogenation of the phenols increases their activity (although to a lesser extent if the halogenation is in the ortho- than in the para-position) and this is even more pronounced when accompanied by the introduction of aliphatic or aromatic groups into the nucleus, e.g. *p*-chloro-

m-cresol (chlorocresol). This increase in antimicrobial activity is, however, paralleled by a decrease in water solubility.

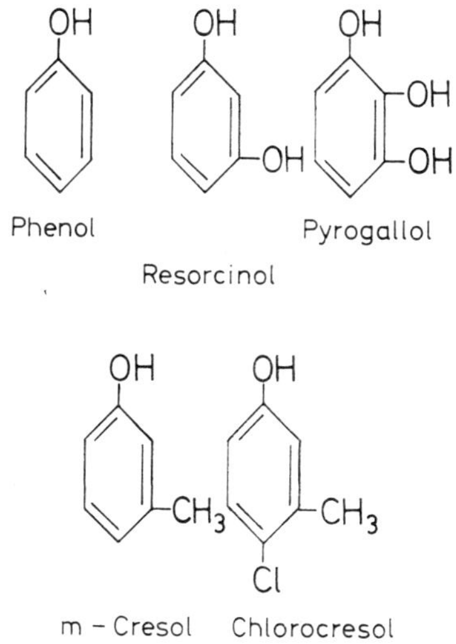

Figure 3

To overcome this decrease in aqueous solubility, various soaps have been used to render water-soluble (solubilize) these substances. However, the effect of soap on the biological efficacy of the phenols depends on two factors, firstly the nature of the soap, and second the proportion of soap to phenol. Solution of Cresol with Soap (Lysol, B.P.), for example, contains 50% of cresol in a saponaceous solvent and has a bactericidal activity which depends on the nature and amount of soap used (Berry and Stenlake, 1942). The ratio of cresols to soap may be critical, the optimal cresol-soap ratio being of the order of 2 : 1. In lysol, the soap content is *c.* 22%, and the ratio is thus 2.2 : 1. The soap solutions are able to solubilize insoluble phenols in the micelles; the critical micelle concentrations (c.m.c.s.) of different soaps vary, and this explains differences in bactericidal action noted above.

Lysol and the so-called "black-fluids," which consist of the lower coal tar phenols, are formulated in sufficient soap so that they are retained in solution when diluted with water. In contrast, the white fluids consist of concentrated emulsions of high boiling phenols stabilized with protective colloids; they can be diluted with hard or soft waters, whereas black fluids dilute to emulsions with soft waters only.

Micelles have been considered as being reservoirs of phenols, so that when the concentrated solution is diluted before use, the phenols are released by dilution below the c.m.c. to give a highly active solution. Two types of system have been investigated experimentally in attempts to assess the exact role of the micelles; these are (a) constant phenol concentration (Alexander and Trim, 1946), and (b) a constant phenol/soap ratio (Berry, 1951; Berry and Briggs, 1956; Berry, Cook and Wills, 1956; Cook, 1960). With a constant phenol system, there is a rapid increase in bactericidal activity below the c.m.c. of the soap which could be the result of an increased uptake of phenol together with an increased permeability of the bacterial surface (Mulley, 1964); however, above the c.m.c. in this system, there is a decrease in bactericidal activity, as the increasing numbers of micelles being formed compete for the phenol with the cell surface. In systems where there is a constant phenol/soap ratio, there is likewise an increase in activity below the c.m.c., and a decrease in activity at the c.m.c. However, in this system the activity again increases at higher soap concentrations; thus, this increased bactericidal activity parallels a saturation of the soap micelles with the phenol. The reasons for this remain unclear, for although the soap itself could contribute to or be responsible for the increased activity, systems in which the soap has a low activity give similar results. The actual role of the micelles is thus difficult to assess. Of the two systems described, it is apparent that the system with the constant phenol/soap ratio is more important from the viewpoint of practical disinfection. It is necessary to use the lowest possible proportion of soap to phenol, thus giving a comparable situation to bactericides in two-phase systems, in which the bactericidal efficiency is related to the concentration in the aqueous phase (Bean and Heman-Ackah, 1965).

Phenol itself is an effective bactericide and antifungal agent, which is used as a preservative in some injections and creams; it is also the standard reference substance in some methods of testing phenolic bactericides (section III). Cresol, a mixture of *o*-, *m*- and *p*-cresol, in which the *meta*-isomer predominates, and of other phenols obtained from coal tar, is highly bactericidal and fungicidal. Apart from being the active constituent of Lysol, it is also used as a preservative in certain injections and creams. Chlorocresol is a powerful antimicrobial agent which, in conjunction with heat, is employed for the sterilization of certain injections.

It is also used as a preservative in certain cosmetic creams and lotions, and is included in Sodium Benzoate and Chlorocresol Solution, which may be used for the storage of sterilized surgical instruments, the sodium benzoate delaying rusting. Chloroxylenol has low water solubility and is solubilised by means of soap, the solution being known as Roxenol. Its antimicrobial activity is markedly reduced in the presence of organic matter. Thymol is a potent bactericide and fungicide which, in the form of Glycerin of Thymol, is employed as an oral antiseptic. Hexachlorophane [2,2′-methylenebis (3,4,6-trichlorophenol)] is most frequently used as a soap containing 2% of the substance; the effectiveness of this soap as a skin disinfectant may depend upon the accumulation of hexachlorophane on the skin. Pentachlorophenyl dodecanoate is extensively used as a fungicide and insecticide in the textile and packing industries (D'Arcy, 1971). Unlike the parent molecule, pentachlorophenol, it is non-toxic to humans.

None of the above compounds can be relied upon to kill bacterial spores at ordinary temperatures.

Another group, related to the phenols, consists of the esters of p-hydroxy-benzoic acid (the parabens). Unlike benzoic acid, the dissociation of which increases with increasing pH with a corresponding decrease in activity, the parabens are active against bacteria and fungi over a fairly wide pH range. Their bactericidal and fungicidal properties increase within the homologous series, but this is paralleled by a decrease in aqueous solubility. The parabens are not sporicidal, and although on their discovery they were heralded as being the ideal preservatives, they are now used (often in combination of two or more) mainly as preservatives in various pharmaceutical and cosmetic products (Russell, Jenkins and Harrison, 1967).

The phenols and parabens are believed to have an effect on the cytoplasmic membranes of bacteria and fungi.

2. Biguanides and bisbiguanides

Biguanides have the general formula

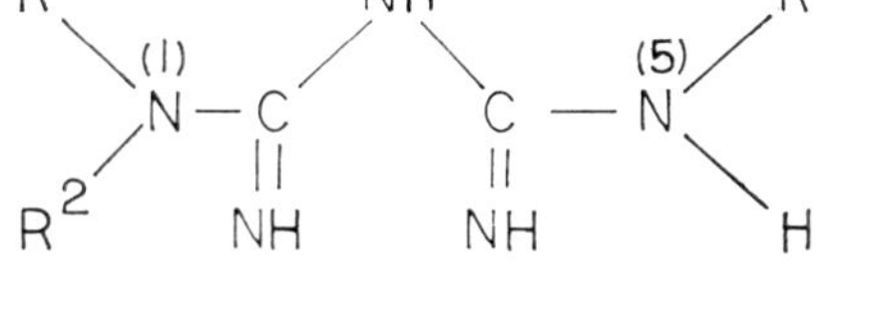

Three distinct antimicrobial actions of N^1, N^5-substituted biguanides have to date been recognized (Weinberg, 1968): germicidal, antiviral and antimalarial. Unfortunately, no one compound is generally active in more than one of these three categories. The requirements for each type of activity are fairly unique, e.g. for maximum broad-spectrum germicidal activity, both N^1 and N^5 should have a halogen–substituted aralkyl substituent.

Bisbiguanides have the general formula

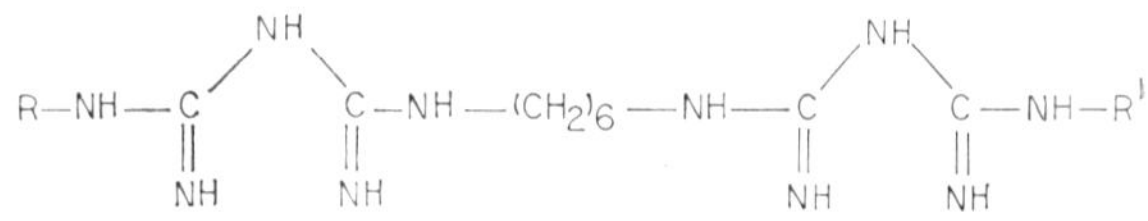

For maximum broad spectrum germicidal activity R and R′ should consist of a halogen–substituted aryl group. A number of bisbiguanides are known to be germicidal, and the most important member of this group is chlorhexidine (R and R^1 are both C_6H_5Cl).

Chlorhexidine was first described in 1954 (Davies *et al.*, 1954), and it is bacteriostatic in low concentrations; higher concentrations are bactericidal to several bacterial species, including strains of the genus *Proteus*, of which *Pr. mirabilis* is the most resistant, and *Pr. rettgeri* and *Pr. morganii* are the most sensitive to chlorhexidine, with *Pr. vulgaris* occupying an intermediate position. Some strains of *Ps. aeruginosa* may be highly resistant. Chlorhexidine also possesses antifungal activity. It is a membrane-active agent.

Chlorhexidine, the active constituent of "Hibitane," is used in surface disinfection, as an antimicrobial agent in eye-drops, and, in the presence of sodium nitrite to prevent corrosion, for the storage of surgical instruments.

3. Chlorine-releasing compounds

It was observed by Dakin (1915, 1916) that the commercial hypochlorites then in use were not of constant composition and contained free alkali and sometimes free chlorine. He thus developed a solution (Dakin's Solution or Chlorinated Soda Solution, Surgical) which is still in use today. The stability of free available chlorine in solution is dependent on a number of factors, in particular on the chlorine concentration, pH of the solution, the presence of catalysts, temperature, the presence of organic matter, and light (Dychdala, 1968).

The types of chlorine compounds which are

frequently used are (1) Hypochlorites. These are cheap and convenient to use, and have a wide antibacterial spectrum (Davis, 1963), but they are less active against spores than against vegetative bacteria and are inactive against Mycobacteria (Report, 1965).

The hypochlorites are moderately effective against animal viruses.

The antibacterial activity of the hypochlorites decreases with increasing pH (Charlton and Levine, 1937; Weber, 1950; Ito *et al.*, 1967; Hays *et al.*, 1967), e.g. whereas 99% of spores of *B. cereus* are killed after 2.5 min at pH 6 by a solution containing 25 ppm available chlorine, nearly 8 hours are required for a comparable kill at a pH of *c.* 13 (Rudolph and Levine, 1941). At constant pH, the time to kill bacteria depends on the concentration of available chlorine. The sporicidal activity of sodium hypochlorite may be potentiated by various compounds, e.g. by the addition of ammonia (Weber and Levine, 1944) or 1.5–4% sodium hydroxide (Cousins and Allan, 1967), notwithstanding the earlier comment about pH. In the presence of bromide, hypochlorite has an enhanced effect in bleaching cellulosic fibres as compared with hypochlorite alone, possibly because of a continuous generation of hypobromite when hypochlorite is in excess. A potentiation of the bactericidal effect of hypochlorite has been achieved by the addition of small amounts of bromide (Farkas-Himsley, 1964).

The antimicrobial activity of hypochlorites is considerably reduced by organic matter. However, the hypochlorites are used in the disinfection of water, dairy equipment and eating utensils. (2) Chloramine-T (sodium *p*-toluene sulphonchloramide). Dakin *et al.* (1916) considered that chloramine-T had a powerful germicidal action. It is bactericidal and sporicidal, although the rate of kill is slower than with the hypochlorites. Its activity is considerably higher at acid than at alkaline pH (Charlton and Levine, 1935; Weber, 1950), and a drop of 10°C in the reaction temperature results in a 3–4 fold increase in the time necessary to kill microorganisms (Weber and Levine, 1944). Chloramine-T is employed as a wound "disinfectant," and as a general surgical disinfectant. It is non-irritant and non-toxic, in contrast to Dichloramine-T (toluene-*p*-sulphon-dichloramide) which although a powerful disinfectant is not used because of its toxicity and instability.

The mode of action of chlorine compounds is unknown, although several proposals have been made, e.g. the formation of chloramines as a result of combination of chlorine with bacterial protoplasm, halogenation or oxidation reactions of chlorine with bacterial cells, changes in cellular permeability and an effect on enzyme systems. It has also recently been found, however (Bernarde *et al.*, 1967) that chlorine dioxide causes a marked and immediate cessation of protein synthesis in growing cells.

4. Iodine and iodophors

Iodine in aqueous or alcoholic solution is considered by most authors (Gershenfeld and Witlin, 1950; Gershenfeld, 1956; Report, 1965; Sykes, 1970) to be a reliable and effective germicide which is lethal to vegetative bacteria, bacterial spores and acid-fast bacilli. Spaulding (1968), however, considers that alcoholic iodine (0.5% iodine in 70% alcohol) possesses good activity against non-sporing bacteria and *M. tuberculoses* but none against bacterial spores, whereas Rubbo and Gardner (1965) state that bacterial spores are moderately resistant to iodine. Viruses are considered by Rubbo and Gardner (1965) to be moderately sensitive to iodine.

Iodine has a high fungicidal or fungistatic activity against yeasts and various moulds, but its antimicrobial properties are to a great extent inhibited in the presence of organic matter, since iodine is a highly reactive element.

Iodine is sparingly soluble in cold water, but more soluble in hot water. Stronger solutions can be made in potassium iodide solution or in aqueous alcohol. Iodine is more effective as a germicide at acid than at alkaline pH, but is less affected by pH than are chlorine compounds. The concentration of iodine to disinfect does not vary greatly with different types of micro-organisms. Various types of iodine solution are used for the first-aid treatment of small wounds and abrasions, and in pre-operative skin "disinfection." Iodine has also been employed for the sterilization of surgical catgut (although this method is now little used) and is nowadays used for the disinfection of drinking and swimming pool water, the disinfection of instruments and of clinical thermometers, and the sanitization of eating and drinking utensils.

Unfortunately, iodine solutions stain fabrics and tissues and tend to be toxic. However, certain non-ionic surface-active agents can solubilize iodine to form compounds, the iodophors (Blatt and Maloney, 1961; Davis, 1962, 1963, 1968) which retain the germicidal activity of iodine, but not its undesirable properties; these iodophors are literally "iodine-carriers." They are active against bacterial spores (Johns, 1954), including pathogenic anaerobic spores (Lawrence *et al.*, 1957; Gershenfeld, 1962). It is the concentration of free iodine in an iodophor which is responsible for its microbicidal action; this has been well demonstrated by Allawala and Riegelman (1953) who made a log-log plot of killing time against amount of free iodine, and found that the 99% killing time of *B. cereus* spores was a function of the concentration of free iodine in the presence or absence of added surface-active agent. The bactericidal properties of the iodophors are increased at low pH values, but their stability is unaffected (cf. hypochlorites). They may thus be employed with acids, e.g. phosphoric acid, to enhance their microbicidal action and also to assist in preventing the formation of film or milkstone (see section VI) in the dairy industry. The iodophors are considered (Davis, 1968) to be powerful detergents, although they do not dissociate protein as readily as do alkalis. The formulation of acidic solutions of iodophors is particularly useful when calcium or magnesium scale is encountered, but they can be corrosive, especially with galvanized iron.

5. Surface-active agents

Surface-active agents have 2 regions in their molecular structure, one a hydrocarbon, water-repellant (hydrophobic) group, and the other a water-attracting (hydrophilic or polar) group. Depending on the basis of the charge or absence of ionization of the hydrophilic group, surface-active agents are classified into (James, 1965):

(a) Anionic agents, which usually have strong detergent, but weak antimicrobial properties, e.g. sodium lauryl sulphate.

(b) Cationic agents, which have strong bactericidal, but weak detergent, properties. The term "cationic detergent" usually signifies a quaternary ammonium compound (QAC, onium compound), but this is not strictly accurate, as the concentration at which a QAC is germicidal is so low that its detergent activity is negligible.

(c) Non-ionic agents, which consist of a hydrocarbon chain attached to a non-polar water-attracting group, which is usually a chain of ethylene oxide units, e.g. cetomacrogols. Non-ionic agents have no antimicrobial properties.

(d) Ampholytic (amphoteric) agents, which are compounds of mixed anionic-cationic character, in which the charge can be similarly positive and negative. They are effective antimicrobial compounds, which are extensively used in the dairy industry.

The QACs are organically substituted ammonium compounds, in which the nitrogen atom has a valency of 5, four of the substituent radicals (R^1–R^4) are alkyl or heterocyclic radicals, and the

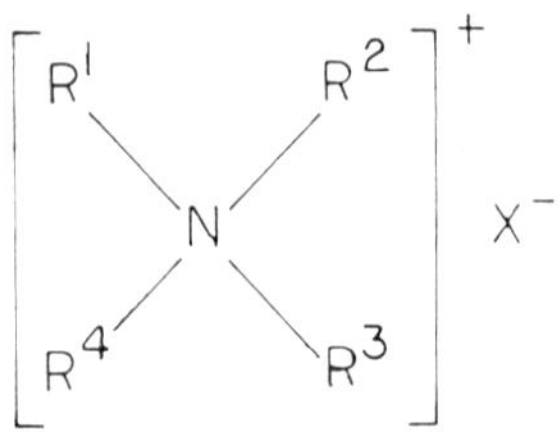

fifth is a small anion. The sum of the carbon atoms in the 4 R groups is more than 10. For a QAC to have a high activity, at least one of the four organic radicals must have a chain length in the range C_8 to C_{18}. The germicidal activities of the QACs were originally recognized in 1916, but they did not attain prominence until Domagk's work in 1935. They are primarily active against Gram-positive bacteria, with concentrations as low as about 1 in 200,000 being lethal; higher concentrations (c. 1 in 30,000) are lethal to Gram-negative bacteria, although *Pseudomonas aeruginosa* is highly resistant. The QACs possess antifungal properties, are sporostatic and not sporicidal (Russell, 1971) and are inactive against the tubercle bacillus. Their antimicrobial activity is markedly affected by organic matter, and they are incompatible with anionic surface-active agents, some of the non-ionic agents (such as Lubrols and Tweens) and the phospholipids, e. g. lecithin and other fat-containing substances. The QACs exert their maximal bactericidal activity under alkaline conditions. Although the QACs are effective bactericidal and fungicidal agents, it has been found (Grossgebauer, 1970) that viruses are rather more resistant than bacteria or fungi.

Although several hundred QACs have been prepared and tested for antimicrobial activity, only a few are regularly used. These are:

1. Cetrimide (cetyltrimethylammonium bromide, CTAB), a mixture of dodecyl, tetradecyl- and hexadecyl-trimethylammonium bromide. In addition to its uses for pre-operative skin disinfection and treatment of seborrhoea of the scalp, cetrimide is also employed, in conjunction with sodium nitrite, which delays or prevents rusting, for the storage of sterilized surgical instruments, although this practice should be discontinued (Rubbo and Gardner, 1965).

2. Benzalkonium chloride (Zephirol, Zephiran) is the active constituent of a general purpose and skin sterilizing agent, "Roccal." It is also used to alleviate or prevent napkin rash caused by urea-splitting organisms, as a preservative in eye-drop preparations and for the disinfection of blankets.

Other important QACs are domiphen bromide and cetylpyridinium chloride.

The value of QACs in the disinfection of woollen blankets has been observed by various authors (Blowers and Wallace, 1955; Frisby, 1957; International Wool Secretariat, 1961), although they cannot be incorporated in the wash with an anionic compound, because of incompatibility. In addition, it must be emphasized that residues of anionic agents on blankets from previous washings could interfere with the action of QACs. To overcome this, blankets may be washed in an appropriate non-ionic detergent, with a final prolonged rinse in a QAC, or they may be treated with a non-ionic detergent incorporating a QAC.

The QACs are also of considerable value as disinfectants in food and dairy plant. If an alkali detergent containing anionic surface-active wetting agents is used prior to sanitizing, then utensils and equipment must be thoroughly rinsed (Barrett, 1969). The QACs at their normally used concentrations are odourless and non-corrosive, but many are not free-rinsing, and undesirable traces may remain on equipment or even be present in dairy food (Clegg, 1967, 1970). However, a combination of a free-rinsing type of QAC and a suitable non-ionic agent may be usefully employed for washing instruments and cutlery, etc. (Barrett, 1969).

The cytoplasmic membrane of bacteria and fungi is the site of action of the QACs. The membrane is composed of lipoprotein, and it is considered (Russell, 1969) that the lipid moiety is involved in the lethal action of these compounds.

There appears to be a clear relationship between the thermodynamic and antibacterial activities of QACs (Weiner *et al.*, 1965), with solutions having equal antimicrobial activity against an organism having surface concentration values of the same order of magnitude. Thermodynamic activities of QACs at two bacterial survivor levels (1% and 0.01%) have been shown to be sufficiently constant (Laycock and Mulley, 1970) to support the Ferguson (1939) principle that compounds with the same thermodynamic activity will have an equal bactericidal activity.

Amphoteric compounds, as already stated, are of mixed anionic-cationic character, and they combine the detergent properties of anionic compounds with the bactericidal properties of the cationic substances; their bactericidal properties remain virtually constant over a wide pH range (Barrett, 1969) and they are less readily inactivated by proteins than are the QACs, (Clegg, 1970). Examples of amphoteric surface-active agents are dodecyl-β-alanine, dodecyl-β-aminobutyric acid and dodecyl-di(aminoethyl)-glycine (Davis 1960a, b; 1968), the last named being a "Tego" compound. The Tego series of compounds have a high molecular weight, and in addition to being recommended for use in pre-operative hand cleansing and pre-operative skin preparation it has also been found that they are suitable for the cleansing of surgical operating theatre floors, walls and equipment and for ward cleansing (Frisby, 1959, 1961). It has, however, recently been shown that Tego 103S in 1% solution is less active than a 0.5% solution of chlorhexidine in 70% alcohol (Kuipers and Dankert, 1970). Amphoteric surface-active agents are inactivated by soaps and other anionic compounds (Frisby, 1959), but they are non-irritating and non-corrosive. Unfortunately, they tend to be expensive.

6. *Aldehydes*

The two most important aldehydes are glutaraldehyde and formaldehyde.

Glutaraldehyde is a dialdehyde which has been

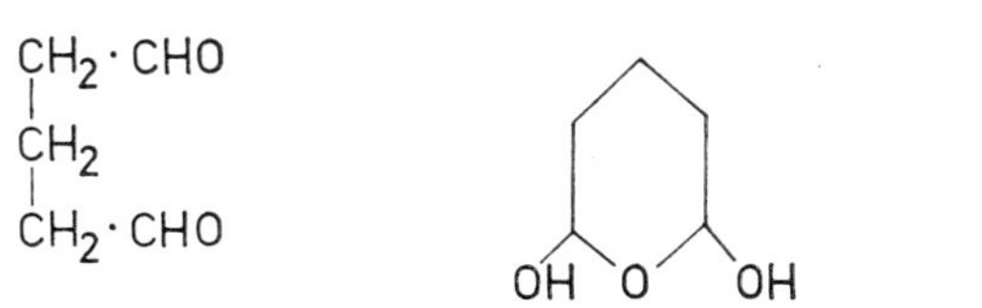

Hydrated Ring Structure

used for several years as a fixative un electron microscopy. Investigations into its antimicrobial activity have been comparatively recent (see Rubbo and Gardner, 1965; Rubbo *et al.*, 1967; Borick, 1968), but they do indicate that this substance has a valuable role to play in chemosterilization. A 2% solution of glutaraldehyde buffered with sodium bicarbonate (0.3% w/v is considered to be the optimum bicarbonate concentration) is effective in killing non-sporing bacteria within 2 min, *M. tuberculosis*, fungi and viruses in 10 min, and spores of *Bacilli* and *Clostridia* in 3 hours. Aqueous solutions of glutaraldehyde are acid, and are considerably less active against micro-organisms than are alkaline ones (Pepper and Chandler, 1963; Stonehill *et al.*, 1963; Borick *et al.*, 1964; Snyder and Cheatle, 1965; Lane *et al.*, 1966; Ross, 1966; Rubbo *et al.*, 1967; Munton and Russell, 1970a, b), but solutions become progressively less stable at pHs above 7. Concentrated solutions of glutaraldehyde (25%) can be purchased, diluted to the required concentration (2%) and "activated" by the addition of sodium bicarbonate. (Alternatively, 2% solutions ready for use when "activated" can also be purchased.) When made alkaline, glutaraldehyde solutions gradually undergo polymerization with a consequent loss of activity, this polymerization proceeding rapidly at pH values above 9. At pH 7.5–8.5, however, activity is maintained for at least 2 weeks.

Serum does not affect the antimicrobial activity of glutaraldehyde, but the dialdehyde is considerably less active in nutrient broth at pH 7.5 than it is at the same pH in buffer (Rubbo *et al.*, 1967; Munton and Russell, 1970a), the reason being that glutaraldehyde discolours the broth as a result of combining with the peptone present in it.

Glutaraldehyde is used as a fixative in the preparation of microbial cells for electron microscopy. It is a useful hospital disinfectant, particularly for articles which cannot be sterilized by physical means (Report, 1965). It has been employed in the sterilization of cytoscopes in urology (Lane, McKeever and Fallon, 1966) and of endoscopic instruments, such as bronchoscopes (Snyder and Cheatle, 1965), as it has no deleterious effect on the cement or lens coating. Glutaraldehyde is also employed as a tanning agent in preference to glyoxal and formaldehyde (Fein *et al.*, 1959), and has been shown to inactivate rapidly influenza virus and a coliphage in mouse tissue blocks (Sabel *et al.*, 1969).

Glutaraldehyde is non-corrosive, and does not affect rubber and plastic articles or the sharpness of cutting instruments; because it does not coagulate protein matter, such as blood and mucus, it does not render the cleaning of blood-covered instruments more difficult. It is obvious, therefore, that glutaraldehyde is a most useful chemosterilizer.

Rubbo *et al.* (1967) have proposed that the microbicidal activity of glutaraldehyde is due to the presence of two free aldehyde groups in the molecule. In solution, glutaraldehyde exists in an equilibrium between the open chain structure and the hydrated ring structure (see above), and there is a complete loss of activity if one or both of the aldehyde groups is altered, whereas a substitution elsewhere in the molecule reduces, but does not abolish, its activity. It is thus essential to have free aldehyde groups, which may react with cell sulphydryl or amino groups. Glutaraldehyde is about 10 times as active as glyoxal, with succinaldehyde occupying an intermediate position (Pepper and Chandler, 1963). Certain bacteria treated with glutaraldehyde become pink in colour (Munton and Russell, 1970b) as a result of cell-aldehyde interaction.

Formaldehyde has long been employed as a disinfectant. Formaldehyde solution is rapidly sporicidal to *B. subtilis* (Ortenzio *et al.*, 1953; cf. Klarmann, 1956, 1959) but not to various *Clostridia* (Ortenzio *et al.*, 1953; Klarmann, 1956, 1959). Ethanol (Rubbo *et al.*, 1967) and methanol (Willard and Alexander, 1964) cannot be recommended as vehicles for formaldehyde, as there is a reduction in antibacterial activity.

Formaldehyde is an important virucidal agent which finds its greatest use in the preparation of certain sterile vaccines, e.g. Poliomyelitis Vaccine (Inactivated). As a result of the experimental evidence accumulated over several years, a considerable amount of information is now available on the kinetics of the virus inactivation by formaldehyde. This particular vaccine consists of poliovirus Types I, II and III, and each is inactivated separately and then blended to give the trivalent vaccine. It is thus essential that formaldehyde treatment be sufficient to destroy the viruses without affecting their antigenicity; prolonged exposure to the aldehyde will, in fact, destroy the antigenic potency also (Morris and Darlow, 1971). The inactivation of poliovirus by formaldehyde has been considered to be a first-order reaction so that extrapolation of the death curve would give a point at which the probability of any infective particles remaining would approach zero. However, first-order kinetics cannot be used with any degree of safety to extrapolate the inactivation curve (Timm *et al.*, 1956; Meier, 1957). Similarly, although the inactivation of SV_{40} virus by formaldehyde has been shown by Sweet and Hilleman (1960) to be linear, it is now known that this linear inactivation is followed by a flattening of the curve indicating the persistence of a residual fraction which resists inactivation (Gerber *et al.*, 1961).

Formaldehyde is rapidly lethal to vegetative bacteria, and is sometimes used for this purpose in the preparation of dead bacterial vaccines.

7. Metals

Because of their antibacterial and antifungal activity, compounds of mercury, silver, copper and tin are of importance from both medical and industrial points of view.

A. Mercury compounds

These are of two types, the inorganic mercuric and mercurous salts and the organic substances. Mercuric salts are primarily bacteriostatic and fungistatic and contrary to earlier findings are not sporicidal (see Russell, 1971). Because of their toxicity, the mercuric salts do not find widespread use in modern medicine, but are extensively employed as industrial preservatives, e.g. in the preservation of wood, leather and paper, and in the control of fungal infections in seeds and bulbs. Mercurous salts have no applications as preservatives.

The most important organic mercury compounds are the phenylmercuric salts (nitrate, acetate and borate) and thiomersal. Phenylmercuric nitrate (PMN) and acetate (PMA) are now mainly employed as preservatives in various pharmaceutical and cosmetic products. PMN is also used as a spermicide in certain contraceptive formulations, as a plant fungicide and for the disinfection of leather and timber. However, because of their lack of sporicidal activity at ordinary temperatures, the organic mercury compounds cannot be recommended as sterilizing agents.

Various sulphydryl compounds, such as cysteine and thioglycollic acid, can reverse mercury-induced bacteriostasis, which led Fildes (1940) to propose that these compounds combined with, and displaced, mercury from its combination with the —SH group of an enzyme (E).

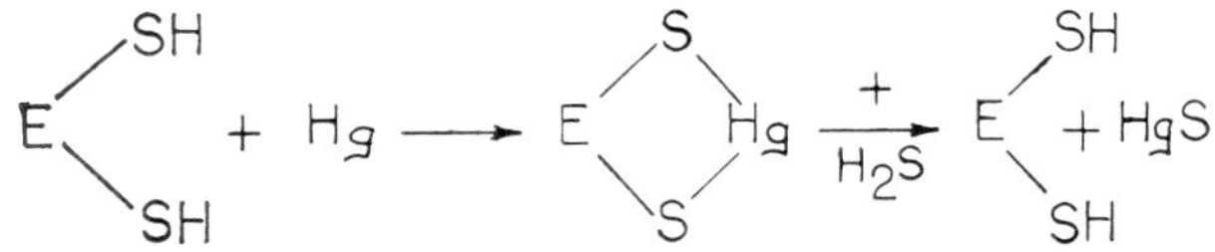

B. Silver compounds

Silver compounds have long been used in medicine for their antimicrobial activity, which extends to Gram-positive and Gram-negative bacteria and fungi. Of the silver compounds available, silver protein and silver nitrate are the most important. The latter, in the form of compresses, is highly effective in preventing the colonisation of burns with *Ps. aeruginosa* and *Proteus* species.

C. Copper compounds

These are bactericidal and fungicidal. They have been used for the latter purpose for more than 200 years, and their sole use nowadays is as industrial preservatives against fungal spoilage. The most frequently used substances are copper naphthenate, oxinate, 1-phenylsalicylate and sulphate; the last-named, in combination with a lime mixture, is known as Bordeaux mixture.

Dialkyldithiocarbamates are considered (Albert, 1963) to be converted into active bactericides and fungicides in the presence of copper. Such salts are highly successful, widely used, agricultural fungicides (Owens, 1969).

D. Tin compounds

Stannous and stannic salts have little antimicrobial activity. However, when tin is coupled with organic radicals, forming what are known as the organotins, potent antimicrobial activity results. If R represents the organic radical linked directly to a tin atom, by C—Sn bond, and X an inorganic or organic radical not so linked, various types of compounds can be obtained, of which $R—SnX_3$ is most active. Gram-negative bacteria are less sensitive than Gram-positive bacteria to organotin compounds. Triphenyltin acetate and hydroxide are important agricultural fungicides.

8. Dyes

The acridines have held a valuable place in medicine for several years, although with the advent of the antibiotics and other chemotherapeutic agents, they are now less widely used than hitherto. The acridines are active against several Gram-positive and Gram-negative bacteria, and have been used mainly in treating infected wounds. Their uses and mode of action have recently been reviewed (Foster and Russell, 1971).

The most important members of the triphenylmethane group of dyes are crystal violet, brilliant green and malachite green. These have mainly been employed for local application to burns and wounds.

Some members of a third group of dyes, the quinones, are important agricultural fungicides. The quinones are natural dyes which impart colour to many forms of plant and animal life. Chemically, the quinones are diketocyclohexadienes, the simplest of which is 1,4-benzoquinone (Fig. 4). Naphtaquinones are the most toxic to bacteria, moulds and yeasts, followed (in this order) by phenanthrenequinones, benzoquinones and anthraquinones (Fig. 4). Antimicrobial activity is increased by halogenation, and two powerful antimicrobial agents employed as fungicides are chloranil (tetrachloro-1,4-benzoquinone) and dichlone (2,3-dichloro-1,4-naphthaquinone).

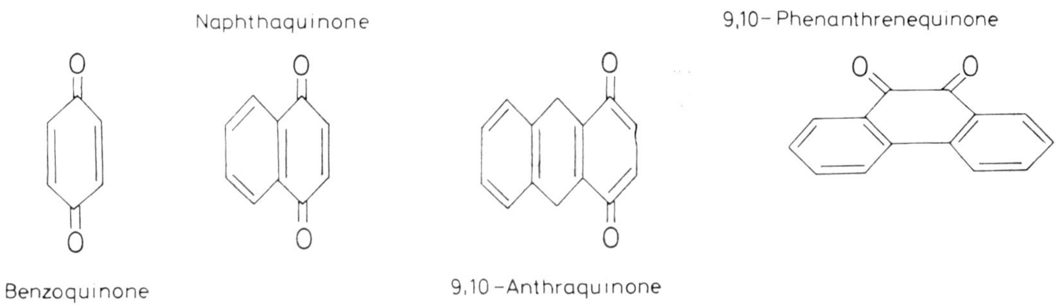

Figure 4

9. Alcohols

Ethyl alcohol, although active against Gram-positive and Gram-negative bacteria, is devoid of lethal activity against bacterial spores, and thus cannot be relied upon as a sterilizing agent (Russell, 1971). Methyl alcohol is likewise not sporicidal. Moreover, alcohols may reduce the sporicidal activity of aldehydes.

V. Gaseous Chemosterilizers

Gaseous chemosterilizers play an important role in sterilization of certain types of medical equipment and of components used in outer space research. The most important substances are:

1. Ethylene oxide

Ethylene oxide (EO) exists as a gas which is soluble in rubber, water, oils and several organic solvents. Chemically, it is $\underset{\diagdown O \diagup}{CH_2{-}CH_2}$. Its inflammability when in contact with air is overcome by using mixtures of EO with carbon dioxide or with fluorocarbon compounds (Phillips and Kaye, 1949; Barwell and Freeman, 1959; Freeman and Barwell, 1960). EO diffuses freely through paper, cellophane, cardboard and some plastics, but less rapidly through polythene; it cannot penetrate crystalline materials (Opfell and Miller, 1965). The antimicrobial activity of EO has been dealt with in many papers and reviews (Phillips, 1949, 1958, 1961, 1968; Phillips and Kaye, 1949; Kaye, 1949, 1950; Kaye and Phillips, 1949; Kaye *et al.*, 1952; Phillips and Warshowsky, 1958; Thomas, 1960; Bruch, 1961; Sutaria and Williams, 1961; Russell, 1965, 1971; Kelsey, 1961, 1967; Sykes, 1970; Hoffman, 1971).

Several factors are known to influence the antimicrobial activity of EO. These are:

(1) Concentration. As would be expected, the higher the concentration of EO (expressed as mg/l, which refers to the actual amount present in the sterilizer) the more rapid is the rate at which micro-organisms are killed. However, even at high concentrations, EO is only slowly lethal, and long periods may, therefore, be needed for sterilization to be achieved.

(2) Temperature. The lethal activity of EO increases with a rise in temperature. It has a temperature coefficient of 2.74 for each 10°C rise in temperature.

(3) Type of organism. EO gas will kill bacteria and their spores, yeasts, moulds and their spores, and viruses (Griffith and Hall, 1938), and resistant strains have not been developed. In contrast to

many other chemical substances which are considerably less effective against spores than against vegetative cells, bacterial spores are only about 2–10 times as resistant as the latter to EO (Toth, 1959; Phillips, 1958, 1968) and in some cases, e.g. *B. stearothermophilus* (Thomas *et al.*, 1969) even less resistant. These results imply that EO can freely penetrate the outer layer(s) of the bacterial spore, although experimental results in support of this contention are sadly lacking.

In addition to its antibacterial and antifungal activity, ethylene oxide is also effective against rickettsiae and viruses (Hoffman, 1971).

(4) Relative humidity (R.H.). Of all the factors which influence the activity of EO, R.H. is probably the most important. The optimum R.H. is considered to be c. 28–33% (Schley *et al.*, 1960), and EO is considerably less microbicidal at high R.H. and in relatively dry air. The correct R.H. may be achieved by prehumidification of the sterilization chamber (Halowell *et al.*, 1958; Ernst and Schull, 1962). However, the moisture content of microorganisms themselves, as well as the R.H. of the environment, is also important. Bacterial cells which have been desiccated and then equilibrated to successively high R.H. values contain less water and are more resistant to EO than cells which have not been desiccated but have instead been allowed to dry naturally until equilibrated to the same R.H. values. To overcome this resistance to EO produced by dehydration, the cells have to be actually wetted (Perkins and Lloyd, 1961; Winge-Heden, 1963; Gilbert *et al.*, 1964.)

Alkylating agents act through alkylation of sulphydryl, amino, carboxyl, hydroxyl and phenolic groupings, with the loss of a hydrogen atom and the production of an alkyl hydroxyethyl group (Phillips, 1952, 1958; Kelsey, 1967), and it seems likely that EO kills micro-organisms by an alkylation of protein molecules (Gilbert *et al.*, 1964). The influence of R.H. on the microbicidal activity of EO is related to this, since the presence of insufficient water prevents alkylation, whereas excess water causes hydrolysis of EO to ethylene glycol, CH_2OHCH_2OH (Wilson and Bruno, 1950).

(5) Effect of drying medium on spore resistance. Organisms dried from salt solution always show some survivors after exposure to EO (Royce and Bowler, 1961; Beeby and Whitehouse, 1965). Moreover, organisms enclosed within crystals are protected from the action of EO (Abbott *et al.*, 1956; Royce and Bowler, 1961) as a result of the inability of the gas to penetrate crystalline materials.

EO has several uses in the pharmaceutical and medical fields. These include the sterilization of ophthalmic instruments (Sheehan *et al.*, 1956), anaesthetic equipment (Snow *et al.*, 1962), heart-lung machines (Bracken *et al.*, 1960), disposable syringes (Rubbo and Gardner, 1968) and hospital blankets (Kaye, 1950), and as a decontamination procedure for articles handled by tuberculous patients (Newman *et al.*, 1955). There may, however, be a "crazing" of disposable syringes, and EO treatment is a slow, costly process, with a "quarantine" period necessary at the end of the process to ensure that complete dissipation of the gas has been achieved.

2. β-Propiolactone

β-Propiolactone (BPL) is a colourless liquid at room temperature. It has been employed as a chemosterilizer in both the liquid and gaseous states,

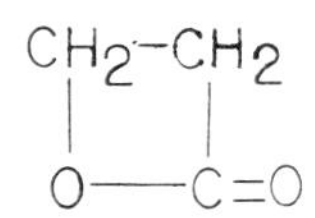

and is microbicidal in both forms (Allen and Murphy, 1960). Its antimicrobial activity depends on its concentration and on the temperature and R.H. at which the vapour is used (Spiner and Hoffman, 1960), a temperature above 24°C being required for optimal activity. As with EO, R.H. is of considerable importance in determining the activity of BPL vapour, although with the latter a R.H. in excess of 70% is required for rapid microbicidal activity, (Hoffman and Warshowsky, 1958). Again, however, it is not necessarily the atmospheric R.H. which is of greatest importance but the location and content of water within the microbial cell. *B. globigii* spores equilibrated to 98% are rapidly killed by BPL at 45% R.H., whereas they are more resistant if preconditioned at 75% R.H. before treatment with BPL at 45% R.H., and a small proportion of spores equilibrated at 1% R.H. are subsequently highly resistant to BPL at 75% R.H. (Hoffman, 1968).

Bacterial spores are more resistant to BPL than vegetative cells, viruses or fungi (Lo Grippo *et al.*, 1955; Trafas *et al.*, 1954; Bruch and Koesterer, 1962; Toplin, 1962) although some strains of

Staph. aureus may be almost as resistant as spores (Hoffman and Warshowsky, 1958).

BPL is also highly active against viruses and rickettsiae (Hoffman, 1971).

BPL has been used for the chemical sterilization of regenerated collagen sutures (Ball *et al.*, 1961), for the decontamination of enclosed spaces (Bruch, 1961 b) and for the sterilization of a variety of instruments contaminated with various sporing and non-sporing bacteria (Allen and Murphy, 1960). However, its reported carcinogenic effects in rats and mice (Walpole *et al.*, 1954) mean that a considerable degree of caution is needed before BPL is employed as a chemosterilizer.

3. Formaldehyde

Formaldehyde vapour may be obtained by evaporating appropriate dilutions of standardised batches of commercial formalin (a 40% solution of formaldehyde in water) with 10% methanol added to prevent polymerization (Report, 1958). Temperature affects the activity of the gas, as does the R.H., there being an increase in activity with increasing R.H. up to 50%, but little further increase in killing rate as the R.H. rises from 50 to 90% (Nordgreen, 1939; Report, 1958). In contrast, gross wetting retards killing.

Bacteria protected by organic matter, such as blood and sputum are less rapidly killed by formaldehyde vapour (Nordgren, 1939; Bullock and Rawlins, 1954). Micro-organisms may also be protected from it when they are included in a crystal mass, in contrast to surface-contaminated crystals (Abbott *et al.*, 1956).

Although bacterial spores are more resistant than vegetative cells to formaldehyde vapour, the degree of difference is not high (Phillips, 1952; Report, 1958; Sykes, 1965). The vapour only has weak penetration, and its application is thus normally limited to surface sterilization (Borick, 1968; Davis, 1968). However, the addition of formaldehyde vapour to steam under subatmospheric pressure at temperatures below 90°C results in deep penetration into fabrics with destruction of heat-resistant micro-organisms (Alder *et al.*, 1966).

Formaldehyde vapour has long been used for the disinfection of blankets, and is considered to be one of the best methods available for disinfecting woollen blankets that have not received a shrink-resist treatment (International Wool Secretariat, 1961). It is also used to decontaminate rooms, buildings and instruments (Hoffman, 1971).

VI. Practical Uses

This final section brings together some of the data presented in the preceding sections, and also provides information in certain specific instances. Disinfectants and chemosterilizers will be considered from two points of view, first their medical, and second their non-medical uses.

A. Medical uses

The use and choice of disinfectants in hospitals have been extensively considered in the last decade in Britain by a special committee (Report, 1965) set up for this purpose. Our comments here are thus based on the recommendations of this report and on the findings presented by Kelsey and Maurer (1967). The Report (1965) recommended that two classes of disinfecting agents were needed, (a) for general disinfection and (b) for surface disinfection of clean objects. Agents for general disinfection should have a wide spectrum, and at appropriate dilutions should remain active in the presence of organic matter; the main purpose of such disinfectants, e.g. phenolic disinfectants based on coal-tar acids, is not necessarily to kill *all* bacteria but to ensure that an object is free of significant numbers of organisms. Chemicals for surface disinfection must be quick-acting, have a wide spectrum, be non-harmful to materials and leave no objectionable odours. Such disinfectants, e.g. hypochlorites, should be used for the rapid disinfection of clean surfaces such as trolley tops, kitchen tables and clinical thermometers.

Kelsey and Maurer (1967) have presented a list of the steps to be taken in drawing up a policy for the use of general purpose disinfectants in hospitals, and among the points they make is the non-usage of disinfectants in certain cases, especially where sterilization is the objective or where other more reliable means are available.

B. Non-medical uses

The main non-medical uses of disinfectants occur in the food, dairy, brewing and fermentation industries (Foster *et al.*, 1958; Frazier, 1967). The maintenance of equipment for use in these industries in a proper sanitary condition cannot be over-emphasized. This therefore means that the cleaning of such equipment is of considerable importance, since the presence of organic matter can reduce or virtually eliminate the effect of many disinfectants (section II, page 166). In the dairy industry, milk stone—resulting from milk drying on equipment, and thus consisting of fat, protein and minerals—and milk film are a well-known problem in disinfection (Clegg, 1967). Chemicals which are of use against micro-organisms in liquid suspension in laboratory tests may be of little use against such organisms on a soiled surface if they are poor detergents (Cousins, Hoy and Clegg, 1960).

To counteract the unwanted effects of organic matter, one of the following two methods may be employed:

(a) detergent first, followed by a disinfectant;

(b) combination of detergent and disinfectant (this corresponds to a sanitizer, or to the detergent-sterilant of Davis (1968), as described in the Introduction).

Harris (1969) stresses the need for using two operations, i.e. the use of cleaning before disinfection. Cleaning is the first essential in the sanitary care of food equipment, and approximate sterility the last (Foster *et al.*, 1958). Steam under pressure is an obvious method of sanitization, but this is limited only to closed systems which can withstand pressure (Frazier, 1967). Theoretically, separate procedures would be expected to give a better result (Clegg, 1970), because of the inactivating effect of organic matter on disinfectants; however, a finding made several years ago (Neave and Hoy, 1947) suggests that a detergent-disinfectant combination would be of greater use, because "the effect of the detergent on milk solids more than outweights the effect of milk residues inactivating the disinfectant" (Clegg, 1970). An effective detergent should dissociate organic and inorganic solids, emulsify, saponify or suspend grease, fats and oils, have good wetting ability, be easily rinsed and be non-corrosive (Olivant and Shapton, 1970). Detergents are thus of considerable importance in this field, because they can also be responsible for the mechanical removal of bacteria (Gilbert, 1969). The most commonly used detergents are strong and mild alkalis, alkali salts, strong acids, anionic alkyl sulphates and aryl sulphonates and non-ionic condensates (Davis, 1968). An excellent descaling agent is nitric acid which can be used hot at concentrations of 0.25–0.5% or cold at 0.5–1%, and which is, in addition, a powerful disinfectant. It is, however, less effective than alkali in the removal of hardened protein films, and is normally employed with a corrosion inhibitor. In actual fact, many detergents are good disinfectants and vice versa, e.g. a detergent such as sodium hydroxide possesses considerable germicidal power (Whitehouse and Clegg, 1963), whereas hypochlorites have a useful detergent effect by disintegrating protein matter (Davis, 1968). Cleaning is an essential first part in high-speed food and beverage processing plant, and considerable economic benefit is achieved as a result of the production of stable liquid detergents and disinfectants which can be delivered by tanker and then distributed to the cleaning areas by pipeline (Hill, 1969). However, even with good pre cleaning, traces of inactivating protein material may remain on equipment, and it is thus important to choose a disinfectant which has a high protein tolerance (Harris, 1969). The design of equipment to facilitate cleaning must also be stressed, and stainless steel is an obvious example, with glass pipelines to give a high degree of visual cleanliness (Harris, 1969).

The disinfection of fermentation laboratories has been described by Darlow (1969). Gaseous disinfectants may be employed, and aerosols appear to have an important role to play. Local disinfection of bench tops, floors, etc. is also a standard practice. The importance of disinfectants in water conservation is emphasised by Fielden (1969).

The control of air-borne micro-organisms is of particular importance in the fermentation industry, in laboratories where strict asepsis is essential (e.g. in the production of various sterile products in the pharmaceutical industry and in hospital pharmacy departments, as well as in the rearing of germ-free animals), in hospital wards to reduce the incidence of cross-infection, and in special wards set aside for patients with a rare disease (hypogammaglobulinaemia) who are particularly sensitive to infection.

This control is normally achieved by the use of special air filters, often in conjunction with ultraviolet lamps to irradiate the upper atmosphere. BPL may be used to decontaminate unoccupied enclosed areas (section V). Disinfectants in the form of aerosols are also of importance in aerial disinfection. To be effective for air disinfection, a chemical should ideally possess the following properties:

(a) be odourless, cheap and stable

(b) be without toxic or irritant properties

(c) be capable of being dispersed in the air, with consequent complete and rapid mixing of infected air and chemical

(d) be capable of maintaining an effective concentration in the air

(e) be highly and rapidly effective against air-borne organisms

(f) be unaffected by relative humidity (R.H.).

Aerosols consist of a very fine dispersed liquid phase in a gaseous (air) continuous phase. The germicide must be nebulized in sufficiently fine spray (aerosol droplets of $< 1\ \mu m$ are the accepted standard) so that it will remain air-borne and thus have ample opportunity to collide with any micro-organisms present in the air. At low R.H., particles are too dry for adequate condensation of the disinfectant on, and diffussion into, the microbial particles, and thus bacteria occurring on dust or on surfaces are much less susceptible to the aerial disinfectant than such organisms enclosed in droplets (Sykes, 1963). The optimum R.H. is usually *c.* 40–60%.

Chemical aerosols are often generated in the following manner (MacKay, 1952): if the chemical is liquid, it may be sprayed directly into the air from an atomiser; if the chemical exists as a solid, it may be dissolved in an appropriate solvent, e.g. propylene glycol, and atomised, or alternatively the solid may be vaporised by heat from a thermostatically-controlled hot-plate.

Chemicals which have been used as aerial disinfectants include hexylresorcinol, lactic acid, propylene glycol (this possesses antimicrobial activity in its own right), hypochlorous acid, formaldehyde gas and sulphur dioxide.

REFERENCES

Abbott, C. F., J. Cockton and W. Jones (1956), *J. Pharm. Pharmac.* **8**, 709.

Albert, A. (1963), *Adv. Appl. Microbiol.* **5**, 1.

Alder, V. G., A. M. Brown and W. A. Gillespie (1966), *J. Clin. Path.* **19**, 83.

Allawala, N. A. and S. Riegelman (1953), *J. Am. Pharm. Ass., Sci. Ed.* **42**, 396.

Allen, H. F. and J. T. Murphy (1960), *J. Am. Med. Ass.* **172**, 1759.

Association of Official Agricultural Chemists (1960). Official Methods of Analysis of the A.O.A.C., 9th Ed., Washington, D.C.

Ball, E. L., A. C. Dornbrush, G. M. Seiger, F. E. Stirn, J. C. Vitucci and J. F. Weidenheimer (1961), *Appl. Microbiol.* **9**, 269.

Barrett, M. (1969), *Process Biochem.* **4**, 23.

Barwell, C. F. and M. A. R. Freeman (1959), *Lancet*, i, 917.

Bean, H. S. (1967), *J. Appl. Bact.* **30**, 6.

Bean, H. S. and A. Das, (1966), *J. Pharm. Pharmac.* **18**, 107 S.

Bean, H. S. and S. M. Heman-Ackah (1964), *J. Pharm. Pharmac.* **16**, 58 T.

Bean, H. S. and V. Walters (1955), *J. Pharm. Pharmac.* **7**, 661.

Beeby, M. M. and C. E. Whitehouse (1965), *J. Appl. Bact.* **28**, 349.

Benarde, M. A. W. B. Snow, V. P. Olivieri, and B. Davidson (1967), *Appl. Microbiol.* **15**, 257.

Bennett, E. O. (1959), *Adv. Appl. Microbiol.* **1**, 49.

Berry, H. (1951), *J. Pharm. Pharmac.* **3**, 689.

Berry, H. and H. S. Bean (1954), *J. Pharm. Pharmac.* **7**, 224.

Berry, H. and A. Briggs (1956), *J. Pharm. Pharmac.* **8**, 1143

Berry, H. and J. B. Stenlake (1942), *Pharm. J.* **148**, 112.

Berry, H., A. M. Cook and B. A. Wills (1956), *J. Pharm. Pharmac.* **8**, 425.

Berry, H., E. Jensen and F. K. Silliker (1938), *Quart. J. Pharm. Pharmac.* **11**, 729.

Blatt, R. and J. V. Moloney (1961), *Surg. Gynaec. Obstet.* **113**, 699.

Borick, P. M. (1968), *Adv. Appl. Microbiol.* **10**, 291.

Borick, P. M. (1970), in *Disinfection.* Ed. M. A. Benarde. Marcel Dekker, New York.

Briggs, A. (1966), *J. Appl. Bact.* **29**, 490.

British Disinfectant Manufacturers Association (1970), Terminology in Sterilization and Disinfection.

British Pharmacopoeia (1968), Pharmaceutical Press, London.

British Standards Institution (1934), British Standard techniques for determining the Rideal-Walker Coefficient of Disinfectants: B.S. 541.

British Standards Institution (1938). British Standard specification for the modified technique of the Chick-Martin Test for Disinfectants: B.S. 808.

Bruch, C. W. (1961 a), *Ann. Rev. Microbiol.* **15**, 245.

Bruch, C. W. (1961 b), *Am. J. Hyg.* **73**, 1.

Bruch, C. W. and M. G. Koesterer (1962), *Appl. Microbiol.* **10**, 123.

Bullock, K. and E. A. Rawlins (1954), *J. Pharm. Pharmac.* **6**, 859.

Charlton, D. B. and M. Levine (1937), *Iowa Eng. Exp. Sta. Bull.* No. 132, Iowa, U.S.A.

Chick, H. (1908), *J. Hyg., Camb.* **8**, 92.

Chick, H. and C. J. Martin (1908), *J. Hyg., Camb.* **8**, 698.

Clegg, L. F. L. (1967), *J. Appl. Bact.* **30**, 117.

Clegg, L. F. L. (1970), in *Disinfection.* Ed. M. A. Benarde. Marcel Dekker, New York.

Cook, A. M. (1960), *J. Pharm. Pharmac.* **12**, 19 T.

Cook, A. M. and B. A. Wills (1954), *J. Pharm. Pharmac.* **6**, 638.

Cousins, C. M. and C. D. Allan (1967), *J. Appl. Bact.* **30**, 168.

Cousins, C. M., W. A. Hoy and L. F. L. Clegg (1960), *J. Appl. Bact.* **23**, 359.

Dakin, H. D. (1915), *Br. Med. J.* **2**, 809.

Dakin, H. D. (1916), *Br. Med. J.* **2**, 318.

Dakin, H. D., J. B. Cohen and J. Kenyon (1916), *Br. Med. J.* **1**, 160.

Darlow, H. M. (1969), *Process Biochem.* **4**, 15.

Davies, G. E., J. Francis, A. R. Martin, F. L. Rose and G. Swain (1954), *Br. J. Pharmac.* **9**, 192.

Davis, J. G. (1960a), *J. Pharm. Pharmac.* **12**, 29 T.

Davis, J. G. (1960b), *J. Appl. Bact.* **23**, 318.

Davis, J. G. (1962), *J. Appl. Bact.* **25**, 195.

Davis, J. G. (1963), *The Sanitarian* October, p. 3.

Davis, J. G. (1968), *Progr. Indust. Microbiol.* **8**, 141.

Dychdala, G. R. (1968), in *Disinfection, Sterilization and Preservation.* Ed. C. A. Lawrence and S. S. Block. Lea and Febiger, Philadelphia.

Ernst, R. R. and J. J. Shull (1962), *Appl. Microbiol.* **10**, 337.

Farkas-Himsley, H. (1964), *Appl. Microbiol.* **12**, 1.

Fein, M. L., E.H. Harris, J. Naghski and E. M. Filachione (1959), *J. Am. Leather Chem. Ass.* **54**, 488.

Ferguson, J. (1939), *Proc. Roy. Soc. B.* **127**, 387.

Fielden, T. B. (1969), *Process Biochem.* **4**, 35.

Finch, W. E. (1958), *Disinfectants, Their Values and Uses.* Chapman and Hall, London.

Flett, L. H., R. C. Haring, A. F. Guiteras and R. L. Shapiro (1945), *J. Bact.* **50**, 591.

Foster, E. M., F. E. Nelson, M.L. Speck, R. N. Doetsch and J. C. Olson (1958), *Dairy Microbiology.* Macmillan and Co. Ltd.

Foster, J. H. S. and A. D. Russell (1971), in *Inhibition and Destruction of the Microbial Cell.* Ed. W. B. Hugo. Academic Press, London and New York.

Frazier, W. C. (1967), *Food Microbiology*, 2nd Ed. McGraw-Hill, New York.

Freeman, M. A. R. and C. F. Barwell (1960), *J. Hyg., Camb.* **58**, 337.

Frisby, B. R. (1959), *Lancet* **2**, 57.

Frisby, B. R. (1961), *Lancet* **2**, 829.

Garrod, L. P. (1934), *J. Hyg., Camb.* **34**, 322.

Garrod, L. P. (1935), *J. Hyg., Camb.* **35**, 219.

Gershenfeld, L. (1956), *Am. J. Pharm.* **128**, 335.

Gershenfeld, L. (1962), *Am. J. Pharm.* **134**, 78.

Gershenfeld, L. and B. Witlin (1950), *Ann. N.Y. Acad. Sci.* **53**, 172.

Gilbert, G. L., V. M. Gambill, D. R. Spiner, R. K. Hoffman, and C. R. Phillips (1964), *Appl. Microbiol.* **12**, 496.

Gilbert, J. (1969), *Process Biochem.* **4**, 37.

Gilireas, F. W. and J. E. O'Brien (1941), *Am. J. Publ. Health* **31**, 143.

Goetchins, G. R. and W. E. Botwright (1950), *J. Milk Technol.* **13**, 63.

Griffith, C. L. and L. A. Hall (1938), *U.S. Patent* 2,189,947.

Grossgebauer, K. (1970), in *Disinfection.* Ed. M. A. Benarde. Marcel Dekker, New York.

Halowell, P., J. T. Murphy and A. B. Mangiaracine (1958), *Anaesthesiology* **19**, 665.

Hare, R., E. Raik and S. Gash (1963), *Br. Med. J.* **1**, 496.

Harris, C. (1969), *Process Biochem.* **4**, 19.

Harris, N. D. (1963), *J. Appl. Bact.* **26**, 387.

Hays, H., P. R. Elliker and W. E. Sandine (1967), *Appl. Microbiol.* **15**, 575.

Hill, R. P. (1969), *Process Biochem.* **4**, 31.

Hoffman, R. K. (1968), *Appl. Microbiol.* **16**, 641.

Hoffman, R. K. (1971), in *Inhibition and Destruction of the Microbial Cell.* Ed. W. B. Hugo, Academic Press, London and New York.

Hoffman, R. K. and B. Warshowsky (1958), *Appl. Microbiol.* **6**, 358.

Hoy, W. A. and L. F. L. Clegg, *Proc. Soc. Appl. Bact.* **16**, 1.

International Wool Secretariat (1961), the Sterilization of Hospital Woollen Blankets. *Wool Science Review*, No. 19 and No. 20.

Ito, K. A., D. J. Seslar and W. A. Mercer (1967), in *Botulism* **66**. Ed. M. Ingram and T. A. Roberts, Chapman and Hall, London.

Jacobs, S. E. and N. D. Harris (1960), *J. Appl. Bact.* **23**, 294.

James, A. M. (1965), *Surface Activity and the Microbial Cell.* Soc. Chem. Ind., London.

Jensen, V. and E. Jensen (1933), *Dansk. Tidss. Farm.* **7**, 77.

Johns, C. K. (1946), *Am. J. Publ. Health* **37**, 1322.

Johns, C. K. (1954), *Can. J. Technol.* **32**, 71.

Kaye, S. (1949), *Am. J. Hyg.* **50**, 289.

Kaye, S. (1950), *J. Lab. Clin. Med.* **35**, 823.

Kaye, S. and C. R. Phillips (1949), *Am. J. Hyg.* **50**, 296.

Kaye, S., H. F. Irminger and C. R. Phillips (1952), *J. Lab. Clin. Med.* **40**, 67.

Kelsey, J. C. (1961), *J. Clin. Path.* **14**, 313.

Kelsey, J. C. (1967), *J. Appl. Bact.* **30**, 92.

Kelsey, J. C. and I. M. Maurer (1967), *Mon. Bull. Min. Health* **26**, 110.

Kelsey, J. C. and Sykes (1969), *Pharm. J.* **202**, 607.

Kelsey, J. C., M. M. Beeby, and C. M. Whitehouse (1965), *Mon. Bull. Min. Health* **24**, 162.

Klarmann, E. G. (1956), *Am. J. Pharm.* **128**, 4.

Klarmann, E. G. (1959), *Am. J. Pharm.* **131**, 86.

Knox, W. E., V. H. Auerbach, K. Zarudnaya and M. Spirtes (1949), *J. Bact.* **58**, 443.

Koch, R. (1881), *Arb. Kaiserl. Gesundh.* **1**, 1.

Kronig, B. and T. Paul (1897), *Z. Hyg. Infekt. Kr.* **25**, 1.

Kuipers, J. S. and J. Dankert (1970), *J. Hyg., Camb.* **68**, 343.

Lamy, P. P. and H. L. Flack (1962), *Am. J. Hosp. Pharm.* **19**, 473.

Lane, V., J. D. McKeever and M. Fallon (1966), *J. Irish Med. Ass.* **58**, 131.

Lawrence, C. A., C. M. Carpenter and A. W. C. Naylor-Foote (1957), *J. Am. Pharm. Ass., Sci. Ed.* **46**, 500.

Laycock, H. H. and B. A. Mulley (1970), *J. Pharm. Pharmac.* **22**, 1515.

Lo Grippo, G. A., P. R. Overhulse, D. E. Szilagyi and F. W. Hartman (1955), *Lab. Invest.* **4**, 217.

Loosemore, M. and A. D. Russell (1963), *J. Pharm. Pharmac.* **15**, 558.

Loosemore, M. and A. D. Russell (1964), *J. Pharm. Pharmac.* **16**, 817.

Mallman, W. L. and M. Hanes (1947), *J. Bact.* **49**, 526.

Mandels, G. R. and R. T. Darby (1953), *J. Bact.* **65**, 16.

Mather, R. L. (1949), *Biometrics* **5**, 127.

Morris, E. J. and H. M. Darlow (1971), in *Inhibition and Destruction of the Microbial Cell.* Ed. W. B. Hugo, Academic Press, London and New York.

Mulley, B. A. (1964), *Adv. Pharm. Sci.* **1**, 87.

Munton, T. J. and A. D. Russell (1970a), *J. Appl. Bact.* **33**, 410.

Munton, T. J. and A. D. Russell (1970b), *Experientia*, **27**, 109.
Neave, F. K. and W. A. Hoy (1947), *J. Dairy Res.* **15**, 24.
Needham, N. V. (1947), *J. Hyg., Camb.* **45**, 1.
Nordgren, G. (1939), *Acta Path. Microbiol. Scand., suppl.* No. 40, 1.
Olivant, D. J. and D. A. Shapton (1970), in *Disinfection.* Ed. M. A. Benarde. Marcel Dekker, New York.
Opfell, J. B. and C. E. Miller (1965), *Adv. Appl. Microbiol.* **7**, 81.
Ortenzio, L. F., L. S. Stuart and J. L. Friedl (1953), *J. Ass. Off. Agric. Chem.* **36**, 480.
Owens, R. G. (1969), in *Fungicides*, Vol. 2. Ed. D. C. Torgeson, Academic Press, New York.
Pepper, R. E. and V. L. Chandler (1963), *Appl. Microbiol.* **11**, 384.
Perkins, J. J. and R. S. Lloyd (1961), in *Recent Developments in the Sterilization of Surgical Materials.* Report of a Symposium, p. 76. Pharmaceutical Press, London.
Phillips, C. R. (1949), *Amer. J. Hyg.* **50**, 280.
Phillips, C. R. (1952), *Bacteriol. Rev.* **16**, 135.
Phillips, C. R. (1958), Becton, Dickinson Lectures on Sterilization: Seton Hall, New Jersey, U.S.A.
Phillips, C. R. (1961), in *Recent Developments in the Sterilization of Surgical Materials.* Report of a Symposium, p. 59. Pharmaceutical Press, London.
Phillips, C. R. (1968), in *Disinfection, Sterilization and Preservation.* Eds. C. A. Lawrence and S. S. Block. Lea and Febiger, Pennsylvania, U.S.A.
Phillips, C. R. and S. Kaye (1949), *Amer. J. Hyg.* **50**, 270.
Phillips, C. R. and B. Warshowsky (1958), *Ann. Rev. Microbiol.* **12**, 525.
Prokop, A. and A. E. Humphrey (1970), in *Disinfection.* Ed. M. A. Benarde. Marcel Dekker, Inc., New York.
Report (1958), Report by the Committee on Formaldehyde Disinfection. *J. Hyg., Camb.* **56**, 488.
Report (1965), Report by Public Health Laboratory Service Committee on the Testing and Evaluation of Disinfectants. *Br. Med. J.* i, 408.
Rittenbury, M. S. and M. E. Hench (1965), *Ann. Surg.* **161**, 127.
Roberts, M. H. and O. Rahn (1946), *J. Bact.* **52**, 639.
Rogers, M. R., J. J. Mather and A. M. Kaplan (1961), *Appl. Microbiol.* **9**, 497.
Royce, A. and C. Bowler (1961), *J. Pharm. Pharmac.* **13**, 87 T.
Rubbo, S. D. and J. F. Gardner (1965), *A Review of Sterilization and Disinfection.* Lloyd-Luke, London.
Rubbo, S. D. and J. F. Gardner (1968), *J. Appl. Bact.* **31**, 164.
Rubbo, S. D., J. F. Gardner and R. L. Webb (1967), *J. Appl. Bact.* **30**, 78.
Rudolph, A. S. and M. Levine (1941), *Iowa Eng. Exp. Sta. Bull.* No. 150. Iowa, U.S.A.
Russell, A. D. (1964), *Lab. Pract.* **13**, 114.
Russell, A. D. (1965), *Mfg. Chem. Aerosol News* **36**, 38.
Russell, A. D. (1971), in *Inhibition and Destruction of the Microbial Cell.* Ed. W. B. Hugo, Academic Press, London and New York.
Russell, A. D. and R. J. Gilbert (1964), *Mfg. Chem. Aerosol News* **35**, 42.
Russell, A. D., J. Jenkins and I. H. Harrison (1967), *Adv. Appl. Microbiol.* **9**, 1.
Russell, A. D. and M. Loosemore (1964), *Appl. Microbiol.* **12**, 403.
Sabel, F. L., A. Hellman and J. J. McDade (1969), *Appl. Microbiol.* **17**, 645.
Salk, J. E., U. Krech, J. S. Younger, B. L. Bennett and L. J. Lewis (1954), *Amer. J. Publ. Health* **44**, 563.
Salton, M. R. J. (1957), *2nd Int. Congr. Surface Activity.* Butterworths, London.
Schley, D. G., R. K. Hoffman and C. R. Phillips (1960), *Appl. Microbiol.* **8**, 15.
Snyder, R. W. and E. L. Cheatle (1965), *Amer. J. Hosp. Pharm.* **22**, 321.
Spaulding, E. H. (1958), *Becton, Dickinson Lectures on Sterilization*, p. 51.
Spaulding, E. H. (1968), in *Disinfection, Sterilization and Preservation.* Ed. C. A. Lawrence and S. S. Block. Lea and Febiger, Philadelphia.
Spiner, D. R. and R. K. Hoffman (1960), *Appl. Microbiol.* **8**, 152.
Stonehill, A. A., S. Krop and P. M. Borick (1963), *Am. J. Hosp. Pharm.* **20**, 458.
Sutaria, R. H. and F. H. Williams (1961), *Pharm. J.* **186**, 311.
Sykes, G. (1939), *J. Hyg., Camb.* **39**, 463.
Sykes, G. (1958), *J. Pharm. Pharmac.* **10**, 40 T.
Sykes, G. (1965), *Disinfection and Sterilization.* Spon, London.
Sykes, G. (1970), *J. Appl. Bact.* **33**, 147.
Sykes, G. and M. C. Hooper (1954), *J. Pharm. Pharmac.* **11**, 235 T.
Ten Cate, L. (1965), *J. Appl. Bact.* **28**, 221.
Thomas, C. G. A. (1960), *Guy's Hosp. Rep.* **109**, 57.
Thomas, C. G. A., B. West and H. Besser (1959), *Guy's Hosp. Rep.* **108**, 446.
Tilley, F. W. (1942), *J. Bact.* **43**, 521.
Toplin, I. (1962), *Biotechn. Bioeng.* **4**, 331.
Toth, L. Z. J. (1959), *Arch. Mikrobiol.* **32**, 409.
Trafas, P. C., R. E. Carlson, G. A. LoGrippo and C. R. Lam (1954), *Arch. Surg.* **69**, 415.
Walpole, A. L., D. C. Roberts, F. L. Rose, J. A. Hendry, and R. F. Homer (1954), *Br. J. Pharmac.* **9**, 306.
Weber, G. R. (1950), *Publ. Health Rep.* **65**, 503.
Weber, G. R. and M. Levine (1944), *Am. J. Publ. Health* **34**, 719.
Wedderbern, D. L. (1964), *Adv. Pharm. Sci.* **1**, 195.
Weinberg, E. D. (1968), *Ann. N.Y. Acad. Sci.* **148,** 587.
Whitehouse, R. L. and L. F. L. Clegg (1963), *J. Dairy Res.* **30**, 315.
Willard, M. and A. Alexander (1964), *Appl. Microbiol.* **12**, 229.
Wilson, A. R. and P. Bruno (1950), *J. Exp. Med.* **91**, 449.
Winge-Heden, K. (1963), *Acta Path. Microbiol. Scand.* **58**, 225.

A. D. RUSSELL
P. J. DITCHETT
University of Wales

E

DRAINAGE-SURFACE: see URBAN RUNOFF

ECOLOGY: see ECOSYSTEM THEORY

EARTH-SPACE ORGANIZATION: see THE TERRESTRIAL SYSTEM

ECOLOGY OF FISH: see POLLUTION EFFECTS ON FISH; THERMAL EFFECTS ON FISH ECOLOGY

ECOLOGY OF HUMANS: see EFFECT OF CHEMICALS

ECOLOGY OF PLANTS

Ecology is the study of the relationship between organisms and the environment and the interrelationships between organisms. The response of an organism to the environment is determined by the effects of the environment on the processes in the organism. Stated in another way, all environmental factors affect the growth and distribution of organisms only by affecting rates of processes in the organisms. The study of the effects of environmental factors on processes in organisms is *physiological ecology*, or in other words, it is the study of the physiological basis of ecological behavior. This field obviously represents a marriage between ecology and physiology, and is sometimes called *environmental physiology*. The field of physiological ecology is extremely broad; thus, this paper is restricted to certain aspects of physiological *plant* ecology.

Environmental Factors

The environment is very complex, consisting of all influences external to an organism. Many systems have been devised to classify the environment but space does not allow a discussion here. The following list of *specific* environmental factors is a satisfactory one although the list could obviously be subdivided in many ways.

1. water
2. temperature
3. radiant energy
4. essential elements
5. aeration
6. toxins
7. wind
8. topography
9. nature of geologic strata

10. altitude
11. organic matter content of soil
12. texture and structure of soil
13. base exchange capacity of soil
14. color of soil
15. plant growth regulators
16. soil organisms
17. parasites
18. plants and animals other than above

Even a cursory examination of this list makes it clear that the effects of all these factors on processes in plants could not be discussed in a brief survey. For example, thorough coverage of the effects of the various essential elements on plant processes would require a sizeable manuscript. I have decided, therefore, to discuss a few factors in sufficient depth to give the reader a feeling for the area.

Interrelationships Between Environmental Factors

In addition to having a direct effect on many plant functions, the water factor affects other factors through the amount and type of precipitation involved. It affects: (1) amount of radiant energy, (2) temperature of air and soil, (3) amount of nitrogen brought down by rain, (4) amounts of minerals leached from the soil, (5) availability of various elements to plants, (6) topography of the region, (7) numbers and types of organisms in the soil, (8) toxins in the soil, (9) aeration of soil, and (10) numbers and types of parasites in a region.

All other environmental factors have many indirect affects also.

Effects of Water Factor

Warming (1909) stated that "no other influence impresses its mark to such a degree upon the internal and external structures of the plant as does the amount of water present in the air and soil." This is due primarily, of course, to the following roles of water: (1) it is important as a solvent system in the soil and in cells of all living organisms on earth, (2) it serves as the dispersion medium for the colloidal systems present in cells, (3) there is a close interaction between the nature of the hydration shell surrounding protein molecules and the physico-chemical properties of the proteins (Klotz, 1958), (4) it is important as a raw material in photosynthesis, (5) it is important as a raw material in all hydrolytic processes, and (6) it has some influence on leaf temperatures through transpirational cooling.

Klotz (1958) suggested that protein molecules in cells have hydration shells of lattice-ordered water which act like ice-shells around the molecules and that the maintenance of an active configuration is due to the shells. The effect of heat is thought to reduce the extent of the ice-shell, whereas lower temperatures increase the thickness of the shell. Urea is thought to be an effective denaturizer of proteins because its strong hydrogen bonding characteristics may break down the frozen structure of the water envelope. The observed effects of water stress on the viscosity of protoplasm suggest that reduced hydration has a definite effect on the ice-shell.

Jacobson (1953) offered considerable evidence for the presence of an ordered-water lattice around DNA molecules also. He suggested that such an ice-shell would make separation of the two chains of DNA during replication possible with low energies.

The models of Klotz and Jacobson have apparently been well received (Stocker 1960, Slatyer 1967). If these models are indeed correct, it is easy to see how water deficiency can have pronounced effects on all sorts of processes, because of the role of DNA in protein synthesis and the role of proteins in enzymes. It is probable that the integrity of specific water-protein structures is necessary for the continuance of organic processes at optimum rates. It is likely that the variations in reactions of different species to water stress are due chiefly to different degrees of sensitivity of essential metabolic systems.

Buckman and Brady (1960) state that a slight lowering of the available water content in the soil below field capacity stimulates growth of plants generally. Blair *et al.* (1950) found that the rate of stem elongation in sunflower (a dwarf type from an inbred line, apparently of *Helianthus annuus*) is markedly reduced before half the available water is used, and that zero growth results during the use of the last one-fourth of the available water. A reduction of 30–40% between field capacity and the wilting point probably slows growth in most plants.

Probably the most direct affect of water on plant growth is its effect on the turgor pressure of the individual cells. Turgor pressure affects cell enlargement and stomatal closure. Closing of the stomates due to reduced turgor pressure in the guard cells reduces both transpiration and photosynthesis and can ultimately reduce growth. Probably, many of the detrimental effects of moderate water deficits are due to stomatal closure. One important effect is that the temperature of leaves can rise to detrimental levels under some conditions when the transpiration rate is lowered markedly (Slatyer 1967).

It appears that cell division is influenced less by water deficits than cell enlargement (Slatyer 1967). Gates and Bonner (1959) found that the amount of DNA in water stressed tomato leaves remained constant during the period of their experiments whereas the amount increased steadily in control leaves. This indicated that chromosomal multiplication was continuing to occur in the control leaves but had ceased in the test leaves. Apparently, cell number is linearly related to DNA content (Slatyer 1967) so it appears that there was a reduction in rate of cell division in the water stressed leaves. Gardner and Nieman (1964) found that water stress in cotyledonary leaves of *Raphanus sativus* (radish) reduced the rate of increase of DNA.

Gates and Bonner (1959) found that water stress prevented a net increase in RNA in *Lycopersicum esculentum* (tomato) leaves also. The water stressed leaves were able to incorporate P^{32}-labeled phosphate into RNA apparently at the same rate as in control leaves. They concluded, therefore, that the rate of destruction of RNA was greater in water stressed leaves than in controls. Kessler (1961) reported that drought stress causes a pronounced decrease in the RNA content of several plants. Despite this, the rate of incorporation of uracil-C^{14} was the same as in control leaves suggesting that RNA is hydrolyzed faster in drought stressed plants. This was confirmed because the activity of RNase was found to increase markedly with water stress.

A reduction in the water content of leaves generally results in a decrease in the rate of photosynthesis (Meyer and Anderson 1939). This may seem strange because less than one per cent of the water absorbed by a plant is used as a raw material in photosynthesis. According to Slatyer (1967) however, there are two main modes of action of water stresses on photosynthesis: (1) stomatal closure and reduced rates of CO_2 exchange can influence the supply of CO_2, and (2) there is probably a direct effect of the water deficit on the biochemical processes involved in photosynthesis. Gates (1964) maintained that the rate of photosynthesis is more closely related to the water content of leaves than to the diffusive capacity of the stomates. Certainly, if Klotz's model of protein hydration and behavior is correct, one would expect Gate's contention to be logical.

Gates (1964) reported that during water stress, the respiration rate is accelerated but later declines. There is an increase in starch hydrolysis and an increase in sucrose which later declines. The respiration rate appears to be correlated with the sucrose content. Protein synthesis is impaired and active proteolysis may occur with prolonged wilting, especially in older leaves. Finally, if the degree of protoplasmic desiccation reaches critical levels, individual cells, tissues and even the entire organism may die.

Flooding lowers the oxygen supply to the roots or roots and tops if the tops are covered with water also. Under these conditions, Crawford and Tyler (1969) found that plants intolerant of flooding have an increased rate of glycolysis including alcohol dehydrogenase induction and the greatest production of acetaldehyde. On the other hand, malic acid accumulates in flood tolerant species and falls in intolerant species. Organic acids can accumulate in cells in much greater quantities than ethanol without damaging the plant. In fact, malic acid helps preserve the electrical neutrality of the cell.

Effects of Temperature

The cardinal temperatures (i.e. minimum, optimum, maximum) often differ for the same function in different plants and different functions in the same plant. They often differ also for the same function at different stages of development, and often in different organs of the same plant at a given stage. Moreover, they may differ with the general physiological condition of the plant, the duration of temperature regime, and changes in other environmental factors.

Daubenmire (1959) states that the minimum temperature for growth of melons (*Citrullus*) and sorghums (*Sorghum*) is in the range of 15 to 18°C; and for peas (*Pisum*), rye (*Secale*), and wheat (*Triticum*), -2 to 5°C. He further states that roots usually have a lower minimum for growth than shoots of the same plants. The optimum temperature for photosynthesis is below that for respiration in all known cases. This is probably the reason peaches, apples, white potatoes and many other plants do not accumulate normal food reserves in low latitudes.

Seeds of many native plants of temperate or colder regions have a low temperature requirement for germination. The value of such a mechanism in seedling survival is obvious. Work in my laboratory at the University of Oklahoma has shown that seeds of *Ambrosia artemisiifolia* (common ragweed), *A. psilostachya* (western ragweed) and *Sporobolus pyramidatus* (drop seed) require exposure to temperatures in the range of 4–7°C for about 6–10 wk in moist conditions to break dormancy. Lane (1965) reported that seeds of *Helianthus annuus* (annual sunflower) require 90 days at 4°C in moist sand to completely break dormancy. Donoho and Walker (1957) reported that Elberta peach (*Prunus persica*) seeds usually require stratification for 60–100 days at 7°C or below to break dormancy.

Overwintering buds of many, if not most, perennial plants of temperate or colder regions also have low temperature requirements for the breaking of dormancy. Weinberger (1956) found that Elberta peach buds require 950 hr at 7°C or below to break dormancy. Daubenmire (1959) stated that peaches in general require 400 or more hours of temperatures at or below 7°C, blueberries (*Vaccinium* spp.) require 800 hr at this level, and apples (*Pyrus malus*) require even more.

The chemical changes which are necessary for the breaking of dormancy in seeds and buds, and which are stimulated by exposure to low tempeatures, are not definitely known. In fact, this is a fertile field for research. There is growing evidence, however, that the breaking of dormancy may be related to a change in balance between growth inhibitors and stimulators. Sondheimer, Tzou, and Galson (1968) reported that abscisic acid inhibits germination of excised nondormant embryos and this can be reversed with a combination of gibberellic acid and kinetin. Natural amounts of abscisic acid in seeds and pericarps of several species of *Fraxinus* (ash) were determined, and it was found that the amount in *F. americana* decreased 37% in the pericarp and 68% in the seed with chilling treatment. They felt that it is unlikely that the amount in the pericarp influences dormancy but that the inhibitor in the seed does appear to play a regulatory role. Junttila (1970) found that dormancy of seeds of *Betula nana* (birch) was effectively broken by low temperatures or treatment with gibberellic acid. Haber, Foard and Perdue (1969), reported 84% germination in dormant lettuce (*Lactuca sativa*) seeds treated with a 5×10^{-4} M gibberellic acid solution, 1.5% germination in seeds treated with a 5×10^{-4} M gibberellic acid solution plus 100 mg/l of abscisic acid, and 0% in water controls.

Donoho and Walker (1957) reported a 98% breaking of dormancy of Elberta peach buds after only 164 hr at 7°C when sprayed with a 4000 ppm gibberellic acid solution, despite the usual requirement of 950 hr at 7°C or below. Aung, Hertogh and Staby (1969) found free gibberellic acid in *Tulipa gesneriana* (tulip) increased by 67% over the initial level in bulbs grown for 4 wk at 18°C. Bound gibberellic acid showed a slight initial increase followed by a rapid decrease. In bulbs treated at 13°C, a marked decline, occurred in free gibberellic acid and a 2-fold increase occurred in the bound form. The rates of floral, shoot and root development were accelerated at 18°C as compared with 13°.C

Cold hardening of plants (i.e. improving the ability of plants to withstand low temperatures without severe damage or death) is probably just as important to plant survival in regions with periods of relatively low temperatures as dormancy. Actually, some of the same internal changes may be responsible for both. Irving (1969) found a buildup during cold hardening in *Acer negundo* (box elder) of an inhibitor similar or identical to abscisic acid which has, of course, been associated with dormancy. He found that it or abscisic acid increased hardiness when applied to branches after a hardening period of 3 wk at 5°C. There are several cellular chemical changes associated with cold hardening (Klages 1947): (1) increase in bound water, (2) decrease in total water, (3) increase in osmotic pressure, (4) change of starch to sugars, (5) increase in pentosans, and (6) change of certain proteins to amino acids. Nothing new has been added to this list since Klages' publication in 1947.

Temperature obviously affects rates of all organic processes in plants, but a comprehensive discussion of such effects is outside the scope of this discussion. The relationship of temperature conditions at the geographic point of origin of plants to the optimum temperatures for photosynthesis and respiration in those plants is very important in explaining their distribution. Mooney and Billings (1961) studied relationships of temperature to photosynthesis and respiration in *Oxyria digyna* (alpine sorrel), a species with a circumboreal distribution of wide latidudinal range. The sources of the populations studied varied from an elevation of 12,300 ft at 39° 40′ N latitude to a zero elevation at 68° 56′ N. They found that plants of northern populations had higher respiration rates at all temperatures than did plants of southern alpine populations. Moreover, they found that plants of northern populations had higher photosynthetic rates at lower temperatures and attained maximum rates at lower temperatures than did plants of the southern alpine populations. Obviously, there are definite temperature ecotypes in this species.

Mooney, Wright and Strain (1964) investigated the gas exchange capacity of plants in relation to elevational zonation in the White Mts. of California. Field measurements over an elevational span of 4000 to 14,000 ft indicated that the ratio of photosynthesis to dark respiration decreased with altitude primarily due to increased respiration rates with increased elevation. When similar measurements were made on plants grown in uniform greenhouse conditions, from seeds collected at different elevations, no differences in respiration rates correlated with elevation were obtained indicating considerable environmentally conditioned plasticity in gas exchange capacity. However, the high elevation plants generally displayed peak photosynthetic activity at temperatures lower than the low elevation plants even though all were grown in the same conditions.

Mooney and West (1964) studied the photosynthetic and respiratory acclimation of five species in the White Mts. of California (Table 1). Seeds were collected at the indicated elevations, planted in pots in the greenhouse, and after establishment of the seedlings, they were taken to stations at various elevations in the mountains for acclimation (Table 2).

They were kept in the same pots and soil and subirrigated at each place. Water was thus not limiting and the soil type was constant. Gas exchange measurements were made after a minimum of 3 wk acclimation. The dark respiration rates of plants of all species from all stations were similar. Plants of all species acclimated to the colder subalpine environment were more efficient in photo-

TABLE 1

Natural Distribution of Experimental Plants

Species	Elev. range (m)	Elev. origin of seeds used (m)
Artemisia tridentata	1829–3200	3078
Chamaebatiaria millefolium	1980–3200	3200
Haplopappus apargioides	3048–3962	3658
Artemisia arbuscula	3200–3719	3475
Encelia virginensis ssp. *actoni*	1372–1829	1402

TABLE 2

Acclimation Stations

Natural vegetation	Elev. (m)	Mean air temp. °C* Max.	Min.
Desert scrub	1402	38.9	11.7
Pinyon woodland	2408	33.4	10.3
Subalpine forest	3094	19.5	−2.4

* Based on weekly maximum and minimum temperatures during period of June 17 to Aug. 12, 1963.

synthesis at colder temperatures than were plants grown at the warmer environment of the lowest elevation. Conversely, plants grown in the desert were more efficient in photosynthesis at high temperatures than were plants grown in the subalpine region. Plants of the two species with the widest natural distributional range in elevation in the White Mts. showed the greatest plasticity in their photosynthetic response as indicated by shifts in the optimum photosynthetic temperatures dependent on treatment.

Miller (1960) compared the effects of temperature on the photosynthetic rate in *Agrostis palustris* (creeping bent) which is adapted to cooler regions of the U.S., and *Cynodon dactylon* (bermuda grass) which is adapted to the warmer regions. The relative rate of apparent photosynthesis in creeping bent increased from 64.6% at 15°C to a maximum of 100% at 25°C, then decreased to 62.2% at 40°C. On the other hand, the relative rate of apparent photosynthesis in bermuda grass increased from 54.9% at 15°C to a maximum of 100% at 35°C, and only dropped to 97.7% at 40°C.

In summary, it is evident that plants vary genetically in their photosynthetic and respiratory responses to temperature changes, and they vary also in such responses depending on the temperature conditions during acclimation.

Effects of Light

According to Johnson (1954) solar radiation at the outer surface of the earth's atmosphere has a virtually constant intensity of 1400 watts m^{-2} or 2 cal cm^{-2} min^{-1}. About 98% of the energy is in the wavelength interval of 0.2–4.5 μ, including about 40–45% in the visible range of 0.4–0.7 μ. About one-third of the energy is reflected back to space by the earth's atmosphere and smaller amounts are absorbed and scattered by various components of the atmosphere. As a result, only about 50% of the total radiation reaches the earth on the average. This figure may be greater than 70% in dry regions with little cloud cover and lower than 40% in tropical rainy areas (Slatyer 1967). Obviously, on very cloudy days the energy reaching the surface of the earth may be very low. Of the solar energy reaching the earth, about 40–45% is still present in the visible range.

Radiation in the visible range is most important in affecting organic processes in plants, so the rest of this discussion will be concerned primarily with visible light. The chief processes in plants which are affected by light are: (1) chlorophyll synthesis, (2) photosynthesis, (3) phototropism, (4) photoperiodism, and (5) transpiration.

Light Intensity Effects. Many plants are able under most conditions to develop well in less than full sunlight. Red kidney beans (*Phaseolus vulgaris*) grow and fruit well in growth chambers in a light intensity of 700–1000 ft-c. Intensity of light has a definite effect on the distribution of plants. Many ferns grow well under trees because such ferns attain their maximum rate of photosynthesis at light intensities considerably below full sunlight. Species which characteristically grow in shaded locations are able to remain alive at very low light intensities. Redwood (*Sequoia sempervirens*) seedlings normally grow in the deep shade of the forest floor and they are able to grow at intensities of as little as 100 ft-c (Bonner and Galston 1952). *Pinus taeda* (loblolly pine) and *Pinus echinata* (shortleaf pine) seedlings are found in great profusion along road cuts and other open areas in the southeastern United States, but they fail to survive under dense forest canopies. Many hardwood seedlings such as those of *Quercus alba* (white oak) and *Quercus rubra* (red oak) survive nicely under such canopies and thus replace the pines in succession. Kramer and Decker (1944) found that photosynthesis in loblolly pine increased with light intensity up to the highest intensity used which was almost that of full sunlight. On the other hand, photosynthesis in white oak and red oak reached a maximum at one-third or less of full sunlight and showed slight decreases at higher light intensities. These results indicate that lack of sufficient light for photosynthesis may be a significant factor in the failure of pine seedlings to become established under forest stands. Seedlings of many other climax forest dominants such as *Fagus grandifolia* (beech) and *Acer saccharum* (sugar maple) have very low light requirements for photosynthesis also (Daubenmire 1959). After a long series of studies, Blackman (1950) concluded that the distribution of *Scilla non-scripta* (bluebell) in English forests is determined primarily by light intensity.

Björkman and Holmgren (1963) investigated the photosynthetic response to light intensity of sun and shade plants of *Solidago virgaurea* (goldenrod). They found that the shade plants had a higher chlorophyll content per unit leaf area in weak light (3×10^4 ergs cm^{-2} sec^{-1}) than in strong light (15×10^4 ergs cm^{-2} sec^{-1}) and more than did sun plants in weak light. On the other hand, the sun plants had lower concentrations of chlorophyll in weak light than strong light. The gross rate of photosynthesis (apparent rate plus respiration rate) in sun plants at light saturation was 25.5 ± 2.0 mg CO_2 dm^{-2} h^{-1} when the plants were grown in strong light (same as above) prior to testing whereas it was only 17.9 ± 0.6 mg CO_2 dm^{-2} h^{-1} in shade plants grown in strong light. The gross rate of photosynthesis at light saturation was higher in the shade plants than in the sun plants when the plants were grown in weak light prior to testing, 2.93 ± 0.08 in shade plants versus 2.31 ± 0.04 in sun plants. The difference of the means in the plants grown in low intensity was highly significant statistically. The authors concluded that the photosynthetic behavior was consistent with light intensities prevailing in the natural environment indicating that the behavior was a result of a genetic adaptation to the habitat. Hybridization experiments supported this conclusion.

Björkman (1966) reported on additional comparative studies of sun and shade plants of several species. Plants of shaded habitats consistently reached a higher percentage of the maximum photosynthetic rate at light saturation, when tested in a low light intensity of 10^4 ergs cm^{-2} sec^{-1} (Table 3). *Plantago lanceolata* (plantain) was found only in open meadows and *Lamium galeobdolon* (henbit) only in dense beech forests. Björkman found that the activity of Photosystem I (far red, 700 mμ) in *Solidago virgaurea* was about the same when grown in weak or strong light, regardless of the original habitat. On the other hand, the action of Photosystem II (>625 mμ used) was impaired in shade plants grown in high light intensity and in sun plants grown in low light intensity. The Emerson enhancement effect (increased photosynthesis from combination red-far red light over sum of photosynthesis in each light quality separately) was drastically reduced in shade plants grown in high light intensities.

TABLE 3

*Photosynthesis at a Low Light Intensity of 10^4 ergs cm^{-2} sec^{-1} in Percent of the Photosynthetic Rate at Light Saturation**

Species	Sun plants	Shade plants
Plantago lanceolata	8	—
Solidago virgaurea	9	17
Rumex acetosa	11	16
Geum rivale	14	16
Lamium galeobdolon	—	25

* From Björkman (1966).

Milner and Hiesey (1964) studied the photosynthetic responses to light intensity of several races of *Mimulus cardinalis* (monkey-flower) collected from an elevation of 45 m to 2220 m in California. They found that the saturating light intensity for the races increased with elevation of their original habitats. At 0°C, the saturating light intensity varied from 290 to 570 ft-c for different races, and at 40°C from 3500 to 5900 ft-c. Light intensity, of course, generally increases with elevation. Mooney and Billings (1961) in their study of ecotypes of *Oxyria digyna* found that the high elevation, low latitude plants attained photosynthetic light saturation at a higher light intensity than did low elevation, high latitude plants even when preconditioned under the same conditions

Light Quality. All visible light is effective in the synthesis of chlorophyll except wavelengths above 680 mμ (Sayre 1928). Chlorophyll a has a maximum absorption peak in the blue-violet (430 mμ) and a secondary maximum in the red (660 mμ), and chlorophyll b has a maximum absorption peak in the blue (455 mμ), and a secondary maximum in the orange-red (640 mμ) (Salisbury and Ross 1969). Hoover (1937) reported that wheat (*Triticum* sp.) has a maximum rate of photosynthesis at 655 mμ in the red and a secondary maximum at 440 mμ in the blue. Warburg and Negelein (1923) gave a general curve for photosynthesis showing a maximum rate in the red and a secondary maximum in the blue. It is now known that photosynthesis actually consists of two photosystems which work together to complete the light reactions which result in the production of ATP and NADPH. Photosystem I has an absorption peak at 700 mμ and Photosystem II has an absorption peak around 672 mμ (Salisbury and Ross 1969).

Phototrophic responses (movements in response to light) are chiefly due to wavelengths in the blue-violet, 400–490 mμ, but there may be some response to U.V. light (Johnston 1934).

The most efficient light quality to interrupt the dark period to prevent a short-day photoperiodic response is in the wavelength range of 620–640 mμ.

Effects of Daylength. The response of organisms to the relative lengths of the light and dark periods is termed photoperiodism. Garner and Allard (1920) discovered this phenomenon while working with Maryland Mammoth Tobacco (*Nicotiana tabacum*). They extended the work to many other species and found that flowering in many plants is regulated by the relative lengths of the light and dark periods. They classified plants into three categories on the basis of their photoperiodic flowering responses: (1) short-day plants—those plants which flower only when the day length is below a *certain critical maximum*, (2) long-day plants—those which flower only when the day length is above a *certain critical minimum*, and (3) day-neutral plants—those in which flowering is not determined by day length. Another category has been added, the intermediate-day plants—those which flower only between two critical day lengths. Larsen (1947) found that the strains of *Andropogon scoparius* (little bluestem) from the area south of 36°N latitude were intermediate-day plants because they failed to flower on a 13-hr photoperiod and also on a 15-hr photoperiod, but flowered well between those daylengths.

It is now known that numerous phenomena in plants are regulated by day length in many species, bulb formation, type of vegetative growth, dormancy, leaf abscission, seed germination, etc. The timing mechanism is still obscure, but it seems to involve a pigment called phytochrome which was discovered and isolated by Borthwick and his associates at the USDA laboratories in Beltsville, Maryland (Borthwick, Hendricks and Parker 1952). They found that phytochrome exists in two forms which can readily be converted from one form to another. One form has an absorption peak in the red at about 660 mμ and is designated as P_r. The other form has an absorption peak at a longer wavelength of red (730 mμ) and is designated as P_{fr}, for far-red. When P_{fr} absorbs light at 730 mμ it is converted to P_r, and when P_r absorbs light at 660 mμ it is converted to P_{fr}. In addition to these rather rapid changes brought about by light, there is a slow conversion of P_{fr} to P_r in darkness. In sunlight and broad spectrum artificial light the conversion of P_r to P_{fr} predominates. The P_{fr} form appears to be inhibitory to the flowering process, so darkness is important to eliminate all or part of this inhibitory form. Salisbury (1958) feels that the plant possibly synthesizes a postulated flowering hormone only during darkness also. Borthwick and Hendricks (1960) reported that the basic difference in long- and short-day plants may be in the ratio of P_r to P_{fr} required to initiate the flowering or other response.

The presence of photoperiodic ecotypes has been well documented in many geographically wide-ranging species of plants. Olmsted (1944) was the first to demonstrate such ecotypes in his work with *Bouteloua curtipendula* (sideoats grama). He found that plants of this species from southern Texas and southern Arizona were generally short-day or intermediate-day plants, with an upper critical photoperiod for flowering between 14 and 16 hours. These strains flowered more vigorously on a 13-hr than on a 9-hr photoperiod. The North Dakota strains consisted largely of long-day plants with a critical photoperiod of about 14 hours. The strains from Oklahoma, Kansas and New Mexico included mostly long-day individuals although the length of the critical photoperiod decreased with decrease in latitude of origin.

Larsen (1947) found that only strains of *Andropogon scoparius* (little bluestem) originating south of 36°N latitude produced visible inflorescences on a 14-hr photoperiod. The northern strains flowered on a 15-hr photoperiod but not the southern strains. The northern strains were apparently long-day plants because they were able to initiate flowers on a long photoperiod but were inhibited from flowering on 13- and 14-hr photoperiods. As was previously indicated at the beginning of this section on effects of day-length, the strains originating south of 36°N latitude were apparently intermediate-day plants.

Vaartaja (1959) collected seeds from 19 genera of trees, 38 species and 82 origins of widely different latitudes in Europe and North America. Plants derived from these seeds were grown in uniform conditions except for photoperiod. Photoperiods of 12, 14, 16 and 18 hours were used with the total light energy being identical in all cases. He found that the farther north the origin of the strain, the longer was the critical daylength below which height growth of seedlings was retarded. Moreover, the more northerly the source, the shorter was the critical dark period which induced dormancy in several species. In some plants from the far north, photoperiods that inhibited elongation permitted great increases in weight of stem and roots, and in numbers of buds. Cambial growth thus would continue later in the growing season than elongation.

Irgens-Moller (1957) collected approximately 100 plants (2–6 yr old) of *Pseudotsuga menziesii* (douglas-fir) from each of seven locations in Oregon at elevations of 60 ft to 4 000 ft. The plants were potted and kept in a cold frame at Corvallis, Oregon during two winters after collection so that all were exposed to the same winter conditions. On Feb. 1 after the second winter, half of the plants from each locality were placed in a 9-hr photoperiod and half on an 18-hr photoperiod with all other conditions the same. There were no effects of photoperiod on time of bud bursting in the low altitude plants, but the long days hastened bud bursting strikingly in high altitude plants as compared with effects of short days. The selection probably results from the killing of seedlings by low night temperatures at high altitudes if they break dormancy in the short days of winter or early spring.

Effects of Plant Produced Toxins

De Candolle first suggested the existence of natural ly-occurring plant growth-inhibitors in 1832 (Bonner 1950). He noted from field observations that

Euphorbia (spurge) was apparently inhibitory to flax (*Linum usitatissimum*), thistles (probably *Cirsium*) to oats (*Avena sativa*) and rye to wheat. Very little research was done on the subject for almost 100 years at which time Cook (1921) described the characteristic wilting of potato (*Solanum tuberosum*) and tomato plants grown near *Juglans nigra* (black walnut). Davis (1928) extracted juglone from mature hulls and roots of black walnut and found it to be very toxic to tomato and alfalfa (*Medicago sativa*) plants when injected into the stems. The juglone was identified as 5-hydroxynaphthaquinone.

Proebsting and Gilmore (1940) related the problem of re-establishing peach (*Prunus persica*) trees in old peach orchards to the presence of toxic substances in the soil. Bonner and Galston (1944) reported that the edge rows in guayule (*Parthenium argentatum*) plantings in California had much larger plants than the center rows and the differences could not be eliminated by heavy watering and mineral application. Experiments indicated that leachates from pots of year-old guayule plants were very inhibitory to guayule seedlings. The inhibitor was subsequently indentified as trans-cinnamic acid.

Muller, Muller and Haines (1964) observed areas free of vegetation around certain shrubs in California grasslands. They found that at least three shrubs, *Salvia leucophylla* (clearleaf sage), *S. apiana* (bee sage), and *Artemisia californica* (California sagebrush) produced volatile materials which inhibited various test plants. Muller and Muller (1964) identified several terpenes which were produced by leaves of three species of *Salvia*, and two of these, camphor and cineole, were found to be very inhibitory to test plants. Muller, Hanawalt and McPherson (1968) related the disappearance of herbaceous species in the open California chaparral during the fire cycle to the production of inhibitory chemicals by chaparral species.

My students and I have obtained good evidence that the rapid disappearance of the pioneer weed stage (2 to 3 years) in our revegetating old-fields in central Oklahoma is due to the inhibition of seedlings of species of that stage by several plants of the pioneer stage (Abdul-Wahab and Rice 1967, Parenti and Rice 1969, Wilson and Rice 1968). Triple-awn grass (*Aristida oligantha*), the dominant of the second stage, is generally not inhibited by species of the pioneer weed stage and can grow well in the infertile soil, so it is able to invade the area.

There is some evidence that antibiotics (substances produced by microorganisms which inhibit other microogranisms) may be important in the growth and distribution of the higher plants as well as microorganisms. Certain soil bacteria inhibit the root-nodule bacteria which are important in adding nitrogen to the soil (Konishi 1931). Iuzhina (1958) reported that many soil bacteria, fungi and actinomycetes are antagonistic to *Azotobacter*, a free living bacterium that adds nitrogen to the soil. Some soil microorganisms have been found to be inhibitory to certain microorganisms that cause diseases of higher plants (Cooper and Chilton 1950).

Another level of inhibition is that of microorganisms against higher plants. There is no question that many, if not most, pathogenic microorganisms produce abnormal symptoms in the host plants through the production of toxins. Very few of these compounds have been identified and this remains an important and fruitful field for research. Microorganisms are sometimes responsible for changing certain non-inhibitory metabolites of higher plants into inhibitory compounds (Börner 1960).

The remaining level of inhibition is that of higher plants against microorganisms. There is considerable evidence that resistance of many plants to fungal, viral, and bacterial diseases may be associated with the production by resistant varieties of inhibitors of the pathogens (Farkas and Kiraly 1962, Hughes and Swain 1960, Schaal and Johnson, 1955).

Ferenczy (1956) found that the seeds or fruits of 52 species and varieties from 19 plant families contained antibacterial compounds and that seeds of some species were attacked intensely by molds whereas others were attacked slightly or not at all. Lane (1965) found that fruits of the native sunflower, *Helianthus annuus*, rotted more frequently during germination tests if certain inhibitors were leached from them before the tests. Those species in which decay of the seeds during germination is less likely to occur would certainly stand a better chance of becoming established.

I have found that many of the low-nitrogen requiring early plant invaders of abandoned cultivated fields produce inhibitors of the nitrogen-fixing bacteria (Rice 1964, 1965b, 1965c). Triple-awn grass, the dominant of the second stage, is one of the inhibitory species. This would give such plant species a selective advantage in competition over plants with higher nitrogen requirements. This probably slows succession in old-fields, because Rice, Penfound and Rohrbaugh (1960) found that the order in which species invade revegetating fields in central Oklahoma, beginning with the second stage, is the same as the order based on increasing nitrogen requirements.

Muller (1965) found that the volatile terpenes from *Salvia leucophylla* reduce cell elongation and cell division in the radicles and hypocotyls of germinating *Cucumis sativus* (cucumber) seeds. Muller *et al.* (1969) reported that two volatile terpenes, cineole and dipentene, which emanate from *Salvia leucophylla* leaves markedly reduce oxygen uptake by suspensions of mitochondria from *Avena fatua* (wild oats) or cucumber. The inhibition appears to be localized in that part of the Krebs cycle where succinate is converted to fumarate, or fumarate to malate. They also reported that the permeability of cell membranes appears to be decreased on exposure to terpenes.

Most of the inhibitors that my students and I have identified from our old-field species are phenolics (Abdul-Wahab and Rice 1967; Parenti and Rice 1969; Rice 1965a, 1965b; Rice and Parenti 1967; Wilson and Rice 1968). Some of the inhibitors identified are: chlorogenic acid, p-coumaric acid, ferulic acid, gallic acid, p-hydroxybenzaldehyde, isochlorogenic acid complex, scopoletin, scopolin, sulfosaliyclic acid, and several tannins.

Sondheimer (1962) reported that chlorogenic acid is a strong inhibitor of several enzyme systems and it is possible that this may be its chief role in growth inhibition. Schwimmer (1958) found that a 10^{-3} M concentration of chlorogenic acid caused a 50% inhibition of potato phosphorylase activity. Mazelis (1962) reported that a peroxidase-catalyzed oxidative decarboxylation of methionine is strongly inhibited by a 5×10^{-6} M concentration of chlorogenic acid. Sondheimer and Griffin (1960) reported that indoleacetic acid oxidase is inhibited by chlorogenic acid and its isomers, and by isochlorogenic acid and dihydrochlorogenic acid.

Zenk and Muller (1963) reported that p-coumaric and ferulic acids increase decarboxylation of indoleacetic acid (IAA), resulting in reduced plant growth. Henderson and Nitsch (1962) found a pronounced inhibiting effect of p-coumaric acid on IAA induced growth also.

Bendall and Gregory (1963) found that phenol oxidases which are important in the formation of lignin-like polymers from coniferyl alcohol and oxidation products of tannins such as quinones react with each other to produce a complex that may make phenol oxidase catalytically inactive, or may produce an active protein of modified properties. Hulme and Jones (1963) and Jones and Hulme (1961) found that tannins caused considerable reduction of Krebs cycle succin oxidase and malic enzyme activity. Gallic and tannic acids were shown also to inhibit IAA induced growth (Zimsmeister and Hollmuller 1964). Hall (1966) and Williams (1963) reported that relatively low concentrations of tannic acid were very inhibitory to pectolytic enzymes. Gallic and tannic acids are very inhibitory to the nodulation of heavily inoculated legumes and to hemoglobin formation in the nodules (Blum and Rice 1969). The hydroxyl groups of tannins are known to have a strong affinity for the peptide linkages of proteins, so it is likely that reactions readily occur between the two compounds when they come in contact at a time when the ice-shell around the protein is appropriately modified. This would undoubtedly change the nature of the protein.

Einhellig *et al.* (1970) found that sunflower (*Helianthus annuus*) and tobacco plants treated with 10^{-4} and 5×10^{-4} scopoletin concentrations had significant increases in scopoletin and scopolin in their tissues. Net photosynthesis in tobacco plants treated with a 10^{-3} M concentration of scopoletin was depressed to as low as 34% of that of the controls. Reduced growth in leaf area over a 12 day experiment correlated well with the significant reduction in the rate of net photosynthesis.

The surface has just been scratched in identifying plant produced inhibitors of other plants. Moreover, the field is wide open for research on the mechanisms of action of the presently known inhibitors.

General Conclusions

I have discussed briefly the presently known physiological basis for some of the ecological responses of plants to the environmental factors of water, temperature, radiant energy, and toxins. It would obviously require most of the present volume to discuss the presently known effects of all environmental factors. Even so, our knowledge of the effects of most factors is superficial. Many of the known physiological effects may be indirect ones also.

In closing, I emphasize that the physiological ecologist is searching for the physiological basis for ecological behavior. It is essential that research in this area be vigorously pursued.

REFERENCES

Abdul-Wahab, A. S., and E. L. Rice, *Bull. Torrey Bot. Club.* **94**, 486 (1967).

Aung, L. S., A. A. de Hertogh, and G. Staby, *Plant Physiol.* **44**, 403 (1969).

Bendall, D. S., and R. P. F. Gregory, "Purification of Phenol Oxidase." In *Enzyme Chemistry of Phenolic Compounds*, MacMillan Co., New York, 1963.

Björkman, O., and P. Holmgren, *Physiol. Plant.*, **16**, 889 (1963).

Björkman, O., *Brittonia* **18**, 214 (1966).

Blackman, G. E., *Ann. Bot.* **14**, 487 (1950).

Blair, G. Y., L. A. Richards, and R. B. Campbell, *Soil Sci.* **70**, 431 (1950).

Blum, U., and E. L. Rice, *Bull. Torrey Bot. Club* **96**, 531 (1969).

Bonner, J., and A. W. Galston, *Bot. Gaz.* **106**, 185 (1944).

Bonner, J., and A. W. Galston, *Principles of Plant Physiology*, W. H. Freeman and Co., San Francisco, 1952.

Bonner, J., *Bot. Rev.* **16**, 51 (1950).

Börner, H., *Bot. Rev.* **26**, 393 (1960).

Borthwick, H. A., S. B. Hendricks, and M. W. Parker, *Proc. Nat. Acad. Sci.* **38**, 929 (1952).

Borthwick, H. A., and S. B. Hendricks, *Science* **132**, 1223 (1960).

Buckman, H. O., and N. C. Brady, *The Nature and Properties of Soils*. The MacMillan Co., New York, 1960.

Cook, M. T., *Phytopathology* **11**, 346 (1921).

Cooper, W. E., and S. J. P. Chilton, *Phytopathology* **40**, 544 (1950).

Crawford, R. M. M., and P. D. Tyler, *J. Ecol.* **57**, 235 (1969).

Daubenmire, R. F., *Plants and Environment*. John Wiley and Sons, Inc., New York, 1959.

Davis, R. F., *Amer. J. Bot.* **15**, 620 (1928).

Donoho, C. W., Jr., and D. R. Walker, *Science* **126**, 1178 (1957).

Einhellig, F. A., E. L. Rice, P. G. Risser, and S. H. Wender, *Bull. Torrey Bot. Club* **97**, 22 (1970).

Farkas, G. L , and Z. Kiraly, *Phytopathologische Z.* **44**, 105 (1962).

Ferenczy, L., *Acta Biol. Acad. Sci. Hung.* **6**, 317 (1956).

Gardner, W. R., and R. H. Nieman, *Science* **143**, 1460 (1964).

Garner, W. W., and H. A. Allard, *J. Agr. Res.* **18**, 553 (1920).

Gates, C. T., *J. Aust. Inst. Agr. Sci.* **30**, 3 (1964).

Gates, C. T., and J. Bonner, *Plant Physiol.* **34**, 49 (1959).

Haber, A. H., D. E. Foard, and S. W. Perdue, *Plant Physiol.* **44**, 463 (1969).

Hall, C. B., *Nature* **212**, 717 (1966).

Henderson, J. H. M., and J. P. Nitsch, *Nature* **195**, 780 (1962).

Hoover, W. H., *Smithsonian Misc. Coll.* **95**, 1 (1937).

Hughes, J. C., and T. Swain, *Phytopathology* **50**, 398 (1960).

Hulme, J. D., and A. C. Jones, "Tannin Inhibition of Plant Mitochondria." In *Enzyme Chemistry of Phenolic Compounds*. MacMillan Co., New York, p. 97–120, 1963.

Irgens-Moller, H., *For. Sci.* **3**, 79 (1957).

Irving, R. M., *Plant Physiol.* **44**, 801 (1969).

Iuzhina, Z. I., *Microbiol.* **27**, 452 (1958).

Jacobson, B., *Nature* **172**, 666 (1953).

Johnson, F. S., *J. Met.* **11**, 431 (1954).

Johnston, E. S., *Smithsonian Inst. Misc. Coll.* **92**, 1 (1934).

Jones, J. D., and A. C. Hulme, *Nature* **191**, 370 (1961).

Junttila, O., *Physiol. Plant* **23**, 425 (1970).

Kessler, B., "Nucleic Acids as Factors in Drought Resistance in Higher Plants." In *Recent Advances in Botany*. Univ. of Toronto Press, Toronto, Canada, 1961.

Klages, K. H. W., *Ecological Crop Geography*. MacMillan Co., New York, 1947.

Klotz, I. M., *Science* **128**, 815 (1958).

Konishi, K., *Mem. Col. Agr., Kyoto Imp. Univ.* **16**, 1 (1931).

Kramer, P. J., and J. P. Decker, *Plant Physiol.* **19**, 350 (1944).

Lane, F. E., "Dormancy and Germination in Fruits of Sunflower." Ph. D. Dissertation, University of Oklahoma, Norman, 1965.

Larsen, E. C., *Bot. Gaz.* **109**, 132 (1947).

Mazelis, M., *J. Biol. Chem.* **237**, 104 (1962).

Meyer, B. S., and D. B. Anderson, *Plant Physiology*. D. Van Nostrand Co., Inc., New York, 1939.

Miller, V. J., *Proc. Amer. Soc. Hort. Sci.* **75**, 700 (1960).

Milner, H. W., and W. M. Hiesey, *Plant Physiol.* **39**, 208 (1964).

Mooney, H. A., and W. D. Billings, *Ecol. Monogr.* **31**, 1 (1961).

Mooney, H. A., and M. West, *Amer. J. Bot.* **51**, 825 (1964).

Mooney, H. A., R. D. Wright, and B. R. Strain, *Amer. Midland Naturalist* **72**, 281 (1964).

Muller, C. H., W. H. Muller, and B. L. Haines, *Science* **143**, 471 (1964).

Muller, W. H., and C. H. Muller, *Bull. Torrey Bot. Club* **91**, 327 (1964).

Muller, W. H., *Bot. Gaz.* **126**, 195 (1965).

Muller, C. H., R. B. Hanawalt, and J. K. McPherson, *Bull. Torrey Bot. Club* **95**, 225 (1968).

Muller, W. H., P. Lorber, B. Haley, and K. Johnson, *Bull. Torrey Bot. Club* **96**, 89 (1969).

Olmsted, C. E., *Bot. Gaz.* **106**, 46 (1944).

Parenti, R. L., and E. L. Rice, *Bull. Torrey Bot. Club* **96**, 70 (1969).

Proebsting, E. L., and A. E. Gilmore, *Proc. Amer. Soc. Hort. Sci.* **38**, 21 (1940).

Rice, E. L., W. T. Penfound, and L. M. Rohrbaugh, *Ecology* **41**, 224 (1960).

Rice, E. L., *Ecology* **45**, 824 (1964).

Rice, E. L., *Physiol. Plant* **18**, 255 (1965a)

Rice, E. L., *Proc. Okla. Acad. Sci.* **45**, 43 (1965b).

Rice, E. L., *Southwestern Naturalist* **10**, 248 (1965c).

Rice, E. L., and R. L. Parenti, *Southwestern Naturalist* **12**, 97 (1967).

Salisbury, F. B., *Scientific Amer.* **198**, 108 (1958).

Salisbury, F. B., and C. Ross, *Plant Physiology*. Wadsworth Publishing Co., Inc., Belmont, Calif., 1969.

Sayre, J. D., *Plant Physiol.* **3**, 71 (1928).

Schaal, L. A., and G. Johnson, *Phytopathology* **45**, 626 (1955).
Schwimmer, S., J. *Biol. Chem.* **232**, 715 (1958).
Slatyer, R. O., *Plant-Water Relationships*. Academic Press, New York, 1967.
Sondheimer, E., and D. H. Griffin, *Science* **131**, 672 (1960).
Sondheimer, E., "The Chlorogenic Acids and Related Compounds." In *Plant Phenolics and Their Industrial Significance*. V. C. Runeckles, ed. Imperial Tobacco Co. of Canada, Montreal, 1962.
Sondheimer, E., D. S. Tzou, and E. C. Galson, *Plant Physiol.* **43**, 1443 (1968).
Stocker, O., *UNESCO Arid Zone Res.* **15**, 63 (1960).
Vaartaja, O., *Ecol. Monogr.* **29**, 91 (1959).
Warburg, P., and E. Negelein, *Zeitschr. Phys. Chem.* **106**, 191 (1923).
Warming, E., *Oecology of Plants; An Introduction to the Study of Plant Communities*. The Clarendon Press, Oxford, England, 1909.
Weinberger, J. H., *Proc. Amer. Soc. Hort. Sci.* **67**, 107 (1956).
Williams, A. H., "Enzyme Inhibition by Phenolic Compounds." In *Enzyme Chemistry of Phenolic Compounds*." MacMillan Co., New York, 1963.
Wilson, R. E., and E. L. Rice, *Bull. Torrey Bot. Club* **95**, 432 (1968).
Zenk, M. H., and G. Müller, *Nature* **200**, 761 (1963).
Zimsmeister, H. D., and W. H. Hollmuller, *Planta* **63**, 133 (1964).

GENERAL REFERENCES

Alexander, M., *Introduction to Soil Microbiology*. John Wiley, New York, 1961.
Bainbridge, R., G. C. Evans, and O. Rackham, *Light as an Ecological Factor*. John Wiley, New York, 1966.
Baker, F. B., and W. C. Snyder, ed., *Ecology of Soil-borne Plant Pathogens*. Univ. of Calif. Press, Berkeley, 1965.
Burges, A., and F. Raw, *Soil Biology*. Academic Press, New York, 1967.
Evans, L. T., ed., *Environmental Control of Plant Growth*. Academic Press, New York, 1963.
Fried, M., and H. Broeshart, *The Soil-Plant System*. Academic Press, New York, 1967.
Geiger, R., *The Climate Near the Ground*. Harvard Univ. Press, Cambridge, 1965.
Gray, T. R. G., and D. Parkinson, ed., *The Ecology of Soil Bacteria*. Univ. of Toronto Press, Toronto, 1968.
Hoagland, D. R., *Inorganic Nutrition of Plants*. Chronica Botanica, Waltham, Massachusetts, 1948.
Kononova, M. M., *Soil Organic Matter*. Pergamon Press, New York, 1966.
Kramer, P. J., *Plant and Soil Water Relationships*. McGraw-Hill, New York, 1949.
Leopold, A. C., *Plant Growth and Development*. McGraw-Hill Book Co., New York, 1964.
Nason, A., and W. D. McElroy, "Modes of Action of the Essential Mineral Elements." In *Plant Physiology, A Treatise, Vol. III*. Academic Press, New York, p. 451 to 536, 1963.
Stewart, W. D. P., *Nitrogen Fixation in Plants*. Athlone Press, London, 1966.
Sutcliffe, J. F., *Mineral Salts Absorption in Plants*. Pergamon Press, New York, 1962.
Treshow, M., *Environment and Plant Response*. McGraw-Hill Book Co., New York, 1970.
Whyte, R. O., *Crop Production and Environment*. Faber and Faber, London, 1946.

ELROY L. RICE
University of Oklahoma

ECOLOGY OF PRIMARY TERRESTRIAL CONSUMERS

Basic Concepts

There are many approaches to the study and appreciation of the natural world. The ecologist looks at it with his interest focussed on the relations between living things and their surroundings. For purposes of quantitative analysis, he finds it useful to think of nature as organized into ecological systems (*ecosystems*), in which the living units interact with their environments to bring about the *flow of energy* and the *cycling of matter* wherever life is found. In this conceptual framework, organisms can profitably be considered according to their major roles in the handling of matter and energy. Thus, living things have been ecologically classified (Thienemann, 1926) as *producers* if they are *autotrophic*, i.e. able to manufacture their own food from simple inorganic substances with energy obtained from sunlight (photosynthetic green plants) or from the chemical oxidation of the inorganic compounds (chemosynthetic bacteria), or as *consumers* if they are *heterotrophic*, i.e. required to depend on already synthesized organic matter as the source of food energy. A special and very important group of consumers are the *decomposers*, which break up the complex organic substances of dead matter, incorporating some of the decomposition products in their own protoplasm and making available simple inorganic nutrients to the producers. Decomposers consist chiefly of fungi and bacteria, which absorb their food through cell membranes and thus differ significantly from the larger consumers, which ingest plant and animal tissue into an alimentary tract. There are, however, many different modes of nutrition, and it has recently been suggested (Wiegert and Owen, 1970) that energy flow and the cycling of matter may be better understood if heterotrophic consumer organisms are classified on the basis of their energy resources rather than in terms of their feeding habits. Thus we may recognize two major groups of consumers: *biophages* if they obtain their energy from living matter, and *saprophages* if they derive their energy from dead and decaying materials. These basic classifications do not accommodate such organisms as Euglena and the Venus fly-trap which are capable of both photosynthesis and heterotrophy, but they provide a reasonable framework for the majority of plant and animal species.

Food Chains and Food Webs

The concept of *food-chain* was developed by Elton (1927) to represent the series of interactions that occur between organisms in their efforts to obtain nourishment. Thus a hierarchy of relationships is formed as green plants or their products are eaten by animals and these are in turn eaten by other animals. Organisms, then, are functionally related to one another as links in a chain, i.e. as successive components in a system for the transfer of energy and matter. An organism can be characterized ecologically in terms of the position it occupies in its food-chain, and consumer organisms that are one, two and three links removed from the producer are referred to as *primary, secondary and tertiary consumers*, respectively. Thus primary consumers are those which subsist on green plants or their products, and are broadly termed *herbivores*, in contrast to secondary and higher-order consumers, which are termed *carnivores*. (Note again, as with the producer-consumer classification, that the herbivore-carnivore dichotomy is not perfect; it does not allow for *omnivores*, which eat both plant and animal tissue, or for those organisms which are herbivorous at one state in their life history and carnivorous at another.)

With each successive transfer, some of the energy that was incorporated by the producer organism (the initial link in the chain) is lost as heat, and for this reason food-chains generally do not involve more than four or five links from the beginning to the ultimate large consumer. A typical terrestrial food-chain consists of a green plant, e.g. grass, eaten by an insect, e.g. a grasshopper, which is in turn eaten by a small bird, e.g. a sparrow, and this in turn by a larger bird, e.g. a hawk. However, food chains are generally linked to other chains at almost every point; the grass is likely to be eaten by numerous species of herbivore, each of which has its own set of predators, etc. The result is that the community ecosystem is made up, not of a set of isolated food-chains but of an interconnected *food-web* whose structure rapidly becomes very complex when more than a few species are considered. This complexity, however, is restricted in part by the limited length of food-chains and in part by the unidirectional flow of energy in these chains. Food-webs are made up of two principal types of food-chains: *grazing food-chains*, which involve the direct consumption and transformation of living tissue, as in the grazing of rangeland by cattle or sheep, and *detritus food-chains*, which involve the disintegration and conversion of dead matter, both plant and animal, by a sequence of decomposer organisms.

Trophic Levels and Ecological Pyramids

The concept of the food-chain permits us to equate, in terms of ecosystem function, all organisms which occupy the same feeding position, i.e. which are the same number of links removed from the producer organism. Organisms which are similar in this respect are said to occupy the same *trophic level*. Such levels form a natural hierarchy arranged from the producer level at the bottom, through the primary consumer (herbivore) level, to one or more successive levels of subsidiary consumers at the top. The trophic level concept was developed by Lindemann (1942) to compare the energy content of the different feeding groups in natural communities and to evaluate the efficiency with which energy is transferred from level to level.

If the numbers of individuals, the total *biomass* (weight of tissue), or the energy content of the organisms of successive trophic levels in a natural community is examined, the quantity is generally found to decrease as one goes upward from producer to ultimate consumer levels. Diagrammatic models of this trophic structure thus tend to be pyramidal and have given rise to the concept of *ecological pyramids*. Comparisons of trophic levels based on numbers of individuals can be misleading, however, especially when species of very different size and rate of growth are involved: herbivorous insects are likely to be much more numerous per unit of suitable habitat than are grazing or browsing mammals (Evans and Lanham, 1960). This difficulty is partly resolved by measuring total weight of organisms present, thus replacing a *pyramid of numbers* with a *pyramid of biomass*. If they are constructed for parasite food-chains (in which one host can support many parasites) or for communities whose producers (such as diatoms and other minute algae) have a more rapid rate of replacement, or *turnover*, than its consumers, such pyramids may appear inverted or may show a very narrow base. Furthermore, because the number of calories varies considerably from tissue to tissue because of differences in composition (fat averages about 9500, and carbohydrate and protein about 4000, calories per gram), biomass may prove a poor indicator of energy content. It is therefore desirable, whenever possible, to replace weights with calorific values and to convert the pyramid of biomass into a *pyramid of energy*. If the total amount of energy utilized at each trophic level over a set period of time is taken into account, the quantity will always be less at each succeeding level and the upright pyramidal shape of the model will be maintained. Thus energy units provide the best basis for comparing the *productivity* (the rate at which energy and matter are stored in the form of organic substances) of different organisms, of different trophic levels, and of different ecosystems. They also offer the best means of evaluating the *efficiency* (the ratio of energy stored or put out in a process to that put in, usually expressed as a percentage) of organisms and ecosystems in carrying out their activities of transferring and transforming energy and matter.

Standing Crops, Production, and Energy Flow

The quantity of living organisms present at a given time may be referred to as the *standing crop* or *stock*. For reasons already explained, this quantity is best expressed in terms of its energy content (in calories). The magnitude of the standing crop will vary from place to place, depending basically on the available quantities of energy and nutrients, and in almost all places from season to season, being influenced by all the factors affecting growth and reproduction. Relatively long-lived consumers like the large herbivorous mammals may accumulate and store considerable quantities of matter and energy in their bodies and thus achieve a large standing crop, but surprisingly high values can be reached by much smaller organisms, such as insects, which feed more efficiently and reproduce more rapidly. Unusually high standing crops are seen at times of population explosions, such as those of defoliating insects and of periodically fluctuating species like snowshoe hares and lemmings, but these levels cannot be sustained for more than a short time. Inverted pyramids of biomass illustrate the possibility of a relatively large standing crop of consumers supported by a small standing crop of producer plants, when the latter are smaller, grow more rapidly, and are replaced more frequently than the consumers.

Because organisms may be eaten or move away from an area, the standing crop often fails to be a good measure of the total quantity of tissue they have produced over a given period of time. This total quantity, or *production*, includes not only the new tissue added as *growth* to the bodies of individual organisms but also that resulting from the reproduction of new individuals. That part of the production that is removed by man (or some other species) is known as the *yield* or harvest. Because of the limited efficiency of the metabolic processes required in the formation of new tissue, more energy and matter are needed by the organism than are stored in production. Thus, to evaluate the complete functioning of an ecosystem, we need a measure of the total energy (or matter) involved in metabolism; for consumer organisms, this is referred to as *assimilation* or *energy flow*. (For producer organisms, the total energy or matter used in their metabolism is called *gross primary production*, while that incorporated as new tissue is called *net primary production*.)

Energy Budgets and Efficiencies

Full understanding of increases or decreases in standing crop or energy flow requires quantitative knowledge of the biological processes involved in the transfer of matter and energy. For consumer organisms, these consist of the processes of *ingestion* or intake, *respiration*—a measure of metabolic activity, and *egestion* (generally taken to include the elimination of both feces and urine). Use is made of the *energy budget* or balance described by the equations

$$\text{Ingestion} = \text{Production} + \text{Respiration} + \text{Egestion}$$

$$\text{Assimilation} = \text{Production} + \text{Respiration}$$

and other expressions to calculate the values for processes whose quantities are not known or to determine the net change of energy transfer. Such calculations enable the ecologist to estimate the efficiency with which these processes are carried out, as for example

$$\text{assimilation efficiency} = \frac{\text{calories assimilated}}{\text{calories ingested}}$$

$$\text{growth efficiency} = \frac{\text{calories in growth}}{\text{calories ingested}}$$

$$\text{yield efficiency} = \frac{\text{calories to man (or other exploiter)}}{\text{calories ingested}}$$

These and other efficiencies provide the means for comparing different trophic levels or other ecological units with respect to their functional roles in the ecosystem. For primary consumers, the ratio

$$\frac{\text{calories consumed by herbivore population}}{\text{calories of available plant food}}$$

measures their *food-chain efficiency*, while the ratio

$$\frac{\text{calories of herbivores consumed by carnivores}}{\text{calories of plant food consumed by herbivores}}$$

measures their *efficiency of energy transfer*.

Diversity of Primary Consumers

All parts of plants—leaves, buds, flowers and their products (seeds, pollen, nectar), stems, sap, roots, bark and wood—are eaten by primary consumers, who have evolved a great diversity of life form and habits in response to their food. Large grazers and browsers, such as elephants and deer, can ingest plant tissue in bulk, but others are specialized to particular products, e.g. the mylabrid weevils that feed on beans and peas, and the sap-sucking aphids and leaf-hoppers. The specialization may extend to particular kinds of plants; the Australian koala "bear," for example, is limited to leaves of *Eucalyptus*, and the larva of the swallowtail butterfly *Papilio marcellus* feeds exclusively on the foliage of the prickly ash (*Xanthoxylum*). Much less is known of the food habits of detritus-feeders, which never attain the large size of some grazers, but Petrusewicz and Macfadyen (1970) point out that "remarkably few species restrict their diets at all narrowly."

All of the principal phyla of land animals have developed forms which are partly or entirely herbivorous, and it is likely that the majority of terrestrial heterotrophic species are primary consumers. About half of the known kinds of insects are plant-eaters (Brues, 1946), and in some habitats the proportion may be considerably higher: Menhinick (1967) found that herbivores made up about 80% of the insect species of a bush clover (*Lespedeza*) stand in South Carolina, and Evans and Murdoch (1968) reported that 85% of approximately 1500 insect species encountered on a 30-year-old abandoned field in Michigan were herbivorous. With such *diversity*, it is unlikely that any species of plant has escaped exploitation by herbivores. Indeed, most plants are host to a wide range of consumers: for example, at least 227 species of herbivorous insects are known to be associated with the oaks (*Quercus robur* and *Q. petraea*) of British woodlands (Elton, 1966, 197).

The most important insect consumers of live vegetation belong to the orders Orthoptera (grasshoppers), Hemiptera (true bugs), Homoptera (leafhoppers, aphids, scales), Coleoptera (beetles), Lepidoptera (moths, butterflies), Diptera (flies), and Hymenoptera (bees, wasps, ants). Insects as a group share dominance as herbivores with the mammals, including such larger types as horses, pigs, deer, antelope, goats, sheep and cattle, and such smaller ones as hares, rabbits, squirrels, marmots, voles and lemmings. (In tropical regions, various monkeys and apes are also important herbivores, as are fruit-eating and nectarivorous bats.) There are significant herbivores among terrestrial birds, e.g. the Galliformes (grouse, quail, pheasants) and the "specialist" seed-eating finches and sparrows, the fruit-eating parrots, and the nectar-sucking hummingbirds, and among land molluscs, such as the "garden" snails and slugs.

Much less is known of the herbivores which feed on roots, underground stems, and other living plant parts in the soil. Studies by Bornebusch (1930), Van der Drift (1951), Cragg (1961) and

Macfadyen (1961) indicate that the most important groups of herbivores in north temperate grassland and forest soils are the parasitic nematode worms, the molluscs, and the larvae of Diptera, Coleoptera, and Lepidoptera. They occur in close association with other primary consumers, the detritus feeders, which include the Oligochaeta (earthworms and enchytraeid worms), Isopoda (woodlice), Diplopoda (millipedes), free-living Nematoda, Collembola (springtails), and Acari (mites). The mechanical comminution of dead matter by these organisms provides a substrate of small particles which can be more easily attacked by fungi, protozoa, bacteria and other micro-organisms (Phillipson, 1966).

Measurement of Primary Consumption

The measurement of primary consumption involves the assessment, in terms of numbers of individuals, biomass, and energy equivalence, of the quantity of herbivorous animals present or produced over a period of time, the *energetic cost* of producing and maintaining that product, and the fate of the matter and energy that becomes incorporated in herbivore tissue. A detailed account of methods and techniques employed for these purposes is beyond the scope of this article, but a brief treatment is presented below.

The mobility, abundance, and small size of many primary consumers make it difficult to assess their numbers. Total counts or *censuses*, as of large mammals by aerial photography (Parker, 1971) or of small ones by intensive trapping (Gliwicz *et al.*, 1968), can only rarely be achieved, and it is general practice to rely on *sample collections* (e.g. Wiegert, 1964, 1965). The marking of individual organisms for subsequent recapture has been employed to estimate population size in such herbivores as grasshoppers (Nakamura *et al.*, 1971), and *indices of relative abundance* based on fecal pellet counts (Southwood, 1966, 229) or on damage to vegetation (Odum and Pigeon, 1970, I-69) are sometimes used. Estimation of change in numbers with time involves a knowledge of birth, death and migration rates, the calculation of which requires considerable *lifetable information*.

Biomass values are commonly derived by multiplying the number of individuals by average weights obtained from sample specimens or calculated from formulas relating weight to body dimensions (Petrusewicz and Macfadyen, 1970, 51). More detailed studies require knowledge of the growth of individuals and the rate of weight gain. Ultimately, biomass should be converted to its energy equivalent by determining its calorific value; this requires the use of a *calorimeter*, such as that developed by Phillipson (1964), and because of the difficulty of obtaining complete combustion, it is generally carried out in the laboratory. Published tables of the caloric content of various plant and animal tissues (e.g. Golley, 1961) show a considerable range of values even for the same species, reflecting differences in life history stage, season, and environmental conditions.

It is also clear that the dimensions of primary consumption cannot be accurately gauged without knowledge of the processes (*ingestion*, *egestion*, *assimilation*, *respiration*) associated with it. It is important, therefore, to investigate the food habits and feeding rates of herbivores, as well as to determine the quality of the food consumed, the proportion rejected as feces, the digestive efficiency, and the rates of respiration (oxygen consumption can be measured by a *respirometer* such as that of Smith and Douglas, 1949; see also the book on manometric techniques by Umbreit *et al.*, 1957). Much of this work has had to be done on animals in confinement, and little is known as yet of activity levels and metabolic rates of herbivores in nature. Recent developments in the use of *radioactive isotope tracers* (Williams and Reichle, 1965) and the *telemetric measurement* of heart rate or other characteristics related to respiration (Adams, 1965) give promise that the study of metabolism under field conditions will eventually be feasible.

Efficiency of Primary Consumers

With careful management and appropriate stocking, large mammalian herbivores can consume a fairly high proportion of the available plant food. In Great Britain, sheep stocked at a density of 10 ewes per hectare on lowland pasture may ingest up to 70% of the annual production of grass (Eadie, 1970), and beef cattle on managed rangeland in the United States, when stocked at the maximum recommended *exploitation rate*, will consume from 30% to 45% of the forage (Lewis *et al.*, 1956). The efficiency of feeding will of course vary with the situation: sheep on hill pastures, where a stocking rate cannot exceed

0.8 ewes per hectare, may utilize no more than 20% of the available food (Eadie, 1970), and Paulsen (1960) indicates that on alpine ranges in the Rocky Mountains the proportion of herbage production removed by sheep was only 7%. This latter figure is close to estimates of feeding efficiency obtained for large mammals in the wild: 10% for the Uganda kob (Buechner and Golley, 1967) and 9.6% for the African elephant (Wiegert and Evans, 1967). Where large, multi-species herds of ungulates occur, as on the savannas of Africa, their combined efficiency may be much higher: estimates of 60% for Uganda grassland and of 28% for Tanganyika grassland were obtained from observations by Petrides and Swank (1965) and Lamprey (1964), respectively.

At ordinary densities, small herbivores, both vertebrate and invertebrate, are much less efficient in their utilization of available food. Golley (1960) reports an efficiency of 1.6% for meadow voles (*Microtus*) in a Michigan grassland, and values of less than 0.5% are estimated for a variety of other small mammals and granivorous birds (Wiegert and Evans, 1967). Similar low efficiencies apparently characterize insect herbivores except when these are present in plague proportions. Wiegert (1965) found that grasshoppers (23 species of acridids and tettigoniids) consumed 1.3% of net plant production in a field of alfalfa, and Smalley (1960) obtained efficiency values of 1.6–2.0% for the meadow grasshopper *Orchelimum* in a *Spartina* salt marsh. Even if all invertebrate herbivores are considered together, their total consumption of net primary production seems rarely to exceed 10%; the following values appear to be representative:

Nature of ecosystem	% net plant production used by invertebrates	References
Bush-clover stand (S.C.)	0.4–1.4	Menhinick, 1967
Festuca grassland (Tenn.)	9.6	Van Hook, 1971
Spartina salt marsh (Ga.)	7.0–9.2	Teal, 1962
mesophytic woodland (Tenn.)	1.5	Reichle and Crossley, 1967
Mature deciduous forest (southern Canada)	1.5–2.5	Bray, 1964

These values do not necessarily indicate the total damage done to the plant crop. For example, Andrzejewska *et al.* (1967) report that grasshoppers may destroy 4.8 times the amount of plant material they ingest, by gnawing the grass blades so that part of the leaf falls off. Such material does not enter the grazing food-chain, however, but drops to the ground and is consumed by detritus feeders.

The nutritive value of most higher plants varies with age of the plant, season and environmental conditions, which also affect the *palatability* of the food. Few herbivores have the ability to digest cellulose without the assistance of *symbiotic bacteria or protozoa*, and much of the food that is eaten fails to be assimilated and is eliminated as feces. Assimilation therefore involves an important split in the flow of matter and energy through the ecosystem.

High assimilation/consumption ratios can be achieved by herbivores which are selective feeders on concentrated foods such as nuts, seeds and fungi; assimilation efficiencies of 85–95% have been recorded for such small mammals as *Clethrionomys*, *Apodemus* and *Microtus* (Drozdz, 1967; Davis and Golley, 1963, 81). Somewhat lower values are reported for large ruminants, ranging from 60–80% for sheep, cattle and deer down to 40% for moose and elephant (Graham, 1964; Dinesman, 1967; Davis and Golley, op. cit.). Interestingly, insect herbivores appear to be considerably less efficient feeders, reaching assimilation levels of 29% for the caterpillar *Hyphantria*, 27% for the grasshopper *Orchelimum*, 36% for the grasshopper *Melanoplus*, and 33% for the spittlebug *Philaenus* (Gere, 1956; Smalley, 1960; Wiegert, 1964, 1965). Detritus feeders seem to have even lower efficiencies, as witness values of 20% for oribatid mites and 10% for the millipede *Glomeris* (Engelmann, 1961; Bocock, 1963). Thus, grazing food-chains and detritus food-chains appear to be characterized by quite different assimilation rates.

Part of the energy assimilated by herbivores is stored as growth or production, while the rest is respired, dissipated as heat in the oxidation of organic matter. These respiratory losses may amount to as much as 98% of the energy assimilated (Petrusewicz and Macfadyen, 1970, 98) but vary greatly, depending on such factors as environmental temperature, level of activity, and age of the individual. In mammals, there is a tendency for the *weight-specific respiration rate* to rise as body weight decreases, and this, in combination with the fact that growth and reproduction rates tend to be greater in small species than in large ones, apparently results in a rather constant low production/assimilation ratio of 1–2% for mammalian populations; values of this magnitude have been calculated for both elephants and mice (Wiegert and Evans, 1967). In contrast, herbivorous insect populations seem to show much higher assimilation efficiencies, on the order of 35–45% and, when only young life stages (larvae, nymphs) are involved, even of 50–60% (Macfadyen, 1967). Data are still too few to permit satisfactory generalizations, but the importance of further studies is evident when it is remembered that energy lost in respiration is irrevocably removed from the ecosystem and cannot be re-cycled.

Synthesis of these process ratios leads to an evaluation of the overall efficiency with which herbivores convert the energy of primary (plant) production into their own tissue. The value of the production/consumption ratio for a fairly broad spectrum of vertebrate and invertebrate herbivores seems to range from less than 1% to 15–20% as a maximum (Petrusewicz and Macfadyen, 1970), with some evidence that the insects are generally more efficient than the non-domesticated mammals; this difference may be due in large part to the necessity for the latter to maintain a steady, high body temperature. When these values are considered along with the proportions of available food ingested (see above), the efficiency of energy transfer of herbivores is rarely found to exceed 15%, and this has been suggested as a likely maximum level for natural ecosystems (Slobodkin, 1962).

Regulatory Mechanisms of Primary Consumers

Although terrestrial herbivores are occasionally so abundant as to deplete the local supply of plant food, as sometimes happens when an introduced species like the Japanese beetle (*Popilia japonica*) or the European gypsy moth (*Porthetria dispar*) enters a new biotic community, such cases seem to be the exception rather than the rule, and, in contrast to many aquatic ecosystems, the bulk of net annual primary production on land is not eaten in the living state by primary consumers but dies and is acted on by decomposer organisms. Thus it appears that, on the whole, land herbivores usually occur at densities well below the level of their available plant food, and the reason or reasons for this are of great ecological interest.

Despite the fact that sound generalizations about abstract concepts such as "trophic levels" are not easily arrived at (Murdoch, 1966), several hypotheses about the *regulation of herbivore numbers* have been suggested. The apparent rarity of obvious depletion of vegetation by herbivores, or of its destruction by meteorological catastrophes, has led Hairston, Smith and Slobodkin (1960) to the belief that herbivores are seldom limited by their food supply; after rejecting weather as an effective control agent, they conclude that herbivores are most often controlled by their predators and/or parasites, interacting in the classical density-dependent manner. These views have been questioned on such grounds as (1) that much green material may often be inedible, unpalatable or even unreachable by the herbivores present, so that food limitation might occur without actual depletion (Murdoch, 1966) or (2) that native herbivores such as forest Lepidoptera and grasshoppers will often increase and cause serious defoliation even in the presence of their predators (Ehrlich and Birch, 1967). Despite these and other criticisms, Slobodkin, Smith and Hairston (1967) find it unnecessary to modify their views in any essential respect.

The possibility that herbivore (and other animal) populations have some capacity for *self-regulation* has not been overlooked. Pimentel (1961, 1968) has suggested the concept of "*genetic feedback*," according to which population density influences the intensity of selective pressure, selection influences the genetic composition of the surviving individuals, and genetic composition influences the subsequent population density. The models thus far proposed to explain how this system works appear to be untenable (Lomnicki, 1971), but the general validity of the concept seems to gain support from the numerous instances of the *co-evolution* of plants and herbivorous animals, e.g. the association between certain orchids and their insect pollinators (Van der Pijl and Dodson, 1966), or that between certain species of *Acacia* and ants (Janzen, 1966), which have been interpreted as involving reciprocal selective interaction.

Another suggested mechanism for self-regulation involves the elaboration of *social behavior patterns*, e.g. territoriality, social hierarchies, and warning displays, which tend to maintain animal populations at relatively low densities, thereby reducing the possibility of depletion of food supplies or other resources (Wynne Edwards, 1962, 1965). The evolution of such a mechanism seems to require that natural selection act on groups rather than on individual organisms, and for this reason

Wynne Edwards' hypothesis has been heavily criticized (Williams, 1966; Maynard Smith, 1964).

The basic importance of self-regulatory mechanisms and other density-dependent interactions in limiting populations of animals has also been questioned. Ehrlich and Birch (1967), for example, deny that "the numbers of all populations are primarily determined by density-regulating factors" and emphasize the significant limiting effects that weather conditions often have on natural populations of herbivores. This view maintains that limiting factors which act on populations without relation to their density commonly maintain such populations at levels where self-regulatory, *density-dependent mechanisms*, if they exist at all, are not called into play.

Wiegert and Owen (1971) suggest that the precise kinds of mechanisms limiting the population of a particular species of herbivore will depend firstly on (a) whether the population, in making use of its plant food resources, directly affects the rate of food supply (either by destroying or by stimulating its capacity to produce new plant tissue), and secondly on (b) the life history characteristics of both the herbivore and its plant food. Most stable terrestrial ecosystems are dominated by relatively large, structurally complex plants that are long-lived, slow-growing and have low rates of population increase, and by herbivores that are smaller, more numerous and faster-growing than their food plants. In contrast, open-water aquatic communities are often composed of small and structurally simple plants (phytoplankton) that are more abundant and multiply more rapidly than their consumers. In such aquatic communities, herbivores can attain high rates of consumption—approaching 100%—of net primary production without danger of completely exhausting their nutritional resources, and their populations are likely to be limited by direct *competition* to levels dictated by the rate at which food is supplied to them. As already noted above, in terrestrial ecosystems, such high rates of consumption by herbivores are rare, and when they occur their influence on the food plants is usually severe. In forest, and to a lesser extent in grassland, communities the herbivores are less likely to be limited by direct competition for food; they tend, rather, to be subject to the effects of predation and parasitism and to have evolved behavioral patterns such as migrations which result in reducing grazing pressure. At the same time, the plants of these systems have developed an array of toxic compounds, unpalatable tissues, thorns and other protective devices to resist grazing. Thus Wiegert and Owen's model of population control stresses the importance of trophic structure (grazing food-chains vs. detritus food-chains) and the biological properties of the interacting populations.

Management of Primary Consumers

Although energy flow studies suggest that man should become predominantly herbivorous in order to make the most effective use of the solar radiation captured by plants, his need for protein, which he can obtain in more concentrated form from animal tissue, is likely to continue to motivate him towards increasing the production of primary consumers such as cattle, chickens, and fish. In fact, the majority of man's domesticated animals are herbivores.

It is clear that if the conversion of solar energy into animal protein for human use is to be maximized herbivores with high growth efficiencies should be cropped. The energy that man himself must spend in order to secure his food must also be assessed, and the time, effort and other costs of cropping have therefore to be taken into consideration. Over the several thousand years in which man has attempted to domesticate plants and animals, his selective efforts have been remarkably successful in developing efficient herbivores for his own use. Modern methods of *animal husbandry*, where animals are kept in specially constructed buildings and are fed specially processed foods, have greatly increased the amount of food energy reaching the animals (Phillipson, 1966). However, these methods require the expenditure of much energy that goes as hidden cost, in the activities involved in the construction and maintenance of facilities and in the production of the processed food. Are there ways of increasing the more direct conversion of plant tissue to animal protein useful to man?

Macfadyen (1964) has indicated that in Great Britain beef cattle raised on grassland commonly consume only one-seventh of the net annual primary production, the rest going to other herbivores and to decomposers. More care in stocking and better management of range are two of the principal ways in which possible improvement can be sought. Less obvious is the extension of husbandry to other herbivorous species, perhaps even to insects, that may be more efficient feeders than the large warm-blooded mammals and that might, through selection, be developed as a productive source of high protein nourishment. The use of faster-growing herbivores, such as rabbits, would yield greater efficiency in terms of meat production per unit of time (Phillipson, 1966). In East Africa, increasing use for human food is being made of antelope, zebra and other native ungulates which are relatively immune to the parasites and diseases that afflict European cattle, and otherwise unproductive areas are being managed for their propagation.

Increased production of plant food as a base for greater herbivore production is also possible. In this effort, man has tried particularly to exploit tropical regions with their higher average temperatures and longer growing seasons. Petrusewicz and Macfadyen (1970) point out, however, that primary productivity does not apparently increase proportionately to the increased light regime of tropical climates, largely due to higher respiration rates in tropical plants and to more rapid rates of decomposition; the tenfold increase in solar radiation experienced by some tropical forests seems not to result in any significant increase in energy available to primary consumers over that found in a temperate deciduous forest. It has long been apparent that agricultural practices developed in temperate regions are poorly adapted for use in the tropics. A great deal more needs to be learned about *tropical ecology* before marked improvement in environmental management can be expected (Owen, 1966).

Concluding Remarks

This article has attempted to outline the current view of nature as an interaction system in which organisms play a variety of roles in facilitating the circulation of matter and the flow of energy within that thin layer of the earth's surface known as the *biosphere*. Despite the relatively small quantities of material and energy that are channeled through primary consumers, as compared with autotrophic plants and decomposer organisms, the former play a vital role in the maintenance of such a system, in that they accumulate and concentrate inorganic nutrients such as nitrogen, phosphorus and potassium, as well as many trace elements, which are otherwise thinly dispersed in the environment. It has not been possible here to present more than a superficial account of the role of primary consumers, and indeed the study of these animals is still in its early stages. But further observation and experimentation in this aspect of ecology are clearly needed, if man is ever to understand and manage wisely the world he lives in.

REFERENCES

Adams, L. (ed.), *Bio. Science* **15**, 83–157 (1965).

Andrzejewska, L., A. Breymeyer, A. Kajak, and Z. Wojcik, pp. 477–495 in *Secondary Productivity of Terrestrial Ecosystems*, edited by K. Petrusewicz, Institute of Ecology, Warsaw, 1967.

Bocock, K. L., pp. 85–91 in *Soil Organisms*, edited by J. Doeksen and J. Van der Drift, Amsterdam, 1963.

Bornebusch, C. H., *Forstl. Forsogskomission* (Copenhagen), **11**, 1–224 (1930).

Bray, J. R., *Ecology* **45**, 165–167 (1964).

Brues, C. T., *Insect Dietary: An Account of the Food Habits of Insects*, Harvard University Press, Cambridge, 1946.

Buechner, H. K., and F. B. Golley, pp. 243–254 in *Secondary Productivity of Terrestrial Ecosystems*, edited by K. Petrusewicz, Institute of Ecology, Warsaw, 1967.

Cragg, J. B., *J. Anim. Ecol.* **30**, 205–234 (1961).

Davis, D. E., and F. B. Golley, *Principles in Mammalogy*, Reinhold Publishing Corp., New York, 1963.

Dinesman, L. G., pp. 261–266, in *Secondary Productivity of Terrestrial Ecosystems*, edited by K. Petrusewicz, Institute of Ecology, Warsaw, 1967.

Drozdz, A., pp. 323–330 in *Secondary Productivity of Terrestrial Ecosystems*, edited by K. Petrusewicz, Institute of Ecology, Warsaw, 1967.

Eadie, J., pp. 7–24 in *Animal Populations in Relation to their Food Resources*, edited by A. Watson, Blackwell Scientific Publications, Oxford, 1970.

Ehrlich, P. R., and L. C. Birch, *Amer. Nat.* **101**, 97–107 (1967).

Elton, C., *Animal Ecology*, Sidgwick and Jackson Ltd., London, 1927.

Elton, C. S., *The Pattern of Animal Communities*, Methuen and Co. Ltd., London, 1966.

Engelmann, M. D., *Ecol. Monogr.* **31**, 221–228 (1961).

Evans, F. C., and U. N. Lanham, *Science* **131**, 1531–1532 (1960).

Evans, F. C., and W. W. Murdoch, *J. Anim. Ecol.* **37**, 259 to 273 (1968).

Gere, G., *Acta Biol. Acad. Sci. Hungary* 7, 43–72 (1956).

Gliwicz, J., R. Andrzejewski, G. Bujalska, and K. Petrusewicz, *Acta Theriol.* **13**, 401–413 (1968).

Golley, F. B., *Ecol. Monogr.* **30**, 187–206 (1960).

Golley, F. B., *Ecology* **42**, 581–584 (1961).

Graham, N. M., *Austral. J. Agric. Res.* **15**, 969–973 (1964).

Hairston, N. G., F. E. Smith, and L. B. Slobodkin, *Amer. Nat.* **94**, 421–425 (1960).

Janzen, D. H., *Evolution* **20**, 249–275 (1966).

Lamprey, H. F., *East African Wildlife J.* **2**, 1–46 (1964).

Lewis, J. K., G. M. Van Dyne, L. R. Albee, and F. W. Whetzal, Bull. S. Dak. Agric. Expt. Sta., 459, 44 (1956).
Lindemann, R. L., *Ecology* **23**, 399–418 (1942).
Lomnicki, A., *Amer. Nat.* **105**, 413–421 (1971).
Macfadyen, A., *Ann. Appl. Biol.* **49**, 216–219 (1961).
Macfadyen, A., pp. 3–20 in *Grazing in Terrestrial and Marine Environments*, edited by D. J. Crisp, Blackwell Scientific Publications, Oxford, 1964.
Macfadyen, A., pp. 383–412 in *Secondary Productivity of Terrestrial Ecosystems*, edited by K. Petrusewicz, Institute of Ecology, Warsaw, 1967.
Maynard Smith, J., *Nature* **201**, 1145–1147 (1964).
Menhinick, E. F., *Ecol. Monogr.* **37**, 255–272 (1967).
Murdoch, W. W., *Amer. Nat.* **100**, 219–226 (1966).
Nakamura, K., Y. Ito, M. Nakamura, T. Matsumoto, and K. Hayakawa, *Oekologia* **7**, 1–15 (1971).
Odum, H. T., and R. F. Pigeon (eds.), *A Tropical Rain Forest*, U.S. Atomic Energy Commission, Washington, D.C., 1970.
Owen, D. F., *Animal Ecology in Tropical Africa*, Oliver and Boyd, Edinburgh, 1966.
Parker, G. R., Occas. Pap. No. 10, Canad. Wildlife Service, Ottawa, 1971.
Paulsen, H. A., Jr., *Iowa State J. Sci.* **34**, 731–748 (1960).
Petrides, G. A., and W. G. Swank, *Proc. Ninth Internat. Grasslands Congress*, San Paulo, Brazil, 1965.
Petrusewicz, K., and A. Macfayden, "*Productivity of Terrestrial Animals: Principles and Methods*," Internat. Biol. Program Handbook No. 13, Burgess and Son, Abingdon, Berkshire, England, 1970.
Phillipson, J., *Oikos* **15**, 130–139 (1964).
Phillipson, J., *Ecological Energetics*, St. Martin's Press, New York, 1966.
Pimentel, D., *Amer. Nat.* **95**, 65–79 (1961).
Pimentel, D., *Science* **159**, 1432–1437 (1968).
Reichle, D. E., and D. A. Crossley, Jr., pp. 563–587 in *Secondary Productivity of Terrestrial Ecosystems*, edited by K. Petrusewicz, Institute of Ecology, Warsaw, 1967.
Slobodkin, L. B., *Adv. Ecol. Res.* **1**, 69–101 (1962).
Slobodkin, L. B., F. E. Smith, and N. G. Hairston, *Amer. Nat.* **101**, 109–124 (1967).
Smalley, A. E., *Ecology* **41**, 672–677 (1960).
Smith, A. H., and J. R. Douglas, *Ann. Ent. Soc. Amer.* **42**, 14–18 (1949).
Southwood, T. R. E., *Ecological Methods*, Methuen and Co. Ltd., London, 1966.
Teal, J. M., *Ecology* **43**, 614–624 (1962).
Thienemann, A., *Verh. deutsch. Zool. Ges.* **31**, 29–79 (1926).
Umbreit, W. W., R. H. Burris, and J. F. Stauffer, *Manometric Techniques*, 3rd ed., Burgess Publishing Co., Minneapolis, 1957.
Van der Drift, J., *Tijdschr. Ent.* **94**, 1–68 (1951).
Van der Pijl, L., and C. H. Dodson, *Orchid Flowers: Their Pollination and Evolution*, University of Miami Press, Coral Gables, Florida, 1966.
Van Hook, R. I., Jr., *Ecol. Monogr.* **41**, 1–26 (1971).
Wiegert, R. G., *Ecol. Monogr.* **34**, 225–241 (1964).
Wiegert, R. G., *Oikos* **16**, 161–176 (1965).
Wiegert, R. G., *Energy Transfer in Ecological Systems*, Patterns of Life Series, Rand McNally and Company, Chicago, 1970.
Wiegert, R. G., and F. C. Evans, pp. 499–518 in *Secondary Productivity of Terrestrial Ecosystems*, edited by K. Petrusewicz, Institute of Ecology, Warsaw, 1967.
Wiegert, R. G., and D. F. Owen, *J. Theoret. Biol.* **30**, 69–81 (1971).
Williams, E. C., Jr., and D. E. Reichle, *Oikos* **19**, 10–18 (1968).
Williams, G. C., *Adaptation and Natural Selection*, Chapter 4, Princeton University Press, Princeton, N. J., 1966.
Wynne Edwards, V. C., *Animal Dispersion in Relation to Social Behaviour*, Oliver and Boyd, Edinburgh, 1962.
Wynne Edwards, V. C., *Science* **147**, 1543–1548 (1965).

FRANCIS C. EVANS
University of Michigan

ECOLOGY-RADIATION: see RADIATION ECOLOGY

ECOSYSTEM THEORY

The concept of the ecosystem is not only the center of professional ecology today, but it is also the most relevant concept in terms of man's environmental problems. Recently the public has seized on the root meaning of ecology, namely "oikos" or "house," to broaden the subject beyond its previously rather narrow academic confines to include the "totality of man and environment," or the whole environmental house, as it were. We are witnessing what is called a historic "attitude revolution" (Odum, 1969, 1970c) in the way people look at their environment for the very simple reason that for the first time in his short history man is faced with ultimate rather than merely local limitations. It will be well for all of us to keep this overriding simplicity in mind as we face the controversies, false starts and backlashes that are bound to accompany man's attempts to put some negative feedback into the vicious spiral of uncontrolled growth and resource exploitation that has characterized the past several decades.

Man has been interested in ecology in a practical sort of way since early in his history. In primitive society every individual, to survive, needed to have definite knowledge of his environment, i.e., of the forces of nature and of the plants and animals around him. Civilization, in fact, began when man learned to use fire and other tools to modify his environment. It is even more necessary than ever for mankind, as a whole, to have an intelligent knowledge of the environment if our complex civilization is to survive, since the basic "laws of nature" have not been repealed; only their complexion and quantitative relations have changed, as the world's human population has increased and as man's power to alter the environment has expanded.

Like all phases of learning, the science of ecology has had a gradual, if spasmodic, development during recorded history. The writings of Hippocrates, Aristotle, and other philosophers of the Greek period contain material which is clearly ecological in nature. However, the Greeks literally did not have a word for it. The word "ecology" is of recent coinage, having been first proposed by the German biologist Ernst Haeckel in 1869. Before this, many of the great men of the biological renaissance of the eighteenth and nineteenth centuries had contributed to the subject even though the label "ecology" was not in use. For example, Anton von Leeuwenhoek, best known as a pioneer microscopist of the early 1700s, also pioneered the study of "food chains" and "population regulation" (see Egerton, 1968), two important areas of modern ecology. As a recognized distinct field of biology, the science of ecology dates from about 1900, and only in the past decade has the word become part of the general vocabulary. Today, everyone is acutely aware of the environmental sciences as indispensable tools for creating and maintaining the quality of human civilization. Consequently, ecology is rapidly becoming the branch of science that is most relevant to the everyday life of every man, woman, and child.

As recently as 1960 the theory of the ecosystem was rather well understood but not in any way applied. The applied ecology of the 1960s consisted of managing components as more or less independent units. Thus we had forest management, wildlife management, water management, soil conservation, pest control, etc., but no ecosystem management and no applied human ecology. Practice has now caught up with theory. Controlled management of the human population together with the resources and the life support system on which it depends as a single, integrated unit now becomes the greatest, and certainly the most difficult, challenge ever faced by human society.

As we have seen, an "anthropocentric" definition of the ecosystem might read something as follows: Man as a part of, not apart from, a life support system composed of the atmosphere, water, minerals, soil, plants, animals and microorganisms that function together to keep the whole viable.

Any unit including all of the organisms (i.e. the "community") in a given area interacting with the physical environment so that a flow of energy leads to a clearly defined trophic structure, biotic diversity and material cycles (i.e. exchange of materials between living and non-living parts) within the system is an *ecological system* or *ecosystem*.

The following formal definition is the one used in the third edition of "Fundamentals of Ecology" (E. P. Odum, 1971)

The word ecology is derived from the Greek *oikos*, meaning "house" or "place to live." Literally, ecology is the study of organisms "at home." Usually ecology is defined as the study of the relation of organisms or groups of organisms to their environment, or the science of the interrelations between living organisms and their environment. Bedause ecology is concerned especially with the biology of *groups* of organisms and with *functional* processes on the lands, in the oceans, and in fresh waters, it is more in keeping with the modern emphasis to define ecology as the study of the structure and function of nature, it being understood that mankind is a part of nature. One of the definitions inWebster's Unabridged Dictionary seems especially appropriate for the closing decades of the 20th century, namely, "*the totality or pattern of relations between organisms and their environment*." In the long run the best defiintion for a broad subject field is probably the shortest and least technical one, as for example, "environmental biology."

To understand the scope and relevance of ecology, the subject must be considered in relation to other branches of biology and to "*ologies*" in general. In the present age of specialization in human endeavors, the inevitable connections between different fields are often obscured by the large masses of knowledge with the fields (and sometimes also, it must be admitted, by stereotyped college courses). At the other extreme, almost any field of learning may be so broadly defined as to take in an enormous range of subject material. Therefore, recognized "fields" need to have recognized bounds, even if these bounds are somewhat arbitrary and subject to shifting from time to time. A shift in scope has been especially noteworthy in the case of ecology as general public awareness of the subject has increased. To many, "ecology" now stands for "the totality of man and environment." But first let us examine the more traditional academic position of ecology in the family of sciences.

For the moment, let us look at the divisions of biology, „the science of life." Morphology, physiology, genetics, ecology, evolution, molecular biology, and developmental biology are examples of such divisions. We may also divide the subject into what may be called "taxonomic" divisions, which deal with the morphology, physiology, ecology, etc., of specific kinds of organisms. Zoology, botany, and bacteriology, are large divisions of this type, and phycology, protozoology, mycology, entomology, ornithology, etc., are divisions dealing with more limited groups of organisms. Thus ecology is a basic division of biology and, as such, is also an integral part of any and all of the taxonomic divisions. Both approaches are profitable. It is often very productive to restrict work to certain taxonomic groups, because different kinds of organisms require different methods of study (one cannot study eagles by the same methods used to study bacteria) and because some groups of organisms are economically or otherwise much more important or interesting to man than others. Ultimately, however, unifying principles must be delimited and tested if the subject field is to qualify as "basic."

Perhaps the best way to delimit modern ecology is to consider it in terms of the concept of *levels of organization* visualized as a sort of „biological' spectrum." Community, population, organism, organ, cell, and gene are widely used terms for several major biotic levels.

Interaction with the physical environment (energy and matter) at each level produces characteristic functional systems. By a *system* we mean just what Webster's Collegiate Dictionary defines as "regularly interacting and interdependent components forming a unified whole." Systems containing living components (biological systems or biosystems) may be conceived at any level in the hierarchy. For example, we might consider not only gene systems, organ systems, and so on, but also host-parasite systems as intermediate levels between population and community.

Historical Review of the Ecosystem Concept

Although the term ecosystem was first proposed by the British ecologist, A. G. Tansley in 1935, the concept is by no means so recent. Allusions to the idea of the unity of organisms and environment (as well as the oneness of man and nature) can be found as far back in written history as one might care to look, and such an idea has been a basic part of many religions (less so in Christian religions as recently pointed out by historian Lynn White; 1967). Anthropologists and geographers have long been concerned with the impact of man on his environment and early debated the question: To what extent has man's continuing trouble with deteriorated environments stemmed from the fact that human culture tends to develop independently of the natural environment? The Vermont prophet George Perkins Marsh wrote a classic treatise on this theme in 1864. He analyzed the causes of the decline of ancient civilizations and forecast a similar doom for modern ones unless man takes what we would call today a "ecosystematic" view of man and nature. In the late 1800s biologists began to write essays on the unity of nature, interestingly enough in a parallel manner in German, English and Russian languages. Thus Karl Möbius in 1877 wrote about the community of organisms in an oyster reef as a "biocoenosis," while in 1887 the American S. A. Forbes wrote his classic essay on "The Lake as a Microcosm." The Russian pioneering ecologist V. V. Dokuchaev (1846–1903) and his disciple G. F. Morozov (who specialized in forest ecology) placed great emphasis on the concept of "biocoenosis," a term later expanded to geobiocoenosis (or biogeocoenosis) (see Sukachev, 1944), which can be considered a synonym of the word "ecosystem."

No one has expressed the relevance of the ecosystem concept to man better than Aldo Leopold in his essays on the land ethic. In 1933 he wrote: "Christianity tries to integrate the individual to society, Democracy to integrate social organization to the individual. There is yet no ethic dealing with man's relation to the land" ... which is "still strictly economic entailing privileges but not obligations." Thus, man is continually striving, with but partial success so far, to establish ethical relationships between man and man, man and government, and, now, man and environment. Without the latter what little progress has been made with the other two ethics will surely be lost. In the context of the 1970 scene Garrett Harden (1968) says it in another way when he points out that technology alone will not solve the population and pollution dilemmas; ethical and legal constraints are also necessary. Environmental science is now being called upon to help determine a realistic level of human population density and rate of use of resources and power that are optimum in terms of the quality of human life, in order that "societal feedback" can be applied before there are serious overshoots. This requires diligent study of ecosystems, and, ultimately, a judgment on the carrying capacity of the biosphere. If studies of natural populations have any bearing on the problem, we can be quite certain that the optimum density in terms of the individual's options for liberty and the pursuit of happiness is something less than the maximum number that can be sustained at a subsistence level, as so many domestic "animals" in a polluted feed lot!

My advanced ecology class recently attempted to determine what might be the optimum population for the State of Georgia on the assumption that someday the state would have to have a balanced resource input-output (i.e. live within its own resources). On the basis of a per capita approach to land use the tentative conclusion was that a density of one person per five acres (2 hectares] represented the upper limit for an optimum population size when the space requirements for quality (i.e. high protein) food production, domestic animals, outdoor recreation, waste treatment and pollution-free living space were all fully considered. Anything less than five acres of life support and resource space per capita, it was concluded, would result in a reduction in the individual person's options for freedom and the pursuit of happiness, and, accordingly, a rapid loss in environmental quality. Since the 1970 per capita density of Georgia is 1 in 8 acres and for the United States as a whole, 1 in 10 acres no more than double the present U.S. population could be considered optimum according to this type of analysis. This would mean that we have about 30 years to level off population growth. The study also suggested that permanent zoning of at least one-third of land and freshwater areas (plus all estaurine and marine zones) as "open space" in urbanized areas would go a long way towards preventing overpopulation, overdevelopment and social decay that is now so evident in many parts of the world today. The results of this preliminary study have been published (E. P. Odum, 1970b). See also the prevocative symposium on "The Optimum Population for Britain" edited by Taylor (1970).

The Two Approaches to Ecosystem Study

G. Evelyn Hutchinson in his 1964 essay, "The Lacustrine Microcosm Reconsidered," contrasts the two long-standing ways ecologists attempt to study lakes or other large ecosystems of the real world. Hutchinson cites E. A. Birge's (1915) work on heat budgets of lakes as pioneering the *holological* (from holos = whole) or *holistic* approach in which the whole ecosystem is treated as a "black box" (i.e., a unit whose function may be evaluated without specifying the internal contents) with emphasis on inputs and outputs, and he contrasts this with the *merological* (from meros = part) approach of Forbes in which "we discourse on parts of the system and try to build up the whole from them." Each procedure has obvious advantages and disadvantages and each leads to different kinds of application in terms of solving problems. Unfortunately, there is something of a "credibility gap" between the two approaches. As would be expected, the merological approach has dominated the thinking of the biologist-ecologist who is species-oriented, while the physicist-ecologist and engineer prefer the "black box" approach. Most of all, man's environmental crisis has speeded up the application of systems analysis to ecology. The formalized, or mathematical model, approach to populations, communities, and ecosystems has come to be known as *systems ecology* which is rapidly becoming a major science in its own right for two reasons: (1) extremely powerful new formal tools are now available in terms of mathematical theory, cybernetics, electronic data processing, etc. (2) Formal simplication of complex ecosystems provides the best hope for solutions of man's environmental problems that can no longer be trusted to trial-and-error, or one-problem one-solution procedures that have been chiefly relied on in the past.

Again we see the contrast between merological and holological approaches in that there are systems ecologists who start at the population or other component level and "model up," and those who start with the whole and "model down." The same dichotomy is evident in the very rewarding studies of experimental laboratory ecosystems. One class of microecosystems can be called "derived" systems because they are established by multiple seeding from nature in contrast to "defined" microcosms which are built up from previously isolated pure cultures. Theoretically, at least, the approaches are applicable to efforts to devise life support systems for space travel. In fact, one of the best ways to visualize the ecosystems concept for students and layment is to consider space travel, because when man leaves the biosphere he must take with him a sharply delimited enclosed environment that will supply all vital needs with sun energy as the only usable input form the surrounding very hostile space environment. For journeys of a few weeks (such as to the moon and back) man does not need a regenerative ecosystem, since sufficient oxygen and food can be stored while CO_2 and other waste products can be fixed or detoxified for short periods of time. For long journeys man must engineer himself into a complete ecosystem that includes the means of recycling materials and balancing production, consumption and decomposition by biotic components or their mechanical substitutes. In a very real sense the problems of man's survival in an artificial space craft are the same as the problems involved in his continued survival on the earth space ship. For example, detection and control of air, and water pollution, adequate quantity and nutritional quality of food, what to do with accumulated toxic wastes and garbage, and the social problems created by reduced living space are common concerns of cities and spacecrafts. For this reason the ecologist would urge that national and international space programs now turn their attention to the study and monitoring of our spaceship earth. As was the case with Apollo 13, survival becomes the mission when the limits of carrying capacity are approached.

The Components of the Ecosystem

From the energy of trophic standpoint an ecosystem has two components which are usually partially separated in space and time, namely, an *autotrophic component* (autotrophic = self nourishing) in which fixation of light energy, use of simple inorganic substances, and the buildup of complex substances predominate; and secondly, a *heterotrophic component* (heterotrophic = other-nourishing) in which utilization, rearrangement and decomposition of complex materials predominate. As viewed from the side (cross section) ecosystems consist of an upper "green belt" which receives incoming solar energy and overlaps, or interdigitates, with a lower "brown belt" where organic matter accumulates and decomposes in soils and sediments.

It is convenient for the purposes of first order analysis and modelling to recognize six structural components and six processes as comprising the ecosystem as follows:

A. Components

1. *Inorganic substances* (C, N, CO_2, H_2O, etc.) involved in material cycles.
2. *Organic compounds* (proteins, carbohydrates, lipids, humic substances, etc.) that link biotic and abiotic.
3. *Climate regime* (temperature, rainfall, etc.).
4. *Autotrophs* or *producers*, largely green plants able to manufacture food from simple substances.
5. *Phagotrophs* (phago = to eat) or *macro-consumers*, heterotrophic organisms, largely animals which ingest other organisms or particulate organic matter.
6. *Saprotrophs* (sapro = to decompose) or *micro-consumers* (also caled osmotrophs), heterotrophic organisms, chiefly bacteria, fungi and some protozoa that break down complex compounds, absorb some of the decomposition products and release inorganic substances usable by the autotrophs together with organic residues which may provide energy sources or which may be inhibitory, stimulatory or regulatory to other biotic components of the ecosystem.

(Another useful division for heterotrophs: *biophages* = organisms that feed on other living organisms; *saprophages* = organisms that feed on dead organic matter.)

B. Processes

1. Energy flow circuits.
2. Foods chains (trophic relationships).
3. Diversity patterns in time and space.
4. Nutrient (biogeochemical) cycles.
5. Development and Evolution.
6. Control (cybernetics).

Subdivision of the ecosystem into these six "components" and six "processes," as with most classification, is arbitrary but convenient, since the former emphasize structure and the latter function. From the holistic viewpoint, of course, components are operationally inseparable. While different methods are often required to delineate structure on the one hand and to measure rates of function on the other, the ultimate goal of study at any level or organization is to understand the relationship between structure and function. It is not feasible

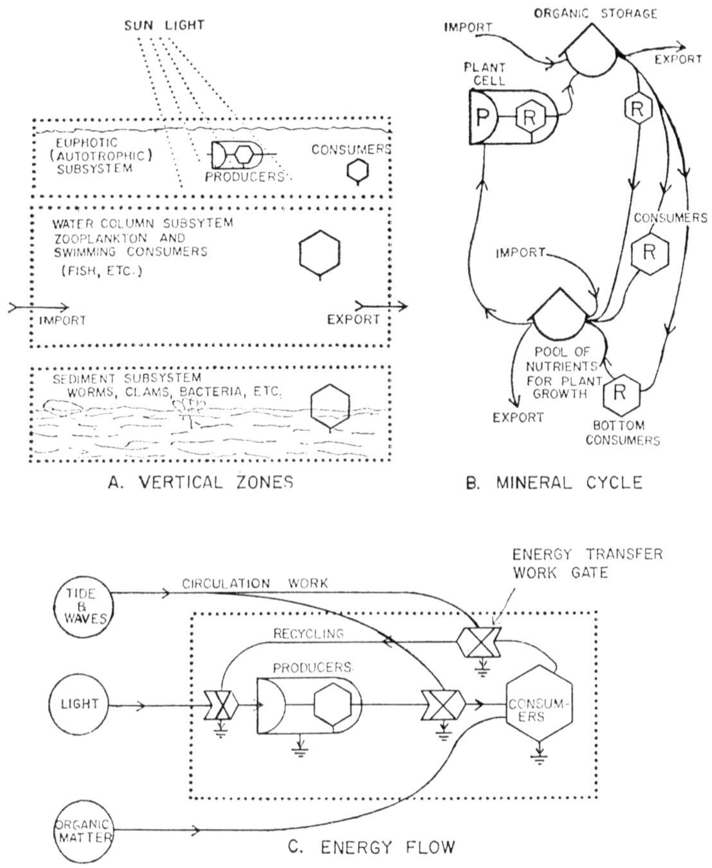

Figure 1 Three aspects of the structure and function of ecosystems as illustrated by an estaurine system. A. Vertical zonation with photosynthetic production above (autotrophic stratum) and most of the respiration and decomposition below (heterotrophic stratum). B. Material cycle with circulation of plant nutrients upward and organic matter food downward. C. Energy flow circuit diagram showing three sources of energy input into the system. The bullet-shaped modules represent producers with their double metabolism, that is P (production) and R (respiration). The hexagons are populations of consumers which have storage, self maintenance and reproduction. The storage bins represent nutrient pools in and out of which move nitrogen, phosphorus and other vital substances. In diagrams B and C the lines represent the "invisible wires of nature" then link the components into a functional network. In diagram C the "ground" symbols (i.e. arrow into the heat sink) indicate where energy is dispersed and no longer available in the food chain. The circles represent energy inputs. The work gate symbols (large X) indicate where a flow of work energy along one pathway assists a second flow to pass over energy barriers. Note that some of the lines of flow loop back from „downstream" energy sources to „upstream" inflows serving various roles there including control functions (saprotrophs controlling photosynthesis by controlling the rate of mineral regeneration, for example). The diagram (C) also shows how auxillary energy of the tide (energy subsidy) assists in (1) recycling of nutrients from consumer to producer and (2) speeding up the movement of plant food to the consumer. Reducing tidal flow by dyking the estuary will reduce the productivity just as surely as cutting out some of the light. Stress such as pollution or harvest can be shown in such circuit models by adding circles enclosing negative signs linked with appropriate heat sinks to show where energy is diverted away from the ecosystem. Both subsidies (+) and stresses (—) can be quantitated in terms of Calories added or diverted per unit of time and space. (From E. P. Odum, 1971 after H. T. Odum, and B. J. Copeland, 1969.)

to go into any detailed discussion of these component-processes in this brief introduction (one can refer to textbooks and review papers), but we can list a few key principles that are especially relevant to human ecology. Fig. 1 is a schematic diagram that may be useful in picturing the basic arrangement and functional linkage of ecosystem components.

1. The living (Items A 4–6 above) and non-living (Items A 1–3) parts of ecosystems are so interwoven into the fabric of nature that it is difficult to separate them, hence operational classifications (B 1–6) do not make a sharp distinction between biotic and abiotic. Elements and compounds are in a constant state of flux between living and non-living states. There are

very few substances that are confined to one or the other state. Exceptions may be ATP which is found only isinde living cells, and humic substances (resistant end-products of decomposition) which are not found inside cells yet are characteristic of all ecosystems.

2. The time-space separation of autotrophic and heterotrophic activity leads to a convenient classification of energy circuits into (1) a *grazing food chain* (where term grazing refers to direct consumption of living plants or plant parts) and (2) an *organic* detritus (from *deterere* = to wear away) food chain which involves the accumulation and decomposition of dead materials. To build up a stable biomass structure there must be negative feedback control of grazing—a need too often neglected in man's domesticated ecosystems.
3. As is well known available energy declines with each step in the food chain (so a system can support more herbivores than carnivores; if man wants to keep his meat-eating option open there will have to be fewer people supported by a given food base). On the other hand, materials often become concentrated with each step in the food chains. Failure to anticipate possible "biological magnification" of pollutants, such as DDT or long-lived radionuclides, is causing serious problems in man's environment.
4. It is becoming increasingly evident that high biological productivity (in terms of calories per unit area) in both natural and agricultural ecosystems is almost always achieved with the aid of *energy subsidies* from outside the system that reduces the cost of maintenance (thus diverting more energy to production). Energy subsidies take the form of wind and rain in a rain forest, tidal energy in an estuary (see Fig. 1), or fuel, animal or human work energy used in the cultivation of a crop. In comparing productivity of different systems it is important to consider *the complete budget*—not just light input.
5. Likewise it is increasingly evident that *both harvest and pollution are stresses* which reduce the energy available for self-maintenance. Man must be aware that he will have to pay the costs of added anti-thermal maintenance, or "disorder pumpout" as H. T. Odum, 1971 (Chapter 5) calls it. It is dangerous strategy to try to force too much productivity, or yield, from the landscape (as is being attempted in the so-called "green revolution") because very serious "ecological backlashes" can result from the following: (1) pollution caused by heavy use of fertilizers and insecticides and the consumption of fossil fuels, (2) unstable or oscillating conditions created by one-crop systems, (3) vulnerability of plants to disease because their self-protection mechanisms have been "selected out," in favor of yield, and (4) social disorder created by rapid shift of rural people to cities that are not prepared to house or employ them (the tragedy here is that industrialized agriculture *can* result in increased food per acre but it *can* also widen the gap between rich and poor so that there are increasing numbers of people unable to buy the food!).
6. While we generally think of production and decomposition as being balanced in the biosphere as a whole, the truth is that this balance has never been exact but has fluctuated from time to time in geological history. Through the long haul of evolutional history production has slightly exceeded decomposition so that a highly oxygenic atmosphere has replaced the original reducing atmosphere of the earth. Man, of course, is tending to reverse this trend by increasing decomposition (burning of fuels, ect.) at the expense of production. The most immediate problem is created by the increase in atmospheric CO_2 since relatively small changes in concentration can have large effects on the heat budget of the earth.
7. Ecological studies indicate that diversity is directly correlated with stability and perhaps inversely correlated with productivity, at least in many situations. It could well be that the preservation of diversity in the ecosystem is important for man since variety may be a necessity, not just the spice of life!
8. At the population level it is now clear that the growth form of the human population will not conform to the simple sigmoid or logistic model since there will always be a long time lag in the effects of crowding, pollution and overexploitation of resources. Growth will not "automatically" level off as do populations of yeasts in a confined vessel where individuals are *immediately* affected by their waste products. Instead, the human population will clearly overshoot some vital resource, unless man can

"anticipate" the effects of overpopulation and reduce growth rates *before* the deleterious effects of crowding are actually felt. Intelligent reasoning behavior seems now to be the only means to accomplish this, as emphasized at the beginning of the article.

9. Some of the most important "breakthroughs" in ecology are in the area of biogeochemical cycling. Since "recycle" of water and minerals must become a major goal of human society the recycle pathways in nature are of great interest; there seem to be at least four major ones which vary in importance in different kinds of ecosystems: (1) recycle via microbial decomposition of detritus, (2) recycle via animal excretion, (3) direct recycle from plant back to plant via symbiotic microorganism such as mycorrhizae associated with roots, and (4) autolysis, on chemical recycle, with no organism involved. Pathway 3 seems to be especially important in the humid tropics which suggests tropical agriculture might be redesigned to include food plants with mycorrhizae.
10. The principles inherent in limiting factor analysis and in human ecology can be combined to formulate the following tentative overview: In an industrialized society energy (power, food) is not likely to be limiting, *but the pollution consequences of the use of energy and exploitation of resources is limiting*. Thus, pollution can be considered the limiting factor for industrialized man—which may be fortunate since pollution is so "visible" that it can force us to use that reasoning power which is supposed to be a special attribute of man.

Ecosystem Development

Principles having to do with the development of ecosystems, that is ecological succession are among the most relevant in view of man's present situation. I have recently reviewed this subject (Odum, 1970); accordingly a brief summary will suffice for this piece.

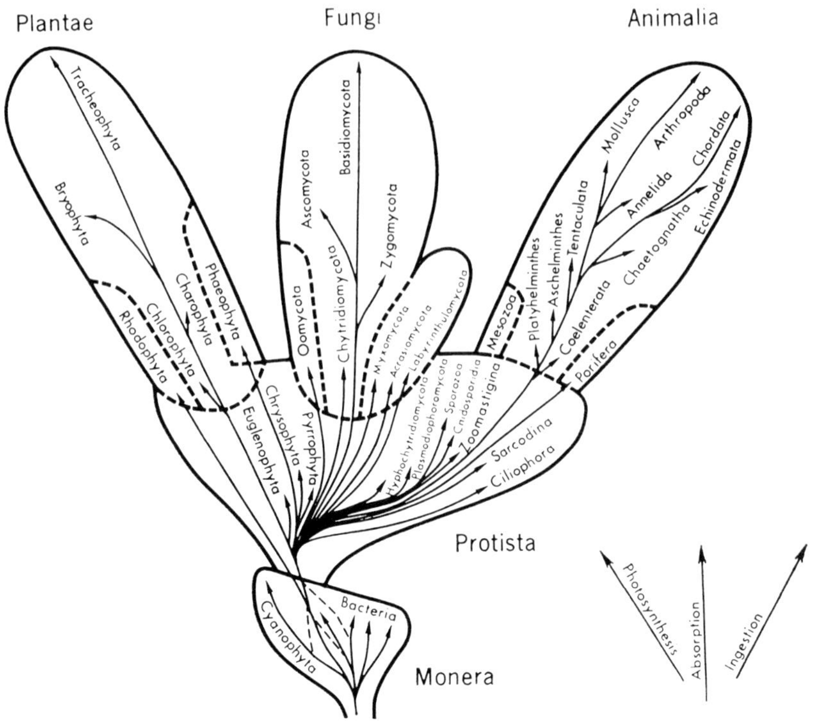

Figure 2 A five-kingdom system based on three levels of organization—the procaryotic (kingdom Monera), apearyotic unicellular (kingdom Protista), and eucaryotic multicellular and multinucleate. On each level there is divvergence in relation to three principal modes of nutrition—the photosynthetic, absorptive, and ingestive. Many ecology and microbiology texts list four kingdoms by combining the "lower Protista" (i.e., Monera) with the higher "Protista" to form "Protista." Evolutionary relations are much simplified, particularly in the Protista. Only major animal phyla are entered, and phyla of the bacteria are omitted. The Coelenterata comprise the Cnidaria and Ctenophora; the Tentaculata comprise the Bryozoa, Brachiopoda, and Phoronida, and in some treatments the Batoprocta. (From Whittaker, 1969.)

In broad view ecosystems develop through a rapid growth stage that leads to some kind of maturity or steady state (climax), usually an oscillating steady state. The early successional growth stage is characterized by a high Production/respiration (P/R) ratio, high yields (net production), short food chains, low diversity, small size of organism, open nutrient cycles and a lack of stability. In contrast, mature stages have a high biomass/respiration (B/R) ratio, food web, low net production, high diversity and stability. In other words, major energy flow shifts from production to maintenance (respiration).

The general relevance of the development sequence to land use planning can be emphasized by the following "mini-model" that contrast in very general terms young and mature ecosystems:

Young	Mature
Production	Protection
Growth	Stability
Quantity	Quality

It is mathematically impossible to obtain a maximum for more than one thing at a time, so one can not have both extremes at the same time and place. Since all six characteristics are desirable in the aggregate, two possible solutions to the dilemma suggest themselves. We can compromise so as to provide moderate quality and moderate yield on all the landscape, or we can plan to compartmentalize the landscape so as to simultaneously maintain highly productive and predominantly protective types as separate units subjected to different management strategies. If ecosystem development theory is valid and applicable to land-use planning (total zoning), then the so-called multiple-use strategy, about which we hear so much will work only through one or both of these cases, because in most cases, projected multiple uses conflict with one another. Some examples and suggestions for implementing compartmental plans are considered in the above mentioned paper.

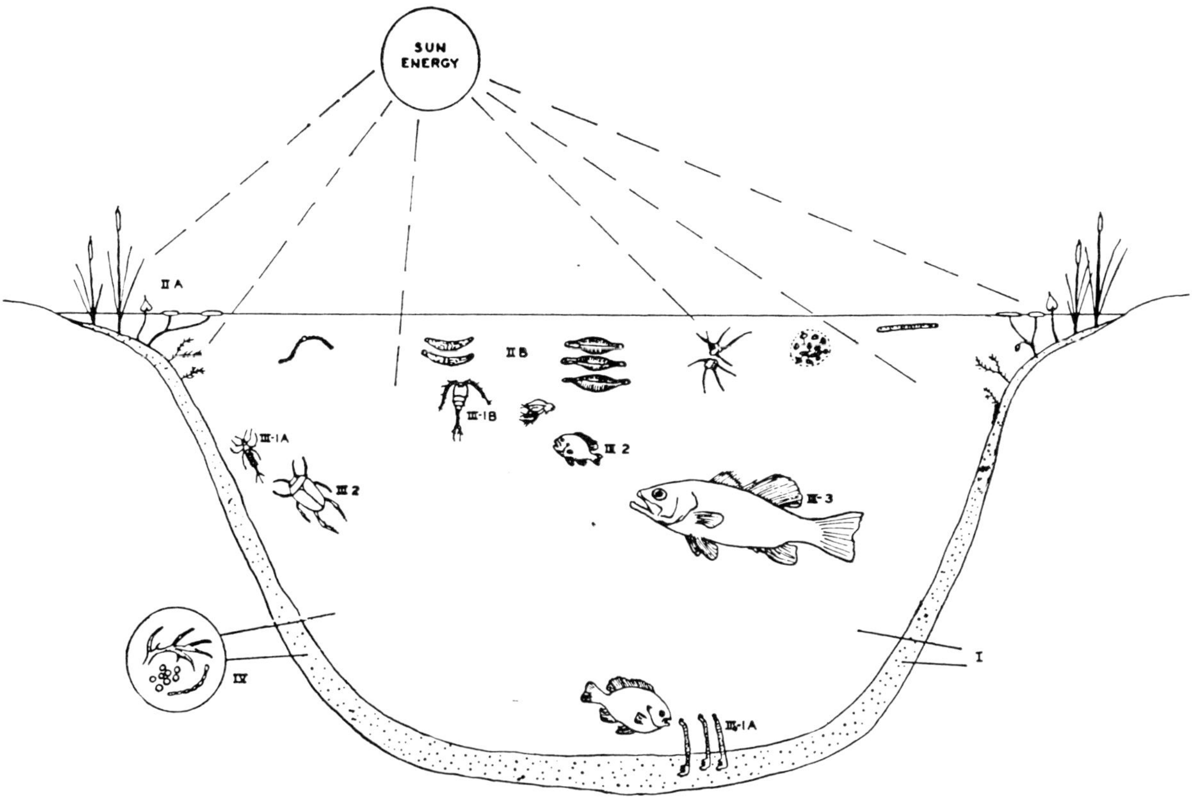

Figure 3 Diagram of the pond ecosystem. Basic units are as follows: I, abiotic substances—basic inorganic and organic compounds; IIA, producers—rooted vegetation; IIB, producers—phytoplankton; III-1A, primary consumers (herbivores)—bottom forms; III-1B, primary consumers (herbivores)—zooplankton; III-2, secondary consumers (carnivores); III-3, tertiary consumers (secondary carnivores); IV, saprotrophs—bacteria and fungi of decay. The metabolism of the system runs on sun energy, while the rate of metabolism and relative stability of the pond depend on the rate of inflow of materials from rain and from the drainage basin in which the pond is located. (From Odum 1971.)

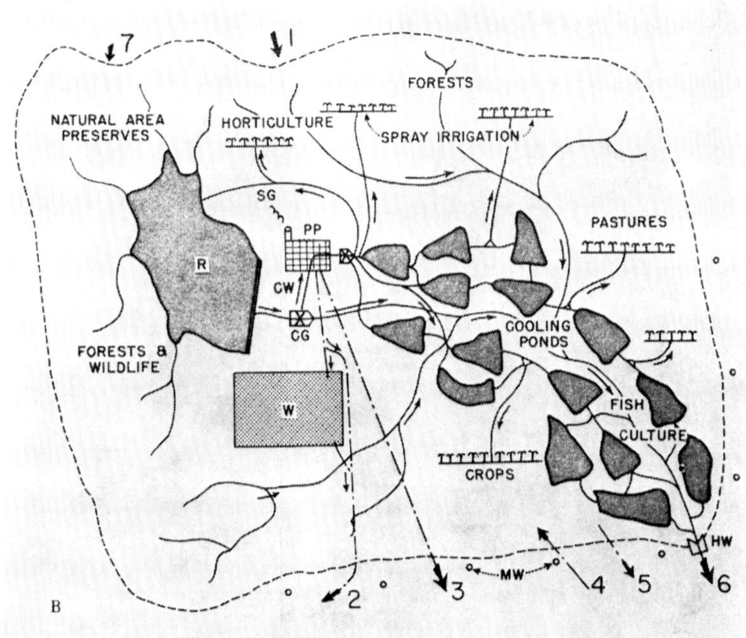

Figure 4 Schematic design for a waste management park for the atomic power plant (PP) of the future which is located in a natural watershed basin (outlined by the dashed lines). Waste heat (i.e., thermal pollution) in the reactor cooling water (CW) flowing from a large storage reservoir (R) is completely dissipated by evaporative cooling from the network of shallow ponds and spray irrigation systems. The warm ponds may be used for fish culture, sport fishing, or other recreation purposes. Irrigation of portions of the terrestrial watershed increases the yield of useful forest or agricultural products, while at the same time water recycles through the "living filter" of the land back into streams, ponds, and ground-water. Low-level nuclear wastes and solid wastes are contained within carefully managed land fill area (W); high-level nuclear wastes in spent fuel elements are exported to a special nuclear burial ground located outside of the management park. Stream flow, ground water, and stack gases are continuously monitored by hydrological weirs (HW), monitoring wells (MW), and stack gas control systems (SG) in order to make certain that no air or water pollution leaves the controlled area. The chief inputs and outputs for this environmental system include (see numbered marginal arrows): 1, input of sunlight and rainfall; 2, export of nuclear wastes to burial grounds; 3, electric power to cities, etc.; 4, input of nuclear and other fuels; 5, output of food, fibers, clean air, etc.; 6, downstream flow of clean water for agriculture, industry, and cities; 7, public and professional use for recreation, education, and environmental research. The size of such a complete waste management park will depend on regional climates and topography, and on the amount of electrical or other energy diverted to power cooling, but something on the order of 10,000 acres for a 2500 megawatt power plant would be the minimum needed to insure 100 per cent pollution control and allow for accidents and mechanical malfunctions as is explained in Chapter 17, p. 466. However, such waste treatment capacity could also support a certain amount of light industry within the park.

Heavy industry should be located within its own waste management park. (From Odum, 1971.)

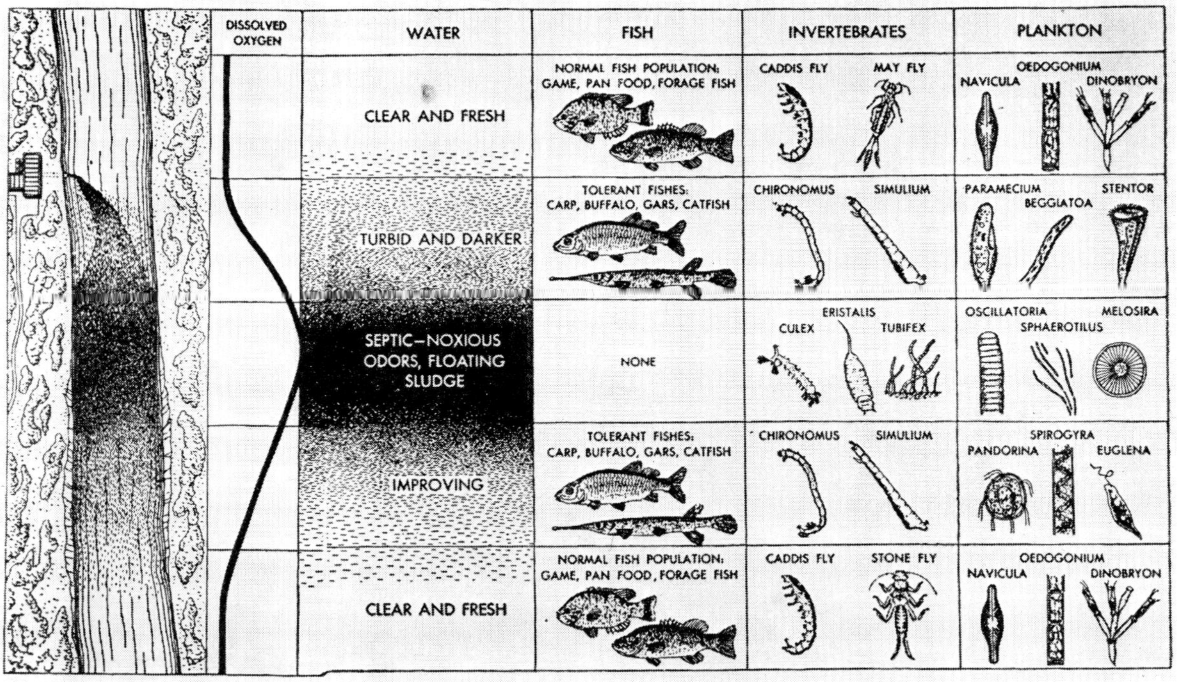

Figure 5 Pollution of a stream with untreated sewage and the subsequent recovery as reflected in changes in the biotic community. As the oxygen dissolved in the water decreases (curve to the left), fishes disappear and only organisms able to obtain oxygen from the surface (as in Culex mosquito larvae) or those which are tolerant of low oxygen concentration are found in zone of maximum organic decomposition. When bacteria have reduced all of the discharged material the stream returns to normal. (After Eliassen, Scientific American, Vol. 186, No. 3, March, 1952.)

REFERENCES

Birge, E. A. (1915), The heat budgets of American and European lakes. *Trans. Wis. Acad. Arts. Lett.* **18**, 166 to 213.

Dokuchaev, V. V., *Teaching about the zones of nature*. (In Russian.) Reprinted 1948. Moscow.

Forbes, S. A. (1887), the lake as a microcosm. Bull. Sc. H. Peoria. Reprinted in: *Ill. Nat. Hist. Surv. Bull.* **15**, 537–550 (1925).

Hardin, Garrett (1968), the Tragedy of the Commons. *Science* **162**, 1243–1248.

Hutchinson, G. Evelyn (1964), the lacustrine microcosm reconsidered. *Amer. Sci.* **52**, 331–341.

Leopold, Aldo (1933), the conservation ethic. *Jour. Forestry* **31**, 634–643.

Marsh, George Perkins (1864), *man and Nature: Or Physical Geography as Modified by Human Action*. Reprinted 1965 by Harvard Univ. Press, Cambridge.

Möbius, Karl (1877), die Auster und die Austernwirtschaft. Berlin, Transl. 1880. Rept. U.S. Fish Comm., 1880, 683 to 751.

Odum, H. T. (1970), *Environment, Power and Society*. John Wiley and Sons, N.Y., 331 pp.

Odum, H. T. (1967), Biological circuits and the marine systems of Texas. In: *Pollution and Marine Ecology* (Olson and Burgess, Eds.), John Wiley and Sons, N.Y., pp. 99–157.

Odum, H. T., and B. J. Copeland, and E. A. McMahan (1969), Coastal Ecological Systems of the United States. A report to the Federal Water Pollution Control Administration, Marine Sci. Inst., Univ. Texas. Mimeograph.

Odum, Eugene P. (1969), the attitude lag. *BioScience*, **19**, 403.

Odum, Eugene P. (1970a), the strategy of ecosystem development. *Science* **164**, 262–270.

Odum, Eugene P. (1970b), Optimum population and environment: a Georgia microcosm. *Current History* **58**, 355–359.

Odum, Eugene P. (1970c), The attitude revolution. In: *The Crisis of Survival*. Scott, Foresman Co., Glenview, Ill., pp. 9–15.

Odum, Eugene P. (1971), *Fundamentals of Ecology*. 3rd Edition. W. B. Saunders, Philadelphia.

Sukachev, V. N. (1944), On principle of genetic classification in bioeoenology. Translated and condensed by F. Raney and R. Daubenmire. *Ecol.* **39**, 364–367. (See also Silva Fennica, **105**, 94 (1960).

Tansley, A. G. (1935), the use and abuse of vegetational concepts and terms. *Ecol.* **16**, 284–307.

Taylor, L. R. (Ed.), *the Optimum Population for Britain*. Academic Press, N.Y., 182 pp.

White, Lynn (1967), the historical roots of our ecological crisis. *Science* **155**, 1203–1207.

EUGENE P. ODUM
University of Georgia

EFFECTS OF ATMOSPHERIC POLLUTANTS

The sole reason for air pollution control is to reduce or minimize the effects of air pollutants on people or their property (private and public). These effects are classified by the Environmental Protection Agency as primary and secondary effects. The former are those affecting health. The latter include injury to animals and vegetation; damage to materials (including soiling); annoyance to the senses of people(eye, nose, and throat irritation, odor, restricted visibility); loss of property values; and disturbing enjoyment of life, including aesthetic and ecological values.

Many of the effects of atmospheric pollutants are difficult to assess much less to quantify. There are two reasons for this situation. Firstly, no two pollutants have identical effects on receptors even when unmixed with other pollutants; the effects of mixtures are complex and usually unknown. Secondly, many of the effects are either subjective or relative with the consequence that specific numerical values may be misleading, even meaningless.

This is not to say physical, chemical, and biological data on the effects of different concentrations and dosages of pollutants on various receptors are unknown. They are known but they cover so wide a range that cutoff points are purely a matter of judgment. Thus one molecule of sulfuric acid will react with a stell plate, but where intolerable corrosion begins is a value judgment.

In any large population there are some individuals just born and some on the brink of death with no sharp division between the infirm and the healthy. The effect of a given dosage of a pollutant will be much greater on some than on others so one has difficulty in establishing a definite value which may be called a health effect. It is somewhat easier to define a biological effect as a biological reaction to a given stimulus, but whether or not the biological change affects health or not may not be known.

A further factor that causes considerable difference of opinion is the value judgment that must be placed on published experimental results by reliable scientists. Many experiments, especially those dealing with people, cannot be duplicated, so to judge the validity of an experiment, one must rely on the statistical soundness of the data and the reputation of the experimentalist. Even when these two factors are acceptable, a third party may extrapolate the results much farther than the experimenter himself would dare to go. Even governmental bodies are not free of criticism in this respect. When the first air quality criteria for sulfur oxides were published by the U.S. Dept. of Health, Education and Welfare in 1967, there was such a hue and cry from nongovernment scientists that the report was withdrawn and later reissued after modification.

The greatest deterrent to assessing health effects of air pollutants is the lack of data from controlled experiments pertinent to the problem. This situation exists because the real problem in the ambient air is not acute health effects from high concentrations of pollutants but the long-term chronic effects of low levels of pollution over a lifetime. Some indications can be derived from experiments on laboratory animals, e.g. rats, hamsters, guinea pigs, etc., but the validity of transferring these data to man is questionable. As a consequence resort is made to epidemiological studies where control of variables is almost impossible or to biological reactions that may or may not be related to health.

The determination of general ecological effects of air pollutants has similar limitations.

With the foregoing limitations in mind, the following discussion of the effects of specific pollutants was prepared. The information is based largely on the air quality criteria booklets prepared by the National Air Pollution Control Administration and its successor, the Air Pollution Control Office of the Environmental Protection Administration. Where considerable controversy exists, the basis for the effect of interest is discussed.

Particulate Matter

The words "particulate matter" cover a broad range of particles from those only slightly larger than gas molecules (>0.01 microns) to those greater than 1000 microns ($^{1}/_{25}$ of an inch). Particulate matter or particles may be either liquids or solids, sometimes both. They may be highly

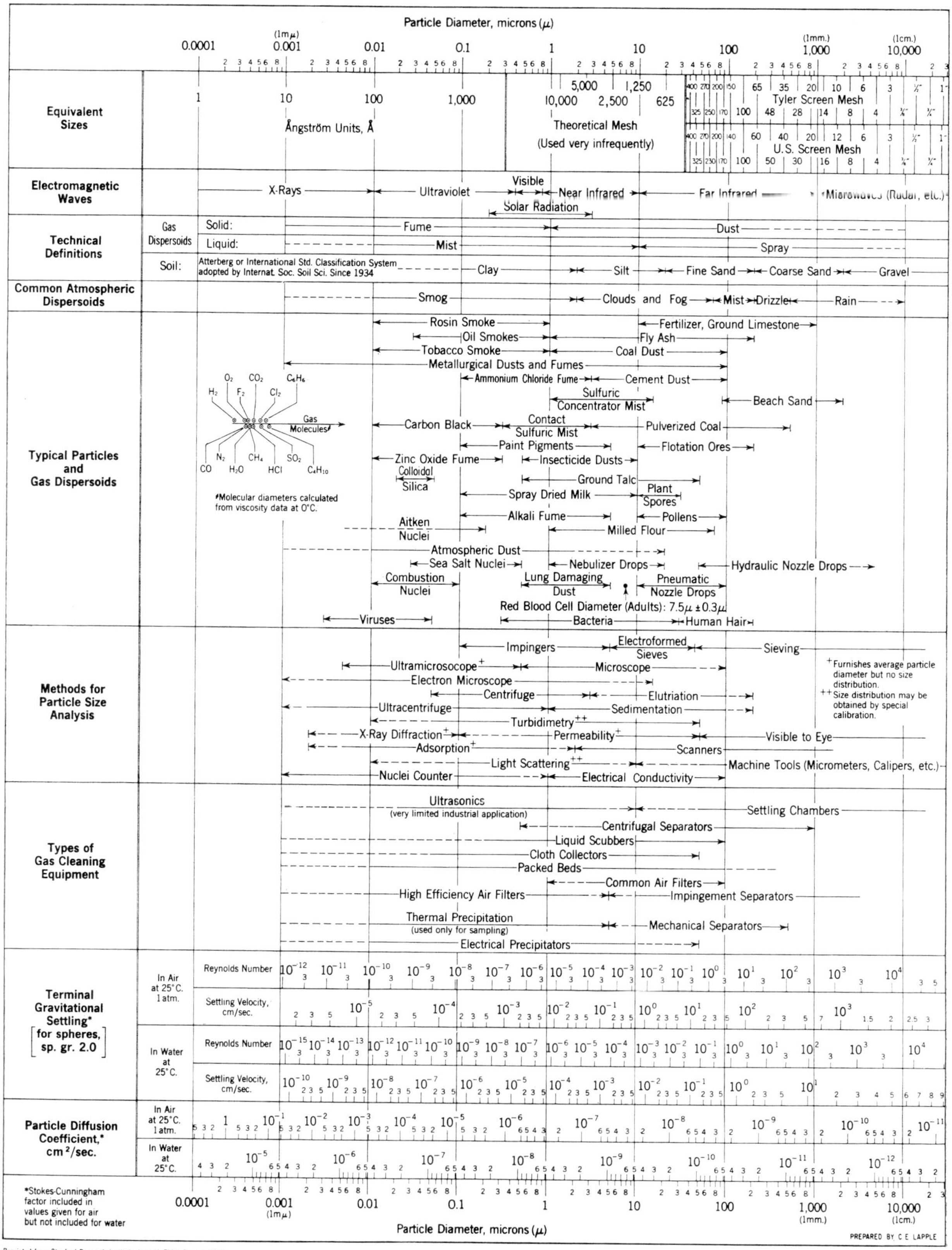

Reprinted from Stanford Research Institute Journal, Third Quarter, 1961.

Figure 1 Characteristics of particles and particle dispersoids.

complex mixtures or relatively pure compounds. Physical and chemical properties may vary widely.

Inasmuch as the most common particulates in the atmosphere are mixtures of dusts, smoke and fumes, the effects of this mixture, known as "suspended particulate matter" is of interest. Legally it is usually defined as the material collected on a high-volume filter of special design over a period of 24 hours.

A special type of suspended particulate matter is smoke, where the particulate is largely soot. Attention is also given to dustfall or settled dust which is the particulate matter which settles from the air on to surfaces in a specified period of time. Dustfall particles are generally 20 microns or greater when their specific gravity is 2.0. Higher gravity particles may be smaller, and vice-versa.

The principal effect of *settled dust* is soiling of surfaces, where the degree of effect varies with location. Thus a heavy industrial area will accept more dustfall than a commercial area or a residential area. The air quality standards of the State of Texas, presumably set below a nuisance level, set as maximum values for dustfall 4 tons per square mile per 30 days in recreational areas, 6 tons in residential areas, 10 tons in commercial areas, and 18 tons per square mile per 30 days in industrial areas.

Settled dust particles are not in themselves a health hazard in that individual particles are so large ($>20\,\mu$) that they will be filtered out in the nose and throat and not enter the lungs. On the other hand, high dustfall values are often found along with high concentrations of respirable dust suspended in the atmosphere and may therefore be an index of health effects.

Leafy vegatation may be affected by excessive amounts of settled dust as witness the poor growth of vegetation along an unpaved highway. Large amounts of dust on leaves may discourage the predators of mites and allow the latter to thrive. Quantitative data are not available.

The chief effects of suspended particulate matter are soiling and reduction in visibility. Soiling arises from the deposition of particles less than 10 microns by impingement on surfaces to thus mingle with settled dust. Thus at some concentration it is a nuisance. Attempts have been made through public opinion surveys to ascertain the atmospheric concentration at which a majority or a given percentage of the population believed an air pollution problem existed. The various surveys indicate values of the annual geometric means between 50 and 200 micrograms per cubic meter are acceptable depending on the specific area sampled. On the other hand the reliability of public opinion surveys allows much to be desired.

A sounder relationship exists between the atmospheric concentration of suspended particulate matter and visibility. Even here a considerable concentration range exists because of varying particle size. Light scattered by a cloud of particles depends not only on the concentration of the particles and path length but also on size distribution, composition and shape of the particles. Charlson developed a formula to relate visibility in miles to concentration of particles in micrograms per cubic meter. In simple form it may be expressed as:

$$L \times M = 1200,$$

where L is visibility in miles, M is concentration of suspended particulates in micrograms per cubic meter. The constant, 1200, actually varies from 600 to 2400. Hence an atmospheric particulate concentration of $100/\mu g/m^3$ would result in 6 to 24 miles visibility depending on particle size. A concentration of $200/\mu g/m^3$ would yield visibilities of 3 to 12 miles.

In the NAPCA publication "Air Quality Criteria for Particulate Matter," suspended particulate matter concentrations are also related to health effects and corrosion. The NAPCA document quotes a variety of studies relating 24-hr. average values of suspended particulate matter to various health effects, e.g. *excess deaths*, *increase in illness*, *increase in mean sputum volume*, acute worsening of symptoms in *chronic bronchitis* patients, increased *absence* of industrial workers due to illness, increased incidence of *respiratory* diseases, and *increased death rates* for persons over 50 years old. The several relationships are derived from epidemiological studies made in communities with comparatively high sulfur dioxide concentrations or sulfation levels. Particulate concentrations were expressed as annual geometric means in micrograms per cubic meter. The lowest values (one study) related to health effects were 80 to 100 $\mu g/m^3$ for particulates (annual geometric means) accompanied by sulfation levels of about 30 mg/cm^2-mo. These values were associated with increased death rates for persons over 50 years of age. No data of significance indicates health is affected by suspended particulate matter except where sulfur dioxide is comparatively high.

The NAPCA criteria publication relates corrosion to suspended particulate matter through a study showing that corrosion of steel and zinc panels in the presence of SO_2 and moisture is 3 to 6 times more rapid in New York City than in State College, Pennsylvania, where the suspended particulate matter loadings were 180 $\mu g/m^3$ and 60 $\mu g/m^3$ respectively (annual geometric means).

A complicating factor, particularly in health effects studies of suspended particulate matter, is the presence of smoke in which the particulate matter consists largely of soot. The several studies relating health effects to suspended particulate matter were carried out in cities where smoke was a major contributor to the atmospheric loading.

It appears then that in those localities where soot-laden smoke contributes significantly to the suspended particulate matter in the air, and the sulfur dioxide is also comparatively high, the resulting mixture has a noticeable effect on the health of some portion of the population when the 24-hr. annual geometric mean of suspended particulate matter exceeds 100 $\mu g/m^3$. Further data are required to relate non-smoke particulate matter to health.

Specific solid and liquid particulates must be considered separately whenever they are toxic or are likely to be major components of atmospheric particulates. Among the more important specific particulates are the sulfuric acid aerosol, fluorides, beryllium, asbestos, polynuclear aromatic hydrocarbons, the "photochemical" aerosol, lead, and arsenic.

The *sulfuric acid aerosol* is of considerable theoretical interest because it is one of the intermediate products in the removal of sulfur dioxide from the air. Thus it is put into the air in the first place as sulfur dioxide, and is removed by reaction with inorganic dusts. The major effects of the sulfuric acid aerosol are corrosion and visibility reduction. In sufficiently high concentrations (highly unlikely) it could be a hazard to health. Inasmuch as the atmospheric aerosol is derived from SO_2, the various effects are discussed in the section on sulfur dioxide.

Similarly *fluoride particulates* have effects similar to gaseous fluorides and are discussed further along in this section. (See Fluorides.)

Beryllium particles, which may exist in the atmosphere in the form of the oxide, BeO, is highly toxic, and can lead to chronic pulmonary disease with a high fatality rate (greater than 35%). In some instances, onset of the disease may not occur until years after exposure to high levels. In 1949, the U.S. Atomic Energy Commission set a community air concentration limit of 0.01 $\mu g/mg$ for beryllium (10–100 fold safety factor).

The breathing of *asbestos* particles in high concentrations over a period of years by asbestos workers has led to specific types of lung tumors. There are no data indicating whether or not the quantities found in community air are sufficient to constitute a health hazard.

Polynuclear aromatic hydrocarbons (PAH), a broad class of hydrocarbons present in combustion effluents, have been of considerable interest for many years because of their carcinogenic activity. Cancers have been produced in tumor-susceptible mice, by painting extracts of PAH-containing material on the skin or by subcutaneous injection. Whether or not inhalation will produce lung cancer in man is not known. A few epidemiological studies have related suspended particulate concentration and atmospheric smoke density to cancer of the bronchia, lung, stomach, and intestines, but the specific carcinogens responsible are unknown.

The "*photochemical aerosol*" is the liquid particulate matter formed by the photochemical reaction between certain organic compounds and nitrogen oxides in the presence of sunlight. Not too much is known concerning the composition of the aerosol except that the liquid phase is organic in nature, and the particles generally contain the elements C, O, H, S, and N. The particles are sub-micron and highly efficient in reducing visibility. No other effects of the aerosol "per se" have been proved, although the complex known as photochemical smog has numerous effects. (See ozone and oxidants in this section.)

Lead is a constituent of most suspended particulate matter found in both urban and rural areas. Generally it is present in the form of an oxide or a complex halide. Because lead compounds are toxic in high concentrations, there has been considerable public pressure to reduce atmospheric concentrations by reducing the use of lead compounds as gasoline anti-knock compounds. A great deal of study has been made of the metabolism of lead entering the human body in various manners, i.e.

through food, drink, and inhalation. Normal levels of lead in the blood vary from less than 0.01 to 0.04 mg/10 g blood; normal urinary levels in the human population are well below the borderline value of 0.08 mg/100 g blood for incipient lead intoxication. Authorities in the field do not believe a health hazard exists, and have suggested that a tentative ambient air quality standard of 10 μg/m^3/30 days would be conservative.

Recent studies indicate that lead inhibits formation of alphaaminolevulinic acid dehydrogenase (ALAD), an enzyme important in heme synthesis, but whether or not this biological reaction produces an adverse health effect or not is controversial. Unfortunately, the normal level of ALA in urine has not been established. A need for more study is indicated.

Domestic animals have contracted chronic lead intoxication from feeding on contaminated forage or pastures, or drinking contaminated water. Toxic doses of lead are 50 to 100 g for ruminants, 500 to 700 g for horses, 20 to 25 g for sheep and goats, and 10 to 20 g for pigs and dogs.

The only effect of lead on vegetation is apparently contamination by fallout. In lead-rich soils, lead absorbed by root vegetables concentrates in the leaves rather than the root.

Arsenic deposited on vegetation near sources of emissions has resulted in arsenic toxicosis to several species of animals and insects. Honey bees are the most susceptible; sheep next. Horses and cattle are most resistant. A dosage of 300 grains of arsenic daily is fatal to horses and cows; 4 to 8 grams for sheep. Wildlife (e.g. red deer and hares) are also susceptible.

Damage to vegetation or to people from airborne arsenic have not been reported.

Gases

In contrast to particulate matter, much more is known about the effects of various concentrations of individual gases in the atmosphere. Analytical methods are simpler and laboratory experiments more convenient. On the other hand, data on mixtures of gases are nearly non-existent. So, once again, epidemiological studies are relied upon in the field of health effects, with their resulting "associations" but few cause and effect relationships.

Sulfur dioxide, SO_2, is a colorless gas with a suffocating odor. Concentrations of 6 to 12 ppm cause instantaneous irritation of the nose and throat. Most people can detect concentrations as low as 0.3 to 1 ppm by taste and about 3 ppm by smell. The maximum allowable concentration in air for 8 hours, as adopted by the American Conference of Governmental Hygienists is 10 ppm.

Concentrations experienced in the ambient atmosphere affect materials, visibility, plants, animals, and man. These effects are complicated by oxidation of SO_2 in the atmosphere to form sulfur trioxide and sulfuric acid, both of which also have effects on various receptors.

Nearly all materials react to sulfur dioxide or to its reaction product, sulfuric acid. In laboratory experiments, *steel panels* dusted with particulate pollutants corroded more rapidly in the presence of traces of SO_2, particularly at relative humidities greater than 70%. In the ambient atmosphere, SO_2 is seldom separated from particulate matter, but considerable corrosion takes place at annual average concentrations of 0.03 ppm to 0.12 ppm SO_2. It is generally accepted that SO_2 has greater influence on corrosion rate than does particulate matter.

Atmospheres containing sulfur dioxide also attack many *non-metallic building materials* (limestone, marble, slate, mortar), and accordingly damage and discolor buildings, walls, statuary and similar works of man. Cotton, rayon and nylon *fabrics* are weakened; dyed fabrics may fade; over a long period of time, *leather* and *paper* become brittle. Concentrations of 1 ppm SO_2 can increase the drying time of certain oil-based *paints* by 50 to 100 per cent; other paints become brittle.

The effect of SO_2 on atmosphere visibility is basically the effect of sulfuric acid particles. These atmospheric particles are most often in the size range that has greatest effectiveness in reducing visibility. The actual visibility may be estimated by use of the equation previously shown (see particulate matter).

Damage to vegetation from sulfur dioxide is a function of many variables: plant species, temperature, humidity, light intensity, soil moisture, nutrient supply and age of the plant tissue. Acute injury results from the rapid absorption of toxic concentrations of sulfur dioxide (high concentrations for short periods of time). On broad-leafed plants injured marginal and intercostal areas dry out and bleach to an ivory color. Acute injury in

Figure 2 Outside of St. Mary's Greek Catholic Church, Taylor, Pa. Shows soiling effects of pollution. Taken August 18, 1966 (Scranton Sunday Times).

pine shows up in bands on the tips of the needles with the injured area becoming reddish brown. Chronic injury, which results from lower concentrations over a period of days or weeks, leads to gradual yellowing or chlorosis of the leaf tissue. In some cases, even though visible injury is not evident, suppression of growth and yield may occur. This "invisible injury" is difficult to detect and document, inasmuch as many other factors, e.g. soil, lack of water, improper plant nutrients, etc., may affect growth. Generally, however, growth suppression is accompanied by chlorosis and excessive leaf drop.

Some varieties of vegetation are more susceptible than others to sulfur dioxide injury. The threshold response of alfalfa to acute injury is 1.25 ppm for 1 hour; privet requires 15 times this concentration. The most sensitive plant species will not show

Figure 3 SO_2 damage to tomato plants (Courtesy Moyer Thomas).

visible injury if SO_2 concentration does not exceed 0.3 ppm for 8 hours.

A specific type of damage to leafy plants has been attributed to sulfuric acid mist. A spotted type of injury to wet leaves may occur at concentrations of 100 $\mu g/m^3$.

The effects of SO_2 on *animals* are, as would be expected, effects on the respiratory system. Only relatively high concentrations of SO_2 or sulfuric acid will produce pathological changes in the lung or mortality in animals.

The *health effects* of sulfur dioscide have been the subject of more debates in the medical field than any other environmental subject. There is no doubt that SO_2 affects the respiratory system and in high concentrations may cause death. Again, as in the case of particulates, most of the data relating sulfur dioxide concentrations to health are from epidemiological studies. The nature of the major conclusions drawn by NAPCA are:

1. At 1500 $\mu g/m^3$ (0.52 ppm) (24-hr. average) and high smoke content, *increased mortality* may occur.—(New York City.)

2. At 715 $\mu g/m^3$ (0.25 ppm) (24-hr. mean) and 750 $\mu g/m^3$ of smoke, *increased daily death rate* may occur.—(London.)

3. At 500 $\mu g/m^3$ (0.19 ppm) (24-hr. mean) and low particulates *increased mortality rates* may occur.—(Rotterdam.)

4. At 300 to 500 $\mu g/m^3$ (0.11 to 0.19 ppm) (24-hr. mean) and low particulates, increased hospital admissions of older persons for respiratory disease may occur; absenteeism from work, particularly with older persons, may also occur.—(New York) (Rotterdam.)

5. At 715 $\mu g/m^3$ (0.25 ppm) (24-hr. mean) (with particulates), a sharp rise in illness rates for patients over age 54 with severe bronchitis may occur.—(Chicago, one-day episode.)

6. At 600 $\mu g/m^3$ (0.21 ppm) (24-hr mean), and 300 mg/m^3 smoke, patients with chronic lung disease may experience accentuation of symptoms.—(London, one-day episode.)

7. At 105 to 265 $\mu g/m^3$ (0.037 to 0.092 ppm) (annual mean) and 185 $\mu g/m^3$ of smoke, *increased frequency of respiratory symptoms and lung disease* may occur.—(Genoa).

8. At 120 $\mu g/m^3$ (0.046 ppm) (annual mean), and 100 $\mu g/m^3$ of smoke, increased frequency and severity of respiratory diseases in school children may occur.—(Great Britain.)

9. At 115 $\mu g/m^3$ (0.04 ppm) (annual mean), and 160 $\mu g/m^3$ of smoke, *increase in mortality* from bronchitis and from lung cancer may occur.—(Great Britain.)

On the basis of these data, the Environmental Protection Administration recommended ambient air quality standards "to protect human health" of 80 μg/m³ (0.03 ppm), annual arithmetic mean and 365 μg/m³ (0.14 ppm), maximum 24-hr. concentration not to be exceeded more than once a year.

Wide acceptance of these values will come only after more studies in American cities, particulalry those with low smoke concentrations.

Hydrogen sulfide is a toxic, disagreeable smelling gas that can readily be detected by smell in concentrations as low as 0.001 ppm. However, the nose soon becomes accustomed to the odor, so this method of detection is not acceptable when the concentration is high or persistent.

The major effects of hydrogen sulfide on materials are discoloration of certain metals and lead-base paints. Tarnishing of copper and silver is noticeable at very low concentrations. In the presence of moisture, lead-base paints will tend to blacken in the presence on 0.05–0.1 ppm hydrogen sulfide over a period of several hours.

Health effects are noticeable only at concentrations considerably above the odor threshold and those that affect paint and metals. At 50 ppm, subacute poisoning, mild conjunctivitis and mild respiratory tract irritation have been experienced after a one-hour exposure. At 700 to 1000 ppm, rapid unconciousness and death may result. However, physiological response tests indicate that a change in light sensitivity of the eye can be detected at 0.07 ppm.

Ambient air quality standards are commonly based on the odor threshold and vary from 0.03 ppm for ½ hour to 0.1 ppm for 1 hour.

Carbon monoxide is a colorless, odorless, highly poisonous gas that is the product of incomplete combustion of carbon and its compounds. It is usually classified as an asphyxiant because of its strong combination with hemoglobin in the blood. Many deaths have resulted from carbon monoxide in coal mines, unventilated rooms with defective heating equipment, and unventilated garages where internal combustion engines have been left running. The maximum allowable concentration for eight hours exposure inside is 100 ppm.

Much work has been done to assess the concentrations of carbon monoxide at which people are affected. Studies of population groups are complicated by smoking habits in that cigarette air streams contain 400 to 475 ppm CO.

When carbon monoxide is breathed it eventually winds up in the blood where it reduces the oxygen carrying capacity of the blood by uniting with hemoglobin to form carboxyhemoglobin (COHb). The elevation of COHb in the blood fortunately is reversible and is reduced when the CO content of the air breathed is lowered. Healthy people can stand up to 100 ppm in the air for many hours, and many of these will not even experience a headache. Extremely high concentrations (3%) may be countenanced for a minute or so. The ill and infirm, particularly cardiac and respiratory cripples, may be considerably affected at much lower concentrations, i.e. those which will result in an increase in normal COHb blood level of 2 to 5%. The variation in "normal" levels makes standard setting difficult. The concentration of CO required to raise the COHb by the amount stated depends, of course, on the length of exposure.

Another effect of CO that has received much publicity is performance impairment, especially impairment in judging time intervals, and in visual acuity. Data are conflicting and indicate CO threshold values varying from 10–15 ppm for 8 hours to more than 50 ppm for longer periods.

Tests with experimental animals also vary widely, with few changes below 50 ppm for 8 hours and often longer periods.

On the basis of a study of the literature, the EPA has recommended as ambient air quality standards:

10 mg/m³ (ca. 8.7 ppm)—maximum 8-hour concentration not to be exceeded more than once per year.

15 mg/m³ (ca. 13.1 ppm)—maximum 1-hour concentration not to be exceeded more than once per year.

Many authorities on carbon monoxide toxicity believe the proposed standards include safety factors of greater than 100 per cent.

Ozone (O_3), is a colorless gas with a characteristic odor. It is made up of 3 atoms of oxygen, has high oxidizing power, and readily decomposes to molecular oxygen (O_2) and atomic oxygen (O). Ozone is seldom emitted into the atmosphere as such, yet is found in small quantities (0.01 to 0.05 ppm) even in unpolluted atmospheres. It is formed naturally by electric discharge, and in the stratosphere by the action of solar radiation; with the maximum concentration (0.20 ppm) occurring 12 miles above the surface of the earth. Under favorable weather

Figure 4 Inversion over the Los Angeles Basin.

conditions (frontal zones) some of the ozone is transferred to ground level. It is also believed that some of the "natural" ozone found over forested land comes from photo-oxidation of hydrocarbons emitted from certain trees. It is formed in polluted atmospheres by a series of reactions between hydrocarbons and oxides of nitrogen in the presence of sunlight. Along with ozone formation, many other substances are formed with oxidizing power greater than molecular oxygen. These substances along with ozone are commonly referred to as a class by the name "photochemical oxidants" or just "oxidants." Some of these oxidants have been identified, e.g. *peroxyacetyl nitrate* (*PAN*). Others include a wide variety of peroxides, hydroperoxides, peroxy compounds etc. Ozone is, however, the major constituent of the photochemical oxidant.

The effects of ozone are primarily related to its strong oxidizing power. Thus at atmospheric concentrations it may cause fading of textile dyes. At 0.2 ppm (40 $\mu g/m^3$) for an hour, it will cause cracking of stretched sheets of vulcanized natural rubber. Its odor may also be detected in less than five minutes at 0.02 ppm.

Long-time exposure of human subjects to 390 $\mu g/m^3$ (0.2 ppm) for 3 hours a day, 6 days a week, for 12 weeks has not produced any apparent effects. Exposure to 980 $\mu g/m^3$ (0.5 ppm) for 3 hours a day, 6 days a week, has caused a decrease in the 1-second forced expiratory volumes of the lungs after 8 weeks. Nasal and throat irritation are experienced with exposures of 0.3 ppm (590 $\mu g/m^3$) for 8 hours. Numerous experiments at 0.6 to 1 ppm for 1 to 2 hours show measurable changes in pulmonary function.

Ozone also affects vegetation in verying degrees depending on variety and growing conditions. For vegetation grown under sensitive conditions, 20% leaf injury will be caused by 0.15 to 0.30 ppm ozone for one hour to some varities of spinach, radish, tobacco, pine and others.

Inasmuch as most experimental work with ozone has been carried out in purified air it appears to be more nearly relevant to examine the effects of the atmospheric oxidant, particularly as it serves as an index to photochemical air pollution.

Plants sensitive to acute injury from the photochemical oxidant include certain varities of tobacco

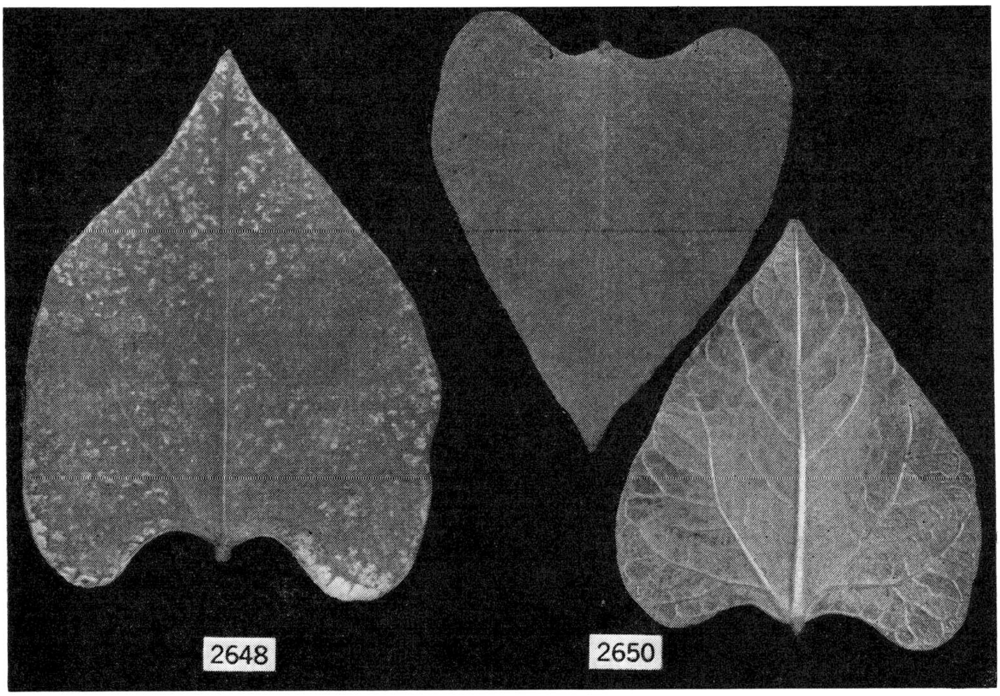

Figure 5 Oxidant damage to leaves of common pinto bean. Damage to the leaf on the left (2648), resulting from ½-hr exposure to 0.5 ppm ozone (KI), is limited to the upper surface. The two leaves under 2650 were exposed to a mixture of ozone (residual 0.15 ppm) and 15 ppm hexene-1. Typical oxidant damage occurs on the lower leaf surface (right); note absence of oxidant injury on upper leaf surface. Small amount of damage on upper leaf surface of middle leaf in photograph is attributable to residual ozone. These results were secured from a 30-minute exposure. (Courtesy University of California, Riverside.)

(0.06 ppm for 1 hour), white pine (0.1 ppm for 4 hours), and the pinto bean (0.15 ppm for a day—moderate injury). One variety of tobacco (Bel-W 3) suffers severe leaf injury with an exposure of 0.03–0.22 ppm for 700 hours. Various plant species affected by ozone and PAN are obviously also affected by "oxidant" but data are sparse. Experiments using PAN alone show that sensitive varieties of petunia show moderate injury at 0.01 ppm for 6 hours. Generally the type of leaf marking shows which oxidant is responsible for the injury. Ozone marks leaves by "stipple" of the top leaf surface. PAN produces a bronzing or silvering of the underside of the leaf. Various combinations have been observed both in the laboratory and in the field.

With respect to the health effects of photochemical smog as measured by oxidant values, there can be little doubt that the major effect is *irritation of the eyes* thought to be caused by formaldehyde, acrolein, PAN, and other constituents of the photochemical smog complex. Not everyone notices eye irritation when peak values exceed 0.1 ppm oxidant, but a sufficient number of people appear to be affected that this value has been selected by the State of California as the point where smog is a nuisance. The eye irritation is not caused by the oxidant; it serves only as an index to describe the severity of the photochemical smog. A few individuals claim to experience eye irritation at 0.03 ppm oxidant. Various studies seeking to relate photochemical smog severity, as measured by oxidant, to helath effects have been made, but they either showed no effect or were undeterminate. One study on aggravation of asthma in Pasadena, California showed an increase in the number of asthmatic attacks in 5 per cent of a group of asthmatic patients when the daily oxidant maximum escceeded 0.13 ppm (ca. 250 $\mu g/m^3$).

A study of high school cross-country runners in a suburb of Los Angeles indicated lack of improvement in performance was associated with oxidant values between 0.03 and 0.30 ppm.

A great deal of laboratory data is available relating ozone, oxidant, and PAN concentrations to effects on animals. Pathological changes were noticed after prolonged exposure to 0.2 to 1.0 ppm ozone; chemical changes in the structural protein of the lung were noticed at 1 ppm ozone for 1 hour.

Review of all the biological and health data on photochemical smog leaves one with little more than suspicion that photochemical smog affects health. It may be that possible chronic effects are subtle but nearly 30 years of intense smog in Los Angeles hasn't clarified the effects.

The one highly noticeable effect of photochemical smog in addition to eye irritation is visibility reduction. Even at oxidant concentrations equivalent to background values, an intense haze may develop under some meteorological conditions and markedly restrict visibility. Much more information will be required to understand the phenomena.

On the basis of the studies outlined in the foregoing the Environmental Protection Administration has recommended an ambient air quality standard for photochemical oxidant of 125 $\mu g/m^3$ (0.064 ppm) not to be exceeded as an hourly average. The value appears to be based on acute damage to sensitive vegetation.

Hydrocarbons consist of a variety of carbon-hydrogen compounds which may be gases, liquids, or solids at normal temperatures depending on molecular weight. The simplest hydrocarbon, methane, CH_4, is a gas (actually the major constituent of natural gas). A typical liquid hydrocarbon is benzene, C_6H_6. The only solid hydrocarbons of much interest in air pollution are the polynuclear aromatic hydrocarbons (see Particulate Matter).

Several classes of hydrocarbons are recognized, based on their chemical structure, and accordingly their chemical reactivity. The important classes are:

1. Paraffins or saturated hydrocarbons, with the general formula, C_nH_{2n+2}. Examples are butane, C_4H_{10}.
2. Olefins or unsaturated hydrocarbons, with the general formula C_nH_{2n}. An example is hexene, C_6H_{12}.
3. Aromatics. The simplest aromatic hydrocarbon is benzene (C_6H_6).

Except under very specialized circumstances hydrocarbons as a general class are not important in air pollution. The exception is photochemical smog formation. The basic chemical reactions taking place in the atmosphere to form photochemical smog are shown under nitrogen oxides (q.v.). The most important energy-absorbing molecules (M) shown in the equations are hydrocarbons. Without hydrocarbons in the air to react with atomic oxygen, ozone, and nitrogen oxides, there would be no photochemical smog and its attendant effects: vegetation damage, high oxidant concentration, eye irritation, and reduced visibility.

The degree to which hydrocarbons must be reduced to eliminate photochemical smog is not known. One complicating factor is the relative *reactivity* of various hydrocarbons. Certain olefins, particularly "dienes" (pronounced *dye-eens*) are the most reactive of all hydrocarbons and lead to the most intense photochemical smog effects. Certain aromatic hydrocarbons, particularly the xylenes and higher molecular weight aromatics also react rapidly and lead to intense eye irritation and probably to the visibility-reducing aerosol. The simplest aromatic hydrocarbon, benzene, is essentially unreactive and therefore an exception to the general statement. Paraffinic hydrocarbons react very slowly and aren't very important. Methane is completely unreactive. Acetylenes and other types of hydrocarbons are also not reactive.

Because of the variation in hydrocarbon reactivity it has been suggested that the easiest way to eliminate photochemical smog is to eliminate olefinic and aromatic hydrocarbons.

In addition to hydrocarbons, various other *organic compounds* in the atmosphere, e.g. alcohols, aldehydes, ketones, esters, etc., may react with oxides of nitrogen under the stimulus of sunlight. Here, again, those species having olefinic or aromatic structure are more reactive than other types.

There are a few specific hydrocarbons which produce undesirable effects in themselves, i.e. in addition to their involvement in the photochemical smog reaction. One of these, the simplest olefin, ethylene, in addition to its photochemical activity, is a plant damaging agent (phytotoxicant). Even at 5 parts per *billion* it injures certain orchid varieties by causing the sepals to wither and die. At 0.05 to 0.10 ppm, it causes epinasty (downward curling of leaves) and inhibits nutation in marigold, sweet pea, and tomato leaves. Various other types of damage have been reported for several species of flowers.

The effects of polynuclear aromatic hydrocarbons have already been discussed (see Particulate Matter).

Figure 6 Effect of naturally occurring air pollutants (possibly ethylene) on roses collected in San Francisco, Calif. (Courtesy University of California, Riverside).

Two *oxides of nitrogen* are important in air pollution, nitric oxide (NO) and nitrogen dioxide (NO_2). The latter is the most important reactant in photochemical smog formation. In fact, the initial reaction in the atmospheric photo-oxidation process is the absorption of ultra-violet light energy by NO_2 [see equation (1)]. The highly energized molecule, NO_2^{+}, then decomposes to nitric oxide and atomic oxygen [equation (2)]. The atomic oxygen quickly reacts with molecular oxygen in the air to form ozone (O_3). Unless an energy-absorbing molecule (*M*) is present the ozone will rapidly decompose. But with a third body (M) present, a stable ozone molecule is formed [equation (3)]. These reactions may be expressed by the following chemical equations:

$$NO_2 + h_v \rightarrow NO_{2*} \quad (1)$$

(quantum of light)

$$NO_{2*} \rightarrow NO + O \quad (2)$$

$$O + O_2 + M \rightarrow O_3 + M \quad (3)$$

Some of the ozone, some atomic oxygen, and nitric oxide react with various hydrocarbons and related organics in the air to form various products and intermediates which may even react further to produce the compounds responsible for the effects of photochemical smog. Obviously both oxides of nitrogen and hydrocarbons (organics) are important, but whether their relative importance are the same or not depends on many unknown factors. There can be little doubt that complete or nearly complete elimination of either nitrogen oxides or reactive organics would eliminate photochemical smog effects. How much *both* must be reduced to accomplish the same goal is moot. There are other effects of nitrogen oxides, particularly NO_2, which may require their control regardless of photochemical smog effects.

Nitric oxide (NO) is the oxide form usually emitted into the atmosphere. It is a colorless, odorless, gas. It is non-toxic in the concentrations normally found in the atmosphere. It oxidizes very slowly to nitrogen dioxide at atmospheric temperatures, more rapidly at higher temperatures or in the presence of reactive hydrocarbons and sunlight. Actually NO is not stable at atmospheric temperatures and tends to decompose to nitrogen and oxygen albeit very slowly except at temperatures above 500°F. Thus the only reason why NO is of interest in air pollution is its liklihood of oxidizing to nitrogen dioxide.

Nitrogen dioxide (NO_2) is a reddish-brown-orange gas with a characteristic pungent odor. It is a strong oxidizing agent, corrosive, irritating and toxic at sufficiently high concentrations. In the presence of water it hydrolyzes to form nitric acid (HNO_3) and nitrous acid (HNO_2).

Insofar as damage to materials is concerned, NO_2 is suspect but no sound data at atmospheric concentrations are available. At 25 ppm (47 mg/m^3) it damages many kinds of plants by producing acute necrotic leaf injury. Continuous fumigation of certain types of citrus with 0.5 ppm NO_2 for 35 hours is said to result in leaf drop and chlorosis, but no actual necrotic lesions.

In man and animals the primary toxic effect of NO_2 is on the lungs. 100 ppm is lethal to most laboratory animals. Exposure of rats to 0.5 ppm for 4 hours or 1.0 ppm for 1 hour has produced degranulation of lung mast cells. Humans are much more resistant. Some can smell NO_2 at about 0.12 ppm. A dosage of 5 ppm for 10 minutes has produced a transient increase in airway resistance. The 8-hour MAC for occupational exposure is 5 ppm. A six-month epidemiological study in Chattanooga, Tennessee related acute bronchitis in infants and school children to average 24-hour concentrations of NO_2 between 0.063 and 0.083 ppm.

On the basis of the foregoing the Environmental Protection Agency has suggested as an ambient air quality standard for total nitrogen oxides (NO_x) an annual arithmetic mean of 100 μg/m^3 (0.053 ppm) and a maximum 24-hr value of 250 μg/m^3 (0.133 ppm). The term NO_x or NOX is commonly used to designate total NO + NO_2 concentration (calculated as if all gas was in the form of NO_2).

In 1965, the State of California adopted an ambient air quality standard for NO_x of 0.25 ppm for 1 hour on the basis that 0.25 ppm NO_2 over a 10-mile path length (on a day of 20-mile visibility) would produce an undesirable brown coloration to objects 10 miles away. There have been many reports of brown atmospheres caused by elevated NO_2 concentrations but these usually have turned out to be red or brown colorations caused by particulate matter.

Pollution from *fluorides* in the atmosphere arises from both gaseous and particulate fluorides. The only toxic atmospheric gas that contains fluorine is *hydrogen fluoride*. Fluorine itself is far too reactive to exist as such for very long or even to be emitted in large quantities. The fluorocarbons used in refrigeration and as an aerosol propellant are non-toxic.

Hydrogen fluorides is an extremely corrosive, poisonous liquid but because of its low boiling-point (19.4°C) it is emitted into the atmosphere as a gas. It is seldom found in the air except in very low concentrations.

The other form of fluoride pollution is particulate matter consisting of inorganic fluorides, chiefly sodium and calcium fluorides.

The most important effect of fluorides is fluorosis of cattle and other farm animals caused by ingestion of vegetable matter that has collected fluorine-containing dusts. If cattle take more than 30 to 50 ppm soluble fluorides in their daily total ration they may develop mottling and abrasion of teeth, osteoarthritis, lameness, and eventual death. Other animals are not so sensitive as cattle. Sheep and hogs can take at least twice the concentration of fluorine in their daily ration.

Gaseous fluorides, particularly HF, are the most important form of fluorine with respect to effects on vegetation. Damage varies with type of vegeta-and with dosage (concentration × time of exposure). Gladiolus is particularly sensitive. Less than 1 part per billion HF for 10 days induces leaf chlorosis and affects flowering. Similar concentrations cause needle tip burn on some conifers. Other sensitive vegetation includes apricot, azalea, corn, sorghum, peach, strawberry, sweet potatoes, and iris.

Man is little affected by atmospheric hydrogen fluoride below 2–3 ppm.

Ambient air quality standards are usually expressed in terms of fluoride concentrations in grasses and plants near known source.

Chlorine and its compounds, particularly *chlotinated organic compounds* are comparatively minor atmospheric pollutants. Chlorine, in the pure state, is a heavy, toxic greenish-yellow gas with a pungent, irritating odor. The maximum allowable concentration for occupational exposure is 3 ppm. It is detectable by odor at 0.05 ppm, is irritating to some people at 1 ppm. As a phototoxicant it is 30 to 50% more toxic than SO_2. The damaging effects on a plant are similar to those of sulfur dioxide. Leaves of sensitive plants first appear spotted; later the spots become straw-colored or brown. The plants most sensitive to injury are peach, coleus, cosmos, buckwheat and hybrid tea rose.

Hydrogen chloride is occasionally emitted into the atmosphere but it is so soluble in water and so reactive that it is seldom found except in minute concentrations. Its principal effect is corrosion. It may damage vegetation at 100–1000 ppm.

The various chlorinated organic compounds that are emitted into the atmosphere are comparatively unreactive and may generally be considered as hydrocarbons. Some chlorinated organics are lachrymators at very low concentrations (e.g. chloroacetophenone); a few are toxic (e.g. phosgene); and some are weed-killer (e.g. 2,4-D).

Ammonia is occasionally found in the atmosphere. It may be noticed by odor at a few ppm. Sensitive vegetation is damaged at 40 ppm for 1 hour, 16.6 ppm for 4 hours. It may react with acidic compounds (sulfuric acid, hydrochloric acid) at very low concentrations to form a visible aerosol.

A great many *sulfur- and nitrogen-containing organic compounds* are important in air pollution because of their odoriferous qualities even at extremely low concentrations. A few of these, with their odor threshold concentration are shown in the following table:

Odorant	Odor threshold concentration (ppm by volume)
Amine, dimethyl	0.047
Amine, monomethyl	0.021
Amine, trimethyl	0.00021
Benzyl sulfide	0.0021
Dimethyl sulfide	0.001
Ethyl mercaptan	0.001
Methyl mercaptan	0.0021
Pyridine	0.021

Atmospheric Mixtures

An unsolved problem in the air pollution field is the effect of *mixtures* of the many gases and particles that pollute the atmosphere and particularly the long-time biological, ecological and economic effects. When two or more pollutants are mixed the effects of one may be increased by the presence of the other, or its effect may be largely counteracted. Rarely will the effects be additive.

There are many cases where *acute* effects appear to be increased in mixtures, notably the disastrous episodes in London, Donora, and the Meuse Valley. Insofar as anyone knows (measurements were not the best) neither the SO_2 or particulate dosages during any of these events was sufficient to cause the experienced mortality and morbidity rates. On the other hand, there may have been constituents present in the smogs to have accounted for the effects if we only knew the true composition of the atmosphere at that time.

The development of chronic disease is even more difficult to relate to atmospheric concentrations of mixtures. Usually many years are required for the development of chronic disease. During the period of time required people undergo many variations in their physical, mental, and social environment, so that it is well nigh impossible to say which variable is responsible for the effect. Even so, it is the concensus of authorities in the field that the chronic effects of many pollutants on *sensitive* individuals are exacerbated in mixtures encountered in urban communities. A large majority, the healthy people in the community, appear to adapt to the mixtures and are not affected as long as no drastic or rapid increase in pollution occurs.

Effects on the highly complex ecological systems that make up our world probably are similar. Adaptation to slow changes in pollution is part of the evolutionary process, so changes in the ecology are masked by time, except for extremely sensitive species. In the latter case, effects are noticeable but the cause is often not apparent, and sometimes the apparent cause is not the real one.

Slow changes in economic effects are subject to the same uncertainties. Decreases in real estate values have been laid to increasing atmospheric pollution, but so many variables affect such values that relative causes may not be separable.

One can only conclude that much remains to be known about the specific effects of air pollutants, particularly the quantitative effects of constituents of mixtures, and the interaction among the atmospheric concentration of various pollutants and other variables affecting the health and welfare of individuals.

REFERENCES

Air Quality Criteria for Sulfur Oxides, EPA Publication No. AP-50 (1969).

Air Quality Criteria for Particulate Matter, EPA Publication No. AP-49 (1969).

Air Quality Criteria for Carbon Monoxide, EPA Publication No. AP-62 (1970).

Air Quality Criteria for Hydrocarbons, EPA Publication No. AP-64 (1970).

Air Quality Criteria for Photochemical Oxidants, EPA Publication No. AP-63 (1970).

Air Quality Criteria for Nitrogen Oxides, EPA Publication No. AP-84 (1971).

Proceedings of the Symposium on Air Quality Criteria. *Journal of Occupational Medicine* **10**, No. 9 (Sept. 1968),

Air Quality Monographs No. 69-1 through No. 69-11, American Petroleum Institute, New York, 1969.

W. L. FAITH
San Marino, Calif.

EFFECTS OF CHEMICALS ON HUMAN ECOLOGY

Problems resulting from technological advances are being uncovered at an alarming rate. While it is possible to predict, with some accuracy, benefits to be expected from improvements and new approaches in applied science, seldom have serious efforts been made to determine adverse effects resulting from these. The general public is increasingly aware of potential hazards from new products and processes and the trend in industrialized countries is toward greater control and regulation of hazardous undertakings. The question as to the more effective approach, gentle persuasion to gain voluntary compliance or strong legislation for strict regulation, has not been satisfactorily answered. Probably there never will be agreement.

The relationship among the air, water and soil is not a static one. Effects of a pollutant may be demonstrated progressively in the various compartments into which the environment is divided. A substance may be initially present, without apparent ill effects, in one component. Later, the same substance, or a demonstrable derivative, may appear in a different part of the environment in a most undesirable way. An excellent example is the high mercury level found in fish. Initially, it was felt that disposal into the marine environment would conveniently remove a bothersome waste. However, by previously unsuspected paths the metal found its way into the systems of game and food fish. Food, thus, has become a secondary distributor of pollutant material.

Chemical pollutants may be divided into four categories:

1) Natural chemicals in excess
2) Naturally occurring toxins
3) Mixtures of air and water pollutants which produce adverse effects but with only partially defined or undefined components
4) Synthetic chemicals.

The first group includes chemicals such as nitrates and nitrites. Nitrates, for example, can cause methemoglobinemis in infants by reduction of the capacity of the blood to carry oxygen. An intermediate reduction is involved. A famous case in New York involved accidental introduction of sodium nitrite into oatmeal and the resulting problems of "Eleven Blue Men." The victims, all heavy users of alcohol, apparently tried instinctively to compensate for low salt in the body. Through an accident, sodium nitrite instead of sodium chloride, was placed in salt shakers in a public eating place. The eleven victims all became cyanotic, with the characteristic blue color giving the name to the episode. Nitrites may also react with secondary amines to form nitrosamines, some of which have been shown to be carcinogenic, teratogenic and mutagenic, all in microgram doses. Oxides of nitrogen, thought to be significant in smog production, can also form nitrosamines.

A description of nitrogen dioxide pollution effect has been described in an article in the Journal of the Air Pollution Control Association.[3]

A unique opportunity to study the effects of nitrogen dioxide without many of the confounding covariables usually present in urban atmosphere existed in Chattanooga, Tennessee. It has been impossible to find similar circumstances elsewhere to permit replication of the study; but as the NO_2 emissions there decrease, we are studying the resultant health implications.

"In the analysis of lower repiratory illness by Pearlman, *et al.*,[4] the information gained from the aerometric measurements made in the two previous studies was used to edlineate the two control areas as 'intermediate' and as 'low' in NO_2 concentrations. This distinction was made in the protocol for the study, and not as a result of the illness rates measured, as seems to be inferred by Heuss, *et al.*[5] Although the highest rates were generally found in the intermediate area, those rates were not significantly different from the high NO_2 area rates. This disparity between the studies might reflect a different threshold for the upper and lower respiratory tracts, with the incidence of lower respiratory infections being a more sensitive index of NO_2 toxicity, as stated by Pearlman, *et al.*

Epidemiological studies are subject to the interaction of many variables, and this interaction tends to cast some doubt on the specificity of the studies. Verification is always desirable; however, this is not a valid argument against taking action on the basis of a 'single' study. As is frequently the case, epidemiological studies are carried out as a result of laboratory evidence of cause-effect relationships. When an epidemiological study can be shown to support the laboratory findings, withholding control becomes the more questionable course of action, especially if the consequences of judgmental error are only economic. There are many examples in medical history wherein a single study later proved to be valid, and delay in its implementation resulted only in needless suffering. The Chattanooga Study was undertaken in response to histological, biochemical, and clinical laboratory studies implicating nitrogen dioxide as a possible potentiator or respiratory disease in man. The epidemiological study in question supports that evidence. In the judgment of the Administrator the the combined evidence justifies the level of the Air Quality Standard."

The second group, natural fungal and plant toxins, usually are introduced into the human ecosystem through accident or carelessness. Conditions of harvesting, storage and processing have been shown to be of possible importance.

An excellent example of the third group has already been mentioned. Although but recently brought to light, the problem of mercury contamination in water is obviously of long term origin. Mercury was discharged to the receiving water as an apparent ultimate solution to a waste disposal problem. The two components of the system, mercury and water, were assumed to be nonreactive. Unfortunately, the two-component system was, in fact, a multicomponent system and no endeavor was made to determine the complete mechanism. The problem was further complicated by the introduction of edible fish into the chain.

The last group, synthetic chemicals, includes pesticides and fertilizers used in agriculture, food additives, compounds containing heavy metals, plasticizers, fuel additives, househould chemicals, industrial chemicals, therapeutic and prophylactic drugs and drugs of abuse. Food additives may be intentional or accidental. Antioxidants and dyes are added routinely to many foods. However, almost any of the aforementioned may be accidentally introduced into food, often with most unpleasant results. There is much controversy concerning synthetic agricultural chemicals. Advantages and disadvantages are numerous and no definite decision has been or soon will be reached concerning continued use of such substances as DDT. NTA (Nitrilotriacetic Acid) as a substitute for phosphates in detergents, is another excellent example of conflicting use. The problem is far from restricted to simple direct physiological effects.

Food additives may be classified as to function. They find use as coloring material, flavor enhancers, shelf life extenders, and in protection of food nutritional value. While valuable, color additives are not always essential. However, many of the foods now enjoyed by modern western society would not be possible, in their present form, without food additives.

It is estimated by the World Health Organization that about one fifth of the food produced in the world is lost by spoilage. Preservation, or retardation of spoilage, can be accomplished by addition of chemical preservatives, or by physical means such as freezing, drying, souring, fermenting, curing or ionizing radiation. There is some concern that irradiation of food may have adverse effects and leave unwanted residues. Little is known about possible chemical chain reactions.

Food additives have been classified by Kermode into five broad groups:

1) Flavours
2) Colours
3) Preservatives
4) Texture agents
5) Miscellaneous

In Table 1 are displayed food additives declared by the United States Food and Drug Administration to be "generally recognized as safe." Not included in Table 1 is a large group of natural flavours and oils. To be on the list an additive must have been in use before 1958 and meet specifications for safety. Materials introduced after 1958 must be tested individually in order to qualify for inclusion on the FDA list. Recent examples of materials formerly listed but now removed are cyclamate sweetners and saccharin. A ban on cyclamtes was ordered by the U.S. government after tests revealed development of bladder cancer in laboratory rats fed on a diet containing cyclamates. Further testing, after the ban, found cancer development in the same test species at dosage rates one sixth as large as those which brought about the ban. Saccharin is now receiving close scrutiny by health authorities in a number of countries.

It is estimated that there are as many as 1400 natural and synthetic flavours available today. Increased use of flavouring has paralled the introduction of new food processing and distribution techniques.

Flavour enhancers are closely related to flavours. The best known of these is monosodium glutamate (MSG). There is evidence linking excessive intake of MSG to Kwok's disease, sometimes known as "Chinese restaurant syndrome." Symptoms include a tightening of the neck and face muscles, at times accompanied by nausea, headache and giddiness. Some countries have placed limits on the amount of MSG in foods and require that the presence of this substance be prominently noted.

Colour additives are utilized primarily for the purpose of giving an appetizing appearance to food. Synthetic dyestuffs are used much more often than natural materials. It can be expected that the synthetic dyes will be produced in a high state of purity. In general, small amounts of dye will suffice to give the desired food colour. A material that gives a colour that is durable is needed by the food processor. Regulations governing the colours that may be added to foods vary markedly from country to country and this causes some difficulties for food processors engaged in international operations.

Some commonly used preservatives are benzoic acid, sodium benzoate, sorbic acid, monosodium phosphate, sodium propionate and sulfur dioxide.

TABLE 1

Additives listed by the U.S. Food and Drug Administration as "generally recognized as safe". (Courtesy, Scientific American).

ANTICAKING AGENTS
Aluminum calcium silicate
Calcium silicate
Magnesium silicate
Sodium aluminosilicate
Sodium calcium aluminosilicate
Tricalcium silicate

CHEMICAL PRESERVATIVES
Ascorbic acid
Ascorbyl palmitate
Benzoic acid
Butylated hydroxyanisole
Butylated hydroxytoluene
Calcium ascorbate
Calcium propionate
Calcium sorbate
Caprylic acid
Dilauryl thiodipropionate
Erythorbic acid
Gum guaiac
Methylparaben
Potassium bisulfite
Potassium metabisulfite
Potassium sorbate
Propionic acid
Propyl gallate
Propylparaben
Sodium ascorbate
Sodium benzoate
Sodium bisulfite
Sodium metabisulfite
Sodium propionate
Sodium sorbate
Sodium sulfite
Sorbic acid
Stannous chloride
Sulfur dioxide
Thiodipropionic acid
Tocopherols

EMULSIFYING AGENTS
Cholic acid
Desoxycholic acid
Diacetyl tartaric acid esters of mono- and diglycerides
Glycocholic acid
Mono- and diglycerides
Monosodium phosphate derivatives of above
Propylene glycol
Ox bile extract
Taurocholic acid

NUTRIENTS AND DIETARY SUPPLEMENTS
Alanine
Arginine
Ascorbic acid
Aspartic acid
Biotin
Calcium carbonate
Calcium citrate
Calcium glycerophosphate
Calcium oxide
Calcium pantothenate
Calcium phosphate
Calcium pyrophosphate
Calcium sulfate
Carotene
Choline bitartrate
Choline chloride
Chopper gluconate
Cuprous iodide
Cysteine
Cystine
Ferric phosphate
Ferric pyrophosphate
Ferric sodium pyrophosphate
Ferrous gluconate
Ferrous lactate
Ferrous sulfate
Glycine
Histidine
Inositol
Iron, reduced
Isoleucine
Leucine
Linoleic acid
Lysine
Magnesium oxide
Magnesium phosphate
Magnesium sulfate
Manganese chloride
Manganese citrate
Manganese gluconate
Manganese glycerophosphate
Manganese hypophosphite
Manganese sulfate
Manganous oxide
Mannitol
Methionine
Methionine hydroxy analogue
Niacin
Niacinamide
D-pantothenyl alcohol
Phenylalanine
Potassium chloride
Potassium glycerophosphate
Potassium iodide
Proline
Pyridoxine hydrochloride
Riboflavin
Riboflavin-5-phosphate
Serine
Sodium pantothenate
Sodium phosphate
Sorbitol
Thiamine hydrochloride
Thiamine mononitrate
Threonine
Tocopherols
Tocopherol acetate
Tryptophane
Tyrosine
Valine
Vitamin A
Vitamin A acetate
Vitamin A palmitate
Vitamin B_{12}
Vitamin D_2
Vitamin D_3
Zinc sulfate
Zinc gluconate
Zinc chloride
Zinc oxide
Zinc stearate

SEQUESTRANTS
Calcium acetate
Calcium chloride
Calcium citrate
Calcium diacetate
Calcium gluconate
Calcium hexametaphosphate
Calcium phosphate monobasic
Calcium phytate
Citric acid
Dipotassium phosphate
Disodium phosphate
Isopropyl citrate
Monoisopropyl citrate
Potassium citrate
Sodium acid phosphate
Sodium citrate
Sodium diacetate
Sodium gluconate
Sodium hexametaphosphate
Sodium metaphosphate
Sodium phosphate
Sodium potassium tartrate
Sodium pyrophosphate
Sodium pyrophosphate, tetra
Sodium thiosulfate
Sodium tripolyphosphate
Stearyl citrate
Tartaric acid

STABILIZERS
Acacia (gum arabic)
Agar-agar
Ammonium alginate
Calcium alginate
Carob bean gum
Chondrus extract
Ghatti gum
Guar gum
Potassium alginate
Sodium alginate
Sterculia (or karaya) gum
Tragacanth

MISCELLANEOUS ADDITIVES
Acetic acid
Adipic acid
Aluminum ammonium sulfate
Aluminum potassium sulfate
Aluminum sodium sulfate
Aluminum sulfate
Ammonium bicarbonate
Ammonium carbonate
Ammonium hydroxide
Ammonium phosphate
Ammonium sulfate
Beeswax
Bentonite
Butane
Caffeine
Calcium carbonate
Calcium chloride
Calcium citrate
Calcium gluconate
Calcium hydroxide
Calcium lactate
Calcium oxide
Calcium phosphate
Caramel
Carbon dioxide
Carnauba wax
Citric acid
Dextrans
Ethyl formate
Glutamic acid
Glutamic acid hydrochloride
Glycerin
Glyceryl monostearate
Helium
Hydrochloric acid
Hydrogen peroxide
Lactic acid
Lecithin
Magnesium carbonate
Magnesium hydroxide
Magnesium oxide
Magnesium stearate
Malic acid
Methylcellulose
Monoammonium glutamate
Monopotassium glutamate
Nitrogen
Nitrous oxide

TABLE 1 (*cont.*)

Papain	Sodium carbonate	Triethyl citrate	Ethyl acetate
Phosphoric acid	Sodium citrate		Ethyl butyrate
Potassium acid tartrate	Sodium carboxymethyl-cellulose		Ethyl vanillin
Potassium bicarbonate		SYNTHETIC FLAVORING SUBSTANCES	Eugenol
Potassium carbonate	Sodium caseinate		Geraniol
Potassium citrate	Sodium citrate	Acetaldehyde	Geranyl acetate
Potassium hydroxide	Sodium hydroxide	Acetoin	Glycerol tributyrate
Potassium sulfate	Sodium pectinate	Aconitic acid	Limonene
Propane	Sodium phosphate	Anethole	Linalool
Propylene glycol	Sodium potassium tartrate	Benzaldehyde	Linalyl acetate
Rennet	Sodium sesquicarbonate	*N*-butyric acid	1-malic acid
Silica aerogel	Sodium tripolyphosphate	*d*- or *l*-carvone	Methyl anthranilate
Sodium acetate	Succinic acid	Cinnamaldehyde	3-Methyl-3-phenyl glycidic acid ethyl ester
Sodium acid pyrophosphate	Sulfuric acid	Citral	
Sodium aluminum phosphate	Tartaric acid	Decanal	Piperonal
Sodium bicarbonate	Triacetin	Diacetyl	Vanillin

Of special interest is the use of sulfur dioxide for inhibition of mold and discolouration in wine. Sulfur dioxide is, of course, a prominent air pollutant. There is anxiety that the acceptable daily intake of sulfur dioxide will be exceeded in countries where wine intake is heavy. Experimental evidence indicates that sulfur dioxide inhibits the growth rate, probably due to destruction of Vitamin B_1 by sulfite.

Antioxidants, added to prevent rancidity in fatty foods, can be put in the classification of preservatives. A second use for antioxidants is developing with the growing practice of display of food in transparent containers and wrapping. Attendant exposure to light causes discolouration, not necessarily detrimental to nutritional value in itself, but detracting from attractiveness. The most commonly used antioxidants are butylated hydroxyanisole, butylated hydroxytoluene, propyl, octyl and dodecyl gallates and tocopherols. Antioxidant effect of a substance can often be increased by addition of a second material, producing a synergistic reaction and allowing more effective product control.

Texture agents include emulsifiers, stabilizers and thickeners. These are the largest single class in terms of total quantity consumed in food. Use of these agents has contributed greatly to the development of the new convenience foods.

The miscellaneous group includes acids, alkalies, buffers, neutralizing agents, anti-glazers, release agents, anti-caking materials, clarifying agents and foaming agents. All of these substances are primarily manufacturing and processing aids. Indeed, without many of these materials the range of modern foods would be much more limited.

Testing of food additives, as with most other chemicals, is not usually done with human subjects. Exposure studies of both limited duration and long term are made on mice, rats, dogs and other laboratory animals. When deviation from normal response is noted, the largest dosage that does not produce the change is used as the base for acceptable intake for humans. Normally, this dosage is reduced by a factor of about 100 in most countries. The allowable, or acceptable, dosage is expressed as milligrams of the substance per kilogram of receptor body weight. It must be emphasized that extrapolation from animals to humans is always a difficult undertaking.

International guidelines have been issued by the World Health Organization and the Joint Expert Committee on Food Additives of the Food and Agriculture Organization. The Codex Alimentarius Commission of the Food and Agriculture Orgnization has published six principles concerning the use of food additives.

1) The use of an additive is justified only when it has the purpose of maintaining a food's nutritional quality, enhancing keeping quality or stability, making the food attractive, providing aid in processing, packing, transporting or storing food or providing essential components for food in special diets. An additive is not justified if the proposed level of use constitutes a hazard to the consumer's health, if the additive causes a substantial reduction in the nutritive value of the food, if it disguises faulty quality or the use of processing and handling

techniques which are not allowed, if it deceives the customer or if the desired effect can be achieved by other manufacturing processes that are economically and technologically satisfactory.

2) The amount of additive should not exceed the level reasonably required to attain the desired effect under good manufacturing conditions.

3) Additives should conform to an approved standard of purity.

4) All additives in use, or proposed, should be subjected to adequate toxicological evaluation and permitted additives should be kept under observation for deleterious effects.

5) Approval of additives should be limited to specific foods for specific purposes and under specific conditions.

6) Use of additives in foods consumed mainly by special groups within the community should be determined by the food intake of that group.

Toxicity is the general term applied to adverse biological effects in man resulting from pollutants. Effects may range from lessened health levels to death. Chronic toxicity is of major interest here. Included under effects of chronic toxicity are carcinogenicity, teratogenicity and mutagenicity. It has been suggested that some behavioral disorders are seated in the effect of toxic substances. This area has not been extensively investigated.

One or more of these types of toxicity may be induced by pollutants in the environment. Pollutants may cause adverse effects alone or may interact synergistically with ordinarily harmless substances to give unexpected results. The mechanism of smog formation is an excellent example, although man is but indirectly affected.

It has been estimated that, over the expected life span of one malformed child, the total costs for care are in the neighborhood of a quarter of a million dollars. This figure, which does not include loss of earnings, is a high price for society to pay.

Widespread, long term genetic effects due to environmental pollution cannot be accurately predicted but the cost is certainly great. Such cost estimates have not, until recently, been applied in matching benefits to cost in evaluation of a potentially hazardous substance. Hazards from a particular substance need not be accepted when another substance of equal worth is available. Mandatory testing is usually looked upon as an unnecessarily bothersome expense by producers of synthetic substances. In fairness, however, it must be said that some chemicals, in use for years, have been unexpectedly indicted as potential hazards and the economic loss has been heavy. Cyclamates are a good example. It is of interest to speculate if such a situation could have been avoided. The basic question concerns the adequacy of existing legislation and difficulty of implementing regulations.

It has become increasingly clear that many diseases formerly regarded as spontaneous are caused by environmental pollutants. These diseases include cancer, birth defects and mutations. The problem is compounded by greatly increased exposure of the population to new synthetic chemicals and their degradation products. Environmental effects of these substances are not usually adequately evaluated.

There is now overwhelming evidence that many human cancers are due to carcinogenic substances in the environment. These are, then, preventable. Studies of epidemiological factors have indicated strongly local environmental factors are of significance in cancer incidence.

Lung cancer has been linked with tobacco smoking. Air pollution is a contributory factor. Ayres[14] describes the cigarette as a private air pollution source. Cancer of the oral cavity in Asia is linked with chewing of tobacco leaves and betel nuts. High incidence of gastric cancer in Chile, Iceland and Japan has been associated epidemiologically with a diet high in fish. It has been suggested that nitrosamines, formed by reaction between nitrites and secondary amines of the fish, may be a significant factor. Nitrites are employed as a preservative of fish. Dietary contamination with aflatoxin is thought to be a causitive effect of liver cancer in the Bantu. Aflatoxin is a fungal carcinogen. Ingestion of azoglucoside carcinogens with Cycad plants is probably the chief cause of liver cancer in Guam. Cancer of the esophagus in Zambians may be related to high nitrosamine content of Kachasu spirits.

There is well documented evidence of occupationally related cancers. These include bladder cancer in the rubber and aniline dye industries, lung cancer in uranium miners, nasal sinus cancer in wood workers, skin cancer in shale oil workers and lung cancer in asbestos workers. Environmental carcinogens can be classed into two categories, potent and weak. Potent carcinogens, such as nitrosamines and aflatoxins, can induce cancers in animals in very low concentrations. Isolation of these substances in food has given rise to endeavors to relate food distribution patterns with local cancer incidence.

Weak carcinogens, such as atmospheric pollutants, some pesticides and food additives, have effects much more difficult to evaluate and thus may pose as great a threat as the potent carcinogens because they are less likely to be recognized as a significant epidemiological factor. Oftern, evaluation must be by indirect means. A causal relationship has been demonstrated to exist between urban air pollution and lung cancer. Lung cancer mortality patterns differ in various sections of the United States and Great Britain. It has been clearly shown that increased mortality due to lung cancer is related to increased urbanization. A survey in the late 1950's found lung cancer rates in the United States to be 39 per one hundred thousand in rural areas and 52 per one hundred thousand in urban areas. Similar data were gathered in Great Britain and the added effect of smoking was evaluated. The conclusion of significance of urban air pollution is inescapable.

An effect of SO_2 and particulate matter pollution on lung function has been described by Chiaramonte, *et al.*[15]

The Long Island College Hospital was well located for the purposes of this study as shown in Fig. 1. Its position was downwind of the Consolidated Edison stacks on the upper East River during November 21 and 22, and on November 23 and 24 the area received atmospheric pollution from Manhattan and New Jersey. Although New Jersey is west of Manhattan, industrial pollutants emitted at high elevations and trapped in the inversion layer are likely to overpass part of Manhattan and come to the ground in Brooklyn.[15] That this actually happened can be presumed from the significantly higher sulfur dioxide levels measured by the Brooklyn Polytechnic Institute as compared with those measured in Harlem (Fig. 1).*

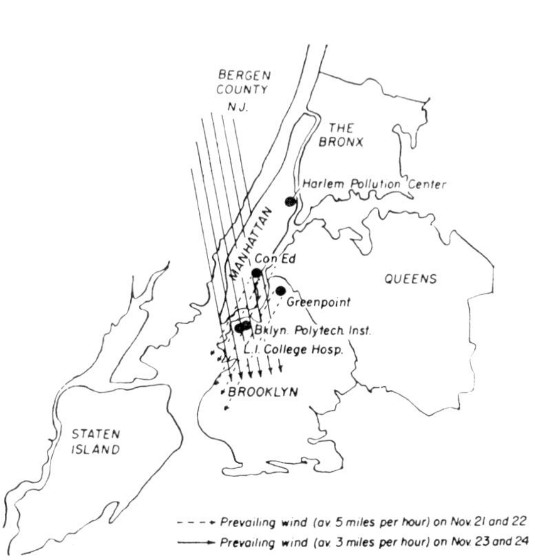

Figure 1 Arrows on map show how direction of winds on November 21 and 22 and November 23 and 24 contributed to movement of atmospheric industrial pollution from Manhattan and New Jersey to vicinity of Long Island College Hospital.

Results

The number of patients visiting the emergency room and the daily sulfur dioxide levels for Manhattan and Brooklyn during the three-week period are shown in Figure 2.

There was no significant difference in the total number of emergency room visits for each of the three weeks. The figures for each day were also approximately equivalent. However, there was a marked rise in the number of patients with respiratory symptoms during the week of the air pollution episode. This increase, when compared with the two control weeks, is statistically significant despite slight rises in the sulfur dioxide during the control weeks ($p = <0.005$).

The increase in respiratory symptoms was most marked on November 26 and 27, suggesting a seventy-two-hour lag between the peak sulfur dioxide level and the increased appearance of symptoms. This fact may reflect the patient's reluctance to seek medical attention over the Thanksgiving holiday. However, such a latent period was also noted by Glasser, Greenburg, and Field in their study of adults during the same pollution episode.[22]

We compared November 26 and 27 with the other days in the control periods and pollution week. The proportion of respiratory patients on those two days was significantly higher than would be expected based on the proportions for the other days ($p = <0.001$ for November 26 and $p = <0.02$ for November 27). During those two days, there was also an increase in the number of visiting children whose respiratory symptoms were

---

* Sulfur dioxide levels for Manhattan in this study were recorded by the New York City Department of Air Pollution Control (now the Department of Air Resources) by their 121st Street station in Harlem and for Brooklyn by the Polytechnic Institute of Brooklyn near Long Island College Hospital.

characterized mainly by expiratory obstruction. Again, the proportions of patients with expiratory wheezing on November 26 and 27 were significantly greater than the proportions of patients with these symptoms on the other days ($p = <0.001$ and <0.02 respectively), as seen in Fig. 2.

Judged on the basis of the various diagnostic criteria used in the allergy evaluations most of the 40 children who completed the evaluations were considered to have allergic constitutions. At least two factors may account for this: (1) asthmatic children tend to have greater difficulty during October and November[14,15] and (2) letters that were sent to recall the patients for evaluation mentioned an interest in allergy and air pollution. Hence, the subjects who returned may represent a biased sampling. Nonetheless, 8 of the 9 children who returned of the first week's group were found to have allergic constitutions, as were 15 of the 18 from the second week and 8 of the 13 from the third week.

Although the percentage of allergic children among those tested with repiratory difficulty was high for each of the three weeks, the total number of those with respiratory difficulty who were found to be allergic was greater for the week of pollution.

One striking difference between the groups from the pollution and control weeks showed up in the analysis of the individual parameters for allergy, specifically the skin test for molds (Table 11). Nineteen (86 per cent) of the 22 children evaluated from the two control weeks showed skin reactivity

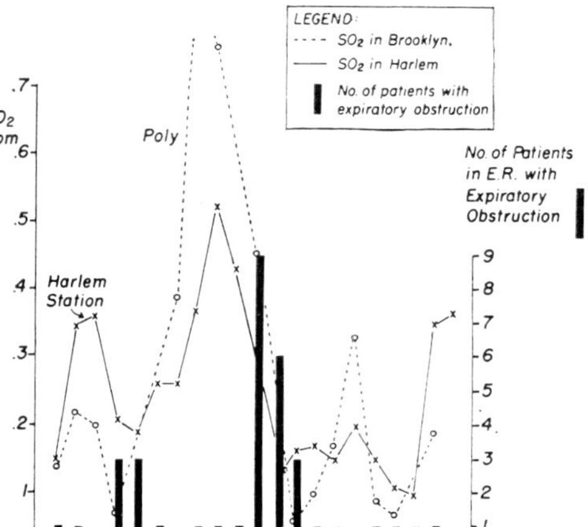

Figure 2 Schematic representation of daily sulfur dioxide levels as recorded in Brooklyn and Harlem from November 16 to December 6 and number of patients visiting emergency room on those days with respiratory symptoms characterized mainly by expiratory obstruction.

to molds in contrast to 8 (44 per cent) of the 18 from the pollution week.

One may hypothesize from these findings that during the week of high pollution, the number of children with positive mold test results were relatively low as compared with the number of those asthmatic patients having difficulty when there was no pollution. Hence, it is reasonable to assume that the high incidence and severity of symptoms during this time occurred in allergic individuals who were reacting to the pollution. More exhaustive studies are necessary to determine the specific noxious agents and, perhaps, the constitutional predisposition of these children.

All of the children with allergic constitutions had histories of respiratory difficulty from asthma, allergic rhinitis, or other continuing nasal symptoms. The frequency of asthma in relation to the age of a populational is bimodal, with peaks under sixteen years and over forty years of age. This study suggests a relationship between the occurrence of severe air pollution and an increase of obstructive respiratory symptoms in children under sixteen years of age. This relationship seems similar to that generally observed in people over forty, despite the absence in children of factors such as smoking and so-called chronic irreversible pulmonary disease.

Summary

"High levels of atmospheric sulfur dioxide occurred in the vicinity of Long Island College Hospital over several days of this study. During a three-week period, including the week preceding and the week following that of the pollution episode, we examined 83 children with respiratory symptoms out of 420 who visited the pediatric emergency room. Forty of the 83 children completed allergy evaluations. Most had allergic constitutions, but those who visited during the period of pollution showed a lower skin reactivity to molds. The study revealed a statistically significant rise in obstructive symptoms during the several days of high sulfur dioxide levels as compared with the days preceding and following the pollution episode ($p = <0.005$). This rise was more evident on two consecutive days seventy-two hours following the peak sulfur dioxide level ($p = <0.001$ and <0.02)."

The first demonstration that environmental pollutants can cause genetic damage came with the discovery that high energy radiation induces mutations. Later, the development of the nuclear industry focussed increased attention on the dangers of unwanted genetic change. Associated with possible genetic damage in this industry was the danger of radiation induced carcinoma. Accordingly, safeguards were developed and limits were set on radiation levels. These were functions of exposure time. Limits were changed periodically in the light of fresh evidence. It is of interest to note that changes were usually downward. Radiation from X-Ray machines and fluoroscopes was treated in the same way. A widely available hazard was the fluoroscope for fitting of shoes. It is impossible to determine possible genetic damage from this source but it was probably considerable. Such installations have all but disappeared. Medical and dental radiation sources are firmly regulated. There was suspicion that some chemicals might induce mutations but only about thirty years ago was the mutagenic effect of mustard gas on fruit flies shown. There is concern in some circles that strongly mutagenic chemicals, their effects not yet recognized, may already be in wide use.

A mutation can be a chemical transformation of a single gene or a rearrangement of a chromosome. The former, called a gene or point mutation, can cause an alteration in function. The latter may be microscopically visible and is known as chromosome aberration. In studies of human subjects the aforementioned changes are not always easily determined.

When genetic function of a cell is changed while reproductive capacity is unaltered the genetic change is transmitted to descendant cells. Mutations in germ cells are most serious in long term effects because changes are transmitted to future generations.

Mutagenic effects are of many types. These range from lethal effects to changes so slight as to remain unnoticed. It is thought that many inherited diseases are based in mutagenic effects. At this time there is no known way to adequately evaluate long term mutagenic effects. It is a cause for serious concern for future generations. Genetic effects over a long term are statistical rather than discrete. As natural selection has had its effect over a long period, so will externally induced mutations make their effect felt over a long period. Dominant mutations appear in the next generation while recessive mutations require contributions from both parents. This characteristic may not appear for many generations.

Most mutations are harmful or neutral. Many mutants which in the past would have produced death or lessened fertility now remain. This is due to higher standards of health care. The former equilibrium, in which old mutations disappeared about as rapidly as new mutants appeared, has been upset in favor of new mutants. Natural selection does not apply as strongly as before. It ha been suggested that medical problems of the future will be more and more due to genetic origin. One estimate places our present health burden as being 25% of genetic nature.

If a mutation causes a gene to have a sterilizing or lethal effect, only one generation is involved. Less severe effects, however, may involve many generations. The less severe the effect, the more people will be exposed. Mild mutational effects, affecting many people, will have a greater public health impact than one severe, or fatal, case. Unfortunately, many of these milder effects may be difficult to detect.

Many chemical mutagens in the environment pose threats which have not yet been adequately evaluated. Mammalian test systems are available but extrapolation from smaller animals to man is indeed difficult and liable to error.

Trimethylphosphate, until recently, was added to gasoline for control of surface ignition and spark plug fouling. This substance has been implicated in chromosome damage in rats. The dosage in rats at which damage occurred was at subtoxic levels. However, extrapolation of dosage levels to man is difficult and man's exposure to the substance in the environment is almost impossible to estimate.

Congenital malformations are those abnormalities which can be recognized at birth or shortly thereafter. The study of these abnormalities, which can cause serious disability or death, is known as teratology. A broader definition can include microscopial, biochemical and functional abnormalities of prenatal origin. It is estimated that about 3% of live births involve congenital malformations, but this is only an estimate. Lack of adequate data at the national level makes such figures rather rough and probably inaccurate.

Three categories of human teratogens have been identified. These are: (1) viral infections, (2) ionizing radiation and (3) chemicals. Thalidomide is the best known of the third category. Experimental evidence of teratogenicity of some compounds had been in existence for many years but requirements of three generation reproductive tests in animals were established only after the sensational thalidomide disaster of 1962. Some questions concerning the validity of such tests have been raised.

Pollutant materials to which humans are exposed must be examined for toxicity. These examinations must also include specific tests for carcinogenicity, mutagenicity and teratogenicity. These effects have been studied in the past by separate disciplines and there has been inadequate interchange of results and ideas. It is to be hoped that the emerging profession of environmental scientist will aid in breaking down these historical and somewhat artificial barriers among disciplines.

There is need for more sensitive and reliable methods of testing effects of single chemicals, degradation products and mixtures. External environmental effects must also be included as system parameters in evaluation of pollutant effects. It is possible that human tests may become necessary, but moral and legal considerations pose serious questions.

Agents under test must be administered at subtoxic, toxic and chronic levels. Effects of possibly significant but normally non-toxic agents on the system containing the toxic agent under investigation must also be evaluated.

Testing procedures must be realistic and reflect the pathways of human exposure. However, the exception to this rule is in sensitivity. Evaluation of carcinogenic effect may require administration to test animals by different means. Normal ingestion in humans might be by inhalation while adequate dosage in rats might require addition of the substance in question to daily food. This is particularly true when dealing with weak environmental carcinogens.

Metabolic compatability between a test species and humans is important for extrapolation of test results. There is seldom a one to one correspondence and most authroties feel that at least two test species should be examined. There is much testing data on rodents, pigs and subhuman primates. Thus, further utilization of these animals is indicated. In special cases, however, a less common species may be necessary in order to gain the desired data.

It is necessary to test at levels much higher than human exposure for carcinogenic, teratogenic and mutagenic substances. It must be noted that, even when large numbers of mice are tested, the number is still small compared to the millions of humans that might be exposed in the everyday environment. Not all humans exposed would respond, in any event, to the dangerous substance. As an example, meclizine, an antihistamine used for treatment of morning sickness, has been found to be teratogenic in the rat but not so in a relatively small number of women tested. The question as to effects on a larger population is unanswered. For thalidomide, humans are found to be 60 times more sensitive than mice, 100 times more sensitive than rats, 200 times more sensitive than dogs and 700 times more sensitive than hamsters. It is obvious, theraefore, that predictions as to teratogenic doses of thalidomide on the basis of animal testing would mean very high and dangerous exposure levels.

Complicating the testing picture is the increased effect of substances when a second, supposedly innocuous, substance greatly increases the undesirable end effect. Such reactions are the basis of the Delaney Amendment of 1968 to the Food, Drug and Cosmetic Act (U.S.). The Amendment requires that no additive will be considered safe if found, under appropriate testing conditions, to induce cancer in man or animal. The question of appropriate testing is a difficult one. It is felt that zero tolerance should also be imposed on occupational carcinogens, such as asbestos.

Often, non-carcinogenic alternatives are available as replacements for questionable materials. There is considerable controversy about the pesticide DDT. Some authorities question the continued utility of DDT for control of cotton insects due to development of resistant strains. This controversy will not soon die down. Cyclamates, banned in recent years, were an intentional food additive and of no value nutritionally. Only after their widespread use for a number of years was there shown a carcinogenic danger.

The argument of economic loss to special interest groups is of no validity. In view of the uncertainty of long term effects of carcinogenic materials, zero tolerance levels must be imposed for these substances.

It is important that chemicals, and their derivatives, suspected or implicated as toxic, carcinogenic, teratogenic and mutagenic be detected and monitored in the environment. Epidemiological studies may show effects but it is necessary to quantitatively establish occurrence of these substances. It has been possible to demonstrate the relationship between cigarette smoking and lung cancer although some special interest groups seriously disputed this. However, in this case there were two dissimilar populations, smokers and non-smokers. In the case of exposure of the general population to a suspect substance, evaluation and epidemiological treatment of data is more difficult. Apart from rubella virus, no known teratogens, such as ionizing radiation, mercury, etc. have been positively identified epidemiologically in the highly industrialized countries. This indicates the great need for better and more comprehensive gathering of data relating to birth defects. It is to be hoped that environmental effects should be segregated from natural and spontaneous mutation effects. It would be necessary to monitor special indicator traits. Such traits would have to be chosen with great care and followed for a considerable period.

Recently it has been suggested that mutation rates could be monitored by means of data on spontaneous abortions. Teratogens may cause chromasome aberrations but these act after fertilization and would be against a normal background.

The question of data on occupational hazards, long the major field of endeavor of the industrial hygienist, must be raised in connection with legislation thought to be inadequate and outdated. The correctness of much of the information on which maximum allowable concentrations are based is in doubt at this time. Probably a critical reexamination will soon come about as a result of increased environmental awareness.

Present toxicological techniques are not sufficiently sensitive to adequately monitor many weak carcinogens. Compounding the problem is the difficulty of recognizing effects of many substances in the general population. Special situations as to sample population at times help to simplify the problem, such as cigarette smoking. But there is certainly a great area in which problems are strongly suspected to exist but statistical validation is not possible at this time. Yet, tolerance and allowable concentration limits must be set. There is a most fruitful area for research here.

Regulatory agencies are often understaffed and underfunded. Attention of existing resources is usually allocated on the "squeaky wheel" principle. They move from current crisis to current crisis and are frequently unable to devote enough time to prevention of new crisis. This is contrary to good public health practice.

Legislation relating to environmental hazards is produced for many reasons but is relatively uncoordinated. The practice of politicians creating great amounts of noise, with attendant publicity, about a specific sensational, but relatively minor problem is all too common. The public is not served. News media would do well to be more selective in their coverage. Rather, more effort and money should be spent in updating legislation as new facts come to light and in streamlining and an reinforcing regulatory machinery. It appears that a step has been taken in the right direction by the United States with formation of the Environmental Protection Agency. There are disquieting signs, unfortunately, that politics may have entered rather heavily in administration of various programs under the agency's jurisdiction. The trend toward establishment of separate Agencies for Environmental Protection at all governmental levels seems to be strong. But, it remains to be seen how much effort will be spent on prevention of future problems as compared to solution of current problems. It is to be hoped that a proper balance will be achieved.

REFERENCES

1. Roueche, Berton, *Eleven Blue Men*. Little, Brown and Co., New York, 1953.
2. Brandt, A. D., *Industrial Health Engineering*. John Wiley and Sons, Inc., New York, 1947.
3. Barth, D. S., *et al., J.A.P.C.A.* **21**, 544 (1971).
4. Perlman, M. E., *et al., Pediatrics* **47**, 391 (1971).
5. Heuss, *et al., J.A.P.C.A.* **21**, 541 (1971).
6. Maxcy, K., Rosenau, *Preventive Medicine and Public Health*. 8th Ed., Appleton-Century-Crofts, Inc., 1956.
7. Faith, W. L., *Air Pollution Control*. John Wiley and Sons, Inc., New York, 1959.
8. Oberle, M. W., *Science* **165**, 991 (1969).
9. Fair, G. M., and J. C. Geyer, *Water. Supply and Waste Water Disposal*. John Wiley and Sons, Inc., New York, 1954.
10. Hopkins, E. S., and W. Schulze, *The Practice of Sanitation*. Williams and Wilkins, Baltimore, Maryland, 1954.
11. Faith, W. L., Effects of Atmospheric Pollutants. *The Encyclopedia of Environmental Science and Engineering*. Vol. 1. Gordon and Breach, New York, 1976.
12. Epstein, S. S., *Nature* **228**, 816 (1969).
13. Gauvain, S. and M. Harrington., Occupational Health. *The Encyclopedia of Environmental Science and Engineering*. Vol. 2. Gordon and Breach, New York, 1976.

14. Aynes, S. M., Conference on Metropolitan Air Pollution. Wagner College, New York, 1967.
15. Chiaramonte, *et al.*, *N.Y.S. J. of Medicine* p. 394 (Feb. 1, 1970).
16. Junge, C. E., *Air Chemistry and Radioactivity*. Academic Press, New York, 1963.
17. Kendall, D. M., A Summary of Panel Recommendations: Report of a Panel on Food Safety to the White House Conference on Food, Nutrition and Health. November, 1969.
18. Solvato, J., Environmental Health. *The Encyclopedia of Environmental Science and Engineering*. Vol. 1. Gordon and Breach, Inc., New York, 1973.
19. Stacks, P., *British Journal of Cancer* **14**, 397 (1960).
20. Federal Register **36**, 8186 (1971).
21. Kermode, G. O., *Scientific American* **3**, 15 (1972).
22. Glasser, M., L. Greenburg, and F. Field, *Arch. Environ. Health* **15**, 684 (1967).

Additiona References on General Gaseous Pollutaants

Air Quality' Criteria for Carbon Monoxide, National Air Pollution Control Administration Publication No. AP-62, March, 1970.

Air Quality Criteria for Nitrogen Oxides, Air Pollution Control Office Publication No. AP-84, January, 1971.

Air Quality Criteria for Hydrocarbons, National Air Pollution Control Administration Publication No. AP-64, March, 1970.

Air Quality Criteria for Photochemical Oxidants, National Air Pollution Control Administration Publication No. AP-63, March, 1970.

Carbon Monoxide

Beard, R. R., and G. A. Wertheim, *Amer. J. Public Health*, **57**, 2012 (1967).

"Effects of Chronic Exposure to Low Levels of Carbon Monoxide on Human Health, Behavior, and Performance," report of National Academy of Sciences and National Academy of Engineering, Washington, D.C., 1969.

(a) Stewart, R. D., J. E. Peterson, E. D. Baretta, R.T. Bachand, M. J. Hasko, and A. A. Herrmann, *Arch. Environ. Health* **21**, 154 (1970); and (b) Stewart, R. D., letter to Environmental Protection Agency commenting on proposed carbon monoxide air quality standards, March 5, 1971.

(a) Mikulka, P., R. O'Donnell, P. Heinig, and J. Theodore, *Toxicol. Appl. Pharmacol.* **18**, 593 (1971); and (b) Theodore, J., R. O'Donnell, and K. C. Back, *J. Occupational Med.* **13**, 242 (1971).

Report to the State of California Air Resources Board by the Technical Advisory Committee, September 30, 1970.

Nitrogen Dioxide

Shy, C. M., J. P. Creason, M. E. Pearlman, K. E. McClain, F. B. Benson, and M. M. Young, *J. Air Poll. Control Ass.* **20**, 539 (1970).

Shy, C. M., J. P. Creason, M. E. Pearlman, K. E. McClain, F. B. Benson, and M. M. Young, *Ibid.* **20**, 582 (1970).

Pearlman, M. E., J. F. Finklea, J. P. Creason, C. M. Shy, M. M. Young, and R. J. Horton, *Pediatrics* **47**, 391 (1971).

Morgan, G. B., C. Golden, and E. C. Tabor, *J. Air Poll. Control Ass.* **17**, 300 (1967).

Purdue, L. J., J. E. Dudley, J. B. Clements, and R. J. Thompson, *Environ. Sci. Technol.*, in press.

Jacobs, M. B., and S. Hochhesier, *Anal. Chem.* **30**, 426 (1958).

Christie, A. A., R. G. Lidzey, and D. W. F. Radford, *Ana-* **95**, 519 (1970).

Eckardt, R. E., *J. Air Poll. Control Ass.* **20**, 832 (1970).

Hydrocarbons

Glasson, W. A., and C. S. Tuesday, *Environ. Sci. Technol.* **4**, 37 (1970).

Dimitriades, B., U.S. Bureau of Mines, *Rept. Invest.* 7433 (1970).

Altshuller, A. P., S. L. Kopczynski, D. Wilson, W. Lonneman, and F. D. Sutterfield, *J. Air Poll. Control Ass.* **19**, 787 (1969).

System Development Corporation Technical Memorandum TM-(1)-4119, Volumes I through VI (1969).

Junge, C. E,, "Air Chemistry and Radioactivity," Academic Press, New York, 1963, p. 94.

Stephens, E. R., and F. R. Burleson, *J. Air Poll. Control Ass.* **19**, 929 (1969).

Cavanagh, L. A., C. F. Schadt, and E. Robinson, *Environ. Sci. Technol.* **3**, 251 (1969).

Went, F. W., *Tellus* **18**, 549 (1966).

Rasmussen, R. A., and F. W. Went, *Proc. Nat. Acad. Sci. U.S.* **53**, 215 (1965).

Rasmussen, R. A., *Environ. Sci. Technol.* **4**, 667 (1970).

Photochemical Oxidants

Schoettlin, C. E., and E. Landau, *Public Health Rept.* **76**, 545 (1961).

Wayne, W. S., P. F. Wehrle, and R. E. Carroll, *J. Amer. Med. Ass.* **199**, 901 (1967).

Koonitz, C. H., as reported by I. T. T. Higgins, and J. R. McCarroll in "Development of Air Quality Standards", A. Atkis son and R. S. Gaines, eds., Merrill Publ. Co., Columbus, Ohio, 1970, p. 90.

"Profile of Air Pollution in Los Angeles County," report published by the Air Pollution Control District, Los Angeles County, January, 1969.

Clean Air for California, California Dept. of Public Health, Berkeley, Initial Report, March, 1955; Second Report, March, 1956; Third Report, 1957.

Hechter, H. H., and J. R. Goldsmith, *Amer. J. Med. Sci.* **241**, 581 (1961).

JOSEPH R. JONES
Windsor, Ontario
EDWARD N. ZIEGLER
JAMES R. PFAFFLIN

ENERGY SOURCES—ALTERNATIVES

Abundant and dependable sources of energy are vital to the social and economic welfare of an industrial nation. Energy resources, including fossil and nuclear fuels as well as solar, tidal, and geothermal energy, may be recovered or captured and converted into other energy forms for a variety of household, commercial, transportation, and industrial applications that enrich the human environment. The economic benefits of energy utilization are best illustrated by the relationship between the per capita GNP of a nation and its per capita energy use shown in Fig. 1, which is based on United Nations data.[1] On the other hand, economic indicators such as GNP are not the sole measure of the quality-of-life and energy sources, particularly in the energy conversion step where resources are consumed to produce power, have been identified as major contributors to air and water pollution and to the despoilment of land; environmental factors that threaten the integrity of the biosphere.

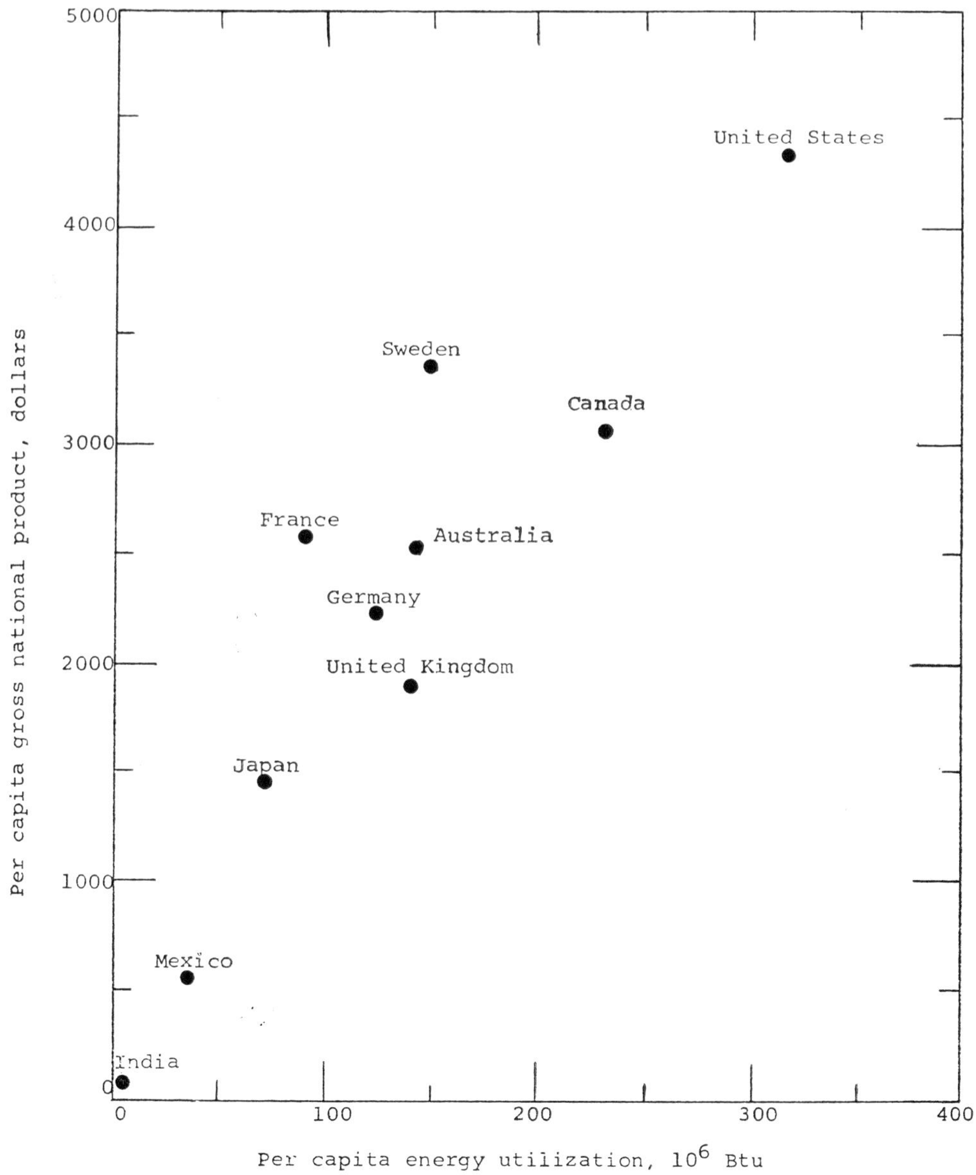

Figure 1 Relationship between per capita GNP and energy utilization for selected countries in 1968.[1,3]

Energy sources have effects on the environment that occur both naturally and as a result of man's efforts to exploit them. Natural effects range from the life sustaining action of solar radiation to the destructive manifestations of geothermal energy, volcanic eruptions, which cause extensive damage and spew particulate matter and gases into the atmosphere at a rate unrivaled by man-made processes. These and other natural phenomena are not now controllable and will not be considered further. Environmental effects encountered in the harnessing of various energy sources depend on the forms of energy in demand and on the resources and systems that are utilized to satisfy that demand. The impacts of this activity are controllable to varying degrees by the use of alternate fuels or resources, and by the proper design and application of the systems that are used to supply and utilize energy. Figure 2 is a flowsheet indicating all of the activities required to deliver a useful form of energy to the consumer. In mining and drilling to recover energy resources, the environmental problems encountered include the despoilment of land, safety of the workers involved, and water pollution. The strip mining of coal and uranium is especially damaging to land areas, and underground coal mining is notoriously hazardous. Acid mine drainage is a significant source of pollution of streams and waterways, and oil drilling in offshore areas carries the possibility of uncontrolled discharges of crude oil that are damaging to the marine environment. Each fuel or resource has unique potential for environmental damage in its shipment and storage and in refining and treatment operations where waste materials are generated and must be disposed of. The major fraction of emissions and wastes however, are generated in the energy conversion process. A distinction is made in this step between centralized energy conversion, where fuels and other resources are consumed and converted to

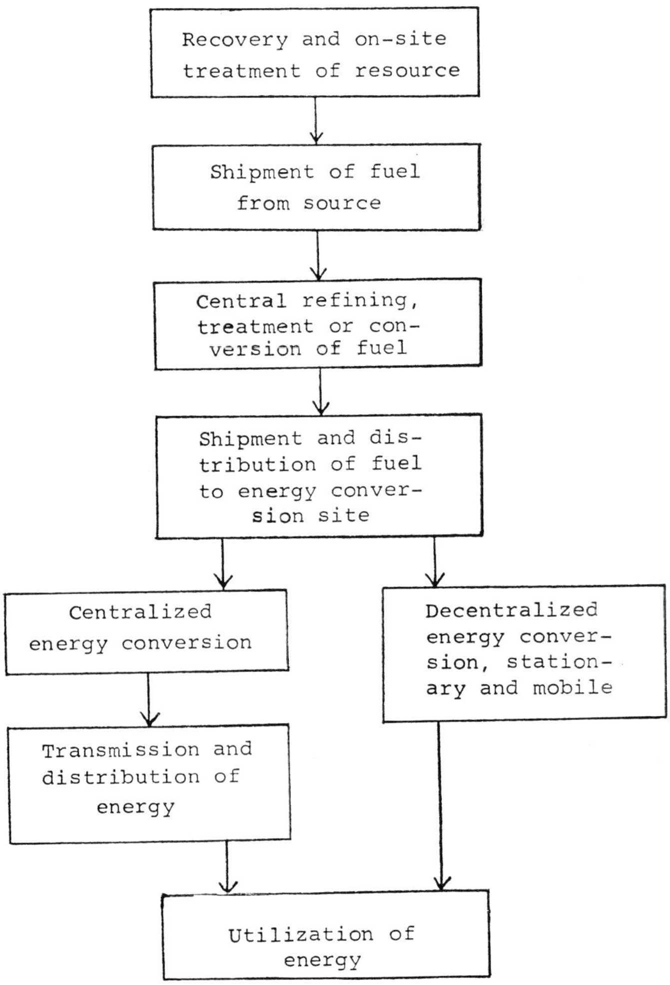

Figure 2 Energy system flowsheet.

other energy forms, usually electrical, for distribution and utilization, and a decentralized operation, where fuels are distributed to the location of the ultimate utilization of energy. Although the centralized conversion system generally has economies of scale which may facilitiate an increased investment in environmental protection systems, the concentrated nature of the activity aggravates emission and waste disposal problems. In certain cases, such as in the disposal of waste heat, dilution of emissions or discharges may be a valid control technique and is feasible with numerous distributed sources while it may not be as practical with large concentrated sources.

The release of wastes and pollutants from energy activities is at a level where significant external costs are being sustained in terms of hazards to health and damage to materials and property. From the economic viewpoint, these wastes and pollutants are residuals which have heretofore been disposed of to the environment by the producing activity with little or no economic penalty. The social and other external costs attributable to these residuals are external to the producing activity and do not affect decisions regarding the utilization and allocation of resources in that activity. Ayres and Kneese[2] have proposed a mathematical framework for tracing the flow of residuals and charging the external costs attributable to them to the producing activity. The materials flow diagram that they have developed is shown in Fig. 3 and addresses the energy conversion operation in relation to industrial activity in the production of goods. The inputs are energy resources and other raw materials which are converted by this industrial activity into final goods and residuals. The residuals, or waste products, identified with the energy conversion operation include carbon dioxide, carbon monoxide, sulfur dioxide, oxides of nitrogen, fly ash, slag, radioactive materials, and waste energy in the form of thermal energy and noise. Control of these residuals may be stimulated by economic measures such as taxation, that lead to the internalization of all of the costs imposed on society by their discharge, or by regulatory action. Regardless of how controls are stimulated, however, the quality and form of residuals and the cost of their control will depend ultimately on the technological alternatives that are available.

Because of the complexity of both the energy system and its impacts on the environment, it will be useful to view the subject from four reference

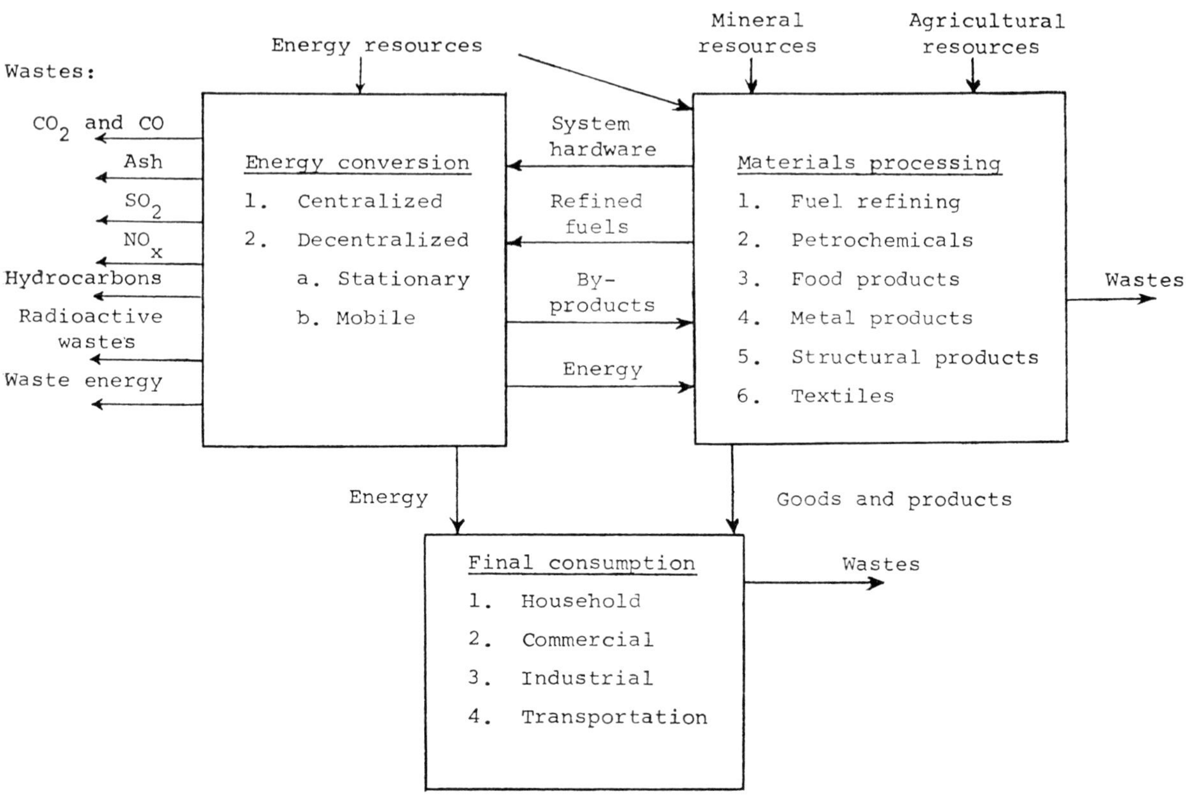

Figure 3 Materials flow diagram in economic activity.[2]

points: *ENERGY DEMAND*, *ENVIRONMENTAL FACTORS*, *ENERGY RESOURCES*, and *ENERGY CONVERSION*. Consideration of the existing and projected patterns of *ENERGY DEMAND*, and a review of all of the *ENVIRONMENTAL FACTORS* involved in satisfying this demand, will serve to define the extent of the conflict between energy and the environment as well as to indicate trends that can aggravate or help to resolve the problem. Measures that can be taken to insure an adequate supply of energy with minimum environmental damage are centered on the availability of alternative *ENERGY RESOURCES* and on *ENERGY CONVERSION* systems. In the short term, fuels may be treated, modified, or substituted, and emission controls can be provided in fuel handling and in energy conversion operations, while in the longer term the development of new resources and conversion systems may also be dictated. Both near- and long-term measures must be discussed to illustrate the range of selection available among alternatives. Statistics on energy demand, emissions, and other environmental factors are given for the year 1968 unless more recent reliable data are available. A variety of units are employed in the discussion of energy demand and resource availability including British thermal units (Btu) as a measure of energy; and tons of coal, barrels of oil, and cubic feet of gas as a measure of resource quantity. Although the energy content of some fuels can vary depending on their source, the following conversion factors may be used as a good approximation:

Coal	
Anthracite:	25.4×10^6 Btu/ton
Bituminous:	26.2×10^6 Btu/ton
Lignite:	14.0×10^6 Btu/ton
Oil	
Crude:	5.67×10^6 Btu/42 gal barrel (bbl)
Residual:	6.29×10^6 Btu/42 gal barrel (bbl)
Distillate fuel oil:	5.83×10^6 Btu/42 gal barrel (bbl)
Gasoline:	5.25×10^6 Btu/42 gal barrel (bbl)
Kerosine:	5.67×10^6 Btu/42 gal barrel (bbl)
Natural gas:	1032 Btu/ft^3 at STP
Fissionable material:	74.0×10^6 Btu/g U^{235} fissioned
Fusion fuel	
Deuterium–tritium reaction:	1.02×10^9 Btu/g D reacted
Deuterium–deuterium reaction:	2.25×10^8 Btu/g D reacted

Energy Demand

Energy is utilized in various forms by the demand sectors of the economy. Household and commercial consumers require thermal energy for comfort and for the preparation of food, and electrical energy for lighting and for the operation of electronic and motor-driven appliances that provide entertainment, communication, and labor-saving functions. Energy in mechanical form is needed for both public and private transportation. Industry has similar needs with added requirements involving thermal and chemical energy for materials processing, refining, and the manufacture of synthetic materials and consumer goods. The use of specific energy resources to satisfy these demands depends on their relative economics and convenience as well as on the systems that are available for conversion into the form of energy that is ultimately required.

In 1968 virtually all of the energy demand in the United States was satisfied by consumption of the fossil fuel resources; coal, oil, and natural gas. Small amounts of power were supplied by hydroelectric and nuclear power plants and even less by solar, geothermal, tidal, and other energy resources. Table I shows the direct consumption of energy resources by demand sector in the United States for the year 1968 with a projection to the year 2000. The consumption of energy resources in the production of electricity represents a centralized conversion, and the ultimate disposition of electrical energy to the various demand sectors is given in Table II, again for the year 1968, with a projection to the year 2000. Since the data presented represent resource consumption, they imply an efficiency of energy conversion and utilization. Comparing the fuel consumption and electrical output of utilities in 1968, an average efficiency of about 32% is indicated. Fuels burned for thermal energy in decentralized household, commercial, and industrial applications are used at efficiencies of 60% to 75% ,while fuels used for transportation are converted first to thermal and then to mechanical energy at an average over-all efficiency of about 25%. The source of the 1968 data is the Minerals Yearbook[3] while the projections are those of Morrison and Readling,[4] who developed a log-linear relationship between the major subcomponents of energy demand and several independent variables that are assumed to influence demand,

TABLE I

Total U.S. Energy Consumption by Fuel and Consuming Sector
(trillions of Btu's in direct fuels used)

		1968[3] (actual)	2000[4] (projected)
Household and commercial:			
Gas		6,451	19,066
Petroleum		6,581	2,000
Coal		568	
	Total	13,600	21,066
Industrial:			
Gas		9,258	17,504
Petroleum		4,474	13,090
Coal		5,616	2,000
	Total	19,348	32,594
Transportation:			
Gas		610	1,000
Petroleum		14,513	41,649
Coal		12	—
	Total	15,135	42,649
Electric utilities:			
Gas		3,245	4,128
Petroleum		1,181	861
Coal		7,130	18,720
Hydro (coal equiv.)		2,359	5,056
Nuclear		130	43,526
	Total	14,045	72,291
Miscellaneous:			
Gas		—	—
Petroleum		295	—
Coal		1	—
	Total	296	
Total consumption		62,424	168,600

TABLE II

Utility Electricity Distribution by Consuming Sector
(trillions of Btu's in electrical form)

		1968[3]	2000[4]
Household and commercial		2,469	20,045
Industrial		2,043	10,793
Transportation		18	100
	Total	4,530	30,938

such as; population, economic indicators, and technological developments. Coefficients of the independent variables were obtained by regression analysis using data for the period 1947 to 1965.

The year 2000 projection of a total energy resource demand of 168.6×10^{15} Btu is a medium range case, and assumes a normal rate of improvement in energy conversion efficiency and the continuation of present trends of competition and substitution in energy markets. Other projections were made by Morrison and Readling on the basis of different assumptions, including high- and low-range cases that reflect changes in energy demand patterns and technological developments anticipated in energy conversion systems. The high-range case forecasts a demand of 209.4×10^{15} Btu in the year 2000 and assumes the establishement of an all-electric economy, where electricity is generat ed in fossil, hydroelectric, and nuclear power plants at an average efficiency of 43% and is consumed in household, commercial, and industrial sectors at 10% efficiency and in transportation, with battery-driven vehicles, at an efficiency of 42%. The low-range forecast is 106.5×10^{15} Btu and assumes that energy inputs to the household, commercial, and industrial sectors are satisfied with electricity from fuel cells operating at an efficiency of 75% with 50% recovery of waste heat, and energy requirements in the transportation sector are also provided by fuel cells but at an efficiency of 50%. This again, is an all-electric economy but no central station power generation is needed under these circumstances. Natural gas was the fuel considered for all sectors in the low-range case, however, alternative fuels suitable for fuel cells could also be provided by hydrogen or methane produced by the gasification of coal, or by hydrogen made by the electrolysis of water. Of course such forecasts are highly speculative since it is not certain that fuel cells will be economical in such applications in the next 30 years. Numerous other forecasts of energy demand have been made and are summarized in a review[5] performed by Battelle-Northwest for the Energy Policy staff of the Office of Sciency and Technology (OST). All of these forecasts neglected supply restraints and thus do not reflect the effect of possible energy cost variations due to shortages of resources, increased requirements for environmental protection, and technological developments. None indicate any significant exploitation of solar, geothermal, or tidal energy resources.

The medium-range forecast for the year 2000, summarized in Table I indicates a shift in energy demand to electricity generated in centralized plants with its share of direct resource usage increasing from about 22% in 1968 to 43% in 2000. This shift is apparent in current trends with the demand for total energy growing at about 4% per year while the demand for electricity is growing at around 7% per year. Nuclear power is expected to account for over half of the electrical generation in 2000. Such changes in energy demand patterns are seen to have profound effects on the amount of resources consumed and, consequently, on environmental effects. For example, in the case of space heating which constitutes a growing market for electrical energy, fossil resources may be used directly for their thermal effect at an average efficiency about twice that involved in the conversion to electricity in central station power plants. This results in a doubling of the amount of fossil fuel that must be burned to provide a given amount of space heat to the consumer. Although this substitution would be expected to result in the production of more wastes and emissions, it may be justifiable on the basis of the economies of scale that are at work in the centralized operation, which permit better control of emissions, and the opportunity it provides to utilize the more abundant nuclear fuels which cannot be applied economically in small decentralized conversion systems.

The demand indicated in Table I for petroleum in 1968 in the various sectors requires more detailed attention since this energy resource is consumed in a variety of product forms. The total demand for petroleum products was 4.9×10^9 barrels and Table III shows how this total breaks down among the various products. The end use of most of these is self-evident and it may be seen that less than 2% of the product was consumed as petrochemical feedstock. Diesel fuel for mobile and stationary power plants is included along with the lighter fuel oils in the distillate fuel oil category. Naphtha-type jet fuel is generally used in military aircraft while the kerosine-type fuel is used in civil aircraft.

TABLE III

Domestic Demand for Crude Oil Products, 1968[3]
(millions of barrels)

Gasoline—motor gasoline	1,925
aviation gasoline	30
Jet fuel—naphtha type	127
kerosine type	222
Special naphthas	27
Kerosine	103
Distillate fuel oil	863
Residual fuel oil	680
Lubricants, wax, coke, asphalt, road oil, and still gas	427
Liquefied gases for fuel and chemicals	386
Petrochemical feedstock	93
Miscellaneous	17
Total domestic product demand	4,900

The non-energy demand for natural gas as petrochemical feedstock and in the production of carbon black for the rubber industry accounted for about 2½% of the total demand for that resource. In the case of coal, about half of the resources used in the industrial sector were consumed as coke, carbonized from coal. About 1% of the total coal demand is consumed as industrial raw materials for the production of crude light oil and coal tars. There is much concern over the depletion of petroleum and natural gas for energy purposes when the demand for these materials, as organic molecular building blocks in the petrochemical industry, is growing rapidly and will expand further as natural mineral resources are depleted and must be replaced in their markets with synthetic materials. This dilemma suggests, for the long term, the implementation of an all-electric economy based on power plants consuming the more abundant nuclear fuels and thereby conserving fossil fuel reserves.

A more detailed breakdown of energy utilization for specific applications within the household, commercial, industrial, and transportation sectors is presented by Schurr.[6]

Environmental Factors

All of the activities involved in supplying energy to the various demand sectors have some impact on the environment. The benefits to the human environment obtained in the utilization of energy have a social and economic basis. These benefits must be acknowledged and are also discussed in some detail by Schurr.[6] Other environmental factors involved in energy related activities may be measured in terms of land use, employment and working conditions in energy industries, air and water pollution, radiation hazards, solid wastes, and noise. These environmental factors are attributable to either the characteristics of the energy resource utilized, or to the technology used in the particular activity that is the causative agent.

Land use

Physical space is required for all energy-related activities, with mining, electric power production, and electric power transmission being the largest users. The strip mining of coal is especially damaging to the landscape and represents an increasing method of recovery of coal, with about 34% of the coal produced in 1968 mined by this technique. In 1965 about 1.3×10^6 acres were involved in strip mining and the reclamation of this land was estimated to cost an average of $ 230/acre for complete restoration and $ 149/acre for partial restoration.[7] Strip mining areas are under study as possible solid waste disposal sites to serve urban areas that are nearby, where the cost of transporting the wastes is not excessive. Uranium mining poses potential problems similar to those of coal mining but on a reduced scale. The effects of oil and gas drilling, barring accidental discharges, are less severe.

Fuel shipments by rail and truck consume space with attendant noise, hazards, and general inconvenience. In 1968 there were 209,478 miles of petroleum pipeline and 861,000 miles of natural gas pipeline in service, including distribution lines.

The electric utilities require substantial areas of land for power generation and transmission. About 400 acres of land are required for a 3000 MW (t) nuclear power plant and 1200 acres are needed for a coal-fired plant of the same size, including space for on-site coal and ash storage. These sites are usually preferred near a source of cooling water, such as a river, lake, or other body of water. There are also over 300,000 miles of high-voltage overhead electric transmission lines in service with an average right-of-way width of 110 ft, for a total land requirement of over 4 million acres. This latter figure does not include local distribution lines although there is a growing tendency to put these, as well as high voltage transmission lines, underground. There is about 2400 miles of underground transmission line in use in the 69 to 345 kV range. Electrical energy cannot be conveniently stored in large quantities and some utilities use pumped storage facilities to satisfy peak electrical demands. Elevated water reservoirs are required at pumped storage sites and, like hydroelectric power plants, they are usually located in mountainous areas where there are natural sources of water. The electric utilities are facing severe problems in the siting of new facilities. Recognizing these difficulties and the growing demand for electric power, the Energy Policy staff of the OST has recommended[8] more long-range integrated planning by utilities, participation in this planning by environmental agencies, and notice to the public of new plant sites at least five years in advance of construction. In addition to calling for an expanded resaerch and development program on pollution controls and improved generation techniques, the report asks for more attention to underground power transmission and advanced siting approaches such as offshore locations and underground nuclear installations. The availability of a transmission grid that could move very large blocks of power over some distance would relieve some siting constraints and could also improve service reliability.

Employment and safety

The direct employment of labor in the various energy industries is a small fraction of the total labor force, however, such statistics do not reflect

TABLE IV

Employment and Safety Record in Selected Energy Industries

Industry	No. of employees, 1969 (thousands)	No. of injuries, 1968		Injury rate per 10^6 man-hours, 1968	
		fatal	non-fatal	fatal	non-fatal
Coal mining	136	311	9460	1.33	40.36
Petroleum and natural gas extraction	282	102	9069	0.10	9.19
Petroleum refining	146	—	—	—	8.6[c]
Electric, gas, and sanitary services	664	—	—	—	6.2[c]
Automotive service stations	575[a]	—	—	—	14.5[c]
Fuel dealers	101[a]	—	—	—	—
Federal government services[b]	3006	—	—	—	6.9[c]

[a] 1967 data.
[b] excluding military.
[c] includes fatal injuries.

(Source: Statistical Abstracts of the U.S., 1970.)

the vital need for energy in manufacturing and industrial processes and to provide comfort, light, and labor-saving functions in all working activities. The employment of labor in some energy industries is listed in Table IV along with the injury rate in each industry. Additional data on the fatal injury rate in each industry. Additiodal data on the fatal injury rate and the number of injuries, both fatal and nonfatal, are given for mining activities. Statistics on injuries in federal government service, excluding military personnel, are also given for comparison purposes. It is apparent that the underground coal mining activity has the poorest safety record of the energy industries listed, the most significant problems involving collapse of mine tunnels and explosions due to the methane released from the coal bed during mining operations. Research and development is in progress in the coal mine safety area, including the study of methods of stabilizing mine tunnels as the coal is extracted and other methods of protecting the workers in the event of a fire or mine failure. An example of one novel scheme to guard against fire in high-methane mines is the provision of an oxygen-free atmosphere in the mine with the workers using portable breathing apparatus.[9]

Air pollutants

A survey[10] performed by the National Air Pollution Control Administration (NAPCA) of the emission of air pollutatns in the United States for the year 1968 identifies energy conversion operations in stationary and mobile power plants as the major source of carbon monoxide, particulates sulfur dioxide, hydrocarbons, and oxides of nitrogen. The total emissions of these pollutants are indicated by source in Table V. The Office of Air Programs in the Environmental Protection Agency (EPA), formerly the NAPCA, has also published a compilation of air pollutant emission factors[11] which give the average quantity of pollutants released in the operation of various fuel-energy conversion system combinations. The emissions are given for uncontrolled sources unless otherwise noted. Control measures that may be used to reduce these emission factors involve treatment of fuels, modification of the energy conversion system, or treatment of effluents depending upon the particular case.

Carbon Monoxide: This poisonous gas is produced by the incomplete combustion of carbon in fuels. When combustion conditions are well controlled and excess air can be provided, combustion is generally complete and CO_2 is produced with only trace amounts of CO. In mobile power plants, such as internal-combustions engines which account for almost two-thirds of the total emissions of this pollutant, operating conditions are quite variable and CO may be in equilibrium in high concentrations with other combustion products at the high temperatures involved. These hot gases are cooled rapidly and there is not sufficient time to convert the CO to CO_2 before they are expelled from the cylinder.

Emission factors for CO from uncontrolled sources at average operating conditions are given in Table IV. Control of these emissions is accomplished, when possible, by proper design of the combustion chamber to prevent regions of insufficient air and by providing residence time and temperature conditions that will allow conversion of all of the CO to CO_2. In cases where combustion conditions cannot be adequately controlled to permit complete conversion, the exhaust gases may

TABLE V

Estimated Nationwide Emissions, 1968[10]
(in millions of tons per year)

Source	Carbon monoxide	Particulates	Sulfur oxides	Hydrocarbons	Nitrogen oxides	Total
Transportation	63.8	1.2	0.8	16.6	8.1	90.5
Fuel combustion in stationary sources	1.9	8.9	24.4	0.7	10.0	45.9
Industrial processes	9.7	7.5	7.3	4.6	0.2	29.3
Solid waste disposal	7.8	1.1	0.1	1.6	0.6	11.2
Miscellaneous*	16.9	9.6	0.6	8.5	1.7	37.3
Total	100.1	28.3	33.2	32.0	20.6	214.2

* Primarily forest fires, agricultural burning, coal waste fires.

TABLE VI

Carbon Monoxide Emission Factors[11]

	Average emissions per unit of fuel burned
Stationary sources	
Coal:	
Household and commercial	10 lb/ton
Industry	2 lb/ton
Utility	1 lb/ton
Fuel oil:	
Household	5 lb/1000 gal
Commercial and industrial	0.2 lb/1000 gal
Utility	0.04 lb/1000 gal
Natural gas:	
Household and commercial	20 lb/million ft^3
Utility and industrial	0.4 lb/million ft^3
Mobile sources	
Gasoline-powered vehicle, urban 1970	2620 lb/1000 gal
Diesel-powered bus and truck	225 lb/1000 gal
Aircraft, long range turbofan	26 lb/engine-flight*
jumbo jet	28 lb/engine-flight*

* Emissions during portion of flight at altitudes less than 3500 ft, including takeoff and landing.

be reacted with air either thermally or catalytically to complete the reaction to CO_2.

Particulates:[12,13] This category of pollutants includes any matter, solid or liquid, in which individual particles are larger than molecular size but smaller than 500 microns. The burning of coal in central station power plants accounts for about 64% of all particulate emissions, even with the widespread use of electrostatic precipitators with efficiencies of up to 99.5%. This particulate release is mostly in the form of fly ash, which is predominantly carbon, silica, alumina, and iron oxide, and when recovered from stack gases, poses a solid waste disposal problem. Markets for fly ash involve construction uses and are growing. Oil has a lower ash content than coal but is also a source of particulate emissions. The particulate emissions from mobile power plants include lead compounds, carbon, and metallic oxides. Lead alkyl additives used in gasoline to raise its octane number are present in concentrations of 2 to 4 g/gal and this lead content is ultimately discharged to the environment. It appears that the automotive and petroleum industries are moving toward reduction or elimination of these additives.

Emission factors for particulates from uncontrolled sources are given in Table VII. The emissions from the combustion of coal are given as a function of the per cent ash content, which depends on the grade of coal but averages about 10%. Wet bottom and cyclone-type burners capture more ash in the furnace in the form of bottom ash, which is a solid waste, and thereby reduce the fly ash particulate matter released with the flue gas. Control of particulate emissions form power plants is provided by cyclone separation, filtering of the flue gases, and electrostatic precipitation, and these techniques are used extensively.

Particulate matter in the atmosphere has adverse effects on materials and health and may have long-term effects on the thermal energy balance of the earth. It provides condensation nuclei which can increase cloud cover and thereby reflect more of

TABLE VII

Particulate Emission Factors[11]

	Average emissions per unit of fuel burned
Stationary sources	
Coal, bituminous	
General	16A[a] lb/ton
Dry bottom	17A lb/ton
Wet bottom	13A lb/ton
Cyclone	2A lb/ton
Spreader stoker	
Greater than 10×10^6 Btu/hr	13A lb/ton
Less than 10×10^6 Btu/hr	2A lb/ton
Hand-fired equipment	20 lb/ton
Residual oil:	
Utility	8 lb/1000 gal
Industrial and commercial	23 lb/1000 gal
Distillate oil:	
Industrial and commercial	15 lb/1000 gal
Household	10 lb/1000 gal
Natural gas:	
Utility	15 lb/million ft^3
Industrial	18 lb/million ft^3
Household and commercial	19 lb/million ft^3
Mobile sources:	
Gasoline-powered vehicle, urban 1970	8 lb/1000 gal
Diesel-powered bus and truck	13 lb/1000 gal
Aircraft, long range turbofan	8 lb/engine-flight[b]
jumbo jet	10 lb/engine-flight[b]

[a] Where letter A is shown, multiply number given by the per cent ash in coal.

[b] Emission during portion of flight at altitudes less than 3500 ft, including takeoff and landing.

the solar energy input to the earth back into space. It has been estimated[14] that a 1% increase in the earth's cloud cover can reduce the average temperature of the earth by 1.4° F and an increase of 5% could lead to the return of an ice age. Volcanic eruptions are believed to account for most of the particulate contamination of the upper atmosphere and in the past have been a source of meteorological changes. Jet aircraft exhausts add water vapor and particulates to the upper atmosphere and can also contribute to meteorological effects.

Sulfur Dioxide:[15,16] SO_2 is produced in the combustion of sulfur-bearing fuels such as coal and residual oil and is released to the atmosphere in the flue gases from power plants using these fuels. Approximately 60% of the SO_2 released to the atmosphere in 1968 was produced in the burning of coal while the combustion of oil accounted for another 14%.

Emission factors are given for various uncontrolled sources in Table VIII as a function of the per cent sulfur content of the fuel. The average sulfur content of coal is about 2.5% by weight while that of crude oil is closer to 0.3%, but these levels can vary considerably depending on the source. In petroleum refining, the sulfur generally remains in the heavier fractions and reaches much higher concentrations in residual fuel oil, for example. Sulfur dioxide emissions may be controlled by fuel selection using either low sulfur fuels or by removing the sulfur from the fuel before burning it. Other techniques have been proposed and are under development to capture sulfur compounds during combustion or to remove them from the flue gas before it is discharged to the atmosphere. Flue gas scrubbing processes under development include the use of MgO, alkalyzed alumina, or limestone to neutralize oxides of sulfur, forming precipitates of sulfites and sulfates that can be collected as solids. A catalytic oxidation process, based on the technology of sulfuric acid production, that captures sulfur emissions as SO_3 in a weak acid solution is also quite promising.

TABLE VIII

Sulfur Dioxide Emission Factors[11]

	Average emissions per unit of fuel burned
tationary sources	
Coal	38 S[a] lb/ton
Fuel oil	157S[b] lb/1000 gal
Natural gas	0.6 lb/million ft^3
Mobile sources	
Gasoline-powered vehicle, urban 1970	5 lb/1000 gal
Diesel-powered bus and truck	27 lb/1000 gal
Aircraft, long range turbofan and jumbo jet	2 lb/engine-flight[c]

[a] S indicates % sulfur in fuel.

[b] This factor is for residual oil. For distillate oil use factor of 142S.

[c] Emission during portion of flight at altitudes less than 3500 ft, including takeoff and landing.

Oxides of Nitrogen,[17] *Hydrocarbons,*[18] *and Photochemical Oxidants:*[19] NO_x and reactive organic substances such as hydrocarbons (HC), when exposed in the atmosphere to solar ultraviolet radiation, form photochemical smog consisting of aldehydes and a group of oxidants including ozone and peroxyacyl nitrates.

Oxides of nitrogen are produced in all combustion processes that use air as the oxidant at a rate that is dependent on both the flame temperature and air supply. Nitric oxide (NO) is formed at high combustion temperatures and normally would convert to the more stable nitrogen dioxide (NO_2) as the gas temperature is reduced. The kinetics of that reaction are slow, however, and the gases cool to near ambient conditions before very much NO_2 can be formed. In the combustion of methane, the peak equilibrium NO_x emission concentration is developed with about 15% excess air while the peak flame temperature occurs at a slightly deficient air fuel ratio.

With different fuels the NO_x emissions will vary depending on the theoretical flame temperatures and the combustion process used. Table IX gives average uncontrolled emission factors for NO_x. These emissions may be reduced, but not eliminated, by the adjustment of the air supply to a fuel-rich condition and operation at as low a combustion temperature as possible by means of two-stage combustion, exhaust gas recirculation, or water injection. Some of the flue gas scrubbing techniques proposed to control SO_2 emissions in power plants may also be effective for the removal of NO_x from the gas stream.

The mains sources of hydrocarbon emissions are evaporation of fuel from vehicles and in fuel handling operations, and unburnt fuel in vehicular exhausts. Emission factors for various sources of hydrocarbons are given in Table X. Blowby of

TABLE IX

Emissionrs Facto for Nitrogen Oxides[11]

	Average emissions per unit of fuel consumed
Stationary sources	
Coal	
Household and commercial	6 lb/ton
Industry	18 lb/ton
Utility	18 lb/ton
Fuel oil	
Household and commercial	12 lb/1000 gal
Industry, tangentially fired unit	40 lb/1000 gal
Utility	105 lb/1000 gal
Natural gas	
Household	50 lb/million ft^3
Commercial	100 lb/million ft^3
Industry	230 lb/million ft^3
Utility	390 lb/million ft^3
Mobile sources	
Gasoline-powered vehicle, urban 1970	183 lb/1000 gal
Diesel-powered bus and truck	370 lb/1000 gal
Aircraft, long range turbofan	5 lb/engine-flight*
jumbo jet	6 lb/engine-flight*

* Emission during portion of flight at altitudes less then 3500 ft, including takeoff and landing.

TABLE X

Hydrocarbon Emission Factors[11]

	Average emissions per unit of fuel burned
Stationary sources	
Coal	
Household and commercial	3 lb/ton
Industry	1 lb/ton
Utility	0.3 lb/ton
Fuel oil	
Household	3 lb/1000 gal
Industry and commercial	3 lb/1000 gal
Utility	2 lb/1000 gal
Natural gas	
Household and commercial	8 lb/million ft^3
Utility and industry	40 lb/million ft^3
Mobile sources	
Gasoline-powered vehicle, urban 1970	330 lb/1000 gal
Diesel-powered bus and truck	37 lb/1000 gal
Aircraft, long range turbofan	17 lb/engine-flight[a]
jumbo jet	3 lb/engine-flight[a]
Gasoline evaporation sources	
Vehicle gas tank and carburetor	75 lb/1000 gal
Filling[b] of automobile tanks	12 lb/1000 gal
Filling[b] of service station tanks	12 lb/1000 gal

[a] Emission during portion of flight at altitudes less than 3500 ft, including takeoff and landing.

[b] Assumes splash fill.

gasoline vapors from the cylinder into the crankcase was a large automotive emission source but has been largely brought under control by positive crankcase ventilation techniques that recycle these vapors back to the combustion chamber. Control of other sources may be provided by better fuel handling techniques, improved combustor design, treatment of exhaust gases and reduction of fuel volatility. The treatment of automotive exhaust gases involves control of both CO and hydrocarbon emissions using thermal oxidation in the 1600°to 1800°F temperature range, or catalytic oxidation with transition metal oxide materials. These catalysts are generally poisoned by lead additives and offer extra incentive to eliminate lead additives in fuels.

Carbon Dioxide: This gas is inevitably produced in the complete combustion of carbon-bearing fuels. It is not normally thought of as a pollutant, however, there is some concern over its long-term effect on the thermal equilibrium of the earth because of the so-called "greenhouse effect." Carbon dioxide is a strong absorber in the long wavelength infrared region of the electromagnetic spectrum, a region where much of the energy that the earth loses by radiation to space is concentrated. The CO_2 absorbs this energy, reradiating it in turn to space at a lower source temperature. This mechanism effectively reduces the heat loss from the earth and causes its temperature to increase.

Bolin[20] has reviewed the carbon cycle mechanisms that are at work in the biosphere. The current annual release to the atmosphere of carbon, as CO_2, attributable to fossil fuel combustion is about 5×10^9 tons while a total of some 200×10^9 tons have been released to the atmosphere by combustion processes during the past century. During this same time period the average concentration of CO_2 in the atmosphere has increased from 290 to 320 ppm, a change that accounts for only about $^1/_3$ of the total CO_2 emissions. Apparently, removal processes are effective and include photosynthesis, solution in ocean waters, and deposition in carbonate sediments. One analysis[14] indicates that an increase in atmospheric CO_2 concentration to 400 ppm, which will be attained in 1990 if fossil fuel consumption increases at an annual rate of 5% and if 60% of the CO_2 emissions remain in the atmosphere, would increase the temperature of the earth by 1,4°F. The growth of vegetation is stimulated by high CO_2 levels but, of course, carbon fixed in organic compounds by photosynthesis is ultimately released again either by combustion or natural decay processes.

Water quality[14]

The energy activities have an impact on water quality in several ways. Severe damage is done to the marine environment in the accidental discharge of oil either from offshore drilling operations or from oil tankers and other vessels that carry petroleum products. In 1969 there were 1007 reported spills of oil in quantities over 100 bbl each. Vessels were identified as the source of 53 of the sespills, 331 were from shore facilities, and 144 were unidentified. Individual discharges can be quite large however. In 1967 the tanker Torrey Canyon ran aground and discharged approximately 400,000 bbl of crude oil to the sea. It has been estimated[21] that the oil well blowout in the Santa Barbara channel in 1969 released about 75,000 bbl of oil by the 100th day of discharge. Smith[22] has outlined methods of dealing with spills after they have occurred that have been applied in the United Kingdom, including skimming, absorption, jelling, sinking, and emulsification.

Acid mine drainage, consisting of sulfuric acid and iron compounds formed by the reaction of water and air with sulfur-bearing minerals, contaminates local streams in coal mining areast. Martin,[23] in 1968, estimated that over 4 million tons of polluted water was released annually in the United States to about 11,000 miles of streams and other receiving waters from both abandonded and active mines. Control measures include eliminating the source of polluted water drainage by sealing off underground mines and restoring the surface of strip mines, and by collection and treatment of the drainage.

The petroleum- and coal-producing industries, in 1964, discharged about 7.3×10^{12} gal of industrial waste, or about 10% of the total discharge of industrial waste water. This waste involved 500×10^6 lb of biochemical oxygen demand (BOD), only about 2% of the total industrial waste water BOD. These industries also required 1.2×10^{12} gal of cooling water in the same year which was 2.4% of the total industrial cooling water demand. Electrical utilities accounted for 81.3% of the total demand by using 40.7×10^{12} gal of cooling water. This water is returned to the environment in a heated condition, which reduces the amount of dissolved oxygen that can be retained and thus has an effect similar to oxygen demanding wastes, but it is not otherwise contaminated. The need for cooling water is discussed in a more general context, along with alternate means of heat rejection, in the following section.

Thermal effects

Heat must be rejected from all thermodynamic power cycles in quantities that depend on the efficiency of the energy conversion operation. The primary sources of waste heat are central station power plants, which can reject heat from their condensers to water or to the atmosphere through cooling towers, and automotive power plants, which discharge their waste energy to the atmosphere in exhaust gases. Atmospheric discharges of thermal energy can lead to local meteorological effects while discharges to water can have effects on the marine ecology.

In a 1000 MW (electrical output) fossil fuel power plant of 40% thermal efficiency, about 1275 MW of thermal energy must be rejected to the condenser coolant while about 225 MW is discharged directly to the atmosphere in the flue gas. A nuclear power plant of the same electrical capacity operating at 33% efficiency must reject all of its waste heat, 2000 MW, to the condenser coolant.

Discharges of waste heat can be controlled by improving the thermal efficiency of power plants, and the severity of local effects may be reduced by distributing the waste energy in large quantities of air or water. Many regulatory agencies have placed restrictions on the allowable temperature difference between inlet water and the water returned to the marine environment. Cooling ponds may be employed in such instances to hold up the effluent water until it has cooled to the allowable discharge temperature. Applications for this "waste" energy have been proposed,[24] including space heating and refrigeration in urban areas, thawing of ice-bound seaways, agricultural use to stimulate growth and to extend the growing season, and in aquaculture to stimulate the growth of algae, shellfish, and other potential marine food sources.

Radiation

There has been little addition to natural radiation levels in the environment as a result of energy related activities, excluding nuclear weapons testing as a source.

In the nuclear industry radiation problems begin with the mining of uranium-bearing ores. The alpha particle emitters, particularly radon gas, in the uranium-238 decay chain that ends with stable lead-206 pose the major health problem. Radon is chemically inert and diffuses from rock surfaces into the mine atmosphere where it and its daughter products can be inhaled by workers. Exposure standards and control measures have been published by the Bureau of Mines.[25] Mine tailings also contain these radiation sources which complicate the disposal of these solid wastes.

The largest source of radioactive materials are the products produced by the fissioning of uranium and plutonium fuel in nuclear reactors. These fission products are formed in the encapsulated fuel elements and must be handled in the reactor waste treatment system only in small amounts that may leak from the elements during reactor operation. Other radioactive materials found in the reactor system are isotopes produced by the irradiation of materials of construction, corrosion products, and coolant additives. The fission products must be handled in the fuel reprocessing plant where uranium and plutonium that has not fissioned is recovered for use in new fuel elements. Table XI indicates the activity of the long-lived radioactive materials that are generated in the fuel elements of a nuclear reactor and that pose a long-term disposal problem. Isotopes with a short half-life may be simply held-up until they decay to a harmless product that may be released to the environment, but the others must be chemically reacted and concentrated for disposal by some other technique. The long half-]ife materials must be sequestered in a maanner such that they cannot be released to the environment for centuries. A most promising approach to longterm disposal under study by the Atomic Energy Commission (AEC) involves the concentration of radioactive wastes and their conversion into an inert solid form by

TABLE XI

Long-Lived Radioactive Materials Generated in Typical Nuclear Reactor Fuel

Material	Half-life (in years)	Activity[a] (megacuries[b]/ 10^{15} Btu of thermal energy produced by fission)
Tritium	12.26	0.146
Krypton-85	10.76	3.4
Iodine-129	1.6×10^7	1.03×10^{-5}
Other fission products		
Cesium-137	30	40.2
Strontium-90	28	28.5
Technetium-99	2.1×10^5	4.2×10^{-3}
Transuranium Isotopes		
Americium-241	458	0.091
Curium-244	18.1	1.34
Plutonium-238	89	1.27
Plutonium-239	2.4×10^4	0.116
Plutonium-240	6760	0.212
Plutonium-241	13	55

[a] After 90 days of cooling time following operation of a typical batch of 2.84% U-235, non-recycle, PWR fuel at a specific power of 40 kW/kg U to a burnup of 35,000 MW-days/long ton.

[b] A curie is the amount of a given radioactive substance which undergoes 3.7×10^{10} radioactive transformation/sec (equivalent to 1 g of radium).

(Derived from data presented in: Safety Analysis Report, Barnwell Nuclear Fuel Plant, Allied Chemical Nuclear Products, Inc., Amendment 2, Docket No. 50-332-4, March 21, 1969.)

calcination or fixation in glassy materials, which would be stored underground in isolated geological formations such as salt domes, shale, or rock.[26] Two long-lived radioactive isotopes that cannot be easily captured and concentrated for such disposal are krypton-85 and tritium. Krypton-85 is a chemically inert fission product with a 10.76-year half-life. Tritium, with a half-life of 12.26 years, is produced both by fission and by neutron irradiation of boron and lithium contained in the reactor coolant and ends up in aqueous form in waste treatment systems. Both of these isotopes are discharged to the environment in liquid and gaseous effluents but are first diluted to permissible concentration levels.

Fossil-fueled power plants release small quantities of radioactive material that are attributable to trace amounts of uranium and thorium in coal and oil.

A study performed by Department of Health, Education, and Welfare personnel[27] compared potential exposures to the population in the vicinity of coal fired and nuclear power plants. The only radiation sources attributable to either type of plant that could be detected above background levels were emissions to theat mosphere. A plant fired with a specific coal containing 9% ash and with 97.5% effective fly ash removal could lead to an exposure of 3.08×10^{-4} millirem of radiation annually per MW of electrical power produced. The exposure potential from a typical pressurized water reactor (PWR) plant was 1.05×10^{-5} millirem on the same basis. A boiling water reactor (BWR) included in the study had a higher exposure value but is not representative of plants of this type that are now being built. In any case, the maximum off-site exposure attributable to any nuclear plant was in the 5 to 15 millirem/yr range. The difference between the BWR and PWR in this study is attributed to the fact that the PWR had the ability to store gaseous wastes for about 60 days prior to release to the atmosphere, while the BWR had only a 30 minute hold-up. Larger volumes of waste gases must be handled in the BWR even though the total quantity of radioactive materials produced are comparable in the two reactor types.

Solid wastes[28]

In 1969 over 30 million tons of fly ash were recovered from power plants burning coal and lignite. About 20% of this is used commercially as a soil conditioner and as an additive to concrete and asphalt mixes. Bottom ash, which accounts for about one-fifth of the ash produced in coal-fired plants, is used in the manufacture of cinder blocks, as road ballast, and on highways in icy weather. Unused portions of these materials represent a solid waste disposal problem and are dumped in spoil areas which occupy land and are unsightly.

Large amounts of solid wastes are generated in coal mining, uranium mining, and shale oil recovery. The residual wastes from the processing of coal are put in culm banks and coal waste piles. These contain a high content of carbonaceous material and besides their unsightly appearance and the landslide hazard that they represent, they can ignite spontaneously and emit air pollutants. A survey has located waste piles from 470 bituminous coal cleaning plants and 800 culm banks each containing 10,000 tons or more of waste material from the cleaning of anthracite coal. In 1964 about 495 refuse deposits located in 15 states and containing about 500×10^6 yds^3 of material were found to be burning.

Overburden material from strip mines is piled in spoil banks which may be thousands of feet long and over 100 feet high. These are a source of acid water, will generally not support vegetation, and render the land unsuitable for recreation or industrial use. A thickness of overburden material averaging about eleven times the coal be thickness was handled in U.S. strip mining activities during 1970.

Control of the solid waste problem in these activities will require increase utilization of ash from coal burning power plants and of mill tailings from coal and uranium ore treatment. An obvious disposal method for the portions of these wastes that cannot be used and for spoil bank material is to place them in depleted strip mining areas and thereby assist in restoration of the land whenever this is feasible.

Noise[14]

Energy conversion devices, particularly those used for mobile applications, are noisy in operation. Noise can impair hearing and has physiological and psychological effects on people. Industrial activities are required by law to limit the noise level to 90 db (A) in working areas.

Traffic noise, mainly from vehicular power plants, can reach 90 db (A) in urban areas and on heavily traveled highways, while a four-engine jet aircraft produces up to 120 db (A) on take-off at a distance of 200 feet. In internal-combustion engines the sources of noise are the rapid combustion of fuels, high-velocity gas discharges through exhaust valves, and engine vibrations. The noise output is held down by muffling of the exhaust gases and by acoustic lining of the engine compartment. Turbine noises from both aircraft engines and stationary power plants arise from high-speed rotating components and high-velocity gas flows, and can be controlled by reducing the fan blade tip speed, the use of acoustic linings, and varying the geometry of the exhaust nozzle.

Energy Resources

Modification of fuel form and the substitution of alternate energy resources are useful procedures for

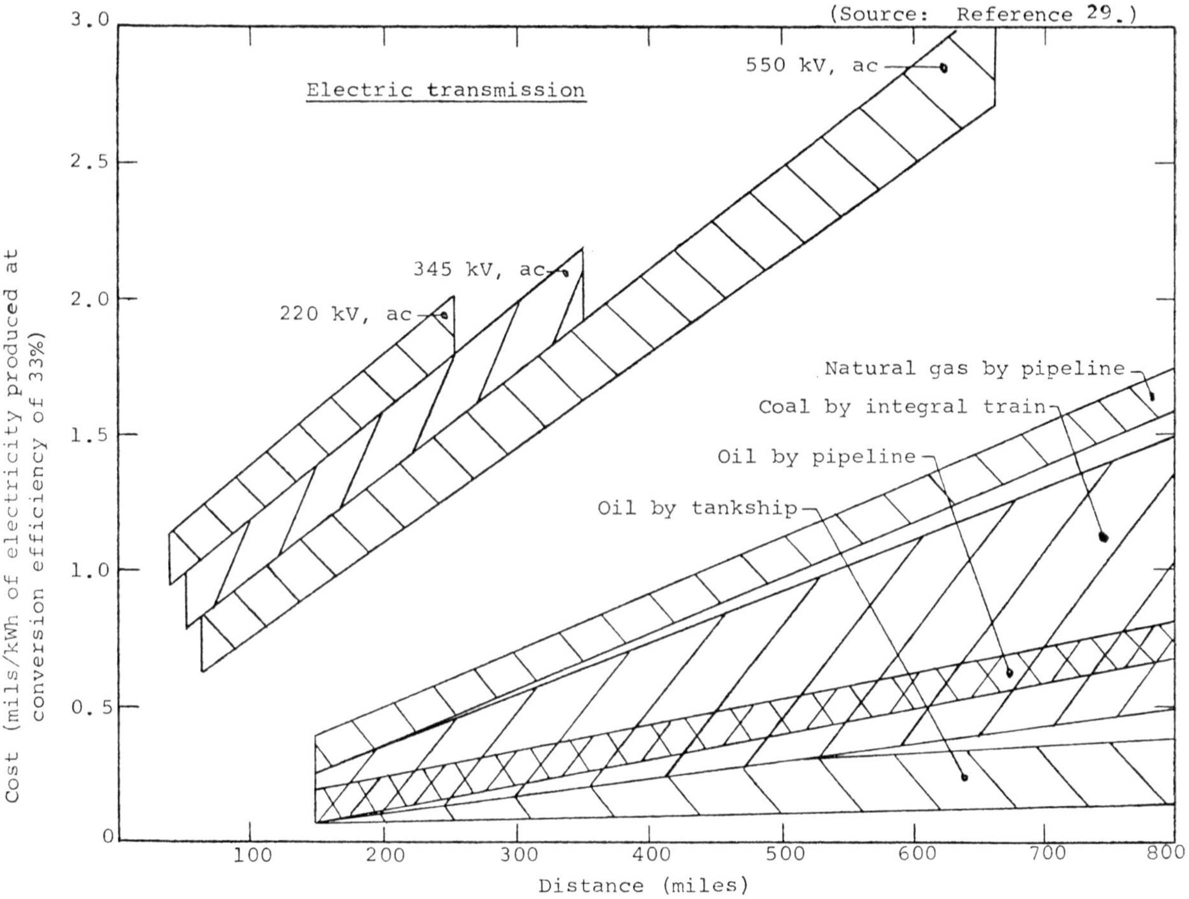

Figure 4 Portion of central station electric cost attributable to the transportation of energy.

controlling environmental effects. The selection among alternative resources depends on their relative abundance, ease of recovery, convenience in handling, and economics. Some resources, such as fossil and nuclear fuels, are present on earth in only finite amounts and are not renewed to any extent, although in the case of nuclear fuels the energy content of known reserves may be quite large. Theother major category of energy resources involve energy that is supplied continually to the biosphere over a geological time span, including geothermal and tidal energy as well as solar energy and its derivatives, wind, waves, and hydropower.

In 1968 the average cost of coal, oil, and natural gas at their source was \$ 4.67 per ton, \$ 2.94 per bbl, and 16.4 ¢ per 1000 ft^3, respectively. These costs have generally escalated and further adjustments may be attributed to the cost of providing environmental protection systems and to governmental policy actions in specific areas such as mine safety, oil import quotas, and natural gas price regulation.

Current prices for nuclear fuel supplied to fission reactors indicate a cost for this form of energy of about 15 ¢ per 10^6 Btu.

The cost of treating, refining, and transporting fuels adds considerably to the price paid by the consumer for energy. Figure 4 shows the portion of the cost of electrical energy, produced in central station power plants, that is attributable to the transportation of coal, oil, and natural gas fuels and to high-voltage electric transmission as a function of distance and transport technique.[29] The spatial separation between the location of resources and the centers of energy demand poses the problem of optimal siting of fuel treatment, refining, and energy conversion activities and has been studied in a geographical format by Manners.[30] Consideration of environmental factors imposes additional constraints on such siting problems.

A number of estimates of the abundance of energy resources are available and have been assembled and refined by Hubbert[31] of the U.S. Geological Survey. The exploitation of an exhaustible energy resource generally begins at a low level and increases to a period of maximum utilization, ultimately returning again to a low level as the resource is depleted or as recovery costs become prohibitive. Of interest with respect to an energy resource is the estimated time of peak annual consumption and the time at which 90% of the resource is expected to be consumed and the resource is expected to be consumed. The summation of annual recovery over the entire recovery time span is the ultimate recovery for that resource, Q_∞. The cost of obtaining a resource increases as that resourc is depleted. In general, it is necessary to qualify any statement concerning resource abundance with a definition of the cost constraints that are assumed. In view of such difficulties, estimates of resource availability vary widely.

The estimated supply of coal and nuclear fuels can provide the projected U.S. needs over several centuries, while Hubbert's analyses indicate that over 90% of the oil and natural gas resources may be consumed by the year 2015. Others are much more optimistic about new discoveries of oil and gas. Homan and Lovejoy[32] have reviewed the methods used in arriving at oil and gas supply estimates and compare the results arrived at by several investigators.

Another pertinent characteristic of a specific fuel or energy resource is its versatility. Some fuels can be modified or treated with relative ease to provide a convenient form and can be utilized in a wide spectrum of energy conversion systems while others may be limited in their periods of availability or in their adaptability. This characteristic accounts more than any other for the great demand for natural gas or oil and its derivative products.

Coal

Coal is formed from plant substances preserved from complete decay producing first peat and then progressing through lignite, subbituminous, bituminous, and anthracite stages. It has been estimated that there is about 1.58×10^{12} tons of coal in the U.S. in seams at least 14 in. thick and at depths of less than 3000 ft and about half of this is considered recoverable at a cost near current levels. These deposits are described by grade and sulfur content in Table XII. On the same basis, the world's coal resources are 8.3×10^{12} tons. The time of peak annual recovery and the time at which about 90% of this resource will be consumed are projected to be the year 2150 and 2440, respectively, for the U.S. and 2150 and 2380 for the world.

TABLE XII

Estimated Remaining Coal Reserves of the United States by Rank and Sulfur Content, 1965
(in billions of tons*)

Sulfur content %	*Coal Rank* Anthracite	Bituminous	Subbituminous	Lignite	Total
<0.7	14	104	257	345	720
0.8–1.0	—	111	131	61	303
1.1–1.5	—	49	—	41	90
1.6–2.0	—	43	1	—	44
2.1–2.5	—	48	—	—	48
2.6–3.0	—	52	—	—	52
3.1–3.5	—	90	—	—	90
3.6–4.0	—	128	—	—	128
<4.0	—	105	—	—	105
Total	14	730	389	447	1580

* Coal in seams at least 14 in thick and less than 300 0ft deep in explored areas. Approximately one-half of these reserves are considered recoverable.

(Data source: Decarlo, J. A., E. T. Sheridan, and Z. E. Murphy, Sulfur Content of U.S. Coals, Dept. of Interior, Bureau of Mines, Information Circular 8312, 1966, Washington, D. C.)

Sulfur is present in coal as pyrite (FeS_2) and organic sulfur in roughly equal amounts. The organic sulfur is difficult to remove while only 30% of the sulfur present as pyrite is removed by the standard coal-cleaning technique of crushing and flotation. Processes are available for the liquefaction and gasification of coal, which provide a sulfur-free fuel in more convenient form. Hydrogenation plants are foreseen that will produce 100,000 bbl/day of synthetic crude oil from 3.24×10^5 tons of coal at a cost after sufficient development of about $ 3.35 per bbl, assuming a coal cost of $ 3.50 per ton.[33] The liquid has a high gasoline content and the process yields a high-sulfur char residue. In the gasification of coal, the sulfur is recovered as H_2S. Carbonization, or pyrolysis, yields a gas, mostly hydrogen, with 15 to 30% of the original energy content of the coal with the balance in the coke that is produced. Hydrogasification processes can produce methane, a pipeline quality gas, at an estimated cost of 30 to 60 $ per million Btu above the cost of the coal feed. The waste products produced in coal conversion include particulates, CO_2 and sulfur compounds but they are generally easier to handle in this step which may also be carried out at a site more remote from population centers than where the fuel is consumed in an energy conversion operation.

Crude oil

Hubbert estimates that the ultimate recovery, QF, of crude oil in the U.S., including Alaska, will be about 200×10^9 bbl, of which about half had been produced by the end of 1969. Other estimates are higher than this by a factor of 2 or 3. The world resources are felt to be between 1350 and 2100×10^9 bbl. Peak annual recovery and the 90% depletion point are projected to be attained by the years 1970 and 2000, respectively, for the U.S. and 2000 and 2030 for the world. Petroleum resources may be extended considerably by the recovery of this fuel from shale oil. About 80×10^9 bbl of oil can be recovered in the U.S. under present conditions, using shale with a content of oil of more than 10 gallons per ton. The total U.S. resources of shale oil in place with more than 10 gallons of oil per ton is figured at 1430×10^9 bbl. The Bureau of Mines has outlined a plan[34] to achieve a production of 35,000 bbl/day by 1973 increasing to 10^6 bbl/day by 1980. It was estimated that the crude could sell for about $ 3.59 per bbl from first generation plants, decreasing to $ 2.72 per bbl from improved plants assuming no credit for by-products. Before these goals could be achieved, however, several prototype plants would have to be constructed.

A wide range of control of product yield is available in the oil refining process. Separation of light and heavy fractions are generally performed by distillation into liquefied petroleum gas (mostly propane and butane), gasoline, kerosine, fuel oil, and residual crude. The normal yield of gasoline from a typical crude is about 15% but this can be increased substantially by thermal and catalytic cracking techniques. In 1968 the average gasoline yield from U.S. refineries was 45% of the crude oil consumption. Petroleum can also be used to produce methane, hydrogen, ammonia, methanol, and other clean fuels that are being studied for various applications including mobile power plants.

The sulfur content of crude oil is generally concentrated by refining operations in the heavier fractions, such as the residual oil used to fuel central station power plants. Table XIII gives the 1966 crude oil production in the U.S. by area and sulfur content of the original crude. Since much crude oil and petroleum product is imported, similar information on foreign production is given in Table XIV. Sulfur may be removed from oil

TABLE XIII

United States Crude Oil Production by Area and Sulfur Content Category – 1966

Area	Crude oil production, % of U.S.	Annual crude oil production, 10^6 bbl					
		sulfur, weight %					
		0.00–0.25	0.26–0.50	0.51–1.00	1.01–2.00	2.00	Total
Alaska	0.47	14.3					14.3
Appalachia	0.68	20.4		0.1			20.5
California	11.34	25.7	40.7	110.1	122.5	45.4	344.4
Gulf Coast	29.52	568.4	328.4	0.4			897.2
Illinois, Indiana, Kentucky	3.02	67.3	24.2				91.5
Michigan	0.47	12.0	0.9	1.2	0.1		14.2
Mid-continent	23.09	215.2	227.9	145.1	79.6	33.8	701.6
Rocky Mountain	9.08	140.4	23.1	47.7	14.0	50.7	275.9
Southeastern U.S.	2.14	13.8	20.0	1.5	7.1	22.5	64.9
West Texas and Southeast New Mexico	20.19	150.2	107.4	90.2	171.3	94.4	613.5
Totals	100.0	1227.7	772.6	396.3	394.6	246.8	3038.0

(Source: McKinney, C. M., and E. M. Shelton, Sulfur Content of Crude Oils of the Free World, U.S. Dept. of the Interior, Bureau of Mines, Report of Investigations 7059, 1967, Washington, D.C.)

TABLE XIV

Foreign Crude Oil Production by Area and Sulfur Content Category

Area and sulfur content range, weight %	1966 production 10⁶ bbl	per cent
Africa:		
0.00–0.25	637	63.7
0.26–0.50	144	14.4
0.51–1.00	216	21.6
1.01–2.00	—	—
>2.00	3	0.3
Canada:		
0.00–0.25	112	35.3
0.26–0.50	18	5.7
0.51–1.00	107	33.8
1.01–2.00	40	12.6
>2.00	40	12.6
Middle East:		
0.00–0.25	—	—
0.26–0.50	—	—
0.51–1.00	—	—
1.01–2.00	1509	44.8
>2.00	1862	55.2
South America:		
0.00–0.25	24	1.6
0.26–0.50	19	1.3
0.51–1.00	52	3.5
1.01–2.00	225	15.2
>2.00	1161	78.4

(Source: McKinney, C. M., and E. M. Shelton, Sulfur Content of Crude Oils of the Free World. U.S. Dcpt. of the Interior, Report of Investigations 7059, 1967, Washington, D.C.)

by several techniques including delayed coking, solvent deasphalting, and hydrogen treatment. Studies made of the desulfurization of fuel oil indicate a cost of about 60 ¢ per bbl to reduce the sulfur content to a level of about 1%. Such studies have been summarized by the National Air Pollutional Contr Administration.[15]

Natural gas and natural gas liquids

The principal component of natural gas delivered to the consumer is methane. As recovered in the field the gas contains hydrocarbons of the composition C_nH_{2n+2} (methane, ethane, propane, butane, and pentane), natural gas liquids, and other gases such as hydrogen sulfide (H_2S) and, in some cases, helium. The H_2S content varies from field to field and is removed to recover and sell the sulfur. Natural gas resources are classified into nonassociated, associated, and dissolved categories depending on whether the gas exists alone, in a gas cap with oil beneath it, or dissolved in oil. About 75% of the gas reserves are in the nonassociated category. Gas associated with oil generally provides pressurization of the oil pool and stimulates its recovery, so the gas is usually left in place until the oil is depleted. Natural gas liquids (NGL) are recoverable from the gas and supplement other liquid fuel resources.

The ultimate recovery, Q_∞, of natural gas in the U.S. is estimated by Hubbert at about 1200×10^{12}ft^3 and for the world at 8000 to 12,000 $\times$ 10^{12} ft^3. Cumulative production through 1969 was about 400 $\times$ 10^{12} ft^3 in the U.S. On the basis of these figures, the year of peak annual recovery will be about 1980 and 90% of the resources will be consumed by the year 2015.

The ultimate supply of natural gas liquids corresponding to the above natural gas recovery is 41 $\times$ 10^9 bbl for the U.S. and 250 to 420 $\times$ 10^9 bbl for the world.

Uranium and thorium

Natural uranium has an isotopic makeup of 0.711% uranium-235, 0.006% uranium-234, and the balance is uranium-238. Only uranium-23 5among the naturally occurring isotopes is fissionable to any extent and natural uranium must be enriched in this isotope to between 2 and 3% to provide fuel for current nuclear reactors. If only the naturally occurring uranium-235 were utilized in fission reactors, the nuclear industry could be limited, for economic reasons, to those U.S. resources of U_3O_8 that are recoverable at a cost of less than $ 10 per lb. These are estimated at 660,000 tons and would furnish only about 232 $\times$ 10^{15} Btu of energy. Fortunately, it is possible to extend the degree of utilization of our uranium resources by converting the abundant uranium-238 isotope into fissionable plutonium-239. Thorium, which is available in large quantities, can also be converted into a fissionable isotope, in this case uranium-233. These conversion processes, involving the capture of neutrons by the fertile isotopes, may be carried out in power producing reactors, which are called breeders if the amount of new fissionable generated is greater than that consumed, or converter if the amount generated is less than that consumed. Breeding is required in order to fully utilize uranium-238 and thorium-232 resources and makes the economics of nuclear power less dependent on the cost of fuel resources. There are estimated to be about 25×10^6 tons of U_3O_8 in the U.S. recoverable at costs of less than $100 per pound and eventual utilization of all of thisna tural uranium, including the uranium-238 isotope after conversion into plutonium-239, would develop some 1.25×10^{21} Btu of energy. Current AEC estimates of uranium and thorium resources recoverable at various price levels are given in Tables XV and XVI.

TABLE XV

Estimated United States Uranium Resources[43]

U_3O_8 price per pound	(cumulative millions of tons of U_3O_8) Reasonably assured	Estimated additional	Total
$ 8.00	0.16	0.26	0.42
10.00	0.32	0.35	0.67
15.00	0.47	0.55	1.02
30.00	0.67	1.00	1.67
50.00	6.00	4.00	10.00
100.00	12.00	13.00	25.00

TABLE XVI

Estimated United States Thorium Resources[43]

ThO_2 price per pound	(cumulative millions of tons of ThO_2) Reasonably assured	Estimated additional	Total
$ 10	0.1	0.5	0.6
30	0.2	0.6	0.8
50	3.2	7.6	10.8
100	11.2	24.6	35.8

Fusion resources

There are several fusion processes that are currently under study as a source of energy but none have been attained in a controlled reactor. These reactions depend on the hydrogen isotope deuterium ,D, in seawater as a primary fuel. One process, the D–D reactionj, involves the feeding of deuterium into a magnetically confined high temperature plasma and has the net result of converting five deuterium atoms into one atom each of helium-4, helium-3, and hydrogen, in addition to two neutrons, releasing 24.9 MeV of energy in the process. An alternate and, in theory, easier reaction to obtain is the D–T reaction which involves the fusion of one atom of deuterium and one atom of tritium to produce an atom of helium-4, one neutron, and 17.6 MeV of energy. The neutrons generated in the D–T reactor must be captured in a lithium-6 blanket producing another atom of helium-4, an additional 4.8 MeV of energy, and most importantly, an atom of tritium which is recovered and fed to the fusion, device along with fresh deuterium to fuel the reactor. Thus, the D–T concept produces a net 22.4 MeV for each deuterium atom supplied, but also requires a continuous supply of lithium-6 and seems to be limited by the availability of this material rather than deuterium.

There are 1.028×10^{25} atoms of deuterium per cubic meter of seawater or about 1.5×10^{43} atoms in all of the oceans. Recovery and utilization of 1% of these in a D–D Reaction would provide 1.14×10^{26} Btu of energy.

The measured and probable resources of lithium in the U.S. amount to about 21×10^6 tons containing 6.54×10^{34} atoms of lithium-6. Since the number of lithium-6 atoms recoverable appear to be less than the deuterium resources, the energy available through the D–T reaction is limited by the availability of lithium-6 to about 2.2×10^{20} Btu, an amount comparable to the energy content of the world's coal resources.

Geothermal

The temperature of earth increases with depth at the rate of about 2°F every 100 ft in rock and at about 3 times that rate near natural steam sources. The thermal energy in deep underground zones can be recovered either by tapping natural steam or hot water sources, or by fracturing hot rock formations and flooding them with water to generate steam for use in a conventional steam-electric power plant. As this energy is withdrawn it is replaced by the normal conduction of heat from below. The installed electrical capacity in 1969 of such power plants was about 82 MW in the U.S. and 830 MW in the world.

Recent surveys indicate a potential electrical output of 20,000 MW from natural steam sources in California's Imperial Valley, with a like amount available in Oregon. The use of nuclear explosives to fracture hot rock formations can extend the availability of this resource considerably and is under study by the AEC.

Solar energy and derivatives

The rate at which solar energy intercepts the diametral plane of the earth is about 17.7×10^{10} MW. At the surface of the earth the average solar power is about 1820 Btu/ft^2-day but this figure can vary widely depending upon the location and, of course, is only available during sunny periods. This energy may be captured and stored in thermal form for space heating, or converted to electricity directly using solar cells.

Water power is derived from the evaporative effects of solar energy on bodies of water, causing subsequent cloud formation and precipitation over higher land areas. The potential energy of the water may be recovered to produce electricity by installing turbine-generator systems at selected locations along the rivers and streams that return the water to the sea. The Federal Power Commission (FPC) estimates the electrical production potential of hydropower sources in the U.S. at 161,000 MW with about 45,000 MW of this developed by 1966. The worldwide potential was estimated at 2.9×10^6 MW with only 7.5% of this developed as of 1964.

The development of hydropower sites generally involves flood control and the creation of recreational areas and may also involve pumped storage facilities.

Wind and wave power are also derived from solar energy as are the temperature gradients between the surface and deeper parts of the ocean. Proposals have been made to utilize these energy sources but they appear to be unsuitable for large-scale exploitation.

Tidal power

This energy resource results from gravitational interaction between the earth and the moon. It may be harnessed by constructing dams in tidal areas to create a difference in elevation between two bodies of water. The potential energy in the higher body of water can then be recovered to produce electrical power by the same techniques used in hydroelectric plants. The maximum energy recoverable is proportional to the area of the basin and to the square of the tidal range.

In North America various sites on the Bay of Fundy have a total potential for the production of about 29,000 MW of electrical power, although little of this can be developed economically.

Refuse

Solid waste products including paper, rags, cardboard, and plastics may be considered as an energy resource. Combustible wastes may be used as a fuel for electric power production and process heat and have a heating value of about 5000 Btu/lb. If fully utilized, these wastes could furnish about 10% of the fuel needs of central station power plants. Environmental problems are encountered in burning this material just as in the incineration of waste, and emissions depend upon the chemical characteristics of the refuse.

An alternate use of refuse, studied by the Bureau of Mines,[35] involves its conversion by pyrolysis into fuels such as combustible gases and light oils as well as other useful by-products. A ton of urban, refuse containing mainly plastics can be converted, by pyrolysis at 900°C, into 382 lb of solid residue, 1.5 gal of tar, 0.5 gal of light oil (mostly benzine), 97.4 gal of aqueous liquor, 31.5 lb of ammonium sulfate, and 18,058 ft^3 of gas consisting of hydrogen, methane, CO, and ethylene. Other processes may be used to produce predominantly liquid fuels rather than a gaseous product.

Agricultural wastes, from animals and vegetation, are more abundant than urban, wastes and constitute a large, though difficult to exploit, energy resource.

Energy Conversion

Most of the air pollutant and waste heat discharges attributable to energy-related activities are released in the energy conversion step. Here, energy resources are consumed to provide other energy forms either for direct utilization or for transmission to centers of energy demand.

The range of technological alternatives available in energy conversion is illustrated in Fig. 5. The columns and rows represent input and output energy forms, respectively, with specific devices or power plants capable of making the transformation indicated in the appropriate block. The most widely used systems involve chemical (fossil fuel) and nuclear energy inputs and the following conversions or series of conversions:

1. For space heat (non-electric);
   chemical → thermal.
2. For the generation of electricity;
   chemical or nuclear → thermal → kinetic → electrical.

To \ From	Electro-magnetic	Chemical	Nuclear	Thermal	Kinetic (mechanical)	Electrical	Gravitational
Electro-magnetic		Chemilumines-cence (fireflies)	Gamma reac-tions (Co^{60} source) A-bomb	Thermal radiation (hot iron)	Accelerating charge (cyclo-tron) Phosphor	Electromagnet-ic radiation Electrolumines-cence	Unknown
Chemical	Photosynthe-sis (plants) Photochemis-try (photo-graphic film)		Radiation catalysis (hydrazine plant) Ionization (cloud cham-ber)	Boiling (water/steam) Dissociation	Dissociation by radiolysis	Electrolysis (production of of aluminum)	Unknown
Nuclear	Gamma-neu-tron reactions ($Be^9 + \gamma \rightarrow Be^8 + n$)	Unknown					
Thermal	Solar absorber (hot sidewalk)	Combustion (fire)	Fission (fuel element) Fusion		Friction (brake shoes)	Resistance-heating (electric stove)	Unknown
Kinetic	Radiometer Solar cell	Muscle	Radioactivity (alpha parti-cles) A-bomb	Power cycles (thermal expan-sion turbines, combustion engines)		Motors Electrostriction (sonar trans-mitter)	Falling object (water)
Electrical	Radio anten-na Solar cell	Fuel cell Batteries	Nuclear bat-tery	Thermoelec-tricity, Therm-ionics, Ther-momagnetism, Ferroelectric-ity	MHD Conventional generator Piezoelectricity		Unknown
Gravitational	Unknown	Unknown	Unknown	Unknown	Rising object	Unknown	

(Source: Adapted from USAEC matrix.)

Figure 5 Energy conversion matrix.

3. For the propulsion of a vehicle (non-electric); chemical → thermal → kinetic.

In the conversion from thermal to kinetic energy a power cycle is used and its efficiency is limited by the second law of thermodynamics to a value less than that of the theoretical Carnot cycle. The Carnot cycle efficiency is equal to the difference between the temperatures at which heat is supplied to and rejected from the cycle working fluid, divided by the absolute temperature at which the heat is added. This theoretical limit on conversion efficiency increases for a given heat rejection temperature as the temperature at which heat is supplied to the cycle working fluid is raised.

A 1968 inventory of prime movers that are involved in the kinetic energy portion of various conversion systems is given in Table XVII. Although automotive engines make up over 90% of the total installed horsepower, this figure is deceptive because of differences in the annual usage of the various power plants. A better indication of their relative use can be obtained from fuel consumption data.

The state-of-the-art in energy conversion systems and associated operations has been reviewed in a study performed for the OST[36] and by the FPC in both its 1964 and most recent 1970 National Power Survey.[29] It is convenient in reviewing alternative power devices that might be used to reduce harmful environmental effects, to maintain

TABLE XVII

Total Horsepower of all Prime Movers, 1968
(10^6 horsepower)

Item	Horsepower
Automotive[a]	16,854
Nonautomotive	
Factories	52
Mines	43
Railroads	58
Merchant ships and vessels	20
Farms[b]	292
Electric central stations	372
Aircraft	137
Total	17,828

[a] Includes cars, buses, trucks, and motorcycles.
[b] Includes about 1.3×10^6 hp in work animals and 22,000 hp in windmills.
(Source: Statistical Abstracts of the U.S., 1970.)

the distinction between ecntralized and decentralized power plants, where the latter may be either stationary or mobile.

Centralized power plants

To take advantage of economies of scale, electrical energy is generated in large centralized facilities for distribution to consuming activities. The demand for elecrtical energy varies on a daily, weekly, and seasonal cycle. Since this form of energy cannot be conveniently stored, thc installed generating capacity must match the peak demand and some of this capacity must be idle during periods of low demand. The more economic plants in the system, which may be fossil or nuclear facilities operating on a steam power cycle or hydroelectric plants, are usually operated as much as possible in a so-called base-loaded condition, while the older and less efficient plants and special gas-turbine installations are utilized only during periods of peak demand. The use of pumped storage can serve to flatten out the demand curve by using excess power produced during periods of low demand to pump water uphill where it can be used during later periods of high demand by recovering its potential energy. Transmission ties between electric utility systems in different regions of the country can be used to take advantage of load diversity since the timing of seasonal and daily peak demands varies depending on location.

Electric generating plants with an output of over 1000 MW are not uncommon and stations of up to 10,000 MW capacity are anticipated by the year 2000, perhaps involving an energy center concept where the power source is integrated on the same site with industrial, agricultural, and/or desalination operations. The concentrated environmental effects of large central station power plants are receiving increased attention and have been the subject of congressional hearings.[37] Natural gas and low-sulfur residual oil may be used in steam-electric plants with minimal air pollution but these fuels are limited in their availability. The provision of control systems for SO_2 and NO_x emissions from steam plants fired with coal or high sulfur oil will increase the cost of power generation. Baron[38] estimates that flue gas scrubbing would add about 7% to the cost of power generation in a 1000 MW output plant using oil with a 1% sulfur content. The provision of dry cooling towers that reject waste heat to the atmosphere with no water usage would increase the cost in the same plant by another 30%. The EPA has recently proposed[39] emission standards for coal-fired steam generators that would limit particulate releases to 0.2 lb per 10^6 Btu, SO_2 releases to 1.2 lb per 10^6 Btu, and NO_x to 0.7 lb per 10^6 Btu heat input.

The trend in central station power plants is towards more efficient fossil-fueled energy conversion systems and systems that can utilize the more abundant nuclear fuels. Improvements in the efficiency of energy conversion reduce the fuel consumption and attendant environmental effects directly for a given electrical output. Steam-cycle power plants operating with fossil fuels attain a thermal efficiency of about 40% with a maximum steam temperature in the power cycle of 1000° to 1050°F, which is well below the combustion temperature of these fuels. Current nuclear power plants operate with a maximum steam temperature of about 550°F and have a lower thermal efficiency. The cycle efficiency increases as the steam temperature is raised but the more severe temperature and pressure conditions would lead, in both fossil and nuclear systems to materials problems and higher plant capital costs that would more than offset the economic gains of increased efficiency. Topping cycles, which are added to the steam cycle and use other thermodynamic working fluids such as mercury or potassium in higher temperature ranges, therefore providing higher efficiency, have been studied for fossil-fueled systems and actually operated in the case of mercury but the gains appear to be marginal. Thermonic topping devices[40] that convert thermal energy directly to electrical energy have also been proposed and promise to increase the over-all efficiency of conventional plants to about 50% performance must be improved before they can be used to advantage. A combined cycle, involving a gas turbine topping unit that would operate at high temperature on gas produced from coal and would reject its heat to a conventional steam cycle, is another promising system and could have an over-all efficiency in the 55 to 60% range.

Some promising alternatives for centralized power production that can utilize fossil resources are fuel cells and magnetohydrodynamic (MHD) generators. The number of nuclear reactor plants being installed by utilities is expanding rapidly and further developments in breeder reactors and fusion systems are also anticipated. Solar power plants are of interest in the longer term as a clean energy source.

Fuel Cells: This direct conversion device is not limited in efficiency by thermodynamic considerations and can convert chemical energy directly into electrical energy without the intermediate conversion to thermal and kinetic energy. Over-all efficiencies as high as 75% have been attained in special purpose applications. With hydrogen and oxygen fuel fed to electrodes submerged in a suitable electrolyte, this device produces 0.7 to 0.85 volts (dc)/cell and many cells must be connected in series to produce useful voltages for transmission. Development work is in progress on cells that could operate on gasified coal or other gaseous fuels with air as the oxidizer and a conceptual design has been completed for such a plant.[41]

Air pollution and waste heat discharges would be reduced in a fuel cell power plant because of the improved efficiency of fuel utilization. If coal were used as the source of a gaseous fuel, the impurities in the coal and particulates would have to be controlled in the gasification plant, but such control would be easier to provide in this operation than in a conventional power plant. In the course of the development of fuel cells, it is more likely that they will be applied first to smaller decentralized power systems.

Magnetohydrodynamics: MHD generators can utilize either a fossil or nuclear thermal energy source. In a fossil-fueled system, hot combustion gases, seeded with potassium or cesium to make the gas conductive, are expanded at high velocity through a magnetic field. The dc current produced in the moving conductive gas is picked up at electrodes embedded in the walls of the gas channel in various geometries depending upon the particular type of generator. The advantage of this system is its high thermal efficiency of 50 to 60% when operated at a gas temperature in the 4000° to 5000 °F range. A panel[42] has conducted a study of MHD for the OST and concluded that additional research and development should be performed before an MHD plant is constructed. Much of this development effort must be directed towards materials research to permit extended operation at high temperatures.

The environmental effects of MHD generation would be similar to those in any fossil-fueled combustion system but would be mitigated because of the improved efficiency of conversion. The seed material in the gases could not be released in any case, for economic as well as environmental reasons, and must be recovered. It is anticipated that recovery of the other pollutants can be accomplished in the same operation.

Fission Reactors: Current light-water cooled nuclear reactor systems used for electric power production operate with fuel slightly enriched in the uranium-235 isotope and have a conversion ratio of about 0.6, i.e., they generate about 60% as much new fuel as they consume by converting fertile uranium-238 into fissionable plutonium-239. These plants have a thermal efficiency of about 33% because of limitations on the temperatures at which they can operate. Alternate reactor systems that are under study, including the molten salt reactor (MSR) and the high temperature gas-cooled reactor (HTGCR), can generate steam at conditions comparable to those of fossil-fueled plants and the molten salt system, when fueled with uranium-233 fuel that is produced from thorium by neutron capture in a reactor, can breed about 5% more fuel than it consumes. All of these systems operate with a thermal (or slow) neutron spectrum in the reactor. By operating with a higher energy neutron spectrum in a fast reactor, better neutron economy can be achieved and more neutrons are available for the conversion of fertile material. Such reactors may be cooled with helium, steam, or liquid sodium. These fast breeder reactors, when developed and integrated into the centralized power system, will greatly extend the nuclear energy resources by fully utilizing the more abundant uranium-238 isotope to produce additional fissionable fuel. An analysis of the various reactor concepts and their potential roles in a nuclear power economy has been performed by the AEC[43] and led to the selection of the liquid metal cooled fast breeder reactor (LMFBR) as the high priority development task.

The various topping and MHD systems discussed previously are also applicable to nuclear power plants and may be useful additions if higher reactor operating temperatures can be reached.

All reactor systems generate similar quantities of radioactive materials for a given thermal energy output and the handling of these materials is closely controlled during all fuel cycle and reactor operations. The amount of such materials produced, and waste heat discharged, for a given electrical output depends, of course, on the plant efficiency. It is important to recognize that in all reactor systems the cooling water that carries away the waste heat is physically separated from the reactor primary coolant whether it be gas, water, or liquid sodium. Because of the large quantities of radioactive materials present in the fuel region of a reactor during operation, there is concern over incidents that might accidently discharge this material to the environment and no nuclear plants have been constructed in regions of high population density. Protection against such discharge is provided by safety systems that minimize the possibility of an accident and by incorporating several physical barriers within the plant to contain the radioactive products in the unlikely event that they are released from the fuel region.

Fusion Reactors: The successful development of a controlled and economic fusion device would make available the tremendously large energy reserves of deuterium in sea water. Current research is directed toward devices in which the combination of plasma temperature, density, and confinement time required for a sustained reaction may be attained. The status of this research and the outlook for fusion power has been summarized by Rose.[44] Some estimate that the feasibility of fusion reactors may be demonstrated by 1980 and that a pilot plant of several megawatts output may be constructed by 1990. The deuterium–tritium (D–T) reaction, wherein these materials are confined as a plasma at a temperature of about 40×10^{6}°C in a high strength magnetic field while they react, releases most of its energy in the form of neutrons that must be captured in a lithium blanket to generate new tritium fuel. The blanket fluid may then be circulated to generate steam for a conventional steam-cycle system. Topping cycles may also be used depending on the operating temperatures and the over-all efficiencies will be competitive with fossil-fired power plants.

The deuterium–deuterium (D–D) reaction appears to be more difficult to attain, requiring confinement at a temperature of some 350×10^{6}°C, but releases much of its energy in the form of charged particles which might be used to generate electricity directly without using a power cycle. Higher operating efficiencies may be achieved in such a system. There are still other fusion reactions that are of interest because of their special characteristics.

It now seems that fusion power will be more economical in very large plants of greater than 5000 MW capacity and such plants will pose substantial waste heat disposal problems unless very high efficiencies are attained. In the D–T concept, tritium must be recovered from the blanket and handled during the fuel cycle. Another source of radioactive products is the structural material that is activated by the neutrons produced in either reaction. In any case, the problems of disposing of radioactive materials appear to be less severe than those in fission reactors and, because of the small amounts of fuel material present in the device at any given time, the accident hazard is apparently minimal.

Solar power plants: Several concepts for utilizing solar energy in a centralized power plant have been proposed. Glaser[45] outlined a conceptual system that uses a satellite with a large array of solar cells placed in a synchronous orbit. The solar cells would convert the solar energy to electricity and this energy would be beamed by microwave to a receiving station on earth. To supply the annual electrical energy demand of the northeast region of the U.S. in 1980, estimated at 1.7×10^{15} Btu, a 105 square mile array of silicon cells would be required on the satellite and the receiving antenna on earth would be of the order of 20 square miles. Another scheme proposed by Meinel[46] involves a terrestrialbased plant that would capture thermal energy from the sun by means of specially coated surfaces to provide a "greenhouse" effect. A working fluid would be heated to a temperature of about, 1,000°F for use in a power cycle to produce electricity. A terrestrial-based system of 1,000 MW electric capacity would require a land area of about 5.5 square miles. The satellite concept, like the utilization of fossil and nuclear resources, does add to the thermal burden that must be dissipated to the environment, while the terrestrial-based solar plant merely implies a redistribution of energy that impinges on the earth in any event.

Decentralized stationary power plants

Energy conversion sytstems currently utilized in this category involve mainly combustion devices for space heat and industrial process heat applications. For economic reasons they are generally not as amenable to as high a level of emission control as larger centralized systems, save for conversion to cleaner fuels such as natural gas and perhaps methane or hydrogen produced from coal. These applications are also candidates for conversion to electrical energy thereby shifting the emission problem from decentralized plants to centralized power generating facility.

Decentralized electric power generating facilities of up to 5 MW output are sometimes installed in shopping centers and housing complexes. These now use diesel generators or gas turbines but could also use fuel cells when these are developed. Waste heat from these power plants is used for space heating and air-conditioning, and it is claimed[47] that up to 85% of the energy in the fuel may be effectively utilized in these systems as compared with only about 40% in central station power plants.

Solar energy has long been studied for individual home use in space conditioning and water heating but involves a substantial capital investment in these applications. Its use in certain areas will probably increase as fossil fuel and electric costs rise.

Nuclear reactors are not competitive in small decentralized applications for economic and technical reasons partially attributable to the requirement for shielding and other auxiliary systems. They are used, like other new devices such as fuel cells, in special military and space applications and may have potential in very large aircraft. Certain isotopes, produced by nenutro irradiation in a nuclear reactor, have also been used as energy sources for small power systems in space and remote terrestrial applications.

Mobile power plants

Energy conversion devices used in motor vehicle, rail, and aircraft applications must provide acceptable performance characteristics in terms of horsepower-weight ratio and specific fuel consumption. The gasoline-fueled, internal-combustion (I–C) engine has dominated the motor vehicle field and the gas turbine has achieved the same level of preeminence in aircraft. An appraisal of automotive transportation in relation to the problem of air pollution was performed by a Department of Commerce panel.[48] Alternate systems were reviewed and a five-year development program was recommended to support innovative developments on energy sources, propulsion systems, and emission control devices. Alternate energy sources were considered further in system studies performed on electric vehicles by A. D. Little[49] and on the whole spectrum of unconventional low pollution-potential power sources by Battelle Memorial Institute.[50] It also is possible to use alternate fuels such as propane, methanol, natural gas, and hydrogen in I–C engines to reduce the level of emissions. Proipulsion devices that may have some future potential include gas-turbines, external combustion systems operating with a steam or organic Rankine cycle or with a Stirling cycle, fuel cells, and battery driven electric motors. The characteristics of some alternative power plants are illustrated in Fig. 6 which indicates the range of specific power in watts/lb and specific energy in watt-hr/lb available with these devices. The propulsion system requirements for a 2000 lb vehicle propelled with a 500 lb power source at speeds of 20, 40, and 60 mph, and for vehicle ranges of 50, 100, and 200 miles are also indicated. From these characteristics it may be seen that the gas turbine and external-combustion engine are the systems most competitive with the current I–C engine. Various types of battery systems are shown and these as well as fuel cell systems may be applicable to limited-purpose vehicles. The sodium–sulfur and lithium–chlorine systems may approach the performance of combustion engines but are still in the early development stages. Introduction of a battery driven vehicle would shift the emission problem from the automobile to a central station power plant where the electrical energy required to recharge the battery would be generated.

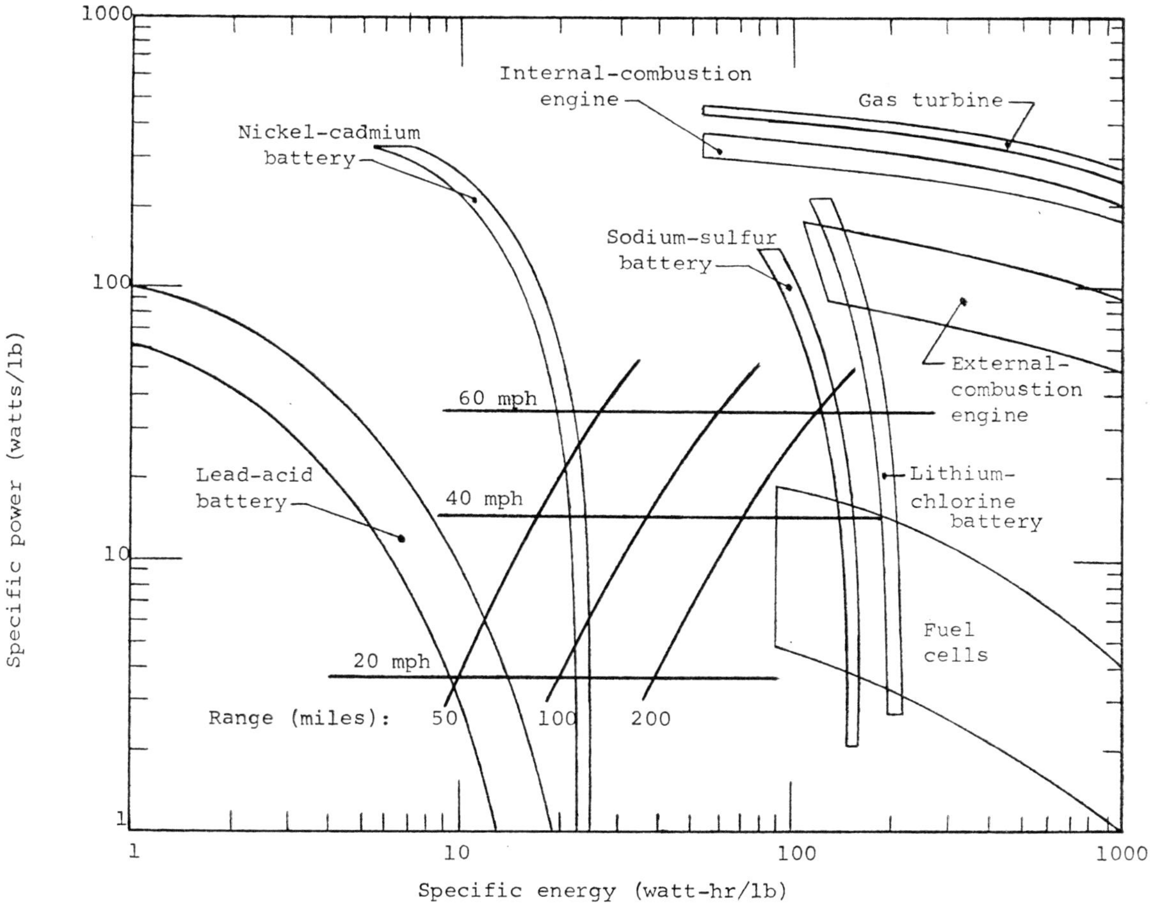

Figure 6 Vehicle requirements, and characteristics of mobile power plants for a 2000 lb vehicle with a 500 lb power source.[48]

TABLE XVIII

Emissions from Alternate Vehicular Power Plants

	Emissions (g/mile)		
	HC	CO	NO_x
California standard, 1971[a]	2.2	23	4
EPA standard for 1975[b]	0.41	3.4	3
EPA standard for 1976[b]	0.41	3.4	0.4
Gasoline I–C engine, no controls	11.6	64	5.6
Natural gas I–C engine[a]	1.5	6	1.5
Diesel engine[c]	3.5	5	4
Regenerative gas turbine[c]	0.32	3.5	1.9
Stirling engine[c]	0.1	1.0	2.6
Steam engine[a]	0.2	1.0	0.4
Electric battery vehicle[d]	0.04	—	4.3

[a] Congressional Record H 903, February 17, 1970.
[b] Federal Register, July 2, 1971, pp. 12658.
[c] Personal communication, R. Tom Sawyer.
[d] Reference 51; emissions in this case are from a coal-fired central power plant and also include about 16.25 g of SO_2/mile and 2.1 g particulates/mile.

Emissions from automotive power plants are variable and depend upon engine, load, level of maintenance, and individual driving habits. To allow for variable engine loads, emission standards are often related to a given driving cycle involving specified warm-up, acceleration, deceleration, cruise, and idle periods. The hydrocarbon, CO, and NO_x emission standards now in effect in California and those established by the EPA for 1974 concurrent are shown in Table XVIII along with emissions from a representative uncontrolled gasoline fueled I–C engine as well as alternative power plants. The pollutants from current I–C engines have been reduced from uncontrolled levels by control of gasoline evaporation sources and by crankcase ventilation systems that recirculate cylinder blow-by gases back to the combustion chamber. Further reductions of CO and hydrocarbon releases will involve high temperature or catalytic reaction of exhaust gases. The advantages of the gas turbine and external combustion engines, which can operate on distillate fuels such as kerosine, over the I–C engine lie in the better control of combustion conditions that can be achieved by separating the fuel combustion chamber from the working parts of the engine. The emission figures for a battery driven-electric vehicle indicated in Table XVIII were developed by Agarwal[51] on the bases that the electrical energy is provided from a central station power plant. It was assumed that 0.95 lb of coal was burned at the power plant for each mile of vehicle travel. In this case the emissions would also include an SO_2 release of 16.25 g/mile if the coal contained 2% sulfur and no control were provided.

Diesel engines used in trucks, buses, and trains are accountable for a small fraction of transportation vehicle emissions but have local objectionable effects such as smoke and odor. Aircraft engines have similar effects but in the long run may be able to effectively utilize alternate fuels such as methane and hydrogen that will reduce emissions.

The expansion of mass transportation facilities to reduce automobile usage, especially in urban areas, is an effective way to utilize energy more efficiently in moving people from place to place. A bus carrying 50 people and traveling at a speed of 60 mph may consume about 0.003 gallons of fuel/passenger mile, while an automobile carrying three peopel will consume about 0.023 gallons of fuel/passenger mile traveling at the same speed with a corresponding increase in emissions and traffic congestion.

Conclusion

The role of energy in the well being and progress of a nation is boviously quite complex, involving many conflicting social, economic, and environmental factors. It is generally recognized that a national energy policy is urgently needed. In 1968, Resources for the Future completed a study for the OST[52] in which the research tasks that should be performed to provide a basis for an energy policy were identified. The President's energy message of June 4, 1971[53] includes recommendations for a reorganization of federal organizations to provide for a more integrated energy policy. In this message, a broad program is outlined to accelerate the development of fast breeder reactors, SO_2 control techniques, coal gasification, and several other energy conversion techniques. The program also includes the development of leasing provisions for geothermal fields and shale oil resources and the acceleration of gas lease sales on the Outer Continental Shelf, steps are that intended to increase the availability of fuels which are or will be in short supply. The balanc eof th eprogram outlined in the energy message includes programs to utilize energy more wisely and specifies increased insulation requirements in Federal Housing Administration standards for federally insured homes.

Increasing attention is being directed toward individual energy utilizing activities, such as space heating and cooling, to determine how more efficient systems and practices can serve to reduce per capita energy demands.

The views of the electric power and petroleum industries regarding future developments in the energy sector of the economy and the need for research and development are outlined in two reports published in the summer of 1971. The National Petroleum Council study[54] estimates a capital investment requirement of some 374 billion dollars in energy systems, including oil, gas, and lectric utility facilities, during the 1971–1985 period. The Electric Research Council report[55] recommends a 32.5 billion dollar R & D program for the 1972 through year 2000 period for the electric utility industry alone.

REFERENCES

1. *United Nations 1969 Statistical Yearbook*, Statistical Office of the U.N., Department of Economic and Social Affairs, New York, 1970.
2. Ayres, R. U., and A. V. Kneese, Production, Consumption and Externalities, *American Economic Review*, June 1969.
3. *Minerals Year-book*, 1968, Vol. I–II, Metals, Minerals, and Fuels, USGPO, Washington, D.C., 1969.
4. Morrison, W. E., and C. L. Readling, an Energy Model for the United States, Featuring Energy Balances for the Years 1947 to 1965 and Projections and Forecasts to the Years 1980 and 2000, U.S. Dept. of the Interior, Bureau of Mines Information Circular 8384, USGPO, Washington, D.C., 1968.
5. A Review and Comparison of Selected United States Energy Forecasts, prepared for the Office of Science and Technology by the Battelle Memorial Institute, December 1969, USGPO, Washington, D.C., 1970.
6. Schurr, S. H., and B. C. Netschert, *Energy in the American Economy, 1850–1975, an Economic Study of its History and Prospects*, the Johns Hopkins Press, Baltimore, Md., 1960.
7. *The Economy, Energy, and the Environment*, A Background Study prepared for the use of the Joint Economic Com-

mittee, Congress of the United States, September 1970, USGPO, Washington, D.C., 1970.

8. *Electric Power and the Environment*, A report sponsored by the Energy Policy Staff, Office of Science and Technology, August 1970, USGPO, Washington, D.C., 1970.
9. Rice, J. K., Coal Mining in an Oxygen-Free Atmosphere, American Society of Mechanical Engineers Winter Annual Meeting, Nov. 1970, New York, Paper No. 70-WA/PID-4.
10. *Nationwide Inventory of Air Pollutant Emissions 1968*, August 1970, U.S. Dept. of HEW, NAPCA Publication No. AP-73, USGPO, Washington, D.C., 1970.
11. *Compilation of Air Pollutant Emission Factors* (revised), U.S. Environmental Protection Agency, Office of Air Programs, Publication No. AP-42, 1972, USGPO, Washington, D.C., 1972.
12. *Control Techniques for Particulate Air Pollutants*, January 1969, U.S. Dept. of HEW, NAPCA Publication No. AP-51, USGPO, Washington, D. C., 1969.
13. *Air Quality Criteria for Particulate Matter, January 1969*, U.S. Dept. of HEW, NAPCA Publication, No. AP-49, USGPO, Washington, D.C., 1969.
14. *Environmental Quality*, the First Annual Report of the Council on Environmental Quality, August 1970, USGPO, Washington, D.C., 1970.
15. *Control Techniques for Sulfur Oxide Air Pollutants*, January 1969, U.S. Dept. of HEW, NAPCA Publication No. AP-52, USGPO, Washington, D.C., 1969.
16. *Air Quality Criteria for Sulfur Oxides, January 1969* U.S. Dept. of HEW, NAPCA Publication No. AP-50, USGPO, Washington, D.C., 1969.
17. Bartok, W., *et al.*, *Systems Study of Nitrogen Oxide Control Methods for Stationary Sources*. Vol. II, November 1969, PB 192,789.
18. *Air Quality Criteria for Hydrocarbons*, March 1970, U.S. Dept. of HEW, NAPCA Publication No. AP-64, USGPO, Washington, D.C., 1970.
19. *Air Quality Criteria for Photochemical Oxidants*, March 1970, U.S. Dept. of HEW, NAPCA Publication No. AP-63, USGPO, Washington, D.C., 1970.
20. Bolin, B., The Carbon Cycle, *Scientific American*, Vol. 223, No. 3, September 1970.
21. McCaull, J., The Black Tide. *Environment*, Vol. 11, No. 9, November 1969.
22. Smith, J. W., Problems in Dealing with Oil Pullution on Sea and Land, *Journal of the Institute of Petroleum*, Vol. 54, No. 539, November 1968.
23. Martin, E. J., adn R. D. Hill, Mine Drainage Research Program of the Federal Water Pollution Control Administration, Preprints of Papers presented before the Second Symposium on Coal Mine Drainage Research, Mellon Institute, May 14–15, 1968, Pittsbuburgh, Pa.
24. Miller, A., *et al.*, Use of Steam Electric Power Plants to Provide Thermal Energy to Urban Areas, ORNL-HUD-14, January, 1971 National Technical Information Service, U.S. Department of Commerce, Springfield, Va.
25. Rock, R. L., and D. K. Walker, Controlling Employee Exposure to Alpha Radiation in Underground Uranium Mines, U.S. Dept. of the Interior, Bureau of Mines, USGPO, Washington, D.C., 1970.
26. Schneider, K. J., Solidification and Disposal of High-Level Radioactive Wastes in the United States, *Reactor Technology*, Vol. 13, No. 4, pp. 387–415.
27. Martin, J. E., E. D. Harward, and D. T. Oakley, Comparison of Radioactivity from Fossil Fuel and Nuclear Plants, Nov. 1969, Appendix 14, Hearings before the JCAE, Congress of the United States, Environmental Effects of Producing Electric Power, October and November 1969, USGPO, Washington, D.C., 1969.
28. Solid Waste Management, Prepared by the Ad Hoc Group for the Office of Science and Technology, May 1969, USGPO, Washington, D.C., 1969.
29. National Power Survey, Federal Power Commission, 1964, USGPO, Washington, D.C., 1964.
30. Manners, G., *The Geography of Energy*, Aldine Publishing Co., Chicago, 1967.
31. Hubbert, M. K., *Resources and Man*, Chapter 8, Energy Resources, W. H. Freeman and Co., San Francisco, 1969.
32. Lovejoy, W. F., and P. T. Homan, *Methods of Estimating Reserves of Crude Oil, Natural Gas, and Natural Gas Liquids*, The Johns Hopkins Press, Baltimore, Md., 1965.
33. DeCarlo, J. A., An Alternative to Oil Imports—Crude Synthetic from Coal, Oil Import Controls, Hearings before the Subcommittee on Mines and Mining of the Committee on Insular Affairs, House of Representatives, 1970, USGPO, Washington, D.C., 1970.
34. Schramm, L. W., Estimated Cost and Producing Capacity of a Government Assisted Shale Oil Industry, 1970 to 1980, Oil Import Controls, Hearings before the Subcommittee on Mines and Mining of the Committee on Insular Affairs, House of Representatives, 1970, USGPO, Washington, D.C., 1970.
35. Sanner, W. S., *et al.*, Conversion of Municipal and Industrial Refuse into Useful Materials by Pyrolysis, August 1970, Dept. of the Interior, Bureau of Mines, report of investigations 7428, USGPO, Washington, D.C., 1970.
36. Energy R & D and National Progress, Prepared for the Interdepartmental Energy Study by the Energy Study Group under the direction of Ali Bulent Cambel, USGPO, Washington, D.C., 1965.
37. Environmental Effects of Producing Electric Power, Hearings before the JCAE, Congress of the United States, October and November 1969, USGPO, Washington, D.C., 1969.
38. Baron, S., Cut Pollution at What Price? *Electrical World*, January 19, 1970.
39. *Federal Register*, August 17, 1971, pp. 15706.
40. Thermionic Topping Converter for a Coal-Fired Power Plant, Office of Coal Research R & D Report No. 52, U.S. Dept. of the Interior, USGPO, Washington, D.C.
41. Final Report, Project Fuel Cell, Office of Coal Research R & D Report No. 57, U.S. Department of the Interior, USGPO, Washington, D.C., 1970.
42. MHD for Central Station Power Generation: A Plan for Action, prepared for the Office of Science and Technology, by the Panel on Magnetohydrodynamics, June 1969, USGPO, Washington, D. C., 1969.
43. U.S. Atomic Energy Commission, Potential Nuclear Power Growth Patterns, WASH-1098, December 1970, USGPO, Washington, D.C.

44. Rose, D. J., Controlled Nuclear Fusion: Status and Outlook, *Science* Vol. 172, No. 3985, 21 May 1971, pp. 797 to 808.
45. Glaser, P. E., Power from the Sun: Its Future, *Science*, Vol. 162, No. 3856, 22 Nov. 1968, pp. 857–861.
46. Meinel, A. B., and M. P. Meinel, *Bulletin of the Atomic Scientists*, October 1971, pp. 32–37.
47. Clark, J. M., Jr., Total Energy, The State-of-the-Art, National Petroleum Refiners Association, National Fuels and Lubricants Meeting, Sept. 17–18, 1969, New York, Paper No. F & L-69-65.
48. The Automobile and Air Pollution, A Program for Progress, U.S. Dept. of Commerce, October 1967, USGPO, Washington, D.C., 1967.
49. George, J. H. B., L. J. Stratton, and R. G. Acton, *Prospects for Electric Vehicles, A Study of Low-Pollution-Potential Vehicles-Electric*, May 15, 1968, Arthur D. Little, Inc., Cambridge, Mass.
50. Hoess, J. A., and R. C. Stahman, Unconventional Thermal, Mechanical, and Nuclear Low-Pollution-Potential Power Sources for Urban Vehicles, International Automotive Engineering Congress, Society of Automotive Engineers, Detroit, Mich., January 13–17, 1969. SAE Paper No. 690231.
51. Agarwal, P. D., Electric Car and Air Pollution, Automotive Engineering Congress, Society of Automotive Engineers, Detroit, Mich., January 11–15, 1971. SAE Paper No. 710190.
52. *Resources for the Future, U.S. Energy Policies*, an agenda for research, Johns Hopkins Press, Baltimore, Md., 1968.
53. Office of the White House Press Secretary, The President's Message on Energy, June 4, 1971, Washington, D.C.
54. National Petroleum Council, *U.S. Energy Outlook, an initial appraisal*, 1971–1985, Vol. 1, July 15, 1971.
55. Electric Utilities Industry Research and Development Goals Through the Year 2000, Report of the R & D Goals Task Force to the Electric Research Council ERC Publication No. 1-71, June 1971.

KENNETH C. HOFFMAN
Brookhaven National Laboratory

ENVIRONMENTAL EDUCATION

Introduction

The planet earth has frequently been described as a space ship. The reason for this of course, is that all of the resources are finite and the only source of energy which permits the recycling and subsequent renewal of resources is the sun. Throughout the history of the earth changes have been taking place continually in terms of climate, atmosphere, the form of the earths crust and the living populations which existed on the earth. The natural system is equipped with a series of checks and balances which has up to now prevented any species from significantly affecting the balance of nature.

Mutations of various types of animals have brought about new species over long periods of time which were better adapted to the environment in which they found themselves. The natural balance had never been threatened by any species of animal until man. The effect which man has had on the planet earth, has in the past been extremely limited, due to the small numbers of humans which existed and their inability to control their natural enemies. But man has, through technology, been able to overcome a number of his natural enemies. He has been able to manipulate the climate, create an environment which was favourable for him to flourish and prosper. The very property which permitted man this freedom is now beginning to threaten his survival. Man's ingenuity or his ability to reason, has been the main force which has allowed man to overcome the hostile environment in which he has found himself.

The same ingenuity has developed a very advanced technology which is now beginning to have some impact upon the environment itself. As with a population of microoranisms the effect which they can exert upon their environment depends very much upon their numbers. Man's numbers are now increasing at an exponential rate. His present impact upon the environment is in reality quite small in a global sense, but it is none the less noticeable.

Fortunately large segments of the population have now begun to express a concern for the environmental deterioration which they have been helping to bring about. It is well known, that biological populations of almost any type of organism follow one of two different types of growth. These two different types of growth are

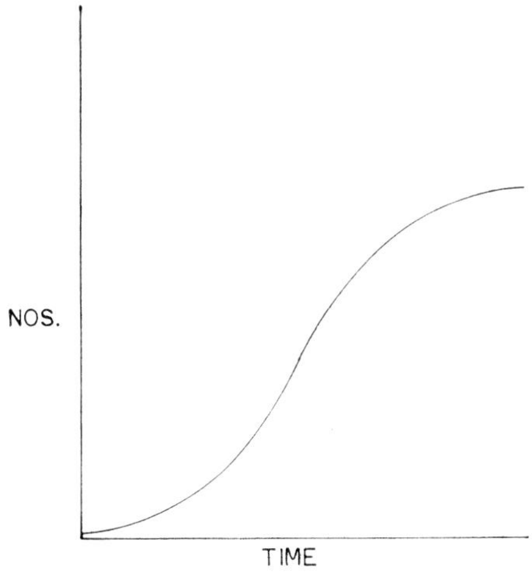

Figure 1

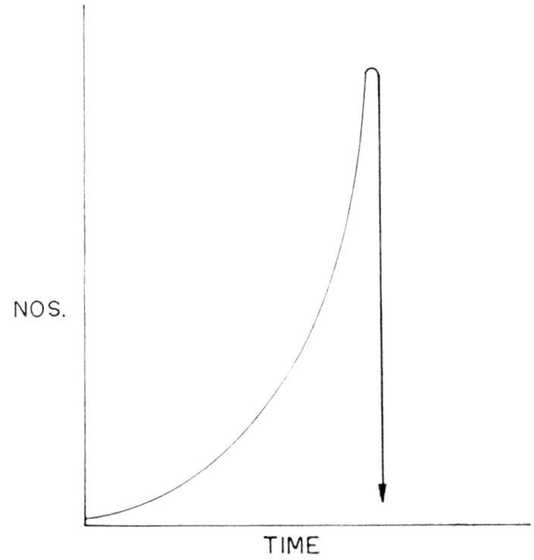

Figure 2

shown in Figs. 1 and 2. Figure 1 shows the traditional S type curve which allows a time lag, while the organism adapts to the environment. It is then followed by accelerated period, leading to an exponential rate of growth. This subsequently gives way to wath is called the stationary stage of growth when there is no net increase in numbers (death equals births). In most living systems, it is understood that the reason for this deceleration in growth may be any one or a combination of three reasons.

The first reason may be that there is physically no space left in which the organism may live. In terms of certain territorial creatures in nature, this stabilization of population occurs as a result of aggression. A further reason may be that there is no further food left to sustain a population any larger than exists. Then of course starvation controls the population level. The third reason, and the one which is most common in the case of microorganisms, is that the environment has been rendered incompatible with life. A death or decline phase follows as a result of a change of this nature. In the case of microorganisms this may be that they have rendered their environment acid and can no longer exist in this unfavourable environment. In the case of the human population, it may very well be that they have polluted the environment to the point where they can no longer exist comfortably.

The second type of curve (Fig. 2) is perhaps the more frightening one, because this indicates that the population grows extremely rapidly until it reaches some certain point and then it crashes down to zero. The reason for this crash may also be any one of the previous three reasons.

In those parts of the world where survival is not considered to be in doubt, in fact is almost assured, the population are starting to look very critically at the environmental degradation which is going on around them. In those other parts of the world, where survival is less certain, the aesthetic values and the long term effects are ignored in favour of the immediate problems of living or even surviving.

The ethics of the great religions of the world have not traditionally dealt realistically with the subject of man's relationship with and dependence upon a naturally balanced environment. In fact many of them have tended to show extreme compassion for organisms of a similar species and domestic animals and have in other ways exerted and abused "man's dominion over the earth." Man of course does not have dominion over the earth, he is a part of an intricate balance of living things. Up until recently, his efforts have had very little effect upon this matter of balance, but now with his advanced technology he has the capability of changing this balance and without the comprehensive wisdom required to predict the outcome, his activities normally produce a bad effect.

Many people are writing books in the early 1970's on the subject of man's survival in an environment which he is attempting to systematically destroy through his technology. This is perhaps encouraging as it does show that at least some percentage of the population are beginning to become aware of the real dangers of man's dominion over the earth. Man's thirst for knowledge of the environmental system is becoming greater year by year, but his frustrations at the obvious complexities and his inability to understand, is also becoming greater year by year. Thus it is particularly important that all of society become more familiar with the effects of their activities and demands up on the natural system.

The laws of thermodynamics dictate that the entropy of the universe tends to increase, and the planet earth being a part of the universe is not exempt from this law. Entropy being loosely defined as a state of confusion or disorder, it is clear that energy is required to maintain order or to avoid further confusion. If man's demands for energy exceed that which can reach the earth and be harnessed from the sun, then the system goes into a rapid state of decline. (Increased entropy.)

Thus environmental education becomes a natural and essential part of education for all people in all walks of life and at all ages. Much of this type of education has in the past been conducted within the family and very much based upon folklore and has largely dealt with the survival of the individual. Such things as knowing which berries one can eat, the dangers of eating poorly cooked pork, and not eating certain shell fish at certain times of the year (unless there is an *R* in the month). The new concern for the environment must be dealt with in a very much more formal manner and will require a fairly rapid and widespread dissemination of those facts of which we feel relatively certain, so that future societal judgements may be wisely made.

Education at Home

The family unit is not normally thought of as being one of the fundamental units of education. Yet it should be realized upon reflection that this is the place where most behavioural patterns are developed and in fact a very considerable amount of the deep rooted understanding that a child develops originates at his home.

It should also be recognized that no longer is this kind of education a one-way pipeline from the parents to the child but in fact there is a good deal of feedback from the children to the parents. This is particularly so after the child has entered school. Thus the home setting, if it is properly used, can be one of the most fundamentally important places for a mutual educational exercise to occur.

In terms of environmental concern and environmental awareness, the exchange of knowledge between parents and children will develop some kind of an ethic. This kind of ethic is sometimes referred to as "upbringing." Thus if individual members of the family are provided with an adequate education of their environmental responsibilities the family unit provides a multiplying effect for this information and attitude. Many bad attitudes are developed at home. We are all familiar with many cases where racial discrimination has permeated a family because of an attitude of one of the members of that family and because of the way he perceived the situation. Similarly, bad environmental habits are inherited from the previous generation.

In addition to having other members of the family as resources for the development of environmental understanding there are the many communication devices which now exist in the home. The television, the radio, the newspapers and magazines and periodicals. Popular type magazines, pseudo-scientific and in fact, the scientific magazines are now quite common in the home. In many homes the work that children bring home from school has a direct impact upon the family.

Of all these media the television appears to have the most significant impact whereas the newspapers and magazines have the most potential for developing the ideas and thoughts which are projected from the television screen or the radio speaker. The message which television provides for the family has developed into quite a dichotomy. Frequently we will see a public service type of program which cries out for conservation and environmental preservation. Between programs we may find commercials which are advertising some industrial product and demanding that you purchase one. Thus we have a message which is saying control your demands of the environment, control your demands of our natural resources and interspersed with these messages we have commercials insisting that we in fact increase our demands. A considerable amount more time is spent suggesting that we buy more goods than appealing to curb these demands. It is perhaps for this reason that the demands for goods are still high, production is high, the depletion of natural resources is high and the pollution of the environment is equally high.

Yet a new danger is developing in the electronic

medium. Many industries are now advertising in a conservation vein and in fact suggesting that in their particular case they are not only doing a lot but seem to have the problem of environmental protection completely under control. This of course is very misleading and can bring about a complacency which is not only unwarranted but in fact can be considered responsible for the unfortunate environmental deterioration which we are now experiencing.

The impact of the radio is diminishing in those very large areas where television reception is either poor or unavailable. By far the greatest impact of the radio must be outside of the family setting and in the case of North America in the mobile setting of a private automobile. A great deal of daytime programming is now being geared to this intimate personal conversation style.

However, both the radio and the television can only provide a very superficial look at any given problem and this look is of necessity only fleeting in nature. The role of explaining in depth the background of the story and the facts of the situation rests almost entirely with the newspapers and magazines. News items dealing with specific situations and developing the story in depth together with feature items explaining background facts to general situations provide the main source of factual information which is available to educate a family.

Here we find a new and contemporary problem which has never previously occurred. In the newspapers and the magazines we will find that eminent scientists are disagreeing with one another, and violently trying to convince society that their point of view is correct. Examples of this type of problem must be familiar to everyone. In a magazine we may have a very eminent and well respected scientist writing a very sober report which indicates that if we continue to burn fossil fuels and create carbon dioxide in the atmosphere at the rate which we presently are, a "greenhouse effect" will result and the temperature of the earth will go up. The result of this will be that the polar ice caps will melt and this in turn will raise the levels of the sea many feet and inundate the major seaports of the world. In this same magazine we will find an equally eminent and well respected scientist stating that if we continue to burn fossil fuels and release the particulate matter, which results from the incomplete combustion of certain materials, into the atmosphere we will develop a screen. This screen will effectively bar the solar energy from reaching the earth and ultimately of course will depress the temperature of the earth and result in another ice age.

This is the type of debate which is currently going on between scientists. This debate has gone on for centuries and in fact has been responsible for bringing about some of the major discoveries known to man and some of the major advances in technology. However, in the past, these debates have taken place behind closed doors, in the hallowed halls of universities and in the booming silence of royal societies. Now for the first time, the doors have been opened and the public have been allowed to see that the scientists in this technocracy, may have feet of clay.

In most governing systems in the world, the decision to do or not to do something does not normally rest with the experts. Public policy or decisions are taken by society as a whole. Society expresses itself through the political system. Thus the provision of conflicting evidence to the family unit allows the family the privilege which they never had before. The privilege to debate amongst themselves and come to a consensus. In a very informal way, this type of debate is extremely valuable and really provides the setting for an extremely useful seminar to occur for the benefit of all.

If the question is asked of any family or any society, are you against pollution?; are you in favour of the restoration and protection of the environment?; the answer of course, must be "yes." However, the other question must also be asked and debated and faced by the individual. That question is how much are you prepared to pay and how clean do you feel the environment should be, bearing in mind the cost? This type of question is very rarely asked. When society is on a witch hunt the very last thing they want to discover is that they in fact are the witches.

Public School

The children of course, have by this stage developed a certain attitude towards many things. These attitudes have been developed at home with their parents family and friends. However, children are not necessarily exclusively influenced at home. Environmental change can be demonstrated even to a kindergarten child. A very simple example of this is almost always readily available. One can take a child to a ravine or an area where a year or two ago he could find bullfrogs and snakes and mice only to find that now an apartment building, a parking lot or a dam has been built there. At this point a very simple discussion can be developed on the subject of choices. It can be shown that the town council had the opportunity to leave that field or ravine for the mice, snakes and frogs to live in or they could have put it to this new use which meant that all the mice and snakes and frogs would either have to go somewhere else or all die. This is a choice which has been made and a little child is quite familiar with the subject of choice.

Even the concept of cost benefit can be implanted in the small child because the youngest of children, even preschoolers, unconciously are continually making cost benefit analyses in their minds. If they are given a dime they go through an analysis which helps them decide whether they will buy a candy bar, an ice-cream cone or keep the dime. They are deciding what the cost is and what the benefit is and they are making a decision based entirely upon which will give them more pleasure. Thus the concepts of environmental protection, cost benefit studies, freedom of choice, social action can all be introduced at very early stages in school.

There is, however, nowadays, a tendency to try to bring environmental studies into a single category. Some school systems insist that environmental studies belong in the science, physics and chemistry department whereas some insist it belongs in the geography department. Others have taken it purely as a biology subject, something similar to nature study. Truely, of course, all of these are quite wrong. Environmental studies have a place in every academic subject which exists. Certainly chemistry, physics, social sciences, geography all have a very real and somewhat more obvious part in environmental studies. However, history and english and mathematics are amongst the academic subjects which all have a very real part to play in the study of the subject. Business and industrial arts on the technical side have a real role to play also. In other words, environmental studies cannot be separated into a physical or a social science, it involves all aspects and should be treated this way.

Many new books and many experimental curricula are being promoted for schools nowadays and many of them are quite good. The public schools have time for trips and visits to a greater extent than has ever been hitherto possible. Groups of small children may visit water treatment plants and sewage treatment plants, go to the council offices where decisions are being made, to the magistrates court where judgements are being passed, and to the local pond or brook to see the effect that their town has had on this natural environment. Considerable good work can be and probably will be done in this area but as in the case of the family, it requires that one of the participants in the exchange be well informed and understand the subject. In the case of the public school this necessarily implies that the teachers understand the subject.

High School

As we move from the family setting to the public school into the high school or technical school we are now dealing with people with a different degree of maturity and in a different state of emotional growth. This must be considered in the approach to educational technique. The very small child has the feeling of family security and has no particular desires for material things except those very small expendable items which give him immediate pleasure. As the young adult develops, an individual identity emerges and the inherited trait of wanting material things, changes his perspective on society.

It is hypothesized that man has a hierarchy of demands from simple survival and the prevention of starvation at one end of the spectrum to the other end where all his basic needs are satisfied and he has achieved a degree of affluence which allows him to demand those things which might be considered aesthetic. It should be pointed out here that many of those things which we might initially consider to be aesthetic in fact, are vital to the functioning of a natural system. This may not be immediately recognized by everyone. Thus teenagers or young adults, in seeking their own identities wish to develop that degree of independence which is demonstrated by having a number of physical durable assets of their own. The type of things we might consider here are televisions, radios, cars and of other major and expensive items. It should be pointed out that this is simply a hypothesis but physical needs satisfied are in effect the most easily influenced and the most sensitive to the problem of environment degradation.

It is thus at this adolescent level of education that the message must be passed along most subtly, it must in fact, permeate stories of history where it can be shown that battles were lost and wars were won because of water pollution bringing upon disease to communities. That various ways of life and cultures were completely wiped out because of environmental pollution. The economic consequences of, ceasing to produce certain goods might also be examined at this level of education the results of eliminating the pulp and paper industry from the Canadian National economy or eliminating the automobile from the American economy.

All of these ideas might be developed within the traditional structure of education but it requires once more personnel who are well informed with a good understanding of the environmental relationships with their subject or speciality. Public meetings, debates, extracurricula clubs, are all methods by which the high school student community can develop a keen environmental awareness and a good understanding of the complex interaction of all parts of the living system both man-made and natural.

Post Secondary Colleges

In many parts of the world the formal education system is finished for most people at the end of public school. For others it is completed at the end of high school, but for many parts of the world a post secondary education is available for a very large percentage of the population. In some parts of the world the only post secondary education which will be available is the university or the technical university. Some parts of the world are developing an intermediate between the highly academic requirements of the university and the high school. These might be described as post secondary colleges.

At all levels of education it is considered desirable to maintain some broad aspects of education as well as the concentrated effort in the field of specialization. This is particularly important when we pass the level of high school and reach into the level of optional education. At this post secondary level it is possible to introduce elements of multi-disciplinary studies at quite a fundamental level. At this point it might be useful to consider courses and seminars taught by a group of instructors with a wide diversity of disciplines and specialities.

Environmental studies may be examined in a series of courses not just utilizing instructors at the institution in question but also professors from neighbouring universities, specialists of various kinds from government and industry with one or two course co-ordinators to insure the continuity of the program. Here the subject of environment can be examined from cultural, historical and religious roots. Population and the politics of population control could be examined together with a demagraphic analysis of food and population. The subject of man playing a role in, but not dominating the natural system could be explained in considerable depth.

The growth ethos should be questioned. The thesis that "bigger is better." The type of training and education which could be made available and should be made available to the post secondary college is very similar to that which will be described in the next section for the university. However, the program should be geared to the intellectual demands which are expected of the particular institution.

University

Most universities are sub-divided into various administrative groups which are frequently further sub-divided into much smaller groups which tend to be based very much on disciplines. This of course was not always so, the original universities were extremely broad but also of course were extremely small. For administrative purposes it is extremely important to set up a series of manageable units. However for the purposes of environmental education these highly specialized water tight compart- ments known as disciplines do tend to inhibit the development of satisfactory programs in environmental studies. As we have already realized environmental studies involve all disciplines.

The disciplines as they are contained in the departments of a university are at their most highly honed and most penetrating level. This of course is as it should be and there is no case to be made for a large number of university professors who are "jacks of all trades and masters of none." The problem exists however, where the depth of interest and expertise which a man may havein his own discipline effectively excludes him from having any understanding of, or sympathy for other disciplines beyond his own. Thus develops the difficulty of communication.

In these highly specialized areas new and foreign "languages" are developed which are known only to the specialists working in these fields. Attempts to bring together multi-disciplinary groups, be they students or staff have frequently ended in frustration because the students and staff in one discipline completely fail to understand the students and staff in another, therefore, dismiss their contribution as gibberish and unimportant. The latter is not true and the difficulty arises only because the professor of discipline X is unable to communicate intelligently with professor of discipline Y.

Thus a somewhat major reorganization is required within most traditional universities to provide the framework for an adequate environmental education. In view of the fact that most of the high school and public school teachers are products of these universities it is extremely important that this type of educational experience is available to these people in order that they may be adequate resource personnel in the lower levels of school.

When we approach the subject from the top we will realize that the multi-disciplinary studies involved in environmental education in fact involves every single discipline within the university. Therefore it is apparent that the resources already exists to provide a satisfactory environmental educational program at any university. Experience however, has shown that the existence simply of resources is not enough. In order that a programme may develop effectively it is first of all necessary to have a small staff of professors each of which is highly skilled in a specialty which is different from his colleagues, and that have a common expressed intent to jointly develop an adequate program in environmental studies which will permeate the entire university and be available for students no matter what their parent department.

In the University's undergraduate areas there are at least two quite separate streams of studies which are identifiable and have different objectives. These are the professional courses such as medicine, law, engineering on the one hand and liberal education devoted to the development of well educated laymen on the other. In the case of the former, the curriculum is usually quite strictly controlled by the faculties in consultation with the professions which they are serving. The bar society or the medical association will prescribe certain areas of study which they consider to be essential for the adequate training and education of professionals in their field. This approach frequently leaves little room for optional courses but fortunately the trend nowadays is towards allowing some freedom for electing to study some non technical subjects.

It is quite evident to many people that the professions and particularly the engineering profession are quite deeply involved in the control and manipulation of the environment. It thus becomes extremely important, in fact vital that the professions be made aware of the environmental consequences of many of their activities. However, although it is extremely important it is still not central to the training of a professional. It becomes thus evident that the professional faculties while having a significant interaction with environmental studies should not provide the central home for these studies.

The liberal arts education on the other hand provides a setting for a broad and general education which may have quite a significant environmental orientation. In the letters arts and sciences divisions there is a good deal more freedom in the choice of subjects which a student may study in pursuit of their undergraduate degree. In this area, a spectrum of possibilities exist from the almost professional approach of, let us say, a fisheries biologist who may take a few courses of an environmental nature, through to the absolute generalist who is simply seeking an education and does not necessarily wish to become trained to undertake some job. In the case of the latter student there is a possibility that an undergraduate degree could almost exclusively be based upon environmental studies. There is still a considerable amount of disagreement on this subject and many people believe that it is impossible to grasp the fundamentals of the environment without first of all, understanding some of the natural sciences which underlie the environmental system. However, it is equally understood that it would be quite impossible for any student during a four year stay at the university to become extremely well versed in all aspects of natural and social sciences. Thus a spectrum of possibilities exist within a university which should be exploited to produce the diversity of expertise which is necessary for today's complex society.

Post Graduate

At the post graduate level of university education an even greater flexibility is required in order to satisfactorily meet the needs of environmental education. At this point it would be almost essential that a student would enter the programme with some area of expertise which had been developed in depth in the undergraduate years. There are some universities with the opinion that a student should receive a general environmental education at the undergraduate level and specialize into one aspect of it at the post graduate level. Presumably both of these approaches have some validity. In the case of the general under-graduate programme the student would attempt to pursue in some depth one area of expertise through one of the traditional disciplined graduate departments. Any difficulties which may arise here would result from a departmental requirement of some demonstrated competence in their field of specialty before admission to the department as a candidate for a graduate degree.

In those other cases where students received a fairly disciplined undergraduate training they might then undertake environmental studies in their post graduate years by a variety of different means. Certainly it would be essential that the student develop an understanding of the "languages" of some of the other disciplines. A student whose background is in say engineering should take some graduate level studies in some social or life science subjects in order to fully understand his colleagues in these other disciplines.

In addition to this the student should be involved in some truly multi-disciplinary educational experiences with graduate students of other disciplines, being taught by a number of professors from a similar variety of disciplines. At the graduate level this type of training can take the form of case studies. The interaction between the students and between the staff in the presence of the students results in a very stimulating type of education. Finally since today's society requires that a person taking a post graduate degree become a "master of something," it is appropriate that such a student should further strengthen his original discipline with advanced instruction in that field presumably with an emphasis on environmental problems.

Any research or project which is undertaken as a part of the degree might very well be jointly supervised by two or more staff members from different disciplines and in fact be integrated into one of the multi-disciplinary research projects undertaken by the academic group as a whole.

The major difficulty with an undertaking such as this is that it does not fit into the traditional university structure. Most universities throughout the world have for many years developed along the lines of the traditional discipline. These may be described as vertical educational institutions. The advent of multi or trans-disciplinary activities introduces a new concept which is a horizontal training process and therefore does not readily fit into the administrative structures of the university. Such problems of course can be overcome with enthusiasm, imagination and the determination to succeed.

Public

Universities, professional associations and private foundations all have a responsibility to provide ongoing education for the public sector. In the case of environmental education there are many different ways in which the public can be broken down for the purposes of education. First of all the popular lectures which are made available as public services through various instiutions can provide a very real impact upon the public. Secondly, there is the grouping which could be made of politicians who are making decisions by virtue of their elected office. These might be members of local town councils, county councils, regional councils, provincial governments, federal government, state governments and any other legislative lay body that is responsible for putting the wishes of the public into action.

Such ventures in educating the public may make use of all the media of communication and could vary. Lecture series offered during the evenings, weekend popular seminars, a series of feature articles in the newspapers and magazines, a series of light and interesting television or radio productions either through commercial or educational television networks. Such undertakings may originate from and be sponsored by any number of groups and it is most certainly not essential that it be the government or universities that sponsor these educational experiences.

The difficulties associated with any form of public education is of course the broad diversity of backgrounds which the participants bring to the meeting place. Whether this meeting place is the printed word in a feature article or a lecture in a local museum. Some attempts at sub grouping is almost essential to ensure that the interest and enthusiasm of all segments are maintained. Clearly a lecture or an article which is prepared for a group with a public school education is not going to have the same appeal for professionals with higher degrees and by the same token educational undertakings designed for people with higher education are going to fail to stimulate those with the limited schooling. Neither of these groups should be ignored as they all represent the combined force of society whose opinions are being sought through the political system.

Environmental education can also be routed through common interest groups such as conservation associations, church groups and professional organizations. In this case professional organizations are the logical vehicles for ensuring the compatibility of the instruction with the needs and capabilities of the audience. Furthermore, at the professional level in matters of environmental concern there could be and perhaps should be a greater degree of interaction between the various professional bodies which exist. Joint meetings of the law society, the medical association and the engineering profession could be extremely fruitful and might provide for an interim solution to the interprofessional communication problem until such time as the new academic generation graduates from university and reaches the degree of prominence in their professions that their voices are heard and their opinions heeded.

Conclusion

At this time in the affairs of man when much of the society of the global village has recognized the dangers of continued growth of population, use of energy, consumption of finite materials and development of technology, the subject of environmental education becomes of paramount importance. This understanding cannot be relegated to a rather pious Sunday activity which in no way affects the Monday, Tuesday and Wednesday activities but in fact it affects all of our activities at all times. Our concern is great but our understanding is little. Perhaps the future of our civilization depends upon an early understanding of our relationship with and dependence upon the natural system. It is said that all animals are aware but only man is aware that he is aware, and with this simple fact goes a very great responsibility which must be understood if our species are to prosper, or even survive.

PHILIP H. JONES
University of Toronto

ENVIRONMENTAL HEALTH

Introduction*

Definitions

"Environmental Health" encompasses what is also known as environmental engineering and sanitation, public health engineering, and sanitary engineering. It is concerned with "the control of all those factors in man's physical environment which exercise or may exercise a deleterious effect on his physical development, health and survival,"[1] with consideration of the physical, economic and social† impact of the control measures applied. Included is the application of engineering principles to the control, modification or adaptation of the physical, chemical, and biological factors of the environment in the interest of man's health, comfort, and social well-being. The concern is not merely with simple survival and prevention of disease and poisoning, although even these are not entirely under control.

* This chapter is based on material in a new edition of a book by the author entitled *Environmental Engineering and Sanitation*, John Wiley & Sons, Inc., New York, N.Y., 1972.

† Includes political, cultural, educational, biological, medical, public health.

In addition, environmental health involves the maintenance of an environment that is suited to man's efficient performance, and to the preservation of comfort and enjoyment of living today and in the future.[2/3]

The environment is defined as the sum of all external influences and conditions affecting life and the development of an organism. The Committee on Environment of the American Public Health Association goes a step further and says that "The Environment is considered the surroundings in which man lives, works and plays. It encompasses the air he breathes, the water he drinks, the food he consumes and the shelter he provides for his protection against the elements. It also includes the pollutants, waste materials, and other detrimental environmental factors which adversely affect his life and health."[4]

Scope

It is apparent from the above that the field of environmental health covers an extremely broad area of man's universe. The World Health Organization, the American Public Health Association, and others have suggested what the scope should be and how the programs should be administered. In all instances, a totality is envisioned with consideration of the impact of man on the environment and the impact of the environment on man, with a balanced appraisal and allocation of available resources to both.

A Report of a WHO Expert Committee[5] considers the scope of environmental health to include or relate to the following:

(1) Water supplies, with special reference to the provision of adequate quantities of safe water that are readily accessible to the user, and to the planning, design, management, and sanitary surveillance of community water supplies, giving due consideration to other essential uses of water resources.

(2) Waste-water treatment and water-pollution control, including the collection, treatment, and disposal of domestic sewage and other water-borne wastes, and the control of the quality of surface water (including the sea) and ground water.

(3) Solid-waste management, including sanitary handling and disposal.

(4) Vector control, including the control of arthropods, molluscs, rodents, and other alternative hosts of disease.

(5) Prevention or control of soil pollution by human excreta and by substances detrimental to human, animal, or plant life.

(6) Food hygiene, including milk hygiene.

(7) Control of air pollution.

(8) Radiation control.

(9) Occupational health, in particular the control of physical, chemical, and biological hazards.

(10) Noise control.

(11) Housing and its immediate environment, in particular the public health aspects of residential, public, and institutional buildings.

(12) Urban and regional planning.

(13) Environmental health aspects of air, sea, or land transport.

(14) Accident prevention.

(15) Public recreation and tourism, in particular the environmental health aspects of public beaches, swimming pools, camping sites, etc.

(16) Sanitation measures associated with epidemics, emergencies, disasters, and migrations of populations.

(17) Preventive measures required to ensure that the general environment is free from risk to health.

The APHA Committee on Environment[4] proposed the following program areas and also the planning considerations, and methods to implement programs.

I. *Environmental Program Area*

- Wastes
  - (1) Air
  - (2) Sewage and liquid
  - (3) Solid
- Water Supply
- Housing and Residential Environment
- Food and Drugs
- Radiation
- Noise
- Accidents
- Occupational and Institutional Hazards
- Vectors
- Recreation

II. *Planning Considerations*

- Health
- Economic
- Demographic and Land Use

Social
Esthetic
Resource Conservation
(Also manpower, facilities and services.)

III. *Methods and Technics*

Research
Demonstration
Education
Standards
Legislation
Inspection
Enforcement
Planning
Evaluation
Incentives
Systems Analysis

The discussions that follow will briefly review and introduce some of the major program areas listed above. Detailed treatment of some of the subject matter can be found in other chapters of this Encyclopedia.

Environmental Health and Engineering Planning

The principle of the total environment

The central community or city is dependent upon the surrounding areas for its human resources and its economic survival. The surrounding areas are dependent in whole or in part on the central community or city for employment, cultural and other needs. Increasing urbanization is erasing the boundary between the two except for the governmental jurisdictions involved and even these are being blended. The health, welfare, library, police, transportation, recreation, drainage, water, sewerage, and superimposed school systems are examples of combined services now provided in some areas. It is self evident that regional problems require regional solutions even though operationally individual control may be temporarily required as a political expedient until mutual confidence is established between adjoining governmental units. This emphasizes however the necessity of interim comprehensive planning so that adjoining governmental units may capitalize on improvements and services provided to the long term mutual benefit of their residents. For no one lives in a vacuum and the interdependency of governmental units must be recognized as a fact of life.

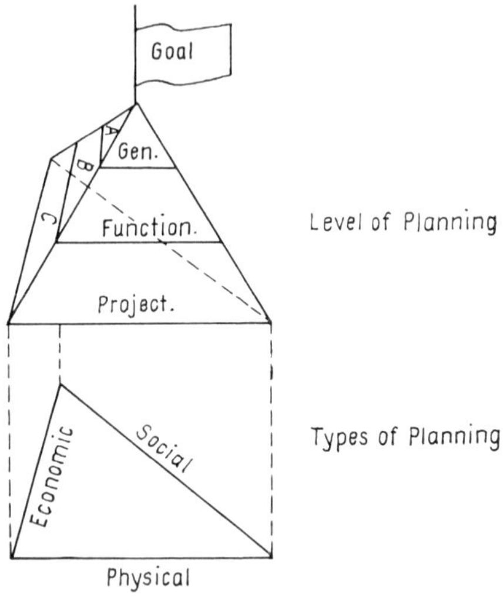

A. *General, Overall Policy Planning*—Identification of goals, aspirations and realistic objectives. Establishment of functional priorities.

B. *Functional Planning*—Such as for transportation, water supply, wastewater, recreation, air pollution, solid wastes, or medical care facilities in which *alternative functional solutions are presented*, including consideration of the economic, social and ecological factors, advantages and disadvantages.

C. *Project Planning*—Detailed engineering and architectural specific project plans, specifications, drawing and contracts for bidding. Plan of action.
Construction, Operation and Maintenance—Plan adjustment as constructed; updating and planning for alterations and new construction.

Figure 1 Types and Levels of planning. (Applicable to National, State, Regional and Local planning.)
From Joseph A. Salvato, Jr., *Environmental Engineering and Sanitation*, John Wiley & Sons, Inc., New York, N.Y., 1972.

Planning and the planning process

It is important that environmental health and engineering planning take place within the context of comprehensive regional or area wide planning. It is equally important however that comprehensive community planning fully recognize the environmental health and engineering functions and needs of the area. As used here, the term planning means the systematic process by which goals (policies) are established, facts are gathered and analyzed, alternative proposals and programs are considered and compared, resources are measured, priorities are established and recommendations are made for the deployment of resources designed to achieve the established goals.[6]

There are many types and levels of planning for the future ranging from family planning to national planning for survival. Figure 1 shows three different

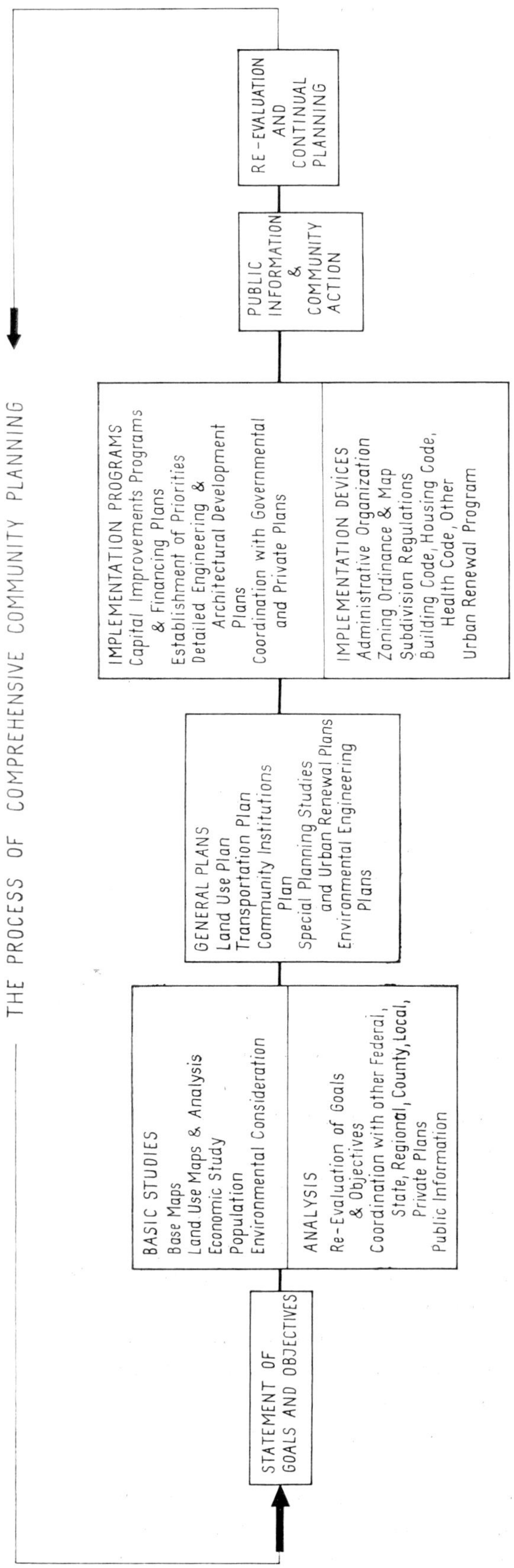

Figure 2 An example of the planning process. From Joseph A. Salvato, Jr., Environmental "Health and Community Planning," *Journal of the Urban Planning and Development Division*, ASCE, Vol. 94, No. UPI, Proc. Paper 6084, August 1968, pp. 22–30.

types of planning and three different levels of planning. These are all interdependent.

Figure 2 gives an example of the *general, overall policy planning process.* It recognizes in the process the environmental health and engineering goals and objectives, considerations, plans and the implementation programs and devices needed. The next level of planning, depending on the functional priorities established, would be specific functional planning such as for recreation, transportation, sewerage, solid wastes, or environmental health.

Environmental health and engineering planning

Since this chapter deals with environmental health, a report outline for that purpose follows.

1. Letter of transmittal to the contracting agency.
2. Acknowledgements.
3. Table of contents.
   a. List of tables.
   b. List of figures.
4. Findings, conclusions and recommendations.
5. Purpose and scope.
6. Background data and analysis, as applicable, including base maps, reports and special studies.

a. Geography, hydrology, meteorology, geology, and ground water levels.

b. Population and characteristics, past, present, future, and density.

c. Soils characteristics; flora and fauna.

d. Transportation and mobility. Adequacy and effects produced, present and future.

e. Residential, industrial, commercial, recreational, agricultural, and institutional development and redevelopment.

f. Land use, present and future; spread of blight and obsolescence; inefficient and desirable land uses.

g. Drainage, water pollution control, and flood control management.

h. Water resources, multi-use planning and development with priority to water supply, environmental impact.

i. Air and water pollution, sewerage and solid wastes management.

j. Public utilities and their adequacy—electricity, gas, oil, heat.

k. Educational and cultural facilities, size, location, effects.

l. Economic studies—present sources of income, future economic base and balance, labor force, markets, industrial opportunities, retail facilities, stability.

m. Sociological factors, characteristics, knowledge, attitudes; behavior of the people and their expectations.

n. Local government and laws, codes ordinances.

o. Special problems, previous studies and findings, background data.

7. Supplementary background environmental health and engineering information.

a. Epidemiological survey including mortality, morbidity, births and deaths, and specific incidence of diseases; social, economic and environmental relationships; also water-, insect- and food-borne diseases; animal and animal related diseases, air-borne and air related diseases, pesticide and chemical poisonings; adequacy of data and programs.

b. Public water supply, treatment and distribution including population served, adequacy, operation, quality control, cross-connection control, storage and distribution protection, operator qualifications. For individual systems—population served, special problems, treatment and costs, adequacy, control of well construction.

c. Wastewater collection, treatment and disposal, adequacy of treatment and collection system, population served, operator qualifications, sewer connection control. For individual systems—population served, special problems, control of installations. Water pollution control.

d. Solid wastes management—storage, collection, transportation, processing and disposal; adequacy.

e. Air resources management and air pollution control including sources, air quality, emission standards, problems and effects, regulation and control program.

f. Housing and the residential environment—control of new construction, housing conservation and rehabilitation, enforcement of housing occupancy and maintenance code, effectiveness of zoning controls and urban renewal. Realty subdivision and mobile home park development and control, also effect of development on the regional surroundings and effect of the region on the subdivision, including the environmental impact of the subdivision.

g. Recreation facilities and open space planning, including suitability of water quality and adequacy of sewerage, solid waste disposal, water supply, food service, rest rooms, safety, and other facilities.

h. Food protection program—adequacy from source to point of consumption.

i. Nuclear energy development, radioisotope and radiation environmental control including fallout, air, water, food and land contamination; thermal energy utilization or dissipation, and waste disposal; naturally occurring radioactive materials; air, water, plant and animal surveillance; federal and state control programs, standards; site selection and environmental impact, plant design and operation control; emergency plans.

j. Planning for drainage, flood control and land use management.

k. Public health institutions and adequacy of medical care facilities such as hospitals, nursing homes, public health centers, clinics, mental health centers, rehabilitation centers, service agencies.

l. Noise and vibration abatement and control.

m. Noxious weed, insect, rodent and other vermin control, including disease vectors and nuisance arthropods; regulation, control and surveillance including pesticide use for control o, aquatic and terrestrial plants and vectors; federalf state and local programs; and effects of water, recreation, housing and other land resource development.

n. Natural and manmade hazards including safety, slides, earthquakes, brush and forest fires, reservoirs, tides, sand storms, hurricanes, tornadoes, high rainfall, fog and dampness, high winds, gas and high tension transmission lines, storage and disposal of explosive and flammable substances and other hazardous materials.

o. Aesthetic considerations, also wooded and scenic areas, prevailing winds and sunshine.

p. Laws, codes, ordinances, rules and regulations.

q. Environmental health and quality protection; adequacy of organization and administration.

8. The comprehensive regional plan.

a. Alternative solutions and plans.

b. Economic, social and ecologic evaluation of alternatives.

c. The recommended regional plan.

d. Site development and reuse plans.

9. Administration and financing.

a. Public information.

b. Administrative arrangements, management and costs.

c. Financing methods—general obligation bonds, revenue bonds, special assessment bonds; taxes, grants, incentives, federal and state aid.

d. Cost distribution, service charges and rates. Capital costs—property, equipment, structures, engineering and legal services; annual costs to repay capital costs, principal and interest, taxes. Regular and special service charges and rates.

e. Legislation, standards, inspection and enforcement.

f. Evaluation, research and re-planning.

10. Appendices

a. Applicable laws

b. Special data.

c. Charts, tables, illustrations

11. Glossary

12. References

Health Aspects

Life expectancy

The life expectancy at birth has varied with time, geography, and with the extent to which available knowledge concerning disease prevention and control could be applied. Table 1 shows the trend in life expectancy with time. The gains in life expectancy between 1900 and 1968 shown in Table la have occurred mostly in the early years, 21 years at birth and 15.5 years at age 5, reducing to 4.8 years at age 45 and 2.4 years at age 70. The life expectancy gains are due to better sanitation and nutrition and to the conquest of the major epidemic and infectious diseases including immunization and chemotherapy.

The vital statistics in Table 2 are of interest in that they show the changes in major causes of death in 1900 related to 1960 and the net reduction in total death rate. Table 2a shows the leading causes of death as of 1967. The leveling off that is apparent in the United States is due to our inability thus far to identify the causes and to control the chronic, non-infectious diseases such as heart disease and cancer.

TABLE 1

Life Expectancy at Birth

Period or year	Life expectancy
Neanderthalman (50,000 B.C.–35,000 B.C.)	29.4*
Upper paleolithic (600,000 B.C.–15,000 B.C.)	32.4*
Mesolithic	31.5*
Neolithic anatolia (12,000 B.C.–10,000 B.C.)	38.2*
Bronze age—Austria	38*
Greece classical (700 B.C.–460 B.C.)	35*
Classical Roman (700 B.C.–200 A.D.)	32*
Roman empire (27 B.C.–395 A.D.)	24
1000	32
England (1276)	48*
England (1376–1400)	38*
1690	33.5
1800	35
1850	40
1870	40
1880	45
1900	48
1910	50
1920	54
1930	59
1940	63
1950	66
1960	68
1970	71

Note. The 1970 life expectancy reported by the United Nations for Sweden was 71.9 for males and 76.5 for females and for the United States 67.0 for males and 74.2 for females.

* E. S. Deevy, Jr., "The Human Population," *Scientific American*, Vol. 203, No. 3, September 1960, p. 200.

Life expectancy figures from 1690 to 1970 are for the United States.

TABLE 1a

*Increase in Life Expectancy Between 1900 and 1968 at Selected Ages**

Age	Years added	Age	Years added
0	21	45	4.8
1	15.5	55	3.6
5	12.0	60	3.1
15	10.4	65	2.7
25	8.8	70	2.4
35	6.7	75	2.0

* *Reference:* The New York Times Encyclopedic Almanac 1971, New York, N.Y., p. 496.

Communicable disease control

Although the communicable diseases as causes of death have been largely brought under control in the more advanced countries, this is not the case in the undeveloped areas of the world. Even in the so-called advanced countries, illnesses associated with contaminated drinking water and food are not uncommon.

Between 1946 and 1960 a total of 228 waterborne outbreaks with 25,984 cases were reported in the United States.[7] An outbreak in California involving *Salmonella typhimurium* affected an estimated 18,000 persons in a population of 130,000.[8] An explosive epidemic of infectious

TABLE 2

Selected Vital Statistics—Deaths per 100,000 in the United States

Cause	Death rate for year 1900	1950–60
Pneumonia and influenza	202	32
Tuberculosis	195	7
Diphtheria	40	0+
Measles	14	0.2
Diarrhea	110	—
Typhoid	31	0+
Malignant neoplasms	64	150
Cardiovascular and renal diseases	345	520
All causes	1719	934

TABLE 2a

*Leading Causes of Death—Deaths per 100,000 Population in the United States**

Cause		Death rate
Diseases of the heart		364.5
Cancer		157.2
Cerebral hemorrhage (stroke)		102.2
Accidents		57.2
Motor vehicle	26.8	
All others	30.4	
Pneumonia and influenza		28.8
Certain diseases of early infancy		24.4
Arteriosclerosis		19.0
Diabetes mellitus		17.7
Other diseases of the circulatory system		15.1
Other bronchopulmonic diseases		14.8
Cirrhosis of liver		14.1
Suicide		10.8
Congenital malformations		8.8
Homicide		6.8
Other hypertensive disease		5.6
Other and ill-defined		88.7
Total of all causes		9.537

* From U.S. Public Health Service, 1967.

hepatitis in India with about 29,300 cases of jaundice was attributed to inadequate chlorination and poor operation control.[9]

Sewage normally contains organisms causing various types of diarrhea, dysentery, infectious hepatitis, salmonella infections, and many other illnesses. It becomes obvious that all sewage should be considered presumptively contaminated, beyond any reasonable doubt, with disease producing organisms. The mere exposure of sewage on the surface of the ground, such as from an overflowing cesspool or septic tank system, or its improper treatment and disposal into a stream or lake, immediately sets the stage for possible disease transmission. The means may be a child's ball, the house fly, or ingestion of contaminated water or food. Sanitary safeguards, including adequate water treatment, must therefore be always maintained if the water- and filth-borne diseases are to be held in check.

It may appear inconceivable, but there are still many urban areas, as well as suburban areas, in the United States and Abroad where the discharge of raw or inadequately treated sewage to roadside ditches and streams is commonplace. Although the disease hazard is ever present, the public pressure for sewage treatment and water pollution abatement stems more from aesthetic, recreational and related economic considerations rather than from actual disease hazard and transmission. As a matter of fact, a critical reappraisal may be in order to review current expenditures and perhaps achieve a better balance in the allocation of public funds for the public good.

Whereas a safe and adequate water supply is taken for granted by most people in the United States, for about 2 billion people, about two thirds of the world's population, this is still a dream. The availability of any reasonably clean water in less developed areas of the world just to wash and bathe would go a long way toward the reduction of such scourges as scabies and other skin diseases, yaws and trachoma, and the high infant mortality. The lack of a safe water makes commonplace high incidences of shigellosis, amebiasis, schistosomiasis, leptospirosis, infectious hepatitis, typhoid and paratyphoid.[10]

Between 1923 and 1937 there were an average of 43 milkborne outbreaks with 1724 cases and 47 deaths reported each year in the United States. Between 1938 and 1956 an average of 24 milkborne outbreaks per year with 980 cases and 5 deaths were reported to the U.S. Public Health Service. Between 1957 and 1960 the outbreaks averaged 9 and the cases 151 per year. There were no deaths reported since 1949. The success achieved in the control of milkborne illnesses can be attributed to the practical elimination of the sale of raw milk, greatly improved equipment and to effective control over the pasteurization of milk and milk products.

Whereas milkborne diseases have been brought

under control, foodborne illnesses remain unnecessarily high. Between 1938 and 1956 there were reported 4647 outbreaks, 179,773 cases and 439 deaths. In 1967, there were still 273 outbreaks reported with 22,171 cases and 15 deaths. The major bacteria related to foodborne illnesses in recent years (1967–68) are Salmonella, *C. perfringens*, and Staphylococcus. Banquets accounted for over half of the illnesses reported with schools and restaurants making up most of the rest. The largest number of outbreaks occurred in the home.[11]

In addition to waterborne and foodborne diseases, consideration should be given to the environmental related respiratory diseases, the insectborne diseases and zoonoses, and the many miscellaneous diseases. These too must be controlled to the extent possible where indicated. The common cold, encephalitides, malaria, rabies, and disabilities related to air pollution are examples.

More effort must be directed to the total environmental pressures and insults to which the body is subjected. The cumulative body burden of all deleterious substances gaining access to the body through the air, food, drink and skin must be examined both individually and in combination. The synergistic effects and neutralizing effects must be learned in order that proper preventive measures may be applied. It has been difficult to determine the effects of the presence or absence of certain trace elements in air, water and food on human health. Some elements such as fluorine for the control of tooth decay, iodine to control goiter, and iron to control iron-deficiency anemia have been recognized as being beneficial in proper amounts. But the action of trace amounts ingested individually and in combination of lead, cadmium, zinc, hexavalent chromium, nickel, mercury, manganese, and other chemicals are often insidious. The effects are extended in time to the point where direct relationship with morbidity and mortality is difficult to prove in view of the many possible intervening and confusing factors. Some may even prove to be beneficial.

Mercury

A brief discussion of one of these metals, namely mercury, is of interest. The poisoning associated with a) the consumption of mercury contaminated fish in Japan between 1953 and 1964; b) bread made from mercury contaminated wheat seed in West Pakistan in 1961, in Central Iraq in 1960 and 1965, and in Panorama, Guatemala in 1963 and 1964; c) pork from hogs fed methylmercury treated seed in Alamagordo, New Mexico; and d) methylmercury treated seed eaten by birds in Sweden, directed worldwide attention to this problem. The discovery of moderate amounts of mercury in tuna and most fresh water fish, and relatively large amounts in swordfish, by many investigators in 1969 and 1970 tended to further dramatize the problem.[12/13/14]

The organic methylmercury forms of mercury are highly toxic. Depending on the concentration and intake, it can cause unusual weakness, fatigue and apathy followed by neurological disorders. Numbness around the mouth, loss of side vision, poor coordination in speech and gait, tremors of hands, irritability and depression are additional symptoms leading possibly to blindness, paralysis and death. The mercury also attacks vital organs such as the liver and kidney. It concentrates in the fetus and can cause birth defects.

Mercury has an estimated biological half-life of 70 to 74 days in man, depending on such factors as age, size and metabolism, and is excreted mostly in the feces at the rate of about one per cent per day. Mercury persists in large fish such as pike one to two years.

Mercury is ubiquitous in the environment. The sources are both natural and manmade. Natural sources are leachings and volatilization from mercury containing geological formations. Manmade sources are waste discharges from chlor-alkali and pulp manufacturing plants, mining, chemical manufacture and formulation, the manufacture of mercury seals and controls, treated seeds, combustion of fossil fuels, fallout, and surface runoff. The mercury ends up in lakes, streams and tidal waters, and in the bottom mud and sludge deposits.

Microorganisms and macroorganisms in water and bottom deposits can transform metallic mercury, inorganic divalent mercury, phenylmercury and alkoxialkylmercury into methylmercury. The methylmercury thus formed, and perhaps other types, in addition to that discharged in wastewaters, are assimilated and accumulated by aquatic and marine life such as plankton, small fish, and large fish. Alkaline waters tend to favor production of the more volatile dimethylmercury, but acid waters are believed to favor retention of the dimethyl form in the bottom deposits. Under anaerobic conditions, the inorganic mercury ions are precipitated to insoluble mercury sulfide in the presence of hydrogen sulfide. The process of methylation will continue as long as organisms are present and they have access to mercury. The form of mercury in fish has been found to be practically all methylmercury and there is indication that a significant part of the mercury found in eggs and meat is in the form of methylmercury.

The amount of mercury in canned tuna fish has averaged 0.32 ppm, in fresh swordfish 0.93 ppm, in fresh water fish 0.42 ppm (up to 1.4 and 2.0 ppm in a few large fish such as Walleyed Pike), and as high as 8 to 23 ppm in fish taken from heavily contaminated waters. The mercury in urban air has been in the range of 0.02 to 0.2μg per cu m, in drinking water less than 0.001 ppm, and in rain water about 0.2 to 0.5 ppb (μg/l). Reports from Sweden and Denmark (1967–69) indicate a mercury concentration of 3 to 8 ppb (ng/gm) in pork chops, 9 to 21 ppb in pig's liver, 2 to 5 ppb in beef filet, 9 to 14 ppb in hen's eggs, and 0.40 to 8.4 ppm in pike.

In view of the potential hazards involved, steps have been taken to provide standards or guidelines for mercury. The maximum allowable concentration for 8 hour occupational exposure has been set at 0.1 mg metallic vapor and inorganic compounds of mercury per cubic meter of air. For organic mercury the threshold limit is 0.01 mg per cu m of air. A maximum allowable steady intake (ADI) of 0.03 mg for a 70 kg man is proposed, which would provide a safety factor of ten. If fish containing 0.5 ppm mercury (the actionable level) were eaten daily, the limit of 0.03 mg would be reached by the daily consumption of 60 gm (about 2 ounces) of fish.[13] The proposed standard for drinking water is 0.002 ppm. A standard for food has not yet been established; 0.05 ppm has been mentioned.

There is no evidence to show that the mercury in the current daily dietary intake has caused any harm, although apparent health does not indicate possible non-detectable effects on brain cells or other tissues. Nevertheless, from a conservative health standpoint, it has been recommended that pregnant women not eat any canned tuna or swordfish. Also implied is caution agains the steady eating of large fresh water fish or other large salt water fish. The general population should probably not eat more than one fish meal per week.

The identification of mercury as coming from manmade and natural sources requires that every effort be made to eliminate mercury discharges to the environment. At the same time the air, drinking water, food, fish and other wildlife, aquatic plants and animals, surface runoff and leachates, precipitation, surface waters, and man himself should be monitored. This should be done not only for mercury, but also for other potentially toxic and deleterious chemicals. Further research and studies are needed to determine the subtle and actual effects of mercury and other metals, as normally found, on man and his environment.

Water Supply

A primary requisite for good health is an adequate supply of water that is of satisfactory sanitary quality. It is also important that the water be attractive and palatable to induce its use; for otherwise, water of doubtful quality from some near by unprotected stream, well or spring may be used. Where a municipal water supply is available, it should be used as such supplies are usually under competent supervision, ample in quantity and also provide fire protection. However, this is not always the case. Because of the excellent water service generally available in the United States, the people and public officials have tended to become complacent and take for granted their water supply. As a result, in some instances, funds have been diverted to other more popular causes rather than to maintenance, operation and upgrading of the water supply system.

Status of water supply

A survey[15] made by the Public Health Service in 1962 is of interest in pointing out the number and type of public water supplies in the United States and populations served. It was reported that there were 19,236 public water supplies serving approximately 150,000,000 people; 75 per cent were ground water supplies, 18 per cent were surface water supplies, and 7 per cent were a combination. Of significance is the finding that 75,000,000 people in communities under 100,000 population were served by 18,873 public water supplies, and 77,000,000 people in communities over 100,000 were served by 399 public water supplies. Also, of the 19,236 supplies, 85 per cent served communities of 5000 or less. This information emphasizes the need for giving at least equal attention to small public water supplies as is given to large supplies. In addition, millions of people on vacation in relatively uncontrolled rural environments depend on small water supplies which often are not under close surrveillance.

The Public Health Service completed a study in 1970 covering 969 small to large public surface and ground water supply systems serving 18,200,000 persons (12 per cent of the total United States population served by public water supplies) and 84 special systems serving trailer and mobile home parks, institutions and tourist accommodations.[16] Although the drinking water supplies in the United States rank among the best in the world, the study showed the need for improvements. Based on the 1962 USPHS Drinking Water Standards it was found that in 16 per cent of the 969 communities surveyed the water quality exceeded one or more of the *mandatory* limits established for coliform organisms (120), fluoride (24), lead (14). It is of interest to note that of the 120 systems that exceeded the coliform standard, 108 served populations of 5000 or less and that 63 of these were located in a state where disinfection was not frequently practiced or was inadequate. An additional 25 per cent of the systems exceeded the *recommended* limits for iron (96), total dissolved solids (95), manganese (90), fluoride (52), sulfate (25), nitrate (19). The study also showed that 56 per cent of the systems were deficient in one or more of the following: source protection; disinfection or control of disinfection; clarification (removal of suspended matter) or control of clarification; and pressure in the distribution system. It was also reported that 90 per cent of the systems did not have sufficient samples collected for bacteriological surveillance; 56 per cent of the systems had not been surveyed by the state or local health department within the last three years; in 54 per cent, cross-connection prevention ordinances were lacking; in 89 per cent, reinspection of existing construction was lacking; in 61 per cent, the operators had not received any water treatment training; in 77 per cent, the operators were deficient in training for microbiological work and 46 per cent of those who needed chemistry training did not have any. Finally, the smaller communities had more water quality problems and deficiencies than the larger ones, showing the advisability of consolidation and regionalizaton where this is feasible.

Water cycle

The movement of water from the atmosphere to the earth and back again to the atmosphere can be best illustrated by the hydrologic or water cycle shown in Fig. 3. The average annual precipitation in the United States is about 30 inches of which 72 per cent evaporates from water and land surfaces and transpires from plants, and 28 per cent contributes to the ground water recharge and stream flow.[17]

Water quality

Water quality is usually determined by the water source, the treatment it receives and its method of distribution. Standards of water quality take into consideration the physical, chemical and microbiological characteristics, the radioactivity, and compliance with the *Public Health Service Drinking Water Standards 1962.* [18] Interpretation of water quality analyses should be based on a sanitary survey. This is defined as "1) An investigation of any condition that may affect public health. 2) A study of conditions related to the collection, treatment, and disposal of liquid, solid, or airborne wastes to determine the potential hazard contributed from these sources to the environment. 3) A study of wastewater discharges on sources of water supply, on bathing or other recreational waters, on shellfish culture, and other related environments."[19]

Surface waters by their very nature are subject to natural and manmade pollution which in many instances is not readily controllable. Hence the selection of a surface water supply source must take into consideration the treatment that may be needed now and in the future to ensure that the finished water will at all times meet the *Drinking Water Standards*. It is obvious that this objective can be met with greater certainty if the source of water is from a protected watershed and reservoir rather than from a stream passing through urban areas which may also be used for wastewater disposal and navigation.

Raw water quality criteria for public water supplies have been developed which recognize the great variability in surface water quality. The criteria are shown in Table 3. The *permissible criteria* are for raw waters which can by conventional rapid sand filtration treatment or less meet the *Drinking Water Standards*. The *desirable*

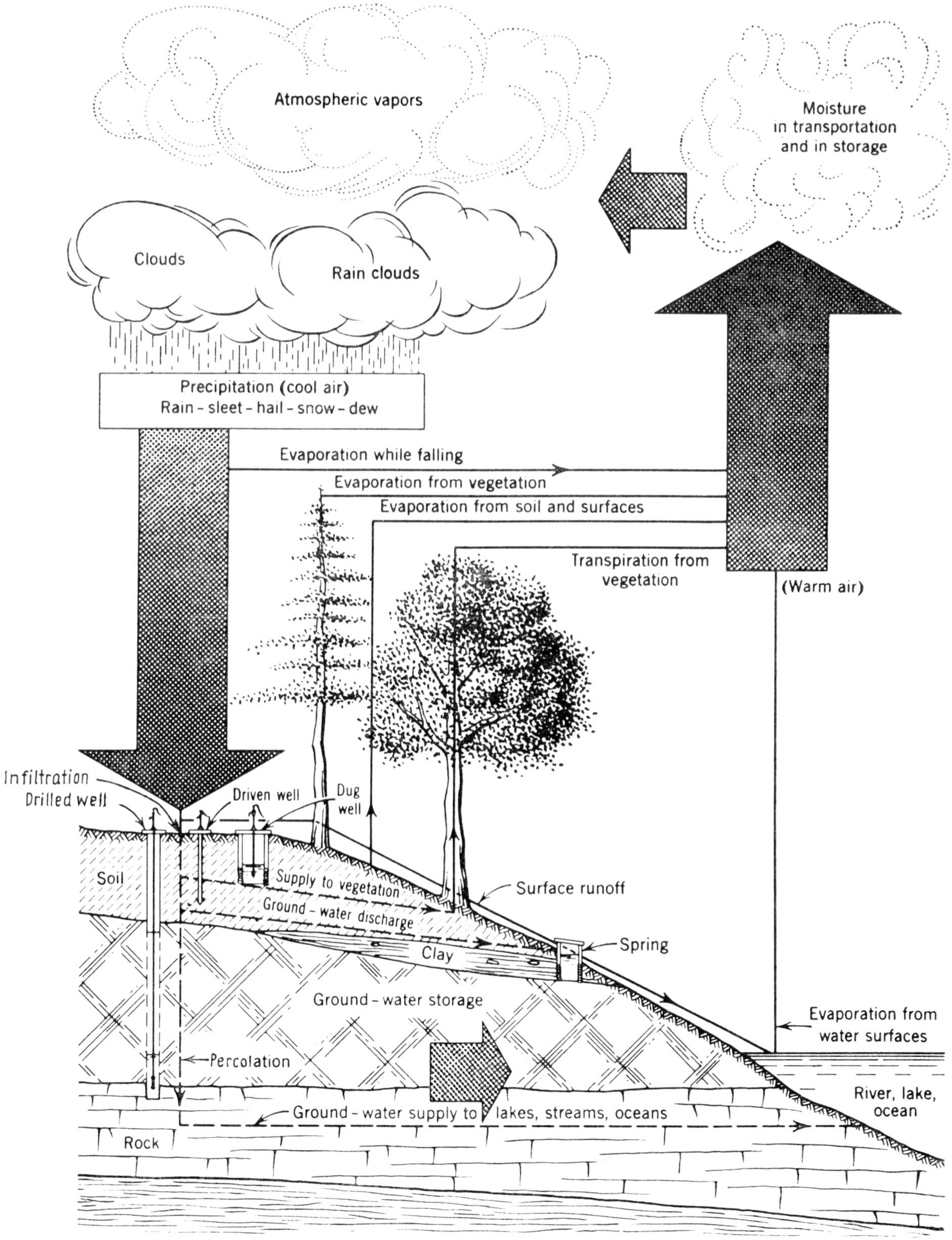

Figure 3 The hydrologic or water cycle.
From Joseph A. Salvato, Jr., *Environmental Engineering and Sanitation*, John Wiley & Sons, Inc., New York, N. Y., 1972.

TABLE 3

Surface Water Criteria for Public Water Supplies

Constituent or characteristic	Permissible criteria	Desirable criteria
Physical:		
Color (color units)	75	<10
Odor	+	Virtually absent
Temperature*	do	+
Turbidity	do	Virtually absent
Microbiological:		
Coliform organisms	10,000/100 ml [1]	<100/100 ml[1]
Fecal coliforms	2,000/100 ml [1]	<20/100 ml [1]
Inorganic chemicals:	(mg/l)	(mg/l)
Alkalinity	+	+
Ammonia	0.5 (as N)	<0.01
Arsenic*	0.05	Absent
Barium*	1.0	do
Boron*	1.0	do
Cadmium*	0.01	do
Chloride*	250	<25
Chromium,* hexavalent	0.05	Absent
Copper*	1.0	Virtually absent
Dissolved oxygen	≥4 (monthly mean) ≥3 (individual sample)	Near saturation
Fluoride*	+	+
Hardness*	do	do
Iron (filterable)	0.3	Virtually absent
Lead*	0.05	Absent
Manganese* (filterable)	0.05	do
Nitrates plus nitrites*	10 (as N)	Virtually absent
pH (range)	6.0–8.5	+
Phosphorus*	+	do
Selenium*	0.01	Absent
Silver*	0.05	do
Sulfate*	250	<50
Total dissolved solids* (filterable residue)	500	<200
Uranyl ion*	5	Absent
Zinc*	5	Virtually absent
Organic chemicals:		
Carbon chloroform extract* (CCE)	0.15	<0.04
Cyanide*	0.20	Absent
Methylene blue active substances*	0.5	Virtually absent
Oil and grease*	Virtually absent	Absent
Pesticides:		
Aldrin*	0.017	do
Chlordane*	0.003	do
DDT*	0.042	do
Dieldrin*	0.017	do
Endrin*	0.001	do
Heptachlor*	0.018	do
Heptachlor epoxide*	0.018	do
Lindane*	0.056	do
Methoxychlor*	0.035	do
Organic phosphates plus carbamates*	0.1 [2]	do
Toxaphene*	0.005	do
Herbicides:		
2,4-D plus 2,4,5-T, plus 2,4,5-TP*	0.1	do
Phenols*	0.001	do
Radioactivity:	(pc/l)	(pc/l)
Gross beta*	1,000	<100
Radium-226*	3	<1
Strontium-90*	10	<2

* The defined treatment process has little effect on this constituent. (Coagulation, sedimentation, rapid sand filtria tion and chlorination.)

\+ No concensus on a single numerical value which is applicable throughout the country. See reference.

[1] Microbiological limits are monthly arithmetic averages based upon an adequate number of samples. Total coliform limit may be relaxed if fecal coliform concentration does not exceed the specified limit.

[2] As parathion in cholinesterase inhibition. It may be necessary to resort to even lower concentrations for some compounds or mixtures. (Permissible levels are based upon the recommendations of the Public Health Service Advisory Committee on Use of the PHS Drinking Water Standards.)

From *Water Quality Criteria*, Report of the National Advisory Committee to the Secretary of the Interior, April 1, 1968, Washington, D.C., p. 20.

criteria are for raw waters which can meet the *Drinking Water Standards* at less cost and with greater factors of safety than is possible with waters meeting the desirable criteria. It is important to point out that the constituents marked with an asterisk in Table 3 are not removed by conventional rapid sand filtration treatment. In some instances additional treatment with coagulant aids or activated carbon may suffice, and in others elimination of the source of the contaminant may be the only practical answer.

Water use

The quantity of water used for domestic purposes will, in general, vary directly with the availability of the water, the habits of the people, the number and type of plumbing fixtures provided, the water pressure, the air temperature, the newness of a community, the type of establishments, and other factors. Wherever possible, the actual water consumption under existing or similar circumstances and the number of persons served should be the basis for the design of a water and sewerage system. Special adjustment must be made for industrial use. Municipal water consumption varies widely dependent on location, industrial usage, metering, size, economic, social and other factors. An average is 155 gpcd.

TABLE 4

Guides for Water Use

Type of establishment	Gallons per day*
Residential:	
Dwellings and apartments (per bedroom)	150
Temporary quarters:	
Boarding houses	65
Additional (or non-resident boarders)	10
Camp sites (per site)	100
Cottages, seasonal	50
Day camps	15–20
Hotels	65–75
Mobile home parks (per unit)	125–150
Motels	50–75
Restaurants (toilets and kitchens)	7–10
Without public toilet facilities	$2^1/_2$–3
With bar or cocktail lounge, additional	2
Summer camps	40–50
Public establishments:	
Boarding schools	75–100
Day schools	15–20
Hospitals (per bed)	250–500
Institutions other than hospitals (per bed)	75–125
Places of public assembly	3–10
Turnpike rest areas	5
Turnpike service areas (10 per cent of cars passing)	15–20
Amusement and commercial:	
Airports (per passenger)	3–5
Country clubs	25
Day workers (per shift)	15–35
Drive-in theaters (per car space)	5
Gas station (per vehicle serviced)	10
Milk plant, pasteurization (per 100 lbs of milk)	11–25
Movie theaters (per seat)	3
Picnic parks with flush toilets	5–10
Self-service laundries (per machine)	400
Shopping center (per 1000 sq ft floor area)	250
Stores (per toilet room)	400
Swimming pools and beaches with bath-houses	10

TABLE 4 (*cont.*)

Type of Establishment	Gallons per day*
Farming:	
Livestock (per animal)	
Cattle	12
Dairy	35
Goat	2
Hog	4
Horse	12
Mule	12
Sheep	2
Steer	12
Poultry (per 100):	
Chickens	5–10
Turkeys	10–18

* Per person unless otherwise stated.

From *Rural Water Supply*, New York State Department of Health, Albany, N.Y., 1966.

Table 4 gives estimates of water consumption at different types of places. Additions should be made for car washing, lawn sprinkling, and miscellaneous uses. If provision is made for firefighting requirements, then the quantity of water provided for this purpose to meet fire underwriters' standards will be in addition to that required for normal domestic needs in small communities.

Sources of water

The sources of water supply are divided into two major classifications; namely, ground water and surface water. To these should be added for completeness rain water and demineralized water. Ground water supply sources include dug, bored, driven, and drilled wells; rock and sand or earth springs, and infiltration galleries. Figures 4–7 show details of well and spring construction and sanitary protection. Standards for well construction are given in *AWWA Standard for Deep Wells*, AWWA A 100-66 published by the American Water Works Association and in *Recommended State Legislation and Regulations*, Department of Health, Education, and Welfare, Public Health Service, Washington, D.C., July 1965. Surface water supply sources include lakes, reservoirs, streams, ponds, rivers, and creeks. The development of surface water supplies is covered in standard sanitary engineering texts and state publications.

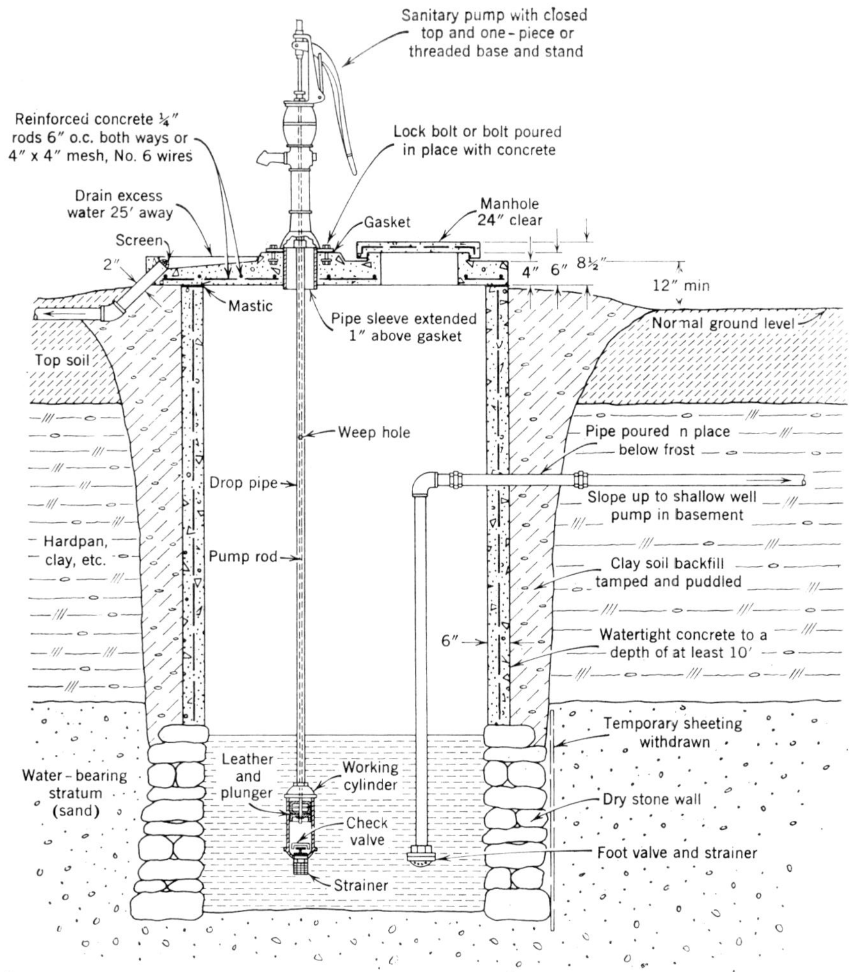

Figure 4 A properly developed dug well.

Reprinted, with permission, from *Environmental Sanitation*, Joseph A. Salvato, Jr., John Wiley & Sons, Inc., copyright, 1958.

Note: 1. *Locations*. Wells should be located at least 100 ft and preferably 200 ft from privies, barnyards, leaching pits, cesspools, tile fields, and other sources of pollution. Wells should not be located downgrade so as to be in the direct line of drainage from sources of pollution. 2. *Concrete*. Mix one bag of cement, $2\frac{1}{2}$ cu ft of sand, and 3 cu ft of gravel. Then add 5 gal of water, for moist sand, and mix again.

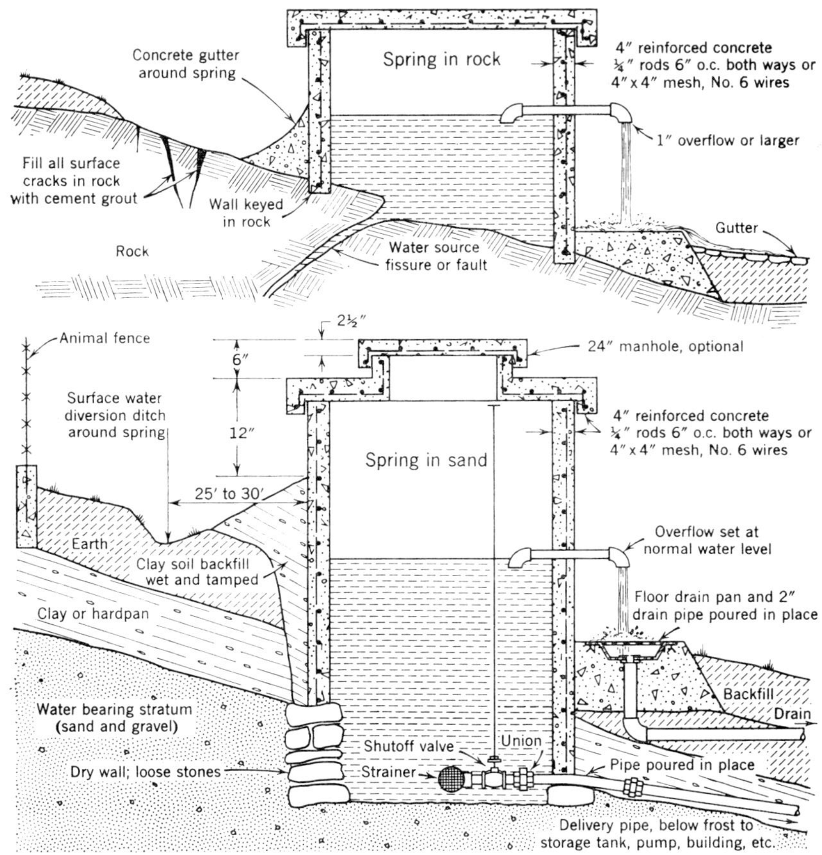

Figure 5 Properly constructed springs
(Reprinted, with permission, from *Environmental Sanitation*, Joseph A. Salvato, Jr., John Wiley & Sons, Inc., copyright, 1958.)

Note: 1. Springs should be located at least 100 ft and preferably 200 ft from privies, cesspools, barnyards, leaching pits, tile fields, etc. Springs should not be located downgrade so as to be in the direct line of drainage from sources of pollution. 2. If a handpump is provided it should be of sanitary design, installed with a watertight connection.

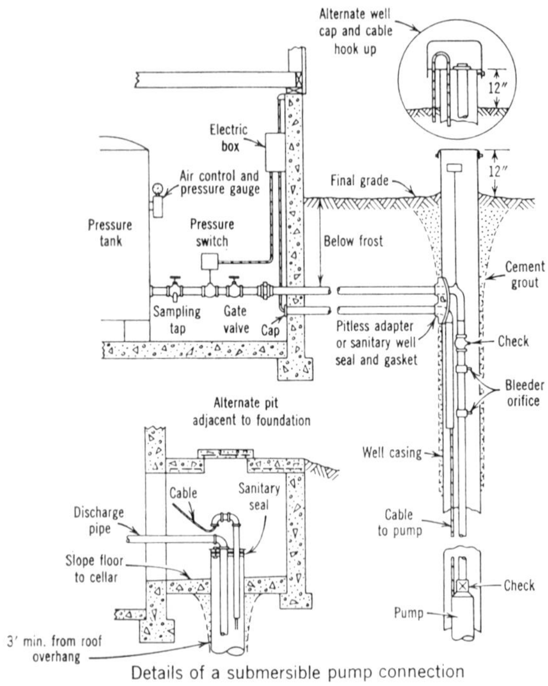

Figure 6 Sanitary well caps and seals.
Reprinted, with permission, from *Environmental Sanitation*, Joseph A. Salvato, Jr., John Wiley & Sons, Inc., copyright, 1958.

Water treatment

As an aid in determining the treatment that should be given water to make it safe to drink, the United States Public Health Service has classified waters into several groups.[20] The treatment required by this classification is based upon the most probable number (MPN) of coliform bacteria per 100 milliliters (ml) of sample. The classification is summarized in Table 5. It needs to be supplemented by chemical, physical and microscopic examinations. For water to be generally acceptable, other treatment may be required in addition to that necessary for the elimination of disease-producing organisms. People expect the water to be safe to drink, attractive to the senses, soft, non-staining, and neither scale-forming nor corrosive to the water system. The various treatment processes employed to accomplish these results are shown in the flow diagram in Fig. 8. The untrained individual should not attempt to design a water treatment plant for life and health will be jeopardized. This is a job for a competent sanitary engineer.

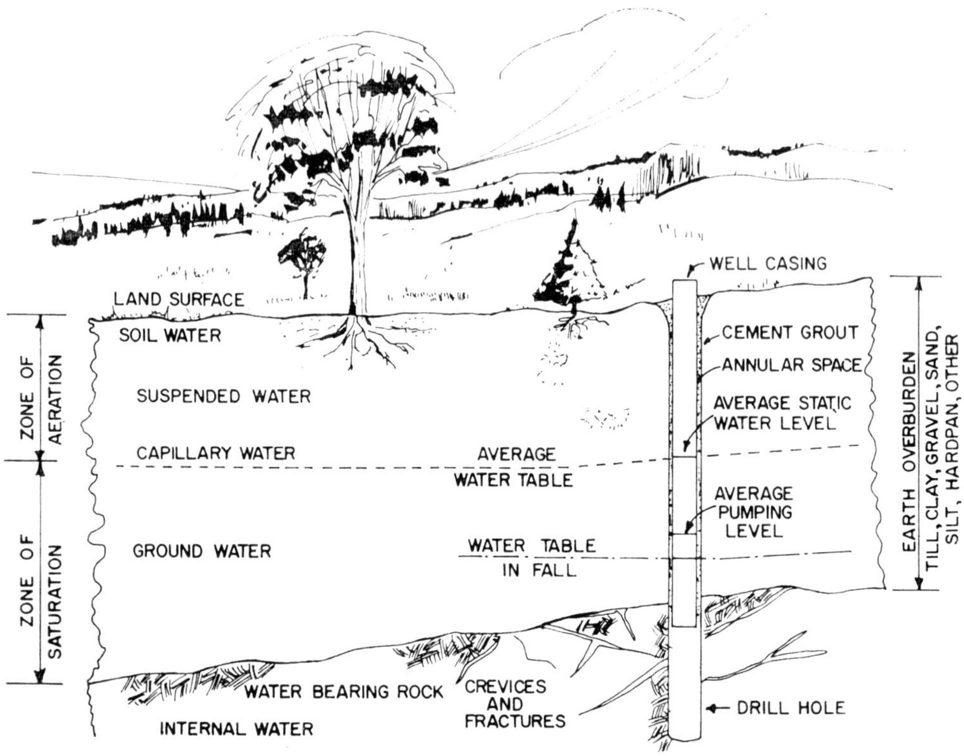

Figure 7 A geologic section showing ground water terms.
From *Rural Water Supply*, New York State Department of Health, Albany, N.Y., 1966.

TABLE 5

A Classification of Waters by Concentration of Coliform Bacteria and Treatment Required to Render the Water of Safe Sanitary Quality

Group No.	Maximum permissible average MPN coliform bacteria per month†	Treatment required*
1	Not more than 10% of all 10-ml or 60% of 100 ml portions positive; not more than 1.0 coliform bacteria per 100 ml.	No treatment required of underground water, but a minimum of chlorination required of surface water, advised for ground water.
2	Not more than 50 per 100 ml.	Simple chlorination or equivalent.
3	Not more than 5000 per 100 ml and this MPN exceeded in not more than 20% of samples.	Rapid sand filtration (including coagulation) or its equivalent plus continuous chlorination.
4	MPN greater than 5000 per 100 ml in more than 20% of samples and not exceeding 20,000 per 100 ml in more than 5% of the samples.	Auxiliary treatment such as 30–90 days storage, presettling, prechlorination, or equivalent plus complete filtration and chlorination.
5	MPN exceeds group 4.	Prolonged storage or equivalent to bring within groups 1–4.

Adapted from *Environmental Sanitation*, Joseph A. Salvato, Jr., John Wiley & Sons, Inc., New York, 1958.

* Physical, inorganic and organic chemicals, and radioactivity concentrations in the raw water and ease of removal by the proposed treatment must also be taken into consideration. See Table 3, *Water Quality Criteria*, Report of the National Technical Advisory Committee, to the Secretary of the Interior, April 1, 1968, Washington, D.C., and *Manual for Evaluating Public Drinking Water Supplies*, U.S. Department of Health, Education, and Welfare, Public Health Service, Environmental Control Administration, Cincinnati, Ohio 1969.

† Fecal coliforms not to exceed 20% of total coliform organisms. The monthly geometric mean of the MPN for group 2 may be less than 100 and for group 3 and 4 less than 20,000 per 100 ml with the indicated treatment.

Wastewater

Definitions

Excreta is the waste matter eliminated from the body, including feces, urine and sweat. *Domestic sewage* is the used water from a home or community and includes toilet, bath, laundry, lavatory, and kitchen-sink wastes. Sewage from a community may include some industrial wastes, ground water and surface water, hence the more inclusive term *wastewater* is coming into general usage. Normal sewage from a private sewage disposal system contains about 99.8 per cent water and 0.2 per cent total mineral and organic solids. Domestic sewage contains less than 0.1 per cent total solids. The strength of wastewater is commonly expressed in terms of 5-day biochemical oxygen demand (BOD), suspended solids and chemical oxygen demand (COD).

The biochemical oxygen demand of sewage, sewage effluents, polluted waters, industrial wastes or other wastewaters is the oxygen in parts per million (ppm) or milligrams per liter (mg/l) required during stabilization of the decomposable organic matter by aerobic bacterial action. Complete stabilization requires more than 100 days at 20°C. Incubation for 5 or 20 days is not unusual but as used in this chapter BOD refers to the 5-day test unless otherwise specified.

Suspended solids are those which are visible and in suspension in water. They are the solids which are retained on the asbestos mat in a Gooch crucible.

The *chemical oxygen demand* (COD) is also used, particularly in relation to certain industrial wastes. The COD is the amount of oxygen expressed in parts per million (ppm) or milligrams per liter (mg/l) consumed under specific conditions in the oxidation of organic and oxidizable inorganic material. The test is relatively rapid. It does not oxidize some biodegradable organic pollutants (pyridine, benzene, toluene) but does oxidize some inorganic compounds which are not measured, i.e., affected by the BOD analysis.

When storm water and domestic sewage enter a sewer it is called a *combined sewer*. If domestic sewage and storm water are collected separately, in a *sanitary sewer* and in a *storm sewer*, then the result is a *separate sewer system*. A *sewage works* or *sewer system* is a combination of sewers and appurtenances for the collection, pumping, and transportation of sewage, sometimes, also called *sewerage*, plus facilities for treatment and disposal of sewage, known as the *sewage treatment plant*.

The *privy*, or one of its modifications, is the common device used when excreta is disposed of without the aid of water. When excreta is disposed of with water, a *water-carriage* sewage disposal system is used; generally all other domestic liquid wastes are included.

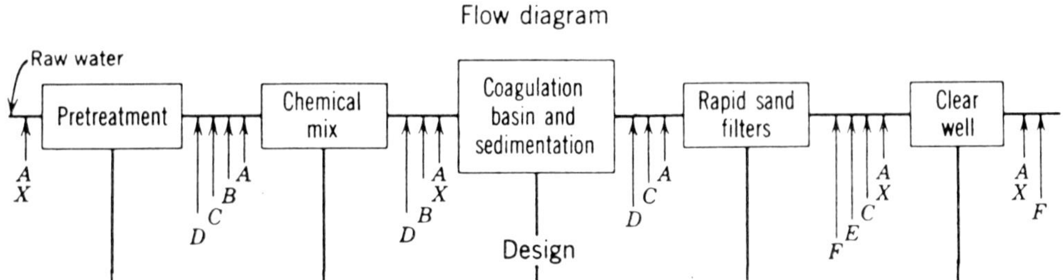

1. Prechlorination 2. Pre-settling 2 to 3 days, or 3. Storage, 30 to 90 days or longer, or 4. Prefiltration (Where needed) 5. Aeration 6. Micro-straining 7. Screening	1. Detention for 20 to 45 min. 2. Velocity 60 to 150 ft. per min. 3. Rapid mix 30 sec. plus slow mix	1. Detention for 4 to 6 hr. 2. Maximum velocity 0.5 ft. per min. 3. Duplicate units 4. Length twice to 3 x width or greater 5. Overflow rate $\ngtr$ 20,000 gpd/ft. weir	1. Rate 2 gal. per sq.ft. per min. 2. Sand depth 24 in. Eff. size 0.35-0.50 mm Unif. coef. 1.5-2.0. For anthracite E.S. = 0.8-2.0 mm U.C. $\ngtr$ 1.7 3. Rate controller 4. Surface wash optional	1. Capacity equal to average days demand 2. Covered, if possible

Plant capacity—50 to 100% greater than average daily demand, with clear well.

Possible Chemical Combinations

A Chlorine.

B Coagulant; aluminum sulfate (pH 5.5–8.0), ferrous sulfate (pH 8.5–11.0), ferric chloride (pH 5.0–11.0), sodium aluminate and or other (activated silica) (polyelectrolytes).

C Alkalinity adjustment; lime, soda ash, or polyphosphate,

D Activated carbon.

E Dechlorination; sulfur dioxide, sodium sulfite, sodium thiosulfate.

F Fluoridation treatment.

X Chlorine dioxide.

Note: The chlorinator should be selected to pre-chlorinate surface water at 20 mg/l and post-chlorinate at 3 mg/l. Provide for a dose of 3 mg/l plus chlorine demand for ground water.

Figure 8 Rapid sand filter plant flow diagram.
From *Environmental Sanitation*, Joseph A. Salvato, Jr., John Wiley & Sons, Inc., New York, 1958.

Stream pollution and recovery

Although the 5-day BOD is the best single measure of wastewater or polluted water organic loading, aquatic organisms when measured with the biochemical oxygen demand, the chemical oxygen demand where indicated, the dissolved oxygen, and the sanitary survey taken all together are the best indicators of water pollution. Other chemical and bacterial analyses provide additional information.

In a freshly polluted water, during the first stage (5 to 20 days at 20°C), mostly carbonaceous matter is oxidized. This is demonstrated by an immediate increase in the stream BOD and oxygen utilization in the area of pollution discharge, followed by the second or nitrification stage where a lesser but uniform rate of oxygen utilization takes place for an extended period of time. This is accompanied by a related characteristic change in the stream biota. The amount of dissolved oxygen in a receiving water is the single most important factor determining the waste assimilation capacity of a body of water.

Stream pollution (organic) is apparent along its length by a zone of degradation just below the source, a zone of active decomposition and, if additional pollution is not added, a zone of recovery. In the zone of degradation the oxygen in the water is decreased, suspended solids may be increased and the stream bottom accumulates sludge. The fish life changes from game and food species to coarse. Worms, snails and other biota associated with pollution increase. In the zone of active decomposition, the dissolved oxygen is further reduced and may approach zero. The water becomes turbid and gives off foul odors. Fish disappear, anaerobes predominate in the bottom mud and sludge worms become very numerous. In the zone of recovery the process is gradually reversed and the stream returns to normal. The zones mentioned may not be discernible or experienced where sewage has been given adequate treatment before discharge.

Small water-borne sewage disposal systems

The provision of running water in a dwelling or structure immediately introduces the requirement for sanitary removal of the used water. Where public sewage works are available, connection to the sewer will solve a major sanitation problem. Where public or central sewage works are not provided or anticipated, such as in predominately rural areas, then consideration must be given to the proposed method of collection, removal, treatment, and disposal of sewage on each lot as shown in Fig. 9. With a suitable soil, the disposal of sewage can be simple, economical, and inoffensive; but careful maintenance is essential for continued satisfactory operation. Where rock or ground water is close to the surface or the soil is a tight clay, it would be well to investigate some other property.

The common system for sewage treatment and disposal at a private home in a rural area consists of a proper septic tank for the settling and treatment of the sewage, and a subsurface leaching system for the disposal of the septic tank overflow, provided the soil is satisfactory. The soil percolation test and soil characteristics are used as means for determining soil permeability or the capacity of a soil to absorb settled sewage. This and the quantity of sewage from a dwelling are the bases upon which a subsurface sewage disposal system is designed. Sand filters, evapotranspiration beds, aeration systems, stabilization ponds or lagoons, recirculating toilets and various types of privies are used under certain conditions.

Large sewage works

The design of sewers and treatment plants should take into consideration the broad principles of comprehensive community planning discussed at the beginning of this chapter. Regional and area-wide sewerage planning (preliminary) should take place within the context of the general overall comprehensive community plan. It recognizes the extent of present and future service areas, the established water quality and effluent standards, and the alternative solutions with their first costs and total annual costs. This information is needed to assist local officials in making a decision to proceed with the design and construction of a specific sewerage system and treatment plant. These are essential first steps to ensure that the proposed construction will meet the community, state and national goals and objectives.

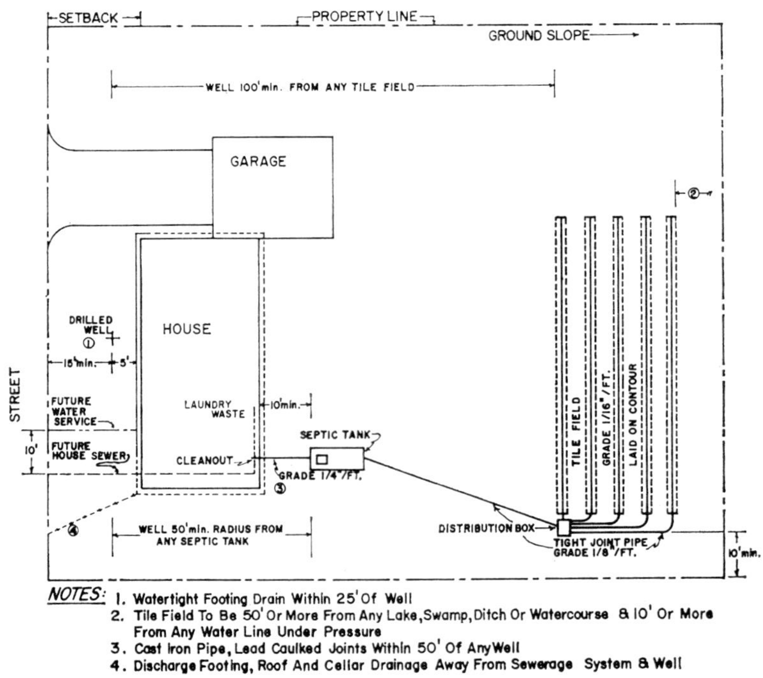

Figure 9 Private water supply and sewage disposal layout.

Water quality standards. The purpose of stream and other surface water quality standards is to protect, improve where needed, and maintain the highest practicable water quality for certain specified best uses. Such uses are usually grouped under "public water supply," "bathing and recreational," "fishing, aquatic life and wildlife," and "natural drainage, agricultural and industrial water supply." In general, sludge deposits, solid refuse, floating solids, oils, grease and scum are not allowed under any uses. Waters are also required to be free from organic and inorganic chemical or radiologic constituents in concentrations or combinations which would be harmful to human, animal, or aquatic life for the specified water use. For sources of drinking water supplies, the sewage and industrial waste treatment before discharge to a stream is such that with conventional (coagulation, sedimentation, rapid sand filtration) water treatment the Public Health Service Drinking Water Standards are not exceeded.

For sources of drinking water with chlorination treatment alone, the maximum permissible average most probable number (MPN) of coliform bacteria per month should not exceed 50 per 100 ml with not more than 20 per cent of the samples exceeding 240. With coagulation, rapid sand filtration and chlorination the average MPN should not exceed 5000 with not more than 20 per cent of the samples exceeding 5000. Fecal coliforms should not exceed 20 per cent of the total coliform counts. For sources of water for bathing and other contact recreation uses the average MPN should not exceed 2400 (or 1000) per 100 ml with not more than 20 per cent (or 10 per cent) of the samples exceeding 5000 (or 2000). The fecal coliform count (log mean) should not exceed 200 with not more than 10 per cent of the samples collected during any 30-day period exceeding 400. In all cases the coliform counts should be interpreted in the light of a sanitary survey and related factors.

Water quality standards also specify a minimum dissolved oxygen content of 5.0 mg/l for trout waters, 5.0 or 4.0 mg/l for non-trout waters and a minimum 2.0 or 3.0 mg/l for any other waters. Similar standards as above are used for marine waters. If waters are used for shellfish harvesting for direct human consumption, the coliform bacteria shall not exceed a median MPN of 70 per 100 ml and not more than 10 per cent of the samples shall ordinarily exceed 230.

Effluent standards. It should be obvious that the realistic compliance with water classification standards requires control over all sewage and industrial wastewater discharges to the classified waters. In Britain, the Ministry of Housing and local government re-affirmed in 1966 the Royal Commission "General Standard" as a "norm" for sewage effluents, i.e., 5-day BOD 20 mg/l and suspended solids 30 mg/l with a dilution factor of 9 to 150 volumes in the receiving watercourse having not more than 4.0 mg/l BOD. A higher effluent standard of 10 mg/l BOD and suspended solids may be required if indicated.[21] In any case, sewage and industrial waste effluents should not contain any matter likely to render the receiving stream poisonous or injurious to fish, or if used for drinking water supply injurious to man.

Wastewater treatment. The need for wastewater treatment can be better justified for aesthetic, economic, recreation and fish reasons rather than for reason of reduction of health hazards. Treatment of sewage and industrial wastes does not usually remove the need for nor reduction in required treatment for drinking water. On the contrary, greater emphasis needs to be placed on the improvement and maintenance of drinking water quality. The increasing urbanization and industrialization has and will continue to result in the discharge of viruses, organic and inorganic chemicals, which are not readily removed by conventional water treatment, and whose long-term health effects are not yet known. Many people may be mislead into thinking that money and effort spent in stream pollution abatement justifies relaxation of effort, money to be spent and support for drinking water treatment. Nothing could be further from the truth!

Wastewater treatment usually involves screening and sedimentation for primary treatment. When followed by oxidation and disinfection it is given secondary treatment. If secondary treatment plant effluent is passed through a sand filter, oxidation pond, or some combination of chemical treatment and filters plus chlorination, in which additional organic and inorganic materials are removed, together with bacteria, protozoa and viruses, then it would be receiving tertiary treatment. Some possible combinations of sewage treatment unit processes are shown in Figs. 10 and 11; their efficiencies are shown in Table 6.

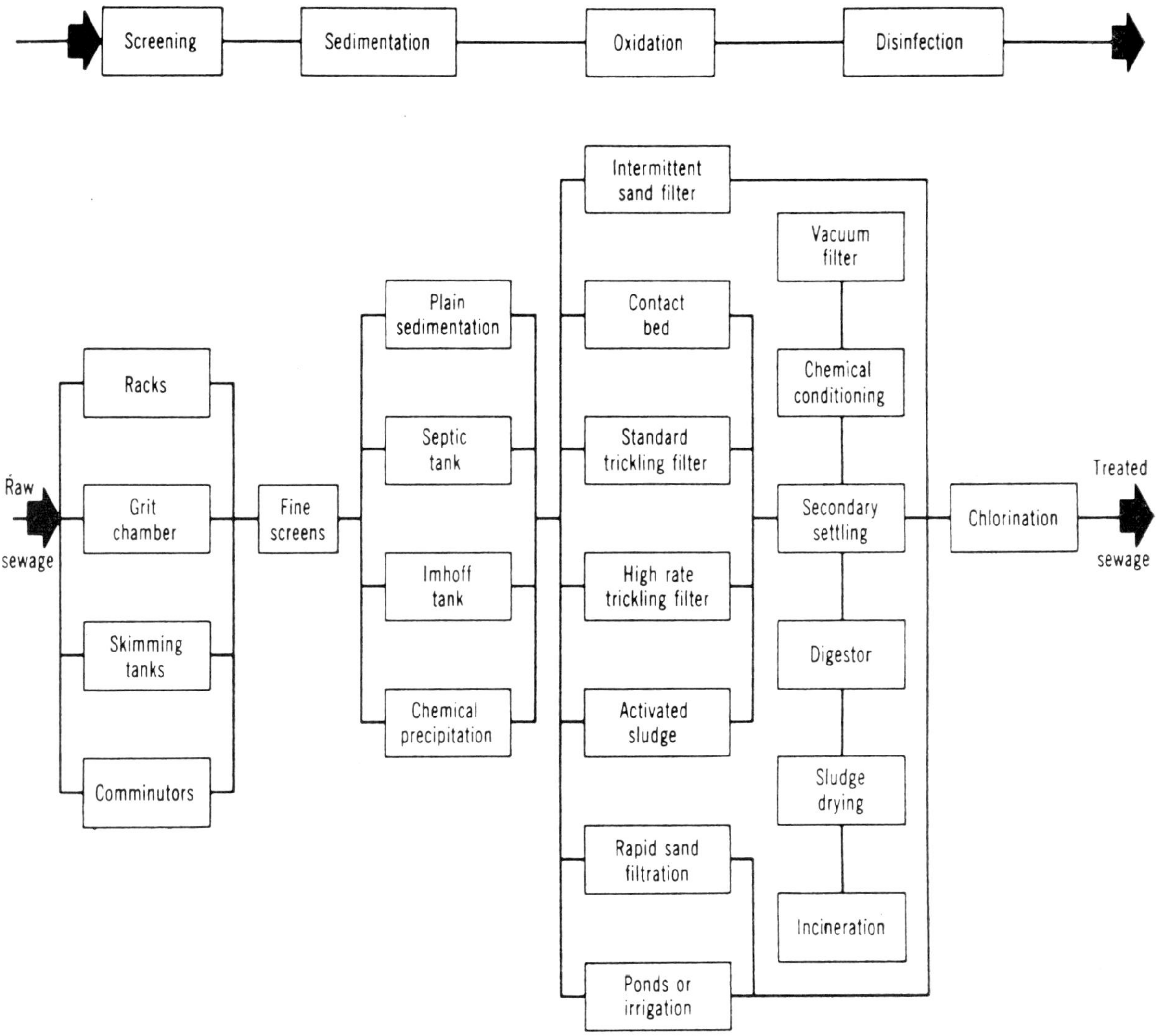

Figure 10 Sewage-treatment processes.
From *Environmental Sanitation*, Joseph A. Salvato, Jr., John Wiley & Sons, Inc., New York, 1958.

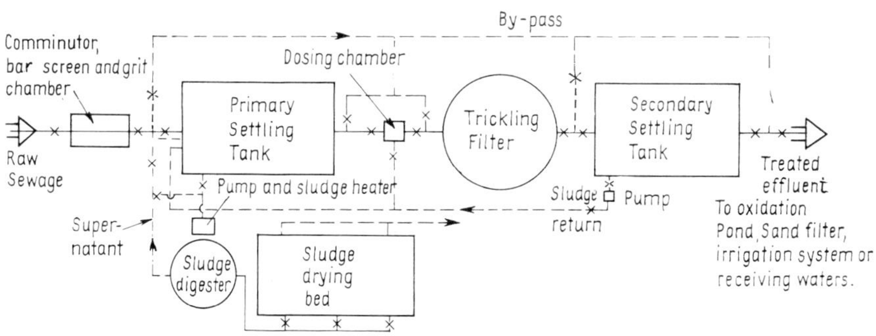

Figure 11 A secondary sewage treatment plant.

TABLE 6

Sewage-Treatment Plant Unit Combinations and Efficiencies

Treatment Plant	Total per cent Reduction—Approximate	
	Suspended Solids	Biochemical Oxygen Demand
Sedimentation plus sand filter	90 to 98	85 to 95
Sedimentation plus standard trickling filter, 600 lb BOD per acre-foot maximum loading	75 to 90	80 to 95
Sedimentation plus single stage high rate trickling filter	50 to 80	35 to 65*
Sedimentation plus two stage high rate trickling filter	70 to 90	80 to 95*
Activated sludge	85 to 95	85 to 95
Chemical treatment	65 to 90	45 to 80
Pre-aeration (1 hr) plus sedimentation	60 to 80	40 to 60
Plain sedimentation	40 to 70	25 to 40
Fine screening	2 to 20	5 to 10
Stabilization, pond aerobic		70 to 90
Anaerobic lagoon	60 to 70	40 to 70

* No recirculation. Efficiencies can be increased within limits by controlling organic loading, efficiencies of settling tanks, volume of recirculation, and the number of stages; however, effluent will be less nitrified than from standard rate filter, but will usually contain dissolved oxygen. Filter flies and odors are reduced. Study first cost plus operation and maintenance.

From *Environmental Sanitation*, Joseph A. Salvato, Jr., John Wiley & Sons, Inc., New York, 1958.

There are two major factors which can determine the success or failure of a sewer and treatment plant construction program once completed. First is the positive exclusion of surface water, ground water, roof leaders and footing drains by ordinance and inspection. Many a sewage treatment plant constructed has been so hydraulically overloaded as to render it almost useless as a stream pollution abatement facility. Next is proper operation of the treatment plant by competent adequately paid personnel. Here the consulting engineers and the regulatory agencies share a joint responsibility to ensure that public officials are properly informed and that the huge investments in water pollution control accomplish the intended objectives.

Solid Wastes

General

Aesthetic, land use, health, water pollution, air pollution and economic considerations make proper solid waste storage, collection and disposal municipal, corporate and individual functions which must be taken seriously. Indiscriminate dumping of solid waste and failure of the collection system in a populated community for two or three weeks would soon cause many problems. Odors, flies, rats, roaches, crickets, wandering dogs and cats, and fires would dispel any remaining doubts of the importance of proper solid waste storage, collection and disposal.

The complexities of proper solid waste management are not readily appreciated. There are numerous sources and types of solid wastes ranging from the home to the farm and from garbage to radioactive wastes, junked cars and industrial wastes. Handling involves storage, collection transfer and transport. Processing includes incineration, densification, composting, separation, treatment, and energy conversion. Disposal shows the environmental inter-relation of air, land and water, and the place of salvage and recycling. All these steps introduce constraints—social, political, economic, technological, ecological, legal, information and communications which must be considered in the analysis of the problem and in coming up with acceptable solutions.

All these factors show the need for comprehensive planning. It is necessary to come up with alternative solutions together with the environmental and social impact of each. In this way one is more likely to come up with a practical and acceptable plan that will have a good chance of being implemented.

Composition, weight and volume

The solid waste of a municipality may consist of garbage, rubbish, ashes, bulky wastes, street refuse, dead animals, abandoned vehicles, construction and demolition wastes, industrial solid wastes, special hazardous wastes, animal and agricultural wastes, sewage treatment plant residues and septic tank cleanings. Garbage is the putrescible organic material discarded or remaining as a result of the storage, processing, preparation and consumption of food. Rubbish includes paper, wood, glass, metal, leaves, plastics, tin cans, stones and dirt. Bulky wastes include stoves, refrigerators, auto parts, furniture, large crates, trees and tree branches.

Various estimates have been made of the quantity of solid waste generated and collected per person per day. The amount of municipal solid waste *generated* in 1968 was estimated to be 7 pounds per capita per day and expected to increase at about 4 per cent per year. The amount collected was found to average 5.32 pounds. The figures given in Tables 7a and 7b are averages, subject to adjustment dependent upon many local factors, including time of the year, habits and economic status of the people, whether urban or rural, education and location. The estimates should not be used for design purpose. Each community should be studied by itself and actual weighings made to obtain representative information.

TABLE 7a

Average Solid Waste Collected, 1968
(pounds per person per day)

Solid waste type	Urban	Rural	National
Household	1.26	0.72	1.14
Commercial	0.46	0.11	0.38
Combined	2.63	2.60	2.63
Industrial	0.65	0.37	0.59
Demolition, construction	0.23	0.02	0.18
Street and alley	0.11	0.03	0.09
Miscellaneous	0.38	0.08	0.31
Totals	5.72	3.93	5.32

Anton J. Muhich, "Sample Representativeness and Community Data," *Proceedings, Institute for Solid Wastes*, APWA, Chicago, Ill., 1968.

TABLE 7b

*Approximate Composition of Domestic Solid Wastes**

Component	Per cent by weight
Food waste	15
Paper products	50
Plastics, rubber, leather	3
Rags	2
Metals	8
Glass and ceramics	8
Wood	2
Garden wastes	5
Rocks, dirt, miscellaneous	7

* Moisture content approximately 30 per cent.

Collection

Collection cost has been estimated to represent about 80 per cent of the total cost of collection and disposal by sanitary landfill and 60 per cent when incineration is used. In 1968 collection cost alone was found to average 10 to 13 dollars per ton.

The frequency of collection will depend upon the quantity of solid waste, time of the year, socio-economic status of area served, and municipal or contractor responsibility. In business districts refuse, including garbage from hotel and restaurants, should be collected daily except Sundays. In residential areas, twice a week refuse collection during warm months of the year and once a week at other times should be the maximum permissible interval. Slum and ghetto areas usually require at least twice a week collection. The receptacle should be either emptied directly into the garbage truck or carted away and replaced with a clean container. Refuse transferred from can to can will invariably cause spilling, with resultant pollution of the ground and attraction of rats and flies. If other than curb pick-up is provided, the cost of collection will be high. Some property owners are willing to pay for this extra service.

Transfer station

The urban areas around cities have been spreading leaving fewer nearby acceptable solid waste disposal sites. This has generally made necessary the construction of incinerators in cities or on the outskirts, or the transportation of wastes longer distances to new sites. However, as the distance from the centers of solid waste generation increases, the cost of direct haul to a site increases. A distance is reached (in terms of time) when it becomes less expensive to construct an incinerator, or a transfer station near the center of solid waste generation where wastes from collection vehicles can be transferred to large tractor-trailers for haul to more distant disposal sites.

Rail haul and barging to sea also involve the use of transfer stations. They may include one or a combination of grinding, compaction to various densities, and baling.

Treatment and disposal of solid wastes

Solid waste treatment and disposal methods include the open dump, hog feeding, incineration, grinding and discharge to a sewer, milling, compaction, sanitary landfill, dumping and burial at sea, reduction, composting, pyrolization, wet oxidation, and anaerobic digestion. The common accetable methods are incineration, sanitary landfill and in some parts of the world composting.

Open dump. The open dump is all too common and needs no explanation. It is never satisfactory as usually maintained. Refuse is generally spread over a large area providing a source of food and harborage for rats, flies and other vermin. It is unsightly, an odor and smoke nuisance, a fire hazard, and often a cause of water pollution. It should be eliminated or its operation changed to a sanitary landfill.

Hog feeding. Where garbage is fed to hogs, careful supervision is necessary. The spread of trichinosis in man, hog cholera, the virus of foot-and-mouth disease, and vesicular exanthema in swine is encouraged when uncooked garbage is fed to hogs. In some instances, tuberculosis, swine erysipelas and stomatitis may also be spread by raw garbage. The boiling for thirty minutes of all garbage fed to hogs will prevent transmission of trichinosis and economic loss to the swine industry due to hog illness and death.

Grinding. The grinding of garbage is fast becoming a common method of garbage disposal. It is highly recommended from a convenience and public health standpoint, but the disposal of other refuse remains to be handled. The putrescible matter is promptly removed thereby eliminating this as a source of odors and food for rats, flies and other vermin. In one system, the home grinder is connected to the kitchen sink drain. Garbage is shredded into small particles while being mixed with water and is discharged to the house sewer. In another system, garbage is collected as before but dumped into large, centrally located garbage grinding stations that discharge garbage to the municipal sewerage system. In small communities, the garbage grinding station may be located at the sewage-treatment plant. The strength of the sewage is increased and additional sludge digestion and drying facilities will be required when a large amount of garbage is handled.

Disposal at sea. Where dumping at sea is practiced, all garbage and other refuse is dumped into large garbage scows or barges. The scows are towed by tugs and the garbage is taken out to sea and dumped a sufficient distance out to prevent the refuse being carried back to shore and causing a nuisance. Bad weather conditions hamper this operation and unless this method is kept under very careful surveillance, abuses and failures will result. Because of the cost of maintaining a small navy and difficulties in satisfactorily carrying out this operation, coastal cities have reverted to sanitary landfill and incineration. In recent years, consideration has been given to the compaction of refuse to a density greater than 66.5 pounds per cubic foot prior to transport and then disposal by burial in the ocean at depths greater than 100 feet, based upon oceanographic conditions, to ensure there will be no mixing with surface water. More research is needed to determine stability of the compacted refuse and effect on marine life. See high density compaction.

Garbage reduction. In the reduction method of garbage disposal, the garbage is cooked under pressure. Fats melt out and are separated from the remaining material. The fat is used in the manufacture of soaps or glycerines and the residue is dried, ground and sold for fertilizer or cattle feed. Odor complaints are associated with this process and, where a solvent such as naphtha is used to increase the extraction of fat, a greater fire hazard exists. The use of synthetic detergents and chemical fertilizers and high operating costs have led to the abandonment of this process.

Composting. Composting is the controlled decay of organic matter in a warm, moist environment by the action of bacteria, fungi, molds and other organisms. This may be an aerobic and/or anaerobic operation. Moisture is maintained at 40 to 65 per cent, 50 to 60 per cent is best. Composition of the refuse, disposal of refuse not composted, demand for compost and salvaged material, odor production and control, public acceptance and total cost are factors to be carefully weighed. Compost is a good soil conditioner but a poor fertilizer. The process is very attractive, but because of the factors mentioned has not met with success in the United States.

The composting operation involves a combination of steps. These may include: a) weighing, b) separation of non-compostables and salvage by hand and by a magnetic separator, c) size reduction to 2 inches or less by means of a shredder, grinder, chipper, rasp mill, hammermill, d) ballistic and magnetic separation, e) biological digestion by any one of a number of composting methods, f) screening and possible standardization of fertilizer value, and g) disposal by bagging for sale, or to a sanitary landfill.

Pyrolization. Pyrolization as applied to solid

wastes (metal and glass removed) is an experimental thermochemical process for conversion of complex organic solids, in the absence of added oxygen, to water, combustible gases, tarry liquids, and a stable residue. Intermediate products may be collected or may be used to contribute heat to support the process. The end products would be carbon, water and carbon dioxide if carried to completion. If the raw material contains sulfur and nitrogen, these oxides will be formed with resultant air pollution unless provision is made for their removal. Temperatures of 900 to 1700°F have been used. In a variation, some oxygen and a temperature up to 2100°F is used. It is a process of destructive distillation similar to that used for making charcoal and for the recovery of organic by-products such as turpentine, acetic acid and methanol from wood.[22] The Lantz converter and the Urban Research and Development Corporation unit are variations of the pyrolytic process. The Lantz system uses a temperature of 1200 to 1400°F with ground refuse.

High temperature incineration. High temperature incineration is carried out at 3400°F (Melt-zit), 2500–2600°F (FLK Slagging Incinerator, Germany), 2600–3000°F (Torrax system), 3000°F (American Design and Development Corporation). These units are in the developmental and pilot stage. Combustibles are destroyed and non-combustibles are reduced to slag or sand-like grit. A 97 per cent volume reduction is reported.

Wet oxidation. Wet oxidation and anaerobic digestion of refuse are in the experimental stage. In wet oxidation, the refuse can be processed under high pressures and temperatures, or the refuse can be ground and aerated while in suspension in a liquid medium. In anaerobic digestion, decomposable material is separated from refuse, ground and then digested at controlled temperature in the absence of air. The resultant gases are mostly methane and carbon dioxide and the residue can serve as a soil conditioner.

Size reduction. Shredders, hoggers and chippers are devices which reduce refuse, bulky items and other solid wastes to a manageable size for disposal by landfill or for processing through an incinerator.

High density compaction. High density compaction of solid wastes is accomplished by compression to a density of more than 66.5 pounds per cubic foot. The resulting bales may be enclosed in chicken wire, hot asphalt, vinyl plastic, plain or reinforced cement, or welded sheet steel depending on the method of disposal or intended use of the bales or blocks. Rogus reports on the Tezuka-Kosan process in Tokyo that liquid release during compression ranged from 2 to 5 per cent by weight and that it will require treatment. The bales had a density of 70 to 109 pounds per cubic foot; they sank in sea water; had good structural cohesiveness; resisted corrosion; were reasonably free of odors, and insect, rodent hazard; and showed no evidence of aerobic or anaerobic decomposition.[23] This system has not yet been adopted in the United States but offers attractive possibilities for solid waste reuse, economic hauling, and more acceptable disposal.

Other methods of achieving high density are subjecting refuse to pressure in a hydraulic pressure chamber and forcing baled refuse through a tube extending 30 to 50 feet below the water surface. Plastic toys and containers do not remain compressed or submerged. The effect of time on decomposition with possible gas formation, disintegration and floating of the refuse is not known. The long-term effect of the submerged material on the marine ecology is also an unknown.[24]

Incineration

Incineration is a controlled combustion process for burning solid, liquid, or gaseous combustible waste to gases and a residue containing little or no combustible material *when properly carried out.* It is a volume reduction process suitable for about 70 per cent of the municipal solid wastes.

Incineration, if properly designed and controlled, is satisfactory for burning combustible refuse provided air pollution standards can be met. Continuous operation six days a week and a controlled temperature are needed for efficiency, prevention of excessive air pollution and odor control.

An operating design temperature range of 1500 to 1800°F is generally recommended. Excessively high temperatures and extreme variations cause cracking and spalling with rapid deterioration of fire tile and brick linings (refractories). Batch feed or one-shift operation promote spalling and loosening of tile linings. Other types of lining and design may permit higher operating temperaures. Actually, temperature in the furnace will range from 2100 to 2500°F. When the gases leave the combustion chamber the temperature should be between 1400 and 1800°F and the gas entering the stack 1000°F or less. The temperature will have to be lowered to 450 to 500°F before the gas is filtered or to 600°F or less if electrical precipitators are used.

Sanitary landfill

A sanitary landfill is an engineering method of disposing solid waste on land by spreading the waste in thin layers, compacting it to the smallest practical volume and covering it with a layer of earth at the end of each day or more frequently in a manner which prevents environmental pollution.[25] The cost of sanitary landfill versus incineration for various populations is shown in Fig. 12.

Social and political factors. An important aspect of refuse disposal site selection, in addition to engineering factors, is the evaluation of public reaction and education of the public so that understanding and acceptance is developed. A program of public information is also needed. Equally important are the climate for political cooperation, cost comparison of alternative solutions, available revenue, aesthetic expectations of the people, organized community support, and similar factors.

Experience shows that where open dumps have been operated, there will be opposition to almost any site proposed for sanitary landfill or an incinerator for that matter. However, local officials will have to study all of the facts and make a decision to fulfill their responsibility and exercise their authority for the public good in spite of any expressed opposition. Usually the critcal factor is convincing the public that a nuisance-free operation can and will in fact be conducted. A private contract may make possible regional disposal, whereas an inter-governmental operation may be politically impractical.

Planning. Local officials can make their task easier by planning ahead together on a county or regional basis for 20 to 40 years in the future, and by acquisition of adequate sites at least 5 years prior to anticipated needs and use. The availability of federal and state funds for planning for collection, treatment and disposal of refuse on an area-wide basis such as a county should be explored. The planning will require an engineering analysis of alternative sites including population projections, volume and characteristics of all types of solid wastes to be handled, cost of land and site preparation, expected life of the site, haul distances from the sources of refuse to the site, cost of operation, and possible value of the finished sites. Consideration would be given to the climate of the

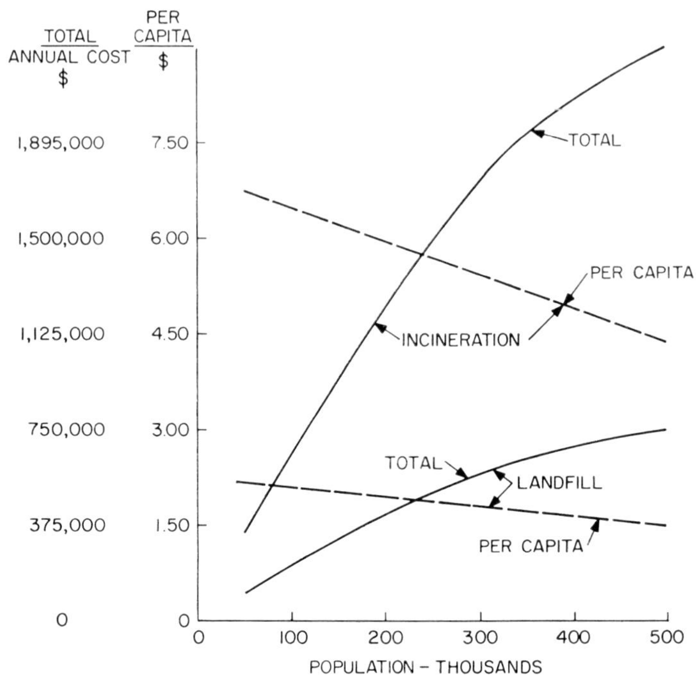

Figure 12 General cost comparison of sanitary landfill and incineration.
From *Municipal Refuse Collection and Disposal* Office for Local Government, Albany, N.Y., 1964. Adjusted to 1970 Costs.

region, prevailing winds, zoning ordinances, geology and topography. Location and drainage to prevent surface water or ground water pollution, access roads to major highways, and availability of suitable cover material are other considerations.

Special attention must be given to the diversion of surface water, to the highest ground water level so as to keep the bottom of the fill well above it, and to the soils available. The type of soil used for cover and its placement should minimize infiltration. The intervening soil between the bottom of the landfill and maximum ground water level should prevent or control leachate travel, or if this cannot be reasonably assured, an artificial membrane or clay soil barrier with an underdrain collection system may be needed.

Once a decision is made it should be made common knowledge and plans developed to show how it is proposed to reclaim or improve and reuse the site upon completion. This should include

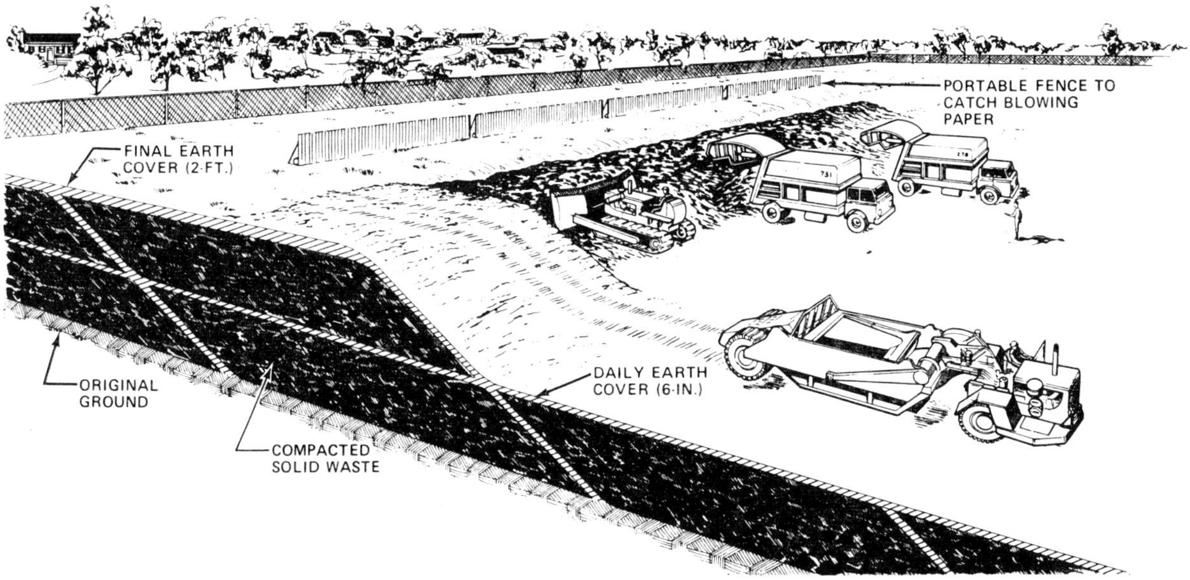

Figure 13 Area method. The bulldozer spreads and compacts solid wastes. The scraper (foreground) is used to haul the cover material at the end of the day's operations. Note the portable fence that catches any blowing debris. This is used with any landfill method.

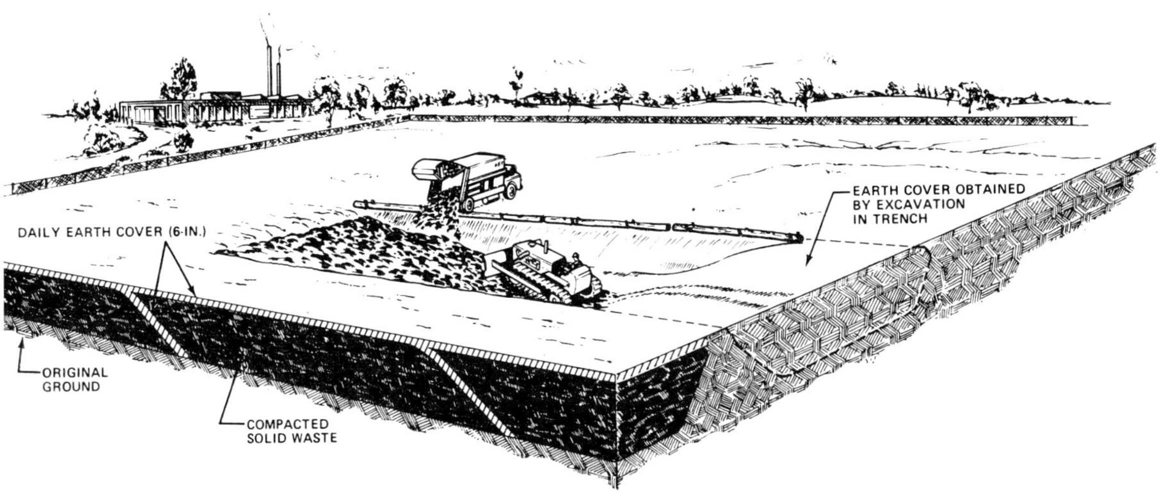

Figure 14 Trench method. The waste collection truck deposits its load into the trench where the bulldozer spreads and compacts it. At the end of the day the dragline excavates soil from the future trench; this soil is used as the daily cover material. Trenches can also be excavated with a front- end loader, bulldozer, or scraper. From *Sanitary Landfill Facts*, Thomas J. Sorg and H. Lanier Hickman, U.S. Department of Health, Education, and Welfare, Public Health Service Publication No. 1792, Superintendent of Documents, Washington, D.C., 1970.

talks, slides, news releases, question and answer presentations, and inspection of good operations. Artist's renderings are very helpful in explaining construction methods and final use of the land. Figures 13 and 14 show the area and trench sanitary landfill methods.

Recreation Areas and Temporary Residences

Bathing beaches and swimming pools

The sanitation of bathing places is dictated by health and aesthetic reasons. Few people would knowingly swim or water ski in polluted water; and insanitary surroundings are not conductive to the enjoyment of "a day at the beach." People are demanding more and cleaner beaches and pools, and a camp, motel, club, or resort without a pool or beach is not nearly as popular as one that has this facility.

Health hazards

From our knowledge of disease transmission, it is known that certain illnesses can be transmitted by improperly operated or located swimming pools and beaches through contact and ingestion of polluted water. Among these are typhoid fever, dysentery, and other gastrointestinal illnesses; conjunctivitis, trachoma, leptospirosis, ringworm infections, schistosomiasis, or swimmer's itch; upper respiratory tract diseases such as sinus infection, septic sore throat, and middle ear infection. The repeated flushing of the mucous coatings of the eyes, ears, and throat, and the excessive use of alum or lack of pH control, expose the unprotected surfaces to possible inflammation, irritation, and infection. Contraction of the skin on immersion in water may make possible the direct entrance of contaminated water into the nose and eyes.

Stevenson[26] reported that "an appreciably higher over-all illness incidence may be expected in the swimming group over that in the nonswimming group regardless of the bathing water quality." It was further stated that "eye, ear, nose, and throat ailments may be expected to represent more than half of the over-all illness incidence, gastrointestinal disturbances up to one-fifth, and skin irritations and other illnesses the balance." Although based on limited data, swimming in a lake water with an average coliform content of 2300 per 100 ml caused "a significant increase in illness incidence ..." and swimming in a river "water having a median coliform density of 2700 per 100 ml appears to have caused a significant increase in such (gastro-intestinal) illness." The study also showed the greatest amount of swimming was done by persons 5 to 19 years of age.

Diesch and McCulloch[27] summarized incidences of leptospirosis in persons swimming in waters contaminated by discharges of domestic and wild animals including cattle, swine, foxes, racoons, muskrats and mice. Pathogenic leptospires were isolated from natural waters, confirming the inadvisability of swimming in streams and farm ponds receiving drainage from cattle or swine pastures.

Joyce and Weiser report that enteroviruses which are found in human or animal excreta, if introduced into a farm pond by drainage or direct flow, can constitute a serious public health hazard if used for recreation, drinking or other domestic purposes.[28]

Many other studies[29,30,31] have been made to relate bathing water bacterial quality to disease transmission with inconclusive or negative results.

British investigators[29] have drawn the following general conclusions:

(i) That bathing in sewage-polluted sea water carries only a negligible risk to health, even on beaches that are aesthetically very unsatisfactory.

(ii) That the minimal risk attending such bathing is probably associated with chance contact with intact aggregates of faecal material that happen to have come from infected persons.

(iii) That the isolation of pathogenic organisms from sewage-contaminated sea water is more important as evidence of an existing hazard in the populations from which the sewage is derived than as evidence of a further risk of infection in bathers.

(iv) That, since a serious risk of contracting disease through bathing in sewage-polluted sea water is probably not incurred unless the water is so fouled as to be aesthetically revolting, public health requirements would seem to be reasonably met by a general policy of improving grossly insanitary bathing waters and of preventing so far as possible the pollution of bathing beaches with undisintegrated faecal matter during the bathing season.

The findings of the Public Health Activities Committee of the ASCE Sanitary Engineering Division [31] are summarized in the following abstract:

"Coliform standards are a major public health factor in judging the sanitary quality of recreational waters. There is little, if any, conclusive proof that disease hazards are directly associated with large numbers of coliform organisms. Comprehensive research is recommended to provide data for establishing sanitary standards for recreational waters on a more rational or sound public health basis. British investigations show that even finding typhoid organisms and other pathogens in recreational waters is not indicative of a health hazard to bathers but is only indicative of the presence of these diseases in the population producing the sewage. The Committee recommends that beaches not be closed and other decisive action not be taken because current microbiological standards are not met except when evidence of fresh sewage or epidemiological data would support such action."

In view of the available information, emphasis should be placed on elimination of sources of pollution (sewage, storm water, land drainage), effective disinfection of treated waste-waters, and on the proper interpretation of bacterial examinations of samples collected from representative locations.

In swimming pools there is a possibility of direct transmission of infection from one bather to another if the water does not have an active disinfectant such as free available chlorine. Proper operation and treatment therefore are of prime importance.

Food Protection

Site location and planning

It is extremely important that a preliminary investigation be made of a site proposed for a new plant to determine the suitability of a site for a given purpose. This should include restrictions related to zoning, sewage and wastewater disposal, solid wastes disposal, air pollution control, and noise.

Time and money spent in study, before a property is purchased, is a good investment and sound planning. For example, a factory that requires millions of gallons of cooling or processing water would not be located too distant from a lake or clear stream, unless it was demonstrated that an unlimited supply of satisfactory well water or public water was available at a reasonable cost. An industry having as an integral part of its process a liquid, solid, or gaseous waste would not locate where adequate dilution or disposal area was not available, unless an economical waste treatment process could be devised.

Such factors as topography, drainage, highways, railroads, watercourses, exposure, swamps, prevailing winds, dust, odors, insect and rodent prevalence, type of rock and soil, availability and adequacy of public utilities, the need for a separate power plant, and water-, sewage-, or waste-treatment works and special air pollution control equipment should all be considered and evaluated before selecting a site for a particular use.

Engineering and architectural considerations

The design of structures and facilities are properly engineering and architectural functions that should be delegated to individuals or firms who have become expert in such matters. The design and construction material used is dependent upon the type of structure, its geographical location and operation. Some materials can be used to greater advantage because of location, availability of raw materials, labor costs, type of skilled labor available, local building codes, climate, and other factors.

In many instances plan approval and permits are required from federal, state or local regulatory agencies. Check with the agency having jurisdiction before any decisions are made or any plans are prepared.

Basic sanitation requirements for food processing and food service establishments

Regulation. The regulation and supervision of food establishments such as restaurants, caterers, commissaries, pasteurizing plants, frozen dessert plants, frozen prepared food plants, vending machine centers, slaughter houses, poultry processing, bakeries, shellfish shucking and packing plants, and similar places is in the public interest. This responsibility is usually vested in the state and local health and agriculture departments, and also in the Food and Drug Administration, the U.S. Department of Agriculture and in the U.S. Public Health Service when interstate traffic of the food is involved. The industry affected as a rule also recognizes its fundamental responsibility.

Industry and regulatory control emphasis needs to be placed at the foci of food collection, processing and distribution and on self-inspection.

See page 310. Proper design, equipment and layout, and construction must build out environmental sanitation problems and simplify sanitary operations. Quality control procedures must also be applied to convenience type food such as prepared frozen dinners. The food catering industries which now serve banquets, hospitals and nursing homes, airlines and other mass feeding facilities are particularly vulnerable. Food preparation is leaving the home and small restaurant and is becoming a centralized production process leaving only final conditioning or heating at the point of service. Such operations make possible more centralized control and certain efficiencies, but also increase the potential of foodborne outbreaks affecting large numbers of people.

General requirements. It is to be noted that certain basic sanitation requirements are common to all places where food is processed. McGlasson[32] has proposed a set of standards under the following headings:

1. Location, construction, facilities, and maintenance. a. Grounds and premises, b. Construction and maintenance. c. Lighting. d. Ventilation. e. Dressing rooms and lockers.

2. Sanitary facilities and controls. a. Water supply. b. Sewage disposal. c. Plumbing. d. Restroom facilities. e. Handwashing facilities. f. Food wastes and rubbish disposal. g. Vermin control.

3. Food-product equipment and utensils. a. Sanitary design, construction, and installation of equipment and utensils. b. Cleanliness of equipment and utensils.

4. Food, food products, and ingredients. a. Source of supply. b. Protection of food, food products, and ingredients.

5. Personal. a. Health and disease control. b. Cleanliness.

A similar set of basic standards were published in the Federal Register in April 1969.[33]

Specific requirements. The general sanitation requirements should be supplemented by specific regulations applicable to a particular establishment or operation. Excellent codes, compliance guides, and inspection report forms are available from regulatory and training organizations.

Food preparation. Temperature control and clean practices should be the rule in kitchens and food-processing plants if contamination is to be kept to a minimum. All food contact surfaces and equipment used in food preparation must be kept clean and in good repair.

Frozen meat, poultry, and other bulk frozen foods should be thawed slowly under controlled refrigeration (36°–38°F is ideal) and *not left to stand at room temperature overnight to thaw.* Frozen vegetables and chops need not be thawed but can be cooked directly. Prepared foods, especially protein types, should be served immediately, kept temporarily at a temperature of 45°F or less, or on a warming table maintained at a temperature above 140°F until served. If not to be served, the food should be refrigerated in shallow pans to a depth of 2 to 3 inches within 30 minutes at a temperature of 45°F. Bulk foods such as roasts should be refrigerated within 30 minutes of preparation and cooled so that the internal temperature is 40°F or less within two hours, unless immediately served. Note that souring of hot foods is prevented by prompt refrigeration.

Food handlers are expected to have hygienic habits. A conveniently located wash basin in the kitchen or work room, and in the toilet room, supplied with warm running water, soap, and individual paper towels are essential.

Microbiological standards. Natural microbial variations in different foods, and the statistical aspects of sampling present considerable difficulties in the establishment of firm standards. In addition, because of normal errors inherent in laboratory techniques, it is practical to allow some leeway in the standards proposed and hence refer to them as guidelines. One might use the geometric mean of say 10 samples or allow one substandard sample out of four. Parameters used include total aerobic count, toxigenic molds, number of coliforms, number of *E. coli*, coagulase-positive staphylococci, salmonellae, shigella, *Clostridium perfringens*, *Clostridium botulinum* and beta-hemolytic streptococci as indicated.

Milk control

The milk industry and the regulatory agencies have a joint responsibility in ensuring that all milk and milk products consistently meet the standards established for protection of the public health. Inspection duplication should be avoided; instead there should be a deliberate synergism of effort. With proper planning and cooperation the industry, local, state, and federal systems can actually strengthen the protection afforded the consumer.

The role of industry and official agencies to accomplish the objective stated is described below.

Certified industry inspection. Industry quality control inspectors are qualified by the official agency (usually health or agriculture) based upon education, experience and examination to make dairy farm inspection pursuant to the milk code or ordinance. Certificates issued are issued for a stated period of time, one to three years, may be revoked for cause, are renewed based on a satisfactory work record, and may require participation in an annual refresher course. Copies of all inspections, field tests, veterinary examinations, and laboratory reports are promptly forwarded to the official agency or to an agreed upon place and kept on file at least one year.

Cooperative State—Public Health Service program for certification of Interstate Milk Shippers. The federal-state program is commonly referred to as the IMS Program. A State milk sanitation rating officer certified by the U.S. Public Health Service makes a rating of a milk supply. The name of the supply and rating is published quarterly by the Public Health Service. If the milk and milk products are produced and pasteurized under regulations which are substantially equivalent to the PHS Milk Ordinance, and is given an acceptable milk sanitation compliance and enforcement rating, the milk and milk products may be shipped to another area of jurisdiction which is participating in the IMS Program.

The procedures for rating a milk supply are carefully designed with detailed instructions to be followed by the industry and the rating officer. Independent evaluations are made by Public Health Service rating officers to confirm ratings given, or changes since the last rating, and ensure reproducibility of results.

Official local program supervision and inspection. The official agency makes regular review of the industry inspection files mentioned above, takes whatever action is indicated, and at least annually makes joint inspections with the industry inspector of a randomly selected significant number of dairy farms, including receiving stations. The quality of work done is reviewed, the need for special training is determined, and recommendation concerning certificate renewal is made to the permit issuing official. A similar review is also made of the sample collection, transportation and the procedures, equipment and personnel in the laboratory making the routine milk and water examinations.

In addition, the local regulatory agency collects official samples as required by the State milk code, advises the industry having jurisdiction of the results and corrective action required, participates in training sessions, and serves as the State agent in securing compliance with the State milk code. The authorized local, city or county agency usually has responsibility for the routine inspection of processing plants, sample collection, and overall program supervision for compliance with the Code. The agency sanitarian serves as a consultant to the industry in the resolution of the more difficult technical, operational and laboratory problems. This whole procedure makes possible better use of the qualified industry inspector, and the professionally trained sanitarian, with better direct supervision over dairy farms and pasteurization plants, and more effective surveillance of milk quality. In some states the local activities are carried out in whole or in part by the state regulatory agency.

Official state surveillance and program evaluation. The state department of health, and in some instances the state department of agriculture, share responsibility for milk sanitation, wholesomeness and adulteration. The responsibility is usually given in state law, sanitary code or milk code, and in rules and regulations promulgated pursuant to authority in the law. Most states have adopted the *Grade "A" Pasteurized Milk Ordinance*, 1965 Recommendations of the United States Public Health Service, or a code which is substantially the same. This makes possible a reasonable basis for uniformity in regulations and interpretation both interstate and intrastate. However short-term and alleged economic factors frequently limit reciprocity and interstate movement of milk. Both industry and regulatory agencies need to cooperate in the elimination of milk codes as trade barriers. Health agencies usually have no objection to reciprocity where milk quality compliance and enforcement is certified under the Public Health Service IMS Program. However not all states or farms participate in the program.

The IMS Program also gives the state regulatory agency a valuable tool to objectively evaluate the effectiveness of the local routine inspection and enforcement activities. The types of additional training needed by the qualified industry inspector, the assistance the dairy farmer should have, and the supervisory training needed by the regulatory agency sanitarian become apparent. Changes in technology and in practices are noted and need for clarification of regulations, laws and policies are made known to the state agency.

A common function of a state regulatory agency is periodic in-depth evaluation of local milk programs. This includes the effectiveness with which the local unit is carrying out its delegated responsibilities as described above, the quality of work done, the staff competency and adequacy, the reliability of the official laboratory work, the record keeping, the equipment and facilities available, the number of inspections and reinspections made of dairy farms and processing plants and their adequacy or excessive frequency. The state agency usually has responsibility for approving all equipment used in milk production and service from the cow to the consumer. The standards recommended by national organizations are generally used as a basis for the acceptance of equipment.

Pasteurization. Pasteurization is the process of heating every particle of milk or milk product to at least 145°F and holding at or above such temperature continuously for at least 30 minutes, or to at least 161°F and holding at such temperature continuously for at least 15 seconds, in approved and properly operated equipment. A temperature of at least 150°F for 30 continuous minutes (155°F for 30 minutes in practice) or at least 166°F for 15 seconds (175°F for 25 seconds in practice) is needed for milk products which have a higher milkfat content than milk or contain added sweeteners. Higher temperatures and shorter times are also being used. The heat treatment should be followed by prompt cooling to 45°F.

The effectiveness of pasteurization in the prevention of illnesses that may be transmitted through milk-borne disease organisms has been demonstrated beyond any doubt. The continued sale and consumption of raw milk must therefore be attributed to ignorance of these facts. Pasteurization does not eliminate pesticide residues, anthrax spores, or toxins given off by certain staphylococci; but the production of toxins is nil when milk is properly refrigerated. The equipment used to pasteurize milk is of three general types. These are the holder, high-temperature short-time, and in-the-bottle pasteurizers. High-temperature pasteurizers include the vacuum type (temperature 162 to 200°F) and the ultra-high temperature type (190 to 270°F).

Quality control. Quality control involves herd health, milk handling, transportation, processing and distribution. Field and laboratory testing coupled with inspection, supervision, education surveillance, enforcement and evaluation are the major methods used.

The tests used to control the quality of milk are explained in detail in *Standard Methods for the Examination of Dairy Products* Twelfth Edition. The major tests are mentioned here. Raw milk quality is determined by temperature, sediment, odor and flavor, appearance, antibiotics test, direct microscopic counts including clumps of bacteria, leucocytes and streptococci; standard plate counts, abnormal milk tests, and thermoduric determination. Tests for brucellosis and animal health are also made. Pasteurized milk quality is indicated by the standard plate count, direct microscopic count, phosphatase test, coliform test, taste, and odor tests. Other common tests are for butterfat, total solids, and specific gravity.

REFERENCES

1. WHO Expert Committee on Environmental Sanitation, Wld. Hlth. Org. techn. Rep. Ser., 1950, No. 10, p. 5.
2. Stead, Frank M., "Levels in Environmental Health," *AJPH*, Vol. 50, No. 3, March 1960, p. 312.
3. Wld. Hlth. Org. techn. Rep. Ser., 1954, No. 77, p. 9.
4. "Environmental Factors in Health Planning," *AJPH*, Vol. 58, No. 2, February, 1968, pp. 358–361.
5. "National Environmental Health Programmes: Their Planning, Organization, and Administration," Wld. Hlth. Org. techn. Rep. Ser., 1970, No. 439, pp. 10–11.
6. *Local and Regional Planning and Development*, Governor's Task Force Report, West Virginia, Regional Review, National Service to Regional Councils, Washington, D.C., January 1970, p. 2.
7. Weibel, S. R., F. R. Dixon, R. B. Weidner, and L. J. McCabe, "Waterborne-Disease Outbreaks, 1946–60," *Jour. AWWA*, Aug. 1964, pp. 947–58.
8. Craun, Gunther F., and McCabe, Leland J., "Review of the Causes of Waterborne-Disease Outbreaks," *Jour. AWWA*, Jan. 1973, pp. 74–84.
9. Dennis, Joseph M., "1955–56 Infectious Hepatitis Epidemic in Delhi, India," *Jour. AWWA*, Oct. 1955, pp. 1288–98.
10. Wolman, Abel, "Water Supply and Environmental Health," *Jour. AWWA*, Dec. 1970, pp. 746–749.
11. Woodward, William E., *et al.*, "Foodborne Disease Surveillance in the United States, 1966 and 1967," *AJPH*, Jan. 1970, pp. 130–37.

12. Lofroth, Goran, "Methylmercury," Redaktionstjansten, Natural Science Research Council, Stockholm, Sweden, March 20, 1969.
13. A special report to the Secretary's Pesticide Advisory Committee, *Hazards in Mercury*, Department of Health, Education, and Welfare, and the Environmental Protection Agency, Washington, D.C., November 1970.
14. Cooke, N. E., and A. Beitel, "Some Aspects of Other Sources of Mercury to the Environment," Symposium Mercury in Man's Environment, Royal Society of Canada, February 15, 1971.
15. *Statistical Summary of Municipal Water Facilities in the United States*, Public Health Service Publ. No. 1039, Washington, D.C., Jauary 1, 1963.
16. McCabe, Leland J., James M. Symons, Roger D. Lee and Gordon G. Robeck, "Survey of Community Water, Supply, Systems," *Jour. AWWA*, Nov. 1970, pp. 670 to 687.
17. *Hydrology Handbook* Manual 28, Am. Soc. of Civil Engrs., N.Y., 1949.
18. *Public Health Service Drinking Water Standards, 1962* PHS Pub. 956, U.S. Department of Health, Education, and Welfare, Washington, D.C.
19. *Glossary Water and Wastewater Control Engineering*, APHA, ASCE, AWWA, WPCF, 1969.
20. *Manual of Recommended Water-Sanitation Practice*, Recommended by the United States Public Health Service, 1946, Public Health Bulletin No. 296.
21. Saunders, W. R., "Water and Wastes, The Public Health Viewpoint," *Royal Society of Health Journal*, Sept./Oct. 1968, pp. 247–250.
22. Hoffman, Donald A., and Richard A. Fitz, "Batch Retort Pyrolysis of Solid Municipal Wastes," *Environmental Science and Technology*, Vol. 2, November 1968, pp.1023–1026.
23. Rogus, Casimir A., "High Compression Baling of Solid Wastes," *Public Works*, June 1969, pp. 85–89.
24. Golueke, Clarence G., and P. H. McGauhey, "Future Alternatives to Incineration and their Air Pollution Potential," *Proceedings: The Third National Conference on Air Pollution*, U.S. Department of Health, Education, and Welfare, Washington, D.C., December, 12–14, 1966.
25. Brunner, D. R., D. J. Keller, C. W. Reid, Jr., and J. Wheeler, *Sanitary Landfill Guidelines-1970*, Public Health Service, Dept. of Health, Education, and Welfare, Washington, D.C., 1970.
26. Stevenson, Albert H., "Studies of Bathing Water Quality and Health," Presented before the second session of the Engineering Section of the American Public Health Association at the 80th Annual Meeting in Cleveland, Ohio, Oct. 23, 1952.
27. Diesch, Stanley L., and William F. McCulloch, "Isolation of Pathogenic Leptospires From Waters Used for Recreation," *Public Health Reports*, Vol. 81, No. 4, April 1966, pp. 299–304.
28. Joyce, Gayle, and H. H. Weiser, "Survival of Enteroviruses and Bacteriophage, in Farm Pond Waters," *Jour. AWWA*, Vol. 59, No. 4, April 1967, pp. 491–501.
29. "Sewage Contamination of Coastal Bathing Waters in England and Wales," Committee on Bathing Beach Contamination of the Public Health Laboratory Service, *The Journal of Hygiene*, Vol. 57, No. 4, December 1959, pp. 435–472.
30. Henderson, John M., "Enteric Disease Criteria for Recreational Waters," *Journal of the Sanitary Engineering Division*, ASCE, SA6, Proc. Paper 6320, December, 1968, pp. 1253–1276.
31. "Coliform Standards for Recreational Waters," Committee Report, *Journal of the Sanitary Engineering Division*, ASCE, SA4, Proc. Paper 3617, August, 1963, pp 57–94.
32, McGlasson, E. D. "Proposed Sanitation Standards for Food-Processing Establishments," Association of Food & Drug Officials of the United States, Quarterly Bulletin, April 1967.
33, Title 21—Food and Drugs, Chapter 1—Food and Drug Administration, Department of Health, Education, and Welfare, Part 128—Human Foods; Current Good Manufacturing, Processing, Packing, or Holding, Federal Register, April 25, 1969.

JOSEPH A. SALVATO, JR.
New York State Department of Health

ENVIRONMENTAL HEALTH ADMINISTRATION

Environmental health administration must be defined in order to set a course of action and to delineate environmental health administration from natural resources or environmental conservation administration. The delineation applies to the area of responsibility of environmental health as contrasted to natural resources or environmental conservation. The basic philosophy of administration is the same for both areas but the goals and objectives vary.

Environmental health is by the World Health Organization's definition: "The control of all those factors in man's physical environment which exercise or may exercise a deleterious effect on man's physical, mental or social well-being." This means we must be concerned with all conditions which *may affect total man.*

Administration means to manage, direct, or superintend the execution, use or conduct of public affairs. Normally, administration is distinguished from policy-making. Environmental health administration means action.

The environment of concern is the total universe. This universe has two basic components; natural and human resources. Natural resources are important only so long as we have human resources to use and enjoy them. Therefore, human resources must have the top priority for a maximum quality of life. A concern for the management and use of our total environment is everyone's responsibility, but those concerned with environmental health must concentrate on the physical factors which directly or indirectly affect man's health where he lives, works or plays and the food he eats.

The universal environment is our natural and human resources. The human resources segregate into environmental health and personal health. The control of those physical factors affecting the environmental health phase of the human resource is different from the prevention and control of individual illness and the preservation of man's personal health. This does not mean that these two important elements are unrelated. It means the basic approach to the solution of the cause of the personal or environmental ill is different, requiring different but interrelated knowledges and skills. Just as the solution of the cause of natural resource ills requires different knowledges and skills.

Natural resources include air, water, land, wild life, forests, minerals, oil and gas. Maintaining, protecting and developing the natural resources is for man's benefit and use. In and of themselves, natural resources normally do not have a direct effect on man's well-being. The man-developed insults are factors induced by man's activities which affect or may affect his well-being. Collectively, these insults cover the area of environmental health administration. Physical factors affecting the human resource are the pollutants such as bacteria, viruses, radiation, chemicals, drugs, noise and pests. Noise and chemicals have multiple insults on man's well-being. Chemicals, viruses, radioactive materials and bacteria are in one instance of concern in municipal water quality. However, a major emphasis must be with the availability of adequate quantities of water. In addition man's well-being includes a responsibility for adequate housing, hospitals and the institutional environment.

The man-developed insults, or environmental health, are interrelated with natural resources and personal health, but each is administered on a different basis of goals. Good administration demands clearly stated and understood goals, objectives and responsibilities. Therefore, the best body of government might mean better cooperation among the many Federal, State and local programs concerned with the total environment (natural and human resources), rather than the integration of the various parts with basically different but interrelated concepts. This fragmentation is pointed out in a paper on "Ecology and Administration" by Community Health Incorporated in May of 1970. Portions of this paper indicate "The ecological perspective toward man and his world has taught us that there are literally thousands of finely articulated sub-systems in all-encompassing ecosystems. Man's actions as a manipulative species causes changes in this environment whose effects may be proximate or distant, anticipated or unanticipated. In the current environmental crisis we are harvesting the fruits of centuries of lack of concern or lack of appreciation of the ecologic consequences of human activity. The cumulative insult to the environment has risen continuously while the response in society has been highly incremental and oriented toward single problems. One result of this incremental, uncoordinated approach to societal programming for the environment has been the development of a multitude of administrative subdivisions in government that deal with one sub-system or another *without efforts to achieve integration.* Environmental control programs have grown out of concern as diverse as preservation of wild life, management of natural resources, protection against communicable disease, and increasing agricultural production. In addition, there are many other governmental programs that are related to environmental problems, either as part of the cause or part of the solution. As a result, we find programs of considerable environmental impact distributed widely within government—in Departments of Commerce, Health, Housing, Conservation, Urban Affairs, Agriculture, and Transportation, to cite a few. Such subdivisions seldom share goals or information and many operate in direct competition. There is obvious need for better coordinating of programs dealing with the environment, the cause of its deterioration, and the means for its enhancement. As government at every level strives to respond to the ecological crisis, the solution emerging tends—more frequently than not —to be an attempt to create some type of ecological superagency. Such agencies—according to their proponents—will unite the fragmented environmental programs that have grown in number in recent years, and create combinations which will be an ecological whole. In our view, creation of such agencies represents an approach that is neither logical nor ecological. There is a real danger that—while appearing to do something to improve environmental programming —such agencies will merely perpetuate fragmentation at a time when *a coordinated response is essential.* If the logic behind proposals for recombining environmentally-related programs into super-agencies is presented as ecological, there are some very real problems. First, where should the line be drawn for inclusion and exclusion of programs? To do less than putting all environmentally-related programs together destroys the logic of recombination. This appears a practical and political impossibility, and none of the realignments proposed or accomplished, even begins to approach this magnitude of change. It is more common to take conservation programs, water quality, air pollution and solid waste under the ecology banner and ignore ecologically-equal activities in other fields. In reality, *the problems of the environment are so pervasive that every agency of government has some responsibility.* Would it not be more efficacious to concentrate on seeing that everyone fulfills their respective responsibilities? *Administratively recombination itself is not a panacea for environmental problems.* Organizational proximity does not necessarily enhance coordination of cooperation."

Administration and management of the natural resources is concerned with the preservation, development and maintenance of the water, air, land, minerals, forests, oil, wild life and fish for man's use.

Administration and managemen tofenvironmental health activities is concerned with man's quality of living in a broad aspect (total well-being) as developed under the definition of environmental health. Environmental health activities involve directly the engineer, the sanitarian, the physician, the planner, the biologist, the nurse, the chemist, technicians, specialists, consultants and related personnel. These activities cover same and varying responsibility for municipal water supplies, water and air pollution control, milk and food, rodent and vector control, pesticide control, interstate carriers, bakeries, restaurants, housing, noise, radiological health, swimming pools, bathing beaches, accidents, hospitals, institutions, camps, private water and sewage and *standard setting.*

Personal health, concerned with the health of the individual's body, is the responsibility of the physician, dentist, nurse, nutritionist, physiotherapist and related personnel. They are concerned with preventing and curing diseases such as cancer, heart disease, tuberculosis, dental caries, smallpox, typhoid fever and other personal illnesses. The personal health group has "good guy" image inasmuch as they provide advice and assistance in keeping an individual healthy, which advice the individual may or may not follow. On the other hand, the environmental health group is concerned with convincing a community, commercial establishment, industry or individual to do something which will cost money and which they may not wish to do but must, therefore creating a different image and developing a need for a different approach.

In many environmental health programs it is necessary to have the input of the physician epidemiologist and toxicologist if a program is to be developed on a sound broad based health cause and effects and economic basis. The physician with scientific input must provide the health effects basis for adding fluoride to municipal water supplies. The environmental health engineer will provide the know-how to the community in installing and operating equipment so that fluoride will be added within specified predetermined ranges. The medical team with scientific back-up determines the level of radioisotopes, such as Iodine 131, which may have an effect on children, through the consumption of milk. Control is the responsibility of the engineer, sanitarian, technician of the environmental health team. See table IV. A successful noise program based on feasibility and economics must be based on a decibel system which has been developed through the medical team's research of the multiple insult of noise on man in the work place, home and community. Therefore, environmental health, in the true concept of that term, must have a close working relationship with the medical team. The concept of environmental health today is considerably expanded from that of yesteryear. Activities in the field of environmental health at one time were controlled only because they were disease-related. Today the concept is to provide a quality of living.

An agency concerned with Natural Resources has as one of its responsibilities the need to plan for and implement the provision of adequate quantities of water for power, irrigation, transportation, recreation, water supply, industry and flood control. They provide for adequate quantities of water at the right place at the right time to meet all needs. Environmental health administration requires responsibility for both the quantity and quality of water available for municipal water supply purposes. Adequate municipal water quality cannot be provided merely through the construction of industrial and municipal waste treatment facilities. These facilities may meet the needs of fish but not man. Municipal water treatment facilities are a necessary adjunct.

The interrelated responsibilities for natural resource administration and environmental health administration is shown by Tables I, II, III, IV.

The administration of environmental health programs includes responsibility for not only legally-delegated programs, but also the responsibility for providing dynamic leadership, guidance and coordination of related environmental activities in other agencies.

Environmental health administration then involves programs which will provide a quality of living through the control of all those factors in man's physical environment which exercise or may exercise a deleterious effect on man's physical, mental or social well-being. In developing and administering environmental health programs, it is necessary to have an understanding of the words used in program planning and administration.

TABLE I

Basic Responsibilities of Natural Resource and Environmental Health Agencies

THE UNIVERSE

↓

MAN'S ENVIRONMENT

Environment — Man

Responsibilities

NATURAL RESOURCES	ENVIRONMENTAL HEALTH
Plan, Develop, Preserve	Plan, Recommend, Enforce
Water	Water
Land	Land
Air	Air
Minerals	Minerals
Oil	Oil
Gas	Gas
Wildlife	Wildlife
Fish	Fish
Forests	Forests
	Noise
	Shelter

TABLE II

Responsibilities of Natural Resource and Environmental Health Agencies for Water

Natural resources	Environmental health
Water	Water
Multipurpose dams	Municipal supply
Power	Quantity
Flood control	Quality
Recreation	bacterial
Water supply	viral
Industry	chemical
Irrigation	physical
	Recreation
	bacterial
	physical

Natural resource administration assumes responsibility to provide adequate quantities of water for all purposes at the right place at the right time.

TABLE III

Environmental Health Concerns for Municipal Water Supply

CONCERNS	
bacterial	WATERSHED
viral	pesticide use
chemical	subdivisions
nitrate	agriculture use
phosphate	camping
fluoride	waste disposal
sodium	animal feed lots
calcium	WATERCOURSE, RESERVOIR
lead	bathing
mercury	boating
selenium	fishing
radioisotopes	waste discharge
physical	TREATMENT, DISTRIBUTION, STORAGE
taste	fluoridation
clarity	cross-connections
odor, color	

TABLE IV

(From Table III concerns)

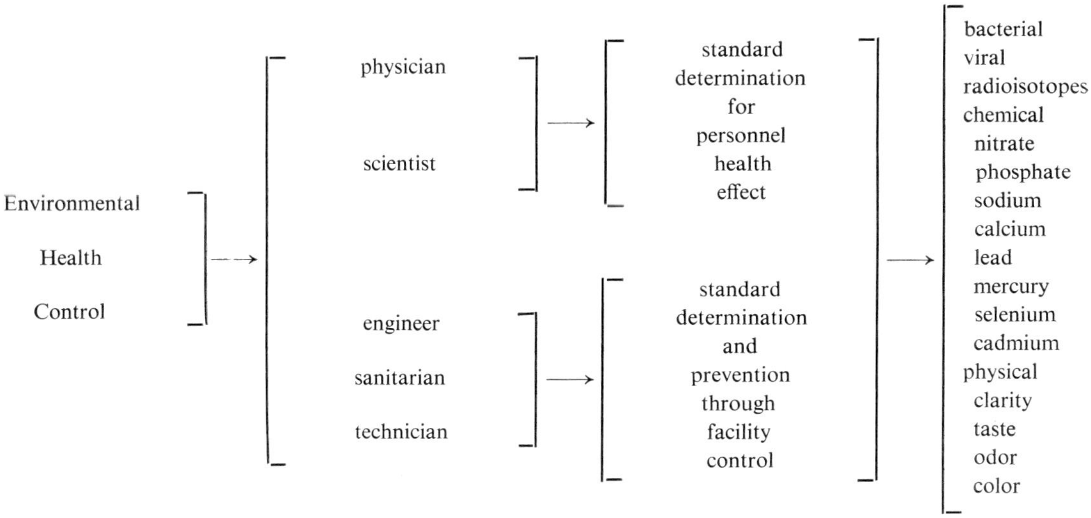

1. *Program.* An organized array of steps or activities intended to achieve an objective.

2. *Activity.* Performance of a specific and purposeful jobrelated work element.

3. *Objective.* A statement of program purpose, expressed in terms of the impact on people or the end product.

4. *Need*—reason for a program, in terms of identified problems.

5. *Program size indicator.* A quantification of activities or objectives of a problem.

6. *Program Effectiveness Indicators.* An epidemiological or analytical tool permitting inferences about achievements of program objectives.

7. *Characteristics of a Problem.* Specific quality of the problem which distinguishes it from other problems, including identificaiton of target group.

The initiation of programs comes from various sources: the Federal, State or local legislative bodies; the Governor, the Mayor, State or local agencies, such as the Health Department or groups of citizens concerned about certain problems, such as radiation from nuclear reactors.

The legislation or mandate for a new program merely provides some basis for the need for a program and usually designates agency responsibility. However, the legislation does not indicate the complete extent of the problem, the organization needed to resolve the problem, or the cost for preventing or resolving the problem. Some steps in the development of any program are:

1. *Determine the need for a program.*

2. *Analyze the collected information and data.*

Does a problem exist? It on the sound basis of existing data a problem does exist, then a program would be developed to abate existing problems and prevent new problems.

3. *Set long and short-range objectives of a proposed program.*

Short-range objectives are those that can be accomplished within a short period of time such as a year, while long-range objectives are those which will take a longer period of time, such as five or more years.

4. *Determine the functions needed in the program to*:

(a) Prevent the problem from recurring.

(b) Resolve existing problems.

5. *Determine specific activities in connection with needed functions.* What specific activities such as inspections and other day by day activities are required in preventing or resolving the problem. See Chart V (b).

6. *Determine available resources.* What existing agencies have related responsibilities for the problem? Determine the number and type of people available, funds available, equipment and other facilities available. Develop an agreement of cooperative activity.

7. *Determine personnel needs.* On the basis of

activities to be carried out and available resources, a preliminary estimate can be made of the number and kind of people needed to carry out the proposed activities and the administration of the program.

8. *Determine facility and equipment needs.* Determine the type of field and office testing equipment needed, laboratory facilities needed, cars needed, and furniture and related equipment needed.

9. *Determine method of evaluation.* Effective evaluation is one of the most important ingredients of any program for its continued effective and economic progress. Therefore develop effective measurements as criteria and standards. In developing basic standards and criteria, it is necessary to be sure that they are based on current scientific and technical knowledge which is arrived at after free

CHART V

(a) The Organization Chart

BUREAU OF PUBLIC WATER SUPPLY

Office of the Director

	Tele. No.	Room No.
1. John C. Bumstead, Director	457-2091	115
Eileen Jones, Sr. Stenographer	457-2092	115
The People ...		
Water Supply Operations Section		
2. Samuel Syrotynski, Chief	457-2974	115
3.		
4. Kenneth Slade, Sr. San. Engr.	457-2973	115
5. Daniel Stone, Jr. San. Engr.	457-2973	115
6. Thomas Brewer, Asst. San. Engr.	457-2973	115
7. Frank McKeefe, Asst. San. Engr.	457-2973	115
8. Ralph DeLuke, Statistical Clerk	457-2973	115
Margaret Nolte, Stenographer	457-3122	115
Yvette Mays, Typist	457-3122	115
Water Supply Design and Construction Section		
9. Gilbert Faustel, Chief	457-3166	115
10. Steven Mayzel, Sr. San. Engr.	457-2094	115
11. George Philip, Asst. San. Engr.	457-2094	115
12. Manoj Ajmera, Asst. San. Engr.	457-2094	115
13. Richard Bonczek, Engr. Tech.	457-2093	115
14. George Schaefer, Sr. Engr. Tech.	457-2093	115
Margaret DeCrescenzo, Stenographer	457-2093	115
Water Supply Planning Section		
15. James McCarthy, Sr. San. Engr.	457-7105	114
16. Paul Schulte, Sr. San. Engr.	457-7105	114
17. Raymond F. McDermott, Asst. San. Engr.	457-7105	114
Ann DeVoe, Stenographer	457-7105	115

CHART V (b)

(b) The Programs and Activities ...

Programs and activities	Responsible person(s)
Annual Inspection of Public Water Supply	6, 2
Bulk and Bottled Water	7, 2
Completed Works Approvals	14
Comprehensive Studies	15, 16, 17
Emergencies	All Bureau Staff
Emergencies: Remedial followup	15, 16
Enforcement	1
Federal Aid: Water Supply Projects	9, 13
Fluoridation	4, 2
Interstate Carrier Water Supplies	7, 2
Monitoring and Surveillance	5, 2
New Processes and Equipment Evaluation	11, 12, 9
NARS—North Atlantic Regional Water Resources Study	15, 16
NEWS—North East Water Supply Study	16
Operational Problems	2, 4, 5, 6, 7
Operator Certification	4, 2
Plan Review and Approval	9
Southeastern New York Temporary Water Commission	16
Special Health Hazard Problems	1,2
Statewide Water Plan	15, 16
Technical Assistance to Field	2, 9
Water Supply Applications	10, 9
Watershed Rules and Regulations	6, 2

and full exchange of views and information among all persons concerned and that they are promulgated and enforced by competent authority both reasonably and firmly.

Development of reports and inspection forms are necessary. (See table VI)

10. *Develop a budget.* Conforming with the program objectives, develop a realistic budget for the first year and a practical as possible budget through the fifth year. Long range budgets will be altered as a program progresses.

11. *Develop an organizational chart.* An organizational chart will show the structure of responsibility for carrying out a program, but is does not show responsibility for specific activities of a program. See Chart V (a)

12. *Determine research and development needs.* This may not be feasible for the first year of a program, but may develop as the program progresses.

13. *Determine training needs.* Provide for various types of training, such as on-the-job training, in-service training, and continuing education train-

TABLE VI (a)

Completion of Form San 255–Report of Annual Inspection of Public Water Supply

General The intent of this form is to provide an annual evaluation of the public water supply program in terms of statutory requirements. The form is distribution systere oriented. Therefore, the complcxity of completion, where multiple sources, treatment facilities, etc., exist, should be minimized.

Completion of the form requires either a designated number entry indicating YES, NO, UNKNOWN, and NOT APPLICABLE or a numerical entry for dates, quantities or sample results. Where not specified, a blank would indicate no report or not applicable.

The form should be completed in four copies: 2-BWWUM, 1-Region, 1-Field.

The information reported on the form can serve as a basis for a letter report to the water supply official upon completion of the annual inspection.

A narrative summary should be attached to the report to explain and clarify elements of the water supply program, if necessary, and is warranted if the program deviates from statutory requirements.

Municipality, Water District, etc. An annual inspection report is required for each Water Authority, County Water District, City, Town, Town Water District, Village, Private Water Company, Corporation and similar community water supply entities. The fact that the water supply serves or receives water from another water supply will be entered on the form as explained herein. Enter the name of the water supply.

City, Town, Village and County Political sub-division(s) in which the distribution system is located.

Water Works in Charge of The person or persons in charge of the operation of the public water supply, i.e., Name of the Authority, Town Board, Water Department, Company or individual. Enter also the proper address and telephone number where the water utility operating head (Authority, Board, etc.) may be reached.

Name of Chief Operator The name, business address and telephone number of the certified chief operator of the water supply.

Population Served by Distribution System Permanent retail consumers served, regardless of political boundary lines, excluding population served by other water supplies.

Ownership Public (Federal, State, County, Municipality, District, etc.)
Private (Private Corporation, Company, Individual, etc.)

Source(s) For each category, enter the number of permanent and auxiliary, not emergency, sources. A well field constitutes one source. For a water supply without a source, i.e., water district, municipality purchasing water, etc., enter purchased, not the ultimate source(s) of the water.

Number of Water Purveyors Served Enter the number of municipalities, districts, companies, etc., served, for which reports will be submitted.

Water Purchased From Enter the name(s) of the municipality, company, etc., from which the water is received or purchased.

Treatment Used Enter each general category of treatment provided, although multiple treatment facilities may exist. Examples: Chlorination does not differentiate between pre and post. The use of pressure diatomaceous earth filters would be noted by checking D.E. only, pressure filtration is a distinct treatment method. If water is received or purchased from another water supply, do not enter the treatment provided by the supplier.

Average Daily Production (MGD) The total plant production and/or amount of water purchased or received from another water supply.

Average Daily Consumption (MGD) Consumption by the population served, excluding water sold or served to other water supplies.

Plant Type (A, B or C) and Operator Grade (Chief) Enter the plant type and operator grade (Chief) required under Part 11 of the Sanitary Code (10 NYCCR 11). If more than one plant exists, enter the information for the largest plant.

Source Storage (gal $\times 10^3$) Phe impounded storage, in gallons, divided by 1000. If ground sources or purchased, no entry is required.

Distribution Storage (gal $\times 10^3$) The available storage of finished water in the distribution system area, (clearwells, underground or low level storage, elevated tanks, standpipes, high level reservoirs, etc.).

Number of Services Number of services, accounts, etc., not meters.

Percent Metered The per cent of services, accounts and connections metered.

Local Water Supply Program This section measures the water supply control program in terms of the requirements of the Sanitary Code, Part 5, Drinking Water Supplies (10 NYCRR 5)

32	Sec. 5.2,	Approval of plans and completed works
33	Sec. 5.3,	Reporting emergency changes
34	Sec. 5.20,	Providing treatment for public water supplies
35	Sec. 5.35,	Protection of equalizing and distribution reservoirs
36	Sec. 5.21 (b),	Operation of a public water supply
37	Sec. 5.32,	Adequacy of distribution system
38	Sec. 5.33,	Physical Connections
40	Sec. 5.10,	Protection and Supervision

Examinations of Samples of Water This section pertains to the requirements to the Sanitary Code, Part 5, Section 5.30 (10 NYCRR 5.30) and the Administrative Rules and Regulations, Part 72 (10 NYCRR 72). Sample results reported shall be indicative of the water supply since the last annual inspection or analysis results of samples collected at the time of the inspection. If any substance listed in Part 72 of the Administrative Rules and Regulations is found to approach the maximum permissible concentration, a routine sampling program and examination for that constituent shall be initiated to determine the maximum concentration that might occur, and that result should be reported.

Laboratories Used Note type of exam done in each lab category (10 NYCRR 5.30 (a) (1).

Bacteriological Examinations

Untreated Reference: 10 NYCRR 5.30 (a) (3)
Distribution Reference: 10 NYCRR 5.30 (a) (4)

TABLE VI (a) (*cont*).

Distribution Bacteriological Examinations
Reference: 10 NYCRR 72.3
Physical Analysis
Source Reference: 10 NYCRR 5.30 (a) (7)
Distribution Reference: 10 NYCRR 5.30 (a) (8)
Distribution Physical Analysis
Reference: 10 NYCRR 72.4
Chemical Analysis
Source Reference: 10 NYCRR 5.30 (a) (5)
Distribution Reference: 10 NYCRR 5.30 (a) (5)
Distribution Chemical Analysis (A blank indicates no report)
Reference: 10 NYCRR 72.5
10 NYCRR 5.30 (a) (6)
References
Public Health Law, Article 11, Title 1, Potable Waters
10 NYCRR 5 State Sanitary Code, Part 5, Drinking Water Supplies
10 NYCRR 72 Administrative Rules and Regulations, Part 72, Drinking Water Standards
Environmental Health Manual, Pure Water, W.S. Item 151, Annual Inspection of Public Water Supplies

Note Enter sample results as reported by the lab, i.e., 1.5 not 01.50. Average of bacteriological results by tube method: 2.2 equals 0.

ing for agency personnel and people involved with a program concerned with industry or utilities.

Action in carrying out these steps does not follow such steps consecutively either at the inception of a program or at any time during the progress of a program. Program planning, budgeting and evaluation is a continuous non-linear action.

Program: Municipal Water Supply Need for the Program:

A. Immediate problem

1594 water supply systems provide drinking water to 15.97 million people. While surveillance shows all municipal water supplies to be delivering safe water, 68% of the systems (representing 72% of the served population) are in need of improvements such as additional sources, treatment facilities, reservoirs, transmission lines and other essential works. Emergency measures were required by 55 systems during 1970. Water use restrictions are common each summer on many supplies. State construction grants are absent and Federal grants have provided insufficient monies for significant improvements.

The achievement of improvements is necessary not only to assure continuing supply of safe water quality, but also to upgrade the quality of water beyond minimum standards and to improve fire protection for life and property.

B. Trend

Potable water use is expected to increase from 150 gallons/capita/day to 190 gcd by 1990. Water quality requirements are for total acceptability in physical, chemical and biological parameters. Continuity of adequate supply and pressure are established criteria.

Although the number of water supply systems is decreasing and all but 9% of the population is served, population growth, industrial growth and migration from central cities to suburban areas will require increased capacity, a real expansion and integration of systems. Competitive water uses result in the utilization of poorer raw water quality. More complex treatment facilities and increasingly exotic pollutants necessitate increasingly skilled and sophisticated operation. National attention is being drawn to the need for improvement of water supply (viz. Strategy for a Livable Environment–Linton Report) and Federal legislation is proposed.

Objectives

To assure safe and adequate potable water supply for all consumers.

A. Short-range (one year)
   1. Obtain completion of intermunicipal water supply planning studies.
   2. Achieve inspection for 68% of the municipal water supply systems, representing all having independent sources of supply.
   3. Obtain compliance with Drinking Water Standards 68% of distribution systems.

B. Long-range (5 years)
   1. Obtain compliance with New York State Drinking Water Standards for 100% of the distribution systems.
   2. Obtain compliance with sanitary code regulations for 100% of the distribution systems.

Methodology to Meet Objectives

A. Legal basis

1. Public Health Law regulation of water supplies.
2. Code "Drinking Water Supplies."

3. Code "Drinking Water Standards."
4. Public Health Law "Potable Waters."
5. Watershed rules and regulations.
6. Public Health Law qualifications of operators.

B. Activities

1. Regulation of project design and construction for water supply development and treatment.
2. Initiation of studies and award of grants for comprehensive intermunicipal water supply planning; project review and approval; obtain implementation.
3. Inspection and evaluation of existing water supplies for compliance with regulations and desirable operating criteria.
4. Coordination and assistance in preventive and corrective measures in water supply emergency situations.
5. Technical consultation; investigation of complaints and special problems.
6. Water treatment plant operator training and certification.
7. Promotion of water supply development and improvement in areas of need.
8. Promulgation of rules and regulations for the protection of sources of public water supplies.

C. Program size indicators

1. Number and value of comprehensive intermunicipal water supply studies underway and completed.
2. Number of water works systems inspected for evaluation.
3. Number of projects approved for construction and completed according to approval.
4. Number of operators certified.
5. Number of water analyses required.

D. Resource requirements

E. Data needs

1. Inventory of water supply facilities for
   a. technical characteristics
   b. management information
   c. emergency operations
   d. source and distributed water quality.
2. Relationship of water supply sources to sources of pollution.
3. Operating data, review and storage; qualitative and quantitative.
4. Relationship to other comprehensive planning studies such as wastewater, solid wastes, vector control, etc.
5. Treatment effectiveness and chemical toxicology.

F. Field responsibility

1. Evaluate project proposals for field implications.
2. Make routine waterworks inspections and reports, recommendations for improvement.
3. Mobilize emergency resources, when necessary; enunciate emergency measures.
4. Provide data for inventories.
5. Refer cases for legal action.
6. Investigate complaints and obtain correction.
7. Provide consultation on new sites and facilities, treatment and protection, emergency operations.
8. Identify existing and potential problems and report on program needs.
9. Promote water supply development and improvement in areas of need.

G. Other agencies

1. Water Resources Commission; approval of the acquisition of sources of water supply; approval of intermunicipal water supply studies.
2. Public Service Commission; regulation of investor-owned utilities, investigation of service complaints.
3. Department of Health, Education and Welfare; certification of interstate carrier water supplies.
4. Department of Urban and Housing Development; construction grants.
5. Department of Agriculture, Farmers Home Administration; construction loans.
6. Civil Defense Commission, Engineering and Public Works Service; release of stockpiles water supply equipment, coordination of emergency services.
7. Division of Laboratories and Research; water analyses; certification of laboratories; special investigations.
8. Office of Environmental Health Manpower; water treatment plant operator training.

H. Research

1. Design economic treatment and disposal of water treatment plant wastewater.

2. Reclamation of wastewaters for water supply use.

Evaluation

A. Effectiveness measures

1. Population receiving drinking water consistent with operational criteria and quality standards; number of supplies represented.
2. Population benefited through implementation of comprehensive intermunicipal water supply planning studies.
3. Areas and population embraced by comprehensive intermunicipal water supply planning studies.
4. Communities and population benefited by development and construction of waterworks and improvements in accordance with established criteria.
5. Per cent waterworks systems having qualified operators.

B. Reports

1. Status of intermunicipal water supply planning studies, monthly.

An important aspect of effective and efficient environmental health administration is program evaluation. The basic purpose of program evaluation is to assess the progress and efficiency of the program and to determine the direction and rate at which to proceed. Therefore, determine the specific data which is needed from various sources in providing guidance toward determining progress in a program. Data is useless and costly unless it serves predetermined needs and uses. The following questions should be answered very clearly:

1. Why is the data needed?
2. Who will use the data?
3. How will the data be used?
4. How will the data be handled and tabulated?
5. How will the data be made available for use?

Before a final determination is made as to data needs, *first* determine the *final* use of the data. Answer all of the above questions. Data can be secured in many ways, such as through inspection reports annual reports, monthly reports and special reports, Some of the basis for needed data is to:

1. Know who is doing what.
2. Know the status of the program at any particular time.
3. Know the location of facilities.
4. Determine unsolved or unsatisfactory specific problems.
5. Determine on the basis of 4 the priority for people, programs and activities.

It is necessary to know who will use the data and for what purpose. Is data for the use of the inspector alone, for his supervisor, for the program director, for the Commissioner, for the legal counsel, for the legislative body, for the Governor, for Federal commitments. Will the data be used primarily for planning work or for determining new or existing problems, determining progress, setting program priorities, developing a budget, determining personnel effectiveness, determining need for personnel training, research needs and related intelligent information or a combination of several or all such needs. How much of what type of data will be developed? Of equal importance is how will the data be tabulated? Will it be done by manual processing, by mechanical processing, by electronic data processing, or by computerization?

Effective continuous ongoing program administration requires determination of the programs and activities to be carried on in subsequent years based on current conditions of programs and activities. To plan for a subsequent year's activity it is necessary to make assumptions based on accurate and inaccurate, complete and incomplete information.

Decisions to be arrived at for planned activities:

1. Will the same programs be operated?
2. Will the same number and type of personnel be available?
3. Will new programs and personnel be added?
4. Will new programs be added but no additional personnel?

At the end of a calendar or fiscal year determine from predetermined statistical data the status of each program and each component part of the program. For instance, for a municipal water supply program know; (a) the number of municipal water supply systems; (b) the total number inspected; (c) which have unsatisfactory conditions; (d) how many meet established water quality standards. On the basis of such data for all programs it is possible to arrive at a reasonable determination of program needs and activity priorities. Next determine the number and type of personnel available to carry out needed activities. A flexible analysis should be developed for each person, showing the activities to which they will be assigned and the

TABLE VI

SUMMARY OF ANNUAL INSPECTION OF PUBLIC WATER SUPPLY

FOR USE OF CENTRAL OFFICE	MUNICIPALITY, WATER DISTRICT, COMPANY, ETC.	CITY, TOWN, VILLAGE / COUNTY
1 [ ] 4 5 [ ] 6 7 [ ] 8 Sequence County DB	WATER WORKS IN CHARGE OF	NAME OF CHIEF OPERATOR

DATE THIS INSPECTION MADE	POPULATION SERVED BY DISTRIBUTION SYSTEM (EXCLUSIVE OF SERVICE TO OTHER WATER PURVEYORS)	OWNERSHIP
9 [ ] [ ] [ ] MONTH DAY YEAR	[ ]	1 = Public 2 = Private [ ] 23

SOURCE(S) - - NUMBER OF EACH	NUMBER OF WATER PURVEYORS SERVED -	WATER PURCHASED FROM	SUPPLIER SEQUENCE NO.
24 [ ] Surface [ ] Ground [ ] Purchased	[ ]		[ ] 32 FOR USE OF CENTRAL OFFICE

TREATMENT USED - - 1 = YES, 2 = NO

33	[ ] Copper Sulfate	[ ] Aeration	[ ] Microscreening	[ ] Activated Carbon	[ ] Lime-Soda Softening 37
38	[ ] Coagulation	[ ] Settling	[ ] Rapid Filtration 2 GPM/SF or less	[ ] Slow Sand Filtration	[ ] Diatomaceous Earth Filtration 42
43	[ ] High Rate Filtration More than 2 GPM/SF	[ ] Pressure Filtration	[ ] Sand Media	[ ] Anthracite Media	[ ] Mixed Media 47
48	[ ] Chlorination	[ ] Iron and/or Manganese Removal	[ ] Fluoridation	[ ] Ion Exchange	[ ] Corrosion Control 52
53	[ ] Potassium Permanganate	[ ] Phosphate Addition	[ ] Chlorine Dioxide	[ ] Chlorine Ammonia	[ ] None 57

AVERAGE DAILY PRODUCTION (MGD)	AVERAGE DAILY CONSUMPTION (MGD)	PLANT TYPE (A, B OR C)	OPERATOR GRADE (CHIEF)
58 [ ]	[ ]	[ ]	Required [ ] Actual [ ] 76

SOURCE STORAGE (GAL x 10^3)	DISTRIBUTION STORAGE (GAL x 10^3)	NUMBER OF SERVICES	% METERED
5 [ ]	[ ]	[ ]	[ ] 31

LOCAL WATER SUPPLY PROGRAM - - 1 = YES, 2 = NO, 3= UNKNOWN, 4 = NOT APPLICABLE

32 [ ] Have plans and specifications for all additions or modifications been submitted to and approved by the State Commissioner of Health since the last annual inspection?

33 [ ] Have there been any emergencies due to water shortage, deterioration of water quality or other reasons since the last annual inspection?

34 [ ] Are all surface water sources disinfected?

35 [ ] Are all distribution reservoirs covered or the water leaving the reservoirs effectively disinfected?

36 [ ] Are complete daily operating records maintained and submitted to the local health unit having jurisdiction?

37 [ ] Is a minimum working pressure of 20 psi continuously maintained in all parts of the distribution system?

38 [ ] Is there a local ordinance or program for Inspection of physical connections to prohibit or eliminate all unsafe or hazardous connections?

39 [ ] --Is the ordinance enforced or the program effective in the elimination of health hazards?

40 [ ] Have rules and regulations for the protection of the water supply been enacted by the State Commissioner of Health under Section 1100 of the Public Health Law?

41 [ ] --Are they enforced?

42 [ ] -- Is an annual watershed enforcement report submitted to the local health unit?

43 [ ] Have uncorrected water quality problems been the subject of consumer complaints since the last annual inspection?

FORM SAN. 255 (5-69)

TABLE VI

1 [ ][ ][ ][ ] 4 Seq. No.

EXAMINATIONS OF SAMPLES OF WATER

LABORATORIES USED - - 1 = YES, 2 = NO

5	Treatment Plant Laboratory	Chemical	Physical	Bacteriological	Microscopic 9
10	Commercial Laboratories	Chemical	Physical	Bacteriological	Microscopic 14
15	Local Government Laboratory	Chemical	Physical	Bacteriological	Microscopic 19
20	State Laboratory	Chemical	Physical	Bacteriological	Microscopic 24

BACTERIOLOGICAL EXAMINATIONS

UNTREATED	Required	Examined	DISTRIBUTION	Required	Examined (Avg)
25 Number of samples per year	[ ][ ][ ]	[ ][ ][ ]	Number of samples per month	[ ][ ][ ]	[ ][ ][ ] 36

DISTRIBUTION BACTERIOLOGICAL EXAMINATIONS

TUBE METHOD FOR COLIFORM ORGANISMS: - - 1 = YES, 2 = NO, 3 = UNKNOWN, 4 = NOT APPLICABLE	MONTH AND YEAR (MO. / YR.)	NUMBER OF DISTRIBUTION SAMPLES
37 Did the arithmetic average of the most probable number (MPN) per 100 ml of all samples collected in any month exceed 1.1 per 100 ml of sample?		
Did a most probable number (MPN) per 100 ml of 8.8 or greater occur in:		
38 More than one sample when less than 20 were examined in the month or more than 5% of the samples when 20 or more were examined in the month?		
39 Two consecutive samples?		
MEMBRANE FILTER METHOD FOR COLIFORM ORGANISMS: - - 1 = YES, 2 = NO, 3 = UNKNOWN, 4 = NOT APPLICABLE		
40 Did the arithmetic average of the membrane filter count per 100 ml of sample examined in any month exceed 1 per 100 ml of sample?		
Did coliform colonies per sample exceed 3 per 50 ml, 4 per 100 ml, 7 per 200 ml or 13 per 500 ml in:		
41 More than one sample when less than 20 were examined in the month or more than 5% of the samples when 20 or more were examined in the month?		
42 Two consecutive samples?		

43 [ ] 1 = Yes 2 = No 3 = Unknown — The bacteriological quality of the drinking water conforms to the standards of the State of New York Administrative Rules and Regulations, Part 72, DRINKING WATER STANDARDS.

PHYSICAL ANALYSIS

SOURCE		DISTRIBUTION	
44 Are the untreated waters of the source(s) sampled at the minimum frequency? Ground 1/yr Surface 1/wk	1 = Yes 2 = No [ ]	Previous physical sample taken: MO. DAY YR.	Date this physical sample taken: MO. DAY YR. 56

DISTRIBUTION PHYSICAL ANALYSIS

57 Turbidity (5 s.u.) ------------ [ ][ ].[ ]	Color (15 s.u.) ------------ [ ][ ].[ ]	Odor (3 s.u.) - ------------------ [ ] 63

64 [ ] 1 = Yes 2 = No 3 = Unknown — The physical characteristics of the drinking water conform with the standards of the State of New York Administrative Rules and Regulations, Part 72, DRINKING WATER STANDARDS.

CHEMICAL ANALYSIS

SOURCE		DISTRIBUTION	
65 Are the untreated waters of the source(s) sampled at the minimum frequency? Ground 1/yr Surface 2/yr	1 = Yes 2 = No [ ]	Previous chemical sample taken: MO. DAY YR.	Date this chemical sample taken: MO. DAY YR. 77

DISTRIBUTION CHEMICAL ANALYSIS (mg/1) (A blank indicates no report)

05 ABS (.5)	Arsenic (.05)	Chromium (hexavalent) (.05)	13
14 Barium (1.0)	Cadmium (.01)	Iron (.3)	23
24 Chloride (250)	CCE (.2)	Manganese (.3)	33
34 Copper (1.0)	Cyanide (.2)	Nitrate (as N) (10.)	43
44 Fluoride (1.5)	Lead (.05)	pH	50
51 Sulfates (250)	Phenols (.001)	Alkalinity (as Ca CO_3)	60
61 Zinc (5.0)	Selenium (.01)	Hardness (as Ca CO_3)	69
70 TDS (500)	Silver (.05)	(Numbers in ( ) are maximum limits in mg/1)	76

77 [ ] 1 = Yes 2 = No 3 = Data Incomplete — The chemical characteristics of the drinking water conform to the standards of the State of New York Administrative Rules and Regulations, Part 72, DRINKING WATER STANDARDS.

Sodium	Nitrite (as N)	Ammonia Free (as N)
Magnesium	Oxygen Consumed	Ammonia alb. (as N)

INSPECTING ENGINEER (NAME AND TITLE) | HEALTH AGENCY

TABLE VII

PROGRAMS and ACTIVITIES
BUREAU of PUBLIC WATER SUPPLY

	CENTRAL OFFICE STAFF - BUREAU OF PUBLIC WATER SUPPLY																			
	Dir.	Planning			Design and Construction						Operations									
Position Title	Prin. San. Engr.	Sen. San. Engr.	Sen. San. Engr.	Asst. San. Engr.	Assoc. San. Engr.	Sen. San. Engr.	Asst. San. Engr.	Asst. San. Engr.	Engr. Tech.	Sen. Engr. Tech.	Assoc. San. Engr.	Sen. San. Engr.	Sen. San. Engr.	Asst. San. Engr.	Asst. San. Engr.	Asst. San. Engr.	Stat. Clerk			
Base Salary April '71	22,691	14,915	14,915	12,103	18,438	14,915	12,103	12,103	6,518	7,729	18,438	14,915	14,915	12,103	12,1C3	12,103	5,520			
	Bumstead	McCarthy	Schulte	McDermott	Faustel	Mayzel	Phillip	Ajmera	Bonczek	Schaefer	Syrotynski	** Vacant	Slade	Stone	Brewer	McKeefe	Delke	Current Total Man-Days per Activity or Program	Percent of Total Bureau Engineering man-days	***Desirable man-days to achieve optimum program implementation
WHAT MUST BE DONE... IMMEDIATELY																			%	
Emergencies	5	1	2	5	2	2	1	1	1	1	5		3	3	3	3	/	36	1.08	40
Executive Office Requests	5	12	2	10	10				20	20	2							81	2.30	100
Governor's Requests	1	3	2	3	3						2							14	0.40	20
Legislative Requests	5	5	2	10	5						2							34	0.97	75
Technical Assistance to Field	1	10	4	10	20	10	6	8	2	4	15		10	10	10	10	10	140	4.00	440
Program Development and Evaluation	30	5	6	5	13						10		5	5	5	5	5	94	2.70	100
Budget	20	12	2	20	20					15	5							94	2.70	140
Personnel	5				10	2					5							22	0.63	50
Purchasing	1				1						2							4	0.11	20
WHAT SHOULD BE DONE... --Routine Programs and Activities--																				
Plan Review and Approval	1				15	18	91	107	69	65								366	10.40	650
Federal Aid: Water Supply Projects	1				5				10									16	0.46	100
Operator Certification	5										5		60				5	75	2.14	220
Water Supply Applications (EnCon)	1	1	1		13	153	12	17	10	10								218	6.18	250
Monitoring and Surveillance	4										17			75	5		15	116	3.30	556
Emergencies: Remedial follow-up	5	8	12								5							30	0.85	100
Enforcement: Preparation and follow-thru	10	1	4	2	5		2	2				40						26	0.74	260
Special Health Hazard Problems *	5										30	40	35	50	20		65	205	5.82	640
Complete Comp. PWS Studies	10	72	60	58	5													205	5.82	220
Implementation: Comp. PWS Studies	20	20	20	10	5	1	10	12	4	4	5							111	3.16	518
Monthly Summary Reports	3	1	2	10	12				15	15	12		10	10	10	10	10	120	3.42	200
Annual Report	2	1	2	20	4	1			10	13	2						5	60	1.70	150
Routine Inspections: PWS Systems SAN 255	1			10							5				30		60	106	3.01	306
Watershed Rules and Regulations	1										10	40			35			46	1.31	306
Develop Statewide Water Plan	5	49	50	20	10				20	10	5						5	174	4.95	220
Fluoridation	1				2		15				5		40				10	73	2.07	75
Bottled and Bulk Water	5										5					55	5	70	2.00	110
Public Water Supply Systems Evaluations	1										5	90	30	20	10	40		106	3.00	876
New Process and Equipment Evaluations	1				8		50	50	28	35								172	4.90	812
Interstate Carrier Water Supply Certification (USPHS)	5										10					60		75	2.14	110
Completed Works Approvals					1				2	6								9	0.26	229
Report of Illnesses *											2		2	2	2	2		10	0.28	110
Training Given and Received	10	7	8	5	7	5	9	3	4	6		10	10	10	10	10	10	114	3.23	344
WHAT CAN BE DEFERRED... --Desirable Support Activities--																				
Information Requests (complaints)	10	1	1	5	3	2	3	6	20	1	5		5	5	10	15	5	97	2.77	150
Physical Connections Program					2				1	3								6	0.17	426
Publications	10	1	3	5	8	1	4	2	2	3	5		5	5	5	5		64	1.82	220
Engineering Reports (Special)		2	6		5		15	10		7	5							50	1.42	220
Data Management Systems	5			5	5						7			5	50		5	82	2.34	200
Research (Proposals for applied research)	5			5	2						5			5	5			27	0.77	467
Inter-agency Requests	5	5	6		10	21					2							49	1.40	50
Special Meetings: Fed., State & Local	10	3	25	2	6	4	2	2	2	2	5		5	5	5	5	5	88	2.5	220
Drinking Water Standards (revisions)	5				3						5			10	5			28	0.79	50
	220	220	220	220	220	220	220	220	220	220	220	220	220	220	220	220	220	3520	100. %	10,340

* Special health hazard problems are a medical responsibility with engineering assistance provided.

** This position is not filled, and its indicated man-days are not being performed. These man-days are not included in "Current Total Man-Days"

All Values in "MAN DAYS" Unless Otherwise Noted

RFMcD

TABLE VII

STATEWIDE RESPONSIBILITIES

	CENTRAL OFFICE STAFF - BUREAU OF PUBLIC WATER SUPPLY																REGIONS													
	Dir.	Planning			Design and Construction						Operations						Albany		Buffalo		Rochester		Syracuse		White Plains		ONYCA			
	Bumstead	McCarthy	Schulte	McDermott	Faustel	Mayzel	Phillip	Ajmera	Bonczek	Schaefer	Syrotynski		Slade	Stone	Brewer	McKeefe	Deluke	Regional Engr.	Water Engr.	Regional Engr.	Water Engr.	Regional Engr.	Water Engr.	Regional Engr.	Water Engr.	Regional Engr.	Water Engr.	Regional Engr.	Water Engr.	District or County Engr.	Local Responsibility
W H A T M U S T B E D O N E ... IMMEDIATELY																															
Emergencies	A	B	B	B	B	B	B	B	B	B	S		B	B	B	B		X	X	X	X	X	X	X	X	X	X	X	X	X	X
Executive Office Requests	S	P	P	P	P				B	B	P							X		X		X		X		X		X		X	
Governor's Requests	S	P	P	P	P						P							X		X		X		X		X		X		X	
Legislative Requests	S	P	P	P	P						P							X		X		X		X		X		X		X	
Technical Assistance to Field	A	P	P		S	[illegible]		[illegible]		B	S		P	P	B	B		X		X		X		X		X		X		X	
Program Development and Evaluation	A	P	P		P						P		B					X		X		X		X		X		X		X	
Budget	S	P	P	P	P					P	P							X		X		X		X		X		X		X	
Personnel	S				P	P					P																				
Purchasing	S				P						P																				
W H A T S H O U L D B E D O N E ... -- Routine Programs and Activities--																															
Plan Review and Approval	A				S	P	P	P	P	P									X		X		X		X		X		X	X	X
Federal Aid: Water Supply Projects	A				S				P									X		X		X		X		X		X		X	X
Operator Certification	A										S		P				B	X	X	X	X	X	X	X	X	X	X	X	X	X	X
Water Supply Applications (EnCon)	A	B	B		S	P	B	B	B	B								X		X		X		X		X		X		X	X
Monitoring and Surveillance	A										S			P	B		B		X		X		X		X		X		X	X	X
Emergencies: Remedial follow-up	S	P	P	P							P								X		X		X		X		X		X	X	X
Enforcement: Preparation and follow-thru	S	B	B	B	P		B	[illegible]				F						X		X		X		X		X		X		X	X
Special Health Hazard Problems*	B										B	F	B	B	B																
Complete Comp. PWS Studies	S	P	P	B	B													X	X	X	X	X	X	X	X	X	X	X	X	X	X
Implementation: Comp. PWS Studies	S	P	P		B	B	B	B	B	B	B							X	X	X	X	X	X	X	X	X	X	X	X	X	X
Monthly Summary Report	A	B	B	P	B				B	B	B						B														
Annual Report	A	B	[illegible]	P	[illegible]					B	B						P	X	X	X	X	X	X	X	X	X	X	X	X	X	
Routine Inspections: PWS Systems - SAN 255	A			S							S				P		B		X		X		X		X		X		X	X	X
Watershed Rules and Regulations	A										S	F			P			X	X	X	X	X	X	X	X	X	X	X	X	X	X
Develop Statewide Water Plan	S	P	P	B	B				B	B	B						[illegible]	X		X		X		X		X		X		X	
Fluoridation	A				P		P				S		P				B		X		X		X		X		X		X	X	X
Bottled and Bulk Water	A										S					P	B	X	X	X	X	X	X	X	X	X	X	X	X	X	X
Public Water Supply Systems Evaluations	A										S	P	B	B	P	B															X
New Process and Equipment Evaluations	A				S		P	P	B	B																					X
Interstate Carrier Water Supply Certification (USPHS)	A										P					B		X	X	X	X	X	X	X	X	X	X	X	X	X	X
Completed Works Approvals					S				B	P									X		X		X		X		X		X	X	X
Report of Illnesses*											P		B	B	B	[illegible]		X		X		X		X		X		X		X	
Training Given and Received	A	P	P	B	P	B	B	B	B	B		B	B	P	B	B		X	X	X	X	X	X	X	X	X	X	X	X	X	X
W H A T C A N B E D E F E R R E D ... --Desirable Support Activities--																															
Information Requests (complaints)	A	B	B	B	B	B	B	B	B	B	P		B	B	P	B	B	X	X	X	X	X	X	X	X	X	X	X	X	X	X
Physical Connections Program					S				B	P	S																				
Publications	S	P	P	P	P	B	B	B	B	B	P		B	B	B	B			X		X		X		X		X		X	X	X
Engineering Reports (Special)	A	P	P		S		P	P		B	S		P	P	P	B															X
Data Management Systems	S			P	[illegible]						B			B	B		P		X		X		X		X		X		X	X	
Research (Proposals for applied research)	A			[illegible]	S						S			P	P																
Inter-agency Requests	S	P	P		P	B					P																				X
Special Meetings: Fed., State & Local	S	P	P	P	P	B	B	B	B	B	P		B	B	B	B	B	X		X		X				X				X	X
Drinking Water Standards (revisions)	S				B						P			B	P																X

* Special health hazard problems are a medical responsibility with engineering assistance provided

LEGEND
A - Administrative Responsibility
S - Supervisory Responsibility
P - Primary Responsibility
B - Backup Responsibility
X - Field or Local Responsibility

man-days assigned to each activity for each person. Such a plan must be flexible so as to change emphasis during the work year if necessary with changing conditions and emergencies. A composite chart should show the total responsibilities on a county-wide or state-wide basis of the personnel for activities and man-day assignments. (See tables VI, VII).

In assigning activities, plan the time of the year when certain activities might be best carried out. Such planning will provide a more efficient use of personnel. Programs concerned with the inspection of farm labor camps usually can be done most effectively in the summer, while the inspection of x-ray facilities in hospitals and physicians' offices can be done equally effectively during the winter months. Thus planning for a year's activities for all personnel is necessary, if all activities are to be carried out most efficiently.

Equally important to assigning activities for efficient operation is consideration to travel routes and facilities. Pre-planning of activities in connection with travel routes will assist in covering varied environmental health activities by one generalist person in a specific geographic area. However, if not planned, such activities may develop the daily overlap and duplication of travel routes by personnel carrying out similar activities. As an illustration, a sanitarian, engineer or technician on an unplanned day by day work basis of operation is assigned a responsibility for the inspection of fairy farms and restaurants in a particular geographic area, while, at the same time, in the same general geographic area, on the same day, another sanitarian assumes the inspection of private water supply and sewage disposal facilities.

This inefficient operation is basically caused by poor or no planning but also because of bad administration in permitting personnel to operate on a day by day basis on the theory that each daily request for service must be handled that day. This philosophy will end a fiscal year's activities with the statistical report showing a great number of daily requests being handled with excessive total travel but also showing unsatisfactory conditions existing at important facilities having major public health significance. Equally important is the inefficient use of time by permitting personnel to come to the office at the beginning and end of each day. Normally there is no need for such daily contact if activities are preplanned.

Program evaluation includes personnel evaluation. Unless there are adequate numbers of personnel who are competent, qualified and efficient, it is not possible to carry out a program which is progressive, effective and efficient. Personnel evaluation must be based on the fact that an agency has a dedicated community responsibility and is interested and concerned in developing the full potential of each person in meeting this responsibility. Each person must understand clearly the requirements of his job. What is specifically expected from each person? Are there any restraints placed on his actions? Is there a clear and concise written statement of responsibilities and activities? Such understanding can be developed only through person to person discussion. Clear understanding is necessary for the supervisor and the supervised. How does the employee view his on-going activities in meeting the job duties? More important is how does the supervisor view the employee's on-going activities. These two viewpoints can be conflicting revolving around not only the activities themselves but also the quantity and quality of work activity. A responsible person in the agency must have a first-hand practical knowledge of the job to be done as understood by the employee and as the supervisor views the employee's meeting the written and explained activities. Such personnel evaluations are important, not only to the individual if the employee is expected to make changes or improvements, but also to the morale and the efficiency and effectiveness of the agency. Personnel evaluations should be made frequently so that any misunderstandings can be corrected before they become of major concern. The end result of an evaluation between a supervisor and an employee should be recorded and the written statement agreed upon by both parties.

Personnel need and recruitment. Determine the type of personnel needed for agency programs and activities. Frequently in developing programs an arbitrary decision is made that only engineers or sanitarians are needed for a specific activity. However, if specific activities for each function are spelled out in detail, it might be shown that the degree of knowledge and ability for effective work varies for each activity. In developing personnel needs for the environmental health aspects of a hospital program, first biased conclusions not based on actual activities might indicate that only the expertise and ability of sanitarians need be used.

However, a detailed study of a hospital and nursing home program and the associated environmental health activities will indicate that some of the routine and follow-up activities can be effectively carried out by technician personnel with less than a college degree, and some of the activities will require the expertise of the engineer.

Program evaluation involves not only the *quantity* of effort expended on various environmental health activities as shown by periodic or annual statistics but also and more important the *quality* of the effort being expended by personnel. A program may consist of several hundred public water supply systems. The data collected statistics may show that the program goal of two complete inspections a year is being met. However, what is the quality of the effort? Such determination of quality requires an evaluation by a competent supervisor of actual field installations with the responsible field Inspector. Such field evaluation should be used not merely to check on the quality of work but to provide an educational opportunity. The dual purpose is effected by the supervisor and inspector each completing a separate inspection report and discussing the variances on the two reports. The field evaluation will substantiate the annual statistics that the facilities are or are not satisfactory.

An additional benefit accrues to the supervisor in that such field evaluation of county-wide or state-wide field activities will provide for equal interpretation, application and enforcement of codes and rules and regulations.

A closing but not the least important aspect of environmental health administration must be effective legal enforcement. Legal enforcement of environmental health programs is an integral part of total administration. Program development, direction, inspection, education, hearings and legal action are continuous and integrated.

Frank P. Grad, LL. B, Columbia Law Schools, points out that "An enforcement program in the environmental health area, regardless of the special field, air pollution, water pollution, food sanitation, housing code enforcement, etc., invariably consists of three components: first, a legal basis in statute and regulations; second, staff, inspectional, clerical and supervisory—and legal; and third, a method, plan or system of administration and enforcement. A good enforcement program is the result of a proper balance of the components mentioned, combined in every instance with another ingredient, namely, the will to enforce."

An effective enforcement program depends upon adequate laws and regulations. The laws and regulations must be clearly understood and be capable of intelligent interpretation by the public and the agency inspector. In addition, enforcement must be by competent personnel both reasonably and firmly.

Effective enforcement depends on effective health protection standards and criteria for the prevention or control of the many insults to man's health. Such standards and criteria to be effective must be based on all the pertinent scientific and technical knowledge available and that they are arrived at after free and full exchange of views and information among all concerned.

Effective and efficient administration of environmental health programs depends on adequate numbers of competent, qualified personnel to carry out and enforce activities based on clearly stated and understood goals and objectives, the effort for which can be evaluated both quantitatively and qualitatively to determine progress and direction.

MEREDITH H. THOMPSON
New York State Department of Health

ENVIRONMENTAL LAW

The environment has become a concern of the law and Environmental Law is now a recognized and established legal discipline.

The body of Environmental Law is growing in response to the felt needs of society which is beginning to acknowledge that the technological advances of this generation are destroying the legacy of the past and the capital assets of the future.

Historic forces forged common law doctrines suited to a world of endless frontiers, where ruthless exploitation of natural resources was a manifestation of the desperate need to subdue all nature as the means of survival in the wilderness of a new world. Such survival techniques are no longer appropriate on an earth whose life support systems are as fragile as those in a space capsule.

As natural resources are being utilized at rates unprecedented in the history of civilization, and the quality of environment deteriorates, there is increasing public demand for legal aid to the environment. Attorneys are being called upon to act as public defenders of the environment and the law is being asked to restore the quality of life.

Environment and the law interface in three significant areas: Legislation, Administration and Executive action, and Judicial determination.

There already exists a substantial body of statutory law at the federal, state and local level dealing with matters of environmental concern, and there has been a perceptible evolution of ecologically sophisticated, environmentally responsible, socially relevant and politically feasible legislation at all operative levels of government.

Administrative agencies characterized by a combination of delegated legislative, executive and judicial powers, functions, and responsibilities affect the environment in the furtherance of their statutory mission.

The administrative agencies are legislative creations. In theory, they exist to effect policy established by the elected legislative representatives of the people. To accomplish this the legislature ceded rule-making power from its legislative mandate under the Constitution, the executive ceded a certain amount of administrative power, and the judiciary ceded certain judicial functions, in particular fact-finding and preliminary hearing. As a result of this tripartite grant of power, administrative agencies represent not a fourth branch of government as some seem to think, but the foundation of all practical government operations: Administrative agencies provide the substantial bulk of bureaucracy.

The judiciary of the United States and that of the several states furnish the forum for environmental litigation. Such litigation involves the interpretation of statutes and the adjudication of liabilities for damages resulting from nuisance, negligence, trespass and other traditional common law torts; as well as general actions seeking declaration of the rights of the people (in the form of class actions for declaratory judgment) and equitable relief (injunction, reparations) based on such ancient common law equitable principles as the Trust Doctrine and the maxim *sic utere to alienam non laedas*—so use your own property as not to injure that of another. Following the trail blazed by the American Labor Movement and retracing many of the judicial steps of the Civil Rights struggle, while seeking judicial vindication of a fundamental human right, to a salubrious environment, as one of those rights "so basic and important to our society that it would be inconceivable that it is not protected from unwarranted in reference" and is a right retained by the People of the United States under the Ninth Amendment of the Constitution of the United States and protected from disparagement by the actions of the federal government by operation of the *due process* and *equal protection* clauses of the Fifth Amendment of the Constitution and protected from disparagement by the actions of any of the several states by operation of the *privileges or immunities*, *due process* and *equal protection* clauses of the Fourteenth Amendment of the Constitution.

Environmental Law and Environmental Litigation became recognized elements of the Anglo-American legal system in the Spring of 1966 when a suburban housewife brought an action on behalf of all the citizens of Suffolk County, New York seeking equitable relief from a toxic insult to the community ecosystem; challenging not merely the local mosquito control commission still routinely using DDT in an increasingly futile attempt to control a mosquito population that had long since become resistant to that pesticide, but the broad-spectrum, persistent chemical biocide, 1,1,1-trichloro-2,2-bis(parachlorophenyl) ethane: DDT itself.

The New York State Supreme Court issued a temporary injunction restraining the County of Suffolk from using DDT for mosquito control on August 15, 1966 and continued this "temporary" injunction until December 6, 1967, finally holding that:

"DDT has, by its inherent chemical stability, become a continuing factor in some ecological life cycles so as to profoundly alter them and the environmental equilibrium.

Thus, it is reasonably apparent that DDT is capable of and actually has to some extent caused extraordinary damage to the resources of this county. If in no other way, the chemical by its very stability has introduced an element of instability in the general ecosystem. For instance, by reducing a food source of some of the larger wildlife and so reducing the over-all larger wildlife population, lesser elements multiply more quickly. These lower forms are presumably more of a nuisance, assuming they in turn survive. Furthermore, DDT affects wildlife directly. Its ingestion, from whatever source has the capability, it seems, to disrupt reproductive processes or even more simply act as a poison. It is fairly apparent then that the application of DDT in Suffolk County has and is continuing to have a demonstrable effect on local wildlife, reducing it slowly but surely, either directly across the board or indirectly from the top down, but reducing it nevertheless.

We have a situation where plaintiff has at least minimally sustained a massive effort to validate the allegation that DDT does in fact do biological harm." [*Yannacone* v. *Dennison et al.* (1967) 55 Misc 2d 468, 471–472, 285 NYS 2d 476.]

Environmental Legislation

a. National Environmental Policy Act of 1969

The principal national legislative statement on the environment is the National Environmental Policy Act of 1969 (National Environmental Policy Act) the purpose of which is.

To declare a national policy which will encourage productive and enjoyable harmony between man and his environment; to promote efforts which will prevent or eliminate damage to the environment and biosphere and stimulate the health and welfare of man; to enrich the understanding of the ecological systems and natural resources important to the Nation; and to establish a Council on Environmental Quality.

The National Environmental Policy Act (NEPA) is a synthesis by a Conference Committee of bills introduced in the Senate by Senator Jackson and in the House by Congressman Dingell.

The declaration of a national environmental policy contained in section 101 of the act speaks for itself.

Sec. 101. (a) The Congress, recognizing the profound impact of man's activity on the interrelations of all components of the natural environment, particularly the profound influences of population growth, high-density urbanization, industrial expansion, resource exploitation, and new and expanding technological advances and recognizing further the critical importance of restoring and maintaining environmental quality to the overall welfare and development of man, declares that it is the continuing policy of the Federal Government, in cooperation with State and local governments, and other concerned public and private organizations, to use all practicable means and measures, including financial and technical assistance, in a manner calculated to foster and promote the general welfare, to create and maintain conditions under which man and nature can exist in productive harmony, and fulfill the social, economic, and other requirements of present and future generations of Americans.

(b) In order to carry out the policy set forth in this Act, it is the continuing responsibility of the Federal Government to use all practicable means, consistent with other essential considerations of national policy, to improve and coordinate Federal plans, functions, programs, and resources to the end that the Nation may—

(1) fulfill the responsibilities of each generation as trustee of the environment for succeeding generations;

(2) assure for all Americans safe, healthful, productive, and esthetically and culturally pleasing surroundings;

(3) attain the widest range of beneficial uses of the environment without degradation, risk to health or safety, or other undesirable and unintended consequences;

(4) preserve important historic, cultural, and natural aspects of out national heritage, and maintain, wherever possible, an environment which supports diversity and variety of individual choice;

(5) achieve a balance between population and resource use which will permit high standards of living and a wide sharing of life's amentities; and

(6) enhance the quality of renewable resources and approach the maximum attainable recycling of depletable resources.

(c) The Congress recognizes that each person should enjoy a healthful environment and that each person has a responsibility to contribute to the preservation and enhancement of the environment.

Section 101 (b) (5) is recognition by Congress that uncontrolled magnitude and distribution of population underlies many of this nation's environmental and resource problems. To insure that a high standard of living is made available to all citizens and that all citizens have the opportunity for aesthetic enjoyment from a quality environment, the intent of NEPA is that the Federal Government must strive to maintain the magnitude and distriment's of population so as not to exceed the environbution capacity to provide such benefits. By section 101 (b)(5) Congress has thus implied that the rights of underprivileged citizens to a high standard of living are not to be sacrificed as a result of the national environmental policy to protect the environment.

The provision of NEPA most debated by the members of Congress was Section 102,

Sec. 102. The Congress authorizes and directs that, to the fullest extent possible: (1) the policies, regulations, and public laws of the United States shall be interpreted and administered in accordance with the policies set forth in this Act, and (2) all agencies of the Federal Government shall—

(A) utilize a systematic, interdisciplinary approach which will insure the integrated use of the natural and social sciences and the environmental design arts in planning and in decisionmaking which may have an impact on man's environment;

(B) identify and develop methods and procedures, in consulation with the Council on Environmental Quality established by title II of this Act, which will insure that presently unquantified environmental amenities and values may be given appropriate consideration in decisionmaking along with economic and technical considerations;

(C) include in every recommendation or report on proposals for legislation and other major Federal actions significantly affecting the quality of the human environment, a detailed statement by the responsible official on—

(i) the environmental impact of the proposed action,

(ii) any adverse environmental effects which cannot be avoided should the proposal be implemented,

(iii) alternatives to the proposed action,

(iv) the relationship between local short-term uses of man's environment and the maintenance and enhancement of long-term productivity, and

(v) any irreversible and irretrievable commitments of resources which would be involved in the proposed action should it be implemented.

Prior to making any detailed statement, the responsible Federal official shall consult with and obtain the comments of any Federal agency which has jurisdiction by law or special expertise with respect to any environmental impact involved. Copies of such statement and the comments and views of the appropriate Federal, State, and local agencies, which are authorized to develop and enforce environmental standards, shall be made available to the President, the Council on Environmental Quality and to the public as provided by section 552 of title 5, United States Code, and shall accompany the proposal through the existing agency review processes;

(D) study, develop, and describe appropriate alternatives to recommended courses of action in any proposal which involves unresolved conflicts concerning alternative uses of available resources;

(E) recognize the worldwide and long-range character of environmental problems and, where consistent with the foreign policy of the United States, lend appropriate support to initiatives, resolutions, and programs designed to maximize international cooperation in anticipating and preventing a decline in the quality of mankind's world environment;

(F) make available to States, counties, municipalities, institutions, and individuals, advice and information useful in restoring, maintaining, and enhancing the quality of the environment;

(G) initiate and utilize ecological information in the planning and development of resource-oriented projects; and

(H) assist the Council on Environmental Quality established by title II of this Act.

NEPA strengthened the right of public access under the Freedom of Information to records and information of Federal agencies Act. In addition, Congress has required all Federal agencies, to the fullest extent possible, to "make available to States, Counties, municipalities, institutions and individuals, advise and information useful in restoring and maintaining, and enhancing the quality of the environment ..."

These provisions enable citizens, under the Freedom of Information Act, to gain access to those records of Federal agencies that are relevant to environment protection in addition to the environmental impact statements required to be filed under Section 102.

NEPA requires all Federal agencies and officials to consider environmental values in reaching decisions or in planning agency action.

Though Congress has enacted numerous laws over the past years that constitute Congressional mandates on various aspects of environmental policy, areas of Federal policy and action exist which have no environmental goals or policies and in which the conflicting operational necessities of different agencies complicate and often frustrate attainment of environmental quality objectives which are in the interest of the entire country. Many of the older operating agencies of the Federal Government, for example, do not, at present, have a mandate within the body of their enabling laws permitting them to give adequate attention to environmental values. In other agencies, especially when the expenditures of funds are involved, the latitude of a public official to deviate from the most economical alternative to reach an environmental goal may be strictly circumscribed by Congressional authorizations which have overlooked existing or potential environmental problems. There is also reason for serious concern over the activities of those agencies which do not feel they have sufficient authority to undertake needed research and action to enhance, preserve, and maintain the quality of the environment in connection with development activities.

The National Environmental Policy Act re-emphasizes the importance of existing statutory programs relating to the environment. Prior to the passage of the National Environmental Policy Act of 1969, existing legislation involving certain areas of Federal activity did not provide "clear authority for the consideration of environmental factors which conflict with other objectives," and many Federal agencies had not given substantial and consistent consideration to environmental factors during decision making in certain areas of their responsibility.

Section 102 (1) remedies these shortcomings in

the statutory foundations of existing agency programs by incorporating the policy and goals set forth in section 101 into the actions and programs of all Federal agencies. Section 102 (2) establishes procedures which will help to insure that the policies enunciated in sections 101 are implemented.

Prior to the passage of the National Environmental Policy Act of 1969, planning and decision-making that might have an effect on the quality of the environment was too often the exclusive province of the engineer and cost analyst. The National Environmental Policy Act requires Federal agencies to consider all relevant points of view and draw upon the broadest possible range of social and natural scientific knowledge and design arts in planning and decision making. In the past, Federal agencies have all too frequently ignored environmental factors in planning and decision-making or omitted them from consideration during the early stages of planning because of the difficult in evaluating such factors as compared with the ease of evaluating economic and technological factors. Unter NEPA, Federal agencies and officials are now required to develop the methodology and techniques necessary to determination the total environmental impact and full course of actions by the Federal government.

One of the most frequent sources of environment litigation involves the construction and application of the National Environmental Policy Act of 1969 and the sufficiency or environmental impact statements prepared pursuant to section 102 (2) (C). Although the initial determination of environmental impact is made by the agency itself, should the agency find that the proposed activity will have a significant effect on the environment, the agency recommendation of report supporting the proposal must make findings with respect to the environmental factors set forth in section 102(2)(C)(i)–(v). If adverse environmental effects cannot be avoided by reasonable alternative actions the agency must demonstrate that such adverse environmental effects are justified by other considerations of national policy, which must be stated in detail. The use of local, short-term resources must be found to be consistent with the maintenance and enhancement of the long-term productivity of the environment. Proposals involving significant commitments of resources that will be irreversable and irretrievable under conditions of known technology and reasonable economics must be found necessary.

The National Environmental Policy Act supplements the statutory authority of most Federal agencies and requires them to follow the policies, goals and procedures of the Act, unless the existing law governing such agency's operations expressly prohibits full compliance, however section 103 provides that no Federal agency is to interpret its existing statutory authority in such a way as to avoid compliance with the directives of NEPA. If the agency finds a clear conflict between its existing statutory mandate and the Act, the agency is required to propose to the President of the United States such measures as will be necessary to bring the statutory authority and mandate of the agency into conformity with the provisions of the National Environmental Policy Act.

The National Environmental Policy Act is a relevant statute within the meaning of the Administrative Procedure Act, thus confering standing upon representatives of the general public seeking judicial review of actions and decisions by federal agencies which may have failed to follow the policies, goals and procedures of the Act, where such action or decision of the agency threatens to have adverse effects upon the ecological system an individual may reside in or use for recreation. Section 101 (C) of the National Environmental Policy Act recognizes that "each person should enjoy a healthful environment and that each person has a responsibility to contribute to the preservation and the enhancement of the environment," indicating that Congress intends to recognize the interests of individual citizens in the protection of the ecosystem they may reside in or use for recreation.

Such an individual citizen is within the class of persons which the National Environmental Policy Act was meant to protect, and is thus a person adversely affected or aggrieved within the meaning of the Administrative Procedure Act and as such a person is entitled to judicial review of the agency action.

The national goals and policies for the protection, establishment and enhancement of this Nation's environment established by the National Environmental Policy Act are in the words of its Senate sponsor, Senator Jackson,

> More than the statement of what we believe as a people and as a nation. It establishes priorities and gives expressions to our National goals and aspirations. It provides a statuatory foundation to which administrators may refer ... for guidance in making decisions which find environmental values in conflict with other values.

What is involved is a congressional declaration, that we do not intend, as a government or as a people to initiate actions which endanger the continued existence or the health of mankind: that we will not intentionally initiate actions which will do irreparable damage to the air, land, and water which support life on Earth.

An environmental policy is an policy for people. Its primary concern is with man and his future. The basic principle of the policy is that we must strive in all that we do to achieve a standard of excellence in man's relationship to his physical surroundings. If there are to be departures from this standard of excellence, they should be exceptions to the rule and policy.

b. Administrative Procedure Act

The Administrative Procedure Act provides both an independent source of federal jurisdiction and a grant of standing to conservation organizations and aggrieved seeking persons to challenge decisions of, or actions by, Federal agencies which may result in environmental degradation.

Senator McCarren, the author of the bill which ultimately became the Administrative Procedure Act, explained that it conferred no administrative powers, but provided definitions of, and limitations upon, administrative action, to be interpreted and applied by the agencies in the first instance, but to be reviewed by the courts in the final analysis.

The scope of judicial review of agency action is governed by the language of the Administrative Procedure Act:

To the extent necessary to decision, and when presented the reviewing court shall decide all relevant questions of law, interpret constitutional and statuatory provisions, and determine the meaning or applicability of the terms of an agency action. The reviewing court shall ...

(2) hold unlawful and set aside agency action, findings and conclusions found to be—

(A) Arbitrary, caprisious, an abuse of discretion, or otherwise not in accordance with the law;

(B) Contrary to constitutional right, power, privilege or immunity;

(C) In excess of statuatory jurisdiction, authority, or limitations, or short of statuatory jurisdiction, authority, or limitations, or short of statutory rights; ...

(F) Unwarranted by the facts to the extent that the facts are subject to trial *de novo* by the reviewing court.... (5 United States Code § 706.)

Under the Administrative Procedure Act any person "suffering legal wrong because of agency action or adversely affected or aggrieved by agency action within the meaning of a relevant statute" may seek judicial review.

Agency action made reviewable by statute and final agency action for which there is no other adequate remedy in a court are subject to judicial review. A preliminary, procedural, or intermediate agency action or ruling not directly reviewable is subject to review on the review of the final agency action. Except as otherwise expressly required by statute, agency action otherwise final is final for the purposes of [the Administrative Procedure Act] whether or not there has been presented or determined an application for a declaratory order, for any form of reconsideration, or, unless the agency otherwise requires by rule and provides that the action meanwhile is inoperative, for an appeal to a superior agency or authority. (5 United States Code § 104.)

The "legal wrong" contemplated by the Administrative Procedure Act is something more than the mere personal damage normally contemplated in the law of torts. The damage necessary to constitute legal wrong under the provisions of the Administrative Procedure Act must produce some illegal effect and be demonstrably contrary to law in either substance or procedure. However, the law considered relevant upon judicial review of agency action under the Administrative Procedure Act is not only the statutory law directly involved with the agency action, but general constitutional law as well. The legal wrong necessary to confer jurisdiction on the court and standing to the party complaining can be the invasion of any legally protected right.

Jurisdiction can also be obtained over agents and officers of the United States Government such as contractors of the Atomic Energy Commission, the Secretary of Agriculture, the Secretary of Interior, and divisions and departments of Federal agencies such as the Division of Wildlife Services, the Bureau of Sport Fisheries and Wildlife, and the Bureau of Land Management of the United States Department of the Interior, the Corps of Engineers of the United States Army of the Department of Defense, and the Federal Aviation Administration and the Bureau of Public Roads of the Department of Transportation.

There is a basic presumption in the law favoring judicial review of agency or administrative action even where the statute prescribing agency action does not on its face provide for judicial review. For the right of judicial review to be denied, the agency must produce clear and convincing evidence of legislative intent to foreclose judicial review. The Supreme Court of the United States [Association of Data Processing Service Organizations, Inc. v. Camp (1970) 397 US 150, 25 h Edzd 184, 90 SCt 827] has reemphasized the presumption of judicial review of agency action in recent litigation which construed the Administrative Procedure Act not grudgingly but as serving a broadly remedial purpose holding that "the mere failure to provide specially by statute for judicial review is certainly no evidence of intent to withhold review."

No presumption in favor of "administrative absolutism" and against judicial review exists. A congressional purpose to preclude judicial review of agency action must be "fairly discernable in the statutory scheme," since the right to judicial review is ordinarily inferred where congressional intent to protect the interests of the class of which the complainant is a member can be found.

Administrative action is not immune from judicial review because it may be committed to agency discretion or simply because the statute authorizes a public official to "prescribe such regulations as he may deem proper to carry out the provisions of [the legislation]."

The right to judicial review of agency action is all that stands between the citizen and government by administrative fiat not subject to correction by the elective process. Although a court usually will not substitute its judgment for that of an administrative agency, it can compel the proper exercise of statutory authority by the agency, since it is not for the administrative agency to finally determine the limits of its own power. That is a judicial function.

Where a party aggrieved has no other adequate remedy following administrative action there is certainly the right, upon proper application, to judicial review of agency action. Some confusion existed in the Federal Courts for a number of years after the passage of the Administrative Procedure Act because the provisions for judicial review apply, "According to the provisions thereof, except to the extent that agency action is committed to agency discretion by law." (5 United State Code § 701).

Some Federal Courts held that agency action involving the exercise of discretion could not be reviewed for any reason under the Administrative Procedure Act, however, the majority of Federal Courts have adopted the rule that any agency action, even that involving the exercise of discretion and dependent upon the special technical and scientific expertise of the agency, is subject to judicial review to the extent provided by the Administrative Procedure Act. The mere fact that the acts of some administrative official, even a cabinet level Secretary, require the exercise of discretion and judgment does not preclude judicial review of official action. Any contrary decision would lead to establishment of a shadow government by Administrative officials not subject to popular election or judicial review and therefore totally outside the systems of checks and balances so carefully developed during the Constitutional Convention. The rights of the public must receive active and affirmative protection at the hands of the administrative agencies.

Much environmental litigation has involved the standing of parties to bring suit against administrative agencies, their officials, employees and agents. Recent decisions have established the right of public benefit organizations whose purposes include protection of the environment, or protection the lives, health, and property of human beings or animals, as well as individuals seeking to proceed in the public interest as "private attorneys general," to bring actions a persons "adversely affected" or "parties aggrieved," within the meaning of the Adminstrative Procedure Act.

Neither economic injury nor a specific individual legal right are necessary adjuncts to standing. A complainant need only demonstrate that it is an appropriate person (corporate or human) to question the alleged failure of an agency to protect those values recognized by law as in the public interest.

C. Environmental Protection Agency

By virtue of the executive authority to reorganize Federal agencies, President Nixon created an Environmental Protection Agency to which was transferred, effective December 2, 1970, significant environmental responsibilities formerly scattered throughout the several executive and administrative branches of government.

Sec. 2. *Transfers to Environmental Protection Agency.* (a) There are hereby transferred to the Administrator [of the Environmental Protection Agency].

(1) All functions vested by law in the Secretary of the Interior and the Department of the Interior which are administered through the Federal Water Quality Administration, all functions which were transferred to the Secretary of the Interior by Reorganization Plan No. 2 of 1966 (80 Stat. 1608), and all functions vested in the Secretary of the Interior or the Department of the Interior by the Federal Water Pollution Control Act or by provisions of law amendatory or supplementary thereof.

(2) (i) The functions vested in the Secretary of the Interior by ... 16 U.S.C. 742d-1 (being an Act relating to studies on the effects of insecticides, herbicides, fungicides, and pesticides upon the fish and wildlife resources of the United States), and (ii) the functions vested by law in the Secretary of the Interior and the Department of the Interior which are administered by the Gulf Breeze Biological Laboratory of the Bureau of Commercial Fisheries at Gulf Breeze, Florida.

(3) The functions vested by law in the Secretary of Health, Education, and Welfare or in the Department of Health, Education, and Welfare which are administered through the Environmental Health Service, including the functions exercised by the following components thereof:

(i) The National Air Pollution Control Administration,

(ii) The Environmental Control Administration:

(A) Bureau of Solid Waste Management,

(B) Bureau of Water Hygiene,

(C) Bureau of Radiological Health,

except that functions carried out by the following components of the Environmental Control Administration of the Environmental Health Service are not transferred: (i) Bureau of Community Environmental Management, (ii) Bureau of Occupational Safety and Health, and (iii) Bureau of Radiological Health, insofar as the functions carried out by the latter Bureau pertain to (A) regulation of radiation from consumer products, including electronic product radiation, (B) radiation as used in the healing arts, (C) occupational exposures to rediation, and (D) research, technical assistance, and training related to clauses (A), (B), and (C).

(4) The functions vested in the Secretary of Health, Education, and Welfare of establishing tolerances for pesticide chemicals under the Federal Food, Drug and Cosmetic Art, as amended, 21 U.S.C. 346, 346a, and 348, together with authority, in connection with the functions transferred, (i) to monitor compliance with the tolerances and the effectiveness of surveillance and nforcement, and (ii) to provide technical assistance to the States and conduct research under the Federal Food, Drug, and Cosmetic Act, as amended and the Public Health Service Act, as amended.

(5) So much of the functions of the Council on Environmental Quality under section 204(5) of the National Environmental Policy Act of 1969 ... as pertains to ecological systems.

(6) The functions of the Atomic Energy Commission under the Atomic Energy Act of 1954, as amended, administered through its Division of Radiation Protection Standards, to the extent that such functions of the Commission consist of establishing generally applicable environmental standards for the protection of the general environment from radioactive material. As used herein, standards mean limits on radiation exposures or levels, or concentrations or quantities of radioactive material, in the general environment outside the boundaries of locations under the control of persons possessing or using radioactive material.

(7) All functions of the Federal Radiation Council (42 U.S.C. 2021(h)).

(8) (i) The functions of the Secretary of Agriculture and the Department of Agriculture under the Federal Insecticide, Fungicide, and Rodenticide Act, as amended (7 U.S.C. 135-135k), (ii) the functions of the Secretary of Agriculture and the Department of Agriculture under section 408(l) of the Federal Food, Drug and Cosmetic Act, as amended (21 U.S.C. 346a(l)), and (iii) the functions vested by law in the Secretary of Agriculture and the Department of Agriculture which are administered through the Environmental Quality Branch of the Plant Protection Division of the Agricultural Research Service.

(9) So much of the functins of the transferor officers and agencies referred to in or affected by the foregoing provisions of this section as is incidental to or necessary for the performance by or under the Administrator of the functions transferred by those provisions or relates primarily to those functions. The transfers to the Administrator made by this section shall be deemed to include the transfer of (1) authority, provided by law, to prescribe regulations relating primarily to the transferred functions, and (2) the functions vested in the Secretary of the Interior and the Secretary of Health, Education, and Welfare by section 169(d)(1)(B) and (3) of the Internal Revenue Code of 1954 (as enacted by section 704 of the Tax Reform Act of 1969, ...); but shall be deemed to exclude the transfer of the functions of the Bureau of Reclamation under section 3(b)(1) of the Water Pollution Control Act (33 U.S.C. 466a(b)(1)).

(b), There are hereby transferred to the Agency:

(1) From the Department of the Interior, (i) the Water Pollution Control Advisory Board (33 U.S.C. 466t), together with its functions, and (ii) the hearing boards provided for in sections 10(c)(4) and 10(f) of the Federal Water Pollution Control Act, as amended (33 U.S.C. 466g(c)(4); 466g(f)). The functions of the Secretary of the Interior with respect to being or designating the Chairman of the Water Pollution Control Advisory Board are hereby transferred to the Administrator.

(2) From the Department of Health, Education, and Welfare, the Air Quality Advisory Board (42 U.S.C. 1857c), together with its functions. The functions of the Secretary of Health, Education and Welfare with respect to being a member and the Chairman of that Board are hereby transferred to the Administrator.

Sec. 6. *Abolitions.* (a) Subject to the provisions of this reorganization plan, the following, exclusive of any functions, are hereby abolished:

(1) The Federal Water Quality Administration in the Department of the Interior (33 U.S.C. 466-1).

(2) The Federal Radiation Council (..., 42 U.S.C. 2021(h)).

Among the procedural landmarks established in the Project Rulison litigation was the right of the plaintiff COSCC to take the depositions of experts prior to the determination of the defendants motion for summary judgment to dismiss the complaint. COSCC argued that the motion for summary judgment could not be decided without considering the relevant testimony of certain experts under the control of the defendants and the Court directed the defendants to produce those experts for pre-trial oral examination. The information obtained in those pre-trial examinations established the need for a full hearing on the merits as a condition precedent to determination of any AEC motion for summary judgment dismissing the complaint.

d. Sovereign Immunity and Popular Sovereignty

Throughout the relatively short history of environmental litigation involving agencies of government, sovereign immunity has been raised as a defense by the government agency asserting that the action is a suit against the sovereign United States of America brought without the consent of Congress and that the relief sought would invade the powers of officers of the executive branch of the Federal Government to whom the actions complained of had been delegated. This defense is usually buttressed by an appeal for judicial restraint under the dictrine of Separation of Powers.

The general doctrine of the immunity of the United States from suit without consent of Congress is a rule propounded in a diction by United States Supreme Court Chief Justice John Marshall in the case of *Cohens* v. *Virginia* [19 US (6 Wheat) 264, 411–412 (1821)]. However, in the earliest discussion of sovereignity by the United States Supreme Court [*Chisholm* v. *Georgia*, (1973) 2 US 419], the court stated,

> ... the term sovereign has for its correlative, subject. In this sense, the term can receive no application; for it has no object in the Constitution of the United States. Under the Constitution, there are citizens but no subjects
>
> ... the people of the United States ... have reserved the supreme power in their own hands; and on that supreme power have made these [governments] dependent, instead of being sovereign ...
>
> Even in almost every nation which has been denominated free the state has assumed a supercilious preeminence above the people who have formed it: hence the haughty notions of State independence, State sovereignity, and State supremacy. In despotic governments, the government has usurped in a similar manner both upon the state and the people: hence all the arbitrary doctrines and pretensions concerning the supreme, absolute and incontrollable, power of government. In each, man is degraded from the prime rank which he ought to hold in human affairs ...
>
> Another instance, equally strong, but still more astonishing is drawn from the British Government as described by Sir William Blackstone and his followers. As described by him and them, the British is a despotic Government. It is a Government without a people. In that Government, as so described, the sovereignty is possessed by the Parliament: In the Parliament, therefore, the supreme and absolute authority is vested: In the Parliament resides that incontrollable and despotic power, which, in all Governments, must reside somewhere. The constituent parts of the Parliament are the King's Majesty, the Lord's Spiritual, the Lord's Temporal, and the Commons. The King and these three Estates together form the great corporation or body politic of the Kingdom. All these sentiments are found; the last expressions are found verbatim, in the *Commentaries Upon the Laws of England.* The Parliament forms the great body politic of England. What then, or where, are the people? Nothing! No where! They are not so much as even the 'baseless fabric of a vision!' From legal contemplation they totally disappear! Am I not warranted in saying, that, if this is a just description; a Government, so and justly so described, is a despotic Government?

Chief Justice Jay also recognized the people of the United States, not the Federal government, as the sovereign:

> From the crown of Great Britain, the sovereignty of their country passed to the people of it ... [T]he people, in their collective and national capacity, established the present Constitution. It is remarkable that in establishing it, the people exercised their own rights, and their own proper sovereignty, and conscious of the plentitude of it, they declared with becoming dignity. "We the people of the United States, do ordain and establish this Constitution." Here we see the people acting as sovereigns of the whole country ... [T]he sovereignty of the nation is [in] the people of the nation ...

Chief Justice Jay noted that in England the doctrine of sovereignty was based on feudal principles that considered the prince as sovereign and the people as his subjects. These feudal principles contemplated the sovereign

> As being the fountain of honor and authority; and from his grace and grant derives all franchises, immunities and pirivileges; it is easy to perceive that such a sovereign could not be amenable to a Court of Justice, or subjected to judicial control and actual constraint. It was of necessity, therefore, that suability, became incompatible with such sovereignty. Besides, the Prince having all the Executive powers, the judgment of the Courts would, in fact, be only monitory, not mandatory to him, and a capacity to be advised, is a distinct thing from a capacity to be sued. The same feudal ideas run through all their jurisprudence, and constantly remind us of the distinction between the Prince and the subject. No such ideas obtain here. At the Revolution, the sovereignty devolved on the people and they are truly the sovereigns of the country, but they are sovereigns without subjects ... and have none to govern but themselves, the citizens of America are equal as fellow citizens, and as joint tenants in the sovereignty.
>
> ... Sovereignty is the right to govern; a nation or State-sovereign is the person or persons in whom that resides. In Europe the sovereignty is generally ascribed to the Prince; here it resides with the people ... Their Princes have personal powers, dignities, and pre-eminences, our rulers have none but official; nor do they partake in the sovereignty otherwise, or in any other capacity, than as private citizens.

A majority of the Supreme Court of the United States, just five years after the adoption of the Constitution of the United States, rejected the idea of the United States as sovereign—and necessarily the rule of sovereign immunity from suit because "suability became incompatible with such sovereignty." The people of the United States are the sovereign not the United States government.

In 1882, the Supreme Court of the United States further said,

Under our system the *people*, who are there [in England] called *subjects*, are the sovereign. Their rights, whether collective or individual, are not bound to give way to a sentiment of loyalty to the person of a monarch. The citizen here knows no person, however near to those in power, or however powerful himself, to whom he need yield the rights which the law secures to him when it is well administered. When he, in one of the courts of competent jurisdiction, has established his right to property, there is no reason why deference to any person, natural or artificial, not even the United States, should prevent him from -using the means which the law gives him for the protection and enforcement of that right. [United States v. Lee (1882), 106 US 196, 205 to 209.]

e. Land Use and Resource Management.

Throughout history, land and natural resources have been considered the property of the sovereign to be used, abused, given or taken at the sovereign whim. The history of civilization, to a certain extent, can be considered a chronicle of wars and revolutions fought over the ownership and utilization land and natural resources.

The Constitution of the United States provides that the rights not explicitly given by the people of the United States to the federal government are retained by the people of the United States, individually, and as collectively assembled in the several states.

The enumeration in the Constitution, of certain rights, shall not be constructed to deny of disparage others retained by the people. (*United States Constitution*, Ninth Amendment.)

The powers not delegated to the United States by the Constitution, nor prohibited by it to the States, are reserved to the States respectively, or to the people. (*United States Constitution*, Tenth Amendment.)

As the individual state constitutions were formulated, the rights of individual property owners were strengthened, but at no time did the sovereign people of the United States relinquish the ultimate right to determine the highest and best use of the land and natural resources subject to the jurisdiction of the United States. The justification for any legal restriction on the private use of land or natural resources is rooted in this concept of sovereignty in the people.

According to the report commissioned by the Council on Environmental Quality as a preparation for consideration of a National Land Use Policy Act,

This country is in the midst of a revolution in the way we regulate the use of our land. It is a peaceful revolution ... a quiet revolution ...

The *ancien regime* being overthrown is the feudal system under which the entire pattern of land development has been controlled by thousands of individual local governments, each seeking to maximize its tax base and minimize its social problems, and caring less what happens to all the others.

and if we are to live in harmony with that which has been given to us from preceding generations, and from the earth before mankind appeared, we must make certain assumptions with respect to every accessible parcel of real property and every natural resource capable of exploitation.

- The area or resource is vulnerable.
- Development, exploitation or utilization of some kind is inevitable.
- The highest and best use of the land or resource as an element of human ecology must be accommodated.
- Development must be consistent with natural ecological constraints.
- Planned achievement of the highest and best use of land and resources is more profitable to mankind than unplanned development.
- The police power of the state, the ultimate sovereignty of the people, and the maintenance of private property concepts are compatible and can result in the harmonious, beneficial development of land and resources.

Recognizing the limited availability of land itself and the place of land as the basic capital asset of civilization, land use historically has been limited by executive, legislative and judicial process.

An early example of land use legislation—zoning—has been upheld by the courts on the grounds that it represents action by an individual community to assure the highest and best use of limited resources for the greatest number of people without undue infringement upon the rights of individuals.

Zoning is not just an expansion of the common law of nuisance. It seeks to achieve much more than the removal of obnoxious gases and unsightly uses. Underlying the entire concept of zoning is the assumption that zoning can be a vital tool for maintaining a civilized form of existence only if we employ the insights and the learning of the philosopher, the city planner, the economist, the sociologist, the public health expert and also the other professions concerned with urban problems ...

This fundamental conception of zoning has been present from its inception. The almost universal, statutory requirement that zoning conform to a "well-considered plan" or "comprehensive plan" is a reflection of that view [See Standard State Zoning Enabling Act, U.S. Dept. of Commerce (1926).] The thought behind the requirement is that consideration must be given to needs of the community as a whole. In exercising their zoning powers, the local authorities must act for the benefit of the community as a whole following a common, deliberate consideration of the alternatives, and not because of the whims of either an articulate minority or even majority in the community ... Rather, the comprehensive plan is the essence of zoning. Without it, there can be no rational allocation of land use. It is the insurance that the public welfare is being served and that zoning does not become nothing more than just a Gallup poll. [Udell v. Hass (1968), 21 NY2d 463, 235 NE2d 897, 288 NYS2d 888.]

The key to successful land use and resource utilization legislation is the community determination of the highest and best use of land and resources in terms of intrinsic suitability and naturally imposed constraints. Of necessity this must be done by a team of individuals trained in the several disciplines necessary to define the environmental parameters of a Regional Ecological System. The community itself, particularly its people, constitute elements of that regional ecological system just as surely as topography, hydrology, and climate. One of the most significant contributions of environmental litigation to the development of Anglo-American jurisprudence has been the broadening of the concept of "expert witness" to include an individual member of a multidiscipline scientific team testifying about the work of the entire team even though the individual scientist may not have performed the work testified about personally. Environmental litigation, particularly that involving such environmental toxicants as DDT and radionuclides has led to legal acceptance of the "environmental scientist" who is permitted to testify generally, upon a showing, to the satisfaction of the Court, of demonstrated competence in some recognized scientific discipline together with regular collaboration with other scientists of diverse disciplines upon general problems of environmental concern.

The adequate determination of the highest and best use of the limited natural resources of any regional ecological system mandates a systems approach supported by modern computer technology in order to determine the relevant boundary conditions and elemental optimizations of the complex, non-linear dynamic relationships that describe a natural system as it actually exists. Courts increasingly are accepting systems concepts as a method of presenting evidence in environmental litigation, and the techniques of general systems analysis have become key factors in the resolution of environmental controversy such as that concerning the continued use of DDT, the distribution of radionuclides from peaceful applications of nuclear energy, and the Cross-Florida Barge Canal.

The acceptance of systems methods in ecology by the courts now means that any land use or resource utilization law which does not fully reflect the system characteristics of the Regional Ecological System in which it will operate is fatally defective in the legal sense. It can not be sustained in the face of sophisticated legal challenge; while any such law which does, in fact, reflect the system characteristics of the region can be sustained in the public interest, even where such a law appears to limit the rights incident to private property ownership.

The current litigation involving the adequacy of statements prepared under the mandate of section 102 of the National Environmental Policy Act (environmental impact statements) demonstrates that any failure to consider the ecological integrity of a region and fully determine the interrelationships among each element of the land, landscape and natural processes diminishes the legal value of the effort; while a full evaluation of any environmentally significant action with respect to its effect upon the overall ecological integrity of the region and the interrelationships among each element of land, landscape, and natural processes of the region can form the basis for legal restraints upon land use and resource exploitation, even where such restraints appear to infringe upon private property rights or government agency prerogatives.

Environmental Litigation

Environmental litigation includes almost all the actions brought before the courts of the United States, the several states, and the various administrative agencies exercising quasi-judicial powers which involve a determination which may affect natural processes, resources or the environment.

a. The Scenic Hudson Preservation case.

One of the landmark cases in environmental law is *Scenic Hudson Preservation Conference* v. *Federal Power Commission* [354 F2d. 608 (1965, CA2), *certiorari denied* 384 US 941, 16 LEd 2d 540, 86 SCt 1462], which established the right of conservation organizations and other public interest groups to challenge the extent of consideration given to historic, scenic, aesthetic, conservation and recreational aspects of power development by the Federal Power Commission. The holding of that case has been extended to permit similar organizations to challenge the consideration of such values by the Secretary of Transportation (in approving federal-aid highways) and the Secretary of the Army and Chief of the Corps of Engineers (in licensing and granting permits for projects on navigable waterways).

The *Scenic Hudson* saga began on January 29, 1963 with Consolidated Edison of New York (the major electric power utility servicing the City of New York) filing an application with the Federal Power Commission (FPC) for a license to construct a pumped-storage hydroelectric facility on the west side of the Hudson River at Storm King Mountain. As information concerning the project spread, the Federal Power Commission directed that a public hearing on the project be held February 25, 1964. The Scenic Hudson Preservation Conference was hastily formed and filed a petition on February 6, 1964 seeking to intervene in the proceeding. The petition was granted on February 14, 1964; intervention was permitted; hearings were held in Washington, D.C.; and the hearing examiner made an initial determination approving the license application on July 31, 1964 subject to review by the entire Federal Power Commission which eventually licensed the project on March 9, 1965.

The pumped-storage plant proposed by Consolidated Edison would generate electric power for use during peak load periods in New York City using hydroelectric units driven by water from a reservoir. The project consisted of three major components—a storage reservoir, a powerhouse, and transmission lines. The reservoir would be located over the power house which was to be located on the shores of the Hudson River at the foot of Storm King Mountain, a mountain of striking aspect rising over 1,360 feet above the river. The reservoir would be connected to the power house by a tunnel 40 feet in diameter. During slack periods of demand for electrical power in New York City, Consolidated Edison's steam generating plants in New York City would provide electric power for the pumps at Storm King to force water through the tunnel and up the mountain into the reservoir. The water would be stored in the reservoir until periods of peak demand for electrical power in New York City at which time the water would be released from the reservoir to rush down the mountain through the tunnel to power the generators. The project was to have a capacity of two gigawatts (billion watts, or million kilowatts), but would be capable of expansion to three gigawatts. At that time, the Storm King Project was to be the world's largest pumped-storage electric generating facility and was estimated to cost an estimated $ 162,000,000.

Subsequently the United States Court of Appeals for the Second Circuit reviewed the determination of the Federal Power Commission holding, that

> if the Commission is properly to discharge its duty in this regard [approve those projects best adapted to a comprehensive plan for improving or developing a waterway], the record on which it bases it determination must be complete. The petitioners and the public at large have a right to demand this completeness. It is our view, and we find, that the Commission has failed to compile a record which is sufficient to support its decision. The Commission has ignored certain relevant factors and failed to make a thorough study of possible alternatives to the Storm King Project.

The court found that the Commission must take into consideration the unique beauty and important historical significance of the Hudson River noting that the Federal Power Act sought to "promote the comprehensive development of the Nations's water resources."

The court directed the Federal Power Commission to reexamine all the issues that the court found had not been properly considered.

> The Commission's renewed proceedings must include as a basic concern the preservation of natural beauty and of national and historic shrines, keeping in mind, that, in our affluent society, the cost of a project is only one of several factors to be considered. The record as it comes to us fails markedly to make out a case for the Storm King Project on, among other matters, cost, public convenience and necessity, and absence of reasonable alternatives. Of course the Commission should make every effort to expedite the new proceeding.

Following the hearings on remand, the hearing examiner made an initial determination on August 6, 1968 recommending that the license be issued. New York City then petitioned to intervene prior to a decision by the Commission in order to introduce evidence of the effects project construction activities might have upon the safety of the Catskill Aqueduct which supplies a substantial part of the water for New York City. The proceedings were reopened and on December 23, 1969 a supplemental decision was issued by the hearing examiner recommending that the licence be granted with some modifications.

On August 17, 1970, the Federal Power Commission issued the license for construction of a pumped-storage storage plant by Consolidated Edison at Storn King Mountain. In a lengthy opinion, the Commission paid close attention to the decision of the Second Circuit Court of Appeals, giving detailed consideration to areas in which the court had found that the previous decision had been deficient, including alternative sources for the needed power, alternative sites for the project, fish protective devices, underground transmission lines, and the scenic, aesthetic, and historical values that would be affected by the project. The Commission found the project to be compatible with the natural environment, and not to have an adverse effect upon the ecology of the area, the water quality of the Hudson River, or the fishery resources of the Hudson River. The area would be replanted after construction; the power house would be totally underground and the Commission found that the project, as modified and as licensed, represented no real impairment to the environment and scenic aspects of the highlands and the natural beauty of the area. The Commission recognized man's already awesome impact on this section of the Hudson River shoreline, by stating that,

> ... it should not be inferred that previous and existing environmental detriment permits additional detriment which would otherwise not be permitted but for existing conditions,

pointing out that Consolidated Edison would rehabilitate the waterfront area in the vicinity of the project site, an area which the Commission euphemistically described as "blighted".

Although the hearings had been conducted prior to the passage of the National Environmental Policy Act, the Commission found that full and careful consideration had been given to all the concerns which the Act embodies and that the licensing of the facility complied with the requirements of that Act.

All parties appealed to the Second Circuit Court of Appeals which affirmed the Commission's issuance of the license. The majority of the Court found that the decision of the Federal Power Commission was supported by substantial evidence and had been based on a consideration of all relevant factors, and had in fact complied with the requirements of the Federal Power Act and the National Environmental Policy Act.

After 8 years of litigation, at enormous cost to the Scenic Hudson Preservation Conference, its supporters, and Consolidated Edison, not to mention the inconvenience and cost to the electric power-hungry people of the City of New York, the project would be constructed, little modified from the initial proposal.

The first decision of the Court of Appeals in the Scenic Hudson Preservation case represented a crucial selection event in the evolution of Environmental Law. If the conservationists were willing to submit the controversy to the Federal Power Commission, then, according to the Court, the Federal Power Commission should hear and consider evidence on natural values in addition to evidence on the economics of electric power generation and distribution. The alternative choice open to the conservationists was to remain before the court and challenge the inherent suitability of the Federal Power Commission as a body capable of resoluting environmental conflict. The Scenic Hudson Preservation Conference, a coalition of preservationists and aesthetically concerned conservationists, chose to yield to the Federal Power Commission the authority to make an ecological judgment binding upon generations yet unborn, cloaking the Federal Power Commission with a mantle of ecological competence. Since the Scenic Hudson Preservation Conference chose to take its "day in court" before an administrative agency rather than a court of equity, the findings of fact made by the Commission were binding upon the conservationists in subsequent review by the Courts.

Fortunately, the principles set forth by the Court of Appeals in its initial review of the *Scenic Hudson* case mandating consideration of all relevant issues including environmental issues, and requiring affirmative action by the Agency to protect the public interest are not limited in application to the Federal Power Commission. Other agencies, including the Corps of Engineers of the United States Army, the Atomic Energy Commission, the Department of Transportation, the Department of the Interior and the Department of Agriculture, must also consider all relevant issues, including environmental issues and take affirmative action to protect the public interest.

b. The Defense of Florissant.

The Florissant fossils, located a short distance west of Colorado Springs, Colorado, are found in more than 6,000 acres of an ancient lake bed where seeds, leaves, plants and insects from the Oligocene period (34 million years ago) are remarkably preserved in paper-thin layers of volcanic shale which disintegrate when left exposed to weather unless properly protected. A number of bills had been introduced over the years in Congress to protect the Florissant fossil beds but they did not receive extensive consideration until the United States National Park Service promulgated a master plan detailing the paleontological and palynogical values of Florissant.

At the time the Florissant fossil beds National Monument bill passed the Senate, Park Land Company, a Colorado Springs real estate group had already contracted to purchase 1,800 acres of the ancient lake bed. While the House of Representatives was deliberating its version of the National Monument bill, Park Land Company announced it would bulldoze a road through a portion of the proposed national monument to open the area for development and immediate sale to anyone interested in recreational housing. A group of Colorado conservationists met with the principals of the Park Land Company in an attempt to persuade them to withhold excavation in the area to be included within the Florissant Fossil Beds National Monument at least until the House of Representatives acted on the bill. This request was refused as was a similar request to confine development activities to the area lying outside the ancient lake bed. The only alternative offered the conservationists was the opportunity to purchase the land—for cash immediately—at a price considerably in excess of any appraised value based on recent land sales in the area.

Faced with the irreparable loss of a substantial portion of these unique and irreplaceable fossil beds, a small group of concerned citizens formed a non-profit, public benefit corporation called the Defenders of Florissant and commenced an action for declaratory judgment and injunctive against the Park Land Company and all the other land owners and contract vendees in the area to be included within the proposed National Monument.

The United States District Court for the District of Colorado heard the Defenders of Florissant application for a temporary restraining order on July 9, 1969, and although the plaintiffs' proof that the proposed excavations for roads and culverts would result in the distruction of some of the most valuable fossil areas in the proposed national monument was uncontradicted and unchallenged the District Court held that there was nothing in the United States Constitution preventing the owners from using their property in any way not prohibited by law. The District Court denied the application for a temporary restraining order and a subsequent application for a stay pending appeal, but did, however, note the importance of preserving the fossil beds.

Following the District Court decision, representatives of the plaintiffs' held an informal conference in the Courtroom with two of the partners in the Park Land Company who agreed to postpone excavation until Monday, July 14, if the plaintiffs gave some assurance of raising the purchase price by that day. Refusing to accept an offer they felt was a form of community blackmail, the Defenders of Florissant appealed to the Tenth Circuit Court of Appeals the following morning. July 10. At the hearing before three judges of that Court in the afternoon, the Court questioned whether it had authority to issue a restraining order in the absence of any statute protecting the fossils. ...

Admitting that Congress, "... in its infinite wisdom, had not seen fit to pass legislation protecting fossil beds in general," plaintiffs' argued: "... if someone had found the original Constitution of the United States buried on his land and then wanted to use it to mop a stain on the floor, is there any doubt ... they could be restrained?"

Legally, plaintiffs claimed that the right to preservation of the unique and irreplaceable Florissant fossils, a national, natural, resource treasure, was one of the unenumerated rights retained by the People of the United States under the Ninth Amendment of the Constitution and protected by the *due process* and *equal protection* clauses of the Fifth Amendment, and the *provileges or immunities, due process and equal protection* clauses of the Fourteenth Amendment. Plaintiffs also asserted that the Florissant fossil beds were subject to judicial protection under the Trust Doctrine and while the defendants could profit from their nominal title to the land and make reasonable use of the area, they were under a duty to maintain that portion of the property vested with the public interest—the 34 million year old fossil shales. Procedurally, the Defenders invoked the federal equity jurisdiction relying on the fundamental equitable maxim, "there shall be no wrong without a remedy."

In summation, counsel for the Defenders of

Florissant picked up a fossil palm leaf that had been uncovered at Florissant, and holding it up to the Court, pleaded:

> The Florissant fossils are to geology, paleontology, paleobotony, palynology and evolution what the Rosetta Stone was to Egyptology. To sacrifice this 34 million year old record, a record you might say written by the mighty hand of God, for 30 year mortgages and the basements of the A-frame ghettos of the seventies is like wrapping fish with the Dead Sea Scrolls.

After a short recess, the Court issued an order restraining the defendants from

> disturbing the soil, subsoil or geological formations of the Florissant fossil beds by any physical or mechanical means ...

After a trial on July 29, 1969, the District Court denied the Defenders application for a preliminary injunction for the same reasons it had previously denied the application for a temporary restraining order, and the Park Land Company announced that its bulldozer would begin excavation that afternoon. Several hours later, the Plaintiffs filed a motion for an emergency stay with the Tenth Circuit Court of Appeals, citing defendants threat, and the Court of Appeals for the Tenth Circuit dramatically issued an order extending the restraining order of July 10 indefinitely until further order.

On July 31, 1969, the House Interior and Insular Affairs Committee, through its Subcommittee on Parks and Recreation favorably reported an amended version of the Florissant Fossil Beds National Monument bill, and floor action by the House of Representatives was scheduled for August 4.

During the argument of the appeal before the Tenth Circuit Court of Appeals, the plaintiffs amplified their legal position, asserting that the Federal Courts had a duty to cooperate with Congress, and that by issuing a preliminary injunction, pending the final deliberation of the Congress of the United States the court was thereby aiding the orderly operations of the Legislative and Executive branches of government. Plaintiffs pursued their original theory that the Trust Doctrine protected the fossil beds, arguing that the land had acquired a public character due to the actions of Congress on the bills pending to dedicate the land as a national monument. The Court reserved decision at the close of the arguments and continued the temporary restraining order. That afternoon the House of Representatives passed its version of the bill as a number of concerned Congressmen from all over the country turned out to suspend the rules and consider the bill out of the regular order because of the pending threat to the fossils. The Senate agreed to the House version of the bill on August 7, and the President signed the bill on August 14, 1969. The preliminary restraining order issued by the Tenth Circuit Court of Appeals remained in effect while the United States of America instituted suit to acquire the Park Land Company land by condemnation.

The court order prohibiting excavation of the fossil beds may have deprived the landowners of the most profitable use of their land, but did not prohibit all uses of the land consistent with the protection of the fossil beds. The landowners were free to develop the land for tourism, scientific research, or other uses compatible with maintenance of the paleontological integrity of the area. Such uses, while perhaps not the most profitable use of the land, would still return a reasonable yield on the defendants' speculative investment.

The mere fact that the landowner might not wish to use the land for this purpose does not make the restraint on the land development an unreasonable taking where the public interest in the land was so great.

Certainly where a natural resource is as unique as the Florissant fossil beds were, the value to the public of protecting such a resource is so substantial as to justify the resultant burden upon the private property interests involved, even if it could have been shown that there was no reasonable expectation of profitable use of the property from tourism or other ancillary commercial development.

The message of the Florrisant litigation is that judicial protection of a unique, national, natural resource treasure such as the 34 million year old Florissant fossil beds warrants restraint upon the rights of private property ownership, particularly during the period of due deliberation by Congress or other legislative body representative of the people.

The mere fact that Congress could not move as fast as the developers' bulldozer does not prevent a federal court of equity from acting to protect a national, natural resource treasure threatened with irreparable damage.

c. Project Rulison

Project Rulison provided the first direct confrontation among the several theories currently urged in support of citizen action to protect the environment from federal agency operations. Three separate suits were filed concerned with the AEC-Austral Oil Co. experiment seeking to stimulate production of natural gas by underground nuclear explosion. The first action was supported by the American Civil Liberties Union and relied on conventional theories to establish standing: Individual plaintiffs alleged direct, personal, private injury and special damage, seeking injunctive relief to protect their own property rights. The A.C.L.U. application sought to restrain detonation of the underground nuclear device as its principle request for relief, and had already been denied when the second action was filed by the Colorado Open Space Coordinating Council. The title of that action in itself indicates the contrasting theories.

COLORADO OPEN SPACE COORDINATING COUNCIL, on behalf of all those entitled to the protection of their health and safety and of the health and safety of those generations yet unborn, from the hazards of ionizing radiation resulting from the distribution of radioactive materials through the permanent biogeochemical cycles of the Biosphere as a result of the defendants conduct of *Project Rulison*, and on behalf of all those entitled to the full benefit, use and enjoyment of the national, natural resource treasures of the State of Colorado without degradation resulting from contamination with radioactive material released as a result of the defendants conduct of *Project Rulison*, and all others similarly situated,

Plaintiffs,

—against—

AUSTRAL OIL COMPANY, INCORPORATED
and
CER GEONUCLEAR CORPORATION,

Defendants

U.S. ATOMIC ENERGY COMMISSION,
BUREAU OF MINES; U.S. DEPARTMENT
OF INTERIOR, and
LOS ALAMOS SCIENTIFIC LABORATORY,

as their several interests may appear.

The shift in emphasis in the COSCC action from an emotional outcry against the underground nuclear blast itself to a reasoned demand for care in the release of radionuclides to the environment led to a Court order restraining the "flaring" of the radioactive natural gas following the blast until the hearing and determination of the action brought by COSCC. By amending their complaint, A.C.L.U., on behalf of the other individuals concerned personally with the blast, remained in the action. Subsequently, the District Attorney of the Ninth Judicial District of the state of Colorado attempted to bring an action in the state court on public nuisance theory, but that action was summarily transferred to the U.S. District Court and consolidated with the COSCC and A.C.L.U. actions at the request of the Atomic Energy Commission.

Equity

The truly unique element of the Anglo-American is the concept of equity. Equity jurisprudence as a system of remedial law evolved from a number of common sources. It can be found in the Talmud and the earliest writings of the Roman law. It can be found today, through it is somewhat less than obvious, in the current system of civil jurisprudence derived from the Code Napoleon and used throughout most of Europe.

In its most elementary form, the fundamental principle of equity jurisprudence is the command: so use your own property as not to injure that of another. And the law, in order to give effect to this right provides for appeal to the ultimate power of society, be it king, parliament, state, or people, with a corollary maxim: equity permits no wrong to be without a remedy! The effective assertion of equitable rights by an individual or group of individuals is limited only by the rule that a party seeking equitable relief must come forward with "clean hands"—the party must be morally right as well as legally justified. There is an additional rule of restraint self-imposed by courts of equity: the relief granted must be commensurate with the injury suffered by the party seeking relief and tempered by the needs of society.

The Origin of Jurisdiction in Equity

Aristotle asserted that all law is universal and thereby cannot admit of exceptions, yet laws promulgated to cover a broad range of human action frequently cause injustice to some innocent individual because a particular case does not appear to be covered by the application of the universal law. Aristotle reasoned that when lawmakers make a law, they make it for the good of the community, therfore some legal remedy must be available to the individual treated unjustly by the particular application of a general law.

In order to determine whether the individual was indeed being treated unfairly by the law, one must look to the intent of the lawmaker and the operation of the law. The remedy for the individual unjustly treated in a particular case by application of a general law was termed by Aristotle *Epicheia*, which translates loosely as "There should by an exception." It was left up to the judge deciding each particular case to determine whether a general or universal law was applicable, and if not, to make the appropriate exception in order to preserve the intention of the lawmaker, which was to ensure the good of the community.

The early Roman lawgivers accepted this concept, named it equity, and used it as a cornerstone for the development of what we now call the Roman Law. The intention of the Roman lawgivers was that one person should not benefit by a law while another person was injured by that law unintentionally and unnecessarily.

During the Middle Ages the heir apparent to the Roman legal system, the Christian church, developed the concept of equity even further, establishing the principle that "for every injury there must be some legal remedy," on the philosophical grounds that if the lawmaker did not provide some remedy for injury to an innocent individual, the law would allow certain injustices to go unpunished while others were punished, and this would be unjust. Since it was then an accepted principle that lawmakers were just, they could not have intended an injustice to follow from their laws, therefore, they intended to provide the legal remedy and the court would simply serve as the means to provide that which they intended to provide all along. Although such circumlocution might have found favor with medieval philosophers, there was, nevertheless, a singularly practical reason for development of the principle. For if there was no procedure for individual relief from the unintended application of a general law, and sufficient individuals were oppressed by the unintended application of that law, then the individuals would tend to look with disfavor on the lawmakers and, in spite of the repression inherent in the feudal system, there might be civil unrest, a condition not conductive to the maintenance of the tenuous existence of the feudal estates which represented civilization in western Europe during the Middle Ages.

Equity jurisprudence developed throughout the ecclesiastical courts following the decline and fall of the Roman Empire, but it was to see its most dramatic development in England following the Norman conquest.

The common law, as distinguished from the customary law of the popular courts, originated in the establishment by Henry II of a national court administering a law for the entire nation, and by the end of the reign of Henry II, we find established a *Curia Regis*, a court of the King, which was a true court of law in the modern sense, administering a national law, common to the entire country, and which had largely displaced the customary laws of the different parts of the country. This continued during the thirteenth century, so that by its close, the common law was definitely established as the law of the nation, displacing the customary law and the local courts which were limited to local petty matters. The law of the *Curia Regis*, which had been the law of the very great, extended and adapted to the needs of the people so as to become the common law of a nation.

So long as the common law remained a flexible system with its field undefined, its power of inclusion unlimited, and its organs undifferentiated, there was no reason for distinguishing between the common law and equity. But soon the common law became so fixed and inflexible and its practitioners so absorbed in nice questions of form and pleading, there was no longer room for equity. By the early fourteenth century, the common law, which had supplanted the ancient customary law, had now, in its turn, become the regular system of remedial justice, but with gaps and defects where sufficient remedies were not provided, and resort was necessary to the ancient power of the King as the fountain of all justice.

Equity did not directly contest the existence of settled legal right. Rather, after recognizing those rights, equity went on to insist that the holder of such legal rights if they were acquired or retained unconsionably, or if they were being used in an unconscionable attempt to interfere with the fundamental rights of others, should be subject to the jurisdiction of the "Chancellor" or the "Keeper of the Conscience of the King" as the source of all legal rights.

Equitable Relief

It is a fundamental principle essential to the very existence of organized society, and civilization as we know it, that every person, in exercising personal rights and in the use of personal property shall respect the rights and properties of others. Every person must so conduct themselves in the enjoyment of the rights and privileges which they may enjoy as individual members of society in such a way that they shall prejudice no one in the possession and enjoyment of their personal rights or the rights they hold in common as members of society. When there is an invasion of the rights or privileges of the public or the rights and privileges of any individual held in common by reason of the existence of civilized society, the absence of exact precedent and the fact that commentators on the law do not discuss the subject is of no material importance in awarding equitable relief.

That the exercise of the preventive powers of a court of equity is demanded in a novel case is not a fatal objection. In social evolution, with the march of the arts and sciences, and in the resultant effects upon organized society, it is quite intelligible that new conditions must arise in personal relations, which the rules of the common law, cast in the rigid mold of an earlier status, were not designed to meet. It would be a reproach to equitable jurisprudence, if equity were powerless to extend the application of the principles of common law, or of natural justice, in remedying a wrong, which in the progress of civilization, has been made possible as the result of new social or commercial conditions.

Equity is the agency by which law is brought into harmony with society. It is one of the factors which operate in judicial evolution. It succeeds legal fictions—those judicial assumptions through which a rule of law is modified in its operation—and it precedes legislation.

Equity has neither fixed boundaries, nor logical subdivisions, and its origin, both in Rome and in England, was that there was a wrong for which there was no remedy at law. As Lord Chancellor Cottenham observed:

> It is the duty of this court, [equity], to adopt its practice and course of proceeding to the existing state of society and not, by a strict adherence to forms and rules, under different circumstances, to decline to administer justice and enforce rights for which there is no other remedy ... If it were necessary to go much further than it is, in order to open the doors of this court to those who could not obtain [justice] elsewhere, I should not shirk from the responsibility of doing so.*

---

* *Wallworth* v. *Holt* 4Myl & C619, 41 Eng. Rep. 238 (1841).

A distinguishing feature of equity jurisdiction is that it will apply settled rules to unusual conditions, and mold its decrees so as to do justice between the parties. Peculiar and extraordinary cases will arise in the complex and diversified affairs of men, which perhaps, cannot be classed under any of the distinct heads of equity jurisdiction, but which must be acknowledged, nevertheless, to come within the legitimate powers of a court of equity because complete justice cannot otherwise be done between the parties. Therefore, when no remedy exists at law, courts of equity, to prevent injustice and in many cases on principles of general policy, will fo far in granting relief.

Such was the law of equity at the start of the eighteenth century. If that rule of law had been developed with vigor much of the human and environment degradation of the Industrial Revolation might have been avoided.

Since that time, two anomalous rules have evolved, interfering with the principle that for every wrong there is a remedy. The first of these anomalies concerns the artificial distinction between public and private nuisance and the second is the doctrine of "sovereign immunity." Both of these doctrines can be attributed to the meddling of Sir Wiliam Blackstone (1723–1780).

Public and Private Nuisance

Until Blackstone there was no distinction made between public and private nuisance. The rule had been well established that any indivudual could apply to a court of equity to abate a nuisance. But during the later part of the eighteenth century, Blackstone created a new rule of law that was to represent a classic manifestation of the obfuscation of simple legal principles by "self-proclaimed" legal scholars. This new rule of law proclaimed by Blackstone was to contribute substantially to the environmental and social crises of today and represents the kind of antisocial perversion of the law that made the common law of England during the seventeenth, eighteenth, and nineteenth centuries such an inviting target for the diverse talents of William Shakespeare, Jonathan Swift, Charles Dickens, and W. S. Gilbert.

Blackstone created a difference between public and private nuisance, and the significance of that difference is to be found in the criteria for abatement of the public nuisance: (1) only a public nuisance may be made the basis for a criminal prosecution, and (2) only the public, through the proper officer, may sue to enjoin or abate a public nuisance, in the absence of special damage to a particular private individual—damage which is substantially greater than that suffered by other individuals members of society—a public nuisance is subject to correction only at the hands of public authority. The mischief done and the disastrous consequences of Blackstone's whim are still evident.

In 1965, an action was brought on behalf of the people of the Town of Brookhaven, in the County of Suffolk, New York, by a group of citizens, to restrain the Long Island duck industry from discharging raw, untreated sewage equivalent to that of a city of one million people into the waters of Great South Bay, which was once one of the finest shellfishing and marine recreation areas on the east coast of the United States. The duck industry defended the action by admitting that the eight million ducklings grown along the estuaries of the Great South Bay did in fact deliver the raw sewage effluent equivalent of a city of one million people into the Bay, but that the Court should dismiss the action on the grounds that such an affront to the public waters was so great that it was a "public nuisance" not a "private nuisance" and as a "public nuisance" it could only be attacked by the Attorney General of the state of New York, not by any private citizens, unless of course those private citizens could establish "special damages" different from the damages sustained by the public at large.

A New York State Supreme Court, the same court that less than a year later in a dramatic reversal of precedent would issue the first injunction against the use of DDT ever granted by a Court, dismissed the case against the duck industry, accepting without question the argument that a public nuisance could not be abated by a private citizen or group of citizens.

How could this strange concept have crept into the law of equity? What is the justification for this strange anomaly in that body of law that holds no wrong may exist without a remedy? In 1858, the Court of Appeals of the State of New York identified the source and expounded the justification for this onerous rule,

> A contrary rule would be productive of very great inconveniences. ... No private person or number of persons can assume to be the champions of the community and in its behalf, challenge the public officers to meet them in the courts of justice to defend their official acts.

The court continued and discussed the theory of the decision,

> The general rule is that for wrongs against the public, whether actually committed or only apprehended, the remedy, whether civil or criminal, is by a prosecution instituted by the state in its political character, or by some officer authorized by law to act in its behalf. ...
>
> The principle is further exemplified in questions respecting nuisances. Common or public nuisances, which are such as are inconvenient or injurious to the whole community in general, are, as all are aware, indictable [the People of State can take action in a criminal proceeding] only, and the not actionable [any citizen can sue]; for as Blackstone [*Blackstone's Commentaries*, Book 4, p. 167] says, "it would be unreasonable to multiply suits by giving every man a separate right of action for what damnifies him in common only with the rest of his fellow-citizens.

Just who was Sir William Blackstone that he should exert such a restraint on the general application of equitable principles? Referring to the eleventh edition of the *Encyclopaedia Britannica*, published in 1911 and drawing on continual revisions from the first English edition in 1771, the following information may be elicited under the entry, "Blackstone, Sir William [1723–1780)":

> ... In 1746 he was called to the bar. Though but little known or distinguished as a pleader, he was actively employed during his occasional residences at the university (Oxford), in taking part in the internal management of his college. In May 1749, as a small reward for his services, and to give him further opportunity of advancing the interests of the college, Blackstone was appointed steward of its manors. In the same year, on the resignation of his uncle, he was elected recorder of the Borough of Wallingford. ... He accepted a seat on the bench, and on the death of Sir Joseph (Yates) succeeded him (in the court of common pleas). He died on the 14th of February, 1780. ...
>
> Blackstone was by no means what would now be called a scientific jurist. He had only the vaguest possible grasp of the elementary conceptions of law. Austin, who accused him of following slavishly the method of Hale's Analysis of the Law, declares that he "blindly adopts the mistakes of his rude and compendious model; missing invariably, with a nice and surprising infelicity, the pregnant but obscure suggestions which it proffered to his attention, and which would have guided a discerning and inventive writer to an arrangement comparatively just.
>
> From the small place which equity jurisprudence occupies in his arrangement, he would scarcely seem to have realized its true position in the law of England.
>
> Bentham accuses him of being the enemy of all reform, and the unscrupulous champion of every form of professional chicanery. Austin says that the truckled to the sinister influences and mischievous prejudices of power, and that the flattered the overweening conceit of the English in their own institutions. "He displays much ingenuity in giving a plausible form to common prejudices and fallacies. ...

For more than a century the opinion of that one man has stood in the way of a proper disciplined application of a fundamental principle of equity jurisprudence, "equity will not suffer a wrong without a remedy."

Just what happened at the time of Blackstone? It was obvious that Blackstone set himself to the task of codifying the laws of England, but in the process of attempting to build a logical and consistent body of legal principles he lost the basic insight of the Anglo-American system of jurisprudence, the common law.

The common law grew with civilization and the practices and customs of society. As the oppressed peasants obtained certain remedies during the Middle Ages, these remedies became a part of the common law. That is why the Anglo-American legal system depends on case law or precedent rather than elaborate codes promulgated by a legislature or king. A system of administering justice based on precedent rather than statute is inherently more flexible and capable of meeting the needs of society because the court is free to interpret the law with reference to, and by analogy with, past cases of similar import, though not precisely identical.

There is no doubt that Blackstone did a reasonable job of tidying-up the law, but in his attempt to make of law an axiomatic science little different from geometry, he lost the essential elements of equity jurisprudence, particularly the rule that for every injury there must be a legal remedy.

Blackstone's justification for striking the principle that for every injury there must be a legal remedy from the law of England arose from his belief that the King could do no wrong. Reasoning that failure to provide an adequate legal remedy for an injury was wrong, and that the King could do no wrong, it was obvious that there was no need for a rule of law providing a remedy for every wrong, because the King had obviously provided the remedy, since the King could do no wrong and failure to provide a remedy for an injury would be wrong. This perfectly circular argument has been exposed by many legal scholars, yet it still appears in decisions of courts that should know better.

Consider for a moment the ideological basis for establishing American independence. The founding fathers of this country were asserting the fundamental equitable principle that no wrong should exist without an adequate legal remedy. They did not deny the right of George III to tax his American colonies, they asserted that certain taxes were the wrong taxes, at the wrong time and for the wrong purposes. Taxation without representation was wrong. In other words, the King could, in fact, under certain circumstances, do wrong, and the equitable rule of law that no wrong should exist without a legal remedy was a fundamental human right.

Environmental Law is a mixture of the new and the old. Affirming the timeless principles of equity jurisprudence and asserting the unenumerated rights reserved to the people by the Ninth Amendment of the Constitution and protected by the *equal protection* and *due process* clauses of the Fifth Amendment and the *privileges or immunities*, *due process* and *equal protection* clauses of the Fourteenth Amendment of the Constitution, the law is newly applying these established principles and traditional legal procedures directly to the environmental crises threatening the Biosphere.

The basic element of environmental jurisprudence is recognition by our courts that the people have an absolute right to a salubrious environment as one of the "fundamental and inherent rights with which all humans are endowed even though no specific mention is made of them in either the national or state constitutions. ... The inherent human freedoms with which mankind is endowed are 'antecedent to all earthly governments; rights that cannot be repealed or restrained by human laws, rights derived from the Great Legislator of the Universe."

[Colorado Anti-Discrimination Com. v. Case, 151 Colo. 235, 380 P. 2d 34 (1962)]

The words of Lord Atkin in an English case are compelling: "When those ghosts of the past stand in the path of justice clanking their medieval chains, the proper course for the judge is to pass through them undeterred.... The common law does not go on the theory that a case of first impression presents a problem of legislative as opposed to judicial power."

[United Australia, Ltd. v. Barclay Bank, Ltd. (1940) 4 All Eng. 20, (1941) A C 1, 29]

"The law will protect a flower or a vine as well as an oak. ... [Such] damges are irreparable too, because the trees and vines cannot be replaced."

[Campbell v. Seaman, 63 N.Y. 568 (1876)]

REFERENCES

Detailed treatment of the material contained in this article, including generally:

Environmental Protection and the Law; The Trust Doctrine, The Ninth Amendment; Nuisance and Other Common Law Remedies; Environmental Legislation; Environmental Toxicants; Radiation; Water Pollution; Air Pollution; Noise; Visual Pollution; Solid Waste; Land Use and Natural Resource Management; Environmental Litigation and Administrative Agencies and the Environment, will be found in the treatise ENVIRONMENTAL RIGHTS & REMEDIES by Victor John Yannacone, jr., Bernard S. Cohen, Steven Gebauer Davison (1971), Lawyers Cooperative Publishing Company, Rochester, New York/Bancroft Whitney Co., San Francisco, California.

Current material in the nature of news accounts, federal, state and significant local legislation as promulgated, and texts of the decisions of courts and selected administrative agencies, is published weekly by the Bureau of National Affairs, Washington, D.C., in their service ENVIRONMENT REPORTER.

VICTOR J. YANNACONE, JR.
Yannacone and Yannacone

ESTUARINE WATER QUALITY

Introduction

Estuarine water quality is a term used to describe the quality or characteristics of the water in estuaries. Although the term implies quality in a physical-chemical sense, its use has been extended to include also the acceptability of water in a socio-economic sence. The term "water quality," like environmental quality and air quality, became well known terms to the people of the United States and indeed the world in the late 1960's and early 1970's. The term has to do with the quality of the water or the environment wherever it is found and wherever it is used or encountered. The high chemical and bacteriological quality of the water supplies has become an almost matter-of-fact part of any American's life, but the quality of the waters in which man swims has become of greater concern with man's awareness of degradation of the quality of the waters around him.

A thoroughly technical description of water quality would require several volumes of text the size of this encyclopedia to cover the physical, chemical, and biological characteristics of water and how these characteristics change in different environments, how they interact, and how they influence the many ways water is used by plants, animals, and especially man. This would be true even though this article is limited to estuaries which are among the most complex natural systems known and which feel the impact of man perhaps more than any other natural aquatic system. Suffice it to say that estuarine water quality will be examined in a broad way only, and the reader is referred to the books and articles cited in the bibliography for further discussions on the topics covered herein.

Use context for water quality

Quality of water may be discussed most usefully in the context of water use. That is, for certain uses of water, whether they be recreation, drinking, transportation, or some other use, some level of water quality is required or desired for that particular use. Some uses, such as drinking, will require a much higher level of water quality than will another use such as swimming, and swimming may require a higher quality of water than transportation on water. The important point is that for desired uses of bodies or areas of water, certain levels of water quality are desired, and if the quality of the water for that particular area is not equal to the quality of water desired or needed to support that use, the use may not be sustained. The concept of use applies not only to man's direct uses of the water, but applies also to biological uses of bodies of water such as fish spawning grounds, shrimp nursery areas, and so forth. Indeed, the history of setting levels of desired water quality for particular uses has shown that following the setting of levels for water quality for swimming, levels of water quality were set for protecting and enhancing the survival of fish in streams. Levels of dissolved oxygen in streams, which are in most state and federal water quality standards, are there to protect fish in those streams.

As uses for bodies of water become more numerous, a competition for use of the water begins to develop. Uses such as transportation or navigation, swimming, fishing, dredging, fish and shellfish nursery areas, and perhaps other uses are not uncommon competing uses for a body of water. The quality of water required to support each of these uses is different as noted above, and because of this, some uses may or may not be sustained, depending on which use is the most "beneficial" of that particular body of water.

The Federal Water Quality Act of 1965 stated that water quality standards were to be adopted by all the states by June, 1967, and in preparation of these standards public hearings were to be held to determine the desired uses of all the waters of the state which were under federal jurisdiction. Although many states had already determined uses of their waters, particularly the streams, this was the first time that a nation-wide effort was made to determine desired uses of waters and to set water quality standards for them. The hearings, the water quality standards developed, and the subsequent implementation and enforcement of the standards have shown the very real problems which arise when competing uses of the water resource become very strong and intense. One or two particular uses become dominant, and the water quality for a particular use is set to meet that use. Other uses which require a higher level of water quality may or may not be sustained, while levels of water quality required for other uses may be more than adequately met. However, there are some uses of these waters for which water quality standards need not be set, for these uses are not dependent on water quality for sustenance. These uses however, may degrade the quality of the water for other uses; these uses include waste disposal, dredging, transportion of oil or hazardous materials which may spill, and other similar uses.

Estuaries

The estuary is one of those bodies of water which is the focus of intense competing uses. Estuaries comprise one of the most important resources of any country for the support of such uses as navigation, recreation, nursery and resting grounds for fowl and wildlife, nursery and spawning aras for fish and shellfish, and even sites for urban growth and the consequences or by-products of urban growth.

Estuaries are semi-enclosed coastal bodies of water having a free connection with the open sea and within which the sea water is measurably diluted with fresh water derived from land drainage (Pritchard, 1967). Along the coasts of the United States alone some 45,832 square miles of estuarine waters exist. Of this total, 17,058 square miles are found along the Atlantic Coast, along the Pacific Coast south of Alaska but including the Pacific Islands, 14,353 square miles exist, 2,760 square miles are found along the Coast of Alaska, and 11,661 square miles are found in the Gulf of Mexico and Caribbean Islands (National Estuarine Study, 1971). Of the total, less than 30 per cent is water less than six feet deep, vulnerable to filling, as well as specially productive of fish, shellfish and wildlife. At least 6.8 per cent of the latter have been obliterated through filling, most in the last 50 years (Stroud, 1971).

Development of estuary shorelines indicate some of the uses of the estuaries. Of the 89,571 statute miles of tidal shoreline in the United States estuaries, some 17,853 miles can be described as recreation shoreline, that is, accessible and useful for recreational pursuits. Of this shoreline, 16,559 miles are privately owned and 1,294 miles are publicly owned; however, only 770 miles may be considered recreation areas. Marine transportation terminal facilities are users of a portion of the shoreline estuaries. In 1966, there were 1,626 marine terminals providing deep water berths in 132 ports on the Atlantic, Gulf, and Pacific Coasts. Industries use estuarine waters for cooling and return a heated effluent. Industries and cities use estuaries as disposal sites for their wastes. With a third of the United States population located in the estuarine zone, the impact of man on estuaries must necessarily be quite high (National Estuarine Study, 1971).

Biological uses of estuaries are also quite high. It has been estimated that nearly 63 per cent of the commercial catch on the Atlantic Coast is made up of species of fish believed to be estuarine dependent. Assuming that this applies equally to the combined catches by foreign nationals as to the U.S. domestic catch, the fisheries yield to the U.S. Atlantic continental shelf and present levels of development of the fishery is equivalent to about 535 pounds per acre of estuaries (McHugh, 1966). Similar but somewhat smaller estimates have been made for the Gulf of Mexico estuaries based on commercial catches in the Gulf of Mexico and for the Chesapeake Bay estuary based on catches within the estuary itself (McHugh, 1967).

Factors influencing estuarine water quality

What are the factors that control the quality of waters in estuaries? The predominant factors are the hydraulic characteristics of the estuary, the inputs or sources of materials which make up elements of the quality of the water, and the sinks present in estuaries—those physical, chemical and biological phenomena which cause materials in the water to change in concentration or to be altered chemically to a different form than when originally introduced.

The hydraulic regime of an estuary is dependent upon three particular factors: the physiography of the bay—its size, area in relation to volume, depth, and shoreline development; the amount and seasonality of river inflow to the estuary; and the wind and tidal mixing which takes place in the estuary on each tidal excursion. The latter factor is dependent upon the tidal range, the configuration of the entrance to the estuary, the volume of the river inflow and the periodicity of the tides. The impact of sources of material to an estuary are dependent upon the character and amount of the material and the location in the estuary where the material enters. Materials which enter with the river inflow will very likely reach broad areas in the estuary due to mixing within the esturay and the fact that the material will pass through the estuary on its way to the ocean. On the other hand, material discharged near the mouth of the estuary will travel only a short distance into the estuary and most likely be transported out of the estuary rather quickly. This generalization does not apply to all estuaries, particularly those which are strongly stratified. This type of estuary will be discussed in more detail later on. The size of the sinks for materials in estuaries is dependent on the conservative or non-conservative nature of the material, that is, whether the material can be broken down into by-products or whether it remains in essentially the same form throughout its history within the estuary. Conservative and nonconservative materials may both be removed from the water column due to flocculation or sedimentation within the estuary, in which case materials may become part of the bottom sediments and lost from the water column unless the sediments are disturbed.

Because of the intimate tie between water quality and estuarine hydraulics, they will be examined below as well as the sources and sinks for materials within estuaries, both natural and man-made materials, before discussing water quality—estuarine use interactions.

Estuarine Hydraulics

A spectrum of hydraulic types may occur or exist in an estuary. These may range from the situation in an estuary in which the river flow dominates to the estuary in which the river flow is negligible and the hydraulic regime is dependent on the tidal mixing. Naturally, in a river flow dominated estuary, the water quality in the estuary is most similar to that of the river, whereas in the tidal mixing dominated estuary, its water quality is more like that of the off-shore waters. The other factor greatly influencing the hydraulic regime in an estuary is the physiography of the estuary which is very greatly dependent on the origin of the estuary and the subsequent events which have taken place in geologic time to modify the original shape of the estuary.

Origins of estuaries

From a geomorphological standpoint, there are four primary subdivisions of estuaries: (1) drowned river valleys; (2) fjord type estuaries; (3) bar-built estuaries; and (4) estuaries produced by tectonic processes (Pritchard, 1967). Each of these types of estuaries is characterized by the fact that at some point in geologic time, it has been inundated with ocean water due to the rise in the sea level. During the last glacial stage, sea level was about 450 feet below its present level, and the shorelines of the continent were at or near the present continental slopes. Within the last 50,000 years, the sea level has risen from that stage to the present with the last changes in sea level occuring about 3,000 years ago (Russell, 1967).

As the name implies, drowned river valley estuaries are river valleys found along a coastline with a relatively wide coastal plain, which were inundated with ocean water as the sea level rose. The Chesapeake Bay is a prime example of this type of estuary. During the last glacial period, the Susquehana River reached the ocean about 180 kilometers seaward of the present shoreline; the York River and the other rivers now entering the bay to the north of the York were then tributaries of the Susquehana River. The rise in sea level flooded the valleys of these rivers to form the present Chesapeake Bay system. These drowned river valleys, or as they are more commonly called, coastal plain estuaries, extend up river to a point approximately where the floor of the river rises above sea level. This is also the point at which a major change in water quality occurs from the ocean and estuary type water quality to that of the river. This geographical point may be downstream from parts of the river which are still influenced by the oscillation of the tidal currents.

The fjord type estuary is that formed by glaciers. These estuaries are generally U-shaped in cross-section, and they frequently have a shallow sill formed by terminal glacial deposits at their mouths. The basins inside these sills are often quite deep reaching depths of some 300 or 400 meters. Most fjords have rivers entering at the head and exhibit estuary features in the upper water layers. The sill depths in Norwegian fjords are often so shallow that the estuarine features develop from the surface to the sill depth while the deeper basin waters remain stagnant for prolonged periods.

Bar-built estuaries are those formed in an off-shore area where sand is deposited as a sand island and sand spit build above sea level and extend between the headlands in a chain broken by one or more inlets. Such bays often occur in areas where the land is emerging geologically. The area enclosed by the barrier beaches is generally parallel to the coast line. Frequently, more than one river enters the estuary, though the total drainage area feeding a bar-built estuary is seldom large. The lower valleys of such rivers have frequently been drowned by the rising sea level, and hence the bar-built estuary might be considered as a composite system, part being an outer embayment partially enclosed by the barrier beaches, and part being a drowned river valley or valleys. Tidal action is usually considerably reduced in such estuaries. These systems are usually shallow, and the wind provides the important mixing mechanism (Pritchard, 1967). Several of the North Carolina estuaries and most of those along the Texas Gulf Coast are examples of this type of estuary.

Estuaries produced by tectonic processes are those formed by faulting or by local subsidence, and they usually have an excess supply of fresh-water inflow. San Francisco Bay is an example of such an estuary.

Circulation in estuaries

Other than the physiography of estuaries, the dominant physical processes associated with movement of water and mixing in an estuary are the wind, tides, and the inflow of river water. The composite actions of these processes produce a variable interaction or interfacing of fresh water from the river and salt water from the ocean. Because these two sources of water have very different densities, the less dense fresh river water will tend to float on top of the dense salt water, and the extent that the two types of water mix is dependent on the strength of the mixing mechanisms. In an estuary with no tides or wind and a steady river inflow, and no friction between the fresh and salt water layers, the fresh water inflow would ride on top of the salt water from sea level in the estuary or river bed to the ocean. Because in a real system friction is present, the fresh water will force salt water some distance downstream from the sea level point in the river and the interface between the salt and fresh water layers will tilt downward in the upstream direction in a wedge shape. The friction between the layers will also cause an exchange or water from one layer to another, generally from the salt water, or "salt wedge," to the fresh water. The amount of exange depends strongly on the mixing mechanisms, wind, tides, and river inflow.

In a wind dominated estuary, wind provides most of the energy for moving and mixing the water. In a tide dominated estuary, turbulence associated with the tidal currents results in mixing between the salt and fresh water, which in turn produces the density gradients associated with the non-tidal circulation pattern. In a river dominated estuary, such as the Mississippi River estuary, water movement is predominantly related to riverflow and mixing is caused mostly by the breaking of unstable interfacial waves at the upper boundary between the fresh river water and the salt water from the ocean.

In an estuary in which a salt wedge occurs distinctly the river flow almost completely dominates the circulation. The salt water extends as a wedge into the river and the interface between the fresh and salt water slopes slightly downward in the upstream direction. The steep density gradient at the interface, amounting to a discontinuity, reduces the turbulence and mixing to a very low level. The effect of the Coriolis force causes the interface to slope downward to the right in the northern hemisphere, looking towards the sea. In the moderately stratified estuary, the dominant mixing agent is turbulence caused by tidal action, rather than velocity sheer at the interface between the salt water and overlying fresh water layer, as in the previous case. With a tide of moderate amplitude, random water movements at all depths occur and turbulent eddies transport fresh water downward and carry salt water upward, in contrast to the dominantly upward advection of salt across the interface which constitutes the vertical flux of salt in the river dominated estuary. The result of this two way mixing is that the salt content of both the upper and the lower layers increases toward the sea. At any given point the bottom layer is always higher in salt content than the lower layer. The boundary between the seaward flowing upward layer and the counter flowing lower layer occurs with a middepth region of relatively rapid increase in salt content with depth, compared to the vertical gradient in either the upper or the lower layers. This type of mixing contributes a greater volume of salt water to the upper, seaward flowing layer than in the salt wedge estuary. The rate of flow in the upper layer of the moderately stratified estuary is therefore much greater in volume than in the highly stratified estuary, necessitating a correspondingly larger compensating up estuary flow in the lower layer.

When tidal mixing is sufficiently vigorous, the vertical salinity stratification breaks down, and the estuary approaches true vertical homogeneity. The type of circulation which exists in a vertically homogeneous system depends upon the amount of lateral homogeneity. Owing to the Coriolis force, the water on the right of an observer looking seaward may be lower in salinity than the water to his left. A cyclonic circulation pattern is developed, with fresher, seaward flowing water concentrated to the right of center and a compensating up estuary flow of higher salinity water to the left of center. Although a vertical salinity gradient is absent in a vertically homogeneous estuary, vertical transfer of salt is not lacking. There is also a strong lateral transfer of salt which represents the dominant circulation pattern in this type of estuary.

Certain vertically homogeneous estuaries, particularly those which are relatively deep and narrow, do not exhibit these cyclonic circulation patterns. The direction of water movement is symmetrical about the longitudinal axis, and fluctuations in velocity are related to the tides and the net flow averaged over several tidal cycles is directed seaward at all depths. There is a tendency for salt to be driven out of the estuary by the action of this advective process. There must be a compensating non-advective longitudinal flux of salt directed toward the head of the estuary (Prichard, 1967).

It is very important to note that the quality or character of the water at any point in the stratified, partially stratified, or vertically homogeneous estuary will be strongly correlated with the salinity content of the water. For example, the high salinity, bottom water in a stratified estuary will have a quality much like that of the offshore ocean water. The water at the geographical midpoint of a vertically homogeneous estuary will be a mixture of river and ocean waters. Also, materials introduced into an estuary will be influenced at any point in time or space by the circulation patterns in the estuary. Estuaries which have not felt man's influence either in the estuarine zone or the fresh waters which flow into it have biological systems adapted to whatever water quality patterns exist. Since these water quality patterns are strongly influenced by the circulation system in the estuary, a change in the circulation patterns and/or introduction or removal of materials will have a beneficial or deleterious effect on the biota of the estuary depending on the extent of the change or the nature of the material introduction or removal. Thus, it is important to examine the circulation patterns of estuaries as well as the materials introduced or removed to understand the water quality and biota of the estuary and the uses which may be made of the estuary.

Estuarine circulation models

Numerous attempts have been made to model the hydraulic processes which occur in estuaries. Originally, these models were developed to determine circulation modifications which might occur because of physical modifications to the estuary. These models have been extended in recent years to include constituents of water and the prediction of their transport and fate in estuaries. One of the first types of models developed for estuaries was the hydraulic model. This type of model is a physical representation of an estuary on a small scale. Such models are usually distorted in the vertical direction so that water depth may be represented on a larger scale than a lateral dimension. For example, if an estuary were modeled to a scale of 1:100, the width of the estuary it it were 10 miles, would be 0.1 miles in the model, but the depth of the water if it were 10 feet would be 0.01 feet which would be not much more than a film of water in the model. To avoid this situation wich would make the model unuseable, the vertical scale is reduced to a lesser extent than the horizontal scale. The vertical scale may be 1:50 or 1:10 of the horizontal scale such that the 10 foot depth of water mentioned above would be 0.2 feet or 1 foot respectively for the scale factors. While the hydraulic models are capable of representing tidal currents, momentum entrainment, and gravitational circulation, they are not able to represent local currents and turbulent eddies. For this reason, there is considerable distortion of diffusive processes in the physical model that makes its utility in quantitative concentration distribution studies dubious (Ward and Espey, 1971). From a qualitative standpoint, the physical model possesses an excellent demonstration capability for the visualization of flow patterns in resultant concentration distributions, and this capability should not be underrated.

The other types of models developed for estuaries are mathematical models which may be intended to model tidal currents, net advective movement, or tidal stage in an estuary, or they may be intended to model the transport of salt in the water or other materials in various chemical forms. Such models may be three dimensional to represent transport of material down the estuary as well as laterally and vertically, or they may be two dimensional to represent transport of material down the estuary and laterally or vertically, or they may be one dimensional to represent transport of material down the estuary. Because the complexity of developing and solving mathematical models decreases as the number of dimensions included are decreased, the one dimensional model has received the widest attention in terms of development and use. This type of model is most advantageously applied to linear type estuaries, that is, estuaries which have little or linear variation in cross sectional area and depth with distance down the estuary. Examples of such models include the model of the Thames River in England, the Delaware River in New Jersey, the Potomac River in Maryland, and the Hudson River in New York.

Recently two dimensional models have been developed. These models almost invariably assume that vertical stratification does not occur in the water column and that lateral stratification does occur. Such models are most appropriately applied to estuaries with large surface areas and shallow waters. Such models have been developed for San Francisco Bay in California, Galveston Bay in Texas (Ward and Espey, 1971), Jamaica Bay in New York (Leendertse, 1970), and most recently for Matagorda and San Antonio Bays along the Texas Coast (Masch and Brandes, 1971). While the need for three dimensional models has been recognized for salt wedge type and moderately stratified estuaries, the numerical solutions for three dimensional mathematical models of real estuaries are at the present time still beyond practical attainment.

Water quality models are usually derived from the following basic three dimensional continuity equation:

$$\frac{\partial C}{\partial t} = \frac{\partial}{\partial x}\left(E_x \frac{\partial C}{\partial x}\right) + \frac{\partial}{\partial y}\left(E_y \frac{\partial C}{\partial y}\right) + \frac{\partial z}{\partial}\left(E_z \frac{\partial C}{\partial z}\right) - \frac{\partial}{\partial x}(uC) - \frac{\partial}{\partial y}(vC) - \frac{\partial}{\partial z}(wc) \pm \sum S$$

where:

E = dispersion coefficient along each of the three axes, x, y, and z

u, v, w = velocity in x, y, or z direction respectively

S = source or sink for material

C = concentration of material

This equation expresses a relationship between the flux of mass caused by circulation and mixing in the estuary and the sources and sinks of mass. In the one-dimensional form in which the assumptions have been made that concentrations of some material are of homogeneous concentration laterally and vertically (the y and z directions respectively) and that the net transport of the material through the estuary is of concern, then the following equation has been used (O'Connor and Thomann, 1971):

$$\frac{\partial C}{\partial t} = \frac{1}{A}\frac{\partial}{\partial x}\left(EA\frac{\partial C}{\partial x}\right) - \frac{1}{A}\frac{\partial}{\partial x}QC \pm S$$

where

A = cross sectional area of estuary
Q = freshwater inflow
E = dispersion coefficient in x direction

and other terms are the same as above. Such models may be used to determine changes in material concentration with time for materials whose rate of entry to the estuary and/or loss from the estuary in a sink are steady or only slightly variable. A further assumption is to select the steady state situation, the condition in which the concentration of the material does not change with time. For this condition $\partial C/\partial t$ in the above equation is set to zero and the equation solved.

In all these equations, the source and sink terms become zero for conservative substances. For non-conservative substances, reactions that take place may usually be represented by first-order kinetics, i.e. the rate of reaction is proportional to the concentration of the material. In some cases the reaction term defines the fundamental reaction mechanism, whereas in other uses it is an empirical approximation to the phenomenon.

While it is concluded that as a general rule the modeling of the hydrodynamic transport of a constituent in an estuary is much further advanced than the modeling of its reaction kinetics, the most commonly unsatisfactory aspect of present water quality models is the specification of the source and sink terms. Many of the physical-chemical processes affecting the concentration of parameters lack adequate formulation. These include sedimenation and deposits of particulate matter, non-linear reaction kinetics, surface exchange of gaseous constituents, and chemical and biological reactions. Modeling of the relation of water quality and estuarine biota is not well advanced at all. Models of phytoplankton production, of nitrogen cycling, and of gross ecological parameters have been attempted, but presently these are at best experimental.

Sources and Sinks

In addition to the hydraulic regime of an estuary, the other factors which have great influence on the water quality of estuaries are the sources and sinks of the materials. The circulation patterns and water movements in estuaries will dictate the distribution of fresh and salt water in the estuary. Superimposed on this distribution is another pattern made up of materials introduced by sources and lost to sinks.

Sources

Sources for the materials which are found in the waters of estuaries include two major sources, river inflow and ocean water inflow. The concentrations (or ranges) of selected chemical constituents of fresh and ocean waters are given in Table 1.

TABLE 1

Quality of Fresh and Ocean Water

(Concentration Units are mg/l)

Constituent	Fresh	Ocean
Chloride	1.0–1,000	18,980
Sodium	1.0–1,000	10,560
Sulfate	1.0–1,000	2,560
Magnesium	1.0–1,000	1,272
Calcium	1.0–1,000	400
Potassium	0.01–10.0	380
Bicarbonate	1.0–1,000	142
Carbonate	0.01–10.0	
Bromide	0.0001–0.1	65
Strontium	0.01–10.0	13
Boron	0.01–10.0	4.6
Fluoride	0.01–10.0	1.4
Aluminum	0.0001–0.1	0.16–1.9
Iodide	0.0001–0.1	0.05
Silicate	1.0–1,000	0.04–8.6
Nitrogen	0.01–10.0	0.03–0.9
Zinc	0.0001–0.10	0.005–0.014
Lead	0.0001–0.1	0.004–0.005
Iron	0.01–10.0	0.002–0.02
Phosphorus	0.0001–0.1	0.001–0.10
Mercury		0.0003

Fresh waters may have large ranges of concentrations of materials due to the soil and rock composition of the lands which they drain. These ranges are quite different from those of oceanic waters. In fresh waters, calcium is usually the most abundant cation and sulfate is the most abundant anion although carbonate may also be quite high in concentration. In sea water, on the other hand, chloride is the most abundant constituent and anion followed by sulfate and bicarbonate. Sodium and magnesium constitute the majority of the cations. Depending on the relative balance of river inflow and the intrusion of seawater brought in by tidal action, the quality of the water in the estuary assumes a composition in proportion to the two

TABLE 2

Quality of Domestic Wastes

(Concentration Units are mg/l)

Constituent	Strong	Medium	Weak	Mass discharge coefficients (lbs/person/day)
Solids, total	1,000	500	200	
Volatile	700	350	120	
Fixed	300	150	80	
Suspended, total	500	300	100	0.23
Volatile	400	250	70	
Fixed	100	50	30	
Dissolved, total	500	200	100	
Volatile	300	100	50	
Fixed	200	100	50	
BOD (5-day, 20 °C)	300	200	100	0.2
Dissolved oxygen	0	0	0	
Nitrogen, total	86	50	25	0.06
Organic	35	20	10	0.035
Ammonia	50	30	15	
Nitrites (NO_2)	0.10	0.05	0	0.025 (Ammonia, Nitrites, Nitrates)
Nitrates (NO_3)	0.40	0.20	0.10	
Chlorides	175	100	15	
Alkalinity	200	100	50	
Fats	40	20	0	0.03
Phosphorus				0.012
Flow				135 gal/person/day

Source: Water Encyclopedia, 1971.

sources. However, the location of constituents from the various sources either laterally in the estuary or vertically in the water column is highly dependent on the circulation patterns existing in the estuary which were discussed earlier.

Although in relation to river and tidal flows, direct precipitation is a small hydraulic input to an estuary, its water quality cannot be ignored. In shallow bays with little river inflow and a restricted opening to the ocean such as bar-built estuaries, rainfall directly on the estuary may be an important source of fresh water.

Waste discharges may exert a dominant influence on the water quality of estuaries depending on the amount of material discharged and its character. Because urbanization typically occurs around estuaries, waste discharges are usually directed to the estuaries since they are the most convenient waste disposal site. Domestic wastes, wastes derived from municipalities and ultimately humans, contains large amounts of organic and nutrient (nitrogen, phosphorus, trace) materials. Some typical concentration values are given in Table 2. The relative amounts of material discharged to estuaries or other bodies of water may be estimated by knowing the population served by a sewerage system and mass discharge coefficients. These coefficients indicate the amount of material discharged per person per day. Such coefficients are also given in Table 2.

Industrial wastes also reach estuaries either as a direct discharge to the estuary, as spills form vessels carrying materials to or from the industries, as the result of dredging activities, as the discharge of heated effluents from power plants and heated effluents from nuclear power plants which also carry radioactive materials, and in other forms. Most industrial activities involve the use of and/or the disposal of water. Such waters usually contain the by-products of the industrial process and are characteristic of the process. For example in manufacturing steel, a certain amount of water is required for cooling and washing purposes. The amount of water used to produce a ton of steel by

TABLE 3

Industrial Wastes Characteristics

Industry (Unit)	Flow (gal/unit)	BOD (lb/unit)	Suspended solids (lb/unit)
Brewery (barrel)	370	1.9	1.03
Cannery (case)	75	0.7	0.8
Dairy (100 lb)			
Butter	410–1,350	0.34–1.68	—
Cheese	1,290–2,310	0.45–3.0	—
Ice Cream	620–1,200	0	—
Milk	200–500	0.05–0.26	—
Meat Packing (100 lb live wt. killed)	1,294	14.4	—
Poultry Proc. (1000 birds)	10,400	26.2	—
Petrol. Ref. (barrel)	100	0.1	—
Pulp & Paper (ton)			
Bleached kraft	45,000	120	170
Bleached sulfite	55,000	330	100
Steel mill (ingot ton)	10,000	—	100
Tannery (100 lb)	660	6.2	13.0
Textile (lb cloth)			
Wool	63	0.30	—
Cotton	38	0.16	0.07
Synthetics	~15	0.07–0.10	0.02–0.07

Source: Malina, 1970.

a given process is fairly consistent and the quality of the water resulting from the process is also fairly consistent. Table 3 lists various types of industrial activities, the amount of water used in the activity, and the pounds of oxygen required to oxidize the organic material in the wastewater as well as the pounds of suspended solids produced in making some unit amount of product.

Another source of waste material frequently overlooked in the past is urban and rural runoff. Urban runoff may consist of storm water runoff from the streets and gutters which is routed to the nearest waterway by storm water pipes, or it may consist of a mixture of storm water runoff and sanitary sewage in what is called a combined sewer system. Such systems are typical of older cities in the United States and other countries which built one pipe to carry both sanitary wastes and storm water wastes. Because of economics the pipe could be built just so big, and at that size it could carry all the domestic wastes during dry weather but only a portion of the wastes during wet weather. During a large storm, the pipe would fill up and the flow would have to be diverted to a waterway to insure that backups did not occur in this sewerage system. For such drainage systems, each large rainfall results in a certain amount of material being washed into the nearest waterway. The amount of material produced is highly dependent on the drainage system itself, on the use of land in the drainage area, and for some systems on the stage of the tide when the overflow occurs. The quality of such wastewater is given in Table 4 as well as some mass discharge coefficients. A more detailed discussion of this type of wastewater is given in another part of this Encyclopedia (see *Urban Runoff*).

Rural runoff, though less innocuous than urban runoff, cannot be disregarded as a wastewater source. Mass discharge coefficients relating to quality of the runoff water to the land use and amount of rainfall are given in Table 5. Miscellaneous waste discharges occur into estuaries which cannot be quantified in the way done for other types of waste discharges. These include such waste as oil spills, spills of toxic or hazardous materials, waste from houseboats or larger vessels with quarters for crew for living purposes, dredging spoils, heat from power or nuclear plants, and other sources.

Another source of material in estuaries is biological recycling. Although biological recycling may also be considered as a sink and will be discussed as such later, recycling of material is ex-

TABLE 4

Combined Sewer Overflow and Urban Storm Runoff Characteristics

Combined Sewer Overflows

Constituent	Flow wtd. conc. (mg/l)†	Mass discharge (lb/acre/in. runoff)††	Coefficients (lb/acre/year)‡
BOD	150	30	125
SS	325	70	600
VSS	200	45	180
HEM	50	10	38
TKN	12	3	13
NH_3–N	5		
NO_3–N	0.2		
Total P	5.0	1	3
Total coli*	5×10^5		
Fecal coli*	0.5×10^5		

Urban Runoff

Constituent	Flow wtd. conc. (mg/l)	Mass discharge (lb/acre/year)†	Mass discharge (lb/acre/in. runoff)†	Coefficients (lb/acre/in. runoff)‡
BOD	18	33	2.0	4.1
SS	77	730	45.6	17.4
VSS	25	160	10.0	5.7
HEM	2.8			
TKN	1.2	9	0.56	0.26
TP	0.3	2.5	0.05	0.07
TC*	9×10^3			
FC*	3×10^3			

* Units are MPN/ml
† Data from Weibel, *et al.*, 1964
†† Data from Spring Creek Project, 1970
‡ Data from San Francisco, 1967

TABLE 5

Quality of Rural Runoff

Source	Total nitrogen (lb/acre/year)	Total phosphorus (lb/acre/year)
Forest runoff	1.3–3.0	0.3–0.8
Surface irrigation	2.45–24.0	0.92–3.88
return flow	38.0–166.0	2.5–8.1
Subsurface irrigation		
return flow		
Urban runoff	8.5	0.8

Source: Fruh, 1968

tremely important in transforming waste materials from man-made waste discharges (or waste discharges from other lower animals or even dead organisms) to a chemical form in which it may be used again by the biological system. For example, organic material in a domestic waste discharge is oxidized at least partially by bacteria to carbon dioxide and water. Carbon dioxide is a necessary constituent for the growth of plants in conjunction with light. The process of photosynthesis using light reduces carbon dioxide to an organic material which is incorporated in the tissues of the plant. Since plants comprise the basic food stuff for all organisms, the recycling of carbon from the organic form to the inorganic form as carbon dioxide is vitally important. Similarly, nitrogen which may be in an organic form in a waste discharge is oxidized to an inorganic form which may then be utilized by plants for growth. In both cases the organic form acts as a sink while concurrently the inorganic form becomes the source.

Sinks

There are several types of water quality sinks in estuaries. Water withdrawal from an estuary is one type of sink; withdrawals may be made for industrial uses such as salt production or for cooling purposes. In the latter case, some water may be returned to the estuary in a heated condition, but the water withdrawn which is lost in the cooling process as evaporation is lost to the estuary. Sedimentation becomes another sink as molecules of material which were in the water column sorb onto the particulate matter which settles out and become part of the bottom sediment. Unless the bottom sediment is disturbed, eroded or dredged, this material is essentially lost to the water column. Precipitation may also occur to tie up several minerals which then settle out on the bottom and as with sedimentation are lost from the water column. Precipitation quite commonly occurs in estuaries which have little fresh water inflow but a high evaporation rate. These estuaries are known as hypersaline estuaries because the salinity content rises to levels above that of normal sea water. In the Laguna Madre Bay of Texas, salinity levels reach two or three times that of normal sea water. In this bay, crystals of gypsum are found on the shores.

Another type of sink which is tied intimately to the biological system of the estuary is the degradation of materials, that is, the change from one chemical form to another by biological action. Above, the process of oxidation of organic material was mentioned. This is one form of oxidation in which organic material is oxidized to smaller molecular material which is further oxidized to carbon dioxide and water as the ultimate inorganic products. Oxygen is consumed in the process and is lost to the water. Another oxidation process is nitrification. In this process, nitrogen in the organic form is oxidized to ammonia which is then further oxidized by the bacterium, *Nitrosomonus* to nitrite and further to nitrate by the bacterium, *Nitrobacter*. Under low oxygen or anaerobic conditions, the nitrate or nitrate may be reduced by bacteria to elemental nitrogen gas which may then be removed from the system according to the solubility of the nitrogen gas.

Quality Needed for Estuary Use Sustenance

Now that some of the hydraulic, biological, and man-made mechanisms which influence water quality in estuaries have been described, the use concept for estuaries and the water quality needed to sustain uses should be considered.

The recent San Francisco Bay-Delta Study Report (1969) listed some of the uses which can be made of estuaries, and these are given in Table 6. These uses are quite varied and in some cases would be competing uses if applied to the same part of an estuary. For combinations of uses for a given area in an estuary, usually one use will require a water quality higher than the other uses, and this use will dictate the water quality needed in that particular area. Some of the uses listed are not really dependent on water quality such as shipping, unless the quality is particularly adverse (very acid water or large floating material), and other uses which may or may not be dependent on water quality may contribute to its degradation which prevent other uses from taking place. It is the latter use category, which includes waste discharges, which must be managed most closely.

The general philosophy of estuary use in particular or resource use in general is that a range of uses may exist bounded by two extreme views. These views are: (1) a resource may be used indescriminantly without regart to the consequences (e.g. total consumption, contamination, etc.) of that use; and (2) a resource should be preserved

TABLE 6

Beneficial Uses to be Protected in San Francisco Bay

Municipal Supply (seasonal)
Industrial Supply
Boiler
Cooling
Rinsing
Processing
Agricultural Supply (some seasonally)
Irrigation
Livestock
Fish and Wildlife Propagation and Aquatic Growth
Fish habitat, migration, spawning
Shrimp and crab habitat
Shellfish habitat
Waterfowl habitat
Mammal rookery
Kelp
Commercial Fishing and Shellfishing
Recreation
Swimming, waterskiing, skindiving, picnicking
Surfacing, beachcombing, sunbathing
Pleasure boating
Fishing
Shellfishing
Hunting
Enjoyment of Esthetic Values
Tidepool and Marine Life Study
Navigation
Resource Extraction

with no interaction with man. Both views, despite being widely held today by some industrialists and environmentalists respectively, are recognized as incompatible, and the concept of sustained use, the designation of some use or uses for a resource area and management of the system to support that use best, is favored. As mentioned at the beginning of this article, the designation of uses for estuarine areas in all states through public hearings in 1967 was a recognition of the sustained use concept. However, subsequent attempts to manage waste inputs to estuarine systems and the estuarine systems themselves have shown how crude our management techniques are. The difficulties lie in the technical aspects of management (the interrelatedness of water quality with estuarine circulation and sources and sinks), the economic aspects (the cost of management systems directly and indirectly), and the socio-political-legal aspects (the interrelatedness of social desires and political decisions and the lack of adequate pollution control laws).

For each of the uses listed in Table 6, a list of physical, chemical, and/or biological properties or constituents may be compiled. This list would indicate the levels of the properties or constituents which have to be provided or cannot be exceeded if that use is to be sustained. For example, the National Technical Advisory Committee (1968) recommended the following criteria for waters used for primary contact (e.g. swimming):

Constituent	Level
Mandatory Criteria	
Fecal coliform	$\leq$ log mean of 200/100 ml for 5 samples in 30 day period, 10 per cent of samples in 30 day period $\leq$400/100 ml.
pH	6.5–8.3 except due to natural variations, $5.0 \leq pH \leq 9.0$ is required at all times.
Desirable Criteria	
Clarity	Secchi Disc depth $\geq$4 ft.
Temperature	$\leq$30 °C except due to natural conditions

Other uses will require a different list of constituents and different levels to be achieved. These constituent lists may change as they are found to be unimportant or are found to be needed, and the levels may change as new knowledge is gained about how the levels more specifically affect the desired use.

Unfortunately at this time, we know very little about the bases for many levels set. For example, we still lack good epidemiological evidence relating fecal or total coliform organism concentrations in swimming areas to disease. A larger gap, however, is the criteria list for the fish and wildlife of estuaries and the organisms they feed on. For these organisms the best information available pertains to temperature, dissolved oxygen, pH and salinity. Beyond these criteria, the levels set for other materials such as toxic substances are suspect. Gross toxicity criteria like that developed by Armstrong, Storrs, and Pearson (1970) for San Francisco Bay are possible ways to circumvent our present lack of knowledge.

The reader may obtain more specific information about water quality in general and estuarine water quality in particular by referring to the bibliography references and to other related material. Of special interest might be the reports of studies on major estuaries such as San Francisco Bay, Galveston Bay, Delaware Bay, and Jamaica Bay. These studies include not only the theory of water quality but the practical techniques for and applications of its management.

REFERENCES

1. Armstrong, N. E., P. N. Storrs, and E. A. Pearson, Proc. Fifth Intern. Wat. Poll. Res. Conf. (1970).
2. "Characterization and Treatment of Combined Sewer Overflows," Report to Federal Water Pollution Control Admin. by City of San Francisco. Prepared by Engineering-Science, Inc., November (1967).
3. Fruh, E. G., in *Advances in Water Quality Improvement*, Eds. E. F. Gloyna and W. W. Eckenfelder, Jr., Univ. of Texas Press, Austin (1968).
4. Geyer, R. A., in *The Biological Significance of Estuaries*, Sport Fishing Institute, March (1971).
5. Leendertse, J. J., RM-6230-RC, The Rand Corporation, February (1970).
6. Malina, J. F., Jr., Report No. CRWR 69, Center for Research in Water Resources, University of Texas, October (1970).
7. Masch, F. D., and R. J. Brandes, Tech. Rept. HYD 12-7102, Hydraulic Engineering Laboratory, University of Texas, August (1971).
8. McHugh, J. L., in *A Symposium on Estuarine Fisheries*, Special Publ. No. 3, 133–154. Amer. Fish. Soc., Wash., D.C. (1966).
9. McHugh, 1967, in *Estuaries*, Publ. No. 83, AAAS (1967).
10. "National Estuarine Pollution Study," U.S. Gov't. Printing Office (1971).
11. O'Connor, D. J., and R. V. Thomann, in *Estuarine Modeling: An Assessment*, Environmental Protection Agency, Project 16070 DZV, February (1971).
12. Pritchard, D. R., in *Estuaries*, Publ. No. 83, AAAS (1967).
13. Russell, R. J., in *Estuaries*, Publ. No. 83, AAAS (1967).
14. "San Francisco Bay-Delta Water Quality Control Program," Final Report to the State of California by Kaiser Engineers and Associated Firms, March (1969).
15. Stroud, R. H., in *The Biological Significance of Estuaries*, Sport Fishing Institute, March (1971).
16. Ward, G. H., and W. H. Espey, in *Estuarine Modeling: An Assessment*, Environmental Protection Agency, Project 16070 DZV, February (1971).
17. "Water Quality Criteria," Report of the National Technical Advisory Committee of the Federal Water Pollution Control Administration to the Department of the Interior, April 1(968).
18. Weibel, S. R., R. J. Anderson, and R. L. Woodward, *J. Wat. Poll. Contr.* **36** (7), 914 (1964).
19. *Water Encyclopedia*, David K. Todd, Water Information Center, Port Washington, N.Y. (1970).

NEAL E. ARMSTRONG
University of Texas

EUTROPHICATION

Introduction

For a considerable time, scientists have been aware of the natural aging of lakes, a process so slow that it was considered immeasurable within the lifetime of human beings. In recent years, however, that portion of the nutrient enrichment or eutrophication of these and other natural bodies of water contributed by man-made sources have become a matter of concern. Many bodies of water of late have exhibited biological nuisances such as dense algal and aquatic weed growths whereas in the past they supported only incidental populations of these plants.

Excessive nutrients are most often blamed in the scientific literature for the creation of the plant nuisances. Among the nutrients, dominant roles have been assigned by most researchers to nitrogen and phosphorus. These elements can be found in natural waters, in soils, in plants and animals, and in precipitation. Man-made sources for these nutrients are in domestic wastes and often in industrial wastes.

This chapter concerns itself with the nature of algae, the environmental factors affecting their growth, the nature of the entrophication problem (sources, relative quantities of nutrients contributed by these sources, threshold limits for the growth of aquatic plants), and various techniques for the removal of those nutrients usually associated with the eutrophication problem.

The Physical Nature of Algae

Most bodies of water which can be considered eutrophic exhibit various predominant forms of algae at different times of the year. Algae that are important to investigators concerned with the eutrophication problem may be classified into four groups which exclude all but a few miscellaneous forms. The four groups are:

1. Blue-green algae (Myxophyceae)
2. Green algae (Chlorophyceae)
3. Diatoms (Bacillariophyceae)
4. Pigmented flagellates (Chrysophyceae, Euglenophyceae)

The basis for this classification is the color of the organism. Blue-green and green algae are self descriptive, whereas diatoms are brown or greenish-brown. Pigmented flagellates can be brown or green. They possess whip-like appendages called flagella, which permit them to move about in the water. It is not inferred by the above list that all algae are restricted to these colors. Rhodophyceae, for example, which are primarily marine algae, are brilliant red.

Aquatic biologists and phytologists do not agree on the number of divisions that should be established to identify algae. Some authorities use as many as nine divisions while others use seven, five and four. Nevertheless, the four divisions as suggested by Palmer[1] will be used as they are adequate for the ensuing discussions.

Blue-green Algae

Blue-green algae as a group are most abundant in the early fall at a temperature range of 70 to 80 degrees F. Data obtained from water sources in the southwestern and southcentral United States indicate that for this section of the country maximum growth occurs at the end of February and throughout much of April, May and June. When blue-green algae become predominant, it frequently indicates that the water has been enriched with organic matter, or that previously there had been a superabundance of diatoms. Blue-green algae are quite buoyant due to the oil globules and gas bubbles which they may contain. For this and other reasons they live near the surface of the water often producing offensive mats or blankets. Since these algae are never flagellated, they are not considered swimmers although a few, such as oscillatoria and spirulina, are able to creep or crawl by body movements. Some of the common blue-green algae are anabaena, aphanizomenon, rivularia, gomphosphaeria and desmonema.

Green Algae

Green algae are most abundant in mid-summer at a temperature range of 60 to 80 degrees F. For water bodies in the southwestern and southcentral

United States, maximum growth occurs during the first half of September with little variation throughout the remainder of the year. Like the blue-green algae, green algae usually contain oil globules and gas bubbles which contribute to the reasons why they are found near the surface of the water. Green algae are distinguished by their green color which comes from the presence of chlorophyll in their cells. Many of the green algae are flagellates and due to their swimming ability they are frequently found in rapidly moving streams. Some of the common green algae are chlorella, spyrogyra, chlosterium, hydrodictyon, nitella, staurastrum and tribonema.

Diatoms

Diatoms are usually most prevalent during the cooler months but thrive over the wide temperature range of from 35 to 75 degrees F. For water bodies in the southwestern and southcentral United States, diatoms thrive best in May, September and October with the maximum growth observed in mid-October. It is generally recognized that many diatoms will continue to flourish during the winter months, often under the ice. The reason for the increase in growth twice a year is due to the spring and fall overturn, in which food in the form of carbon, nitrates, ammonia, silica and mineral matter, is brought to the surface where there is more oxygen and a greater intensity of light.[2]

Diatoms live most abundantly near the surface, but unlike the buoyant green and blue-green algae, they may be found at almost any depth and even in the bottom mud. Diatoms may grow as a brownish coating on the stems and leaves of aquatic plants, and in some cases they grow along with or in direct association with other algae. In rapidly moving streams they may coat the bottom rocks and debris with a slimy brownish matrix which is extremely slippery. Lastly, diatoms are always single-celled and non-flagellated.

Pigmented Flagellates

No classification of algae has caused more disagreement than that of the pigmented flagellates. The difficulty arises from the fact that they possess the protozoan characteristics of being able to swim by means of flagella, and the algae characteristic of utilizing green chlorophyll in association with photosynthesis. Thus they could be listed either as swimming or flagellated algae in the plant kingdom, or as pigmented ot photosynthetic protozoa in the animal kingdom. One of many attempts to resolve this problem has been the proposal to lump together all one-celled algae and all protozoa under the name "Protista".[3] This method, however, has not met with general acceptance.[4] For the sanitary engineer the motility of the organism is of lesser importance than its ability to produce oxygen. The pigmentation characteristic associated with green chlorophyll and oxygen production is sufficient criteria for separating these organisms into a class by themselves. Thus a distinction is made between pigmented flagellates (algae) and non-pigmented flagellates (protozoa).

Pigmented flagellates are more abundant in the spring than at any time of the year although there is generally considerable variation among the individual species. Apparently flagellates are dependent on more than temperature.[5] They are found at all depths, but usually are more prevalent below the surface of the water than at the surface.

For present purposes pigmented flagellates can be divided into two groups: euglenophyceae which are grass-green in color and chrysophyceae which are golden-brown.

Euglenophyceae are usually found in small pools rich in organic matter, whereas chrysophyceae are usually found in waters that are reasonably pure.

Some of the more common pigmented flagellates are euglena, ceratium, mallomonas, chlamydomonas, cryptomonas, glenodinium, peridinium, synura and volvox.

Motility

Of additional value in the classification of algea are their means of motility. Three categories have been established, namely:

Nekton — algae that move by means of flagella.
Plankton — algae that have no means of motility.
Benthic algae — algae that attach themselves to a fixed object.

Nekton

Nekton are the most active algae and are often referred to as "swimmers." Due to their activity they use more energy and in turn release more oxygen during the daylight hours. Their cells are supplied with one, two, or more flagella which extend outward from the front, side or back of the cell. These flagella enable the organisms to move about freely in the aquatic environment and to seek food which, in the case of turbulent water, is constantly changing in location.

In general nekton have the most complex structure of the three categories and come nearest to being simple animals. Nekton are the predominant algae found in swiftly-moving rivers and streams.

According to Lackey, results of tests performed on rivers of the Ohio River Valley show that certain nekton are the only algae that provide reliable clear-cut responses to the presence of pollution and thus are true indicator organisms.[6] Five flagellates have been singled out on the genus level as being common and easily recognized. They are (1) cryptomonas, (2) mallomonas, (3) synura, (4) uroglenopsis, and (5) dinobryon. Dinobryon is perhaps the most easily recognized due to its unique shape which resembles a shaft of wheat. Samples taken from several rivers indicate that these algae react adversely to the presence of sewage and are found in abundance only in clean water. Unfortunately not all experts agree on what constitutes clean water and what algae serve as indicator organisms. Patrick states that the "healthy" portion of a stream contains primarily diatoms and green algae.[7] Rafter states that the absence of large amounts of blue-green algae is an indicator of clean water.[8] Palmer lists 46 species which have been selected as being representative of "clean-water algae," and these consist of diatoms, flagellates, green algae, blue-green algae and red algae.[9] In addition Palmer lists 47 species of algae condensed from a list of 500 prepared from reports of more than 50 workers, as being representative of "polluted-water algae." These consist of blue-green algae, green algae, diatoms and flagellates.[10]

Plankton

Plankton are free-floating algae which are most commonly found in lakes and ponds, although they are by no means limited to these waters. Most species are unicellular; however, they tend to become colonial when their numbers increase, as in the formation of a heavy concentrated growth known as a "bloom." An arbitrary definition of a bloom is that concentration of plankton that equals or exceeds 500 individual organisms per ml. of raw water.[11] Blooms usually show a predominance of blue-green algae although algae from other classes can also form blooms.[11]

An algae bloom often becomes sufficiently dense as to be readily visible on or near the water surface, and its presence usually indicates that a rich supply of nutrients is available. Other environmental factors also may stimulate the formation of blooms, and a bloom of the same organism in two bodies of water may or may not result from identical favorable environmental conditions.[12] These growths are extremely undesirable in bodies of water, in general, and in potential water supply sources in particular for the following reasons:

1) The are very unsightly.
2) They interfere with recreational pursuits.
3) When the water becomes turbulent, fragments of the mat become detached and may enter a water treatment system clogging screens and filters.
4) When the algae die (as a result of seasonal changes or the use of algicides), decomposition occurs, resulting in foul tastes and odors.
5) They may act as a barrier to the penetration of oxygen into the water which may result in fish kills.
6) They may reduce the dissolved oxygen in the water through decay or respiration within the bloom.
7) Some blooms release toxic substances that are capable of killing fish and wild life.[13]
8) They may cause discoloration of the water.
9) They attract waterfowl which contribute to the pollution of the water.[14]

Some of the common blue-green algae that form blooms are anabaena, aphanizomenon, oscillatoria, chlorella and hydrodictyon. Synedra and cyclotella are common diatoms that form blooms and synura, euglena and chlamydomonas are common flagellates that form blooms. Filamentous green plankton, such as spirogyra, cladophora and zygnema form a dense floating mat or "blanket" on the surface when the density of the bloom becomes sufficient to reduce the intensity of solar light below the surface. Like blooms, these blankets are undesirable, and for the same reasons cited earlier. However, in addition, blankets also serve as a breeding place for gnats and midge flies, and after storms they may wash up on the shores where they become offensive. In many cases hydrogen sulfide and other gases which are able to spread disagreeable odors considerable distances through the air are liberated. In large amounts, hydrogen sulfide has been known to seriously discolor the paint on lakeside dwellings.[15]

Benthic Algae

Benthic algae are those algae which grow in close association with a supply of food. That is, they seek out an aquatic environment where nutrients are adequate, then attach themselves to a convenient stationary object such as a submerged twig or rock. They may be found in quiet ponds and lakes or in fast-moving rivers and streams. In some cases they break away from their attachments and form unsightly surface mats, or they may re-attach themselves somewhere else. Chlamydomonas is such an organism, where in one growth phase it may be found attached to a fixed object, and in another phase it may be dispersed throughout the water.[16]

Benthic algae include diatoms, blue-green algae, green algae and a few species of red fresh-water algae. None of the pigmented flagellates are benthic. Most attached algae grow as a cluster of branched or unbranched filaments or tubes and are fastened at one end to some object by means of an anchoring device. Others take the shape of a green felt-like mat (gomphonema), a thin green film or layer (phytoconis), or a soft fragile tube (tetraspora). Some of the most common benthic algae are cladophora, chara, nitella, ulothrix, cymbella, vaucheria and gomphonema.

Environmental Factors Affecting Growth of Algae

The effects of certain environmental factors on the growth of the aforementioned forms of algae have been fairly well defined. The most important parameters to be considered in the growth pattern are light intensity, temperature, pH and nutritional requirements.

Light Intensity

Light is essential to all organisms which carry on photosynthesis; however, requirements or tolerance levels differ greatly with the organism. For example, terrestrial species of vaucheria grow equally well in fully-illumniated soil and densely-shaded soil, while a number of blue-green algae grow only in shaded habitats.[17] In addition, some algae are unable to endure in the absence of sunlight caused by several consecutive cloudy days,[18] whereas certain submerged algae are unable to withstand exposure to full sunlight.[17] Thus, an algae kill may be noted during a drought where shallow water prevents depth-dwelling algae from escaping the intensity of the sunlight penetration. Often muddy rivers are virtually algae-free due to the lack of penetration of the sunlight, the Missouri and the Mississippi being two such rivers.

The distribution of algae at the various depths in a body of water is directly correlated with the intensity of illumination at the respective depths. This distribution would be difficult to express in general terms when dealing with algae on the species level. In addition, the depths at which these species would be found would change with such variables as growth phase of the organism, temperature and the absorptive and reflective characteristics of the water. It can be stated, however, that certain fresh-water red algae and blue-green algae are found only at considerable depths and that some diatoms exist in the bottom mud. In the most general terms it can also be stated that algae are found at all levels, but most commonly near the surface.

The vertical distribution may also be related to the division of light rays into various spectral colors. This division varies with the concentration of dissolved color material, plankton and particulate matter, with the seasons, and with the depth.[19] In colored water the violet-blue end of the spectrum is absorbed more readily. As depth increases light rays divide differently with greater absorption occurring at the red end of the spectrum.[17]

The depth to which light penetrates has a direct influence on photosynthetic activity. The seasonal variation in this light and the resulting availability of certain dominant wave lengths may be the reason for fluctuations in the composition of the algal population from spring to fall. Much more work is needed in this area.

Temperature

In general, temperature is not the key factor in determining the nature of the algal flora.[17] Most species are able to grow and reproduce if other environmental conditions are favorable. According to Patrick,[20] however, the above statement is not true in the case of diatoms, where temperature changes are more important than any other environmental factor in influencing their rate of growth. Additional work in this area by Cairns indicates that certain diatoms grow best only at a specific temperature, and that at some temperatures they will not grow at all.[21]

Most algae are not affected by minor changes in pH brought about by the seasonal variations, growths of carbon-dioxide producing organisms, etc. Large changes, such as would be caused by the introduction of industrial wastes or acid mine waters, will greatly affect algae, usually causing a decrease in population.

The majority of algae thrive when the pH is near 7.0. Some blue-green algae prefer high pH$^{\underline{S}}$. Anacystis and coccochloris are found at about pH 10.0 with little or no growth below pH 8.0.[22] Other algae such as eugleny mutabilis, cryptomonas erosa and ulothrix zonata prefer low pH$^{\underline{S}}$.[23]

Nutritional Requirements and Toxic Elements for Algae

Calcium

Calcium is not an essential element for most algae, although some cannot develop without it.[17]

Calcium and magnesium

As bicarbonates they are a supplemental supply of carbon dioxide for photosynthesis. This accounts for the greater abundance of algae in hard-water lakes than in soft-water lakes.[17]

Iron

Most algae grow best when the ferric oxide content of the water is between 0.2 to 2.0 mg per liter. Above 5 mg per liter there is a toxic effect unless it is overcome by the buffering action of organic compounds or calcium salts.[17] Certain diatoms (eunotia and pinnularia) are found in iron-rich water.[24] Effluent from steel mills may be toxic to most algae if the resulting iron concentration exceeds the toxic limitation.

Copper

Copper is extremely toxic to algae in the range of 0.1 to 3.0 ppm as copper sulfate; the sulfate form being used as an algicide.[25] Some algae are able to tolerate large amounts of copper ion and are considered copper-sulfate resistant. Protococcus, for example, is not destroyed by 10 ppm of copper sulfate.[25]

Phenol

At a concentration of up to 1.9 mg per liter, phenol apparently has no toxic effect on diatoms.[26]

Nitrates, phosphates and ammonia

These are essential food elements necessary for growth. Nitrogen may be obtained from nitrates, nitrites or simple ammonia compounds. The primary source of these nutrients is from sewage treatment plant effluents, although nitrogen may be derived from the atmosphere, land runoff, etc. (See section on EUTROPHICATION.) In general as little as 0.3 to 0.015 ppm of nitrates and phosphates will produce blooms of certain species of algae, other conditions being favorable.[27]

Oil

Streams polluted with oil are usually low in algae. One variety of diatom may be dominant in such waters.[24]

Salinity

Increases in salinity up to about one per cent do not affect the algae population.[28] Significant increases, such as caused by salt-brine wastes, may destroy most of the algae present, however. Certain fresh-water algae may become adapted to water with slowly-increasing salinity.

Hydrogen sulfide

At a concentration of 3.9 ppm, hydrogen sulfide is toxic to most diatoms.[29] Some resistant species are achnanthes affinis, cymbella ventricosa, hantzschia amphioxys and nitzschia palea.

Silica

Silica is necessary for the growth of diatoms whose cell wall is composed of silica. Presently no limits have been determined (to the author's knowledge).

Vitamins

Several vitamins in small quantities are a requisite to growth in certain species of algae. Chief among these vitamins are vitamin B-12, thiamine and biotin.[30] These vitamins are supplied by bottom deposits, soil runoff and by the metabolites produced by other organisms.

Micronutrients

Substances such as manganese, zinc, molybdenum, vanadium, boron, chlorine, cobalt, etc. are generally present in water in the small concentrations sufficient for plant growth.

Carbon dioxide

Carbon dioxide is necessary for respiration. If it is deficient, algae may remove carbon dioxide from the atmosphere.

Chlorine

Chlorine is toxic to most algae and is used as an algicide in the range of 0.3 to 3.0 ppm[31]. It is used as an algicide in the treatment plant and distribution system. Some algae, cosmarium for example, are resistant to chlorine.[32] Protococcus, which is resistant to copper sulfate, is killed by 1 ppm of chlorine.[33] Therefore, algae resistant to the copper ion not may be resistant to the chlorine ion and vice versa.

Calcium hydroxide (lime)

An excess of lime in the water, as may be introduced during pH adjustment for coagulation, results in the death of certain algae. Five ppm of lime with an exposure of 48 hours has been lethal to melosira, nitzschia and certain protozoa and crustacea.[31]

The Eutrophication Problem

Of the factors previously discussed which promote the growth of algae, that factor which man has altered is the nutrient concentrations in many of the natural waterways.

In simplest terms, eutrophication is the enrichment of waters by nutrients from natural or man-made sources. Of the many nutrients which are added to the waters by man-made sources, nitrogen and phosphorous are most often cited by researchers as being the key nutrients responsible for the promotion of algae growth. In nearly all cases when the nitrogen and phosphorus level of a body of water increases, there will be a corresponding increase in the growth of algae and aquatic plants. Such growth greatly speeds up the aging process whereby organic matter invades and gradually displaces the water until eventually a swamp or marsh is formed. Unfortunately, the process of eutrophication is often difficult to reverse in bodies of water such as large lakes where the flushing or replacement time for these waters can be of the order of years.

The following sections provide the relative magnitude of natural and man-made sources of nutrient material associated with plant growth.

Sources of Nutrients

While it is recognized that certain algae require a number of chemical elements for growth, it is also known that algae can absorb essential as well as superfluous or even toxic elements. Although every essential element must be present in algae, this does not mean that every element is essential. On the other hand, the absence of certain nutrients elements will prevent growth. Nutrients may be classified as (1) "absolute nutrients," which are those which cannot be replaced by other nutrients, (2) "normal nutrients," which are the nutrients contained in the cell during active growth, and (3) "optimum nutrients," which are the nutrients permitting maximum growth.[1] It may also be well to assign a broad meaning to the work "nutrient" and define it as anything that can be used as a source of energy for the promotion of growth or for the repair of tissue.[2]

In evaluating the effects of nutrients on algae, care must be exercised to consider the interaction between nutrients and other physical, chemical or biological conditions. Rapid growth of algae may be stimulated more by factors of sunlight, temperature, pH, etc., than by an abundance of nutrient material. Tests performed with nitzschia chlosterium, in order to study the interaction of environmental factors, showed that two identical cultures of the organism, when supplied with a reduced nutrient level, had a lower optimum light intensity and optimum temperature for maximum growth.[3] Thus light intensity and temperature data should accompanay data on nutrient concentration and growth rate.

Of all the possible nutrients, only nitrogen and phosphorus have been studied in depth both in the field and in the laboratory. This is because of the relative difficulties associated with the study, analysis and measurement of trace elements, compounded by the minute impurities present in the reagents and distilled water. In addition, nitrates and phosphates have a long history of use in agricultural fertilizers where determination of their properties have been essential to their economical use.

The following are the most common sources of nitrogen and phosphorus in bodies of water:

(1) Rainfall—Based on experimental data, it has been found that rainwater contains between 0.16 and 1.06 ppm of nitrate nitrogen and between 0.04 to 1.70 ppm of ammonia nitrogen.[4] Computations based on the nitrogen content of rainwater show that for Lake Mendota, Wisconsin, approximately 90,000 pounds of nitrogen are available each year as a result of rainfall.[5] Thus it can be seen that rainfall plays a significant role in building up the nitrogen content of a lake or reservoir especially if the surface area is large.

An examination of the phosphorus content of rainwater of different countries shows that a number of concentrations may exist ranging from 0.10 ppm to as little as an unmeasurable trace, the latter reported in the Lake Superior region of the United States.[6] In view of the wide variation in the determinations, little can be stated at present regarding the degree of phosphorus build-up in impoundments resulting from rainwater.

(2) Groundwater—Studies conducted on subsurface inflows to Green Lake, Washington, show that this water contains approximately 0.3 ppm of phosphorus.[7] Other reports, however, claim that the amount of phosphorus in groundwater is negligible.[8]

Investigations into the nitrate content of groundwater produced variable results; however, it can be stated that 1.0 ppm is a reasonable figure.[9] The results of the above studies on both nitrogen and phosphorus can be summarized by stating that groundwater should not be discounted as a possible source of nutrients and that quantitative values should be obtained for the specific locality in question.

(3) Urban Runoff—Urban runoff contains storm water drainage, overflow from private disposal systems, organic and inorganic debris from paved and grassed areas, fertilizers from lawns, leaves, etc. In view of the variable concentration of the above material, precise figures cannot be obtained on the phosphorus or nitrate content that would be meaningful for all areas.

Studies conducted in 1959 and 1960 by Sylvester on storm water from Seattle street gutters showed the following nutrients.

Organic nitrogen—up to 9.0 ppm
Nitrate nitrogen—up to 2.8 ppm
Phosphorus—up to 0.78 ppm soluble and up to 1.4 ppm total

(4) Rural Runoff—Rural runoff for the purposes of definition may be considered as runoff from sparsely-populated, wooded areas with little or no land devoted to agriculture. Investigations by Sylvester showed that the phosphorus content of drainage from three such areas in the state of Washington contained 0.74, 0.77 and 0.32 lbs./acre/year, or a total concentration of 0.069 ppm.[10] The corresponding nitrate nitrogen concentration and organic nitrogen concentration amounted to 0.130 and 0.074 ppm, respectively.

(5) Agricultural Runoff—Agricultural runoff is one of the largest sources of enrichment material and may be derived from two sources—wastes from farm animals and the use of nitrogen and phosphorus-containing fertilizers.

Farm-animal wastes add both large quantities and high concentrations of nutrients to adjacent streams and rivers. The large concentrations are due primarily to the practice of herding animals in relatively confined areas. A comparison on the nutrient value of animal wastes and human wastes has been made in a study by the President's Science Advisory Committee.[11] According to the findings, a cow generates the waste equivalent of 16.4 humans, a hog produces as much as 1.9 humans and a chicken produces as much as 0.14 humans.

The use of chemical fertilizers in the United States has grown almost 250 per cent in the decade from 1953 to 1963.[12] In 1964 the use of phosphorus-containing fertilizers and the use of nitrogen-containing fertilizers reached approximately 1.5 and 4.4 million tons, respectively, per year.[13] Most all of this fertilizer is distributed to soil already high in natural-occurring nitrogen. When nitrogen fertilizer and natural soil nitrogen combine, there is a great increase in crop production, but also a greater opportunity for loss of this nitrogen in runoff. This loss will increase if the fertilizer is not properly applied, if it is not completely utilized by the crops, if the crops have a short growing season (the land being non-productive for a time), if the land is irrigated, and if the land is sloped.

The addition of nitrogen-bearing fertilizers also increases the quantity of mineral elements in the soil runoff which are necessary for the growth of aquatic plants and algae. When applied, the nitrogen in the fertilizer is converted into nitric acid which combines with the minerals in the soil, such as calcium and potassium, rendering them soluble and subject to leaching.[14]

(6) Industrial Wastes—The nutrient content of industrial waste effluents is variable and depends entirely upon the nature, location and size of the industry. In some cases the effluents are totally free of nitrogen and phosphorus.

The meatpacking industry is one of the chief producers of nitrogen-bearing wastes. The greatest producer of phosphate-bearing wastes is most likely the phosphate-manufacturing industry itself.[15] Most phosphate production in the United States is concentrated in Florida and as a result many severe localized problems of eutrophication have resulted in that state.[15]

Fuel processing industries and petroleum refineries discharge vast quantities of nitrogen into the atmosphere both in the gaseous state and solid state as particulate matter. This nitrogen is then washed from the atmosphere by the rain and carried back to earth. In 1964, the 500 billion tons of coal used in the United States released about 7.5 million tons of nitrogen into the armosphere, most of which has returned to be combined with the soil.[15] This greatly exceeds the use of nitrogen in the form of fertilizers which, as previously stated, amounted to 4.4 million tons for that year. Thus, through the atmosphere we are bringing more nitrogen into the soil than we are taking out, and much of this excess ultimately gets washed out into our waterways.

(7) Municipal Water Treatment—The water treatment plants themselves are to a degree responsible for adding to the eutrophication problem as approximately 33 per cent of the municipal water in the United States is treated with compounds containing phosphorus or nitrogen.[16] Some of the commonly used nutrient-bearing chemicals or compounds are ammonia (in the use of chloramines) organic polyelectrolytes, inorganic coagulant aids, sodium hexametaphosphate, sodium tripolyphosphate, and sodium pyrophosphate.

(8) Waterfowl—It has been estimated that wild ducks contribute 12.8 pounds of total nitrogen/acre/year and 5.6 pounds of total phosphorus/acre/year to reservoirs or lakes.[17] A number of studies have been conducted on waterfowl, but it may be concluded that, although there may be some bearing on localized eutrophication, in general the overall effect is negligible.[15]

(9) Domestic Sewage Effluent—Undoubtedly the greatest contributor toward the eutrophication of rivers and lakes is the discharge from sewage treatment plants. Conventionally treated domestic sewage usually contains from 15 to 35 ppm total nitrogen and from 6 to 12 ppm total phosphorus.[18] In addition there are a large number of minerals present in sewage which serve as micronutrients for algae and aquatic plants.

Phosphorus in domestic sewage may be derived from human wastes, waste food (primarily from household garbage-disposal units), and synthetic detergents. Human wastes have been reported in domestic sewage at the rate of 1.4 pounds of phosphorus/capita/year.[19] The largest source of phosphorus, however, is from synthetic detergents which amounts to approximately 2.1 pounds/capita/year.[13] Sawyer indicates that detergent-based phosphorus represents between 50 and 75 per cent of the total phosphorus in domestic sewage.[20] It should be noted that both the use of household garbage-disposal units and detergents is fairly recent, and accordingly they may be considered as contributing strongly to the development of the recently magnified eutrophication problem.

Not all the phosphorus entering a sewage treatment plant will leave the plant since chemical removal does occur during the treatment process. Calcium and metallic salts in large concentrations form insoluble phosphates which are readily removed. Very often phosphate-precipitating agents are present in waters containing industrial wastes, and when these agents are received at the plant, removals in the neighborhood of 60 per cent may be realized.[21]

Nitrogen in domestic sewage is derived from human wastes and from waste food primarily from household garbage-disposal units. Human wastes, the major source of nitrogen, contributes an average of about 11 pounds of nitrogen/capita/year.[22] Some reduction in the nitrogen also takes place during the treatment of the sewage. Many plants treat the sludge anerobically which permits significant release of the nitrogen. In general the removal amounts to between 20 and 50 per cent.[23] The higher percentage of removal occurs when fresh wastes are given complete treatment with no return of sludge nutrients to the effluent.

Eutrophication Studies

In recent years a considerable number of studies have been made on eutrophication and related factors. Most of the studies can be grouped into the following categories:[24]

(1) nutrient content of runoff, rainwater, sewage effluent, bottom mud, etc.

(2) nutrient analysis and physical distribution of nutrients in bodies of water before and/or after enrichment

(3) methemoglobinemia (illness in infants due to drinking high nitrate-content water)

(4) toxicological and other effects on fish of high nitrate/high phosphate-content water

(5) the chemical composition of plants in both eutrophied and non-eutrophied waters

(6) the nutrient values of various fertilizers, manures and other fertilizing elements

(7) the nutrient value of various soils

(8) the effects of eutrophication on aquatic plants, animals and fish

(9) studies on specific algae under either controlled laboratory conditions or in a particular body of water, using artificial or natural environmental conditions

(10) methods for the removal or reduction of nitrogen and phosphorus

(11) nutrient thresholds for growth of algae and aquativ weeds

(12) the effects of eutrophication on the oxygen balance.

Of the above list, only studies conducted in the area of (11) and (12) will be presented below. Work done in regard to (1) has already been presented. The removal or reduction of nitrogen and phosphorus (10) will be discussed separately as part of the subject matter in "CONTROL METHODS".

Nutritional Thresholds for the Growth of Algae

Studies conducted by Chu indicate that for growth on artificial media most planktonic algae flourish if the total nitrogen content ranges from 1.0 to 7.0 ppm and the total phosphorus content ranges from 0.1 to 2.0 ppm.[25] If the nitrogen is reduced below 0.2 ppm and the phosphorus below 0.05 ppm, the growth of algae appears to be inhibited. The same inhibiting effect is created when the nitrogen or phosphorus content is raised above 20.0 ppm. The lower limit of the optimum range of nitrogen varies with the organism and with the type of nitrogen. For ammonia nitrogen the optimum range varies from 0.3 to 5.3 ppm and for nitrate nitrogen the optimum range falls between 0.3 and 0.9 ppm.[26] Below these values the growth rate decreases as the concentration of nitrogen decreases.

Apparently the use of the various forms of nitrogen by algae is not constant throughout the year. Tests conducted at Sanctuary Lake in Pennsylvania (1965) indicate that the order of preference for the three forms of nitrogen—ammonia-nitrogen, nitrate-nitrogen, and nitrite-nitrogen—are defined by three seasonal periods, which are:[27]

Spring	(1) Ammonia nitrogen
	(2) Nitrate nitrogen
	(3) Nitrite nitrogen
Midsummer	(1) Ammonia nitrogen
	(2) Nitrite nitrogen
	(3) Nitrate nitrogen
Fall	(1) Ammonia nitrogen
	(2) Nitrate nitrogen
	(3) Nitrite nitrogen

The amount of nitrogen in the aquatic environment is important to algae because it determines the amount of chlorophyll that may be formed. Too much nitrogen, however, inhibits the formation of chlorophyll and limits growth.

Laboratory studies on algae conducted by Gerloff indicate that of all the nutrients required by algae, only nitrogen, phosphorus and iron may be considered as limiting elements, and of these three, nitrogen exerts the maximum limiting influence. Approximately 5 mg of nitrogen and 0.08 mg of phosphorus were necessary for each 100 mg of algae produced.[28] The corresponding nitrogen/phosphorus ratio is 60 to 1.

Hutchinson cites phosphorus as being the more important element since it is more likely to be deficient.[29] When phosphorus enters a body of water, only about 10 per cent is in the soluble form readily available for algal consumption. During midsummer total phosphate may increase greatly during the formation of algal blooms, while soluble phosphate is undetectable due to rapid absorption by the growing algae. Very often during warm weather these blooms are stimulated by the decomposition and release of soluble phosphates from the bottom sediments, deposited by the expired blooms of previous seasons. Thus when phosphates are added to a lake, only a portion of the phosphates are used in producing blooms. The blooms thrive and consume phosphates for only a short time, and a significant amount finds its way to the bottom sediments where it will be unavailable to further growth of aquatic vegetation.

Prescott examined a number of algae and concluded that most blue-green algae are highly proteinaceous.[30] Aphanizomenon flos-aquae, for example, was shown to contain 62.8 per cent protein. Green algae were found to be less proteinaceous. Spirogyra and cladophora, for example, contain 23.8 and 23.6 per cent respectively. Thus it can be conclused that the nitrogen requirement (for the elaboration of proteins) depends on the class of algae, and that blue-green algae would require more nitrogen than green algae.

Provasoli examined 154 algal species to determine the requirements for organic micronutrients. He found that although 56 species required no vitamins, 90 species were unable to live without vitamins such as B_{12}, thimin and biotin, either alone or in various combinations.[31] He concluded that these vitamins are derived from soil runoff, bottom muds, fungi and bacterial production (B_{12}), and from a natural residual in the water.

Ketchum and Pirson conducted a series of examinations on the inorganic micronutrient requirements of algae and concluded that a number of elements are necessary for growth.[32] No numerical values were assigned to the requirement levels. Those elements shown to be essential were C, H, O, P, H, S, Mg, Ca, Co, Fe, Me, K and Mo. Those elements which may be essential (subject to further study) were Cu, An, B, Si, Va, Na, Sr, and Rb.

In summation, absolute values on nutrient thresholds cannot be set at this time because too little is known regarding the requirements of individual species. It might be stated in general terms, however, that nitrogen and phosphorus are two essential nutrient elements related to the production of blooms, and that if they are present in the neighborhood of 0.2 ppm and 0.05 ppm, respectively, algal growths will increase significantly.

Nutritional Thresholds for the Growth of Aquatic Plants

Studies conducted by Harper and Daniel indicate that submerged aquatic plants contain 12 per cent dry matter of which 1.8 per cent are nitrogen compounds and 0.18 per cent are phosphorus compounds.[33] Hoagland indicates that when the nitrate content of water is high, nitrates may be stored in aquatic plants to be reduced to the usable ammonia nitrogen form as required.[34] Subsequent investigations show that ammonia nitrogen can be substituted for nitrate nitrogen and used directly. Light apparently is not a necessary factor in the reduction of the nitrogen.

Muller conducted a number of experiments on both algae and submerged aquatic plants, and concludes that excessive growths of plants and algae can be avoided in enriched waters if the concentration of nitrate nitrogen is kept below 0.3 ppm, and if the concentration of total nitrogen remains below 0.6 ppm.[35]

Oxygen Balance

Recently, attention has been given to the effect of the intense growths of algae on the oxygen balance of natural waterways. It has been established that the dissolved oxygen concentrations may exhibit wide variation throughout the course of the day. This variation is attributed to the ability of algae to produce oxygen during the daylight hours, whereas they require oxygen for their metabolic processes during the hours of darkness.

In addition, since algae are organic in nature, they exert a biochemical oxygen demand (BOD) on the stream oxygen resources as does other materials which are organic.

Extensive tests were run on the Fox River in Wisconsin by Wisniewski in 1955 and 1956 to examine the influence of algae on the purification capacity on rivers.[36] In the most general terms, the studies indicate that algae increase the B.O.D by adding organic matter capable of aerobic bacterial decomposition and by the respiration of the live cells which utilize oxygen during the absence of light. In the presence of light, algae produce oxygen and as a result may cause a "negative" B.O.D. or a production of oxygen in excess of that required for the normal B.O.D. requirements or aerobic bacterial stabilization. In addition to the above, the following specific conclusions were drawn from the tests:

(1) The oxidation rate resulting from the respiration of live algae was much lower than that obtained by the biological oxidation of the dead algae.

(2) The ultimate B.O.D. of live algae was practically the same as for dead algae.

(3) A linear relationship was found to exist between the five-day B.O.D. of suspended matter and volatile suspended solids concentration.

(4) The B.O.D. increases with increases in suspended solids, the latter consisting largely of algae.

Additional work was done in this area and reported in 1965 by O'Connell and Thomas.[37] They note that the oxygen produced by photosynthetic plants is affected greatly by changes in the availability of light due to cloud cover, turbidity in the water, etc., and therefore it may be too variable to be used as a reliable factor in evaluating the oxygen resources of a river. Another variable may be the loss of oxygen to the atmosphere during the daylight hours, caused by excess oxygen production and localized supersaturation.

An important consideration is the type of photosynthetic plants which are prevalent in a river. According to the above authors, if benthic algae and/or rooted aquatic plants are predominant (in lieu of phytoplankton), there will be little beneficial effect on the oxygen balance. In addition nighttime absorption of oxygen through respiration can seriously reduce daily minimum concentrations of dissolved oxygen. Determination of the effects of the benthic algae oscillatoria along a five-mile stretch of the Truckee River in Nevada indicated that on the average the organism produced 72.5 pounds/acre/day of oxygen through photosynthesis. Oxygen uptake for these same organisms amounted to an average of 65.4 pounds/acre/day.

An examination of the oxygen profiles indicated that the oxygen variation throughout the day ranged from 2 (at night) to 13 (during daylight) parts per million.

It is dissolved oxygen variations such as the above which has been responsible for the disappearance of high quality game fish in many of our natural waterways.

Control Methods to Prevent Eutrophication

There are a number of methods which attempt to limit the amounts of nutrients in bodies of water once the point of eutrophy has been reached. Some of these include dredging and removing bottom sediments with an inert liner, harvesting the algae, fish, aquatic weeds, etc., and diluting the standing water with a water of lower nutrient concentration. Although these methods may have their proper application, if eutrophication is to be decelerated, nutrient removal must start before wastes are permitted to enter the receiving waters.

Regarding the specific nutrients necessary to be removed, most researchers have placed the blame of eutrophication in waters to the inorganic forms of phosphorus and nitrogen. A smaller number of researchers[38,39,40] are claiming that the algae-bacteria symbiosis relationship might be responsible for the rapid growth of blooms and that the amount of algae present in natural waters is in direct balance with the amount of carbon dioxide and/or bicarbonate ions in the waters. They further argue that an external supply of the above elements is necessary for the growth of algae populations. Since neither theory has been proved conclusively to date, the control methods given will be for the removal of nitrogen and phosphorus since it is these nutrients which most researchers lay to the blame of eutrophication and which have been therefore subsequently studied in detail.

Nitrogen Removal

1. Land application

It has been found that nitrogen-bearing waters, when percolated through soil are subject to physical adsorption and biological action which removes the nitrogen in the ammonium form.[41] It appears, however, that the nitrate form of nitrogen remains unaffected. At present this process is only at the theoretical stage, and to the author's knowlege no full-scale application has been attempted. Considerable land area would be involved which may prove a deterrent.

2. Anaerobic denitrification

In this process, the nitrate present in sewage is reduced by denitrifying bacteria to nitrogen and nitrous oxide gases which are allowed to escape into the atmosphere. In order to satisfy the growth and energy requirements of the bacteria, methanol in excess of 25 to 35 per cent must be added as a source of carbon.[42]

The removal efficiency ranges from 60 to 95 per cent. The major advantage to anaerobic denitrification is that there are no waste products requiring disposal. This process is still primarily in the experimental stage at this date.

3. Ammonia stripping

Ammonia stripping is an aeration process modified by first raising the pH of the wastewater above 10.0. At this pH the ammonia nitrogen present is readily liberated as a gas and is absorbed into the atmosphere. Aeration is usually accomplished in a packed tray tower through which air is blown.

This process is suited to raw sewage where most of the nitrogen is either in the ammonia form or may be readily converted to that form. In secondary treatment processes the conversion of ammonia nitrogen to nitrate nitrogen can be retarded by maintaining a high organic loading rate on the secondary process.

Efficiency of nitrogen removal by ammonia stripping is excellent with 80 to 98 per cent reported.[42] There is also the advantage that there are no waste materials which must be disposed of.

Phosphorus Removal

1. Chemical Precipitation

Precipitation of phosphorus in wastewater may be accomplished by the addition of such coagulants as lime, alum, ferric salts and polyelectrolytes either in the primary or secondary state of treatment, or as a separate operation in tertiary treatment. In general, large doses in the order of 200 to 400 ppm of coagulant are required.[42] However, subsequent coagulation and sedimentation may reduce total phosphates to as low as 0.5 ppm, as in the case of lime.[43] Doses of alum of about 100 to 200 ppm have reportedly reduced orthophosphates to less than 1.0 ppm.[44]

Phosphorus removal by chemical coagulation generally is efficient with removals in the order of 80 to 95 per cent reported.[39] Additional benefits are gained in the process by a reduction in B.O.D. to a value of less than 1.0 ppm.[39] Both installation and chemical costs are high, however, and the sludges produced are both voluminous and difficult to dewater.[45]

2. Sorption

Sorption is the process of passing wastewater downward through a column of activated alumina whereby the common form of phosphate are removed by ionic attraction. Regeneration of the media is accomplished by backwashing with sodium hydroxide followed by acidification with nitric acid.

Contrary to alum treatment, this process has the advantage in that sulfate ions are removed and thus the sulfate concentration is not increased. Since no salts are added, the pH and the calcium ion concentration remain unchanged. The process is efficient with more than 99 per cent removal reported.[46] The process should be limited to wastewater with a moderate amount of solids so as not to clog the media.

Removal of Nitrogen and Phosphorus

1. Biological (secondary) treatment

In the secondary method of sewage treatment, bacteria utilize soluble organic materials and transform them into more stable end products. In the process nitrogen and phosphorus are removed from the wastes, utilized to build new cellular material, and the excess is stored within the cell for future use. For each pound of new cellular material produced, assuming the material to be in the form of $C_5H_7NO_2$, about 0.13 pounds of nitrogen and about 0.026 pounds of phosphorus would be removed from the sewage.[42] In the actual operation of this process not all of this nitrogen is removed unless additional energy material in the form of carbohydrates is added.[42] Although it may be possible to eliminate all the nitrogen, a considerable amount of soluble phosphorus may remain, possibly because of the high ratio of phosphorus to nitrogen in sewage, attributable to synthetic detergents. Much of this phosphorus can be removed by absorption on activated sludge floc when it is later separated and removed.

This process offers a 30 to 50 per cent removal of nitrogen and about a 20 to 40 per cent removal of phosphorus without the special addition of carbohydrates.[39,42]

2. Reverse osmosis

The process of reverse osmosis consists of passing wastewater, under pressures as high as 750 psi, through a cellulose acetate membrane. The result is the separation of water and all ions dissolved therein. In actual practice the process has been plagued with difficulties primarily due to membrane fouling or premature failure of the membrane. In addition some nitrate and phosphate ions escape through the membrane.

Removal efficiency ranges from between 65 to 95 per cent (for nitrogen).[42]

3. Electrodialysis

Like reverse osmosis, electrocialysis is a non-selective demineralization process which removes all ions which would include the nitrate and phosphate ions. Essentially an electric current is used in conjunction with a membrane inserted in the line of current flow to separate the cations and anions.

The problems that have developed in the operation of this process include membrane clogging and precipitation of low-solubility salts on the membrane. Acidification of the water and removal of some of the solids prior to treatment has been effective in minimizing these problems, although it adds to the cost.

Removal efficiency ranges from between 30 to 50 per cent (for nitrogen).[42]

4. Ion exchange

In the ion exchange process wastewater is passed through a media bed which removes both anionic phosphorus and anionic nitrogen ions and replaces them with another ion from the media. Regeneration of ion exchangers is commonly accomplished with inexpensive sodium chloride, and frequently the salt is salvaged by recycling the backwash water.

Difficulties in the process may be caused by fouling of the exchange resin due to organic material and reduction in the exchange capacity due to sulfates and other ions. The former may be reduced by removing the organic matter from the resin with sodium hydroxide, hydrochloric acid, methanol and bentonite.[47]

The efficiency and cost of nitrogen and phosphorus removal by ion exchange depends largely on the degree of pretreatment and/or the quality of the water to be treated. Removal of nitrogen range from between 80 to 92 per cent.[42]

A number of ion exchange resins are available for nitrogen removal alone. These include zeolites, strong base anion resins (Amberlite IRA-410) and nuclear sulfonic cation resins (Nalcite HCR and Amberlite IR-120).[48]

5. Algae harvesting

A two-phase process which involves (1) growing algae in special shallow wastewater ponds where they feed on and absorb nutrients, and (2) removing the algae which then contain the nutrients within their systems. Algal predominance will depend upon the type of nutrients available and the concentrations. Frequently the flagellates euglena and chlorella will predominate where the nutrient concentration is high, and filamentous green algae such as spirogyra, vaucheria and ulothrix will predominate where the nutrient concentration is more moderate.[49] The most desirable algae would be those that are large, those that would grow rapidly and those that would require vast quantities of food for energy, such as the swimming algae. In addition to removing nutrients, algae produce oxygen which reduces the B.O.D., and certain flagellates which ingest inorganic solids, are able to stabilize some of the organic material.

One of the major difficulties experienced with this process involves the harvesting procedure. A number of methods have been tried which include screening, settling, centrifuging and chemical coagulation. All have been found to present some form of difficulty, although it appears from the standpoint of performance and economy, the screening method may be the least unsatisfactory.[50] Another problem is that complete nitrogen removal is seldom achieved unless a carbon source, such as carbon dioxide or methanol, is supplied. Still another problem is the need for disposing of a huge, sloppy mass of slimy and odoriferous dead algae.

Pilot studies show that a high-rate continuous-flow process is feasible when light is not limiting, and that orthophosphate concentrations can be reduced 90 per cent to less than 1.0 ppm within 6 to 12 hours.[50] Nitrogen removal is variable with estimates ranging from 40 to 90 per cent efficiency depending upon the feed rate, pond design, and climatic conditions.[39] The major drawbacks to this process aside from those mentioned, are the large land requirements needed and the necessity to rely on climatic conditions. In the latter case, artificial illumination may prove of value depending on power rates in the aera.

Conclusions and Observations

It might be concluded after examining a vast array of material that the bio-physical and bio-chemical factors which affect algae are extremely complex, and it is difficult to predict, with exacting certainty, future events relating to algal growth. The complexity is greatly magnified by the interaction which apparently exists between the numerous factors themselves. When these factors are examined and evaluated, the conclusions reached by one observer are not always in complete agreement, or may disagree entirely, with the conclusions reached by another observer. Commonly, specific organisms may produce different reactions because of their phase of life, seasonal changes, or because of other complex and little understood metabolic functions.

In few areas is there less accord then in the literature written about algae. Authorities vary sharply in their opinions as to classification, physical descriptions, toxicity thresholds, etc. When this disagreement is added to the previously-expressed uncertainties, it can be seen that our current knowledge about algae is deficient and subject to various interpretations. Thus it is evident that there is a need for more work; work to develop new and useful information which will receive universal acceptance and which will clarify and expand our present knowledge.

Acknowledgements

My sincere appreciation is extended to Mr. Robert G. Wieland who helped in the preparation of the manuscript and to the Research Foundation of the Newark College of Engineering for their aid in the typing of the manuscript.

REFERENCES

The Physical Nature of Algae

1. Palmer, Mervin C., *Algae in Water Supplies*, Robert A. Taft Sanitary Engineering Center, Cincinnati, Ohio, Public Health Service Pub. No. 657, 1959, p. 8.
2. Babbitt, H. E., and J. J. Doland, *Water Supply Engineering*, McGraw-Hill, N.Y., 1955, p. 457.
3. Kudo, R. R., *Protozoology*, Chas. C. Thomas Publisher, 3rd Ed., 1950.
4. Palmer, Mervin C., and W. M. Ingram, "Suggested Classification of Algae and Protozoa in Sanitary Science," *Sewage and Industrial Wastes*, Vol. 27, 1955, pp. 1183 to 1188.
5. Silvey, J. K. G., and A. W. Roach, "Studies on Microbiotic Cycles in Surface Waters," *Jour. of the Amer. Water Works Assn.*, Jan. 1964, p. 61.
6. Lackey, James B., "Two Groups of Flagellated Algae Serving as Indicators of Clean Water," *Jour. of the Amer. Water Works Assn.*, June 1941, p. 1000.
7. Patrick, Ruth, "A Proposed Biological Measure of Stream Conditions," Proc. of the 5th Industrial Waste Conference, Purdue, Univ. Engineering Bull. No. 34, 1950, pp. 379–399.
8. Rafter, G. W., *The Microscopial Examination of Potable Water*, Van Nostrand Co., N.Y., 1960.
9. See Reference 1. above, pp. 41 and 42.
10. See Reference 1. above, p. 38.
11. Lackey, J. B., "Plankton as Related to Nuisance Conditions in Surface Waters," Amer. Assn. for the Advancement of Science, 1949, pp. 56–63.
12. Krauss, Robert W., "Fundamental Characteristics of Algal Physiology," Seminar on Algae and Metropolitan Wastes, Robert A. Taft Inst., Cincinnati, Ohio, Apr. 27–29, 1960, p. 41.
13. Ingram, W. M., and G. W. Prescott, "Toxic Fresh-Water Algae," *Amer. Midland Naturalist*, Vol. 52, 1954, pp. 75–87.
14. Derby, R. L., and D. W. Graham, "Control of Aquatic Growths in Reservoirs by Copper Sulfate and Secondary Effects of Such Treatment," *Proc. Amer. Soc. of Civil Engineers*, Vol. 79, 1953.
15. Borchardt, J. A., "Eutrophication, Causes and Effects," *Jour. of the Amer. Water Works Assn.*, June, 1969, pp. 272–275.
16. Thomas, S., and J. W. Burger, "Chlamydomonas in Storage Reservoirs," *Jour. of the Amer. Water Works Assn.*, July, 1933.
17. Smith, Gilbert M., *The Fresh-Water Algae of the United States*, McGraw-Hill, N.Y., 1933, p. 13.
18. Mackenthum, K. M., and W. M. Ingram, *Limnological Aspects of Recreational Lakes*, U.S. Public Health Service Pub. No. 1167, 1964, pp. 18–19.
19. Lauff, George H., "The Role of Limnological Factors in the Availability of Algae Nutrients," Contrib. No. 22, Univ. of Georgia Marine Inst., Sapelo Is., Ga.
20. Patrick, Ruth, "Factors Effecting the Distribution of Diatoms," *Botanical Review* No. 14, 1948, pp. 473–524.
21. Cairns, J., Jr., "Effects of Increased Temperatures on Aquatic Organisms," *Industrial Wastes*, Vol. 1, 1956, pp. 150–152.
22. Gerloff, G. C., *et al.*, "The Mineral Nutrition of Coccochlores Peniocystis," *Amer. Jour. of Botany*, Vol. 37, 1950, pp. 835–840.
23. Lackey, J. B., "Aquatic Life in Waters Polluted by Acid Mine Waste," U.S. Public Health Service Report No. 54, 1939, pp. 740–746.
24. Claassen, P. W., "Biological Studies of Polluted Areas in the Genesee River System," N.Y. State Dept. of Conservation, Supplement to the 16th Annual Report for 1926, 1927.
25. *Water Quality and Treatment*, Amer. Water Works Assn., N.Y., 1951, p. 119.
26. See Reference 1. above, p. 57.
27. Lackey, J. B., "Algal Density as Related to Nutritional Thresholds," Seminar on Algae and Metropolitan Wastes, Robert A. Taft Inst., Cincinnati, Ohio, Apr. 27–29, 1960, p. 59.
28. See Reference 17 above, p. 15 (see reference to Moore, 1917, and Moore and Carter, 1923).
29. See Reference 17 above, p. 57.
30. Provasoli, Luigi, "Micronueri nts and Heterotrophy as Possible Factors in Bloom Production in Natural Waters," Seminar on Algae and Metropolitan Wastes,

Robert A. Taft Inst., Cincinnati, Ohio, Apr. 27–29, 1960, p. 48.
31. See Reference 25 above, pp. 112–113.
32. See Reference 1. above, p. 65.
33. See Reference 25. above, p. 111.

Eutrophication

1. Fruh, Gus E., "Biological Responses to Nutrients—Eutrophication: Problems in Fresh Water," Water Resources Symposium No. 1, Univ. of Texas, Austin, 1968.
2. Ketchum, B. H., "Mineral Nutrition of Phytoplankton," *Annual Review of Plant Physiology*, No. 5, 1954, p. 55.
3. Maddux, W. S., and R. F. Jones, "Some Interactions of Temperature, Light, Intensity and Nutrient Concentration During the Continuous Culture of Nitzschia Closterium and Tetraselmis sp.," *Jour. of Limnology and Oceanography*, Vol. 9, 1964.
4. Carrol, D., "Rainwater as a Chemical Agent of Geological Progress—A Review," Geological Survey Water—Supply 1535-G, 1962.
5. Fruh, Gus E., and G. F. Lee, "The Effect of Eutrophication upon the Water Resources Management of the Yahara River Basin," Report of the Water Chemistry Program, Univ. of Wisconsin, Madison, Wisc. (in press in 1968).
6. Putnam, H. D., and T. A. Olson, "An Investigation of Nutrients in Western Lake Superior," School of Public Health Report, Univ. of Minnesota, 1960.
7. Sylvester, R. O., and G. C. Anderson, "A Lake's Response to its Environment," American Society of Civil Engineers, Sanitary Engineering Div., Vol. 90, 1964.
8. Sawyer, C. N., "Causes, Effects and Control of Aquatic Growths," *Jour. of the Water Pollution Control Fed.*, Vol. 34, 1962, p. 279.
9. Reference 1. above, p. 51.
10. Sylvester, R. O., "An Engineering and Ecological Study for the Rehabilitation of Green Lake," Univ. of Washington, Seattle, Wash., 1960.
11. "Restoring the Quality of the Environment," Report of the Environmental Pollution Panel, President's Science Advisory Committee, The White House, Wash., D.C., 1965.
12. Sauchelli, V., *Chemistry and Technology of Fertilizers*, ACS Monograph Series No. 148, Reinhold Pub. Co., N.Y., 1960.
13. "Sources of Nitrogen and Phosphorus in Water Supplies," Task Group Report, *Jour. of the Amer. Water Works Assn.*, Mar. 1967, p. 347.
14. Andrews, W. R., *The Response of Crops and Soils to Fertilizers and Manures*, State College, Mississippi, 1947.
15. Reference 13. above, p. 349.
16. Reference 13. above, p. 350.
17. Paloumpis, A. A., and W. C. Starrett, "An Ecological Study of Benthic Organisms in Three Illinois River Flood Plain Lakes," *Amer. Midland Naturalist*, Vol. 64, 1906.
18. Hume, N. B., and C. E. Gunnerson, "Characteristics and Effects of Hyperion Effluent," *Jour. of the Water Pollution Control Fed.*, Vol. 34, 1962, p. 15.
19. Hawk, P. B., *et al.*, *Practical Physiological Chemistry*, 13th Ed., The Blakiston Co., Phila., Pa., 1955.
20. Sawyer, C. N., "Problems of Phosphorus in Water Supplies," *Jour. of the Amer. Water Works Assn.*, Nov. 1965, p. 1431.
21. Hurwitz, E., *et al.*, "Phosphates—Their Fate in a Sewage Treatment Plant Waterway System," *Water and Sewage Works*, Vol. 112, 1965, p. 84.
22. "Standard Practice in Separate Sludge Digestion—Committee Report," *Proc. of the Amer. Soc. of Civil Engineers*, Vol. 63, 1937, p. 39.
23. Reference 13. above, p. 348.
24. Mackenthun, Kenneth M., *Nitrogen and Phosphorus in Water, An Annotated Selected Bibliography of Their Biological Effects*, U.S. Public Health Service, 1965.
25. Chu, S. P., "The Influence of the Mineral Composition of the Medium on the Growth of Planktonic Algae—Part I," *Jour. of Ecology*, Vol. 30, No. 2, 1942, pp. 284 to 325.
26. Chu, S. P., "The Influence of the Mineral Composition of the Medium on the Growth of Planktonic Algae—Part II," *Jour. of Ecology*, Vol. 31, No. 2, 1943, pp. 109 to 148.
27. Dugdale, V. A., and R. C. Dugdale, "Nitrogen Metabolism in Lakes III; Tracer Studies of the Assimilation of Inorganic Nitrogen Sources," *Limnology and Oceanography*, Vol. 10, No. 1, pp. 53–57.
28. Gerloff, G. C., and F. Skoog, "Nitrogen as a Limiting Factor for the Growth of Microcystis Aereginosa in Southern Wisconsin Lakes," *Jour. of Ecology*, Vol. 38, No. 4, 1957, pp. 556–561.
29. Hutchinson, G. I., *A Treatise on Limnology*, John Wiley, N.Y., 1957.
30. Precsott, G. W., "Biological Disturbances Resulting from Algal Populations in Standing Waters, The Ecology of Algae," Spec. Pub. No. 2, Pymatuning Laboratory of Field Biology, Univ. of Pittsburgh, 1959, pp. 22–37.
31. Provasoli, Luigi, "Micronutrients and Heterotrophy ad Possible Factors in Bloom Production in Natural Waters," Algae and Metropolitan Wastes, Trans. of the 1960 Seminar, Taft Sanitary Engineering Center, Cincinnati, Ohio.
32. Reference 2. above, pp. 55–74.
33. Harper, H. J., and H. R. Danial, "Chemical Composition of Certain Aquatic Plants," *Botanical Gazette*, Vol. 96, 1939, p. 186.
34. Hoagland, D. R., "Lectures on the Inorganic Nutrition, of Plants," Chronica Botanica Co., Waltham, Mass., 1944.
35. Muller, W., "Nitrogen Content and Pollution of Streams," Gesundheitsing., *Water Pollution Abstracts*, Vol. 28, No. 2, Abs. No. 454, Vol. 74, p. 256.
36. Wisniewski, Theodore F., "Algae and Their Effects on D.O. and B.O.D.," Parts I–III, Water and Sewage Works, June–Aug., 1958.
37. O'Connell, R. L., and N. A. Thomas, "Effects of Benthic Algae on Stream Dissolved Oxygen," *Jour. of the Sanitary Engineering Div.—Amer. Society of Civil Engineers*, June, 1965, pp. 1–16.
38. Lange, Willy, "Effect of Carbohydrates on the Symbiotic

Growth of Planktonic Blue-Green Algae with Bacteria," Nature, Sept., 1967.
39. Kuentael, L. E., "Bacteria, Carbon Dioxide and Algal Blooms," *Jour. of the Water Pollution Control Fed.*, October, 1969.
40. King, D. L., "The Role of Carbon in Eutrophication," *Jour. of the Water Pollution Control Federation*, Dec., 1970.
41. Herbert, C. P., and G. S. Schroepfer, "The Travel of Nitrogen in Soils," *Jour. of the Water Pollution Control Fed.*, Vol 39, No. 1, Jan. 1967.
42. Eliassen, Rolf, *et al.*, "Removal of Nitrogen and Phosphorus," 23rd Purdue Industrial Waste Conference, Purdue Univ., Lafayette, Ind., May 8, 1968.
43. Missingham, G. A., "Occurrence of Phosphorus in Surface Waters and Some Related Problems," *Jour. of the Amer. Water Works Assn.*, Feb., 1967, p. 207.
44. Culp, R. L., "Wastewater Reclamation by Tertiary Treatment," *Jour. of the Water Pollution Control Fed.*, Vol. 35, No. 6, June, 1963.
45. Reference 20. above, p. 1438.
46. Yee, W. C., "Selective Removal of Mixed Phosphates by Activated Alumina," *Jour. of the Amer. Water Works Assn.*, Feb., 1966, p. 246.
47. Bennett, G. E., *et al.*, "Water Reclamation Study Program," Water Quality Control Lab., Stanford Univ., Stanford, Conn., Aug. 1968.
48. Rolich, G. A., "Chemical Methods for the Removal of Nitrogen and Phosphorus from Sewage Treatment Plant Effluents," Algae and Metropolitan Wastes, Trans. of the 1960 Seminar, Taft Sanitary Engineering Center, Cincinnati, Ohio.
49. McKinney, Ross E., *Microbiology for Sanitary Engineers*, McGraw-Hill, N.Y., 1962, p. 243.
50. Bogan, R. H., "The Use of Algae in Removing Nutrients from Domestic Sewage," Algae and Metropolitan Wastes, Trans. of the 1960 Seminar, Taft Sanitary Engineering Center, Cincinnati, Ohio.

ROBERT DRESNACK
Newark College of Engineering

F

FISH ECOLOGY: see POLLUTION EFFECTS ON FISH; THERMAL EFFECTS ON FISH ECOLOGY

FRESH WATER: see EUTROPHICATION, WATER-FRESH

G

GASEOUS POLLUTANT CONTROL

The toxic gases produced during combustion and other chemical processes may be removed by destructive disposal, dispersive dilution or as recoverable side products. The removal path chosen is at the present time motivated primarily by economics, but public pressure and a growing awareness of our environmental problems are beginning to influence the choices.

This section will concern itself with destructive disposal and the various recovery processes, the subject of dispersion being capably handled in the section on *Air Pollution Meteorology* and *Urban Air Pollution Modelling*. The main emphasis will be on principles of gaseous reaction and removal with some description of the equipment used for air pollution abatement. Problems specifically concerned with the automobile can be found under *Automotive Pollution Control*. Although the control principles to be described below are general, it is usually necessary to design equipment for each installation because of variations in physical and chemical properties of effluents. Also, in general, the cost of adding pollution devices to an existing unit will be higher than if they were placed in the original design, because of operational downtime, for example.

Although the majority of effluent material from combustion occurs in the gaseous state, it is important to characterize the total effluent stream for control purposes. For example, the effluent may be condensible at operating temperature (a vapor) or noncondensible (a gas), but is usually a mixture of the two. Particulate matter (solids) and mists (liquids) are often suspended in the gaseous stream; if the particles do not separate upon settling they are called aerosols. The considerations in this section deal with gas or vapor removal only and not with liquid or solid particle removal.

To get some concept of the magnitudes involved in pollution control work it is useful to review the units commonly employed for gas handling. At standard conditions (s.c.) of 0°C and 1 atm, one liter of any gaseous phase contains 0.0446 gram-moles of that gas (this would correspond to 1.285 grams of air). Air pollutant (P) concentrations are usually measured in fractional parts per million by weight or by volume and in milligrams per cubic meter. At s.c. since the amount of clear air is virtually constant compared to the pollutant,

$1 \text{ ppm (wt.)} = 1.285 \times 10^{-6} \text{ gms P/liter}$
$= 1.285 \text{ mg P/m}^3$
or
$1 \text{ ppm (vol.)} = 4.46 \times 10^{-8} \text{ g-mol P/liter}$
$= 4.46 \times 10^{-2} \text{ mg-mol P/m}^3.$

Thus at s.c., 0.15 mg/m^3 of SO_2 is $0.15/1.285 = 0.117$ ppm (wt.) and $(0.15/64)/4.46 \times 10^{-2} = 0.052$ ppm (vol.).

For other temperatures and pressures, gas law corrections may be sued to correct the conversion factors above since the volume will vary from the s.c. value. For an ideal gas the conversion factors above should be multiplied by $273\,P/(T + 273)$ with P, pressure in atmospheres and T, temperature in °C. Gases whose molecular weights are heavier than that of air (vis. >29) will have ppm (wt.) > ppm (vol.) and vice versa. The number of instances of available literature not mentioning whether ppm is by wt. or vol. is unfortunately very high.

Sulfur dioxide, SO_2, and trioxide, SO_3

The sulfur oxides are found in particularly high concentrations in the vicinity of power plants (several ppm (vol.) SO_2), incinerators, paper mills and chemical facilities using sulfuric acid. Coal and oil fired power both will produce SO_2 during the combustion process. A large power station could discharge 1000 tons of sulfur oxides daily. For a 3% coal fired system, SO_2 might typically be 1.5% in the flue gas. The SO_2 molecules combine readily with mists and aerosols, thus compounding the removal problem. The U.S.S.R. standards call for daily SO_2 maximum levels of 0.055 ppm (vol.) at ambient temperatures with 0.180 ppm (vol.) at any one time. The New York City standards are more liberal using 0.10 ppm (vol.), 0.20 ppm and 0.40 ppm for alert, warning and emergency, respectively. These values are useful in control design, establishing targets to shoot for.

Sulfur dioxide reacts slowly with a large excess of oxygen in the presence of sunlight to form the trioxide. Gerhard[1] showed that the process occurs with O_2 at a rate of 0.1–0.2% per hour and Cadle[2] with ozone, O_3, at <0.1% per day. Niepenburg[3] illustrates the effects of oxygen in the waste gas during combustion of oil.

The conversion of SO_2 to SO_3 is believed to be possible at realistic rates because of the presence, on divers surfaces, of Fe_2O_3 which acts as a catalyst. The trioxide has a short lifetime since it readily combines with water vapor in the atmosphere to form sulfuric acid.

TABLE 1

National sulfur oxides air quality standards (1972)

Standard	Concentration µg/cm	ppm	Description
Primary	80	0.03	Annual arithmetic mean
	365	0.14	24 hr maximum not to be exceeded more than once per year
Secondary	60	0.02	Annual arithmetic mean
	260	0.1	24 hr maximum not to be exceeded more than once per year
	1,300	0.5	3 hr maximum not to be exceeded more than once per year

Oxides of nitrogen, NO_x

Oxides of nitrogen are produced in all combustions which take place using air as an oxygen supply and in those chemical industries employing nitric acid. Concentrations of almost 1 ppm (vol.) have been reported at ground level in urban atmospheres. Over half the total NO emissions of about 20 million tons eminate from stationary sources. Nitrogen dioxide, NO_2, is the most common, followed by nitric oxide, NO; nitrous oxide, N_2O, and nitrogen pentoxide, N_2O_5. The oxides are very soluble and will readily combine with mists and aerosols. Ermenc[4] found that at high temperatures nitrogen and oxygen combine to form both NO and NO_2. The yield of NO increases from 0.26% @ 2800°F to 1.75 @ 3800°F. If the temperature is reduced slowly the reverse reaction will take place, but if the products are quenched by rapid heat exchange, the reverse reaction rate becomes small and the oxides remain in the exhaust stream. In a more recent paper[5] several NO formation charts have been prepared as a function of oxygen concentration and temperature. The oxide NO can usually be oxidized to form NO_2 according to

$$2\,NO + O_2 \rightarrow 2\,NO_2$$

The reaction is second order with respect to NO and first order with respect to O_2 and independent of temperature. The rate is slow enough so that nearly complete conversion might take more than 25 hours at ambient temperatures. The dioxide also reacts with oxygen to form ozone, another oxidant capable of complexing with exhaust hydrocarbons forming still other pollutants. Mechanistic models of both the oxide reactions and the photochemical formation of smog[6,7] have been used in estimations of the effects of temperature, concentration and burner design on the pollutant

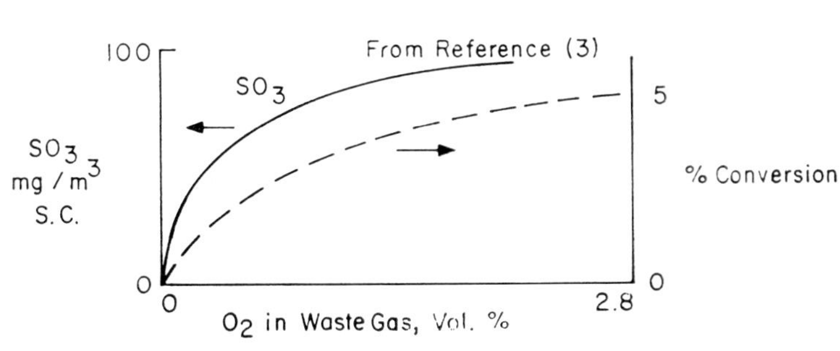

Figure 1

removability of various control devices. The existence of nitrogen trioxide at low concentration in polluted atmospheres is postulated[8] to form by the reaction with ozone

$$NO_2 + O_3 \rightarrow NO_3 + O_2$$

and remain in equilibrium with N_2O_5

$$N_2O_5 \rightleftarrows NO_2 + NO_3$$

with equilibrium constant $K = 1.33 \times 10^{-10}$ moles/liter. The NO_3 formation reaction only takes place after NO is substantially depleted in the atmosphere and O_3 begins to appear.

Some typical NO_x exhaust concentrations are as follows; 50–150 ppm for incineration, 100 to 500 ppm in boilers and heaters, and 400–700 ppm in power plants of both gas and oil fired types[9]. Automobile exhausts have been estimated to produce 5.3 grams NO_x or about 1300 ppm per vehicle mile, but federal standards are calling for 0.4 g/mile, or about 100 ppm by 1980[10,11]. The U.S.S.R. ambient standards for NO_x (taken as NO_2) are 0.08 ppm (vol.) daily and 0.27 ppm (vol.) at one time. The British and American ambient standards are generally becoming more restrictive, setting 0.05 ppm as a maximum permissible level for 1975.

Although the nitrogen oxides are generally present in the atmosphere at about the same level as SO_2, they are potentially more difficult to remove, since they cannot be controlled by treating the fuel alone.

Oxides of carbon

The carbon dioxide, CO_2, concentration threshold for humans is 5% (5000 ppm) for an 8-hour exposure. This compares with a normal atmospheric CO_2 concentration of 0.03% (300 ppm). With a perfect stoichiometric combination of pure carbon in air, a CO_2 concentration of about 21% could be attained. Considering the usual dispersion of combustion gases it would take an unusual isolation to produce a CO_2 health hazard. A more detailed description of CO_2 consequences may be found under *Effects of Atmospheric Pollutants.*

Incomplete combustion of fuels is the more serious problem, since carbon monoxide, CO, will form. This rarely happens in stationary furnaces for which efficiencies of combustion are high and oxygen is available in excess of the theoretical requirements.

It has been estimated[12] that slightly more than 100 million tons of CO are emitted annually in the U.S.A., of which the major sources are automobiles

TABLE II

Properties of Gaseous Pollutants

Name	Formula	Molec. WT	Sense Properties	Boiling PT, °C	Solubility, CC per 100 GMS Cold H_2O	Warm H_2O	Other
Ammonia	NH_3	17.03	Colorless Pungent	−33.4	very soluble	>1000 (99°)	
Carbon monoxide	CO	28.01	Colorless Odorless	−192	3.5 (0°)	2.32 (20°)	Alcohol, Cu_2Cl_2
Chlorine	Cl_2	70.91	Gn-yellow Pungent	−34.6	310 (10°)	177 (30°)	Aq. NaOH or KOH
Fluorine	F_2	38.00	Gn-yellow	−187	decomposes		
Hydrochloric acid	HCl	36.47	Colorless	−85	very soluble	>1000 (99°)	Alcohol, Ethylether
Hydrafluoric acid	HF	20.01	Colorless	19.4	very soluble	>1000 (99°)	
Hydrogen sulfide	H_2S	34.08	Colorless Decay odor	−59.6	437 (0°)	186 (40°)	Alcohol, CS_2
Mercury	Hg	200.61	–	356.9	–	–	–
Nitric oxide	NO	30.01	Colorless	−151	7.34 (0°)	0 (100°)	Alcohol, H_2SO_4
Nitrogen dioxide	NO_2	46.01	Red.-brown	21.3	decomposes		HNO_3, H_2SO_4, CS_2
Ozone	O_3	48.00	Faint blue	−112	0.494 (0°)	0 (50°)	Oil turp., oil cinn.
Sulfur dioxide	SO_2	64.07	Colorless Choking	−10	8000 (9°)	1600 (50°)	H_2SO_4, alcohol, acetic acid
Sulfur trioxide	SO_3	80.06	Colorless	44.6	decomposes		H_2SO_4

(59%), various open burnings (16%), chemical industry (10%) and other transportation means (5%). New York City, with its acute urban traffic problem, has established a first alert at 15 ppm of CO over an 8-hour period. High temperatures favor the equilibrium dissociation of CO_2 to CO, with the latter being very stable at high temperatures.

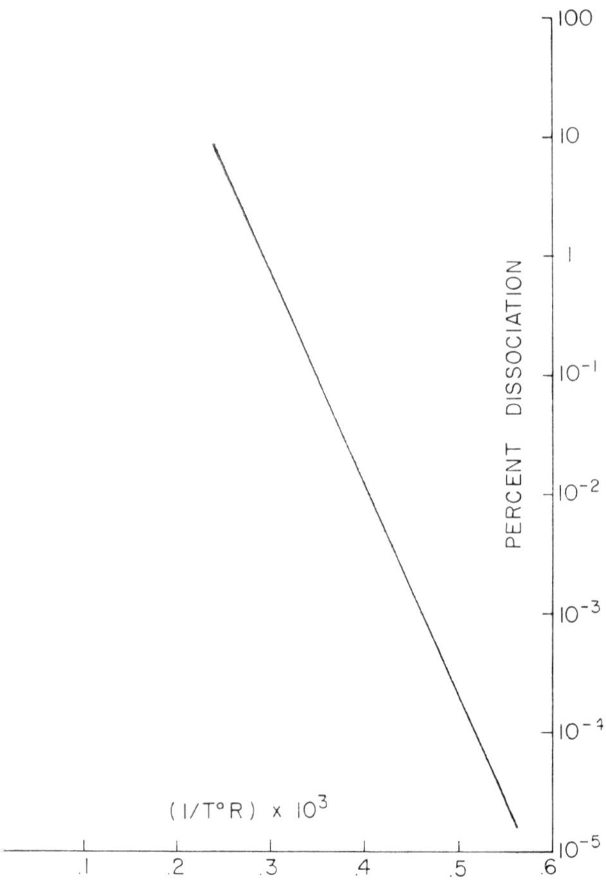

Figure 2

Thus if a CO_2–CO mixture is quenched from its high temperature zone the % CO may remain high, since at lower temperatures longer times are required to reach equilibrium. Rich fuel-air mixtures favor the formation of CO over CO_2. A complete description of CO control methods may be found in the section *Automative Pollution Control*.

Miscellaneous gases

Compounds of *fluorine* are inown to have negative effects (fluorisis) at concentrations as low as 5×10^{-3} ppm. They are generated as waste gases of fertilizer aluminum and ceramic processes, but are present to a lesser extent in most flue gases. A concentration of 0.1 ppm (vol.) of fluorine has been set as a maximum permissible value by the American Conference of Governmental Industrial Hygrenists; U.S.S.R. standards are roughly one tenth as stringent.

Ozone, O_3, is one of the strongest gaseous oxidants and is formed naturally from oxygen during electrical discharges in the atmosphere and at the high temperatures of combustion.

$$O_2 \rightarrow 2O\cdot$$

$$O_2 + O\cdot \rightarrow O_3$$

Taken as oxidants, New York City classifies an ozone level above 0.03 ppm as unsatisfactory and above 0.07 ppm as unhealthy over a 6 hour period. Eye irritation commences at concentrations of about 0.1 ppm. Interestingly enough ozone in the lower stratosphere affords part of the protective shield against ultraviolet radiation from the sun which could destroy land vegetation.

$$O_3 \rightarrow O_2 + O\cdot$$

Some scientists are concerned that nitric oxide formed by supersonic jets may deplete the ozone supply in the lower stratosphere eroding the barrier to the destructive rays.

High temperature processes involving metal recovery from ores emit *mercury* vapor in addition to sulfur dioxide. Mercury is available at concentrations up to a few hundred ppm[13] during zinc sulfide ore processing for example. *Hydrochloric* and hydrofluoric acids also appear in the roaster gases of such processes. No danger levels for mercury vapor have been officially established in ambient air quality standards.

A few limits have been established for less common pollutants of the process industries in U.S.S.R. standards given below.

	Daily		Instantaneous	
	mg/m^3	ppm/wt	mg/m^3	ppm/wt
Cl_2	0.03	(0.024)	0.10	(0.081)
H_2S	0.01	(0.0081)	0.03	(0.024)
CS_2	0.15	(0.122)	0.50	(0.406)
P_2O_5	0.05	(0.049)	0.15	(0.122)
Phenol	0.10	(0.081)	0.30	(0.24)

The U.S. ambient air quality standards call for *hydrocarbon* concentrations below 160 mg/m^3 (0.24 ppm) between 6–9 am.

Aldehydes and other *oxygenated hydrocarbons* are formed by the action of ozone on unburned hydrocarbons in the presence of sunlight. For example,

O_3 + 1–3 Butadiene → Acrolein + Formaldehyde

In the above reaction both products have been linked to the severe eye irritation encountered in urban environments.

Transport of Pollutants

The feed and waste materials of any combustion or chemical process travel through ducts or pipes. Control devices may be placed at various stages of the process, depending on the separation technique to be employed. It will be valuable to review the flow and transport behavior for fluids and then the separation methods. The important process variables to be considered are the mass flow rate of the waste gas, its temperature, pressure and composition. The raw material feed rate variables may also be of significance, as in the desulfurization of fuel oil. Control devices may broadly be classified according to the physical separation process being used, adsorption; absorption; extraction; distillation; or to the chemical process, homogeneous or heterogeneous catalytic reaction. In each instance, equations which account for the transport of material and energy must be developed.

In a sense almost any process may be considered as taking place in a pipeline. The simplest model of flow is called plug flow and assumes that no mixing takes place along the axis of the pipeline, but that lateral mixing is complete. Also, this assumes a flat velocity profile exists at each longitudinal position, or that the average velocity is the same at each lateral position.

Continuity

If the density, ϱ, of the fluid at a distance along the pipe, Z, of cross section, S, changes, the velocity must also change as seen by an elemental mass balance across dZ distance.

$$\text{Mass Accumulation Rate} = \text{Mass Net Input Rate} + \text{Mass Generation}$$

$$\frac{\partial(\varrho S\, dZ)}{\partial t} = \frac{\partial(\varrho v S)}{\partial Z} dZ + 0$$

$$\boxed{\frac{\partial \varrho}{\partial t} = \frac{1}{S}\frac{\partial(\varrho v S)}{\partial Z}} \qquad (1)$$

For steady state results, $\varrho v S = \text{const.} = W_0$ and the mass flow rate becomes the same at all axial positions. If the fluid is incompressible, $\varrho = \text{const.}$, as for most liquids, vS, the volumetric flow rate, does not vary with position even during transient conditions.

Motion

In a comparable manner an elemental momentum (force) balance may be made over length dZ.

$$\text{Momentum Acc. Rate} = \text{Net Input Rate of Momentum} + \text{Sum of Forces Acting}$$

$$\frac{\partial(\varrho v S\, dZ)}{\partial t} = -\frac{\partial(\varrho v v S)}{\partial Z} dZ - S g_c \frac{\partial p}{\partial Z} dZ + \left[\frac{g_z}{g_c}\varrho S - (F + F_w)\right] g_c\, dZ \qquad (2)$$

This equation describes the relationship between velocity and pressure along the pipe. The quantities F and F_w are the magnitudes of skin frictional

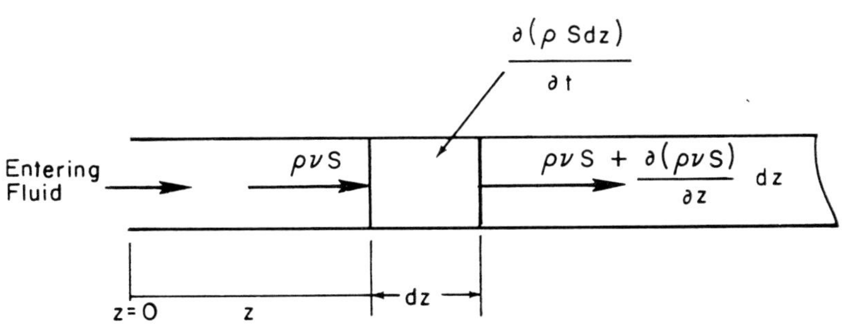

Figure 3

force and force doing work on external surfaces, respectively, both per unit length of pipe.

At steady state the equation simplifies to

$$W_0\, dv + S g_c\, dp = [-g_c(F + F_w) + \varrho g_z S]\, dZ \quad (3)$$

or

$$\frac{dv^2}{2g_c} + \frac{dp}{\varrho} = \left[-\frac{F + F_w}{\varrho S} + \frac{g_z}{g_c}\right] dZ \quad (4)$$

For incompressible flow

$$\frac{\partial v}{\partial t} = \frac{1}{2}\frac{\partial v^2}{\partial Z} - \frac{g_c}{\varrho}\frac{\partial p}{\partial Z} - \frac{g_c(F + F_w)}{\varrho S_0} + g_z \quad (5)$$

For both steady and incompressible flow

$$\frac{dp}{dZ} = \left[\frac{F}{S_0} - \frac{\varrho g_z}{g_c}\right] = \text{const.} \quad (6)$$

Energy

The First Law of Thermodynamics may be written for the differential element of length, dz, at steady state.

$$\frac{dH}{dz} + \frac{g}{g_c} + \frac{v}{g_c}\frac{dv}{dz} = \delta Q - \delta W_s \quad (7)$$

For unsteady behavior where temperature gradients are desired the equation of thermal energy may be applied assuming a uniform temperature at any cross-section and no axial conduction.

$$\varrho C_v \frac{\partial T}{\partial t} = -T\left(\frac{\partial p}{\partial T}\right)_\varrho \left(\frac{\partial v}{\partial z}\right) + \delta q - \delta w_s - \frac{\partial(\varrho c_v v T)}{\partial z} \quad (8)$$

in which q and w_s are the volumetric thermal energy input rate (produced for example by an electrical or chemical phenomenon) and the work output rate, respectively. For a constant density fluid equation (8), left hand side, represents the accumulation of internal energy, and the right hand terms represent the influence of pressure on the energy transport rate, the combined energy input rate per unit volume by generation and forces and the net energy input rate by flow (force convection), respectively.

Component balance

The equations of continuity, motion and energy often may be applied to describe the situation in stacks of power plants, in the flow of fuels and effluents, and in the analysis of material, momentum and energy requirements of a pollution producing process. To analyze the concentrations of pollutant it still remains to make component material balances on $n - 1$ species within the system (for which n components exist). For separation processes an additional phase equation is usually required transfer of pollutants between the rich and lean phases.

The mass balance on a particular species may be found for component A by examining the imaginary stationary differential element of thickness dz. Assuming plug flow we may derive an expression for C_A:

Accum. rate of A	Net Input Rate of A	Generation Rate of A

$$\frac{\partial c_A}{\delta t} = -\left(\frac{\partial F_A}{\partial Z}\right) + r_A \quad (9)$$

c_A – molar concentration
F_A – flux of A at position Z, moles A flowing/(time) (cs. area)
r_A – production rate of component A, by chemical or nuclear reaction, moles A formed/(time) (vol.)

If axial diffusion is small in comparison to the bulk flow, the flux may be described by

$$F_A = v c_A$$

where v is the average fluid velocity. Substituting for F_A and r_A in terms of c_A equation (9) reduces to

$$\frac{\partial c_A}{\partial t} = -\frac{\partial(v c_A)}{\partial Z} + r_A \quad (10)$$

in which r_A is usually some empirical function of c_A such as $-kc_A^n$ for an irreversible decomposition reaction of nth order.

Dispersion model

It is known that mixing in the tube can appreciably alter the component material balance since it tends to reduce the concentration gradients along the direction of the flow.

The subject of dispersion of substances is of interest in various control units in which mixing occurs, as well as in the atmosphere for establishing pollutant concentrations over a distributed vicinity. Taylor[14] analyzed the dispersion of soluble matter in a fluid flowing through tubes for various Reynold's number regimes.

In the laminar flow regime a conductive tracer

injected into a flowing water stream was found to spread out almost symmetrically from a point which moves with the mean speed of the fluid. This is not obvious in that the velocity is distributed over the radial direction and the laminar flow material balance on a tracer is in differential form

$$D_m\left(\frac{\partial^2 c}{\partial r^2}+\frac{1}{r}\frac{\partial c}{\partial r}+\frac{\partial^2 c}{\partial z^2}\right) = \frac{\partial c}{\partial t} + v_0\left[1-\left(\frac{r}{a}\right)^2\right]\frac{\partial c}{\partial z} \quad (11)$$

The observations which were similar to those found in an earlier experiment with fluorescent tracer by Griffiths[15] led Taylor to attempt a model which considers the transfer to (from) a differential element moving with the average speed of the main stream and to determine its region of validity. The model called the dispersion model allows for a transfer to the differentail element by a diffusion (Fick's Law) mechanism. For the non-reactive situation, a net input flux to the system is given by

$$\left(-D\frac{dc}{\partial z}\right)-\left[-D\frac{\partial c}{\partial z}+\frac{\partial}{\partial z}\left(-D\frac{\partial c}{\partial z}\right)dz\right]$$

Thus, for a tracer, Eq. (9) will now include dispersion

$$\boxed{\frac{\partial c}{\partial t_m}=\frac{\partial}{\partial z}\left(D\frac{\partial c}{\partial z}\right)} \quad (12)$$

and for chemical reaction with dispersion

$$\boxed{\frac{\partial c}{\partial t_m}=\frac{\partial}{\partial z}\left(D\frac{\partial c}{\partial z}\right)+r_A} \quad (13)$$

The quantity D is called the dispersion coefficient and has units comparable to a diffusivity, length^2/time. Its value is for convenience usually considered independent of concentration, and can be factored out of the derivative. Taylor showed that if equation (12) is to be a valid representation of equation (11) for laminar flow with $D = \dfrac{a^2v^2}{48\,D_m}$, where D_m is the molecular diffusivity, it is necessary that the Peclet number for molecular diffusion

$$Pe = \frac{av}{D_m} \gg \sqrt{48} \approx 6.9 \quad (14)$$

For turbulent flow equation (12) was found to be a valuable empirical method for determining the dispersion coefficient from tracer data. It was pointed out that Taylor's model is not strictly correct for turbulent flow for some initial period after the injection of tracer, but then is valid for later periods[16].

Gas adsorption

Solid materials may be used to remove pollutants from a gas stream by adsorption. Adsorption is the process by which solid surfaces attract fluid phase molecules and form a chemical or physical bond with them. The attraction for a specific gas phase component will depend on properties of both the solid adsorbent and the gas phase species. The concentration of the gas phase component, the total surface area of adsorbent, the temperature, polarity of the component and adsorbent, and similar properties of competing gas molecules are the most important of these properties.

Adsorption is usually used in air pollution processes to concentrate the species of interest for future disposal or recovery. For chemisorption, properties which affect reaction kinetics will also come into play see section on *Gas Reaction*.

Activated carbon, and silicon and aluminum oxides make up the majority of commercially significant adsorbents. Activated carbon is the least affected by humidity and physically adsorbs nonpolar compounds since it has no great electrical charge itself. On the other hand alumina and silica materials preferentially adsorb polar compounds.

The Langmuir isotherm[17] is popularly used to express chemisorption equilibria. If it is assumed that A_1 reacts with an unoccupied site X_0 to form adsorbed component X_1,

$$A_1 + X_0 \underset{k_{-1}}{\overset{k_1}{\rightleftharpoons}} X_1$$

the equilibrium adsorption concentration, c, is obtained in terms of the gas phase concentration c_1 and the total absorption site concentration, c_0.

$$\frac{C}{C_0}=\frac{K_1C_1}{1+K_1C_1} \quad (15)$$

Here, K_1 is the adsorption equilibrium constant which varies only if temperature varies.

Diatomic molecules such as chlorine might be expected to simultaneously adsorb and dissociate on adjacent sites. Such an adsorption might be described symbollically by

$$A_1 + 2X_0 \underset{k_{-1}}{\overset{k_1}{\rightleftharpoons}} 2X_1$$

in which case the equilibrium isotherm expression can readily be shown to be

$$\frac{C}{C_0} = \frac{(K_1C_1)^{1/2}}{1 + (K_1C_1)^{1/2}} \tag{16}$$

If more than one pollutant is being adsorbed, each component, Ai, undergoes

$$A_j + X_0 \rightleftarrows X_j$$

The equilibrium for single site adsorption of component A_j becomes

$$\frac{C}{C_0} = \frac{K_1C_1}{1 + \sum_{j=1}^{n} K_jC_j} \tag{17}$$

The latter may be referred to as competitive adsorption. Figure 4 depicts the different dependence on gas phase concentration for the adsorption types described thus far.

Another isotherm finding wide use, particularly in multilayer adsorption, is that of Brunauer, Emmett and Teller (18), the B.E.T. equation

$$\frac{C_s}{C_{max}} = \frac{Q_2X}{(1 - X)[1 + (Q_2 - 1)X]} \tag{18}$$

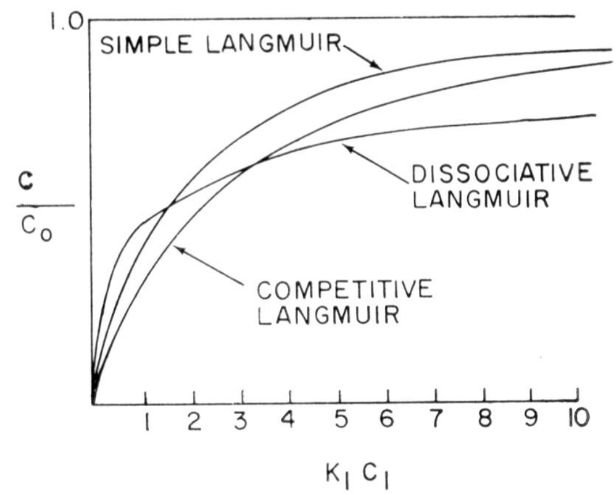

Figure 4

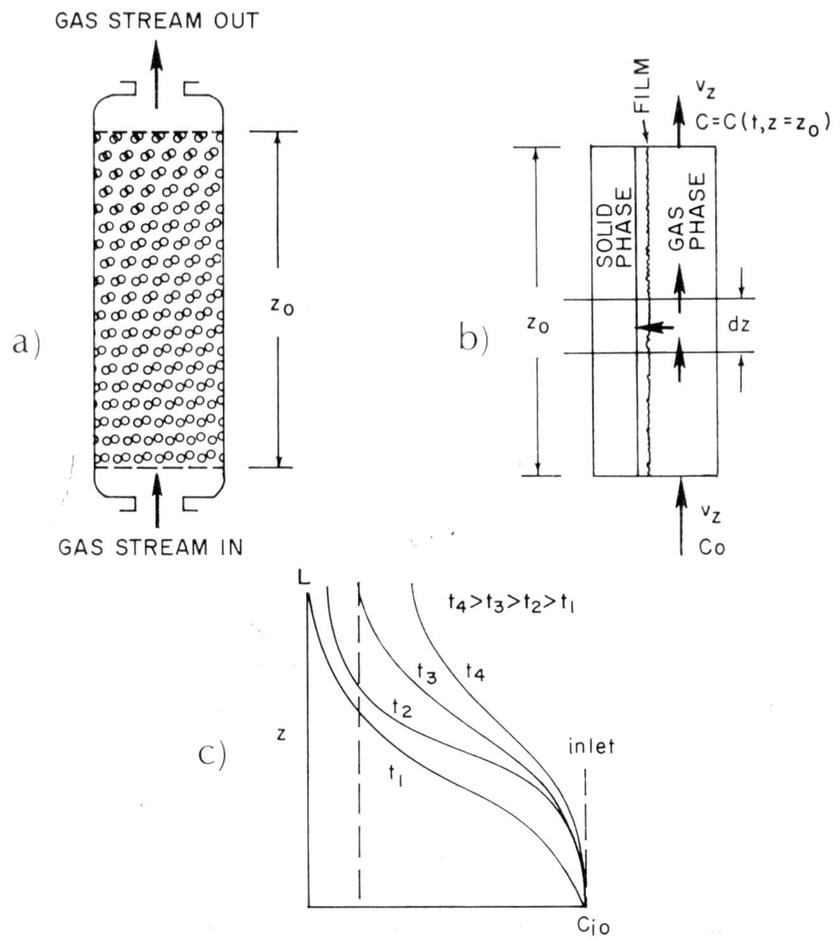

Figure 5 (a) Pictorial representation. (b) Schematic model showing a differential element over which a mass balance is made. (c) Pollutant concentration as a function of time at various heights.

in which: C_s, C_{max} – amount of gas absorbed per gm. of solid, and maximum amount, respectively

Q_2 – term exponentially dependent on heat of adsorption

X – ratio of equilibrium gas phase concentration to saturation value.

If neither the Langmuir or B.E.T. equations are satisfactory, a polynomial fit to adsorption data may be required.

For adsorption the particles are usually dumped into a column as a packed bed (Fig. 5).

The time t_3 in which breakthrough (C_{ig} at the exit equals the permissable set amount) occurs may be established by analyzing the differential component mass balance assuming that the transport of material again is governed by diffusion through a film.

$$N_i = k_g(C_{ig} - C^*_{ig}) \tag{19}$$

in which $C^*_{i_g}$ is the gas phase concentration in equilibrium with the absorbed concentration C_{is} at the same elevation.

The equilibrium concentrations have been determined for most commercially available adsorbents and typical pollutants and are presented as either langmuir or B.E.T. isotherms. In general such equations take the form

$$C_{is} = f(C^*_{ig}) \tag{20}$$

The generation term is excluded, as it is assumed that chemical reaction is not taking place in the system. To develop an expression for the net rate of accumulation of the mass of A_i in the column, i.e., the rate expression for the adsorption process, one also needs the function of the concentration of A_i in the fluid phase.

Considering a differential column segment and writing the continuity relationship for pollutant A_i in each phase for this differential section, one gets for the solid phase in this section,

$$\frac{1}{M_a} P_s(1 - \varepsilon)\left(\frac{\partial C_{is}}{\partial t}\right) - k_g(C_t - C^*) \tag{21}$$

where:

k_g : is the gas phase mass transfer coefficient in units of (time^{-1}).

M_A: is the molecular weight of A_1.

P_s : is the averaged density of the solids.

ε : is the fractional voidage in the bed.

For the gas phase in this differential segment,

$$\varepsilon \frac{\partial C_{if}}{\partial t} = - V_z \frac{\partial C_{if}}{\partial z} - k_g(C_i - C^*_g) \tag{22}$$

where:

V_Z: is the superficial velocity of the fluid.

These partial differential equations (20)–(22) may be solved simultaneously by numerical analysis using difference formulas to approximate the partial derivatives. In such a way the breakthrough curves of hazardous organic vapors may be predicted for a given absorbent.

The real computerized results were plotted on Fig. 6 for five different compounds having Langmuir type behavior inactivated carbon under the same hypothetical operating conditions. If one wishes to attain a 90% removal of certain organic vapor, one

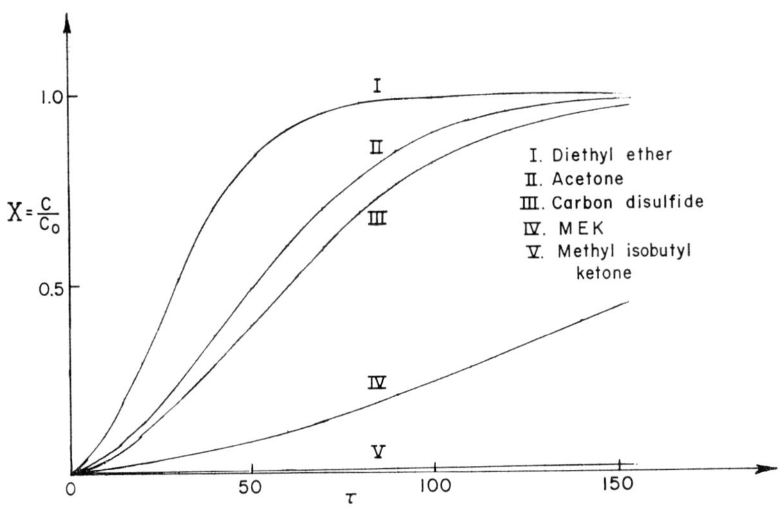

Figure 6 Break through curves for various compounds at 20°C and 1 atm with C_0 = 0.00548 mole/liter.

could easily see from Fig. 6 that diethyl ether requires the shortest re-cycle time and methyl isobutyl ketone the longest among the five materials on the graph.

Properties of adsorbents

Figures A and B are adsorption isotherms for activated carbon with nitrous oxide and carbon dioxide respectively. A more sophisticated correlation of adsorption data is presented in Fig. C–E for pure CO, C_2H_4 and CO_2 gases. Here $(RT/V_s) \ln f_s/f_g$ is plotted versus N_s (in which: R – gas constant, T – temperature, °K, V_s – molar volume of adsorbate, cc/mole, f_s and f_g – fugacities of adsorbate and gas and N – amount of gas adsorbed, g – moles/gm. adsorbent. Hydrocarbons and SO_3 adsorb

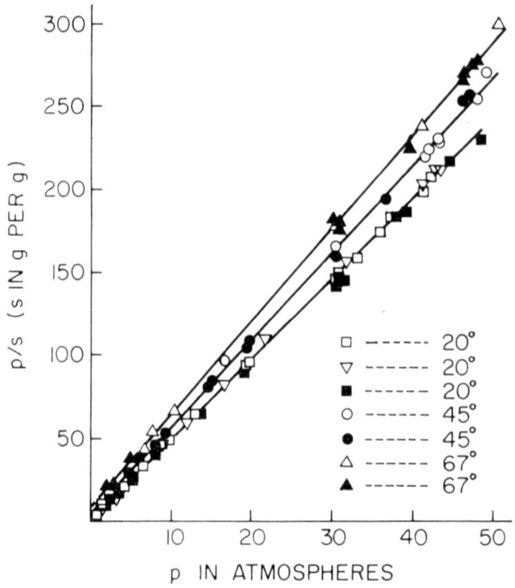

Figure A The system nitrous oxide-carbon McBain and Britton: *J. Am. Chem. Soc.* **52**, 2217 (1930) (Fig. 14).

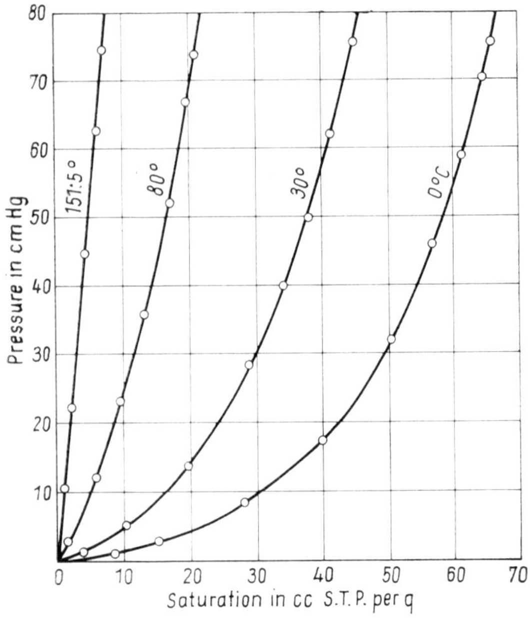

Figure B Adsorption isotherms for the system CO_2–carbon (note that $t_c = 31°C$).

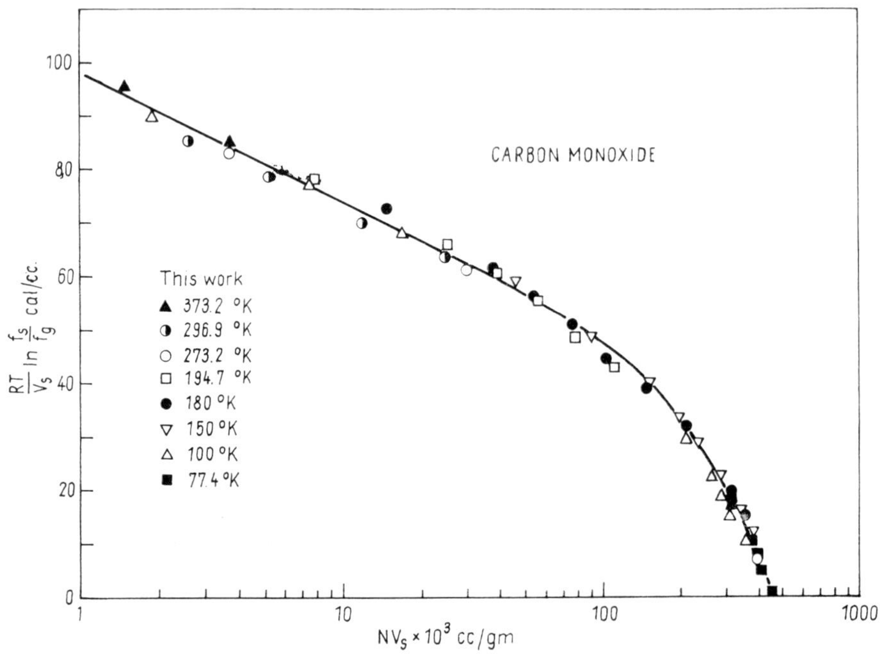

Figure C

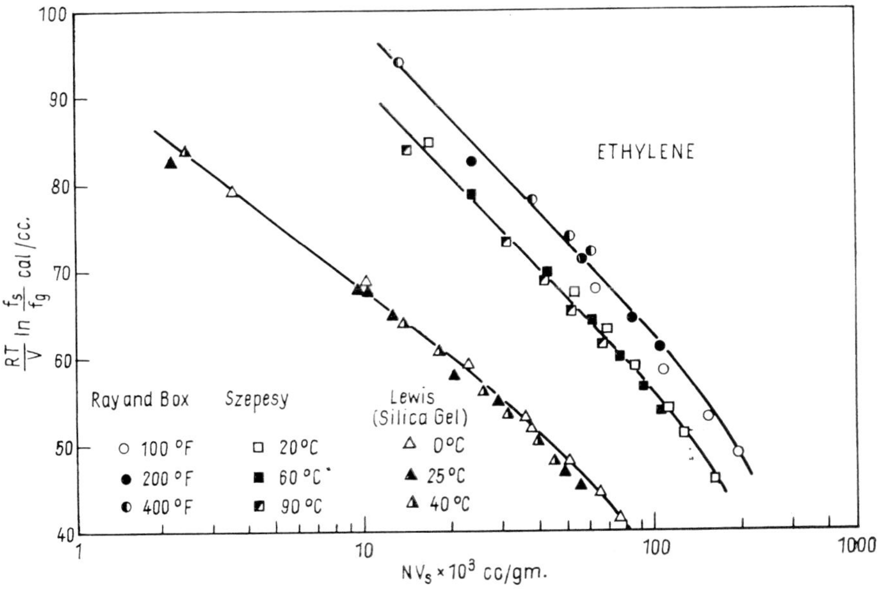

Figure D

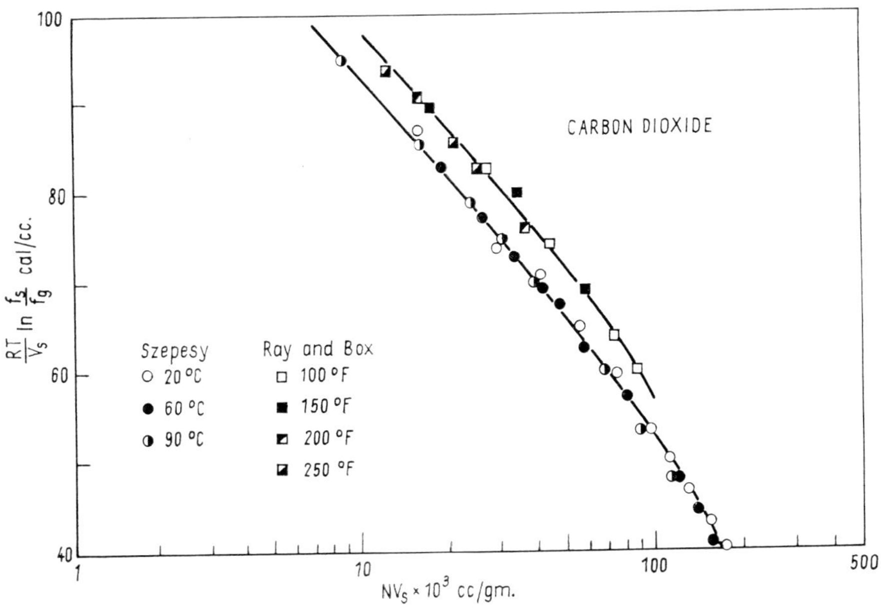

Figure E

readily on activated carbon. SO_2 has a maximum retention of 10 wt.% on carbon at 20°C, 760 torr. Ozone decomposes to oxygen on carbon[32].

Figure F has comparable results plotted for CO_2 adsorption on silica gel. Activated carbon has significantly better equilibrium properties than does silica gel (vis Fig. B vs. Fig. F).

Other results for activated carbon and zeolites may be found in the book by Strauss[35]. Basic facts about adsorption properties of activated charcoal, system types and components and applications are discussed by Lee[44]. He tabulated data on the air purification applications for inexpensive, nonregenerative, thin bed adsorbers and for regenerative systems and discusses the design of a solvent vapor recovery system.

"Activated carbon filters were used to concentrate atmospheric mixtures of acrolein, methyl sulfide, and *n*-propyl mercaptan. Removal efficiency and carbon capacity for each of the odor compounds were investigated using two different carbones, Cliffchar (4–10 mesh) and Barnebey–Cheney (C–4).

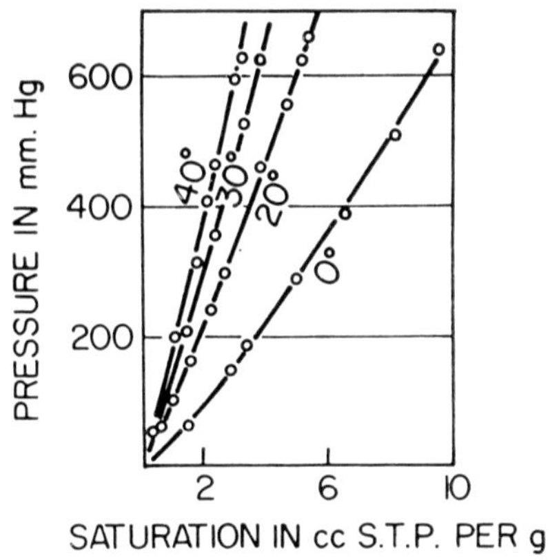

Figure F Adsorption isotherms for the system CO_2–silica gel (note that $t_c = 31°C$) Patrick, Preston and Owens: *J. Phys. Chem.* **29**, 421 (1925) (Fig. 1).

A closed system was devised to establish a known atmospheric odor concentration for each filter run. Solvent extraction techniques were employed to desorb and recover the odor compounds from the carbon filters. All quantitative analyses were conducted with gas liquid chromatography utilizing the hydrogen flame ionization detector. The removal studies conducted indicate that the efficiency of removal of a carbon filter is essentially 100 per cent up to the point of filter breakthrough. This breakthrough point is governed by the filter's capacity for a particular compound. This study indicated that the filter capacity is dependent both on the type of carbon employed and the particular odor compound adsorbed. Solvent recovery of the odor compounds from the carbons varied from 0 to 4.5 per cent for the mercaptan up to 96 to 98 per cent for acrolein. Per cent recovery was found to vary for a given odor compound with different carbons and for a given carbon with different odor pollutants."[44a].

Gas absorption

One may wish to use the selective solibility of gaseous pollutants towards certain liquid solvents as an advantageous means of cleaning a gas stream. Such a process referred to as scrubbing or absorption is usually carried out countercurrently by bubbling the vapor through the liquid, permitting the liquid to pour down along the walls of a container with the vapor simultaneously rising through the center of the column or by using a padring to increase the surface of contact of the falling liquid.

When a gas containing soluble components is brought into contact with a liquid phase an exchange of the soluble components (gas absorption) will occur until equilibrium in a batch system or steady state in a flow system is attained. The comparable situation (extraction) exists for a separation between two imniscible liquids.

A physically realistic model which describes this type of transport, called the *penetration theory*[19] envisions pockets or eddies of fluid leaving a bulk or well mixed portion of the phase and arriving at the interface. During their residence at the interface, molecules are exchanged between the phases after which the eddies return to the bulk phase. Although this theory is appealing, it has certain disadvantages associated with estimating parameters needed for adsorption calculations.

The film theory approach described below is of more practical value and usually the only method applied in pollution control problems.

At equilibrium, for gaseous species of low or moderate solubility, the partial pressure of the component is related to its liquid phase mole fraction according to Henry's law

$$P_i = Hx_i,$$

Both partial pressures and mole fraction may be related to concentration.

$$P_i = C_{ig}RT \quad \text{and} \quad X_i = C_{il}/C_{\text{tot}}$$

At constant temperature, $C_{ug} = H'C_{il}$, where $H' = H/C_{\text{tot}}RT$.

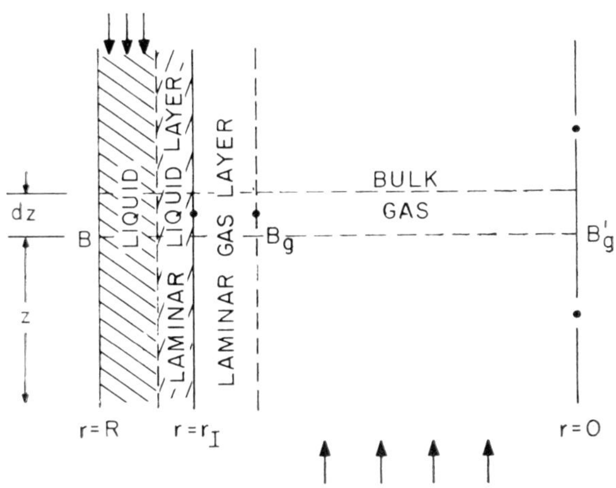

Figure 7

A typical absorption of gas phase component *A* into a liquid film is depicted in the Figure of a wetted wall column of radius *R*. The bulk gas and liquid phases are assumed to be in plug flow, with concentration at each *Z* elevation being constant except for a narrow region B_1–B_g in which surface drag limits the flow rate of the gas and in which the concentration is assumed to be linearly related to distance from the interface *I*. The mass transfer flux across the phase boundary at dz is defined as

$$N_i = k_2(C_{ig} - C_{igI}) = k_1(C_{il} - C_{igI})$$

Usually one or the other k values is large and only one of these equalities need be used. In a wetted-wall tower with an appropriate solvent, the liquid mass transfer coefficient for a soluble pollutant will be large, so that $C_{ilI} = C_{il}$. If it si assumed that the interface concentrations are at equilibrium, the molar flux of A_i for this system may be described over an element dz by

$$N_i = k_g(C_{ig} - C_{ig}^*) \quad \text{(equil.)}$$

and absorption is said to be gas phase controlled; C_{ig} is the gas phase concentration in equilibrium with C_{il}.

The correlation of Gilliland and Sherwood[20] may be used to estimate the value of k_g for pollutants

$$\frac{2k_gR}{D_r} = 0.023Re^{0.83}\,Sc^{0.44}$$

A liquid phase correlation of Stephens and Morris[21] for a disk column coefficient is

$$\frac{k_1}{D_r} = 21.0\left(\frac{4L}{\mu}\right)^{0.7} Sc^{0.5}$$

L – liquid flow per unit width of surface, lbs/hr/st

"The Counter Current Tower, Figure 8, has its primary application in the removal of an undesired gas or mist component. The gas stream moves upwards through a packed bed and the absorbing or reacting liquor is injected at the top of the packing. This counter current flow provides the maximum uniform driving force throughout the packed bed. The most contaminated gas at the bottom of the tower is in contact with contaminated liquor – the clean gas at the top of the tower with clean liquor. These towers are ideal for cooling gas since the coolest gas is in contact with the coolest liquor."

For packed towers using Raschig rings, Berl saddles or Intalox saddles, the following correlation may be used to determine gas phase and liquid phase transport coefficients[22].

Gas phase:

$$\frac{k_g}{a_tD_r} = \alpha\left(\frac{G}{a_t\mu_r}\right)^{0.7}\left(\frac{\mu G}{\varrho_rD_r}\right)^{1/3}(a_td_p)^{-2.0} \pm 40\%$$

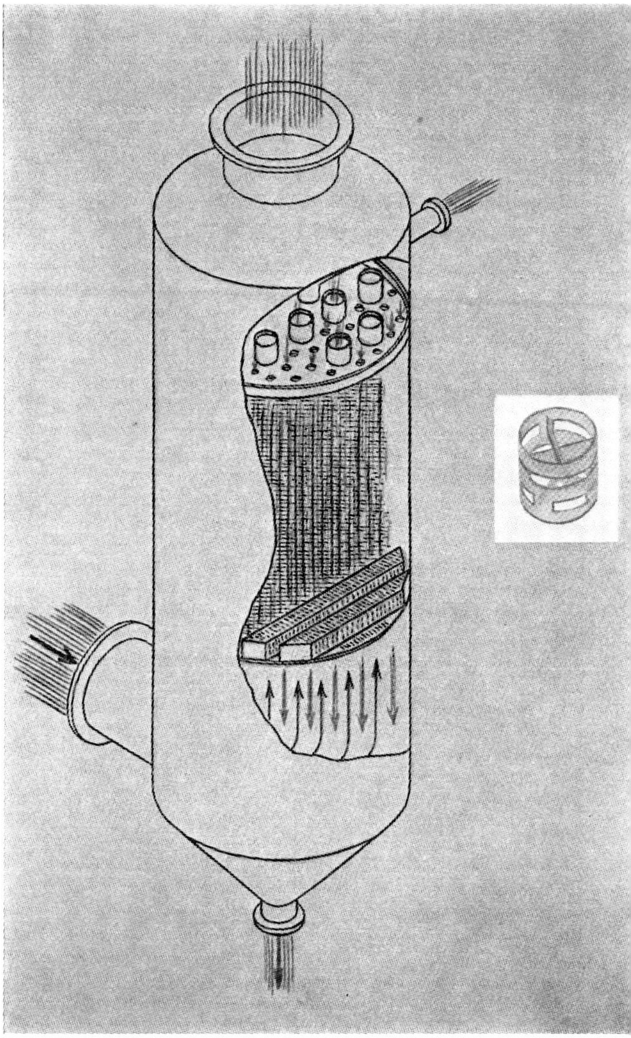

Figure 8 Counter current packed tower (Courtesy Koch Engineering).

$\alpha = 5.23$ if d_p — 15 mm, $\alpha = 2.0$ if $d_p < 15$ mm.

a_t – total surface area of packing per unit volume of bed.

d_p – normal direction of packing pieces.

Liquid phase:

$$\frac{k_1a}{D_1} = \alpha\left(\frac{L}{\mu}\right)^{1-n}\left(\frac{\mu}{\varrho D_1}\right)^{0.5}$$

$\alpha = 80$, $n = 0.22$ – 2 in Raschig rings

$\alpha = 100$, $n = 0.22$ – 1 in Raschig rings

$\alpha = 550$, $n = 0.46$ – 3/8 in Raschig rings

$\alpha = 160$, $n = 0.28$ – avg for Berl saddles

$\alpha = 110$, $n = 0.28$ – 3 in spiral tile

Tray type towers have also been used successfully. Bubble cap plates correlations have been proposed by Andrew

$$k_g = 7u^{1/4}S^{-1/2}D_r^{1/2} \text{ can/sec}$$

$$k_1 = 11u^{1/4}S^{-1/2}D_1^{1/2} \text{ can/sec}$$

$$a = 0.7u^{1/2}S^{5/6}$$

in which: S = effective liquid area on plate can^2
u = superficial gas velocity can/sec
a = interferial area/unit area of plate

A conservative estimate for the liquid phase coefficient in a sieve plate may by obtained from the Equation of Claderbank and Moo–Young[24].

$$k_i = 0.31(gv_1)^{1/3}(D_1/v_1)^{2/3} \text{ can/sec}$$

where v_1 is the kinematic liquid viscosity. For CO_2 and water at ordinary temperature, $k_1 \approx \approx 0.01$ can/sec. Typical sieve plate values are about 10^{-4} g mole/can^2-sec-atm.

"Each tray of Figure 9 has a series of drawn orifices fitted with a cage and cap. The orifice has a flared entrance. This reduces the dry pressure drop allowing a greater percentage of the work expended to be utilized for scrubbing. The floating cap maintains the scrubbing efficiency even with variations in gas flow as wide as 40–110% of capacity.

Gas enters the vessel flowing upward through the valve trays. Liquid is introduced on the top tray and flows across each tray over a weir and then to a sealed downcomer to the next lower tray. A level is maintained over each tray by the weir. The upward flowing gas is given a horizontal component by the cap causing atomization of the liquid on the tray. The froth formed consisting of a myriad of small droplets traps the particles and absorbs or reacts with acid or alkaline vapors. The liquid agitation on the tray surface prevents build-up. This has been demonstrated by the many successful applications in the Petrochemical Industry involving tarry solids and liquids.

Each tray has a $1\frac{1}{2}$ W.G. pressure drop. A typical installation with four (4) trays would require less than 8″ W.G. pressure loss."

An excellent review of gas phase absorption may be found in the work of Danckwerts[25]. The molecular diffusivities in the vapor phase D_r and in the liquid, D_1 may be found from existing correlations, for example see (26). Unlike the solid in adsorption the liquid solvent in absorption usually leaves the system where it can be regenerated. Hence a steady state plug flow analysis in either phase in

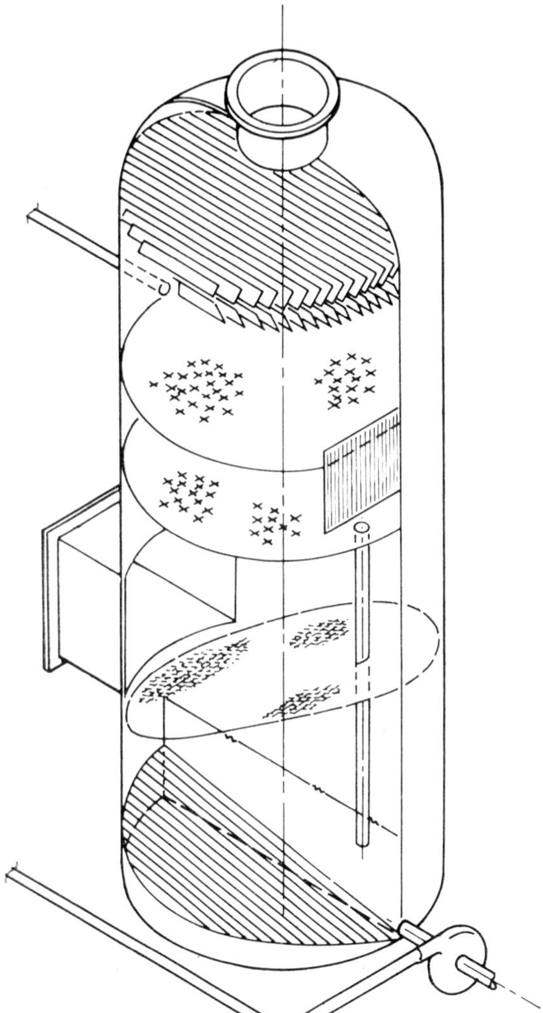

Figure 9 Flexitray Scrubber (Courtesy of Koch Engineering).

terms of overall coefficients is possible.

$$L de_{il} = G de_{ig} = K_g a(C_{ig} - C_{ig}^*)\, dz$$
$$= K_1 a(C_{il}^* - C_{il})\, dz$$

The required tower height is then given by either of the following:

$$Z = \frac{G}{K_g a}\int_{C_{go}}^{C_{ge}} \frac{dC_g}{C_g - C_g^*} = \frac{L}{K_1 a}\int_0^{C_{le}} \frac{dc_1}{C_1^* - C_1}$$

in which: a is the surface area of contact per unit volume of bed; L and G are per superficial mass velacities.

Properties of absorbents

Henry's law constants for CO, CO_2, NO and H_2S, are presented in Table A. Lower values of H such as those for H_2S correspond to higher solubility values. Table B contains specific wt. fraction absorbed at equilibrium vs. gas partial pressure for both ammonia and SO_2 in water; Table C has similar material for HCl.

Highly soluble materials have absorption rates which are controlled by diffusion through the gas phase (see Table D).

Reaction

Processes exist for catalytically removing gaseous pollutants by either forming harmless products or products are more amenable to recovery. The behavior of most catalytic reactions requires a more substantial analysis that homogeneous systems because of the presence of at least two phases. One of the primary differences is that in heterogeneous reactions require the adsorption of a fluid phase component on the catalyst surface. In addition to adsorption, diffusion rates mag be significant for both main stream to surface gas transfer and for transport inside catalyst pores. The analysis is often simplified by lumping the parameters of the system and assuming plug flow.

Thus for a packed bed of catalyst as in Fig. 10 the catalyst mass is lumped as a single substance and the gas phase as another, the latter in plug flow.

As in the p.f.r. model, a differential volume of material is examined at steady state for component A_i.

$$0 = -Sd(u\bar{c}_i) + \bar{P}_i(\bar{c}_i)\,\varrho_B S\,dz$$

u = superficial velocity based on total cross-sectional area S

$u = Eu_{act.}$, $u_{act.}$ is based on average cross-section occupied by fluid

ϱ_B = bulk density, mass of solids/unit volume

$P_i(\bar{c})$ = production rate, moles of A_i formed/(mass solids) (time)

TABLE A

Henry Law constant for slightly soluble pollutants, $H \times 10^{-4}$ atm/mole-fraction

T, °C	CO	CO_2	NO	H_2S	SO_2
0	3.52	0.0728	1.69	0.0268	0.0016
20	5.36	0.142	2.64	0.0483	0.0033
40	6.96	0.233	3.52	0.0745	0.0062
60	8.21	0.341	4.18	0.103	
80	8.45	–	4.48	0.135	
100	8.46	–	4.54	0.148	

TABLE B

Solubility data for ammonia and sulfur dioxide in water

Ammonia

Mass NH_3 per 100 Masses H_2O	Partial Pressure NH_3, mm Hg						
	0°C	10°C	20°C	30°C	40°C	50°C	60°C
100	947						
90	785						
80	636	987					
70	500	780					
60	380	600	945				
50	275	439	686				
40	190	301	470	719			
30	119	190	298	454	692		
25	89,5	144	227	352	534	825	
20	64	103.5	166	260	395	596	834
15	42.7	70.1	114	179	273	405	583
10	25.1	41.8	69.6	110	167	247	361
7.5	17.7	29.9	50.0	79.7	120	179	261
5	11.2	19.1	31.7	51.0	76.5	115	165
4		16.1	24.9	40.1	60.8	91.1	129.2
3		11.3	18.2	29.6	45.0	67.1	94.3
2			12.0	19.3	30.0	44.5	61.0
1					15.4	22.2	30.2

Sulfur Dioxide

Mass SO_2 per 100 Masses H_2O	Partial Pressure of SO_2, mm Hg							
	0°C	7°C	10°C	15°C	20°C	30°C	40°C	50°C
20	646	657						
15	474	637	726					
10	308	417	474	567	698			
7.5	228	307	349	419	517	688		
5.0	148	198	226	270	336	452	665	
2.5	69	92	105	127	161	216	322	458
1.5	38	51	59	71	92	125	186	266
1.0	23.3	31	37	44	59	79	121	172
0.7	15.2	20.6	23.6	28.0	39.0	52	87	116
0.5	9.9	13.5	15.6	19.3	26.0	36	57	82
0.3	5.1	6.9	7.9	10.0	14.1	19.7		
0.1	1.2	1.5	1.75	2.2	3.2	4.7	7.5	12.0
0.05	0.6	0.7	0.75	0.8	1.2	1.7	2.8	4.7
0.02	0.25	0.3	0.3	0.3	0.5	0.6	0.8	1.3

From Sherwood, T. K., *Ind. Eng. Chem.*, *17* (1925), 745.

The mass of catalyst required to attain a given conversion may be calculated if the dependence of u and $\bar{P}_i$ on c_i is known.

$$M = \varrho_B V = S \int_{c_{io}}^{c_i} \frac{d(u\bar{c}i)}{\bar{P}_i}$$

TABLE C

Equilibrium Data for Hydrogen Chloride Gas and Water[36]

Mass of HCl per 100 Masses H_2O	Partial pressure of HCl, mm Hg (Torr)				
	10°C	30°C	50°C	80°C	110°C
78.6	840	—	—	—	—
66.7	233	627	—	—	—
56.3	56.4	188	535	—	—
47.0	11.8	44.5	141	623	—
38.9	2.27	9.90	35.7	188	760
31.6	0.43	2.17	8.9	54.5	253
25.0	0.084	0.48	2.21	15.6	83
19.05	0.016	0.106	0.55	4.66	28
13.64	0.00305	0.0234	0.136	1.34	9.3
8.70	0.000583	0.00515	0.0344	0.39	3.10
4.17	0.000069	0.00077	0.0064	0.095	0.93
2.04	0.0000117	0.000151	0.00140	0.0245	0.280

TABLE D

Molecular diffusivities and Schmidt numbers for gaseous pollutants in air at 0°C and 1 atm.

		D_v		$Sc = \dfrac{\mu}{\varrho D_v}$
Typical Sc	SO_2	0.40 ft²/hr	1.11×10^{-3} *lb mole/ft-hr*	1.28
High Sc	n-octane	0.196	0.55	2.62
Low Sc	NH_3	0.836	2.33	0.61
	CO_2	0.535	1.49	0.96
	CO	0.713	$(\mu/\varrho)_{air} = 0.512$ ft²/hr	

Prandtl numbers for gases	$C_\varrho \mu/k$
Air	0.69
CO_2	0.75
CO	0.72
NO_2, NO	0.72

Dependence of gaseous diffusion coefficient on temperature: $D \propto T^{3/2}_{\text{absolute}}$. Values for many gases can be estimated from Hirschfelder Curtis & Bird [37] and also Reid & Sherwood [38].

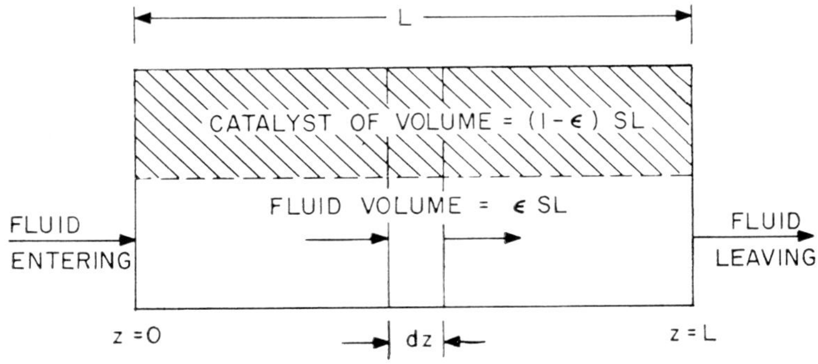

Figure 10

TABLE 3

Combustion Constants of Pure Gases (27)

No. Substance	Formula	Mol wt[a]	Lb/ft^3 [b]	ft^3/lb[b]	Sp. Gr. Air = 1.000[b]	Heat of Combustion[c] Btu/ft^3 Gross	Btu/ft^3 Net[d]	Btu/lb Gross	Btu/lb Net[d]
1 Carbon	C	12.01						14,093[g]	14,093[g]
2 Hydrogen	H_2	2.016	0.005327	187.723	0.06959	325.0	275.0	61,100	51,623
3 Oxygen	O_2	32.000	0.08461	11.819	1.1053				
4 Nitrogen (atmos.)	N_2	28.016	0.07439[e]	13.443[e]	0.9718[e]				
5 Carbon monoxide	CO	28.01	0.07404	13.506	0.9672	321.8	321.8	4,347	4,347
6 Carbon dioxide	CO_2	44.01	0.1170	8.548	1.5282				
Paraffin series (C_nH_{2n+2})									
7 Methane	CH_4	16.041	0.04243	23.565	0.5543	1013.2	913.1	23,879	21,520
8 Ethane	C_2H_6	30.067	0.08029[e]	12 455[e]	1.04882	1792	1041	22,320	20,432
9 Propane	C_3H_8	44.092	0.1196[e]	8.365[e]	1.5617[e]	2590	2385	21,661	19,944
10 *n*-butane	C_4H_{10}	58.118	0.1582[e]	6.321	2.06654	3370	3113	21,308	19,680
11 Isobutane	C_4H_{10}	58.118	0.1582[e]	6.321	2.06654[e]	3363	3105	21,257	19,629
12 *n*-pentane	C_5H_{12}	72.144	0.1904[e]	5.252[e]	2.4872[e]	4016	3709	21,091	19,517
13 Isopentane	C_5H_{12}	72.144	0.1904[e]	5.252[e]	2.4872[e]	4008	3716	21,052	19,478
14 Neopentane	C_5H_{12}	72.144	0.1904[e]	5.252[e]	2.4872[e]	3993	3693	20,970	19,396
15 *n*-hexane	C_6H_{14}	86.169	0.2274[e]	4.398[e]	2.9704[e]	4762	4412	20,940	19,403
Olefin series (C_nH_{2n})									
16 Ethylene	C_2H_4	28.051	0.07456	13.412	0.9740	1613.8	1513.2	21,644	20,295
17 Propylene	C_3H_6	42.077	0.1110[e]	9.007[e]	1.4504[e]	2336	2186	21,041	19,691
18 *n*-butene (Butylene)	C_4H_8	56.102	0.1480[e]	6.756[e]	1.9336[e]	3084	2885	20.840	19,496
19 Isobutene	C_4H_8	56.102	0.1480[e]	6.756[e]	1.9336[e]	3068	2869	20,730	19,382
20 *n*-pentene	C_5H_{10}	70.128	0.1852[e]	5.400[e]	2.4190[e]	3836	3586	20,712	19,363
Aromatic series (C_nH_{2n-6})									
21 Benzene	C_6H_6	78.107	0.2060[e]	4.852[e]	2.6920[e]	3751	3601	18,210	17,480
22 Toluene	C_7H_8	92.132	0.2431	4.113[e]	3.1760[e]	4484	4284	18,440	17,620
23 Xylene	C_8H_{10}	106.158	0.2803[e]	3.567[e]	3.6618	5230	4980	18,650	17,760
Miscellaneous gases									
24 Acetylene	C_2H_2	26.036	0.06971	14.344	0.9107	1499	1448	21,500	20,776
25 Naphthalene	$C_{10}H_8$	128.162	0.3384	2.955[e]	4.4208[e]	5854[f]	5654[f]	17,298[f]	16,708[f]
26 Methyl alcohol	CH_3OH	32.041	0.0846[e]	11.820[e]	1.1052	867.9	768.0	10,259	9,078
27 Ethyl alcohol	C_2H_5OH	46.067	0.1216[e]	8.221[e]	1.5890[e]	1600.3	1450.5	13,161	11,929
28 Ammonia	NH_3	17.031	0.0456[e]	21.914[e]	0.5961[e]	441.1	365.1	9,668	8,001
29 Sulphur	S	32.06						3,983	3,983
30 Hydrogen sulphide	H_2S	34.076	0.09109[e]	10.979[e]	1.1898[e]	647	596	7,100	6,545
31 Sulphur dioxide	SO_2	64.06	0.1733	5.770	2.264				
32 Water vapor	H_2O	18.016	0.04758[e]	21.017[e]	0.6215[e]				
33 Air		28.9	0.07655	13.063	1.0000				

* From *Gaseous Fuels*, Ed. by L. Shnidman (New York: American Gas Association, 1954).

All gas volumes corrected to 60 °F and 30 in. Hg dry. For gases saturated with water at 60 °F, 1.73% of the Btu value must be deducted.

[a] Calculated from atomic weights, *Am. Chem. Soc.*, February 1937.

[b] Densities calculated from values given in grams per liter at 0 °C and 760 mm in the International Critical Tables allowing for the known deviations from the gas laws. Where the coefficient of expansion was not available, the assumed value was taken as 0.0037 per °C Compare this with 0.003662 which is the coefficient for a perfect gas. Where no densities were available the volume of the mol was taken as 22.4115 liters.

[c] Converted to mean Btu/lb (1/180 of the heat per pound of water from 32 °F to 212 °F.) from data by Frederick D. Rossini, National Bureau of Standards, letter of April 10, 1937, except as noted.

ft³/ft³ of combustible						lb/lb of combustible						Experimental error in heat of combustion %
Required for combustion			Flue products			Required for combustion			Flue products			
O_2	N_2	Air	CO_2	H_2O	N_2	O_2	N_2	Air	CO_2	H_2O	N_2	
						2.664	8.863	11.527	3.664		8.863	0.012
0.5	1.882	2.382		1.0	1.882	7.937	26.407	34.344		8.937	26.407	0.015
0.5	1.882	2.382	1.0		1.882	0.571	1.900	2.471	1.571		1.900	0.045
2.0	7.528	9.528	1.0	2.0	7.528	3.990	13.275	17.265	2.744	2.246	13.275	0.033
3.5	13.175	16.675	2.0	3.0	13.175	3.725	12.394	16.119	2.927	1.798	12.394	0.030
5.0	18.821	23.821	3.0	4.0	18.821	3.629	12.074	15.703	2.994	1.634	12.074	0.023
6.5	24.467	30.967	4.0	5.0	24.467	3.579	11.908	15.487	3.029	1.550	11.908	0.022
6.5	24.467	30.967	4.0	5.0	24.467	3.579	11.908	15.487	3.029	1.550	11.908	0.019
8.0	30.114	38.114	5.0	6.0	30.114	3.548	11.805	15.353	3.050	1.498	11.805	0.025
8.0	30.114	38.114	5.0	6.0	30.114	3.548	11.805	15.353	3.050	1.498	11.805	0.071
8.0	30.114	38.114	5.0	6.0	30.114	3.548	11.805	15.353	3.050	1.498	11.805	0.11
9.5	35.760	45.260	6.0	7.0	35.760	3.528	11.738	15.266	3.064	1.464	11.738	0.05
3.0	11.293	14.293	2.0	2.0	11.293	3.422	11.385	14.807	3.138	1.285	11.385	0.021
4.5	16.939	21.439	3.0	3.0	16.939	3.422	11.385	14.807	3.138	1.285	11.385	0.031
6.0	22.585	28.585	4.0	4.0	22.585	3.422	11.385	14.807	3.138	1.285	11.385	0.031
6.0	22.585	28.585	4.0	4.0	22.585	3.422	11.385	14.807	3.138	1.285	11.385	0.031
7.5	28.232	35.732	5.0	5.0	28.232	3.422	11.385	14.807	3.138	1.285	11.385	0.037
7.5	28,232	35.732	6.0	3.0	28.232	3.073	10.224	13.297	3.381	0.692	10.224	0.12
9.0	33.878	42.878	7.0	4.0	33.878	3.126	10.401	13.527	3.344	0.782	10.401	0.21
10.5	39.524	50.024	8.0	5.0	39.524	3.165	10.530	13.695	3.317	0.849	10.530	0.36
2.5	9.411	11.911	2.0	1.0	9.411	3.073	10.224	13.297	3.381	0.692	10.224	0.16
12.0	45.170	57.170	10.0	4.0	45.170	2.996	9.968	12.964	3.434	0.562	9.968	
1.5	5.646	7.146	1.0	2.0	5.646	1.498	4.984	6.482	1.374	1.125	4.984	0.027
3.0	11.293	14.293	2.0	3.0	11.293	2.084	6.934	9.018	1.922	1.170	6.934	0.030
0.75	2.823	3.573		1.5	3.323	1.409	4.688	6.097		1.587	5.511	0.088
						0.998	3.287	4.285	SO_2 1.998		3.287	0.071
1.5	5.646	7.146	SO_2 1.0	1.0	5.646	1.409	4.688	6.097	SO_2 1.880	0.529	4.688	0.30

[d] Deduction from gross to net heating value determined by deducting 18,919 Btu per pound mol of water in the products of combustion. Osborne, Stimson, and Ginnings, *Mechanical Engineering*, p. 163, March 1935, and Osborne, Stimson, and Fiock, National Bureau of Standards Research Paper 209.

[e] Denotes that either the density or the coefficient of expansion has been assumed. Some of the materials cannot exist as gases at 60 °F and 30 in. Hg pressure, in which case the values are theoretical ones given for ease of calculation of gas problems. Under the actual concentrations in which these materials are present their partial pressure is low enough to keep them as gases.

[f] From Third Edition of "Combustion."

[g] National Bureau of Standards, RP 1141.

The dependence of the production rate on concentration is usually determined assuming either a Langmuir-Hinshelwood or a Langmuir Rideal mechanism, if diffusion rate constants are large.

The *Langmuir-Hinshelwood*-reaction sequence is a mechanism based on the combination of adsorbed species.

$$A_1 + A_2 \rightleftarrows A_3 + A_4 \quad \text{(overall)}$$

$$A_1 + X_0 \underset{k_{-1}}{\overset{k_1}{\rightleftarrows}} X_1 \qquad \text{Langmuir chemisorption}$$

$$A_2 + X_0 \underset{k_{-2}}{\overset{k_2}{\rightleftarrows}} X_2$$

$$X_1 + X_2 \underset{k_{-s}}{\overset{k_s}{\rightleftarrows}} X_3 + X_4 \quad \text{Surface reaction}$$

$$X_3 \underset{k_{-3}}{\overset{k_3}{\rightleftarrows}} A_3 + X_0 \qquad \text{Langmuir desorption}$$

$$X_4 \underset{k_{-4}}{\overset{k_4}{\rightleftarrows}} A_4 + X_0$$

Here the X_i represent an adsorbed component A_i on an active site X_0.

The *Langmuir-Rideal* sequence assumes that a fluid phase reactant bombines with a surface adsorbed molecule. For the same overall reaction as above the sequence is

$$A_1 + X_0 \underset{k_{-1}}{\overset{k_1}{\rightleftarrows}} X_1 \qquad \text{Langmuir adsorption}$$

$$A_2 + X_1 \underset{k_{-2}}{\overset{k_2}{\rightleftarrows}} X_3 + A_4 \qquad \text{Precursor—surface reaction}$$

$$X_3 \underset{k_{-3}}{\overset{k_3}{\rightleftarrows}} A_3 + X_0 \qquad \text{Langmuir desorption}$$

Most any conceivable mechanism might be formulated as a combination of these two sequences. Some typical variations are dissociative dual-site adsorption, or higher order surface reaction.

Usually, to simplify the analysis all the equations save one are assumed to be at equilibrium and the unsteady equation is said to control the rate. Thus for the Langmuir-Hinshelwood equation if

a) Surface Reaction Controls

$$P_3 = \frac{(k_s K_1 K_2 C_1 C_2 - k_{-s} K_3 K_4 C_3 C_4)\, C_0^2}{\left(1 + \sum_{j=1}^{4} k_j C_j\right)^2}$$

b) Chemisorption of A_1 Controls

$$P_3 = \frac{[k_1 C_1 - C_3 C_4 / K C_2]\, C_0}{\left[1 + \sum_{j=2} K_j C_j\right] + \frac{K_3 K_4}{K_2 K_2} \frac{C_3 C_4}{C_2}}$$

c) Desorption of A_3 Controls

$$P_3 = \frac{k_3 [k_s K_1 K_2 C_1 C_2 - k_{-s} K_3 K_4 C_3 C_4]\, C_0}{k_{-s} K_4 C_4 [1 + K_1 C_1 + K_2 C_2 + K_4 C_4] + k_s K_1 K_2 C_1 C_2}$$

The denominators indicate that each component competes for adsorption sites on the catalyst surface. The rate constants and equilibrium constants are denoted by k and K, respectively, with subscripts denoting surface reaction and j referring to adsorption of component A_j. The concentration of each component can be put in terms of C_3 by stoichiometry considerations.

Catalyst properties

A typical conversion described by Langmuir adsorption followed by chemical reaction is that for SO_2 removal over vanadium oxide catalyst[33].

$$SO_2 + \partial V^{5+} + O^{2-} \rightleftarrows SO_3 + \partial V^{4+}$$

95% Yield a 450 °C

$$r = k p_{O_2} \frac{K p_{SO_2} / p_{SO_3}}{[1 + (K p_{SO_2} / p_{SO_3})^{1/2}]^2}$$

k = rate constant, K = equilib. constant for above reaction

Similarly for the oxidation of ethylene

$$r = \frac{k p_{C_2H_4}^2 p_{O_2}}{[1 + K_1 C_{C_2H_4} + K_2 C_{CO_2}^2]^2}$$

The simplifications in the above expressions come about because of differences in the order of magnitudes of various rate and equilibrium constants.

One other place where reaction occurs frequently is in the *combustion* process. Oxygen in the air usually combines with a fuel containing carbon, hydrogen, sulfur, hydrogen sulfide and hydrocarbons. In addition, the nitrogen in the air will also react with oxygen at the elevated temperatures of a combustion furnace. In Table 3 many gaseous properties including the heat liberated during combustion are presented.

An excess of air is typically used over the theoretical amount of oxygen actually required for combustion; see Table 4.

Typical curves for combustion of coal, where combustion is to CO_2 and H_2O.

A combustion calculation, carried out for a bituminous coal is presented in Table 5 below.

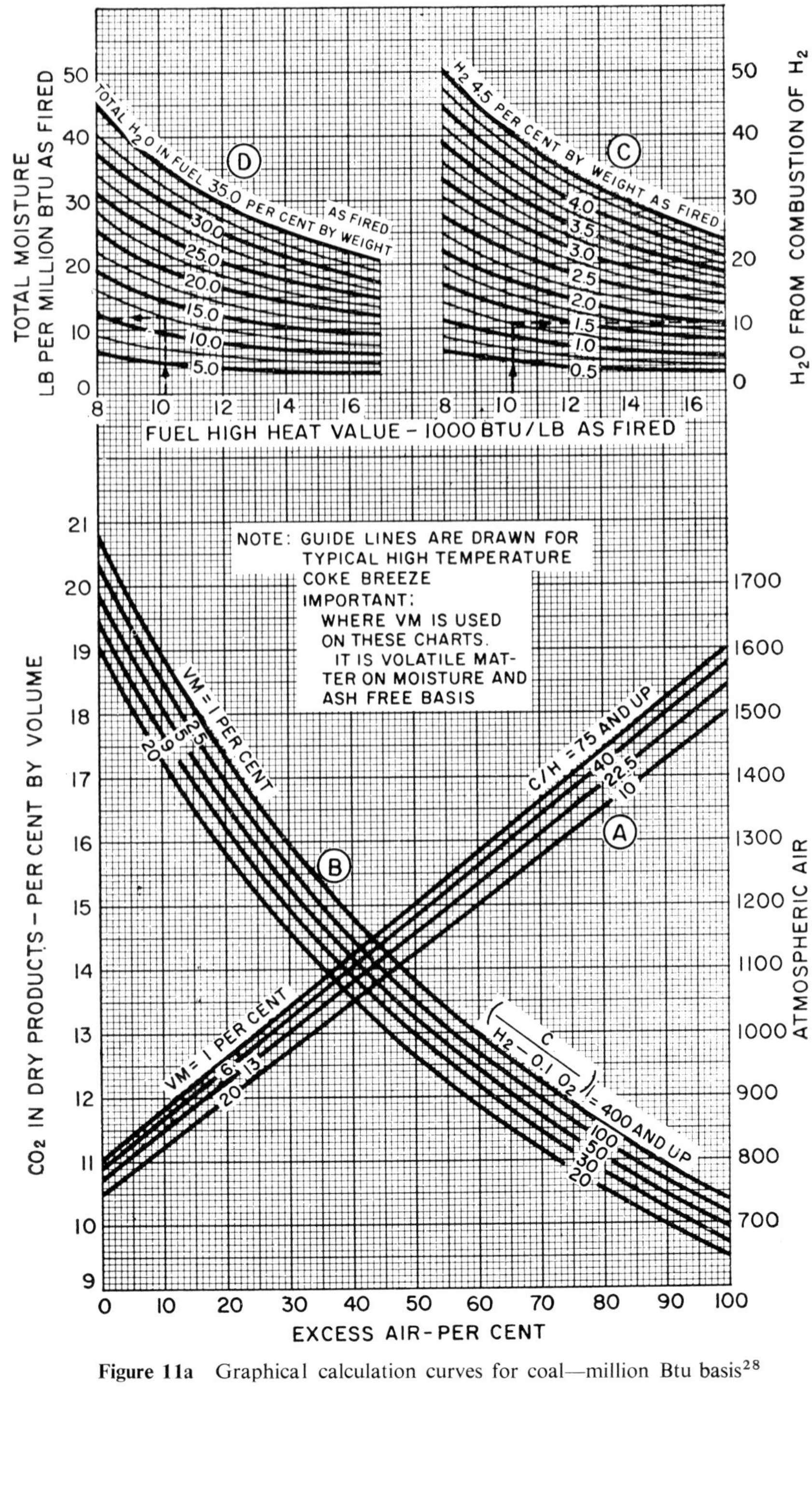

Figure 11a Graphical calculation curves for coal—million Btu basis[28]

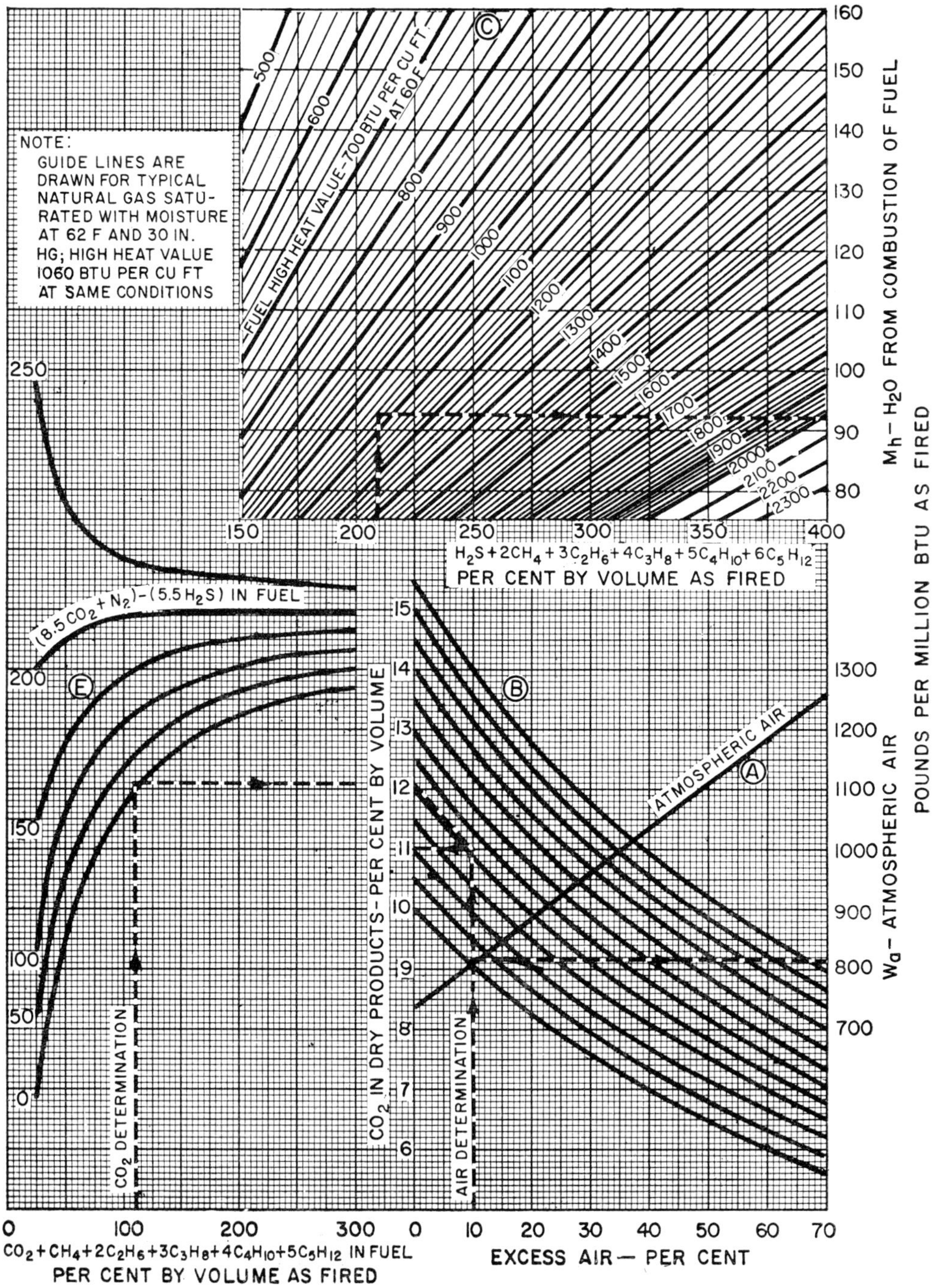

Figure 11b Graphical calculation curves for natural gas—million Btu basis[28]

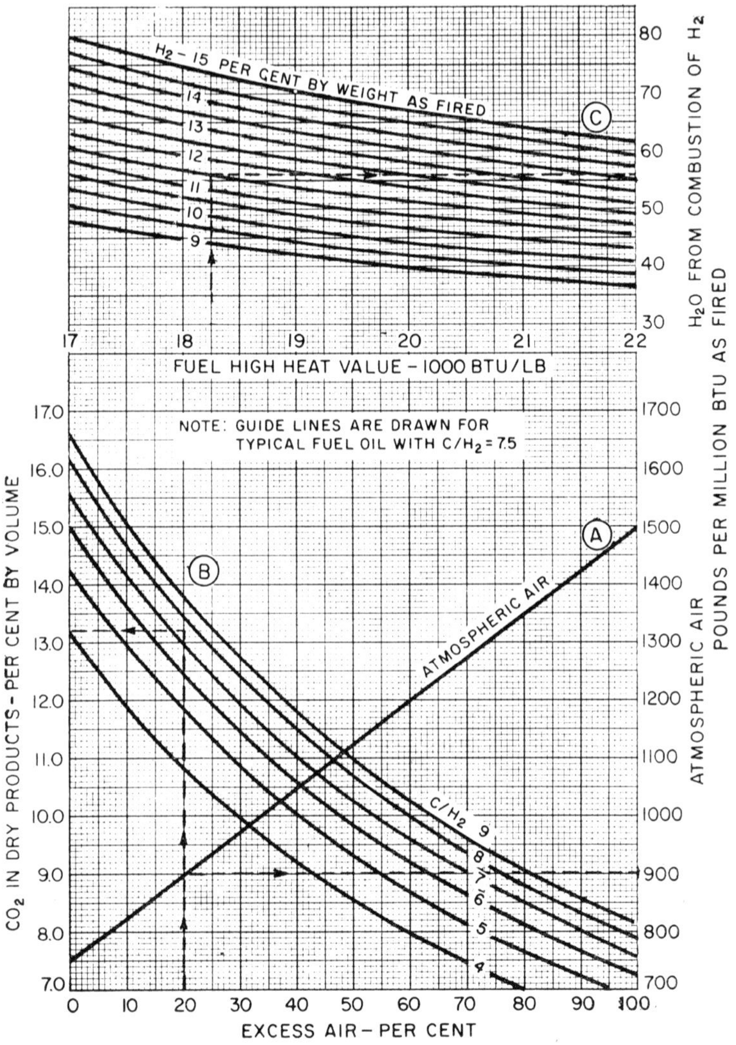

Figure 11c Chart for fuel oil[28]

TABLE 4

(28) *Excess air at furnace outlet* (29) *At inlet*

	Fuel	Excess air, %	Gas fuels	air/gas (vol)
Solid fuels	Coal	10–40	Natural	9.8
	Coke	20–40	L.P.	29.5
	Wood	25–50	Manufacturing gas	4.9
	Bagasse	25–45		
			Coke oven	4.9
Liquid fuels	Oil	3–15	Blast furnace	0.7
Gaseous fuels	Natural gas	5–10	Producer gas	1.0
	Refinery gas	8–15	Sewage gas	6.3
	Blast furnace gas	15–25		
	Coke oven gas	5–10		

TABLE 5

Typical complete combustion of bituminous coal

Gas	Weight/ 100 lb coal	Molar Basis	Volume fraction
SO_2	3	3/64 = 0.047	0.015
CO_2	207	207/44 = 4.70	0.145
O_2	40	40/32 = 1.25	0.040
N_2	750	750/28 = 24.8	0.800
	1000 lb	= 30.8	1.000

In addition to furnace combustion waste gases are often burned. Hydrogen cyanide and hydrocarbons are often flared at the exit of a stack to produce less harmful products. The addition of steam to stacks often inhibits the smoke formation of aromatic hydrocarbons by setting up the water gas reaction producing H_2 and CO. Sometimes a closed combustion chamber is used. In both instances safety considerations are important for preventing explosive mixtures. The detonation limits for various pure gases with air are presented in Table 6 below.

TABLE 6

Detonation limits

	g/m^3 with air
C_2H_2	26–844
NH_3	106–198
CoH_6	45–217
CO	145–863
C_2H_4	36–373
H_2	3–62
CH_4	35–93
H_2S	61–645

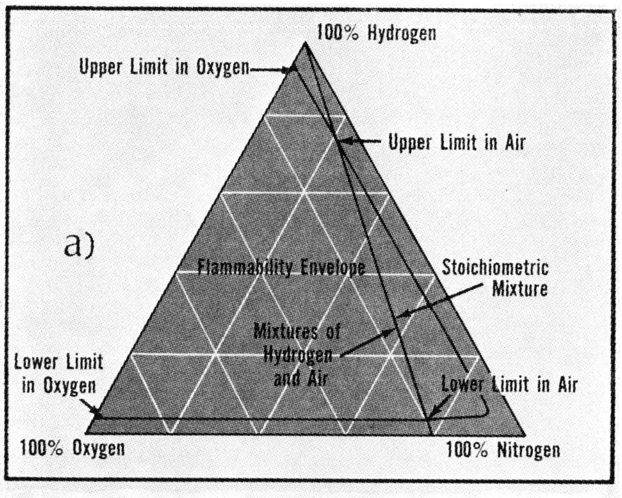

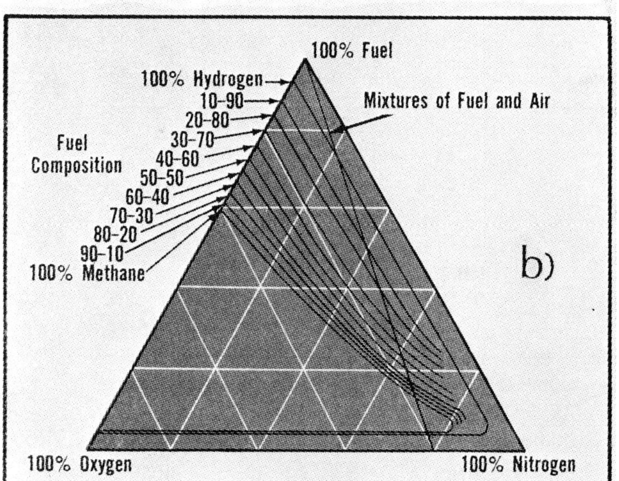

Figure 12 Flammability diagram. a) Hydrogen-oxygen-nitrogen 1 atm 20°C. b) Fuel-oxygen-hydrogen 1 atm 20°C (Talmage)[30].

Talmage[30] describes a flammability envelope diagram which must be considered for handling flammable vapors, Fig. 12a, b. By appropriately adding inerts or other fuel, it is possible to operate outside of such an envelope.

Reaction Rate Parameters:

Table 1 is a compilation of bimolecular reaction rate constants involving typical pollutants.

The decompostion of ozone takes place with a rate constant[37]

$$k = 4.6 \times 10^{15} e^{-2{,}400/RT} \quad \text{cc mole}^{-1}\ \text{sec}^{-1}$$

Pollutant control methods

Gases containing compounds of sulfur such as SO_2, SO_3, H_2S and mercaptans, have received the widest attention for the purpose of control of all noxious gases. For this reason the subject has been treated separately in the over view article entitled *Sulfur Removal*. In this section we shall consider the removal of NO_x, ozone, HF, HCl and mercury vapor. Additional information on NO_x, HC and CO removal may be found in the section entitled *Automotive Pollution*.

Nitrogen oxides

Several techniques for combustion control of NO_x emissions are possible. The most obvious is to reduce the amount of excess air used. This suggestion, among others, is proposed by Ermenc[4]. However, it must be realized that as the oxygen availability is reduced to the stoichiometric level, CO and unburned hydrocarbon emissions will increase. This will depend to a great extent on the type of fuel used.

Burner modification seems to be the best route to reduced NO_x emissions. A recent report from the Southern California Edison Co.[39] indicates that modified burner designs and improved gas flow control may reduce NO_x emissions by over 90%. Burner design can be improved by using two-stage combustion. In the first stage, fuel is mixed with substoichiometric amounts of air and burned. In the second stage fuel burn-out is completed by injecting secondary air into the stage. Formation of NO is thereby limited in the first stage because of the low air level. By using interstage cooling, temperatures can be held down in the second stage where the excess air is injected. This kinetically limits NO_x formation. Flue gas recirculation also reduces NO_x formation.

TABLE 7

Kinetic parameters for some bimolecular reactions[37]

Reaction	Activation energy, kcal per mole	Logarithm of frequency factor, cc $mole^{-1}$ sec^{-1}			Reference
		Observed	Calculated by absolute rate theory	Calculated by simple collision theory	
$NO + O_3 \rightarrow NO_2 + O_2$	2.5	11.9	11.6	13.7	*a*
$NO + O_3 \rightarrow NO_3 + O$	7.0	12.8	11.1	13.8	*b*
$NO_2 + F_2 \rightarrow NO_2F + F$	10.4	12.2	11.1	13.8	*c*
$NO_2 + CO \rightarrow NO + CO_2$	31.6	13.1	12.8	13.6	*d*
$2\,NO_2 \rightarrow 2\,NO + O_2$	26.6	12.3	12.7	13.6	*e*
$NO + NO_2Cl \rightarrow NOCl + NO_2$	6.9	11.9	11.9	13.9	*f*
$2\,NOCl \rightarrow 2\,NO + Cl_2$	24.5	13.0	11.6	13.8	*g*
$NO + Cl_2 \rightarrow NOCl + Cl$	20.3	12.6	12.1	14.0	*h*
$F_2 + ClO_2 \rightarrow FClO_2 + F$	8.5	10.5	10.9	13.7	*i*
$2\,ClO \rightarrow Cl_2 + O_2$	0	10.8	10.0	13.4	*j*

[a] H. S. Johnston and H. J. Crosby, *J. Chem. Phys.*, **22**, 689 (1954).
[b] H. S. Johnston and D. M. Yost, *ibid.*, **17**, 385 (1949).
[c] R. L. Perrine and H. S. Johnston, *ibid.*, **21**, 2200 (1953).
[d] H. S. Johnston, W. A. Bonner, and D. J. Wilson, *ibid.*, **26**, 1002 (1957).
[e] M. Bodenstein and H. Ramstetter, *Z. Physik. Chem.*, **100**, 106 (1922).
[f] E. C. Freiling, H. S. Johnson, and R. A. Ogg, *J. Chem. Phys.*, **20**, 327 (1952).
[g] G. Waddington and R. C. Tolman, *J. Am. Chem. Soc.*, **57**, 689 (1953).
[h] P. G. Ashmore and J. Chanmugan, *Trans. Faraday Soc.*, **49**, 270 (1953).
[i] P. J. Aynoneno, J. E. Sicre, and H. J. Schumacher, *J. Chem. Phys.*, **22**, 756 (1954).
[j] G. Porter and F. J. Wright, *Discussions Faraday Soc.*, **14**, 23 (1953).

Another burner design improvement is to use tangentially fired burners, where the furnace is used as the burner. Conventional horizontally fired burners create a great deal of turbulence and have higher peak flame temperatures. Tangential fired burners can cut NO_x emissions in half.

To get a better picture of how basic parameters such as temperature, kinetics, excess air, burner design, etc. effect NO_x formation, a kinetic model must be developed. One model has been developed by the Esso Research and Engineering Company[7] and contains several important burner design parameters such as specified flow velocity, pressure, flow area and heat transfer profiles. Two stage air injection and flue gas recirculation was also simulated. The mathematical kinetic model therefore contains detailed descriptions of mass, heat, and momentum transport occurring in the burner.

Naturally, several simplifications concerning the model are made. For example, the fuel (natural gas) and NO rate equations are assumed to be uncoupled. Nevertheless, the model yielded excellent results and closely matched the data obtained from the Southern California Edison (SCE) pilot plant tests[45].

The data presented demonstrate the effectiveness of off-stoichiometric combustion and reduced combustion temperature as techniques for controlling NO emissions from utility boilers. Off-stoichiometric combustion has been implemented on most of Southern California Edison's boilers and product gas recirculation has been implemented on the remaining units having a capability of recirculating the gas in the combustion air. Implementation of these techniques has resulted in a substantial reduction in total NO emissions. The techniques of off-stoichiometric combustion and gas recirculation are being combined on two large units presently under construction for maximum control in NO emissions. The modified burner pattern and degree of active burner fuelrichness is dependent upon the type of unit tested however, as is the minimum level of NO emissions achieved.[45]

When going from single stage to two stage

combustion (at 5% excess air) the model predicts a reduction in NO concentration of from 1200 ppm to 200 ppm. A similar reduction of NO is obtained when lowering the amount of excess air from 5% to 0%. Obviously, this idealized simulation has some flaws. The actual reduction in NO formation is probably less than that predicted. Surprisingly, the model predicts a slight drop off in NO emission when going from 5% to 10% excess air. This contradicts the data cited by Ermenc on the Wisconsin Process, which did not consider the combustion of fuel. However, the results agree with the SCE data. The authors, nevertheless, admit that the model does not take into account mixing limitations. In spite of this directional trends of NO formation can be accurately predicted by varying any one of a number of important design parameters.

One major drawback of the model is that it does not indicate the effect of varying excess air, temperature, etc. on CO emissions. Undoubtedly, CO emissions would markedly increase in fuel oil or coal burners at low excess air levels.

Burner design modifications presently can be applied only to newly constructed or updated facilities. Existing plants must therefore resort to flue gas treatment techniques to reduce NO_x emissions. Even in new plants containing the previously described burner modifications, NO_x emission levels will still probably not meet future legal standards. Thus, flue gas treatment techniques must be developed to treat NO_x emissions.

Before discussing various combustion gas treatment techniques one should look at the basic problem. The major disadvantage in flue gas treatment is the handling of massive volumes of gas—over one million SCFM for a 750 MW plant. Another problem is the low NO concentration level in the combustion gases—about 1000–2000 ppm. At low NO concentration levels it takes longer for NO to oxidize to NO_2, which is more readily absorbed. For example, it takes 0.175 min for half of 20,000 ppm of NO to oxidize in an environment of pure air and 35 min for half of 100 ppm to oxidize[40]. On the plus side it should be noted that the source of pollution is stationary. Unlike auto emissions, which probably can be handled by small catalytic devices, stationary emissions can be treated by any of the known unit operations.

In treating any waste gas the removal of all pollutants is important. From power plants, in particular, SO_2 and NO_x emissions must both be controlled. Therefore, only those treatment techniques which remove both pollutants can be considered. Basically, removal methods can be classified into three areas:

1 Catalytic decomposition
2 Adsorption
3 Absorption

The first method, catalytic reduction or decomposition, is probably the most effective but least economical method of removing NO_x and SO_2 from combustion gases. Catalysts, mostly noble metal and heavy metal oxides, have been developed to reduce NO_x and SO_2. SO_2, however, is in most cases a catalytic poison and may have to be removed first by other means. Although capital investments are comparatively low for catalytic reduction, operating costs, especially for regeneration, are prohibitive. As research continues in the area of automotive converters, catalytic removal methods may be found that are applicable to stationary sources as well.

The second possibility to consider is adsorption. The use of common adsorbants such as activated carbon and alumina show low adsorption capacity at typical NO combustion gas concentrations. Molecular sieves[41] have also been investigated in conjunction with removing NO_2 tail gas from nitric acid plants. The severe operating conditions of high temperature and high flow rates of combustion gases make most adsorption methods still technically unfeasible.

By far the most practical and economical method for handling large volumes of flue gas is absorption. Scrubbing is a well established engineering operation and can be studied without extensive R& D background information.

The earliest efforts in NO_x pollution were concerned with the removal of NO_x emissions from nitric acid plants. Initially, simple aqueous absorption was studied[42,43] and found to be adequate in removing NO and NO_2 and recovering HNO_3. Catalytic processing of tail gas and reduction with NH_3 has also been studied and found to be evonomically attractive. Nitric acid plant tail gas, however, contains mostly NO_2 impurities (rather than NO) which are present in much higher concentrations than in flue gases. In addition, no SO_2 is found in HNO_3 tail gas. When applied to combustion gases, the methods used to remove NO_x from tail gas, are economically unsound.

Slightly lower NO_x concentrations are encountered in the off-gases from metal etching processes. M. First and F. Viles, Jr.[40] have studied absorber efficiency using NaOH scrubbing of these off-gases. For a 16 stage absorber using a 4.6% NaOH solution, 90% of the nitrogen oxides (95% NO_2, 85% NO) can be removed. Although not applicable to flue gas treatment the article demonstrates how existing operations can be modified to handle efficiently and economically off-gases containing NO_x contaminants.

Probably the most comprehensive analysis of absorption systems has been carried out by Esso as reported by Bartok, et al[7]. Esso has singled out two processes as being the most promising. Both processes remove SO_2 and NO_2 from the combustion flue gas.

One of the processes, using sulfuric acid, has been developed by Tyco Loboratory under NAPCA contract. In this process sulfuric acid is used to scrub the flue gas. In addition, NO_2 is recycled back into the flue gas to oxidize SO_2 to sulfuric acid mist. However, in order to obtain efficient absorption NO and NO_2 must be in equimolar amounts. This is because the combined oxide, N_2O_3, is more readily absorbed than either NO or NO_2. The recycling of NO_2 is accomplished by oxidizing the NO stream of the spent acid absorbent (before regeneration). The requirements of SO_2 oxidation and NO_2 recycle into the flue gas places highly exact and difficult to maintain requirements on the scrubbing system. This, naturally, increases cost.

A new magnesium hydroxide scrubbing system has been proposed by Esso. In this process $Mg(OH)_2$ is used to scrub the flue gas. The scrubber effluent is then sent to a settler where $Mg(NO_2)_2$, $MgSO_3$ and recycle $Mg(OH)_2$ are separated. The $Mg(NO_2)_2$ is then sent through a nitrite decomposer and reactor. Ammonia and nitric acid are added to the reactor forming ammonium nitrate byproduct and recycle $Mg(OH)_2$. NO exiting from the decomposer is oxidized to form the HNO_3 used in the reactor and the NO_2 recycled to the flue gas. The $MgSO_3$ is also decomposed into SO_2 and recycle $Mg(OH)_2$. The SO_2 can be used to make sulfuric acid or other by-products. The process is shown schematically on the next page.

The advantage of this scrubbing system is twofold. Firstly, no exacting requirements are placed in the scrubber, as is the case with the Tyco acid process. Secondly, NO_2 and SO_2 are separated in the form of useful by-products. Another important factor is that magnesium hydroxide is easily regenerated in both the nitrite and sulfite decomposer.

Cost analyses show that an actual net savings can be obtained with the $Mg(OH)_2$ scrubbing system. With $Mg(OH)_2$ scrubbing, a capital investment of \$12 million is needed for a 1000 MW coalfired power plant. Operating costs, however, are −6−/ton of coal because of by-product recovery. Limewater scrubbing, which is similar to $Mg(OH)_2$ scrubbing, costs \$8.5 million with operating costs at 50−/ton of coal. The Tyco H_2SO_4 scrubbing process requires an investment of \$17.3 million with operating costs estimated at 17−/ton of coal. Previously mentioned catalytic reduction processes have operating costs around \$1.00/ton of coal. Thus, only the $Mg(OH)_2$ scrubbing process can be considered economically favorable.

The ammonia burner emits a gas which contains approximately ten per cent NO_x. Because these compounds are valuable intermediates in the process, an ideal facility would emit no NO_x, but would rather trap the gases in an absorber for eventual sale as Nitric Acid. However, because the absorption of NO_x in water is an equilibrium process, the ideal can be only approached using a very high pressure system and a very large water flow rate. High pressure will force the most soluble gas (NO_2) into solution, so as to releive the stress, and high water flow rate will keep the vapor pressure of the gas already in solution low.

It is here that the other constraint, cost, presents itself, the one which is the primary subject of Newman's[46] paper. His data shows that the operating point where increased revenues from making a higher percentage of NO_x product (i.e. more severe operating conditions) vs. the increased capital and operating costs necessary to maintain these conditions safely is 135 psia and 1500 ppm NO_x effluent concentration.

Generally, this is insufficient control. It is impossible to determine the most economical approach to abatement until the statutory effluent limit is known. There is the choice of driving the absorber harder, even though it is less economical (barring fines) to do sok, or of building a separate catalytic oxidation abator. Newman says 700 ppm NO_x is the approximate break point, below which the separate unit becomes more attractive. This operating point is developed under the assumption that one is building a new plant. The economics of updating an old plant may be quite different, since it may be necessary to remove old pumps and compressors, replacing them with higher pressure models and bigger motors. Down time costs enter into the decision, too.

Adlhart[47] *et al.* mention two types of catalytic units. The first they call a "cosmetic unit." This decolorizes the tail gas (Makes NO), rather than abates it (makes N_2, H_2O, and CO_2). This technology is the cheapest solution, but NO reaching the atmosphere recombines with oxygen, producing the same old problem.

Once a catalytic abator has been chosen, there is at present only one way to turn the money spent on it into a cash return. Because of the high temperature at the exit port of the catalyst bed, the heat can be recovered as high Btu steam. Andersen *et al.*[48] present the operating data on this plant. There are, however, many variables which can drive the cost of operation much higher than it need go.

Adlhart, *et al.*[47] deal primarily with these problems. Temperature is the most significant problem, because efficient operation requires at least 750°C for the abatement reaction to proceed at a reasonable rate. Furthermore, natural gas must be preheated to 450°C before it will burn, and once it is burning, the temperature will go way above the 750–800°C operating range if the oxygen content of the gas coming from the absorber is too high. This occurs because all oxygen must be burned, thus leaving a reducing atmosphere, before any destruction of nitrogen oxides will take place, and its reaction enthalpy must be absorbed as sensible heat.

Therefore, the consideration which will limit the use of a single catalyst bed flow scheme is simply the oxygen content of the tail gas. Otherwise a flow scheme using an interstage steam boiler or a bypass system in which part of the feed tail gas is not reacted and at interstage is used for cooling must be used.

In the two stage system another problem encountered is consumption of the catalyst. If simple Palladium metal is used, the life at a space-velocity of 100,000/hr. is only ten weeks at abating conditions (i.e. the second bed). "Stabilization," the act of combining the catalyst with a refractory, can keep the abative bed from degrading for up to a year. This is considered a satisfactory catalyst life, and can be maintained as long as the design space-velocity is not exceeded.

There is no mention, whatsoever, as to the breakdown of costs for a catalytic combustor. This is a reasonable objection, because in all the papers, the authors were generally referring their comments to building a new plant. Anyone not directly connected with the nitric acid production business would have no way of judging profitability from any of the data presented on costs, because it is presented as dollars vs. process variables. There are no equipment costs, nor anything to indicate how the $60,000–90,000 compares to the investment in equipment and operation of the 500 ton/day nitric acid plant. A figure required in a meaningful economic balance is the price increase of the product acid necessary to maintain a constant profit margin.

One can assume that a significant reduction in costs can be obtained if less palladium catalyst need be employed. Unfortunately, decreased performance precludes this possibility. The technology described is not really new and has been proven to work efficiently. The best path of improvement is to find a cheaper, longer lasting catalyst.

Fluorides

Not all fluoride emissions are gases. The system employed must have the capability of handling particulates, dispersed in huge quantities of air. Phosphoric acid production is a typical fluoride producer.

The process begins with the grinding of rock phosphate (typically $Ca_{10}F_2(PO_4)_6$). This is naturally a source of particulates, as is the drying of the pulverized phosphate. Acidulation with sulphuric acid produces the gaseous pollutants associated with this process, most notably hydrogen fluoride, and silicon tetrafluoride. Both of these gases can be scrubbed in aqueous medium, but each one causes a severe handling problem, even after scrubbing. The HF is highly corrosive, and the SiF_4 combines with water to form a gelatinous mass which can clog an improperly designed scrubber.

Another constraint is the extremely low effluent concentrations allowed by law, about 0.02 mg/ft^3. Since each unit in the process stream is a source, there are at least 250,000 ft^3/min which must be scrubbed down from 200–500 mg/ft³. This is a reasonable requirement, considering the toxicity of the pollutants, but it is still a difficult objective to attain.

Teller[49] claims to solve all these problems at once with a cross flow scrubber and a high water flow rate. The device is a baffled tank with wetted baffles so that any solids which do not immediately dissolve will just drop to the bottom after colliding and not clog the air ducts. The large water flow rate keeps the fluorine concentration low, so that the same efficiency (4.8 transfer units) typical of a countercurrent flow scrubber can be maintained.

Cost factors cause the designed to look very hard at the size of this system and its pumping costs. It is obvious that a scrubber which has a greater than necessary number of transfer units will be uneconomical. Even with a tight design on both size and pressure drop, the enormous volumes of gas can still make this system undesirable. A way must be found to decrease the volume.

The best way to do this is by taking advantage of the fact that not all the units produce pollutants at the same concentration level. Since the total flow is large enough that multiple scrubbers would probably be necessary anyway, the first scrubbers in the line take the dirtiest gas and clean it to the level of the next stream. The two are mixed and sent to the next scrubber, and so on. Only the last scrubber need be large enough to accommodate the entire flow. This should solve the air pollution problems to a sufficient degree.

There are two problems which deserve mention: these are water pollution from the dissolved fluorides and solid waste sludges which slough off the baffles. Both of these are highly contaminated and must ultimately lead to a disposal problem.

Teller suggests the water problem be solved by the use of a closed cycle holding pond. Byproduct recovery would appear to be a possible way to solve this problem, but the situation requires further study. There is mention of neutralization of HF, but this will give a CaF_2 precipitate, which would eventually fill even a large holding pond. Regardless of the method chosen, the water must be cleaned of excess fluoride before it can be recycled.

Efficiency of fume control in the aluminum smelting industry has been the subject of a recent Alcoa study[50]. They note that the recovery process may be divided into two categories: "interception of total particulate which includes both fluorine compounds and relatively innocuous aluminum oxide.

Hydrogen fluoride is by far the dominant gaseous fluorine compound in pot gas. Recovery of HF is a function of surface area of alumina, the HF concentration in pot gas delivered to the reactors, quantity of alumina through the system, and contacting efficiency between pot gas and alumina. Reaction between HF and alumina is rapid and with proper reactor design, contacting is no problem. With a sufficient feed rate of alumina to the reactor, reaction efficiency remains high even at very low bed levels. On the other hand, insufficient feed rate or total cessation of feed for a period of time can impair the efficient recovery of HF as the bed becomes saturated. Similarly high concentration of HF in pot gas due to poor pot operating conditions also can overload the sorptive capacity of available alumina. Provided an adequate alumina supply, the recovery of HF gas is invariably above 99% and frequently above 99.9%.

The interception of particulate fluoride compounds takes place primarily in the fluid bed of the reactor. Fluoride concentration in the bed is found to be about $3\frac{1}{2}$%, nearly twice the concentration of hydrogen fluoride sorbed on alumina.

The final interception of particulate, both alumina and fluoride, is a mechanical filtration step which depends upon bag condition. Old or worn bags or poor connection to the tube sheet cause particulate loss. Bag fabric quality, a highly variable factor in its own right, also weighs heavily on particulate recovery performance.

Solids efficiency tends to be less than gas efficiency, although this is not always the case. The two are not related because of factors previously discussed. As we are more vulnerable to particulate losses than to gas phase fluoride losses, so are we more vulnerable to fluoride loss in particulate than to alumina loss. About 10% of total particulate loss is fluoride compounds.

The Alcoa 398 system, or any other treatment system, can only capture fluorides from pot fume collected and delivered to the system. Hooding and fume duct, therefore, become important parts of the control system. In newer potlines, hooding is designed to achieve 95% interception.

Measurements of losses through roof vents help to determine overall fume control performance as summarized in Table III.

"The recovery of fluorine in a directly reusable form in the process results in reduced fluoride additions to smelting installations. In addition to the performance measurements taken from operating reactors and potroom monitors, patterns of fluorine consumption by individual pots, potlines, and entire plants were examined. Observation of the cells served by the prototype installation of Alcoa 398 Process revealed a reduction in aluminum fluoride consumption from 101 lb/ton aluminum produced to 32 lb/ton aluminum. Since AlF_3 is added to bath neutralize Na_2O incoming with alumina as well as to compensate for loss of fluorides to potlining and atmosphere, the fluorine cycle in smelting cells is complex, and part of this reduction in aluminum fluoride consumption was compensated by additional cryolite requirements. A more general measure of fluoride recovery was derived by evaluation of calcium fluoride sludge collected in ponds serving wet scrubber installations. All of the HF and solids which end in a sludge pond, and a little more besides, are returned directly to the pots with this process. On a fluorine basis, 30 lb of fluorine are retrieved per ton of aluminum produced."

Packed towers have also been used successfully[51], particularly in phosphate fertilizer manufacture.

Use of the packed tower principle has been twofold:

1. To handle high SiF_4 gas loadings in a single, long vessel. When high SiF_4 gas loadings are scrubbed, the plugging effects can be disastrous unless there are provisions for a spray chamber to reduce the concentration of SiF_4 and for the subsequent removal of silica. A working design includes a crossflow packed section for optimum removal of SiF_4 and then a counterflow packed section.
2. To handle very weak SiF_4 concentrations (less than 0.5 mg as F/std. cu. ft.) in a simplified counterflow tower or to serve as a positive entrainment eliminator.

"Generally, the cyclonic scrubber or packed tower requires 7 to 10 gal/M cu. ft. for every 30 mg as F reduction. Depending on the severity of service, liquid pressure at the cyclonic and spray chamber may be 50 to 100 lb/sq in. gauge, while the liquid pressure at the packing area, depending on the method of distribution, may be 5 to 20 lb/sq in. gauge."

An interesting example of a multiple gas cleaning operation involving SO_2, mercury, selonium, chlorides and fluorides has been described elsewhere[52].

REFERENCES

1. Gerhard, E. R. in *Air Pollution Handbook*, Magill *et al.* edit., McGraw Hill (1956).
2. Cadle, R. D., ibid.
3. Niepenburg, H., *J. Fuel Heat Technol.* **14**, 34 (1966).
4. Ermenc, E. D., *Chem. Eng. Prog.* **52**, 149 (1956).
5. Ibid., *Chem. Engrg.* **77** No. 12, 193–6 (1970).
6. Haagen-Smit, A. J., *Ind. Eng. Chem.* **44** 1342 (1952).
7. Bartok, W., A. R. Crawford, and A. Skopp, *Chem. Eng. Prog.* **67**, No. 2, 64–72 (1971).
8. Hanst, P. L., *J. of A.P.C.A.* **21**, 269 (1971).
9. Tow, P. S., *Jnl. Air Pol. Control Ass.* **7**, 234–240 (1957).
10. Nat. Air Pol. Contr. Admin., U.S. Dept. of H.E.W., Report AP-66, March 1970.
11. Bernstein, L. S., *et al.*, Soc. of Auto. Engrs., Paper 710014 Presented at Automotive Engineering Congress Detroit, Mich. Jan. 1971.
12. Nat. Air Pol. Contr. Admin., U.S. Dept. of H.E.W., Report AP-66, March 1970.
13. Kangas, J. *et al.*, *Chem. Engrg.* Sept. **6**, 55–57 (1971).
14. Taylor, Sir Geoffrey, *Proc. Roy. Soc.* **A 219** 186 (1953), **A 223** 446 (1954) and **A 225** 473 (1954).
15. Griffiths, A., *Proc. Phys. Soc.* London **23**, 190 (1911).
16. Fischer, H. B., *Air and Water Pollut. Int. Jl.* **10**, 443 (1966).
17. Langmuir, I., *J. Amer. Chem. Soc.* **40**, 1361 (1918).
18. Brunaner, S., P. H. Emmett, and E. Teller, *J.A.C.S.* **60**, 309 (1938).
19. Higbie, Ralph, *Trans. Am. Inst. Chem. Engrs.* **31**, 365 (1935).
20. Gilliland, E. R., and T. Sherwood, *Ind. and Eng. Chem.* **26**, 516 (1934).
21. Stephens, E. J., and G. A. Morris, *Chem. Eng. Prog.* **47**, 232 (1951).
22. Onda, K., H. Takeuchi, and Y. Okumoto, *J. Chem. Eng.* (Japan) **1**, 56 (1968).
23. Andrew, S. P. S., *Alta Technologia Chimica*, Roma p. 153 (1961).
24. Calderbank, P. H., and M. B. Moo-Young, *Chem. Eng. Sci.* **16**, 39 (1961).
25. Danckwerts, P. V., *Gas Liquid Reactions*, McGraw-Hill (1970).
26. Bird, R. *et al.*, *Transport Phenomena*, Wiley (1960).
27. Schmidt, A. X., and H. List, *Material and Energy Balances*, Prentice Hall (1962).
28. Fryling, G. R. ed., *Combustion Engineering*, C. E. Inc. (1966).
29. Stern, A. C. ed., *Air Pollution* Vols. 1–3, Academic Press (1968).
30. Talmage, W. P., *Chem. Tech. P.* 117, Feb. 1971.
31. Cook, W. H., and D. Basmadjion, *Can. J. of Chem. Eng.* **42**, 146 Aug. 1964.
32. Ray, G. C., and E. O. Box, Jr., *Ind. and Eng. Chem.* **42**, 1315 (1950).
33. Mars, P., and J. G. H. Maessen, *J. Catal.* **10**, 1 (1968).
34. Falkins, and E. Miller, *Ind. and Eng. Chem.* **29**, 241 (1957).

35. Strauss, W., *Industrial Gas Cleaning*, Peraman Press Ltd. (1966).
36. *Int. Critical Tables*, Ed. Washburn, E. W., McGraw-Hill, N.Y. (1926).
37. Laidler, K. J., *Chemical Kinetics*, McGraw-Hill, N. Y. (1965).
38. Reid, R., and T. K. Sherwood, *Properties of Gasses and Liquids*, McGraw-Hill, N.Y. (1968).
39. Weissmantel, G. E., *Chem. Eng.*, **77**, No. 23, 82 (1970).
40. First, M. W., and F. J. Viles, *J. A. P. C. A.* **21**, 122 (1971).
41. Sunderason, B. B., C. I. Harding, and P. P. May, *Env. Sci. Tech.* **1**, 151 (1967).
42. Peters, M. S., and C. P. Ross, *A. I. Ch. E. J.* **1**, 105 (1955).
43. Koval, E. J., and M. S. Peters, *Ind. Eng. Chem.* **52**, 1011 (1960).
44. Lee, D. R., *Heat Piping and Air Cond.* **42**, No. 4, 76 (1970).
44a. Brooman, D. L., and E. Edgerley, Jr., *J. A. P. C. A.* **16**, 25 (1966).
45. Bagwell, F. H. *et al.*, *J. A. P. C. A.* **21**, 702 (1971).
46. Newman, D. J., *C. E. P.* **67**, 79 (1971).
47. Adlhart, O. J., S. G. Hindin, and R. E. Kenson, *C. E. P.* **67**, 73 (1971)
48. Andersen, H. C., W. J. Green, and D. R. Steele, *I. and E. C.* **53**, 199 (1961).
49. Teller, A. J., *C. E. P.* **63**, 74 (1967).
50. Cook, C. C., G. R. Swany, and J. W. Colpitts, *J. A. P. C. A.* **21**, 479 (1971).
51. Specht, R. C., and R. R. Calaceto, *C. E. P.* **63**, 78 May (1967).
52. Kangas, T. *et al.*, *Chem. Eng.* p. 55, Sept. 6 (1971).

EDWARD N. ZIEGLER
Polytechnic Institute of New York

GROUNDWATER

The earth has been likened to a spaceship in which one carries all of his resources with him. The particular resource of present interest is water. The quantity of the earth's waters have been estimated to be in excess of three hundred million cubic miles. If nature and man did not continually recycle this supply it would be quickly depleted. Recycling allows man to supply his need for water for all his uses. This recycling may continue to supple his needs only so long as the quality of the resource is sufficiently high to satisfy his needs. As man continues to pollute, the quality must continue to degrade. While the situation may not yet be of catastrophic importance it is beginning to take on increasing and often alarming significance

The waters of present interest are those which are found in consolidated or unconsolidated sedimentary deposits, fissures in igneous rocks, limestone caverns and other water bearing sources. These waters are below the groundwater table which may lie near the surface or may be hundreds of feet beneath the surface.

In order to establish an intuitive feel for the importance of groundwater it is necessary to describe its relation to the total water resource. The oceans, which of course are saline, cover approximately 71% of the earth's surface and contain about 97% of the earth's waters. The remaining 3% represents the total fresh water on the earth at any one time. This fresh water is found in the atmosphere as water vapor, on the land as water, dew, snow, ice, and in the ground usually in the form of water.

This total supply of fresh water (estimated at about 12 million cubic miles) vast as it may be, would be quickly depleted if it were not for its endless movement from the ocean and back.

This movement of the earth's waters in shown schematically in Fig. 1.

Since the total supply is constant we can easily conclude that on a long term average basis the budget equation for fresh water is simply

Total precipitation = evaporation + transpiration.

Table 1 indicates that at any one time approximately 99.5% of all the earth's water is in a form of location which renders it unavailable for use, leaving but 0.5% of the earth's water to fill man's needs.

It is obvious that the largest amount of available water is in the form of groundwater. This is fortunate for several reasons including the fact that in regions where surface supplies may be nonexistent groundwater is usually available. To emphasize this, it is noted that for about 13 states, groundwater constitutes 25 per cent or more of their total supplies. The average % groundwater use for the U.S. as a whole is approximately 20%. Further it should be noted that groundwater is not as susceptible to degradation by man's various activities

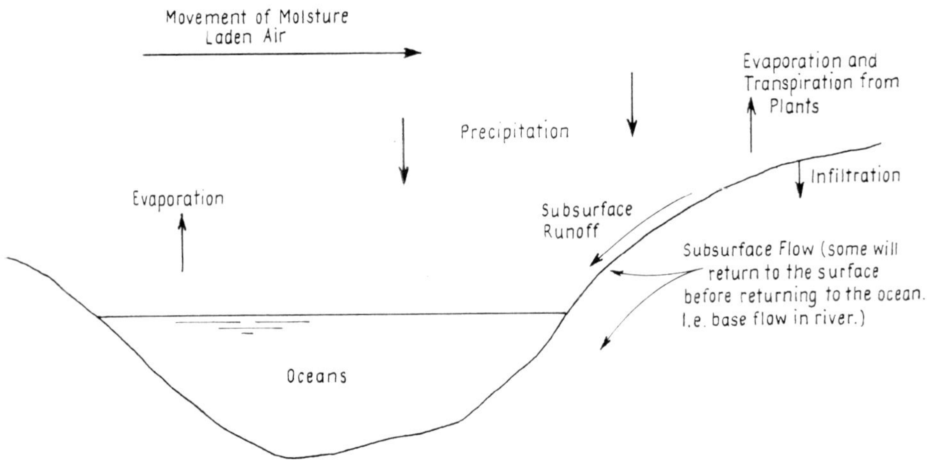

Figure 1 The hydrologic cycle.

TABLE 1

*Average Distribution of Earth's Waters**

Source	Volume in Cubic Miles	Volume in Cubic Feet	Percentage of Total Supply
Oceans	328×10^6	5×10^{19}	97%
Polar Ice Caps and Glaciers	7.1×10^6	1.5×10^{18}	2.1%
Ground water (2,500 to 12,500 ft) (below the surface)	1.3×10^6	2×10^{17}	0.39%
Ground water (0 to 2,500 ft) (below the surface)	1.05×10^6	1.5×10^{17}	0.31%
Fresh water lakes	2.7×10^4	3.8×10^{15}	0.08%
Soil Moisture above the ground water table	5.4×10^3	7.7×10^{14}	0.017%
Rivers	2.7×10^3	3.85×10^{14}	0.008%
Atmosphere	3.1×10^3	4.4×10^{14}	0.010%
			99.915%

* These figures are gross estimations only. Therefore the percentages will not add up to 100%.

and hence somewhat protected from undesirable contamination.

Another interesting figure leads to a gross estimate of the rate of replenishment.

The total annual precipitation for the earth as a whole is about 34 inches per year. Since the earth's surface is roughly 196 million square miles, this precipitation figures out to be about 95,000 cubic miles while from the table above the total moisture in the atmospheres is replenished about once every week and a half or a little more often.

In general the hydrologist is concerned with two main problems.

1. The determination of the long range *yield* from watersheds or aquifers, i.e., how much water can we expect to get from a watershed over an indefinite period?

In general, yield is a function of the time storage capacity of the watershed and the supply which fluctuates with time. Drought periods, of course, are the critical times and the problem is to ensure enough supply to last through such times. It is not enough to supply storage capabilities if supply isn't there.

2. The second problem of concern is just the opposite of the above, i.e., the very short high flow periods of flooding.

While flooding may sometimes be a problem in free surface aquifers, the rate of movement is relatively slow allowing time to provide for its effects. It might be pointed out, however, that since its presence cannot usually be seen until it presents some problem, in making any change or contemplating any change in a drainage basin one should give thought to this aspect of groundwater. For the most part, however, the main problem in groundwater is supply, i.e., the determination of yield and the development of drainage basins. Typical means of groundwater removal are shown in Figure 2. Qualitatively the flow rate in rivers is on the order of several miles per day, the flow rates in phreatic of "free surface" aquifers may be several feet per day, while flows in confined or artesian aquifers are much less. (It has been estimated that it takes five hundred years for rain seeping into the ground in Pennsylvania to reach lower New

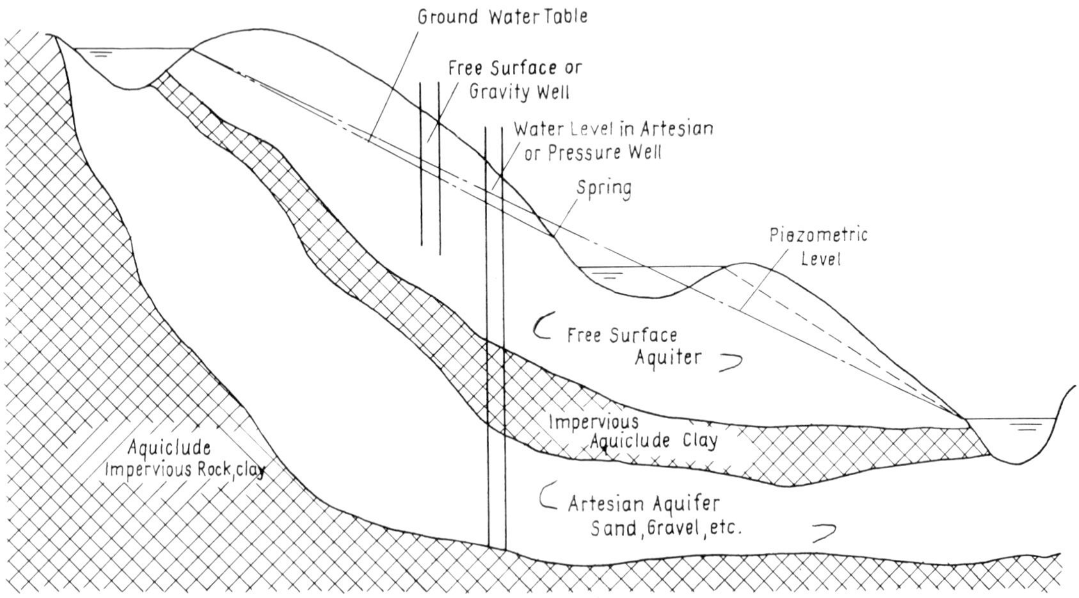

Figure 2 Types of groundwater withdrawal.

Jersey, and perhaps more astounding that it takes from 500,000 to a million years for water to travel a hundred miles in the Albian Formation beneath the Sahara Desert.)

The development of groundwaters which discharge through springs represents the relatively simple problem of improving the point of egress. While this may often prove technically difficult, it seldom represents any dislocation or disruption in the general hydraulics of the system. Such is not the case when subsurface waters are extracted through wells or borings. Since these are not natural points of egress, consideration must be given to all essential factors and the degree to which they upset the natural "equilibrium" of the system. The key factors which determine the effect of well extraction are:

1. The characteristics of the replenishment supplies.
2. The distance of wells from the recharge zone.
3. The distance of wells from the natural discharge zone, and
4. The character of each well's cone of depression.

While the quantity of available groundwater and the study of the movement of groundwater ("pure" groundwater hydrology) is still of ever increasing interest to engineers and scientists, an increasing interest in the quality of groundwater has been evidenced recently. This interest is hardly surprising. The study of groundwater has its major emphasis on water supply. In most portions of the world, even some of the most arid or heavily populated and industrialized regions the problem of supply is not one of having enough water, but rather one of having fresh, clean, usable water, where and when it is needed. Vast quantities of water are of little use if that water is polluted to an extent which makes it unfit for its intended use.

In order to discuss the problems of groundwater quality, it is first necessary to define what is meant by *pollution*. Pollution is a difficult word to define exactly. However, it is essential to the discussion of quality to have a working definition of pollution. It is extremely difficult to obtain chemically pure water, that is H_2O. For a drinking water source one would not want pure H_2O even if it could be obtained. Pure water is tasteless or seems to have a flat taste. Thus in a strict sense all the water that is used is contaminated, i.e. it contains something other than H_2O. It is hard to draw a line as to when a water is contaminated and when it is polluted. The average drinking water contains approximately 0.03% by weight material other than H_2O. but an average domestic sewage contains only 0.06% by weight material other than H_2O. It is not at all difficult to find drinking water supplies which contain 0.1% contamination by weight nor is it difficult to find domestic sewages which contain 0.04% contaminants by weight or even less. It is obvious then, that the amount of contaminants cannot be used to define when a water is polluted. One good definition is that a water may be considered polluted when it is contaminated to such an extent that its intended use is affected. Thus, 0.1% contaminants may not affect a water as a drinking water supply, but 0.04% contaminants may render a water unfit for any use.

Another tact may be taken in defining *pollution*. Botanists define a weed as "a plant out of place." The National Academy of Sciences, National Research Council has proposed a definition of pollution as "a resource out of place." Chemical fertilizers with their phosphates and nitrates are a valuable resource when spread on farming land, but they are pollutants when they are dissolved in the waters of a lake and the nitrogen in the form of nitrates is not only a pollutant but a health hazard when dissolved in drinking water supplies.

Once pollution has been defined it is possible to discuss water quality. For the purposes either of the above definitions will suffice. Of interest to groundwater hydrologists is the quality of existing groundwater resources and the effect that man has on the quality of groundwater. Man effects groundwater quality in many ways, recharge of aquifers, often with reclaimed wastewaters, the leaching of contaminants into groundwater supplies from agricultural operations, wastewater discharges, sanitary landfills, and others, and the intrusion of salt water into fresh groundwater supplies due to lowering of water table levels by withdrawal of water for use.

History

The history of groundwater hydrology, like the history of science in general, does not show discrete jumps in knowledge, but rather a continum of learning from broad attempts to qualitatively describe the relationships observed in nature to the detailed quantitative descriptions of particular phenomena.

A long time passed in the history of man before the concept of the *hydrologic cycle* was established. Homer in the *Iliad* mentioned the origin of groundwater. Herodotus described in very rough terms the relationship between evaporation, rainfall and runoff. Also before the birth of Christ, Anaxagoras qualitatively described the relationships between evaporation, rainfall and groundwater recharge. Plato wrote about rainfall runoff and the infiltration of surface water as the source for springs. Aristotle recognized the hydrologic cycle including evaporation, condensation, streamflow, infiltration and the reappearance of groundwater as springs.

In the XVIth Century, Leonardo da Vinci recognized the existence of artesian aquifers and Bernard Palissy described the correct relationships between wells and rivers, the hydrologic cycle, and variations of water levels in wells. The XVIIth Century brought the work of Perrault who showed that the quantity of rainfall is more than sufficient to account for the flows of rivers and springs, Mariotte who showed that the water in springs and wells is supplied by infiltration and Halley who showed that evaporation was sufficient to account for the return of water flowing into the sea. In 1791 La Metherie described the fate of rainfall. He accounted for direct runoff, evapotranspiration and percolation.

Between 1856 and 1967 developments established the principles needed to make the study and evaluation of groundwater resources a quantitative subject. In 1856 Darcy studying the flow of water in sand, developed *Darcy's Law* ($V = Ki$) which expressed the velocity of percolation as a function of permeability and hydraulic gradient. Darcy's Law provides the basis for many quantitative methods used in groundwater evaluation. In this period most of the developments centered on the problems of well hydraulics and well-field design. This is probably due to the fact that the most pressing problem in the field of groundwater hydraulics has been that of water supply. Since 1968 the tendency in groundwater research has been a renewed interest in Darcy's Law. Attempts are being made to define exactly the meaning of the permeability constant, hydraulic head and hydraulic gradient. Attempts are also being made to make much wider application of *Darcy's Law* and to determine if other general laws exist which also govern groundwater movements.

Future research, while not abandoning the question of groundwater hydraulics, will move into the areas of groundwater resource management including such topics as economical use and management of a groundwater resource and environmental considerations in groundwater management. The problem of recharge of aquifers using reclaimed wastewaters has been and will continue to be studied. Multipurpose use of groundwater resources will be studied. One of the major areas of study in the future will be the study of man's effect on the groundwater resource.

The history of groundwater research belongs to fluid mechanics, hydrologists, mathematicians, engineers and geologists. The future belongs to these people but with the addition of sanitary engineers, microbiologists, ecologists, planners and management specialists.

Quality

The quality of any water is a relative term. Water which is of good quality for an industrial cooling water supply may be of unacceptable quality as a drinking water source. Generally, water quality may be considered to be a function of chemical, biological and radiological content, temperature and intended use.

As water moves through the *hydrologic cycle* its quality is altered. Even if, for the moment, the effect of man and his activities on water quality is ignored, large changes in water quality occur during its movements. Even the "purest" rainwater contains substances other than H_2O. Rainfall by the time it reaches the surface containes dissolved gasses (CO_2, O_2, N_2, NO_2, NH_3, CO, CO_2, SO_2, H_2S, etc.) dust particles, smoke particles, bacteria, virus, radionucleides, dissolved vapors, dissolved ions (SO_4, NH_4, etc.), plus various kinds of dissolved organic materials. The vast majority of these substances are of little or no importance in biological processes, or are present in such small concentrations that they cause very few pollutional problems.

As water reaches the soil further changes in its quality occur. As water percolates through the biologically active soil mantle, organic material, both suspended and dissolved, is removed due to aerobic bacterial action. Filtration occurs as does ion exchange and adsorption. Thus, by the time water has reached the level of rock, a great deal of the organic material, effectively all of the suspended material, and some of the dissolved inorganic material has been removed. However, inorganic materials may also have been added to the water, due to ion exchange reactions and chemical equilibriums. For example, ammonia dissolved in the rainwater or present in the soil will be oxidized by soil bacteria to nitrate and the nitrate will be

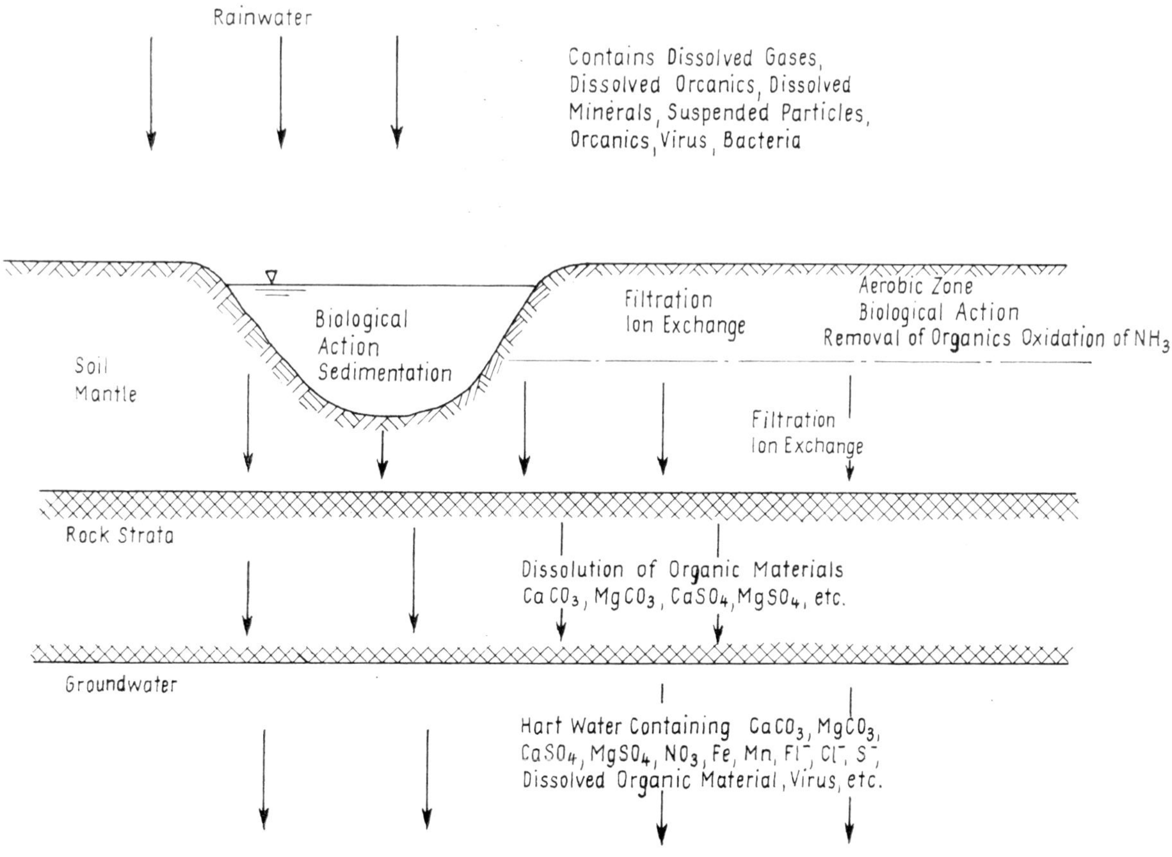

Figure 3 Effect of water movement on water quality.

released into the water and carried downward to the water table. Since bacteria do not remain in the percolating water, groundwater will be free of bacteria, but any organic material remaining dissolved in the water will not be removed. As the water moves through rock strata, it picks up inorganic matter by dissolution. It is this dissolving action which is responsible for the relative hardness of most groundwaters. It is also this action which is responsible for caves and caverns, especially in limestone ($CaCO_3$) strata. The natural changes in groundwater quality may be seen summarized in Figure 3.

Well hydraulics

The quantity $\frac{\text{(Water/Unit Area)}}{\text{Unit Change in Head}}$ is the amount of water released from a column of an artesian aquifer having unit cross sectional area when the head acting on the aquifer is reduced by 1 foot. This quantity, S_c, is dimensionless and is usually called the *Storage coefficient* or sometimes referred to as the specific yield and has the form:

$$S_c = \varrho g m \theta \left(\frac{1}{E_w} + \frac{c}{\theta E_s} \right)$$

Where: ϱ = Density
g = Gravitational Acceleration
θ = Porosity of the aquifer
E_w = Bulk modulus of elasticity of water
E_s = Bulk modulus of elasticity of the aquifer
c = Relative area at the top of the aquifer over which the pressure acts.

To illustrate some of the inadequacies of the expression, it should be pointed out that in its derivation it is assumed that the soil above the aquifer is void of water. This could only be true if the aquifer was solid rock. It is much more likely that the overlying aquiclude will be clay. Clay is a highly porous material, although it has a very low permeability, and hence over the long period of time through which such aquifers are formed would undoubtedly have absorbed much water. Secondly, it assumes that release occurs instantaneously and that any compaction occurs only in the aquifer. If the aquifer is a limestone, such compaction is not too likely. On the other hand, clay will compact quite readily when it loses its moisture content. Before controlled withdrawals were put into effect, the ground levels of Mexico City subsided several feet. It is believed that most or much of this subsidence was caused by the compaction of the clays overlying the limestone formation from which water is drawn. In any case if compaction does occur, S_c will vary. Since the variation in S_c is rather slow, however, S_c can be considered to be a property of the soil for most purposes.

Since we are concerned with the extraction of water from underground sources, we will begin with the derivation of the *non-equilibrium drawdown* equation for an artesian aquifer. Although the derivation is based on a highly idealized situation, as we shall see later, it is itself a useful tool and with slight modifications illuminates many factors which need to be understood in order to evaluate and interpret field data which is necessary in order to properly develop any groundwater basin. Figure 4 illustrates the assumptions of the derivation.

Assumptions:

1. The artesian aquifer is horizontal and infinite in extent in all directions.
2. There is no leakage from above or below. All water comes from the aquifer. This is another way of saying that all flow is radial.
3. The non-pumping piezometric surface is horizontal, i.e., initially there is no flow in the aquifer.
4. S_c is constant, i.e. soil properties are uniform (homogeneous).
5. The well screen penetrates the entire aquifer. This says that all streamlines are horizontal.
6. There is no lag time in S_c, i.e. we start pumping at a rate equal to Q instantaneously.

In addition we made use of *Darcy's Equation* which is:

$$Q = KiA$$

Where: Q = discharge or flow rate
K = coefficient or permeability
i = slope of the piezometric surface or gradient
A = area through which flow occurs

From the basic mass balance equation it can be shown that the resulting expression is as follows:

(Supply) − (Outflow) = Change in storage

OR

$$\frac{1}{r}\frac{\partial s}{\partial r} + \frac{\partial^2 s}{\partial r^2} = \frac{S_c}{T}\frac{\partial s}{\partial t}$$

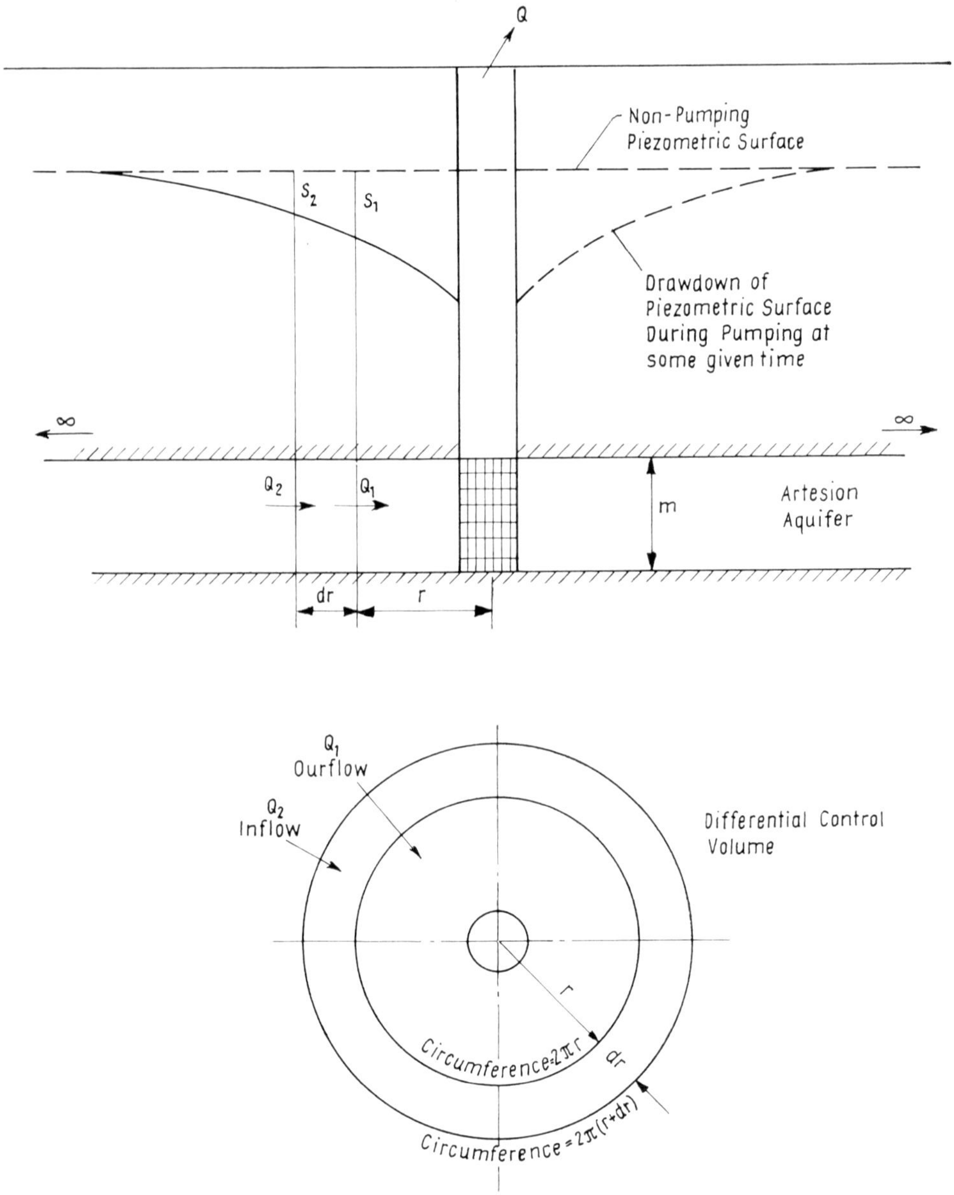

Figure 4 Definition sketch for the derivation of the nonequilibrium Theis equation.

where T is $= Km =$ coefficient of transmissibility cfd/ft. The pertinent initial and boundary conditions are:

At $t = 0$ $\quad S = 0$ and $Q_1 =$ pumping rate $= Q$

$R = Rw$

At $r = \infty$ $\quad Q_2 = 0$

$$\frac{\partial s}{\partial r} = 0$$

With these conditions the above equation can be solved. The final solution is:

$$S = \frac{Q}{4\pi T} \int_{-\mu}^{\infty} \frac{e^{-\mu}}{\mu} d\mu = \frac{Q}{4\pi T} W(\mu)$$

where $W(\mu) =$ Well Function which is a definite integral for which values have been tabulated and

$$\mu = \frac{r^2 Sc}{4Tt}$$

This equation is sometimes called the *Theis Equation* and is the expression for the drawdown for radial flow to a well which has *not* reached equilibrium. It is perhaps the most widely used equation (at least in the U.S.) for the field determination of T and S_c. However, due to the mathematical difficulties involved, the direct usage of the equation is seldom attempted. Rather one of several approximate but readily applied techniques are used.

The Theis method

This is simply the solution of the above equation by graphical means and is as follows:
The Theis equation may be rewritten in the form

$$W(\mu) = \frac{4\pi T}{Q} S$$

which for constant pumping rate becomes:

$$W(\mu) = K_1 S$$

further:

$$\mu = \frac{r^2 S_c}{4Tt} = K_2 \frac{r^2}{t}$$

Since S_c and T are soil properties, i.e., constants, it follows that:

$$\frac{W(\mu)}{\mu} \doteq \frac{S}{\frac{r^2}{t}}$$

For any given well which is pumped at a constant rate Q, the values of the drawdown, S, versus r^2/t could be plotted on log-log paper. For reasons which will become apparent later the drawdown values in the well being pumped cannot be used but rather the value of some observation well located at a distance, r, from the well being tested must be used.

If a curve of $\frac{W(\mu)}{u}$ vs $\frac{S}{r^2/t}$ is also plotted on log-log paper and these curves are overlapped with their co-ordinate axes parallel as shown in Figure 5, the values of $W(\mu)$, μ, s, and r^2/t may be obtained for the center of the matching segments.

Since

$$T = \frac{0}{4\pi S} W(\mu) \qquad T \text{ can be found.}$$

and then,

$$S_c = \frac{\mu\, 4T}{r^2/t}$$

and now S_c can be calculated.

Second method—the method of Jacob

This method involves a simplification of the Well Function itself. Hence is has certain disadvantages. However, in the method above, as the curves becomes flatter the length of the matching curves usually becomes greater, and more difficult to determine. Hence the accuracy of determination decreases as the curves flatten out. The present method does not suffer quite as badly.
The definite integral called the well function may be expanded into a convergent power series which is as follows:

$$W(\mu) = \int_{-\mu}^{\infty} \frac{e^{-\mu}}{\mu} d\mu$$

$$= \left[-0.5772 - \ln\mu + \mu - \frac{\mu^2}{2.2!} + \frac{\mu^3}{3.3!} - \frac{\mu^4}{4.4!} + \cdots \right]$$

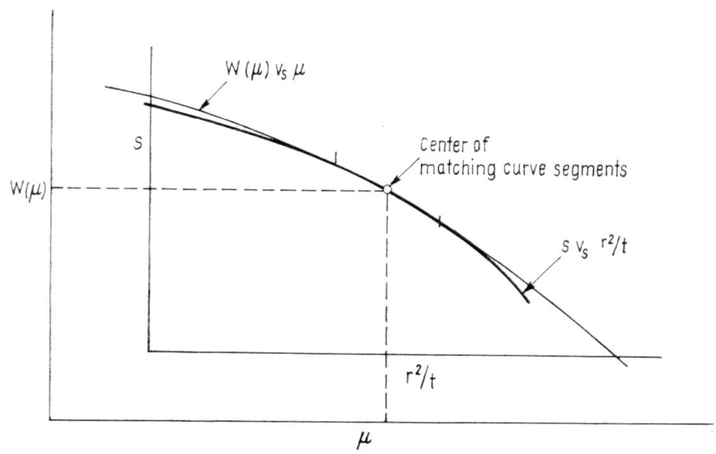

Figure 5 Application of Theis method.

If μ is small, the well function may be approximated by the equation:

$$W(\mu) = 0.5772 - \ln \mu = \ln \frac{1}{\mu} - \ln 1.78$$

since

$$\mu = \frac{r^2 S_c}{4Tt} = K_2 \frac{r^2}{t}$$

$$\mu \rightarrow \text{small} \quad \text{if} \quad r \rightarrow \text{small} \quad \text{or} \quad \text{if } t \rightarrow \text{large}$$

Hence for small μ the *Theis Equation* becomes

$$S = \frac{Q}{4\pi T}\left[\ln \frac{4Tt}{r^2 S_c} - \ln 1.78\right]$$

or

$$S = \frac{Q}{4\pi T}\left[\ln \frac{4Tt}{1.78 r^2 S_c}\right] = \frac{Q}{4\pi T}\left[\ln \frac{2.25Tt}{r^2 S_c}\right]$$

Up until now the units of the various terms of the equation that were being used have been consistent and are:

$$Q = \frac{\text{cubic feet}}{\text{day}} \qquad r = Ft$$

$$T = \frac{\text{cubic feet}}{\text{day}} \Big/ \text{Ft} \qquad t = \text{days}$$

Now if it is desired to convert gallons rather than cubic feet and to logs to the base 10 rather than natural logs, the equation becomes:

$$S = \frac{2.3Q}{4\pi T} \log_{10} \frac{0.3Tt}{r^2 S_c}$$

This is the form of the approximate solution best suited for field test purposes.

Assume that a test well is being pumped at a rate Q. Further assume that observation wells are located at r_1 and r_2 in a radial direction from the pumped well. As the test well is pumped the drawdowns are observed, as well as the time of drawdown, for each of the observation wells. On the basis of the approximate equation, it can be said that:

$$S_2 - S_1 = \frac{2.3Q}{4\pi T}\left(\log_{10} \frac{0.3Tt_2}{r_2^2 S_c} - \log_{10} \frac{0.3Tt_1}{r_1^2 S_c}\right)$$

where the subscripts refer to the observation wells. This equation can, of course be simplified to:

$$S_2 - S_1 = \frac{2.3Q}{4\pi T} \log_{10} \frac{r_1^2/t_1}{r_2^2/t_2}$$

Now assume that there is only one observation well, i.e., $r_1 = r_2$ but subscripts now refers to time at which observations are made and recorded. Then the above equation reduces to:

$$S_2 - S_1 = \frac{2.3Q}{4\pi T} \log_{10} \frac{t_2}{t_1}$$

This equation is now in a form which can be used to evaluate T from field data. The data is plotted on semi-log paper, reasons for which are obvious or will become obvious.

From the observation readings in the observation well with the test well being pumped at a constant rate Q, drawdowns versus time of drawdown can be plotted on semi-log paper as shown in Fig. 6.

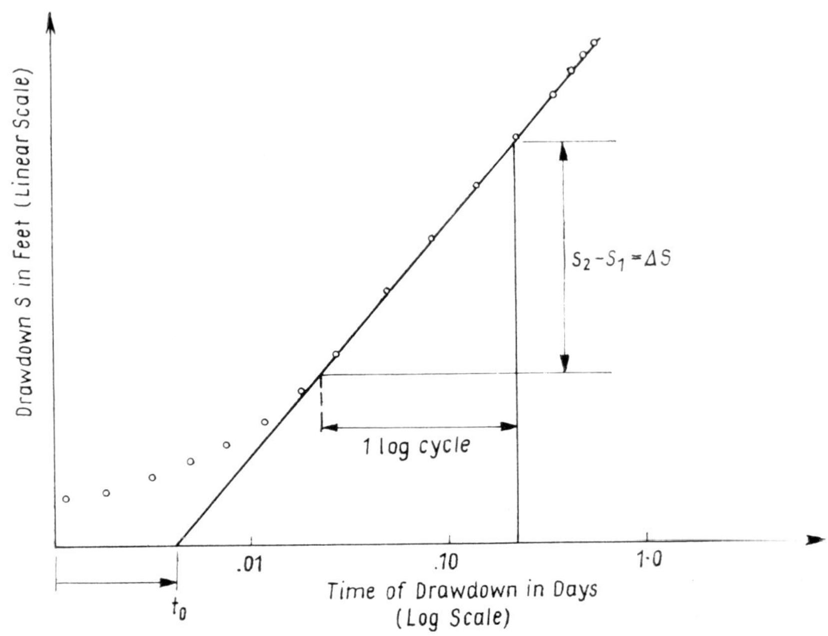

Figure 6 Application of semi-log drawdown curve—method of Jacob.

If points on the linear portion of the curve one log cycle apart are chosen we can simplify the calculations since then:

$$\frac{t_2}{t_1} = 10 \quad \text{and} \quad \log_{10}\frac{t_2}{t_1} = 1$$

Hence the equation now becomes simply:

$$\Delta S = S_2 - S_1 = \frac{2.3Q}{4\pi T}$$

or

$$T = \frac{2.3Q}{4\pi(S_2 - S_1)} \qquad \text{Hence } T \text{ is known.}$$

In this case Q and T are in GPD units i.e. GPD and GPD/Ft respectively. Now all that is needed is the value of S_c. Returning to the original approximate form of the drawdown equation which was:

$$S = \frac{2.3Q}{4\pi T}\log_{10}\frac{0.3Tt}{r^2 S_c}$$

Note that in simplifying the *Theis Equation*, it was assumed that μ was small. However, as t becomes small the approximation becomes invalid. The effect of this is shown in Fig. 6. Note that at small t the observed data (the small circles) deviate from the straight line, i.e. the approximation is in error in this region. However, if it is assumed that drawdown began at some time, t_0, and followed the linear curve, we have a means of evaluating S_c, i.e. extend the linear portion of the curve until it crosses the abscissa at t_0 when $S = 0$.

Then

$$S = \frac{2.3Q}{4\pi T}\log_{10}\frac{0.3Tt_0}{r^2 S_c}$$

But $S = 0$

Now since Q or T are not zero, S can be zero only if the log term is zero.

Hence

$$\log_{10}\frac{0.3Tt_0}{r^2 S_c} = 0$$

Therefore

$$\frac{0.3Tt_0}{r^2 S_c} = 1.0$$

or

$$S_c = \frac{0.3Tt_0}{r^2} \qquad \text{and } S_c \text{ is now known.}$$

A third approach that can be used involves *intermittent pumping* and *recovery*. When a well is being pumped the rate of pumping, Q, is sometimes somewhat erratic due to variations in the power source, vibrations etc., but when the pump is shut down, the water in the aquifer is still subject to the gradation of pressure of the cone of depression. Hence the flow continues towards the well after pumping ceases and at a rate Q which is, if anything, more nearly constant since it is dependent only on the pressure gradient. Hence data observation during recovery can be most useful. Further it is unlikely that the well would be pumped continously for any great length of time. Hence expressions for drawdown due to intermittent pumping are of interest in determining the long range effects of such pumping.

Intermittent pumping

To begin with assume that a test well is pumped at a rate Q for one day, shut the pump off during the next day and on the third day pump again at the same rate Q. This sequence would look as shown in Fig. 7(a). In order to approximate this situation mathematically, it is assumed that the actual pumping conditions were as shown in Fig. 7(b). I.e., the first day the well was pumped at a rate Q, the second day there were two imaginary pumps, one pumping at a rate Q, the other at $-Q$, the third day the original pump is operating plus the two imaginary pumps. The sum total is the same as the actual situation, but our situation is now mathematically continuous over the *w* whole period, and hence amenable to analysis. The drawdown for out idealized situation is shown in Fig. 7(c).

The appropriate expression during *recovery* is:

$$S = \frac{2.3Q}{4\pi T}\log_{10}\frac{t}{t - t_1}$$

where $t_1 < t < t_2$

If during *recovery* observations are taken out, and plotted on semi-log paper as in Fig. 8, and values are then taken over 1 log cycle so that the log term = 1:

$$T = \frac{2.3Q}{4\pi\,\Delta S}$$

where T is known and S_c can be found by the previous method using the observed values during recovery, i.e. find t_0 and evaluate as before. Since the determination of the *yield* of any particular basin and the manner in which the development of that basin is determined will depend on how readily the basin yields its wealth as well as on its total amount, this knowledge is quite important.

If the alternate pumping cycle was continued for say N (an even number of days) days, during the

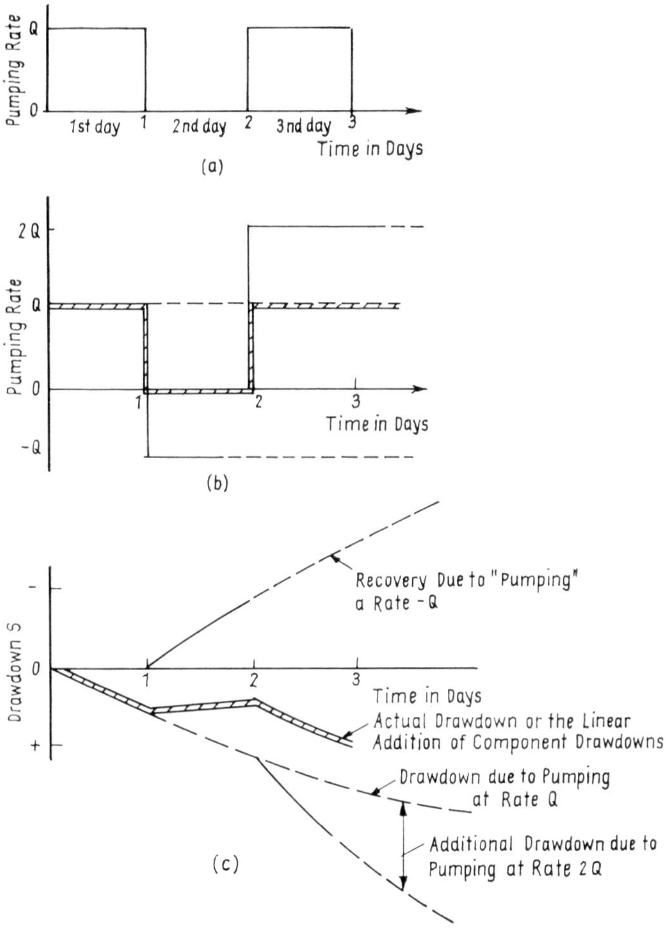

Figure 7 Pumping and drawdown curres—intermittent pumping.

Nth day the drawdown equation (using the approximate form again) would be as follows:

$$S_w = \frac{2.3Q}{4\pi T}\left[\log_{10}\frac{0.3Tt}{r_w^2 S_c} - \log_{10}\frac{0.3T(t-t_1)}{r_w^2 S_c}\right.$$
$$+ \log_{10}\frac{0.3T(t-t_2)}{r_w^2 S_c} - \log_{10}\frac{0.3T(t-t_3)}{r_w^2 S_c}$$
$$+ \cdots + \log_{10}\frac{0.3T(t-t_{N-1})}{r_w^2 S_c}$$
$$\left. - \log_{10}\frac{0.3T(t-tN-1)}{r_w^2 S_c}\right]$$

Since there are an even number of days there are an even number of + log terms, hence the factors $\frac{0.3T}{r_w^2 S_c}$ will all cancel when terms are combined and simplified to the following:

$$S_w = \frac{2.3Q}{4\pi T}$$
$$\log_{10} \times \left[\frac{t(t-t_2)(t-t_4)(t-t_6)\ldots(t-t_{N-2})}{t(t-t_1)(t-t_3)(t-t_5)\ldots(t-t_{N-1})}\right.$$

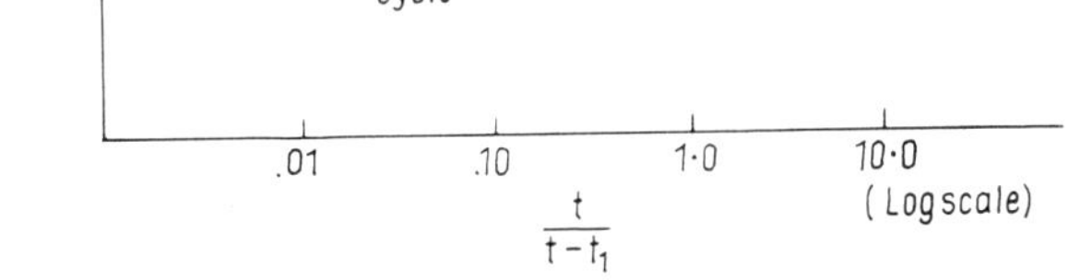

Figure 8 Plot of Drawdown vs. Time during recovery.

If we generalize this slightly to say that the periods of pumping and shut down are not necessarily equal but only that their sum is cyclical, we arrive at a somewhat more useful expression. In this light then let us assume that the period of pumping is p and that cycle is one day...p is a fraction i.e., some portion of the day. (It isn't necessary that the cycle be a day, but merely some consistent period.)

If we wish to determine the drawdown at the end of the *recovery* period of the $\frac{N\text{th}}{2}$ day, where N is now the number of on *and* off periods of pumping. Then

$$t = N/2 \text{ days} = d \text{ days}$$

$$t_2 = 1 \text{ day}$$

$$t_4 = 2 \text{ days}$$

$$t_6 = 3 \text{ days}$$

$$t_{N-2} = \frac{N}{2} - 1 \text{ day}$$

or $t\,(t - t_2)(t - t_4)\ldots(t - t_{N-2}) = d\,(d - 1)(d - 2)\ldots 1 - d$; also

$$t_1 = p \qquad t - t_1 = d - p$$

$$t_3 = 1 + p \qquad t - t_3 = d - (1 + p)$$

$$t_5 = 2 + p$$

$$t_{N-1} = d - (1 - p) \qquad t - t_{N-1} = d - d - (1 - p) = 1 - p$$

and

$$\ldots(t - t_1)(t - t_3)\ldots(t - t_{N-1})$$

$$= (d - p)(d - (1 + p)\ldots(d - (d - (1 - p)))$$

or

$$(1 - p)(2 - p)\ldots(d - p)$$

and substituting into the equation we have

$$S_R w = \frac{2.3Q}{4\pi T}\log_{10}\left[\frac{d!}{(1 - p)(2 - p)\ldots(d - p)}\right]$$

where d is in days or whole number of cycles. And p is *a* fraction of days or the cycle period.

Strictly speaking $S_R w$ only locates the upper peaks of the intermittent drawdown curve as shown in Fig. 9.

$S_R w$ is really an expression for the loci of points on the drawdown curve at the ends of the recharge periods.

With similar reasoning an expression for an odd number of pumping *and recovery* periods will result in the following:

$$S_w = \frac{2.3Q}{4\pi T}\left[\log_{10}\frac{0.3Tt'}{r_w^2 S_c} - \log_{10}\frac{0.3T(t' - t_1)}{r_w^2 S_c} + \cdots + \log_{10}\frac{0.3T(t' - t_{N-3})}{r_w^2 S_c} - \log_{10}\frac{0.3T(t' - t_{N-2})}{r_w^2 S_c} + \log_{10}\frac{0.3T(t' - t_{N-1})}{r_w^2 S_c}\right]$$

In this case the number of periods of pumping *and* recovery is now an odd number. And if it is assumed that t' is now one pumping period later that of t as assumed in deriving $S_R w$ then

$$t' = \frac{N}{2} + p \quad \text{or} \quad d + p \text{ days}$$

also

$$\log K_1 a - \log K_1 b + \log K_1 c = \log\frac{K_1 a}{K_1 b}K_1 c = \log\frac{K_1 ac}{b}$$

i.e. the constant term $\frac{0.3T}{rw^2}S_c$ no longer cancels.

Simplifying the equation we have

$$S_p w = \frac{2.3Q}{4\pi T}\log_{10} \times \left[\frac{t'(t' - t_2)(t' - t_4)\ldots(t' - t_{N-1})\,0.3T}{(t' - t_1)(t' - t_3)\ldots(t' - t_{N-2})\,r_w^2 S_o}\right]$$

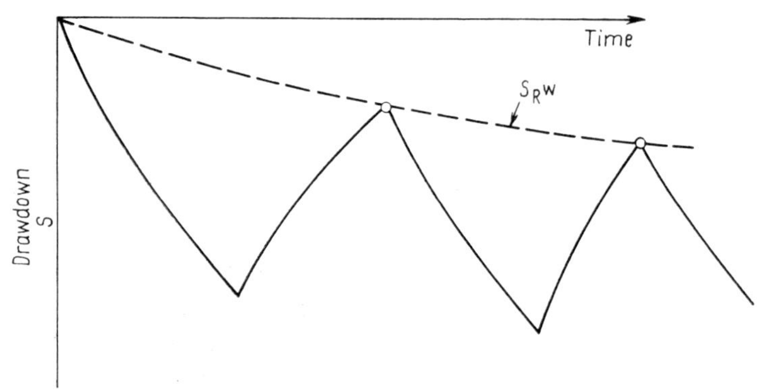

Figure 9 Drawdown vs. Time—intermittent pumping.

but now

$$t' = d + p \qquad t' - t_1 = d + p - p = d$$

$$t_1 = p \qquad t' - t_3 = d + p - (1 + p)$$

$$= d - 1$$

$$t_3 = 1 + p$$

$$t_{N-1} = d \qquad t' - t_{N-1} = d + p - d = p$$

$$t_{N-3} = d - 1 \qquad t' - t_{N-3} = d + p - (d - 1)$$

$$= 1 + p$$

Also

$$t' - t_2 = d + p - 1$$

$$t' - t_4 = d + p - 2$$

$$t' - t_{N-2} = d + p - d = p$$

$$\therefore \quad (t' - t_1)(t'_w - t_3) \dots (t' - t_{N-2})$$

$$= d(d - 1) \dots (1) = d!$$

or

$$t'(t' - t_2)(t_w - t_4) \dots (t' - t_{N-2})$$

$$= (d + p) \dots (2 + p)(1 + p) P$$

$$S_{Pw} = \frac{2.3Q}{4\pi T} \log_{10} \times \left[\frac{0.3T}{r_w^2 S_c} \frac{p(1 + p)(2 + p) \dots (d + p)}{d!} \right]$$

This expression is the loci of points on the drawdown curve at the ends of the pumping periods, i.e. points of greatest drawdown.

The value of this pair of equations is in determining the danger of drawing the well dry, i.e., is S_{Pw} in danger of reaching the bottom of the well. Further it is useful in determining the long term effects of *intermittent pumping* i.e. S_{Rw} becomes how severe?

So far we have dealt with rather idealistic situations. Seldom are field conditions so nice. In order to account for some of the different situations normally encountered in the field, the basic equations are modified slightly. Figure 10 illustrates a more nearly "normal" field experience. The losses indicated in Fig. 10 are assumed to be due to turbulent flow conditions. Hence.

$$\text{Well Friction Losses} = kQ^2$$

The form KQ^n is sometimes used but n is almost always 2 and the refinement seldom practical. From Fig. 10 then

$$S''_w = S'_w + kQ^2$$

but

$$S'_w = CS_w$$

where

$$S_w = \frac{2.3Q}{4\pi T} \log \frac{0.3Tt}{r_w^2 S_c}$$

And C is a coefficient which is a gross measure of the degree of well screen penetration, and hence flow curvature, the radius of the well itself, and the anisotropy of the permeability of the aquifer, i.e.

$$C = f\left(r_w, m, \frac{m'}{m}, \frac{K_H}{K_V}\right)$$

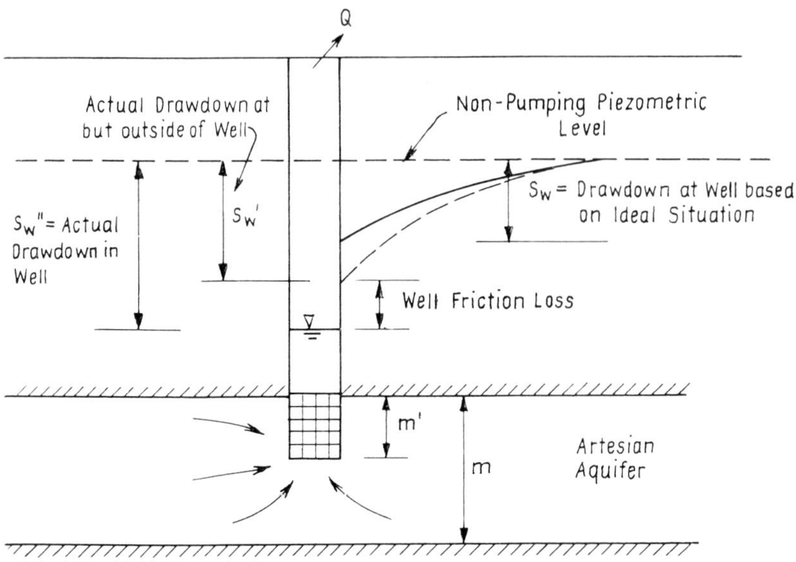

Figure 10 Definition sketch for drawdown from partially penetrating artesian well.

Where:

K_H = Horizontal coefficient of permeability

K_V = Vertical coefficient of permeability.

In general, C is too complex for analysis, and besides the unknowns cannot be too easily, if at all, determined with any great degree of precision. However, C, can be found by making field tests at the side.

Now,
$$S''_w = \frac{2.3CQ}{4\pi T}\log_{10}\frac{0.3Tt}{r_w^2 S_c} + kQ^2$$

The unknowns are T, S_c plus the new ones C and k.

Since the effects of C and k on the cone of depression are confined to the near vicinity of the well, the drawdown experienced by observation wells follows theory quite qell. Hence S_c and T may be found by the methods of Theiss or Jacob discussed previously. The problem, the, becomes one of how to evaluate C und k.

During pumping at a constant rate Q, observation readings of the drawdowns within the test well are taken. At two different times the readings are:

$$S''_{w2} = \frac{2.3CQ}{4\pi T}\log_{10}\frac{0.3Tt_2}{r_w^2 S_c} + kQ^2$$

$$S''_{w1} = \frac{2.3CQ}{4\pi T}\log_{10}\frac{0.3Tt_1}{r_w^2 S_c} + kQ^2$$

and their difference is.

$$S''_{w2} - S''_{w1} = \frac{2.3CQ}{4\pi T}\log_{10}\frac{t_2}{t_1}$$

Plotting the observed drawdowns versus time of dradown on semi-log paper, a straight line plot is obtained. If values are taken over 1 log cycle ($\log \frac{t_2}{t_1} = 1$) and:

$$S''_{w2} - S''_{w1} = \frac{2.3CQ}{4\pi T}$$

or

$$C = \frac{4\pi T}{2.3CQ}(S''_{w2} - S''_{w1})$$

and C is known.

To find k, the values determined for T, S_c, and C are substituted into the expression (at some time t with its corresponding S''_w)

$$S''_w = \frac{2.3CQ}{4\pi T}\log_{10}\frac{0.3Tt}{r_w^2 S_c} + kQ^2$$

from which k may be evaluated.

However, k is not a constant and may change with time due to plugging of the screen or the area adjacent to the screen by limestone deposits, corrosion, etc. It is sometimes possible to unplug such conditions by dropping dry ice into the well and redissolving the limestone. Surging i.e., over pumping, intermittently may gradually work the well free and redevelop it.

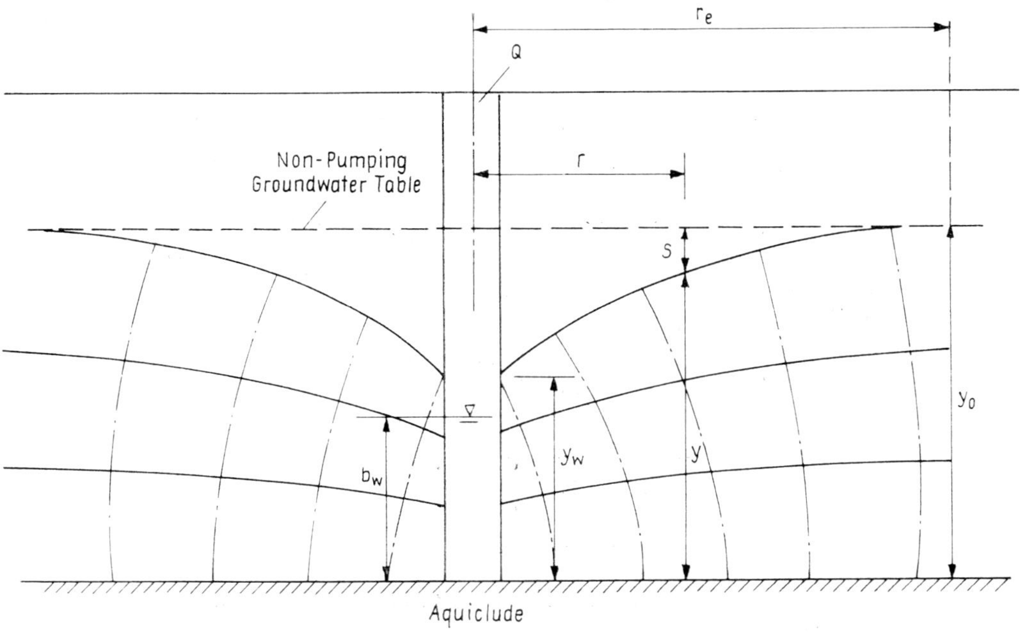

Figure 11 Definition sketch for free surface well.

Up until now our concern has been with the non-equilibrium or time dependent aspects of artesian wells. To conclude the investigations of *non-equilibrium* conditions let's now take a look at *phreatic* or *free surface* wells.

With the use of potential flow theory an exact expression for the drawdown curve can be derived (Hantush). Although the results are presented here the derivation has been omitted. The definition sketch for free surface wells is presented as Fig. 11.

Hantush's solution is

$$2y\bar{y} - y^2 = y_0^2 - \frac{2.3Q}{4\pi T}\log\left(\frac{r_0}{r}\right)$$

where $\bar{y}$ = depth of water in an observation well at a distance r, which is fully screened throughout the aquifer, i.e., this is the average hydraulic head in the observation well. Both theory and experiment indicate that if $r > \frac{3}{2}\, y_0; \bar{y} \to y$. At such distances Hantush's Equation becomes

$$y^2 = y_0^2 - \frac{2.3Q}{\pi K}\log\left(\frac{R_0}{R}\right)$$

Since time does not appear in this result the observed values of r_p, y_0, r and y must be made simultaneously, if this expression is to be used during the *non-equilibrium* period of pumping.

Another approach is to modify the artesian formula as follows:

For two observation wells, the difference in drawdown using the artesian approximate formula is

$$S_2 - S_1 = \frac{2.3Q}{4\pi T}\log_{10}\frac{(r_1^2/t_1)}{(r_2^2/t_2)}$$

Now comes the modification.

If T is now defined such that $T = \bar{y}K_p$.

Where K_p = effective permeability,

$$\text{and } \bar{y} = \frac{y_1 + y_2}{2}, \text{ the average head.}$$

Then when $t_1 = t_2$, i.e. for instantaneous observations this equation can be expressed as:

$$Q = \frac{\pi K_p(y_1^2 - y_2^2)}{2.3\log_{10}(r_1/r_2)}$$

Comparing this with the results of Hantush, it may be seen that when $r > \frac{3}{2}y_0$ the equations are identical.

This expression does *not* apply near the well, however, either Hantush's wholly penetrating screen or a well which penetrates to the bottom of the aquifer and hence responds to the piezometic pressures on the bottom of the aquifer will yield field-theory agreements on the order of 1 or 2%.

This completes the presentation of the *non-equilibrium* aspects of well hydraulics. It should be pointed out however, that the last pair of equations, i.e. the free surface equations do not include C and k factors. In the field evaluation of the aquifer constants adjustments for C and k may be necessary.

This series of equations which are really merely modifications and or extensions of the basic *Theis equation* enable the field determination of the characteristics of aquifers which are necessary for any evaluation of the *yield* of and developments of underground water resources.

In all of the foregoing discussions, stated and implied is the assumption that the spread of the cone of depression or the drawdown near the well will continue to spread indefinitely so long as pumping continues.

Since we would not normally be too likely to want to continue pumping a test well for more than a few days, it is not likely that we would achieve *equilibrium* conditions except for the free surface wells which may intercept a point of recharge in a few days or perhaps hours. Hence these *non-equilibrium* techniques are valuable. However, since the main concern is with the development of such underground reservoirs it *is* important that we understand *equilibrium* and how it might be reached under a variety of conditions.

True equilibrium between the rate of withdrawal and the rate of replenishment is seldom, if ever, achieved. However, a state of quasi-equilibrium can be attained through several means. Replenishment may be by a stream located near the point of withdrawal or, in the case of artesian aquifers by rains falling on an exposed portion of the aquifer miles away, or by rain waters infiltrating to the groundwater in the near vicinity of the well. Several of the ways in which equilibirum may be reached are shown schematically in Fig. 12.

Quality Considerations

Man's influence on the quality of groundwater results primarily from those activities which produce wastes. There are some rather large exceptions to this statement such as nitrates in waters due to

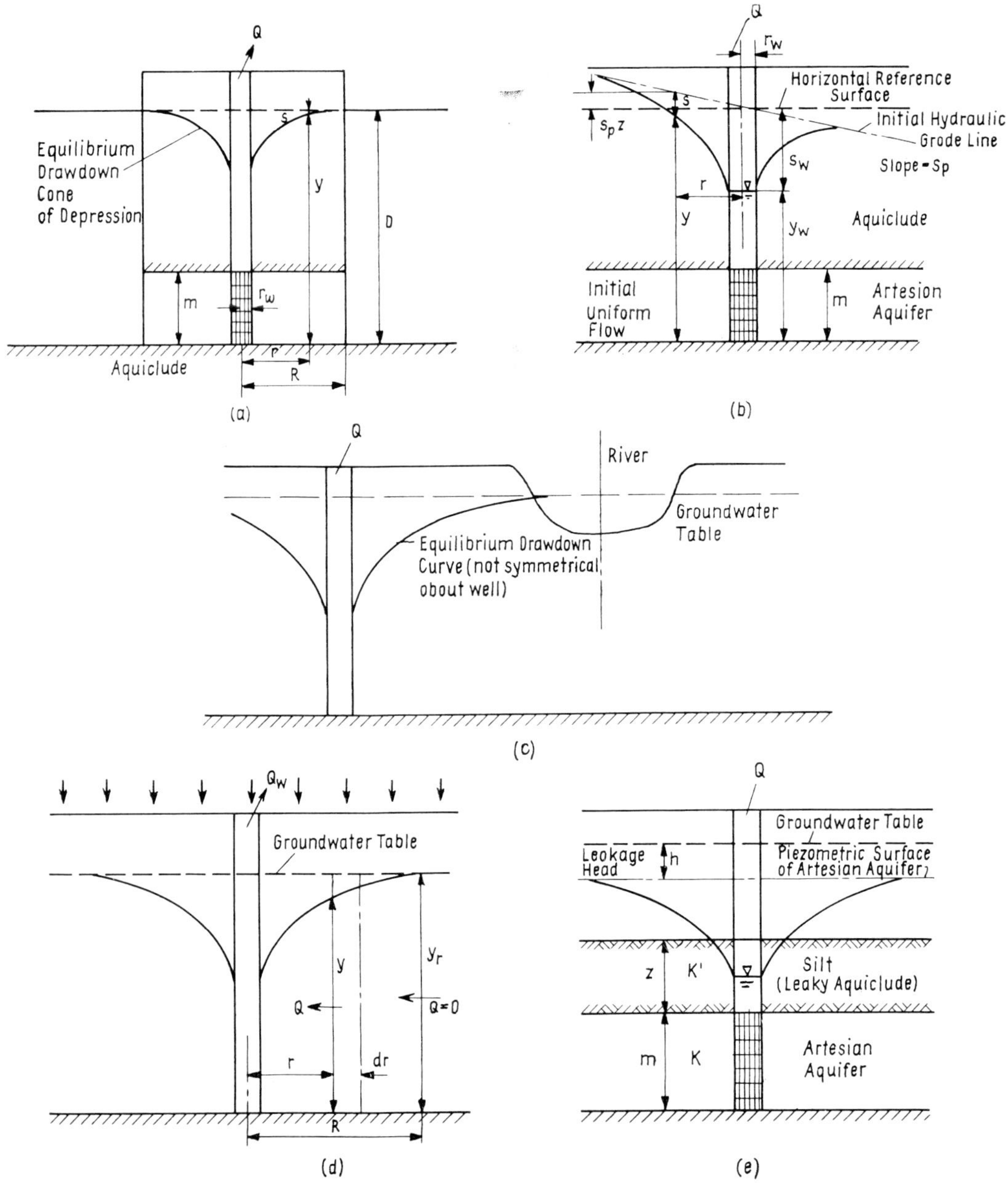

Figure 12 Composite sketch showing conditions under which equilibrium may exist.

fertilizers, and salt water intrusion due to pumping of aquifers, but primarily the major effects on groundwater quality may be attributed to pollutants from man's waste.

Pollutants have been broadly classified into eight categories by the U.S. Public Health Service.

1. Oxygen demanding wastes

These include domestic and industrial wastewaters, leachates from sanitary landfill and any other wastes containing biodegradable organic material. These wastes are stabilized by microorganisms and will remove dissolved oxygen from water. Oxygen demanding wastes may enter groundwater through infiltration. As the water pases through the biologically active soil mantle, much of the organic (oxygen demanding) material will be removed by biological action and filtration. However, organic material which is not removed in this zone will remain in the groundwater until the water again reaches a biologically active area.

2. Infective agents

Bacteria, virus, protozoa and other infectious agents may be present in domestic wastewaters. Those which reach drinking water or even recreational water supplies represent a health hazard to man. Fortunately, groundwater supplies are protected from the bulk of pathogenic organisms. Filtration of water through the soil removes all of the bacteria and other relatively large microorganisms. Virus may pass through the soil but the low rate of movement of groundwaters allows sufficient time in most cases for the virus to become inactive. There have been cases of illness due to contaminated groundwater sources, primarily caused by poor placement or construction of wells and septic tanks.

3. Plant nutrients

Mineral substances, particularly nitrates and phosphates stimulate the growth of plants. While plant nutrients as such are of little interest in groundwater, nitrates particularly in large concentrations as found in groundwaters infiltrating through heavily fertilized soil can be a public health hazard.

4. Organic chemicals

Chemicals such as insecticides, pesticides and detergents which are toxic at low concentrations, cause significant changes in groundwater quality. Some insecticides and pesticides are removed from infiltrating water by adsorption on clays. Detergents may cause severa problems in groundwaters used as drinking water supplies since under anaerobic conditions there is no breakdown of the ABS or LAS molecule. Too little is known about the long term effects of these chemicals to predict their exact dangers. However, it is known that they do present a hazard of undefined magnitude.

5. Other minerals and chemicals

This group consists primarily of materials found in industrial wastes, including chemical residues, petrochemicals, salts, acids, etc. New chemical substances are being created at the rate of 400 to 500 compounds per year. Many of these substances are known to be toxic although very little is known about their biological action. Almost nothing is known about possible long term effects of these chemical compounds. Industrial wastes containing these compounds may find their way into groundwater through deep well disposal, spreading of effluents on land, or indirectly through infiltration of streamflow to which the waste has been discharged. Very little removal of soluable material will occur during percolation and there is no reason to expect inactivation during groundwater storage.

6. Sediments from land erosion

These materials do not effect groundwater quality since they would be filtered out in the first stages of percolation.

7. Radioactive substances

Expected increases in the use of nuclear powered reactors for power generation indicate that radioactivity may be a serious problem in the near future. The extremely long half life of some radioactive materials make the disposal of radioactive wastes difficult. Some adsorption of radionuclides by clays during percolation may be expected however, once a radionuclide reaches the groundwater it remains a hazard until decay is complete.

8. Heat

Heat from industry and power generation is considered a serious pollutant. Its effect on groundwater is limited to reducing the value of a groundwater source for cooling water.

In addition to the eight categories of pollutants used by the U.S. Public Health Service a nineth category could be considered to be oil. *Oil spills* in coastal waters have found their way into groundwater supplies near the coast.

Water quality as stated previously, is a relative term. The desired quality of a water is a function of its intended use. Groundwater has and is primarily used as a water source for public and private drinking water supply, industrial process and cooling water, and agricultural supply. The acceptability of a groundwater source for use depends on the standards for quality for that use. Most drinking water sources in the United States conform to the Standards for Drinking Water Supply, U.S. Public Health Service (1962). These standards are shown in Table 2.

The quality of water required for industrial process uses varies widely with the type of industry and the type of process used. For example, the recommended total solids concentration in boiler feed waters varies from 50 to 3000 mg/l depending on the design of the boiler and the pressure of operation.

TABLE 2

U.S. Public Health Service Drinking Water Standards (1962)

Substance	Concentration which should not be exceeded if other supply is, or can be made available (Mg/l)	Grounds for rejection (Mg/l)
Alkyl benzene		
Sulfonate (ABS)	0.5	–
Arsenic (As)	0.01	0.5
Chloride (Cl)	2.50	–
Copper (Cu)	1.0	–
Carbon Chloroform		
Extract (CCE)	0.2	–
Cyanide (CN)	0.01	0.2
Iron (Fe)	0.3	–
Manganese (Mn)	0.05	–
Nitrate (NO_3)	45	–
Phenols	0.001	–
Sulfate (SO_4)	250	5
Total Dissolved	500	–
Solids (TDS)	5	–
Zinc (Zn)		
Barium (Ba)	–	1.0
Cadmium (Cd)	–	0.01
Chromium (Cr^{+6})	–	0.05
Lead (Pb)	–	0.05
Selenium (Se)	–	0.01
Silver (Ag)	–	0.05

The quality of water used for agricultural operations again varies considerably. Range cattle use water containing 5,000 mg/l dissolved solids but low concentrations of sodium, magnesium and sulfate are desirable. The quality required for irrigation water depends on the soil, composition, climate, type of crops, amount of groundwater used, method of application, location of groundwater table and local topography.

Wastewater Reuse Through Groundwater Recharge

Population increase and increased industrialization has led to a greatly increased demand for water supply. At the same time these factors have led to the lowering of water quality. Surface waters are receiving large quantities of untreated and partially treated domestic and industrial wastes. Groundwater quality is adversely affected by infiltration of low quality waters, leachates from sanitary landfills, deep well disposal and other waste discharges. In addition pumping of wells in coastal areas may lead to intrusion of salt water into the fresh groundwater supplies. These combined pressures have led to the reuse of treated wastewaters to supplement water supplies. The potential for increased water supplies from wastewater reuse is enormous. In the United States, one cycle of wastewater recovery and reuse could make available for use, by 1980, 224 billion gallons per day of essentially new water.

Recharge of groundwater with reclaimed wastewater is now being practiced both in the United States and abroad. Public acceptance of recycled waters has been extremely good.

At Whittier Narrows, California since 1962, 12 mgd of renovated wastewater has been recharged into the groundwater table using injection wells. The wastewater is treated by standard secondary treatment (Activated Sludge) which is followed by foam separation. The groundwater to which the reclaimed wastewater is recharged is part of the potable water supply of Los Angeles County. The drinking water of Los Angeles County has shown no ill effects from the recharging operation. A similar recharge operation is also conducted at the Hyperion Wastewater Treatment Plant of Los Angeles County. It is predicted that the Whittier Narrows Plant alone could reclaim 50 mgd of wastewater.

The city of Tel Aviv, Israel uses wastewater renovation and groundwater recharge as a method of increasing its water supply. Following an extensive testing program in the early 1960's, it was decided that lagooning of the wastewater would provide sufficient treatment if the wastewater was then placed on appropriately spaced seepage basins. The water recovered from the wells is of potable quality and the lagoon wastewater treatment removes sufficient solids so there is no danger of clogging or diminution of the infiltration rate in the recharge basin.

In Nassau County, New York a wastewater renovation plant and injection wells were placed in operation at the Bay Park Treatment Plant in 1968. The wastewater is treated at Bay Park by the conventional activated sludge treatment process. After the conventional secondary treatment the activated sludge effluent is then treated by lime coagulation, flocculation and sedimentation. Ammonia is removed in a combination water cooling, ammonia stripping tower. The wastewater is then recarbonated and the pH is lowered using carbon dioxide. Alum and polyeletrolytes are added to aid in coagulating the finely divided calcium carbonate recarbonation sludge. The remaining solids are removed on a reverse graded filter. Activated Carbon adsorption on granular activated carbon is used to remove the soluble "refractory" organics. The effluent from this tertiary process is then chlorinated before injection into the groundwater table. The Bay Park Plants is the first of a series of planned injection plants which will inject reclaimed wastewater to form a barrier against salt water intrusion. The injection wells will allow renovated wastewater to flow into the ocean, rather than the existing groundwater supplies flowing into the ocean. It is expected that the system will add 27 mgd to the Nassau County water supply by 1988 and will add 58 mgd to the County's water supply by the year 2010.

The reuse of reclaimed wastewater by injecting the treatment plant effluents into the groundwater table will in all probability be limited to water which will be used to supplement drinking water supplies. The high cost of injection facilities added to the cost of treatment required to produce the high quality water needed to protect the existing groundwater supplies will probably preclude the use of this method for purposes other than supplementing the drinking water supply.

REFERENCES

Todd, David K., *Ground Water*, John Wiley and Sons, Inc., New York (1959).

MacKichan, K. A., "Estimated use of water in the United States," 1955, *JAWA* **49**, 369–391 (1957).

Walton, W. C., *Groundwater Evaluation*, McGraw-Hill, New York (1970).

National Academy of Sciences—National Research Council, "Waste Management and Control," Washington, D.C. (1966).

Stephan, D. G., and L. W. Weinberger, "Water Reuse—Has It Arrievd," *JWPCF* **40**, 4, 529 (1968).

Parkhurst, J. D., and W. E. Garrison, "Whittier Narrows Water Reclaimation Plan—Two Years of Operation", *Civil Engineering* **34**, 9, 60 (1969).

Amramy, A., "Wastewater for Groundwater Recharge", *JWPCF* **36**, 3, 296 (1964).

Peters, J. H., and D. Cuming, "Water Conservation by Barrier Injection", *Water and Sewage Works* **114**, 2, 63, (1967).

ERICK R. GIDLUND
Polytechnic Institute of New York
BRUCE A. BELL
Evirotech Corporation

H

HEALTH PHYSICS: see MANAGEMENT OF RADIOACTIVE WASTES; OCCUPATIONAL HEALTH; RADIATION ECOLOGY

HYDROLOGY

The Purposes of Hydrological Studies

Hydrology is concerned with all phases of the transport of water between the atmosphere, the land surface and subsurface, and the oceans, and the historical development of an understanding of the hydrological processes is in itself a fascinating study[6]. As a science, hydrology encompasses many complex processes, a number of which are only imperfectly understood. It is perhaps helpful in developing an understanding of hydrological theory to focus attention not on the individual physical processes, but on the practical problems which the hydrologist is seeking to solve. By studying hydrology from the problem solving viewpoint, we shall see the interrelationship of the physical processes and the approximations which are made to represent processes which are either imperfectly understood or too complex for complete physical representation. We shall also see what data is required to make adequate evaluations of given problems.

A prime hydrological problem is the forecasting of stream-flow run-off. Such forecasts may be concerned with daily flows, especially peak flows for flood warning, or a seasonal forecast may be required, where a knowledge of the total volume of run-off is of prime interest. More sophisticated forecast procedures are required for the day to day operation of flood control reservoirs, hydropower projects, irrigation and water supply schemes, especially for schemes which are used to serve several purposes simultaneously such as hydropower, flood control, and irrigation.

Hydrologists are also concerned with studying statistical patterns of run-off. A special class of problems is the study of extreme events, such as floods or droughts. Such maximum events provide limiting design data for flood spillways, dyke levels, channel design, etc. Minimum events are important, for example, in irrigation studies and fisheries projects. A more complex example of statistical studies is concerned with sequential patterns of run-off, for either monthly or annual sequences. Such sequences are important when testing the storage capacity of a water resource system, such as an irrigation or hydropower reservoir, when assessing the risk of failing to meet the requirements of a given scheme. A specially challenging example of sequential flow studies concerns the pattern of run-off from several tributary areas of the same river system. In such studies it is necessary to try to maintain not only a sequential pattern but also to model the cross-correlations between the various tributaries.

The question of land use and its influence on run-off occupies a central position in the understanding of hydrological processes. Land use has been studied for its influence on flood control, erosion control, water yield and agriculture, with particular application to irrigation. Perhaps the most marked effect of changed land use and changed run-off characteristics is demonstrated by urbanization of agricultural and forested lands. The paving of large areas and the influence of buildings has a marked effect in increasing run-off rates and volumes, so that sewer systems must be designed to handle the increased flows. Although not so dramatic, and certainly not so easy to document, the influence of trees and crops on soil structure and stability may well prove to be the most far-reaching problem. There is a complex interaction between soil biology, the crop and the hydrological factors such as soil moisture, percolation, run-off, erosion and evapotranspiration. Adequate hydrological calculations are a prerequisite for such studies.

A long term aim of hydrological studies is the clear definition of existing patterns of rainfall and run-off. Such a definition requires the establishment of statistical measures such as the means, variances and probabilities of rare events. From these studies come not only the design data for extreme events but also the determination of any changes in climate which may be either cyclical or a long-term trend. It is being suggested in many quarters that air pollution may have a gradual effect on the Earth's radiation balance. If this is true we should expect to see measurable changes in our climatic patterns. Good hydrological data and its proper analysis will provide one very important means of evaluating such trends and also for measuring the effectiveness of our attempts to correct the balance.

A Brief Note on Statistical Techniques

The hydrologist is constantly handling large quantities of data which may describe precipitation, streamflow, climate, groundwater, evaporation and many other factors. A reasonable grasp of statistical measures and techniques is invaluable to the hydrologist. Several good basic textbooks are referenced[1,2,3,8,9], and *Facts from Figures* by Moroney, is particularly recommended for a basic understanding of what statistics is aiming to achieve.

The most important aspect of the nature of data is the question of whether data is independent or dependent. Very often this basic question of dependence or independence is not discussed until after many primary statistical measures have been defined. It is basic to the analysis, to the selection of variables and to the choice of technique to have some idea of whether data is related or independent. For example, it is usually reasonable to assume that annual flood peaks are independent of each other, whereas daily streamflows are usually closely related to preceding and subsequent events: they exhibit what is termed *serial correlation*.

The selection of data for multiple correlation studies is an example where dependence of the data is in conflict with the underlying assumptions of the method. Once the true nature of the data is appreciated it is far less difficult to decide on the correct statistical technique for the job in hand. For example, maximum daily temperatures and incoming radiation are highly correlated and yet are sometimes both used simultaneously to describe snowmelt.

In many hydrological studies it has been demonstrated that the assumption of random processes is not unreasonable. Such an assumption requires an understanding of statistical distributions and probabilities. Real data of different types has been found to approximate such theoretical distributions as the binomial, the Poisson, the normal distribution or certain special extreme value distributions. Especially, in probability analysis, it is important that the correct assumption is made concerning the type of distribution if extrapolated values are being read from the graphs.

Probabilities and return periods are important concepts in design studies and require understanding. The term "return period" can be somewhat misleading unless it is clearly appreciated that a return period is in fact a probability. Therefore when we speak of a return period of 100 years we imply that a magnitude of flow, or some other such event, has a one percent probability of occurring in any given year. It is even more important to realize that the probability of a certain event occurring in a number of years of record is much higher than we might be led to believe from considering only its annual probability or return period. As an example, the 200 year return period flood or drought has an annual probability of 0.5%, but in 50 years of record, the probability that it will occur at least once is 22%. Figure 1 summarizes the probabilities for various return periods to occur at least once as a function of the number of years of record. From such a graph it is somewhat easier to appreciate why design floods for such critical structures as dam spillways have return periods of 1000 years or even 10,000 years.

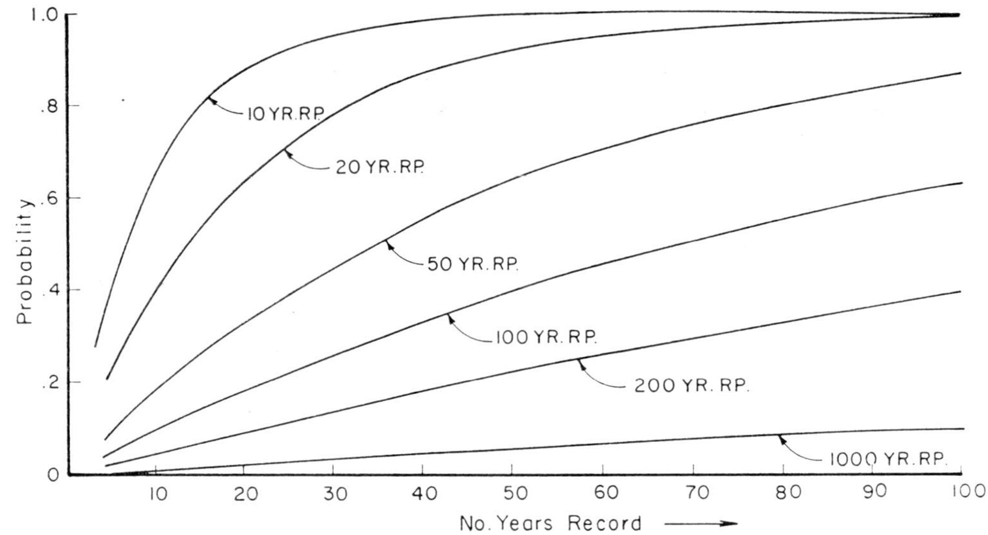

Figure 1 Probability of occurrence of various annual return period events as a function of years of record.

Analysis of Precipitation Data

Before analyzing any precipitation data it is advisable to study the method of measurement and the errors inherent in the type of gauge used. Such errors can be considerable (Chow[1], and Ward[5]).

Precipitation measurements vary in type and precision, and according to whether rain or snow is being measured. Precipitation gauges may be read manually at intervals of a day or part day. Alternatively gauges may be automatic and yield records of short-term intensity. Wind and gauge exposure can change the catch efficiency of precipitation gauges and this is especially true for snow measurements. Many snow measurements are made from the depth of new snow and an average specific gravity of 0.10 is assumed when converting to water equivalent.

Precipitation data is analyzed to give mean annual values and also mean monthly values which are useful in assessing seasonal precipitation patterns. Such figures are useful for determining total water supply for domestic, agricultural and hydropower use, etc.

More detailed analysis of precipitation data is given for individual storms and these figures are required for design of drainage systems and flood control works. Analysis shows the relationship between rain intensity (inches per hour) with both duration and area. In general terms, the longer the duration of storm, the lower will be the average intensity of rainfall. Similarly, the larger the area of land being considered, the lower will be the average intensity of rainfall. For example, a small catchment area of, say, four square miles may be subjected to a storm lasting one hour with an average intensity of two inches per hour while a catchment of two hundred square miles would only experience an average intensity of about one inch per hour. Both these storms would have the same return period or probability associated with them. Such data is prepared by weather agencies like the U.S. Weather Bureau and is available in their publications for all areas of the country. Typical data is shown in Fig. 2. The use of these data sheets will be discussed further in the section on run-off.

Winter snowpacks represent a large water storage which is mainly released at a variable rate during spring and early summer. In general, the pattern of snowfall is less important than the total accumulation. In the deep mountain snowpacks, snowtube and snowpillow measurements appear to give fairly reliable estimates of accumulated snow which can be used for forecasts of run-off volumes as well as for flood forecasting. On the flat prairie lands, where snow is often quite moderate in amounts, there is considerable redistribution and drifting of snow by wind and it is a considerable problem to obtain good estimates of total snow accumulation.

When estimates of snow accumulation have been made it is a further problem to calculate the rate at which the snow will melt and will contribute to stream run-off. Snow therefore represents twice

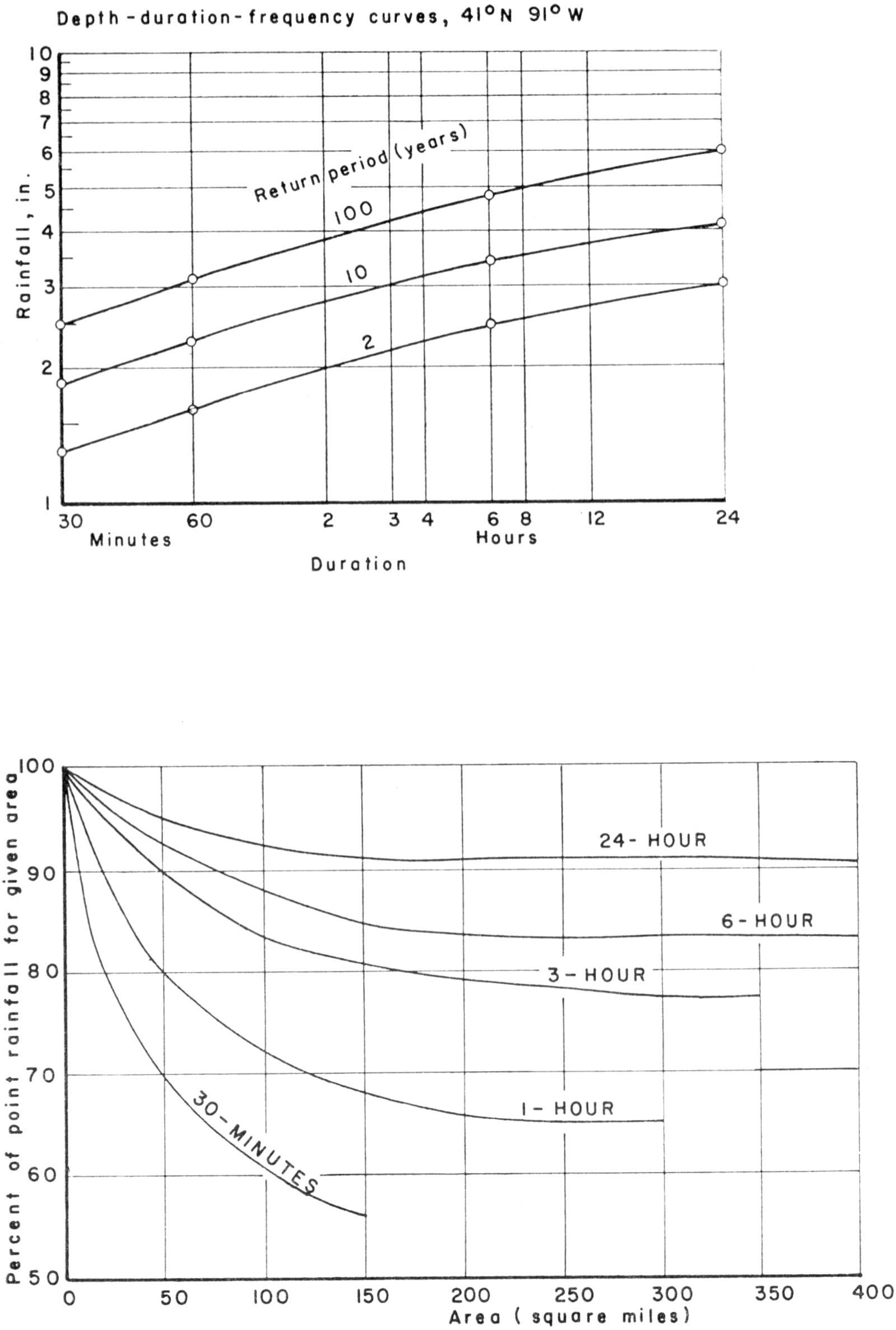

Figure 2 Rainfall depth-duration and area-frequency curves (U.S. Weather Bureau, after Chow[1]).

the problem of rain, because firstly we must measure its distribution and amount and secondly, it may remain as snow for a considerable period before it contributes to snowmelt.

Evaporation and Evapo-Transpiration

Of the total precipitation which falls, only a part finally discharges as streamflow to the oceans. The remainder returns to the atmosphere by evaporation. Linsley[2] points out that ten reservoirs like Lake Mead could evaporate an amount equivalent to the annual Colorado flow. Some years ago, studies of Lake Victoria indicated that the increased area resulting from raising the lake level would produce such an increase in evaporation that there would be a net loss of water utilization in the system.

Evaporation varies considerably with climatic zone, latitude and elevation and its magnitude is often difficult to evaluate. Because evaporation is such a significant term in many hydrological situations, its proper evaluation is often a key part of hydrological studies.

Fundamentally, evaporation will occur when the vapor pressure of the evaporating surface is greater than the vapor pressure of the overlying air. Considerable energy is required to sustain evaporation, namely 597 calories per gram of water or 677 calories per gram of snow or ice. Energy may be supplied by incoming radiation or by air temperature, but if this energy supply is inadequate, the water or land surface and the air will cool, thus slowing down the evaporation process. In the long term the total energy supply is a function of the net radiation balance which, in turn, is a function of latitude. There is therefore a tendency for annual evaporation to be only moderately variable and to be a function of latitude, whereas short term evaporation may vary considerably with wind, air temperature, air vapor pressure, net radiation and surface temperature.

The discussion so far applies mainly to evaporation from a free water surface such as a lake, or to evaporation from a saturated soil surface. Moisture loss from a vegetated land surface is complicated by transpiration. Transpiration is the term used to describe the loss of water to the atmosphere from plant surfaces. This process is very important because the plant's root system can collect water from various depths of the underlying soil layers and transmit it to the atmosphere. In practice it is not usually possible to differentiate between evaporation from the soil surface and transpiration from the plant surface, so it is customary to consider the joint effect and call it evapo-transpiration. This lumping of the two processes has led to thinking of them as being identical, however we do know that the evaporation rate from a soil surface decreases as the moisture content of the soil gets less, whereas there is evidence to indicate that transpiration may continue at a nearly constant rate until a plant reaches the wilting point.

To understand the usual approach now being taken to the calculation of evapo-transpiration, it is necessary to appreciate what is meant by *potential evapo-transpiration* as opposed to *actual evapo-transpiration*. Potential evapo-transpiration is the moisture loss to the atmosphere which would occur if the soil layers remained saturated. Actual evapo-transpiration cannot exceed the potential rate and gradually reduces to a fraction of the potential rate as the soil moisture decreases. Various formulae exist for estimating potential evapo-transpiration in terms of climatic parameters, such as Thornthwaites method, or Penman or Turk's formulae. Such investigations have shown that a good field measure of potential evapo-transpiration is pan evaporation from a standard evaporation-pan, such as the Class A type, and such measurements are now widely used. To turn these potential estimates into actual evapo-transpiration it is commonly assumed that actual equals potential after the soil has been saturated until some specific amount of moisture has evaporated, say two inches or so depending on the soil and crop. It is then assumed that the actual rate decreases exponentially until it effectively ceases at very low moisture contents. In hydrological modeling an accounting procedure can be used to keep track of incoming precipitation and evaporation so that estimates of evapo-transpiration can be made. The potential evapo-transpiration rate must be estimated from one of the accepted formulae or from pan-evaporation measurements, if available. Details of such procedures are well illustrated in papers by Nash[17] and by Linsley and Crawford[44] in the Stanford IV watershed model.

Runoff: Rain

It is useful to imagine that we start with a dry catchment, where the groundwater table is low, and the soil moisture has been greatly reduced, perhaps almost to the point where hygroscopic moisture alone remains. When rain first starts much is *intercepted* by the trees and vegetation and this *interception* storage is lost by evaporation after the storm. Rain reaching the soil *infiltrates* into pervious surfaces and begins to satisfy soil moisture deficits. As soil moisture levels rise, water percolates downwards towards the fully saturated water table level. If the rain is heavy enough, the water supply may exceed the vertical percolation rate and water then starts to flow laterally in the superficial soil layers towards the stream channels: this process is termed *interflow* and is much debated because it is so difficult to measure. At very high rainfall rates, the surface infiltration rate may be exceeded and then direct surface runoff will occur. Direct runoff is rare from soil surfaces but does occur from certain impervious soil types, and from paved areas. Much work has been done to evaluate the relative significance of these various processes and is well documented in references (1, 2, 3).

Such qualitative descriptions of the run-off process are helpful, but are limited because of the extreme complexity and inter-relationship of the various processes. Various methods have been developed to by-pass this complexity and to give us usable relationships for hydrologic calculations.

The simplest method is a plot of historical events, showing run-off as a function of the depth of precipitation in a given storm. This method does not allow for any antecedent soil moisture conditions or for the duration of a particular storm.

More complex relationships use some measure of soil moisture deficiency such as cumulative pan-evaporation or the antecedent precipitation index. Storm duration and precipitation amount is also allowed for and is well illustrated by the U.S. Weather Bureau's charts developed for various areas (Fig. 2). It is as well to emphasize that the antecedent precipitation index, although based on precipitation, is intended to model the exponential decay of soil moisture between storms, and is expressed by

$$I_N = (I_0 k^M + I_M)\, k^{(N-M)}$$

where I_0 is the rain on the first day and no more rain occurs until day M, when I_M falls. If k is the recession factor, usually about 0.9, then I_N will be the API for day N. The expression can of course have many more terms according to the number of rain events.

Before computers were readily available such calculations were considered tedious. Now it is possible to use more complex accounting procedures in which soil moisture storage, evapo-transpiration, accumulated basin run-off, percolation, etc. can all be allowed for. These procedures are used in more complex hydrological modeling and are proving very successful.

Runoff: Snowmelt

As a first step in the calculation of runoff from snow, methods must be found for calculating the rate of snowmelt. This snowmelt can then be treated similarly to a rainfall input. Snowmelt will also be subject to soil moisture storage effects and evapo-transpiration.

The earliest physically-based model of snowmelt was the degree-day method which recognized that, despite the complexity of the process, there appeared to be a good correlation between melt rates and air temperature. Such a relationship is well illustrated by the plots of cumulative degree-days against cumulative downstream flow, a rather frustrating graph because it cannot be used as a forecasting tool. This cumulative degree-day versus flow plot is an excellent example of how a complex day to day behaviour yields a long term behaviour which appears deceptively simple. Exponential models and unit hydrograph methods have been used to turn the degree-day approach into a workable method and a number of papers are available describing such work (Wilson[38], Linsley[32]), Arguments are put forward that air temperature is a good index of energy flux, being an integrated result of the complex energy exchanges at the snow surface (Quick[33]).

Light's equation,[31], for snowmelt is based on physical reasoning which models the energy input entirely as a turbulent heat transfer process. The equation ignores radiation and considers only wind speed as the stirring mechanism, air temperature at a standard height as the driving gradient for heat flow and, finally, vapour pressure to account for condensation–evaporation heat flux. It is set up for 6 hourly computation and requires correction for the nature of the forest cover and topography. It is interesting to compare Light's equation with the U.S. Corps equation[36] for clear weather to see the magnitude of melt attributed to each term.

By far the most comprehensive studies of snowmelt

have been the combined studies by the U.S. Corps of Engineers and the Weather Bureau (U.S. Corps of Engineers[36,37]). They set up three field snow laboratory areas varying in size from 4 to 21 square miles and took measurements for periods ranging from 5 to 8 years. Their laboratory areas were chosen to be representative of certain climatic zones. Their investigation was extensive and comprehensive, ranging from experimental evaluation of snowmelt coefficients in terms of meteorological parameters, to studies of thermal budgets, snowcourse and precipitation data reliability, water balances, heat and water transmission in snowpacks, streamflow synthesis, atmospheric circulations, and instrumentation design and development.

A particularly valuable feature of their study appears to have been the lysimeters used, one being 1300 sq. ft. in area and the other being 600 sq. ft. (Hildebrand and Pagenhart[30]). The results of these lysimeter studies have not received the attention they deserve, considering that they give excellent indication of storage and travel time for water in the pack. It may be useful to focus attention on this aspect of the Corps work because it is not easy to unearth the details from the somewhat ponderous Snow Hydrology report. Before leaving this topic it is worth mentioning that the data from the U.S. studies is all available on microfilm and could be valuable for future analyses.

It is perhaps useful at this stage to write down the Light equation and the clear weather equation from the Corps work to compare the resulting terms.

Light's equation[31] (Simple form in °F, inches of melt and standard data heights)

$$D = U(0.001\,84T_a \cdot 10^{-0.0000156} + 0.005\,78(e^{-6.11}))$$

where

U = average wind speed (m.p.h.) for 6 hr. period

T = air temperature above 32 °F for 6 hr. period

e = vapour pressure for 6 hr. period

h = station elevation (feet)

D = melt in inches per 6 hr. period

The U.S. Corps equation is[36]

$$M = k'(0.00508I_i)(1 - a) + (1 - N) \times (0.0212T_a - 0.84) + N(0.029T_c) + k(0.0084U)(0.22T_a + 0.78T_d)$$

M = Incident Radiation + incoming clear air longwave + cloud longwave + [Conduction + Condensation]

k' and k are approximately unity.

N = fraction of cloud cover.

I_i = incident short wave radiation (langleys/day).

a = albedo of snow surface.

T_a = daily mean temperature °F above 32 °F at 10′ level.

T_c = cloud base temperature.

T_d = dew point temperature °F above 32 °F.

U = average wind speed—miles/hour at 50′ level.

Putting in some representative data for a day when the minimum temperature was 32°F and the maximum 70°F, incoming radiation was 700 langleys per day and relative humidity varied from 100% at night to 60% at maximum temperature, the results were:

Light equation

D = Air temp melt and Condensation melt
= 1.035 + 0.961 inches/day
= 1.996 inches/day

U.S. Corps equation

M = incoming shortwave + incoming longwave + air temp + Condensation radiation
= 1.424 − 0.44 + 0.351 + 0.59
= 1.925 inches/day.

Note the large amount attributed to radiation which the Light equation splits between air temperature and radiation. It is a worthwhile operation to attempt to manufacture data for these equations and to compare them with real data. The high correlations between air temperature and radiation is immediately apparent, as is the close relationship between diurnal air temperature variation and dew point temperature during the snowmelt season. Further comparison of the formulae at lower temperature ranges leave doubts about the influence of low overnight temperatures.

There is enough evidence of discrepancies between real and calculated snowmelt to suggest that further study may not be wasted effort. Perhaps this is best illustrated from some recent statements made at a workshop on Snow and Ice Hydrology[39]. Meier indicates that, using snow survey data, the Columbia forecast error is 8 to 14% and occasionally 40 to 50%. Also these errors occured in a situation where the average deviation from the long term mean was only 12 to 20%. For a better comparison of errors it would be interesting to know the standard error of forecast compared with standard "error" of record from the long term mean. Also, later in the same paper it is indicated that a correct heat exchange calculation for the estimation of snowmelt cannot be made because of our inadequate knowledge of the eddy convection process. At the same workshop the study group on Snow Metamorphism and Melt reported: "we still cannot measure the free water content in any snow cover, much less the flux of the water as no theoretical framework for flow through snow exists".

Although limitations of data often preclude the use of the complex melt equations, various investigators have used the simple degree-day method with good success, (Linsley[32] and Quick and Pipes[40,46,47]). There may be reasonable justification for using the degree-day approach for large river basins with extensive snowfields where the air mass tends to reach a dynamic equilibrium with the snowpack so that energy supply and the resulting melt rate may be reasonably well described by air temperature. In fact there seems to be no satisfactory compromise for meteorological forecasting; either we must use the simple degree-day approach or on the other hand we must use the complex radiation balance, vapour exchange and convective heat transfer methods involving sophisticated and exacting data networks.

Computation of Run-off – Small Catchments

Total catchment behaviour is seen to be made up of a number of complex and inter-related processes. The main processes can be reduced to evapo-transpiration losses, soil moisture and groundwater storage, and flow of water through porous media both as saturated flow and unsaturated flow. To describe this complex system the hydrologist has resorted to a mixture of semi theoretical and empirical calculation techniques. Whether such techniques are valid is justified by their ability to predict the measured behaviour of a catchment from the measured inputs.

The budgeting techniques for calculating evapo-transpiration losses have already been described. From an estimation of evapo-transpiration and soil moisture and measured precipitation we can calculate the residual precipitation which can go to storage in the catchment and run-off in the streams. A method is now required to determine at what rate this effective precipitation, as it is usually called, will appear at some point in the stream drainage system. The most widely used method is the unit hydrograph approach first developed by Sherman in 1932[16].

To reduce the unit hydrograph idea to its simplest form, consider that four inches of precipitation falls on a catchment in two hours. After allowing for soil moisture deficit and evaporation losses, let us assume that three inches of this precipitation will eventually appear downstream as run-off. Effectively this precipitation can be assumed to have fallen on the catchment at the rate of one and a half inches per hour for two hours. This effective precipitation will appear some time later in the stream system, but will now be spread out over a much longer time period and will vary from zero flow, rising gradually to a maximum flow and then slowly decreasing back to zero. Figure 3 shows the block of uniform precipitation and the corresponding outflow in the stream system. The outflow diagram can be reduced to the unit hydrograph for the two hour storm by dividing the ordinates by three. The outflow diagram will then contain the volume of run-off equivalent to one inch of precipitation over the given catchment area. For instance, one inch of precipitation over one hundred square miles will give an area under the unit hydrograph of 2690 c.f.s. days.

When a rainstorm has occurred the hydrologist must first calculate how much will become effective rainfall and will contribute to run-off. This can best be done in the framework of a total hydrological run-off model as will be discussed later. The effective rainfall hydrograph must then be broken down into blocks of rainfall corresponding to the time interval for the unit hydrograph. Each block of rain may contain P inches of water and the corresponding outflow hydrograph will have ordinates P times as large as the unit hydrograph ordinates. Also, several of these scaled outflow hydrographs will have to

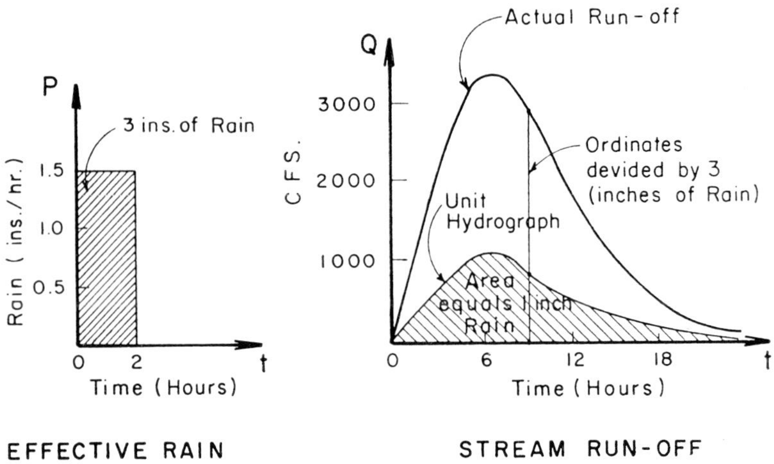

Figure 3 Hydrograph and unit hydrograph of run-off from effective rain.

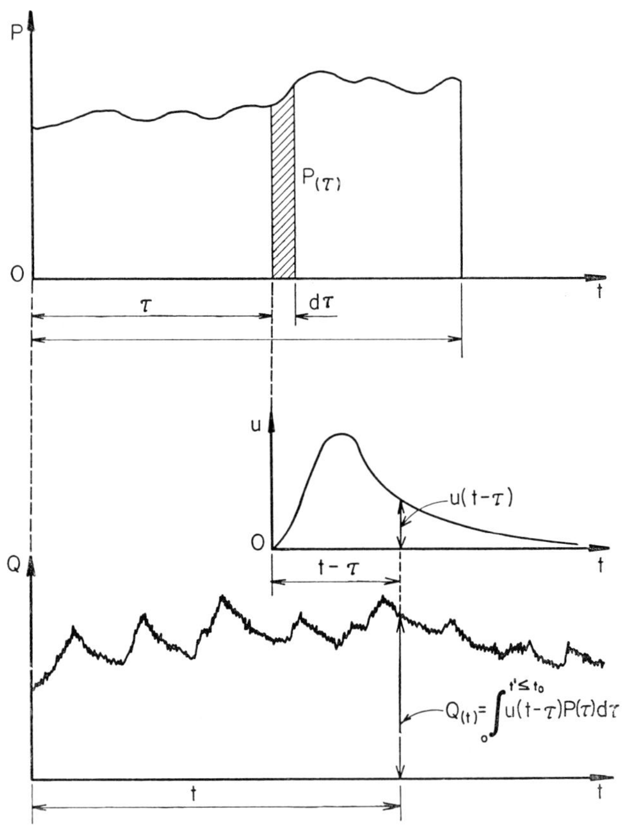

Figure 4 Determination of streamflow from precipitation input using an instantaneous unit hydrograph

be added together. This process is known as convolution and is illustated in Fig. 4 and 5.

The underlying assumption of unit hydrograph theory is that the run-off process is linear, not in the trivial straight line sense, but in the deeper mathematical sense that each incremental run-off event is independent of any other run-off. In the early development, Sherman[16] proposed a unit hydrograph arising from a certain storm duration. Later workers such as Nash[17 and 23] showed that Laplace transform theory, as already highly developed for electric circuit theory, could be used. This led to the instanteneous unit hydrograph and gave rise to a number of fascinating studies by such workers as Dooge[18], Singh[19], and many others. They introduced exponential models which

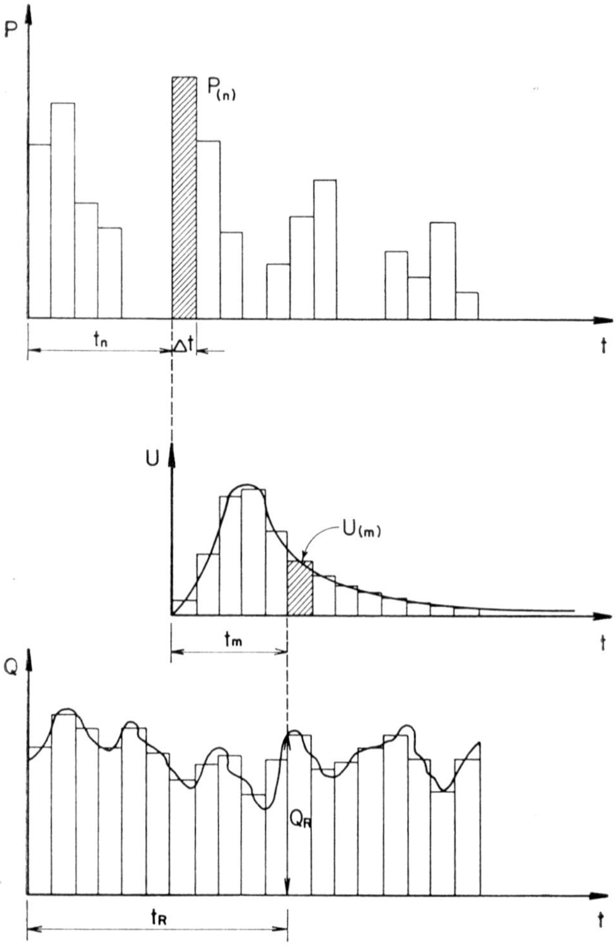

Figure 5 Convolution of precipitation by unit hydrograph on a finite difference basis.

are interpretable in terms of instantaneous unit hydrograph theory. Basically, however, there is no difference in concept and the convolution integral, equation (1) can be arrived at by either the unit hydrograph or the instantaneous unit hydrograph approach. The Convolution integral can be written as:

$$Q(t) = \int_0^{t<t_0} u(t-\tau)\, P(\tau)\, d\tau \qquad (1)$$

Figure 4 shows the definition diagram for the formulation is only useful if both P, the precipitation rate, and u, the instantaneous unit hydrograph ordinate are expressible as continuous functions of time. In real hydrograph applications it is more useful to proceed to a finite difference form of equation (1) in which the integral is replaced by a summation, equation (2), and Fig. 5.

$$Q_R = \sum_1^M u(m)\, P(n)\, \Delta t \qquad (2)$$

Where M is the number of unit hydrograph time increments, and m, n and R are specified in Fig. 2. It should be noted that from Fig. 5,

$$m + n = R + 1 \qquad (3)$$

Expanding equation (2) for a particular value of R,

$$Q(10) = u(10)\,P(1) + u(9)\,P(2) + \ldots u(1)\,P(10) \qquad (4)$$

The whole family of similar equations for Q may be expressed in matrix form (Snyder[20])

$$\begin{Bmatrix} Q_1 \\ Q_2 \\ Q_3 \\ \cdot \\ \cdot \\ \\ \\ \\ \\ Q_R \end{Bmatrix} = \begin{bmatrix} P_1 & 0 & 0 & \cdot & \cdot & \cdot & \cdot & \cdot & 0 \\ P_2 & P_1 & 0 & & & & & & 0 \\ P_3 & P_2 & P_1 & 0 & & & & & 0 \\ \cdot & \cdot & \cdot & \cdot & & & & & \cdot \\ \cdot & \cdot & \cdot & \cdot & & & & & \cdot \\ P_n & & & & & & & & \\ 0 & P_n & & & & & & & \cdot \\ 0 & 0 & P_n & & & & & & \\ 0 & 0 & 0 & & & & & P_1 & 0 \\ \cdot & \cdot & \cdot & & & & & P_2 & P_1 \end{bmatrix} \begin{Bmatrix} u_1 \\ u_2 \\ u_3 \\ \cdot \\ \cdot \\ u_m \end{Bmatrix} \qquad (5)$$

$\leftarrow m$ columns $\rightarrow$

Or more briefly

$$\{Q\} = [P]\,\{u\} \qquad (6)$$

Equation (6) specifies the river flow in terms of the precipitation and the unit hydrograph. In practise Q and P are measured and u must be determined. Some workers have guessed a suitable functional form for u with one or two inknown parameters and have then sought a best fit with the available data. For instance, Nash's series of reservoirs yields[17,23]

$$u(t) = \frac{1}{K(n-1)!}\,\frac{t^{n-1}}{K}\,e^{\frac{-t}{k}} \qquad (7)$$

in which there are two parameters, K and n. Another approach is to solve the matrix equation (6) as follows (Snyder[21])

$$\{u\} = [P^T P]^{-1} P^T \{Q\} \qquad (8)$$

It has already been demonstrated that $R = m + n - 1$ so that there are more equations available than there are unknowns. The solution expressed by equation (8) therefore automatically yields the least squares values for u. This result will be referred to after the next section.

Multiple Regression and Streamflow

The similarity of the finite difference unit hydrograph approach to multiple regression analysis is immediately apparent. The flow in terms of precipitation can be written as

$$\begin{aligned} Q_R &= a_1 P_t \quad + a_2 P_{t+1} + a_3 P_{t+2} \cdots + a_n P_{t+n-1} \\ Q_{R+1} &= a_1 P_{t+1} + a_2 P_{t+2} + a_3 P_{t+3} \cdots + a_n P_{t+n} \qquad (9) \\ Q_{R+2} &= a_1 P_{t+2} + \text{etc.} \ldots \end{aligned}$$

Again we can write

$$\{Q\} = [P]\{a\} \qquad (10)$$

The similarity with equation (6) is obvious and may be complete if we have selected the correct precipitation data to correlate precipitation at 6 a.m. with downstream flow at 9 a.m. when we know that there is a 3 hour lag in the system. Therefore, using multiple regression as most hydrologists do, the method can become identical with the unit hydrograph approach.

Lake, Reservoir and River Routing

The run-off calculations of the previous sections enable estimates to be made of the flow in the headwaters of the river system tributaries. The river system consists of reaches of channels, lakes and perhaps reservoirs. The water travels downstream in the various reaches and through the lakes and reservoirs. Tributaries combine their flows into the main stem flow and also distributed lateral inflows contribute to the total flow. This total channel system influences the flow in two principal ways, first the flow takes time to progress through the system and secondly, some of the flow goes into temporary storage in the system. Channel storage is usually only moderate compared with the total river flow quantities, but lake and reservoir storage can have a considerable influence on the pattern of flow.

Calculation procedures are needed which will allow for this delay of the water as it flows through lakes and channels and for the modifying influence of storage. The problem is correctly and fully described by two physical equations, namely a continuity equation and an equation of motion. Continuity is simply a conservation of mass relationship while the equation of motion relates the mass accelerations to the forces controlling the movement of water in the system. Open channel fluid mechanics deals with the solution of such equations, but at present the solutions have had little application to hydrological work because the solutions demand detailed data which is not usually available and the computations are usually very complex, even with a large computer.

Hydrologists resort to an alternative approach which is empirical; it uses the continuity equation but replaces the equation of motion with a relationship between the storage and the flow in the system. This assumption is not unreasonable and is consistent with the assumption of a stage-discharge relationship which is widely utilized in stream gauging.

Reservoir Routing

The simplest routing procedure is so-called reservoir routing, which also applies to natural lakes. The continuity equation is usually written as;

$$I - O = \frac{dS}{dT}, \qquad (11)$$

where

I = Inflow to reservoir or lake

O = Outflow

S = Storage

The second equation relates storage purely to the outflow, which is true for lakes and reservoirs, where the outflow depends only on the lake level. The outflow relationship may be of the form: –

$$O = KBH^{3/2} \qquad (12)$$

if the outflow is controlled by a rectangular weir, or:

$$O = K'H^n \qquad (13)$$

where K' and n depend on the nature of the outflow channel.

Such relationships can be turned into outflow—storage relationships because storage is a function of H, the lake level. The equations (12) and (13) can then be rewritten in the form

$$O = K''S^m \qquad (14)$$

Alternatively, there may be no simple functional relationship, but a graphical relationship between O and S can be plotted or stored in the computer. The continuity equation and the outflow storage relationship can then be solved either graphically or numerically, so that, given certain inflows, the outflows can be calculated. Notice the assumption that a lake or reservoir responds very rapidly to an inflow, and the whole lake surface rises uniformly.

Channel Routing

The assumptions of reservoir routing no longer hold for channel calculations. The channel system takes time to respond to an input. Also, storage is a function of conditions at each end of the length of channel being considered, rather than just the conditions at the outflow end.

The simplest channel routing procedure is the so-called Muskingum method developed on the Muskingum river (G. T. McCarthy[24], Linsley[2]).

Channel routing is also based on continuity, equation 11. Before it can be utilized this equation must be rewritten in finite difference form

$$\frac{I_1 + I_2}{2} + \frac{O_1 + O_2}{2} = \frac{S_2 - S_1}{t} \tag{15}$$

Also, some assumption must be made from which storage can be computed. The Muskingum method linearizes the problem and assumes that storage in a whole channel reach is completely expressible in terms of inflow and outflow from the reach, namely

$$S = K[xI + (1 - x)\,O] \tag{16}$$

Substituting from (16) in (15), following Linsley[2], the result obtained is,

$$O_2 = c_0 I_2 + c_1 I_1 + c_2 O_1 \tag{17}$$

where c_0, c_1 and c_2 are functions of K, x and t and $c_0 + c_1 + c_2 = 1$. The influence of x is illustrated in Fig. 6. Extending this equation to N days:

$$O_N = c_0 I_N + c_1 I_{N-1} + c_2 c_0 I_{N-1} + c_1 c_2 I_{N-2} + c_2^2 c_0 I_{N-2} + c_2^2 c_1 I_{N-3} + \cdots + c^{N-2} c_1 I_1 + c_2^{N-1} O_1 \tag{18}$$

Usually, because c_0, c_1 and c_2 are less than one, terms in c^3 and higher are ignored, but whatever the simplification, the result is of the form

$$O_N = a_1 I_N + a_2 I_{N-1} + a_3 I_{N-2} + a_4 I_{N-3} + a_5 I_{N-4} \text{ etc.} \tag{19}$$

Once again the set of equations for O_1 to O_R can be written as

$$\{O\} = [I]\,\{a\} \tag{20}$$

(N.B. The use of I for inflow is unfortunate and leads to confusion with the identity matrix. Also O is confusable with the null matrix.) It must be admitted that, from the theory, only a_1, a_2 and a_3 are independent, but in practice the precision with which a solution can be obtained for c_0, c_1 and c_2 does not justify calculating a_4, a_5 etc. from the first three. The best approach at this stage is once more to resort to least squares fitting, recasting equation (20) in the form of equation (8).

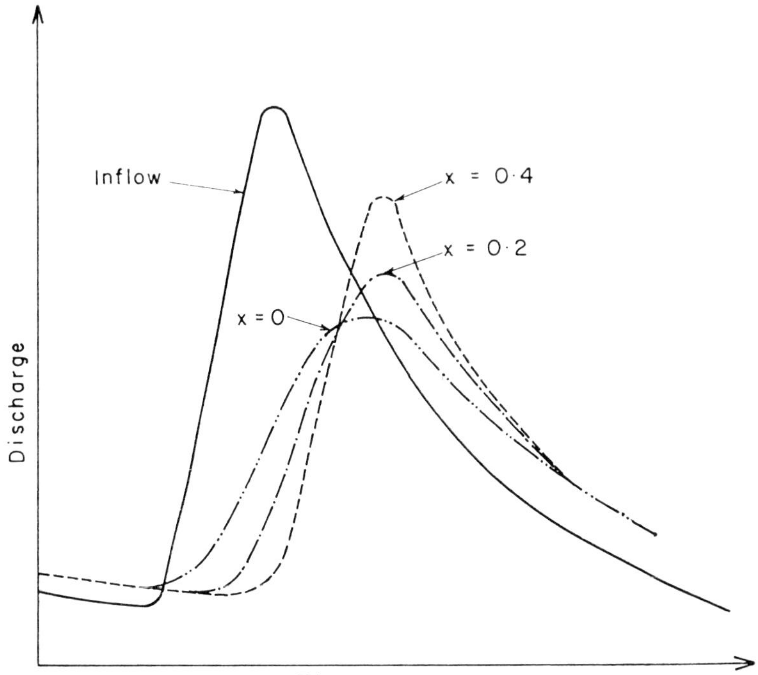

Figure 6 Muskingam routing: to illustrate the influence of x on outflow hydrograph (after Linsley[2]).

Many hydrologists calculate the routing coefficients by evaluating K and x in the traditional graphical method.

Many assumptions are made in the Muskingum method, such as the linear relationship between storage and discharge, and the implied linear variation of water surface along the reach. In spite of these assumptions the methods has proved its value.

Another difficulty which occurs with the Muskingham method usually occurs when the travel time in the reach and the time increments for the data are approximately equal, such as when the travel time is about one day and the data available is mean daily flows. From the strict mathematical viewpoint, the time Δt should be a fraction of the travel time K, otherwise flow gradients such as $\frac{dO}{dt}$ are not well represented on a finite difference basis. However, it is still possible to use time increments Δt greater than K if the flow is only changing slowly, but caution is necessary. The reason for caution is that the C values in equation (17) are a function of $\frac{\Delta t}{K}$ and are not constant.

Solving for the C's in terms of K, x and Δt:–

$$C_0 = -\frac{x + 0.5\left(\frac{\Delta t}{K}\right)}{1 - x + 0.5\left(\frac{\Delta t}{K}\right)} \tag{21 a}$$

$$C_1 = \frac{x + 0.5\left(\frac{\Delta t}{K}\right)}{1 - x + 0.5\left(\frac{\Delta t}{K}\right)} \tag{21 b}$$

$$C_2 = \frac{1 - x - 0.5\left(\frac{\Delta t}{K}\right)}{1 - x + 0.5\left(\frac{\Delta t}{K}\right)} \tag{21 c}$$

Hence

$$C_0 + C_1 + C_2 = 1 \tag{21 d}$$

To illustrate the influence of Δt and K, some synthetic data was used to construct Fig. 7. Values

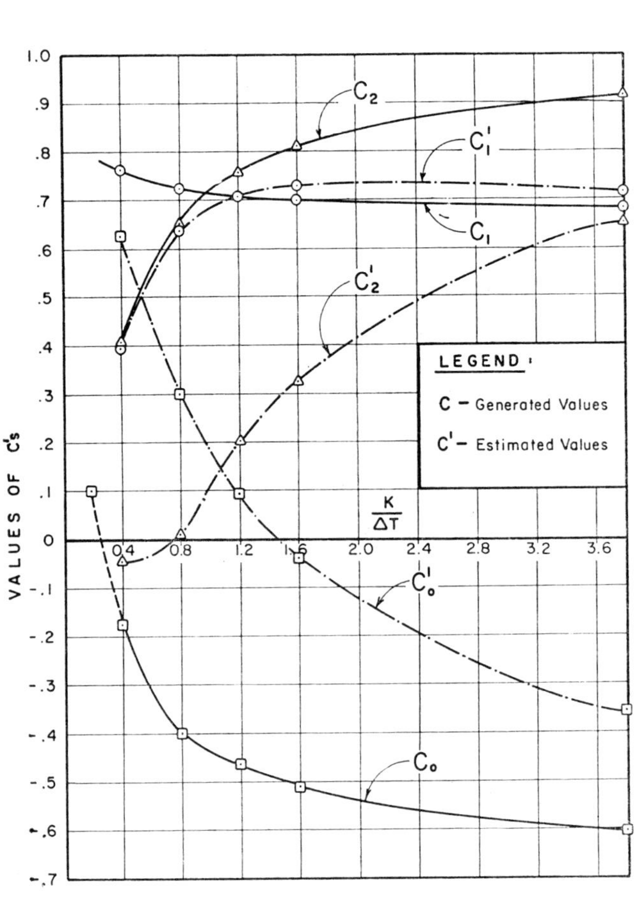

Figure 7 Muskingum coefficients as a function of storage and routing period.

for K and x were chosen and then values of C_0, C_1, and C_2 were calculated for different values of Δt. In addition, an assumed inflow was routed using the K and x values and using a time interval Δt which was small compared with K. These resulting inflows and outflows were then reanalyzed for C_0, C_1 and C_2 using time increments five times greater than the original Δt. Such Δt values exceeded K. The new estimates of C_0, C_1 and C_2 are also plotted in Fig. 7. Note the general agreement in shape of the C_0 and C_0' curves, etc. Exact agreemnt is lost because of the diminishing accuracy of the $\frac{dO}{dt}$ terms, etc. as Δt increases. Note also how rapidly the C values change when Δt is aproximately equal to K.

These remarks are not necessarily meant ot dissuade hydrologists from using the Muskingum method. The intention is to illustrate some of the pitfalls so that it may be possible to evaluate the probable validity or constancy of the coefficients (Laurenson[22]). The worst situation appears to be when $\Delta t \simeq K$, because in real rivers K decreases with rising stage and the C values are very sensitive to whether Δt is just less than or greater than K.

A very real problem in the application of the Muskingum method, and in fact of any channel routing procedure, is the problem of lateral inflow to the channel reach. Given the inflows and outflows for the reach as functions of time, it is necessary to separate out the lateral flows before best fit values of K and x can be determined. The lateral flows will, in general, bear no relationship to the pattern of main stream flows. Sometimes it is possible to use flow measurements on a local tributary stream as an index of total lateral flow. The cumulative volume of lateral flow can be determined by substracting summed inflows from summed outflows. The measured tributary flow can then be scaled up to equal the total lateral flow and these flows can then be subtracted from the reach outflows. This residual outflow can be used in the determination of the Muskingum coefficients.

Sometimes it is possible to find periods of record where lateral flows are small or perhaps have a more predictable pattern, such as during recession periods. Also, the routing coefficients can be refined by an iterative procedure and by using various sets of data, although not infrequently it is found necessary to define Muskingum coefficients for different ranges of flow, because real stream flow is not linear as is assumed in the model.

Kinematic Wave Theory

It has been known for many years that flood wave movement is much slower than would be expected from the shallow water wave theory result of $\sqrt{gy}$. Seddon[25] showed as early as 1900 that flood waves on the Mississippi moved at about 1.5 of river velocity. The theoretical reasons for this slowness of flood wave travel have only been realized during the last twenty years. The theory describing these waves is known either as monoclinic wave theory or kinematic wave theory.

Kinematic wave theory has suffered from neglect because of its usual presentation in the literature. The common approach is to start with the continuity equation and a simplified Manning equation. These equations are combined to yield Seddon's Law that flood waves travel at about 1.5 times the mean flow velocity. Such an approach restricts the method to an unalterable flood wave translating through the channel system. Although such a simple model has the benefit of approximating the real situation, much more useful and general results are available by setting up the two equations in finite difference form and carrying out simultaneous solution on a computer. In particular the method handles lateral inflow with great ease and also calculates discharge against time, eliminating the guess work from the time calculation.

Kinematic wave theory can be seen to be a simplification of the more general unsteady flow theory in open channels, for which the equation of motion and the continuity equation can be written (Henderson and Wooding[4,26])

$$S_0 - \frac{\partial y}{\partial x} - \frac{v}{g}\frac{\partial v}{\partial x} - \frac{1}{g}\frac{\partial v}{\partial t} = \frac{v^2}{c^2 R} \qquad (22)$$

Bed Slope	Depth Slope	Convective Acceln.	Local Acceln.	Friction term

$$\frac{\partial Q}{\partial x} + B\frac{\partial y}{\partial t} = \delta i \text{ (lateral inflow)} \qquad (23)$$

In different situations various terms in equation (22) become negligible. In tidal flow the friction and bed-slope terms are sometimes omitted, although this simplification is not strictly necessary. In this way the positive and negative characteristics are obtained with propagation velocities of $u \pm \sqrt{gy} = \frac{dx}{dt}$. At the other end of the spectrum is kinematic wave theory in which the bedslope term is assumed to overpower the other three terms on the left hand side, namely, depth slope, convective and local acceleration. Hence kinematic waves are friction controlled.

Equation (22) is therefore reduced to

$$S_0 = \frac{v^2}{c^2 R} \qquad (24)$$

To combine equations (22) and (24) into the simplest characteristic form, it is assumed that R, the hydraulic radius, may be replaced by y. Then equation (24) yields

$$Q = vBy = B \cdot Cy^{3/2} S_0^{1/2} \qquad (25)$$

Substituting for $\frac{\partial Q}{\partial x}$ in (23) by differentiating equation (25),

$$\frac{3}{2} BCS_0^{1/2} y^{1/2} \frac{\partial y}{\partial x} + B \frac{\partial y}{\partial t} = \delta i \qquad (26)$$

The left hand side is the total differential $\frac{dy}{dt}$ if the term

$$\frac{3}{2} BCS_0^{1/2} y^{1/2} = \frac{dx}{dt} = \text{wave velocity} \qquad (27)$$

The important fact that monoclinal waves can only propagate down-stream with a wave velocity given, in the simple case, by equation (27), is well known, as is best discussed by Lighthill and Whitham[27]. As previously mentioned, as an empirical result, equation (27) dates back to Seddon[25] and his study of flood wave movement.

Comparing kinematic wave theory with channel routing, equation (23) can be recast in exactly the same form as equations (15) and (16) if the lateral in flow term is ignored. The difference in the two methods depends on the second equation chosen, which is basically a stage-discharge relationship. Channel routing, as illustrated by the Muskingham method, assumes a linear relationship between stage and discharge to arrive at equation (13) for storage. Kinematic theory uses a simplified equation of motion which is usually either the Chezy or Manning equation.

In a further development of kinematic theory researchers such as Hyami[29] have shown that the influence of channel storage can be included and this results in an extra term which is effectively a diffusion term: —

$$\frac{\partial y}{\partial t} + c \frac{\partial y}{\partial x} = K \frac{\partial^2 y}{\partial t^2} \qquad (28)$$

Solutions of this equation are approximately of the form: —

$$y = y_0 e^{-ax} \cos(bt - ax) \qquad (29)$$

This solution assumes a sinusoidal wave profile, and the solution is seen to be delayed as it moves downstream and also it decays in amplitude. It is seen that effectively an exponential damping term is introduced by the inclusion of diffusion.

Such relationships are complex and cumbersome, demanding detailed data on the channels which often is not available. Inspection of the resulting equations reveals a similarity to the old lag and route method. The author has used a simplified method based on these equations which has proved itself to be very successful[46]. The method uses a variable travel time calculated from the velocity-stage curves for the upstream end of the given channel reach. The kinematic result of approximately 1.5 times channel velocity is used, although a best fit result can be found by optimization. The flow is then routed through a simple linear reservoir, the size of which is related to the channel storage. This reservoir allows a check to be kept on continuity and also it produces an exponential decay term as demanded by the diffusion term in the full kinematic theory. For relatively steep rivers like the Fraser River in British Columbia, one linear reservoir is adequate. Less steep streams may require two reservoirs, but two reservoirs should be adequate in that a two parameter exponential will fit almost any curve imaginable.

This simple channel routing procedure has been found to be very powerful because it approximates the true physical behavior very closely. It yields a simple non-linear behavior and models the well documented flood wave subsidence. On the Fraser River system the method proved very successful in separating out the very large lateral flows which are characteristic of that particular river system, and this separation was really the key to the modelling of the whole system behavior.

Artificial Generation of Streamflow

If streamflows can be considered to be statistically distributed, then it may be possible to establish statistical measures of their distribution. Such statistical measures will be the mean flow, the variance and perhaps some correlation between successive flows. It may also be possible to determine the nature of the distribution, such as normal, or log normal. If such parameters can be found, then these parameters can be used to regenerate data with the same statistical patterns.

The reason for such data generation is not apparent until we examine the purposes for which such data can be used. If we are given some 30 years of stream flow data, it is possible to make reasonable estimates of the statistical parameters of, say, monthly or annual flows. The 30 years of record will contain only a very few extreme events, such as floods or prolonged droughts, and will therefore not impose very severe tests on any water resource system which is storage dependent, such as a hydro system or an irrigation project. Generated data could presumably be constructed for very long periods of time and would, if the procedure is justified, reproduce many more extreme flows and consecutive runoff patterns which will test the proposed resource system more thoroughly. The statistical risk of failure can then be evaluated.

This is the aim of streamflow generation studies. Whether the methods currently in use are adequate or valid is still under debate .The aims are sound and the methods promising, if used with caution and an awareness of the intrinsic assumptions.

It is worthwhile comparing the artificial generation methods with the approaches previously used. It has been customary to construct a test period or sequence from the recorded data. As an example, it was not unusual to use the three driest years of record and assume that they occur in successive years. This became the design drought. Without care to evaluate just how extreme an event this might be, such a method might lead to either over or under design. Hopefully, streamflow generation will lead to more balanced designs, especially as experience with the techniques and the techniques themselves improve.

To consider the generation techniques, it is first of all important to appreciate that streamflow forms a time series which can be highly dependent or almost entirely independent according to the time interval chosen. Daily flows are usually highly correlated with each other; monthly flows may still have a reasonable serial correlation, but annual flows may be almost totally independent. Certainly it is usual to assume that annual flood peaks are independent.

The simplest streamflow model assumes that there is no long term trend in the run-off pattern, so that the mean flow for the period of record is a reasonably good estimate of the long term mean. Also it is assumed that the standard deviation from the mean is a good measure of the dispersion of flows around the mean. The model expresses:

Generated flow = Mean flow + Random Component

i.e.

$$Q_i = \bar{Q}_j + t_i\sigma_j \tag{30}$$

i refers to synthetic sequence, j to historic sequence.

Q_i = synthetic streamflow for month i

$\bar{Q}_j$ is mean flow for same month

t_i is random normal deviate with mean zero and variance unity (Available in most computer libraries)

σ_j is standard deviation of flows for month j

The random component is seen to be made up of the calculated standard deviation multiplied by a random variable t_i which is generated by a random number generation, and has a statistical distribution which is the same as the normal random error. The variable t_i therefore redistributes σ_j in a random Gaussian fashion.

A more sophisticated model by Thomas and Fiering[42,43] includes the memory that this month's flow has of last month's flow. This memory is described by serial correlation:

Generated Flow = Mean Flow + Serial Component + Random Component

i.e.

$$Q_{i+1} = \bar{Q}_{j+1} + b_j(Q_i - \bar{Q}_j) + t_i\sigma_{j+1}(1 - r_j^2)^{1/2} \tag{31}$$

The additional symbols are

b_j is regression coefficient for serial correlation of j and $j+1$ month

r_j is correlation coefficient between j and $j+1$ month

It will be noticed that when a serial component is introduced, the random component must be correspondingly decreased by the factor $(1 - r_j^2)^{1/2}$ so that there is not a net increase in flow.

Many other synthetic streamflow methods have been tried or suggested but at the present time only the simple models [equation (30) and (31)] Thomas and Fiering's model[43], seem to be in use by practising hydrologists. Also, the reader should appreciate, that the methods are not without their problems, for generated flows may go negative, so that modelling of droughts is questionable. However, a study of the statistical frequency of failure may still be better than the old methods.

So far, generation techniques have been discussed only for a single streamflow record. If a whole river basin with several reservoirs on different tributaries and on the main stem is being studied, then a much more difficult problem arises. It is not difficult to generate artificial data for each tributary or for the main stem. But such data will have none of the cross-correlation which exists in real data. That is to say, with the real data, if flows are high in one tributary, then probably the other tributaries are also reasonably high. This cross-correlation is difficult to preserve in generating techniques. It is discussed at great length by Fiering but the method suggested for preserving cross-correlations is very demanding on computer storage and time.

The writer wonders whether use of a physically based computer model of a river basin, coupled to a random generation of precipitation events both for one area of storm and intensity of storm might not produce data more economically and with the correct statisticalinterdependence.

Hydrological Simulation

Simulation is a very broad term covering many types of mathematical and physical procedures. The streamflow generation techniques already discussed are one type of simulation. Analogue computer models and physical flow models are also simulations. Digital computer models for simulation may be based on statistics entirely or on the physical behaviour of various components of the system. Some models may be a mixture of both physical and statistical methods.

Although the types of simulation models vary widely, their aims are similar, namely to describe the behaviour of a complex system so that predictions of system behaviour can be made from some specified input. These predictions may be short term flood flow predictions to give flood warning, especially for operation of flood protection schemes. Alternatively, predictions may be needed to evaluate water resource schemes and to examine the influence of storage, diversion, consumptive use, etc. on the system operation.

Statistical Techniques

Many simulations, particularly run-off forecasts, have been based on multiple regression analysis. Flow has been treated as the dependent variable and many factors such as precipitation, antecedent precipitation, sunshine hours, etc. have been used as the independent variables[13]. Such methods have had at least moderate success. The difficulties with the method have been that many of the so-called independent variables are in reality somewhat related, for example, rainfall is inversely correlated with sunshine hours. Also, the lack of true functional relationship usually means that coefficients for one year's data vary for other years or for periods of higher or lower flow, and therefore forecast precision is inherently limited and evaluation of changes to the system, such as introduction of reservoirs is not feasible.

The deficiencies of multiple regression analysis have been somewhat overcome, according to some investigators, by the use of multivariate analysis. The principle component techniques select the most significant relationship even if the data is interrelated. However, this type of analysis is still blind to the physics of the system. Examples of this technique and the theory are given in Kendall, Snyder and Wallis[10,14,15] and the reader can consult these for more details.

Monte Carlo methods, of which the stream generation is a simple example, have been proposed and used. Such methods have had staggering success in some fields of physics, such as nuclear studies. The classic example was the original study of critical mass for the atomic bomb. Such success led people to hope for more general applications. But even the keenest exponents of such methods agree that they are no substitute for an understanding of the physical behavior of the system. The writer considers that such statistical techniques are a reasonable approach if there is no physically based alternative. Experience with constructing physically based simulation models shows immediately how valuable it is to incorporate a functional relationship which bears some resemblance to the physical system. An example of this statement is the unit hydrograph approach which can be considered as a simple simulation model. The actual physical behavior of a catchment seems to be well described by a unit hydrograph. As soon as such a unit hydrograph is introduced, it becomes a relatively simple matter to relate cause and effect, which in that case is precipitation and run-off. A similar results can be achieved if the correct data is used in a multiple regression analysis, but such data selection pre-supposes a knowledge of the system behavior.

Physical Computer Simulation Models

Qualitatively, we can describe the behavior of a hydrologic system. The catchment soil layers have storage which decreases and delays run-off and permits evapo-transpiration to occur. The lake and channel system further delays flows and modifies the shape of the outflow hydrograph. Groundwater supplies a highly damped outflow which is significant during dry spells. Quantitatively we may know the precipitation input and the run-off output, although there may be data error, especially in the precipitation. There may also be data from which estimates of potential evapo-transpiration can be made. Presumably data will also be available of such matters as lake areas, stage-discharge relationships, catchment areas and elevations, and the details of the streamflow network.

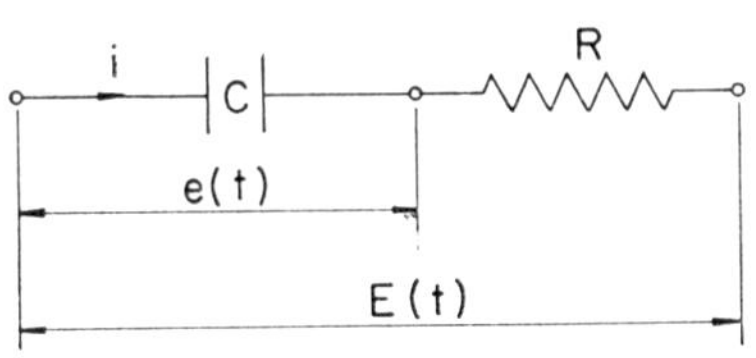

$$E(t) = e(t) + Ri(t)$$

$$i(t) = C\frac{de}{dt}$$

$$\therefore e(t) = \frac{1}{1 + CRD} E(t)$$

Figure 8 Simple system demonstrating non-separability of R and C (after Nash).

It is important to realize what can be discovered about a system simply from studying the input to the system and the output from the system, as has been very well domonstrated by Nash[23]. He considers a simple electrical system made up of a capacitor and a resistance as shown in Fig. 8. Then if $E(t)$ is considered to be the input and $e(t)$ the output it can be shown that

$$e(t) = \frac{1}{1 + RCD} E(t) \tag{32}$$

where D is the differential operator. Hence we can solve for $e(t)$ if $E(t)$ is given. Alternatively if $e(t)$ and $E(t)$ are given we can solve for RC as a lumped term, but we can never find the separate values for R and C. Therefore when we construct a simulation model, we may be able to correctly model the output from the input, although the parameters used in the model may be lumped terms describing various factors in the real system.

Hydrological systems are considerably more complex than the above example and it is important to realize that we cannot start to "fit" the model to the data until we have simulated the total system behavior. Also, in general, the system will not be linear like the simple electrical network, so that response becomes a function of flow.

In constructing a simulation model we must first of all decide the factors or processes which should be included to correctly and adequately describe the system. Each factor must then be approximately fitted using any data or knowledge which we may have. It is a basic rule that these factors, such as evaporation, unit hydrograph, soil moisture storage, etc. should be modeled with as few parameters as possible, as long as adequate description of a process is not jeopardized. This minimization of parameters has been well named as the "principle of parsimony". The final optimisation of the model parameters is done when the whole model is complete. One parameter at a time is adjusted in a step by step process until best fit values are obtained.

To illustrate the process we shall consider a simple model, Quick and Pipes[47].

It will be assumed that precipitation data is available in sufficient detail to provide information on intensity, duration and effective catchment area influenced by rain. Downstream run-off data is also available.

Simple models will be constructed to describe the following:

1. Soil storage and surface infiltration
2. Infiltration to groundwater
3. Evapo-transpiration
4. Fast component of run-off by unit hydrograph (Surface flow and interflow)

5. Slow component of run-off by unit hydrograph (Groundwater flow)
6. Channel routing, lateral flows and lake storage

It will appear that items 1 to 5 are closely interrelated and fitting of the model can only be achieved when the whole model for at least items 1 to 5 has been constructed. Item 6 can also be difficult as far as lateral flows are concerned.

The parameters used in a simulation model may be typified in the following list:—

Catchment parameters

$P(J, I)$	Precipitation. J = day, I = precipitation gauge number
$A(I)$	Catchment area associated with precipitation gauge
S	Soil storage
PS	Precipitation to soil moisture storage
SD	Soil moisture deficit
SI	Surface infiltration
GI	Infiltration to groundwater
EP	Potential evapo-transpiration
K	Actual evapo-transpiration recession factor
EPA	Actual evapo-transpiration
$SUMEPA$	Summed EPA
QF	Fast component of run-off
QS	Slow component of run-off
QT	Total run-off

Unit hydrograph ordinates for day N:

$UF(N)$	Fast component unit hydrograph
$US(N)$	Slow component unit hydrograph

Channel routing parameters

$QI(N)$	Inflow to reach
$QO(N)$	Outflow to reach
$QL(N)$	Lateral inflow to reach
TL	Lag time (function of $QI(N)$)
SR	Storage factor for linear reservoir

The computer program is written so that evaluations are made sequentially of the various run-off processes. First, if $P(J, I)$ exceeds SI then surface run-off will occur, although in most areas this is rare. Secondly, $P(J, I)$ enters soil storage S, which is limited and an amount GI percolates to groundwater. The remaining $P(J, I)$ is routed to the channel system as fast run-off component. The groundwater component is routed to the streams by the slow run-off hydrograph. Both these routing procedures utilize the convolution equation (2).

Evapo-transpiration is assumed to draw on the soil storage S. The simplest model might assume that EPA equals EP until one inch of soil moisture has evaporated. EP must be supplied to the model from either pan data or calculations based on one of the recognized formulae. From then on the EPA decreases by using a simple model such as:

$$EPA = EP*EXP(-K*SD)$$

The model implies that EPA gradually diminishes as a function of soil moisture deficit.

$$SD = \sum EPA - \sum PS$$

The EPA returns to the EP rate following any rain and continues at this rate until all the rain or until 1 inch of moisture, whichever is the greater, has evaporated. The EPA then depends on the residual SD value. Similar models, but generally more complex, are used by the Stanford models and by Nash[42,43].

The role of the EPA calculation is seen to be primarily that of determining how much of the incoming precepitation is stored and how much runs off, either as fast run-off or as slow, groundwater run-off.

Once the computer program has been written, approximate values are assigned to the various coefficients, and best fit values of these coefficients must then be achieved. It is assumed that basic precipitation data, temperature data and run-off data for the stream headwaters are available for this process. The parameters can be varied sequentially by a small amount and correspondence between the input, precipitation and the output, stream run-off, can be evaluated by a type of hill-climbing technique. Good common sense is very valuable in such a process. As an example, it may be possible to select a wet winter period when the soil is permenantly saturated and evaporation is low. For this period a good fit may be made of the fast component unit hydrograph parameters. A recession period may be used for the slow component hydrograph parameters. Precipitation occurring after a prolonged dry spell may give estimates of soil storage requirements before run-off commences. Blind use of optimisation procedures may never achieve separation of these inter-related coefficients.

The stream channel routing and lake routing portions of the model also offer a challenge to the model builder. For large lakes it may be necessary to allow for lake evaporation and data will be required for variations in lake level as well as lake out-flow if the lake outflow is controlled. For a lake outflow which has a constant stage-discharge relationship, either stage or discharge is adequate. Given this data, lake or reservoir routing offers little problem.

Channel routing is more complex, especially if there is much lateral inflow which is ungauged. Ungauged lateral inflow must be separated from the main stem flow, and yet this can only be achieved either for long term periods when routing is insignificant or for short term periods if the routing coefficients are already known. This problem has been discussed when the Muskingham method was presented. The author[46,47] has had good success with a modified lag and route procudere. The lag time or monoclinic travel time was estimated from stage velocity curves. This travel time is therefore a function of flow, or river stage, which is true of a real river system. Estimates of storages for a linear reservoir were made from channel characteristics. Using this two parameter routing model, separation of lateral flows can be achieved. The lateral flows can be modeled from precipitation data for several time periods. Systematic errors due to error in the routing model then become apparent and the model parameters can be adjusted to minimize such errors. This is a time consuming process which can be greatly helped by careful selection of study periods, such as when lateral flows are in a recession pattern or at a minimum.

In summary, simulation techniques should be kept as simple as possible, provided that representation of reality is not jeopardized. As much fitting of each element of the model should be done during the model construction, and this should be done by use of good common sense. The value of graphical fitting should be utilized because, by use of graphical representation, the mind is able to handle large quantities of data simultaneously. Final fitting of the model is done when the main simulation blocks have been assembled, and at this point it is useful to have a simple quantitative measure of the goodness of fit. Several measures aave been proposed, but Nash's[23] is a sgood as any. This measure is calculated by subtracting the recorded output from the simulated output for each time increment. These differences are squared and summed to give a *residual variance*. The optimization can then proceed to minimize this residual variance.

Representative Design Applications

(*a*) *Minor design problems*

Such structures as small dam spillways, road culverts, small bridges, etc. require estimation of a design flood. The approach by which such a design flood can be estimated is shown diagrammatically in the flow diagram Figure 9.

First it must be decided whether rain floods or snowmelt floods are likely to be limiting and it may be necessary to calculate each before a conclusion can be reached.

Secondly, the response time of the catchment must be considered and this is typically characterized by the unit hydrograph, especially the time from the start of run-off to the peak.

Thirdly, the precipitation records for rain events must be examined. In many instances data for the catchment may not exist, but representative data from adjacent areas may be available. Such data must be analyzed to yield duration-intensity plots for given return periods, or probabilities and for the catchment area in question.

The unit hydrograph data and the precipitation data must now be combined to determine the critical period of run-off. As an example, starting with the 1 hour unit hydrograph, it is easy to superimpose several 1 hour unit hydrographs to obtain 2 hour, 3 hour, 4 hour, etc. unit hydrographs. These longer duration unit hydrographs have relatively lower peaks as is illustrated in Figure 10. However, the greater volume of rain in longer duration storms may give a higher resultant peak and such a situation is illustrated in the figure.

A similar calculation for snowmelt would yield a maximum flow for run-off. Note that, for small catchments, the diurnal temperature variations impose limitations on the maximum snowmelt run-off because the high temperatures only last for a limited number of hours each day[30,31,32].

Once the peak flow has been determined for a given rain or snow event probability, a first

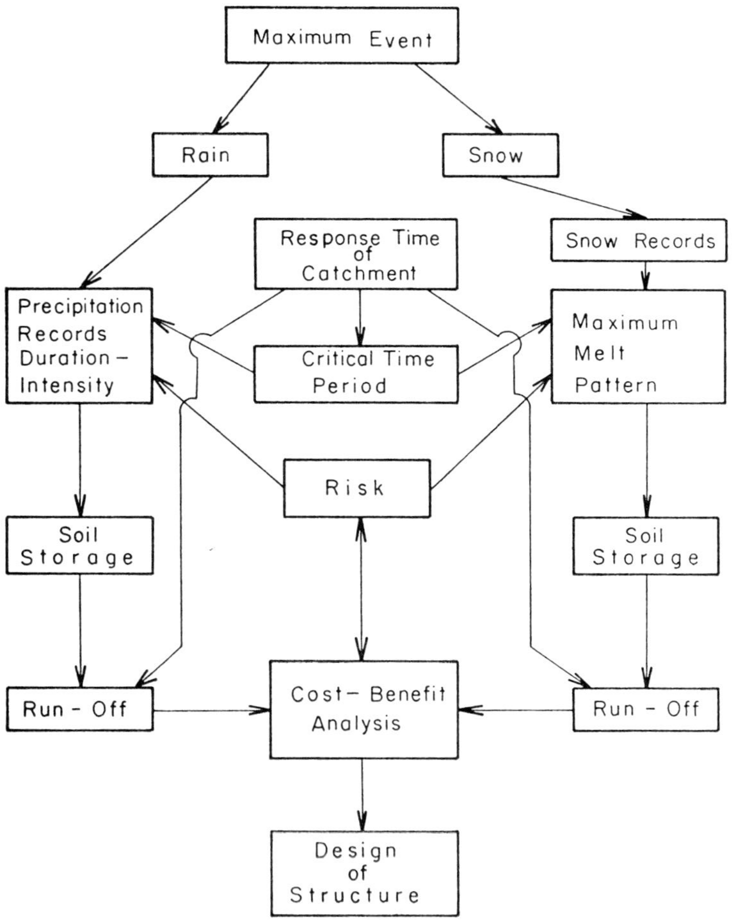

Figure 9 Estimation of design flood for a small spillway or road culvert, etc.

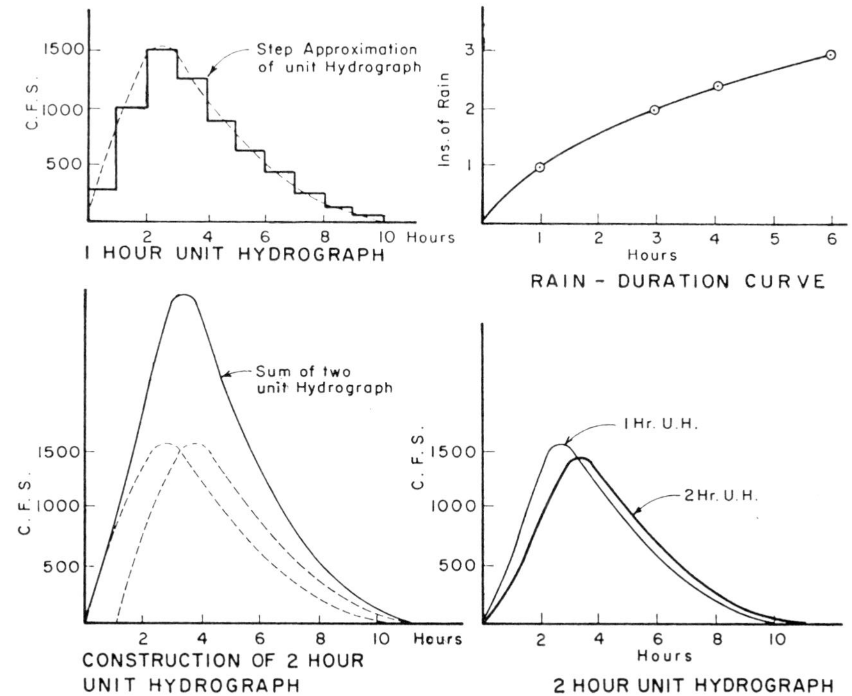

PEAK RUN-OFF

	RAIN	U.H. PEAK	RUN-OFF - RAIN x U.H. PEAK	
1 Hour Storm	1.0 inch	1540	1540 C.F.S.	
2	1.7	1290	2190	
3	2.0	1290	2580	
4	2.5	1125	2810	Max.
5	2.75	990	2720	

∴ Critical Duration is 4 Hour Storm of 2.5 ins. of Rain, **which produces maximum Run-off of 2810** C.F.S.

Figure 10 Determination of critical storm duration for a given catchment.

design of the structure should be made. The costs can be estimated and also damages resulting from higher floods can be at least approximated. At this point the cost of added protection can be considered and the assumed risk can be re-evaluated. Experience of similar structures is very useful at this stage, especially in evaluating such added constraints as possible log jams or the influence of siltation or erosion. (Bureau of Reclamation[50])

(b) Major spillway design floods

Large dams are a particularly challenging design problem from many viewpoints. In particular, the consequences of failure due to inadequate spillway design has given rise to numerous extensive hydrological studies and the development of suitable techniques. For such a major structure it is usual to calculate design floods by several methods, as discussed in such references as Kuiper and Linsley[48,49].

The first method is based on extreme probability theory as previously discussed in the statistics section. The analysis of extreme independent time series is based on the work of Fisher and Tippett[11] and is well illustrated by Gumbel[12]. The big question is what level of risk to assume, and probabilities of 1 : 1000 to 1 : 10,000 are not uncommon for such projects. The economic arguments reveal that the spillway cost is often small compared with the havoc that could occur if the dam failed. Hence it is quite usual for spillways to be designed for much more rare events than the design flood used for the rest of the system. Clearly the determination of the spillway design flood involves the incorporation of a large safety factor, but it is still necessary to make as good an estimate as possible of the true probability of such an event.

The second method for estimation of design floods is the *maximum probable flood*[1] approach. The physical processes involved in producing a flood are studied. In some areas, such as the North-Western part of North America, snowmelt is the principle factor for large rivers. Elsewhere rain floods may be limiting. It is necessary to study the weather patterns which produce either high snowmelt or high rainfall intensities. Storm patterns, durations and movements can be studied. The storm movement can be crucial for a large river for it is not unknown for storms to move downstream at a rate similar to the flood wave movement, so that considerable reinforcement of the flood can take place. On northern rivers like the Mackenzie, which flows to the north, snowmelt and ice breakup occur from south to north, resulting in build up of ice jams and floods. There is also the question of combination of rare events such as rain on snow or several consecutive storms. Concerning rain on snow, it is noticeable that the snow-covered areas are the ones which mainly contribute to run-off, probably because the snowmelt has already primed the soil and also because the 100% humidity of the warm air during rain can produce considerable condensation melt.

The calculation of flood run-off from such storm or melt patterns requires a run-off simulation calculation of the type previously discussed. The resulting flood flow may be modified by the reservoir storage until maximum permissible reservoir stage is reached. From then on the spillway must be capable of passing the total flood, less any flow passed through the turbines or diverted to irrigation.

A third approach, which is often of comfort to the designer, is to compare the design flood with that used on other similar rivers. Careful comparison and cross-correlation may help to confirm the reasonableness of the final design values.

Standard project floods[1] can be determined by similar processes and used for the design of other aspects of a scheme, such as downstream diking and channel protection schemes. In this case the resulting damage from failure may not be so widespread and may not endanger life. It is not unusual for such design floods to be only a fraction of the value of a spillway design flood, perhaps a half.

(c) Complex systems

Usually a complex water resource system involves the operation of storage or even several storages. Such operation aims to minimize flood risk at one extreme and drought risk at the other end of the scale. The scheme must satisfy various complex demands for such uses as hydro-power, irrigation, fisheries requirements, water supply, water quality and navigation. Recreation is also now a demand of increasing importance and imposes limitations regarding water levels on beaches or in marinas. Water intakes and waste-water effluent lines set additional limitations. Water quality also sets demands on minimum flows, water temperature, lake levels, etc.

An important aspect of river basin behavior which has long been neglected is the effect of the alteration of run-off pattern on the ecological balance and erosion patterns downstream. Hydrological studies can supply the biological ecologists with information on how flood levels and seasonal run-off patterns will change with given schemes so that harmful effects can be recognised before damage is done.

Added to this complexity is the question of run-off forecast accuracy which can greatly influence the efficiency of water utilization. In the evaluation of such schemes hydrology is seen in its most necessary and challenging role. The hydrologist is called on to develop accurate forecasts of run-off. He also must simulate the total system behavior and then subject the system to sequences of run-off patterns generated from the characteristics of the recorded data. The resulting performance of the system is then analysed from the economic viewpoint and, frequently, projections must be made into the future to estimate eventual demands and trends and system capability. In this type of problem the whole technical capability of the hydrologist is utilized and as techniques and data improve so the conflicting requirements of water use, water quality, recreation and flood control can be better reconciled.

REFERENCES

Textbooks and reference books

1. Chow, V. T., *Handbook of Hydrology* (McGraw-Hill).
2. Linsley, Kohler, and Paulhus, *Hydrology for Engineers* (McGraw-Hill).
3. Bruce, and Clark, *Introduction to Hydrometeorology* (Pergamon).
4. Henderson, *Open Chennel Flow* (MacMillan).
5. Ward, *Principles of Hydrology* (McGraw-Hill).
6. Meinzer (Editor), *Hydrology* (Dover).
7. Maas *et al.*, *Design of Water Resource Systems* (Harvard).

Statistics

8. Moroney, M. J., *Facts from Figures* (Pelican).
9. Parl, B., *Basic Statistics* (Doubleday College Courses), Refs. 1, 2, 3.
10. Kendall, M. G., *A course in Multivariate Analysis.*
11. Fisher, and Tippett, Limiting Forms of the Frequency Distribution, *Cambridge Phil. Soc.* Vol. 24, 1927–28.
12. Gumbel, Applications of Extreme Statistics, Lecture 4, Applied Mathematics Series 33, U.S. Nat. Bureau of Standards.
13. Matalas, N. C., and M. A. Benson, Effect of Interstation Correlation on Regression Analysis. *J. of Geophys. Res.* Vol. 66, No. 10, pp. 3285–3293, Oct. 1961.
14. Snyder, W. M., Some possibilities for Multivariate Analysis in Hydrological Studies, *J. of Geophys. Res.* Vol. 67, No. 2, pp. 721–729, Feb. 1962.
15. Wallis, J. R., Multivariate Statistical Methods in Hydrology—A comparison using data of known functional relationship. Water Resources Research, Vol. 1, No. 4, pp. 447–461, 1965.

Runoff

16. Sherman, L. K., Streamflow from rainfall by the unit-graph method, *Eng. News. Rec.* Vol. 108, pp. 501–505, April 7, 1932.
17. Nash, J. E., The form of the instantaneous unit hydrograph, Intern. Assoc. of Sci. Hydrology, Pub. 45, Vol. 3, pp. 114–121, 1957.
18. Dooge, J. C. I., A general theory of the unit hydrograph, *J. Geophys. Research* Vol. 64, No. 1, pp. 241–256, 1959.
19. Singh, K. P., A non-linear approach to the instantaneous unit hydrograph, Ph. D. thesis, University of Illinois, 1962.
20. Snyder, W. M., Hydrologic studies by electronic computers in TVA, A.S.C.E. HY 2, pp. 1–10, Febr. 1960.
21. Snyder, W. M., Matrix operations in hydrograph computations, Tennessee Valley Authority, Res. Paper No. 1, 10 pages, Knoxville, Tennessee, Dec. 1961.
22. Laurenson, E. M., Storage analysis and flood routing in long river reaches, *J. Geophys. Res.* Vol. 64, No. 12, pp. 2423–31, Dec. 1969.
23. Nash, J. E., Lectures on Parametric or Analytical Hydrology, Ontario Committee for the Hydrologic Decade, PR 28, 1969.
24. McCarthy, G. T., The Unit Hydrograph and Flood Routing, Conference of North Atlantic Division, U.S. Corps of Engineers, June 1938.
25. Seddon, J. A., River Hydraulics, *Trans. A. S. C. E.* Vol. 43, p. 179, 1900.
26. Henderson, F. M., and R. A. Wooding, Overland Flow and Interflow From Limited Rainfall of Finite Duration. *J. Geophys. Res.* Vol. 69, No. 8, p. 1531, April 1964.
27. Lighthill, and Whitham, On kinematic waves: I-Flood movement in long rivers, *Proc. Roy. Soc.* (London) A., Vol. 229, No. 1178, p. 281, May 1955.
28. Dooge, and Harley: Linear Routing in Uniform Open Channels, Intern. Hydrology Symposium, Fort Collins, Sept. 1967, 8.1, pp. 57–63.
29. Hyami, S., On the propagation of flood waves, Bulletin No. 1, Disaster Prevention Research Institute, Kyoto University, Japan (Dec. 1951) (Summarized by Henderson (4) p. 384).

Snow Hydrology

30. Hildebrand, C. E., and T. H. Pagenhart, (1955) Lysimeter studies of snowmelt. U.S. Corps Research Note.
31. Light, P. (1941), Analysis of high rates of snowmelting. *Trans. Am Geophys. Union* Pt. I. Vol. 22, pp. 195–205.
32. Linsley, R. K. (1943), A simple procedure for the day to day forecasting of run-off from snowmelt, *Trans. Am. Geophys. Union* Pt. III, pp. 62–67

33. Quick, M. C. (1965), River Flood Flows, *Journal of the Hydraulics Division*, *ASCE*, No. HY 3, Vol. 91, pp. 1–18.
34. SIPRE (1959), Report No. 12, Annotated Bibliography of Snow and Ice.
35. Snow and Ice Hydrology, Proceedings of Workshop, 1969, Colorado State University.
36. U.S. Corps of Engineers (1956), Report No. PB 151660 "Snow Hydrology", U.S. Corps of Engineers, Dept. of Commerce, Washington, D. C.
37. U.S. Corps of Engineers (1960), Run-off from Snowmelt, EM 1110-2-1406, U.S. Govt. Printing Office, Washington, D. C.
38. Wilson, W. T. (1941), An outline of the thermodynamics of snowmelt. *Trans. Am. Geophys. Union* Pt. I, pp. 182 to 195.
39. Snow and Ice Hydrology, Proceedings of Workshop 1969, Colorado State University.
40. Quick, M. C., "Experiments with Physical Snowmelt Models" National Research Council Hydrology Symposium No. 8: Runoff from Snow and Ice, Quebec City, May 1971.
41. Ouick, M. C., "Forecasting Runoff—Operational Practices" invited theme paper: International Symposium on the Role of Snow and Ice in Hydrology, Banff 1972. UNESCO, WMO, IHD, Published by National Research Council, Ottawa·

Artificial streamflow generation

42. Maass *et al.*, *Design of Water Resource Systems* Chapter12 (Harvard).
43. Fiering, M. B., *Streamflow Synthesis*.

Simulation

44. Crawford, N. H., and R. K. Linsley, "Digital Simulation in Hydrology: Stanford Watershed Model IV", Tech. Report No. 39, Dept, Civil Eng., Stanford University, Stanford, California, 1966.
45. O'Connell, Nash, and Farrell, River Flow Forecasting Through Conceptual Models II, *Journal of Hydrology*, 1970, p. 322.
46. Quick, M. C., and A. Pipes, UBC Fraser Run-off Model for Forecasting and Planning Studies Dept. of Civil Eng. Water Resource Series, Universtiy of British Columbia 1971.
47. Quick, M. C. and A. Pipes, "Daily and Seasonal Forecasting with a Water Budget Model", Proceedings of the International Symposium on the Role of Snow and Ice in Hydrology, Banff 1972, UNESCO, WMO and IHD, Published by National Research Council, Ottawa.

Designing

48. Kuiper, *Water Resources Development* (Butterworths).
49. Linsley and Franzini, *Water Resources Engineering* (McGraw-Hill).
50. Bureau of Reclamation: Design of Small Dams U.S. Govt. Printing Office, Washington, 1965.

MICHAEL C. QUICK

University of British Columbia

I

INCINERATION: see MANAGEMENT OF SOLID WASTE

INDUSTRIAL HEALTH: see OCCUPATIONAL HEALTH

INDUSTRIAL HYGIENE: see OCCUPATIONAL HEALTH

INDUSTRIAL WATER REUSE: see WATER REUSE

L

LAKES: see AQUATIC PRIMARY PRODUCTION, WATER—FRESH

LEGAL ASPECTS: see ENVIRONMENTAL LAW

LIMNOLOGY: see WATER—FRESH

LIQUID WASTES OF INDUSTRY*

Introduction

The author believes that solving waste treatment problems should demand not an ability to memorize details, but rather the application of scientific principles and judgement to the solution of practical problems. Although the reader may have to search the literature for the detailed information he needs, the material provided in this chapter should suffice as a suitable beginning for all potential wastewater engineers.

Industrial wastes contribute at least 2/3 of the organic matter entering our watercourses in the United States. Unless we understand more about the industries which discharge these wastes amelior-

* For a more complete and detailed coverage of this subject the reader is referred to the new text, *Liquid Wastes of Industry* by Nelson L. Nemerow, published by Addison-Wesley Company, Reading, Mass. 1971.

ation of the overall problem is extremely difficult. One cannot legislate or regulate without a firm knowledge of what can be accomplished feasibly.

The newly created Environmental Protection Agency is currently making important inroads in this area. Each industry is being studied by a separate contractor in order to ascertain reasonable levels of treatment and residual wasteloads that each can be expected to achieve at any cost if necessary. The E.P.A. intends to establish minimum wasteloads of each contaminant for each industry. Presumably state and local governments may require more treatment than this minimum in case local water quality is not maintained by the industrial treatment. Therefore, both effluent and stream standards may be vital in future industrial waste treatment situations.

1.1 Effects on Streams

All industrial wastes affect, in some way, the normal life of a stream. When the effect is sufficient to render the stream unacceptable for its "best usage," it is said to be polluted. Best usage means just what the words imply: use of water for drinking, bathing, fishing, and so forth. A more detailed description of these uses is given in Section 2.

Streams can assimilate a certain quantity of waste before reaching a polluted state. Generally speaking, the larger, swifter, and more remote streams that are not much used are able to tolerate a considerable amount of waste, but too much of any type of polluting material causes a nuisance. To call a stream polluted, therefore, generally means that the stream contains an excessive amount of a specific pollutant or pollutants. The following are examples of materials which can cause pollution:

Inorganic salts	Heated water
Acids and/or alkalis	Color
Organic matter	Toxic chemicals
Suspended solids	Microorganisms
Floating solids and liquids	Radioactive materials
	Foam-producing matter

In the following table the source of these materials as well as some major detrimental effects are given.

TABLE 1

Contaminants—Source and Effect

Type of contaminant	Industrial source (some examples)	Some major effects
Inorganic salts	Oil refinery desalination plants munitions manufacturing	Interferes with industrial usage (i.e. insulating paper) municipal (drinking water), and acriculture (irrigation water)
Acids and/or alkalies	Chemical manufacturing	Corrosion of pipelines and equipment kills fish
Organic matter	Tanneries, canneries, textile mills, etc.	Food for bacteria and thus depletes oxygen
Suspended matter	Paper mills, canneries, etc.	Suffocates fish eggs, degrades stream appearance
Floating solids and liquids	Slaughterhouse, refinery oil	Unsightly, odorous, interferes with oxygen transfer
Heated water	Cooling waters from most industries and power plants	Accelerates bacterial action lowers total oxygen saturation level
Color	Textile, metal finishing, and chemical plants	Objectionable appearance
Toxic chemicals	Munitions mfging., metal plating, steel mills, etc.	Alters stream biota
Microorganism	Tanneries, combined municipal-industrial plants	Unsafe for drinking
Radioactivity	Nuclear power plants, chemical laboratories	Concentrates in fish, harmful for drinking
Foam-producing matter	Glue mfging., slaughterhouse, detergent manufacturing	Aesthetically objectionable

1.2 Effects on Sewage Plants

It is only natural for industry to presume that its wastes can best be disposed of in the domestic sewer system, and municipal officials often feel that it is their responsibility to accept any wastes flowing into their city's disposal system. However, city authorities should not accept any waste discharges into the domestic sewer system without first learning the facts about the characteristics of the wastes, the sewage system's ability to handle them, and the effects of the wastes upon all components of the city disposal system. Institution of a sewer ordinance, restricting the types or concentrations of waste admitted in the sewer leading to a treatment plant, is one means of protecting the system.

To remove pollution from industrial wastes, a sewage-treatment plant must have sufficient capacity and be of the proper type. Theoretically, a sewage treatment plant could be designed to handle any types of industrial waste, but present plants fall short of this ideal. Joint treatment of municipal and industrial wastewaters which are amenable to treatment may offer greater removal efficiencies, but economics will usually be the deciding factor.

The pollutional characteristics of wastes having readily definable effects on sewers and treatment plans can be roughly classed as follows: (1) biochemical oxygen demand (BOD); (2) suspended solids, (3) floating and colored materials; (4) volume; and (5) other harmful constituents. Table 1.2 presents a comparison of domestic-sewage pollutional characteristics and those of some industrial wastes.

TABLE 1.2

Domestic Sewage–Industrial Wastes
A General Comparison

Type Waste	BOD (ppm)	Suspended solids (ppm)
Domestic sewage	250	300
Pulp and papermill waste		
—no chemical pulping	150	350
—with chemical pulping	5,000	2500
Tannery	1,200	2000
Cannery	150–1000	250–1500

BOD of an industrial waste affects mainly the biological or secondary treatment units of a sewage treatment plant. If these units are large enough and if the industrial waste can be biodegraded as rapidly as domestic sewage, it can be handled by the plant. *Suspended* solids, if readily setteable, can be disposed of along with domestic sludge as long as the industrial sludge is either or both readily drainable and decomposable and provided extra sludge treatment unit capacity is provided. Special treatment plant design and operation must be provided in order to take care of floating and/or colored material and other 'harmful' constituents. Excess industrial volume shortens treatment plant retention time, thus enlargement of most units will be necessary if and when detention time is a significant factor in overall plant effectiveness.

Stream Protection Measures

In general the quality of receiving waters can be preserved by either or both of two methods: stream standards or effluent standards. Although the ultimate goal is to raise the quality of the river or lake to the optimum for its best usage, it is sometimes easier to accomplish this by requiring each polluter to discharge only a given quantity of contaminant or a given concentration with a stipulated total volume of waste water. Many state agencies ascertain the best usage of a stream and assign certain quality standards to that water use. Any polluter found adding contaminants in such quantities as to contravene these standards is cited as a violator and must then abate the pollution. A more recent trend followed by some agencies is to establish the receiving water quality desired and attempt to maintain this quality by controlling each waste discharge to the minimum contaminant units per unit of production or per capita. The latter is generally determined on an industry-wide basis from an analysis of effective treatment potential on an economically feasible basis. An example of the stream standards system is shown in Table 2 as used in New York State, while an illustration of the effluent standard system as used by Pennsylvania is shown in Table 3.

Determination of Allowable Organic Loadings on Receiving Waters

If a regulatory agency decides to allow an industry to use a portion of the assimilating capacity of the receiving water, the agency must resort to some method of predicting an acceptable organic loading. The agency orders a decision usually based upon presentation of a certain minimum concentration of dissolved oxygen in the stream. Many factors determine the consumption of dissolved oxygen in streams. Some of these include (1) nature of the organic matter—whether easily decomposable for normal stream biota, (2) amount of receiving water—as a diluting effect as well as an oxygen provider, (3) physical nature of the stream—swift or sluggish, pools or riffles, (4) biological nature of the stream—whether containing adapted flow, slime growths, benthal deposits, algae growth,

TABLE 2

New York State Classes and Standards for Fresh Surface Waters

	Water standards†				
Class and best use*	Minimum dissolved oxygen, ml/liter	Coliform bacteria median no/100 ml	pH	Toxic wastes, deleterious substances, colored wastes, heated liquids, producing substances‡	Floating solids, settleable solids, oil, and sludge deposits
AA Source of unfiltered public water supply and any other usage	5.0 (trout) 4.0 (nontrout)	Not to exceed 50	6.5–8.5	None in sufficient amounts or at such temperatures as to be injurious to fish life or make the waters unsafe or unsuitable	None attributable to sewage, industrial wastes or other wastes
A Source of filtered public water supply and any other usage	5.0 (trout) 4.0 (nontrout)	Not to exceed 5000	6.5–8.5		None which are readily visible and attributable to sewage, industrial wastes or other wastes
B Bathing and any other usages except as a source of public water supply	5.0 (trout) 4.0 (nontrout)	Not to exceed 2400	6.5–8.5		
C Fishing and any other usages except public water supply and bathing	5.0 (trout) 4.0 (nontrout)	Not applicable	6.5–8.5	None in sufficient amounts or at such temperatures as to be injurious to fish life or impair the waters for any other best usage	
D Natural drainage, agriculture, and industrial water supply	3.0	Not applicable	6.0–9.5	None in sufficient amounts or at such temperatures as to prevent fish survival or impair the waters for agricultural purposes or any other best usage	

* Class B and C waters and marine waters shall be substantially free of pollutants that: unduly affect the composition of bottom faune: unduly affect the physical or chemical nature of the bottom; interfere with the propagation of fish. Class D and SD (marine) will be assigned only where a higher water use class cannot be attained after all appropriate waste-treatment methods are utilized. Any water falling below the standards of quality for a given class shall be considered unsatisfactory for the uses indicated for that class. Waters falling below the standards of quality for Class D, or SD (marine), shall be Class E, or SE (marine), respectively and considered to be in a nuisance condition.

† These Standards do no apply to conditions brought about by natural causes. Waste effluents discharging into pubiic water supply and recreation waters must be effectively disinfected. All sewage-treatment plant effluents shall receive disinfection before discharge to a watercourse and/or coastal and marine waters. The degree of treatment and disinfection shall be as required by the state pollution control agency. The minimum average daily-flow for seven consecutive days that can be expected to occur once in ten years shall be the minimum flow to which the standards apply.

‡ Phenolic compounds cannot exceed 0.005 mg/liter; no odor-producing substances that cause the threshold-odor number to exceed 8 are permitted: radioactivity limits are to be approved by the appropriate state agency, with consideration of possible adverse effects in downstream waters from discharge of radioactive wastes, and limits in a particular watershed are to be resolved when necessary after consultation between states involved.

TABLE 3

Pennsylvania raw-waste standards for pulp and paper mills

Type of product or process	Population equivalent, per ton of product, based on 5-day BOD		Pounds of suspended solids per ton of product	
	3-day average*	8-hr average	3-day average	8-hr average
Group A				
Tissue paper	75	80	40	50
Glassine paper	25	30	15	20
Parchment paper	40	45	20	30
Miscellaneous papers	25	30	5	10
Flax papers—Condenser	375	415	300	350
Group B (speciality group)				
Fiber paper	800	850	200	235
Asbestos paper	125	185	290	350
Felt paper	210	230	60	65
Insulating paper	2250	2500	325	350
Speciality papers	1000	1200	135	160
Group C (coarse paper)	90	120	35	50
Group D (integrated mills)				
Wood preparation	80	100	40	50
Pulp (sulfite)	3000	3500	35	40
Pulp (alkaline)	300	350	20	35
Pulp (groundwood)	115	130	80	85
Pulp (deinked unfilled stock)	500	650	375	500
Pulp (deinked filled stock)	400	500	600	800
Pulp (rag cooking)	1400	1550	475	500
Bleaching (long-fiber stock, multisingle-stage bleaching and short-fiber stock, single-stage bleaching)	60	70	3	6
Bleaching (Short-fiber stock, multistage bleaching)	165	185	30	35
Paper-making	100	125	75	85

* All averages are for periods of consecutive operation.

(5) chemical nature of the stream—whether containing toxic chemicals, nutrients for promoting biological growth, (6) environmental character of stream—temperature and pH of the water. It is not possible to discuss all of these parameters in this treatise so the reader is referred to Chapter 3 in the writer's previously mentioned textbook for a more complete coverage of the subject.

In general, industry should determine the ultimate (20 day) biochemical oxygen demand, BOD_μ, of its waste and the quantity (gallons) discharged per day. For the reaction of the receiving water the engineer must compute the stream deoxygenation rate, k_1, and its reaeration rate, k_2, as well as the travel time of the industrial waste downstream from its entrance. From an evaluation of the stream assets and liabilities one can predict with some degree of approximation the resulting dissolved oxygen sag curve below the industry as shown in the following Fig. 4.

The Streeter-Phelps Sag curve formula is generally used

$$D_t = \frac{k_1}{k_2 - k_1} L_a(10^{-k_1 t} - 10^{-k_2 t}) + D_a^{-k_2 t}$$

where D_t = dissolved oxygen deficit at waste travel time, t, downstream and $L_a = BOD_\mu$ in the receiving water after the industrial waste has mixed with the stream water.

Several simplifications of the above formula have been proposed as well as several methods for arriving at true values of the stream reaction rates, k_1 and k_2. Variations in values of the reaction rates can result in vastly different oxygen sag curves with correspondingly different degrees of waste treatment required. Considerable cost differences develop

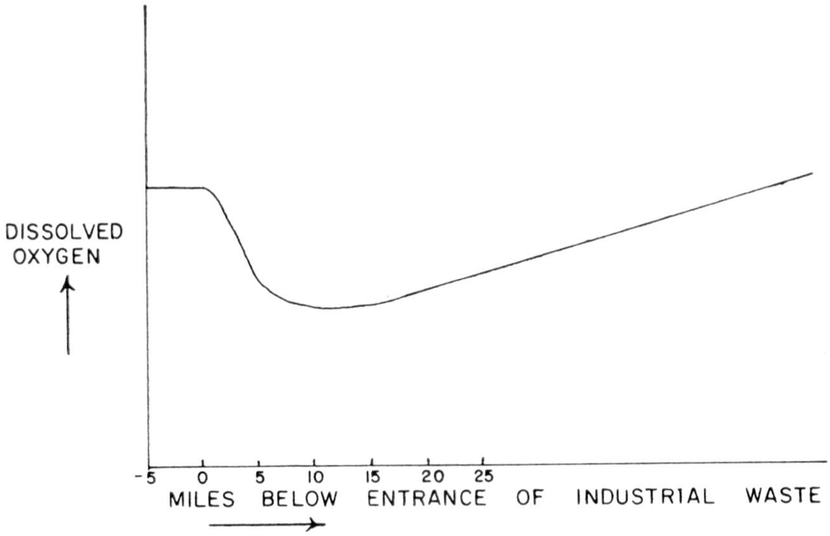

Figure 4

from these situations. Industry often decides to overtreat its wastes (at greater costs) to protect itself against unpredictable nature of the stream and conservative decisions of the regulatory agencies.

Stream Sampling

Receiving streams should be sampled under the most critical conditions likely to be encountered. This generally means sampling at the lowest streamflow, highest river temperature, and maximum flow of industrial waste. Under most situations it is practically impossible to sample under these three extreme conditions simultaneously. As a result the stream analyst has to compromise and sample the stream under the most critical conditions existing. Usually this means a short, intensive sampling during the hot, dry summer period. It becomes necessary then to project the findings to critical design conditions. This, in itself, is a procedure which can cause many hours of debate among stream analysts.

One should sample a receiving water as follows:

1 – as frequently as possible in consecutive stretches below the waste entrance—at least every 1/4 mile and less if the stream character changes suddenly

2 – below and above every major source of contamination or dilution

3 – at appropriate sections across the entire stream

4 – at every significant pool or riffle area

6 – at just before daylight and at midday when algae effect in the stream is suspected

Analyses for dissolved oxygen, temperature, B.O.D., and streamflow should be made at each sampling point. Biological analyses for algae, protozoa, fish, etc., will often yield information of pollution not readily obtained from chemical data.

Economics of Industrial Waste Treatment

While capital costs of waste treatment have been variously estimated at $ 100–300 per pound of BOD per day, 1% of production cost, or 8 to 30 cents per 1000 gallons, much discrepancy in both methods and results as well as from industry to industry exists. What is more important is the fact that cost of treatment increases geometrically with degree of removal of contaminants as shown in Fig. 5.

It is imperative that industry as well as the regulatory industries develop some dependable method for knowing how much an industrial plant can actually afford for waste treatment and still remain competitive. Obviously costs of waste treatment, if assumed by a plant, must eventually be passed on to the public consumer. It is he alone who will finally make the decision; for if he finds the price of the product too high for his needs, he can exercise his option to refrain from buying the product. Industry may then choose one of several alternatives: cease production, curtail some production, seek government subsidy, change type of production, or merge with a competitor in order to be able to afford waste treatment.

Waste Water Volume Reduction

Although reduction of waste volume per se may not result in any less total pounds of contaminants

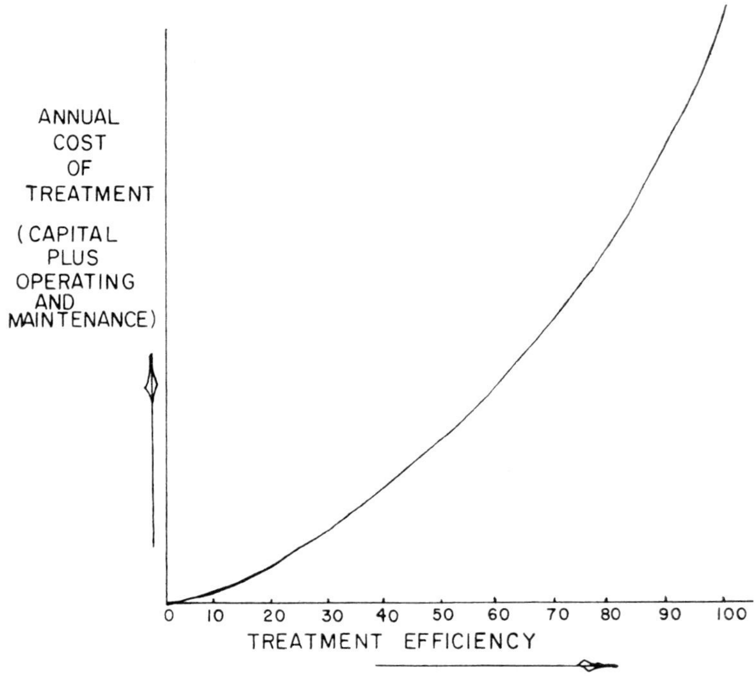

Figure 5

in wastewater, it will contain the pollution in a smaller volume. Usually this means less capital expense for waste treatment. Sometimes it is possible when reducing the volume of wastes to also reduce the total quantity of contaminants. This is accomplished by incorporating more waste matter in the final manufactured product rather than allowing it to be wasted to streams. Another procedure for accomplishing the same result is to "dry-collect" as much waste material from manufacturing machines and operating floors rather than "hosing" down the same matter into drains. The author points out, however, that this results in an increased solid waste disposal problem. Likewise burning of reclaimed wastes may create an air pollution situation. One must consider all environmental impacts of manufacturing and process change decisions such as those involved with a reduction of wastewater volume. Most beneficial changes to affect this reduction can be achieved by a careful study and evaluation of manufacturing processes by an industrial waste engineer.

Waste Strength Reduction

This usually makes a wastewater effluent stream less pollutional in nature, a desirable result. It can be accomplished by (1) changing mechanics or operation of the process, (2) eliminating a step in (or an entire) process, (3) substituting for the raw material or a process chemical. Segregating small volumes of "strong" wastes will give an industry a large volume of relatively "weak" waste and a very small volume of "strong" waste. Both wastes may be treated in different manners and sometimes more economically and effectively than where combined as one waste. Unless either economics or reducing total pollution is effected by waste strength reduction the writer does not advocate the step as an aid in waste treatment. Figure 6 depicts both types of methods of waste strength reduction.

Neutralization

Adjustment of the reaction pH is required for many reasons in numerous instances. Some of these are listed below:

1 – protect sewer lines and plant structures from corrosion
2 – prevent requiring of excessive chemicals for coagulation as a treatment method
3 – protect optimum bacterial activity in biological treatment of wastes
4 – prevent odors which may emit at low or high pH values
5 – protect receiving waters from detrimental effects of wastes with either less than 6.0 or greater than 9.5 pH values.

pH adjustment to correct values can be obtained by (1) adding acid (usually sulfuric) or base (usually sodium hydroxide or lime) (2) mixing acid wastes with alkaline wastes in proper proportion (3) acidifying with flue gas (4) acidifying with submerged combustion (burning of fuel under water) (5) passing acid wastes through limestone beds.

Neutralization costs money and should be used where (1) justified economically (2) it is required by receiving water quality standards (3) it can eliminate a nuisance to the environment.

Equalization and Proportioning

Often when industrial wastes are introduced into a domestic sewage treatment system in a uniform volume and strength in accordance with sewage flows more effective combined treatment results. Equalization involves the holding or retention of industrial wastes for a period of time sufficient to produce at the discharge end of the basin a waste more uniform in quality. This treatment is illustrated in terms of BOD in the following graph (Fig. 7)

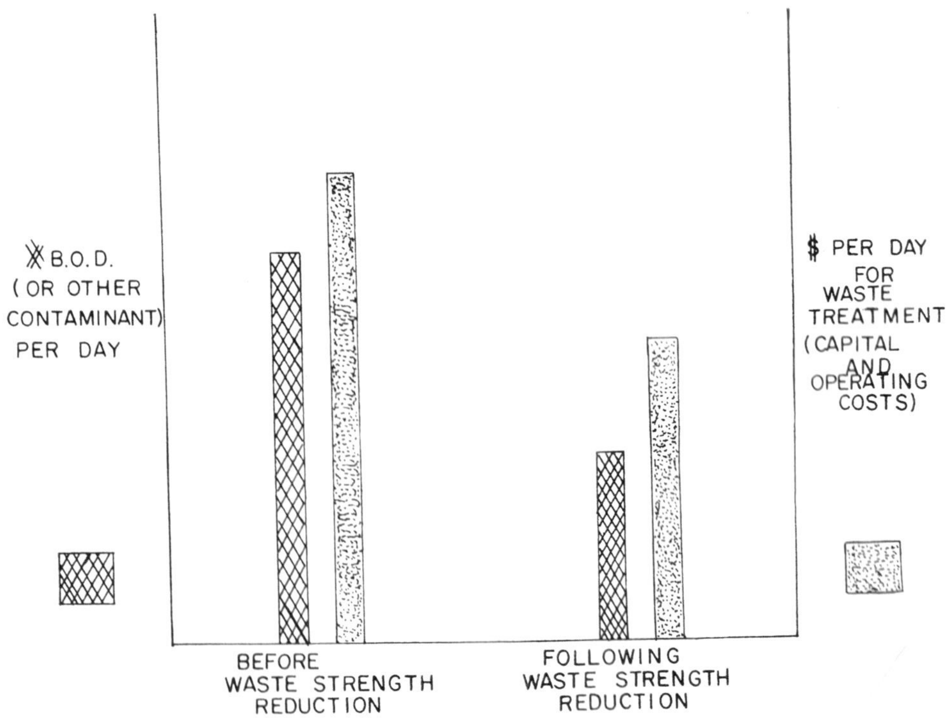

Figure 6

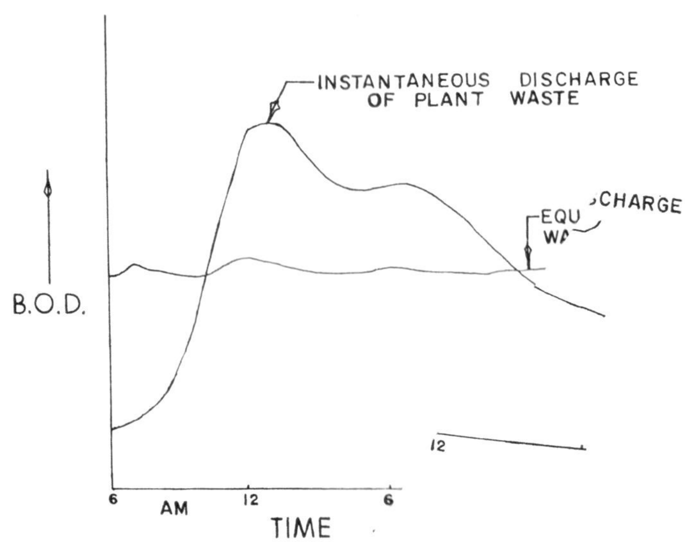

F

The optimum retention time depends upon the nature and repetition of various industrial manufacturing processes. For example, a process resulting in a high BOD waste as shown above occurring at 12 noon and only once every 24 hours may require a 24 hour holding period.

The plant waste may also be discharged in proportion to the sewage flow at the treatment plant or to the river flow into which the waste flows. Proportioning wastes assures that neither treatment plants nor receiving streams are overloaded with "slug" loads of industrial wastes. Proportioning can be accomplished by pumping or discharging industrial plant wastes at previously contemplated or at instantly signaled rates. The system requires flow measurement, signal transmission, and valve responses at plant waste pumping devises. An example of one type of proper proportioning is shown in Fig. 8.

Proportioning of plant wastes can also be practiced to prevent overloading of a contaminant in the system at any one time. This often means revising our previous concept and reducing industrial plant waste flows when municipal flows and contaminants are high.

Removal of Suspended Solids

Many industrial wastes such as cannery, pulp and paper, and tannery contain suspended matter. In general, these solids are usually removed by either differential sedimentation or flotation. Since suspended matter in wastes are generally more dense than water, they can be separated by providing ample detention time to settle. In many instances these types of solids may contain sufficient surface area relative to their density so that they may stratify, be buoyed up, and subsequently floated to the surface and skimmed off. Air dissolved in wastewater under pressure and then released at atmospheric pressures is generally used to accomplish effective floatation of suspended solids. Industrial wastes of the kind which contain suspended solids normally contain from 1 to 20 per cent of these solids by volume. Either sedimentation or flotation produces a sludge with from 2 to 12 per cent solids on a dry weight basis.

These two major methods of suspended solids removal generally reduce the solids by approximately 60 to 90 per cent. In settling removal methods an optimum retention time of one to three hours is required while only 20 to 30 minutes are necessary in flotation solids removal. A typical sedimentation removal curve is shown in Fig. 9.

Removal of Colloidal Solids

These solids which vary in size from 1 to 200 mμ border in characteristics between the suspended and dissolved solids. They are small enough to exhibit stability by virtue of the slight residual electrical charge (generally negative); but large enough to interfere with the passage of light and therefore cause turbidity. They will not settle physically unless destabilized, coagulated, and flocculated into larger masses with sufficiently

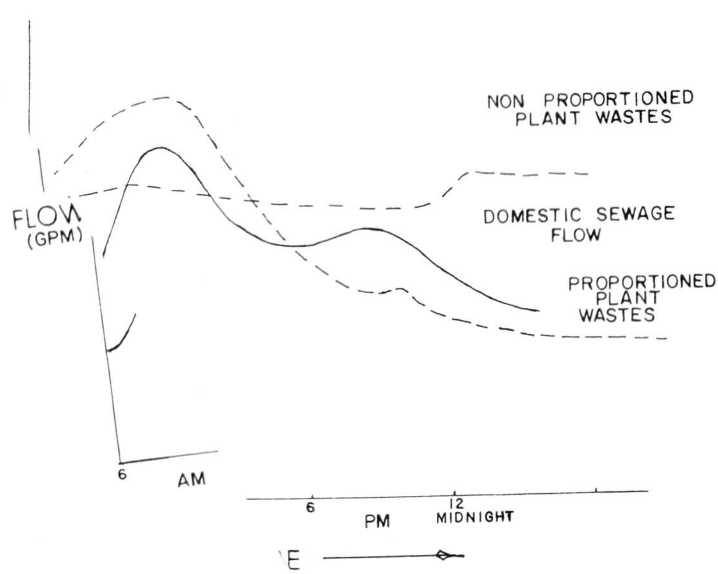

Figure 8

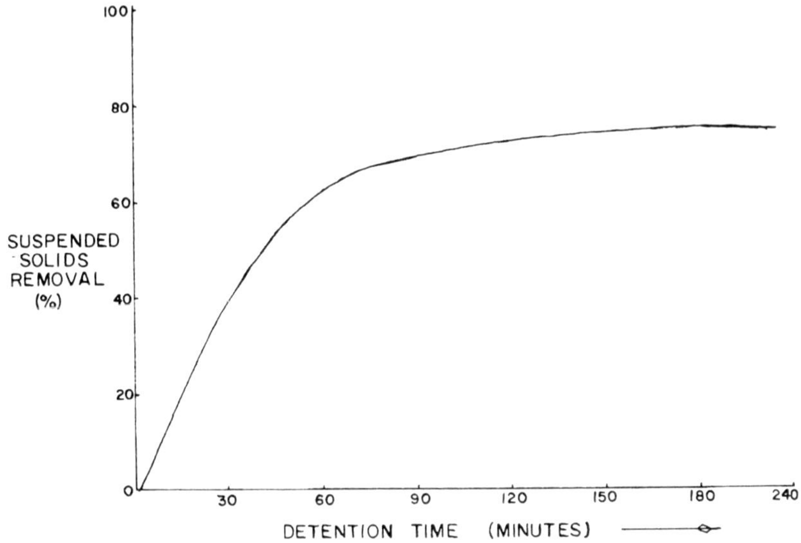

Figure 9

greater densities than water. Generally it follows that this is precisely the treatment that is used to remove them from industrial wastewater. Coagulants employed are normally those electrolytes possessing strong positive charges when dissolved in waters; for example salts of Fe^{+++} or Al^{+++}. The coagulants appear to react simultaneously with the negative hydroxyl ions and the negative colloidal impurities in the wastewaters. Both are reduced substantially by the reaction. An example of such a reaction is shown below:

$$\text{coagulant} + \text{water} +$$

$$Al_2(SO_4)_3 + 6\,H + 6\,OH + \begin{matrix}\text{stable negative}\\ \text{colloidal impurity}\end{matrix} \rightarrow$$

$$\left[2\,Al(OH)_3 \cdot \begin{matrix}\text{destabilized}\\ \text{colloidal impurity}\end{matrix}\right]$$

In many instances colloidal solids may represent from 1/3 to 1/2 of the total oxygen demand of an industrial wastewater. On the other hand, little benefit will be gained by attempting to employ chemical coagulation to remove solids which are not truly colloidal —for larger ones could be removed by settling and smaller ones will not be affected by this method of treatment.

Removal of Inorganic Dissolved Solids

Salts of nitrates, phosphates, carbonates, chlorides, sulfates are often contained in industrial water. Seldom are they removed before discharge to the receiving waters because (1) it is quite costly to do so (2) until recently it has not been considered essential and (3) society has taken the view that it is the downstream user's responsibility to remove them. It turns out that unless the downstream user finds it absolutely essential to remove all inorganic salts from the water he does not do so. Consequently, in general, each subsequent downstream use tends to add more of these salts to our streams. Inorganic salts can cause various diseases and discomforts if taken internally; can ill-effect an industrial product if used by certain manufacturers; can cause scale in pipelines and on equipment of any user; can accelerate algal growth in the waters; can require an excessive amount of soap use in households; can enhance corrosion of metals. It is possible to remove inorganic salts by ion exchange, permselective membranes, evaporation, and by certain biological treatments such as growth of algae. The writer anticipates that such treatment (as well as others) will be used much more in the near future than they are at present.

Removal of Organic Dissolved Solids

Manufacturing of textiles, paper, canned goods, leather as well as preparing many food products such as milk, cheese, and meat results in the wastage of certain amounts of dissolved organic matter. This matter is readily available to microorganisms ever present in streams as a source of food resulting in a diminution of dissolved oxygen in the watercourses. Because these solids are amenable to bacterial degradation waste treatment methods are largely patterned after biological treatment. In effect biological treatment is that which is occurring in streams but only differs in that it is accelerated by creation of optimum environmental conditions for bacterial growth. Some types of biological treatment used on industrial wastes containing a relatively high percentage of dissolved organic include activated sludge and its many modifications, trickling filtration and its variations, as well as oxidation pond treatment. Activated sludge treatment consists of aerating biological flocculent growths within the industrial wastes. Surface for biological oxidation takes place on the flocculent growths. With sufficient detention time for biological reaction, ample supply of oxygen, and a readily degradable organic matter to serve as bacterial food activated sludge treatment will be very effective. In trickling filtration the surface for biological oxidation consists of solid media such as rocks, plastic, or glass. The surface must provide maximum area for biological growth, be resistant itself to deterioration, be relatively light in weight, and allow for sufficient pore or void space for heavy volumetric loadings of industrial waste. Oxygen for biological oxidation comes from the void area in the media. As the growth builds up on the media some sloughs off and is removed by final clarification. Some seed material after filtration is usually returned to the filter influent to keep the "bed" active. Both activated sludge and trickling filtration are capable of removing from 80 to 90 per cent of the B.O.D. Oxidation pond treatment is less effective, requires much land area and detention time, can yield odors unless properly designed or aerated, but is much less expensive to build and to operate. Some ponds develope an algal growth especially in shallow basins which assist in providing oxygen to the bacteria as well as in removing inorganic nutrients. 50 to 75 per cent B.O.D. reduction can usually be easily obtained with these wastes and one to three days detention.

Disposal of Sludge Solids

Screening, settling, and flotation, biological oxidation, and coagulation all produce (in addition to an improved effluent) various quantities and types of sludges. Often the immediate and continual removal and subsequent effective ultimate disposal of these sludges controls the efficiency of the entire treatment process. For if these sludges tend to build up in treatment basins and are not removed they decompose and/or rise and find their way into the plant effluent—thus causing deterioration of its quality.

When sludge is removed it usually still contains from 90 to 95 per cent water. Thus, the first step is one which involves concentration of the solids to lower the water content. Filtration and sand drying beds are popular methods for accomplishing concentration, although centrifugation and lagooning are also used for this same purpose. These processes produce sludges of from 15 to 35 per cent solids. They can then be readily handled for ultimate disposal.

Ultimate disposal can be by incineration, landfill, high temperature—high pressure oxidation, fertilization of farmlands. Although these methods are an end in themselves to the industrial sludge problems, they often cause further deterioration of the air and/or land environment.

Reuse of the sludge solids is being practiced in a few industries such as in the paper and tannery plants and offers a very promising potential for future solutions of sludge disposal problems.

Joint Treatment with Domestic Sewage

It is often advantageous to an industrial manufacturing plant to deliver its entire wastewater—under control—to a municipal sewer and sewage system. Certain state and federal legislation encourages industries to combine their wastes with those of its nearby municipality. Incentives are usually given in terms of ten advantages to the industries.

The many advantages of joint treatment of municipal and industrial wastes are as follows:

1. Responsibility is placed with one owner, while at the same time the cooperative spirit between industry and municipality is increased, particularly if division of costs is mutually satisfactory.
2. Only one chief operator is required, whose sole obligation is the management of the treatment plant; i.e., he is not encumbered by the miscellaneous duties often given to the industrial employee in charge of waste disposal, and the chances of mismanagement and neglect, which may result if industrial pro-

duction men operate waste-treatment plants, are eliminated.

3. Since the operator of such a large treatment plant usually receives higher pay than separate domestic plant operators, better-trained people are available.
4. Even if identical equipment is required, construction costs are less for a single plant than for two or more. Furthermore, municipalities can apply for state and/or federal aid for plant construction, which private industry is not eligible to receive.
5. The land required for plant construction and for disposal of waste products is obtained more easily by the municipality.
6. Operating costs are lower, since more waste is treated at a lower rate per unit of volume.
7. Possible cost advantages resulting from lower municipal financing costs, federal grants, and municipal operation can be passed on to the users and may permit higher degrees of treatment at lower removal levels.
8. Some wastes may add valuable nutrients for biological activity to counteract other industrial wastes that are nutrient-deficient. Thus, bacteria in the sewage are added to organic industrial wastes as seeding material. These microorganisms are vital to biological treatment when the necessary BOD reduction exceeds approximately 70 per cent. Similarly, acids from one industry may help neutralize alkaline wastes from another industry.
9. The treatment of all waste water generated in the community in a municipal plant or plants enables the municipality to assure a uniform level of treatment to all users of the river and even to increase the degree of treatment given to all waste water to the maximum level obtainable with technological advances.
10. Acceptance of the joint treatment project and relinquishment of individual allocations would give the municipality full control of the river's resources and permit it to use the capacity of the river to the best advantage for the public at large. The municipality has greater assurance of stream protection, since it has the opportunity for closer monitoring of effluent quality.

Among the many problems arising from combined treatment, the most important is the character of the industrial waste water reaching the disposal plant. Equalization and regulation of discharge of industrial wastes are sometimes necessary to prevent rapid change in the environmental conditions of the bacteria and other organisms which act as purifying agents, to ensure ample chemical dosage in coagulating basins, and to ensure adequate chlorination to kill harmful bacteria before the effluent is discharged to a stream.

In recent years, two factors in particular have focused attention on the subject of combined treatment for sewage and industrial wastes: the increased interest in stream-pollution abatement and the phenomenal growth of industry in the postwar years, with the subsequent increase in demand for water.

Since most sewage plants use some form of biological treatment, it is essential for satisfactory operation that extremes in industrial waste characteristics be avoided and the waste mixture be: (1) as homogeneous in composition and uniform in flow rate as possible and free from sudden dumpings (shock loads) of the more deleterious industrial wastes; (2) not highly loaded with suspended matter; (3) free of excessive acidity or alkalinity and not high in content of chemicals which precipitate on neutralization or oxidation; (4) practically free of antiseptic materials and toxic trace metals; (5) low in potential sources of high BOD, such as carbohydrates, sugar, hydrolyzed starch, and cellulose; and (6) low in oil and grease content.

If the industrial-waste characteristics are such that the waste can be treated safely and effectively in the municipal sewage plant, there still remain two major considerations: a municipal ordinance which protects the treatment plant from any individual or industrial violation and sewer-rental charges which enable the municipality to defray the increased costs of construction and operation resulting from acceptance of the industrial wastes.

Combined municipal and industrial waste treatment is the most desirable arrangement and at the same time it is the most difficult to achieve. It is the author's contention that the difficulty is usually not a scientific one but rather one of human compatability and understanding. In numerous cases, the economics of the situation were overwhelmingly in favor of combined treatment and yet separate waste treatment installations were finally used, primarily because of personality clashes and lack of sympathetic understanding.

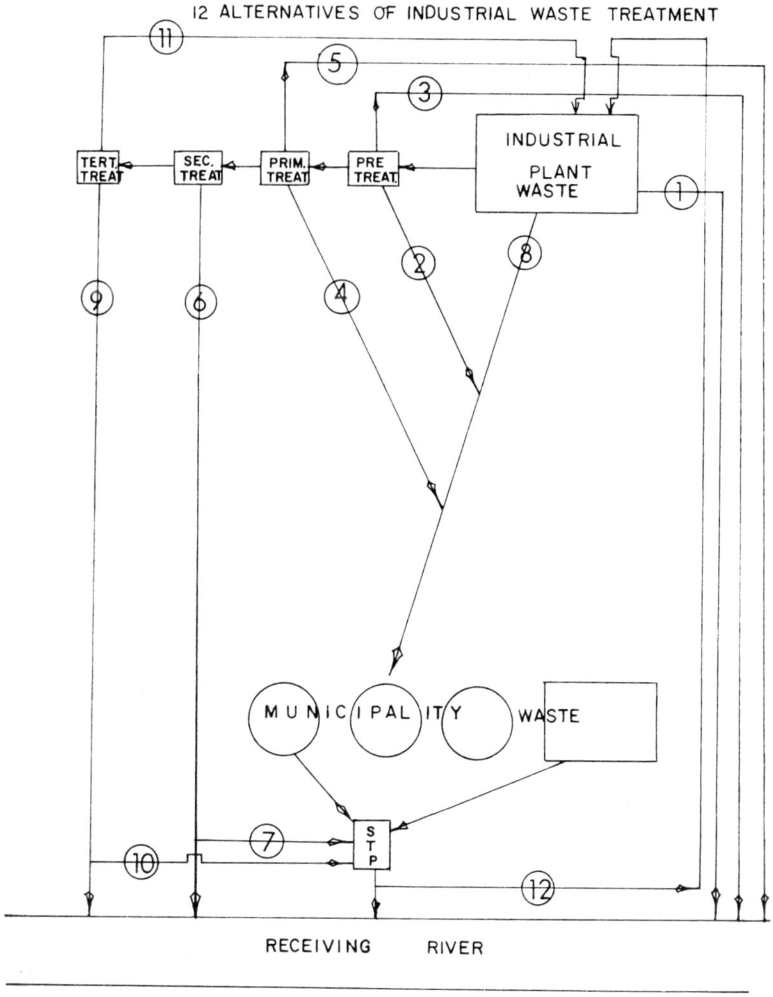

Figure 10

Alternatives of Systems for Industrial Waste Treatment

Industrial wastes problems should be solved by the use of "systems engineering." Systems engineering when applied to industrial wastes implies combining the artful and scientific factors toward optimum solutions to treatment problems. When an engineer considers only the scientific or solely the art factors involved in any waste treatment problem, his solution is less than ideal. The more informed engineer utilizes as much of both art and science as possible in solving these problems. Even so, his result may also be less than an optimum solution because of interdependencies of many factors. Todays enlightened and progressive waste treatment engineer takes all of both the social and physical factors into consideration.

These factors can usually be applied to the major question of "which path to follow." As you can see from the following diagram, Fig. 10, industry must decide whether to treat its own waste or to contract with the municipality to accept, treat, and dispose of its wastes. Twelve (12) potential combinations of alternatives, or side paths, are also available to the industry. For example, it may decide to partially treat its waste (alternates 2, 4, 7 or 10), and then deliver the residual waste volume and load to the municipal treatment plant for final treatment and disposal. It is vital, then, for us to point out and discuss many factors of both types which will assist us in selecting the correct alternative path to follow and hence the overall solution to these problems.

A. Art in industrial wastes

1. Precedent
2. Social relationship of industry and municipal officials

3. Political compatability of industry and municipality
4. Sewer service charge
5. Location of industry-especially in relation to the municipal plant and receiving stream
6. Competence of municipal plant operator
7. Permanence of industrial production

B. Science in industrial wastes

1. *Type of municipal* sewage treatment
2. *Character of industrial* waste
3. Required water quality of the receiving stream
4. *Volume* of industrial waste *in relation* to that of the municipality
5. Economics of the alternatives

I suggest that we consider the art or social factors first, since they have been less publicized and are perhaps less well-known and understood by enginers.

A 1—Precedent

People are most inclined and even biased towards that which they are most familiar with or that which has been done rather uniformly and continuously in the past. There is a form of security in knowing that what is being suggested is in keeping with past practice and its effects have been experienced previously. If in the past the practice has been to include all types of industrial wastes regardless of volume or character in the municipal treatment system, this practice may be expected to be repeated in the future. On the other hand, if separate treatment of industrial wastes and municipal sewage has been encouraged, it will be difficult to change in the future. Any change in policy will require considerable effort, education, and enthusiasm of all parties. The engineer must recognize that even with proper theoretical justification, it may be impossible to overcome and alter that which has become practice by precedent.

A 2—Social Relationship of Industry and Municipal Officials

The consulting engineer must be aware of and appraise the social state of the relationship between the appropriate members of the industrial and municipal community. For one reason or another this relationship may have been enhanced by a previous event which makes future compatibility for joint treatment possible. Likewise, some past experience, sometimes slight or seemingly insignificant may have alienated the mutual social relationship and made future reasoning and negotiations for joint treatment rather difficult. A designing engineer must develop an art of detecting and evaluating this compatability at the onset of his project planning. An oversight at this stage or even an improper consideration or evaluation can be disastrous in terms of the search for an overall optimum path of solution to the joint problems.

A 3—Political Compatability of Industry and Municipality

When both industry and the influential major municipal officials are of the same political leaning, cooperative solutions to waste treatment are facilitated. The engineer must determine this relationship and he must consider the likelihood of a change in municipal officials in an election year during the progress of his project.

A 4—Sewer Service Charge

Municipalities are composed of elected officials each possessing his own philosophy on the subject of industrial patronage or subsidy for local industry. Some prefer to accept all wastes at a minimum charge or a flat fee or on a water use basis. Others believe that industry should solve its own environmental problems and should it wish to have the municipality treat its waste, it should be ready and willing to pay a sewer service charge which correctly and adequately covers a capital as well as operating costs. The engineer should ascertain which type of feeling prevails in the municipality prior to any establishment of a charge. Industry usually does not object to a reasonable charge, but often objects to the philosophy of a comprehensive, complicated, or involved charge which may lead to many unnecessary confrontations. It might prefer to treat its own waste regardless of the "on the surface" economics rather than face potential disagreements with public officials.

A 5—Location of Industry

The location of the industrial plant relative to both municipal sewers and treatment facilities as well as to the final receiving stream plays a vital part in influencing industry towards or against combined treatment.

At first consideration one might presume that

economics alone is effected by plant location. True, cost of connecting to and transmitting through existing or potential municipal sewers for a given distance directly effects costs. However, accessibility to sewers and proximity to treatment plants and receiving streams is important from a psychological viewpoint as well. Questions such as why have two treatment plants so close to each other, or why pump and maintain private pumping station and connecting sewer systems when we can treat our own wastes and dispose of them by gravity nearby, are difficult to answer only in terms of dollars. Proximity enhances joint solutions regardless of the economic consideration.

A 6—Competence of Municipal Plant Operator

The experience, congeniality, and prominence of a supervisor and/or operator of a municipal treatment facility will enhance or hinder cooperative arrangements of industry and government. These qualities of the operating personnel are especially important in the initial stages of decision-making. For example, industry usually considers combined treatment feasible and even desirable when the municipal operator is held in high esteem by his associates on a state or national level. The basis for this feeling is deeply rooted from a psychological standpoint, but most probably stems from a feeling of security in knowing that the wastes will be treated in an optimally-effective manner. An "easy-to-get-along-with" operator will enhance initial as well as continued discussions and negotiations. Combined treatment always manifests problems and difficulties of one type or another. Most can be solved with some effort, ingenuity, and cooperation. When industry knows a priori that it is dealing with a man or men possessing these characteristics, combined treatment becomes practical. An experienced operator usually knows what plant operating changes to make during critical situations in order to avoid malfunctioning. This is reassuring to both industrial and municipal officials. Conversely, the lack of those operator characteristics demands greater consideration of separate industrial waste treatment.

A 7—Permanence of Industrial Production

Industry must show a profit for its owners in order to stay in business. The future of any given industrial facility depends primarily upon the whims of the purchasing public for its product. When whims change, a given industrial plant has three alternatives open to it: (1) *reduce its production*—hopefully cutting costs so as not to allow net profits to deteriorate (2) *change type of production* to meet the need of society or (3) *cease all production* to avoid deficit financing. In all three alternatives the type and/or quantity of waste is altered considerably. Many industrial operations have found it necessary to resort to any or all of these alternatives over a 20–40 year period (during which times bonds for waste treatment facilities are usually amortized). Industry has begun to recognize this as "a fact of industrial life". Because our technology is changing so rapidly and with it the needs and demands of its customers industry hesitates to enter into long-term contractual committments especially concerned with waste treatment facilities. Thus, industry is now recognized as being relatively impermanent when compared to municipal operations. Certain industries are more permanent or less apt to cease production entirely (as evidenced from past records only). For example, DuPont Chemical Corporation or U.S. Steel's main production facilities could be expected to continue its operation-with some modifications-with greater assurance-for longer periods into the future than a small leather tannery or a peach cannery. The former types of industries would be more likely to consider joint ownership of a treatment facility than the latter. The cannery or tannery would be inclined towards private ownership of its separate facilities. There are exceptions and even variations within these latter industries mostly depending upon the size and type of ownership of the industry, as well as specific past experiences it may have had in waste treatment problems.

Some of the more important scientific factors associated with industrial waste treatment include the following:

B 1—Type of Municipal Sewage Treatment

A secondary biological treatment plant—if adequately sized—can best be utilized to treat a readily-decomposible, organic-laden industrial waste. Typical examples of these types include dairies, canneries, slaughter houses, and tanneries. I hasten to make clear, however, that each of these wastes—as well as other typical organic wastes—contain contaminants which can interfere with effective treatment when combined with domestic sewage. For example, dairy wastes often turn acid extremely fast and the lowered pH can affect biological oxidation, many canneries contain an extremely alkaline lye peel waste which when discharged in slug loads can also hamper biological oxidation, slaughterhouse wastes contain slug loads of grease and blood which could hinder physical and biological processes, while tannery wastes contain chromium, sulfides, and lime, which are not compatible with normal sewage treatment. Proper pretreatment and plant operation, however, can remedy these problems.

On the other hand, little would be gained by either party if an industrial waste, such as from a dairy or textile mill, were treated in a municipal treatment plant consisting solely of primary treatment.

Industry must make an assessment of the objectives of treatment required and then examine the municipal plant available or being planned. Mutually compatible objectives and treatment facilities would be conducive to combined treatment.

B 2—Character of Industrial Waste

When considering the treatment of the wastes from a tissue paper mill, industry needs a municipal plant which concentrates its equipment units in the removal of finely divided suspended solids area of waste treatment. It is of little benefit to the tissue mill of the municipal plant possesses only a high-rate trickling filter primarily designed for BOD removal.

Along similar lines a metal plating waste also would not be a welcome addition to the high-rate trickling filter plant because of its potentially toxic metals and acids as well as its lack of organic matter and hence need for BOD reduction.

The waste engineer must carry out a complete analysis of his industrial waste to ascertain its compatability for treatment by the varied possible methods. Some analyses often overlooked by the traditional sanitary engineer include the k value, the waste deoxygeneration rate; ultimate oxygen demand, toxic chemicals and metals, temperature, grease content at reduced temperatures, refractory organic matter, phosphates and nitrates, and other algal nutrients, etc. These and other important characteristics of certain industrial wastes should signal the key to its eventual successful treatment.

B 3—Receiving Stream Water Quality

It is a foregone conclusion that a stream which must be maintained in a high water quality state requires the maximum offshore waste treatment. Generally this means a minimum of the equivalent to secondary treatment. But what is secondary treatment to an industry whose waste contains many varied types of contaminants. Often the conventional biological treatment system will not adequately remove sufficient amounts of the contaminants. Sometimes specific treatment such as chemical precipitation followed by adsorption on activated carbon may remove more industrial contaminents than a secondary type trickling filter plant. Industry has inherited the moral, if not legal, obligation of treating its waste in a manner so as to maintain the highest possible quality of water in the receiving stream. This cannot always be accomplished by following established state or federal rules. In many instances ingenuity must be used and sacrifices must be made in order to remove the proper amount of specific contaminants. In one waste problem it was necessary for management to render a decision to completely eliminate one of its three raw materials. This particular raw material was relatively inexpensive and from its use the industry was able to produce a satisfactory product and considerable profit. However, the waste resulting from processing with this particular raw material accounted for more than 80 per cent of the total plant contamination by conventional treatment proved extremely costly and after much deliberation the industry decided reluctantly to discontinue the use of this particular raw material.

It is even more disconcerting for industry to be faced with a legal decision to install so-called complete treatment of its wastes when the receiving stream—even under critical conditions—shows little deterioration from the wasteload. In this situation industry should endeavor to prove to regulatory agencies its dilemma in order that it operate at optimum economic efficiency. The latter is essential for the long-term benefit of society as long as valuable river resources are not degraded in the process.

B 4—Volume Ratio of Industrial to Municipal Wastes

A relatively small volume of industrial waste can usually be assimilated in a municipal sewage treatment system regardless of its contaminants. This fact does not always depend upon rational reasoning, but is often based on results of the very fact that an attempt to handle the wastes is made. In other words, municipal plant operators generally react optimistically towards *small volumes* of any industrial wastes, agree to try to treat them, and end up accepting them with or without certain pre conditions. If the ratio of waste volume to sewage had been greater, an attempt may never have been made to handle the waste, regardless of the potential of acceptability or treatability. Thus, in general, when ratios are high, industry usually builds its own treatment plant despite the potentially favorable economics or the potential compatability for joint treatment. There are exceptions to this generality; but they usually exist when a new facility is being contemplated by both municipal and industrial officials simultaneously.

B 5—Economics of Alternatives

I would like to dwell upon this aspect of industrial waste treatment a little longer than the others because of its importance and relative newness to the sanitary engineer. To begin with, industry tends to select the least costly alternative—especially when other conditions are equal. How does industry select the least costly system? Usually industry prefers to compare alternative system costs on the basis of total capital expenditures. The least expensive capital outlay is often preferred. I would point out that this method can often be misleading and even erroneous over the long term. To start with, it does not take into consideration the annual cost of operation and maintenance required for effective treatment. Often the least costly alternative can be the most expensive to maintain especially when perfection is required in efficiency of operation. Furthermore, the cost of obtaining money for capital expense is not taken into account when considering only capital costs. This is significant during high money cost periods and when one or more alternative entails public borrowing rather than private borrvowing. It is well-known that municipalities can usually borrow money at lower interest rates than private corporations. And lastly, the financial rating of the borrower is a significant and over-looked factor in comparing capital costs only. Certain public authorities may possess such poor financial ratings that many prominent, profitable industrial firms may be able to borrow money as cheaply or cheaper than the municipality. What I'm suggesting, then, is that economic alternatives be based upon net annual costs which include operation, maintenance as well as amortization of capital costs. The latter shall be selected on a realistic basis of the financial rating of the appropriate borrower as well as for a reasonable bond amortization period depending upon the ability to repay of the borrower and the expected life of the specific equipment required in each alternative.

I strongly suggest that the waste engineer consider another matter of great economic importance. The question of whether the optimum alternative selected is justified on *any* economic basis is one which up to now has defied our engineers. We have blindly presumed that what is required in the way of waste treatment by state or federal edict is required at any cost. This may be so, but the case may be strengthened by a proper economic evaluation. I'm suggesting that we should compute the total net benefits to the surrounding, immediate society of the specific waste treatment alternative selected. Heretofore we have lamented that these benefits were intangible and hence immeasurable. In a research effort for the Department of Interior we were able to identify some of the major benefits of pollution abatement for elevating receiving water quality. In addition, we have devised a procedure for quantifying in terms of dollars the incremental benefits of changes in water quality resulting from waste treatment. Briefly, these benefits as effected by water quality include (1) recreational (2) land use (3) water withdrawal (4) waste treatment and (5) in-place water uses. We have previously developed a means of measuring overall stream water quality by use of a Stream Pollution Index. We just completed a study of establishing an administrative body to measure benefits, make proper water related decisions, and to allocate valuable stream resources on an economic basis.

Forty-Three Industrial Wastes

It is impossible for the writer to present (in the short space allocated him in this chapter) even the major facts about each of these important wastes.

After more than 20 years of experience the writer believes it best for each reader to study certain wastes (which are significant to him) in great depth and detail. To be a real master of any one industrial waste may require a lifetime of serious effort on the part of a qualified waste engineer. Many persons have done this: Trebler in the dairy industry; Souther in textiles; Gehm in pulp and paper; Trubnick in yeast manufacturing; Hoak and Kinney in steel making, and Mercer in canning. The writer suggests that a detailed study of all 43 wastes simulaneously will lead to confusion at best and discouragement of the overall task at the least. Rather the reader should master the art, philosophy, and scientific methods underlying industrial waste treatment. In addition, the writer suggests the reader study a production book for each industry he is reviewing. The language of each industry may be entirely different and the production methods not similar to other industries. For more details on each waste as well as all significant references the reader is directed once again to the writers full length textbook as referenced on the first page of this chapter.

The following table provides the reader with a brief summary of the major liquid wastes, their origin, characteristics, and current methods of treatment. It should be useful as a quick reference, but in no way can it be considered complete for any specific industry.

The author has divided industrial wastes roughly into five major classifications—apparel, food and drugs, materials, chemicals, and energy.

Summary of Industrial Waste: Its Origin, Character, and Treatment (*Courtesy Nemerow**)

Industries producing wastes	Origin of major wastes	Major characteristics	Major treatment and disposal methods
Apparel			
Textiles	Cooking of figers; desizing of fabric	Highly alkaline, colored, high BOD and temperature, high suspended solids	Neutralization, chemical precipitation, biological treatment aeration and/or trickling filtration.
Leather goods	Unhairing, soaking, deliming, and bating of hides	High total solids, hardness, salt, sulfides, chromium, pH precipitated lime, and BOD.	Equalization, sedimentation, and biological treatment.
Laundry trades	Washing of fabrics	High turbidity, alkalinity, and organic solids	Screening, chemical pricipitation, flotation and adsorption.
Food and Drugs			
Canned goods	Trimming, culling, juicing, and blanching of fruits and vegetables	High in suspended solids colloidal and dissolved organic matter	Screening, lagooning, soil absorption of spray irrigation
Dairy products	Dilutions of whole milk, buttermilk, and whey	High in dissolved organic matter, mainly protein, fat, and lactose	Biological treatment aeration, trickling filtration, activated sludge
Brewed and distilled beverages	Steeping and pressing of grain; residue from distillation of alcohol; condensate from stillage evaportaion	High in dissolved organic solids, containing nitrogen and fermented starches or their products	Recovery, concentration by centrifugation and evaporatoin, trickling filtration; use in feeds digestion of slops
Meat and poultry products	Stockyards; slaughtering of animals; rendering of bones and fats; residues in condensates; grease and wash water; picking chickens	High in dissolved and suspended organic matter, blood, other proteins, and fats	Screening, settling and/or flotation, trickling filtration
Animal feedlots	Excreta from animals	High in organic suspended solids and BOD	Land disposal and anaerobic lagoons
Beet sugar	Transfer, screening, and juicing waters; drainings from lime sludge; condensates after evaporator; juice and extracted sugar	High in dissolved and suspended organic matter, containing sugar and protein	Reuse of wastes, coagulation, and lagooning

* Nemerow, N. L., *Liquid Wastes of Industry*, published by Addiso-Wesley Company, Reading, Mass. (1971).

(Summary of Industrial Wastes, continued)

Industries producing wastes	Origin of major wastes	Major characteristics	Major treatment and disposal methods
Pharmaceutical products	Mycelium spent filtrate, and wash waters	High in suspended and dissolved organic matter, including vitamins	Evaporation and drying; feeds
Yeast	Residue from yeast filtration	High in solids (mainly organic) and BOD	Anaerobic digestion, trickling filtration
Pickles	Lime water; brine, alum and turmeric, syrup, seeds and pieces of cucumber	Variable pH, high suspended solids, color, and organic matter	Good housekeeping, screening, equalization
Coffee	Pulping and fermenting of coffee bean	High BOD and suspended solids	Screening, settling, and trickling filtration
Fish	Rejects from centrifuge pressed fish; evaporator and other wash water wastes	Very high BOD, total organic solids, and odor	Evaporation of total waste; barge remainder to sea
Rice	Soaking, cooking, and washing of rice	High BOD, total and suspended solids (mainly starch)	Lime coagulation-digestion
Soft drinks	Bottle washing; floor and equipment cleaning syrup-storage-tank drains	High pH, suspended solids, and BOD	Screening, plus discharge to municipal sewer
Bakeries	Washing and greasing of pans; floor washings	High BOD, grease floor washings, sugars, flour, detergents	Amenable to biological oxidation
Water production	Filter backwash; lime-soda sludge; brine; alum sludge	Minerals and suspended solids	Direct discharge to streams or indirectly through holding lagoons
Materials			
Pulp and paper	Cooking, refining, washing of fibers, screening of paper pulp	High or low pH, color, high suspended colloidal, and dissolved solids, inorganic fillers	Settling, lagooning, biological treatment, aeration, recovery of by-products
Photographic products	Spent solutions of developer and fixer	Alkaline, containing various organic and inorganic reducing agents	Recovery of silver, discharge of wastes into municipal sewer
Steel	Coking of coal, washing of blast-furnace flue gases, and pickling of steel	Low pH, acids cyanogen phenol; ore, coke, limestone, alkali, oils, mill scale, and fine suspended solids	Neutralization, recovery and reuse, chemical coagulation
Metal-plated products	Stripping of oxides, cleaning and plating of metals	Acid, metals, toxic, low volume, mainly mineral matter	Alkaline chlorination of cyanide; reduction and precipitation of chromium; lime precipitation of other metals
Iron-foundry products	Wasting of used sand by hydraulic discharge	High suspended solids, mainly sand; some clay and coal	Selective screening, drying of reclaimed sand
Oil fields and refineries	Drilling muds, salt, oil and some natural gas; acid sludges and miscellaneous oils from refining	High dissolved salts from field; high BOD, odor, phenol, and sulfur compounds from refinery	Diversion, recovery, injection of salts; acidification and burning of alkaline sludges
Fuel oil use	Spills from fuel-tank filling waste auto crankcase oils	High in emulsified and dissolved oils	Leak and spill prevention, flotation
Rubber	Washing of latex, coagulated rubber, exuded impurities from crude rubber	High BOD and odor high suspended solids, variable pH, high chlorides	Aeration, chlorination sulfonation, biological treatment
Glass	Polishing and cleaning of glass	Red color, alkaline nonsettleable suspended solids	Calcium-chloride precipitation
Naval stores	Washing of stumps drop solution, solvent recovery, and oil-recovery water	Acid, high BOD	By-product recovery, equalization recirculation and reuse, trickling filtration
Glue manufacturing	Lime wash, acid washes, extraction of nonspecific proteins	High COD, BOD, pH, chromium, periodic strong mineral acids	Amenable to aerobic biological treatment, flotation chemical precipitation

(Summary of Industrial Wastes, continued)

Industries producing wastes	Origin of major wastes	Major characteristics	Major treatment and disposal methods
Wood preserving	Steam condensates	High in COD, BOD, solids, phenols	Chemical coagulation, oxidation pond and other aerobic biological treatment
Candle manufacturing	Wax spills, stearic acid condensates	Organic (fatty) acids	Anaerobic digestion
Plywood manufacturing	Glue washings	High BOD, pH, phenols, potential toxicity	Settling ponds, incineration
Chemicals			
Acids	Dilute wash waters; many varied dilute acids	Low pH, low organic content	Upflow or straight neutralization, burning when some organic matter is present
Detergents	Washing and purifying soaps and detergents	High in BOD and saponified soaps	Flotation and skimming, precipitation with $CaCl_2$
Cornstarch	Evaporator condensate or bottoms when not reused or recovered, syrup from final washes, wastes from "bottling-up" process	High BOD and dissolved organic matter; mainly starch and related material	Equalization, biological filtration, anaerobic digestion
Explosives	Washing TNT and guncotton for purification, washing and pickling of cartridges	TNT, colored, acid odorous, and contains organic acids and alcohol from powder and cotton, metals, acid, oils, and soaps	Floatation, chemical precipitation, biological treatment, aeration, chlorination of TNT, neutralization, adsorption
Pesticides	Washing and purification products such as 2,4-D and DDT	High organic matter, benzenering structure, toxic to bacteria and fish, acid	Dilution, storage activated carbon adsorption, alkaline chlorination
Phosphate and phosphorus	Washing, screening, floating rock, condenser bleedoff from phosphate reduction plant	Clays, slimes and tall oils, low pH high suspended solids, phosphorus, silica and fluoride	Lagooning, mechanical clarification coagulation and settling of refined waste
Formaldehyde	Residues from manufacturing synthetic resins and from dyeing synthetic fibers	Normally high BOD and HCHO, toxic to bacteria in high concentrations	Trickling filtration, adsorption on activated charcoal
Plastics and resins	Unit operations from polymer preparation and use; spills and equipment washdowns	Acids, caustic, dissolved organic matter such as phenols, formaldehyde, etc.	Discharge to municipal sewer reuse, controlled discharge
Energy			
Steam power	Cooling water, boiler blowdown, coal drainage	Hot, high volume, high inorganic and dissolved solids	Cooling by aeration storage of ashes, neutralization of excess acid wastes
Coal processing	Cleaning and classification of coal, leaching of sulfur strata with water	High suspended solids mainly coal; low pH, high H_2SO_4, and $FeSO_4$	Settling, froth flotation, drainage control, and scaling of mines
Nuclear power and radioactive material	Processing ores; laundering of contaminated clothes; research-lab wastes; processing of fuel; power-plant cooling waters	Radioactive elements, can be very acid and "hot"	Concentration and containing or dilution and dispersion

REFERENCES

1. Bozick, T. A. (1959), The Toxicity of Metals on the Activated Sludge Process. Masters Thesis, Case Institute of Technology.
2. Coburn, S. E. (1949), Limits for Toxic Wastes in Sewage Treatment. *Sewage Works Jour.* Vol. 21, p. 519.
3. Doudoroff, P., and M. Katz (1950), Critical Review of Literature on the Toxicity of Industrial Wastes and Their Components to Fish, I. Alkalies, Acids, and Inorganic Gases. *Sewage and Industrial Wasts* Vol. 22, p. 1432.
4. Doudoroff, P., and M. Katz (1953), II. The Metals, as Salts. *Sewage and Industrial Wastes* Vol. 25, p. 802.
5. Doudoroff, P. *et al.* (1951), Bio-Assay Methods for the Evaluation of Acute Toxicity of Industrial Wastes to Fish. *Sewage and Industrial Wastes* Vol. 23, p. 1380.
6. Doudoroff, P. *et al.* (1966), Acute Toxicity to Fish of Solutions Containing Complex Metal Cyanides, in Relation to Concentrations of Molecular Hydrocyanic

Acid. *Transactions of The American Fisheries Society*, Vol. 95, p. 6.
7. Ellis, M. M. (1937), Detection and Measurement of Stream Pollution. In *Biology of Water Pollution*, FWPCA, 1967.
8. Garrett, J. I. (1957), Toxicity Investigations on Aquatic and Marine Life. *Public Works*, Vol. 88, p. 95.
9. Kugelman, I., and P. L. McCarty (1964), Cation Toxicity and Simulation in Anaerobic Waste Treatment II Daily Feed Studies 19th Ind. Waste Conf., Purdue.
10. Malany, G. W. *et al.* (1959), Toxic Effects of Metallic Ions on Sewage Microorganisms. *Sewage and Ind. Wastes*, Vol. 31, p. 1309.
11. McKee, J. E., and H. W. Wolf (1963), Water Quality Criteria. Calif. State Water Quality Control Bd., Pub. No. 3–A.
12. Moulton, E. Q., and K. S. Shumate (1962), Toxic Effects of Heavy Metals on the Biological Treatment Processes Rpt. No. EES-197-1, Eng. Exp. Station Ohio State Univ.
13. U.S. Dept. of Interior (1965), Interaction of Heavy Metals and Biological Sewage Treatment Processes. Pub. Health Service Pub. No. 999-WP-22.
14. U.S. Dept. of Interior (1968), Report of the Committee on Water Quality Criteria, FWPCA.
15. U.S. Dept. of Interior (1968), A State-of-the Art Review of Metal Finishing Waste Treatment, FWQA—Water Pollution Control Research Series, 12010 EIE 11/68.
16. Report of the Council on Environmental Quality on Ocean Dumping, 1970. Govt. Printing Ofc.
17. Warren, Charles (1971), *Biology and Water Pollution Control*. W. B. Saunders Co.
18. Wilber, C. G. (1969), *The Biological Aspects of Water Pollution*. Charles C. Thomas Pub.

Selected Readings

1. *Industrial Waste Treatment*, E. B. Besselievre. McGraw-Hill 1952.
2. *Waste Treatment*, Peter C. G. Isaac. Pergamon Press 1960.
3. *The Treatment of Industrial Wastes*, E. B. Besselievre. McGraw-Hill 1969.
4. *Aqueous Wastes from Petroleum and Petrochemical Plants*, M. R. Beychok. John Wiley and Sons 1967.
5. *Principles of Industrial Waste Treatment*, C. F. Gurnham. John Wiley 1955.
6. *Industrial Wastewater Control*, ed. by C. F. Gurnham. Academic Press 1965.
7. *Industrial Waste Disposal*, ed by R. D. Rose. Reinhold Pub. Co. 1968.
8. *Industrial Water Pollution Control*, W. W. Eckenfelder. McGraw-Hill Co. 1966.
9. *Treatment and Disposal of Industrial Waste Waters*, B. A. Southgate. H. M. S. O. London 1948.
10. Ohio River Survey Supplement D, U. S. P. H. S. 1943.
11. *Industrial Waste Treatment*, ed. by W. Rudolfs. Reihold Pub. Co. 1953.

NELSON L. NEMEROW
Syracuse University

M

MANAGEMENT OF RADIOACTIVE WASTES

1. Introduction

(a) Philosophy

The group of people engaged in management of radioactive wastes has evolved from a small body of operators who, originally with little or no expert knowledge, were engaged in day-to-day solution of unpleasant problems. They now form a recognized profession, extending from whole-time research scientists to field workers who in some countries are conducting a profit-making industry.

The members of the profession came mainly from Health Physics and brought with them the caution and "conservative" attitude to radiation hazards characteristic of Health Physicists. They regard their mission as being to ensure that members of the public, as well as workers in the field of nuclear energy, will not be harmed by the radioactive material for which they are responsible. With their Health Physics background this sometimes leads to an attitude which industry regards as over-restrictive, although recent controversies have tended to cast them ironically in the role of particularly dangerous polluters of the environment.

It is clear that any human activity that involves conversion of something into something different must produce waste. Conversion of energy from one form to another is no exception. It is sometimes possible for an industry to re-cycle its waste products and to convert part of them to a useful form, but there is always some minimal residue which cannot be retained within the system. This must find some place within the environment. Usually the cheapest procedure is to discharge it in some way that will ensure a sufficient dilution to make it innocuous. If this is impracticable for technical or political reasons it must be confined, but usually the more effective the confinement, the higher the cost. To say that a process must be conducted with no waste is equivalent to saying that the process may not be conducted at all, and to demand a certain level of confinement or restriction of wastes implies an acceptance of the cost of the waste management system as a necessary part of the cost of the process.

Discharge of potentially noxious materials into the environment involves some risk, which may or may not be measurable. Within very broad limits research in nuclear hazards enables us to forecast the effects of exposure of large groups of people, for extended periods, to low doses of radiation. We can also estimate, with less accuracy, the probability that an individual will suffer some harm from such exposure, and we can say with much greater confidence what will happen if an individual is exposed to larger doses—say 50 rem* and upwards—in a single dose. The nuclear industry, then, can provide some information on the probable consequences of environmental contamination extended over a lifetime, and better information on the probable consequences of a major nuclear accident which leads to high radiation exposure. In other words we can, within rather broad limits, estimate the risks.

The situation is different in most other industries. The consequences of acute doses of cyanide, lead, fluoride or carbon tetrachloride are well known, and there is some evidence for the effects from low doses received over a lifetime, but who knows what effect to expect in humans from benzpyrene or nitrous oxide emitted from smoke stacks or

* Rem = unit of radiation exposure: does of radiation that causes the same biological damage in the human as 1 rad (100 ergs per g) of 250 keV X-rays.

from the low levels of polychlorodiphenyls and mercury compounds that are liberated into the environment? They affect every age group in the population and are a potential life-long hazard. But nothing is known about the probability that they will eventually do harm, and it is difficult to see how such knowledge could be obtained in a human population.

Every human activity is associated with some risk, however small. Normally we do not solemnly calculate the risk, weigh it against the benefit we expect to obtain, and then decide for or against the activity. Yet to decide to do something—such as driving a car, getting up in the morning, or going mountain climbing—must involve some sort of conscious or unconscious weighing of risk against benefit.

In deciding upon a particular waste management system, or in deciding to license a particular kind of nuclear power station, a much more deliberate weighing of cost vs benefit must be undertaken. There is, however, a fundamental difficulty which up to now has made it impossible to express such a judgement in numbers. It is characteristic of a ratio that the numerator and the denominator must be in the same units. It should be possible to express most of the benefits of nuclear power, for example, in dollars, but if we regard part of the cost of nuclear power as an increase in the probability that people will develop cancer or that they will experience a shortened life-time, how can that be expressed in dollars?

One benefit of nuclear power is the difference between death and injury among uranium miners and processors and the corresponding figure for equivalent energy production by the coal mining industry. This, again, cannot be expressed in dollars. To work out a true COST/BENEFIT ratio is thus little better than a dream, and the people responsible for approving a waste management system or a new power station are therefore faced in the last analysis with a value judgement, which is at least to some extent subjective. It is not a scientific decision. In the broadest sense, the decision is political.

(*b*) *Controls*

The responsibility for making decisions on matters related to "dealing in"—i.e. having anything to do with—radioactive materials, machines capable of producing electromagnetic radiation (except for medical purposes) and certain scheduled materials such as heavy water, usually rests with a national atomic energy authority. Typically, regulations are issued by the authority that have the force of law. Assistance is given to the authority in assessing hazards of reactors and other installations—including waste management systems—by an independent advisory committee which can call on the services of an expert staff.

In most countries regulations lay down the maximum permissible exposure to radiation for workers in nuclear industry and also for the general population (Table I). Maximum permissible doses (MPDs) have been recommended by the International Commission on Radiological Protection (ICRP), which have received worldwide acceptance as the fundamental basis for national regulations. The ICRP has derived from the MPDs a list of maximum permissible concentrations (MPCs) in air and water on the basis that if workers were to breathe air, or drink water, at the MPC for any particular radionuclide† over a life-time they would not suffer any unacceptable harm.

"Unacceptable" means "detectable", in the sense that it could reasonably be regarded as caused by the radiation. The ICRP has also laid down rules for calculating the MPC for mixtures of more than one radionuclide.

The MPDs are constantly under review by the ICRP, which consists of people who have devoted their professional lives to assessment of radiation hazards. They draw upon the work of large numbers of scientists throughout the world, many of whom are actively engaged in research on somatic and

TABLE I

*Radiation Received Compared with Maximum Permissible Dose**

Dose actually received (N. America):	mRem/year
Natural radiation	125
Medical sources	55
Weapons fallout	1.5
Miscellaneous (including nuclear industry, TV sets luminous dials, etc.	2.0
Maximum dose permissible in excess of natural + medical radiation	500
Permissible dose for atomic energy workers	5000

* Adapted from J. P. Blakely, *Necluar Safety* 1971 (12) 56–58.

† Radionuclide = radioactive isotope.

genetic effects of radiation. Changes have been made time to time in details of the ICRP recommendations but it is remarkable that in such a rapidly developing field the necessary changes have been so few.

The ICRP has consistently emphasized that the MPD and its associated MPCs are *maximum permissible* figures. The Commission has made another recommendation equal in force and status to those on maximum permissible doses. This states that exposure to radiation must always be held down to the lowest PRACTICABLE dose. The word "practicable" was carefully chosen, after considerable debate. If "possible" had been used it could have been claimed that a single contaminated rat must be buried in a platinum box. It is our mission to see that all practicable steps are taken to protect mankind from exposure to radiation, and we can do that very effectively.

2. Sources of Wastes

(*a*) *Uranium mining and milling*

Apart from the normal hazards associated with hardrock mining, the workers in uranium mines are exposed to radon and the decay products which arise from the radium content of the ore. These hazards can be controlled by sealing old workings and general "good housekeeping" but more particularly by installation of an efficient ventilation system and, where necessary, the use of respirators. The ventilation air contains radioactive material and dust, some of which can be removed if necessary by filtration, but the radon remains. The large volume of air used for mine ventillation is ejected at high velocity from a stack, which ensures adequate dilution into the atmosphere.

The end products of the mill are uranium oxide and "tailings". The tailings, together with mine drainage water, contain most of the radium originally present in the ore. Radium is one of the most toxic of all radionuclides, and presents a serious potential hazard. Various methods of treatment, such as co-precipitation with barium, render most of the radium insoluble, but the water draining from tailings ponds often contains more radium than is permissible in drinking water.

Proper design of outfalls into suitable bodies of water can ensure adequate dilution, but vigilance is necessary to prevent rupture of the tailings ponds or improper practices that will nullify or bypass the treatment system. A monitoring system for analysis of downstream water and fish is common today, but in the early days of the industry the dangers were little understood or ignored, with the result that lakes and streams in uranium mining areas became contaminated.

In Canada the existence of a problem was recognized in time to avert a public hazard, but the Report of a Deputy Minister's Committee[1] showed that action was necessary to protect the environment in the Elliott Lake and Bancroft areas. This was particularly urgent as greatly increased activity in uranium mining was anticipated within a few years.

The size of the problem can be judged from the fact that a Congressional Hearing was told in 1959[2] that 12,000,000 gallons of water containing nearly 10 g of radium was discharged daily to the tailings ponds of American uranium mills.

(*b*) *Processing of uranium oxide*

The crude (70%) U_3O_8 produced by the mills may be converted to metal, to UO_2 or to UF_6. The hexafluoride is used in separation of ^{235}U* from ^{238}U. A serious waste problem would result from nuclear fission if a critically large amount of ^{235}U were to accumulate accidentally in one place. This is a rare event, but is not impossible. Otherwise, the wastes consist of uranium chips and fines, contaminated clothing and respirators and dust accumulated in air-cleaning systems. The uranium at this stage is practically free from radium so it is hardly a radioactive hazard. The toxicity of natural uranium or ^{238}U is that of a toxic metal rather than of a radionuclide.

Uranium metal is produced by converting the dioxide to tetrafluoride which is then reduced to the metal at high temperature with magnesium. The waste from this process—magnesium fluoride slag and uranium metal fines from trimming the ingots—is a normal slag disposal problem since it is sparingly soluble in water.

(*c*) *Fuel fabrication*

There are many different kinds of fuel elements, but their manufacture produces little waste beyond dust and faulty pellets or fuel pins. This material is usually recycled, particularly if it contains added ^{235}U.

* ^{235}U = fissionable isotope of U. Abundance in natural U = 0.7%.

(d) Reactor wastes

An operating reactor contains a very large inventory of fission products. A 500 MW* (thermal) reactor, after operating for 180 days, contains four hundred million curies of fission products, measured one day after shut-down. This is equivalent to the activity of about 400 metric tons of radium. The fission products decay rapidly at first, leaving 80 million curies at the end of a week, and more slowly later. After a month, the inventory is reduced to about 8 million curies.

Nuclear power stations rated at 1000 MW (electrical)—i.e. 3000 to 5000 MW thermal—are not unusual. At first sight it would seem that these plants would be enormous potential sources of radioactive wastes, but in practice this is not the case (Fig. 1). In an operating power reactor the fuel is contained within a non-corrodible cladding—usually zirconium or stainless steel—and the fission products cannot get out unless the cladding is ruptured.

It is possible to operate the reactor with defects in a few fuel elements, but these sources of leakage make the primary cooling circuit radioactive. It is impracticable to operate a station in the presence of high radiation fields, so the primary coolant is continually purified by ion exchangers. Again, it is a practical necessity to renew the ion exchangers after they have developed a certain level of radiation. The net result of these considerations is that for reasons of operator safety and economics the presence of more than a small proportion of ruptured fuel in a reactor will require its removal.

Fuel removed from the reactor is normally stored on site for a considerable time to permit desay of short-lived radioactivity. Storage facilities are usually deep tanks filled with water, which acts simutaneously as coolant and radiation shield. If defective fuel is present the water will rapidly become contaminated, but even if there are no defects in the cladding the water in cooling ponds does not remain free from radioactive material. This is because the cladding and the reactor structure contribute neutron activation products (or corrosion products) to the cooling water and the cladding itself always contains minute traces of uranium which undergoes fission in the reactor. Hence, the pond water must be purified, usually by resin ion exchangers, so these resins also become a waste.

If resins are regenerated, the regenerants (acids, alkalis or salts) will appear as a liquid waste for disposal. Otherwise, the resin will be handled within its original container or as a powder or slurry.

The radioactive content of gaseous effluents from reactors depends upon the design of the reactor. If air passes through the core very large amounts of argon-41 may be emitted from the stack. Although ^{41}Ar is a hard gamma emitter it has a short half-life (about two hours) so its effects are only noticeable within or very near to the plant. Radioactive isotopes of nitrogen and oxygen decay so rapidly that they do not reach the stack in appreciable amount and the long-lived carbon-14 is not produced in sufficient amount to

* MW = megawatt. 1 MW = 1000 kilowatts.

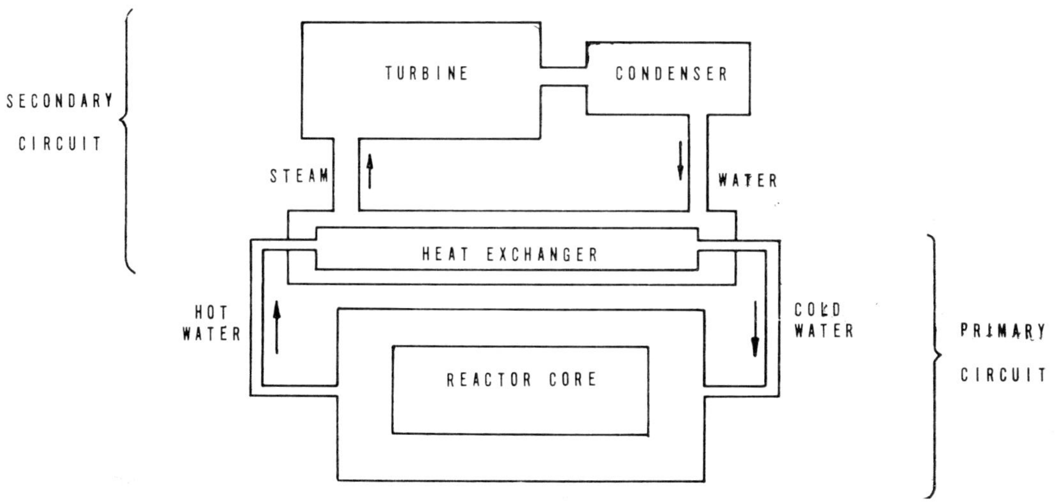

Figure 1 Schematic diagram of processes in nuclear power station. Nearly all radioactive material remains inside the fuel, which is inside the core, which is inside the primary circuit.

be hazardous at the present scale of nuclear power generation. Some concern has, however, been expressed that by the end of this century the build-up of ^{14}C in the atmosphere might become a significant source of radiation within the biosphere.

More concern attaches to radioactive krypton, ^{85}Kr, with a half-life of 10.4 years. This, in contrast with ^{41}Ar and ^{14}C, is a fission product. It is liberated via fuel defects and by diffusion through fuel cladding. It is not a hazard from any single plant, but with increasing numbers of nuclear power stations it might become an ubiquitous source of low-level radiation, though the source of most of the ^{85}Kr would be spent fuel processing plants rather than power stations.

Similar concern has been expressed regarding tritium, the radioactive isotope of hydrogen, which is produced within the fuel and by neutron activation of the heavy hydrogen in ordinary water or the D_2O coolant and moderator of heavy-water reactors. It is also formed by neutron activation of lithium, sometimes used as a neutralising agent in reactor coolants, or of boron which functions as a "poison" in some reactor control systems.

Sometimes the significance of a "source" of radioactive waste depends on whether one is considering the safety of people within the plant, or the public outside. For example, ruptured fuel elements or ordinary day-to-day type mechanical failures can produce air-borne radioactive iodines and other fission products which are a nuisance to operators because they have to work in plastic suits and respirators. The ventilation filtration system and the high dispersion capability of the atmosphere combine to make sources of this kind insignificant beyond the boundary of the exclusion area. However, they may reduce efficiency and disrupt work schedules within the station very seriously, and give rise to significant disposals in the form of clean-up solutions, contaminated clothing, mop-heads and metal scrap.

A noteworthy source of this nature is the tritium which builds up in the coolant and moderator of heavy-water reactors. In a 1000 MW (electrical) power station the equilibrium tritium concentration in the moderator is about 50 Ci/litre. This leads to stack discharges which are quite negligible, but any leaks in pump seals, valves or pipe joints within the station would produce operating problems for those responsible for the radiation safety of the staff. On the other hand, material sent for waste disposal would be no problem, partly because heavy water is recovered for economic reasons and partly because the maximum permissible concentrations of tritium in air and water are much higher than those of most other radionuclides.

In summary, in spite of the enormous potential source of radionuclides within an operating power station the amount of waste generated is small compared with that arising from a research and development establishment, and minute in comparison with a spent fuel processing plant. This statement covers normal operation, including the ordinary accidents and malfunctions expected in any well-designed plant. It does not include the consequences of the "Maximum Credible Accident" which is, in fact, so improbable* that designers of waste management systems do not normally make provision for it.

(*e*) *Spent fuel processing*

Wastes arising from processing of spent fuel account for more than 99.9% of the "waste disposal problem". Fuel which has been enriched with ^{235}U must be treated for recovery of unburned ^{235}U because the fission product load of spent fuel reduces its efficiency as a source of energy. It ceases to be economic as fuel long before the expensive ^{235}U is exhausted.

After removal from the reactor, and storage for sufficient time for decay of short-lived fission products, the fuel is de-sheathed and dissolved, usually in strong nitric acid (Fig. 2). Uranium and plutonium are extracted into an organic solvent, and the acid solution of fission products left behind forms the high level or primary waste. Washing of the organic extractant produces Medium Level wastes, whereas Low Level waste consists of further washings, cooling water, scrubber water and liquids from other sources too numerous to catalogue.

As long ago as 1959 fifty million gallons of High Level wastes were stored in stainless steel tanks at Hanford (USA) alone. It has been forecast that by 1980 stored high-level wastes will contain 10^{10} Curies of activity, 8% of which will be ^{90}Sr. The radionuclides in solution generate so much decay heat that many of the tanks boil, making the provision of elaborate off-gas cleaning systems necessary. Some high level waste tanks have ruptured,

* Usually quoted as around one in a million per year.

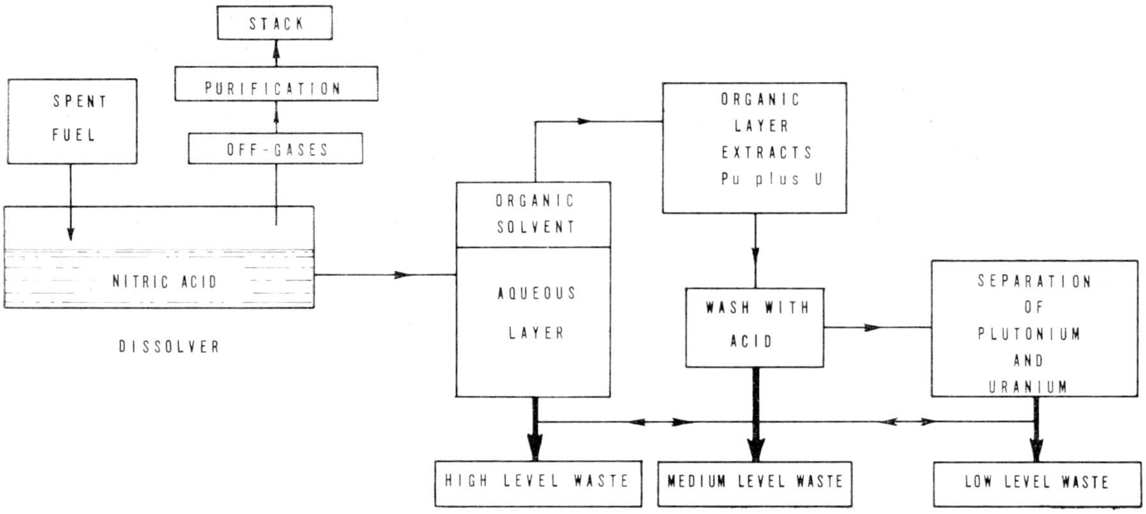

Figure 2 Schematic diagram of fuel processing plant. Showing origins of main waste streams. Reactor fuel contains over 99.95% of the total radionuclides eventually disposed of as waste.

but since they are constructed on a cup-and-saucer principle, with adequate monitoring for spills, and spare tankage is kept available, no unexpected contamination problems have arisen.

Gases from the dissolvers and storage tanks contain tritium, bromines, iodines, xenon, krypton and smaller amounts of less volatile elements such as ruthenium and cesium. After storage for decay, scrubbing and filtration, off-gases can be liberated from a tall stack. As mentioned in the section on reactors, proliferation of fuel processing plants in the future might conceivably lead to local or even eventual world-wide atmospheric contamination if improved containment is not provided in time at spent fuel processing sites.

Solid waste may include glasses or ceramics, used as a means for fixing the activity in high-level liquid wastes, and bitumen or concrete blocks containing less active material. Products of waste processing such as sludges, evaporator bottoms, incinerator ash, adsorbers, filters and scrap fuel cladding are usually in the medium level category. Worn and failed equipment such as pipes, tanks and valves, unserviceable protective clothing, cleanup materials and even whole buildings may have a variety of levels of contamination, by numerous different radionuclides, which defies quantitative assessment. This is not a serious difficulty, except for administrative and recording purposes when quantitative reports have to be made, because most of these wastes have to be contained in some way and none of them are dumped into the environment.

The most difficult problem for the fuel processing industry is not high or medium level waste, off-gases or heterogeneous contaminated scrap. The real problem is very low level liquid waste, because it arises in such enormous volume. Coming from numerous different sources—e.g. cooling and final wash waters, laundry and decontamination center effluents, floor drainage from cleanup operations, personnel shower drainage and effluent from the final stages of liquid waste purification plants—low level and "essentially uncontaminated but suspect" waste adds up to billions of gallons per year. Although some countries (Sweden and Japan, for example) evaporate such effluents on a large scale they are usually discharged by some route into the environment.

(*f*) *Research and development*

A wide variety of wastes arises in such research establishments as Brookhaven (USA), Chalk River (Canada) or Harwell (UK) and they include many of the types mentioned under the heading of fuel processing. In addition the research reactors usually produce very large quantities of radioisotopes which may be processed on-site. However, the quantities involved are very much lower, especially in the high level category, and elaborate waste processing systems are seldom needed even at large research centers unless they are situated in built-up areas or immediately over important aquifers.

At Chalk River the whole of the radioactive effluent reaching the Ottawa River, including drainage from the waste management areas, amounts (1970) to less than 10% of the load of fission products (derived from nuclear weapons fallout) carried by the river from the upper watershed. This is achieved mainly by ground disposal, for Chalk River has no waste processing plant. At the Argonne National Laboratory (US) only 0.06% of the liquid waste was sufficiently active to be passed through the purification plant[3].

(*g*) *Hospitals and biological laboratories*

Organic material and excreta make wastes from these institutions difficult to handle. The radioactive content is usually small, and limited to a restricted list of radionuclides. Those used as sealed sources seldom appear as waste, and the rest are practically confined to ^{131}I, ^{32}P, ^{59}Fe, ^{51}Cr, ^{35}S and ^{24}Na. Other nuclides may be used in small amounts for special purposes such as specific location in certain organs. The nature and amount of radionuclides used in these institutions are such that a high proportion of the waste can be handled safely by the municipal sewage and garbage systems.

Sealed sources are, however, a very different matter. While they remain sealed they are usually within heavy shielding in teletherapy machines*, which are only operated by competent people, or they are in the form of needles and plaques for implantation, or instrumental standard sources used by specialists. However, the time comes when such sources have decayed to the point where they are no longer useful. Sufficient activity remains for them to be highly dangerous to the unwary, so they are dealt with in special ways, usually after return to the supplier.

(*h*) *Isotope production plants*

These facilities are often associated with large reactors, and wastes are similar to those generated in Research and Development plants. Processing of very large sources of volatile elements such as iodine and tellurium necessitates an elaborate ventilation cleaning system. Manufacture of large sources of ^{90}Sr, ^{137}Cs or the tran-suranic elements† as power sources may call for sophisticated remote handling equipment in heavily shielded cells. But the waste problems are different only in scale from those encountered in an *R* and *D* plant.

Some people have considered the separation of ^{90}Sr and ^{137}Cs from fuel processing wastes as a helpful step in their management. Removal of these nuclides leaves a mixture which, during 20 years' storage, would decrease in activity by a factor of about 30,000. However, an industry handling the fission products from 50 tons of ^{235}U burned in one year would have to deal with 500,000,000 Curies of separated ^{90}Sr and about the same amount of ^{137}Cs. It might be difficult to find a market for sources of this scale unless they were cheap, and it must be remembered that they would eventually come back as "waste."

(*i*) *Industrial applications*

Use of radioisotopes in industry is not a significant source of wastes. Most industrial sources are sealed, and nearly all unsealed sources are short-lived.

(*j*) *Transportation*

Ships are the only form of transportation using nuclear reactors as a source of power. They include naval ships, ice breakers and merchant vessels. They contain large amounts of fission products within the reactors, but as a source of waste they are not important, expect possibly in some harbours and inshore waters.

During start-up of the reactor the secondary coolant expands and the limited space in submarines necessitates the dumping of this expansion water. In common with landbased reactor coolant it contains radioactive corrosion products and tritium. The coolant is maintained at a low level of activity by means of ion exchangers, which become waste eventually. Normally this material is disposed of on land, although it has been shown by the Brynielsson Panel of the International Atomic Energy Agency[4] that resin from a fleet of as many as 300 nuclear ships could be dumped safely if this were done only on the high seas.

Apart from these sources wastes from nuclear shipping consist of clean-up solutions, laboratory wastes, laundry effluent and other minor sources common to all reactor operations. Except in submarines, practically all wastes can if necessary be retained on board for disposal ashore.

* For example, the "cobalt bomb".

† Coming after uranium in the Periodic Table, *e.g*, americum and curium.

3. Disposal Principles

There are two main procedures available for disposal—Concentration and Confinement: or Dilution and Dispersion.

(a) If wastes are truly confined, in the sense that in no credible circumstances could they be liberated into the environment, then the only additional requirement is "perpetual custody" to ensure that the confinement is never broken. This is easier said than done. In the field of high level wastes when we say "perpetual" we are speaking in terms of thousands of years. Few private firms go back for 100 years, political regimes have seldom lasted for at long as 500 years, and there are few civilizations shat have survided for 2000 years. In our own day ofrecasters tend to regard dates beyond 2000 A.D. as being in the distant future. What, then, can we do about "perpetual custody" of wastes containing, for example, plutonium with a half-life of 24,000 years?

This is not a fanciful dilemma. A story from Chalk River will illustrate the point. When the Canadians decided to concentrate on natural uranium heavy water reactors for power production it became apparent that processing of spent fuel would be uneconomic until the price of uranium or plutonium rose considerably. Processing was therefore stopped, but the wastes accumulated during the pilot plant operation had to be disposed of. A considerable volume of medium level waste was mixed with cement in steel drums and enclosed within solid concrete monoliths below ground in the waste management area (Fig. 3). The question then arose "What if some archeologist digs this structure up 1000 years from now and thinks it is an ancient temple or tomb?" Eventually someone suggested that its true nature should be inlaid in non-corrodible metal on the top of the monolith. Dr. A. J. Cipriani, who had listened to the debate in silence, then asked "In what language?".

The implications of this question are profound. Some of the wastes for which we are responsible will still be radioactive after our present civilisation has disappeared and perhaps been forgotten. So far

Figure 3 Pouring a concrete monolith. Steel drums filled with waste, solidified by mixing with cement, were stacked on concrete slabs surrounded with forms. The forms were filled with concrete. The monoliths were about 2 m below ground level. (Courtesy Atomic Energy of Canada Limited, Chalk River, Ontario).

as we know there is no practicable solution to the problem. The best we can do is to ensure than the nature, amount and location of all major disposals are recorded in the nearest approximation we have to a perpetual repository of archives—a government department. Beyond that we can only rely on folk memory. After all, farmers in Europe have been ploughing around neolithic tumuli and prehistoric roads for thousands of years for no good reason known to them, except that it was accepted to be the right thing to do.

(b) Dilution and dispersion is the traditional method that men have always used for dealing with their wastes. Until recently it seemed to work fairly well unless populations became very concentrated, but it is now becoming clear that there are so many people that the system is showing signs of breaking down. It depends upon the capacity of the environment to dilute or detoxify the wastes to a level that is innocuous to man and to organisms of interest to man. We are still a very long way from contaminating our environment with radioactivity to a point where radiation effects are observable, even in close proximity to nuclear enterprises, but we must maintain vigilance to ensure that slow and subtle changes do not occur which escape our notice until it is too late.

Safety in discharge to the environment depends upon three factors—(1) Dispersion by such means as atmospheric dilution, mixing into big bodies of water, or spreading through large volumes of soil. (2) Fixation of radionuclides on soil minerals and organic detritus. (3) Decay of radionuclides, dispersed or fixed, before they are able to affect man.

During the 20-year history of nuclear industry there has been no recorded case of injury to a member of the public from radioactive wastes—indeed the nuclear power industry has been completely free from incidents that have had effects on the public beyond the loss of some thousands of gallons of milk which were destroyed after the Windscale reactor accident in 1957.

The principle of dispersion has one logical trap into which regulatory bodies have sometimes fallen. In some countries the discharge of liquid and gaseous wastes is limited by the concentration in the effluent pipe or the concentration at the stack mouth. This is based upon the assumption that if the concentration is limited to the maximum permissible value, all will be well. However, the "dilution capacity" of a river is a function of the number of Curies per day put into the river, divided by the daily flow of water. If an operator wishes to dispose of double the amount of waste, and he is limited only by the concentration in the effluent pipe, he need simply double the amount of water flowing in the pipe. But the down-stream effect will be a doubling in the concentration, unless he has doubled the flow in the river.

For this reason limitations must be made in Curies per unit time, not in micro-curies per millilitre, and account must be taken of volume of river flow if this is seasonally variable. Regulations set on the basis of concentration at the point of discharge only protect people close to the discharge point.

4. Disposal Practices

(a) Gases

Radioactive gases arise mainly in reactors, spent fuel processing, isotope production, and research and development facilities. The general principles are the same for all procedures that depend upon dispersion into the atmosphere.

If we have a stack that is emitting Q Curies/sec., the concentration C at a given distance downwind will be KQ. The parameter K is a very complex function which depends upon wind speed and direction, weather conditions, stack height, topographical features, variability of temperature with height, velocity and buoyancy of the effluent and other conditions. Values of K for a range of conditions can be calculated from equations proposed by Sutton[5], Pasquill[6] and Holland[7]. These equations have been used to calculate the permissible emissions from stacks by inserting appropriate numbers and parameters applying to unfavorable weather conditions likely to obtain at the site. The permissible emission rate has been set at a value which would ensure that populations downwind would not be exposed to more than an agreed maximum radiation dose rate.

The classical equations have been based on statistical theory with empirical values for the diffusion parameters being obtained from experimental work which has sometimes had little relation to real emissions from actual stacks.

Returning to the superficially simple equation $C = KQ$, it is apparent that if we could observe, over a long period of time, the maximum value of

C ever attained per unit emission rate, we could define a figure K_{max} which was not likely to be exceeded. With a sufficient number of observations of C and Q, extended over a sufficient variety of weather conditions, we could estimate the probability that our value K_{max} could ever be exceeded.

When a maximum permissible concentration is set for a noxious substance the decision really depends upon a belief that the probability of damage is so low that it is acceptable. If, then, C is set at the *MPC* at a given distance from the stack, and K_{max} is known for that distance, then Q_p, the maximum permissible release rate, is determined.

It has been shown by Barry[8] that K_{max} is not very dependent upon topography or climate, because it depends mainly on rather large-scale behaviour of the atmosphere, and the frequency of most adverse conditions normally experienced do not vary grossly from one place to another.

The maximum permissible emission rate—or in some cases the *MPC* at the stack mouth—is given in the regulations governing the plant or laboratory. It is then the responsibility of the operator to ensure that emissions are kept as for below the permissible level as may be practicable. Numerous methods are available, other than variation of stack height, for achieving this end (Fig. 4).

(*i*) *Filtration* It is advisable to filter contaminated air near to the source of the activity. This reduces the amount of air to be filtered and also cuts down the "plating-out" of radionuclides on the ductwork, which can be a source of radiation fields within the plant.

Filters must be suitable for the job they are supposed to do. They should be made of non-flammable material such as glass or asbestos fibre and should be tested before and after installation. If fine (e.g. "Absolute") filters are used it is often necessary to precede them with a coarse filter to avoid rapid clogging with dust.

Filters must be very efficient to be adequate for fuel processing plants and incinerators burning highly active waste. For example, a sand filter at Hanford[9] capable of passing 10,000 m^3/min had an efficiency of more than 99.5%, but this was inadequate. The necessary efficiency of 99.99% was attained with a bed of glass fibres 100 cm thick.

(*ii*) *Electrostatic precipitators* Small airborne particles are usually electrically charged. The charge can be increased by passing the air through a corona discharge, or through a charged fabric screen. The particles are attracted to a surface carrying the opposite charge, from which they can be removed mechanically. It is possible to use the same principle by imposing a charge on filters.

(*iii*) *Steam ejector nozzles* The most efficient air cleaning device other than "Absolute" filters consists of a nozzle in which the air is mixed with steam and expelled into an expansion chamber where the steam condenses on the particles. After passing through a second constriction into another

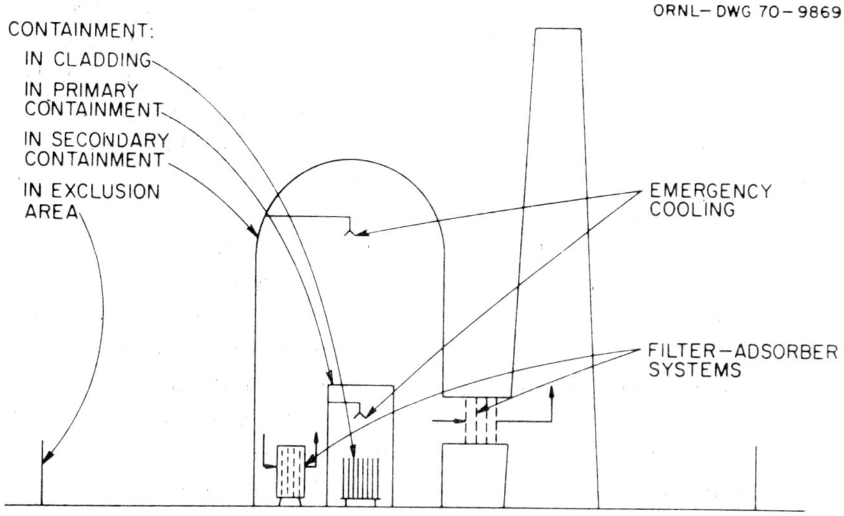

Figure 4 Reactor containment system (Diagrammatic). Any leakage from fuel must pass through the cladding, the primary containment and either the secondary containment or the stack filters. Contamination within the building can be removed by sprays and/or filters.
(Courtesy R. P. Shileds, Report Conf. 700816, Vol. 2, p. 771, 1970).

expansion chamber, where the air is scrubbed with water jets, the removal efficiency for 0.3 micron particles is 99.9%[10].

(*iv*) *Incinerator off-gases* The hot gas from an incinerator carries with it fly ash, tars and water vapour as well as particles. Tars may be removed and the gases cooled by water scrubbing devices. Water droplets must then be eliminated by reheating or passage through a "cyclone". This is a cylinder with a conical bottom. Gas injected tangentially at the top sets up a vortex which causes deposition of particles on the sides.

In smaller incinerators the gases are cooled and some fly ash is removed by passage through a cooling chamber fitted with baffles. After this stage a roughing or "bag" filter is used, followed if necessary by Absolute or charcoal filters.

(*v*) *Processing plant gases* The devices required for cleaning gaseous effluent depend on the nature of the process. Off-gas from boiling high level wastes must be passed through condensers and scrubbers to recover nitric acid as well as to remove volatile radionuclides. However, these and other air cleaning equipment previously mentioned will not remove gases such as ^{85}Kr, nor hold back all of the radioactive halogens.

Radioactive iodine in molecular form is fairly easily absorbed by alakaline scrubbers and copper or silver mesh filters, but in the form of methyl iodine it can only be arrested by an activated charcoal filter. These filters have to be kept cool, not only to remove the decay-heat of adsorbed halogens but also because ^{85}Kr is absorbed much more powerfully by cold charcoal. This is the only practical means we have for removal of radioactive noble gases.

The very large dispersive capacity of a high stack usually makes it unnecessary to remove ^{14}C (as $^{14}CO_2$) or tritium (mainly $^{3}H^{1}HO$) because their toxicity is very low. However, the coolant CO_2 in a gas-graphite reactor does contain enough ^{14}C to require alkaline scrubbing, which removes radioiodine as well.

(*b*) *Liquids*

(*i*) *Storage* The necessity for long-term storage of very large quantities (many millions of gallons) of high level, strongly acid waste has led to the development of tankage and pipeline systems which have stood up to severe conditions for many years. Failures have occurred, but good design and carefully selected materials have prevented environmental contamination.

Tanks are constructed from material, often stainless steel, which will not be corroded by the solutions to be stored. Secondary containment is provided by catch tanks or drip trays and sufficient spare tankage is kept available for rapid emptying of a ruptured tank. Leakage is detected by a monitoring system which alarms immediately radioactive liquid appears in the catch tank (Fig. 5). Movement of active liquid is effected by pumping rather than by gravity to ensure that it is the result of deliberate action rather than accident.

(*ii*) *Evaporation* The most straight-forward and apparently the simplest method of treatment for radioactive liquid wastes is evaporation. In a carefully designed evaporator with an efficient droplet de-entrainment system the radionuclide content of the distillate can be about one millionth of that in the pot. There is little about the design that is specifically related to radioactivity except that shielding may have to be provided for the operator, and off-gases must be monitored and possibly treated in some way. Unfortunately, evaporation is expensive because it consumes a large amount of energy and the end product—the concentrate—is still a radioactive liquid waste. Evaporation to dryness or to the point of crystallization has been practiced, but the residue is so soluble in water that without further processing it is not suitable for disposal.

Where discharge of a large volume of low-level waste into the environment is unacceptable the cost of evaporation may be justified by its many advantages. Practically all liquid wastes are treated by evaporation in Denmark and Sweden, and it is also widely used in Japan.

Residues from evaporation may be mixed with cement, fused with glass frit or various ceramic mixtures, or incorporated with melted bitumen. The product is then handled as a solid waste.

(*iii*) *Flocculation and precipitation* The cheapest and simplest process for treatment of radioactive liquids is removal of the activity on some kind of precipitate, either as an integral part of the precipitated material, or adsorbed on its surface.

In most waste tanks a sludge settles out which may contain up to 90% of the activity, and a copious precipitate of metallic hydroxides is formed

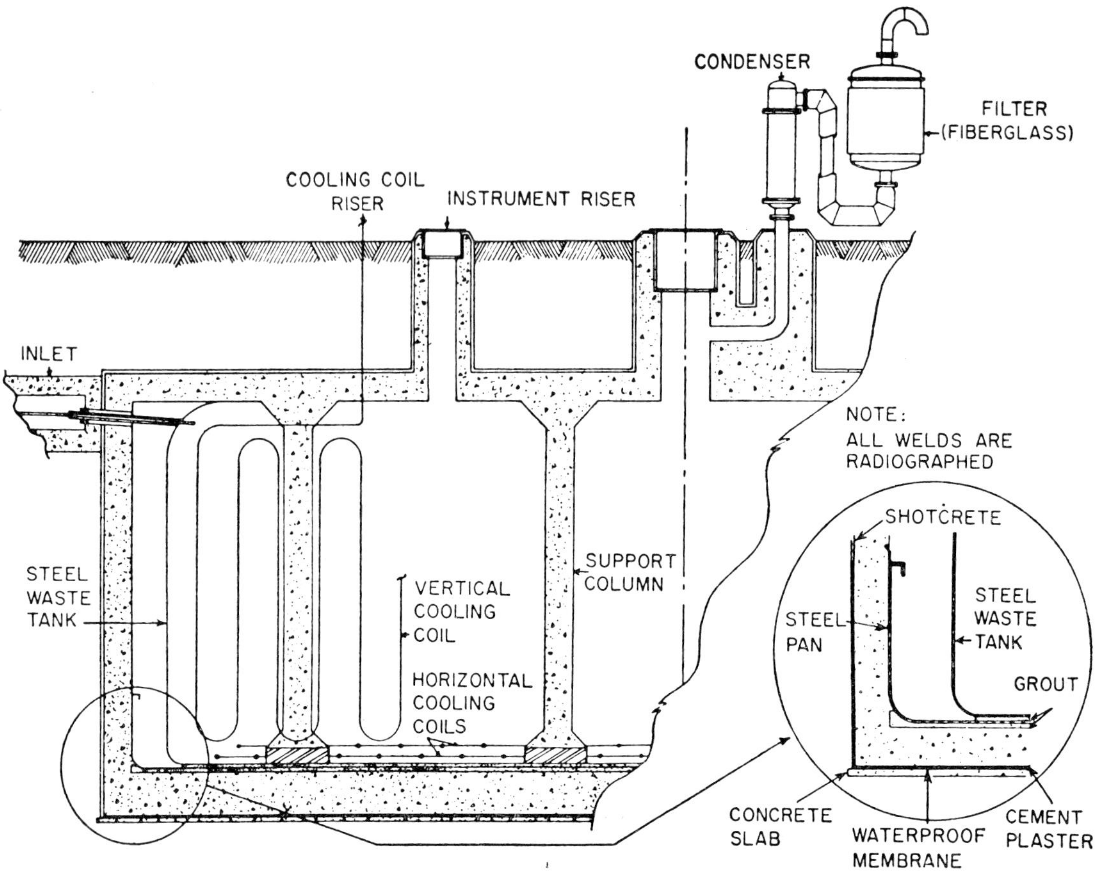

Figure 5 Structure of high level waste tank at Savannah River (Courtesy US Government Printing Office, *Hearings on Industrial Radioactive Waste Disposal*, Vol. 1, 1959, Fig. 4, p. 834).

on neutralisation which may carry down up to 90% of the remainder. Further purification of the clear effluent after separation of these sludges can be achieved by addition of lime and sodium carbonate. Up to 99% of the remaining activity can sometimes be removed by this treatment. Treatment with lime and sodium phosphate is also very effective (Fig. 6).

The treatment used depends upon the particular radionuclides present in the waste, and also on its gross composition e.g. the pH and salt content of the solution. In some cases ferric chloride, clay or other additives are introduced at carefully chosen points in the process. The selection of the process, and modifications introduced as the composition of the waste changes, require constant analysis and control by specialised chemists.

One problem common to all flocculation processes is how to deal with the sludge. The floc settles very slowly and after it has been drained through filters or separated by centrifugation it is in the form of a thick cheese-like solid which, in spite of its appearance, still contains 80 to 90% of water. In a successful British process[11] the sludge is repeatedly frozen and thawed. The separation of pure ice crystals leaves behind a concentrated salt solution which coagulates the small particles of floc into a form which settles more rapidly and is less likely to clog vacuum filters.

(*iv*) *Ion Exchange* The effluent from a flocculation process may still contain too much activity for discharge to public waters. It can then be passed through ion exchangers, which are expensive but very efficient. They cannot be used economically on a solution with a high salt content because their ion-exchange capacity would rapidly be exhausted by adsorbing the dissolved salts.

The effluent from a well-controlled flocculation process has a low total-solids content and after filtration to remove traces of floc it can be passed through a cation exchanger or mixed-bed resin suitable for removal of the radioactive contaminants. If properly chosen such a resin will remove 99.9% of most radionuclides (Fig. 7). Certain mi-

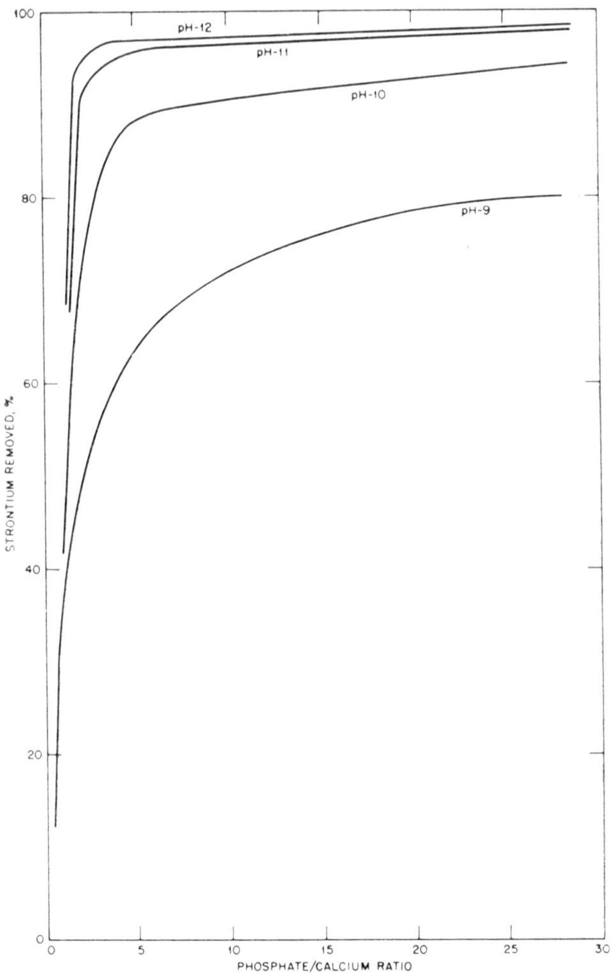

Figure 6 Decontamination with lime and phosphate—effect of pH and lime/phosphate ratio on removal of strontium-90. (Courtesy C. P. Straub, *Low Level Radioactive Wastes*, US AEC, 1964, p. 185).

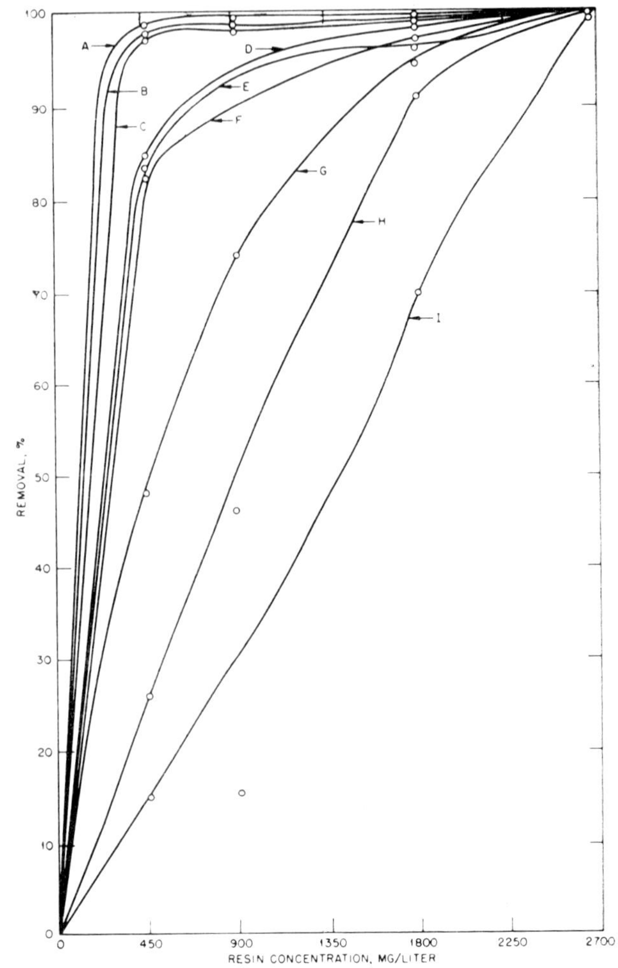

Figure 7 Decontamination with ion-echange resin— efficiency for various radionuclides: A, ^{182}Ta; B, ^{144}Ce–^{144}Pr; C, ^{95}Zr–^{95}Nb; D, ^{140}Ba–^{140}La; E, ^{131}I; F, fission product mixture; G, ^{32}P; H, ^{115}Cd; I, ^{137}Cs–^{137m}Ba. (Courtesy C. P. Straub, "Low Level Radioactive Wastes", p. 175).

nerals—clinoptilolite, greensand and vermiculite are examples—are also efficient ion-exchangers. They are much cheaper than synthetic resins but they require longer contact times for maximum effectiveness.

(*v*) *Glass* The very high-level "self heating" wastes—the primary wastes held in stainless steel tanks—are too active to be treated by flocculation or ion exchange. Storage in liquid form is seldom regarded as a permanent solution—somehow these wastes must be fixed in a non-leachable solid form which can be stored safely without danger of leakage or constant maintenance costs. One of the most promising ways to fix high level waste is to incorporate it into a glass.

Glass is a leach-resistant material which can be made from simple ingredients. Its quality varies with composition but is not usually sensitive to changes in minor constituents. Its low melting point makes it convenient for casting in various shapes and sizes for different disposal procedures. Glasses are supercooled, very viscous, liquid solutions of silicates. Soda glass is made by melting together silica, calcium carbonate and sodium carbonate. Other varieties contain potassium or potassium plus lead instead of sodium, and phosphate or borate in place of part of the carbonate. Metallic oxides are incorporated to form coloured glasses. Such a mixture might well be suitable for fixing the radioactive metallic oxides which form

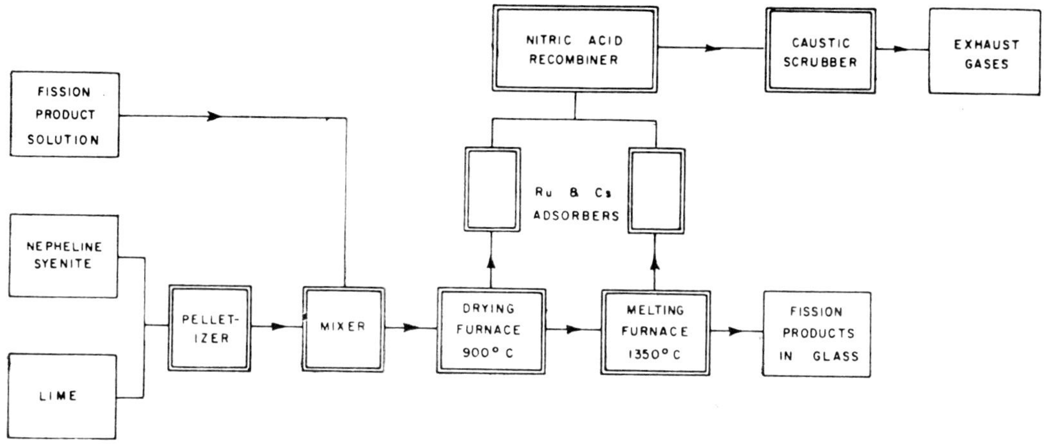

Figure 8 Fixation of fission products in glass. Fission product solution is added to pelletized nepheline syenite + lime, dried and melted at 1350°C.

the major proportion of "mixed fission products", after the nitric acid has been removed and the residue ignited.

Successful fixation of radionuclides in glass has been reported from the USA[12], UK[13] and Canada[14]. British and American practice has concentrated on borate and silicate glasses, or fusing the waste oxides with glass frit, whereas the Canadians have used a natural silicate, nepheline syenite, instead of a glass mix (Fig. 8).

Glass fixation is now being done on quite a large scale at the Pacific Northwest Laboratory (Hanford) Washington, USA. By the end of July 1970 nineteen million Curies had been solidified, representing waste from about ten tons of irradiated fuel.

In Canada nepheline syenite glass containing 1100 Ci mixed fission products has been buried in a swamp, below the water table, since 1960[15]. Careful studies of leaching, by sampling of the soil and ground water down-stream from the disposal, have shown that fission products equivalent to the dissolving of 10^{-10} g of glass per cm^2 per day are being removed from the disposal. Less than 1 mCi has been dissolved in ten years from 1100 Ci. This suggests that burial of active glass in dry soil, or even disposal into a big body of water, would be acceptable for quite large quantities of wastes.

(*vi*) *Calcination* Several methods have been developed for evaporation and subsequent calcination of wastes. Oxides are often soluble in water, so materials are usually added that will bind the oxides into soluble complexes. The calcination process is done in a heated steel container[16], a fluidized bed[17] or a spray calciner[18]. The Hanford pilot plant had put four million Curies through the pot calciner and 25 million through the spray plant in 1970.

The pot calciner is essentially an expendable piece of steel pipe heated in an electric furnace. The waste, mixed with flass-forming fluxes such as borax or lead oxide, is heated to about 900°C.

The spray calciner is a heated steel cylinder with a nozzle at the top through which the waste is sprayed. At a temperature of 875°C a fine powder is produced which must be stored in a dry place as it is leachable by water.

It is characteristic of all waste fixation methods involving evaporation, sintering and fusion that elaborate off-gas treatment systems are required to prevent environmental contamination by dust and volatile radionuclides. The concentrating equipment itself is essentially simple and often not expensive to build, but the gas purification plant is always sophisticated, complex and expensive. However, it is also very effective.

(*vii*) *Rock fracturing* The oil industry has developed methods for creating fissures in rock in order to encourage movement of oil or gas through a formation towards a well. This process has been adapted to disposal of medium-level wastes[19].

A horizontally bedded formation—shale has been used up to now—is drilled to several thousand feet. A high pressure jet of sand and water cuts through the well casing and penetrates between the strata near the bottom of the hole. The well is then sealed and water forced down under very high pressure, splitting the rock between the bedding planes. The water is followed up by the waste, mixed with cement, sugar and other additives. The mixture spreads out in a thin horizontal sheet, which solidifies after several hours. Typically the sheet is about a half inch thick.

The method has been used for disposal of a very large volume of waste at Oak Ridge, Tennessee. The equipment, including large bins for ingredients of the cement mix, mixing apparatus, a drilling rig and a very powerful pump is expensive, but the method is suitable for large-scale operation because successive sheets can be injected at intervals of a few feet through the depth of the bedded rock formation.

(*viii*) *Salt mines* The hazard that must be met by most radioactive waste management systems is contamination of public waters leading directly or indirectly to intake of radionuclides by man. An ideal situation for disposal would therefore be one where public access was impossible and contact with water incredible. The nearest approach to these conditions is found in a deep salt mine[20]. The presence of the salt guarantees that water has been absent for millions of years, and geological study can produce assurance that water is not rapidly penetrating into the salt bed. The excavated galleries of salt mines are large and stable tunnels, suitable for storage and roomy enough for safe work with active loads.

Major disposals of "solidified" waste are being made (1970) in a salt mine in Kansas, where detailed investigation has shown that eventuelly the creep characteristics of the salt will seal the disposal sites of heat-producing wastes. Work on disposal of liquids into cavities cut in the salt suggests that hot liquid waste could also be placed in such a site[21].

(*c*) *Solids*

As with liquid wastes, the most intractable problem is the safe management of the high-volume, low activity waste. The high activity waste is at first sight more dangerous, but although safe custody may be expensive it is not technically difficult.

Low-level waste consists mainly of "garbage"—contaminated clothing, equipment and structural material; broken glassware, cleanup materials such as cloths and mops; and a large amount of "potentially contaminated" material such as packing and paper which must be treated as active simply because it originates in an active area.

Much of this material can be reduced in volume by incineration or baling under high pressure. Fumes and smoke from incinerators and the dusty air from baling plants are cleaned up by methods dealt with under Gases (pp. 376–405), but the ash and baled waste remain to be dealt with.

In some countries geographical or legal circumstances restrict the possibility of burial of radioactive material in the ground. Elsewhere, ground burial is regarded favorably. In the latter case bales and non-combustible waste are likely to be buried in sparsely populated regions. Where land is chep, low-level wastes may be buried without any volume-reducing process.

(*i*) *Conditioning* Pre-treatment of waste before final disposal is called "conditioning". The aim is usually immobilization of radionuclides together with, if possible, volume reduction. There is very wide variation in practice from one country to another. For example, in France quite low level solid wastes are put into concrete containers[22] which are then filled with cement mixture so that the end product is a large concrete block. These blocks are stored, under a roof, on a concrete floor. In Canada, on the other hand, similar wastes are put into open trenches at Chalk River and covered with the local sandy soil. Practical measurements seem to show that both procedures are equally safe in the local circumstances.

A very effective conditioning process is fixation in bitumen or asphalt. Bitumen is very resistant to radiation, has a low melting point, is impermeable to water and has some mechanical flexibility[23]. Radionuclides enclosed in, or even mixed with, bitumen leach very slowly into water. Sludges are dewatered when mixed with melted bitumen, which helps considerably in restricting the volume of the disposals. In general, bitumen is beginning to be favored over concrete as the method of choice for "fixing" otherwise mobile waste radionuclides.

(*ii*) *Ground disposal* In some countries direct burial of contaminated material in the ground is forbidden at any level, whereas in others the amount and nature of ground disposals is left to the discretion of the operator.

Nearly all cations move through soil more slowly than the ground water although some anions—ruthenate and iodide for example—are retarded very little. In the case of the average "mixed fission products" usually of concern in waste management the fastest moving radionuclide is ruthenium, usually followed by Sr, Cs and Ce in that order. Relative rates of movement are affected by the nature and pH of the soil and the ground water, but even in acidic sandy soil ^{90}Sr moves through the soil at only 1/25 to 1/100 of the rate of movement of the ground water.

If the site of the waste management area is selected with care in relation to potable water supplies, so that the time of transit between the point of disposal and the point of human consumption is prolonged in relation to the half-life of the citical radionuclides, direct ground disposal of low level waste is effective and safe. There are a great many places where knowledge of the rate and direction of movement of the ground water, together with the distribution coefficients of radionuclides between water and soil, make it apparent that no significant discharge into the environment would be credible as a result of direct disposal into the ground.

When simple burial is unacceptable, disposal trenches and areas can be drained, with processing of the drain-water, or the area can be covered with asphalt and protected from encroachment of ground water by circumferential drainage.

A further step in the direction of safety is the "engineered enclosure". This is a structure built like a concrete house basement. It usually takes the form of a long concrete-lined trench divided into sections by concrete cross-walls. The section in use

Figure 9 Concrete Trench. Double trench, for medium-level solid wastes, is covered with light roof when in use. The filled trench is levelled with sand and a concrete roof is poured. Note galvanized steel seals for joint. (Courtesy Atomic Energy of Canada Limited, Chalk River, Ontario).

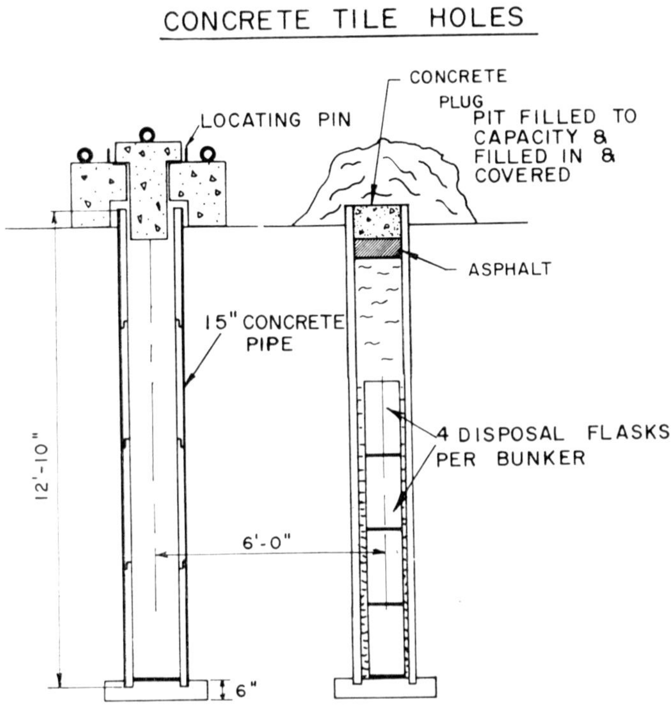

Figure 10 Concrete Holes, constructed from drain pipe painted with bitumen, provide well-shielded receptacles for high-level solid waste. (Courtesy Atomic Energy of Canada Limited, Chalk River, Ontario).

is covered by a temporary roof (Fig. 9). The object of the structure is to prevent the ingrers of water, so joints in the concrete are water-sealed and the base of the work is laid well above the maximum height of the water table. Since the facility will be used for reception of quite high-level waste the approaches must be suitable for trucks and mobile cranes.

Concrete trenches are unsuitable for reception of small, intensely radioactive objects such as spent teletherapy sources because of the inconvenience of scattered radiation fields. They can be accommodated in concrete-lined holes fitted with removable shiedling plugs. Canadian practice is to construct these from sections of concrete drain pipe, painted on the outside with bitumen, which is also used to seal joints between sections[24] (Fig. 10).

There are many different versions of the types of disposal facility just described. Some are in the open, some within buildings, but all are designed to prevent access of water to the contents.

It is often convenient to delay the passage of radionuclides contained in high-volume low-level wastes before discharge into the environment in order to take advantage of radioactive decay. If the local soil and ground water regime are suitable this can often be done by discharge into seepage pits. In 1969 the Hanford (USA) laboratories discharged 5000 million gallons of low-level waste into the ground, containing nearly 4000 Ci of radionuclides. More than 99.9% of this activity is held on the sediments immediately below the disposal facilities[25].

5. Methods for "Small users"

The International Atomic Energy Agency has issued a code of practice[26] on management of radioactive wastes by hospitals, research institutes and industry when no special facilities are available on the site. It gives a review of the scope and nature of the necessary control, particularly in the establishment of permissible limits for discharge into the environment. These institutions rely heavily upon the public sewers and garbage disposal systems, depending upon the fact that in practice there are levels of radioactivity below which things are not regarded as radioactive. Sometimes this is the level at which measurement becomes practicable and sometimes legal limits exist.

The ICRP and the IAEA both recognise 10^{-4} μCi/ml as the concentration in the sewer of an institution below which no restrictive action is required, irrespective of the nature of the radionuclide. This assumes that since large dilutions will occur before ingestion by the public, the individual at risk is the sewer worker. Plumbers can also encounter hazards in traps and filters, and they must be made aware of the situation. Similarly, although very low-level discharges to public disposal areas are usually acceptable, the waste management authorities must know of the practice so that they can warn their staff or undertake special procedures such as tip-and-fill operations. It is particularly important that scavenging should be prevented, because a very small source, normally innocuous, can be hazardous if carried for a long time in a pocket.

Apart from the use of public facilities, radioactive waste disposal for the "small user" does not differ in principle from the methods available to the larger producer of waste. A good deal of common sense and sense of proportion are required in dealing with the problem, aided by technical advice such as that in the IAEA report. It is also useful to remember that an ordinary illuminated wristwatch gives a count of several thousand per minute on a Geiger counter from the face side and almost zero from the back.

Sea Disposal

The public news media are convinced that vast quantities of radioctive wastes are dumped into

Figure 11 Sea Disposal of radioactive waste, solidified with concrete, inside steel drum. (Courtesy European Nuclear Energy Agency, Paris).

the sea particularly by the nuclear power industry. It is difficult to understand how this misunderstanding arose, because except in very special circumstances sea disposal is prohibitively expensive. The British discharge fairly low-level liquids via long pipelines into the Irish Sea off Windscale and into the English Channel off Winfrith, and the European Nuclear Energy Agency has made two experimental disposals of packaged solid wastes by ship (Fig. 11) into the deep waters of the Atlantic[27]. American disposal operations on the Pacific and Atlantic coasts, now abandoned, have been confined to low-level garbage-type wastes. The practice was given up because it was too expensive.

None of these operations has been concerned with more than a few tens of thousands of Curies in any one year, the monitoring has been very good and the potential hazards have been studied in detail. Contrasting with this situation is the fact that tens of *millions* of Curies of strontium-90 and similar amounts of other fission products have fallen into the oceans of the world from nuclear weapons fall-out. Yet this man-made radioactivity in sea water is difficult to detect against the background of natural potassium-40.

The standard work on radioactive waste disposal into the sea is still the Brynielsson Report (Safety Series No. 5, 1961) of the International Atomic Energy Agency[28]. This, with the ENEA report on their Alantic disposals should be consulted for further information.

REFERENCES

1. Hilliard, T. R. *et al.*, *Radiological Water Pollution in the Elliott Lake and Bancroft Areas*, Government of Ontario, Toronto 1965.
   Ontario Water Resources Commission, *Interim Report on Radiological Water Pollution in the Elliott Lake and Bancroft Areas*, OWRC, Toronto 1966.
2. Holaday, A. D., *Hearings on Industrial Radioactive Waste Disposal*, Vol. 1, p. 82. U.S. Gov't Printing Office, Washington, D.C.1959.
3. Rodger, W. A., Report ANL-6233 (1960).
4. International Atomic Energy Agency, Safety Series No. 5, *Radioactive Waste Disposal into the Sea*, IAEA, Vienna 1961.
5. Sutton, O. G., *Quart. J. Roy. Meteorol. Soc.* 73, **252** and **426** (1947).
6. Pasquill, F., *Meteorol. Mag.* **90** (1961).
7. Holland, J. L., Report ORO-99 (1953).
8. Barry, P. J., *Atmospheric Env.* **3** (1969).
9. See Reference 3 above.
10. Schauer, P. J., *Ind. Eng. Chem.* **43**, 1532 (1956).
11. DeBruyn, J., and K. W. Pearce, Report AERE-M 713 (1960).
12. McElroy, J. L., J. N. Hartley, and K. J. Schneider, Report BNWL-1185 (1970).
13. Elliott M. N., and J. R. Grover, Report AERE-R 4844 (1965).
14. Watson, L. C., A. M. Aikin, and A. R. Bancroft, *Disposal of Radioactive Wastes*, Vol. 1, pp. 373–399, IARA, Vienna 1960.
15. Merritt, W. F., *Disposal of Radioactive Wastes into the Ground*, pp. 403–408, IAEA, Vienna 1967.
16. Elliott, M. N., R. Gayler, J. R. Grover, and W. H. Hardwick, *Treatment and Storage of High Level Radioactive Wastes*, pp. 465–506, IAEA, Vienna 1963.
17. Lakey, L. T., G. E. Lohse, and W. P. Palica, Report IDO-14608 (1965).
18. Hartley, J. N. and W. R. Bond, Report BNWL-1494 (1970).
19. De Laguna, W., *Nuclear Safety* **11**, 391 (1970).
29. Galley, J. E. (ed.), Subsurface Disposal in Geologic Basins, American Assoc. Petrol. Geologists, Tulsa 1968.
21. Bradshaw, R. L. *et al.*, Report ORNL-4584 (1970), pp. 16–36.
22. Cerre, P., *Disposal of Radioactive Wastes*, Vol. 1, pp. 226 to 234.
23. International Atomic Energy Agency, Technical Reports Series No. 116, *Bituminization of Radioactive Wastes*, IAEA, Vienna 1970.
24. Mawson, C. A., *Management of Radioactive Wastes* pp. 117–124, Van Nostrand, Princeton, N.J., 1965.
25. Beard, S. S., and W. L. Godfrey, *Disposal of Radioactive*, Wastes into the Ground, pp. 123–134, IAEA, Vienna 1967.
26. International Atomic Energy Agency, Safety Series No. 12, *Management of Radioactive Wastes Produced by Radioisotope Users*, IAEA, Vienna 1965.
27. European Nuclear Energy Agency, *Radioactive Waste Disposal Operation into the Atlantic* 1967, OECD, Paris 1968.
28. International Atomic Energy Agency, Safety Series No. 5, *Radioactive Waste Disposal into the Sea*, IAEA, Vienna 1961.

COLIN A. MAWSON
Atomic Energy of Canada Limited

MANAGEMENT OF SOLID WASTE

Any discussion of solid waste nearly divides into three categories: (a) source and composition, (b) collection and (c) disposal (or hopefully, recycle). This section will deal primarily with the first and last, leaving the important problem of collection to a special section.

Solid wastes may be defined as any solid matter which is discarded as no longer being useful in the economy. It consists of organic and inorganic matter in a wide variety of forms, as is discussed below.

Solid wastes are one of three major interacting pollution vectors. The others are air and water pollutants. Solid matter if improperly handled will be a source of land, air and water pollution. Open dumps for example pollute the land from an aesthetic as well as use point of view. Solids dumped into the ocean can create "dead" zones where neither plants nor animal matter survive. Waste water from incinerators, if not properly treated can pollute the receiving waters. Similarly stack gases from incinerators will add to the air pollution problem if not properly treated to remove particulate matter and trace noxious gases.

The triangle is completed when the potential effects of air and water pollutants are considered on the solid waste sphere. The particulate matter removed from the air is a solid that must be disposed of if it cannot be exonomically utilized. About 70,000 tons per year of particulates are released above New York City. Some of these can be used in concrete or wood aggregate, but what of the remainder? Cleaning up an air pollution problem can create a solid waste disposal problem. The removal of solid and dissolved impurities from waste water also creates a solid waste disposal problem; some of the matter can be burned or processed to produce a soil additive but some will end up as a solid waste.

Solid waste is also an aesthetic pollutant. It mars the landscape in the form of litter and improperly operated disposal sites. This can be counteracted to a degree by individual action and community enforcement of proper disposal procedures. However, as the solid waste problem increases we will find that other attacks must be pursued vigorously.

Solid wastes contain significant amounts of valuable materials; 40 to 50% of urban waste is paper which if recycled can replace virgin stock equivalent to about nine trees per person per year. Not all paper is recyclable, but even if only 50% were recycled it would be equivalent to about 500 million trees annually. Exploration of methods for recycling solid wastes is still in its infancy but is and must gain momentum.

Sources of Waste

Solid waste differs from air and water pollutants in that it comes in discreet quanta and is very heterogeneous in nature. Both composition and rate vary significantly from day to day as well as from otherwise similar sources.

The solid waste production in the United States is in excess of four billion tons/year, and is expected to increase to five billion tons by 1980.[1] Table 1 breaks this down for the year 1967 by major source. The estimate for 1980 is probably conservative because loads resulting from more stringent enforcement of the removal of air and water pollutants will result in significantly greater solid waste loads. This is indicated by a comparison of the expected particulate emissions by the steel industry; 0.14 mil-

TABLE 1

Major Sources of Waste Matter
United States 1967[1]

Source	Solids Generated	
	lbs/Cap/Day	Million tons/yr
Urban		
Domestic	3.5	128
Municipal	1.2	44
Commercial	2.3	84
Sub total	7.0	256
Industrial	3.0	110
Agricultural		
Vegetation	15.0	552
Animal	43.0	1563
Sub total	58.0	2115
Mineral	30.8	1126
Federal	1.2	43
Total	100.0	3650

lion tons/year of particulate matter will be emitted in 1975 instead of 1.42 million tons because of the more stringent requirements of the 1969 Air Quality Control Act.[2] The difference will end up as a potential solid waste. In addition the trend, unless reversed, to use more convenience products will increase the per unit output of solid waste.

Urban waste

Urban waste collected is between 4 and 8 lbs per person per day, with typical values lying between 4.5 and 5.5 lbs per day. This differs from the amount generated because of self and private disposal. The major wastes included in this category are tabulated in Table 2, which includes a summary of disposal trends. One should be careful in the terminology because often domestic and municipal are used interchangeably to indicate the total refuse picked up from residential (domestic), institutional, small business and light industrial sources.

Some further definition of terms may be useful at this point. In general usage many of the terms have been used interchangeably. However an effort to standardize the terminology is being made. This effort is being made by the Institute for Solid Waste of the American Public Works Association and the Bureau of Solid Waste Management of the Environmental Protection Administration.[3] The standard usage of terms detailed by these groups is summarized here.

- Refuse is all solid waste matter.
- Garbage is the animal and vegetable wasteresulting from the preparation of food.
- Rubbish is the waste from homes, small businesses, etc., excluding garbage.
- Trash is used equivalent to rubbish.
- Litter is street refuse.
- Industrial waste is specialized refuse from manufacturing plants, and usually excludes rubbish.

Domestic waste composition will vary seasonally, as well as with locale and economic status. Typical analyses for domestic plus municipal refuse are shown in Table 3. As can be seen in a comparison of the data, the composition has not changed drastically with time except for a significant reduction in ash because of the change from coal as a home heating fuel. Location variations noted are as great or greater. A study of seasonal variations

TABLE 2

Composition of Wastes From Urban Sources[1]

Urban sources	Waste	Composition	Disposal, present
Domestic	Garbage	Wastes from preparation, handling and sale of food.	Landfill
	Rubbish, trash	Paper, wood, excelsior, rags, yard trimmings, metals, dirt, glass, crockery, minerals.	Incineration
	Ashes	Residue from fuel and combustion of solid wastes.	
	Bulky wastes	Furniture, appliances, rubber tires.	Dumping
Commercial, Institutional	Garbage	Same as domestic.	Landfill
	Rubbish, trash	Same as domestic.	Inceneration
	Ashes	Same as domestic.	
	Demolition wastes, urban renewal, expressways	Lumber, pipes, brick masonry, asphaltic material and other construction materials	Dumping Landfill
	Construction wastes	Scrap lumber, pipe, concrete, other construction materials.	Dumping Landfill Open burning
	Special wastes	Hazardous solids and semiliquids, explosives, pathologic wastes, radioactive wastes	Burial, Incineration, Special
Municipal, streets, incinerators, sewage treatment plants, septic tanks	Street refuse	Sweepings, dirt, leaves.	Fill
	Dead animals	Cats, dogs, horses, etc.	Bury or Incinerate
	Abandoned vehicles	Unwanted cars and trucks.	Reclaim
	Fly ash, incinerator residue, boiler slag	Boiler house cinders, metal scraps, shavings, minerals	Land Fill Or Dump
	Sewage treatment residue	Solids and sludge.	Landfill

TABLE 3

Urban Refuse
Typical Compositions

Source time	Hempstead, N.Y. 64		Hempstead, N.Y. 6/66	2.67	Chicago 56–58 average	Chandler Arizona 1953	St. Clara county, Calif. 1967	Berkeley California 1952	1967
Material									
Paper and paper prod.	56.01	53.5	32.71	53.33	56.5	42.7	50	69.0[c] compostable material	69.7[c]
Wood	2.82	–	1.22	1.46		2.3	2		
Grass, leaves, etc.	7.56	9.14	33.33	0.26	9.6	1.3	9		
Rubber	0.42	0.38				0.7	1		
Plastic	3.50	0.76	2.45	3.54		0.4	1	–	1.9
Oil, paint	0.84	0.76					–	–	–
Dirt	2.52	2.29	a	a			–	–	–
Rags	0.84	0.76	3.00	2.24		1.9	2	1.5	1.1
Miscellaneous	0.52	0.38					8	7.6	7.4
Rubbish									
Garbage	9.24	6.11	9.58	16.70	4.8	21.8	12	d	d
Fat		2.29				11.3			
Metal	7.53	6.85	7.96	10.60	14.8[b]	9.8	8	10.6	8.7
Glass, ceramics	8.50	7.73	9.75	11.87		7.8	7	11.4	11.3
Ash					18.7	–	–	–	–
Reference	(4)	(5)	(6)	(6)	(7)	(8)	(9)	(10)	(10)

[a] Included in glass and leads.
[b] Glass averaged 6.4% range 3.5–9.3%.
[c] Includes garbage.
[d] Included in compastable material.

made in 1939 for New York City also showed greater variations: the ranges were garbage, 44 to 3.5%; and metal, 11.6 to 3.1%.[11] Base data have been difficult to obtain because of the many variabilities in the base. The most significant variables include the economic level of the area, the ratio of commercial to residential property, the type of commercial establishments and the housing density and age. The entire picture on obtaining accurate data on urban and/or domestic refuse is further complicated by the sampling problem. A discussion of this problem is beyond the scope of this work; the reader is referred to some basic work in this area by Carruth.[12] An excellent review of sampling and testing has been prepared by the Institute of Solid Wastes.[13]

The ultimate chemical composition of municipal refuse has been examined by a number of investigators. Table 4 give the range of values to be expected. Recently 0.3 to 0.5% chloride has been found in refuse independent of the presence of polyvinyl chloride; this is due to the presence of salt primarily.[16]

Density of municipal refuse varies with the load applied to it. Typically household refuse has a density of 3 to 10 pounds per cubic foot. Transfer stations and/or landfill operations can compact it to between 20 and 30 pounds per cubic foot,

TABLE 4

Municipal Refuse[14]
Ultimate Chemical Analysis

Constituents	Per cent by weight (as received)
Proximate analysis	
Moisture	15–35
Volatile matter	50–65
Fixed Carbon	3–9
Noncombustibles	15–25
Ultimate Analysis	
Moisture	15–35
Carbon	15–30
Oxygen	12–24
Hydrogen	2–5
Nitrogen	0.2–1.0
Sulfur	0.02–0.1
Chloride	0.3–0.5 (16)
Noncombustibles	15–25
Heating Value, gross	3,000–6,000 Btu/lb.

depending upon the material and conditions. The effect of compression on density for the Chandler, Arizona refuse is shown in Fig. 1. High pressure compaction (see Compaction) can increase the density to 80 to 90 pounds per cubic foot.

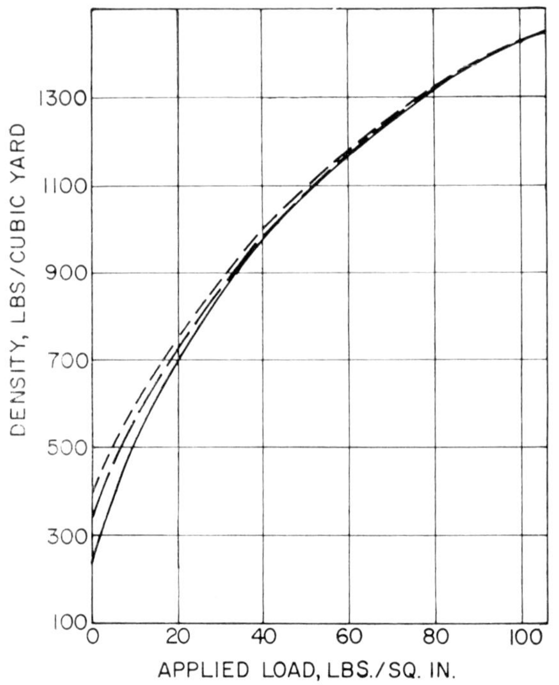

Figure 1 Refuse density. Household Refuse, Chandler, Ariz., 1954. Crdit: APWA, Municipal Refuse Disposal; 1966.

Industrial wastes

Industrial wastes amount to about 115 million tons annually. They include any discarded solid materials resulting from an industrial operation or establishment with the exception of dissolved or suspended solids in domestic or industrial waste waters. The composition and quantity of industrial solid wastes vary significantly from location to location, as well as between industries and within a given industry. Table 5 lists the type of wastes to be expected from the various SIC Industrial Groups. A large fraction of the wastes are generally common to most industries, and are listed on Table 6.

Data on the amounts of waste generated by or collected from various industries is very limited. Industry, quite naturally, has considered this type of data confidential in that it often reveals significant process and economic information. Average data, even if available, are of limited value because wide variations can result from process differences, process efficiences and direct recycle, as shown in a study based on detailed interviews. The results of this study giving total waste by industry are summarized in Table 7. Industry waste production on a unit per employee basis vary widely[17] and are summarized for large and small companies in Tables 8 and 9.

Increased efficiency as well as new uses for present industrial waste streams will alter both the quantity

TABLE 5

Sources and Types of Industrial Wastes

S.I.C. Group classification	Waste generating processes	Expected specific wastes
Plumbing, heating air conditioning Special trade contractors	Manufacturing and installation in homes, buildings, and factories	Scrap metal from piping and duct work; rubber, paper, and insulating materials, misc. construction and demolition debris
Ordonance and accessories	Manufacturing and assembling	Metals, plastic, rubber, paper, wood, cloth, and chemical residues
Food and kindred products	Processing, packaging, and shipping	Meats, fats, oils, bones, offal vegetables, nuts and shells, and cereals
Textile mill products	Weaving, processing, dyeing, and shipping	Cloth and fiber residues
Apparel and other finished products	Cutting, sewing, sizing, and pressing	Cloth and fibers, metals, plastics, and rubber
Lumber and wood products	Sawmills, mill work plants, wooden container, misc. wood products, manufacturing	Scrap wood, shavings, sawdust; in some instances metals, plastics, fibers, glues, sealers, paints, and solvents
Furniture, wood	Manufacture of household and office furniture, partitions, office and store fixtures, and mattreses	Those listed under Code 24, and in addition cloth and padding residues

TABLE 5 (cont.)

S.I.C. Group Classification	Waste generating processes	Expected specific wastes
Furniture, metal	Manufacture of household and office furniture, lockers, bedsprings, and frames	Metals, plastics, resins, glass, wood, rubber, adhesives, cloth, and paper
Paper and allied products	Paper manufacture, conversion of paper and paperboard, manufacture of paperboard boxes and containers	Paper and fiber residues, chemicals, paper coatings and fillers, inks, glues, and fasteners
Printing and publishing	Newspaper publishing, printing, lithography, engraving, and bookbinding	Paper, newsprint, cardboard, metals, chemicals, cloth, inks, and glues
Chemicals and related products	Manufacture and preparation of organic chemicals (ranges from drugs and soups to paints and varnishes, and explosives)	Organic and inorganic chemicals, metals, plastics, rubber, glass, oils, paints, solvents and pigments
Petroleum refining and related industries	Manufacture of paving and roofing materials	Asphalt and tars, felts, asbestos, paper, cloth, and fiber
Rubber and miscellaneous plastic products	Manufacture of fabricated rubber and plastic products	Scrap rubber and plastics, lampblack, curing compounds, and dyes
Leather and leather products	Leather tanning and finishing; manufacture of leather belting and packing	Scrap leather, thread, dyes, oils, processing and curing compounds
Electrical	Manufacture or electric equipment, appliances, and communication apparatus, machining, drawing, forming, welding, stamping, winding, piaintng, plating, baking, and firing operations	Metal scrap, carbon, glass, exotic metals, rubber, plastics, resins, fibers, cloth resides
Transportation Equipment	Manufacture of motor vehicles, truck and bus bodies, motor vehicle parts and accessories aircraft and parts, ship and boat building and repairing motorcyles and bicycles and parts, etc.	Metal scrap, glass, fiber, wood, rubber, plastics, cloth, paints, solvents, petroleum products
Professional, scientific controlling instruments	Manufacture of engineering, laboratory, and research instruments and associated equipment	Metals, plastics, resins, glass, wood, rubber, fibers, and abrasives
Miscellaneous manufacturing	Manufacture of jewelry, silverware, plated ware, toys, amusement, sporting and athletic goods, costume novelties, buttons, brooms, brushes, signs, and advertising displays	Metals, glass, plastics, resins, leather, rubber, composition, bone, cloth, straw, adhesives, paints, solvents
Stone, clay, and glass products	Manufacture of flat glass, fabrication or forming of glass; manufacture of concrete, gypsum, and plaster products; forming and processing of stone and stone products, abrasives, asbestos, and misc. nonmineral products	Glass, cement, clay, ceramics, gypsum asbestos, stone, paper, and abrasives
Primary metal industries	Melting, casting, forging, drawing, rolling, forming, and extruding operations	Ferrous and nonferrous metals scrap, slag, sand, cores, patterns, bonding agents
Fabricated metal products	Manufacture of metal cans, hand tools, general hardware, nonelectric heating apparatus, plumbing fixtures, fabricated structural products, wire, farm machinery and equipment, coating and engraving of metal	Metals, ceramics, sand, slag, scale, coatings, solvents, lubricants, pickling liquors
Machinery (except electrical)	Manufacture of equipment for construction, mining, elevators, moving stairways, conveyors, industrial trucks, trailers, stackers, machine tools, etc.	Slag, sand, cores, metal scrap, wood, plastics, resins, rubber, cloth, paint solvents, petroleum products

TABLE 6

Solid Wastes Common

Parking materials	fiber
	metal
	paper
	plastic
	wood
Maintenance materials	paints
	metal
	grease
	plastic
	rags
General housekeeping	waste aper
	fires
	glass
	solvents
	industrial chemicals

TABLE 7[16]

Industry	Waste for disposal thousand tons/yr.
Saw mills	33,000
Demolition	20,000
Food	7,200
Paper	5,000
Automobile and Aerospace	1,600
Rubber	1,500
Chemical	1,400
Printing and publishing	1,300
Glass	1,400
Electronics	1,000
Wood products	3,000
Tanning	400
Paints	160

and composition of the material for disposal in the next decade. For example, saw mill waste is being reprocessed into composition board and this utilization could essentially eliminate this waste stream. Only limited projections can and have been made and these show only a reduction in saw mill wastes.[20] Conversely, enforcement of air pollution statues will increase the amount of potential solid wastes significantly. Greater purification of industrial waste water will also affect the solid waste load.

Agriculturql wastes

Agricultural wastes are principally organic as indicated in Table 10. The exceptions are chemicals used in various facets of farming such as pesticides, containers, and small amounts of miscellaneous waste matter resulting from maintenance and general housekeeping.

Most crop waste is either plowed back into the soil or composted. Some open burning takes place. In some special cases such as bagasse (sugar cane stalks) industries have been established to utilize the waste material. Essentially none of this material finds its way into the usual disposal facilities.

TABLE 8

Waste Generation Typical for Large Firms[18]

Industrial Classification — Title	Employment 1[a]	Annual Wastes Vol. cu/yd 2[b]	Annual Wastes per Employee cu/yd 3[c]
Ordnance and accessories	29,356	131,404	4.476
Canning and preserving[d]	11,389	102,238	8.977
Other food processing (except 203)	2,012	17,545	8.720
Tobacco	e	e	e
Textiles	e	e	e
Apparel	601	1,248	2.077
Lumber and wood products	e	e	e
Furniture and fixtures	e	e	e
Paper and allied products	250	9,360	37.440
Printing, publishing, and allied	968	7,020	7.252
Chemicals and allied	e	e	e
Petroleum refining	e	e	e
Rubber and plastics	481	9,069	18.854
Leather	e	e	e
Stone, clay, glass, and concrete	1,258	6,617	5.260
Primary metals	e	e	e
Fabricated metal products	3,565	47,078	13.206
Nonelectrical machinery	8,872	101,153	11.401
Electrical machinery	7,807	57,252	7.333
Transportation equipment	4,100	100,776	24.580
Instruments	e	e	e
Miscellaneous manufacturing industries	e	e	e

[a] Column 1: Data on employment were obtained for those large firms which were surveyed and included in the wastes calculation from the research department of the Association of Metropolitan San Jose (Greater San Jose Chamber of Commerce).

[b] Column 2: FMC report, Solid Waste Disposal System Analysis (Preliminary Report), Tables 10 and 11, 1968. [3]

[c] Column 3: Column 2/Column 1

[d] For Canning and Preserving (SIC 203), no individual firm data were available. The industry total developed for the county as a whole was divided by the total employment in the industry (specially tabulated) to arrive at the multiplier. See text for further explanation.

[e] Data not available.

TABLE 9

Waste Generation Typical for Small Firms[19]

Industrial classification — Title	Weekly wastes Vol. per firm cu/yd 1[a]	Annual wastes Vol. per firm cu/yd 2[b]	Average employment per firm 3[c]	Annual waste Vol. per employee cu/yd 4[d]
Ordnance and accessories	2.500	130.00	[e]	[e]
Canning and preserving		(not surveyed)		
Other food processing (except 203)	10.875	565.50	26.979	20.961
Tobacco		N.A.		
Textiles		N.A.		
Apparel	4.000	208.00	5.882	35.360
Lumber and wood products	16.083	836.33	17.247	48.492
Furniture and fixtures	23.000	1,196.00	13.767	86,877
Paper and allied products	44.650	2,321.80	35.479	65.442
Printing, publishing, and allied	6.448	335.29	13.289	25.230
Chemicals and allied	6.506	338.31	18.439	18.348
Petroleum refining	[e]	[e]	[e]	[e]
Rubber and plastics	5.275	274.30	9.596	28.583
Leather	[e]	[e]	[e]	[e]
Stone, clay, glass, and concrete	9.415	489.60	16.747	29.235
Primary metals	2.000	104.00	23.409	4.443
Fabricated metal products	5.284	274.75	12.951	21.214
Nonelectrical machinery	4.450	231.40	12.921	17.909
Electrical machinery	6.733	350.13	21.036	16.645
Transportation equipment	4.550	236.60	16.490	14.348
Instruments	3.600	187.20	20.933	8.943
Misc. manufacturing industries	1.250	65.00	10.931	5.946

[a] Column 1: Data obtained and calculated for each SIC on the basis of small firm questionnaire responses supplied by FMC.
[b] Column 2: Weekly average in Column 1 multiplied by 52.
[c] Column 3: Average size of small firm estimated from the distribution of firms by employment size, supplied by the California Department of Employment (Research and Statistics), San Francisco Office.
[d] Column 4: Column 2/Column 3.
[e] Data not available.

Animal wastes pose a different problem because much is produced in very concentrated areas such as feed lots or poultry farms. The disposal of these wastes is posing a greater problem than crop waste, but may be more easily solved because it is concentrated and therefore susceptible to processing without collection. Average waste yields for a variety of domestic animals are summarized on Table 11.

Mineral wastes

Mineral wastes including solids generated in mining, milling and processing industries are expected to reach between two and four billion tons per year in 1980. The 1965 the waste was 1.4 billion tons, as summarized in Table 12.

TABLE 10[20]

Agricultural Waste (1966)

Waste	Composition	Amount (million tons/yr)
Crop residue	Corn stalks, grain stubble, cull, fruit and vegetable, vines, rice hulls, bagasse, tree prunings etc.	552
Animal manure (paunch manure)	Organic matter, protein, fat, carbohydrates etc. nitrogen, phosphorous	1,532[a]
Poulty manure	Same as animal manure	30[b]
Animal carcasses	–	–
Forest operations	–	25
Pesticides, insecticides etc. residue and containers	Chlorinated hydrocarbons organo-phosphorous compounds, other organics and inorganics, e.g. sulfur, lead arsenate etc.	–

[a] Excludes horse and mule manure.
[b] Excludes ducks etc.

TABLE 11[21]

Unit Generation Rates

Animal	Waste (tons/unit/yr)
Cattle	12.0
Cows, milk	10.6
Hogs	8.0
Sheep	3.0
Chicken, broilers	0.0045
Turkeys	0.025
Chicken, layers	0.047

TABLE 12

Generation by Type of Solid Wastes from the Mineral and Fossil Fuel Industries (1965)

Industry	Mine waste	Mill tailings	Washing plant rejects	Slag	Processing plant wastes	Total (thousands of tons)
Copper	286,600	170,500	–	5,200	–	466,700
Iron and steel	117,599	100,589	–	14,689	1,000	233,877
Bituminous coal	12,800	–	86,800	–	–	99,600
Phosphate rock	72	–	54,823	4,030	9,383	68,308
Lead-zinc	2,500	17,811	970	–	–	20,311
Aluminium	–	–	–	–	5,350	5,350
Anthracite coal	–	–	2,000	–	–	2,000
Coal ash	–	–	–	–	24,500	24,500
Other	–	–	–	–	–	229,284
Totals	419,571	288,900	144,593	23,919	40,233	1,146,500

TABLE 13

Solids Waste Management Methods

Type	Present usage	Relative cost	Items disposed of	Principle benefits
Sanitary landfill	Most used (80%) decreasing[a]	Low[b]	All except hazardous materials	Low intial cost takes all items may reclaim land
Central incineration	Increasing second largest method (9%)	High	All burnable except special items and over sized items	Reduces volume clean product can produce by product items
Local incineration	Decreasing used in multiple dwellings and special situations	High	All burnable except large items	Reduces volume at site
Open burning	Illegal in many areas	Low	Construction wastes, leaves, agriculture waste	
Compaction high pressure	3 plants under construction Medium		All except hazardous materials	Produces dense, essentially inert blocks for fill
Composting	Very few	Medium-high	Organic only. No tires, large pieces	Provides soil conditioner
Garbage grinding	Large number home units	High	Organic only	Reduces domestic collections
Dumping	Not legal in most areas	Lowest	Non-putrescibles	–
Recycling	Only for selected materials and areas, increasing	High	Selected. Depends on process	

[a] Many landfill are not sanitary but are included in this classification.
[b] Low under $ 3/ton
Medium 3 to 6 $/ton
High 6 + $/ton

Processing Methods

A variety of processing methods, as summarized in Table 13, are available at present for handling solids wastes. Most have been in use in some modification for at least the last 50 years. The choice of processing method will depend not only on the type of waste but also on location, sources, quantity of waste, method of collection, public opinion and ultimately economics.

Solid waste management was a 4.5 billion dollar industry in 1968. It is only in recent years that the public has begun to worry about disposal of solids, prior to that it was "out-of sight, out-of-mind." With ever growing amounts of solid waste as detailed in the discussion on sources, public pressure is becoming a factor in any decision on waste management.

The major disposal methods in use are landfill and incineration. Of potential interest in the United States are high pressure compaction and reclamation by recycling. Composting is practiced in Europe, but has not been successfully applied in the United States although it does have potential. There are new processes and techniques appearing for waste disposal and for the first time an organized research and development effort has been mounted to look at solid waste disposal.

Disposal methods could be discussed from the point of view of source: a brief summary of the most used methods for a variety of sources may be found in Table 14. This discussion will instead focus on the disposal methods most commonly in use today, landfill and incineration, followed by discussion of compaction, composting and some of the newer disposal techniques.

TABLE 14

Waste Source and Disposal Methods

Source	Methods recently used	Methods for future consideration
Municipal	Landfill (80%)	Compaction
	Incineration (9%)	Composting
		Recycle and reclamation
		Chemical processing
Industrial	Landfill	Recycle and reclamation
	Incineration	Chemical processing
	Recycle	
Demolition	Dump	Reclamation
	Open burning	Incineration
Construction	Dump	Reclamation
	Open burning	Incineration
Sewage	Landfill	Incinerate
	Dump	Compost
		Chemical processing
Agriculture	Landfill (plowback)	Compost
	Incinerate	Chemical processing
	Open burning	
	Dump	

The oldest method of disposal is dumping either on land or at sea. Here dumping is distinguished from Sanitary Landfill (see below). Dumping costs
The oldest method of disposal is dumping either on land or at sea. Here dumping is disting aished from Sanitary Landfill (see below). Dumping costs
between $ 1 and $ 2 per ton and has been used for all waste materials. It is totally unsatisfactory for putrescible materials such as food wastes and unsatisfactory from a public health as well as aesthetic and land use viewpoint even for non-putrescible materials such as demolition waste.

Open burning is often used for demolition waste, tree branches and stumps, and similar items; it is unacceptable because of the air pollution it creates. Neither dumping nor open burning have a place in the modern waste disposal scheme and are illegal in many places.

Sanitary landfill

Landfill is the most widely used method of waste disposal. There were 8,900 authorized sites (about half publicly operated) used by the 6,300 communities surveyed in 1968.[23] There appeared to be an equal number of unauthorized dumps. Unfortunately only six percent of the sites were considered to be "truly" sanitary. The remainder fell either into Category B or C on the U.S. Public Health Service Classification Scale, summarized in Table 15.

TABLE 15

U.S. Public Health Service Landfill Classification

A: *Sanitary Landfill* Operated without public nuisance or public health hazard; covered daily and adequately; no deliberate burning practiced.

B: Operated without public nuisance or public health hazard, but location permits modification of "A" such as burning of certain types of waste at site, or covering of fill only three times weekly.

C: Operating techniques permit development of public nuisance and potential health hazards, such as fly breeding, rodent substance, and odors.

Sanitary landfill is a perfectly acceptable method of disposal of solids and provides for the ultimate disposal of many types of waste; exception are non-degradable materials such as plastic and aluminum which are placed in landfills. Other items material, toxic chemicals, are not allowed in landfills for safety. Where land is plentiful, or marginal areas are available for reclamation, sanitary landfills offer a number of advantages over other disposal methods including low initial and operating costs. Other advantages and disadvantages are summarized in Table 16.

Sanitary landfill is basically the dumping of wastes followed by compaction and the daily

TABLE 16

Sanitary Landfill–Advantages and Disadvantages

Advantages

1. Most economical method when land is available.
2. Low initial investment.
3. Complete and final disposal.
4. Short period of time from need to full operation.
5. Flexible daily capacity with same working force.
6. Reclamation of marginal land for recreational and other uses.
7. All types of waste are acceptable.

Disadvantages

1. Lack of closeby suitable land in urban areas may make uneconomical.
2. Public opposition in, or near, residential areas.
3. Settling after completion means continued maintenance.
4. Public nuisance and health hazard if not properly operated.
5. Products of decomposition, methane and other gases, may create hazard.
6. Requires special practices for construction on completed fill.

Source, U.S. Dept. HEWPHS. Pub. No. 1792.

application of an earth cover. Several techniques are available, some of which are depicted in Fig. 2, depending on the type of site available. The one constant in all operations is the daily earth cover, preferably a sandy loam, amounting to, usually, one part earth for every four parts refuse.

Proper site selection is as critical to a satisfactory landfill as is sound operation. Selection criteria include proper ground and surface water drainage to prevent pollution of the ground water table. Location in a drainage basin or near streams or lakes present special problems and should be avoided where possible. Accessibility of cover material is an important consideration. The use of tidal areas and marshes presents special problems and is frowned upon, now, as a misuse of this type of natural resource. Dry pits, abandoned quarries and certain types of canyons or depressions are satisfactory landfill sites.

The size of landfills is often restricted by the amount of land available. The capacity can be estimated with a fair degree of accuracy. Refuse on arrival may vary in density from 100 to 800 pounds per cubic yard, depending on the delivery method. Typically the density in the "fill," of the initial compaction with a typical crawler tractor will be 1,000 lbs/yd for a single lift (layer) with a depth of 20 feet or less. For miltiple lifts the initial density

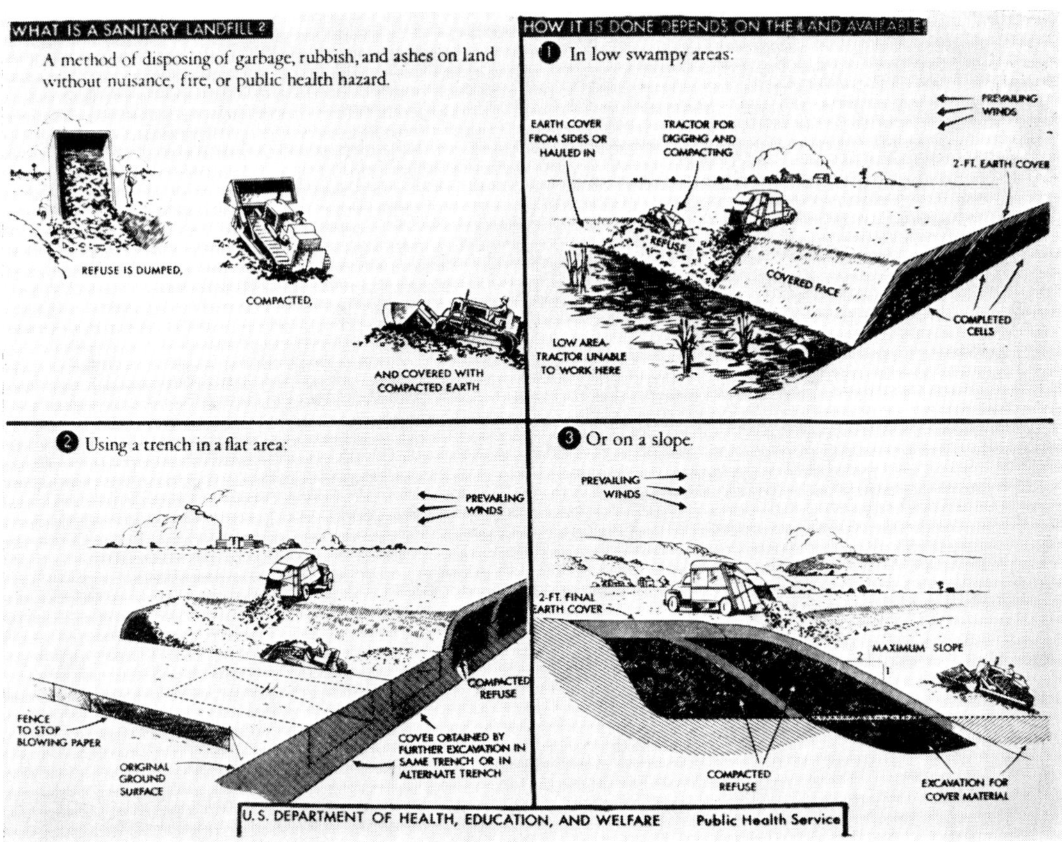

Figure 2 Sanitary land fill operations. Credit: U.S. Dept. of HEW, Public Health Service.

can reach 1,250 lbs/yd. This initial loading increases by as much as 50% over a period of time as further compaction and decomposition take place.[24]

Much of the material in the sanitary landfill decomposes over a period of between three and ten years depending on climate, permeability of the cover, composition of the refuse and degree of compaction. The decomposition in sanitary landfills is anaerobic as compared to aerobic degradation often found in other types of fill. Temperatures typically reach 120°F in the fill as a result of the degradation. The principal gas products are carbon dioxide and methane. The greatest gas production takes place in the first two years, according to a study made at the University of Washington. Ammonia and hydrogen sulfide are not problems in sanitary landfills although small amounts of these gases are produced. Odors resulting from the decomposition of putrescible material can be controlled by observing good operating practices; that is, covering the fill continuously and sealing surface cracks. Fire hazards and insects and vermin are not a problem, as compared to dumps, in a properly operated sanitary landfill although chemical control of the latter two is simetimes required.

Completed landfills are suitable for use as recreational facilities, airfields and parking areas; light industrial buildings may be erected on landfill. Building of residential structures on fill requires special precautions because of the potential hazards associated with the evolution of methane and other decomposition gases.

The cost of operating a sanitary landfill makes it an attractive means of disposal where land is available. Costs for a sanitary fill will vary between $ 1 and $ 5 per ton, depending on location and size of the fill. Small fills, handling less than 50,000 tons per yeark, will have a unit cost of $ 1.50 to $ 6.00 per ton. A large urban fill more typically shows costs of $ 1 to $ 3 per ton. The wide variation is a result of location differences which include differences in land acquisition costs, labor costs and operating differences due to local surface conditions and requirements.

The use of landfill will continue; however, its future, particularly in densely populated urban areas, is in doubt. Land is at a premium for this type of application close to urban centers. What land is available must be preserved for noncombustible material and ashes. For example, one urban county in New Jersey has less than five years landfill capacity available and in portions of Long Island no more land for landfill is available. Hauling costs to, as well as public resistance in, more rural areas is making landfill less attractive for urban areas such as Metropolitan New York. Finally, landfill does not provide for maximizing the value of refuse as a source of raw materials.

Recent studies to find alternatives to traditional landfill practices include a demonstration of shredding prior to filling. Only domestic refuse was shredded; the product was a superior fill compared to "raw" refuse. It could be left incovered with satisfactory sanitary and aesthetic results and was easier to dump and compact. Flies and rats did not breed on the shredded refuse. The compacted, uncovered fill also had better weathering and load bearing characteristics. This can be achieved at a cost of about $ 2.75 per ton in a 65,000 ton per year operation.[25] The method has some attractive features and some commercial facilities including one in Providence, R. I., which incorporates some recycle, are being designed using this principle. However operating and investment costs do appear to be higher than the more traditional method of filling "raw," as collected, refuse.

Baling of refuse may be particularly attractive where landfill sites are not locally available. A feasibility study was carried out in Chicago which showed that this method overcomes many of the present objections to landfill. The Japanese have been leaders in this area using high pressure presses to provide solid cubes suitable for use in building new land in tidal areas. More details may be found in the discussion of compaction.

Incineration

Incineration is essentially a method for reducing residue volume and at the same time producing an inert, essentially inorganic, solid effluent from material which is largely organic. Typical feed analyses are shown in Table 4. In addition to the solid product a gas is produced consisting mainly of CO_2, H_2, O_2 and N_2 but containing other gaseous components in trace quantities depending on the type of material burned and the operating conditions. Incineration is not an ultimate disposal method in that the solid residue which is primarily an ash containing some metal must still be disposed of, usually as landfill. The primary advantage is that it reduces the volume to be disposed of and results in a "clean" inert fill. For every 100 tons of material fed to the incinerator approximately 20 tons of residue result. The volume reduction is even more significant, often resulting in a 90% lower solids volume for organic materials.

The theory of incinerator operation is very simple. A unit is designed to expose combustible material to sufficient air at high temperature to achieve complete combustion. Combustion is usually carried out in deep fuel beds to ensure good contact of air and refuse. Several types of configurations are used to achieve contact; these include cocurrent flow of fuel and air-underfire, countercurrent flow of fuel and air-overfire, flow of fuel and air at an angle to each other—crossfeed; and combinations of these. The combustion is basically the same for all methods in that at the ignition front oxygen is rapidly consumed in the reaction $O_2 + C \rightarrow CO_2$ and if oxygen is depleted $CO_2 + C \rightarrow 2\,CO$. Therefore sufficient oxygen must be available to obtain complete combustion; usually this is provided by adding additional air in the chamber above the fuel. Incinerators are typically operated with about 50 to 150% excess air in order that the gas temperatures do not drop below that required for good odor-free combustion; this is usually in the 1700–2300°F range. Recent trends have been to go to the higher part of this range while old units often operate at 1500°F or below. The effect of excess air on gas composition is summarized in Table 17 for a typical refuse. A detailed discussion of typical air requirements and their effect on the thermal balance may be found in Principles and Practices of Incineration.[26]

Trace components in the incinerator-start gas include some SO_2 and NO_x. The former depends on the sulfur in the refuse and is typically around 0.01 to 0.02 per cent. Nitrogen oxide is generally formed in combustion processes and depends on the amount of excess air and to some degree the operating temperature of the incinerator. Typical values of two pounds of equivalent NO_2 per ton of refuse have been reported;[27,28] this is equivalent to 500 to 1000 ppm of NO_x in the off-gas depending on the refuse composition and the amount of excess air. Other trace components can be found in the off-gas and are summarized in Table 18. Their presence or absence is very much dependent on the type of refuse incinerated and the operating conditions.

TABLE 17

Incinerator Effluent Gas Composition[d]

Feed		Combustion Product		Products Gas Composition, Vol. %
Component	Wt. %	Excess Air Component	0%	50%[e]
Carbon	30.0	CO_2	16.0	9.86
Hydrogen[a]	6.1	H_2O	19.5	12.04
Oxygen[a]	43.3	O_2	–	7.94
		NO_x	–	~700 ppm[b]
Nitrogen	0.5	N_2	64.5	70.15
Sulfur	0.1	SO_2	0.02	0.01
Non-combustible	20.0[c]	–	–	–

[a] Includes moisture of 20%.
[b] Not computed.
[c] No combustion of metal assumed, in actuality same takes place.
[d] Excluding particular matter.
[e] Based on theoretical required less O_2 contained in feed.

TABLE 18

Incinerator Stack Gas Contaminents

Component	Amounts reported
Organic acids	
Formic	25–133 ppm (29)
Palmitic	0.6 lbs/ton of refuse (30)
Acetic (all organics)	40–600 mg/m³ (31)
Esters	
Methyl acetate	5–137 ppm (30)
Ethyl acetate	
Aldehydes	
Acetaldehyde	2.8×10^{-4} (31)
Formaldehyde	1.1 lbs/ton refuse (30)
Hydrocarbons	4 μgm/gm of particulate (27[a])
Halogenated hydrocarbons (depends on plastics and aerosols)	6–120 mg/m³ (31)
	0.44–10 ppm (29)
	0.3 lbs/ton refuse (30)
Ammonia	0.15–1.5 ppm (29)
Nitrogen dioxide	0.15–5 mg/m³ (31)
Nitrogen trioxide	4–100 mg/m³ (31)
HCl	300–1200 mg/kg refuse (31)
	30–350 mg/m³ (31)
SO_2	0.25–1.2 ppm (29)
	1.9 lbs/ton refuse (30)

[a] When burning rubber.

Particulate matter is also present in the stack gas and is removed by the usual techniques discussed in the section on Air Pollution. Particulate loadings of 3 to 25 pounds per ton of refuse burned have been reported.[22,34,35] Typically, particles range from 5 to 350 microns in size with 30% by weight under 10 microns and 75% less than 200 microns in size.

Solids residue from incinerators will vary widely with the type of feed and incerator operating conditions. Typical residues have been examined by the Bureau of Mines. The results of this work are summarized in Table 19. A typical ash and slag chemical analysis may be found in Table 20. This residue can be ulilized in road fill or separated (see Reclamation).

TABLE 19

Incinerator Residue Composition Ranges[36]

	Wt. Per cent
Moisture	24–40
Components, dry Wt. basis	
Tin cans	16–22
Iron, all types	9–14
Nonferrous metals	0.1[a]–3.7
Stones and bricks	0.8–1.9
Ceramics	0.6–1.5
Unburned paper and charcoal	4–12[b]
Partially burned organics	0.1–1.3[b]
Ash	12–18
Glass	37–50

[a] After hand picking.
[b] High temperature operation will decrease this markedly.

TABLE 20

Incinerator Ash and Slag Analysis[37]

SiO_2	46%	Na_2O	3%
Al_2O_3	21%	K_2O	1%
Fe_2O_3	8%	P_2O_5	2%
TiO_2	3%	BaO	0.6%
CaO	10%	SO_3	0.3%
MgO	3%	ZnO	0.5%

Incineration can effectively be divided into local, onsite, and cetral methods. The basic principles are the same but the applications vary considerably. Central incineration facilities handle refuse from many sources and a wide variety of feeds. Local incinerators handle either special feeds, onsite, such as industrial or hospital wastes, or serve a particular small location such as an apartment house. Size is not necessarily a criterion although generally central incineration facilities have capacities in excess of 100 tons per day.

At the present time there are approximately 400 central incineration facilities in use[38,39] making this type of waste reduction facility the most prevalent one. Central incineration handles about 15 million tons of waste annually and is concentrated in the northeastern part of the United States. It is also widely practiced in Europe. The practice of incineration of wastes is growing as land for fill, particularly in urban areas, becomes scarcer and technological improvements provide more efficient and cleaner systems.

A typical incineration facility will have a capacity ranging from 100 to 1200 tons per day with individual furnaces usually limited to a 300 ton per day rating. Most large incinerators today are continuous-feed rather than batch design because operation is more controlled and easier. In addition the absence of the heating and cooling cycle results in lower maintenance and a higher capacity per investment dollar. Air pollution control is improved significantly in continuous-feed incinerators as compared to batch plants.

A large central incineration facility is schematically shown in Fig. 3. It can be divided into five areas: (1) the receiving section which includes the weigh station, storage hopper and bucket crane; (2) the furnace—which includes the charging hopper, stokers, furnace chamber and air feed system; (3) the effluent gas treating facilities; (4) the ash handling system; and (5) the cooling water system. The particular system shown does not have provision for waste-heat recovery; very few systems incorporate this at present.

For mixed refuse a typical refractory-wall incinerator will have 12.7 cubic feet in the primary furnace chamber and 18.5 cubic feet in the secondary chamber per ton of refuse per 24 hours with a grate loading of 77 pounds per square foot per hour. Volume and loading requirements will vary with the type of feed as well as furnace configuration.[40] Typically the values quoted correspond to a 12,500 Btu per hour per cubic foot heat release. A detailed discussion of furnace design is beyond the scope of this work and the reader is referred to an excellent work by Richard C. Corey.[41]

Incineration in the past has received a bad reputation because of poor control of gaseous effluents and sloppy handling of solid and liquid effluents. With proper design and operation an

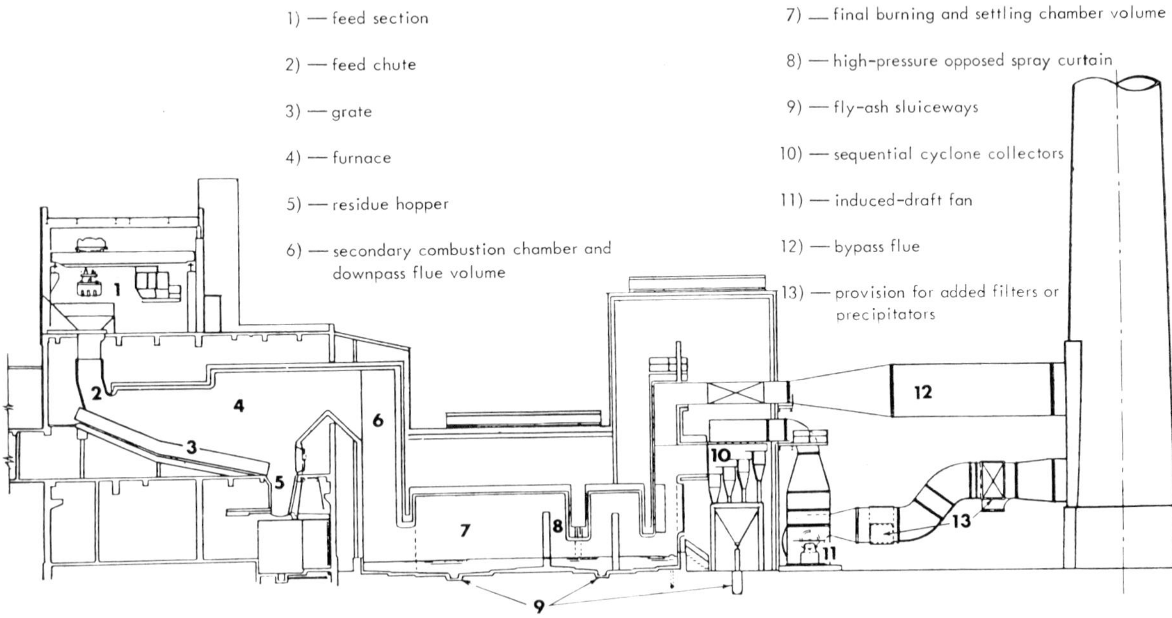

Figure 3 Large incinerator schematic.

incinerator can meet or exceed requirements on all effluent discharges. A modern central incinerator is a more complex operation than a large commercial steam boiler. It therefore requires skilled operating, maintenance and supervisory personnel to ensure efficient operation.

At the present time control of particulate matter in the effluent gas is the most critical problem in incinerator design and operation. A typical modern facility will include either a wet scrubber or a spray chamber followed by solid separation in in a multicyclone separator for primary off gas cleaning. Often this is followed by passage of the gas through either a baghouse filter or electrostatic precipitator. These methods can achieve up to 99% removal of particulate matter[42] which will meet the code requirement of 0.1 pound particulate per 1000 pounds of flue gas at 50% excess air in almost all cases. The costs for this required deanup are significant and can be as high as $ 2,000 per daily ton of refuse capacity[43] (see Air Pollution Control). Odor control is achieved by providing adequate time (0.5 sec) in the combustion chamber at temperatures above 1500°F. As incineration temperatures in modern units are between 1800 and 2200°F this poses no problem.

Waste heat recovery has been practiced to only a very limited degree in the United States. Less than 10% of the incinerators surveyed in 1966[39] had waste heat recovery facilities. This is contrary to the practice in Europe where waste heat recovery is practiced in a large number of units. Those U.S. plants that do recovery of heat incorporate waste heat boilers using the sensible heat of the off gas rather than the usual water wall principle used in furnaces; again this compares to the European practice where water wall incinerators are common. One water-wall incinerator (600 T/D) has been in operation since 1967 at the U.S. Navy Base, Norfolk, Va.[44] Several Reasons have been advanced for the lack of heat recovery in the United States. These include adverse economics. In addition much more severe slagging of the walls (due apparently to differences in waste composition) seems to have held back the use of water walls. With new technology, and a tighter fuel picture, waste heat recovery in incinerators will become commonplace in the next decade. In general municipal service a 1,200 ton/day facility in Montreal went on stream in 1970 and produces 100,000 pounds of steam per hour and a 1,600 ton per day facility in Chicago started up in 1971.

The investment and operating costs for incinerators are high and to date have been one of the major deterrents to wider use. Typical installations in the 600 to 1000 ton per day range require an investment of $ 10,000 to $ 12,000 per ton of installed daily capacity (1970 basis) depending in part on the air pollution control devices which can account for 20 per cent of the total cost as well as

Figure 4 Compact incinerator. "Combustopak Waste Disposal System". Credit: Combustion Engineering Corp.

size. Water-wall installations typically will run 20% more than refractory lined incinerators; the 1,600 ton per day Northwest Chicago plant cost about $ 16,000 per ton, which is unusually high. Operating costs incuding amortization of the investment will vary between $ 5 and $ 8 per ton. Actual values for generated steam have not been published but one estimate for Washington D.C. indicates a potential recovery of $ 2 to $ 3 per ton of refuse for the steam sold.[45]

In addition to the installation of units with waste heat recovery facilities the trend will be to larger installations to take advantage of the economics of size. However, small skid mounted units with capacities in the range of 20 to 50 tons per day will become available at a cost which will make central incineration economic for smaller communities. The first of these units which will meet most air pollution codes (Fig. 4) has been tested and is being marketed by Combustion Engineering at present. Another development will be the installation of more combined sewage sludge-refuse incineration facilities using the Nichols Oven or similar installations.

Incineration in fluidized beds is being developed by Black and Clawson in a facility in Franklin, Ohio, as well as by Combustion Power Co. which is combining a fluid bed combustor with a gas-turbine generator to produce 15,000 KW of power from 400 tons per day of refuse.

On-site incinerators have a number of distinct applications. They are of particular value in reducing the volume of domestic refuse that must be collected and have therefore been very popular in larger apartment houses as well as more rural areas where collections are infrequent or non-existent. In New York it was estimated that if on-site incineration would service about two million people the city could save 18 million dollars annually in collection costs. Unfortunately the savings are not without cost.

The use of on-site incinerators for reducing domestic refuse has been discouraged recently because of the potential pollution problems associated with inadequate design and operation. The outdoor or backyard incinerator is essentially an open refuse burner producing smoke, odors and large amounts of airborne particulate matter. It is inexpensive and therefore is still used in rural areas. In urban areas, outdoor burners are toally unsatisfactory and their use is illegal in most areas.

Indoor home burners have in general proved almost as unsatisfactory as the outdoor incinerator. Most indoor burners today use auxiliary fuel to improve their operation. There are generally three types of auxiliary fuel fired home incinerators: (a) high-BTU input, (b) dehydrating and (c) after-burner. The first two units essentially provide auxiliary fuel so that sufficient heat is available to dry and ignite the wet refuse. They tend to have smoky gas discharges, with the typical odors associated with the burning of garbage, because their gas flue temperatures are usually below the 1300°F minimum required for odor-free combustion. The after-burner incinerator provides a secondary flame in the stack to ensure proper combustion of the off gases; this will usually reduce odor and smoke. The use of indoor single dwelling incinerators is not necessarily economic: annual operating costs are in the range of $ 10 to $ 20 and it is doubtful if the collection costs are reduced by this amount. In many urban areas this type of incineration is no longer allowed because of the poor performance characteristics of the units with respect to air pollution.

On-site incineration has been used for years in apartment houses. There are a number of design variations but essentially the incinerator consists of a chute down which the refuse is conveyed from the upper floors into the combustion chamber. The off gases pass either through the same chute or a parallel flue. This type of incinerator creates the same type of nuisances as do the smaller single dwelling units. It can, however, be modified to provide overfire and afterburners to ensure adequate combustion of the off gases so that smoke and odor are eliminated. Newer units are designed with fly ash removal facilities such as washers, scrubbers or precipitators. These units now will all be multi-chamber design with afterburners. A properly designed and operated unit can be almost as satisfactory as a large central incineration unit. However, experience in New York City has shown that proper operation and maintenance is difficult to achieve because building operators are not trained to handle complex equipment such as wet-scrubbers.

Numerous on-site incinerators are operated with satisfactory results for the reduction of industrial wastes. These facilities are usually specially designed to handle one type of refuse. Typical materials that are incinerated include plastics, rubber, wood scrap and paper. The economics of waste recovery are changing so that often these materials are no longer burned. For example wood chips and sawdust at sawmills are often sent to papermills or composition board producers as feed, where formerly they were burned. Insulated copper wire and automobiles are incinerated to remove the organic components prior to recovery of the metal. Special liquid wastes such as water containing organics are incinerated. The special applications are numerous, including the incineration of radioactively contaminated wastes.

In addition to the more traditional incinerators, whether rectangular or cylindrical, special designs are employed in industrial waste disposal. For example shredded plastic[46] as well as "white water" from papermills[47] is incinerated in a fluidized-bed combustion chamber. Industrial sludge is being burned in a rotary kiln[48] by Kodak. Solid cyanide waste in automobile plants is put into solution and then burned while aluminum chloride sludge from a petrochemical operation can be burned to produce HCl and alumina.

Although hauling to landfill sites is the present disposal method for many industries, on-site incineration of industrial wastes will receive wider use for waste disposal where recycle is not possible and volume is sufficient, in excess of 500 lb/day, to justify an installation. Proper design and operation of an incineration facility in an industrial setting is easier than in a dwelling. Usually trained people are available to maintain the equipment, enabling the operation to meet pollution control requirements. Special, centralized, privately owned and operated, incineration facilities are being built to compete with on-site facilities. These operations will have the exonomy of size in their favor but haulage and variability of feed may more than negate this advantage.

Hospital wastes are now commonly being disposed of in onsite incinerators. To eliminate the possibility of spreading infection wastes should be promptly incinerated. This is best done in an on-site facility. The average load for hospital incinerators

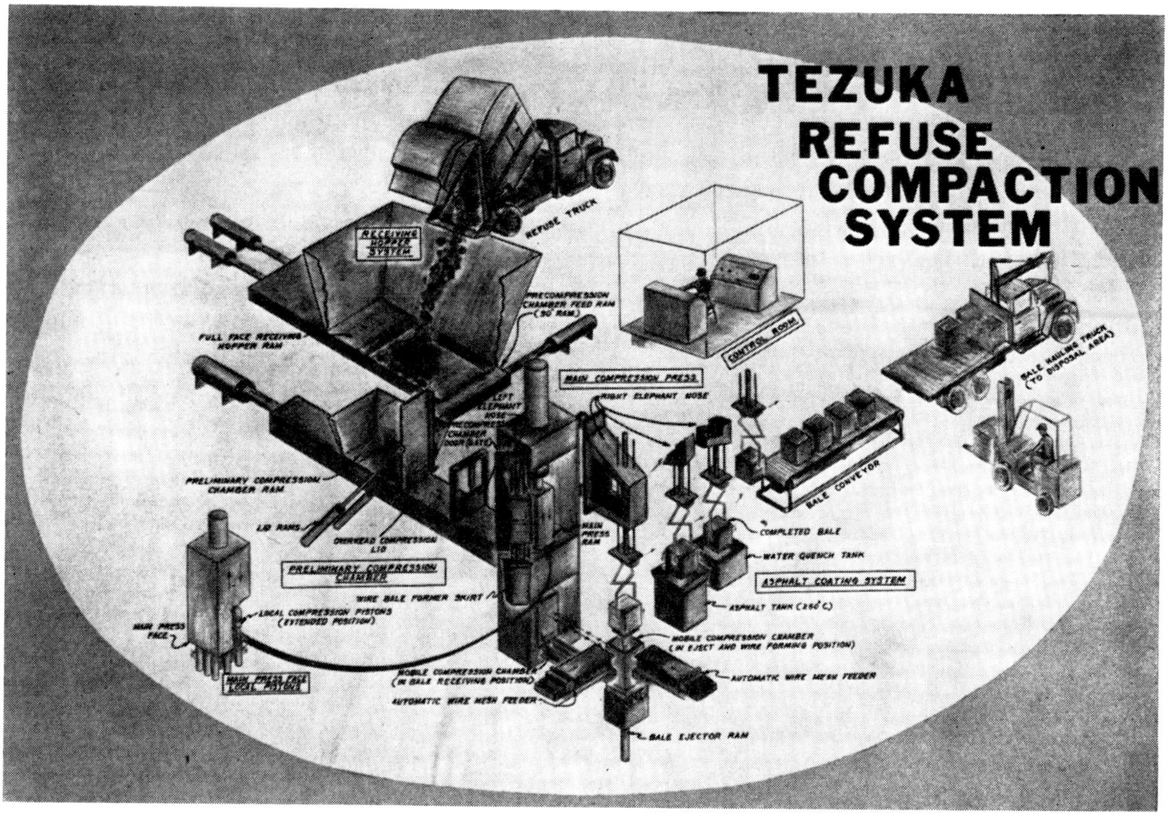

Figure 5 Refuse compaction. Tezuka Refuse Compaction System. Credit: Waste Management Compaction Systems, Inc.

is about 20 pounds per day per patient with a very high fraction of garbage and paper and plastic throw-away products.[49] Provision must also be made to handle pathological wastes, therefore combustion temperatures should be in excess of 2000°F and adequate residence time for the gases at 1500°F should be provided.

Compaction

The reduction of waste volume is receiving considerable attention in an effort to reduce collection costs; compaction is one of the favored methods to achieve this reduction. High pressure compaction has been developed by Tezuka Kosan of Japan to provide a high density product suitable as an essentially inert fill or even as a building material.[50] Using this product as a base covered with a minimal earth cover, the Japanese have reclaimed land from tidal areas having a water depth of 10 feet.[51]

The Japanese process shown in Fig. 5 collects refuse and subjects it to three stages of compression with the final main press exerting 3000 psi on the refuse. The resulting bale is usually wrapped in chicken wire and coated with asphalt for ease of handling and to prevent crumbling and/or leakage. The bales have a density of between 1900 and 2300 pounds per cubic yard and result in a volume reduction of about 90%. This compares to densities of about 1200 to 1500 pounds per cubic yard achieved in lower pressure compaction. The product bale is inert and such bales have survived exposure in Tokyo Bay for three years to date without visible signs of degradation.

Studies by the Japanese indicate that the high pressure squeezing and resultant elevated temperatures decrease the BOD from 6,000 in the raw refuse to 200 in the product. Similarly the COD of 8,000 (which compares to about 14,000 in U.S. refuse) was reduced to about 150. Inspection of the interior of the bale shows a homogenous, plastic like mass. The bale will not support vermin, rodents or insects and is essentially odor free even if it is not protected by an asphalt coating. The only other product of the compaction is waste liquor which amounts to 5% of the feed in Japan and will probably be about 3% with U.S. refuse because of its lower moisture content.

There are a number of these facilities operational in Japan, and the first U.S. facility started up in late 1970. A large facility with a single train can handle 150 tons per 8-hour day and requires an investment of about $ 2 iillion. Its expected operating cost will be about $ 5.00 per ton including amortization if operated on a one-shift basis and about $ 4.00 on as three shift basis (amortization over 20 years). To this should be added the cost of operating the land reclamation site which can amount of $ 0.40 to $ 0.60 per ton for this type of operation.[54]

This system appears to be of particular interest where long hauls are required to dispose of waste. In addition it is attractive where it is desirable to reclaim tidal lands or obtain prompt use of landfill areas because the highly compacted fill eliminates the usual hazards and nuisances associated with the usual landfill operations.

Composting

Composting is the biochemical degradation of organic material (see: COMPOSTING) to yield a sanitary soil supplement. Anaerobic composting has been practiced in Asia and is the process by which sanitary landfills degrade refuse. Modern composting has been practiced in Europe for over 50 years using aerobic microorganisms. The practice has been attempted on a commercial scale in the United States but with very limited success. The unattractiveness of composting is primarily the result of American agriculture's orientation to chemical fertilizers. This has made the large scale marketing of compost difficult. Composting while not economical now could prove more attractive as public opinion moves toward an attitude which requires that wastes returned to the earth be compatible with the environment.

Composting has one overriding advantage; it is the only process which provides for recycling of organic residue. The process can handle garbage and other organic refuse (but not plastics) as well as sewage sludge and industrial waste from certain operations such as saw and paper mills. The primary disadvantages are cost, the need for fairly large areas for final outdoor curing, a slight odor associated with a composting plant, and lack of a market for the product.

Composting is practiced in several forms. Traditionally rows of refuse, shredded or ground, four to six feet high, are exposed to the environment and turned regularly. This is known as the "windrow" method and is still used. Complete composting can be achieved in 10 to 14 days if seeding with compost is employed but often takes four to six weeks.

Mechanically aided aerobic composting is carried out in a number of processes. Among the more prevalent are the Dano process, the Earp-Thomas Multi-Bactor comopst tower, and a number of cell-type stated tower systems. Decomposition takes place under aerobic conditions with the micro-organisms supplied by seed compost. Typical operating temperatures reach 130 to 140°F. Material is held in the unit from one to six days depending on the process. This is usually followed by an open air curing. A new plant a Schweinfurt, Germany, using the Caspari-Brikollare process, produces briquettes in which form the compost is stored until it is to be used.

Raw materials suitable for aerobic composting will be finely ground (coarse for windrowing) and have a maximum carbon to nitrogen ratio of 50 to 1. It is important that good dispersion of air be achieved and that the moisture level be maintained between 50 and 60 per cent. Recycling of between 1 and 10 per cent of active compost enhances the composting process by minimizing the time required for sufficient microorganisms to develop. The yield from composting is about one volume for every three volumes of feed; the weight yield is between 30 and 40 per cent.

In Europe compost is utilized as an organic soil conditioner in luxury agriculture such as vineyards, hotbed vegetable farming, flower and seed production, fruit farming and the improvement of recreational land. It has found little or no use in basic agriculture, nor is it used for erosion control. In Germany less than one per cent of the domestic refuse is composted and in Holland only about 15 per cent is so treated. There appears to be no increase in composting operations because of a lack of additional marketing opportunities.

The investment for a composting facility varies widely depending on size and process. Investments of between $ 1,500 and $ 6,000 per daily ton of capacity have been reported; no valid average figures can be reported because there are so few operational plants. Operating costs, including the cost of capital, will vary between $ 6 and $ 8 per ton of refuse (assuming labor at $ 12,000 per man year) on a U.S. basis. European investments appear to be as low as $ 1,000 per daily ton and operating costs in 1964 were between $3.20 and $6.60 per ton, with an average of $4.50 per ton of refuse.[53] Part of this cost was recovered by sale of salvage (16¢ per ton) and compost; the average recovery amounted to $1.17 per ton resulting in a net average cost of $3.38.[54]

The economics of waste disposal

Economic considerations have and will, of course, continue to play a significant role in the choice of waste management method. Table 21 summarizes both operating and investment costs for principal waste processing methods. It must be realized that values can vary widely depending on local conditions. Technological improvements can also alter the price structure. Most important, however, is the fact that long range implications of waste management, environment and resource considerations cannot readily be reduced to a quantitative cost, and these factors should weigh heavily in the choice of a waste processing method.

TABLE 21

Waste Processing Investment and Operating Costs

(1970 basis)

Method	Investment $/ton Daily capacity	Operating costs %/ton
Sanitary landfill	0.50–1.00[a]	1.00–5.00
Landfill or dump	0.50–0.75	0.50–2.00
Central incineration		
No waste heat recovery	10,000	5–8
Waste heat recovery	12,000	5.50–7.50[b]
Onsite incineration	2000–6000[c]	10–20
Composting	1000–6000	6–8[e]
Compacting, high pressure	4,500[d]	4–5

[a] Per ton of capacity, based on 2–20 foot lifts.

[b] No allowance for value of steam which will be between $1.00 and $2.00 per ton.

[c] Actual capacity of home incinerator is about 100 lbs of refuse; an apartment house incinerator will vary between 1/4 and 8 tons per day based on 8 hours of operation. No costs are developed for industrial units because of special design requirements.

[d] 3 shift operation 450 tons/day.

[e] No credit taken for sale of compost.

Reclamation, reuse and conversion

There has been a salvage industry as long as there has been waste. The intensity of this effort has been limited however and varies very much with location as well as the economic situation at the time. No concentrated effort, except perhaps in wartime, has ever been made to recover and recycle a high percentage of waste. True landfill can be considered a form of reclamation but it is a very low grade use of refuse. This is not to say that the salvage industry is small: the latest figures indicate that it has sales in excess of $7 billion per year.[55]

In certain areas such as paper with 11 million tons recycled in 1968 (25% of consumption), aluminum with 700,000 tons recycled (30% of production), copper with 1.5 million tons recovered, and iron and steel salvage is an important source of raw material.[55] Unit salvage values are summarized in Table 22. Automobiles are regularly recovered; it is economically a break-even operation once the autobody is delivered to the salvage operation.

TABLE 22

Value of Reclaimed Waste
Average 1970 Prices

	$/ton
Paper	
Newsprint	10–15
Other sorted	up to 50[a]
Glass	10–15
Cotton	4–20
Metals	
Iron and steel	20–60
Copper and brass	270–800
Aluminum	160–250
Lead	70–160

[a] Depends on type, color, etc.

Mixed waste such as municipal refuse has not been a major source of this salvage and contains large amounts of additional metals as well as other potentially valuable materials. However it is only recently that essentially total recovery has been considered. That is, "waste" is being looked at as a potential "natural resource." As an example, one company, Industrial Services of America, is operating a separations plant for industrial solid wastes where odor is no problem and manual separation is feasible.[56]

One of the major problems of recycle and reuse has been separation. Most ferrous metal is easily separated by magnetic means, but other waste separation requires hand picking, which is very costly. This method can also create public nuisance in that the odors from such operations can be

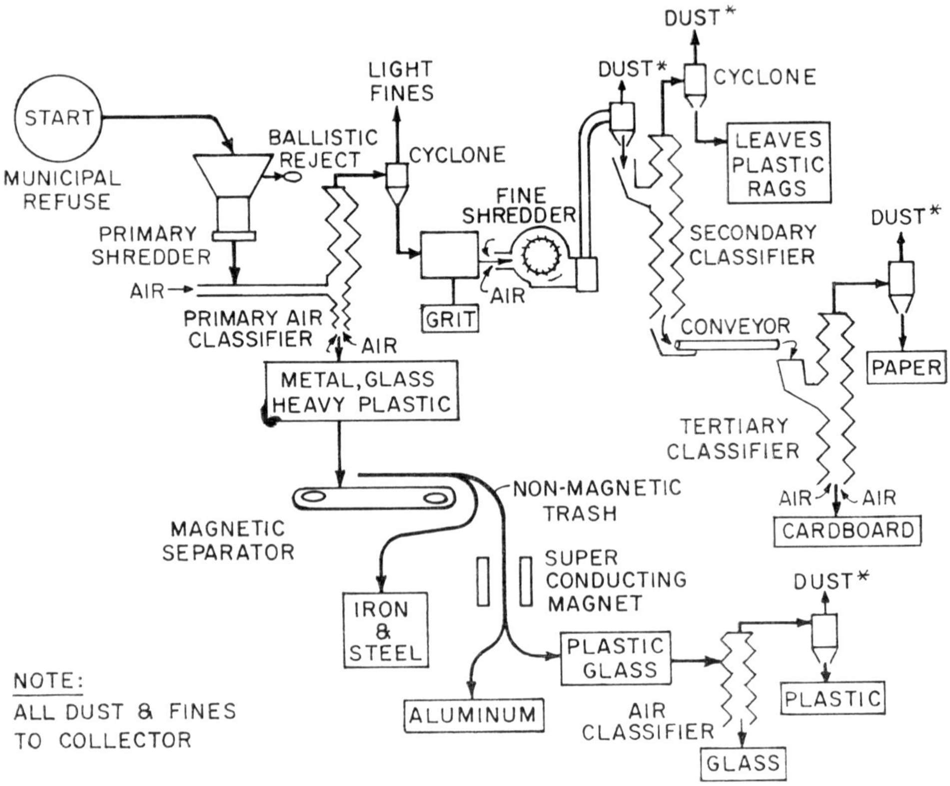

Figure 6 Proposed waste separation system, using SRI Air Classifier.

significant and have resulted in the closure of several such operations. Recent studies by the Bureau of Mines, Stanford Research Institute and others are beginning to provide potential alternates for waste separation.

The work at SRI is particularly interesting because it is intended as a total separation process; waste is shredded and then classified into components by density in an air classifier.[57] While the process has not achieved separation of materials with densities that are close togehter (and much refuse is in this situation) the concept as illustrated in Fig. 6 is in the right direction and should lead to promising results.

Similar work has been piloted using a hydrapulper to separate organic, glass and cellulose from metal. Hydrapulping as developed by Black Clawson Company costs about \$ 6 per ton of feed.[58] A demonstration unit (Fig. 7) with a 70 T/day capacity is operating in Franklin, Ohio using the Hydraposal system (of which Hydrapulping is a part). Installed costs are about \$ 14,000/daily ton for a 300 T/day plant with net operating cost of about \$ 4.50/ton. About 400 pounds/ton of Fiber are reclaimed along with glass cullet, metals and steam.[71] A material balance is shown in Fig. 8. However a dry separation such as that proposed by SRI may have significant advantages over wet separation because the removal of water from cellulose is a very costly operation. The fiber can be used as feed to a kraft paper mill.

The aluminum industry, for example, as shown in Fig. 9, has suggested a process for taking municipal refuse and separatiog it into aluminum, other nonferrous metals, ferrous metals, glass and other waste. Dr. James Etzel of Purdue has piloted a process, based on hydrapulping, which handles sewage and solid waste and yields metals and a slurry containing fine particles of glass and organics which can be further treated or used as a soil supple ment.[60] Such techniques require considerable additional development and refining but will be one of the key solid waste management tools of the future.

Chemical processing of solid waste will become increasingly important in the next decade. Some of these techniques will depend heavily on the separation techniques to provide suitable feed but others will handle mixed refuse.

The most promising process for handling mixed

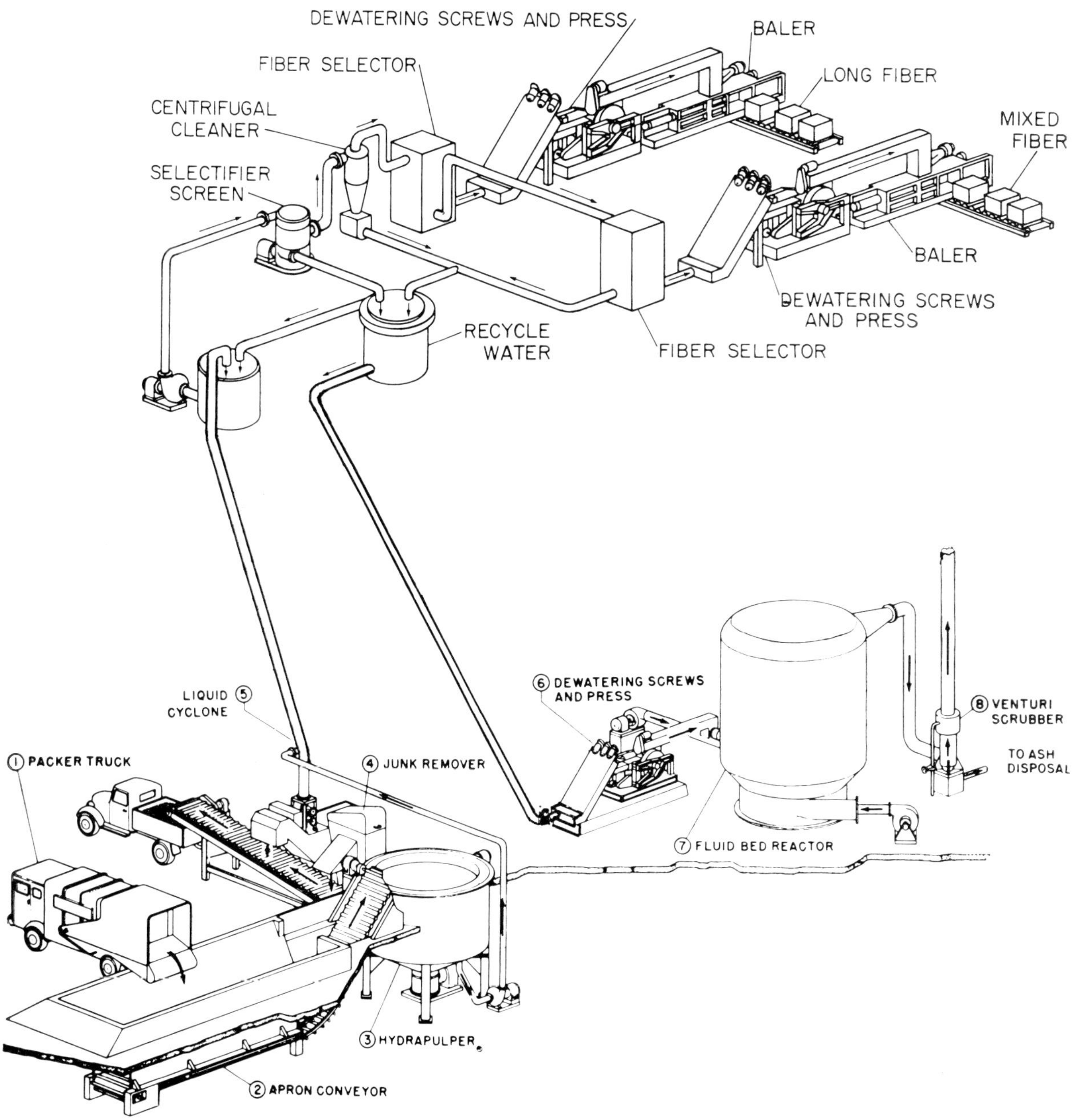

Figure 7 Black Clawson Hydraposal-Fiberclaim. Waste recycling system. Credit: Black Clawson Co.

waste is pyrolysis (also known as destructive distillation). Pyrolysis is the thermochemical degradation of complex organic molecules into low molecular weight molecules. The process is carried out in the absence of added oxygen (or with very limited oxygen), and with the addition of heat, at temperatures in the range of 900–1700°F. The process will produce a fuel gas, oil and carbon. A study in San Diego indicates that about 50% of the refuse is susceptible to pyrolysis.[61] This study yielded low BTU gas, char and oxygenated hydrocarbons. Cities Services has evaluated a similar process and believes it to be potentially economic in plants with capacities of 5000 tons per day which are subsidized at the rate of $ 2 to $ 8/ton.[62] Pilot plant studies by the Bureau of Mines indicate that both industrial and municipal refuse yield large amounts of gas and solid, as shown in Table 23.

A process using pyrolysis has been piloted for tires, which present some particularly difficult disposal problems in incineration and landfill.[64] Similar processes can be effective in recovering chemicals from plastics but have not been developed because of separation problems.[65]

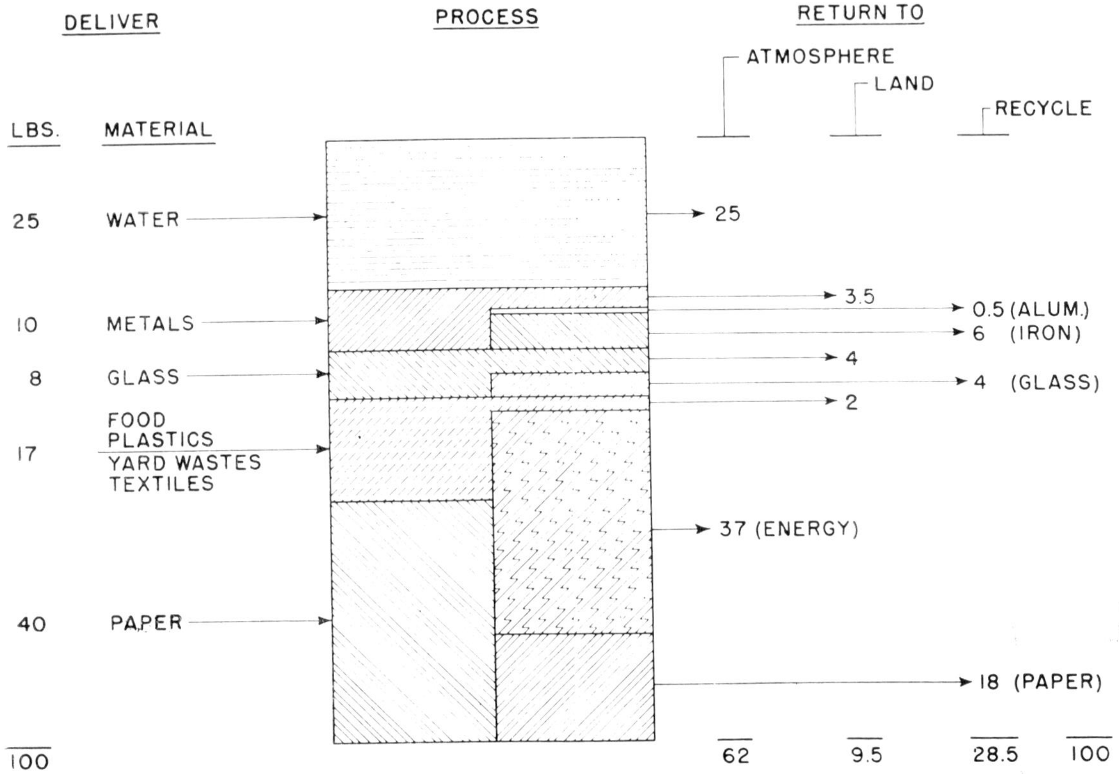

Figure 8 Black Clawson Hydraposal-Fiberclaim. Material balance. Credit: Black Clawson Co.

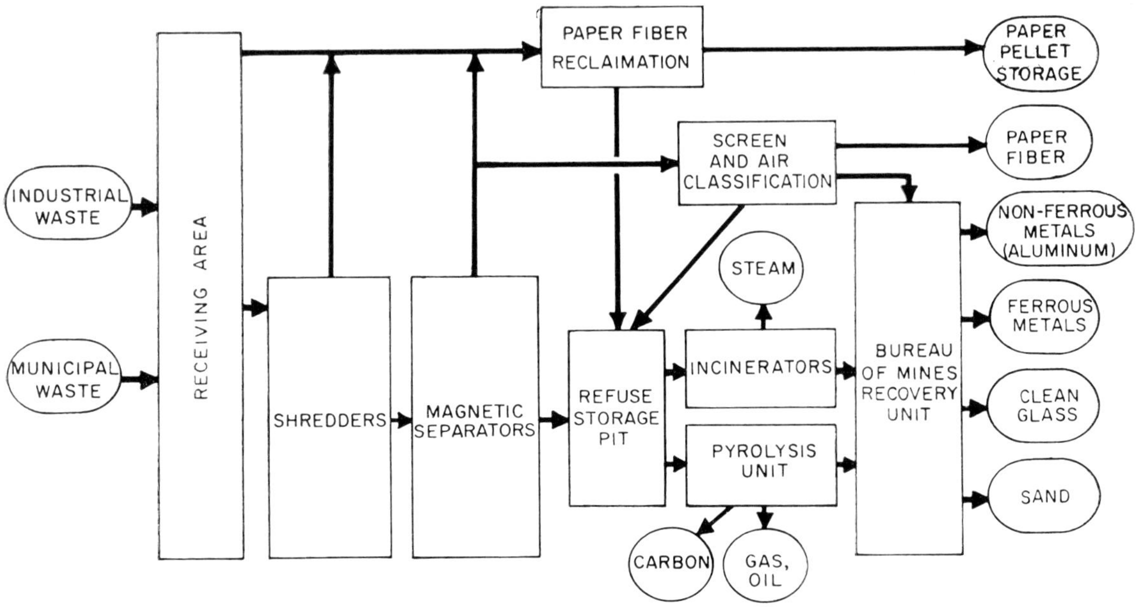

Figure 9 Refuse recycling plant. (Aluminum Association of America).

TABLE 23

Pyrolysis of Refuse Typical Yields[63]

	Source	
	Raw material	Heil milled industrial
Feed, million btu		
Available per ton dry basis	17.09	11.29
Pyrolysis temp	900°C	900°C
Yield, wt % of refuse		
Residue	7.7	38.8
Gas	39.5	29.4
Tar	0.2	0.2
Light oil	–	0.6
Free NH_3	0.3	0.04
Sour liquor	47.8	21.8
Yield per ton of refuse		
Gas, cu ft	17,741	12,318
$(NH_3)SO_4$ lbs	25.1	21.7
Heating Value		
Gas, btu/ft^3	447	498
Residue, btu/lb	5,260	2,180

TABLE 24

Reduction of Garbage by Hydro-oxynation[66]

1 BBL of oil/ton of garbage (50% cellulose)

Conditions	
1500 psig initial pressure, 5000 psig operating pressure 350°C.	
Addition of $CO + H_2O$	
Yield, wt %	
Oil	40%
Residue	10%
H_2O	27–36%
CO	15–20%
Other	2%
Oil analysis: C 83%, H_2 7.8%, O_2 7.8%, N_2 1.9%, S 0.13%	

The U.S. Bureau of Mines has also piloted a high pressure process where refuse free of glass and metal is reacted in the presence of water and carbon monoxide (hydro-oxynation) at 1400 psig and 500 to 700°F to yield oil, gas and carbon.[66] This process appears to have very favorable oil yields. Process variable studies have shown that conversions as high as 90% can be obtained with a 40% yield of oil; a typical yield is shown in Table 24. These yields indicate a potential oil production rate of 2×10^8 tons annually as compared to U.S. crude production of 5×10^8 tons per year. Similar studies using hydrogen showed lower yields and conversions. The use of CO does present some significant operating problems as well as economic debits.

Several systems using pyrolysis are ready for commercialization. Hercules in Delaware is planning a unit to pyrolyze industrial waste. The details of the process are not available at this time. The Monsanto Envirochem is ready to build and guarantee pyrolysis units using the LANDGARD process (Fig. 10). This process emphasizes waste reduction (with recovery of ferrous metal), rather than recovery of variable by-products; it has been piloted in a 35 T/day semi-works facility. The process reduces the solid waste, typically, by 90%; a typical stack gas analysis is given in Table 25 and indicates the very low particulate matter in the effluent. The investment appears to be lower than incineration, averaging $ 9,000 per daily ton for a 1,000 to per day plant. Operating costs are about $ 7 per ton with no credits for the residue or cost of capital included.[72]

In contrast to the Monsanto-Landgard system, a pyrolysis process emphasizing recovery of valuable products has been developed by Garrett Research and Development Co. This process is

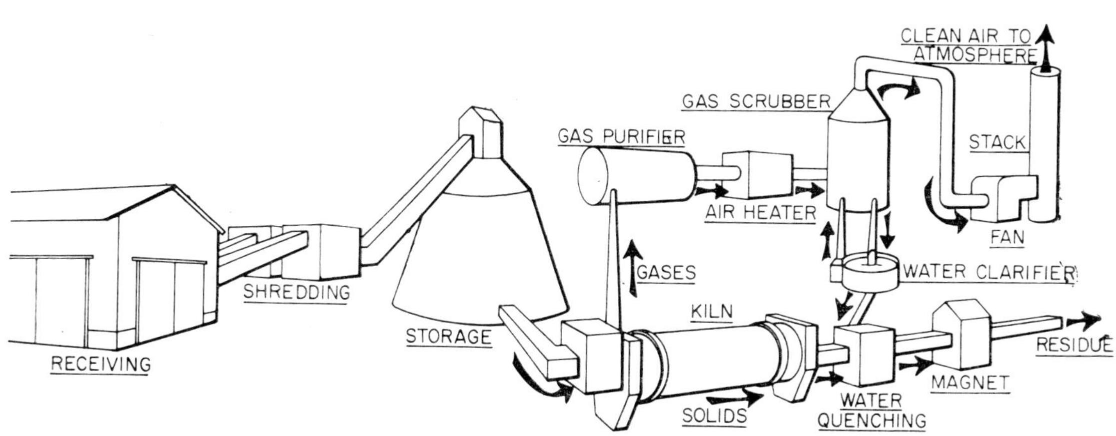

Figure 10 Monsanto-Landgard waste pyrolysis process. Credit: Monsanto Enviro-Chem Systems Inc.

just being piloted at a 4 T/day level. It consists of shredding, air classification, pyrolysis and pyrolysis product separation steps. Product recovery is similar to that obtained by the Bureau of Mines. Based on bench scale experiments investments in the order of \$ 6,000 per daily ton at a level of 2,000 T/day are anticipated. Garret believes that operating costs will be about \$ 4.50 per ton and product values \$ 5.00 per ton.[73]

TABLE 25

Landgard Pyrolysis System

Typical Stack Gas Analysis[72]

Component	Average value
N_2	40%
O_2	3%
CO_2	7%
H_2O vapor	50%
Combustibles	none
NO_x	50 ppm (vol.)
SO_2	100 ppm (vol.)
Chlorides	10 ppm (vol.)
Particulates	0.06 grains per SCF dry gas corrected to 12% CO_2

One may ask why solid refuse should be subjected to complex processes such as hydropulping or pyrolysis. Where actual wastes such as paper or cellulose fiber can be recovered, hydropulping is certainly attractive. On the other hand where waste is utilized as fuel the possibility of using it directly rather than converting it to fuel oil (Bureau of Mines process) must be considered. Pyrolysis, hydro-oxydation or similar processes should be competitive with direct fire waste-heat recovery incineration where clean fuel is desired or fuel is to be used at locations other than where the refuse is available. The recovery of chemical raw materials from refuse using these processes may also become attractive as the process technology is refined.

The recovery of valuable products from incinerator effluent is not, however, being neglected. The Bureau of Mines has demonstrated technology to recover metals and glass from incinerator residues using high-intensity magnetic sorting as well as chemical techniques at a cost of about \$ 4 per ton.

ITT Research Institute has shown that refuse may be selectively separated into a ceramic and metal slag when incinerated at temperatures between 2800 and 3200°F.[67] While technically feasible, this separation incurs some severe economic costs; it does yield some interesting products including tiles, pipe and structural shapes. In addition to the newer methods being studied, the use of incinerator fly ash and even slag for road fill and concrete aggregate appears to be attractive in some situations.

Specific processes to convert refuse fractions are also receiving attention and will be utilized in limited situations where the economics are attractive. An engineering study has shown that the organic portion of raw refuse can be economically converted to sugar by acid hydrolysis.[68] This can then be a raw material for alcohol production. One of the limitations to this and similar processes is a limited market for the product and competition with other sugar sources as well as other sources of alcohol; for example conversion of three per cent of the refuse to ethanol would saturate the normal market.

Though fuel use of alcohols is possible, only limited amounts of refuse can be converted using these processes unless major shifts occur in our economy. Process development to convert citrus waste to citric acid is being conducted at the University of Florida. Other conversion studies are being carried out, but all face the difficulties inherent in processing a heterogeneous, complex, often variable, mixture.

Of more than passing interest are studies being carried out at the University of Maryland to obtain a protein concentrate for human and animal consumption from food processing wastes. Similar studies at Louisiana State University, in the pilot plant stage, has shown that agricultural cellulose wastes can be broken down by selected microorganisms to yield a low cost, high protein food. Yeast can also be produced from cellulose wastes.[69]

The use of waste as a "resource material" is still in its infancy. It is gathering momentum quickly and the next decade should see significant changes in waste management.

Some special problems

Radioactive solid wastes create special problems and are discussed elsewhere in a section on Radioactive Wastes.

Industrial wastes, as mentioned previously, has been dumped as a general rule. Because of the highly specialized nature of industrial wastes, it is impossible to discuss them in a general way. Total recycling of many industrial wastes will become a more frequent practice. Mine tailings will find their way into construction material, fill or may be recycled into the mines. Slag from steel mills should become less of a problem as different processes are used but will still remain a significant contributor; slag can be used in special concrete and efforts in this area will continue.

Sewage sludge (see: SEWAGE) presents some special problems. To date it has generally been dumped. Composting (see: COMPOSTING) should be a major process for handling sewage sludges. Several attempts to sell composted sludge have only been marginal because of the lack of markets; however this should change in the future. Material that, for economic reasons, cannot be composted, can be burned to recover waste heat in specially designed incinerators.

More of a problem will be spent solids from water treatment facilities. These solids are high in carbonates and often have a foul odor due to entrained organic material. They are now being filled and this practice will have to continue unless chemical recovery methods which produce a useful product are found; this is not likely as carbonates are in oversupply.

The management of solid wastes will undergo dramatic changes in the next decade. From a "cottage industry" it will emerge as a major process industry recovering many vital materials and converting others into valuable products. Landfill and incineration will continue to play a role with the former decreasing in importance and the latter coming into greater prominence. However new process technology, only some of which is now in development, will play an ever more important role in total solid waste resource management.

REFERENCES

1. Eliassen, R., "Solid Waste Management", Off. of Science and Tech. Exec. Off. of the Pres., Washington (1969).
2. Eliassen, R., *C. and E. News*, **48**, No. 13, p. 30 (Mar. 30, 1970).
3. Amer. Pub. Works Ass'n Inst. for Solid Wastes, "Municipal Refuse Disposal", Public Administration Service, Chicago, p. 11–20 (1970).
4. Kaiser, E. R. *et al.*, *Proc. 1968 Nat'l Incinerator Conf.*, ASME, New York, p. 142 (1968).
5. Kaiser, E. R., *Proc. 1966 Nat'l Incinerator Conf.*, ASME, New York (1966).
6. Kaiser, E. R. *et al.*, *Proc. 1968 Nat'l Incinerator Conf.*, ASME, New York, p. 142 (1968).
7. Amer. Pub. Works Ass'n, Comm. on Solid Wastes, "Municipal Refuse Disposal," 2nd Ed., Public Administration Service, Chicago, p. 47 (1966).
8. Ibid. p. 52.
9. Golueke, C. G., and P. H. McGauhey, "1st Ann. Report, Comprehensive Studies of Solid Waste Management," U.S. Dept. HEW, PHS Report No. 2039, Washington, p. 20 (1970).
10. Ibid., p. 34.
11. Amer. Pub. Works Ass'n, Comm. on Solid Wastes, Op. Cit. p. 47.
12. Carruth, D. E., and A. J. Klee, "Analysis of Solid Waste Composition—Sample Size," U.S. Dept. HEW, PHS Report #SW-19ts, Washington (1969).
13. Amer. Pub. Works Ass'n Inst. for Solid Wastes, Op. Cit. p. 389.
14. De Marco, J. *et al.*, "Incinerator Guidelines—1969," U.S. Dept. HEW, PHS Pub. No. 2012, Washington, p. 6 (1969).
15. Eliassen, R., Op. Cit. p. 21.
16. Kaiser, E. R., Private Communication.
17. Ibid., p. 8.
18. Golueke, C. G., and P. H. McGauhey, Op. Cit. p. 221.
19. Ibid., p. 222.
20. Eliassen, R., Op. Cit. p. 32–33.
21. Ibid., p. 33.
22. Ibid., p. 35.
23. Ibid., p. 59.
24. Amer. Pub. Works Ass'n, Inst. for Solid Wastes, Op. Cit. p. 94–133.
25. Reinhardt, J. J., and G. Rohlich, "City of Madison Pilot Plant Demonstration Project—An Interim Report," U.S. Dept. HEW, PHS, Cincinnati (1968).
26. Corey, R. C. (ed.), "*Principles and Practices of Incineration*, Wiley, New York, p. 9–33 (1969).
27. Hangebrauck, R. P. *et al.*, *J. APCA* **14** (No. 7,) p. 267 (1964).
28. Hovey, H. H. *et al.*, *J. APCA* **16** (No. 7), p. 362 (1966).
29. Jacobs, M. B. *et al.*, Proc. *APCA*, 51st Meeting, Paper No. 2464, Philadelphia (1958).
30. Jacobs, M. B. *et al.*, "The Smog Problem in Los Angeles County," Western Oil and Gas Ass'n, Los Angeles (1954).
31. Kondo, K. *et al.*, *Trans. SHASF Japan* **7**, p. 95 (1969).
32. Stenburg, R. L. *et al.*, *J. APCA* **12**, No. 2, **83** (1962).
33. Flood, L. P., *J. APCA* **9**, No. 1, 63 (1959).
34. Lenehan, J. W., *J. APCA* **12**, No. 9, 414 (1962).
35. Stenburg, R. L. *et al.*, *J. APCA* **8**, No. 2, 297 (1959).
36. Kenahan, C. B. *et al.*, "Comp. and Characteristics of Municipal Incinerator Residues," U.S. Bureau of Mines, R. I. 7204, Washington (1968).
37. Kenahan, C. B. *et al.*, Municipal Incineration, M. H. Detrich Co., Chicago (1967).
38. Eliassen, R., Op. Cit. p. 60.
39. Eliassen, R., *Proc. 1966 Nat'l Incinerator Conf.* Appendix II, ASME, New York (1966).
40. Amer. Pub. Works Ass'n, Comm. on Solid Wastes, Op. Cit., p. 167 (1970).
41. Corey, R. C., Op. Cit., p. 163–209 (1969).
42. De Marco, J. *et al.*, Op. Cit.
43. Fife, J. A., and R. M. Boyer, Jr., *Proc. 1966 Nat'l Incinerator Conf.*, ASME, New York, p. 89 (1966).
44. Parisius, P. S. *et al.*, 68th Nat'l AIChE Mtg. Paper No. 35c, Houston (1971).

45. Day and Zimmermann, "Special Studies for Incinerators," U.S. Dept. HEW, PHS Pub. 1748, Cincinnati (1968).
46. Sauter, W. E., Proc. SPE Retech, Cherry Hill, N.J., Oct. 27, 1970.
47. Overall, J. E., *Amer. Paper Industry* **48**, p. 51 (1966).
48. Meile, R. L., *Proc. 1966 Nat'l Incinerator Conf.*, ASME, p. 202 (1966).
49. Amer. Pub. Works Ass'n, Inst. for Solid Wastes, Op. Cit. p. 231.
50. Rogers, C. A., Public Works, June 1969, p. 85 (1969).
51. Caver, P. T., *Proc. 3rd ANERAC Conf.*, Univ. of R. I., Kingston, R. I., (1970).
52. Liner, E. H., Waste Management and Compaction Systems, Private Communication.
53. Hart, S. A., "Solid Waste Management-Composting," U.S. Dept. HEW, PHS Pub. No. 1826, Cinnati (1968).
54. Kupchick, G. J., *Bull. World Health Organ.* **34**, No. 5, p. 798 (1966).
55. Nat'l Ass'n of Secondary Materials Industries.
56. Schroering, J. B., *Proc. 3rd ANERAC*, Univ. of R. I., Kingston, R. I., p. 176 (1970).
57. Boettcher, R. A., *Proc. 2nd Mineral Waste Utilization Symp.*, IITRI, Chicago, p. 271 (1970).
58. Abrahams, J. H. Jr., *Proc. 2nd Mineral Waste Utilization Symp.*, IITRI, Chicago, p. 363 (1970).
59. Aluminum Ass'n, News Release, Jan. 25, 1971.
60. Aluminum Ass'n, *C. and E. News* **48**, No. 12, p. 44 (1970).
61. Hoffman, D. A., ASME Ann. Mtg., 1969, ASME Pub. No. 69-WA/PID-10, (1969).
62. Hoffman, D. A., *C. and E. News* **48**, No. 10, p. 38 (1970).
63. Sanner, W. S. *et al.*, "Conversion of Municipal and Industrial Refuse by Pyrolysis," U.S. Bureau of Mines, R. I. 7428, Washington (1970).
64. Wolfson, D. E. *et al.*, "Destructive Distillation of Scrap Tires," U.S. Bureau of Mines, R. I. 7302, Washington (1969).
65. Lederman, P. B., Proc. SPE Retech., Cherry Hill, N. J., Oct. 27, 1970.
66. Appell, H. R. *et al.*, *Proc. 3rd ANERAC*, Univ. of R. I., Kingston, R. I., p. 225 (1970).
67. Boetz, S. A., and A. G. Pincus, *Proc. 2nd Mineral Waste Utilization Symp.*, IITRI, Chicago, p. 327 (1970).
68. Fagen, R. D. *et al.*, *Proc. 3rd ANERAC*, Univ. of R.I., Kingsten, R. I., p. 124 (1970).
69. Meller, F. H., "Conversion of Organic Solid Waste into Yeast," U.S. Dept. HEW, PHS Pub. No. 1909, Rockville, Md. (1969).
70. Stanczyk, M. H., "Continuous Beneficiation of Metals—in Incinerator Refuse." *Proc. 2nd Mineral Waste Utilization Symp.*, IITRI, Chicago, p. 256 (1970).
71. Herbert, W., Private Communication from Black Clawson Co.
72. Buss, T. F., Private Communication, Monsanto Enviro-Chem. Systems Inc.
73. Mallan, G. M., "A Total Recycling Process ...," I and EC Div., 161st ACS Mtg., Los Angeles (1971).

SELECTED BIBLIOGRAPHY

General

1. Amer. Public Works Ass'n, Inst. for Solids Wastes, "Municipal Refuse Disposal," Public Admin, Service, Chicago (1970).
2. Eliassen, R., "Solid Waste Management, Off. of Science and Tech., Exec. Off. of the Pres., Washington (1969).
3. Frey, D. N. (Chairman), "Policies for Solid Waste Management," U.S. Dept. NEW, PHS Pub. 2018 (1970).
4. Train, R. E. (Chairman), "Environmental Quality—1st Ann. Report of the Council on Environmental Quality," Washington, Chapter VI (1970).
5. Hanks, T. G., "Solid Wastes/Disease Relationships," U.S. Dept. HEW, PHS Pub. No. 999-UIH-6, Cincinnati (1967).
6. Cooke, L. M. (Chairman), "Cleaning Our Environment, The Chemical Basis for Action," ACS, Washington (1969).
7. Guinstead, R. R., *Environment* **12**, 10, p. 2 (1970).

Sources

1. Golueke, C. G., and P. H. McGauhey, "Comprehensive Studies of Solid Waste Management," U.S. Dept. HEW, PHS Report No. 2039, Washington (1970).
2. Amer. Public Works Ass'n, Inst. for Solid Wastes, Op. Cit.
3. Copp, W. R. *et al.*, "Technical-Economic Study of Solid Waste Disposal Needs and Practices," Vol. I—Municipal Inventory, Vol. II, Industrial Inventory, U.S. Dept. HEW, PHS Pub. No. 1886, Washington (1969).
4. Darnay, A. J. Jr., and W. E. Franklin, "The Role of Packaging in Solid Waste Management," U.S. Dept. HEW, PHS Pub. No. 1885, Washington (1969).
5. Black, R. J. *et al.*, "The National Solid Wastes Survey—An Interim Report," U.S. Dept. HEW, Washington (1968).

Disposal methods, general

1. Amer. Pub. Works Ass'n, Inst. for Solid Wastes, Op. Cit.
2. Eliassen, R., Op. Cit.
3. Golueke, C. G., and P. H. McGauhey, Op. Cit.
4. Siebel, J. E. *et al.*, "Technical-Economic Study of Solid Waste Disposal Needs and Practices," Vol. IV, Technical-Economic Overview, U.S. Dept. HEW, PHS Pub. No. 1886, Washington (1969).
5. Jensen, M. E., "Observations of Cont. European Solid Waste Management Practices," U.S. Dept. HEW, PHS Pub No. 1880, Washington (1969).

Disposal methods, sanitary landfill

1. Sorg, T. J., and H. L. Hickman, Jr., "Sanitary Landfill Facts," U.S. Dept. HEW, HPS Pub. 1792, 2nd Ed., Washington (1970).
2. Steiner, R. L., and R. Kantz, "Sanitary Landfill; a Bibliography," U.S. Dept. HEW, PHS Pub. N. 1819, Washington (1968).
3. Lambia, J. A. (Proj. Dir.), "Development of Construction and Use Criteria for Sanitary Landfills," U.S. Dept. HEW, PHS Grant Do. 1-UI-00046, Cincinnati (1969).

Disposal methods, incineration

1. Corey, T. C. (Ed.), *Principles and Practices of Incineration*, Wiley, New York (1970).
2. DeMarco, J. *et al.*, "Incinerator Guidelines—1969", U.S. Dept. HEW, PHS Pub. 2012, Washington (1969).

3. Day and Zimmerman, "Special Studies for Incinerators—for the Government District of Columbia," U.S. Dept. HEW, PHS Pub. 1748, Cincinnati (1968).
4. Combustion Power Co., "Combustion Power Unit-400," U.S. Dept. HEW, PHS, Rockville, Md. (1969).

Disposal methods, composting

1. Hart, S. A., "Solid Waste Management/Composting," U.S. Dept. HEW, PHS Pub. No. 1826 (1968).
2. Gainesville Municipal Waste Conversion Authority, "Gainesville Compost Plant," U.S. Dept. HEW, Cincinnati (1969).
3. Shell, G. L., and J. L. Boyd, "Composting Dewatered Sewage Sludge," U.S. Dept. HEW, PHS, Washington (1969).

Reutilization

1. Drobny, N. L., H. E. Hull, and R. F. Testin, "Recovery and Utilization of Municipal Waste," U.S. EAC, Solid Waste Management Office, Washington (1971).

Bibliographies

1. Golueke, C. G., "Solid Waste Management Abstracts and Excerpts from the Literature, Vol. 1 and 2," U.S. Dept. HEW, PHS Pub. 2038, Washington (1970).
2. Black, R. J. *et al.*, "Refuse Collection and Disposal 1954–61," U.S. Dept. HEW, PHS Pub. No. 91, Sup. B, C, D, E and F, Washington D. C. (1966).

PETER B. LEDERMAN
Environmental Protection Agency

MARINE SPILLAGE — SOURCES AND HAZARDS

1.0 Introduction

1.1 Scope and limitations

Pollution of the world's navigable waterways as the result of the operation of commercial and naval vessels may occur from normal service or from casualties such as collision and stranding. Fuel oil as well as liquid cargo systems are involved in both modes of pollution and will be considered in this chapter. Sewage and other waste disposal from shipping is of comparatively small magnitude, compared to waterway pollution from shoreside disposal systems, and will not be discussed. The emphasis will be on the definition and description of the pollution problem and its sources and on means of prevention. Consideration of retention, vollection and cleanup of spills is considered elsewhere in this text.

1.2 Problem classification

1.2.1 Normal operations—The primary source of pollution of the world's waterways is the intentional discharge of oily ballast water into the sea during normal ship operations. For most services, including virtually all seagoing operations, ballast water must be taken on for a portion of the voyage in order to obtain sufficient draft and trim for propeller immersion and an acceptable condition of seaworthiness, and to satisfy operational and regulatory requirements for standards of stability. In the event that insufficient volume is available within the hull to satisfy ballasting requirements with tanks designated only for ballast, it becomes necessary to ballast empty fuel oil tanks, or liquid cargo tanks in the case of tank vessels. This procedure results in the creation of large quantities of contaminated ballast water which must be disposed of before taking on bunkers and cargo. The implication of these requirements on the operations of the major ship types is discussed in the following paragraphs.

1.2.2 Casualties—Spills from ship casualties receive the greatest attention in the world press. Spills may occur from operational mishaps in the pumping of fuel oil bunkers and liquid cargoes. Incidents of spills that occur from collision and grounding are accompanied by associated dangers to personnel and are likely to involve the largest quantities of pollutant. The magnitude of such spills is clearly far greater in the case of a loaded tank vessel than in the event of grounding and rupture of the double bottom fuel tanks of a dry cargo vessel. The implications of ship type and characteristics on the mode and magnitude of such spill incidents is considered in the following paragraphs.

16,000 DWT PARIA LIMPA, built in the 1940's alongside the 326,000 DWT UNIVERSE PORTUGAL.
Courtesy, *The Motor Ship*.

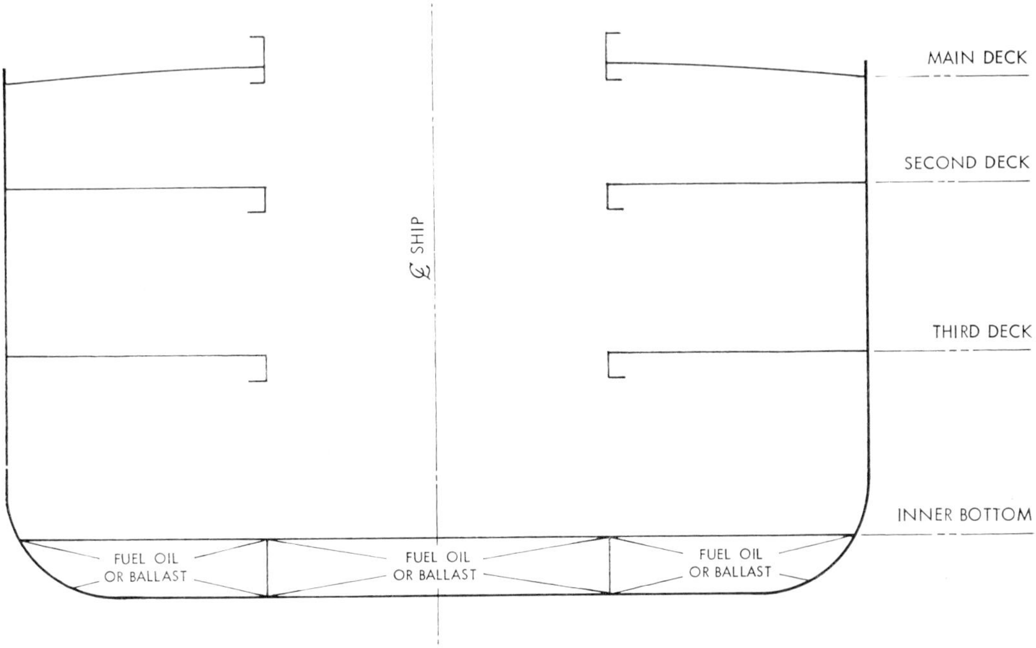

Figure 1 Outline midship section through cargo hold, typical break bulk dry cargo vessel.

2.0 Ship Classification and Description

The arrangements and general characteristics of the various merchant ship types are well described in such standard texts as Reference (1) and in the technical literature. Accordingly, the following discussion will be restricted to characteristics pertinent to the pollution problem, e.g., arrangement of tank spaces.

The following standard abbreviations have been used throughout for convenience:

DWT = deadweight = total displacement − light weight
= cargo + consumables
mld. = molded
fbd. = freeboard
B.P. = between perpendiculars

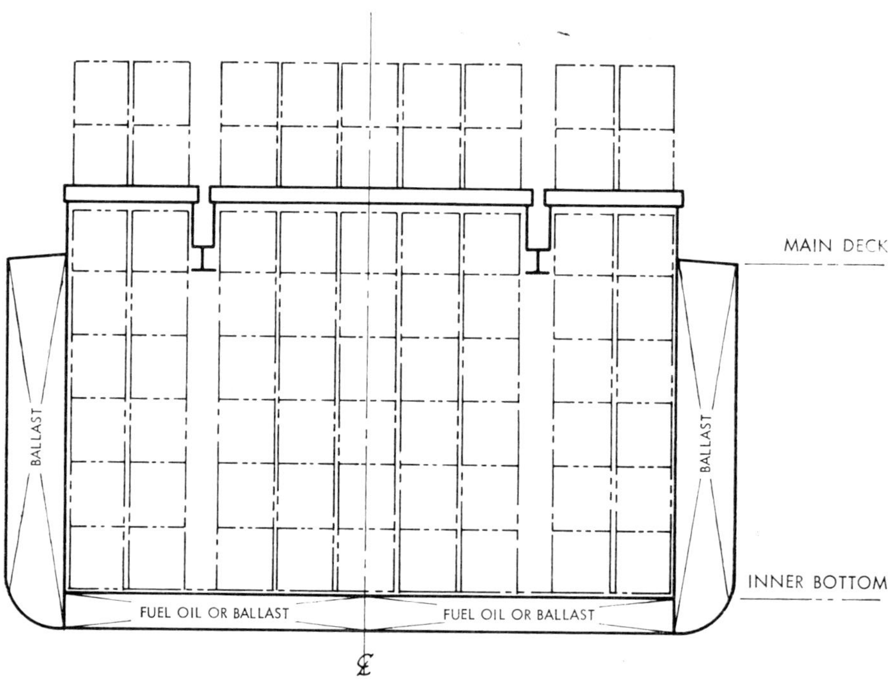

Figure 2 Outline midship section through container hold, typical container ship.

2.1 Break bulk vessels

The largest number, and greatest variety of seagoing vessels are in this category which includes the ordinary general cargo vessel operating in tramp or liner trades and carrying a great variety of dry products in raw material as well as finished and packaged form. An outline sketch of the cross-section through a typical cargo hold, showing hatchway, tween decks and fuel oil or ballast spaces, is shown in Fig. 1. Fuel oil is commonly carried in the double bottoms, as indicated but may also be carried in deep tanks, particularly outboard of shafting and in the vicinity of the machinery spaces. Except for settling and daily service tanks, all bunker spaces are normally piped for fuel oil or ballast. The availability of cubic capacity for tankage assigned only to ballast service is limited in such vessels and frequent use of fuel tanks for ballasting is likely in most operations.

2.2 Unitized cargo carriers

Ships in this category are usually designed for the exclusive transport of standard containers or wheeled trailer vans, and, to a lesser extent as hybrid carriers to handle containers, wheeled vehicles and general break-bulk cargo. In the case of container ships, as illustrated in the typical hold section, Fig. 2, the cellular nature of the cargo stowage requires some "squaring off" of the hold spaces, with the result that considerable wing space is available for ballast tanks. As a result such vessels are able to operate with clean ballast only and no ballasting of fuel tankage is normally required.

Roll-on/roll-off or trailer ships are similarly "squared off" internally and, in addition, may have extensive deep tank spaces available below the lowest vehicle deck. As with the container ships, tankage is likely to be available in sufficient quantity to permit full clean ballast operations.

2.3 Dry bulk carriers

Dry bulk carriers are engaged primarily in the transport of such commodities as coal, grain and ores. Two general configurations exist, as shown in Fig. 3. For light weight, high cubic cargoes, such as coal and grains, the hold configuration is such that water ballast capacity, in the amount of 35 per cent to 40 per cent of cargo deadweight, is available for clean ballast service, as shown in

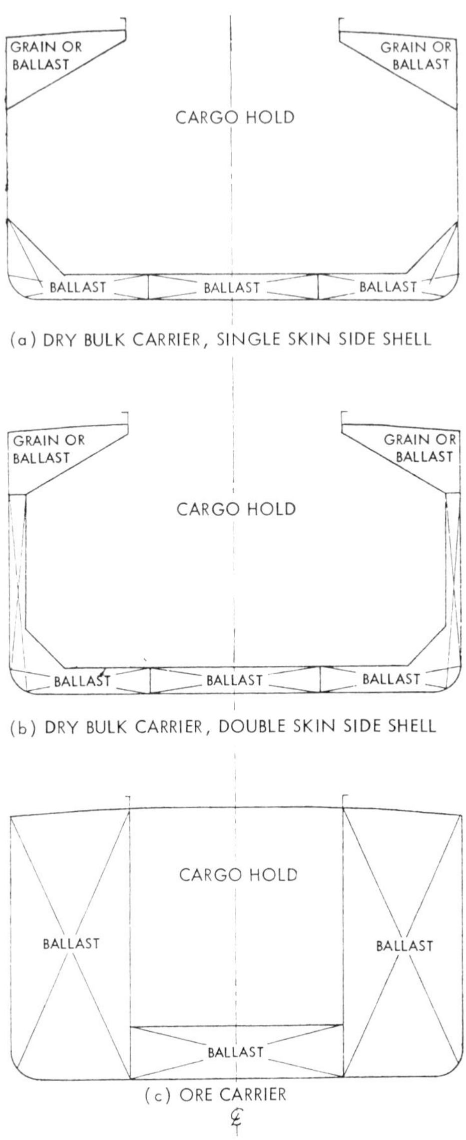

Figure 3 Outline midship sections through cargo holds, typical dry bulk carriers
(A) Dry bulk carrier, single skin side shell.
(B) Dry bulk carrier, double skin side shell.
(C) Ore carrier.

Fig. 3A and 3B. Fuel oil bunkers are generally confined to deep tanks within the machinery spaces or to portions of the wing and double bottom tanks ajacent to the machinery spaces. Clean ballast operation is generally feasible.

Vessels designed specifically for heavier cargoes such as ores are generally arranged with comparatively small cargo holds and large surrounding tank spaces as shown in Fig. 3C. Clean ballast operation is readily accomplished under all loading conditions.

2.4 Liquid bulk tank vessels

The modern tank vessel has evolved from the standard 16,000 deadweight ton (DWT) "T2" tanker of World War II to modern tank vessels exceeding 300,000 DWT capacity. The transport of liquid cargoes, predominantly petroleum crudes and refined petroleum products, is the single largest category of waterborne commerce and represents the greatest potential pollution hazard with respect to normal operations as well as casualties. Accordingly, characteristics of vessels in the liquid bulk trades will be considered in somewhat greater detail than other ship types.

2.4.1 Petroleum crude and products carriers—With the exception of the steam turbo-electric main propulsion machinery and electric drive cargo pumps, the World War II T 2 tanker is, in general arrangement, a parent of the tanker designs developed during the early post-war years. Typical characteristics of these vessels include:

(1) Cargo section divided by a pair of longitudinal bulkheads into port, center and starboard tanks.
(2) Relatively short cargo tanks, on the order of 40 feet in length, regardless of ship size.
(3) Poop, bridge and forecastle superstructures with navigating bridge located amidships.
(4) Forward and after fuel bunkers.
(5) Forward and after pump rooms.
(6) Relatively long, single screw, main propulsion machinery, with separate boiler and engine rooms.

Characteristics of American flag ships of the period are well described in two comprehensive papers, References (2) and (3). The range of sizes extends from 18,000 DWT to the 106,000 DWT MANHATTAN. The characteristics listed above are generally typical of these ships. The one outstanding exception is the MANHATTAN, fitted with twin screw main propulsion machinery with a total output of 43,000 SHP. This ship remains the highest powered tanker in operation today.

From the late 1950's until the present, tanker design evolved through a number of important changes, all directly related to reduced cost of construction and operation. These changes include increased ship size, reduction in complement, simplification of tank and cargo pumping arrangement and changes in traditional ship proportions. These developments, and their implication with respect to pollution, will be considered in later discussions.

Characteristics of the World War II "T-2" tankers are compared in Table I with corresponding characteristics of representative existing and planned tankers of the current generation. Of particular interest is the comparison between the CITIES SERVICE BALTIMORE of 1956 vintage and the more recent OVERSEAS ALICE class tankers, which are identical in hull form and power plant but differ significantly in internal arrangement, extent of superstructure, outfitting and complement. Schematic arrangements of these two vessels are shown in Fig. 4.

Also included in Table I are characteristics of the NISSEKI MARU described in detail in Reference (4), currently the largest tanker in operation, and preliminary information for the 470,000 DWT tanker currently under construction in Japan.

In summary, the recent evolution of tank vessel design has followed the general trends listed below:

(1) Increase in size to over 300,000 DWT capacity, with corresponding increases in dimensions and operating drafts.

(2) Simplification of arrangement, particularly by reduction in number of cargo tanks. Typical modern crude oil tankers are arranged with five center tanks and ten wing tanks. Secondary arrangement changes have included elimination of superstructure and houses amidships, location of all accommodation and navigation spaces aft and elimination of forward pump room and fuel bunkers.

(3) Speed has remained within the 15 to $16\frac{1}{2}$ knot range.

(4) Crew size has been reduced substantially, to as low as 25–27 men on U.S. flag as well as foreign tankers.

(5) Propulsion system power levels have increased with size, reaching as high as 40,000 SHP on a single screw. Centralized pilot house control of all propulsion machinery is a state-of-the-art development available to operators of diesel as well as steam turbine machinery.

(6) Cargo pumping systems are generally similar to those of the post World War II period, except

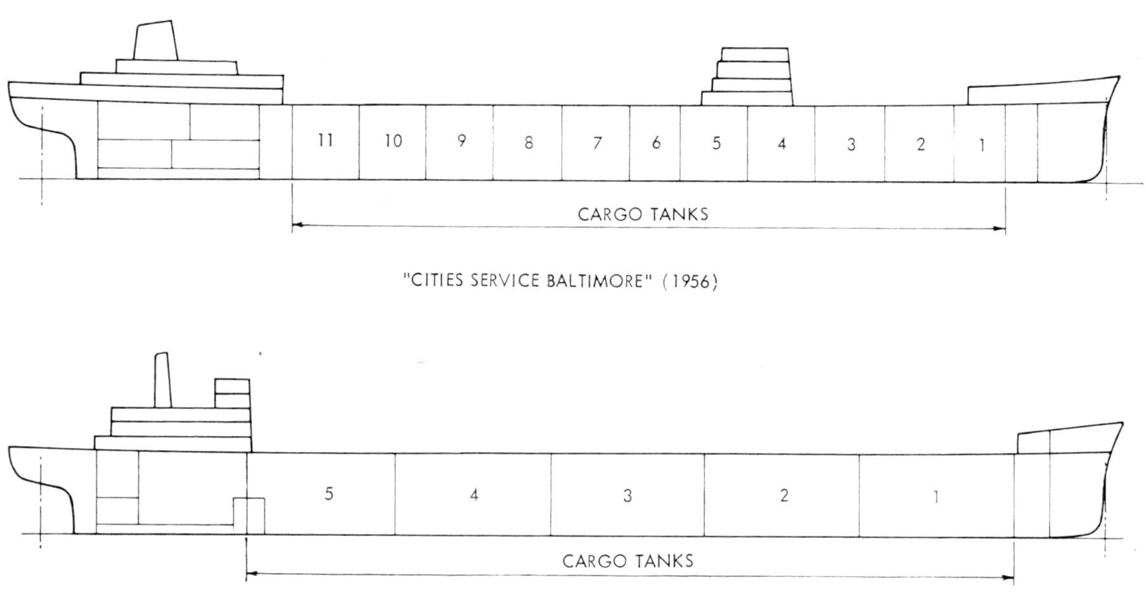

Figure 4 Outline inboard profiles, typical tank vessels.
(A) Cities service baltimore class (1956).
(B) Overseas alice class (1968).

TABLE I

Representative Tank Vessels—World War II to the Present

Name/Designation	T 2-SE-A 1	Cities service Baltimore Class	Overseas Alice Class	Esso Houston Class	Man-hattan*[1]	"CMX" Proposal	Universe Kuwait Class	Nisseki Maru	470,000*[4] DWT
Nominal deadweight	16,700	32,000	37,000	67,000	106,000	125,000	326,000	367,000	470,000
Nationality	U.S.	U.S.	U.S.	U.S.	U.S.	U.S.	Liberia	Japan	U.K.
Date built	W.W. II	1956	1966	1964	1962	1970*[3]	1968	1971	1973
Length, B.P.	503′–0″	630′–0″	630′–0″	763′–0″	895′–0″	900′–0″	1075′–11″	1082′–5″	1181′–0″
Breadth, mld.	68′–0″	90′–0″	90′–0″	116′–0″	132′–0″	147′–6″	174′–10″	178′–9″	203′–6″
Depth, mld.	39′–3″	45′–3″	48′–9″	54′–6″	67′–6″	63′–6″	105′–0″	114′–10″	118′–0″
Draft, summer fbd., keel	30′–1″	35′–7½″	36′–7½″	40′–4″	49′–1½″	48′–6″	81′–5″	88′–7″	91′–11″
Displacement, total, tons	21,880	42,866	46,243	81,780	137,200	147,400	375,811	–	–
Deadweight, total, tons	16,770	32,825	37,814	67,880	106,600	125,200	326,848	366,800	469,500
Cargo capacity, bbls.	141,000	273,000	335,000	582,800	910,000	937,000	2,513,000	–	–
No. of wing tanks	18	20	10	12	30	12	16	–	–
No. of center tanks	8	10	5	5	15	6	8	–	–
Length, wing tanks	36′–6″	40′–0″	90′–0″	Various	40′–0″	105′–0″	105′–3″	–	–
Length, center tanks	36′–6″	40′–0″	90′–0″	Various	40′–0″	105′–0″	105′–3″	–	–
Shaft horsepower, max.	6,600	15,000	15,000	19,000	43,000*[2]	–	37,400*[2]	40,000	45,000
Sustained sea speed, knots	14½	16½	16¼	16¼	17½	15½	15½	14½	15
Crew	–	50	33	27	73	28	47	43	–

——U.S. coastwise service——

*[1] Prior to conv. to icebreaking tanker.
*[2] Twin screw.
*[3] Proposal to U.S. maritime Administration.
*[4] Preliminary data.

for elimination of the forward pump room and increase in pumping rate with ship size.

2.4.2 Special products—A great variety of liquid products are carried in specially constructed tankers. While the quantities carried are small compared to the volume of petroleum crudes and refined products, transport of these commodities may involve unique containment problems and associated hazards. Reference (5) provides technical background on design and transport requirements for major categories of special products carriers.

Special products carriers may be classified in the following manner, according to nature of cargo:

(1) Liquefied natural gasses (LNG) and liquefied petroleum gasses (LPG).

(a) Low temperature—ambient pressure containment—The most exacting containment requirements are in this category, with cargo carried at about −260°F for liquefied natural gas (LNG). The magnitude of this trade is increasing rapidly. The largest vessels in operation are the POLAR ALASKA class, with capacities of about 71,500 cubic meters each. The largest LNG carriers on order at this time will have capacities of about 125,000 cubic meters and will be delivered in 1974. Reference (6) contains a comprehensive review of the design and operation of LNG carriers.

LPG transport includes the carriage of such gasses as propane, butane and ethylene, with propane, carried at about −50°F, the most common.

In a typical LNG or LPG carrier, cargo is carried in an independent, insulated tank or membrane liner. Double bottom and wing tank spaces are normally assigned to salt water ballast and fuel oil is carried in a relatively small portion of the double bottom and in deep tanks within the machinery spaces. Geometry is similar to that of a container ship and clean ballast operation is accomplished with no difficulty.

(b) High pressure—ambient temperature—LPG may be carried in pressure vessels, designed to the A.S.M.E. Code for Unfired Pressure Vessels. While this mode of containment has been generally superceded by low temperature transport for international trade, a considerable amount of LPG and similar cargoes is carried in this manner on the inland waterways of the United States and Europe and in smaller coastwise vessels and barges. The limiting design condition is usually for propane, in cylindrical tanks designed for 250 psig. In general, vessels carrying cargoes in pressure vessels have sufficient cubic capacity to permit clean ballast operation.

(2) Miscellaneous liquefied gasses—Anhydrous ammonia is carried in significant quantities in U.S. inland and coastwise waters. This commodity may be carried at low temperature or under pressure, in containment designed for the transport of propane.

Chlorine gas is commonly transported by barges in U.S. waters, primarily in pressure vessel containment. Other commodities of importance are primarily petro-chemicals, including butadiene, ethane, ethyl chloride, propylene and vinyl chloride.

(3) High temperature commodities—The transport of molten sulfur at about 275°F in heated independent insulated tanks has become the most common high temperature commodity carried on international and inland waters. Internal hull geometry resembles that of the low temperature LPG or LNG vessel, with double bottom and wing tank spaces available for clean ballast.

The transport of asphalt and bitumen in the molten state is lesse xacting than the case of molten sulfur transport and cargo is normally carried in conventional integrated tanks.

Sulfur and bitumen cargoes are relatively dense and clean ballast operation should be expected. The use of cargo tanks for ballast service is not feasible, except for emergency situations.

(4) Toxic and corrosive chemicals—A great variety of hazardous cargoes are carried in relatively small quantities in a variety of containment systems. References (7) and (8) contain a hazardous cargo classification and data for specific commodities, with particular respect to marine transportation.

References (9), (10) and (11) contain U.S. Coast Guard requirements for transport of various hazardous cargoes as well as normal petroleum products. Reference (12) is generally applicable to all U.S. flag ship construction but tends to be less exacting than the U.S. Coast Guard requirements, with respect to transport of LPG, LNG and other hazardous cargoes.

Virtually all hazardous cargo carriers will be built with sufficient ballast tank capacity, in the form of integral double bottom or wing tanks. It is unlikely that the use of cargo tanks for salt water ballast would be permitted, except for emergency conditions.

2.5 *Combination bulk carriers*

In order to improve the overall utilization of conventional dry or liquid bulk carriers, combination bulk carriers have been developed to permit transporting dry and liquid cargoes within the same cargo hold spaces. A typical voyage, for example, would involve carrying crude oil from the Persian Gulf to Maine, ballast from Maine to Hampton Roads, coal from Hampton Roads to Japan, Japan to Persian Gulf in ballast, etc. Cargo operations of this type involve unique cargo handling and hold cleaning problems, with associated potential pollution problems.

Two general configurations exist, the ore-oil carrier and the more common ore-bulk-oil (OBO) carrier. These are analogous in function and similar in geometry to ore carriers and general bulk carriers, respectively, illustrated in the section sketches in Fig. 3. A common modification in the latter case is the provision of a double skin side shell to facilitate hold cleaning.

Referring to Fig. 3, the ore-oil carrier will be equipped to carry cargo oil in the wing tanks as well as the center cargo hold. The double bottom space is normally reserved for clean ballast. The degree to which an ore-oil carrier can maintain a clean ballast operation, when operating as a tanker, will depend on the relation of cargo density to cargo volume available. The OBO will be operated with dry and liquid cargoes restricted to the main hold spaces, hence such vessels will normally operate with clean ballast, as a conventional bulk carrier. In both cases, however, hold cleaning between cargoes is a major operational problem that will be considered in later discussions.

References (13) and (14) contain detailed discussions of the design and operation of combination carriers. The relative importance of such vessels is shown, in References (14), to be increasing rapidly, with sizes of ore-oil carriers reaching 150,000 DWT to 250,000 DWT capacity and OBO carriers

75,000 DWT to 150,000 DWT capacity. At the time of this writing, the largest OBO in operation is of about 150,000 DWT capacity and numerous vessels of 100,000 DWT to 151,000 DWT are under construction. The largest ore-oil carrier currently under construction is of about 270,000 DWT capacity.

2.6 Naval vessels

Characteristics of naval vessels are considered here briefly in the context of combatant vessels. Auxiliary supply vessels are conversions or adaptations of equivalent commercial vessels and may be considered with those categories.

The salient requirement for a combatant vessel that is related to pollution problems is the necessity for maintaining an acceptable margin of stability under all operating conditions. For combatant vessels, this condition is difficult to maintain without generating contaminated ballast water because of the following inherent characteristics:

a) Naval vessels are cubic as well as weight controlled with respect to design and operation, i.e., little weight and space is available for the assignment of tankage exclusively for ballast.
b) Fuel may constitute as much as 20 per cent to 25 per cent of the total ship displacement in the full load condition.

Because of these characteristics, it is usually necessary to reballast each fuel tank promptly following total consumption of the fuel, thus generating contaminated ballast.

The condition described above is rapidly improving with the change from use of residual to distillate fuel. The latter is readily separated from sea water by conventional methods.

Several modern classes of destroyers and destroyer escorts are fitted with fully compensating fuel systems wherein seawater is allowed to flow into the tanks under the fuel oil, thus maintaining a near constant condition of displacement and stability. Such systems have long been standard on submarines and the current experience indicates that pollution problems can be minimized.

2.7 Miscellaneous commercial vessels

The great variety of miscellaneous and floating craft that could be sources of pollution are too numerous to consider here. In general, all can be considered, with respect to pollution, in one of the categories considered earlier. One particular case, of current interest, however, is the recent development of large, unmanned seagoing barges for the ocean transport of dry and liquid bulk commodities. Tank barges of 31,000 DWT are in service and active consideration is being given to barges as large as 45,000 DWT capacity. The geometry of a tank barge resembles that of an austere crude oil tanker of comparable deadweight, with five center tanks and 10 wing tanks. Operational as well as casualty pollution hazards are comparable to those of a self-propelled tanker, with the added complication that no personnel are aboard when the vessel is underway.

Service craft that dispose of waste products at sea are not being considered in these discussions. Sludge vessels and dredges that discharge in specified offshore locations are special vessels that operate under direction differing from conventional commercial or naval shipping.

3.0 Pollution from Normal Operations

3.1 Ballasting and tank cleaning

3.1.1 Break bulk vessels—The major source of pollution from break bulk general cargo vessels is in the intentional discharge of dirty ballast. As consumables, primarily fuel, are expended, displacement, draft and stability changes and may reach the condition that the addition of water ballast may be required. Some tankage may be available for clean ballast, but, in general, ballasting of fuel tanks will probably become necessary at some point beyond the expenditure of one half the consumables on board.

Figure 5 is a stability diagram of a recent design of a U.S. flag general cargo vessel. The diagram is a plot of draft against corresponding metacentric height, GM. The magnitude of the latter, a measure of initial stability in the upright condition, is generally accepted as one criterion of intact stability. A thorough discussion of ship stability may be found in any of the standard texts on naval architecture, such as Reference (15).

The diagram contains a grid plot of GM available as a result of nine loading conditions, ranging from full cargo to no cargo, and full consumables to 10 per cent consumables. A plot of minimum GM required, to meet prudent operating as well as regulatory stability requirements, intersects the GM available grid in a region below the one half

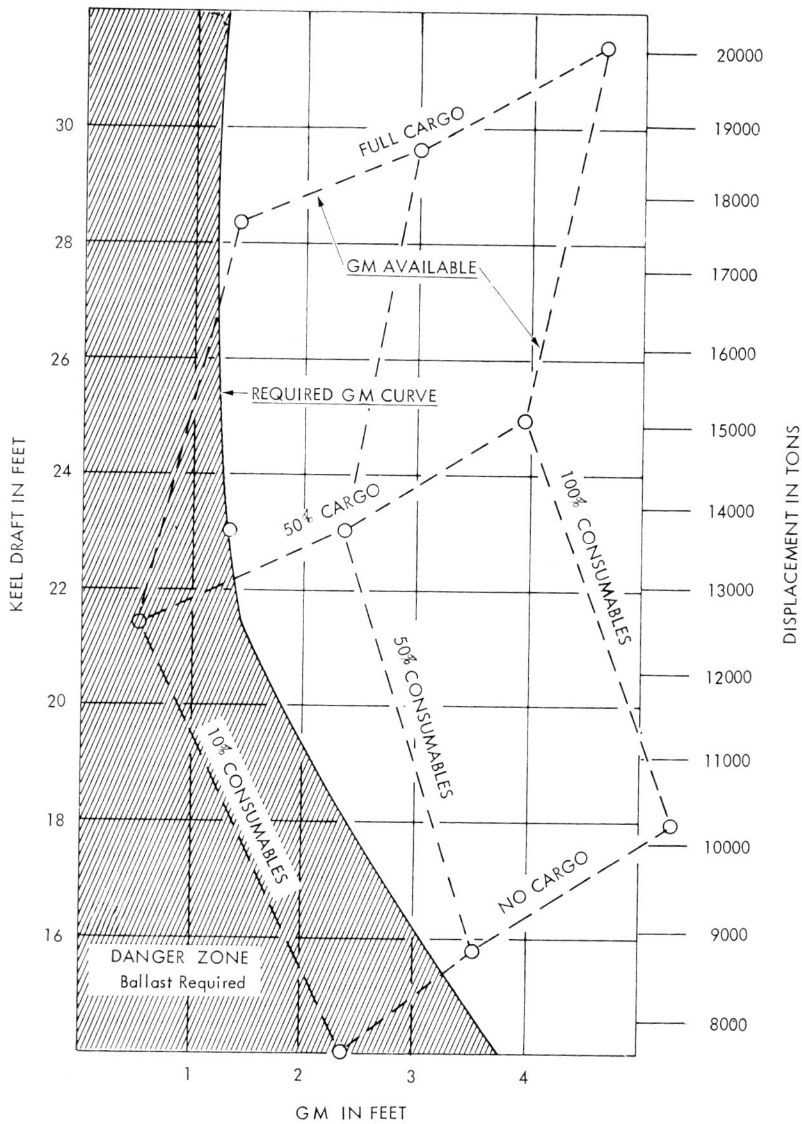

Figure 5 Stability diagram.

consumables condition. All loading conditions to the left of the GM-required curve have inadequate stability and ballasting is required until the metacentric height, GM, is equal to or greater than the minimum required value. The particular vessel represented by Fig. 5 is unusually stable and ballast requirements are moderate. For many general cargo vessels, however, the ship must be ballasted for any condition below one half con sumables, i.e., the GM-required curve crosses the GM-available grid in the general area of the one half consumables condition.

Disposition of the oily ballast at sea is a major pollution source and retention of the ballast for disposal at a shore facility imposes an economic and operational penalty. Further, the operating crews of most ships are reluctant to take on seawater in the fuel tanks to avoid possible contamination of the fuel system. The operator, accordingly, is faced with several poor choices:

a) Pump ballast at sea prior to entering coastal waters. This procedure is likely to violate existing or pending anti-pollution regulations and will place the ship in an unsafe stability condition until arrival in port.
b) Avoid taking on ballast. This procedure avoids the dirty ballast problem as well as any possibility of fuel contamination but may result in operation in unseaworthy condition.
c) Take on ballast as required and retain oily ballast for disposal into port receiving facilities. This procedure may not be feasible if no dockside

storage tanks or sludge barges are available in the particular port. At best, this procedure imposes an economic penalty on the operator, while maintaining standards of stability.

An obvious solution to the problem of disposal of oily ballast would be the provision of oil-water separators of a level of effectiveness sufficient to meet current and proposed pollution standards. Unfortunately, for the typical high specific gravity residual oil fuels normally carried in ship's bunkers, no effective separation system is currently available. Considerable development effort continues to be expended by public and private agencies and there is hope that a practical oil and water separator will be developed for this application.

3.1.2 Tank vessels, crude and refined petroleum products—The greatest source of intentional discharge of contaminated ballast into the sea is from the operation of tank vessels transporting petroleum crudes and products. Tankers are normally one way product carriers and return voyages to the cargo source are in ballasted condition. The weight of the light ship plus fuel and other consumables, on the order of 20 per cent the total displacement, is small compared to the weight of cargo. Accordingly, on the return voyage, the tanker must take on some water ballast to obtain sufficient operating draft for propeller immersion and to avoid slamming of the hull forebody in a seaway. It should be noted that ballast is not required for stability. The tanker normally suffers from excessive stability in the light or ballast condition.

The class of tankers designed for crude oil transport includes the largest hull sizes, the greatest total tonnage, and, accordingly, the greatest potential for oily ballast pollution. Typically, a crude oil tanker would be designed to carry a full deadweight cargo of about 40° A.P.I., or 0.825 specific gravity. Total tankage within the cargo section is generally sufficient to carry full deadweight cargoes of considerably lower specific gravity, on the order of 0.71 to 0.75. Accordingly, excess tankage can be assigned permanently to clean ballast service. This is the case with the typical crude carrier. Referring to Fig. 4, for example, two pairs of wing tanks could be piped exclusively for clean ballast in the case of the CITIES SERVICE BALTIMORE class and a single pair of the longer wing tanks for the more recent OVERSEAS ALICE class. The clean wing ballast cubic capacity amounts to about 10 per cent of the total volume of the cargo tank portion of the hull. As much as an additional 15 per cent of the equivalent cargo deadweight in clean ballast could be available in tankage outside the cargo section, including peak tanks and fore and aft deep tanks. To obtain a total ballast capacity at least equal to 40 per cent of the cargo deadweight would require using some cargo oil tank space, thus generating some oily ballast. In the ideal case considered here, this amounts to at least 15 per cent of the existing cargo tank volume, more if less clean ballast space is available or if greater ballast draft is required.

Tank vessel operators have developed various techniques for the management of the dirty ballast problem, including the "load-on-top" system currently in use for a large number of vessels. Referring to Fig. 4, the load-on-top procedure may be briefly summarized by the following sequence:

a) The ship departs the oil discharge port with clean salt water ballast only in permanent clean ballast spaces, e.g., peak tanks, and no. 3 wings for the modern austere tank arrangement.
b) While departing for the sea voyage, additional cargo tanks to be used as ballast spaces are cleaned by the conventional Butterworth system, using heated salt water under pressure. Tank washings, containing an oil-water mixture, are pumped to spaces designated as slop tanks, located aft of the cargo tanks.
c) At sea, after a reasonable time for the oil-water mixture to settle, the oil will tend to float to the top and the clean sea water is pumped overboard, from under the oil layer.
d) At the oil receiving terminal, all clean ballast is pumped overboard before taking on cargo oil. Cargo oil is loaded over the oil-water mixture in the slop tank, hence the term "load-on-top." Refineries are prepared to take the resulting mixture from the slop tank, provided the amount of sea water contaminant is within certain specified limits.

For tank vessels with no permanently designated clean ballast spaces within the cargo tank length, the procedure is somewhat more complicated. Again referring to Fig. 4, the ship may depart with dirty ballast in the no. 3 cargo wings, and carry out the following procedure:

a) During departure, additional cargo tanks to be used for ballast, e.g. nos. 2 and 4 wings, are cleaned, and the washings pumped to the slop tank. Clean ballast is then taken aboard nos. 2 and 4 wings.

b) Seawater from no. 3 wings is pumped overboard, below the oil layer floating on top, and the remaining oil is pumped to the after slop tanks.

c) At the oil receiving port, clean ballast is discharged and cargo oil is taken aboard as required to obtain a full deadweight cargo, including loading into the after slop tanks.

3.1.3 Tank vessels, special products carriers—The special products carriers described in earlier discussions are predominantly clean ballast vessels. Sea water will rarely be pumped into cargo tanks and sufficient tankage is normally provided, in the form of double bottoms and wing tanks, to serve as cargo tank protection as well as clean ballast tankage.

3.1.4 Dry bulk carriers—The typical dry bulk carrier operates in ballast over a significant portion of the operating life. Many trade situations exist in which return cargoes are not available and ballasting is required for adequate propeller immersion and seaworthiness. Bulk carriers are inherently stable and ballasting is not required for this purpose. As discussed earlier, sufficient cubic capacity is normally available in the wing and double bottom spaces to permit clean ballast operation.

3.1.5 Combination carriers—The most common of the combination carriers, the OBO, has the same general configuration as the dry bulk carrier and, accordingly, is generally capable of clean ballast operation. Hold cleaning between voyages with incompatable cargoes, however, is an additional souce of pollution. In the example given in earlier discussions, the OBO discharges crude oil at Portland, Maine, proceeds in ballast from Maine to Hampton Roads and takes on coal at Hampton Roads for delivery to Japan. During the ballast voyage from Maine to Hampton Roads, the holds are cleaned by conventional means and the dirty oil washings discharged into slop tanks located in a pair of wings immediately aft of the cargo holds. In the normal tanker operation, the clean sea water would be discharged overboard from the slop tanks, after allowing sufficient time for settling. For this particular example, the coastal voyage in ballast requires only about two days and the hold washings are currently pumped at sea in unrestricted zones. As pollution regulations become more restrictive such procedures will be illegal and the operator will face the alternatives of discharging to a port facility or violating the law.

A more complex situation arises for the ballast voyage from Japan to the Persian Gulf, after discharging coal. Several days of manual labor, are required to remove coal residue, followed by Butterworth cleaning to remove the fine coal powder remaining. The solids and wash water are discharged at sea in unrestricted zones.

From the above examples, it is clear that unique operational pollution problems exist with the combination carriers. Assuming time permits, oily ballast or cargo oil wash water can be handled by the "load-on-top" procedures. No simple solution exists at this time for the disposal of the solid wastes.

In general, combination carriers can be changed over from liquid to solid cargoes in a comparatively short time, say a period of 18 hours to two days. The reverse procedure may require more time consuming cleanup procedures. Operators, accordingly, will tend to prefer maintaining a given ship in a single cargo trade for seasonal periods, if permitted by the economics of the trade conditions. Operational pollution problems are reduced in complexity when the occasions for cleaning between incompatable cargoes are minimized.

3.1.6 Naval vessels—The earlier discussions, in Section 2.6, indicated the current status of naval vessels, with respect to fuel oil and ballast requirements for combatant vessels. It is expected that the change to distillate fuels will minimize pollution from oil ballast.

3.2 Cargo transfer, loading and unloading

3.2.1 Liquid cargoes—Some pollution inevitably occurs as a result of fuel and cargo oil transfer between ship and terminal or between ship and lighter alongside. The majority of such incidents results, directly or indirectly, from human failure. Typical incidents include overflow through tank vents and hose failures. In most cases, proper monitoring or automatic control of cargo transfer and fueling operations will minimize the probability of oil spillage. Normal inspection routines should permit anticipating most equipment failures.

The rapid development of the large crude oil tankers has been accompanied by the parallel development of offshore terminals to accommodate the deep draft vessels. The tankers moor to large "monobuoy" single point mooring buoys which are anchored permanently to the sea floor. Oil pipelines are led to the underside of the monobuoy along the sea floor, from the shore tanks. Flexible hoses are led from the buoy to the midship pumping station on the tanker, usually by a tending launch. Means for mooring the tanker and connecting up to the oil hoses are under constant development and are reaching a high level of reliability. It is not expected that operations of the offshore terminals, under proper control, will represent a major source of pollution.

A simple description of the offshore terminal operation is given in Reference (16).

3.2.2 Dry bulk cargoes—Earlier discussions of combination operations included mention of solids pollution from hold washings when converting from dry bulk to liquid bulk operations. Of far greater importance is the harbor pollution occurring at dockside from the simple transfer, by grabs or similar mechanical devices, of dry bulk products between ship and shore storage facility. Over a long period of time, dry bulks spilled between ship and pier accumulate and become a local, but significant source of harbor pollution. While many of the commodities are inert, others, including coal and some ores have an adverse effect on the ecology. The cargoes involved are of low value and command low freight rates, hence there is little incentive to control spillage of small, but accumulative, quantities into harbor waters.

Advanced cargo handling methods, including shipboard self-unloader systems and the recently developed MARCONAFLO slurry system for iron ore transport, Reference (17), are in use in a few cases and are likely to minimize this source of pollution.

4.0 Pollution from Casualties

The magnitude of a particular oil spill, or other pollution casualty, is a function of ship type, ship size and nature of the incident. Tank vessel collisions or groundings involve the greatest magnitude of pollution resulting from individual casualties, and, accordingly, will be considered in some depth in these discussions. Other vessel types will be considered briefly.

4.1 Break bulk vessels

Pollution resulting from rupture of fuel tanks, as a result of collision or grounding, is the only significant casualty of this class of shipping likely to result in pollution. Less important is the potential rupture of deep tanks carrying various special cargo oils, primarily edible oils. Figure 1, showing a typical section through a cargo hold, indicates that double bottom tanks are normally assigned to fuel oil or ballast service. The largest general cargo vessel operating under the U.S. flag has a maximum fuel oil capacity, including all double bottoms, deep tanks and settling tanks, of about 3700 tons. About 83 per cent of this tankage, or about 3000 tons, is in the double bottoms. This represents about $9\frac{1}{2}$ per cent the maximum displacement, or about $13\frac{1}{2}$ per cent of the maximum deadweight of the ship. In general, the operator will tend to carry a minimum weight of fuel oil in order to maximize cargo deadweight, hence somewhat less than the 3700 tons of fuel oil is likely to be aboard.

The largest double bottom fuel tank in this particular vessel holds about 280 tons. A one-compartment damage collision, assuming damage from the side to the centerline, could expose 465 tons of fuel oil to the sea. Two-compartment damage, again from the side to the centerline, could expose about 900 tons of fuel capacity to the sea. A grounding incident, in which the bottom shell is opened to the sea for a considerable portion of the ship's length, could expose as much as 2/3 the double bottom fuel capacity, or about 2000 tons, to the sea. These values represent the maximum quantities likely to be exposed following a casualty. A considerable portion of the fuel would be released to the sea in any of these incidents. Magnitudes of such spills are significant but small relative to the catastrophic effects of a comparable tank vessel incident.

4.2 Unitized cargo

Container, roll-on/roll-off and unitized cargo combination vessels in liner service are larger and higher powered than the break-bulk vessels, hence carry greater quantities of fuel oil. Arrangements of fuel, ballast and cargo oil tanks are varied and fuel oil may be located in bottom, wing or deep tanks. In general, the mode of release of fuel oil to the sea would be as discussed for break bulk vessels, with the quantities somewhat greater. A considerable portion of the wing and bottom tankage of unitized cargo vessels is piped only for ballast, thus lessening the probability that only fuel tanks would be breached in the event of a collision or grounding.

4.3 Tank vessels

Fuel tanks are of relatively minor importance in the case of tank vessels involved in casualties. Fuel is generally confined to two or three deep tanks and settling tanks and represents a small portion of total tankage exposed to the sea following a casualty.

The evolution of tanker design since the early 1950's, with respect to pollution from collision and grounding, is considered briefly in earlier discussions and in greater depth in References (18) and (19).

4.3.1 Effects of major design and construction changes—The major design and construction changes that have generally increased the potential pollution hazards resulting from individual cases of collision and stranding are briefly summarized:

(a) Ship size increases—A consequence of size increase is the greater probability of grounding because of greater operating drafts and an increase in the quantity of oil cargo released in the event of collision or stranding. With tank vessel speeds essentially unchanged over the past 20 years, the energy to be absorbed in stopping the ship is directly proportional to the displacement, a quantity which has increased more than ten-fold during this period. Further, the average speed of dry cargo and many special products carriers has increased from about 15 to 16 knots to well over 20 knots. On the credit side is the likelihood that a lesser number of collisions and strandings will occur as fewer, larger size, tankers are built, replacing obsolete smaller tonnage.

(b) Arrangement changes—For a given ship size, the adoption of tank lengths on the order of twice the tank lengths of 10 to 15 years ago, results in the following:

(1) Larger quantity of cargo release in the event of tank rupture. A typical wing tank of a modern austere crude oil tanker will hold about 1/20th the total cargo capacity. In the case of a 320,000 DWT tanker, the wing tank size is equal to the total cargo deadweight of a World War II T2 Tanker.

(2) Less flexibility in control of cargo, i.e., simpler piping systems, fewer valves and controls, during normal operations as well as in the event of rupture of tank or piping.

(3) The adoption of cargo section lengths forming a greater proportion of ship length, while beneficial in the reduction of bending stresses, reduces to some extent the protection of cargo in the event of end collision.

(c) Hull proportions—The trend toward designing for minimum values of length/depth has necessarily increased the tendency toward greater drafts, thus increasing the probability of grounding.

(d) Bridge location—The general trend toward locating the bridge aft has been a matter of some controversy. Some pilots maintain that the bridge amidships location is preferable for safe navigation. It should be noted, however, that events have demonstrated the aft bridge location to be a far safer place for crew members in the event of collision, explosion or fire.

(e) Structural design—The partial elimination of transverse oil-tight bulkheads in the wings and the general reduction of tank vessel scantlings, based on more rational design procedures and use of protective coatings, results in the existence of less energy absorbing structure in the event of collision or grounding.

(f) Machinery simplification—The elimination of redundancy, while based on rational design practices, does necessarily reduce the ability to cope with emergencies to some degree. Specific examples are the installation of a single main boiler and reduced standby generator capacity.

(g) Cargo pumping—Simplification of cargo pumping systems reduces the flexibility in cargo handling and may limit the ability to meet certain emergency conditions. A particular example is the elimination of cargo oil stripping systems in some installations.

(h) Manning—The general reduction in manning levels may limit the remedial action which could be taken in the event of damage or fire.

(i) Subdivision and damaged stability—Requirements for load line assignment and damaged stability are more specific today than 20 years ago. However, with the reduced degree of subdivision, modern tank vessels are less likely to survive certain types of collision and grounding damage.

4.3.2 Special hazards in tanker operations—In addition to collision and grounding, the following hazards to tank vessel operation should be considered as potential sources of pollution:

(a) Fire—Of the many possible disasters which may occur to a tank vessel in petroleum trades, fire is among the most common. Should rupture of a tank occur incidental to the fire, the cargo may be burned, thus preventing extensive pollution. However, gross structural failures may occur as a consequence of fire damage, resulting in uncontrolled release of cargo that may not be consumed by the fire.

(b) Explosion—Despite the many regulations and operational precautions, explosions occur when an obscure source of ignition takes place in the gaseous atmosphere above the liquid cargo level. Some of the most disastrous incidents have taken place when tankers in ballast, with empty tanks not gas-free, have been involved in minor collisions.

Of particular interest is the series of major explosions which occurred recently on a group of large crude oil carriers in the 200,000 DWT size range. Each vessel was in ballast and undergoing tank cleaning operations at the time of the incident. Other than conjecture regarding static electricity and obscure sources of ignition, no definitive explanation of these disasters has been published at this time. Discussions of these and related incidents are covered in References (20) and (21), which also give consideration to problems of tank cleaning and proposed tank inerting systems. It is clear that disasters of this sort, while not necessarily involving collision or stranding, will result in gross structural failure and resulting release of oily ballast.

(c) Structural failure—The occurrence of major structural failures, hull fracture in longitudinal bending particularly, is well documented for ships built during the World War II period. Improper loading, notch sensitivity of steels, and poor structural details are all causes of past failures. The potential release of cargo oil to the sea is a clear probability in these cases. Further, total fracture of the hull may separate part of the cargo section from the power source and pumping equipment. Fortunately, with improvement in design and quality control, such failures occur infrequently.

4.3.3 Unmanned seagoing barges—The development of large, unmanned, seagoing barges is a recent phenomenon, particularly in U.S. coastwise trade. The regulatory bodies have been aware of the potential hazards, since it is clear that control problems in extreme weather, and the succeeding events which could occur in event of collision or stranding of either the tug or barge, represent major pollution hazards. Certain automatic controls and safeguards have been provided voluntarily, e.g., radio-controlled release of the anchor and control of ballast pump operations and radio-controlled steering in the case of certain conversions from ship to barge operations. In general, however, the ability to control release of cargo oil from a barge as far as two thousand feet from the tug will be limited, particularly in heavy weather.

General discussions of seagoing barge design and operations are given in References (22), (23) and (24).

4.3.4 Special products carriers—Casualties to the special products carriers described in earlier discussions involve unusual hazards and pollution problems. The responsible regulatory agencies were active in studying and regulating various special hazardous cargoes long before oil pollution was recognized as a major hazard. Accordingly, safety codes and regulations have been formulated and effectively enforced for a long period of time. Regulations of the U.S. Coast Guard, for example, which apply to various hazardous cargoes, as well as liquefied petroleum and natural gasses, are covered in References (9), (10) and (11).

Hazardous cargo transportation is given special treatment by the designer and the cognizant regulatory agencies by consideration of the following:

(a) Materials—Where indicated by the corrosive or otherwise restrictive nature of the cargo, special materials, subject to rigorous testing requirements, are specified. Liquefied gas containment, for example, requires special steels or other materials to avoid brittle fracture at low temperature. Similar restrictions hold for piping and cargo pumps.

(b) Containment—In all cases where special hazards exist, it has generally been the practice to require transport of cargoes inboard of shell plating, i.e., cargo is contained within an inner hull formed by an inner bottom and longitudinal bulkheads. Further restrictions in some cases require that the cargo containment be structurally independent of the ship's main hull structure. This is the case with pressure vessels and rectangular tanks for carrying liquefied gasses at low temperature and molten sulfur at high temperature. The requirement for secondary containment, i.e., the ability to contain the cargo for indefinite periods of time in the event of rupture of the primary tank, has long been established as a basic requirement for the transport of liquefied gasses at low temperature.

(c) Cargo handling—Elaborate safety precautions, to avoid spillage and sources of ignition are inherent in established design practices and regulatory requirements.

The above brief discussion indicates a comparatively high level of design for safety in the case of hazardous cargoes. In the event of a low energy collision or stranding there is a reasonable probability that the primary containment system will remain intact. For the liquefied gasses, the greatest tonnage category of special hazardous cargoes, pollution hazards are limited, or at least of short duration, since the cargoes are volatile and would be dissipated in a short period of time in the event of primary tank rupture. For the other cargoes, toxic materials in particular, the pollution hazard remains. Cargoes are shipped in relatively small quantities, however, and effective containment is readily achieved with modern construction methods.

4.4 *Dry bulk carriers*

Fuel oil is likely to be carried primarily in deep tanks, and, to a lesser extent, in selected double bottom and wing tanks. Accordingly, fuel bunkers comprise a relatively small portion of total tank space and do not represent an unusual hazard in the event of collision or grounding.

4.5 *Combination bulk carriers*

When transporting oil cargoes, combination carriers function as tankers. The internal arrangement of ore/oil carriers, Fig. 3, resembles the arrangement of a tanker, with respect to location of longitudinal bulkheads at least 20% of the breadth from the side shell. Accordingly, the extent of cargo oil pollution from side collision is comparable to the case of the tanker, assuming comparable compartment lengths.

The ore/bulk/oil, or OBO, carrier, however, is designed with high capacity cargo holds in order to transport low density dry bulk cargoes as well as oil. As a result, the boundary of the center holds is well outboard the 20% breadth penetration line, for either the single skin or smooth skin side shell arrangement. In the event of a side collision in way of a single compartment of a nine-hold vessel about 1/9 the oil cargo will be exposed to the sea, compared to about 1/20 the total cargo in the case of a modern crude tanker. This amounts to almost 17,000 tons of oil the case of a homogeneously loaded 150,000 DWT vessel. In the case of a low energy collision, the OBO with double skin side shell will offer a somewhat greater resistance to penetration than the more common single skin arrangement.

The double bottom arrangement of combination carriers does provide for some grounding protection, particularly in way of the center holds. The lower outboard sloping wing tanks of some vessels, however, are regularly used as cargo oil tanks.

4.6 *Naval vessels*

High fuel requirements for combatant vessels necessitates the assigning of a relatively high proportion of cargo volume to fuel tankage, hence significant quantities of fuel oil would likely be released in event of a collision. On the credit side, however, is the relatively small size of naval combatant vessels, except for the few large aircraft carriers, and the high level of subdivision and damage control which, relative to commercial vessels, may increase the probability of survival in the event of collision and damage.

5.0 Prevention

Improvements in maritime safety standards have historically followed some major disaster at sea. This was the case with respect to loss of the TITANIC, the 1960 SOLAS convention with respect to the sinking of the ANDREA DORIA, and now the accelerated IMCO activity in the wake of the TORREY CANYON incident. Radical changes in regulatory requirements should not be expected, but rather an intensification of studies and a slow evolution of modification to existing requirements.

It must be emphasized that economic incentives are the fundamental basis for the increased potential for pollution due to collision and stranding. The trend to larger tank vessels and toward simplification of the ship design, all directed to substantial economies in construction and operation, will not easily be reversed. Yet this trend necessarily increases the pollution hazard occurring from an individual incident. In the light of the overwhelming short term economic incentives, it is clear that all improvements must necessarily come from the imposition of requirements from outside the industry; i.e., from the deliberations of regulatory bodies at an international level and perhaps just as effectively, from the pressures of the insurance industry.

Some suggested directions for study in improvement of design and operation are included in the following discussion:

5.1 Operational measures

5.1.1 Dirty ballast—To a great extent, the ship operator is dependent on the design characteristics of the ship when ballasting to obtain a required draft or stability condition. Tank cubic capacity, hydrostatic characteristics and arrangement of hold and tank spaces are built into the ship and are beyond his control. A skilled operator can minimize ballast requirements within these limits by exercising control over loading plans and loading sequences, including disposition of cargo and fuel oil. Until such time as effective oil and water separators are available, the prudent operator will take every precaution to minimize the generation of dirty ballast.

Continued use and improvement of "load-on-top" and similar tank cleaning and ballasting techniques for petroleum carriers will contribute toward minimizing operational pollution. Continued improvements are necessary to minimize the proportion of pollutant remaining in the discharged ballast. Of particular value will be the development of methods of accelerating the separation of oil and sea water in the slop tanks, to permit decanting ballast water at sea during ballast voyages of short duration.

As anti-pollution regulations become more demanding with respect to prermissible levels of contamination, the ship operator will require reliable instrumentation to monitor the quality of overboard discharges. At this time there is no reliable sensor to provide the basis for such instrumentation.

5.1.2 Fuel and cargo oil transfer—Fuel and cargo transfer operations are, to a large extent, under control of shore personnel. The ship's crew is responsible for tank loading sequences and monitoring the loading operation to avoid overflow through tank vents. The general adoption of cargo automation systems involving existing and developmental hardware should minimize this source of pollution.

5.2 Ship design, arrangement and structure

Broad reviews of ship design, with respect to pollution from normal operations and casualties, are included in References (18), (19), (25) and (26). Dillon, in Reference (26), has extended such studies to include the added capital costs required to obtain certain pollution prevention characteristics. The following discussion, taken in part from the above references, briefly covers those areas of design where significant gains in pollution control could be realized.

5.2.1 Clean ballast operation—An obvious solution to operational dirty ballast problems is the construction of new vessels with sufficient cubic capacity to operate with 100% clean ballast at all times, i.e., single-purpose ballast tankage would be provided with sufficient volume to meet all ballast requirements. For all break-bulk dry cargo vessels and most tank vessels this would require a significant increase in ship size, with a consequent increase in construction cost. In Reference (26) Dillon reports on studies made by the U.S. Maritime Administration to estimate the cost penalties imposed by clean ballast design. For the case of the conventional break bulk freighter, the S/S SCHUYLER OTIS BLAND was selected as a conservative example because of the marginal stability characteristics of this ship. The study indicated that an increased freight rate of from 3.4% to 7.9% would be required to compensate for loss in cubic capacity and/or increase in construction cost. The higher figure corresponds to an increase in ballast tank volume, at the expense of cargo colume, with no change in ship dimensions. The lower figure corresponds to a small increase in beam, to obtain greater cubic capacity, in addition to an increase in ballast tankage.

In a more comprehensive study of tank vessels, using a given 250,000 DWT reference design, Dillon shows an increase in required freight rate of about $3\frac{1}{3}$% to $12\frac{1}{2}$%, as clean ballast capacity is increased to obtain a ballast displacement varying from 30% to 60% of full load displacement. The studies were made assuming a constant deadweight and draft, with suitable adjustments in breadth and depth to accommodate the increased ballast tankage. The variations in internal geometry of the tanker, assumed for the increased clean ballast tankage, were made in association with shell protection against collision and grounding.

5.2.2 Protection against collision and grounding—Emphasis is placed here on the protection of cargo tanks of liquid and combination bulk carriers from the effect of collision and grounding. The direct approach to collision protection for a conventional tanker would be provision of a structural barrier at the ship's side in way of the cargo tanks. This outer space, while lost to cargo capacity, would be available for clean ballast. For a given set of hull dimensions, the resulting loss in cubic capacity would be important when light cargoes such as gasoline are carried. Dillon's study, Reference (26), assumed a corresponding increase in breath and/or depth to maintain capacity. One effort in the direction of combining side protection with clean ballast tankage is described in References (27) and (28), with respect to the "Duokleen" tank arrangement. Transverse sections through the cargo tanks of the FORTUNA and conventional tanker are compared in Fig. 6A and 6B. Longitudinal bulkheads of the conventional tanker are located within the 20% breadth line, the assumed depth of collision penetration contained in the rules of most regulatory bodies. The structural and subdivision arrangement of larger tankers would probably include the provision of a third longitudinal bulkhead on the centerline. The effectiveness of the "Duokleen" arrangement was demonstrated in an incident in which one ship of the class survived a broadside collision without breaching the cargo tanks.

Grounding protection is normally provided by fitting of a double bottom in conventional dry cargo vessels. Except for the case of combination ore/oil or OBO carriers, where a double bottom is installed inboard of the longitudinal bulkheads, innerbottoms are not provided in tank vessels. One recent exception is the 212,000 DWT tanker MOBIL PEGASUS shown in a diagrammatic section in Fig. 6C. The vessel, described in detail in Reference (29), was built with a continuous double bottom tank, assigned exclusively to clean ballast. While the primary motivation was to obtain a clean ballast operation, it is evident that this arrangement provides for substantial cargo tank protection against grounding. The arrangement of sloping longitudinal bulkheads reflects a structural design preference and is incidental to the double bottom concept. The tanker operators acknowledge a higher construction cost for this arrangement, but note a reduced operating cost, particularly in the reduction or elimination of tank cleaning costs.

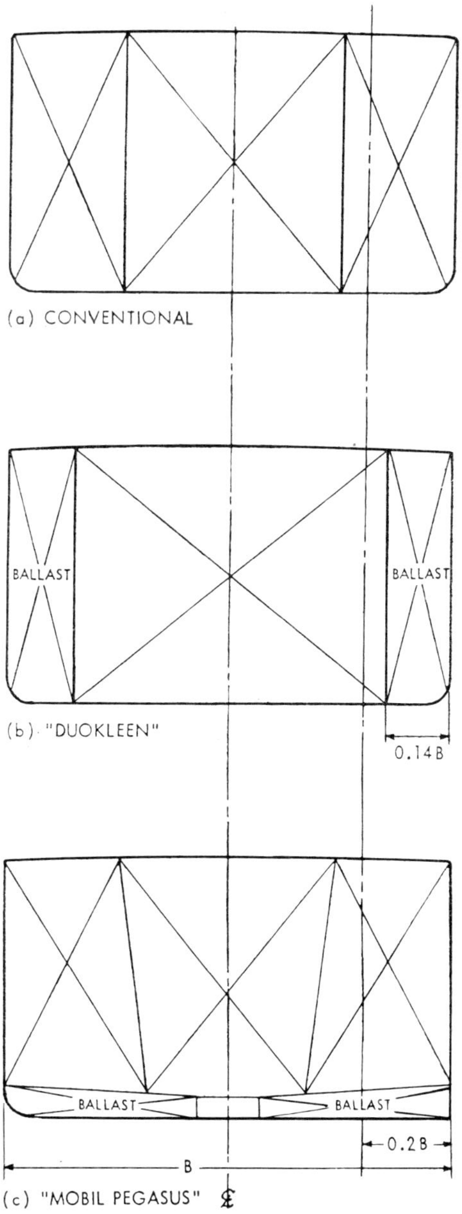

Figure 6 Outline midship sections of tank vessels.
(A) Conventional
(B) "Duokleen".
(C) MOBIL PEGASUS

A combination of side and double bottom protection has been proposed in various studies. The arrangement, analogous to that of the OBO carrier with double skin side shell, provides for clean ballast operation as well as collision and grounding protection, at a significant increase in construction cost.

At the domestic as well as the international level

studies have been made of the consequences of the trend toward long cargo tanks and the probability of losing large quantities of cargo oil in event of collision or grounding. Some existing published data demonstrate that the highest probability of collision damage exists in the forward half of the ship hull. Accordingly, further studies may indicate the desirability of providing comparatively short tanks forward and permitting longer tanks in the after portion of the cargo section. Similar studies could also be made to establish minimum distance of the cargo tank section forward bulkhead from the forward end of the ship, and a minimum number of bulkheads between stem and cargo tanks, to provide a minimum standard of collision protection. Such standards already exist in the rules of the classification societies, but modification may be justified following appropriate study.

An important action in this direction was taken recently by the Maritime Safety Committee of the International Maritime Consultative Organization (IMCO), an agency of the United Nations. In an effort to minimize outflow of oil in the event of tank rupture, formulae have been established which effectively limit cargo oil tank sizes to 22,500 cubic meters for wing tanks and 50,000 cubic meters for center tanks. It is intended to apply these limitations to new construction in which the building contract is placed after January 1, 1972.

Among the structural measures which may warrant further study are the following:

a) Structural design to increase collision protection. Outboard wings could be designed to absorb collision energy and limit penetration. Techniques are available to study various arrangements empirically, as in the case of collision protection structure provided outboard of the N. S. SAVANNAH reactor space.
b) Associated with the above study, potential gains from use of higher strength steels could be considered.
c) Mandatory installation of stress measuring instrumentation at strategic locations in the hull structure, with indicators located on the bridge.

5.3 *Ship design, performance and maneuverability*

Maneuverability is considered here as a primary means for avoiding collison and grounding in emergency conditions. Accordingly, the following observations and studies are pertinent to the overall problem.

5.3.1 Backing power—Data given in Reference (2) demonstrates that steam turbine propelled single screw U.S. flag tankers, with displacement ranging from 25,000 to 77,000 tons, can be expected to stop in $8\frac{1}{2}$ to 9 ship lengths from the full ahead condition and periods of $5\frac{1}{2}$ to 9 minutes will be required to execute this maneuver from the time the crash stop order is given. The 144,000 DWT, combination oil/bulk vessel CEDROS, described in Reference (30), required $10\frac{1}{2}$ ship lengths and over 11 minutes to stop from an initial speed of 18 knots.

The geared steam turbines installed in the above ships are fitted with astern elements with a capability of about 80 per cent full ahead torque and 50 per cent full ahead RPM, i.e., about 40 per cent full ahead power. The time required to obtain full astern revolutions, from the full ahead condition, is on the order of one minute.

Model experiments reported in Reference (31) indicate that crash stop characteristics of the diesel ship may be somewhat better, probably because full astern power is available.

It has been shown that tanker sizes have grown at a far greater rate than the installed power, hence, the stopping ability has been reduced significantly, with the obvious reduction in ship maneuverability. Accordingly, the following studies have been suggested in References (18), (19) and (25):

a) Study of backing characteristics of existing vessels, particularly with respect to known casualties.
b) Establishment of minimum recommended or mandatory standards, defining acceptable values of ahead reach and time to execute the crash stop maneuver, as measured on standardization trials, directed toward establishment of minimum acceptable values of astern powers for the various types of power plants.
c) Study of effects of propeller type on stopping characteristics—The model studies discussed in Reference (30), for example, indicate that the controllable pitch propeller is superior in this characteristic to fixed pitch propellers.
d) Study of relative characteristics of various transmission systems—Where geared or electric drive systems are applicable, various reversing methods are available. In the extreme example, the 20,000 SHP diesel-electric plant of a modern icebreaker can be stopped from full ahead to 0 RPM in 10 seconds and full ahead to full astern in about 20 seconds. The potential gains from the various

systems should be evaluated by an analysis of the RPM-torque-thrust characteristics over the entire crash stop time sequence.

5.3.2 Auxiliary braking devices—Study of auxiliary braking devices, including the following, is proposed:

(1) Shaft brake systems, mechanical or other.

(2) Hydrodynamic—Various schemes have been considered in the past, including a Japanese rudder design with hinged leaves which open to form flat braking surfaces at 90° to the direction of motion. Other concepts may be considered, including adaptation of thruster systems, active rudders, etc.

5.3.3 Steering and directional control—Table II summarizes turning circle data for a few tank vessels ranging from the early 19,000 DWT tankers to the CEDROS and NISSEKI MARU, each over 100,000 DWT. Turning circles of tank vessels have remained essentially constant in this limited sample. However, in view of the greatly increased sizes and relative reduction in power, consideration should be given to improving maneuvering characteristics at sustained sea speeds as well as at reduced speed conditions. Accordingly, the following studies have been suggested:

a) Extensive study of maneuvering characteristics of existing ships, with particular reference to casualty records.
b) Establishment of minimum standards for control, including turning rate, turning, circles, etc., as determined by standardized trials procedures.
c) Study of rudder design—For the large single screw tanker particularly, consideration should be given to rudder designs other than the current standard types in order to improve characteristics at high rudder angles and at low maneuvering speeds. The various configurations that might be studied include active rudders, biplane and triplane rudders, flapped rudders, kort nozzle rudders, and dynamic systems that relate optimum braking ability to periodic rudder reversal during the stopping maneuver.
d) Thruster design—Various bow and stern thruster systems have been developed, some well over 1,000 HP in capacity. Applicability to the large tanker, as an augment to the steering system, should be considered.

5.3.4 Mooring gear and ground tackle—A last resort menas of stopping the ship, and a normal tool for low speed maneuvering, is the use of the ship's anchors. Progress in the development of ship's anchors has not kept pace with general ship technology, and studies have been suggested in the following areas:

a) Develop more rapid means for release of anchor and chain. Centralized control of the mooring gear from the bridge has been provided in a few recent designs and is a step in this direction.
b) Study advisability of requiring the fitting of stern anchors, particularly in case of larger vessels, as an emergency stopping and maneuvering device, and to facilitate maneuvering in congested waters.

5.4 Ship design, safety requirements

Regulatory agencies such as the U.S. Coast Guard actively upgrade national shipping regulations to suit changes in international regulations and changes in the technical nature of marine transportation. The most recent example is given in the

TABLE II

Typical Turning Circles

Vessel	CEDROS	Newportt News* TKR "C"	Newport News* TKR "A"	ORIENTAL GIANT	ESSO SUEZ	NISSEKI MARU
Length, B.P.	940′–0″	685′–0″	535′–0″	804′–0″	600′–0″	1082′–8″
Breadth	142′–0″	93′–0″	75′–0″	108′–0″	82′–6″	178′–9$\frac{3}{4}$″
Depth	81′–0″	48′–7″	40′–5″	–	42′–6″	114′–10″
Draft	54′–0″	36′–8″	31′–9″	–	31′–10″	88′–7″
Displacement, tons	175,940	50,200	25,500	92,980	34,640	–
Shaft horsepower	27,500	26,500	15,000	–	12,500	40,000
Speed, knots	18.3	19.5	18.8	17.6	16	14$\frac{1}{2}$
Turning cirde, feet	3,105	2,380	3,200	2,590	2,280	3,530
Turning circle/LBP	3.31	3,47	5,97	3.22	3.8	3.26

* From reference (2).

U.S. Coast Guard agenda, Reference (32), containing proposed changes in existing rules for manning, marine engineering, subdivision and stability, for tank vessels and dry cargo vessels.

Specific areas of ship safety which have been suggested for active study include the following:

5.4.1 Inerting—Establishment of mandatory inerting requirements for normal use as well as for emergency measures. Recent discussions of the cognizant IMCO subcommittee resulted in recommendations for mandatory provision of inerting systems within the cargo tanks of petroleum tankers.

5.4.2 Liquid cargo systems—Study cargo pumping methods and piping systems with the object of establishing mandatory emergency cargo pumping systems, including mandatory installation of standardized fittings at each tank, to facilitate pumping to an attending ship. Consider increasing requirements for provision of emergency power sources, to provide limited capability for cargo pumping in event of loss of the main power plant.

5.4.3 Navigation—The development of improved radar, echo ranging and depth sounding and recording is under continuous development and is well documented in the literature.

Installation of simple instrumentation to provide accurate forward, midship and after draft readings to the deck officers. These values are continuously changing during a voyage and such information becomes more important as ship sizes continue to increase.

The installation of visible signals to complement the audible whistle signals has been suggested as an aid to avoiding collisions and is, in fact, common practice on the U.S. Great Lakes.

The development of an effective vessel traffic safety system is under serious consideration and is the subject of legislation pending in the U.S. Congress at the time of this writing.

REFERENCES

1. D'Arcangelo, "Ship Design and Construction," The Society of Naval Architects and Marines Engineers, 1969.
2. Long, Stevens, and Tompkins, "Modern High Speed Tankers," Transactions of the Society of Naval Architects and Marine Engineers, 1960.
3. Nickels, Rubin, and Danielson, "Some Aspects of Large Tanker Design," Transactions of the Society of Naval Architects and Marine Engineers, 1960.
4. "NISSEKI MARU—A Fine Ship in Every Respect," ZOSEN, November 1971.
5. Abrahamsen, E., "The Carriage of Special Liquid Cargoes," *European Shipbuilding*, No. 6, 1964; No. 1, 1965.
6. Thomas, W., and A. Schwendtner, "LNG Carriers: The Current State of the Art," Transactions of the Society of Naval Architects and Marine Engineers, 1971.
7. National Academy of Sciences, National Research Council, "Evaluation of the Hazard of Bulk Water Transportation of Industrial Chemicals—A Tentative Guide," A Report to the U.S. Coast Guard, prepared under contract Tcg-15559-A, Washington, D. C. 1970.
8. "Chemical Data Guide for Bulk Shipment by Water," U.S. Coast Guard, CG-388.
9. U.S. Coast Guard, Dept, of Transportation, "Rules and Regulations for Tank Vessels," publication CG-123.
10. U.S. Coast Guard, Dept. of Transportation, "Rules and Regulations for Cargo and Miscellaneous Vessels," publication CG-257.
11. U. S. Coast Guard, Dept. of Transportation, "Marine Engineering Regulations and Material Specifications," publication CG-123.
12. American Bureau of Shipping, "Rules for Classing and Building Steel Vessels," published annually.
13. Dorman, W. J., "Combination Bulk Carriers," *Marine Technology*, October 1966.
14. "The Combination Bulk Carrier," *Surveyor*, August 1970.
15. Comstock, J. P. (editor), "Principals of Naval Architecture," The Society of Naval Architects and Marine Engineers.
16. "The Single Point Mooring," *Marine Engineering/Log*, July 1968.
17. Thompson, N. J., and T. B. Thomas, "Marine Transportation of Mineral Slurries," *Marine Technology*, April 1971.
18. Roseman, D. P., "The Design of Tank Vessels with Respect to Pollution from Collision and Stranding," *Hydronautics*, Incorporated Technical Report 822-1, November 1967.
19. Battelle Memorial Institute, "Oil Spillage Study; Literature Search and Critical Evaluation for Selection of Promising Techniques to Control and Prevent Damage," Report No. AD-666289, Defence Documentation Center.
20. "Explosions in Tankers—Some Methods of Prevention ...," *Shipping World and Shipbuilder*, March 1970.
21. "Explosions in Tankers—International Chamber of Shipping Report ...," *Shipping World and Shipbuilder*, November 1970.
22. "Barges on the High Seas," Surveyor (publication of the American Bureau of Shipping), November 1967.
23. American Bureau of Shipping, "Rules for Building and Classing Steel Barges for Offshore Service."
24. Jones, R. M., and C. S. Smith, "Large Seagoing Unmanned Barge/Tug Developments," Southeast Section, Society of Naval Architects and Marine Engineers, 23 October 1970.
25. Price, R. I., "Anti-Pollution Measures—IMCO Subcommittee on Ship Design and Equipment," *Marine Technology*, January 1971.

26. Dillon, E. S., "Ship Design Aspects of Oil Pollution Abatement," presented at the Chesapeake Section, Society of Naval Architects and Marine Engineering," Washington, D. C., 17 March 1971.
27. "FORTUNA, New Duokleen Tanker Construction," *Shipping World and Shipbuilder*, August 1966.
28. "Duokleen Construction, an Added Safety Factor," *Shipbuilding and Shipping Record*, 8 June 1967.
29. "MOBIL PEGASUS, Japanese Built Tanker with a Double Bottom," *Shipping World and Shipbuilder*, March 1970.
30. "CEDROS, Combination Oil/Bulk Cargo Carrier," *Shipping World and Shipbuilder*, January 1967.
31. Hooft, J. P., and J. D. van Manen, "The Effect of Propeller Type on the Stopping Abilities of Large Ships," *Shipping World and Shipbuilder*, August 1967.
32. Merchant Marine Council Hearing Agenda, March 29, 1971, U.S. Coast Guard publication CG-249.

DONALD P. ROSEMAN
Hydronautics, Inc.

MATERIAL RECYCLING: see RECYCLING OF WASTE MATERIALS

MERCURY: see POLLUTION EFFECTS ON FISH

MICROBIOLOGY

Introduction

Microbiology is the study of organisms which are small enough to require the aid of a microscope to be seen. In a few cases, organisms are included in this group which can be seen by the unaided eye because these organisms are clearly related to the smaller ones. Microorganisms include viruses, bacteria including rickettsia, mycoplasma, fungi (yeasts and molds), most algae, protozoa and, if one interprets "micro" broadly, certain tiny multicellular plants and animals. The study of cells and tissues from higher plants and animals (*tissue culture*) uses techniques common to the microbiologist and is frequently considered part of modern microbiology.

Cells in general vary greatly in size but have many similarities in internal organization. Among the most primative types of cells, it is impossible to clearly distinguish whether they are distinctly "plants" or "animals" since they may have some of the properties of each type. Viruses, on the other hand, are *not cells* at all. Instead of arguing endlessly about whether a microorganism is more plant-like or more animal-like and worrying how to assign viruses, many scientists have divided organisms in general into those which have 1. only animal characteristics, 2. only plant characteristics and 3. the *Protista* (Table 1) which have some characteristics of both plants and animals. Some protists, viruses, may have characteristics not shared by either plants or animals, *i.e.* crystallizability and ability to reproduce only by infecting some cell and using the cell's manufacturing machinery.

Physical characteristics of microorganisms

Protists vary greatly in size, shape and internal architecture. Protists are subdivided into *prokaryotes*, and *eukaryotes. Prokaryotes* do not have their genetic material (chromosomes) separated from the rest of the cell by a membrane whereas eukaryotes have a true nucleus (*eu*—true, *karyo*—nucleus) separated from the rest of the cell by a nuclear membrane. Viruses (virions) are usually included among the prokaryotes. There are 9 types of prokaryotes.

Prokaryotes

1. *Viruses* are the *smallest protists*. They range in size from about 30–300 nm. The smallest viruses can only be visualized with an electron microscope

TABLE 1

Characteristics of the Protista

Name	Size Range	Shape	Genetic material	Method of reproduction
Virion (virus)	30–300 nm	icosahedron, hollow cylinder icosahedral head + tail	RNA, DNA	requires partipation of host machinery
Mycop asma	100–300 nm	pleomorphic prokaryotes	DNA	fission
True bacteria	250–3000 nm	spherical, rod, spiral rods. prokaryotes	DNA	fission
Higher bacteria	500–5000 nm	spherical, rod, spiral rods, filamentous. prokaryotes	DNA	fission, budding
Prokaryotic algae	500–5000 nm	spherical, rods in chains, spiral rods in chains	DNA	fission, internal septation, gonidia
Eukaryotic algae	500 nm to macroscopic	unicellular or multi-cellular, filamentous, leafy	DNA in nucleus, chloro-plasts, mitochondria	asexual or sexual simple fission to complex life cycles
Protozoa	500–500,000 nm	unicellular or colonial various forms	DNA in nucleus, mitochondria	asexual or sexual simple fission to com-plex life cycles

while the largest can be seen with a light microscope. Viruses are composed of two general molecular types a) *only one* nucleic acid, either *ribonucleic acid* (RNA) or *deoxyribonucleic acid* (DNA), and b) a group of proteins also called *protein subunits* or *capsomeres*, which surround the nucleic acid and form a protective coat or *capsid*. The smallest viruses appear spherical but magnification in the order of 150,000–700,000 x reveals that they are *icosahedrons* (20 triangular faces and 12 corners) *e.g.* wart virus. Other viruses, *e.g.* tobacco mosaic virus (TMV), the first virus crystallized in 1935 by Wendell Stanley, is grossly rodlike. Tobacco mosaic virus is composed of a central, spirally-arranged RNA to which capsomeres are attached to form the outside of a cylinder. The center of the RNA spiral of TMV is hollow. Structurally, the most complicated viruses are some which attack bacteria and blue-green algae.

These complicated viruses are composed of an icosahedral head, containing DNA, a protenaceous tail and sometimes accessory tail structures which are important for the attachment of the virus to its host cell.

2. *Mycoplasma* are prokaryotes which overlap viruses in size. They range from 100–300 nm in size. They are highly *pleomorphic*: they do not have one typical shape but rather can appear coccoid, filamentous, or highly branched. Unlike most other prokaryotes, they do not have cell walls external to their cell membranes. Their cell membranes usually contain sterols which are thought to lend strength to these cell-limiting membranes (See also Table 2).

3. The *true bacteria* or *Eubacteriales* are prokaryotes which are built on three general geometric forms: *spheres* or cocci, *rods*, and *spirals* (including spiral helices). All true bacteria have rigid cell valls. They are either permanently immotile or move by means of one to many flagella. They may be aerobes or anaerobes. Some of the anaerobes are photosynthetic. Their sizes and shapes are usually constant except among the rods in which rapidly-multiplying cells may be somewhat smaller than usual. When the cells divide, they often remain attached to each other and form characteristic, multicellular clusters. The shape of the cluster is determined by the number of division planes.

When *cocci* divide in only one plane, they form chains which may be as much as 20 cells long. *Diplococcus pneumoniae* forms chains only two cells long while *Streptococcus* is an example of the long-chain-forming type. Onthe other hand, cocci which divide along two planes, at right angles to each other, form sheets of cells, and cocci which divide in three planes from cube-shaped packets. If there is no regular pattern of the orientation of successive division planes, a randomly-shaped

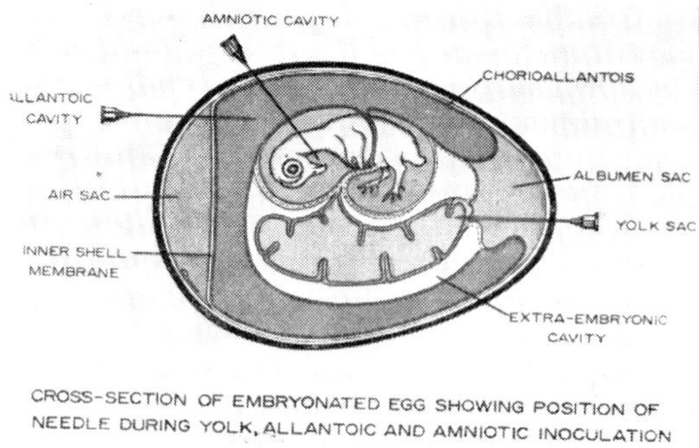

Figure 1 A. Animal viruses are often grown in embryonated eggs. The position of the hypodermic needles indicate three common inoculation places.
B. Lesions on the chlorioallantoic membrane caused by vaccinia pox.

Table 2

Some Characteristics of Prokaryotic ynd Eukaryotic Cells

Structure	Prokaryote	Eukaryote
Weight	0.001–1.0 pg	10–10,000 pg
Chromosome	one, single circular DNA double helix not complexed with histones	several linear DNA double helices (several chromosomes usually complexed with histones)
Nucleus	No true nucleus. Chromosome not separated from cytoplasm by a membrane	True nucleus. Chromosomes enclosed in a nuclear membrane
Reproduction	Usually asexual, conjugation takes place rarely, no mitosis or meiosis	Asexually by mitosis or sexually after meiosis.
Membranes	Only cell limiting membrane present. Usually lacks sterols (except for mycoplasma)	Cell limiting membrane plus membrane limited organelles present. Composition includes sterols
Organelles	None	Many including mitochondria, chloroplasts (plants only), Golgi apparatus, lysosomes, etc.
Apparatus for protein synthesis	Ribosomes, 70 S type usually not associated with membranes	Ribosomes, 80 S type in cytoplasm associated with endoplasmic reticulum. 70 S type in mitochondria and chloroplasts not associated with membranes

cluster is formed. *Staphylococcus* is an example of a coccus which forms random clusters. A typical coccus is in the size range of 0.5–1.5 μm in diameter.

Rods always divide in only one plane. They mya appear as single cells or groups of only two when they separate rapidly. The common intestinal bacterium *Escherichia coli* (size 0.5 × 2.0 μm) is an example of this type. Frequently rods form long chains or *streptobacilli. Bacillus megaterium* (size 1.35 × 3.0 μm), the organism responsible for the "bloody bread" of ancient times, is an example of a chain-forming rod. Some basically rod-shaped bacteria are either curved or helical rods. Their sizes range from almost as small as the smallest straight rod shaped form to close to twice the length of the largest straight rod.

True bacteria always divide by binary fission after their single circular chromosome replicates in a semiconservative fashion.

Some true bacteria have complicated life cycles which includes *spore-formation.* Spore-formers are all rods but belong to diverse genera. They are ecologically related in that they are found primarily in soil. Since that natural environment is rather variable in that it can range from very hot to very cold and from very wet to very dry, the heat- and cold-resistant dormant spores offer the bacteria a means of surviving adverse environmental conditions for months or even years. Many important pathogens and commercially-important organisms are spore formers, *e.g. Bacillus anthracis* which causes anthrax, *Clostridium tetani* which causes tetanus and *Clostridium acetobutylicum* which can ferment corn or potato mash into acetone, ethanol and butanol.

Corynebacteria are also rod-shaped bacteria but they are *pleomorphic* and often look club-shaped.

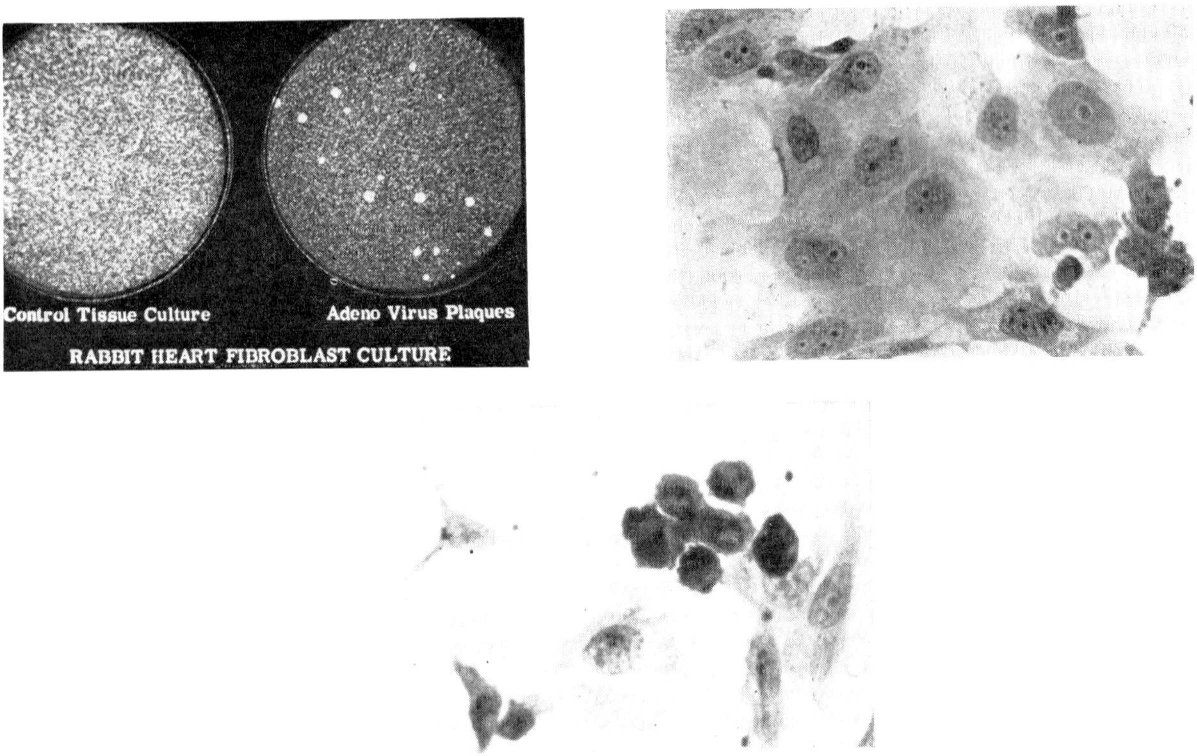

Figure 2 A. Effect of adenovirus on rabbit heart fibroblast cells grown in petri dishes on a solid substrate. The cells are too small to see individually. They only appear as dark specks. Adenoviruses form plaques or holes eaten out of the confluent lawn of cells as a consequence of viral destruction of the infected fibroblast.
B. Photomicrograph of normal rabbit fibroblasts in culture.
C. Photomicrograph of the same culture after injection with adenovirus. The darkened clump is the remains of 6 cells which have been attacked by the virus. The clear area is the edge of a plaque from A. A few relatively normal cells are still present.

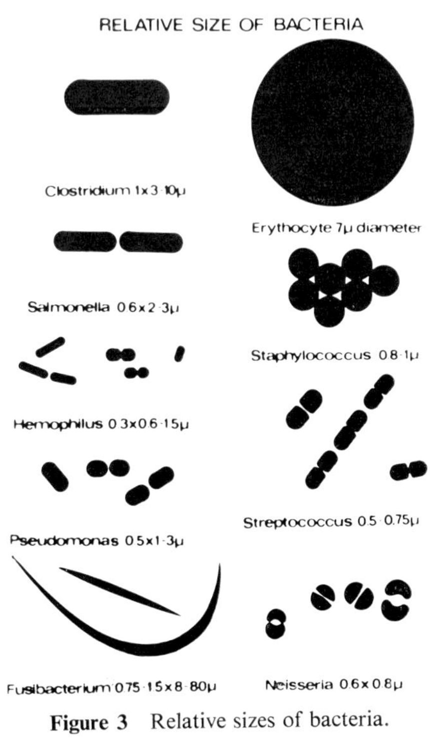

Figure 3 Relative sizes of bacteria.

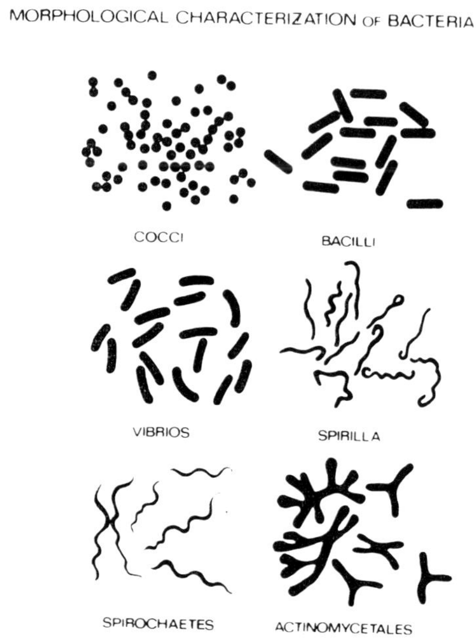

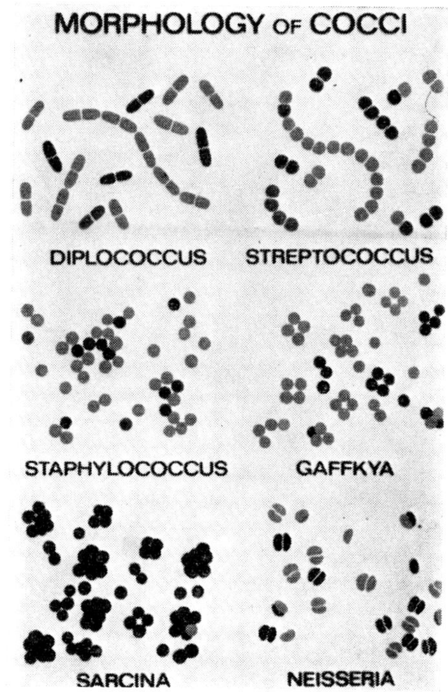

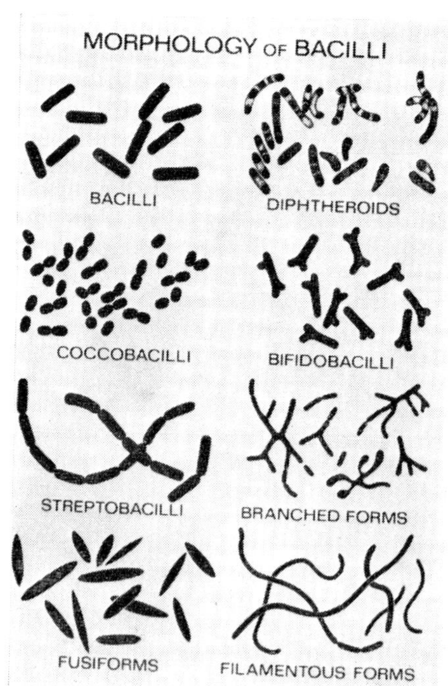

Figure 4 A. General morphological characteristics of bacteria.
B. Variety of morphological types among the cocci.
C. Variety of morphological types among the bacilli (rods).

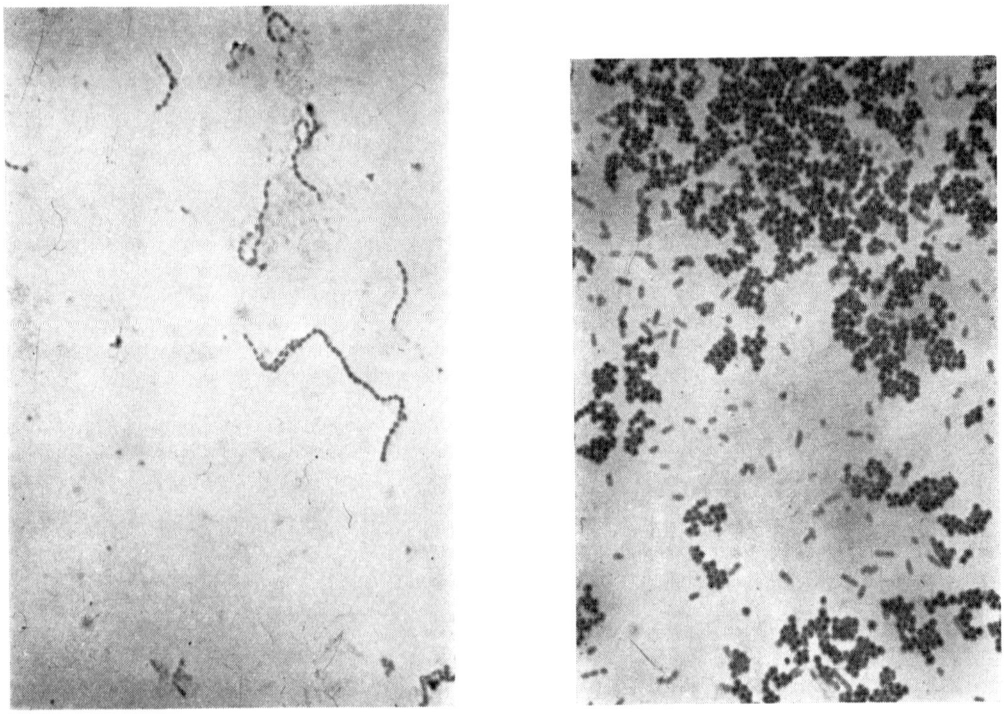

Figure 5 A. *Streptococci* spp.
B. *Staphylococcus* spp. A small number of bacilli are also visible.

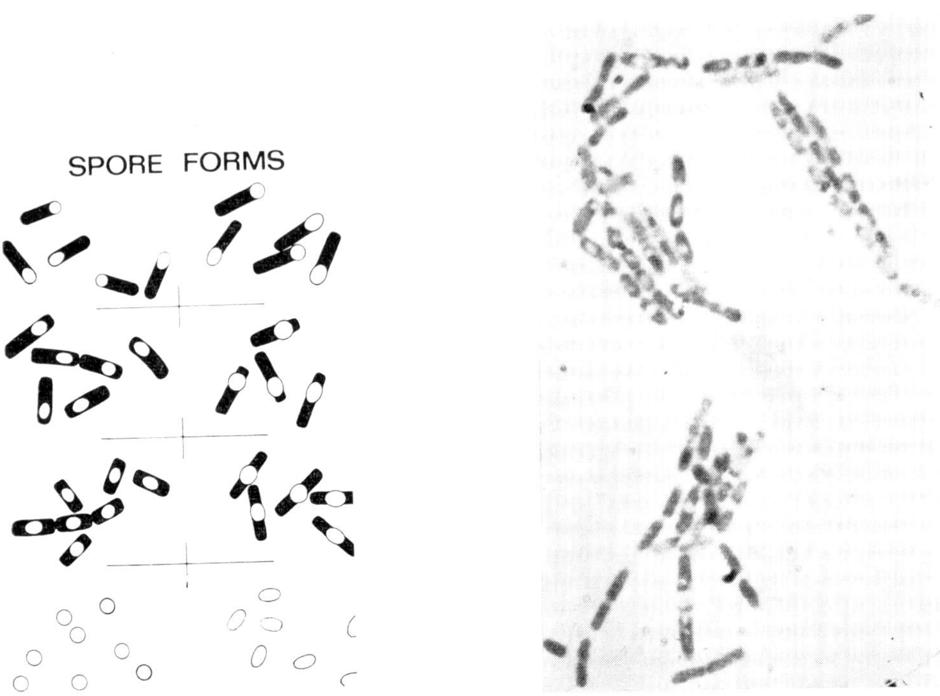

Figure 6 A. Diagramatic representation of spores (clear areas) inside rod-shaped bacteria. Note (bottom row) that free spores may be ball or egg-shaped.
B. Streptobacilli treated with a spore stain.

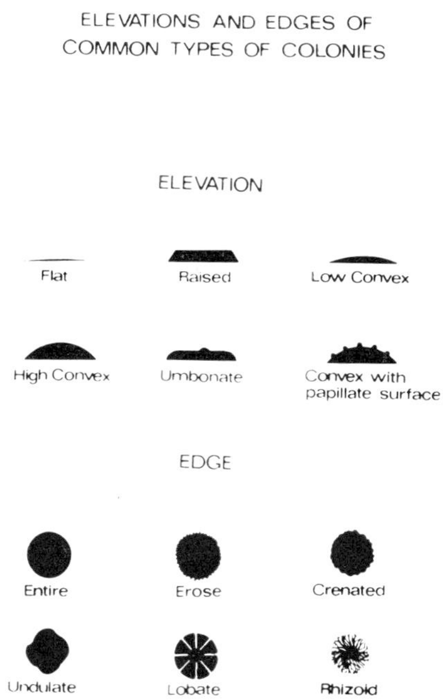

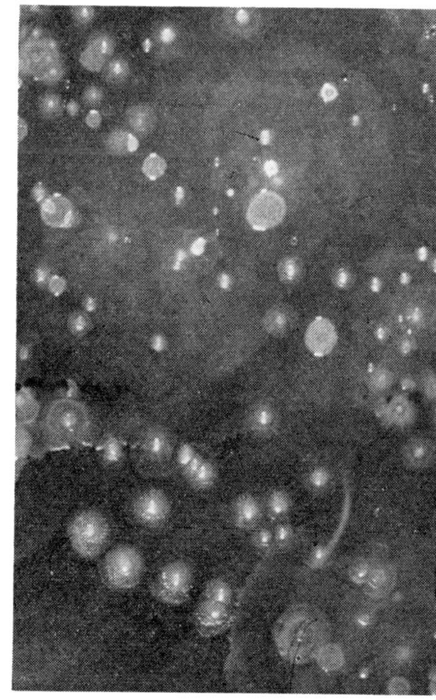

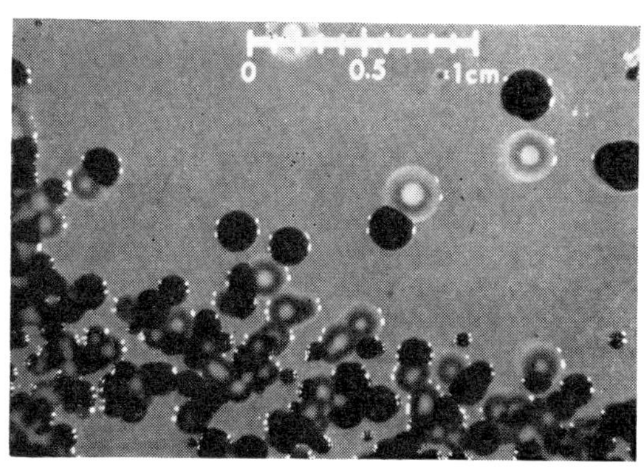

Figure 7 A. Diagramatic representation of types of bacterial colonies. These shapes are specific for individual types and are therefore quite useful as a step in the process of identification of unknown organisms.
B. Mixed colony types on the surface of an agar plate. Note that some colonies have spread over the surface of several others.
C. Colonies of the usually pigmented *Seratia marcescens* showing several colonies with altered pigment. Each individual colony arose from one bacterium. Each colony with altered pigment represents the progeny of a mutant *S. marcescens*.

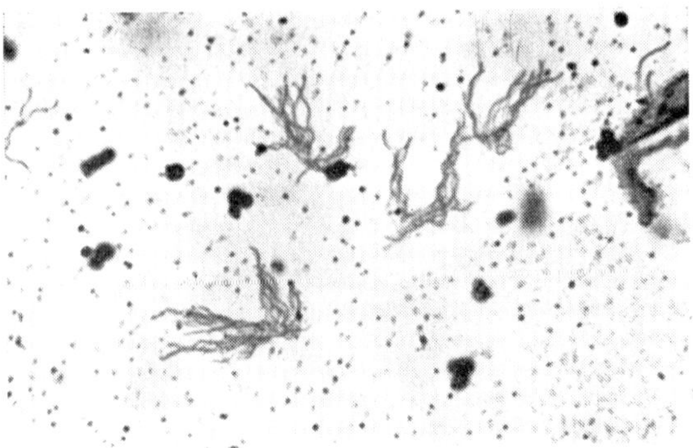

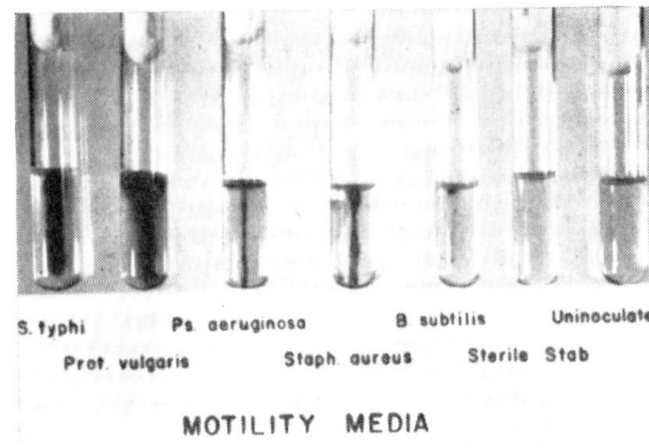

Figure 8 Bacterial mobility.
A. Bacteria are motile by means of flagella. The flagella of *Salmonella typhosa* are shown here in a stained preparation.
B. Motility is tested by stabbing an inoculated needle into a tube of very viscous growth medium. The motile organisms (*S. typhi* and *P. vulgaris*) grow away from the stab mark.

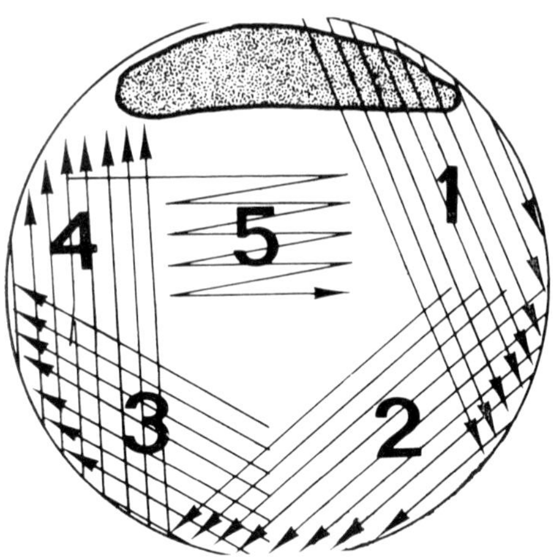

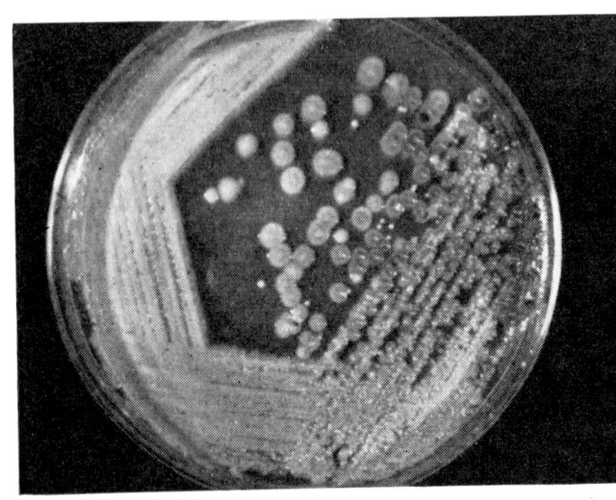

Figure 9 Isolation of single bacterial colonies on agar plates by dilution streaking.
A. Diagramatic representation of method if streaking inoculated needle across nutrient-containing plate. Stippled area is the primary inoculation. The inoculating needle is then flamed to sterilize and is then drawn across the stippled area as indicated for area 1. The needle is then resterilized and drawn across area 2, etc.
B. Actual plate prepared as indicated in A. Note that the areas are arranged here *counterclockwise*.

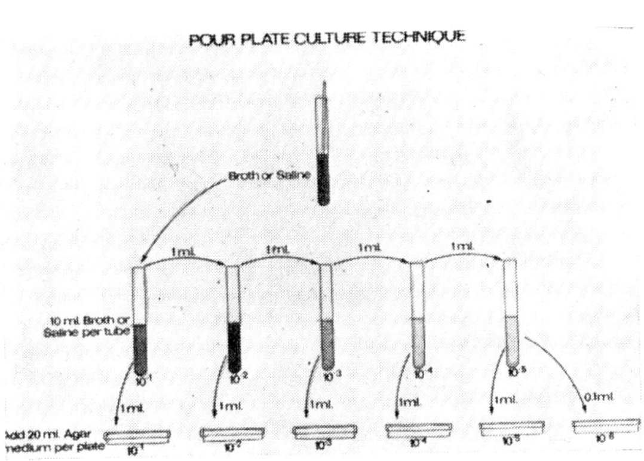

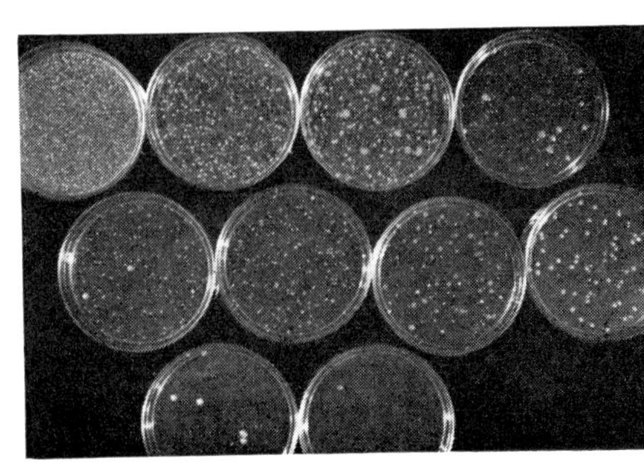

Figure 10 Isolation of single colonies by pour plate technique.
A. Diagramatic representation of the technique by this.
B. Actual plates prepared method. The most concentrated inoculum was placed in the plate at the upper left, the least concentrated in the two plates at the bottom.

One of the best known members of the genus is *C. diphtheriae* which causes diphteria. Other members of the genus are commerically important as producers of the vitamin folic acid.

Arthrobacter species are found widely in soil and water. Depending upon the nutrients supplied, they can appear as cocci or pleomorphic rods.

4. *Spirochetes* are NOT true bacteria though they resemble Eubacteriales in that they are spirally-curved, unicellular and multiply by binary fission. They differ from eubacteria by the *absence of a rigid cell wall* which allows them to be quite flexible. They are all motile by means of *axial filaments* attached at the cell poles and spirally-wrapped around the cell. The smallest spirochete is $0.1 \times 5\,\text{nm}$ while the largest is $3.0 \times 120\,\mu\text{m}$. One of the most important spirochetes is *Treponema pallidum* which causes syphilis.

5. *Actinomycetes* are NOT true bacteria. Rather, they are naturally-branching, filamentous, spore-forming organisms which have a *mycelial structure* similar to that of filamentous fungi. Many actinomycetes, especially those from the genus *Streptomyces*, are commercially important sources of antibiotics.

6. *Mycobacteria* are rods which can form a rudimentary mycelium which resembles actinomycetes but they differ in that their cell walls are particularly rich in waxes which allows them to retain stain imparted by such dyes as basic fuchsin even after treatment with dilute acid. This property, called *acid fastness*, is characteristic of mycobacteria. Many species occur in soil but the best known are the human pathogens *M. tuberculosis* and *M. leprae* which cause tuberculosis and leprosy respectively.

7. *Budding bacteria* are NOT true bacteria. They possess a complicated life cycle which includes *multiplication by budding* rather than binary fission. Their type of budding can be readily distinguished from that of true fungi such as yeast. The budding bacterium *Hyphomicrobium* exists for part of its life cycle as a flagellated, slightly curved rod. For multiplication, the flagellum is lost, the chromosome replicates, and one chromosome migrates to one end of the cell where a hypha-like lengthening takes place. When the hyphal extension ceases, it becomes a rounded bud which contains the chromosome. The bud grows in length and diameter until it reaches the size of the mother cell, grows a new flagellum, and separates from the hyphal extension.

8. *Gliding bacteria* are a diverse group of prokaryotes which are motile without having flagella. They have very close affinities to blue-green algae although gliding bacteria are not themselves photosynthetic. They may be unicellular rods, helical or spiral-helical, or filamentous.

9. *Blue-green algae* or *Cyanophyta* are the only prokaryotic algae. They are a diverse group that include both unicellular and filamentous forms. They have cell walls that resemble Gram-negative bacteria but their photosynthesis more closely resembles that of eukaryotes in that it is aerobic rather than anaerobic (as in photosynthetic bacteria) They are among the most complex prokaryotes. Even though they lack defined organelles, *e.g.* they lack *chloroplasts*, many species have complex membranous or vesicular substructures which are continuous with the cell membrane.

Some filamentous forms contain specialized structures such as *gas vacuoles*, *heterocysts*, or *resting spores* (*akinetes*). Gas vacuoles are frequently found in *planktonic species*, *i.e.* those which live in open water. These vacuoles are thought to provide the algae with a means of floating and sinking to the depth most appropriate to support photosynthesis. *Heterocysts* arise from vegetative cells and are thought to function in N_2 fixation. Some blue-green algae show gliding motility. None are flagellated. They are very widely distributed either in terrestrial or aquatic habitats from the arctic to the tropics. Some forms are found in hot springs. Other *Cyanophyta* are *symbionts* in a variety of plants and animals. For example a species of *Anabaena* fixes N_2 for its host the water fern, *Azolla*. Many blue-green algae form especially luxurient mats of growth called *blooms* which clog waterways and limit their use for navigation, etc.

The prokaryotic blue-green algae, *Cyanophyta*, are usually divided into 5 groups:

Chrooccocales are unicellular. They sometimes occur in irregular packets or colonies. Cells multiply by binary fission.

Chamaesiphonales are unicellular, filamentous, or colonial epiphytes or lithophytes. Cells show distinct polarity from apex to base. The base usually has a holdfast which permits attachment to the substrate. Cells multiply by internal septation or by formation of spherical cells (*gonidia*) at the ends of filaments.

Pleurocapsales are filamentous with differentiation into aerial and nonaerial elements. Cells multiply by crosswall formation or by internal septation.

Nostocales are filamentous without differentiation into aerial and nonaerial elements. They are unbranched or falsely branched and frequently have pale, empty-looking cells called *heterocysts* and resting spores (*akinetes*). Reproduction is by liberation of a short filament only a few cells long, called a *hormogonium*, which then elongates.

a) *Nostacaceae* are unbranched and produce heterocysts. They frequently produce akinetes.

b) *Rivulariaceae* are unbranched or falsely branched. Filaments taper from base to tip. Heterocysts are usually present at the base. There is some akinete formation.

c) *Scytonemataceae* are falsely branched. Heterocysts are frequently found at branch points.

d) *Stigonematalis* are filamentous with aerial and nonaerial differentiation. Hormogonia and heterocysts are present. They often show true branching and have pit connections between cells. Akinetes are rare.

Eukaryotes

Eukaryotic microorganisms include all the algae (except the *Cyanophyta*), all the protozoa, and most fungi. All are microscopic in size.

The eukaryotic algae are separated into nine *divisions* based upon their pigment and carbohydrate reserves (Table 3). They are all photosynthetic and, like higher plants, evolve oxygen during photosynthesis. Many algae are *obligate phototrophs*. That is, they are completely dependent upon photosynthesis: they can not use exogenously supplied organic compounds for growth in either the dark or light. Some algae are *facultative phototrophs*: they are able to utilize organic compounds for growth in the dark but fix carbon dioxide photosynthetically in the light.

Occasionally algae, especially unicellular forms, permanently lose their chloroplasts by exposure to any one of several adverse conditions, *e.g.* heat or

TABLE 3

Divisions and Characteristics of the Eukaryotic Algae

Division	Pigments							Reserve material	
	chlorophyl					biliproteins		Starches (α-1,4-glucans)	
	a	b	c	d	e	phycocyanin	phycoerythrin	True starch	Floridian starch
Chlorophyta (green algae)	+	+	—	—	—			+	
Euglenophyta	+	+	—	—	—	—	—		
Xanthophyta	+	—	—	—	+	—	—		
Chrysophyta	+	—				—	—		
Phaeophyta (brown algae)	+	—	+	—	—	—	—		
Bacillariophyta (diatoms)	+	—	+	—	—	—	—		
Pyrrophyta	+	—	+	—	—	—	—	+	
Cryptophyta	+	—	+	—	—	+	+	+	
Rhodophyta (red algae)	+	—	—	?	—	+	+		+

Division	Reserve material (cont'd.)					Sugar alcohols	Lipid
	β-1,3 glucans			Sugars			
	Laminarin	Paramylon	Chrysolaminarin	Floridoside	Sucrose	Mannitol	
Chlorophyta (green algae)					+		
Euglenophyta		+					
Xanthophyta		+					+
Chrysophyta							
Phaeophyta (brown algae)	+					+	+
Bacillariophyta (diatoms)			+				+
Pyrrophyta							+
Cryptophyta							
Rhodophyta (red algae)				+			

chemicals. If the organism had been a facultative phototroph, before the loss of the chloroplasts, it has the enzymatic machinery necessary to survive except that now, in its chloroplastless state, it is indistinguishable from certain other unicellular organisms more commonly called protozoa. The ease with which an organism at this primative level of evolution may be interchanged between groups containing a preponderance of plant-like or animal-like attributes underlines the need for the term *protist* rather than plant or animal to describe them. Indeed both botanists and zoologists claim the protists. Some algae *e.g. Euglena* spp., normally only form chloroplasts when they grow in the light while others *e.g. Chlorella* spp. form chloroplasts regardless of the presence or absence of light. There is great diversity in size, shape, presence or absence of life cycles, type of multiplication, motility cell wall chemistry, and chloroplast structure. Although these parameters are of great assistance in defining affinities among alage, there are still groups whose proper place is debated.

Many algae are important as sources of food, chemical intermediates of industrial and medical importance, and research tools. Others are nuisances which clog waterways or poison other aquatic life with their potent toxins.

Eukaryotic algal groups

The 8 groups are:

1) *Chlorophyta* (green algae) are either marine or freshwater forms. This large and diverse group includes forms which are either unicellular, colonial, filamentous, *tetrasporal* (cells separated but held together in groups of four in a mucilaginous material), *coenobial* (cells more or less attached to each other in an aggregate), or *siphonaceous* (simple, nonseptate filaments). They frequently have life cycles which include motile, flagellated stages. Both asexual and sexual reproduction occurs.

2) *Euglenophyta* differ from the other algae by possessing a rather flexible cell wall which allows considerable plasticity of form. They are either fresh water or marine forms. They all have two flagella but in some genera the second flagellum is often rudimentary. Many forms are *phagotrophic* (can ingest particles). Chloroplastless forms are fairly common. Multiplication is only by asexual means.

3) *Xanthophyta* are mostly freshwater forms. They may be unicellular, colonial, filamentous or siphonaceous. Some forms have life cycles which include both asexual and sexual reproduction. Motile anteriorly flagellated cells are found.

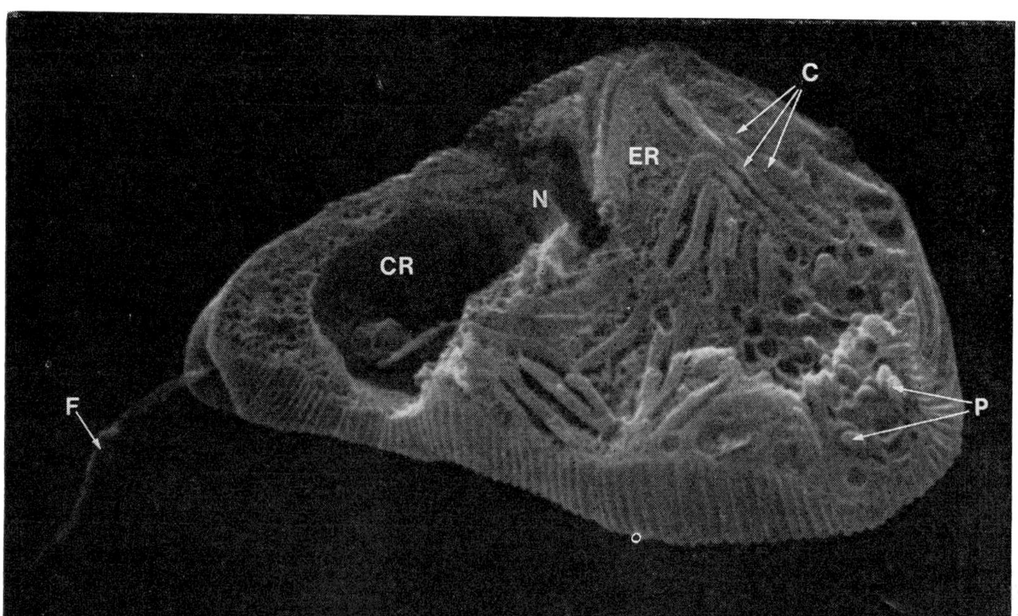

Figure 11 A scanning electron micrograph of a fractured cell of the alga *Euglena gracilis*. The stringbean-shaped chloroplasts (C) and carbohydrate storage granules, paramylon (P) are embedded in spongy endoplasmic reticulum (ER). The edge of the nucleus (N) can be seen below the cavity of the cytostomal reservoir (CR). The flagellum (F) arises from beneath the base of the cytostomal reservoir and emerges through a narrow canal (which is not visible in this picture).

4) *Chrysophyta* are mainly freshwater forms but important marine forms are known. Most genera are unicellular but there are some colonial forms. Cell walls often composed of siliceous or calcareous plates. Some form siliceous cysts. They are mainly found in fresh water but some important marine forms exist. Reproduction is asexual.

5) *Phaeophyta* (diatoms) are unicellular or colonial forms with distinctly patterned siliceous cell walls. Both asexual and sexual multiplication is found. Freshwater, marine, soil and aerial forms exist.

6) *Pyrrophyta* are unicellular flagellates with cellulose cell walls which are sometimes formed in plates. Reproduction is asexual. Sexual reproduction is rare.

7) *Cryptophyta* are unicellular, usually flagellated forms which produce asexually.

8) *Rhodophyta* (red algae) are unicellular, filamentous or leafy forms with complex sexual cycles. Most are marine but there are a few freshwater forms.

2. Fungi

The "true" fungi or *Eumycota* are eukaryotes which are related to both protozoa and algae. They are divided between microscopic and macroscopic groups. In general, they have rigid cell walls, lack chlorophyll, and are usually immotile. Most fungi reproduce asexually or sexually by means of spores though important budding groups such as yeasts are well known. Since fungi are classified by the pattern of their sexual reproduction and the structure of their sexual structures, fungi whose sexual stages are unknown are placed into a group called *Fungi Imperfecti* and assigned genera on the basis of their asexual structures. They are further subdivided into the so-called lower and higher fungi.

The lower fungi, *Phycomycetes*, are also called water molds but not all are aquatic (*e.g.* black bread molds). Some species multiply by means of flagellated gametes or flagellated spores *i.e.* more like certain green algae than other fungi. Most, but not all, Phycomycetes have *hyphae*, micro-

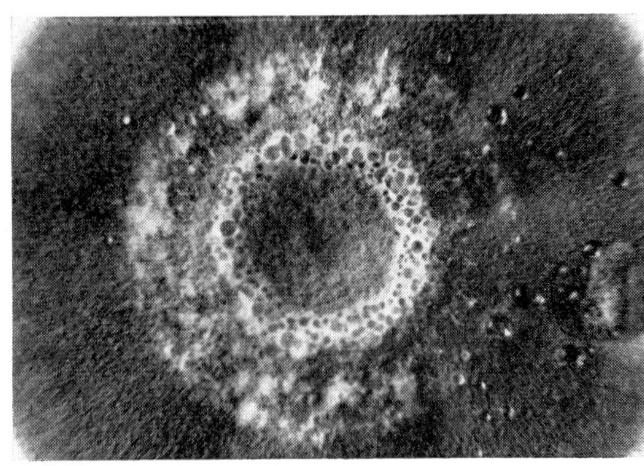

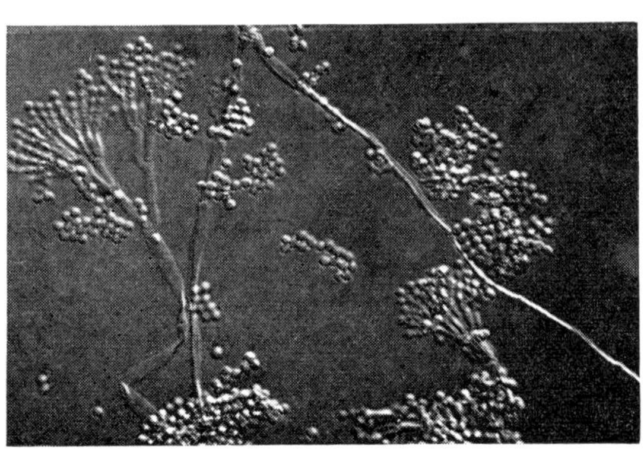

Figure 12 A. One large colony of the mold *Penicillium* is seen in the petri dish on the left. The edge of an uninoculated petri dish is shown on the right.
B. Closeup view of a *Penicillium* colony shown the antibiotic-containing juice being expressed from the colony.
C. Photomicrograph of *Penicillium* cells showing cellular arrangement and the many globular spores.

scopic cytoplasm-filled tube-like branches (lacking crosswalls), which together make a felty mat called a *mycelium*. Individual hyphae are microscopic but the mycelium, equivalent to a bacterial colony, is macroscopic. Growth takes place by extension of the hyphae. Specialized spore-containing bodies called *sporangia* can form at the ends of some hyphae. Sexual reproduction requires fusion of hyphae from two different mycelia to form a specialized *zygospore*.

It is more common now to discard the term Phycomyetes and instead subdivide the group into 4 classes in which affinities are much clearer. However, at present, the literature is divided in its use of the older and newer terminology. As with bacteria, chemical analyses of structures and metabolic pathways followed are important in defining the classes. These 4 classes are:

1) *Chytridiomycetes* lack true mycelia. They are aquatic, have posteriorly uniflagellated zoospores and cell walls composed of chitin.

2) *Hyphochytridiomycetes* have true mycelia. They are aquatic, have anteriorly uniflagellated zoospores and cell walls composed of chitin.

3) *Oomycetes* have true well developed mycelia and cell walls composed of cellulose.

a) *Saprolegniales* are generally aquatic and have asexual spores on specialized mycelear structures. Only male gametes are motile.

b) *Peronosporales* are generally terrestrial. Sporangia either produce asexual zoospores or may germinate directly to form hyphae. Both gametes are nonmotile.

4) *Zygomycetes* are terrestrial and have large and well developed mycelia and nonmotile spores. Asexual spores are produced in sporangia. Cell walls are made of chitosan or chitin.

There are two classes included in the higher fungi.

1) The *Ascomycetes* are the best known and largest class of fungi. Ascomycetes have hyphae divided by porous crosswalls. Each of these hyphal compartments usually contains a separate nucleus. Asexual spores called *conidia*, form singly or in chains at the tip of a specialized hypha. The sexual structure called *ascus*, is formed at the enlarged end of a specialized fruiting structure and usually contains eight *ascospores*. Some important microscopic members of this group include yeasts, mildews, the common red bread mold and many species which produce antibiotics. On the other hand macroscopic forms include *Morchella esculenta* or morels which are highly regarded as a delicacy by gourmets.

2) The *Basidiomycetes* are entirely macroscopic and are commonly known as mushrooms and toadstools.

3. Slime molds

The slime molds, *Myxomycetes*, are at times classified with either true fungi or protozoa or, as here, treated separately. They produce vegetative structures which look like ameboid protozoa and fruiting bodies which produce spores with cell walls like fungi. There are two major subdivisions, a) Cellular and b) Acellular. They both primarily live on decaying plant material and can ingest other microorganisms, such as bacteria, phagocytically. Both have life cycles, but that of the acellular slime molds is more complicated.

Cellular slime molds have vegetative forms composed of single ameboid cells. Cyclically, ameboid cells aggregate to form a slug-shaped *pseudoplasmodium* that begins to form *fruiting bodies* when the slug becomes immotile. Spores are finally produced by the fruiting bodies.

Acellular slime molds have vegetative forms called *plasmodia* which are composed of naked masses of protoplasm of indefinite size and shape and which travel by ameboid movement (protoplasmic streaming). Two kinds of nesting structures are produced: *fruiting bodies* (part of the sexual cycle) and *sclerolia*.

4. Protozoa

The last major group of microorganisms are the protozoa. As already stated, it is very hard to distinguish plants from animals at this primative stage in evolution where organisms have some attributes of each. Most workers therefore are less interested in whether protozoa should be claimed by botanists or zoologists as they are in studying the group as the root of a phylogenetic tree which gave rise to clearly separable plants and animals. Protozoa range in size from that of large bacteria to just visible without a microscope. They have a variaety of shapes, multiplication methods and associations which range from single cells to specialized colonies. They are variously found in fresh water, marine, terrestrial, and occasionally, aerial habitats. Both freeliving and parasitic forms are included. Most are motile but there are also important nonmotile forms. The protozoa are divided into four subphyla (I–IV).

I. *Sarcomastigophora* include forms which have either flagella, pseudopodia or both. Usually a single-type of nucleus (though opalinids contain multiples of this one type) is present except in developmental stages of a few forms. Asexual reproduction by binary fission is common. One whole class contains chloroplasts and are claimed by both protozoologists and algologists (they are considered here in detail with the eukaryotic algae). Many important parasites of diverse animal and some plant groups are found here. Sexual reproduction is present in a few forms.

The Sarcomastigophora are divided into 3 superclasses.

A. *Mastigophora* (flagellates) are further subdivided into *Phytomastigophorea* or plant-like flagellates (see eukaryotic algae) and *Zoomastigophorea* or animal-like flagellates which are divided into 9 orders.

1) *Choanoflagellida* have a single anterior flagellum surrounded posteriorly by a collar. Some forms are attached to substrates. They are solitary or colonial and are all free-living.
2) *Bicosoecida* have 2 flagella (one free, the other attached to the posterior of the organism). They are free-living.
3) *Rhizomastigida* have pseudopodia and 1–4 or more flagella. Most species are free-living.
4) *Kinetoplastida* have 1–4 flagella and all have a kinetoplast (specialized mitochondrion). Many important pathogens (*e.g.* trypanosomes) some free-living genera are included.
5) *Retortamonadida* have 2–4 flagella. The cytostome is fibril-bordered. All are parasitic.
6) *Diplomonadida* have 2 karyomastigonts, each with 4 flagella and sets of accessory organelles. Most species are parasitic.
7) *Oxymonadida* have one or more karyomastigonts, each with 4 flagella. All species are parasitic.
8) *Trichomonadida* have mastigont systems with 4–6 flagella. Some have undulating membranes. Many important pathogens (*e.g. Trichomonas*) are included.
9) *Hypermastigida* have mastigont systems with numerous flagella and multiple parabasal apparatus. All are parasitic. Some forms reproduce sexually.

B. *Opalinata* are an intermediary group related to both ciliates and flagellates and are entirely parasitic. Opalinids have many cilia-like organelles arranged in oblique rows over their entire body surface. They lack cytostomes (oral openings). They have multiple nuclei (ranging from 2 to many) which divide acentrically. The whole organism divides by binary fission. Life cycles are complex.

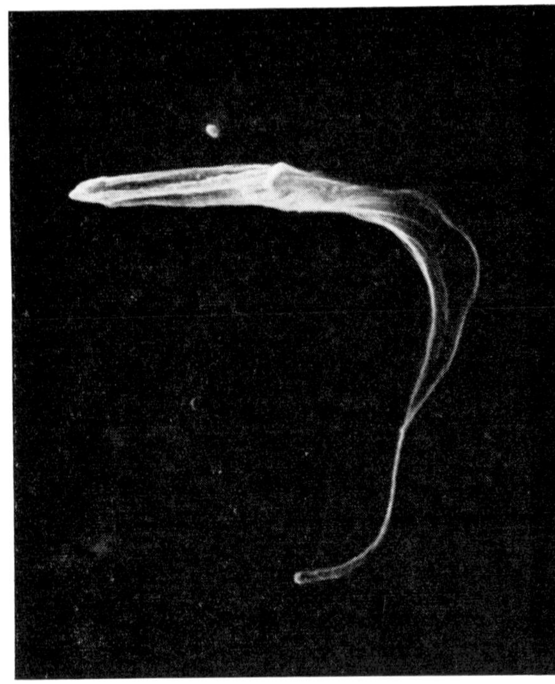

Figure 13 A scanning electron micrograph of the protozoan *Trypanosoma mega* showing the undulating membrane.

C. *Sarcodina* or ameboid organisms have Pseudopodia which are typically present but flagella may be present during certain restricted developmental stages. Some forms have external or internal *tests* or skeletons which vary widely in type and chemical composition. All reproduce asexually by fission but some also reproduce sexually. Most species are free-living (in both aquatic and terrestrial habitats) but some are important pathogens *e.g. Entameba histolytica* which causes amebic dysentary. The sarcodinids are further divided into three classes.

1) *Rhizopodae*, a free-living, mostly particle-eating (phagotrophic) group which includes both naked and shelled species. The specialized pseudopodia are called lobopodia, filopodia, or reticulopodia.

2) *Piroplasmea*. These parasitic small, piriform, round, rod-shaped or ameboid organisms do not form spores, flagella or cilia. Locomotion is by body-flexing or gliding. They reproduce by binary fission or schizogony.

3) *Actinopodea* are free-living, spherical, typically floating forms with typically delicate and radiose pseudopodia. Forms may be naked or have membraneous, clutenoid, or silicated tests. Both asexual and sexual reproduction occurs. Gametes are usually flagellated.

II. *Sporozoa* typically form spores without polar filaments and lack flagella or cilia. Both asexual and sexual reproduction takes place. All species are parasitic. Some have rather complicated life cycles.

The Sporozoa are divided into 3 classes:

A. *Teleosporea* can reproduce sexually or asexually, have spores, move by body flexion or gliding and generally do not have pseudopodia.

B. *Toxoplasmea* reproduce asexually, lack spores, pseudopodia or flagella, and move by body flexion or gliding.

C. *Haplosporea* reproduce asexually and lack flagella. They have spores and may have pseudopodia.

III. *Cnidospora* have spores with one or more polar filaments and one or more sporoplasms. All species are parasitic. There are 2 classes.

IV. *Ciliophora* have simple cilia or compound ciliary organelles in at least one stage of their life cycle. They usually have two types of nucleus. Reproduction is asexually by fission or sexually by various means. Most species are free-living but parasitic forms are known.

Energy and carbon metabolism

All cells require a source of chemical energy and of carbon for building protoplasm. Regardless of whether the cell type is prokaryote or eukaryote or whether it is more plant-like or more animal-like, this basic requirement is the same. The most basic division relates to the source of *carbon* used to build protoplasm. Organisms which can manufacture all their carbon-containing compounds from originally ingested inorganic carbon (CO_2) are called *autotrophs* while those which require ingestion of one or several organic compounds for use in the manufacture of cellular carbon compounds are called *heterotrophs*. Some organisms are nutritionally-versatile and may operate either as autotrophs or heterotrophs and are therefore referred to as *facultative-autotrophs* or *facultative-heterotrophs* (depending upon which mode of nutrition usually predominates).

Autotrophs are further divided according to the manner in which they obtain energy. *Chemoautotrophs* (also called chemotrophs or *chemolithotrophs* oxidize various inorganic compounds to obtain energy while *photoautotrophs* (also called phototrophs or *photolithotrophs*) convert light to chemical energy *via* the absorption of ligth energy by special pigments (chlorophylls and carotenoids). In both cases, chemical energy is stored in the form of chemical bond energy in the compound adenosine triphosphate (ATP).

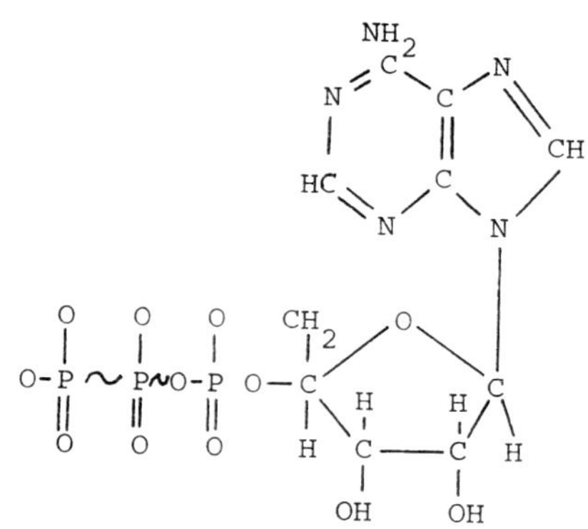

Adenosine triphosphate (ATP)

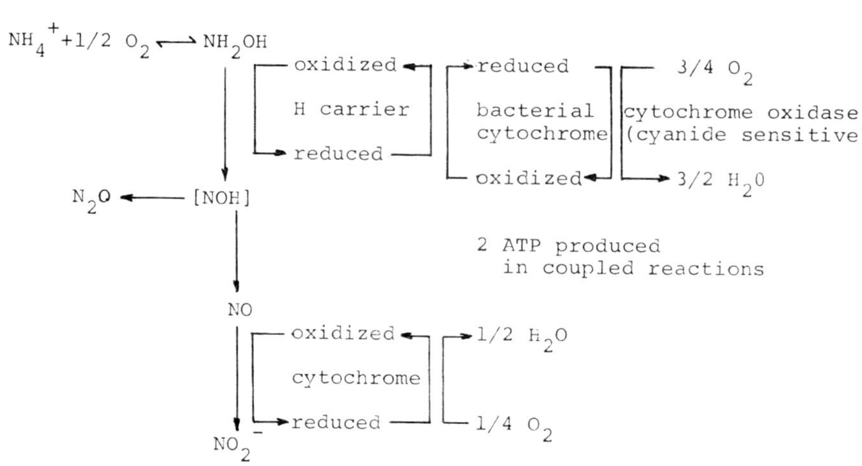

When bonds of ATP indicated by ~ are broken, a considerable amount of energy is released. This ~ bond cleavage energy operates the biological engines: it is the universal chemical power which operates in all cells, autotroph or heterotroph.

Chemolithotrophic nutrition is only used by certain true bacteria. These bacteria are of ecological importance in that they are used to convert one form of nitrogen to another (*i.e.* in the nitrogen cycle) or industrially to oxidize low grade metallic or non-metallic ores. There are six bacterial groups which are chemolithotrophic.

(1) The ammonia oxidizers such as *Nitrosomonas, Nitrosococcus, Nitrosocystis, Nitrosogloea* and *Nitrosospira*.

One scheme for ammonia oxidation has hydroxylamine as an obligate intermediate has been proposed for *Nitrosomonas*. (See p. 548)

(2) The nitrite oxidizers such a *Nitrobacter* and *Nitrocystis*. One proposed scheme for nitrite oxidation for *Nitrobacter* is:

$$NO_2^- \xrightarrow[\text{reductase}]{\text{cytochrome C}} \text{cytochrome C} \xrightarrow[\text{oxidase}]{\text{cytochrome}} O_2$$

ADP ↗ ↘ ATP

(3) Hydrogen oxidizers *Hydrogenomonas*. One proposed hydrogen oxidation scheme is:

$H_2 \rightarrow 2\,H^+ + 2e \rightarrow$ unknown compound $\rightarrow$ flavoprotein

$\rightarrow$ ubiquinone $\rightarrow O_2$
cytochrome
b complex

Nicotinamide adenine → menadione → cytochrome C → cytochrome a → O_2 dinucleotide (NAD)

(4) Ferrous compound oxidizing bacteria such as *Ferrobacillus* and *Thiobacillus ferroxidans*.

One proposed ferrous oxidizing scheme for *F. ferrooxidans* is:

$$4FeCO_3 + O_2 + 6H_2O \rightarrow 4Fe(OH)_3 + 4CO_2$$

(5) Methane oxidizers such as *Methanomonas methanooxidans* and *Pseudomonas methanica* are common in the upper layers of marine sediments and soil.

Methane is oxidized in the following manner:

$$CH_4 \rightarrow CH_3OH \rightarrow HCHO \rightarrow HCOOH \rightarrow CO_2$$

(6) The sulfur-compound oxidizing bacteria *Thiobacillus*.

Four pathways for oxidation of thiosulfate ($S_2O_3^{+2}$) by different *Thiobacillus* species are known. These are:

a) $6Na_2S_2O_3 + SO_2 \rightarrow 4Na_2SO_4 + 2Na_2S_4O_6$

$2Na_2S_4O_6 + 6H_2O + 7O_2 \rightarrow 2Na_2SO_4 + 6H_2SO_4$

b) $Na_2S_2O_3 + 2O_2 + H_2O \rightarrow Na_2SO_4 + H_2SO_4$

c) $5Na_2S_2O_3 + H_2O + 4O_2$

$\rightarrow 5Na_2SO_4 + H_2SO_4 + 4S$

$2S + 3O_2 + 2H_2O \rightarrow 2H_2SO_4$

d) $2Na_2S_2O_3 + H_2O + 1/2O_2 \rightarrow Na_2S_4O_6 + 2NaOH$

Photolithotrophic nutrition is used by photosynthetic bacteria, blue green algae and eukaryotic algae. The general reaction in which both utilization of CO_2 (carbon dioxide fixation) and energy generation is summarized is:

$$CO_2 + H_2A \xrightarrow{h\nu} (CH_2O) + 2A + H_2O$$

Where A is either oxygen for all eukaryotic algae and the prokaryotic blue-green algae ($H_2A = H_2O$), or sulfur for green sulfur bacteria (*Chlorobacteriaceae*) and purple sulfur bacteria, *Thiorhodaceae* ($H_2A = H_2S$) or any one of several organic compounds for nonsulfur purple bacteria, *Athiorhodaceae* (H_2A = H_2-organic compound which is oxidizable).

Both green and purple sulfur bacteria are obligate anaerobes whereas the non-sulfur purple bacteria are facultative anaerobes (they are anaerobic when growing photosynthetically but aerobic when growing heterotrophically). In all cases, photosynthetic organisms operate by the initial transduction of light to chemical energy. In this transduction, chlorophyll + light quanta Chl^+ (excited chlorophyll) + e^- (electron driven off of Chl). Many such events take place simultaneously and electrons released during these reactions migrate through the photosynthetic unit to the reaction center and transfer energy to a special *reaction-center chlorophyll*. At the reaction center, a charge separation of the oxidant and reductant occurs. Electron flow after this event differs in photosynthetic bacteria as compared with algae and higher plants (Figs. 14 and 15). In addition, differences in photosynthetic ability exist among organisms based upon the absorption maxima of their light-transducing pigments (primarily chlorophylls). The combination of light intensity, wavelength of available light, wavelength of operation of principal energy

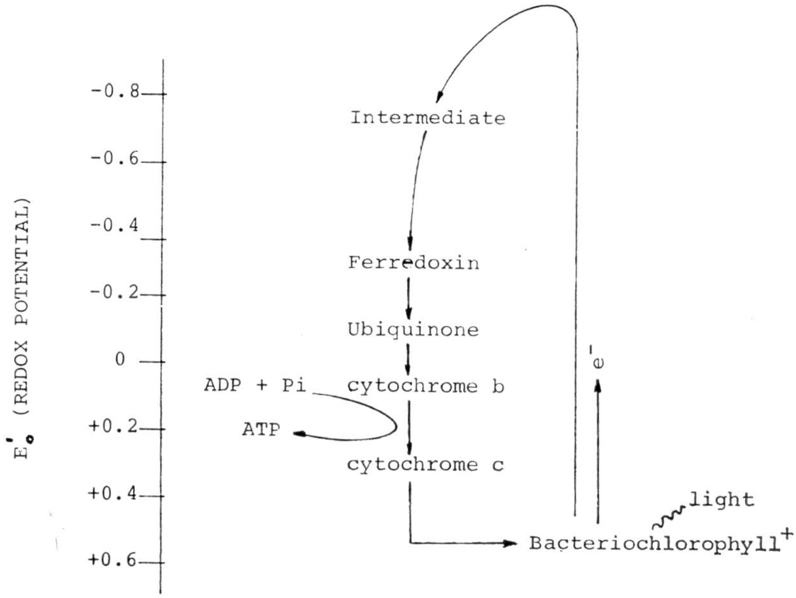

Figure 14 Electron flow in bacterial photosynthesis.

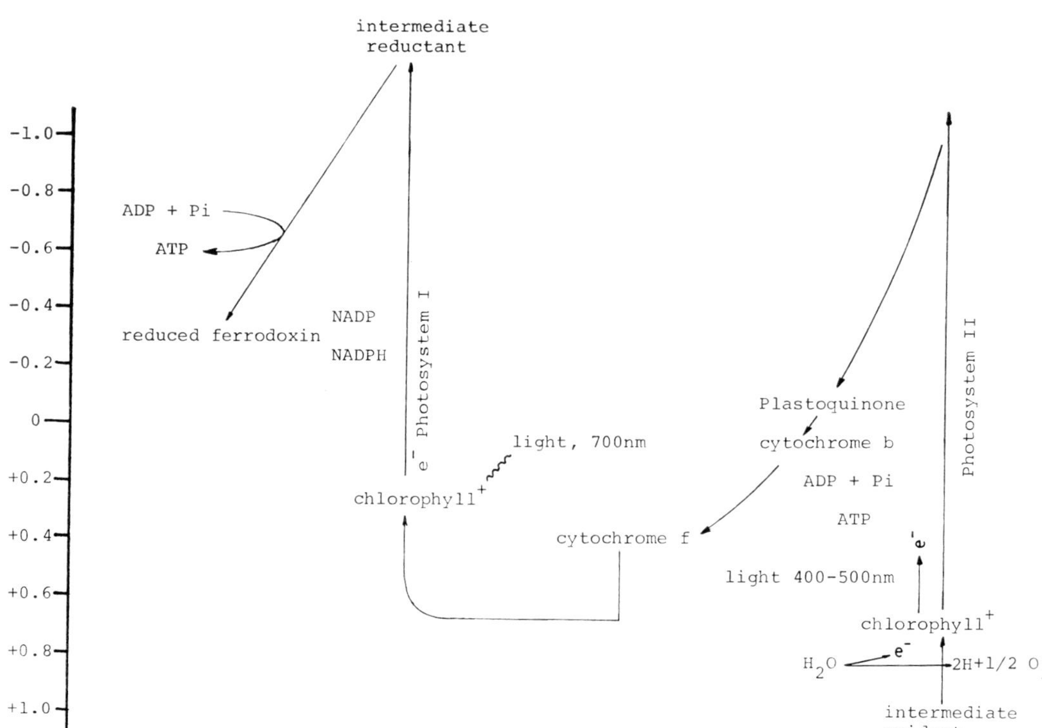

Figure 15 Electron flow in algal and higher plant photosynthesis.

transducing pigment, degree of aerobiasis, and availability of oxidizable compound (H_2O, H_2S, or H_2-organic compound) all influence the efficiency of photosynthesis. These factors should be borne in mind when one looks for the ecological niche occupied by these various organisms.

TABLE 4

Absorption Maxima of Chlorophylls from Various Sources

Organism	Chlorophyll type	Principal absorption maxima in nm
Green sulfur bacteria	Bacterial Chl_c	660
	Bacterial Chl_d	650
Purple sulfur bacteria	Bacterial Chl_a	820
	Bacterial Chl_b	1025
Non sulfur purple bacteria	Bacterial Chl_a	820
	Bacterial Chl_b	1025
Green algae and Euglenids, higher plants	Chl_a	683
	Chl_b	650
Diatoms, brown algae, Pyrrophyta	Chl_a	683
	Chl_c	620
Xanthophyta	Chl_a	683
	Chl_e	
Cyano-, Chryso- and Rhodophyta	Chl_a	683

Ecology of microorganisms

One should understand the physiological requirements of microorganisms *before* investigating the effects of environmental changes on the distribution and activity of diverse microbial types, their interactions and their relationships to higher plants and animals. It is important to note that the "natural" balance may be undesirable. Thus studied efforts to change these balances would be quite desirable. The most important precaution to observe relates to the ancillary consequences of these changes, *i.e.* do the changes produce side effects which may be as unappetizing as the original condition.

The development of techniques which form the bases for studying microbial ecology comprise an important chapter in classical microbiology. These laboratory methods, pioneered by Winogradsky (1856–1953) and Beijerinck (1851–1931) and refined by others, utilize specialized, restrictive, physical and chemical conditions to select and cause to predominate one or few types of organisms from a highly diverse mixture. Hence this method is termed *selective enrichment culture*. Once one understands how to manipulate these laboratory systems, it is easier to analyze field observations in which specific conditions which result in microbial changes can be recognized and, if necessary, altered. We will first explain the principles of selective enrichment techniques and then look at the natural distribution of microorganisms and their relationship to higher plants and animals. Highlights of microbial characteristics which are useful taxonomically have been described in the various sections listed under *Physical Characteristics of Microorganisms.*

Selective enrichment methods

To determine whether a given sample of soil, water, or air contains microorganisms capable of living under a particular set of conditions, one prepares a growth medium which is selective for a property peculiar to those conditions. For example, if organisms which can fix atmospheric nitrogen are sought, all non-atmospheric sources of nitrogen (such as nitrites, nitrates, ammonia, amino acids) are eliminated from the growth medium. If organisms which obligately fix carbon dioxide are desired, all nonatmospheric (organic) sources of carbon are eliminated from the growth medium and, frequently, additional CO_2 is bubbled through the medium.

On the other hand, it might be of particular interest to determine if a certain weed-killer is *biodegradable* before it is used under field conditions. Many different approaches to this important problem are possible. Thus more than one mode of attack is described. The approach which is closest to the general principle of revealing (selectively enriching for) a minor population of desired microbial type among a multitude of undesirable organisms makes use of an enrichment medium in which the weed-killer is used as either a) the only source of organic carbon and nitrogen, b) the only source of organic carbon though other sources of nitrogen are present, or c) the only source of nitro gen though other sources of organic carbon are present. Subsequent microbial growth indicates biodegradability. The organisms may be isolated and used as seed cultures for a percolation system which is fed by the aqueous runoff from a field which was treated with the weed-killer. Thus the more public waterways fed by aqueous effluents from treated fields would not be polluted by potentially-toxic agricultural chemicals.

In order to prepare selective enrichment media, one needs to provide the microorganisms with all their nutritional requirements in proper proportions. Insufficient quantities will not support growth and excesses are frequently toxic. In addition conditions must be biased in some way to insure that most of the undesired organisms will not grow at all or will grow appreciably slower than the desired organisms. It should be recognized that it is rare for any single enrichment to select out only one microbial species. Thus further purification steps are required if one wishes to isolate only one species uncontaminated with other living things. An uncontaminated, single membered culture is called a *pure* or *axenic* (a = absence of, xenos = strangers) culture. The common nutritional requirements of microorganisms, the quantities in which they must be supplied, and the biological uses of each substance is shown in Table 5. The selective enrichment techniques to be described are most frequently used for the isolation of bacteria, yeast, and certain prokaryotic algae. An outline of selecting properties is given in Table 6.

A demonstration of factors involved during the natural selection which takes place under field conditions is shown in Fig. 16. These models are called *Winogradsky columns*. The variety of microbial life which develops over a period of 2–10 days

TABLE 5

Nutritional requirements of typical heterotrophic microorganisms with limited synthetic capacity.

Type compound	Example	Quantity in typical medium
Energy source	Glucose, sucrose	0.1–2.0%
	glutamic acid, succinic acid	0.1–0.5%
Synthesis of protein and fat	sodium acetate	0.01–0.1%
	Lecithin	0.001–0.005%
	amino acids	0.002–0.1%
Synthesis of nucleic acids	purines and pyrimidines	0.0005–0.002%
Coenzymes	vitamins	0.0001–0.005 mg%
Major inorganic requirements	PO_4^-, Mn^{++}, Mg^{++}, Na^+, K^+	0.01–0.05%
Minor inorganic requirements	Ca^{++}, Co^{++}, Zn^{++}, Fe^{+++}, Cu^{++}, Cl^-, NH_4^+ or NO_3^-	0.001–0.01 mg%
Water	All the above are prepared in aqueous solution	

is determined by a) the degree of acidity or alkalinity of these natural growth media, b) the nutrients contained in the liquid and solid phases, and, of course, c) the initial populations of microorganisms. In the examples shown, light is provided to ensure growth of photosynthetic organisms.

If the basic principles of the Winogradsky column are to be used to reveal the microbial population in a particular soil or water sample, then the column and all its components are first sterilized and then inoculated with a *nonsterile* soil or water sample. The microbial population of the sample will develop in the portions of the column which provide the proper physical and chemical conditions.

An important application of Winogradsky columns can be made for testing various chemicals for potential ecological changes. The chemical agent is percolated through a soil column or is simply added to a predominantly liquid column either when the column is started or after its microbial population has developed. Significant changes in the column's normal population (diversity or population density) is indicative of toxicity to one or more types of microorganisms. The profound changes in populations of higher plants and animals due to disruption of the balance of microbial life can be readily appreciated when the cyclic nature of nitrogen and sulfur dissimilation is considered (Figs. 17 and 18). These interdependences emphasize the key role played by microorganisms in maintaining the balance of soil nutrients.

Use of the Winogradsky column for testing biodegradability capacity of a natural soil or water body

In the last section, an example was given in which one of these bodies with little capacity for biodegradation of a week-killer can be controlled so that the body in question does not spread the potential pollutant to a bordering body. We now consider a method for pretesting for biodegradation capacity. One would hope that this or a parallel test would become standard before new agricultural chemicals are marketed. That is, chemicals which are not decomposed *before* they leave the immediate land or water body in which they are used would not be marketed or would only be marketed *after* controls against accumulation of the chemical had been worked out.

The Winogradsky column test for ability of a particular soil or water body to degrade a potentially dangerous chemical such as a weed-killer consists of

TABLE 6

Selecting factors for the isolation specific types of microorganism

- Organic compounds *present* in medium: HETEROTROPHIC ORGANISMS (See 6B)
- Organic compounds *absent* from medium: AUTOTROPHIC ORGANISMS (See 6A)

6A *Selecting Factors for Autotrophic Organisms*

- Anaerobiasis
  - Light absent, S or $S_2O_3^{=}$ oxidized for energy: anaerobic chemoautotrophic bacteria such as *Thiobacillus denitrificans*
  - Light present
    - grows in high sulfide concn: Green sulfur bacteria
    - grows in low sulfide Concn: Purple sulfur bacteria
- Aerobiasis
  - Light and sulfide absent
    1. NH_4^+ is N source. *Nitrosomonas* (bacterium)
    2. NO_2^- is N source, *Nitrobacter* (bacterium)
    3. NH_4^+ plus special atmosphere composed of 85% H_2; 10% O_2; 5% CO_2 at pH 7 Hydrogen bacteria
    4. NH_4^+ plus $S_2O_3^{=}$ at pH 7. Certain *Thiobacillus* spp. (bacteria)
  - Light present and sulfide absent
    - N_2 can serve as sole nitrogen source: Blue-green algae
    - Nitrogen must be present in combined form: Green algae

6B *Selecting Factors for Heterotrophic Organisms*

- Anaerobiasis
  - Light present, neutral pH: Non-sulfur purple bacteria
  - Light absent
    1. Acid pH, fermentable sugar present Yeast spp.
    2. Neutral pH
       - fermentable sugar
         - N_2 as sole nitrogen source *Clostridium* species (bacteria)
         - Presence of combined nitrogen source. Numerous microbial types.
       - non-fermentable substrate, electron acceptor is 1, 2 or 3
         1. NO_3 Denitrifying bacteria
         2. CO_2 Methane bacteria
         3. $SO_4^{=}$ *Desulfovibrio* (bacterium)
- Aerobiasis (see below)

- Aerobiasis
  - Light present. Many algae will grow here, pH aids in further differentiation as does the precise selection of organic components
  - Light absent
    - N_2 is sole nitrogen source *Azotobacter* (bacterium)
    - Presence of combined nitrogen source. Depending upon the complexity of the medium and pH, most aerobic bacteria yeast, protozoa and some algae could grow. Specific texts should be consulted for further details.

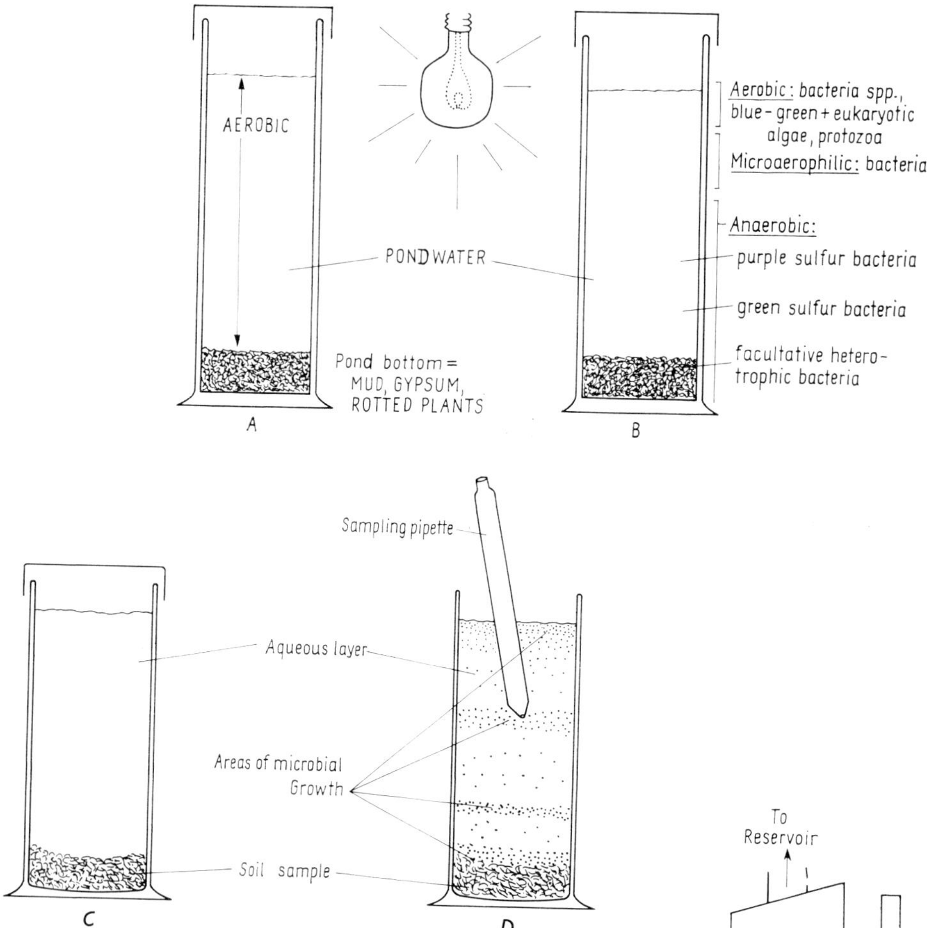

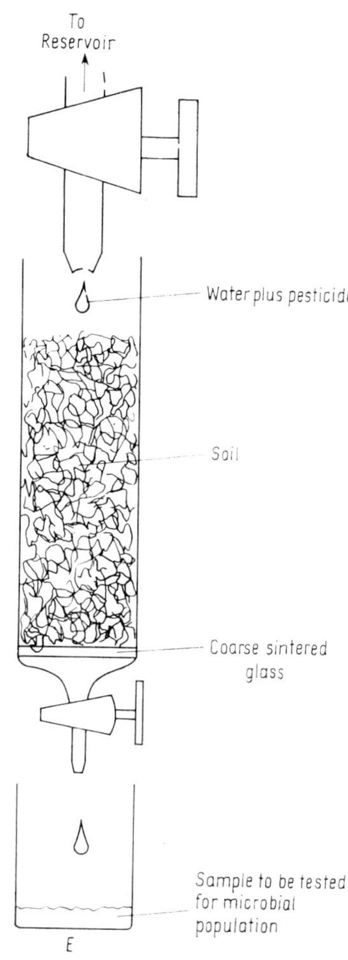

Figure 16 The Winogradsky column. A laboratory model for microbial population development in natural environments.

A and B. Show a model of a stagnant pond. A 1–2 liter graduated cylinder may be used for the column. The liquid and solid phases are drawn from the pond itself.

C and D. Population development in a restricted soil sample. A soil sample is overlaid with sterile water or nutrient-containing water. (C) After days to weeks growth can be observed by the appearance of turbidity. Motile organisms will remain suspended at the level of the column compatible with their natural degree of aerobiases (or Redox potential) whereas only anaerobic non-motile organisms in the soil sample will grow since the soil does not circulate through the aqueous column. To tap the column, a sterile pipette is immersed to the desired depth, and a sample withdrawn (D).

E. Effluent population in watered soil and the effect of pesticides: a dynamic model. A column with an outflow valve (see E) is packed with soil and water, water plus nutrients or water with pesticides are trickled through the column. Continuous population dynamics are studied by sampling the effluent and testing for the desired organisms. Kinetics of population development also may be tested by saturating the column (without outflow) and then washing out the column at intervals. Models such as these also can be used for measuring decomposition of pesticides (recoverable from the effluent) by the developing microorganisms.

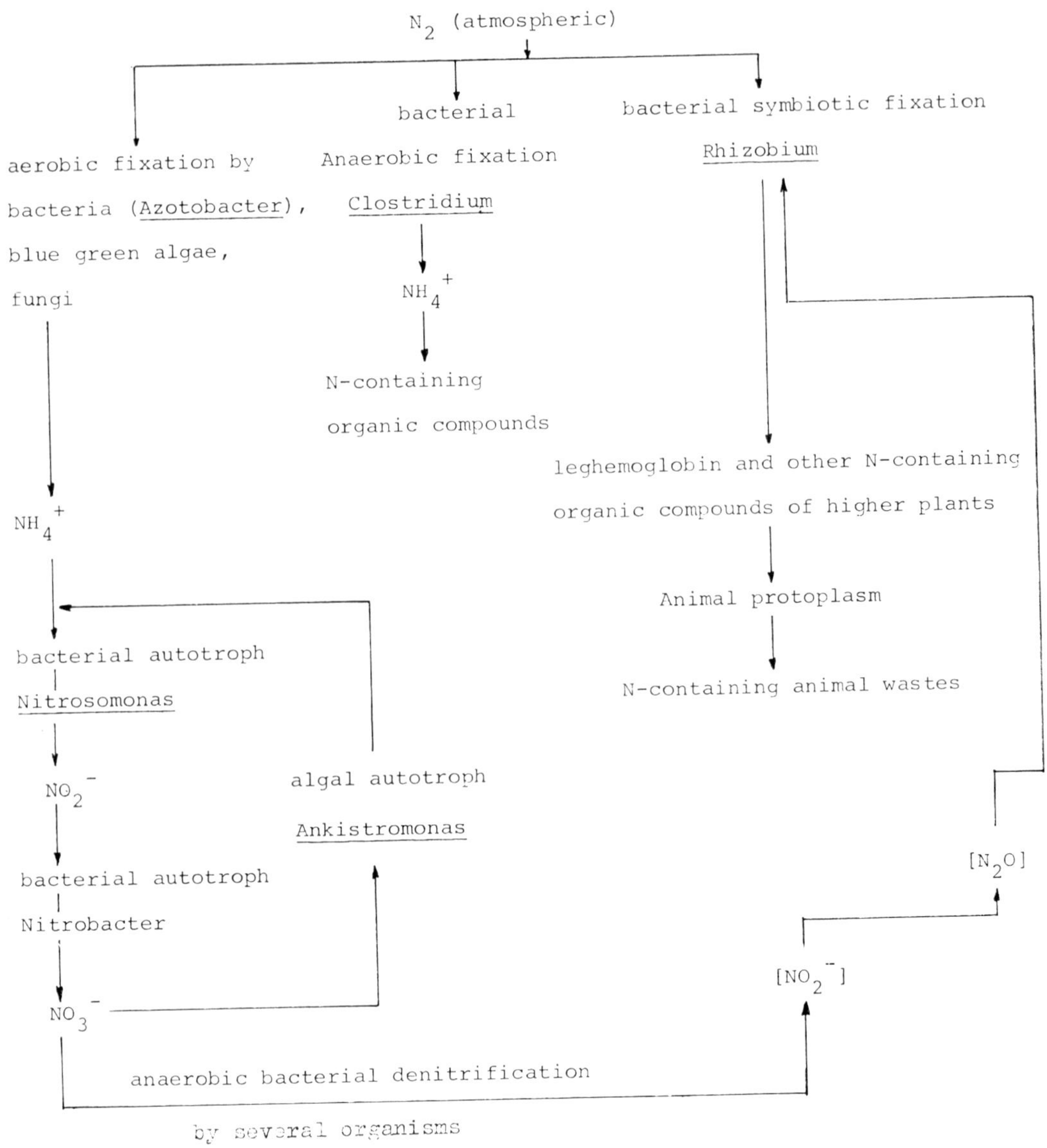

Figure 17 The nitrogen cycle presented in a generalized way so that the role of various microorganisms are indicated as well as the relationship to higher plants and animals. The precise arrangement will vary accoring to several physical parameters of the natural environment.

a) preparing a standard column composed of soil or water from the body in question and, b) after the column has been allowed to develop its natural population c) an isotopically-labelled version of the weed-killer can be added in the concentration (and 10X the concentration) that the weed-killer it to be used. After a time equivalent to that in which the weed-killer is expected to remain in the natural body (*i.e.* account for flushing time from rain or water currents), the column is tapped by elution with water or buffer, and the effluent is analyzed for per cent undegraded weed-killer as well as *the nature of the degradation products*. The latter point is particularly important since the degradation products may themselves be noxious. Thus, the only degradation products which are acceptable are those which may enter normal metabolism or are labile enough to be further degraded to metabolizable compounds by the physical conditions in the body in question.

Determination of gross populations of microorganisms

Quantitative sampling techniques are required for the determination of the microbial population in air, water, or soil samples. The method used for sample collection must ensure against a) loss of

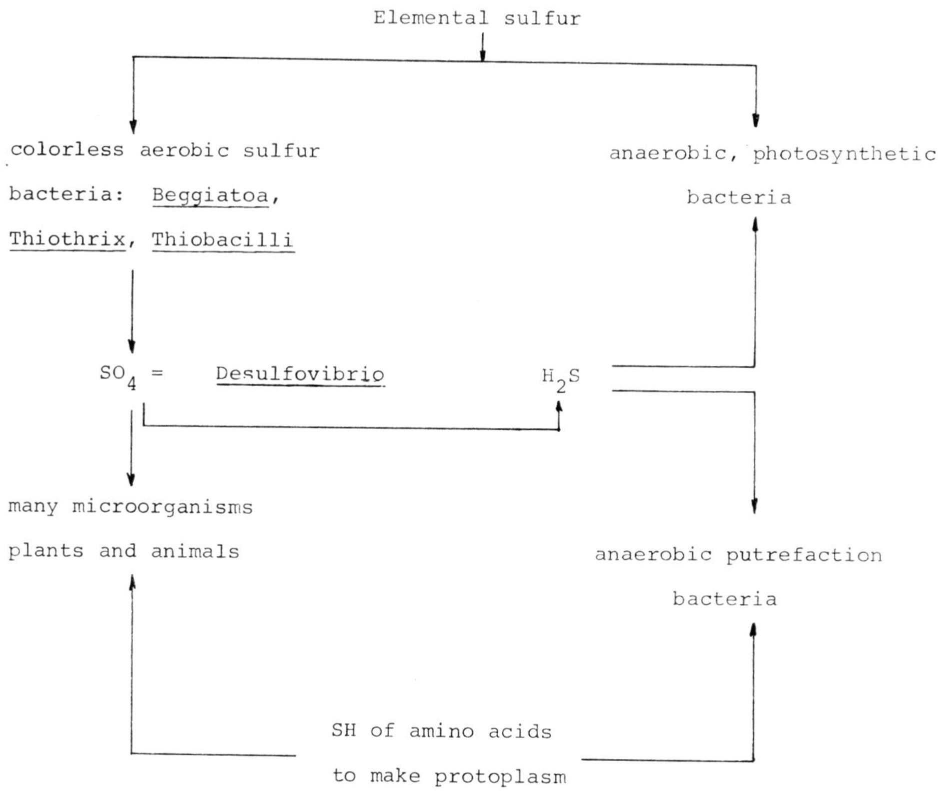

Figure 18 The sulfur cycle. Microorganisms are becoming increasingly useful for processing low-grade sulfur-containing ores. Sometimes overlooked is the natural cyclic redistribution of sulfur-containing compounds in which microorganisms play principle roles.

more than a trivial number of microorganisms and b) cross-contamination from other sources during sample transport and laboratory manipulation. Once the sample has arrived in the laboratory, selective enrichment techniques can be used to reveal the diversity of microorganisms or mixed populations can be counted by using some variation of the pour plate technique (see Fig. 10).

A. *Air sampling techniques*: Non-spore forming organisms are rarely found in air samples because they are too delicate to survive for long in the generally dehydrating conditions of atmospheric transport. A convenient sampler consists of a sterile membrane filter connected to a metering vacuum pump. At the beginning of the sampling period, the membrane is exposed and the desired amount of air passed through it. Air-borne microorganisms are trapped on or in the membrane. Some discrimination among organisms can be made during sampling by using membranes with graded porosities. In this way organisms are segregated according to their cross-sectional dimensions. To reveal the organisms impinged upon the membrane, the membrane is treated as inoculum for the various selecting media.

B. *Water sampling techniques*: Any container which can be sterilized and can be opened and closed by remote signal can be used. It should be recognized that pressure changes after retrieval may influence viability of organisms collected at great depths.

C. *Soil sampling techniques*: A sterilized coring device is usually used to ensure against contamination with airborne organisms. The soil samples are handled aseptically and weighed amounts are tested for their microbial population.

PHOTOGRAPH CREDITS

Figures 1–10, 12: Dr L. LeBeau, Dept. of Microbiology, University of Illinois Medical Center.
Figures 11, 13: Dr H. N. Guttman, Dept. of Biological Sciences, University of Illinois at Chicago Circle.

HELENE N. GUTTMAN
University of Illinois at Chicago Circle

MINE DRAINAGE: see POLLUTION FROM MINE DRAINAGE

MODELING OF RIVER AND ESTUARY WATER QUALITY

I. Introduction

Scope

It is well known that, in recent years, alarm has been expressed concerning the growing use of water in the United States, and the possibility that, far from being in endless supply, the water resources of this nation are severely limited. Our increasing demands for electric power are placing a greater and greater thermal load on our natural water bodies. Our use of vast amounts of detergents and agricultural fertilizers pours increasing quantities of nutrients into our streams. Proliferating suburbs must draw their domestic supplies from farther and farther afield.

Not surprisingly, therefore, water figures largely in the thoughts of planners at all levels of government. At the lowest level, the purity of water in a particular river or lake is of great significance to the local population, primarily in terms of local water use, both domestic and industrial. Secondarily (sometimes unfortunately so), the demands for recreational facilities must be met, i.e., for swimming, boating and sport fishing.

At the next higher level, the effective utilization of water throughout an entire river basin, or lake watershed, requires that attention be given not only to problems of pollution control but also to matters of flood control, navigation, irrigation, power generation and commercial fishing.

Finally, when one comes to the planning for the economic development of a river basin, as a whole, the availability and quality of water is only one factor in the equation, in which others are natural resources besides water (timber, minerals, etc.), population shifts and trends, communications and transportation, and so on.

These higher levels of concern are beyond the scope of this article. For those who may be interested, a short reading list is appended to the list of references. We will concern ourselves here only with the first category of problems, the control of water quality.

The remainder of this introduction will be devoted to describing and discussing in general terms the analytical tools available to the planner.

It should be made clear that the systems covered herein are for the most part bounded by the river bank. Stipulations may sometimes be made as to the flow rate of a waste stream input, and its prior degree of treatment, but the mathematical modeling of waste treatment plants is specifically excluded.

For discussions on the subjects of optimization, probabilistic approaches, thermal pollution and the modeling of lakes and reservoirs, see "Systems Modeling for Water Pollution Control", in Volume 2 of this Encyclopedia.

Systems analysis and mathematical modeling

Needless to say, the definition of Systems Analysis has received considerable attention (Armstrong, 1970; Deininger, 1969; Hetling and Metzler, 1969; Thomann, 1969; Toebes, 1969) from workers in the fields of water quality and use. Perhaps the most pointed definition is due to Hetling and Metzler, 1969. Systems analysis comprehends a critical examination of a problem in an abstract rational manner, delineation of all possible solutions, evaluation of the consequences of each solution, rationally selecting the solution which best solves the problem, and then *presenting the results to a decision-maker*. The point resides in the last clause, i.e., it is not the analyst who makes the final choice. It is always to be borne in mind that the analysis of water systems takes place against a background of political boundaries and governmental structures. The analyst can only provide the best basis he can for whatever agency he works for to come to reasonable decisions.

Armstrong, 1970, makes an interesting distinction

between two kinds of systems analysis. He defines a "descriptive" model and a "prescriptive" (or normative) one. The descriptive model, he says, attempts to predict how the system *will* behave under given conditions, whereas the prescriptive model tells how the system *should* work. He adds that, to determine what are the proper "settings" of the controllable variables that maximize the system objectives, a prescriptive model needs an objective or benefit function. This writer would, perhaps, reserve the title "Systems Analysis" only for the prescriptive model, so defined, and call the predictive model simply that, a "Model."

Systems analysis is a methodology for viewing large-scale complex problems in an orderly, logical way. It is oriented to the optimal management of a resource, that is the evolution of management strategies that produce maximum benefits. A system is some collection of resources in the broadest sense—materials, machines, money, men—all assembled in some fashion to carry out some objective, and it can be defined by a list of variables and parameters. To analyze these, we need to understand how they relate to each other and how the system meets its objectives. This understanding, of course, leads to the construction of a model.

Four major classes of variable and parameter are used to describe a system:

1. Effectiveness measures.
   These describe how well a system is achieving its objectives, e.g., dissolved oxygen concentration.
2. Performance measures.
   These describe how the system components are performing with respect to meeting the objectives, e.g., release or diversion flows from reservoirs.
3. Environmental variables.
   These are uncontrollable within the system, but affect its behavior, e.g., stream reaeration rate.
4. Control variables or management strategies.
   These specify how the system elements shall interact to attain the objectives.

One of the major tasks in any comprehensive analysis of a water body is to provide a means or framework for articulating and specifying the goals or objectives of the program. This may take several forms, including a rigorous quantitative specification of water quality levels (such as dissolved oxygen) or relatively simple qualitative descriptions of the kind and degree of water use that is desired. The articulation of the latter water use descriptions should be sought after by planners and analysts and incorporated into the formulation of any variable environmental control scheme.

After some initial specification of goals, the process continues by analyzing the consequences of the interim goal in terms of:

a) technological feasibility
b) costs of environmental control, and
c) effects of the control program in terms of inproved water use, i.e., the benefits of the interim goal.

Ideally, the results of this first analysis are made known and communicated to those concerned with water use for their further deliberations.

Based on the results of this first analysis, some changes may be made in the initial goal specifications or an entirely new and different goal may be established.

The new goals are then subjected to a similar technical analysis and results again transmitted to the "goal setters."

This feedback process aids to sharpen and identify areas of technical difficulty, policy constraints and strong water using interests. It is not always possible a priori for analysts to know all aspects of a water

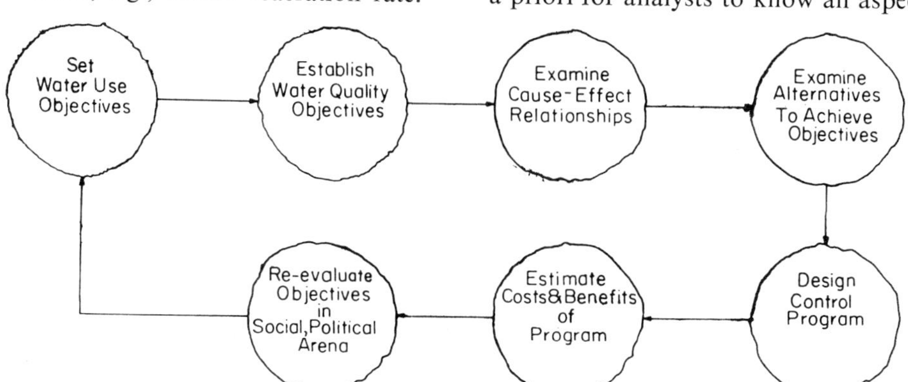

Figure 1 Systems Design—Simplified. (Thomann, 1969)

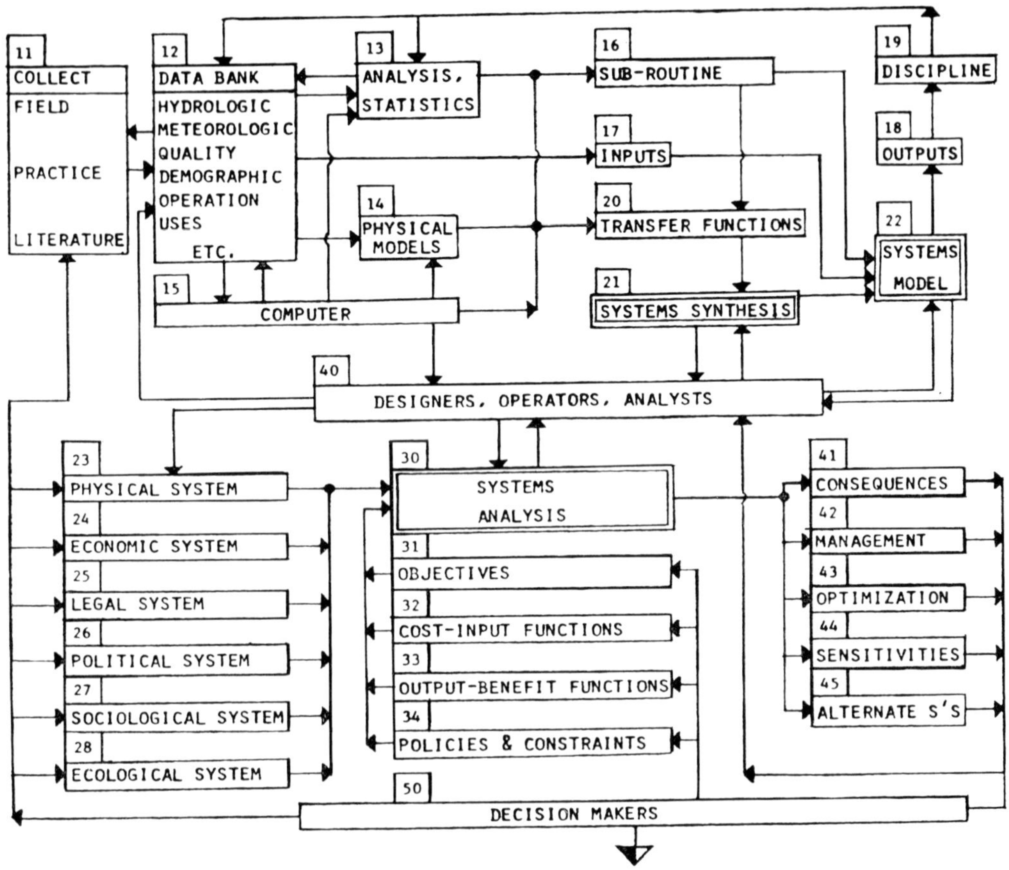

Figure 2 Systems Design. (Toebes, 1969).

quality program and especially to know those particular aspects which will serve later as rallying points leading to success or failure of the program.

A very simple illustration of the process of system design is Fig. 1, taken from Thomann, 1969. Much more detail can be found in Fig. 2 which is owed to Toebes, 1969. The concept of systems design, as portrayed in Fig. 2, emphasizes optimization, systems sensitivity and alternatives. Optimization has, of course, always been the distinguishing characteristic of much physical engineering design. However, the objectives to be implemented and optimized now have expanded greatly. In addition to giving more sophisticated weight than heretofore to the economics and to the welfare economic consequences connected with any large-scale physical design, some elements of the legal, political, sociological and ecological systems (blocks 23 through 28) are taken into account explicitly. Translated into systems functions (blocks 31 through 34), they are increasingly regarded as problem variables rather than as fixed information.

The results obtained from systems analysis (block 30) include:

a) estimates of long-range consequences in case the plans represented by the model are realized;
b) general resource management and operation schedules;
c) optimality of the proposed scheme;
d) model sensitivity to component level variation, input statistics or operating policies;
e) comparison of alternative systems for similar objectives.

The construction and utilization of mathematical models is concerned with the technical aspects of describing cause and effect relationships between waste discharges and the water environment, or designing rigorous analytical optimal control programs. However, it is important to note that, while this is important, it serves primarily as a framework for the comprehensive planning process, and is only a part of the overall planning investigation.

Historically, systems analysis has been used in the operation of water resource systems to examine the effects of various control works (dams, levees, etc.). It has dealt primarily with quantity relations, using fairly simple equations. It has not yet been fully utilized in planning water resource systems, particularly in areas concerned with water quality.

Sensitivity analysis is a critical function that planners must be able to carry out. It uses the model to investigate the relative importance of the variables and parameters, many of which are used in even the simplest model. Under a wide range of operation, which is the most important? Data taking is obviously ciritical, and usually expensive. Sensitivity analysis may yield clues to the relative importance of particular variables. This can then guide the data gathering effort.

Systems analysis helps decision makers make better decisions by clarifying, and making explicit, the conceptual framework in which the decision is to be made. It does this by defining alternative objectives, by exploring the consequences of alternative courses of action (quantitatively where appropriate), taking note of future uncertainties.

Cost functions

A distinction was drawn, previously, between descriptive and prescriptive models. The former describe how a system performs, and can be represented by a set of equations of some kind, differential, integral, algebraic. The latter start with such a set of equations, and then proceed to establish the actual values of certain important variables in order that the system achieve a predetermined objective, or optimum.

The criteria for optimum performance are defined by certain functions whose parameters are usually economic, and such functions have been variously named by different authors the objective function, the benefit function and the net-benefit function, depending on the type of problem.

In its simplest form, the objective function can be the minimum of the sum of costs to be expended on various control activities in a system, in order that a desired end be attained. To cite an example, Smith and Morris, 1969, have studied the problem of upgrading waste treatment at a number of points of pollution discharge in a stretch of the Delaware River.

There are M polluters, and the kth of these removes f_k lbs/day of BOD from his effluent. It costs him C_k dollars to remove one lb/day of BOD. His total cost is therefore $C_k f_k$ dollars.

The entire expenditure on waste treatment by all polluters is

$$\sum_{k=1}^{M} C_k f_k$$

It is desired to minimize this sum. Hence, for this system the objective function is

$$\text{MIN} \sum_{k=1}^{M} C_k f_k$$

The benefit to be derived, in this instance, was an improved river water quality, as determined by the dissolved oxygen concentration profile. However, the authors placed no economic value on this improvement.

A more sophisticated problem would entail applying some cost formulation to benefits achieved. Then expenditures could be deducted from benefits, and the net benefit could be maximized.

In this simple example, it was assumed that the cost of waste treatment is a scalar multiple of the amount of BOD removed per day. Needless to say, such a convenient linear relationship is rarely the case, either for costs or benefits. The establishment of realistic cost functions is essential, however, for a reasonably optimum allocation of resources. A number of authors who have dealt with these problems as applied to water use are Hamilton *et al.*, 1969, Henderson, 1969, Maass *et al.*, 1962, Miernyk, 1969, Stillson, 1969, Sweet, 1970, Tybout, 1969, Whipple, 1969.

As examples of the complexity of cost functions, Fig. 3 gives curves and a nomogram for costs of dams, and Fig. 4 illustrates the savings in reduction of flood damage after the installation of flood control works.

One area of water resource management where economic values are nebulous is the social one. For example, what dollar value can one place on esthetically pleasing living surroundings? James, 1969, and Sweet, 1970, among others, have tried to deal with this subject.

Methods of systems analysis

These consist largely of operations research techniques, such as:

Network Theory
Linear Programming
Non-Linear Programming
Dynamic Programming

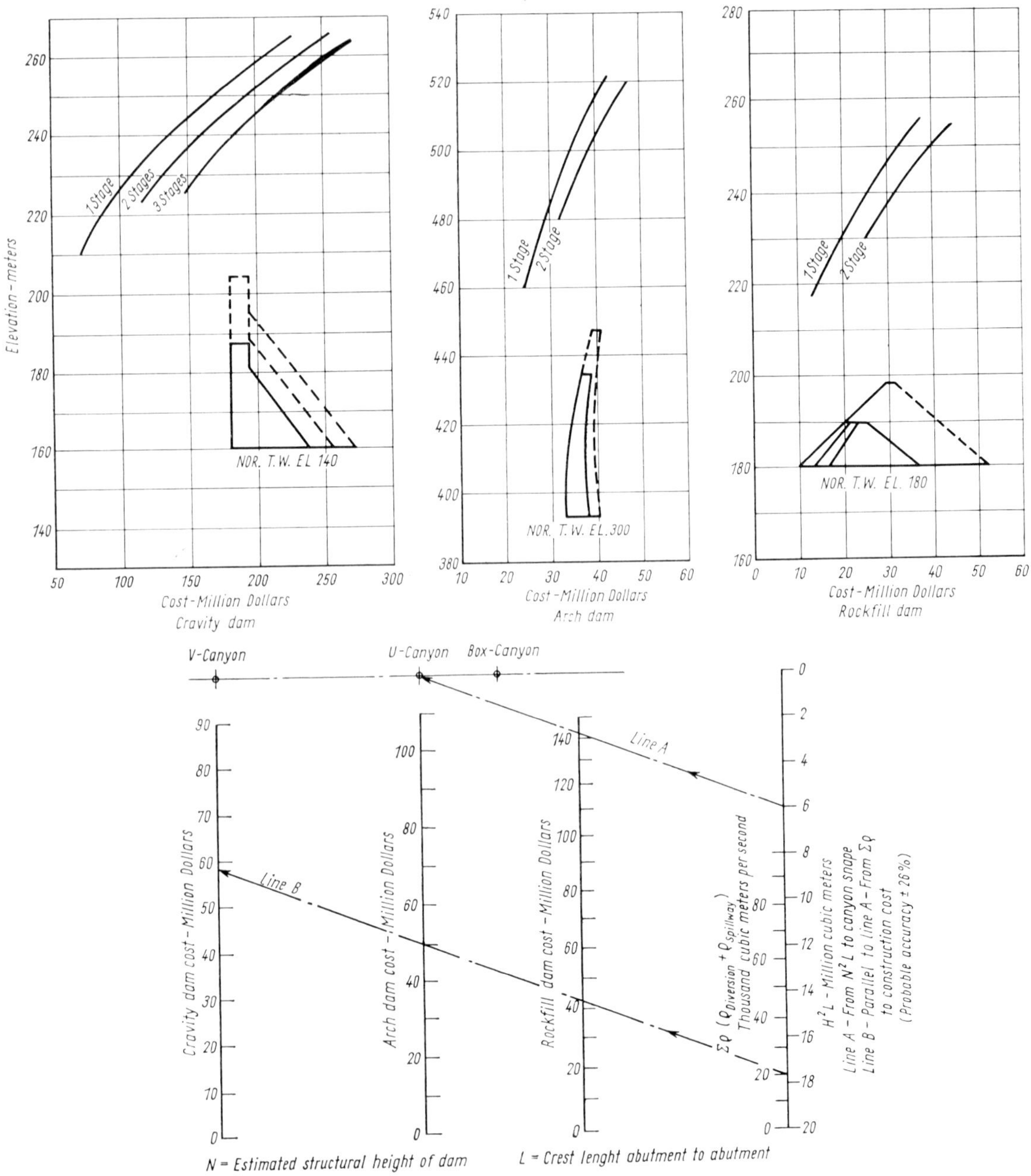

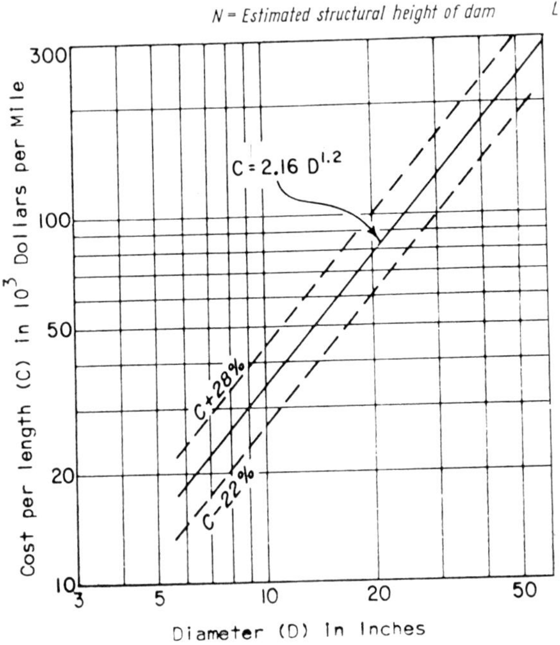

Figure 3 Cost Input Functions. (Toebes, 1969)

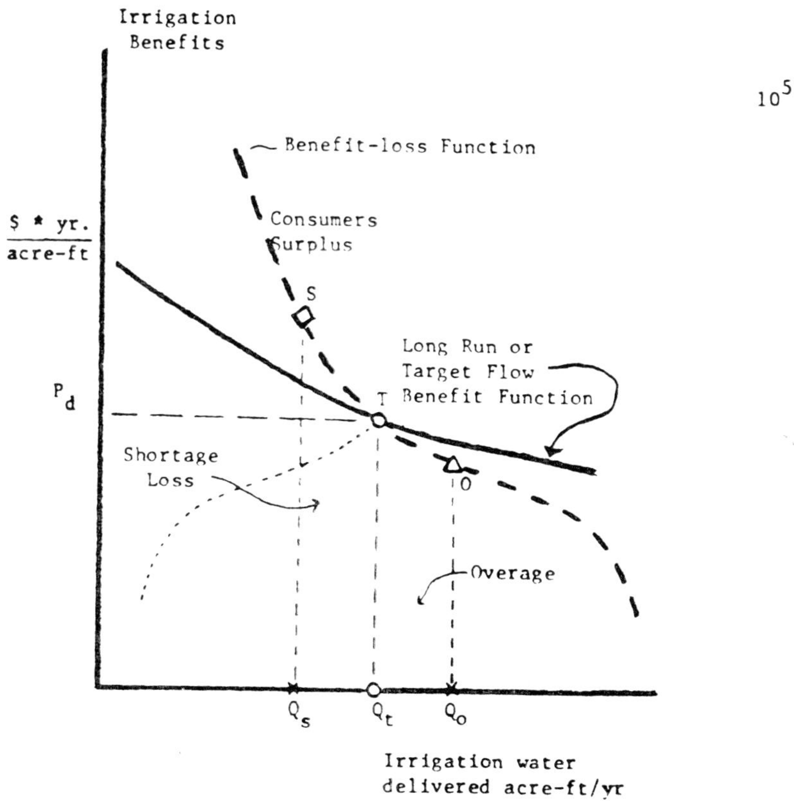

Figure 4 Output-Benefit Function for Irrigation. (Toebes, 1969)

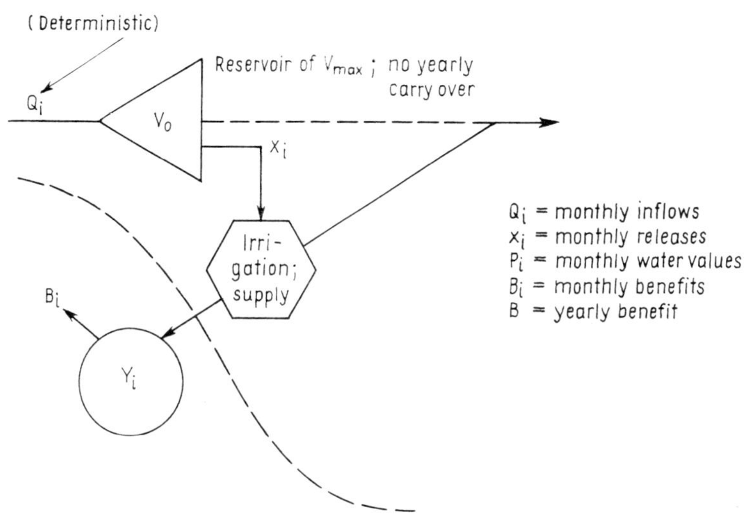

Figure 5 Single Component Policy Optimization Problem. (Toebes, 1969)

Integer Programming
Stochastic Programming
Simulation Techniques

Most of what follows is drawn from Toebes, 1969, who describes some of these methods briefly.

Linear programming

Linear Programming (LP) is designed to identify the optimal allocation of a scarce resource among competing activities. In conventional LP, all analysis functions should be linear. A remarkable variety of problems can be recast into the LP framework. Although its application to a single-purpose reservoir water allocation problem is rather trivial, it may illustrate the recasting of a simple situation in LP terms.

Consider a fixed 12-month inflow distribution Q_i into a reservoir that each January 1 has a filling V_0 (see Fig. 5). What is the optimal set of 12 releases or decision variables, x_i, so that their benefit is optimized? The corresponding objective function to be maximized is:

$$B = \sum_{i=1}^{12} P_i x_i \tag{I.1}$$

The problem can be cast in LP format by noting that, in terms of water quantity, equation I.1 is subject to $p = 23$ constraints, namely:

$$\sum_{i=1}^{12} x_i = \sum_{i=1}^{12} Q_i \tag{I.2}$$

$$\sum_{i=1}^{12} x_i \leqq \sum_{i=1}^{n} Q_i + V_0; \quad 1 \leqq n \leqq 11 \tag{I.3}$$

$$\sum_{i=1}^{n} Q_i + V_0 - \sum_{i=1}^{n} x_i \leqq V_{max}; \quad 1 \leqq n \leqq 11 \tag{I.4}$$

Equation I.2 says that the yearly release must equal the yearly inflow. Equations I.3 state that, up to a month n, the sum of all releases cannot be greater than the sum of all inflows plus V_0. Equations I.4 indicate that, up to a month n, the sum of all inflows plus V_0 less the sum of all outflows, cannot exceed the maximum reservoir volume. Equations I.3 and I.4 can be transformed into linear ones by introducing 22 slack variables S_i and S_j. Together with the x_i, these slack variables define a problem in $q = 34$ variables. A fundamental theorem says that, if there is an optimal solution, then there is also a solution in which $q - p = 11$ variables are zero. The approach is, therefore, to find some solution involving only p variables, to substitute this solution into equation I.1 and then to test whether B has attained a maximum. This is done by incrementing variables, replacing each one that gives $+\Delta B$ by one of the 11 left out, until all $\Delta B < 0$.

In the case of only two decision variables, x_1 and x_2, the geometric representation of this problem is merely that of a parallelepiped, parallel to the B-axis and perpendicular to the x_1 and x_2 axes, which is cut by a plane representing the objective function. The formal process of incrementing and substituting variables corresponds to moving along the intersection until the largest B-coordinate is reached. The corresponding x_i coordinates then represent the optimal x_i-set or "optimal policy."

No real constraints are 100% linear. Also, working with deterministic inputs can be unacceptable. There are LP variants to partially alleviate this. Nevertheless, for large systems, difficulties multiply. For 20 reservoirs, devoted to 10 purposes, a minimum of 2400 constraints would arise. In the informal algebra of dynamic programming, however, constraints are less of a problem. They may be nonlinear and the parallelepiped may have a concave or even discontinuous surface.

Dynamic programming

Dynamic programming (DP) is more an approach than a fixed mathematical scheme. A simple example is shown in Fig. 6. The problem is how to optimally allocate a variable quantity of, say, reservoir water, q, among three users or systems components. Let $q_{max} = Q$. In response to an allocation, each user develops to a commensurate

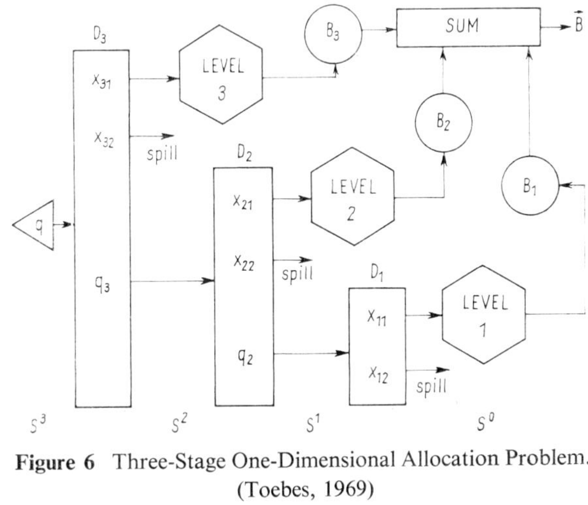

Figure 6 Three-Stage One-Dimensional Allocation Problem. (Toebes, 1969)

"level" having an associated (possibly non-linear) output-benefit function. The systems objective is to maximize the vector of these functions, B.

A trial and error approach would be to divide q into, say, 100 units and to inspect B for every possible x_{ij} division (i = component index; $j = 1$ = applied to purpose of component i; $j = 2$ = spillage). Since there are 100^6 possible divisions, a methodology is clearly needed to reduce the numerical work. The DP strategy is to recast the problem as one in sequential decision making and then to start with the last possible decision rather than the first (*i.e.* the most comprehensive) one. Thus one assumes that only component 1 is to receive part of q. Suppose that x_{11} is the optimal allocation decision D_1, so that $q - x_{11} = x_{12}$ must be spilled (see Fig. 6). Repeating this, for various q's, determines the optimal one-stage process allocation function:

$$f_1(q_2) = \text{MAX}\{B_1(x_{11})\}; \qquad \text{(I.5)}$$
$$0 \leqq x_{11} \leqq q; \quad 0 \leqq q \leqq Q$$

Next assume that components 1 and 2 are to receive parts of q. If $x_{21} + x_{22}$ is alloted to component 2 (i.e., decision stage D_2), the optimal benefit of further allocation (to "1") is already known, namely, $f_1(q - x_{21} - x_{22})$. It remains, therefore, to compute, for various q's the x_{21}'s that determine the optimal two-stage process allocation function:

$$f_2(q_3) = \text{MAX}\{B_2(x_{21})\}; \qquad \text{(I.6)}$$
$$0 \leqq x_{21} \leqq q; \quad 0 \leqq q \leqq Q$$

In the same fashion one can determine for decision stage 3:

$$f_3(q_4) = \text{MAX}\{B_3(x_{31})\}; \qquad \text{(I.7)}$$
$$0 \leqq x_{31} \leqq q; \quad 0 \leqq q \leqq Q$$

Now turning around, one can start with the externally imposed "state variable" q and, using equations (I.7), (I.6) and (I.5) recursively, one can determine the optimal allocation decision sets x_{11}, x_{22}, x_{33}. Collecting these sets in a "policy" table constitutes the "solution."

The dimensionality of the DP problem grows with the number of state variables so that many problems are too large to solve, even when using the refined DP variants developed to overcome this.

The analytical models, tailored to be handled by mathematical programming techniques, have several strong points. One is that the construction of such models forces the analyst to abstract the essence of a problem. Also they permit a direct pursuit of optimality. However, these models have fundamental limitations. One is that feedback in systems is very difficult, if not impossible, to treat. The second is that systems must be rather small and/or their inputs fairly simple. An example of the latter is the simplified deterministic hydrology of the LP illustration of Fig. 5.

Simulation

The objective in many water systems is to transform the stochastic input of natural rainfall into a regulated output by means of impoundments. Consequent-

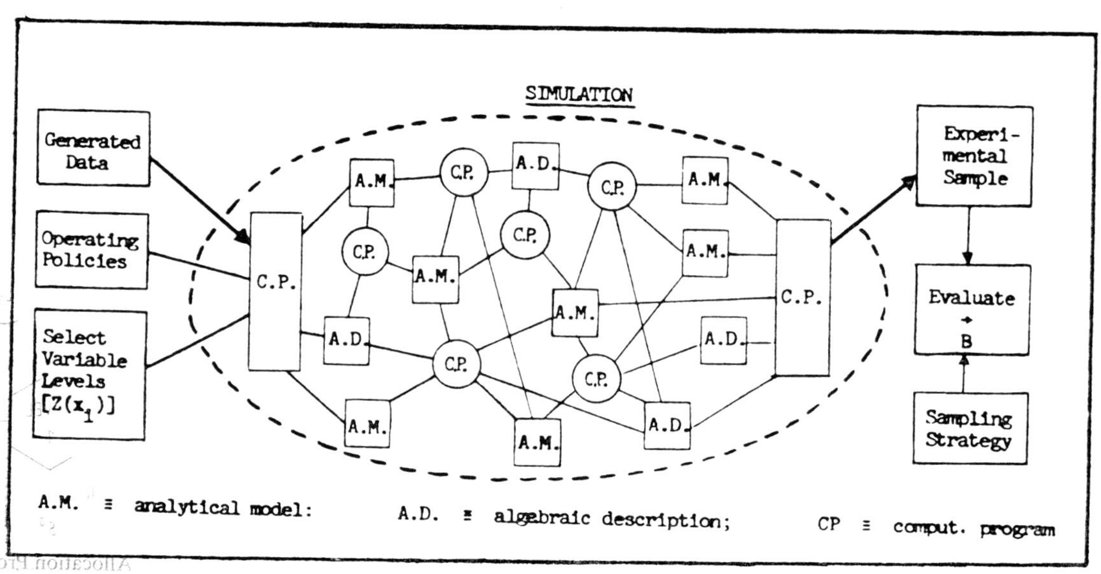

Figure 7 Simulation Model. (Toebes, 1969)

ly, to use simplified, deterministic hydrologies defeats the purpose of the study. Furthermore, for large storage-runoff ratios, the transform "stochastic input-regulated output" can be adequately approximated only by generated inputs of longer duration than most hydrological field records. Using the field records, such long term stochastic inputs can be generated by a number of probabilistic techniques, of which the Monte Carlo and Markov methods are typical.

The simulation model itself can be regarded as a collection of plain algebraic descriptions and small mathematical models, interconnected in order of time and logic by a structure of computer programs (see Fig. 7). The whole is "run" by generated data. If N = number of decision variables plus systems parameters, the quantifiable commensurate systems output is a (N + 1)-dimensional hyper-surface. The location of maxima for large systems is possible only by the use of random and/or systematic sampling strategies.

Realizing furthermore that the operating policy also requires optimization, it is clear that, for large systems, the outputs of each run are essentially statistical samples of systems performance. In other words, one engages in paper experimentation; a direct approach to systems optimality, as in mathematical programming, is not possible. On the other hand, simulation affords great freedom in modeling. Systems sensitivity to a particular input, transfer function, or policy feature is readily ascertained.

However, even though simulation may represent the only avenue towards systematic study of large systems, the modeling cost can be enormous and the completion time is often substantial. Therefore, simulation is most justified for comprehensive and long-range planning efforts. When approximate answers are needed quickly, other approaches should be tried first.

For further reading in the various programming methods and system simulation, consult such texts as Gordon, 1969, Wilde and Beightler, 1967, and Zahradnik, 1971.

II. Basic Modeling

The first systematic study of pollution in rivers to be expressed in mathematical form was that done by Streeter and Phelps on the Ohio River (Streeter and Phelps, 1925). Arising from that work was the now classic "oxygen-sag" equation, upon which most subsequent mathematical modeling has been based.

One of the authors, in a book written some years later (Phelps, 1944), has presented a lucid discussion of the principles involved. Referring to Fig. 8 (itself a classic), we suppose that a stream, initially saturated with oxygen, receives a definite pollutional load at time zero. Assuming, further, that the stream

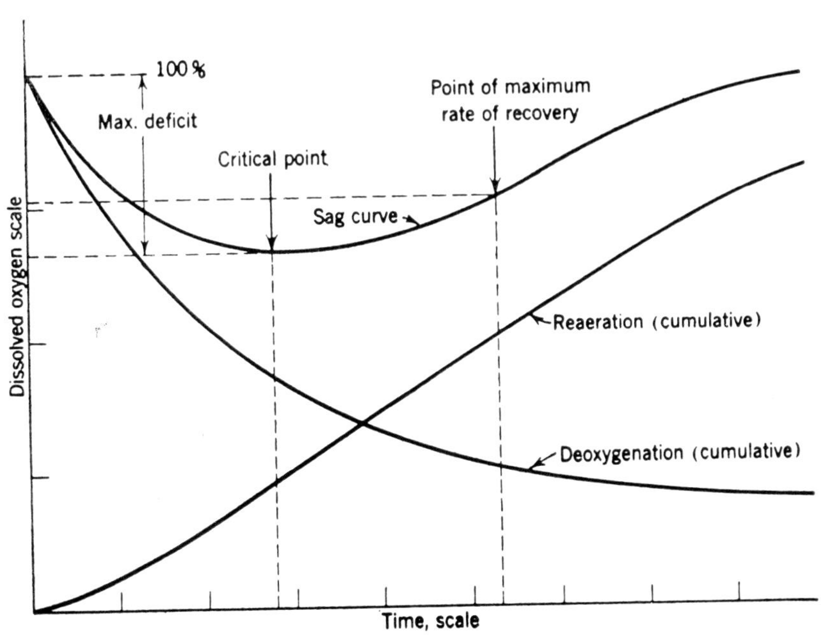

Figure 8 Deoxygenation, Reaeration and Sag Curves. (Phelps, 1944)

flow regime is plug flow (uniform velocity across the entire cross-section), the figure traces out the variation of the dissolved oxygen with time as the water moves downstream. Three curves are shown, the deoxygenation curve, the reaeration curve and oxygen sag curve. Deoxygenation takes place at a rate directly proportional to the instantaneous BOD concentration. Reaeration takes place at a rate directly proportional to the difference between the instantaneous DO concentration and the saturation concentration at the same prevailing conditions, in other words the "deficit."

It can been seen, from Fig. 8, that the rate of withdrawal of oxygen (i.e., the slope at a given point on the deoxygenation curve) starts at a maximum and diminishes continuously to zero. Bear in mind that the ordinate at any point on this curve represents the pollutional concentration in the water, expressed as the concentration of oxygen required to oxidize it.

On the other hand, reaeration starts at zero rate, since initial saturation was assumed. Being always proportional to the DO deficit, it increases progressively for a while, as oxygen depletion increases. With a decreasing rate of loss and an increasing rate of gain, a point must be reached in time at which the two are equal and beyond which the gain exceeds the loss. This represents the lowest point to which the dissolved oxygen will be depleted. It is called the "critical point," and its coordinates are the minimum dissolved oxygen, the critical deficit, and the time, downstream, at which this minimum is reached, the critical time.

Beyond the critical time, reaeration assumes control. The DO profile tends upward and continuously approaches saturation. The total effect of deoxygenation and aeration is shown in the oxygen sag curve.

Mathematically, the sag curve is defined by the differential equation,

$$\frac{dD}{dt} = K_1 L - K_2 D \tag{II.1}$$

where D and L are the instantaneous concentrations of DO deficit and BOD, respectively, and k_1 and k_2 are reaction rate constants of deoxygenation and reaeration. The time of travel or reaction is t.

Furthermore, the variation of BOD with time is given by

$$\frac{dL}{dt} = -K_1 L \tag{II.2}$$

On integration, these equations yield

$$L = L_a e^{-k_1 t} \tag{II.3}$$

$$D = \frac{k_1 L_a}{k_2 - k_1} [e^{-k_1 t} - e^{-k_2 t}] + D_a e^{-k_2 t} \tag{II.4}$$

where L_a and D_a are the initial BOD and oxygen deficit. Note that the exponential terms could be based on common, instead of natural, logarithms. In that case, the rate constants k_1 and k_2 would have different numerical values.

Clearly, the above equations are not a true representation of real streams. Apart from questions of stream geometry and flow regime, they ignore a number of mechanisms whereby BOD and DO concentrations are raised and lowered. A number of authors have subsequently developed more detailed versions. As one example only, Camp, 1963, has added terms representing the resuspension of BOD from the benthos and the production of oxygen by photosynthesis. For tidal streams, in addition, he has introduced terms involving the turbulent diffusion coefficient.

To review the development of the mathematical representations of stream water quality, reference is made to a number of authors (O'Connor and Thomann, 1966; O'Connor, 1967; Thomann *et al.*, 1968; Q.L.M., 1970; O'Connor and DiToro, 1970).

A body of water can be described by a set of partial differential equations with variable coefficients. Each term corresponds to a basic physical characteristic of the water body. For initial and/or boundary conditions, the system has inputs of flow, temperature and wastewater. The system response is the pollutant concentration, distributed in time and space.

The development of the model begins by writing a mass balance on the rate of movement of pollutant through any small segment of the river or estuary.

The volume of this segment is permitted to approach an infinitesimal size. This procedure yields a differential equation which describes the net rate of pollutant transport at any point within the system.

This equation is then integrated to obtain the "general solution." Here it is recognized that the differential equation of stream mass transport describes the behavior for many different stream systems. Its general solution contains the particular solution for any of these; i.e., the general solution contains two as yet undetermined integration constants. These constants, when evaluated, will yield the particular solution for a given system.

For a particular system, these constants are determined by fitting the general solution to the boundary conditions which describe that system. The number of boundary conditions required is equal to the number of integration constants; the "fit" is obtained by substituting the general solution into the boundary conditions and solving a set of simultaneous equations for the integration constants.

The resulting particular solution will describe the distribution of pollutants, such as BOD, or water quality characteristics, such as DO, in the particular waterway being investigated. Since these profiles depend on waste loading and river characteristics, the behavior for various loadings and river conditions may be obtained.

Two basic problem areas exist. One is to draw up a mass balance that realistically models the system. The other is to solve the differential equations that result.

Dissolved oxygen

Factors affecting DO concentration in a natural stream fall into two categories. The first category includes the geophysical characteristics of the drainage area, e.g., freshwater flow, variations in stream geometry, local temperature, etc. The second category includes the biochemical and physical characteristics of the stream iteself, i.e., sources and sinks of DO, both natural and man-made.

Temperature influences oxygen solubility, and rates of both physical and biochemical reactions. Freshwater flow dilutes polutants and affects velocity. Variations with distance of stream cross-section, depth and width play their part in determining the reaeration rate.

Sinks of DO are respiration by bacteria and algae, both free and trapped in the benthal deposits, and chemical oxidation. Sources of DO are atmospheric reaeration, (natural or artificial), photosynthesis and advection.

Biochemical oxygen demand

Organic matter is removed by chemical and biological oxidation. However, it is also removed without use of oxygen by sedimentation, scour and gas transfer from volatization. Biochemical oxygen demand has been shown to occur in two stages, and, in fact, is referred to as "First-stage BOD" and "Second-stage BOD." In any given waste sample, second-stage demand can appear anywhere from one to ten days after first-stage. The two types are also differentiated chemically. First stage BOD is carbonaceous (hence the term "CBOD"), and second-stage BOD is nitrogenous ("NBOD"). Short lags occur when the inputs are biologically treated effluents. Long lags occur when the inputs are raw untreated wastewaters. Secondary DO sags may occur downstream, when the CBOD has been nearly entirely removed and the NBOD starts reacting.

CBOD occurs in both suspended solid and dissolved forms, and NBOD only as dissolved matter not subject to sedimentation.

It is possible to differentiate between the two forms of BOD in the standard equations by including two initial concentrations and two reaction rate constants.

Sinks of DO

In addition to BOD there is benthic demand and plant respiration. If a river has appreciable dissolved oxygen, then the surface layer of the benthos is aerobic. Benthic demand is usually expressed as mass per unit area of stream bed. Multiplying this by the ratio of river volume to river bottom area yields a volumetric rate, with dimensions consistent with the other terms in the oxygen sag equation.

The respiration of plant growths in the river is proportional to temperature, mass of growth, concentration of nutrients and distribution in space. It is also conveniently expressed as a volumetric rate.

Sources of DO

Advection is, natural aeration, artifical aeration and photosynthesis of green plants. Advection is introduced as a flux term or initial condition. Natural aeration is introduced as a source term, a rate constant multiplied by the DO deficit.

Photosynthesis is affected by many factors, sunlight, temperature, the mass of algae, the presence of nutrients, etc.

The discussion so far has dealt specifically with DO and BOD. However, the mass balance approach can be applied to any component whatever in the river or estuary system, both conservative (e.g., salinity) or non-conservative (e.g., radio-activity).

The point form of the mass-balance equation

(often called the "continuity equation") for any component is:

$$\frac{\partial c}{\partial t} = \nabla \cdot j \pm \sum s \tag{II.5}$$

c = concentration of component under study.

t = time

s = sources and sinks of the component

j = mass flux of the component

∇ = "Del" operator (the sum of the differentials with respect to distance in the three coordinate directions of a three-dimensional space).

j is a vector consisting of three elements, one for each of the three coordinate directions. In direction "n"

$$j_n = E_n \frac{\partial c}{\partial n} - U_n c \tag{II.6}$$

where $E_n \frac{\partial c}{\partial n}$ is the "diffusion flux," i.e., the concentration gradient multiplied by the diffusion coefficient E_c (also called "eddy diffusion," "turbulent diffusion" and "dispersion"), and $U_n c$ is the "advection flux," i.e., the point concentration multiplied by the velocity U_c in direction "n".

In non-tidal streams, E is affected by the freshwater flow, in estuaries by tidal translation. E, U and s may be functions of both time and space.

If equation (II.6) is substituted into equation (II.5), the latter will provide a dynamic three-dimensional profile of the component under study. However, the resulting equation would present enormous difficulties for solution. The first simplification universally made is to assume uniform concentration over any given river cross-section. This yields an equation for the concentration as a function only of time and longitudal distance. However, the various coefficients will still be variables also with time and distance. Thus, for rivers:

$$\frac{\partial c}{\partial t} = \frac{1}{A} \frac{\partial}{\partial x}\left(EA \frac{\partial c}{\partial x}\right) - \frac{1}{A} \frac{\partial}{\partial x}(QC) + \sum s \tag{II.7}$$

Q is the freshwater flow.

A is the river cross-section.

E is the diffusion coefficent.

In order to reduce the equations to their simplest possible form, with all coefficients constant, the common technique is to divide a river into segments, in each of which E, Q and A can be considered fixed.

Furthermore, in order to define the various source terms (in the light of previous discussion), simplicity requires that the various reaction rate coefficients be constant. Here again, segmentation provides the means.

Finally, we obtain the following equations for BOD and DO deficit:

$$\frac{\partial L}{\partial t} = E \frac{\partial^2 L}{\partial x^2} - \frac{Q}{A} \frac{\partial L}{\partial x} - k_1 L(x, t) - k_3 L(x, t) + S_c(x, t) + S_n(x, t) \tag{II.8}$$

$$\frac{\partial D}{\partial t} = -E \frac{\partial^2 D}{\partial x^2} + \frac{Q}{A} \frac{\partial D}{\partial x} + k_1 L(x, t) - k_2 D(x, t) - P(x, t) + R(x, t) + B(x, t) \tag{II.9}$$

where the following are constant for the specific river or estuary segment:

E = diffusion coefficient

Q = freshwater flow

A = river cross-section

k_1 = BOD reaction rate coefficient

k_2 = atmospheric reaeration rate coefficient

k_3 = BOD sedimentation rate

and the following vary with time and distance along the segment.

S_c = CBOD input rate

S_n = NBOD input rate

P = oxygen generation rate by photosynthesis

R = algal respiration sink

B = benthic bacterial respiration sink

D = DO deficit concentration

L = BOD concentration

O'Connor and DiToro, 1970, have shown how to account for the diurnal fluctuation in photosynthesis. The term P can be made into a sinusoidal term, such as:

$$P(t) = \begin{cases} P_m \sin\left[\frac{\pi}{p}(t - t_s)\right], & t_s \leqq t \leqq t_s + p \\ 0 & , t_s + p \leqq t \leqq t_s + 1 \end{cases} \tag{II.10}$$

P_m = maximum oxygenation rate, mg/l · day

t_s = time at which the oxygen source begins, days

p = fraction of the day in which the source is active, days.

Equation (II.10) states that, for the day that begins at time t, oxygen is generated for a fraction p, and no oxygen is generated for a fraction $(1 - p)$.

In order to define $P(t)$ for more than one day, it might be simpler to assume that it fluctuated sinusoidally continually and not intermittently. However this would entail the negative half wave, and would imply that as much oxygen was destroyed at night as was generated by day. With more realism, the function of equation (II.10) can be transformed into a continuous one in terms of a Fourier series. Then, for simplification, the series could be truncated at the desired term.

For estuaries, the coefficient of the second term on the right of equation (II.7) and (II.8) becomes $\left(\frac{Q}{A} + U_t\right)$, where U_t is the oscillating tidal velocity. For simplicity, conditions are often considered only at slack tide, at which time $U = 0$. In that case, time t must be reckoned in tidal cycles.

The dispersion term is significant in deep, slow streams and in the non-saline parts of estuaries where there are tidal fluctuations but no flow reversals. In most streams, however, the advective term overwhelms the dispersion term and it is neglected.

It is evident that the Streeter-Phelps equations, (II.1) and (II.2), are a special case of equations (II.8) and (II.9), i.e., with a number of terms neglected, notably dispersion. Various investigators have simplified equations (II.8) and (II.9) to a greater or lesser extent and obtained solutions in a number of ways. (See Section IV.)

QLM, 1971, describes the procedure for solving the steady-state form of the equations for a segmented river. The BOD and DO profiles can be computed, reach by reach from upstream to downstream. Terminal values of one reach are initial values of the next. NBOD steady-state equations also can be written. These are simpler than the CBOD equations because NBOD is exclusively soluble material and there is therefore no sedimentation term. All of the equations are linear, ordinary differential equations of the first order, and are readily solved by the integrating factor method.

If diurnal variation of photosynthesis is of interest, an unsteady-state DO model can be constructed to describe the temporal and spatial distribution of DO within a reach. The unsteady-state model is obtained by replacing the average value of P with a photosynthetic oxygen production rate that is a function of time. Differential mass balances are then written in which the accumulation term is NOT equated to zero.

The other parameters, such as rate coefficients, cross-sectional areas, algal respiration and benthic demand, are constant for a given reach. Carbonaceous BOD concentration L, and nitrogenous BOD concentration N, are functions only of distance. River flow can be either constant or vary linearly with distance.

The resultant equations for DO concentration is a quasi-linear, partial differential equation of the first order. It can also be expressed in terms of DO deficit. The method of solution is as follows:

1. The partial differential equation is transformed into a linear, ordinary differential equation by applying Laplace Transforms.
2. The linear ordinary differential equation is solved by the integrating factor method.
3. The Laplace transforms in the resulting algebraic equation are inverted to yield the solution of the original partial differential equation.

The procedure for solving the time-variant equations is necessarily more complicated. Dresnack and Dobbins, 1968, point out that equations (II.8) and (II.9) are similar to equations for heat conduction in solids, but include additional lower order terms. Many heat conduction problems have been solved analytically, but the BOD and DO equations, having lower order terms, plus complicated boundary conditions usually, are not amenable to analytical solution. Furthermore, numerical procedures that work for the heat conduction problem are often unsatisfactory for the pollution problem because of these same lower order terms.

Equations (II.8) and (II.9) are second-order, parabolic, partial differential equations. Solutions are L and D as functions of x and t. Initial conditions specify values of L and D as functions of x along the stretch at time zero. Boundary conditions specify values of L and D as functions of t at the beginning and end of the stretch.

The source and sink terms often vary diurnally. Analytical solutions of equations (II.8) and (II.9) are obtainable only by making simple assumptions about the variable inputs. Numerical solutions, however, can be obtained at some small loss of accuracy in computation, but with some gain due to the retention of otherwise neglected terms. Such a solution requires setting up finite-difference equations. Derivative terms in the differential equations are replaced by their finite-difference equivalents. A grid is then set up, each point in which has a location in time and space, and for each grid point an algebraic equation is obtained for the dependent variable (BOD or DO concentration) in terms of neighboring grid point values. By this means, a set of simultaneous linear algebraic equations is formed, which are then solved by iteration.

Lee *et al.*, 1971, describes the technique for segmenting a river having n waste inputs and m water intakes. Segment boundaries would be located just below an input or intake, yielding $(n + m + 1)$ segments. For each segment a pair of equations of the form of (II.8) and (II.9) could be written, i.e., $2(n + m + 1)$ equations. At points of entry of waste, mixing is assumed instantaneous and complete.

Sufficient boundary conditions can be determined in order to solve for the required number of arbitrary constants in the analytical solutions. The final form of solution is very complicated and is represented by complex matrix equations.

Using this analytical approach, the authors performed a series of simulation studies and obtained the following qualitative results:

1. An increase in the dispersion coefficient reduces the peak BOD concentration at input and reduces the magnitude of the oxygen sag downstream. It also shifts the critical point (minimum DO) in the upstream direction.
2. An increase in the deoxygenation rate coefficient k_1 (or the BOD decay rate coefficient) causes an increase in the rate of dissolved oxygen consumption and an increase in the magnitude of the oxygen sag. The system was highly sensitive to k_1, implying strongly that a good estimate of this parameter is important.
3. Considerable interaction was manifested between the dispersion coefficient E and the deoxygenation and reaeration rate coefficients k_1 and k_2. As k_1 was increased, the input BOD peak was reduced (as for E). As k_2 was increased, the magnitude of the DO sag fell and the minimum point shifted upstream (as for E).
4. If, instead of one BOD input point, the same total amount was distributed along the reach, then the maximum value of BOD in the reach was reduced, as was the amount of the DO sag.
5. Studies also indicated that, even apart from the improved aeration, artificial mixing was beneficial by virture of the improvement in dispersion.

QLM, 1971, have elaborated on the technique of stream segmentation, and have developed the concept of the "Generalized Reach" for constructing one-dimensional, steady-state BOD and DO models.

The generalized reach is one in which all possible boundary conditions and all possible sources and sinks affecting the levels of BOD and DO are included. (Refer to Fig. 9.)

1. Dams, waste discharges and tributaries are located at the upstream end of the reach.
2. Nitrogenous BOD loadings with zero time lag are considered as point inputs at the upstream end of the reach.
3. Groundwater, surface runoff and distributed waste are input uniformly along a reach.
4. Diversion flows are located at the downstream end of a reach.
5. All the geometrical, hydrological, biological, chemical and physical parameters which can affect the material balance are constant within a reach.

Owing to the lag time of nitrogenous oxidation, the point where a NBOD loading is discharged into a waterway is not the same as the point at which it starts to undergo nitrification. The segmenting of the waterway is therefore arranged so that, at a point downstream equal in travel time to the time lag, a new segment is begun. The nitrogenous load is then assumed to enter the stream at this point with zero time lag.

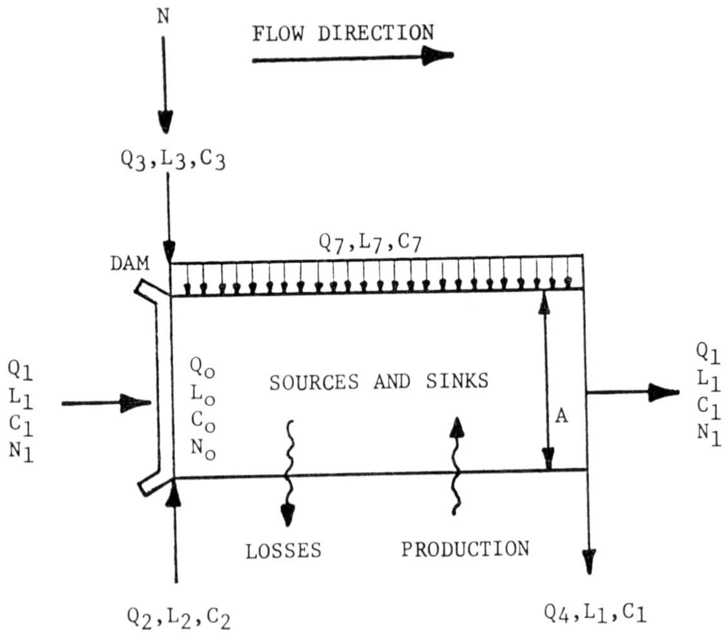

LEGEND

Q_1, L_1, C_1, N_1	=	FLOW, CARBONACEOUS BOD, DO, NITROGENOUS BOD, IN THE WATERWAY (COULD BE EITHER INPUT OR OUTPUT OF A REACH)
Q_2, L_2, C_2	=	FLOW, CARBONACEOUS BOD, DO IN PLANT WASTE DISCHARGE
Q_3, L_3, C_3	=	FLOW, CARBONACEOUS BOD, DO IN A TRIBUTARY
Q_o, L_o, C_o, N_o	=	FLOW, CARBONACEOUS BOD, DO, NITROGENOUS BOD AT THE BEGINNING OF A REACH
Q_7, L_7, C_7	=	FLOW, CARBONACEOUS BOD, DO IN THE UNIFORMLY DISTRIBUTED FLOW
Q_4	=	DIVERSION FLOW (THE BOD AND DO CONCENTRATIONS ARE EQUAL TO THOSE IN THE OUTPUT OF A REACH)
A	=	CROSS-SECTIONAL AREA
N	=	NITROGENOUS BOD LOADING WITH ZERO LAG TIME

Figure 9 Boundary Conditions of the Generalized Reach. (QLM, 1971)

The general outline described here is the one followed by most workers in this field. However, it is not the only approach to modeling BOD and DO river profiles.

Consider the finite-difference method. The classic approach, as described above, has been to derive partial differential equations for the BOD and DO concentrations in a stream, then to put these differential equations into finite-difference form and obtain approximate solutions.

Bella and Dobbins, 1968, eliminate the intermediate abstraction of the differential equations and model the stream by a mass balance approach directly in finite-difference terms.

The stream is divided into segments of length ΔX, in each of which concentration is assumed uniform. Mechanisms of transfer are set up to simulate dispersion, convection and decay. Pollutant sources are simulated as slug loads into a particular segment or segments. Then procedures are discussed for determining DO profiles. Thus one suggested procedure is to determine BOD values for all segments for one time interval, by adding pollutant, dispersing the BOD, convecting it and "decaying" it. The BOD values are then used to compute the DO profile for one time interval. The two steps are repeated for each time interval.

Another suggested procedure is to add pollutant, disperse both BOD and DO, convect both BOD and DO, and finally decay the BOD and add other sources and sinks of DO.

A number of different procedures were tried and their results were found to differ little.

Most mathematical models consider only tidal

cycles, not variations within the cycle. Such models ignore a number of things. Large volumes of water move back and forth across a continuous point source of pollution, thus spreading it. The amount of pollutant discharged into a passing water segment is inversely proportional to the segment velocity, which varies from zero to a maximum. The effect of boundary conditions will vary within a tidal cycle.

However, the model of Bella and Dobbins, 1968, is applicable to to the analysis of profiles within a tidal cycle.

Another discontinuous model approach was described by Thomann, 1963. The general case considers a body of water in the two horizontal dimensions, i.e., vertical mixing is assumed to be complete. From this, the special one-dimensional river case is derived. This method has been applied in the Delaware River estuary. (See Section IV.)

In a hypothetical body of water, in which there is no preferred flow direction, it is assumed that n segments can be defined, all interconnected. Each segment is assumed to be a linear system, and the time rate of change of DO in any segment depends on:

a) advection and diffusion of DO in and out.
b) sources and sinks of DO within the segment.

Consider segment r, having a neighbor r'.

Advection of DO from r' to r is:

$$A_{r'r} = Q_{r'r}[\xi_{r'r}c_{r'} + (1 - \xi_{r'r})c_r] \qquad \text{(II.11)}$$

$Q_{r'r}$ = volumetric transport rate of water from r' to r.

c_r = DO concentration in r.

$c_{r'}$ = DO concentration in r'.

$\xi_{r'r}$ = fraction of $c_{r'}$ advected into r together with $(1 - \xi_{r'r})c_r$.

The value of $\xi_{r'r}$ lies between zero and one and varies with the physical situation of the particular segment, e.g., near the boundaries of the water body.

For estuaries

$$\xi_{r'r} = 0.5$$

Diffusive exchange of DO from r' to r is:

$$D_{r'r} = E_{r'r}(c_{r'} - c_r) \qquad \text{(II.12)}$$

$E_{r'r}$ is the eddy exchange coefficient $\left(\text{with dimensions } \frac{L^3}{T}\right)$, analogous to the eddy diffusion coefficient in a continuous model. Equation (II.12) is valid as long as no strong DO gradients exist within each segment.

In this case, it is assumed that the advective and diffusive exchanges are net over a tidal cycle, and do not model the process *during* a tidal cycle.

Note also that:

$$\left.\begin{aligned} Q_{r'r} &= -Q_{rr'} \\ E_{r'r} &= E_{rr'} \end{aligned}\right\} \qquad \text{(II.13)}$$

The primary sources and sinks of DO are:

a) atmospheric reaeration
b) removal of DO by BOD
c) respiration and photosynthesis of plants

For segment r, the complete DO equation is:

$$\begin{aligned} V_r \frac{dc_r}{dt} = \sum_{r'} \{Q_{r'r}[\xi_{r'r}c_{r'} + (1 - \xi_{r'r})\, c_r] \\ + E_{r'r}(c_{r'} - c_r)\} + V_r k_{ar}[c_{sr} - c_r] \\ - V_r k_{dr} L_r + S_r + M_r \end{aligned} \qquad \text{(II.14)}$$

Where

V_r is the segment volume

c_{sr} is the saturation DO concentration

L_r is the concentration of ultimate BOD in the segment

k_{ar} is the reaeration rate constant

k_{dr} is the deoxygenation rate constant

S_r is the generation rate of DO by photosynthesis

M_r is the net generation rate of DO by all other sources and sinks.

All the quantities listed are time-variant. Equation (II.14) can be rearranged, simplified and rewritten as:

$$\begin{aligned} a_{r1}c_1 + a_{r2}c_2 + \cdots + a_{rr}c_r + \cdots + a_{rn}c_n \\ = k_{ar}c_{sr} - k_{dr}L_r + S_r + M_r = f_r(t) \end{aligned} \qquad \text{(II. 15)}$$

All coefficients are constants, except for a_{rr} which contains the first-order differential operator.

The right-hand side of equation (II.15), i.e., $f_r(t)$, is the forcing function for segment r. The left-hand side is the stationary linear system for c_r.

For the entire system of n segments, equations (II.15) form the matrix equation:

$$\mathbf{Ac} = \mathbf{f} \qquad \text{(II.16)}$$

where:

A is an $n \times n$ matrix, whose diagonal terms contain the first-order differential operator,

c is the vector of segment DO concentrations, and

f is the vector of segment forcing functions.

In general, for a forcing function applied in segment i, the response $c_j(t)$ of the DO concentration in any other segment j is given by:

$$D_A c_j(t) = M_{ij} f_i(t) \qquad \text{(II.17)}$$

where D_A is the determinant of matrix **A**, and M_{ij} is the signed minor of D_A that results from cancelling row i (in which the forcing function is applied) and column j (in which the desired $c_j(t)$ appears).

An important special case: the one-dimensional estuary

In this case, each segment interacts with only two others, its neighbors upstream and down. Consequently, matrix **A** becomes a sparse one, with all elements zero except for those on the diagonal and those on either side of the diagonal.
If the following assumptions are made:

a) streamflow constant, Q.
b) eddy exchange coefficient constant, E.
c) advection factor set to 0.5, ξ.

then the elements of **A** are given by:

$$\left.\begin{aligned} a_{ii} &= \frac{d}{dt} + k_{ai} + \frac{2E}{V_i} \\ a_{ij} &= \frac{1}{V_i}(0.5Q + E) \\ a_{ji} &= -\frac{1}{V_i}(0.5Q + E) \end{aligned}\right\} \qquad \text{(II.18)}$$

III. Parameter Evaluation

This section is a very brief description of the methods adopted by some investigators to obtain numerical values of the various constants and coefficients used in the water quality equations. Reference is made to Harleman *et al.*, 1968, Whipple *et al.*, 1969, Thackston and Krenkel, 1969, and Q.L.M., 1970.

The mass balance equations contain source and sink terms, and it is usually possible to fit almost any mathematical model to observed data by suitably adjusting coefficients. Results may be valid for that particular reach, but it would be dangerous to apply the same coefficients to other estuaries. Also, the predictive ability of the model, even for the original estuary, is questionable. Consequently, it is of great importance to ensure that the values used are as accurate as possible. On the other hand, most rivers and streams have such complex geometries that field data may not be readily applicable. And, of course, laboratory data must be treated with caution when being applied to a real stream.

The U.S. Geological Survey maintains streamflow metering stations in nearly every river in the country, and in many rivers at several points. Empirical correlations have been determined for shallow streams between streamflow on the one hand and stream depth and velocity on the other. They are usually of the form:

$$\left.\begin{aligned} \text{Depth } H &= aQ^b \\ \text{Velocity } U &= cQ^d \end{aligned}\right\} \qquad \text{(III.1)}$$

where Q is the streamflow and the other letters represent numerical constants.

Mean depths in deeper rivers are obtained by sounding, either manually or by fathometer. River width can be scaled from USGS quadrangles, if necessary. Thus cross-sectional areas can be calculated, and hence stream velocities.

Other parameters are discussed under individual headings.

Longitudinal dispersion coefficient

Difficulties arise in determining the value of the longitudinal dispersion coefficient. Sometimes, values determined in the salinity intrusion section of an estuary are used in studies of the upstream freshwater (but still tidal) section, with no regard paid to the different mixing mechanisms in each. On other occasions, dispersion coefficients have been employed which where obtained from dye injection tests on hydraulic models, ignoring the lack of concentration similitude in such models.

The value of the longitudinal dispersion coefficient at any point within the salt-intruded reach of the river can be conveniently obtained by analysis of salinity profiles. The limiting form of the mass transport equation for the case of a conservative substance such as salt, and non-constant values of

Q, A, and E, is:

$$\frac{1}{A}\frac{\partial}{\partial x}\left[EA\frac{\partial c}{\partial x} - Qc\right] = \frac{\partial c}{\partial t} \qquad \text{(III.2)}$$

If the variation of salinity with x and t is known, the derivatives $\partial c/\partial x$ and $\partial c/\partial t$ may be obtained graphically or numerically. Equation III.2 can then be used to compute the value of E at any point within the saline reach of the river.

This procedure requires that a number of profiles be available so that the time derivative, $\partial c/\partial t$ can be computed, and also requires that the values of Q, a time and distance dependent function which is controlling the intrusion, be known. The problem can be avoided by recognizing that drought flows in a river remain relatively constant for extended periods of time; Q, and therefore U, are known and the steady Q gives rise to steady salinity profiles during these periods. Under these conditions, the net flux of the salt in the river must be zero, since there is no sink or source of salt within the estuary.

Equation (III.2) then reduces to:

$$E\frac{\partial c}{\partial x} - Uc = 0 \qquad \text{(III.3)}$$

In the freshwater region of an estuary, the value of E is much smaller than those obtained in the saline region. It has been approximated by the expression:

$$E = \alpha n U R_H^{\beta} \qquad \text{(III.4)}$$

where

n = Manning roughness factor

U = instantaneous longitudinal velocity averaged over the stream cross-section, fps.

R_H = hydraulic radius, ft.

E = dispersion coefficient, ft/sec

α, β = numerical constants

Deoxygenation rate coefficient, k_1

The modelling of oxygen consumption in a river or estuary requires the use of ultimate BOD values rather than the more commonly measured 5-day BOD (BOD_5). The conversion of 5-day BOD to ultimate BOD (BOD_u) requires determination of the reaction velocity constant, k_1, controlling oxygen consumption in long-term BOD studies.

For an estuary, samples can be collected at four instants in the tidal cycle, i.e., low water slack, maximum ebb, high water slack and maximum flow. The samples, unseeded and undiluted are aerated to saturation and then incubated at 20°C. Plotting all the data from each sample point on the same graph provides enough points for confident curve fitting. The line through these points represents the tidal averaged reaction for the particular sampling point. The reaction rate is then computed from the fitted curves.

When the data show convincingly that a plateau of BOD exertion is reached around the 6th to 9th day of incubation, then 2nd stage demand is clearly present and two-stage curves can be used.

BOD sedimentation rate, k_3

$(k_1 + k_3)$ is the measure of the rate at which BOD is being removed from a stream. If BOD is removed by biochemical oxidation alone, k_3 is zero.

Sedimentation can occur during periods of low flow, and scour of bottom deposits, with consequent re-suspension, can occur during high spring flow. BOD removal by sedimentation is often neglected.

Reaeration rate constant, k_2

Several investigators have devised empirical expressions for k_2 as a function of some of the following factors:

- Diffusivity of oxygen in water
- Stream depth
- Stream velocity
- Froude number
- Slope of the stream bed
- Temperature
- ... and so on.

Thackston and Krenkel, 1969, conveniently summarize those due to O'Connor-Dobbins, Krenkel, Chirchill, and Dobbins alone. The most widely used is the first:

$$k_2 = \frac{(D_L \cdot U)^{1/2}}{2.3H^{3/2}} \qquad \text{(III.5)}$$

in which:

D_L = oxygen diffusivity, ft²/hr

U = mean stream velocity (which, in an estuary, varies cyclically with the tide) ft/hr.

H = stream depth, ft.

Whipple *et al.*, 1969, used the simple relation:

$$k_2 = 21.62\frac{U^{0.67}}{H^{1.68}} \qquad \text{(III.6)}$$

Temperature correction for rate constants

Experimental determination of rate constants is always made at 20°C. To determine the value at some other temperature T, the following expression is used:

$$k_T = k_{20}\theta^{(T-20)} \qquad \text{(III.7)}$$

For k_1, the literature reports values for θ from 1.026 to 1.065. For k_2, the reported values range from 1.015 to 1.047.

Oxygen production by algae

The photosynthetic production of oxygen by algae is well known and at times accounts for marked degrees of oxygen supersaturation in lakes and streams. The respiration of the organisms simultaneously depletes oxygen, and at night or during a series of dark cloudy days, may exceed algal oxygen production and result in a measurable drain in the oxygen resources of the water. Oxygen production and depletion by algae are temperature-dependent phenomena, both being highest during those months in which the pollutional load of a stream makes the DO critical. Most species of algae are found in higher concentrations in summer months when water temperatures are highest.

Q.L.M., 1970, working on the Hudson River, have observed that DO values in the summer months appeared to be higher than could be accounted for by reaeration, but that, during the rest of the year, photosynthesis had no measurable effect. They also found that the photosynthesis rate at a depth of one foot was greater than that at four feet. Furthermore, photosynthesis at the one foot depth continued actively until later in the day than at four feet. This was to be expected in such turbid water as the Hudson. Data showed that, on the average, photosynthetic activity was extinguished in the top six feet.

Whipple *et al.*, 1969, working in the much shallower Passaic River, found that the eutrophic zone was to be found only in the top 50 cm. The net production rate in this zone (oxygen production minus algal respiration) was experimentally determined to be 4.6 mg/l/day. Averaged over a total depth of 200 cm, this gave 1.15 mg/1/day. In the Hudson, on the other hand, the production in the six feet layer was found to be 0.72 mg/l during 12 hours of daylight. Averaged over a total depth of 23 feet, the mean daily production was 0.1 mg/1/day, from which algal respiration had to be deducted.

IV. Some Actual River Models

Considerable literature has been published concerning mathematical modeling of the Delaware River Basin. A commission for the Basin was set up in 1961, and has been active in sponsoring projects aimed at upgrading the water quality in the streams within its jurisdiction. Selzer, 1970, describes briefly the scope of the "Delaware Estuary and Bay Water Quality Sampling and Mathematical Modeling Project," and D.R.B.C., 1970, is the Commission's Final Report on it.

A number of approaches were examined for developing mathematical models of the river. As O'Connor and Thomann, 1966, have described, a finite-difference approach may be appropriate for the non-steady state, or if the system cannot be characterized by any regular function. Thus, the continuous derivatives in the partial differential equations (see Section II) may be replaced by finite approximations. Pence *et al.*, 1968, provide an example of the approach, resulting in a pair of ordinary differential equations for each segment of a segmented estuary.

For the kth segment:

$$\left.\begin{aligned} \frac{dL_k}{dt} &= f(t, L_{k-1}, L_k, L_{k+1}) \\ \frac{dc_k}{dt} &= g(t, c_{k-1}, c_k, c_{k+1}) \end{aligned}\right\} \qquad \text{(IV.1)}$$

These can then be integrated by the 4th-order Runge-Kutta method, and the newly computed L_k and C_k values at the end of the kth segment become the initial values for the $(k+1)$th segment.

Alternatively, a material balance can be written around each finite river segment (Thomann, 1963).

In the latter approach, and for the one-dimensional case, one can write for the ith segment of n river segments:

$$\begin{aligned} V_i \frac{dc_i}{dt} = {} & Q_{i-1,i}[\xi_{i-1,i}c_{i-1} + (1 - \xi_{i-1,i})\, c_i] \\ & - Q_{i,i+1}[\xi_{i,i+1}c_i + (1 - \xi_{i,i+1})\, c_{i+1}] \\ & + E_{i-1,i}(c_{i-1} - c_i) + E_{i,i+1}(c_{i+1} - c_i) \\ & + k_{2i}V_i(c_{si} - c_i) + k_{1i}V_iL_i + \sum s_i \end{aligned} \qquad \text{(IV.2)}$$

Here, single subscripts refer to the particular segment. Double subscripts have the form $(a, a+1)$

and refer to the effect of a segment on its next neighbor downstream.

c = DO concentration

c_s = DO saturation concentration

V = segment volume

ξ = dimensionless mixing coefficient

If all terms in c in equation (IV.2) are grouped on the left hand side, the n segments yield the matrix equation (II.16). Repeating it here:

$$\mathbf{Ac} = \mathbf{f} \qquad \text{(IV.3)}$$

where $\mathbf{A}$ is an $n \times n$ matrix, whose elements are functions of $\left(\frac{d}{dt}, Q, E, \xi, V, k_1, k_2\right)$. $\mathbf{c}$ and $\mathbf{f}$ are column vectors, where $\mathbf{f}$'s elements are

$$f_i = k_{2i}V_ic_{si} - k_{1i}V_iL_i + \sum s_i \qquad \text{(IV.4)}$$

and constitute the forcing functions for the system.

Each of the n equations is linear.

Solving equation (IV.3), we obtain

$$\mathbf{c} = \mathbf{A}^{-1}\mathbf{f} \qquad \text{(IV.5)}$$

$\mathbf{A}^{-1}$ is thus the river transfer function, whereby we determine the DO concentration c_i of the ith segment as a sum of the effects of the set of forcing function $\mathbf{f}$ on all n segments.

O'Connor *et al.*, 1968, have developed an analytical method for solving the steady-state version of the basic material balance equations. Thus, for BOD:

$$\frac{d}{dx}\left(-EA\frac{dL}{dx} + QL\right) = -k_1AL \qquad \text{(IV.6)}$$

where k_1 is the removal rate coefficient.

This equation has been solved for constant A, i.e., an idealized river channel of constant cross-section.

For a more realistic river channel, e.g., one whose cross-section increases linearly with distance, let

$$A(x) = mx. \qquad \text{(IV.7)}$$

In such a case, when E, Q and k_1 are constant, the BOD concentration profile is:

$$L = Bx^vI_v\left(x\sqrt{\frac{k_1}{E}}\right) + Cx^vK_v\left(x\sqrt{\frac{k_1}{E}}\right) \qquad \text{(IV.8)}$$

I_v and K_v are modified Bessel functions of the first and second kind, of order v, where $v = Q/2mE$.

B and C are arbitrary coefficients determined from boundary conditions.

The steady-state mass balance for DO, expressed in terms of the deficit concentration D, is:

$$\frac{E}{A}\frac{d}{dx}\left(A\frac{dD}{dx}\right) - \frac{Q}{A}\frac{dD}{dx} - k_2D + k_1L(x) + k_nN(x) + S - P + R = 0 \qquad \text{(IV.9)}$$

k_2, k_1 and k_n are reaction rate coefficients for reaeration, deoxygenation by CBOD and deoxygenation by NBOD.

S = benthic oxygen demand.

P = average daily oxygen production by photosynthesis.

R = average daily algal respiration.

The complementary solution for equation (IV.a) is

$$D = Gx^vI_v\left(x\sqrt{\frac{k_2}{E}}\right) + Hx^vK_v\left(x\sqrt{\frac{k_2}{E}}\right) \qquad \text{(IV.10)}$$

G and H are arbitrary coefficients.

The particular integral for the CBOD source (NBOD is analogous) is:

$$\frac{k_1}{k_2 - k_1}\left[Bx^vI_v\left(x\sqrt{\frac{k_1}{E}}\right) + Cx^vK_v\left(x\sqrt{\frac{k_1}{E}}\right)\right]$$

The particular integral for benthal demand is

$$\frac{S}{k_2}$$

Particular integrals for the remaining sources and sinks are formed analogously.

The authors tested this model against the finite difference model of Thomann, 1963, for the Delaware River from Trenton to the mouth. Four zones were established, each with a different value of k_2, Q and m in equation (IV.7). In addition, the Schuylkill River was designated as a separate zone. For each year, certain periods were selected when flow and temperature remained relatively uniform, as would be required for modeling a steady-state. Agreement was good, and the authors concluded that:

1) The model is a satisfactory engineering tool for assessing the efficacy of proposals for water quality improvement.
2) Principal factors affecting the water quality in the Delaware River are oxidation of CBOD and

NBOD, benthal oxygen demand and "background" pollution.

3) Analysis highlights the importance of nitrification. Benthos and CBOD depress the water quality, and are responsible for delay in oxidation of NBOD during summer.
4) As waste treatment in the Basin improves, reduction in CBOD will create an environment more suitable for the organism responsible for NBOD stabilization. This may foster nitrification where it is now absent.
5) Winter data indicates that the only active demand in that season is CBOD. It is likely that NBOD and benthal demand cease then entirely.
6) Sensitivity analysis indicates that low-flow augmentation is not feasible. However, the location of the points of input of waste loading has a pronounced effect on water quality.

The various models developed for the Delaware River estuary have been applied to a number of water quality management problems in that stream, e.g., Thomann and Sobel, 1964, Smith and Morris, 1969 (see Section I of "Systems Modeling for Water Pollution Control," in Volume 2 of this Encyclopedia), and MacEwen and Tortoriello, 1970.

Areas of interest investigated with the aid of the models include:

1) Waste treatment alternatives

To determine the effect of upgrading existing municipal and industrial waste treatment plants in the estuary where necessary, to comply with effluent standards.

The heaviest loading along the river occurs in the 22 mile reach comprising segments 12 to 18, and the present summer DO level in the reach is low, approximately 1.0 mg/l. It is desired to improve this figure to 4.5 mg/l.

The reach in question runs through Philadelphia, and two municipal treatments plants are located within the reach. A third plant is located about 3 miles upstream.

The steady-state model was used to determine which of the three plants would produce the greatest DO improvement when upgraded. Since they were not the same size, they were rated on "units of river improvement" per 1000 lbs additional BOD removed. "Units of river improvement" was determined by measuring the change in area under the river DO profile (DO concentration plotted against distance). By this criterion, the plant upstream of the reach proved best.

In winter, the DO concentrations in the river are high because of the low temperatures and low BOD reaction rates. However, it is feared that, if higher waste discharges are permitted then, that some will settle out and create large DO deficits when the water got warmer.

The dynamic model was used to determine the effect of permitting such seasonal variations in effluent standards owing to the reduction of waste plant biological efficiencies during cold weather. It was also used to determine the effect of a winter overload on the estuary during the late spring months.

Two model runs were made:

a) with existing loads for the entire year.
b) using loads increased by two-thirds in the upper thirty miles.

Results showed that increased loads in the months January to April caused a small decrease in DO, but that no after-effect was noticeable when the overloading was halted.

2) Salt water intrusion

In the summer of 1964, the average freshwater flow fell to a near record low for several months, and the salt front (i.e., a chloride concentration of 250 mg/l) advanced as far as Camden.

A modified version of the dynamic DO model was used to predict chloride profiles as a function of freshwater flow, and the predicted profile was quite close to the observed one.

For conservative wastes, or those with a very low decay rate, the steady state model can be used to determine which of a number of loading schemes has the least damaging effect on the river.

3) Regionalization

To determine the effect on the river of combining a number waste treatment plant discharges into one.

It is probable that the economies of scale obtained in building large plants is more important, but this was not examined. Two model runs were made:

a) Loads in segments 17, 18 and 19 were combined and discharged into segment 19.
b) As previous, except that segment 21 received the discharge.

The difference was small. The second case showed an improvement in the DO curve at its lowest point, but there was a drop further downstream in a reach where it was relatively high. It was concluded that there was little to choose between the two cases.

Thomann *et al.*, 1968, have verified the techniques developed for the Delaware River by using them to model water quality profiles in other streams. For example:

1. The monthly variation of DO in the East River estuary in New York City.
2. The diurnal variation of DO in the freshwater Grand River.
3. The seasonal salinity profile in the Potomac estuary.

Another example of large-scale river modeling is provided by the Hudson River. In September 1966, Quirk, Lawler and Matusky Engineers were authorized to commence a three year study of the Hudson River's water quality and waste assimilative capacity. Q.L.M., 1970, is their final report, and it describes in detail the approach they followed in constructing a river water quality model. The aim was to make it general enough to study any pollutant whatsoever.

The development began by segmenting the estuary into a finite number of segments. Segment numbering increased in the downstream direction. Two additional segments (segment 1 and segment M) were added at the upstream and downstream ends for the sake of convenience as boundary segments possessing known concentrations or temperatures.

Next, the law of conservation of mass for BOD, dissolved oxygen, salt and dye, or the law of conservation of energy for temperature, was applied over each segment.

The individual quantities involved in the balance are discussed below, and are then brought together in the overall balance equation (IV.11).

Rate of mass or energy input

The rate of mass or energy input is broken into two contributions:

1. advective input: the transport of mass or energy via the net flow; i.e., via freshwater runoff. This is equal to the product of net flow and area-averaged entity concentration at the upstream end of the segment.
2. longitudinal mixing (dispersion): the transport of mass or energy above that transported into the segment via the net flow. This additional contribution is due to salinity-induced circulation, tidal mixing, etc.

The advective input to segment n is:

$$[Q_{n-1,n}]\,[c_{n-1,n}]$$

in which:

$Q_{n-1,n}$ = the net flow at the boundary between segments $n-1$ and n (Volume/Time)

$c_{n-1,n}$ = the area-averaged, tidal-smoothed concentration of the component at the boundary between segments $n-1$ and n (Mass/Volume or Temperature)

The dispersion of material from segment $n-1$ into segment n is:

$$[-E_{n-1,n}]\,[A_{n-1,n}]\,(dc/dx)_{n-1,n}$$

in which:

$E_{n-1,n}$ = the longitudinal dispersion coefficient at the boundary of segments $n-1$ and n (Length^2/Time)

$A_{n-1,n}$ = the cross-sectional area at the same boundary (Length^2)

$(dc/dx)_{n-1,n}$ = the rate of change of entity concentration with respect to distance (Mass/Volume/Length or Temperature/Length)

Dispersive transport always moves in the direction of decreasing concentration. If the concentration is decreasing in the downstream direction, dispersive transport at the upper boundary of the segment should appear as input to the segment, i.e., its numerical value should be positive. Since the gradient (dc/dx) in this case is negative, the entire term will be positive as required.

Conversely, if concentration is increasing in the downstream direction at the boundary between segments $n-1$ and n, dispersive transport should be out of segment n into segment $n-1$. The gradient (dc/dx) is positive, the entire term is negative, dispersive transport is out of the segment, as required.

Rate of mass or energy output

Advective and dispersive output terms are written similar to the input terms developed above.

The advective output from segment n is:

$$[Q_{n,n+1}]\,[c_{n,n+1}]$$

in which:

$Q_{n,n+1}$ = the net flow from segment n into segment $n+1$.

$c_{n,n+1}$ = the concentration at the boundary of segments n and $n+1$.

Similarly, the dispersive output is:

$$[-E_{n,n+1}]\,[A_{n,n+1}]\,(dc/dx)_{n,n+1}$$

Rate of mass or energy loss

Two general mechanisms may be present which can decay the entity being modeled.

1. Removal of the entity due to reaction at the surface or the bottom of the segment. This is the case of removal of oxygen by benthal deposits and of removal of heat by surface heat transfer.
2. Reaction of the entity within the segment. Coliform dieoff and biochemical decay of organic matter falls under this heading.

Consumption of oxygen by benthal deposits can be expressed as:

$$\int_0^{V_n} \frac{fK_s}{R_H}\,dV$$

in which:

f = the fraction of the bottom surface area covered with sludge.

K_s = oxygen uptake rate of the bottom sludge

R_H = hydraulic radius

V = volume

V_n = volume of segment n

Note that the product dV/R_H is equal to the incremental bottom surface area.

For decay throughout (or within) the segment, the decay term can be expressed as:

$$\int_0^{V_n} K\,dV$$

The term K is the removal rate, in mass per unit volume per unit time. If the reaction is first order with respect to the concentration of the entity itself, K is replaced by $K_r c$ and the loss term is written:

$$\int_0^{V_n} K_r c\,dV$$

Decay of coliforms and BOD follow this mechanism.

Decay of the entity in question can also occur as a function of another constituent of the mixture. The case of first-order kinetics with respect to the concentration of a second constituent is written:

$$\int_0^{V_n} K_D(G)\,dV$$

in which:

K_D = first-order reaction coefficient measuring decay with another constituent G.

G = concentration of the decay dependent constituent.

V_n = volume of segment n.

Consumption of oxygen due to the decay of organic matter (BOD) follows this mechanism.

Rate of mass or energy production

The rate of mass or energy production can be broken into two contributions.

1. A direct external input. This is usually the waste loading and is introduced to the segment across its surface or channel boundaries.
2. A concentration dependent term. The concentration-dependent term arises if the entity is present external to the system and soluble in segment n, such as is the case of atmospheric reaeration of oxygen, or net heat transfer from the air into the water.

The direct external input is expressed simply as "W_n," the rate of addition of the pollutant. "W" is expressed as mass per unit time.

The concentration dependent production term may be written:

$$\int_0^{V_n} K_a(c_s - c)\,dV$$

in which:

K_a = rate coefficient for the transfer of the entity into the segment.

c_s = the saturation value of the entity in segment n.

c = concentration of the entity in the segment.

V_n = volume of the segment.

In connection with this latter term, it should be recognized that the true mechanism is actually a boundary transfer phenomenon. However, it appears to be volumetric production, since it is expressed as a growth function over the segment volume. This is somewhat similar to the case of the benthal deposits. In that case, the role of bottom area appeared explicitly in the integral, since the quantity "dV/R_H" is equal to the incremental bottom surface area.

In the case of atmospheric reaeration and surface heat transfer, the actual boundary transfer mechanism is replaced by an equivalent volumetric production mechanism. In the case of oxygen transfer, for example, the correct expression for transfer rate across the surface is:

$$\int_0^{A_s} K_L(c_s - c_B)\, dA_s$$

in which:

K_L = the liquid film coefficient.

c_B = the average DO across the river surface width.

A_s = the water surface area of the segment.

Since the vertical gradient of oxygen in a well-mixed estuary is usually rather small, the quantity c_B is replaced by c, the area-averaged concentration. The quantity $K_L\, dA_S$ is replaced by $K_2\, dV$, in which K_2 is the traditional reoxygenation coefficient. Since dA_S/dV is equal to $1/D$, the reaeration coefficient K_2 is simply equal to the liquid film coefficient, K_L, divided by the mean depth D. In this rather simplified approach, the influences of surface turbulence, waves, surfactant chemical, etc., on the mass transfer rate, appear in the coefficient K_L.

In a similar manner, the surface heat transfer coefficient, usually designated $\bar{K}$, is converted to a so-called temperature exchange coefficient, usually designated K_r. This situation is usually more complex, however, because, in the presence of a thermal discharge, considerable surface stratification usually occurs, and the vertical temperature gradient cannot be ignored.

Accumulation term

The accumulation of the entity within the given segment is the time rate of increase within that segment. This is written:

$$\frac{d}{dt}\left(\int_0^{V_n} c\, dV\right)$$

in which:

t = time from some initial condition.

$\int_0^{V_n} c\, dV$ = mass of constituent, c, in segment n.

This completes the evaluation of each of the terms in the mass or energy balance. Bringing them all together yields:

$$(Q_{n-1,n} \cdot c_{n-1,n}) - E_{n-1,n} \cdot A_{n-1,n}\left(\frac{dC}{dx}\right)_{n-1,n}$$
$$- \left\{(Q_{n,n+1} \cdot c_{n,n+1}) - E_{n,n+1} \cdot A_{n,\,n+1}\left(\frac{dc}{dx}\right)_{n,n+1}\right\}$$
$$- \int_0^{V_n} \frac{fK_s\, dV}{R_H} - \int_0^{V_n} K_r c\, dV - \int_0^{V_n} K_D G\, dV + W_n$$
$$+ \int_0^{V_n} K_a(c_s - c)\, dV = \frac{d}{dt}\int_0^{V_n} c\, dV \qquad \text{(IV.11)}$$

The integrals over volume which appear in equation (IV.11) are equivalent to the products of average concentrations, decay coefficients and areas times the distance increment (segment length) Δx.

Equation (IV.11) is written over each segment from segment 2 to $M - 1$. This procedure yields $M - 2$ equations. There are $M - 2$ unknown boundary entity concentrations. However, substantial modification of the forms of the terms which appear in equation (IV.11) is necessary before simultaneous solution of the $M - 2$ equations can occur.

The approach employed involves assuming that the variation of entity concentration is linear between segment midpoints. This permits evaluation of the integrals and space derivatives which appear in equation (IV.11). The $M - 2$ segment equations then reduce to $M - 2$ first-order linear ordinary differential equations relating the dependence of $M - 2$ segment midpoint entity concentrations to the time variable, t. These equations are then solved by numerical methods.

The linear approximation simply means that the segment length must be chosen sufficiently small, so that the actual concentration variation between segment midpoints can be represented by a straight line.

Thus, for each segment from 2 through $M - 1$, a first-order, linear differential equation can be derived which relates the change in segment midpoint concentration with time to known river values (dispersion coefficients, segment lengths, etc.) and unknown quantities (segment midpoint concentrations).

Functional form of equation set

$$\left.\begin{aligned} &\text{Segment 2:} && \frac{dc_2}{dt} = f(c_2, c_3) \\ &\text{Segment 3 to } (M-2)\text{:} && \frac{dc_n}{dt} = f(c_{n-1}, c_n, c_{n+1}) \\ &\text{Segment } (M-1)\text{:} && \frac{dc_{M-1}}{dt} = f(c_{M-2}, c_{M-1}) \end{aligned}\right\} \tag{IV.12}$$

Since there are $M - 2$ equation and $M - 2$ unknowns, the set of equations has a unique solution. The variations of the known quantities with time and with space, in general, cannot be represented by simple analytical functions, and the set of equations cannot be solved analytically (via Laplace Tansforms or other methods). Numerical solution techniques must be used.

Use of the above equations for salt, dye, BOD, DO and temperature is summarized below.

Application to salt

In applying equations (IV.12) to the time variation of salinity the reaeration, decay, and benthal deposit terms are dropped.

Application to dye, radioactive decay, coliforms

In applying equations (IV.12) to the time variation of dye, radioactive decay, coliforms, or similar substances, the reaeration and benthal deposit terms are dropped. One decay term is kept to account for removal by adsorption, radioactive decay, or coliform dieoff, all of which are assumed to follow first-order kinetics. Decay dependent on another entity is dropped.

Application to BOD

In applying equations (IV.12) to the time variation of BOD, the reaeration and benthal deposit terms are dropped. One decay term (the K_R term) is kept to account for BOD removal dependent on BOD concentration.

Application to dissolved oxygen

In applying equations (IV.12) to the time variation of dissolved oxygen all terms are kept. The resulting equations must be solved simultaneously with the BOD equations because of the BOD dependent term in the oxygen equation.

Application to temperature

In applying equations (IV.12) to the variation of temperature rise above ambient-temperature with time, the reaeration and benthal deposit terms are dropped. One decay term is kept to account for removal of temperature to the surrounding atmosphere. Decay dependent on another entity is also dropped.

V. Two-Dimensional Models

The models described in prior sections have all been unidimensional. That is to say, it has been assumed in all cases that pollutant concentrations in rivers and estuaries are uniform across the entire stream cross-section, and vary only with distance along the stream. This was done, of course, to simplify the mathematics, and has proved to yield useful and valid results in comparatively narrow and fast-flowing estuaries. However in broad, shallow estuaries and deltas, the geometry does not lend itself to such an assumption. This section describes two current projects attempting to model such estuaries.

Feigner and Harris, 1970, describe a model developed specifically for the San Francisco Bay-Delta Estuary, but applicable elsewhere. It models the two-dimensional flow and dispersion characteristics of any estuary where stratification is absent or negligible. Hydrological parameters of tidal flow and stage are computed at time intervals ranging from 0.5 to 5.0 mins and at distance intervals ranging from several hundred to several thousand feet. Predictions of quality levels are computed on the same space scale, but on an expanded time scale, ranging from 15 to 60 mins.

The model is thus truly dynamic in character. It predicts fluctuating tidal flows and computes tidally varying concentrations of constituents, in contrast to a non-tidal model based on the net flow through the estuary such as that developed for the Delaware estuary. It can also accommodate both conservative and non-conservative constituents.

Figure 10 is a map of Suisun Bay, showing how channels are depicted by single lines and open areas by grids. For well-defined channels, the model network follows the prototype configuration. Because of the exigencies of numerical integration,

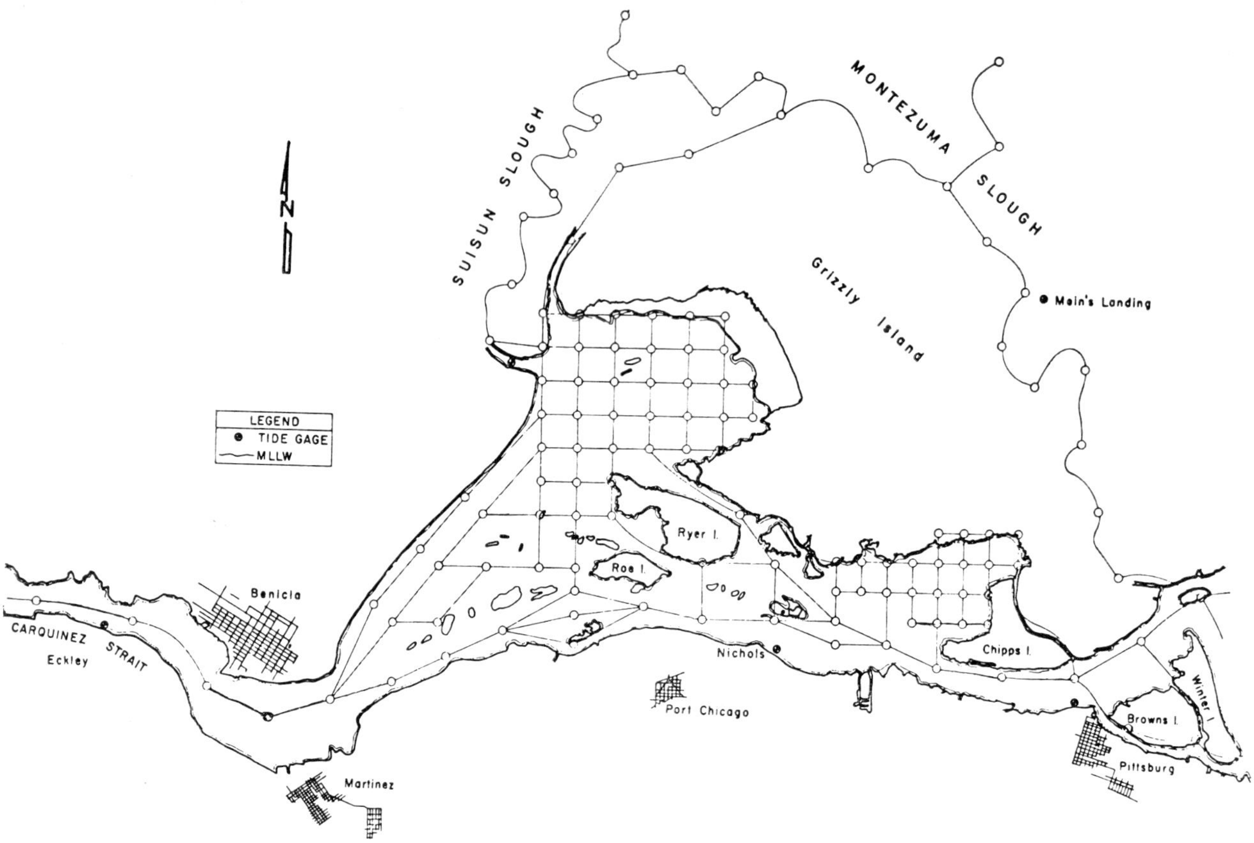

Figure 10 Suisun Bay Network. (Feigner and Harris, 1970)

the desired network scale may dictate channel element lengths, and a prototype channel may have to be divided into a series of channel elements in the model network. Nodes or "junctions" occur in three ways:

a) Joining consecutive channel elements.
b) At the intersection of two or more channels.
c) In bays, at the grid points.

Figure 11 illustrates typical channel and junction elements. Figure 11b shows how the grid lines in the bay areas represent "artifical channels," and how their effective flow widths are determined. Figure 11c indicates how the surface areas associated with junctions are determined.

First, the hydraulic behavior of the estuary is modeled. Having established channel directions, both in the actual prototype channels and (artificially) in the bay areas, the authors use one-dimensional equations. These are based on the following assumptions:

a) Acceleration normal to the x-axis is negligible.
b) Coriolis and wind forces are negligible.
c) The channel is straight.
d) The channel cross-section is uniform throughout its length.
e) The wave length of the propagated tidal wave is at least twice the channel depth.
f) The bottom of the channel is level.

Equations of motion and continuity are, respectively

$$\frac{\partial u}{\partial t} = -u\frac{\partial u}{\partial x} = K\,|u|\,u - g\frac{\partial H}{\partial x} \qquad \text{(V.1)}$$

and:

$$\frac{\partial H}{\partial t} = -\frac{1}{b}\frac{\partial}{\partial x}(uA) \qquad \text{(V.2)}$$

where:

u = velocity along the x-axis

x = distance along the x-axis

H = water surface elevation

g = acceleration of gravity

K = frictional resistance coefficient

t = time

b = mean channel width

A = cross-sectional area of the channel

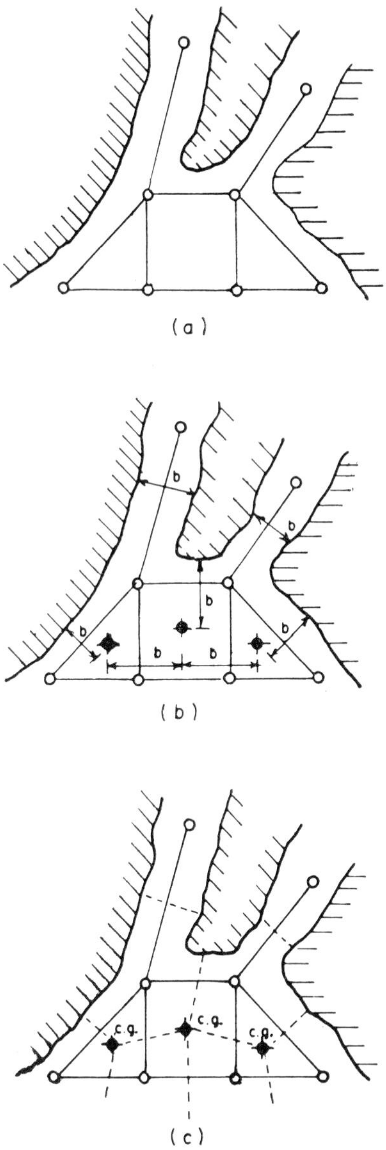

Figure 11 Typical Channel and Junction Elements. (Feigner and Harris, 1970)

The terms on the right hand side of equation (V.1) are, in sequence, the rate of momentum change by mass transfer, the frictional resistance (with the absolute value sign to assure that the resistance always opposes the direction of flow) and the potential difference between the ends of the channel element.

In equation (V.2), the right hand side represents the change in storage over the channel length per unit channel width. To minimize computation, equation (V.1) is applied to the channel elements and equation (V.2) to the junctions.

Both equations are rendered into partial difference form and solved for each channel element and junction, using a modified Runge-Kutta procedure. The results comprise the predicted channel velocities, flows and cross-sectional areas and the predicted water surface elevations at each junction for each time interval. This data is then part of the input to the water quality component of the model. The latter is the common one-dimensional form of the amplified Streeter-Phelps equations, taking into account advection, eddy diffusion, the decay of non-conservative substance and reaeration. The equations are put into finite-difference form and solved to give the concentration of the substance at each junction.

This technique was used to model both San Francisco and San Diego bays. It was found that model predictions and the prototype observations differ somewhat in a number of instances for both locations. These discrepancies result to a great extent from the type of comparison made. Several sources of these differences should be noted.

1. The model concentration is the average for a reach perhaps 3000 to 5000 feet in length which includes the prototype sampling location. This is typically near one shore and is not necessarily representative of the cross-section, much less a long reach.
2. The model used a mean tide repetitively, whereas the prototype tide was continously changing.
3. The number of model junctions for which initial concentrations are known is a trivial fraction of the total number requiring initial concentrations.
4. The hydraulic conditions in the model are defined exactly, whereas flow conditions in the prototype are largely unknown at any time.

Another on-going project is that for Jamaica Bay, New York City. Leendertse, 1970, developed a water-quality simulation model for well-mixed estuaries and coastal seas (i.e., no stratification). Leendertse and Gritton, 1971, have extended the model to include the transport of several dissolved waste constituents in the water, including any interactions among them. The changing tide level influences the location of the land-water boundaries in the shallow areas of coastal waters. To simulate this process, procedures were developed in the model to allow for time-dependent boundary changes.

Large amounts of numerical data are generated by the computer program developed from the simulation model. To assist the investigator in extracting important and meaningful results from these data, machine-made dawings are used to graphically present the results of the computation.

The basic mass-balance equation for 2-dimensional transport of waste constituents in a well-mixed estuary (uniform concentration in the vertical direction) is given in Leendertse, 1970, as:

$$\frac{\partial}{\partial t}(HP) + \frac{\partial}{\partial x}(HUP) + \frac{\partial}{\partial y}(HVP) - \frac{\partial}{\partial x}\left(HD_x \frac{\partial P}{\partial x}\right) - \frac{\partial}{\partial y}\left(HD_y \frac{\partial P}{\partial y}\right) - HS_A = 0 \tag{V.3}$$

where P is the integrated average over the vertical of the waste constituents mass concentration.

U and V are the vertically averaged fluid velocity components in the x (eastward) and y (northward) directions respectively.

S_A is a source function.

D_x and D_y are dispersion coefficients.

H is the instantaneous depth at a point.

The generalized mass-balance equation for n eonstituents is written in matrix notation as:

$$\frac{\partial}{\partial t}(H\bar{P}) + \frac{\partial}{\partial x}(HU\bar{P}) + \frac{\partial}{\partial y}(HV\bar{P}) - \frac{\partial}{\partial x} \times \left(HD_x \frac{\partial \bar{P}}{\partial x}\right) - \frac{\partial}{\partial y}\left(HD_y \frac{\partial \bar{P}}{\partial y}\right) + [K]\,H\bar{P} + H\bar{S} = 0 \tag{V.4}$$

where

$\bar{P}$ = mass-concentration vector with n elements.

$[K]$ = reaction matrix

$\bar{S}$ = source and sink vector.

The reaction matrix $[K]$ in its most general form can give rise to a non-linear transport equation. This occurs because the individual elements of the matrix can be defined as functions of their own concentration, or that of other constituents, or both.

Since the elements of $[K]$ are multiplied by the elements of the concentration vector, such non-linear terms imply kinetics of an order higher than first.

The present study is restricted to first-order reactions, and hence a linear transport equation is obtained. As an example, consider the modeling of coliform bacteria, which can be assumed to have a first-order reaction decay and to come from three independent groups of sources. For this model, the reaction matrix is a single diagonal matrix and $\bar{P}$ is a column vector with 3 elements.

$$[K] = \begin{bmatrix} K_{11} & 0 & 0 \\ 0 & K_{22} & 0 \\ 0 & 0 & K_{33} \end{bmatrix} \tag{V.5}$$

$$\bar{P} = \begin{bmatrix} P_1 \\ P_2 \\ P_3 \end{bmatrix} \tag{V.6}$$

K_{11}, K_{22} and K_{33} are the first-order decay constants, in most cases equal, and P_1, P_2 and P_3 are the 3 coliform concentrations.

For BOD and DO, the model has the form:

$$[K] = \begin{bmatrix} K_{11}/H & K_{12} \\ 0 & K_{22} \end{bmatrix} \tag{V.7}$$

$$\bar{P} = \begin{bmatrix} P_1 \\ P_2 \end{bmatrix} \tag{V.8}$$

$$\bar{S} = \begin{bmatrix} -\dfrac{K_{11}c_{\text{sat}}}{H} \\ 0 \end{bmatrix} \tag{V.9}$$

K_{11} = reaeration coefficient of oxygen (ft/sec)

$K_{12} = K_{22}$ = first-order BOD reaction coefficient (sec^{-1})

P_1 = DO concentration (mg/l)

P_2 = BOD concentration (mg/l)

c_{sat} = saturation concentration of oxygen.

Point sources of BOD and DO, such as occur at the location of sewage discharges into the estuary, are simulated by adding delta-function source terms to the source vector.

For two-dimensional flow in a well-mixed estuary, vertical integration of the momentum and continuity equations yields the following basic equations for the flow model.

$$\frac{\partial U}{\partial t} + U\frac{\partial U}{\partial x} + V\frac{\partial U}{\partial y} - fV + g\frac{\partial \zeta}{\partial x} + g\frac{U(U^2+V^2)^{1/2}}{C^2H} - \frac{1}{\varrho H}\tau_x^s = 0 \tag{V.10}$$

$$\frac{\partial V}{\partial t} + U\frac{\partial V}{\partial x} + V\frac{\partial V}{\partial y} + fU + g\frac{\partial \zeta}{\partial y} + g\frac{V(U^2+V^2)^{1/2}}{C^2H} - \frac{1}{\varrho H}\tau_y^s = 0 \tag{V.11}$$

$$\frac{\partial \zeta}{\partial t} + \frac{\partial}{\partial x}(HU) + \frac{\partial}{\partial y}(HV) = 0 \tag{V.12}$$

where:

f = Coriolis parameter.

g = Acceleration of gravity.

C = Chezy coefficient.

τ_x^s = Component of the wind stress in the x direction.

τ_y^s = Component of the wind stress in the y direction.

ϱ = Water density.

ζ = water level elevation relative to the reference plane.

The wind stress components (from another source) are given by:

$$\tau_x^s = \theta \varrho_a W^2 \sin \psi \qquad \text{(V.13)}$$

$$\tau_y^s = \theta \varrho_a W^2 \cos \psi \qquad \text{(V.14)}$$

where:

θ = wind stress coefficient ≈ 0.0026

ζ_a = atmospheric density

W = wind velocity

ψ = angle between the wind direction and the y axis.

In the finite difference approximations of equations (V.4), (V.10), (V.11) and (V.12), the discrete values of the variables are described on a space-staggered grid, as shown in Fig. 12. The position and time coordinates (x, y, t) are represented on the finite grid by $(j\Delta x, k\Delta y, n\Delta t)$, for $j, k, n = 0, \pm\frac{1}{2}, \pm 1, \pm\frac{3}{2}, \ldots$

Water levels and pollutant concentrations are computed at integer values of j and k (x and y directions). Water depths, obtained from a field survey, are given at half-integer values of j and k.

Velocity component U (x directed) is computed at half-integer values of j and integer values of k.

Velocity component V (y directed) is computed at integer values of j and half-integer values of k.

The set of finite difference equations used to approximate the momentum and mass-balance equations are then presented at two adjacent time levels, n and $(n + \frac{1}{2})$.

Numerical computation of the reaction matrix terms in the mass-balance equations is accomplished by a sequential use of forward and backward information. If M constituents are transported, then for constituent i $(1 \leqq i \leqq M)$, in the first operation at time level n (going from t to $t + \frac{1}{2}\Delta t$), we use information in the reaction matrix terms on the level $(t + \frac{1}{2}\Delta t)$ for all constituents for which the sequence number m is smaller than i. Information at the level t is used for all constituents for which $m > i$. In this step, the constituents are computed in ascending order, from 1 to M.

In the second operation, at time level $n + \frac{1}{2}$ (going from $t + \frac{1}{2}\Delta t$ to $t + \Delta t$), the constituents are computed in descending order, M to 1. Information on the level $t + \frac{1}{2}\Delta t$ is used for all constituents whose $m < i$, and information on the level $t + \Delta t$ is used for all constituents whose $m > i$.

This procedure centers the reaction matrix information of the mass-balance equations within the time interval t to $t + \Delta t$. The reaction matrix terms which involve the ith constituent itself are taken centered over each half timestep.

The sequential use of finite-difference approximations for the continuity equations at n and $n + \frac{1}{2}$ results in alternating forward and backward differences. This means that over a full timestep the terms are either central in time or averaged over the time inteval.

In the first operation at time level n (going from

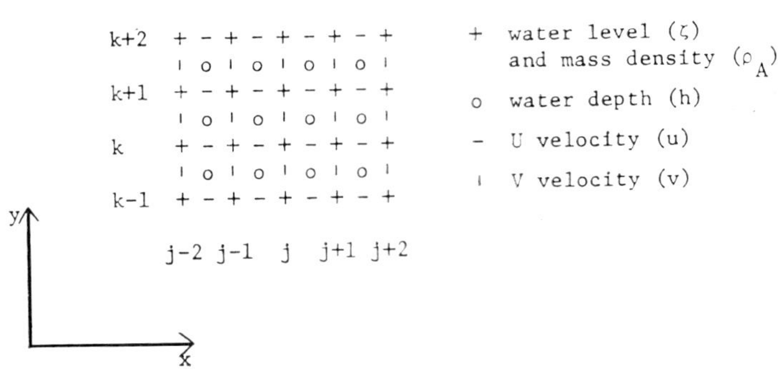

Figure 12 Space-Staggered Grid. (Leendertse and Gritton, 1971)

t to $t + \frac{1}{2}\Delta t$), the momentum and continuity equations are solved first for the water levels and x-directed velocities at time level $n + \frac{1}{2}$. The information generated is then used in the mass balance equations to obtain the constituent concentrations at time level $n + \frac{1}{2}$.

The results of this first operation are then used at time level $n + \frac{1}{2}$ to determine the unknowns in the second half timestep, going from $t + \frac{1}{2}\Delta t$ to $t + \Delta t$. Again, the momentum and continuity equations are solved first, but this time the water levels and y-directed velocities at time level $n + 1$ are obtained. This new information is then used in the mass balance equation to obtain pollutant concentrations at time level $n + 1$.

This procedure is then repeated for each succeeding full timestep. The model can be used to investigate the influence of wind on flow and circulation in the area covered, together with its effect on water levels and distribution of pollutants. This is the first time that real wind effects can be investigated in detail.

REFERENCES

Armstrong, J. M., *Proceedings 1st Annual Ohio Water Resources Conference*, March 1970.

Bella, D. A., and W. E. Dobbins, *Journal Sanitary Engineering Div., ASCE* **94**, 995 (1968).

Camp, T. R., *Water and Its Impurities*, Reinhold Book Corp., New York, (1963).

Deininger, R. A., *Proceedings 1969 Water Resources Seminar*, "Natural Resource Systems Models in Decision Making," Water Resources Center, Purdue University.

Delaware River Basin Commission, Final Progress Report—Delaware Estuary and Bay Water Quality Sampling and Mathematical Modeling Project, May 1970.

Dresnack, R., and W. E. Dobbins, *Journal Sanitary Engineering Div., ASCE* **94**, 789 (1968).

Feigner, K. D., and H. S. Harris, "Documentation Report—FWQA Dynamic Estuary Model," U.S. Department of the Interior, FWQA, 1970.

Gordon, G., *System Simulation*, Prentice-Hall, Inc., Englewood Cliffs, N. J., 1969.

Hamilton, H. R. *et al.*, *Systems Simulation for Regional Analysis—An Application to River Basin Planning*, The M.I.T. Press, Cambridge, Mass. (1969).

Harleman, D. R. F., C. H. Lee, and L. C. Hall, *Journal Sanitary Engineering Div., ASCE* **94**, 897 (1968).

Henderson, J. M., *Proceedings 4th Symposium on Water Resources Research*, Water Resources Center, Ohio State University, Oct. 1969.

Hetling, L. J., and D. Metzler, *Proceedings 4th Symposium on Water Resources Research*, Water Resources Center, Ohio State University, Oct. 1969.

James, L. D., *Proceedings 1969 Water Resources Seminar*, "Natural Resource Systems Models in Decision Making," Water Resources Center, Purdue University.

Lee, E. S., *Quasilinearization and Invariant Imbedding*, Academic Press, New York (1968).

Lee, E. S., L. E. Erickson, and L. T. Fan, The Kansas Water Resources Research Institute, Kansas State University, Jan. 1971.

Leendertse, J. J., Rand Corp. Publication RM-6230-RC, Febr. 1970.

Leendertse, J. J., and E. C. Gritton, The New York City Rand Institute, R-708/9-RC, July 1971.

Maass, A. *et al.*, *Design of Water Resource Systems*, Harvard University Press, Cambridge, Mass (1962).

MacEwen, P. K., and R. Tortoriello, National Symposium on Data and Instrumentation for Water Quality Management, Annual Meeting, Madison, Wisconsin, July 1970.

Miernyk, W. H., *Proceedings 4th Symposium on Water Resources Research*, Water Resources Center, Ohio State University, Oct. 1969.

O'Connor, D. J., *Journal Water Pollution Control Fed.* **34**, 905 (1962).

O'Connor, D. J., *Water Resources Research* **3**, 65 (1967).

O'Connor, D. J., and D. M. DiToro, *Journal Sanitary Engineering Div., ASCE* **96**, 547 (1970).

O'Connor, D. J., and R. V. Thomann, *Proceedings IBM Scientific Computing Symposium on Environmental Sciences*, Nov. 1966.

O'Connor, D. J., J. P. St. John, and D. M. DiToro, *Journal Sanitary Engineering Div., ASCE* **94**, 1225 (1968).

Pence, G. D., J. M. Jeglic, and R. V. Thomann, *Journal Sanitary Eng. Div., ASCE* **94**, 381 (1968).

Phelps, E. B., *Stream Sanitation*, John Wiley and Sons, New York (1944).

Quirk, Lawler and Matusky Engineers (QLM), "Hudson River Water Quality and Waste Assimulative Capacity Study," State of New York, Department of Environmental Conservation, Final Report, Dec. 1970.

Quirk, Lawler and Matusky Engineers (QLM), "Systems Analysis for Water Pollution Control," Commonwealth of Massachusetts Water Resources Commission, Division of Water Pollution Control, June 1971.

Selzer, S. D., ASME Winter Annual Meeting, 1970, 70-WA/PID-6.

Smith, E. T., and A. R. Morris, *Journal WPCF* **41**, 1635 (1969).

Stillson, R. T., *Proceedings 4th Symposium on Water Resources Research*, Water Resources Center, Ohio State University, Oct. 1969.

Streeter, H. W., and E. B. Phelps, U.S. Public Health Service, Public Health Bulletin No. 146, Febr. 1925.

Sweet, D. C., *Proceedings 1st Annual Ohio Water Resources Conference*, March 1970.

Thackston, E. L., and P. A. Krenkel, *Journal Sanitary Engineering Div., ASCE* **95**, 65 (1969).

Thomann, R. V., *Journal Sanitary Engineering Div., ASCE* **89**, 1 (1963).

Thomann, R. V., *Proceedings 4th Symposium on Water Resources Research*, Water Resources Center, Ohio State University, Oct. 1969.

Thomann, R. V., and M. J. Sobel, *Journal Sanitary Engineering Div., ASCE* **90**, 9 (1964).

Thomann, R. V., D. J. O'Connor, and D. M. DiToro, *Chemical Engineering Prog. Symposium Series No.* **90**, **64**, 21 (1968).

Toebes, G. H., *Proceedings of 1969 Water Resources Seminar*, "Natural Resource Systems Models in Decision Making," Water Resources Center, Purdue University.

Tybout, R. A., *Proceedings 1969 Water Resources Seminar*, "Natural Resource Systems Models in Decision Making," Water Resources Research Center, Purdue University.

Whipple, W. Jr. *Proceedings 1969 Water Resources Seminar*, "Natural Resource Systems Models in Decision Making," Water Resources Research Center, Purdue University.

Whipple, W. Jr. *et al.*, "Instream Aeration of Polluted Rivers," Water Resources Research Institute, Rutgers University, Aug. 1969.

Wilde, D. J., and C. S. Beightler, *Foundations of Optimization*, Prentice-Hall, Inc., Englewood Cliffs, N. J., 1967.

Zahradnik, R. L., *Theory and Techniques of Optimization*, Barnes and Noble, New York, 1971.

READINGS IN WATER RESOURCE MANAGEMENT

Beightler, C. S., "Use of New Analytical Methods in Water Resource Management," University of Texas at Austin, Sept. 1970.

Dixon, N. P., D. W. Hendricks, A. L. Huber, and J. M. Bagley, Utah Water Research Laboratory Report PRWG-67-1, Utah State University, June 1970.

Goodman, A. S., and W. E. Dobbins, *Journal Sanitary Engineering Div., ASCE* **92**, 1 (1966).

Goodman, A. S., and R. T. Tucker, FWPCA, Dept. of the Interior, Water Pollution Control Research Series ORD-6, July 1969. (NTIS PB 188494).

Graves, G. W., A. B. Whinston, and G. B. Hatfield, University of Cal. Grad. School of Business Administration, EPA Publicn. 16110 FPX 08/70, 1970.

Hall, C. R., and L. J. Hetling, *Chemical Engineering Prog. Symposium Series No. 90*, **64**, 72 (1968).

Hamilton, H. R., *et al.*, *Systems Simulation for regional Analysis—An Application to River Basin Planning*. The M.I.T. Press, Cambridge, Mass., (1969).

Hufschmidt, M. M., and M. B. Fiering, *Simulation Techniques for Design of Water Resource Systems*, Harvard University Press, Cambridge, Mass. (1966).

Jaske, R. T., M. H. Karr, and C. J. Touhill, U. S. A. E. C. Report BNWL-SA-2738, Febr. 1970 (B).

Jaworski, N. A., W. J. Weber, and R. A. Deininger, *Journal Sanitary Engineering Div., ASCE* **96**, 1 (1970).

Loucks, D. P., *Proceedings 4th Symposium on Water Resources Research*, Water Resources Center, Ohio State University, Oct. 1969.

Maass, A., *et al.*, *Design of Water-Resource Systems*, Harvard University Press, Cambridge, Mass. (1962).

Ortolano, L., and H. A. Thomas, *Chemical Engineering Prog. Symposium Series No. 90*, **64**, 10 (1968).

Pisano, W. C., "River Basin Simulation Program," U.S. Dept. of Commerce Document No. AD 673564, Aug. 1969.

Pyatt, E. E., J. P. Heaney, G. R. Grantham, and B. J. Cartert, "A Model for Quantifying Flow Augmentation Benefits," University of Florida, Final Report to Southeast Region, FWPCA, Atlanta, Ga., Sept. 1969.

Shumate, K. S., *Proceedings 1st Annual Ohio Water Resources Conference*, March 1970.

Tedrow, A. C., *Proceedings 1969 Water Resources Seminar*, "Natural Resource Systems Models in Decision Making," Water Resources Center, Purdue University.

Worley, J. L., M. S. Thesis, Oregon State University, June 1963.

Worley, J. L., F. J. Burgess, and W. W. Towne, *Journal Water Pollution Control Fed.* **37**, 659 (1965).

ISRAEL WILENITZ
Quirk, Lawler and Matusky, Engineers

MONITORING OF STACK EFFLUENTS: see STACK SAMPLING

MUNICIPAL WASTEWATER

Sewage is the spent water supply of a community. Because of infiltration of groundwater into loose sewer pipe joints, the quantity of wastewater is frequently greater than the quantity of water initially consumed. Sewage is about 99.95% water and 0.05% waste material.

A weaker (more dilute) sewage results from greater per capita water consumption. Industrial wastes will contribute to sewage strength. Sewage flow varies with time of day and will be higher during daylight hours. Larger cities will exhibit less variation with time than will smaller cities. Many small communities will have a flow in the late night hours that is due almost entirely to infiltration.

Per capita production will vary from less than 100 gallons per day for a strictly residential community to about 300 gallons per day for a highly industrialized area. The concept of "population equivalent" is frequently applied in evaluating industrial waste contributions to sewage flow and planning for hydraulic, solids and biochemical oxygen demand loadings.

Wastewater treatment facilities have high initial capital costs due to the need for large tanks, equipment and land area. The designed life of a treatment plant is normally equal to the life of the bonded indebtedness of the project. It is expected that capacity will be just reached at the end of this period. In a community with combined sanitary and storm sewers it is often found necessary to bypass wastewater flows during periods of heavy rain or, in low lying areas, during high tides. Excess water may be held in detention basins until normal flow resumes and then treated in the plant. Bypassed flow can be screened and chlorinated before discharge. This subject is receiving increasingly close attention.

Strength of sewage is usually expressed in terms of the following parameters: total solids, settleable solids, suspended solids, volatile solids (sometimes called volatile suspended solids, VSS), filterable solids and biochemical oxygen demand (BOD). Many methods are available for determination of the foregoing. However, in order that operating and research data reported by workers in widely seperated areas be comparable, it is necessary to have analytical methods which are simple, give consistent results and are widely accepted. This need is filled by Standard Methods for the Examination of Water and Wastewater. Contained in each edition are methods for analysis of water and wastewater which have been accepted by committees of experts in various fields. Standard Methods is revised about every five years.

Analysis of a typical american sewage is given below.

Total solids	600 mg/l
a) suspended solids	200 mg/l
1) settleable solids	120 mg/l
2) colloidal solids	80 mg/l
organic	60 mg/l
mineral	20 mg/l
b) Filterable solids	400 mg/l
1) colloidal solids	40 mg/l
organic	30 mg/l
mineral	10 mg/l
2) dissolved solids	360 mg/l
organic	100 mg/l
mineral	260 mg/l
c) BOD (5 day 20 °C)	
1) suspended	42 gms/cap. × day
settleable	19 gms/cap. × day
non-settleable	23 gms/cap. × day
2) dissolved	12 gms/cap. × day
Total	54 gms/cap. × day

Assuming an average daily flow of 100 gallons per capita, 54 gms/cap. × day = 0.12 # BOD/cap. × day. This is a strictly domestic sewage. Per capita BOD values can vary from about 0.10 #/day to above 0.25 #/day. A commonly accepted value used for estimation is 0.17 #/cap. × day.

In the above analysis, the determination of solids, mineral and organic, gives an indication of the loadings to be placed on the plant. Volatile solids give the organic loading and mineral solids are thosewhich must be accommodated by sedimentation equipment.

BOD is a measure of the loadings placed on the oxygen resources of the receiving waters. Several methods have been used over the years for determination of the amount of oxygen necessary for stabilization of the waste. Permanganate and other oxidizing agents were formerly used. One method still finding application, but only as a general indication, is the putrescibility, or methylene blue, test. Methylene blue, a dye, decolorizes in the absence of dissolved oxygen. This test is only quasi-quantitative, but finds use in day to day operational control of wastewater treatment plants. The method used today is an incubation and dilution method. Dissolved oxygen is determined at the beginning and end of a 5 day period by the Alsteberg azide modification of the Winkler method. Water can hold about 10 mg/l dissolved oxygen. Since BOD values can vary from less than 1 mg/l for a clean stream to many thousands of milligrams per liter for some industrial wastes, it is necessary to dilute stronger wastes.

The dissolved oxygen concentration of a seeded nutrient dilution water is determined. A measured quantity of waste is then added to a known volume of dilution water. Total sample volume is usually about three hundred milliliters. The mixture is then incubated at 20°C for five days. At the end of this period the dissolved oxygen concentration of the mixture is again measured. BOD is then equal to the reduction in dissolved oxygen divided by the dilution factor.

Figure 1 shows the exertion of BOD over an extended period (more than 50 days). The trajectory displayed is one which might be followed when waste is added to a relatively clean stream and the watercourse acts as an incubation bottle. Such a complex reaction is of little utility for day to day control. In order to make the concept of BOD more useful, a simplified model is developed. The path followed in the idealized model is displayed in Fig. 2.

The lag phase is eliminated by use of seeded dilution water. In this phase, microorganisms which utilize the waste material for food are becoming acclimatized and assuming dominance in the system. In the carbonaceous phase the dominant organisms are feeding upon the more easily degraded compounds. The nitrogenous phase, in which the character of the food and the microbial population change, does not usually begin for at least ten days. This portion is of interest in research on kinetics but can be ignored in the following. It is hypothesized that a simple monomolecular rate constant can describe the carbonace-

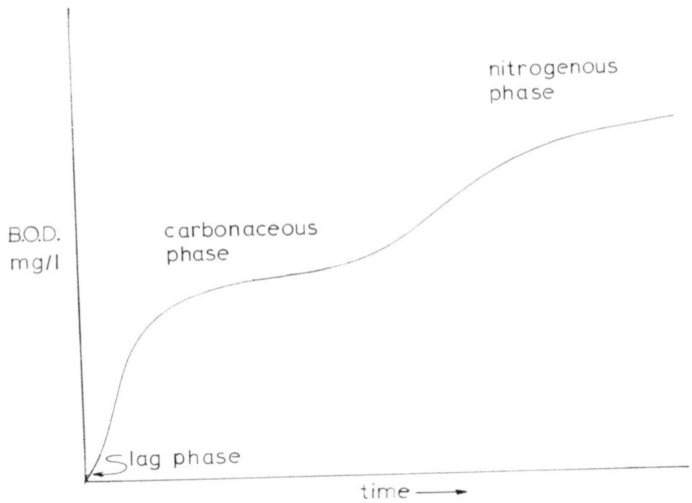

Figure 1

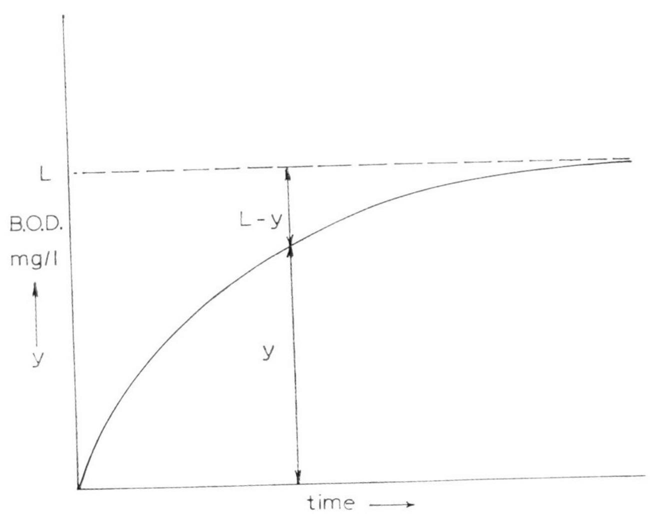

Figure 2

ous phase and the reaction is asymptotic to a limiting value L at infinite time. L is known as the ultimate BOD and is a mathematical artifact. The rate of reaction is assumed proportional to the BOD yet remaining.

$$\frac{dy}{dt} = k'_1(L - y)$$

Integration and taking of proper limits gives

$$y = L(1 - e^{-k'_1t}) = L(1 - 10^{-k_1t})$$

Phelps' relation for decolorization of methylene blue with time

$$\frac{y}{L} = 1 - .794^t$$

is equivalent to the model equation when k_1 equals 0.10/day.

In practice, the 5 day 20 C BOD is used to describe the strength of a waste. Simple enumeration of the BOD value tells nothing about the path by which the number was reached since L and k_1 are not specified. Sewers and treatment plants are not isothermal entities but it is necessary that a common basis be established. One of the reasons advanced for the choice of the five day period is the fact that almost all rivers in England are within five days flow of the sea. Waste once discharged to the the infinite ocean is no longer of interest. Times have changed and there is great concern for effects of waste on the "infinite ocean."

In the literature reference to BOD means 5 day 20°C BOD (BOD_5) unless specifically stated otherwise. Recently, some popular writers have erroneously used the term "biological oxygen demand." There is no such parameter.

The reaction rate constant k_1 actually describes a series of complex microbiological reactions. In the initial stages of biological stabilization of sewage, carbonaceous material is oxidized to carbon dioxide and nitrogenous material is oxidized first to nitrite and then nitrate. One of the measures of degree of stabilization of organic matter in an effluent is the nitrate concentration. However, recent theories concerning eutrophication have raised serious questions concerning the desirability of high effluent nitrate concentrations.

Sewage contains the waste of a city and its inhabitants. It is possible to find almost any microorganism in sewage. In fact, in the sewer system can be found quite unexpected creatures. About 1960, sewer workers in New York City found a number of live and exceedingly unfriendly alligators in that city's sewers. These animals, gifts to city residents, had been disposed of through household toilets. The warm, wet environment of the sewers, rich in food, was excellent for rapid growth.

Traditionally, public health practice applied to water sanitation has placed major emphasis on prevention of typhoid fever, the causative organism of which is S. typhosa. This disease is spread by the pathway of anus to mouth. S. typhosa is a fairly delicate organism and is usually not found in high concentrations. It is felt that if pollution arising from human intestinal discharges is removed to a high degree, then S. typhosa will also be removed. As BOD is used as an indicator of of pollutional loading on oxygen resources, so is the indicator organism E. coli used as an indication of pollutional loading due to human intestinal waste. E. coli is a normal inhabitant of the human intestinal tract. It is excreted in huge numbers and the presence of this organism in receiving waters may be evaluated quantitatively. The lactose broth-brilliant green test is commonly used to determine the presence of E. coli. A common soil organism, A. aerogenes, gives a false positive test and further confirmatory tests are necessary. Serial dilutions of the water being tested are prepared and, on the assumption that one organism is responsible for a positive test, the Most Probable Number (MPN) of organisms is determined. The MPN is based on statistical reasoning. Work by Kupchik, using the enzyme urecase to determine the presence of uric acid has shown a high degree of correlation between the concentration of this acid in wastewater and the MPN.

Most pathogenic organisms are not hardy and do not compete well for food. The use of E. coli as an indicator organism is in the way of a margin of safety. This is consistent with Phelps' concept of multiple barriers.

Microorganisms form the basis of secondary, or biological, wastewater treatment. Stated briefly, microorganisms establish themselves on trickling filter slime or in activated sludge liquor and feed on waste material in the sewage.

Large particles are removed in the primary settling portion of treatment. The larger particles, grease, etc. are in some ways not as objectionable as the colloidal or truly dissolved materials. The larger particles are, of course, objectionable from an aesthetic point of view, but the smaller particles place more of an immediate load upon the receiving waters. In almost all areas of environmental engineering surface area is one of the controlling parameters. Microorganisms, in carrying out their vital processes, utilize the waste as food and the smaller particles offer greater surface area per unit volume. Microbial activity is correspondingly greater and the oxygen required is also greater.

Microorganisms require a readily available source of oxygen. Aerobic conditions are said to exist when the oxygen is in the form of dissolved molecular oxygen or nitrates. At the point of exhaustion of nitrates and nitrites the system is in the anaerobic state and the oxygen sources are then sulfates, phosphates, borates, etc. Reduction of sulfates to give mercaptans (HS^-) carries with it characteristic vile odors. Secondary treatment is an essentially aerobic process while conventional sludge digestion is an anaerobic process.

In the trickling filter and activated sludge processes microorganisms extract their food from the flowing wastewater. By means of extracellular enzymes large molecules are broken down so that they may pass through the microbial cell wall by diffusion. The food is further broken down for cell synthesis and energy by means of intracellular enzymes. End products are largely carbon dioxide and water. The waste material, now part of the cell mass, is removed in the final stage of treatment.

The primary reason for treating any waste is the need for avoiding nuisance and dangerous conditions in the environment. It is necessary to remove some of the waste so that the remaining can be discharged to the surroundings. This is, in effect, disposal by dilution. Discharge standards are usually based on concentration and total quantity.

Sewage purification works were formerly constructed for reasons based primarily on public health. More modern thinking has expanded the original rationale to include protection of oxygen resources of the receiving waters. If the second criterion is satisfied, the first will almost always be also satisfied. Demand for clean waters is increasing even more rapidly than population. Water will be used many times in its passage to the sea. Lakes are essentially a closed system. Leisure time is increasing and the question of water pollution has reached the point where aesthetics is now a significant parameter in planning. It is no longer possible to treat a watercourse as a seperate entity. The approach now finding wider and wider application is that of basin management. This is the systems approach. An excellent example of this is the Ohio River Sanitation Commission. Here it has been possible to rise above local jealousies and self interest. The results are most gratifying and should serve as an example to other areas. It is unfortunate, but true, that one heavy pollution source can undo the efforts of many groups with social responsibility and foresight.

Because objectives in waste management can change drastically at political boundaries, it has occasionally been necessary for intervention by representatives of larger political entities when pollution problems affect, for example, several states. Federal agencies in the United States and other countries have come to play an increasingly larger role in waste management, particularly when problems do not respect political boundaries. The discussion presented here involves primarily technology of wastewater treatment and the underlying philosiphy will not be extensively covered.

It has been mentioned that larger particles in wastewater can be removed by physical means. Removal of colloidal and dissolved components requires other methods and this must be accomplished economically. Treatment is classified as primary if it is without biological basis. Secondary treatment is generally accepted as biological treatment. A few regulatory authorities have endeavored to classify treatment degree on the basis of degree of removal. This relegates some biological treatment to primary status. Many experts do not agree with this change in definition. A widely accepted definition of tertiary treatment is the use of any process, in addition to conventional secondary treatment, for the purpose of further removals. From the foregoing it should be noted that chemical treatment, popular in the past, is a primary treatment process.

Water carriage of waste was practiced in the Minoan civilization of Crete. It may be said that sanitation practices reflect the level of a civilization. Certainly, sanitary practices of the Middle Ages was of a rather low order. Methodical control of waterborne diseases was not attained until the 19th century. It is of historical interest to note that knowledge of the use of creosote for odor control at the Carlisle, England sewage purification works gave inspiration to Joseph Lister for the birth of asepsis in surgery.

The processes by which wastewater is purified can best be understood by following the waste from its source through the collection system and treatment plant. Organic material discharged to a watercourse will eventually undergo stabilization. This is accomplished by natural processes (unit operations) and a wastewater treatment plant basically sets up, under controlled conditions, the processes which act in the river. Indeed, it has often been said that a sewage treatment plant is a river in miniature. In the river heavy particles settle out and lighter particles float to the surface. Biological decomposition takes place. Oxygen present is used by organisms that accomplish decomposition. Some of the settled material will be resuspended, increasing the organic loading. While oxygen is being withdrawn by BOD this resource is being replenished at a rate proportional to the deficit. The oxygen deficit is the difference between the amount that can be held at saturation (about 10 mg/l) and the amount actually present.

Stated mathematically, the oxygen concentration in a river as a function of time is

$$\frac{dD}{dt} = k_1 y - k_2 D.$$

This expression considers only the effects of BOD and atmospheric reaeration. A more complete equation can be written but effects of oxygen production by algae and oxygen reduction by benthal (bottom) deposits are not of major significance here. Solving the above gives

$$D_t = \frac{k_1 L}{k_2 - k_1}\{10^{-k_1 t} - 10^{-k_2 t}\} + D_0 \times 10^{-k_2 t}.$$

This expression is commonly known as the oxygen sag equation and is displayed in Fig. 3. It is of interest here because it illustrates the rationale underlying waste treatment requirements. Waste is treated so that undesirable conditions do not develop in the receiving waters. In effect, a limit has been placed on the allowable oxygen deficit. Regulatory authorities usually require that a minimum dissolved oxygen level be maintained. Normally, this will be stated as a percentage of dissolved oxygen saturation. This is the largest pemissible critical deficit. Games fish may require a minimum of 5 mg/l D. O. while scavengers can survive in a much lower quality water.

The critical deficit is given by

$$D_c = \frac{k_1 L \times 10^{-k_1 t_c}}{k_2}.$$

Once the maximum deficit is specified, the BOD loading on the stream can be immediately estimated. The allowable BOD loading will be that impressed on the watercourse by the wastewater treatment plant effluent. It is of interest to observe the effect of the condition of the river at the point of discharge. A river in poor condition will have a large initial deficit, D_0. This can raise the treatment requirements. It can be seen that it is necessary to integrate the efforts on a basin wide basis. Parameters other than just dissolved oxygen must be controlled by the treatment processes. This is accomplished in some, or all, of the following steps.

Decomposition of the waste begins in the collection system. Ease, or difficulty, of treatment depends to a large degree on the condition of the sewage when it reaches the plant. Some substances

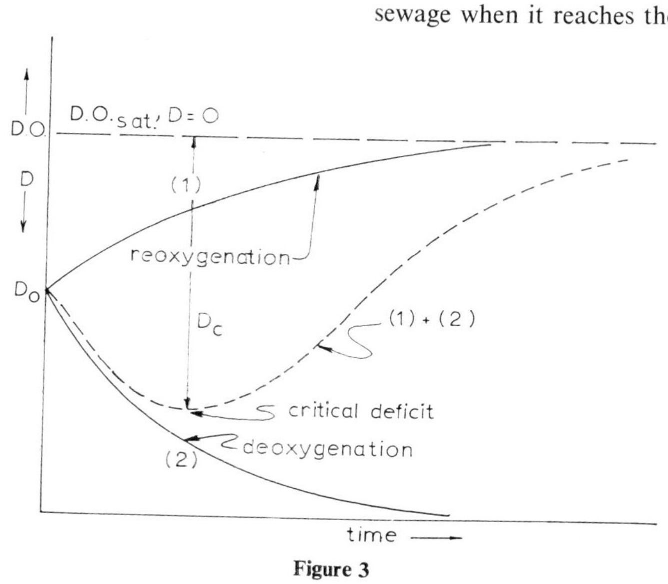

Figure 3

are not permitted in the sewage system. Gasoline and other flammable substances, oil, hexavalent chromium are examples of prohibited substances. These can damage either the collection system or the treatment plant and processes. The legal vehicle by which such materials are excluded is known as a sewer ordinance.

It is most economical to collect sewage by gravity flow. If topography does not permit, pumping will be necessary in order to cross high points and to avoid excessively low flow velocities and deposition of waste material in the pipes. A combined system will, at times, exceed the hydraulic capacity of the treatment plant and flow must be either bypassed or held in detention tanks until heavy flow has subsided.

Multiple treatment units are provided and treatment is not interrupted during periods of maintainence or repairs. Protection must be provided for pumps against large objects, such as floating pieces of wood. Coarse racks, with clear openings of more than 2 inches may be placed at the entrance to the plant. Racks to be placed in advance of grit chambers and settling tanks will have clear openings of 1 to 2 inches. In smaller plants racks are cleaned by hand while larger plants have mechanically cleaned racks. Disposal is by burial, incineration or digestion with sewage sludge. Mechanically cleaned racks have smaller clear openings because head losses are lower with continuous cleaning. Racks with clear openings of 1 to 2 inches can be expected to give from 20 to 100 ft^3 of solids per 1000 people annually. Comminutors macerate floating material into sizes sufficiently small to be easily handled by centrifugal pumps. Racks and screens with very small openings have been almost completely replaced by comminutors.

Rate of flow into the plant will vary over a wide range during any 24 hour period with smaller plants exhibiting greater variation. Flow is measured by Parshall flume, Venturi meter or Sutro (keyhole) weir. The Parshall flume, sometimes called the open channel venturi, is most commonly used. The device operates on the principle of critical flow and measurement of water depth upstream of the flume throat. The governing equation is of the form $Q = cWH^{3/2}$, where Q is the discharge per unit time, W is the throat width. H is the water depth and c is a constant. While c changes with throat width, it is closely constant for a constant throat width. Flume liners of reinforced fibreglass are all but replacing steel and concrete liners. Ease of fabrication, close tolerance and corrosion resistance are advantages cited. The true Venturi operates on closed pipe (pressure) flow and is usually found in larger plants. The Sutro, or keyhole, weir is shaped as its name suggests. Its principal advantage is maintainence of a constant upstream velocity over a wide flow range, but it does have a high energy loss and metering is lost when the opening is submerged.

It is necessary to remove grit in order to protect pumps against excessive wear and to maintain capacity of sludge digesters. It has been found that digesters in plants serving low lying sandy areas can, if grit removal is not efficient, lose up to a third of capacity in but a few years. Grit chambers must operate in a fairly narrow velocity range of from 0.75 to 1.25 ft/sec. Above this range deposited material is scoured back into suspension and below the lower value organic material settles out. The resulting material, called detritus, is unsuitable for landfill uses due to its highly putrescible nature. Grit chambers are usually designed to remove particles with a specific gravity of 2.65 and a mean diameter of 0.02 cm. Because flow variation with depth follows a parabolic function, $Q = cWH \cdot H^{1/2}$, the grit chamber is often given an approximately parabolic shape and better velocity control is attained. The amount of grit collected per million gallons flow is found to vary from 1 to 12 cubic feet. Grit is removed manually in small plants and continuously by mechanical means in larger plants.

Settling tanks are provided for removal of larger, heavier organic particles, oil and grease. Oil, grease and other materials lighter than water are skimmed continuously from the surface and led to digestion. Both circular and rectangular surface configurations are used. Rectangular tanks of the flow through variety have length to width ratios of 4/1 to 6/1. Circular tank size is usually limited by structural requirements of trusses carrying skimming devices. Tank depths vary from 7 to 15 feet. Bottoms are sloped about 1% in rectangular tanks and about 8% in circular tanks to facilitate sludge removal.

Design is on the basis of hydraulic loading. A commonly used figure is 1000 gal/day × ft^2 surface area. It can be expected that a BOD removal of 30% will be achieved in a well operating primary sedimentation unit.

If treatment includes only screening, sedimentation and chlorination of effluent, the treatment is classed as primary. Primary treatment, while inadequate for most areas, is better than no treatment. The adequacy of secondary treatment is now being seriously questioned. Nonetheless, it fits the economics of the situation. Only in the 1960's did wastewater treatment become of interest to any but a small number of people. Sanitary engineers were wont to say "It may be sewage to you but it is bread and butter to me." Theirs was not a profession to which much glamour was attached. Financing bodies were reluctant to invest adequate sums in waste treatment facilities. Hopefully, this has now changed.

There are two main processes utilized for biological (secondary) treatment. These are 1) the trickling filter and 2) acitvated sludge.

The trickling filter is not a true filter. It can best be described as a pile of stones, or other coarse material, which sewage flows. This is the most widely used biological treatment process. Present day biological treatment technology is a logical development from sewage farms (irrigation areas) to intermittent sand filters to contact (fill and draw) beds to trickling filters and activated sludge units. Numerous modifications of the basic processes have evolved but the underlying principles remain unchanged. In biological treatment a suitable environment is provided so that microorganisms may thrive under controlled conditions. The suitable environment is one rich in food and maintained in the aerobic state. The zoogleal mass remains fixed on the filter media in the trickling filter while the sewage flows past. In the activated sludge process the sewage and organisms flow together. In both cases the microorganisms come from the sewage itself. Traditional secondary treatment plants operate in the declining growth phase.

Irrigation by sewage provides water return and some waste stabilization but this means of sewage disposal is in conflict with sound public health practice and ought not be used where there is a possibility that sewage can pass with little change into the groundwater table. Irrigation is best applied in arid regions. When it is utilized for food growing areas, care must taken so that edible plants and fruit are not contaminated. Odors are a problem and removals decline markedly in cold weather.

Intermittend sand filters are much like the slow sand filters used for potable water production. The sewage is applied to a sandy area and allowed to percolate downward. Raw sewage may be applied at rates as high as 80,000 gal/acre × day and secondary effluents at rates as high as 800,000 gal/acre × day. Application of the secondary effluent would be considered tertiary treatment. Biological films that form on the sand grains undergo continuous stabilization. It is necessary to rest the bed between dosings so that objectional conditions do not develop. Surface accumulations of solids must be periodically removed. This method is not recommended for areas underlain by fissured limestone.

Fill and draw beds operate as the name indicates. A tank, packed with coarse material, is filled with sewage and allowed to stand full. It is then drained and allowed to rest. Air is drawn into the bed during filling and emptying. Loadings are about 200,000 gal/acre × ft × day. There is little application of this method today in treatment of municipal wastewater but it does find use in industrial waste treatment. Fill and draw beds are a batch operation and the trickling filter is a continuous operation.

Sewage is distributed over a trickling filter by slowly revolving arms equipped with nozzles and deflectors. Some earlier plant had fixed nozzles but this is no longer done. Revolving arms are driven by hydraulic head. Sewage discharged is allowed to flow slowly downward through the bed. Air is drawn into the bed by temperature differential, thus maintaining a supply of oxygen for the process. Filter media is usually stone. Sizes are in the range of 1 to 4 inches. Packing of this size permits air to be drawn into the bed and the bed is not clogged by biological slime. There appears to be a trend toward more use of plastic filter media. Filter depths range from 3 to 14 feet. A common depth is 6 feet. After passage through the filter the sewage is collected in tile underdrains. These underdrains serve two purposes 1) collection of filter effluent and 2) circulation of air into the filter. The underdrains discharge to a main collection channel which, in turn, discharges to the final settling (humus) tank. The importance of the function of the final settling tank can be seen by an examination of what occurs in the filter itself.

A new filter is "broken in" by applying sewage as in normal operation. After a time the microbial (zoogleal) mass establishes itself on the filter media and carries on the work of waste stabilization.

Waste material in the flowing sewage (food) is first absorbed into the zoogleal mass and then assimilated by the microorganisms. Much of the organic waste material has, at this point, been utilized for cell synthesis and energy. There must be continuous removal of filter slime or the process becomes sluggish due to a lower feeding rate of old organisms. Since waste material is now a part of the filter slime there must be a means provided for removal of sloughed off organisms or the waste material, now in different form, would still appear in the plant effluent and little constructive would have been accomplished. The required removal is carried out in the secondary settling tank. A rate of application lower than that of the primary tank is necessary here because of the different character of the material to be removed. Rates in this portion of the system are in the range of 600 gal/ft^2 × day.

A portion of the effluent is recirculated, as shown in Fig. 4. This is done in order to 1) smooth out flow, 2) keep the food concentration more constant, 3) lower the film thickness and, thus, control the psychoda fly and 4) reseed the applied sewage with acclimatized organisms. The psychoda, of filter, fly is a very small insect which breeds in trickling filter slime. It does not bite but can be extremely bothersome because it does get into the nose and mouth. The range of flight is short but the creature can be carried great distances by the wind. Control of the fly in its developmental phase can be achieved by flooding the filter periodically or through chlorination of influent.

Trickling filters can be classified on the basis of 1) hydraulic loading per unit area and 2) applied pounds of BOD per 1,000 ft^3 of filter volume.

The low rate trickling filter, with hydraulic loadings of 2 to 4 million gallons per acre per day (mgad) and 10 to 20 pounds BOD per 1,000 ft^3, is usually found in use in smaller plants. With proper operation, BOD removals of 80 to 85% can be routinely expected. Raising the applied sewage to 10 mgad produced greater BOD removals per unit filter volume but the effluent organic concentration was found to be high. Influent organic concentration was reduced by greater effluent recirculation and lower effluent organic concentration was realized. Units that operate in the 10 to 40 mgad range are called high rate trickling filters. BOD loadings are up to 90 pounds per 1,000 ft^3, but removals to be expected are in the range of 65 to 75%.

In the 4 to 10 mgad range operational difficulties were frequently encountered and this range was avoided for many years. It appears that, in this range, the hydraulic application was inadequate to keep the filter slime from attaining excessive thickness. Many plants had operated well in this range, but other plants had many problems. The solution seems to have been reached with use of relatively large, 2 to 4 inch, filter stones.

Experimental plants using plastic media have recently achieved very high removal efficiencies (97%) at hydraulic loading rates of 100 mgad. Much of the microbial mass is in the recirculated effluent and these plants are, in effect, modifications of the activated sludge process. Organic loadings are in the region of 100 pounds BOD per 1,000 ft^3 filter volume.

Modifications of the trickling filter process over the years have dealth with improvements in media, air circulation and loadings. One of great interest is that proposed by Ingram. Sewage is introduced at various levels in a very deep filter in an attempt to distribute the load more uniformly over the whole filter depth. Hydraulic loadings up to 500 mgad have been successfully achieved.

Activated sludge serves the same function as trickling filter slime. The major difference lies in the filter slime being fixed to the filter media while the activated sludge is carried along with the flowing

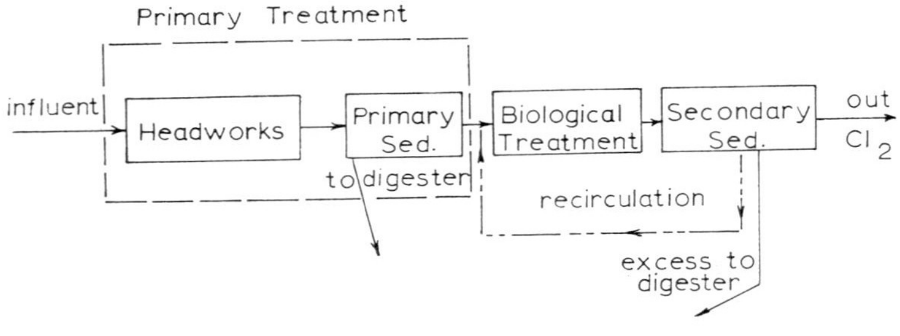

Figure 4

purify sewage by blowing air into it. It was observed that after prolonged aeration, flocs composed of voraciously feeding organisms developed. This floc settled after aeration was stopped. Addition of fresh sewage to tanks containing the settled sludge produced high purification in a practical time. The name activated sludge was assigned this means of waste treatment. At first, this was operated as a fill and draw system. Research showed that continuous operation could be practiced and this is the present means of operation. The process involves

a) return of activated sludge to the aeration tank influent and discharge of excess sludge to digestion.
b) aeration of the sludge-sewage mixture to attain purification and
c) settling of the aeration tank effluent to remove floc before final discharge.

Step (c) is necessary for the same reason as the comparable portion of the trickling filter process-removal of waste material transferre dto the microbial cell mass. Floc is formed in the tank through aerobic growth of unicellular and filamentous bacteria. Protozoa and other organisms will also be found in the floc. This is a strictly aerobic process and air requirements are high. Two aeration systems are used- 1) diffused air units and 2) mechanical aeration. Air diffusers are more commonly used in North America but mechanical aeration systems may be found in plants of less than 1 million gallons per day (mgd) capacity. Both methods of of aeration perform three functions- 1) transfer of oxygen to the mixture and maintainence of aerobic conditions, 2) intimate mixing of floc and sewage and 3) keeping the floc in suspension.

Aeration tanks are normally rectangular in cross section, 10 to 15 feet deep and 30 feet wide. Length to width should be greater than 5 to 1 in order to avoid short circuiting. Detention periods are from 4 to 6 hours. Air is introduced from diffusers in such a way as to set up a spiral flow pattern, thus aiding in mixing of floc and sewage and helping to prevent dead spaces in the tank. It was found that oxygen requirements decreased as the waste proceeded through the tank. The number of diffusers was, therefore, increased at the beginning of the unit and decreased at the effluent end. This is now the common practice and is known as tapered aeration. Mechanical aeration has the same function as air diffusers but is accomplished by rotating paddles or brushes. Peripheral velocity is about 2 ft/sec.

Floc returned to the aeration tank has the purpose as trickling filter slime but floc concentration can be varied as operational needs dictate. Returned sludge varies from about 10 to 30%. Mixed liquor suspended solids (MLSS) will vary from 600 mg/l to 4,000 mg/l, on a dry weight basis. An important parameter in routine process control is the ratio of the volume of MLSS to the dry weight of MLSS. This is known as the sludge volume index (SVI) and is in the range of 50 to 100 in well operating plants. When the value approaches 200 operating difficulties can be expected. Factors which promote or inhibit microbiological growth are important and these include pH, temperature and oxidation-reduction potential (ORP). Hydrogen ion potential, pH, will have a great effect on the dominant species of organisms. Bacteria predominate above pH of 6.5 while fungi assume greater importance below this value. There must be adequate buffering capacity if metabolic products are acidic.

Modifications of the basic activated sludge process have come about for solution to specific operating problems.

The municipal treatment plant at Peoria, Illinois received a waste high in carbohydrates. The resulting nitrogen deficiency caused a light and poorly settling activated sludge floc with attendant poor waste stabilization. Kraus, for whom the modification is named, aerated digester supernatant. This, added to the influent, gave a nitrifying activated sludge. The result was a readily settleable sludge with improved organic removals.

New York City has plants scattered throughout the five boroughs, treating more than 1 billion gallons of sewage per day. A major modification resulting from experimentation with plant operation has come from this city. In conventional plants sewage was added at one end of the aeration tank and allowed to flow through. This gave a high initial microbial food supply and correspondingly high oxygen requirement. The New York City modification involves introduction of sewage at intervals along the tank. This smooths out the food supply and lowers the oxygen requirements. The sewage is added at discrete steps along the unit and the name applied is step aeration.

A low mixed liquor suspended solids concentration of 200 to 500 mg/l is maintained in the high rate process. This gives a high food to microbial mass ratio. This keeps the floc in the active growth phase, but excess food will be discharged in the effluent. BOD removals are only 50 to 60%, but in some areas this is acceptable. New York City applied this method successfully because of a weak sewage and low temperatures. Philadelphia and Los Angeles met with indifferent success because of stronger sewage or higher temperatures.

A process originated simultaneously and independently by Smith and Eckenfelder made use of a phenomenon observed by many researchers but dismissed as experimental error. When activated sludge and raw sewage are mixed together in an aeration vessel there is noted a reduction in BOD, followed by a rise and then another reduction. The first decrease had been ignored by most research workers. Smith and Eckenfelder found that this was due to adsorption of waste material on to the activated sludge floc. This came from material desorbed from colloidal particles. Plants at Austin, Texas and Bergen County, New Jersey were converted from overloaded to underloaded by changing to the Biosorption process.

Putrescible material collected from the primary settling tanks and excess sludge from humus tanks must be disposed of cheaply and efficiently. This material is highly unstable and a potential nuisance source. Because it is putrescible it can be stabilized by biological means, serving as food and energy sources for microorganisms naturally found in the sludge.

Raw sludge is about 95% water, but the water is not easily removed. As the sludge is broken down the water content is lessened, and the volume is markedly reduced. A rough rule is that sludge volume is reduced by half when water is lowered from 95% to 90%, and by two thirds when reduced from 95% to 85%. Fresh sludge has a gray color and can be easily pumped. Its odor is most disagreeable, being due principally to mercaptans. Digested sludge is black in color, granular and has a slight tarry odor.

Sludge digestion is carried out in order to reduce the volume of sludge to be handled, and reduce the number of pathogens. Sludge is usually withdrawn at regular intervals from primary and secondary tanks and lead by gravity to a sludge well. It is then pumped to the digester.

Mixing is very important for efficient sludge digestion. Temperature is equally important.

Since destruction of sludge is carried on by microorganisms, kinetics of their life processes will be temperature dependent. It has been found that sludge temperature of about 95°F will give acceptably short detention times. Even shorter detention times for the same quality of digested sludge can be achieved with temperatures of about 125–130°F, but this temperature range is not widely used for reasons of economics. Above 95°F an increase in detention time is noted, up to 110°F, and then again a decrease. The reason for this is the changing character of the predominant organisms.

Heating of sludge for efficient digestion is carried out in one of two ways. The older installations have hot water coils in the periphery of the tank, and heat is transmitted to the digesting sludge. Mixing was felt to be adequately effected by turbulence due to gas generation. Mechanical mixers have been used. It was found, however, that mixing was not sufficient. In addition, heating of entire tank contents was not achieved due to "baking" of sludge in the vicinity of the heating coils. A second method of sludge heating and mixing was developed, involving the use of external heat exchangers. Sludge is pumped from the digestion tank thru a heat exchanger and returned to the tank. Two objectives are accomplished (1) efficient mixing of sludge, thereby reducing the amount of inadequately digested sludge, (2) more uniform temperature throughout the tank, thus reducing digestion time. The use of external heat exchangers has almost completely supplanted heating coils and internal mixers in new plant design.

Sludge gas generated during digestion is approximately 72% methane and 28% carbon dioxide. Hydrogen and H_2S are present in trace amounts. Gas thus generated has a calorific content of about 600 BTU/ft^3. About 10 ft^3 of gas are produced per cubic foot of raw sludge digested. Generally, the amount of sludge gas produced is sufficient to provide heat used in maintaining digesting sludge at the required temperature, heating plant buildings, provide hot water and incineration of digested sludge, when practiced, and fuel for generators.

Volatile acids, reported as acetic acid, are perhaps the most important parameter in control of sludge digestion. Volatile acids below 1000 mg/l occur in a healthy digestion process. Volatile acids of 6000 mg/l indicate a malfunctioning process. pH values of 6.8 to 7.2 are optimum. Values less than 6.8 usually are due to excessive volatile acid production. In the past liming of malfunctioning tank contents was practiced in an effort to adjust pH to about 7.0. However, the change in volatile acids production was due to changing dominant process microorganisms. The lowered pH and high volatile acids concentrations were a sign of a sick process, rather than the cause.

Digested sludge is reasonably inert, but still has high water content. It is necessary to further dewater, and this is accomplished by (1) drying beds (2) mechanical dewatering (3) wet air oxidation (Zimpro) and (4) ocean disposal.

Digested sludge can be further dewatered by drying on sand beds. These are of two types (a) open and (b) covered.

Open sand beds are exposed to the open atmosphere and drying is accomplished by both downward drainage and evaporation. Covered beds resemble a greenhouse. Temperatures are rather high as a result. In either case sludge is allowed to flow over beds of sand and then left to stand for the required time. The dried sludge is then scraped from the beds for ultimate disposal.

Sludge may be treated by vacuum filtration. Filter drums rotate slowly, picking up wet sludge at the bottom. During rotation a slight vacuum is applied and the water drawn off is returned to the plant stream. At the end of the cycle sludge the vacuum is released and sludge removed by a scraper.

Disposal of dried sludge from drying beds may be by incineration, use as landfill or as fertilizer. Unfortunately, sewage sludge has not found wide use as fertilizer. This is due partly to esthetic reasons and partly to the comparatively low quality of the sludge as fertilizer.

In some cities digested sludge is incinerated at elevated temperatures (1800–2000°F). This practice has fallen into disuse due to attendant air pollution problems.

A relatively new process (1959), known as Wet Air Oxidation or the Zimmerman Process (Zimpro) shows promise of solving some of the problems of sludge disposal. The process is based on the principle that any organic compound can be oxidized in an aqueous solution if sufficient energy, in term of heat and pressure, is supplied.

Water in the operating temperature range of the process has a heat capacity eight times that of the flame gases typical of dry combustion. Thus, the temperature is greatly depressed.

The basic steps of the process operation are as follows:

1) Temperature must lie between 212°F and the critical temperature of water (705°F).

2) Air requirements for complete oxidation depend on the fuel value of the material oxidized. Air input enters into the process economics. COD (chemical oxygen demand) is a measure of the heating value of an aqueously dispersed fuel and a sludge with a high COD will require more air. However, COD of a sludge decrease with increasing temperature.

3) Air supplied to the process becomes saturated with steam from contact with the water. The greater the quantity of air supplied, the more steam is carried from the reactor. There is, however, an upper limit to the amount of air input since all water must not be carried away.

4) Pressure affects the amount of water vapor. The pressure must be increased in order to obtain a temperature increase if water loss is to be kept essentially constant.

Capital costs for process equipment are high, but operational savings have been found to be sufficiently high to indicate that in the future the Zimmerman Process may find wider application.

Disposal of sewage sludge by barging to sea has been practiced by a number of large sea coast cities for many years. The most notable example is New York City and the city maintains a fleet of sludge barges. The future of this means of disposal is in serious doubt. Studies conducted by the Sandy Hook Marine Laboratory have shown that the disposal area used by New York City, located off the New Jersey coast, is apparently a biological wasteland.

It is common practice to chlorinate effluent for bacterial control. Regulations vary from state to state, but most regulations require chlorination to a specified residual. Requirements usually vary from season to season, the most stringent rules governing the swimming season.

A phenomenon not yet fully understood is that of aftergrowth, wherein bacterial count is fairly low immediately after effluent discharge but then suddenly rising to a high figure.

In some plants chlorination of the influent is practiced for the purpose of odor control.

Chlorination of storm water overflow is commonly practiced. In some cases storm water overflow is subjected to simple sedimentation and/or screening, storage and chlorination, then discharged after cessation of the storm.

One commonly used definition of tertiary treatment is any treatment in addition to secondary (biological) treatment.

Tertiary treatment is practiced when an effluent of much higher quality is required than is attainable with conventional biological treatment. The type process used will depend on the final effluent quality necessary and the economics of the total process.

Commonly used tertiary treatment processes are listed below:

a) Sand filtration
b) Microstrainers
c) Oxidation ponds
d) Foam separation
e) Activated carbon adsorption
f) Chemical clarification and precipitation
g) Ion exchange

Disposal of human and kitchen wastes in areas not served by sewers and waste water treatment plants presente unique problems.

Disposal must be in the immediate vicinity of the source of the wastes. If adequate controls are not exercised a closed system may result. An example is Suffolk County (Long Island), New York. Septic tanks are widely used, and there is strong local opposition to the considerable expense of installing sewers. Effluent from the septic tanks found its way into the ground water which is the supply for much of the county. Eventually, the problem was graphically pointed up by the appearance of foaming detergents in water issuing from the tap.

In more primitive societies waste disposal is a matter of convenience. A "cat hole" or communal straddle trench is utilized and covered when capacity is reached. This is the same as the practice with privvies and cesspools.

Privvies, as the name implies, are simply open pits with a structure to provide privacy. Human excreta is deposited into the pit and is slowly stabilized. Stabilization is slow, due principally to the presence of urine. Pits may be open earth or concrete vault. Drawbacks for both types are odors and fly problems. For the unlined pit there is the additional problem of ground water pollution. Older privry construction allowed access to flys around the edge of the pit. For proper protection against flys there must be a tight seal around the edge of the pit and adequate screening of openings in the privry structure itself.

In the 1930's a large number of the older privvies were replaced by concrete vault types. Today, such methods of waste disposal are found only in the smaller rural communities where there is no municipal collection system.

Cesspools are simply pits into which waste is allowed to flow. The term leaching pit is sometimes used. Water seeps into the ground, leaving solid matter in the pit. Construction is of two types. A pit may be unlined, or it may be lined with sewer pipe laid on end. Almost nowhere in the United States are cesspools permitted by health authorities.

Septic tanks are widely used in smaller towns and outlying suburbs of larger cities. They are a combination sedimentation tank and anaerobic digester. Sanitary and kitchen wastes flow into the tank and grease and light material rise to the top. Heavier particles settle to the bottom where anaerobic stabilization occurs. Deflector plates are provided at inlet and outlet in order to minimize short circuting. Effluent flows to a tile field where disposal is into the earth. The tile field is composed of perforated field tile fed by a manifold. The tile is underlain with granular material, usually gravel. Care must be taken that the earth does not become clogged by material carried over from the septic tank. Septic tanks are being replaced as more and more areas are served by municipal systems. Health authorities do not look with favor on septic tanks. Capacity will be a function basically of the number of persons or units served. Some experts feel that, in no case, should capacity be less than 1500 gallons. Lesser volumes are permitted in many codes and the thought that 1500 gallons ought be the minimum permitted arouses home builders and land developers. Periodically it is necessary to employ a scavenger service for emtying the tank of accumulated solids. Solids thus collected may be discharged to a convenient collection system or directly to a wastewater treatment plant.

The Imhoff Tank has not been discussed earlier because it is similar in many respects to the septic tank. Use is generally confined to small communities and isolated installations. Operation is a combination of sedimentation and anaerobic digestion. This tank was invented by Karl Imhoff, who first used them in the Essen District in Germany in 1907. The tank is composed of two chambers, one above the other. Surface configuration may be circular, square or sectangular. Depth is 25′ to 35′.

Sewage flows thru the upper chamber at a low

velocity (about 1 fps). Solids settle out and slide thru a slot into the bottom chamber. Detention period is about 2 hours.

Solids accumulating in the bottom, or digestion, chamber have an initial water content of 85 to 95%. After proper digestion of about 60 days the water content is reduced to about 50% and the volume is greatly reduced. Gases produced during digestion are vented to the atmosphere by gas vents located at the tank sides. Solids buoyed up by gas are prevented from escaping to the upper tank by deflector plates.

Attempts were made to hasten digestion by heating the lower compartment but were of limited success due to overturning of the tank contents.

Some rectangular tanks are arranged so that the direction of flow can be reversed, with outlets becoming inlets and vice versa.

In recent years manufactureres of waste treatment equipment have endeavored to supply complete treatment plants for small communities or developments and isolated installations. Basically, these plants, called package plants, supply primary treatment and sometimes some biological treatment on a small scale without requiring extensive operating supervision. It is felt that such treatment is to be preferred to septic tanks or only primary treatment (Imhoff Tanks, for example), but such installations are not the ultimate solution.

REFERENCES

Andern, E., and W. T. Lockett, *J. Soc. Chem. Ind. London* **33**, 523, 1914.

Bewtra, J. K., Biological Treatment of Wastewater. *Encyclopedia of Environmental Science and Engineering*. Vol. 1, Gordon and Breach, Inc. New York, 1976.

Bewtra, J. K., and H. I. Ali, Wastewater: Physical and Chemical Treatment. *Encyclopedia of Environmental Science and Engineering*. Vol. 2, Gordon and Breach, Inc. New York, 1976.

Camp, T. R., Trans. ASCE, Vol. 111, p. 895, 1942.

Cecil, L. K., Water Reuse. *Encyclopedia of Environmental Science and Engineering*. Vol. 2. Gordon and Breach, Inc. New York, 1976.

Cecil, L. K., Water Reuse and Disposal. *Chemical Engineering*, May 5, 1969.

Disposal of Municipal Sewage. House Report No. 204, 12th Report by Committee on Government Operations, House of Representatives, March 24, 1965.

Dobbins, W. E., *J. of the Sanitary Engineering Division, ASCE* Vol. 90, No. SA 3, Proc. Paper 3949, June 1964.

Eckenfelder, W. W., and D. J. O'Connor, *Biological Waste Treatment*, Pergamon Press, Ltd. London, 1961.

Fair, G. M., J. C. Geyer, and D. A. Okun, *Water and Wastewater Engineering* Vols. 1 and 2. John Wiley and Sons, Inc., New York, N.Y., 1968.

Guttman, H. N., Microbiology. *Encyclopedia of Environmental Science and Engineering* Vol. 1. Gordon and Breach, Inc., New York, 1976.

Imhoff, K., *Taschenbuch der Stadtentwässerung*, R. Oldenbourg Verlag, Munich, 1960.

Imhoff, K., and G. M. Fair, *Sewage Treatment* 2nd. Ed. John Wiley and Sons, Inc., New York, N.Y., 1956.

McKinney, R. E., *Microbiology for Sanitary Engineers*, McGraw-Hill Book Co., New York, N.Y., 1962.

Pfafflin, J. R., Water: Sewage. *The Encyclopedia of Chemical Technology* Vol. 22, Interscience Publishers, Inc., New York, N.Y., 1970.

Pfafflin, J. R., Water: Reuse. *The Encyclopedia of Chemical Technology* Vol. 22, Interscience Publishers, Inc. New York, N.Y., 1970.

Phelps, E. B., *Stream Sanitation*, John Wiley and Sons, Inc. New York, N.Y., 1944.

Salvato, J., *Environmental Sanitation*, 2nd Ed. John Wiley and Sons, Inc., New York, N.Y., 1972.

Sawyer, C. N., and P. L. McCarty, *Chemistry for Sanitary Engineers*, 2nd. Ed., McGraw-Hill, Book Co., New York, N.Y., 1967.

Standard Methods for the Examination of Water and Wastewater, 13th Ed., APHA, 1971.

Streeter, H. W., and E. B. Phelps, A Study of the Pollution and Natural Purification of the Ohio River. Public Health Bulletin 146, U.S. Public Health Service Washington, D. C., 1925.

CAMERON MACINNIS
University of Windsor
JAMES R. PFAFFLIN
Stevens Institute of Technology

MUNICIPAL WATER REUSE: see WATER REUSE

MUNICIPAL WATER SUPPLY: see WATER TREATMENT
